Smart Cyber-Physical Power Systems

IEEE Press
445 Hoes Lane
Piscataway, NJ 08854

Published by John Wiley & Sons, Inc., Hoboken, New Jersey.
Published simultaneously in Canada.

Library of Congress Cataloging-in-Publication Data:

Names: Parizad, Ali, editor. | Baghaee, Hamid Reza, editor. | Rahman, Saifur, editor.
Title: Smart cyber-physical power systems : challenges and solutions Volume 1/ edited by Ali Parizad, Hamid Reza Baghaee, Saifur Rahman.
Description: Hoboken, New Jersey : Wiley-IEEE Press, [2025] | Includes index.
Identifiers: LCCN 2024048504 (print) | LCCN 2024048505 (ebook) | ISBN 9781394191499 (cloth) | ISBN 9781394191505 (adobe pdf) | ISBN 9781394191512 (epub)
Subjects: LCSH: Cooperating objects (Computer systems) | Electric power systems–Automation. | Artificial intelligence.
Classification: LCC TJ213 .S485 2025 (print) | LCC TJ213 (ebook) | DDC 006.2/2–dc23/eng/20241214
LC record available at https://lccn.loc.gov/2024048504
LC ebook record available at https://lccn.loc.gov/2024048505

Cover Design: Wiley
Cover Image: © metamorworks/Shutterstock

Set in 9.5/12.5pt STIXTwoText by Straive, Chennai, India

Smart Cyber-Physical Power Systems

Fundamental Concepts, Challenges, and Solutions

Volume 1

Edited by

Ali Parizad
Virginia Tech
United States

Hamid Reza Baghaee
Tarbiat Modares University
Iran

Saifur Rahman
Virginia Tech
United States

IEEE Press Series on Power and Energy Systems
Ganesh Kumar Venayagamoorthy, Series Editor

IEEE Press
WILEY

To my parents, whose unwavering support and guidance illuminate my journey at every step.
To my beloved wife, whose love, patience, and encouragement have been my greatest source of strength and inspiration.

To my professors and colleagues, from whom I have learned immensely, whose wisdom and mentorship have profoundly shaped my professional path.

And to those envisioning a future where sustainable living, smart cities, and the pioneering spirit of artificial intelligence converge to create a world where technology harmoniously enhances our environment and society, fostering an era of unparalleled freedom and possibilities.

– Ali Parizad

To my beloved family: my parents, whose unwavering support has been my foundation; my wife, who has stood by me at every step; my children, who bring joy to my life; and my entire family for their constant encouragement. I also dedicate this work to my esteemed professors for their valuable supports and to researchers in this field for their dedication to advancing knowledge. This two-volume work, "Smart Cyber-Physical Power Systems: Challenges and Solutions," is a humble reflection of your support and inspiration.

– Hamid Reza Baghaee

I dedicate this book to my parents Mr. Serajur Rahman and Mrs. Sahara Rahman for their deep affection and love and for injecting deep moral values in me. These have laid the foundation on which my life's achievements stand.

– Professor Saifur Rahman

Contents

About the Editors

Ali Parizad, Postdoctoral Associate, Virginia Tech, Advanced Research Institute (ARI), Virginia, USA

Ali Parizad is a Postdoctoral Associate at Virginia Tech's Advanced Research Institute. His tenure at Virginia Tech involves leveraging machine learning (ML) to enhance energy efficiency within smart grids, under the mentorship of Professor Saifur Rahman, IEEE President 2023. Ali's academic foundation was laid at Southern Illinois University, where he obtained his PhD from the Electrical and Computer Engineering Department in 2021. His doctoral research, which was honored with the Dissertation Research Award for the 2020–2021 academic year, focused on pioneering solutions for modern power systems and smart grids. Specifically, he developed innovative software for Ameren Electric Company, aimed at optimizing distribution system planning with an emphasis on distributed energy resources (DERs) to boost the performance of electric distribution networks. His PhD dissertation emphasized the application of machine/deep learning algorithms for load forecasting, alongside exploring cyber-security and false data detection methods within power systems.

Before embarking on his PhD, Ali joined MAPNA Electric and Control Engineering and Manufacturing Company, Iran's premier power company, as a Power Systems Analysis Engineer in 2010. His roles expanded to include Energy Management System and Supervisory Control and Data Acquisition (SCADA) engineer, as well as Commissioning Supervisor in substation and power plant projects in collaboration with ABB and SIEMENS companies. His innovative work in the realm of real-time simulators culminated in the registration of a patent for a real-time islanded simulator for industrial power plants.

Ali's research interests are extensive, covering the application of artificial intelligence, deep learning, big data, information theory techniques in modern power systems and smart grids, distributed generation, renewable energies, and the operation and control of power systems. He has also explored the potential applications of real-time simulators in enhancing power system operations.

His contributions to the field are substantial, with three books, two book chapters, a patent, and numerous papers in reputable power systems journals to his name. Ali is a valued peer reviewer for several prestigious academic journals, including *IEEE Transactions on Power Delivery, IEEE Transactions on Power Electronics*, and *IEEE Access*, among others. His work not only contributes to the academic community but also to the advancement of practical solutions for power systems and smart grid challenges.

As a Senior Data Scientist in the Information and Data Analytics (IDA), Data Science & Machine Learning department at Shell Energy, Ali applied his profound expertise to develop and implement advanced data science solutions for energy demand forecasting and electric vehicle charging station analysis. This role underscored his commitment to leveraging data analytics and machine learning to solve complex challenges in the energy sector, marking his transition from academia to a leading role in industry innovation. Continuing on this path, he holds the position of Staff Power Systems Machine Learning Engineer at Thinklabs AI, where he is dedicated to furthering his impact by addressing critical power systems challenges through state-of-the-art AI technologies.

Hamid Reza Baghaee, Faculty of Electrical and Computer Engineering (ECE) at Tarbiat Modares University (TMU), Tehran, Iran

Hamid Reza Baghaee (SM' 2008, M' 2017) received his PhD in Electrical Engineering from Amirkabir University of Technology (AUT) (Center of Excellence in Power Engineering and the most prestigious university of Iran in electrical power engineering) in 2017. From 2007 to 2017, he was a teaching and research assistant in the Department of Electrical Engineering at AUT. He is the author of three books, three published book chapters, 85 ISI-ranked journal papers (mostly published in IEEE, IET, and Elsevier journals), 70 conference papers, and the owner of one registered patent. Additionally, he has presented 20 workshops and 15 invited talks at national and international conferences and scientific events. His book entitled *Microgrids and Methods of Analysis* was selected as the best book of the year in the power and energy industry of Iran by the technical committee of the Iran Ministry of Energy (MOE) in November 2021 and the winner of the Distinguished Author of the International Books Award in the AUT in December 2021. He has many HOT and HIGHLY-CITED papers in his journal and conference papers, based on SciVal and Web of Science (WoS) statistics. His special fields of interest are micro- and smart grids, cyber-physical power systems, power system cyber security and cyber-resiliency, application of artificial intelligence (AI) and machine learning (ML) and big data analytics in power systems, real-time simulation of power systems, distributed generation, and renewable energy resources, FACTS, HVDC and custom power devices, power electronics applications in power systems, Power Electronics-Dominated Grids (PEDGs), power quality, real-time simulation of power systems, and power system operation, control, monitoring, and protection.

Dr. Baghaee is also the winner of four national and international prizes, as the best dissertation award, from the Iran Scientific Organization of Smart Grids (ISOSG) in December 2017, the Iranian Energy Association (IEA) in February 2018, Amirkabir University of Technology in December 2018, and the IEEE Iran Section in May 2019 for his PhD dissertation. After pursuing his post-doctoral fellowship in AUT (October 2017–August 2019), in August 2019, he joined AUT as an Associate Research Professor in the Department of Electrical Engineering. He is the Project Coordinator of the AUT pilot microgrid project, one of the sub-projects of the Iran grand (National) Smart Grid Project. He has been a co-supervisor and consulting professor of more than 15 PhD and 20 MSc students since 2017. In 2022, he joined the Faculty of Electrical and Computer Engineering (ECE) at Tarbiat Modares University (TMU), Tehran, Iran, where he is now an Assistant Professor. In December 2023, has was selected as a distinguished researcher at TMU for the reputation and citations of his research among papers and patents. He also was a short-term scientist with CERN and ABB Switzerland. Besides, Dr. Baghaee is a member and Vice-Chairperson of the IEEE Iran Section Power Chapter (since 2022), a member and secretary-chair of the IEEE Iran Section Communication Committee (from 2020 to 2023), and a member of the IEEE, IEEE Smart Grid Community, IEEE Internet of Things Technical Community, IEEE Big Data Community, IEEE Smart Cities Community, and IEEE Sensors Council. Since August 2021, he has been elected as a member of the board and chairperson of the committee on publication and conferences at the ISOSG, the Vice-Chairperson and international representative of CIGRE Iran C6 working group on "Active distribution systems and distributed energy resources," a member of the IEE Transmission and Distribution (TD) Committee, IEEE PES Transmission Sub-Committee and its working groups of Reliability impacts of Inverter-based Resources, Generation and Energy Storage Integration, Voltage Optimization, and Transmission Power System Switching, and also IEEE PES Subcommittee on Big Data Analytics for Power Systems, and IEEE PES Task Force on Application of Big Data Analytic on Transmission System Dynamic Security Assessment, IEEE PES Task Force on Resilient and Secure Large-Scale Energy Internet Systems (RSEI), and IEEE Task Force on Microgrid Design. He is also the reviewer of several IEEE, IET, and Elsevier journals, and Guest Editor of several special issues in IEEE, IET, and Elsevier, MDPI, and a scientific program committee member of several IEEE conferences. Since December 2020, he served as an Associate Editor and Energy Section Editor of the IET Journal of Engineering. He has also been selected as the best and outstanding reviewer of several journals, such as IEEE Transactions on Power Systems (Top 0.66 of reviewers, among more than 8000 reviewers in 2020), Elsevier Control Engineering Practice (in 2018, 2019, and 2020), Wiley International Transaction on Electrical Energy Systems in 2020, and the Pablon best and listed among top 1 of the reviewers in Engineering (in 2018) and both Engineering and Cross-Field (in 2019). He was selected as the Star Reviewer of the IEEE JESTPE and IEEE Power Electronics Society (PELS) in 2020, commemorated and presented during the IEEE ECCE 2021 conference in Vancouver, Canada. He has also been listed in 2020, 2021, and 2022 editions of the top 2% of scientists in the field of Energy, Electrical Engineering, and Enabling and Strategic Technologies according to the Science-Wide Citation Indicators (reported by Stanford University, USA), and mentioned among World's top 1% of Elite Scientists according to Web of Science (WoS) and Essential Science Indicators (ESI) ranking since 2020.

Prof. Saifur Rahman, Director, Virginia Tech Advanced Research Institute, Virginia, USA
2023 IEEE President and CEO

Professor Saifur Rahman is the founding director of the Advanced Research Institute at Virginia Tech, USA, where he is the Joseph R. Loring professor of electrical and computer engineering. He also directs the Center for Energy and the Global Environment at the University. He is a Life Fellow of the IEEE and an IEEE Millennium Medal winner. He was the 2023 IEEE President and CEO. He was the IEEE Power and Energy Society (PES) President in 2018 and 2019. He is the founding Editor-in-Chief of the *IEEE Electrification Magazine* and the *IEEE Transactions on Sustainable Energy*. He has published over 160 journal papers and has made over 700 conference and invited presentations. In 2006 he served on the IEEE Board of Directors as the Vice President for publications. He also served on the Virginia Governor's Executive Committee on Energy Efficiency. He currently serves as a Senior Technical Expert of the Global Energy Interconnection Development Cooperation Organization (GEIDCO). He has a Ph.D. in electrical engineering from Virginia Tech.

Dr. Rahman joined Virginia Tech in 1979 as an assistant professor after serving on the faculty at Texas A&M University from 1978 to 1979, later on becoming a full professor of electrical engineering at Virginia Tech in 1987. In 2005 he was named Joseph R. Loring professor of electrical and computer engineering at the university. In 1992–1993 he spent a year with the Tokyo Electric Power Company in Japan as a research engineer in their Artificial Intelligence Laboratory. Upon his return to the university, he was named the director of the Center for Energy and the Global Environment at Virginia Tech in 1994. Two years later, Dr. Rahman joined the U.S. National Science Foundation as the program director in the Engineering Directorate in charge of the Energy Systems Program, a position he held until September 1999. He served as the chair of the U.S. National Science Foundation Advisory Committee for International Science and Engineering from 2010 to 2013. He led the IEEE delegation to the United Nations Framework Convention on Climate Change Conference COP27 and COP28 in Egypt and Dubai, respectively, in 2022 and 2023. He is the General Chair of the IEEE International Symposium on Achieving a Resilient Climate (ISARC) to be held in Geneva in December 2024.

List of Contributors

Fahim Ahmed
Cyber Security Engineering Department
George Mason University
Fairfax, VA
USA

A. Ahmed
Intel Corporation
Hillsboro
OR
USA

Amir Ameli
Department of Electrical and Computer Engineering
Lakehead University
Thunder Bay, Ontario
Canada

Md Tanvir Arafin
Cyber Security Engineering Department
George Mason University
Fairfax, VA
USA

Matti Aro
Smart Energy and Built Environment
VTT Technical Research Centre of Finland
Espoo
Finland

Mohammadmahdi Asghari
Department of Electrical and Computer Engineering
Lakehead University
Thunder Bay, Ontario
Canada

Hamid Reza Baghaee
Faculty of Electrical and Computer Engineering
Tarbiat Modares University
Tehran
Iran

Masoud Barati
Electrical and Computer Engineering Department
University of Pittsburgh
Pittsburgh, PA
USA

S. Basumalik
New York Power Authority
Albany
NY
USA

Stephen Bayne
Department of Electrical and Computer Engineering
Texas Tech University
Lubbock, TX
USA

Mohammed Ben-Idris
Department of Electrical and Computer Engineering
Michigan State University
East Lansing, Michigan
USA

Manohar Chamana
National Wind Institute
Texas Tech University
Lubbock
USA

Mohsen Chegnizadeh
Department of Electrical Engineering
Sharif University of Technology
Tehran
Iran

Ziyu Chen
South China University of Technology
School of Electric Power Engineering
Guangzhou
China

Sabrieh Choobkar
Faculty of Information and Communication Technologies (ICT) Research Group
Niroo Research Institute (NRI)
Tehran
Iran

Tapadhir Das
Department of Computer Science
University of the Pacific
Stockton, California
USA

Mehdi Davoudi
Department of Energy Engineering
Sharif University of Technology
Tehran
Iran

Sanchari Deb
School of Engineering
Newcastle University
Newcastle
UK

Sajjad Fattaheian Dehkordi
Department of Electrical Engineering and Automation
Aalto University
Espoo
Finland

Elahe Doroudchi
Department of Electrical Engineering
School of Technology and Innovations
University of Vaasa
Vaasa
Finland

Alireza Fereidunian
Faculty of Electrical Engineering
K.N. Toosi University of Technology
Tehran
Iran

Mahmoud Fotuhi-Firuzabad
Department of Electrical Engineering
Sharif University of Technology
Tehran
Iran

Marco Galici
University of Cagliari
Deptartment of Electrical and Electronic Engineering
Via Marengo
Cagliari
Italy

Alireza Ghadertootoonchi
Department of Energy Engineering
Sharif University of Technology
Tehran
Iran

Mohsen Ghafouri
Concordia Institute for Information Systems Engineering (CIISE)
Montreal, Quebec
Canada

Emilio Ghiani
University of Cagliari
Deptartment of Electrical and Electronic Engineering
Via Marengo
Cagliari
Italy

Georgios Giannakopoulos
Intelligent Electrical Power Grids
Department of Electrical Sustainable Energy
Faculty of Electrical Engineering
Mathematics and Computer Science
Delft University of Technology
The Netherlands

Ali Golriz
System & Sector Development
Innovation and R&D at Independent Electricity System Operator
Toronto, Ontario
Canada

Arcadio Perilla Guerra
Intelligent Electrical Power Grids
Department of Electrical Sustainable Energy
Faculty of Electrical Engineering
Mathematics and Computer Science
Delft University of Technology
The Netherlands

Chenke He
South China University of Technology
School of Electric Power Engineering
Guangzhou
China

Tran The Hoang
University of Auckland
Auckland
New Zealand

Rabih A. Jabr
American University of Lebanon
Beirut
Lebanon

Mahmood Jamali
School of Electrical and Electronic Engineering
University of Sheffield
Sheffield
United Kingdom

Zejia Jing
Advanced Research Institute (ARI)
The Bradley Department of Electrical and Computer Engineering
Virginia Tech
Arlington, VA
USA

Hossein Jobran
Faculty of Electrical Engineering
K.N. Toosi University of Technology
Tehran
Iran

Jaesung Jung
Department of Energy Systems Research
Ajou University
Suwon
South Korea

Mladen Kezunovic
Department of Electrical and Computer Engineering
Texas A&M University - TAMU
College Station, TX
USA

Rabia Khan
School of Electrical Engineering & Computer Science
Washington State University
Pullman, WA
USA

Mohaddeseh Koochaki
Department of Energy Engineering
Sharif University of Technology
Tehran
Iran

Jonatas Boas Leite
Electrical Engineering Department
Sao Paulo State University – UNESP
Sãao Paulo, Ilha Solteira
Brazil

Yizheng Liao
Department of Electrical
Computer and Energy Engineering
Arizona State University
Tempe, AZ
USA

Meisam Mahdavi
Department of Electrical Engineering
University of Brasilia
Brasilia
Brazil

Zahra Iranpour Mobarakeh
Faculty of Electrical Engineering
K.N. Toosi University of Technology
Tehran
Iran

Susanna Mocci
University of Cagliari
Deptartment of Electrical and Electronic Engineering
Via Marengo
Cagliari
Italy

Moein Moeini-Aghatie
Department of Energy Engineering
Sharif University of Technology
Tehran
Iran

Moein Moeini-Aghtaie
Energy, Water and Environmental Institute
Sharif University of Technology
Tehran
Iran

Amr A. Mohamed
Electrical and Computer Engineering Department
Toronto Metropolitan University
Toronto, Ontario
Canada

Omar Mohamed
School of Technology and Innovations
University of Vaasa
Vaasa
Finland

Sergio Motta
Department of Electrical Engineering
Aalto University
Espoo
Finland

Mohammad Ghafourian Nasiri
Faculty of Industrial and Systems Engineering
Tarbiat Modares University
Tehran
Iran

Firdous Ul Nazir
Glasgow Caledonian University
Glasgow
United Kingdom

Luciane Neves
Department of Electromechanical and Power Systems
Federal University of Santa Maria
Santa Maria
Brazil

Tung Lam Nguyen
New York Power Authority
Albany, NY
USA

Ha Thi Nguyen
University of Connecticut
Storrs, CT
USA

Nadya Noorfatima
Department of Energy Systems Research
Ajou University
Suwon
South Korea

Mahdi Nozarian
Faculty of Electrical Engineering
K.N. Toosi University of Technology
Tehran
Iran

Kaikai Pan
College of Electrical Engineering
Zhejiang University
Hangzhou, Zhejiang
China

Bikash C. Pal
Imperial College London
London
United Kingdom

Peter Palensky
Intelligent Electrical Power Grids
Department of Electrical Sustainable Energy
Faculty of Electrical Engineering
Mathematics and Computer Science
Delft University of Technology
Delft, South Holland
The Netherlands

S. Pandey
Smart Grid and Technology
ComEd
Oakbrook Terrace, IL
USA

Ali Parizad
Advanced Research Institute (ARI)
The Bradley Department of Electrical and
Computer Engineering
Virginia Tech
Arlington, VA
USA

Fabrizio Pilo
University of Cagliari
Deptartment of Electrical and Electronic
Engineering
Via Marengo
Cagliari
Italy

Alfan Presekal
Intelligent Electrical Power Grids
Department of Electrical Sustainable Energy
Faculty of Electrical Engineering
Mathematics and Computer Science
Delft University of Technology
Delft, South Holland
The Netherlands

Saifur Rahman
Advanced Research Institute (ARI)
The Bradley Department of Electrical and
Computer Engineering
Virginia Tech
Arlington, VA
USA

Vetrivel Subramaniam Rajkumar
Intelligent Electrical Power Grids
Department of Electrical Sustainable Energy
Faculty of Electrical Engineering
Mathematics and Computer Science
Delft University of Technology
Delft, South Holland
The Netherlands

Hossein Ranjbar
School of Electrical and Mechanical Engineering
University of Adelaide
Adelaide
Australia

Hossein Saber
Department of Electrical Engineering
Sharif University of Technology
Tehran
Iran

Carlos Sabillon
Northland Power Inc.
Toronto, Ontario
Canada

Mahdieh S. Sadabadi
Department of Electrical and Electronic Engineering
University of Manchester
Manchester
United Kingdom

S.K. Sadanandan
DEWA R&D
Dubai
UAE

Amir Safdarian
Volue Oy
Helsinki
Finland

Talal Saleh
School of Technology and Innovations
University of Vaasa
Vaasa
Finland

Konrad Schmitt
Department of Electrical and Computer Engineering
Texas Tech University
Lubbock, TX
USA

Noel N. Schulz
School of Electrical Engineering & Computer Science
Washington State University
Pullman, WA
USA

Shamik Sengupta
Department of Computer Science and Engineering
University of Nevada
Reno, Reno
Nevada
USA

Miadreza Shafie-khah
Research and Innovation Division
Nowocert
Dublin
Ireland

Raj Mani Shukla
School of Computing and Information Science
Anglia Ruskin University
Cambridge
UK

A.K. Srivastava
Smart Grid Resiliency and Analytics Lab
West Virginia University
Morgantown, WV
USA

and

Smart Grid Demonstration and Research Investigation Lab
Washington State University
Pullman
WA
USA

Alexandru Ştefanov
Intelligent Electrical Power Grids
Department of Electrical Sustainable Energy
Faculty of Electrical Engineering
Mathematics and Computer Science
Delft University of Technology
Delft, South Holland
The Netherlands

José Luis Rueda Torres
Intelligent Electrical Power Grids
Department of Electrical Sustainable Energy
Faculty of Electrical Engineering
Mathematics and Computer Science
Delft University of Technology
The Netherlands

Riccardo Trevisan
University of Cagliari
Deptartment of Electrical and Electronic Engineering
Via Marengo
Cagliari
Italy

Mohammad N. Uddin
Department of Electrical and Computer Engineering
Lakehead University
Thunder Bay, Ontario
Canada

Bala Venkatesh
Electrical and Computer Engineering Department
Toronto Metropolitan University
Toronto, Ontario
Canada

Yu Wang
Chongqing University
Chongqing
China

Yang Weng
Department of Electrical
Computer and Energy Engineering
Arizona State University
Tempe, AZ
USA

Wanli Wu
South China University of Technology
School of Electric Power Engineering
Guangzhou
China

Chenhan Xiao
Department of Electrical
Computer and Energy Engineering
Arizona State University
Tempe, AZ
USA

Seyed Farhad Zandrazavi
School of Technology and Innovations
University of Vaasa
Vaasa
Finland

Jizhong Zhu
South China University of Technology
School of Electric Power Engineering
Guangzhou
China

Foreword (John D. McDonald)

Foresight is a virtue.

In the case of rapidly evolving electric power systems, foresight – *the ability to understand what will be needed in the future* – is also an imperative.

Stakeholders in electric power systems must gain at least a rudimentary understanding of the concepts and technologies that are destined to shape the way forward. The very culture of electric power system operations must adopt a holistic, horizontal approach that demolishes organizational and operational silos. Every technology investment must serve the broadest enterprise and operational needs and goals across the entire organization, while making sense in the context of foreseeable, future investments. To attract support, technological solutions require a sound business case. The more foresight applied, the stronger the business case.

Today's emerging technologies will change everything about how electric power is sustainably and securely generated, distributed, bought, and sold. This book covers an eye-opening array of future possibilities for cyber-physical power system operations and management, ranging from familiar yet still-evolving concepts and practices (e.g., Demand Response, Microgrids, Integration of Distributed Energy Resources, Big Data, Cybersecurity) to still-esoteric applications (e.g., artificial intelligence, machine learning, digital twin) to concepts that remain at the outskirts of practical implementation (e.g., Transactive Energy Systems, Blockchain-based Energy Trading, Quantum Computing).

To acknowledge that components of many electric power systems today still lack situational awareness and automation, while accepting that survival demands a clear path to becoming a cyber-physical power system, helps to frame the challenge posed by the complex and accelerating changes that are upon us. Every electric power system must find its place and future direction on this potentially daunting path. In advising stakeholders how to navigate among uncertainties, I've characterized this path as a continuum and dubbed it, "The Journey to Digital Transformation." This concept provides a simple, practical framework that may assist stakeholders in understanding and making use of the ideas and technologies so expertly articulated in this book.

In my simple construct, The Journey to Digital Transformation has five levels. Today, many power systems remain in a reactive mode (Level 1) with low situational awareness; customers must call in to report outages. Yet others have moved to a responsive mode (Level 2) with improved situational awareness and fault location that enables automated restoration and greater efficiencies. Leading power systems are now moving to a predictive mode (Level 3) in which they can more accurately assess actual demand to guide generation and predict the impact of weather and the risk of asset failure. On the not-too-distant horizon beckons a prescriptive mode (Level 4) that relies on an AI-driven application for the optimization and orchestration of all power system functions, from edge to cloud, to prevent and minimize the extent of outages. Rapidly coming into view is an autonomous mode (Level 5) that enables cyber-physical power systems to self-heal and self-provision to support operations with limited human intervention.

This book's authors have provided an eminently readable, practical approach that lights the way on this journey into the future of cyber-physical power systems. As a nonacademic engineer myself, I believe this accessible approach will broaden this book's readership and, thus, maximize its welcome influence.

Quite recently, I actually wrote that achieving Levels 4 and 5 on the Journey to Digital Transformation is "conceivable but dependent on further technological advancements." This may be true – at the moment. Yet with the present work, *Smart Cyber-Physical Power Systems*, it would appear we are on the very cusp of implementing the concepts and technologies that will realize the ultimate goals of our Journey to Digital Transformation. I suggest that readers devour this book with a sense of urgency. It describes our collective future.

John D. McDonald
Founder/CEO, JDM Associates, LLC, Duluth, GA, USA
IEEE Life Fellow, Member National Academy of Engineering
CIGRE Honorary Member

Foreword (Massoud Amin)

Energy systems are undergoing a transformation of historic significance. Once centralized and dependent on fossil fuels, these systems now face demands for decentralization, flexibility, and resilience. Electricity demand is projected to grow by 25% by 2030, while achieving net-zero emissions requires reducing energy-related CO_2 emissions by over 40% within the same period. These challenges are compounded by growing cybersecurity threats and the complexities of integrating renewable energy resources into aging infrastructure. This two-volume series, *Smart Cyber-Physical Power Systems: Challenges and Solutions*, provides a clear and actionable roadmap for addressing these issues, offering both foundational insights and forward-looking solutions.

Since 1998, I have had the privilege of working closely with Professor Saifur Rahman, initially during his tenure at the National Science Foundation and my leadership at the Electric Power Research Institute (EPRI). Professor Rahman's ability to integrate advanced research with practical solutions has been exemplary. Together with his coauthors, Dr. Ali Parizad and Dr. Hamid Reza Baghaee, he has created a series that combines academic depth with real-world applications. This series offers valuable insights for policymakers, engineers, researchers, and industry leaders.

Key Areas Addressed

Foundational Challenges

Volume 1 introduces the principles and challenges of transitioning to cyber-physical power systems (CPPS):

- System Architecture and Data Integration: Modern grids must manage decentralized energy resources (DERs), such as rooftop solar, battery storage, and electric vehicles, while ensuring operational stability. This volume examines the role of real-time analytics and advanced control frameworks in maintaining system balance and enhancing efficiency [1, 2].
- Cybersecurity and Resilience: The 2015 cyberattack on Ukraine's power grid exposed the vulnerabilities in interconnected energy systems. This volume outlines strategies for mitigating such risks, including hybrid anomaly detection, encryption protocols, and decentralized architectures [3].
- Self-Healing Grids: Building upon decades of research, including my work at EPRI, the authors explore self-healing grid technologies that can autonomously detect, isolate, and recover from faults to minimize disruptions caused by natural disasters or cyberattacks [4].

Advanced Solutions and Applications

Volume 2 explores emerging technologies and their applications in modernizing CPPS:

- Artificial Intelligence (AI) and Machine Learning (ML): AI and ML are transforming grid operations, enabling predictive maintenance, resource optimization, and load forecasting. These technologies reduce equipment downtime and enhance grid stability, as demonstrated by real-world applications [5, 6].
- Blockchain and Transactive Energy: Blockchain technology facilitates secure and decentralized energy trading systems, empowering communities while improving reliability. Case studies from Europe and Australia highlight the successful deployment of these systems [7].
- Quantum Computing: Quantum algorithms, such as the Quantum Approximate Optimization Algorithm (QAOA), address complex grid optimization problems that traditional computing cannot solve, thereby enhancing the integration of renewable energy and improving overall grid performance [8].

Resilience and Policy Integration

Energy resilience is critical as climate events, cyber threats, and geopolitical risks increase. This series offers strategies to ensure system adaptability:

- Localized Microgrids: Microgrids provide uninterrupted power during grid outages, making them essential for critical infrastructure such as hospitals and emergency response centers. This book explores their deployment, operation, and integration within broader energy systems [9, 10].
- Global Standards and Policy Harmonization: The development of CPPS requires harmonized technical standards and regulatory frameworks. International collaboration is vital for aligning policies and enabling cross-border energy systems [11, 12].

Emerging Trends and Innovations

The authors present a forward-looking perspective, addressing:

- IoT in Smart Buildings: IoT-enabled systems in buildings can reduce energy waste by up to 30%, offering significant sustainability and cost-efficiency benefits [13].
- Digital Twins: Virtual models of energy systems allow operators to simulate scenarios, optimize performance, and predict outcomes, improving reliability and resilience.
- Decarbonization Pathways: This series explores practical strategies for transitioning to renewable energy systems, ensuring grid stability while meeting ambitious climate targets.

Economic and Equity Impacts

The transition to CPPS offers significant economic benefits, including cost savings from predictive maintenance, reduced outage expenses, and the creation of jobs in renewable energy and smart grid technologies. Furthermore, this transformation must address energy equity. CPPS technologies help reduce energy poverty by expanding access to clean and reliable power for underserved communities, as demonstrated by microgrid projects in rural Africa and rooftop solar installations in low-income neighborhoods across the United States.

Lessons from Past Failures

Grid failures, such as the Texas power crisis in 2021, underscore the importance of resilience and proactive planning. This series addresses how CPPS can prevent similar failures by integrating predictive analytics, fault-tolerant designs, and decentralized systems that reduce reliance on centralized grid infrastructure.

Why This Series Matters

This two-volume series bridges the gap between theory and practice, offering practical solutions for modernizing energy systems. It addresses the challenges of increasing energy demand, climate change, and cybersecurity threats, while providing a framework for building systems that are adaptive, resilient, and secure. By including lessons from past failures, real-world case studies, and actionable strategies, this work ensures its relevance to both current challenges and future innovations.

Professor Rahman and his coauthors have produced a resource of exceptional quality, reflecting decades of dedicated expertise. This series is indispensable for anyone seeking to understand and shape the future of energy systems. It is my privilege to support this important endeavor.

Massoud Amin, DSc
CTO, Renewable Energy Partners (REP) | https://renewablenrgpartners.com
President & Chairman, Energy Policy & Security (EPS) Associates | https://eps-associates.com
Professor Emeritus, former Director & Honeywell H.W. Sweatt Chair in Technological Leadership | University of Minnesota | cse.umn.edu/ece
Fellow, IEEE and ASME

References

1 International Energy Agency (2023). Global Electricity Demand Outlook. *World Energy Outlook*.
2 National Renewable Energy Laboratory (2020). Distributed Energy Resource Integration. NREL Report No. 73801, 2020.
3 Cybersecurity and Infrastructure Security Agency (2016). Ukraine Cyberattacks 2015–2016: Lessons for Grid Resilience. CISA Report.
4 Amin, S.M. and Wollenberg, B.F. (2005). Toward a smart grid: power delivery for the 21st century. *IEEE Power and Energy Magazine* 3 (5): 34–41.
5 U.S. Department of Energy (2020). AI in energy systems. DOE Report, 2020.
6 IBM (2021). *Quantum Computing in Energy Systems*. IBM Research Publications.
7 Energy Blockchain Consortium (2023). Blockchain in energy applications. Consortium Report, 2023.

8 National Renewable Energy Laboratory (2022). Quantum computing and energy systems. NREL Report, 2022.

9 World Economic Forum (2022). Harmonizing global energy policies. WEF Report, 2022.

10 Amin, S.M. (2005). Critical infrastructure resilience: strategies for secure energy systems. *Proceedings of the IEEE* 93 (5): 861–875.

11 U.S. Department of Energy (2021). IoT for smart buildings. DOE Report, 2021.

12 National Renewable Energy Laboratory (2021). Digital twins in energy systems. NREL Report No. 77424, 2021.

13 Amin, S.M. (2012). Smart grid security, privacy, and resilient architectures: Opportunities and challenges. In *Power and Energy Society General Meeting, 2012 IEEE*, IEEE, pp. 1–2.

Preface for Volume 1: Smart Cyber-Physical Power Systems: Fundamental Concepts, Challenges, and Solutions

The integration of cyber-physical systems (CPSs) into power systems is heralding a new era of energy management, where intelligent, efficient, and resilient technologies are transforming the landscape of power generation, distribution, and consumption. Volume 1 of this book embarks on a comprehensive journey through the foundational concepts and challenges that shape smart cyber-physical power systems (CPPSs), which are at the forefront of this revolution.

In this volume, we begin by establishing the core theoretical foundations that underpin the integration of digital technologies into traditional power systems. From the basic principles of smart grids and microgrids to the emerging concept of the Internet of Energy, the chapters delve into the essential frameworks that support the evolution of modern power infrastructures. At the same time, the book addresses the multifaceted challenges that come with this transformation. These challenges include grid modernization, cybersecurity vulnerabilities, system resiliency, and the integration of renewable energy sources – all of which demand innovative solutions.

The chapters in this volume provide a structured exploration of the obstacles faced by practitioners and researchers in the field. Through detailed discussions on topics such as global demand response strategies, microgrid architectures, and advanced distributed control, we seek to equip readers with the necessary insights to understand and navigate the complexities of CPSs. By drawing from the latest advancements in digital technologies, this volume serves as both a comprehensive guide to the core concepts of smart power systems and a critical resource for understanding the challenges that lie ahead.

As the world pivots toward more sustainable and efficient energy systems, this volume provides the foundational knowledge needed to tackle the myriad challenges facing the power sector today. With contributions from experts around the world, it presents a compelling vision for how smart cyber-physical power systems will define the future of energy. Through this volume, we invite you to explore the building blocks of these transformative systems, setting the stage for the innovations and solutions that will shape the energy landscape in the years to come.

Join us on a journey through the landscape of Smart Cyber-Physical Power Systems (CPPSs), where foundational concepts are explored and structural challenges are addressed. Volume 1 lays the groundwork for understanding the integration and optimization of modern energy systems, setting the stage for the advanced solutions detailed in Volume 2. This volume introduces essential building blocks, paving the way for transformative technologies such as AI/ML, Blockchain, and IoT to revolutionize efficiency, security, and sustainability in modern power systems.

January 2025

Ali Parizad
California, USA

Hamid Reza Baghaee
Tehran, Iran

Saifur Rahman
Virginia, USA

Acknowledgments

We would like to extend our gratitude to the Wiley-IEEE Press staff, especially Nandhini Karuppiah, Victoria Bradshaw, and Mary Hatcher, for their invaluable support and dedication throughout the publication process of this book. Their expertise, patience, and commitment have been essential in bringing the contents of this work to light.

We also extend special thanks to Deenadayalu Govindanagaraj and Sathishwaran our Content Refinement Specialists, for their meticulous copyediting and proofreading efforts. Their precision and dedication have significantly enhanced the clarity and quality of this publication.

Their combined effort in coordinating the review process, managing the production stages, and ensuring the highest standards of quality has been instrumental in crafting a publication that we are all proud to be associated with. Their tireless work behind the scenes has made this book possible, and for that, we are immensely thankful.

Thank you for your perseverance, attention to detail, and unwavering support in navigating the complex pathways towards publication.

Ali Parizad
Hamid Reza Baghaee
Saifur Rahman

1

Overview of Smart Cyber-Physical Power Systems: Fundamentals, Challenges, and Solutions

Ali Parizad[1], Hamid Reza Baghaee[2], and Saifur Rahman[1]

[1]*Advanced Research Institute (ARI), Virginia Tech, National Capital Region, Arlington, VA, USA*
[2]*Faculty of Electrical and Computer Engineering, Tarbiat Modares University, Tehran, Iran*

1.1 Introduction

This first chapter serves as an introductory guide to the world of smart cyber-physical power systems (CPPSs), laying the groundwork for the comprehensive discussions in the subsequent 46 chapters of this book. It presents an accessible overview of how modern power systems are evolving, integrating advanced technologies to meet the demands of a changing energy landscape.

At the heart of this chapter is an exploration of modern power grids replacing traditional systems. These smart grids bring together a range of technologies, such as artificial intelligence (AI), machine learning (ML), and sophisticated data analysis methods, to improve how we generate, distribute, control, and use electricity. The chapter also briefly introduces concepts like blockchain, which adds security to energy transactions, and big data, crucial for handling the vast amounts of information these systems produce.

Furthermore, the chapter discusses how the Internet of Things (IoT)—a network of interconnected devices—plays a vital role in making these power systems more efficient and responsive. It also touches on the role of information theory in understanding these complex systems and quantum computing's potential to solve complex problems more efficiently.

This chapter aims to provide readers with a clear understanding of smart power systems' current state and future potential. It offers insights into these systems' challenges and the innovative solutions being developed. This is an essential read for anyone interested in the evolving field of energy systems, offering a blend of foundational knowledge and a glimpse into future advancements.

1.2 Structural Overview and Roadmap of the Book

This book is structured into four distinct sections, each encompassing a series of chapters that delve deep into the various facets of smart CPPSs, as illustrated in Figure 1.1. This visual representation serves as an overarching guide, showcasing elements such as microgrids (MGs), IoT, blockchain, cloud computing and cyber layers, big data analytics, ML, and AI, among others. This division of the book is designed to provide a comprehensive and coherent journey through the expansive and dynamic field, offering detailed insights and expert analyses on a range of pertinent topics. Each section is thematically organized to address different aspects of the field, from fundamental concepts to advanced solutions, ensuring a well-rounded understanding for the readers. The

Smart Cyber-Physical Power Systems: Fundamental Concepts, Challenges, and Solutions, Volume 1, First Edition.
Edited by Ali Parizad, Hamid Reza Baghaee, and Saifur Rahman.

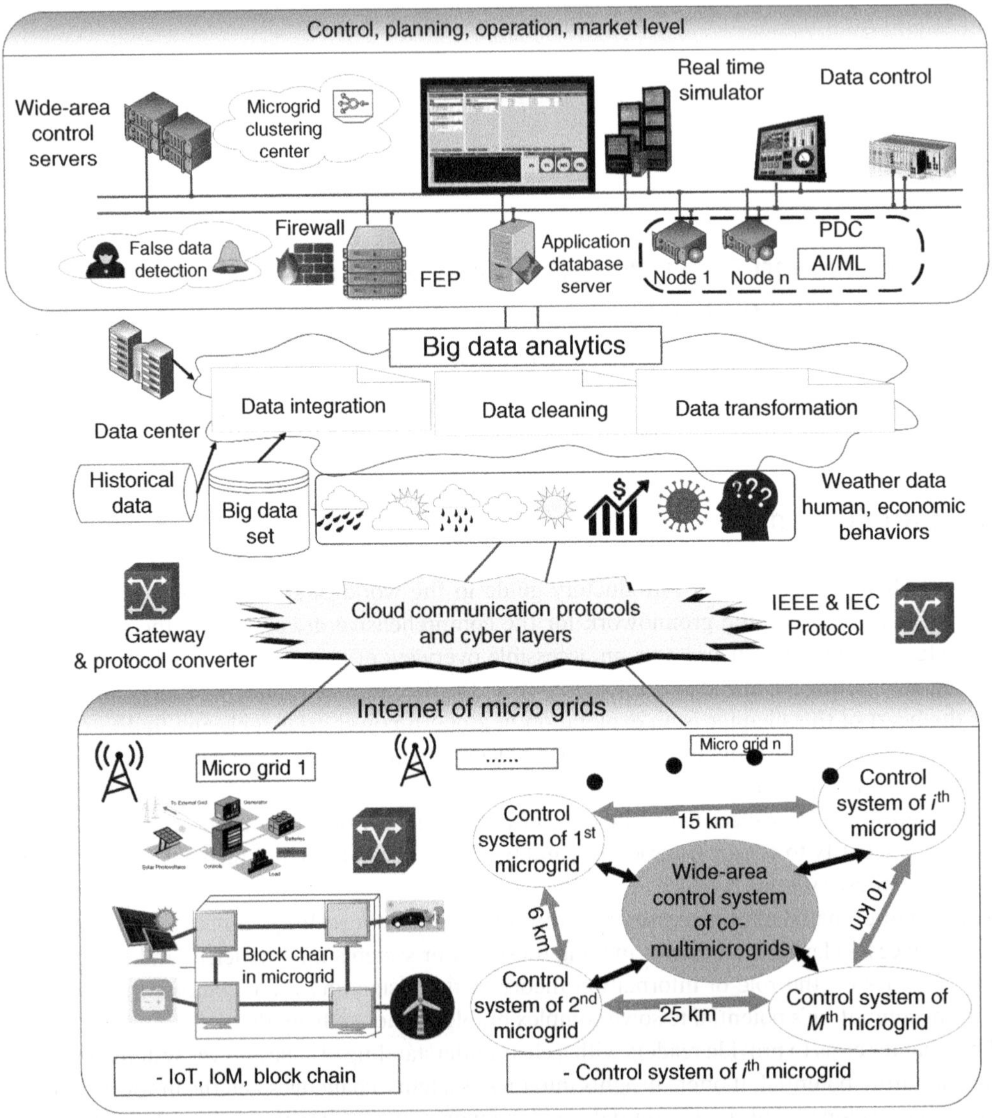

Figure 1.1 An overview of a smart CPPS.

figure visually encapsulates the breadth and interconnectivity of the topics covered. It enhances the reader's comprehension of how these diverse technologies and concepts interplay within smart CPPSs.

Section I: Introduction and Fundamental Concepts

This initial section lays the foundational groundwork, introducing readers to the core concepts and global perspectives underpinning smart CPPSs. This includes an overview of the fundamentals, challenges, and solutions and a deep dive into global demand response dynamics.

Chapter 1, titled "Overview of Smart Cyber-Physical Power Systems; Fundamentals, Challenges, and Solutions," sets the stage by elucidating the pivotal role of power systems in the economic and public safety sectors of countries. This chapter presents an insightful exploration of how the technological revolution has reshaped modern power systems, introducing paradigms such as the smart grid ecosystem (SGE) and cyber-physical systems (CPSs). The integration of information and communication technologies (ICT) into power systems, transforming them into advanced CPSs, is examined in detail. This transformation is presented as a double-edged sword: on the one hand, bolstering national economies and enhancing public safety; on the other hand, introducing a set of new challenges that must be navigated with care and expertise.

Following this comprehensive introduction, Chapter 2 delves into "Global Demand Response Status: Potentials, Barriers, and Solutions," offering a deep dive into the world of AI and ML in the context of global demand response. This chapter provides a critical analysis of the potential and barriers of these technologies, exploring how they can be leveraged for improved efficiency and effectiveness in managing demand response programs globally.

Together, these chapters lay a solid foundation for the book, setting the context for the intricate discussions that follow. They collectively aim to provide the reader with a clear understanding of the current landscape of smart power systems, their challenges, and the innovative solutions that are being developed to address them.

Section II: Cyber-Physical Structure of Power Systems

In Section II, the focus shifts to the intricate structure and operational dynamics of modern power systems, emphasizing their evolution into sophisticated cyber-physical entities. This section delves into various aspects of power systems, encompassing advanced management strategies, infrastructural developments, and innovative control mechanisms. Topics range from smart energy management in MGs and smart city infrastructures to integrating electric vehicles (EVs) and energy efficiency in smart buildings. This section elucidates the architectural and operational nuances of these advanced systems. Chapter 3, titled "Smart Power/Energy Management and Optimization in Microgrids," introduces a ML-based approach for efficient energy management within MGs. This chapter addresses the complexities of managing energy resources in a MG, highlighting the potential of ML algorithms in optimizing power distribution and consumption. Following this, Chapter 4, "Smart City Energy Infrastructure as a CPS of Systems: Planning, Operation, and Control Processes," presents an integrated perspective on smart city energy infrastructure. This chapter explores the multifaceted aspects of planning, operating, and controlling the energy systems within smart cities, demonstrating how they form a complex system of systems (SoS). Chapter 5, "Metaverse Local Energy Market in Smart City: A Descriptive Model and Strategic Development Analysis," brings to the forefront the innovative concept of the Metaverse in shaping local energy markets in smart cities. The chapter provides a descriptive model and conducts a strategic analysis of this emerging technology's role in energy market dynamics. In Chapter 6, the book introduces "Cooperative and Distributed Control Strategies of Microgrids," which discusses the significance of distributed control strategies in AC MG operation. This chapter underscores the importance of cooperative approaches in ensuring the stability and efficiency of MGs.

Further expanding the discussion, Chapter 7, "Interconnected Microgrid Systems: Architecture, Hierarchical Control, and Implementation," delves into the design and control strategies of interconnected MG systems. This chapter examines the architectural complexities and hierarchical control mechanisms pivotal to the successful implementation of these systems. Chapter 8, "Internet of Energy and Internet of Microgrids (IoE, IoM)," presents a forward-looking view of the IoE and the internet of MGs. This chapter examines the concept of interconnecting energy systems at various scales, from local MGs to broader energy networks, highlighting how digital communication technologies are revolutionizing energy distribution and management. In Chapter 9, "Voltage Regulation and Reactive Power Optimization for Integration of Distributed Energy Resources into Smart

Grids," the focus is the technical challenges and solutions associated with integrating distributed energy resources (DERs). This chapter discusses advanced strategies for voltage regulation and reactive power optimization, essential for maintaining grid stability in the presence of diverse energy sources. Chapter 10, "The Role of Data Analysis in Hosting Capacities of Distribution Power Systems for Electric Vehicles," explores the integration of EVs into existing power systems. This chapter covers different methods and research regarding forecasting power demands of EVs, along with uncertainties and their impact on the power grid are discussed. Following this, Chapter 11, "Energy Efficiency in Smart Buildings through IoT Sensor Integration: Implementation of BEMOSS™ Platform," delves into the role of IoT in enhancing energy efficiency in smart buildings. The chapter illustrates the implementation of the BEMOSS Platform, showcasing how IoT sensor integration can lead to significant improvements in building energy management. Concluding Section II, Chapter 12, "Optimal Dispatch of Smart Energy System Based on Cyber-Physical-Social Integration," presents a holistic approach to energy system management. This chapter discusses the integration of CPSs with social aspects to achieve optimal energy dispatch, highlighting the importance of considering human factors in energy system design and operation.

Through these chapters, Section II offers a comprehensive view of the cyber-physical structures in power systems, covering a wide spectrum from the IoE to the integration of EVs and smart building technologies. Each chapter contributes to a deeper understanding of the complexities and innovations in the field, setting the stage for the discussions on challenges and solutions in the subsequent sections.

Section III: Challenges

This section addresses the myriad challenges faced in smart CPPSs. It covers crucial areas such as self-healing in power distribution, system resiliency, cybersecurity threats, anomaly detection, and the evolving landscape of transactive energy systems. This section is pivotal in understanding the vulnerabilities and dynamic complexities of modern power systems. Chapter 13, "Power Distribution Systems Self-Healing," begins this section by addressing one of the most crucial aspects of modern power systems: the ability to recover from unplanned power outages. This chapter explores critical technologies and strategies that enable power systems to detect, diagnose, and rectify faults autonomously, thereby enhancing their resilience and reliability. In Chapter 14, "Resiliency, Reliability, and Security of Smart Cyber-Physical Power Systems," the discussion shifts to a broader perspective, focusing on the escalating reliance on electricity. This chapter examines the vulnerabilities of these systems and the importance of fortifying them against various threats via ML techniques to ensure their continuous and secure operation. Chapter 15, "Cyber Attacks on Power Systems," offers a comprehensive analysis of the state-of-the-art and emerging trends in cybersecurity for power systems. It discusses the different types of cyber threats and the methodologies to detect and mitigate them, emphasizing the criticality of cybersecurity in modern power systems.

Following this, Chapter 16, "Vulnerabilities of ML Algorithms to Adversarial Attacks for Cyber-Physical Power Systems," explores the intersection of AI and cybersecurity. This chapter delves into the vulnerabilities of AI techniques in power systems, particularly focusing on how ML algorithms can be susceptible to adversarial attacks. Chapter 17, "Synchrophasor Data Anomaly Detection for Wide-Area Monitoring and Control in Cyber-Power Systems," then shifts focus to synchrophasor technology, a key component in wide-area monitoring and control. The chapter discusses the challenges and methodologies for anomaly detection in synchrophasor data, a critical element for ensuring the stability and security of power systems. The exploration of challenges in Section III continues with Chapter 18, "Application of State Observers and Filters in Protection and Cyber-Security of Power Grids." This chapter delves into advanced techniques for enhancing the protection and cybersecurity of power systems. It discusses the use of state observers and filters, crucial tools in detecting and mitigating potential cyber threats, and ensuring the integrity of power grid operations. Chapter 19, "Anomaly Detection and Mitigation in Cyber-Physical Power Systems Based on Hybrid Deep Learning and Attack Graphs," addresses the complex task of identifying and countering anomalies in power systems. The chapter explores hybrid deep learning models and attack graphs as innovative methods for detecting unusual patterns that may indicate cybersecurity threats, thereby contributing to the fortification of power systems against cyberattacks. In Chapter 20, "Attack Detection and Countermeasures at Edge Devices," the focus is on the security of edge devices in power systems. It examines the vulnerabilities of these devices and presents strategies for detecting attacks and deploying countermeasures, emphasizing the importance of securing all components of the cyber-physical infrastructure. Chapter 21, "Privacy-Preserving Outage Detection in Modern Distribution Grids: Challenges and Opportunities," brings to light the challenges of maintaining privacy while detecting outages in power distribution networks (DNs). It outlines the balance between efficient outage management and the preservation of consumer privacy, highlighting the potential solutions and opportunities in this domain. Chapter 22, "Transactive Energy Management and Distribution System Reform Using Market Concepts," shifts the discussion to the emerging concept of transactive energy. It explores how market-driven approaches can reform energy management and distribution systems, presenting challenges and solutions for implementing transactive energy models. The discussion then moves to Chapter 23, "Transactive Energy Systems in Decentralized Autonomous Renewable Energy Communities." This chapter delves into the role of transactive energy systems in managing decentralized, autonomous renewable energy communities, discussing the challenges and potential of these innovative systems in enhancing energy efficiency and sustainability. In Chapter 24, "Transactive Coordination Paradigm for Efficient Charging Management of Plug-in Electric Vehicles in Future Distribution Networks," the focus is on integrating EVs into the power grid. The chapter discusses the challenges and solutions related to the efficient management of EV charging within the transactive energy framework, highlighting its importance in future DNs. Chapter 25, "Optimal Peer-to-Peer Energy Trading Using ML: Architecture, Strategies, and Algorithms," explores the application of ML in peer-to-peer energy trading. This chapter presents the architectural design, strategies, and algorithms that enable optimal energy trading among peers, showcasing the potential of AI in revolutionizing energy market dynamics.

The narrative of challenges in smart CPPSs extends into Chapter 26, "Optimal Peer-to-Peer Power Sharing in DC Islanded Microgrids." This chapter focuses on the intricacies of energy sharing in isolated MG environments, discussing optimal power dispatch strategies that enable efficient and sustainable peer-to-peer energy transactions in islanded systems, where the MGs operate independently from the power grid. Chapter 27, "Blockchain-Based Energy Trading Employing Hyperledger and Anomaly Detection Algorithms," explores the revolutionary integration of blockchain technology in the energy sector. This chapter presents an in-depth analysis of how blockchain, specifically the Hyperledger platform, can be employed to enhance the transparency, efficiency, and security of energy trading. Additionally, it delves into the role of anomaly detection algorithms in safeguarding these transactions, ensuring a reliable and secure trading environment. In Chapter 28, "Optimal Coordination of VSC-Interfaced Subsystems to Safeguard the Frequency Performance of Cyber-Physical Power Systems," the discussion shifts to the technical aspects of maintaining frequency stability in power systems. This chapter examines the challenges and solutions related to the coordination of voltage source converter (VSC) interfaced subsystems. It highlights the importance of optimal coordination strategies in preserving the frequency performance and overall stability of CPPSs, which is crucial in the face of increasing integration of renewable energy sources and dynamic load conditions.

With these chapters, Section III thoroughly examines the vast array of challenges facing modern CPPSs. From the complexities of energy sharing in isolated and DC MGs to the cutting-edge applications of blockchain in energy trading, each chapter contributes to a deeper understanding of the obstacles and hurdles in the field, paving the way for exploring solutions in the final section of the book.

Section IV: Solutions and Tools

In the final section, we explore a range of solutions and tools designed to counter the challenges identified earlier. This section is rich with innovative approaches and cutting-edge research findings, from the application of AI and ML, digital twin technologies, and quantum computing to data-driven methods for ensuring system stability and security. In Chapter 29, "Information Theory and Gray Level Transformation Techniques in Detecting False Data Injection Attack on Power System State Estimation," the focus is on a novel semi-supervised learning approach, underlining the use of information theory and gray-level transformation in cybersecurity. The journey continues with Chapter 30, where "AI and ML Applications in Modern Power Systems" are dissected to reveal AI/ML's impact on operational awareness and decision-making in challenging power system conditions. In Chapter 31, "Physics-Informed Deep Reinforcement Learning-Based Control in Power Systems," the exploration centers on integrating physics-informed neural networks within deep reinforcement learning (RL), highlighting grid control's evolving landscape. "Digital Twin Approach toward Modern Power Systems," presented in Chapter 32, unveils the potential and integration of digital twins, a testament to digitalization's potency in modern power systems. Chapter 33, "Application of AI and ML Algorithms in Power System State Estimation," addresses the pressing challenge of state estimation in DNs, emphasizing deep learning's transformative role.

Moving to Chapter 34, "ANN-Based Scenario Generation Approach for Energy Management of Smart Buildings," the narrative shifts to artificial neural networks in generating renewable energy management scenarios, a critical aspect of smart building operations. The discourse in Chapter 35, "Protection Challenges and Solutions in Power Grids by AI/ML," pivots to the transformative impact of AI/ML in addressing modern power grid protection systems' complexities. Chapter 36, "Deep and RL for Active Distribution Network Protection," introduces an advanced deep RL-based system for MG protection, spotlighting fault detection. The essential role of big data in power system management is meticulously unfolded in Chapter 37, "Handling and Application of Big Data in Modern Power Systems for Planning, Operation, and Control Processes," and further elaborated in Chapter 38, "Handling and Application of Big Data in Modern Power Systems for Situational Awareness and Operation," emphasizing its application in power flow sensitivity analysis. Chapter 39's "Data-Driven Methods in Modern Power System Stability and Security" introduces methods to maintain and enhance stability in power systems, integrating stochastic modeling and control theory. Quantum technology's revolutionary potential is the crux of Chapter 40, "Application of Quantum Computing for Power Systems," highlighting its capacity to transform computational challenges in smart CPPSs.

The narrative of load forecasting is split between Chapter 41, "High-Resolution Building Level Load Forecasting - Part 1: Principles and Concepts," and Chapter 42, "High-Resolution Building Level Load Forecasting - Part 2: Simulation and Experimental Results," offering both theoretical foundations and practical applications. In Chapter 43, "PV Energy Forecasting Applying ML Methods Targeting Energy Trading Systems," the focus is on using ML for PV energy forecasting, enhancing energy market participation. Chapter 44, "An Intelligent RL-Based Load Shedding to Prevent Voltage Instability," explores RL's application in optimizing load-shedding procedures for voltage stability. "Deep Learning Techniques for Solving Optimal Power Flow Problems," presented in Chapter 45, introduces learning-based paradigms for efficient DER operation engaging DNN to predict the solutions of an OPF. Chapter 46, "Research on the Intelligent Prediction of Spatial-Temporal Dynamic Frequency Response and the Performance Evaluation," delves into frequency prediction in power systems, utilizing LSTM and average system frequency (ASF) analysis. Concluding the section, Chapter 47, "The Emerging Technologies and the Future Trends in Cyber-Physical Power Systems: Towards a New Era of Innovations," synthesizes the current state and prospective advancements in AI, ML, blockchain, IoT, and information theory within the context of CPPSs.

Each chapter of these four sections, authored by experts in their respective fields, offers in-depth analysis and unique perspectives, providing readers with a comprehensive understanding of smart power systems' current state and future potential. This book aims to offer insights into the challenges of smart power systems and the cutting-edge solutions being developed, making it an indispensable resource for anyone interested in this rapidly evolving domain.

1.3 General Concepts of the Cyber-Physical Systems

1.3.1 The Emerging Technologies

Before discussing CPS, let's talk about the hype Gartner chart. Hype cycles and priority matrices provide a quick overview of the relative market promotion and perceived worth of innovations. They identify overhyped areas, predict the maturation dates of inventions and trends, and offer practical guidance to assist enterprises in making adoption decisions. Gartner's hype cycle is one of the most well-known and significant consultant models for counseling big businesses on their technology strategy. The original hype cycle model, which practitioners and researchers widely use to support and validate specific technological investment or non-investment decisions, has not yet undergone a thorough academic examination concerning its theoretical underpinnings, methodological approach, or empirical validity. Furthermore, three technologies' hype cycles have been tested empirically and contrasted with Gartner's forecasts and assessments. A technology (or related service and discipline innovation) passes through several stages on its path to productivity (see Figure 1.2) [2]:

- **Innovation trigger (formerly called technology trigger)**: The hype cycle starts when a breakthrough, public demonstration, product launch, or other event generates press and industry interest in a technology innovation.
- **Peak of inflated expectations**: A wave of "buzz" builds and the expectations for this innovation rise above the current reality of its capabilities. In some cases, an investment bubble forms, as happened with the web and social media.
- **Trough of disillusionment:** Inevitably, impatience for results replaces the original excitement about potential value. Problems with performance, slower-than-expected adoption, or a failure to deliver financial returns in the anticipated time lead to missed expectations and disillusionment sets in.
- **Slope of enlightenment**: Some early adopters overcome the initial hurdles, begin to experience benefits, and recommit efforts to move forward. Organizations draw on the experience of the early adopters. Their understanding grows about where and how the innovation can be used to good effect and, just as significantly, where it brings little or no value.

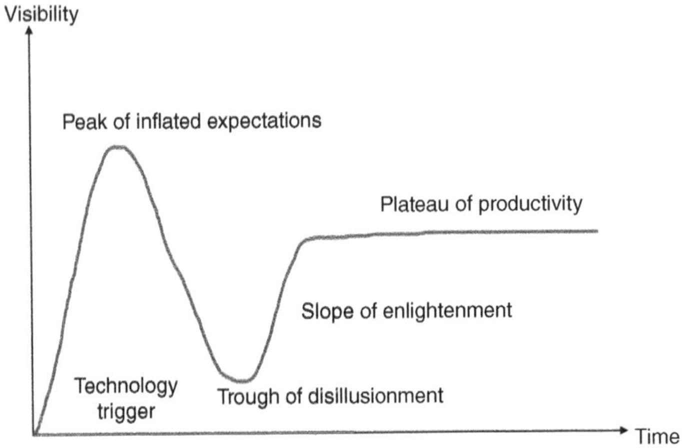

Figure 1.2 The stages of the changes in the hype gartner chart (Source: [1]/Wikimedia/Public domain).

- **Plateau of productivity:** With the real-world benefits of the innovation demonstrated and accepted, growing numbers of organizations feel comfortable with the now significantly reduced levels of risk. A sharp rise in adoption begins (resembling a hockey stick when shown graphically), and penetration accelerates rapidly due to productive and useful value.

Let's look at the hype Gartner chart for emerging technologies in 2018. The 35 must-watch technologies represented on the Gartner Inc. Hype Cycle for Emerging Technologies, 2018 revealed five distinct emerging technology trends that will blur the lines between humans and machines. Emerging technologies, such as AI, play a critical role in enabling companies to be ubiquitous, always available, and connected to business ecosystems to survive in the near future. "Business and technology leaders will continue to face rapidly accelerating technology innovation that will profoundly impact how they engage with their workforce, collaborate with their partners, and create products and services for their customers," said Mike J. Walker, research vice president at Gartner. "CIOs and technology leaders should always be scanning the market along with assessing and piloting emerging technologies to identify new business opportunities with high impact potential and strategic relevance for their business." The hype cycle for emerging technologies report is the longest-running annual Gartner hype cycle, providing a cross-industry perspective on the technologies and trends that business strategists, chief innovation officers, R&D leaders, entrepreneurs, global market developers, and emerging technology teams should consider in developing emerging technology portfolios. The hype cycle for emerging technologies is unique among most Gartner hype cycles because it garners insights from over 2000 technologies into a concise set of 35 emerging technologies and trends. This hype cycle specifically focuses on the set of technologies showing promise in delivering a high degree of competitive advantage over the next 5 to 10 years (see Figure 1.3).

Based on the hype cycle for emerging technologies, 2018, there are five emerging technology trends (that we follow in the themes of this book with an emphasis on CPPSs) such as democratized AI, digitalized ecosystems, do-it-yourself biohacking, and transparently immersive experiences:

- **Democratized AI:** AI technologies will be virtually everywhere over the next ten years. While these technologies enable early adopters to adapt to new situations and solve problems that have not been encountered previously, these technologies will become democratized to the masses. Movements and trends like cloud computing, the "maker" community, and open source will

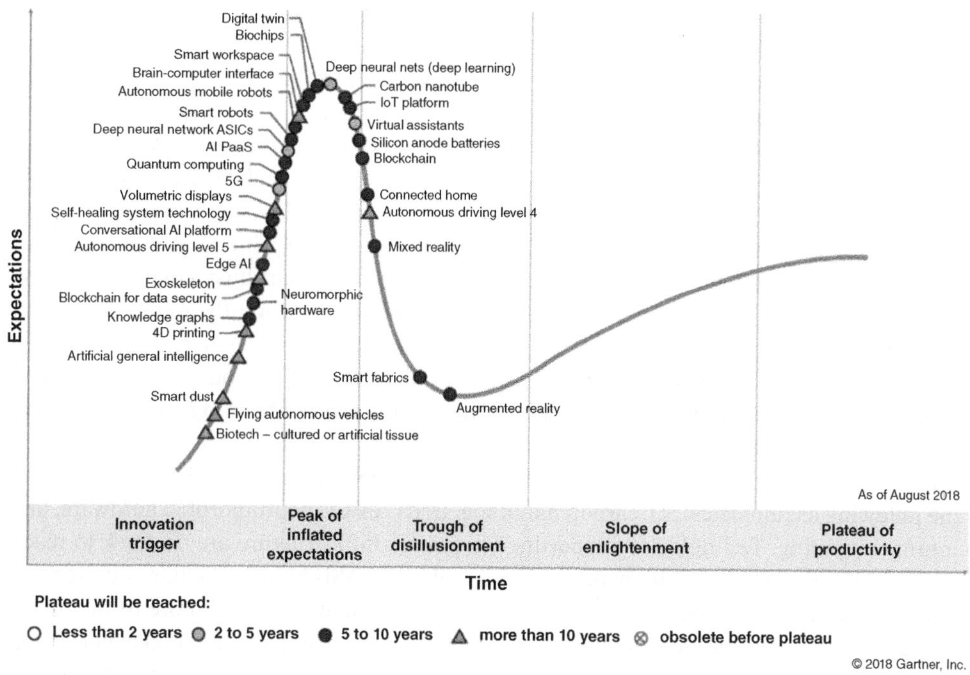

Figure 1.3 Hype cycle for emerging technologies, 2018 (Source: [3]/Gartner).

eventually propel AI into everyone's hands. The following technologies enable this trend: AI Platform as a Service (PaaS), artificial general intelligence, autonomous driving (Levels 4 and 5), autonomous mobile robots, conversational AI platform, deep neural nets, flying autonomous vehicles, smart robots, and virtual assistants. "Technologies representing democratized AI populate three out of five sections on the hype cycle, and some of them, such as deep neural nets and virtual assistants, will reach mainstream adoption in the next two to five years," said Mr. Walker. "Other emerging technologies of that category, such as smart robots or AI PaaS, are also moving rapidly through the hype cycle, approaching the peak and will soon have crossed it."

- **Digitalized ecosystems:** Emerging technologies require revolutionizing the enabling foundations that provide the volume of data needed, advanced computing power, and ubiquity-enabling ecosystems. The shift from compartmentalized technical infrastructure to ecosystem-enabling platforms is laying the foundations for entirely new business models, forming the bridge between humans and technology. This trend is enabled by the following technologies: blockchain, blockchain for data security, digital twin, IoT platform, and knowledge graphs. "Digitalized ecosystem technologies are making their way to the Hype Cycle fast," said Walker. "Blockchain and IoT platforms have crossed the peak by now, and we believe they will reach maturity in the next 5 to 10 years, with digital twins and knowledge graphs on their heels."
- **Do-it-yourself biohacking:** Over the next decade, humanity will begin its "transhuman" era. Depending on lifestyle, interests, and health needs, biology can then be hacked. Biohacking falls into four categories: technology augmentation, nutrigenomics, experimental biology, and grinder biohacking. However, questions remain about how far society is prepared to accept these applications and what ethical issues they create. This trend is enabled by the following technologies: biochips, biotech—cultured or artificial tissue, brain-computer interface, augmented reality,

mixed reality, and smart fabrics. Emerging technologies in do-it-yourself biohacking are moving rapidly through the hype cycle. Mixed reality is reaching the trough of disillusionment, and augmented reality has almost reached the bottom. Those pioneers will be followed by biochips, which have just reached the peak and will have moved on to the plateau in 5 to 10 years.

- **Transparently immersive experiences:** Technology will continue to become more human-centric to the point where it will introduce transparency between people, businesses, and things. These technologies extend and enable smarter living, work, and other spaces we encounter. This trend is enabled by the following technologies: 4D printing, connected home, edge AI, self-healing system technology, silicon anode batteries, smart dust, smart workspace, and volumetric displays. "Emerging technologies representing transparently immersive experiences are mostly on their way to the peak or—in the case of silicon anode batteries—just crossed it," said Mr. Walker. "The smart workspace has moved along quite a bit and is about to peak in the near future." Ubiquitous infrastructure is no longer in the way of obtaining an organization's goals. The advent and mass popularity of cloud computing and its many variations have enabled an always-on, available, limitless infrastructure computing environment. This trend is enabled by the following technologies: 5G, carbon nanotube, DNN ASICs, neuromorphic hardware, and quantum computing. Technologies supporting ubiquitous infrastructure are on track to reach the peak and move fast along the hype cycle. 5G and DNN ASICs, in particular, are expected to reach the plateau in the next two to five years. Gartner clients can read more in the report "Hype Cycle for Emerging Technologies, 2018." This research is part of the Gartner Trend Insight Report, "2018 Hype Cycles: Riding the Innovation Wave". With profiles of technologies, services, and disciplines spanning over 100 hype cycles, this Trend Insight Report is designed to help CIOs and IT leaders respond to the opportunities and threats affecting their businesses, take the lead in technology-enabled business innovations, and help their organizations define an effective digital business strategy. Additional analysis on emerging technologies will be presented during the Gartner Symposium/ITxpo, the world's most important gathering of CIOs and other senior IT executives. IT executives rely on these events to gain insight into how their organizations can use IT to overcome business challenges and improve operational efficiency. Follow news and updates from the events on Twitter using #GartnerSYM.

The disruptive technologies in this hype cycle will affect business and society through 2033. Technology innovation leaders can use them to harness emergent AI, enhance developer experience (DevX), exploit the pervasive cloud, and deliver human-centric security and privacy programs. They fit into four main themes: emergent AI, DevX, pervasive cloud, and human-centric security and privacy. The 2023 Gartner hype cycle (Figure 1.4) identifies 25 must-know emerging technologies designed to help enterprise architecture and technology innovation leaders: Evaluate the business impact of emerging technologies. Examine and explore potentially transformative technologies. Strategize how to benefit from these technologies. These technologies are expected to impact business and society significantly over the next 2 to 10 years. They will especially enable CIOs and IT leaders to deliver on the promise of digital business transformation.

Four Gartner hype cycle themes to think about in 2023 and beyond theme (that are mostly covered in this book with emphasis on CPPSs, too) are [4]:

- ***Emergent AI***
  These technologies provide opportunities for sustainable differentiation and greater workforce productivity. While generative AI can enable competitive differentiation, several other emerging AI techniques offer immense potential to enhance digital customer experiences, make better business decisions, and distinguish yourself from your competition. An example of emergent

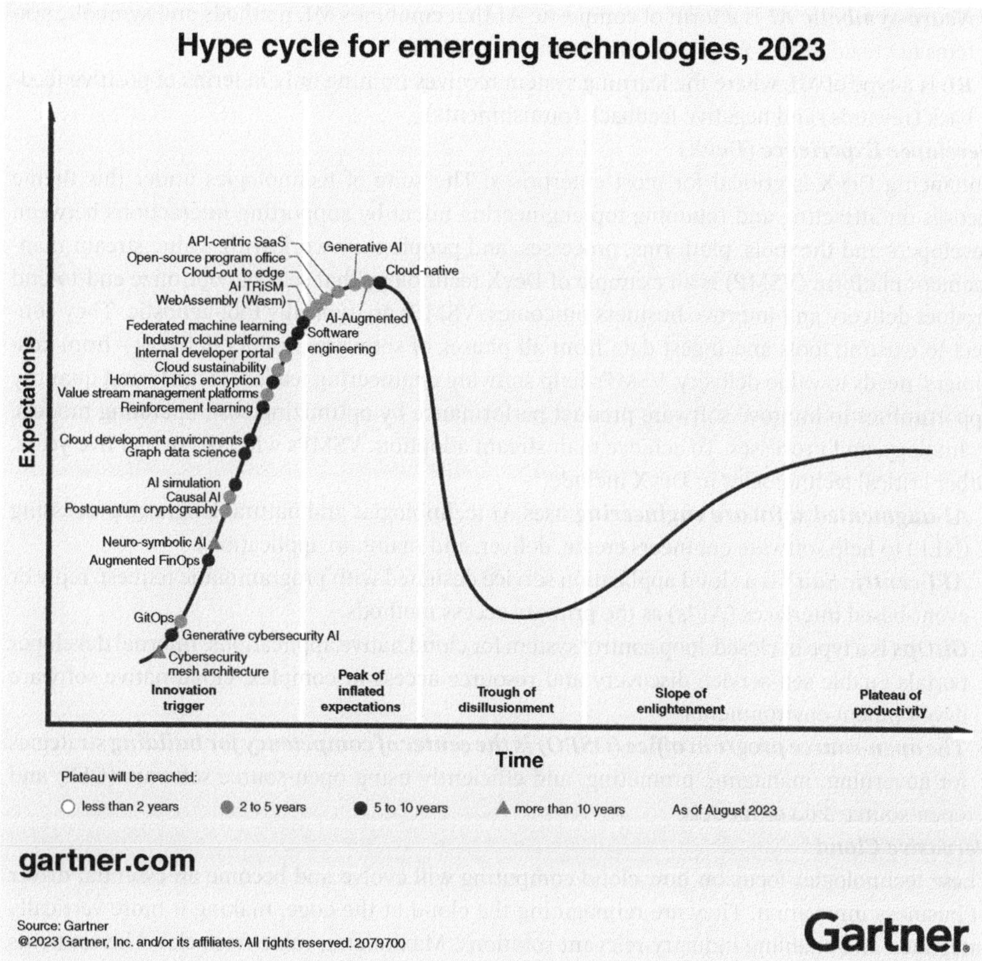

Figure 1.4 Hype cycle for emerging technologies, 2023 (Source: [4]/Gartner).

AI is that generative AI can generate new derived versions of content, strategies, designs, and methods by learning from large-source content repositories. It will continue to have profound business impacts, including content and product development, automation of human work, and enhancement of customer and employee experiences as it reaches mainstream adoption in two to five years. Other critical technologies in emergent AI include:

- ***AI simulation*** is the combined application of AI and simulation technologies to jointly develop AI agents and the simulated environments in which they can be trained, tested, and sometimes deployed.
- ***Causal AI*** identifies and uses cause-and-effect relationships to go beyond correlation-based predictive models and toward AI systems that can prescribe actions more effectively and autonomously. Federated ML aims to train a ML algorithm without explicitly sharing data samples, enabling better privacy and security.
- ***Graph data science (GDS)*** is a discipline in which data science techniques are applied to graph data structures to identify behavioral characteristics that can be used to build predictive and prescriptive models.

- ***Neuro-symbolic AI*** is a form of composite AI that combines ML methods and symbolic systems to create more robust and trustworthy AI models.
- ***RL*** is a type of ML where the learning system receives training only in terms of positive feedback (rewards) and negative feedback (punishments).

- ***Developer Experience (DevX)***

  Enhancing DevX is critical for most enterprises. The suite of technologies under this theme focuses on attracting and retaining top engineering talent by supporting interactions between developers and the tools, platforms, processes, and people they work with. Value stream management platform (VSMP) is an example of DevX technology that seeks to optimize end-to-end product delivery and improve business outcomes. VSMPs are typically tool-agnostic. They connect to existing tools and ingest data from all phases of software product delivery—from customers' needs to value delivery. VSMPs help software engineering leaders identify and quantify opportunities to improve software product performance by optimizing cost, operating models, technology, and processes. To achieve mainstream adoption, VSMPs will take two to five years. Other critical technologies in DevX include:
  - ***AI-augmented software engineering*** uses AI technologies and natural language processing (NLP) to help software engineers create, deliver, and maintain applications.
  - ***API-centric SaaS*** is a cloud application service designed with programmatic request/reply or event-based interfaces (APIs) as the primary access methods.
  - ***GitOps*** is a type of closed-loop control system for cloud native applications. Internal developer portals enable self-service discovery and resource access in complex, cloud-native software development environments.
  - ***The open-source program office (OSPO) is the center of competency for building*** strategies for governing, managing, promoting, and efficiently using open-source software (OSS) and open-source data or models.

- ***Pervasive Cloud***

  These technologies focus on how cloud computing will evolve and become an essential driver of business innovation. They are reimagining the cloud at the edge, making it more vertically integrated and enabling industry-relevant solutions. Maximizing value from cloud investments will require automated operational scaling, access to cloud-native platform tools, and adequate governance. Industry cloud platforms exemplify pervasive cloud and address industry-relevant business outcomes by combining underlying SaaS, PaaS, and IaaS services into a product offering with composable capabilities. These typically include an industry data fabric, a library of packaged business capabilities, composition tools, and other platform innovations. IT leaders can use the composability of these platforms to be adaptable and agile in response to accelerating disruption. They will take 5 to 10 years to reach mainstream adoption. Other critical technologies in the pervasive cloud include:
  - ***Augmented FinOps*** applies the traditional DevOps concepts of agility, continuous integration and deployment, and end-user feedback to financial governance, budgeting, and cost optimization efforts.
  - ***Cloud development environments (CDEs)*** provide remote, ready-to-use access to a cloud-hosted development environment with minimal effort for setup and configuration.
  - ***Cloud sustainability*** uses cloud services to achieve sustainability benefits within economic, environmental, and social systems.
  - ***Cloud-native*** refers to something created to optimally leverage or implement cloud characteristics that are part of the original definition of cloud computing and include capabilities delivered as a service.

- **_Cloud-out to edge_** is an architectural construct where a centrally managed cloud environment, typically a hyperscale cloud, provides cloud service capabilities extended to edge environments. WebAssembly (Wasm) is a lightweight virtual-stack machine and binary code format designed to support secure, high-performance applications on webpages.

- **_Human-Centric Security and Privacy_**
  The technologies in this bucket focus on how organizations can become resilient by implementing human-centric security and privacy programs. They enable enterprises to create a culture of mutual trust and awareness of shared risks in decision-making between many teams. AI trust, risk, and security management (AI TRiSM) is an excellent example of human-centric security and privacy. It ensures AI model governance, trustworthiness, fairness, reliability, robustness, efficacy, and data protection. It includes solutions and techniques for model interpretability and explainability, data and content anomaly detection, AI data protection, model operations, and adversarial attack resistance. It will take two to five years to achieve mainstream adoption. Other critical technologies in human-centric security and privacy include:
  - **_Cybersecurity mesh architecture (CSMA)_** is an emerging approach for architecting composable, distributed security controls that improve overall security effectiveness.
  - **_Generative cybersecurity AI_** generates new derived versions of security-related and other relevant content, strategies, designs, and methods by learning from large source data repositories.
  - **_Homomorphic encryption (HE)_** uses algorithms to enable computations with encrypted data and allows businesses to share data without compromising privacy.
  - **_Postquantum cryptography (PQC)_**, called quantum-safe cryptography, is an algorithm designed to secure against classical and quantum computing attacks.

As a big picture, these strategic technology trends will factor into business and technology decisions over the next three years. Gartner urges you to evaluate the impacts and benefits of each technology trend to determine which innovation—or strategic combination—will significantly impact your organization's success. Every trend is related to one or more of the major themes in business, which include generating value for the evolving needs of internal and external customers, safeguarding and maintaining previous and future investments, and creating the appropriate solutions for the right stakeholders at the right time.

1) AI Trust, Risk, and Security Management (AI TRiSM)
2) Continuous Threat Exposure Management (CTEM)
3) Sustainable Technology
4) Platform Engineering
5) AI-Augmented Development
6) Industry Cloud Platforms
7) Intelligent Applications
8) Democratized Generative AI
9) Augmented Connected Workforce
10) Machine Customers

Although we have not fully addressed this book's 2023 and 2024 technology trends, we try to approach these issues more customized for CPPSs.

1.3.2 Cyber-Physical Systems

In 2006, Helen Gill of the US National Science Foundation first used the term "CPSs." CPSs are engineered systems built from and depend upon the ***seamless integration of computation and physical components***. The capabilities, flexibility, scalability, resilience, safety, security, and usability that these vital systems will be able to provide will be made possible by advancements in CPS. Similar to how the Internet changed how people interacted with information, CPS technologies are changing how people engage with designed systems in different application domains such as energy and power systems, industrial IT, agriculture, aeronautics, building design, civil infrastructure, energy, environmental quality, healthcare and personalized medicine, manufacturing, and transportation.

There are some contradictions in CPSs, including but not limited to contradictions in CPS adaptability vs. repeatability, high connectivity vs. security and privacy, high performance vs. low energy, asynchrony vs. coordination/cooperation, scalability vs. reliability and predictability, laws and regulations vs. technical possibilities, economies of scale (cloud) vs. locality (fog), open vs. proprietary, and algorithms vs. dynamics.

CPS and CPPS are new multidisciplinary areas of research that necessitate having enough knowledge about,

- Energy and Power Engineering
- Control Engineering
- Communication and Information Engineering
- Electronics and Embedded Systems Engineering
- Computer Science, Data Science, and AI Engineering

1.3.3 The Organization of Different Concepts of This Book in the Context of CPPS

To learn about the CPSs, we are trying to provide a scientific, structured approach for designing and implementing embedded systems to realize CPEPS, including understanding the elements, layers, and components of CPEPS, with enough focus on model-based system design on embedded hardware and software, the interdependencies of different layers of the CPPSs, and interaction and overall performance analysis.

Suppose we liken the topics in the CPPS's subset to the football team members. In that case, we can consider the combination of different topics as shown in Figure 1.5. In this team, the general topics, basics, and definitions (part of what is presented in the first chapter of this book) will be the gatekeeper, which is presented in Section I of the book. The defense line of this team includes essential and fundamental topics regarding their structure, architecture, infrastructure, smart ecosystem, and smart city, which are discussed in Section II of the book. Major challenges such as telecommunication infrastructure, management, control, monitoring, and protection, the electricity market in the cyber-physical environment, cyber security and flexibility, reliability, and cyber-physical resilience are considered the middle line players of this team as midfielders, which in Section III of the book is discussed in different chapters. Finally, the attack line of this team as solutions and new tools to solve the challenges of these systems based on AI, data science and data mining, information theory, blockchain, digital twin, and other emerging technologies in Section IV of the book is different.

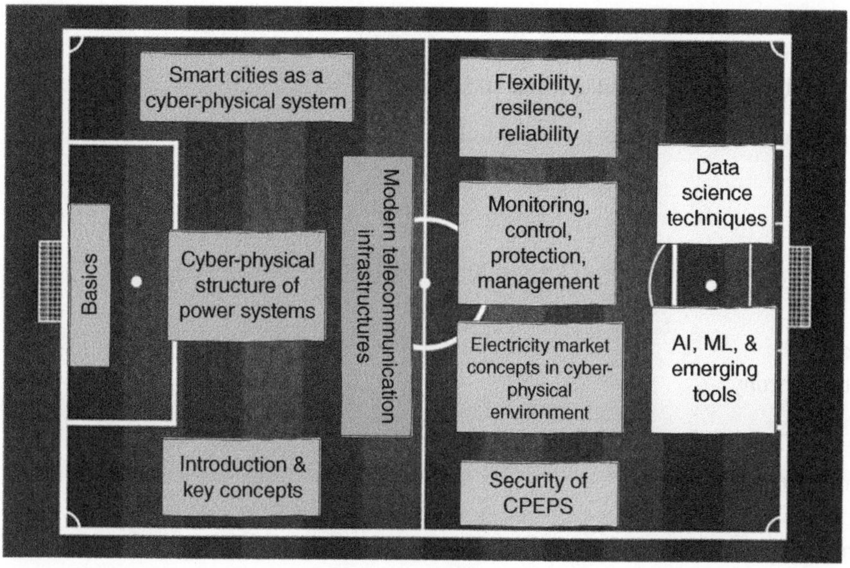

Figure 1.5 The classification of different concepts of the CPPs as the arrangement of a football team.

1.3.4 What Are We Looking for in This Book?

CPSs are the integration of physical systems and processes with networked computing. In CPSs, computations and communications are deeply embedded in and interact with physical processes to equip physical systems with new capabilities that cover a wide range of scales (pacemakers to national power grid). CPSs are physical and engineered systems whose operations are monitored, coordinated, controlled, and integrated. This intimate coupling between the cyber and physical differentiates CPS from other fields.

CPS will transform how we interact with the physical world, just like the Internet transformed how we interact with one another. In this book, we try to address the following questions:

- What is the definition of CPS, CPSS, and cyber-physical-social power systems (CPSPS)?
- What are CPS's security issues, and how do they differ from those in traditional information systems?
- To what extent can a CPS be secured against cybercrime?
- Are some fundamental design principles that should be used when designing or upgrading a CPS?
- What are the curricular ramifications of CPS security?

Based on the definition of the CPSs provided by Helen Gill from the NSF, there are some hallmark characteristics:

- Cyber capability in every physical component
- Networked at multiple and extreme scales
- Complex at multiple temporal and spatial scales
- Constituent elements are coupled logically and physically
- Dynamically reorganizing/reconfiguring; "open systems"

- High degrees of automation, control loops closed at many scales
- Unconventional computational and physical substrates (such as bio, nano, chem, …)
- The operation must be dependable and certified in some cases.

CPSs are faced with different challenges. They should tolerate

- CPS must tolerate
- Failures
- Noise
- Uncertainty
- Imprecision
- Security attacks
- Lack of perfect synchrony
- Scale
- Openness
- Increasing complexity
- Heterogeneity
- Disconnectedness

1.3.5 Cyber-Physical Systems of Systems (CPSoS)

Due to the increase of human interactions with electricity DNs, due to new technologies such as P2P exchanges, IoT, EVs and CEVs, transactive energy, the emergence of prosumers, home solar cells, etc., DNs became dependent on human factors. This dependence on social-human behaviors and factors has turned cyber-physical DNs into CPSDSs. Based on definitions, CPSDSs are, in a way, a SoS, so we can define these systems as CPSSoS. As stated [5], "A SoS is a set of components that may be individually considered as a system and has two additional properties: the operational and management independence of the components." SoS is simply a merge of subsystems. However, SoS integrates a finite number of independent and operable constituent systems that are networked together for some time to achieve a specific higher goal [6]. SoS is a collection of task-oriented or dedicated systems that pool their resources and capabilities together to create a new, more complex system that offers more functionality and performance than simply the sum of the constituent systems. SoS are large-scale concurrent and distributed systems, the components of which are complex systems. A linking overlap that connects the CPSs to the SoSs and can practically realize the CPSoS is IoT.

1.4 Cyber-Physical Energy and Power Systems (CPEPSs)

Cyber-physical energy and power systems (CPEPS) combine computation, communication, and control technologies with physical power systems and realize the efficient fusion of power, information, and control. CPEPS is a power and information-integrated system. Information and communication technologies play an essential role in CPEPS. The communication network connects the devices of sensor, computation, and control units to realize the information sharing in whole systems, and together with distributed computation technology, the physical system's identification, optimization, and control can be performed. The power system's safety, reliability, and efficiency can be improved by the above fusion of power, information, and control, providing physical entities with computation, communication, accurate control, coordination, and autonomy. Moreover,

CPEPS can cooperate with other social systems, like transportation and the environment, to realize a green economy and sustainable development.

The power grid consists of generation units, a transmission network (TN), a DN, and a wide-area network (WAN) (including WAMS, WACS, and WAMPAC), and more importantly, the following systems:

- Wide-Area Monitoring System and Control (WAMS and WAMC)
- Wide-Area Protection and Control System (WAMPAC)

The objective of the power system is to provide a sustainable supply for consumers while minimizing operation costs.

We need the communication system as a backbone to

- Extend exchange between miscellaneous agents
- Enabling central/coordinated/hierarchical control and monitoring schemes to ensure reliability while optimizing the performance

Considering dissimilar time scales between the physical power system and communication rate, cyber and physical systems are traditionally studied separately. However, the power grid is evolving into a complicated multi-layer CPS with the integration of power assets with communication networks and the IoT.

Some reasons for this transmission include, but are not limited to

- high-resolution spatial and temporal measurements such as smart meters and phasor measurement units (PMUs),
- high penetration of DERs and distributed generation (DG) enable MG level operation and
- modern decentralized and distributed peer-to-peer control schemes.

The traditional approach to analyzing the power system does not capture the interdependency of the cyber and physical power systems. Cyber system attributes, such as latency and availability, can affect the cyber-physical grid. In addition, communication and IoT are prone to cyber threats and introduce CP vulnerability in smart grids, which might lead to physical ramifications.

1.4.1 Multi-Layer Representation of the CPPSs

The traditional approach to analyzing the power system does not capture the interdependency of the cyber and physical power systems. Cyber system attributes, such as latency and availability, can affect the cyber-physical grid. In addition, communication and IoT are prone to cyber threats and introduce CP vulnerability in smart grids, which might lead to physical ramifications.

Different models have been proposed for analyzing the CPPSs. Some models have considered two cyber and physical layers that are almost inefficient in technical challenges like reliability or security analysis. By the way, a good model for CPPSs has been offered by CENELEC [7] (Figure 1.6). However, some other efficient models, such as the four-layer model shown in Figure 1.7, have more adoption with the IEC standard.

In the following, we briefly describe the different layers of CPPSs.

1.4.1.1 The Physical Layer

The physical layer, which consists of all physical components (e.g., generators, power lines, transformers, circuit breakers, power electronic devices, energy storage, loads, smart appliances, etc.), is connected to the communication and networking layer through state awareness (sensors) and

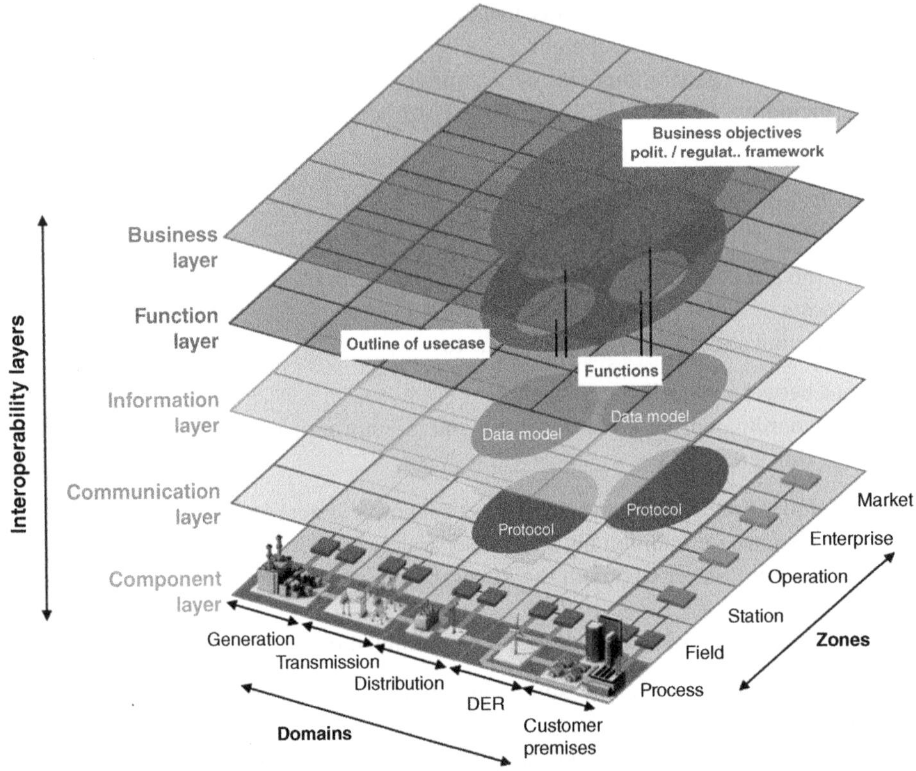

Figure 1.6 The smart grid architecture model (SGAM) and CEN-CENELEC-ETSI smart grid coordination group (2012) [7].

command execution devices. It consists of a generation system that involves several large generators generally located away from load centers. DG may feed into the grid at the sub-transmission or distribution level. The power generated by these generators is delivered to the end-user or customers through the transmission and distribution systems. The electric grid is undergoing several changes, including generation mix, load types, electricity markets, difficulty building new transmission lines, and environmental constraints.

1.4.1.2 The Interface (Control and Protection) Layer

According to the U.S. Department of Energy (DOE) report, the average demand for electricity in the past two decades has been increasing at 2.5% annually. Several blackouts in the past indicated the requirement for sustained improvement on the existing electric power grid. Energy storage requirements, better visibility and situation awareness, automated control, and sustainable energy are some critical factors that have generated the push toward developing a smarter grid. Analysis of past blackouts in the North American grid has shown that the lack of visibility and the unavailability of high-resolution information to make critical decisions were the leading causes of the blackouts.

The smart grid significantly enhances the electric grid infrastructure for improved efficiency, reliability, and safety, efficiently integrating renewable and alternative energy sources through automated control and advanced communication technologies. With the evolution of sensing devices, such as PMUs, smart meters, digital relays, etc., progressively powerful computation capability, and

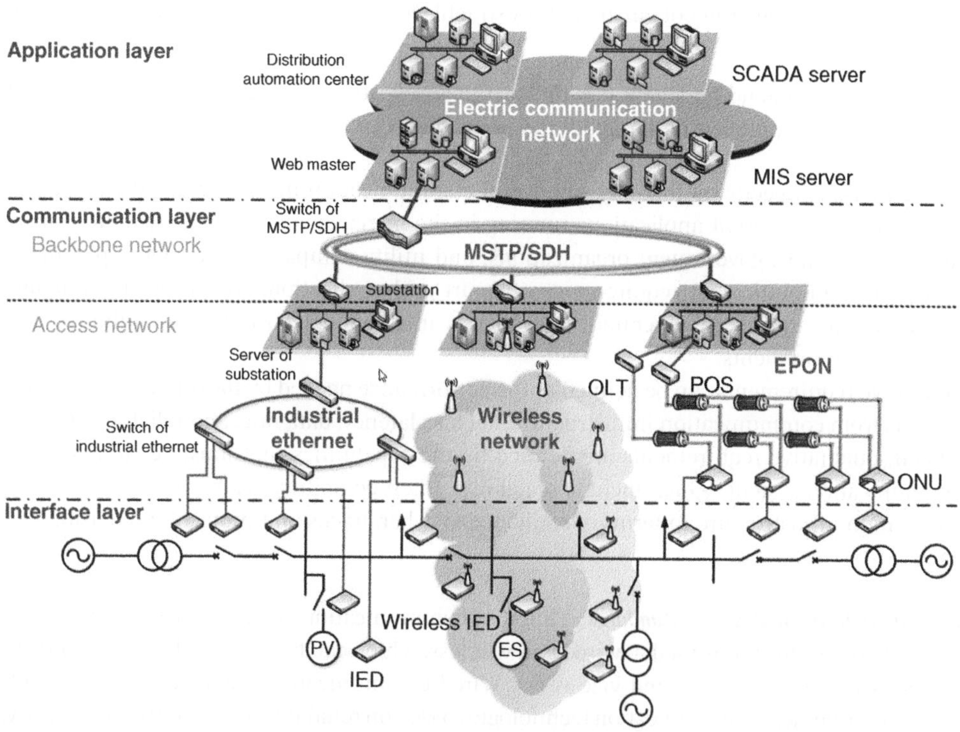

Figure 1.7 A four-layer representation of CPPSs (IEC standard) (Source: [8]).

central communication and networking foundation, the future power grid has moved to a very complex cyber-physical infrastructure with two-way communication of both energy and information across the network. However, such strong couplings between cyber and physical systems significantly increase vulnerabilities in the power grid with potential cyberattacks. Therefore, the electric power grid's reliable operation depends highly on the grid component's efficient functioning and the associated cyber system's security. Hence, monitoring, maintaining, and controlling the power grid should consider the interdependence between cyber and physical systems.

1.4.1.3 Communication and Networking/Information Layer

The communication and networking layer consists of remote terminal units (RTUs), intelligent electronic devices (IEDs), communication networks, substation automation systems, and control centers. In this layer, the interface devices carry out the physical power system layer measurements and the control or decision command to the power system layer. The communication network, which connects the interface devices, consists of various communication devices and the links between them. The failure or malfunction of interface devices and communication networks will impact the decision-layer functions' reliability and accuracy. Therefore, modeling the impact of the communication and coupling layer is critical for effectively operating a CPPS. Several essential issues need to be considered while designing a communication network, such as:

- Which communication technologies need to be used for establishing links between different components?
- Which communication topologies can be used to have better performance?

- Which communication protocols are incredibly suitable for meeting the needs of smart grid communications?

The common understanding is that it is difficult to answer the above questions because smart grids will operate in different environments and use cases.

System Requirement Specifying that any smart grid communication infrastructure should satisfy the requirements of different applications running in the management layer is essential. Hence, different research labs, government organizations, and utility companies have developed many reports on the communication requirements of smart grid applications. Based on the available document, we can divide the essential requirements into two categories: (i) quantitative and (ii) qualitative requirements.

Quantitative requirements can be defined as the performance needed by the different smart grid applications from communication infrastructure, such as latency, data rate, and reliability. On the other hand, qualitative requirements are defined by the capabilities that must be supported by the communication system, which include scalability, interoperability, flexibility, and security of communication infrastructure in terms of resilience to cyberattacks and protection of customers' privacy.

Communication Technologies and Standards Different communications technologies are supported by two main communications media: wired and wireless, which can be used for data transfer from smart meters to control centers and vice versa. Wired communication technologies are considered superior to wireless communication technologies based on reliability, bandwidth, and security. The physical cables are more reliable and secure from external interference, and equipment and maintenance costs are much cheaper than wireless technologies. On the other hand, wireless communication technologies have advantages over wired technologies, such as low-cost infrastructure and ease of connection to rugged or isolated areas. Many communication technologies support smart grid applications. The wired communication technologies include traditional twisted-copper phone lines, fiber optic cable, power line carriers, and broadband over power lines. Wireless communications include cellular, satellite, microwave, WiMAX, and short-range in-home technologies such as WiFi and ZigBee. The smart grid application can be built on communication technologies such as home area networks (HAN), neighbor area networks (NAN), and wide area networks (WAN).

Communication Technologies Two possible communication topologies are mesh topology and star topology (point-to-point). Mesh topology comprises communication lines along the same right-of-way as the power transmission lines. Each node in the network connects to multiple nodes, efficiently routing data from/to the substation and control center. This leads to a more efficient use of the available infrastructure and increases the redundancy with various communication paths. But communication performance might be a little less because of multi-hop networks, where a packet must be routed through multiple nodes before reaching its destination.

On the other hand, star topologies refer to a network where all substations directly communicate with the control center with just one hope. This type of topology is simple and straightforward. Also, the processing time in this topology is shorter since each substation directly transmits the data to the control center. However, start topologies have their drawbacks, such as (i) being very costly to build for large systems, (ii) not being able to keep the communication function integrity under communication contingency, and (iii) in control center unexpected waiting/queuing delays or/and cause the receiving buffer overflow.

Communication Protocol Communication protocols describe the "RULES" by which devices on a network can communicate. They outline the structure of data packets transmitted on the network and other necessary information, such as the device's unique ID/address and the start and end of a data message, and check for transmission errors by the receiving device. Typically, a communication network system has four layers (TCP/IP model): application layer, transport layer, network layer, and network interface layer, from top to bottom. Each communication layer has multiple choices for protocol selection.

For example, data can be transferred in the transport layer using user datagram protocol (UDP) or transmission control protocol (TCP). The choice for each layer is described as follows:

Application Layer: This layer interacts with an application program to access the network components. Depending on the application, different protocols are available. For example, the IEEE C37.118 protocol is designed for PMU data communication similar to DNP3, IEC 61850, or the GOOSE protocol for SCADA measurements. Other commonly used protocols are Hypertext Transfer Protocol (HTTP), File Transfer Protocol (FTP), Terminal Network (TELNET), Domain Name System (DNS), etc.

- **Transport layer:** The transport layer is responsible for the reliability, flow control, and correction of data sent over the network. The two protocols used in the transport layer are UDP and TCP.
- **Network layer:** This layer's responsibility is to transmit the packets from any network, and irrespective of the route the data takes, it should arrive at the destination. Internet protocol (IP), Address Resolution Protocol (ARP), and Internet Control Message Protocol (ICMP) are the different protocols used in this layer, and it is the most significant part of the entire TCP/IP suite.
- **Network interface layer:** This layer comprises the physical and data link layers defined in the OSI reference model. Protocol is selected depending on the communication type, i.e., wired or wireless communication. For example, wired networks Ethernet protocol will be chosen and implemented in this layer.

Note that for the information layer, the best practice is to analyze it based on the open systems interconnection (OSI) model shown in Figure 1.8.

1.4.1.4 Management and Control Layer

The management and control layer consists of functions like automatic generation control, renewable generation control, energy management system, demand response, etc., which are essential for the power system's efficient operation. The real-time measurement obtained from PMUs and RTUs is used to estimate the real-time operating state of the power system. It is continuously monitored by the control room operator on a human machine interface (HMI) to take corrective actions if needed.

1.5 From Conventional Distribution Networks to Smart Grids

1.5.1 Conventional Power Distribution Systems

Conventional power distribution systems have been the backbone of electricity distribution for many years and continue to serve as the primary method of delivering power to consumers in many regions. However, technological advancements and the growing integration of renewable energy sources drive the development of modern smart grid systems that are more efficient, reliable,

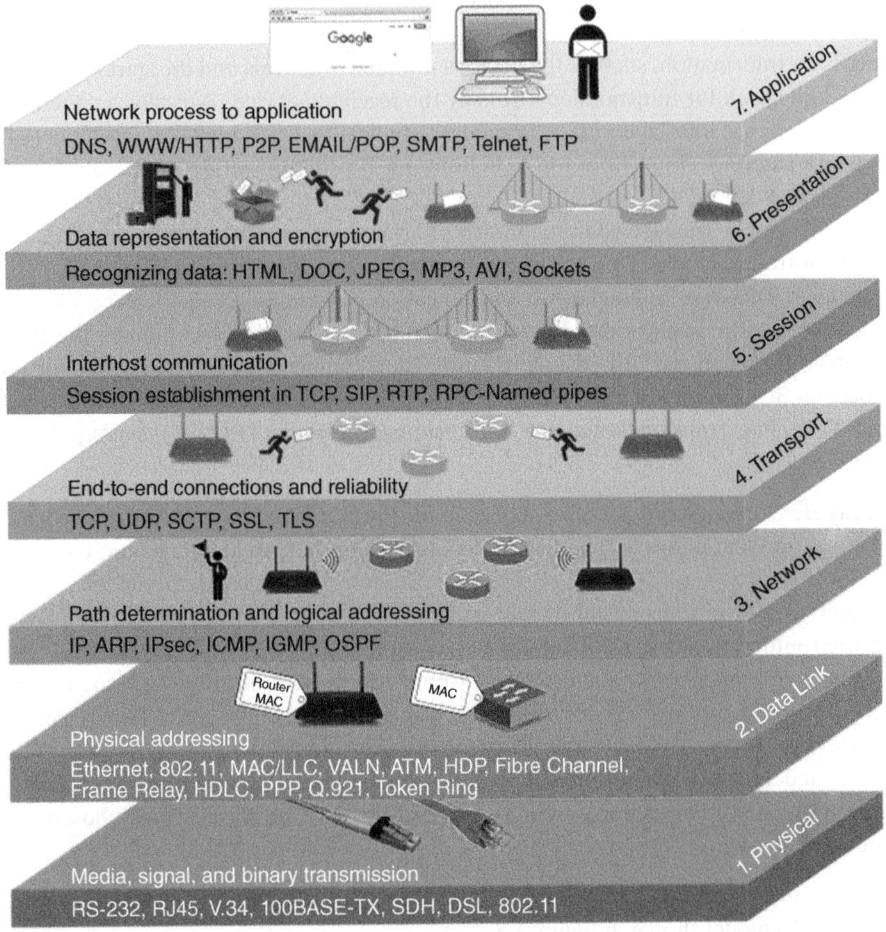

Figure 1.8 Open systems interconnection model (OSI).

and flexible. Conventional power distribution systems typically have the following characteristics [9–13]:

1) **Centralized generation:** Power is generated at centralized power plants and distributed to end users through a transmission and distribution line network.
2) **High voltage transmission:** Electricity is transmitted over long distances at high voltages to reduce energy losses during transportation.
3) **Step-down transformers:** At various points along the DN, step-down transformers reduce the voltage to levels suitable for commercial, industrial, and residential consumers.
4) **Radial configuration:** Conventional distribution systems often have a radial or tree-like configuration, where power flows from the substation to the end users in a unidirectional manner.
5) **Reliability:** These systems often employ redundant components and backup systems to ensure reliability and minimize downtime.
6) **Limited Automation:** Traditional distribution systems may have limited automation and remote control capabilities compared to modern smart grid technologies.

7) **Limited visibility:** Operators may have limited visibility into real-time system conditions and rely on periodic manual inspections for maintenance and operation.
8) **Voltage regulation:** Voltage regulation devices such as capacitors and voltage regulators maintain voltage within acceptable limits as power is distributed across the network.

These characteristics are typical of conventional power distribution systems, but modern advancements in smart grid technologies are transforming how electricity is distributed and managed.

1.5.2 Active Distribution Networks (ADNs)

1.5.2.1 Integration of DERs

The challenges conventional power systems face include low energy efficiency, environmental contamination, and the slow depletion of fossil fuel supplies. A new trend in power generation that uses renewable and non-conventional energy sources, such as natural gas, biogas, wind power generation, solar photovoltaic (PV) cells, fuel cells (FC), combined heat and power (CHP) systems, microturbines (MTs), and Stirling engines, and integrates them into the utility DN. Dispersed generation (DG) is the name used to describe this kind of power production, while DERs are the energy sources. The phrase "Distributed Generation" separates this generation idea from the traditional, centralized generation. With the integration of DG, the DN becomes active, thus the name "active DN [14]."

A DER is a localized, small-scale power-generating device linked to a larger power grid at the distribution level (Mv or LV). It may thus function in grid-connected/on-grid or islanded/autonomous/isolated/off-grid modes.

DERs may have a controlled load and be DG, ESS, EV, or PHEV. Among them are PV units, tiny natural gas-fueled generators, synchronous-based distributed generators (SDG), EVs, and controlled loads, such as air conditioning and water heaters. One of DER's key characteristics is that the energy it generates is often used close to the source.

DER may be either directly coupled or electronically coupled/inverter-interfaced/inverter-based (ECDER/IBDER), dispatchable (DDER) (like FC) or non-dispatchable (NDDER) (e.g., WG or PV), and renewable-based (RER) (e.g., PV or biomass) or non-renewable (e.g., SDG). Renewable energy resources (RERs) may either be non-variable (NVRER) or variable (VRER).

In contrast to controllable renewable energy sources like biomass or dammed hydroelectricity or relatively constant sources like geothermal power, variable RERs, also known as intermittent RERs or VRERs, are RERs that cannot be dispatched because of their intermittent or fluctuating nature. Examples of such sources include wind and solar power.

The intermittent nature of certain resources when employing RERs necessitates using several renewable resources and connecting, managing, and storing their output.

Energy storage, such as battery energy storage systems (BESSs) and flywheel energy storage systems (FESSs), are necessary for hardware such as wind and other turbine types, solar panels, and tidal generating units. Electronic management equipment, such as inverters, and software like storage distributed resource schedulers (SDRS) must closely monitor various power sources and storage devices to optimize energy production. In the residential, commercial, and industrial sectors, DERs are often utilized to control a variety of smaller power-generating and storage techniques. Utility companies and people may utilize them as backup power sources or for generating and storing renewable energy. Technologies, such as smart grids, are essential to increasingly sophisticated power networks.

1.5.2.2 Concept of ADN

Today's energy networks are significantly shifting from reliable passive DNs that carry power in a single direction to active DNs (ADNs) that transport electricity in two directions.

DG units are incorporated in DNs, making them inactive in their absence. It turns on when DG units are added, and clients get electricity from the national grid that flows in both directions via the distribution system.

To facilitate this shift, rich nations should take on the financial and technical difficulties of changing DNs, while poorer nations should prioritize creating sustainable electrical infrastructure [14–17].

Distributed intelligent systems must be integrated with flexible and intelligent control in ADNs. ADNs should also use future technologies that lead to smart grid or MG networks to harvest sustainable energy from renewable DERs. Controllable loads and a large number of DERs make up ADNs. Whereas a MG is a small-scale ADN that may be operated either in islanding or grid-linked modes of operation. Several factors are in favor of the evolution of ADNs, e.g.,

- Pressing customer expectations of high-quality, reliable power distribution,
- Increasing desire of policymakers for the accommodation of RERs with ESSs,
- Carbon commitment to reducing emissions by 50% by 2050,
- Motivating the DN operators (DNOs) toward better asset utilization,
- Management by deferring the replacement of age-old assets, etc.

The focus of the research should be mainly on the following areas:

- Wide area active control,
- Adaptive protection and control,
- Network management devices,
- Real-time network simulation,
- Advanced sensors and measurements,
- Distributed pervasive communication,
- Knowledge extraction by intelligent methods and,
- Novel design of transmission and distribution systems.

1.5.3 Microgrid

Factors of changes in power systems can be summarized as:

- Increasing the participation of consumers in the network (in the sectors of energy generation and storage and demand-side management)
- Less carbon production (use of renewable energy and energy production with high efficiency and less pollution)
- Integration of DERs (DGs, ESSs, and EVs)
- Necessity of investment and end of network lifetime
- The need for congestion management considering electricity market issues
- Advancement of ICT and its application in energy systems
- Increasing development of telecommunication systems
- Stability, reliability, security, and adaptability
- Developing small networks inside the power system, which are called MGs.

These factors result in the development of small networks inside the power system called MGs. A MG is a small-scale power grid in low (or medium) voltage (including some components like electrical and heating loads, ESSs [BESS, FESS, HESS, SC, etc.], DGs, and FDGs) that must be able to locally solve energy issues (using a unified control system) and enhance the flexibility and can operate either in grid-connected or islanding (autonomous) mode of operation.

Because a MG consists of DG systems and various loads at the distribution voltage level, it is an AND. In a MG, electricity is produced at distribution voltage by integrating renewable and non-conventional DERs, micro-sources, or generators (MSs) [14, 16].

Power electronic interfaces (PEIs) and controls are necessary from an operational standpoint to provide the MSs the flexibility to guarantee that they operate as a single aggregated system and maintain the designated power quality and energy output. Because of its control flexibility, the MG could function as a single, controlled unit that satisfies the security and reliability requirements of the local energy market, as seen by the main utility power system.

The U.S. DOE defines a MG as "a group of interconnected loads and DERs within clearly defined electrical boundaries that act as a single controllable entity concerning the grid. A MG can connect and disconnect from the grid, enabling it to operate in both grid-connected or island-mode."

The main advantage of MGs is that they provide higher service quality and reliability.

The key differences between a MG and a conventional power plant are as follows:

- Micro-sources are of much smaller capacity than the large generators in conventional power plants.
- Power generated at distribution voltage can be directly fed to the utility DN.
- Micro-sources are normally installed close to the customers' premises so that the electrical/heat loads can be efficiently supplied with satisfactory voltage and frequency profiles and negligible line losses.

From a grid perspective, the primary benefit of a MG is that it is seen as a regulated entity within the electricity system. One aggregated load may be used to run it. This guarantees its simple controllability and adherence to grid policies and guidelines without jeopardizing the electricity utility's dependability and security.

MGs are advantageous from the client's perspective for addressing their local heating and electricity needs. They may lower feeder losses, increase local dependability, provide uninterrupted power, and sustain local voltage. From an environmental perspective, MGs use low-carbon technologies to lessen global warming and environmental damage.

Before MGs may become widely used, various technical, legal, and financial challenges must be settled to ensure a stable and secure operation. DER production's erratic and climate-dependent character is one issue that would need careful consideration. The low energy content of the fuels and the absence of standards and regulations for running the MGs in synchronism with the power utility. Such concerns would need a large amount of offline and real-time investigation. The summary of the concepts mentioned above is presented in Figure 1.9 [16].

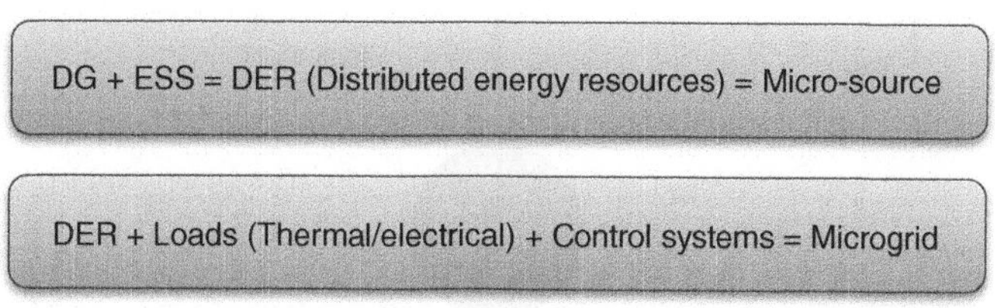

Figure 1.9 Two main concepts: The maximum capacity is normally restricted to approximately 10 MVA as per IEEE recommendation (in some references, 100 kW–Multiple Megawatts).

Definition: A MG is a small-scale power grid in low voltage that must be able to locally solve energy issues and enhance flexibility and can operate either in grid-connected or islanded (autonomous) modes of operation [15, 16, 18]. We should discriminate between three basic definitions:

- **Power grid:** Power grids are larger conventional and spread-out grids with high-voltage power transmission capabilities.
- ***MG*** technology can be applied to weak grids, making the network more robust.

MG: DERs and loads that can be operated in a controlled, coordinated way either connected to the main power grid or in "islanded" mode. MGs are low- or medium-voltage grids without power transmission capabilities and are typically not geographically spread out.

- **Nanogrid**: Low voltage grids that typically serve a single building.

The most important IEEE standards in the field of MGs are

- **IEEE 2030 series:** "IEEE Guide for Smart Grid Interoperability of Energy Technology and Information Technology Operation with the Electric Power System (EPS), End-Use Applications, and Loads"
- IEEE 1547 Series of "DERs Interconnection and Interoperability Standards, Guides, and Recommended Practices"
- IEEE-P2800 "Standard for Interconnection and Interoperability of Inverter-Based Resources Interconnecting with Associated Transmission Electric Power Systems"

IEEE 2030.7 standard reduces MG complexity to two steady state (SS) operating modes and four types of transitions (T) (Figure 1.10):

- SS1—Steady State Grid Connected
- SS2—Stable Island
- T1—Transition from Grid Connected to Steady State Island (Planned)
- T2—Grid Connected to Steady State Island (Unplanned)
- T3—Steady State Island Reconnects to Grid
- T4—Black Start into Steady State Island

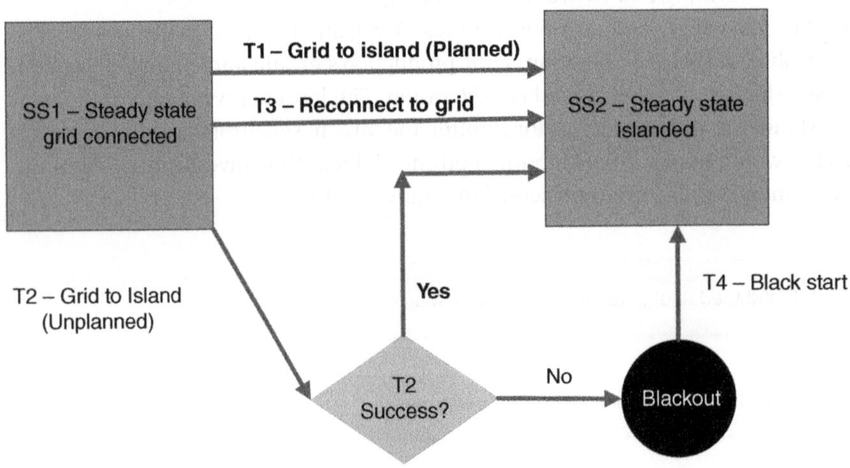

Figure 1.10 The simplified MG control flow diagram is based on the IEEE Std. 2030 (Source: [19]/National Rural Electric Cooperative Association).

Black Start Capability means the ability of a generating unit or station to go from a shutdown condition to an operating condition and start delivering power without assistance from the power system. Black Start Capability means the capability to provide electricity to the customer in the event of an outage.

The main function of a micro-source controller (MC) is to independently control the micro-source's power flow and load-end voltage profile in response to any disturbance and load changes. Here, 'independently' implies no communications from the central controller CC. MC also participates in economic generation scheduling, load tracking/management, and demand side management by controlling storage devices. It must also ensure that each micro-source rapidly picks up its generation to supply its share of the load in stand-alone mode and automatically returns to the grid-connected mode with the help of CC [14].

The most significant aspect of MC is its quickness in responding to the locally monitored voltages and currents, irrespective of the data from the neighboring MCs. This control feature enables micro-sources to act as plug-and-play (P&P or, in some references, PnP) devices. It facilitates the addition/outage (or outage and re-connection) of new micro-sources at any point of the MG without affecting the control and protection of the existing units. Two other key features are that

1) an MC will not interact independently with other MCs in the MG and that
2) it will override the CC directives that may seem dangerous for its micro-source.

The CC executes the overall MG operation and protection control through the MCs. Its objectives are

- to maintain specified voltage and frequency at the load end through power-frequency (*P*–*f*) and voltage control and
- to ensure energy optimization for the MG.

The CC also performs protection coordination and provides the power dispatch and voltage set points for all the MCs. CC is designed to operate automatically, providing manual intervention as and when necessary. Two main functional modules of CC are

- Energy Management Module (EMM) and
- Protection Co-ordination Module (PCM).

1.5.3.1 Energy Management Module

EMM provides the set points for active and reactive power output, voltage, and frequency to each MC. This function is coordinated through state-of-the-art communication and AI techniques. The values of the set points are decided according to the operational needs of the MG. The EMM must see that [14].

a) Micro-sources supply heat and electrical loads to customer satisfaction.
b) MGs operate satisfactorily as per the operational a priori contracts with the main grid.
c) MGs satisfy their obligatory bindings in minimizing system losses and emissions of greenhouse gases and particulates.
d) Micro-sources operate at their highest possible efficiencies.

1.5.3.2 Protection Co-ordination Module

PCM responds to MG and main grid faults and loss of grid (LOG) scenarios to ensure correct protection coordination of the MG. It also adapts to the change in fault current levels during the changeover from grid-connected to stand-alone mode. To achieve this, proper communication between the PCM, the MCs, and upstream main grid controllers exists. For the main grid fault, PCM immediately switches the MG to stand-alone mode to supply power to the priority loads at a significantly lower incremental cost [14].

However, for some minor faults, the PCM allows the MG to ride through in the grid-connected mode for some time, and it continues if any temporary fault is removed [20–24].

Besides, if the grid fault endangers the stability of the MG, then PCM may disconnect the MG fully from all main grid loads. In that case, effective utilization of the MG would be lost in exporting power. Suppose a fault occurs within a portion of the MG feeder. In that case, the smallest possible feeder zone is eliminated to maintain supply to the healthy parts of the feeder.

Under-frequency and undervoltage protection schemes with bus voltage support are normally used to protect sensitive loads. PCM also helps to resynchronize the MG to the main grid after switching to the grid-connected mode of operation through suitable reclosing schemes.

The functions of the CC in the grid-connected mode are as follows:

- Monitoring system diagnostics by collecting information from the micro-sources and loads.
- Using collected information, performing state estimation and security assessment evaluation, economic generation scheduling, and active and reactive power control of the micro-sources and demand side management functions.
- Ensuring synchronized operation with the main grid, maintaining the power exchange at priori contract points.
- The security assessment.
- Economic generation planning.
- Controlling of the active and reactive powers of micro-sources.
- Demand-side management.

The functions of the CC in the stand-alone mode are as follows:

- Performing active and reactive power control of the micro-sources to maintain stable voltage and frequency at load ends.
- Adopting load interruption/load shedding strategies using demand-side management with storage device support for maintaining power balance and bus voltage.
- Initiating a local black start to ensure improved reliability and continuity of service.
- Switching over the MG to grid-connected mode after the main grid supply is restored without hampering the stability of either grid.

1.5.3.3 Microgrid's Control Hierarchy

The interconnected power system is spread over a large geographical span. This intricate system can be controlled through either centralized control or decentralized control. Fully centralized control relies on the data gathered in a dedicated central controller and requires extensive communication between the controller and other units. In a fully decentralized control, each unit is controlled by its local controller, unaware of system-wide disturbances and independent of other controllers. The hierarchical control scheme compromises between fully centralized and fully decentralized control schemes.

The hierarchical control scheme in power systems includes three control levels: primary, secondary, and tertiary.

These control levels differ in their

i) speed of response and the time frame in which they operate, and
ii) infrastructure requirements, e.g., need for communication.

This control hierarchy can also be implemented for the MG control. The principles of operation and control of a MG can be best described in two distinct grid-connected and islanded modes of operation and are described in the rest of this section.

In the grid-connected mode, the voltage of the point of common coupling (PCC) of the MG is dominantly determined by the host grid, and the main role of the MG is to accommodate.

i) the real or reactive power generated by the DER units and
ii) the load demand.

Reactive power injection by a DER unit can be used for

i) power factor correction,
ii) reactive power supply, or
iii) voltage control at the corresponding point of connection (PC).

The DER units with limited power generation capacity cannot practically assist a strong utility network in its voltage and/or frequency regulation. In the grid-connected mode, the host utility may not permit regulation or control of the PCC voltage by the DER units to avoid interaction with the same functionality performed by the grid. Therefore, the DER units in the proximity of the PCC (determined by the electrical distance and SC of the grid) should not actively implement a voltage control scheme. In the islanded mode, the MG operates independently and must provide voltage and frequency control and real and reactive power balance. For example, if the load demand exceeds the total generation, the MG central controller should decrease the net generated power. This is accomplished by assigning new set points to the DER units. On the other hand, if the power generated within the MG cannot meet the load demand, either non-critical load shedding or activation of storage units must be considered [16, 20, 25–27].

Primary Control Primary control is the 1st control level in the control hierarchy and features the fastest response. Primary control responds to system dynamics and ensures that the system variables, e.g., voltage and frequency, track their set points. Primary control mostly employs conventional linear control methods and is performed locally based on locally measured signals. Because of their speed implications, islanding detection and the subsequent change of controller modes lie in this control level [16, 17, 26–29].

Secondary Control Secondary control is the next level of control. It is responsible for ensuring power quality and mitigating longer-term voltage and frequency deviations by determining the set points for the primary control. While this is a common task between a secondary controller and an energy management strategy, the latter lacks (i) the use of communication between the MG components and (ii) the use of possible distributed storage (DS) units such as spinning reserves. Secondary control operates on a slower time frame than the primary control, e.g., it has a settling time in the order of a minute in a conventional grid so that the primary controller mostly handles the initial transients of the MG. The primary control loop reaches its steady state before the secondary controller updates the set point. This assists in (i) decoupling secondary control from primary control and (ii) reducing the communication bandwidth, as the secondary control uses sampled measurements of the MG variables [16, 26, 29, 30].

Tertiary Control Tertiary control is the highest level of control. It sets the long-term set points depending on the requirements of an optimal power flow, e.g., based on the information received about the status of the DER units, market signals, and other system requirements.

Each primary, secondary, or tertiary control layer can be realized based on centralized, decentralized, or distributed control structures. The hierarchical control structure is realized when two or more control layers (e.g., primary and secondary) are hierarchically coupled. Each layer can be independently realized based on a centralized/decentralized/distributed control structure (Figures 1.11 and 1.12) [16].

The basic hierarchical control structure of the MGs was first proposed in [13] (Figure 1.9). However, in [14], different MG structures and control techniques at different hierarchical levels were proposed. The power converters' basic operation modes and control structure for MG applications

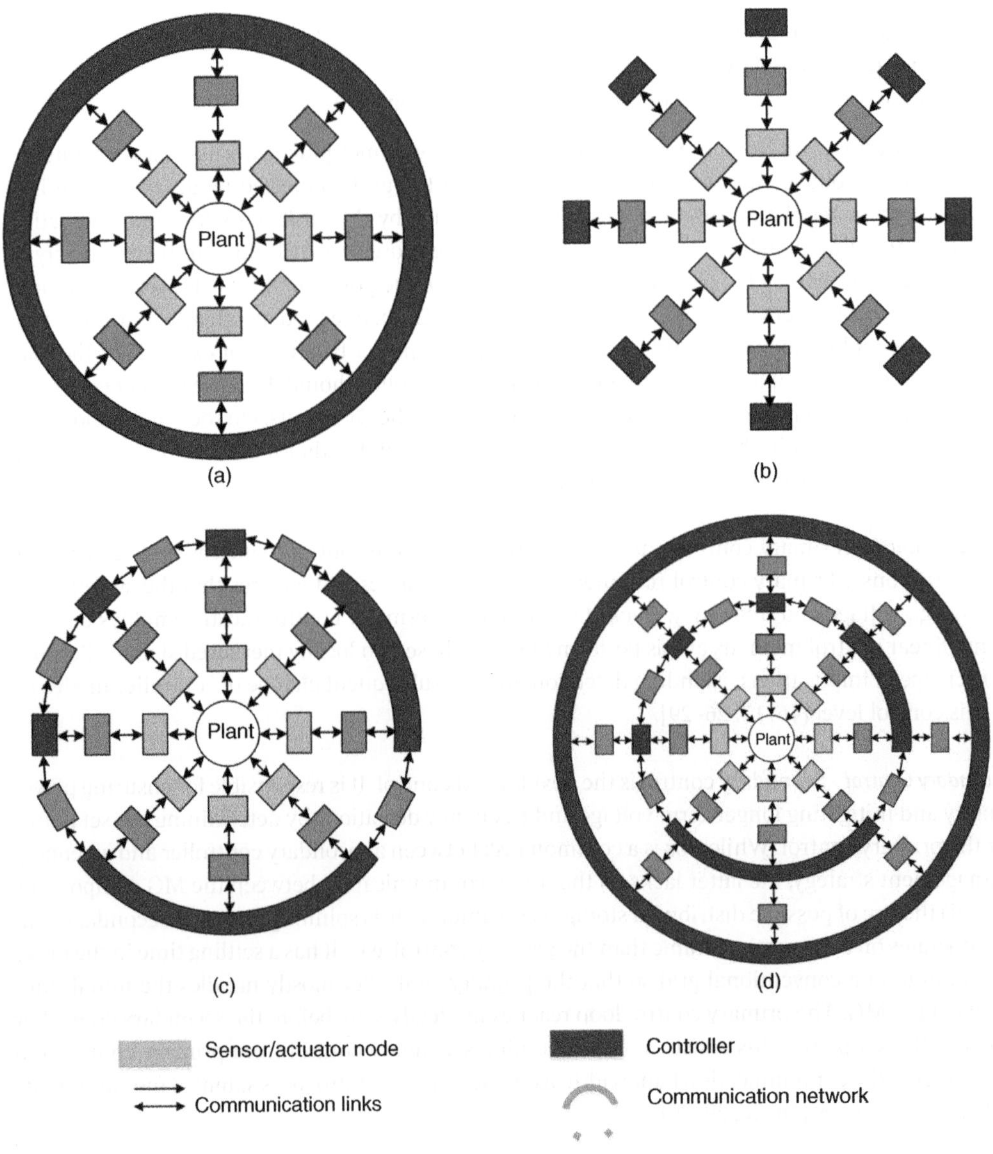

Figure 1.11 The basic control structures are (a) centralized control, (b) decentralized control, (c) distributed control, (d) hierarchical control schemes, and (e) legend.

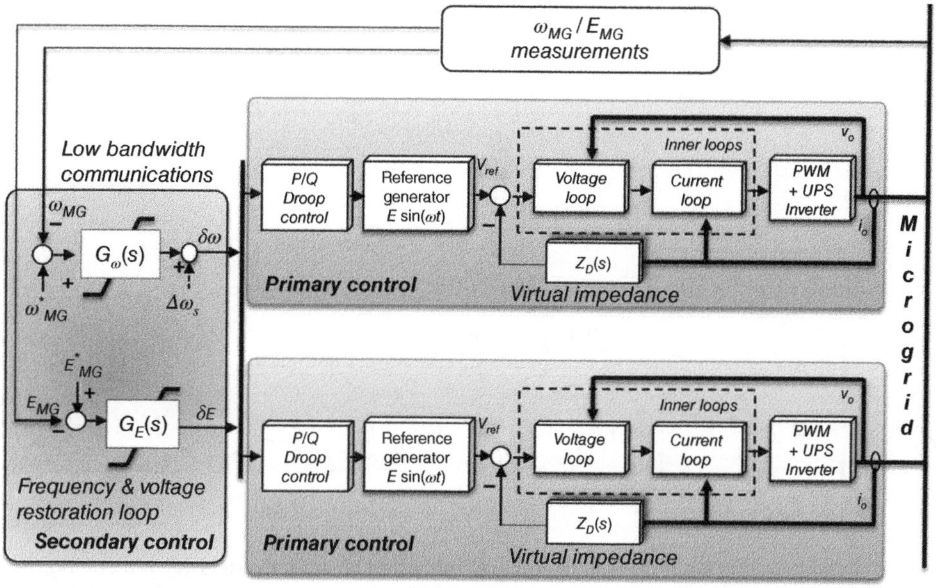

(a)

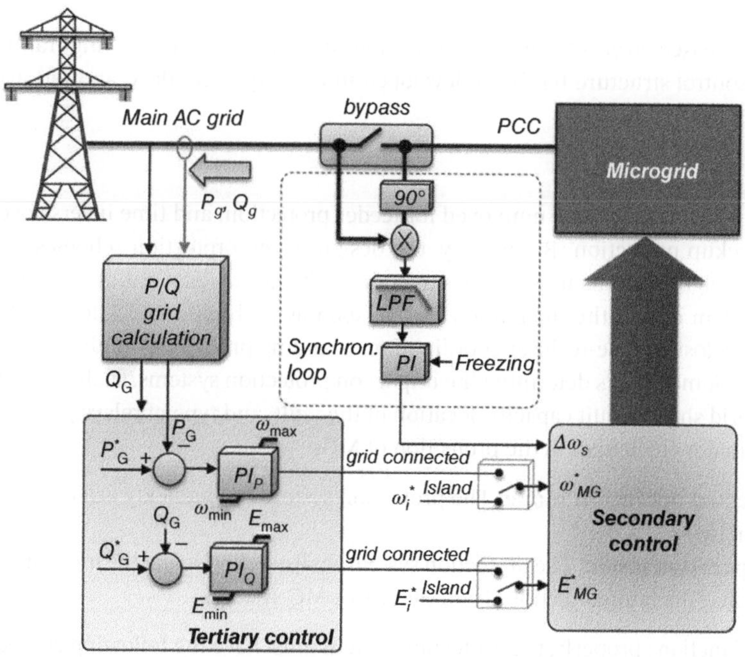

(b)

Figure 1.12 Block diagrams of the hierarchical control of an AC MG. (a) Primary and secondary controls of an AC MG. (b) Tertiary control and synchronization loop of an AC MG (Source: [26]).

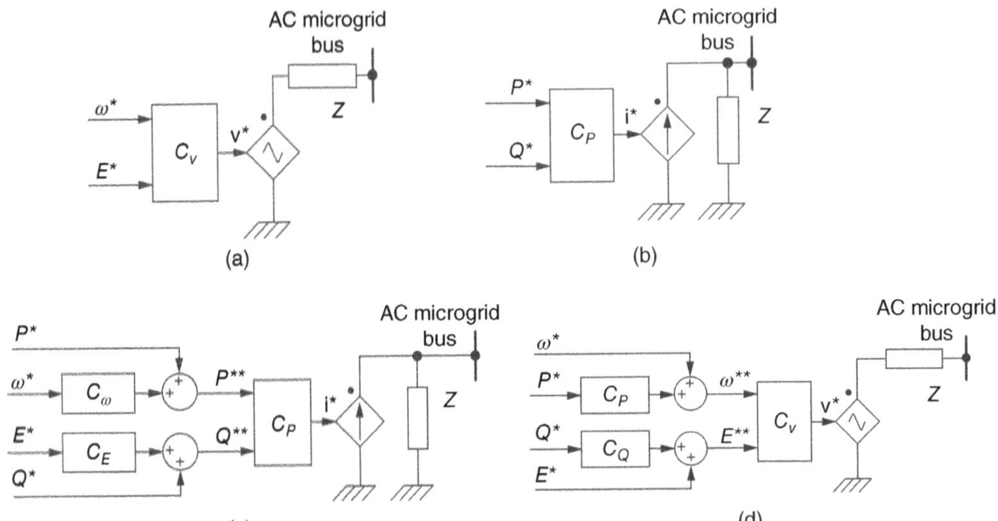

Figure 1.13 Circuit/block-diagram representation of grid-connected power converters: (a) gridforming, (b) grid-feeding, (c) current-source-based grid-supporting, and (d) voltage-source-based grid-supporting (Source: [31]).

have been classified based on grid-forming, grid-feeding, and grid-supporting configurations. Then, the hierarchical control structure has been developed in primary, secondary, and tertiary control levels (Figure 1.13 and Figure 1.14).

1.5.3.4 Microgrid Protection

Traditionally, overcurrent relays are employed for feeder protection, and time inverse overcurrent relays provide backup protection. By the way, the design of the protection schemes has moved from traditional design methods to adaptive protection schemes on account of some impacts on the protection system due to the integration of DG, such as prohibition of automatic reclosing, unsynchronized reclosing, fuse-recloser coordination, islanding problems, blinding of protection, and false tripping. Some factors determine the impact on protection systems, such as location, size and type of DG, grid short circuit capacity, location of the fault, and type of relays [16, 22, 32].

There are two major challenges in the protection of MGs

- **In grid-connected operation mode**: The protection system shall quickly isolate the MG from the upstream network.
- **In islanded operation mode**: The protection system shall quickly isolate the minimum possible section of the MG (the faulted section) from the whole MG system.

To perform its functions properly, the protection system must have the following characteristics: speed, selectivity, reliability, and sensitivity.

- **Sensitivity:** Sensitivity refers to the characteristic of a protective relay that operates reliably when required in response to a fault that produces the minimum short-circuit current flowing through the relay.
- **Speed:** The protection system's speed refers to the protection relays' operating times. The potential damage to the faulted element depends on the length of time the short-circuit currents are allowed to flow. The speed of clearing or isolating the faulted system component also affects the

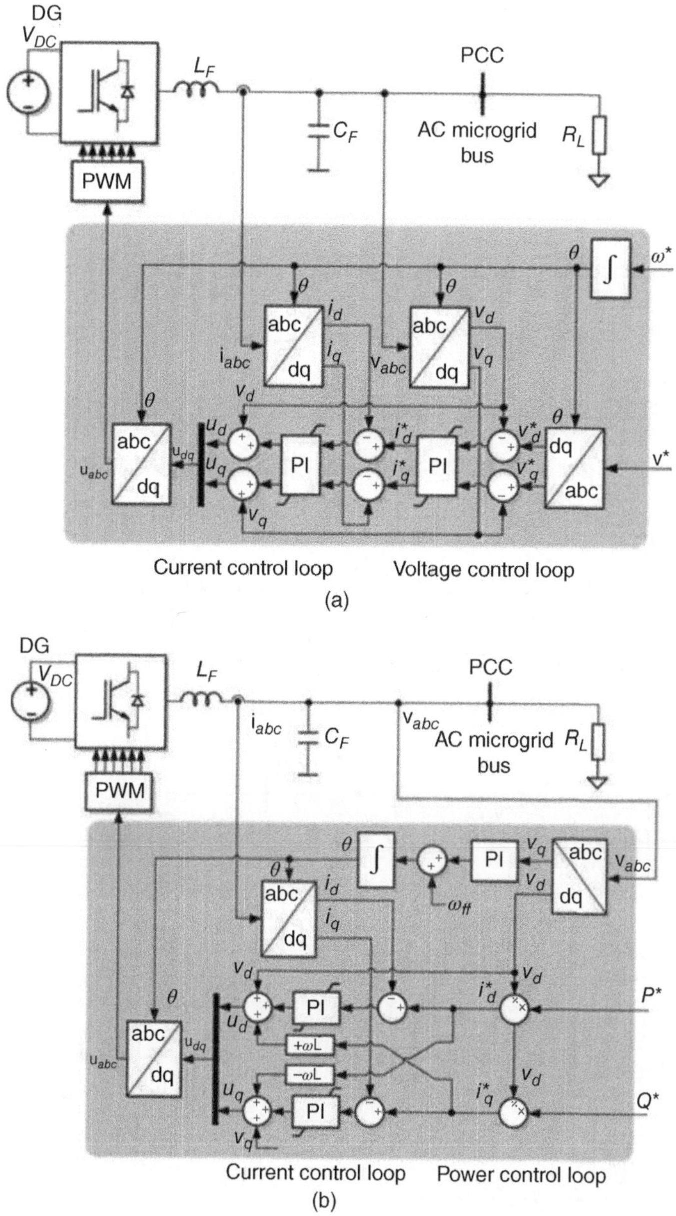

Figure 1.14 Basic control structure in (a) a three-phase grid-forming and (b) a three-phase grid-feeding power converter, (c) a grid-supporting power converter operating as a current source, and (d) a grid-supporting power converter operating as a voltage source (Source: [31]).

stability of the whole system. Protective relays may be instantaneous with an operating time of about 0.10 seconds or as high speed with a working time of less than 0.05 seconds. Solid-state or static relays can have operating times as low as one-quarter of a cycle.

- **Reliability:** The reliability of the protection system is its ability to operate upon the occurrence of any fault for which it was designed to protect. In other words, the protection system should operate when it is supposed to and not operate when it is not supposed to.

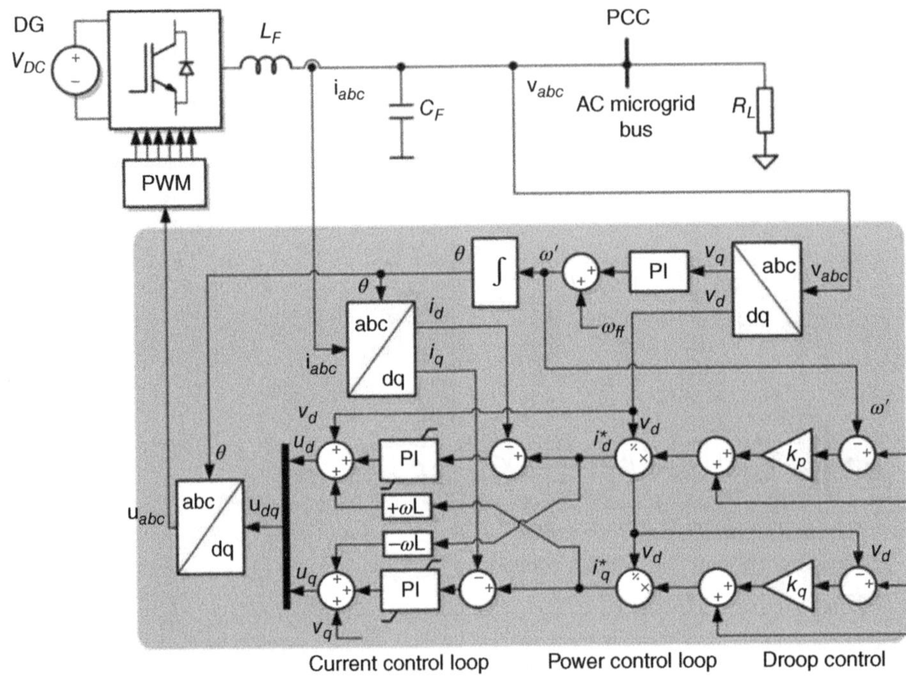

(c)

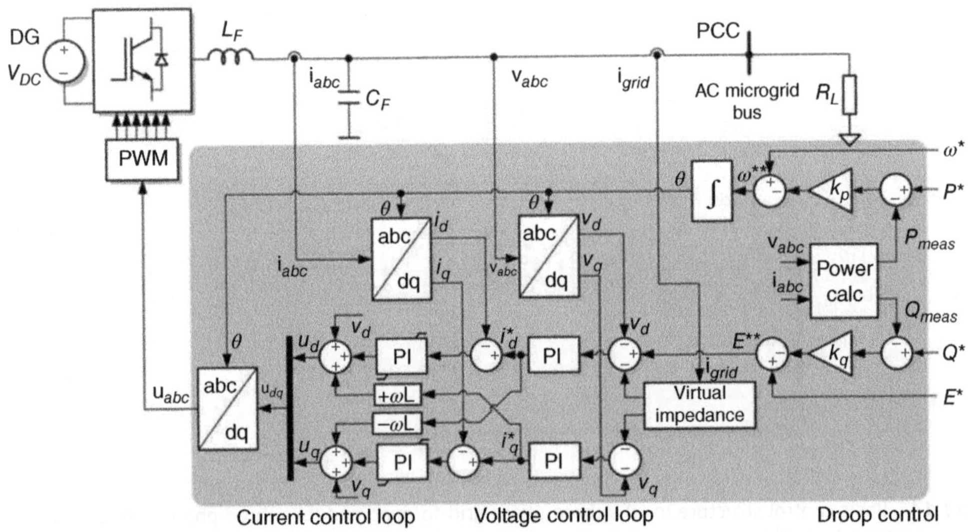

(d)

Figure 1.14 *(Continued)*

- **Selectivity**: Selectivity is the ability of the protection system to detect a fault, identify the point at which the fault occurred, and isolate the faulted circuit element by tripping the minimum number of circuit breakers. Selectivity of the protection system is obtained by properly coordinating the protective relays' operating currents and time delays.

Besides, traditional networks are almost radial and unidirectional (one-way feeding); consequently, they have simple coordination of protective equipment. By the way, in modern distribution systems in the form of ADNs, MGs, multi-feed networks, and networked MGs, new challenges like generation uncertainties, inverter-interfaced DERs, and the possibility of changing the operation mode are added that make coordination of protective equipment complicated.

Basically, we can categorize the protection conflicts caused by the integration of DERs into the DNs as follows:

- Sympathetic tripping
- Protection blinding
- Reclosing operation issue
- Change in short circuit level
- Coordination complexity
- Islanding and subsequent protection problems
- Low short circuit current in island mode

Finally, the future trends of the research in the field of MG protection include, but are not limited to, the following items:

- Communication infrastructure
- Fault location/fault detection
- Protection coordination
- Hierarchical protection strategies
- Adaptive protection schemes
- Monitoring and control protection schemes
- Interaction of protection system and control structure
- Blockchain-based modern protection strategies
- Real-time simulation test
- Fault current limiting strategies
- …

1.5.4 Virtual Power Plants (VPPs)

A virtual power plant (VPP) is a cloud-based distributed power plant that aggregates the capacities of heterogeneous DERs' capacities to enhance power generation, trade or sell power on the electricity market, and demand side options for load reduction. DER assets in a VPP can include PV solar, energy storage, EV chargers, and demand-responsive devices (such as water heaters, thermostats, and appliances), with examples of VPPs existing in the United States, Europe, and Australia. A VPP integrates several power sources to give a reliable overall power supply. The sources often form a cluster of dispatchable and non-dispatchable, controllable, or flexible load (CL or FL) DG units controlled by a central authority. It can include micro-CHPs, natural gas-fired reciprocating engines, small-scale wind power plants (WPP), PV, run-of-river hydroelectricity plants, small hydro, biomass, backup generators, and ESS. Benefits: the ability to deliver peak load

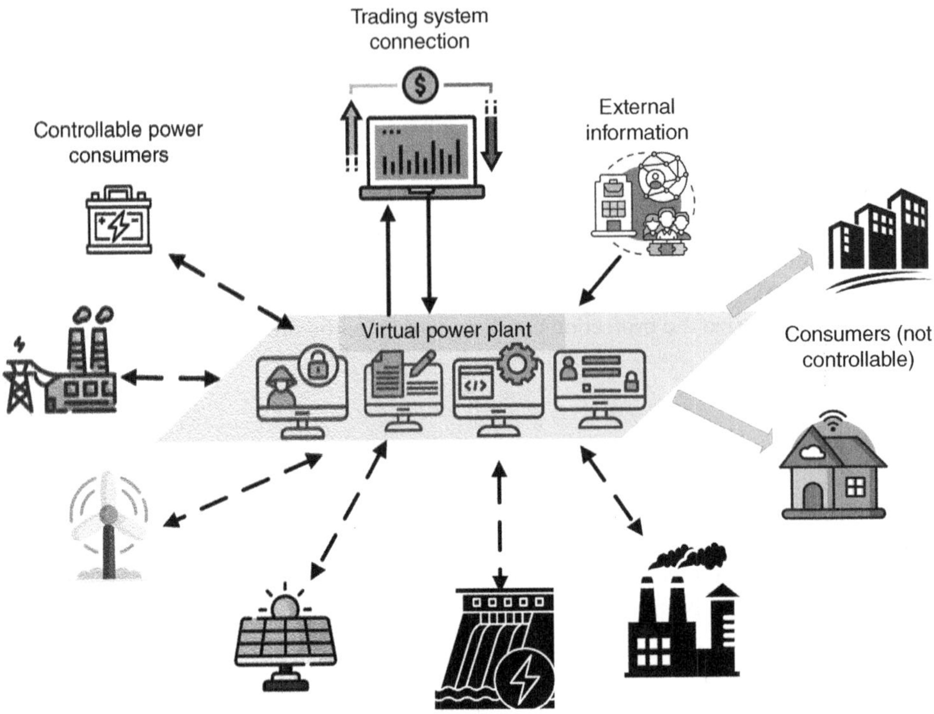

Figure 1.15 A block-diagram representation of the VPP.

electricity or load-following power generation on short notice (Figure 1.15). VPP can replace a conventional power plant while providing higher efficiency and more flexibility, which allows the system to react better to load fluctuations. The drawback is the higher complexity of the system, which requires complicated optimization, control, and secure communications [33–39].

"Virtual power plants represent an 'Internet of Energy'," said senior analyst Peter Asmus of Pike Research. "These systems tap existing grid networks to tailor electricity supply and demand services for a customer. VPPs maximize value for the end user and the distribution utility using sophisticated software-based systems. They are dynamic, deliver value in real-time, and can react quickly to changing customer load conditions." VPPs can also provide ancillary services to grid operators to help maintain grid stability. Ancillary services include frequency regulation, load following, and providing operating reserve. These services primarily maintain the instantaneous balance of electrical supply and demand. Power plants providing ancillary services must respond to signals from grid operators to increase or decrease load on the order of seconds to minutes in response to varying levels of consumer demand [33–39].

Future carbon-free electrical networks that include large proportions of solar and wind power will need to depend on additional sources of controlled power production or consumption since controllable fossil-fuel generators usually supply ancillary services. A prominent illustration of this is the Vehicle to Grid (V2G) technology (Figure 1.16).

In this scenario, a collection of dispersed, grid-connected electric cars may be coordinated to function as a single VPP. Through selective regulation of each car's charging rate, the grid observes a net injection or energy consumption as if a large-scale battery were providing this service.

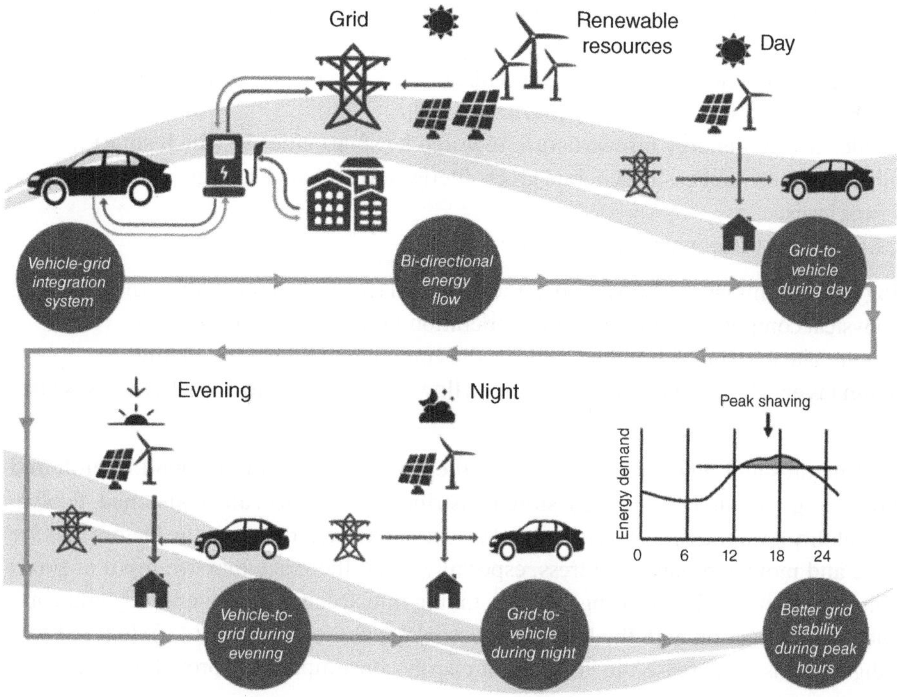

Figure 1.16 The configuration of vehicle-to-grid technology (Source: [40]/MDPI/Public Domain).

1.5.5 The Internet of Energy and Internet of Microgrids

1.5.5.1 Internet of Energy

The IoT is a name for the aggregate collection of network-enabled devices, excluding traditional computers like laptops and servers. Types of network connections can include Wi-Fi connections, bluetooth connections, and near-field communication (NFC). The IoT includes devices such as "smart" appliances, like refrigerators and thermostats; home security systems; computer peripherals, like webcams and printers; wearable technology, such as Apple Watches and Fitbits; routers; and smart speaker devices, like Amazon Echo and Google Home. The IoT promises to transform a wide range of fields. For example, connected devices can help medical professionals monitor patients inside and outside a hospital. Computers can then evaluate the data to help practitioners adjust treatments and improve patient outcomes. To know how the IoT works, first we should note that these devices use IP. The goal behind the IoT is to have devices that self-report in real-time, improving efficiency and bringing important information to the surface more quickly than a system depending on human intervention. The term "IoT" is attributed to Kevin Ashton of Procter & Gamble, who, in a 1999 article, used the phrase to describe the role of RFID tags in making supply chains more efficient [41–46].

As a definition, Energy Internet (EI) refers to a combination of advanced power and electronics technology, information technology (IT) and intelligent management technology, and a large number of new power networks, petroleum networks, natural gas networks, etc., which are composed of DG/DERs, distributed energy storage systems (DESs), and various types of loads. The IoE refers to the upgrading and automating of electricity infrastructures for energy producers and manufacturers. This allows energy production to move forward more efficiently and cleanly with the least waste. The term is derived from the increasingly prominent market for IoT technology, which has helped develop the distributed energy systems that make up the IoE. As noted above, it reduces inefficiencies, making energy transmission much more productive. There are also significant savings in money and a great reduction in energy waste. This, in turn, can be passed down to consumers or end users, who will also see a cost saving [39, 42, 47–50].

The EI provides the info-energy infrastructure required by smart cities in which smart homes, factories, utilities, EVs, etc. can easily generate green energy and exchange energy and information on a peer-to-peer basis.

Generally, EI includes three main layers (Figure 1.17):

1) **Technology layer**: This layer, also known as the physical layer, is the main core of the EI, which includes physical components such as loads, generation units, physical devices, etc.
2) **Market layer**: The market layer mainly reflects the commercial side of the EI and
3) **Information layer**: This layer is related to the data flow in different system levels. In EI systems, the energy.

Regulatory structures and new energy services in the smart grid seem to respond to profound changes in how energy should be generated, stored, distributed, managed, and consumed. Ecological and sustainability awareness, comfort-oriented consumer behavior, etc. are the factors that are becoming more and more complex to address, especially when there is a massive input of smart devices combined with multiple operating systems, taking into account interoperability and compatibility issues. In conjunction with the Fourth Industrial Revolution (Industry 4.0) (Figure 1.18), the IoT paradigm seems to increase the visibility of energy consumption by providing intelligent

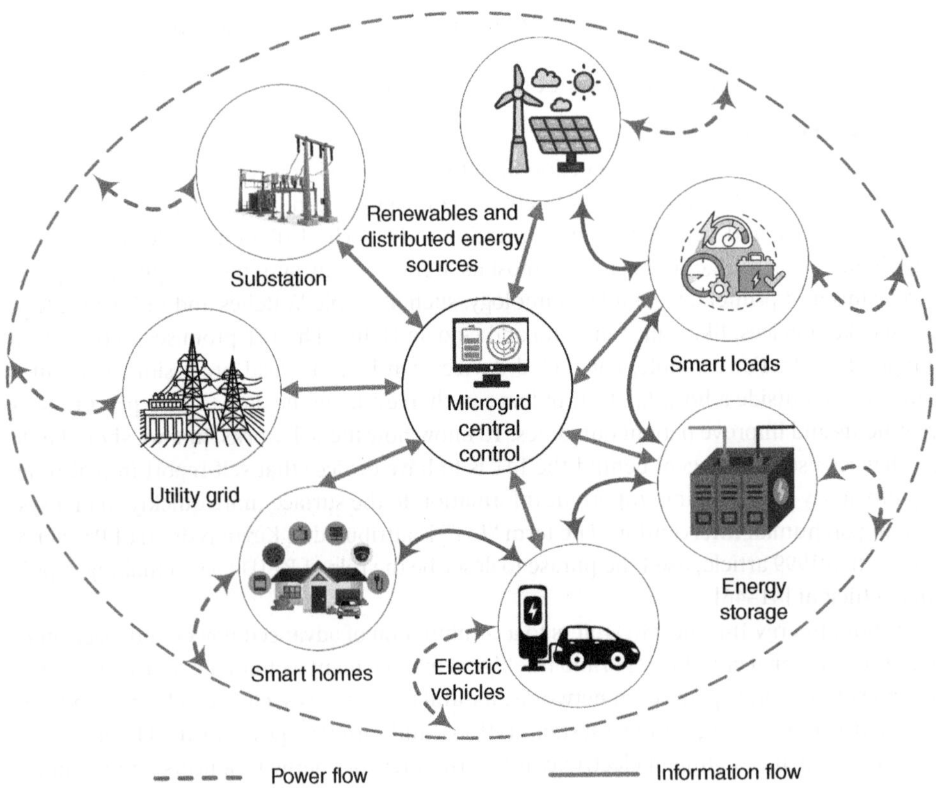

Figure 1.17 The Internet of Energy.

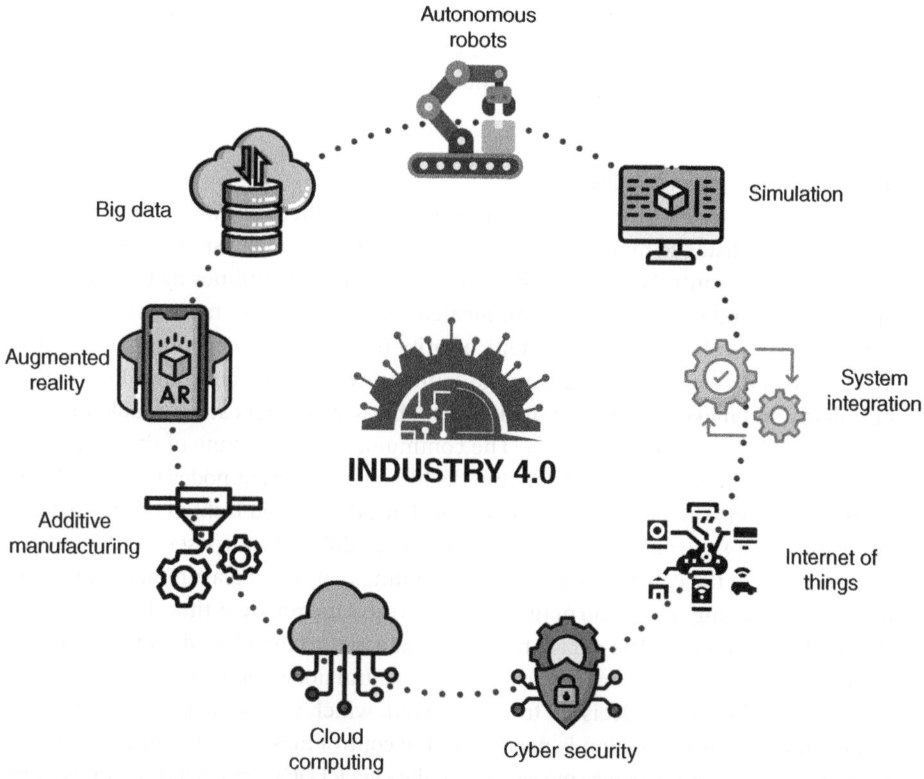

Figure 1.18 The elements of the industry 4.0.

and automated systems to improve comfort and energy efficiency through comprehensive solutions for connectivity, manageability, and security in future smart grids.

Also, thanks to the distributed allocation of smart sensors and smart meters, real-time energy consumption data can be collected easily and monitored in a user-friendly interface to improve energy-aware decision-making. We should focus on models, methods, and techniques/tools that will enable a new level of intelligence and effectiveness in IT infrastructures for a user-interactive energy system, which can also be called IoE. More specifically, advanced IoT-based structures, algorithms, and functionalities will be developed to initially interact with the surrounding environment in terms of sensing and metering (e.g., electricity, heat, and water in the case of Denmark), then process the information to extract knowledge for more efficient data management and transmission and eventually optimize energy efficiency in both supply and demand sides.

1.5.5.2 Energy Internet of Microgrids (EIoM)

The networked information system and communication infrastructure of a MG are the core components of its control structure because distributed/networked control systems (DCS/NCS) architectures intrinsically rely on tight coupling between the cyber and physical layers of MGs or co-multi-MGs. Smart energy systems are an essential open area of research for applications of IoT; however, the emerging EI and IoM systems are so complicated regarding operational requirements, communication infrastructure, and data interoperability systems. For developing a pilot and deploying a reference architecture for massive multi-party data exchange, management and governance, and real-time processing in the energy sector and translating this reference architecture into an open, modular data analytics toolbox for the safe and effective operation of grids and provision of innovative energy services, integration of relevant digital technologies like the IoT, AI/ML, cloud, and big data services should be enabled, which necessitates a unified data-driven EIoM framework.

1.5.5.3 Wide-Area Control System (WACS)

The wide-area control system (WACS) is introduced to respond to the real-time requirements of the power system. A large distance may separate the control centers and substations, so the control actions generated at the control center will be communicated to the utilities by the wide-area communication network, and the status is communicated back. The control messages in a smart grid can be of different levels of criticality. Therefore, priority-based routing, depending on the criticality of messages, is essential in controlling message communication. It can improve the system's reliability and stability and make good power quality available to customers at affordable rates by enabling a bidirectional flow of power and data. The communication backbone of the smart grid is the WAN, which will connect the small-area networks located at different nodes in a grid. Measurement and control actions should take place on the grid in a distributed fashion (Figure 1.19).

The communication in a smart grid can be mainly for three different purposes: (i) communicating the meter readings to the control station, (ii) distributing real-time measurement data, and (iii) controlling message passing. The cloud network receives the data from real-time data collection units (RTDCUs) and smart meters. The decision will be made in the cloud based on the collected data, and the control decision will be communicated to the central node (data collector [DC]). The central node will receive the control decision from the cloud, which will be delivered to the control devices. Bidirectional communication is needed for the control message communication in a smart grid on a real-time basis as it receives measurement data via RTDCU and meter readings from smart meters. A cloud is a collection of computers with databases housing RTDCU and smart meter data (or AMR or PMU), which will collect information from the grid, process the information, and generate the control decision (Figure 1.20).

1.5.5.4 Big Data Processing

After the technological revolution in power systems, large volumes of data are incorporated into the processes of generation companies and customers. Big data refers to a large volume of received data from gateways (with distinct protocols) consisting of different categories (weather, power grid, MG, etc.) and needs a framework to process all the information effectively. As shown in the middle part of Figure 1.20 (big data processing), all these data are imported to a cloud via data gateways. Then, they should be managed through some steps, i.e., data integration, data cleaning, data transformation, and false data detection. As a case in point, we detected duplicated data and removed it (e.g., keep the last measurement). The next process is data cleaning and missing data (e.g., because of smart meter error), and abnormal values (e.g., bad measurement) should be detected to avoid AI and ML algorithms such as long short-term memory (LSTM), gated recurrent units (GRU), etc. from miss-training. In the third step (i.e., data transformation), we applied data analysis methods such as standardization. In the last step, false data should be detected by the implemented technologies and algorithms.

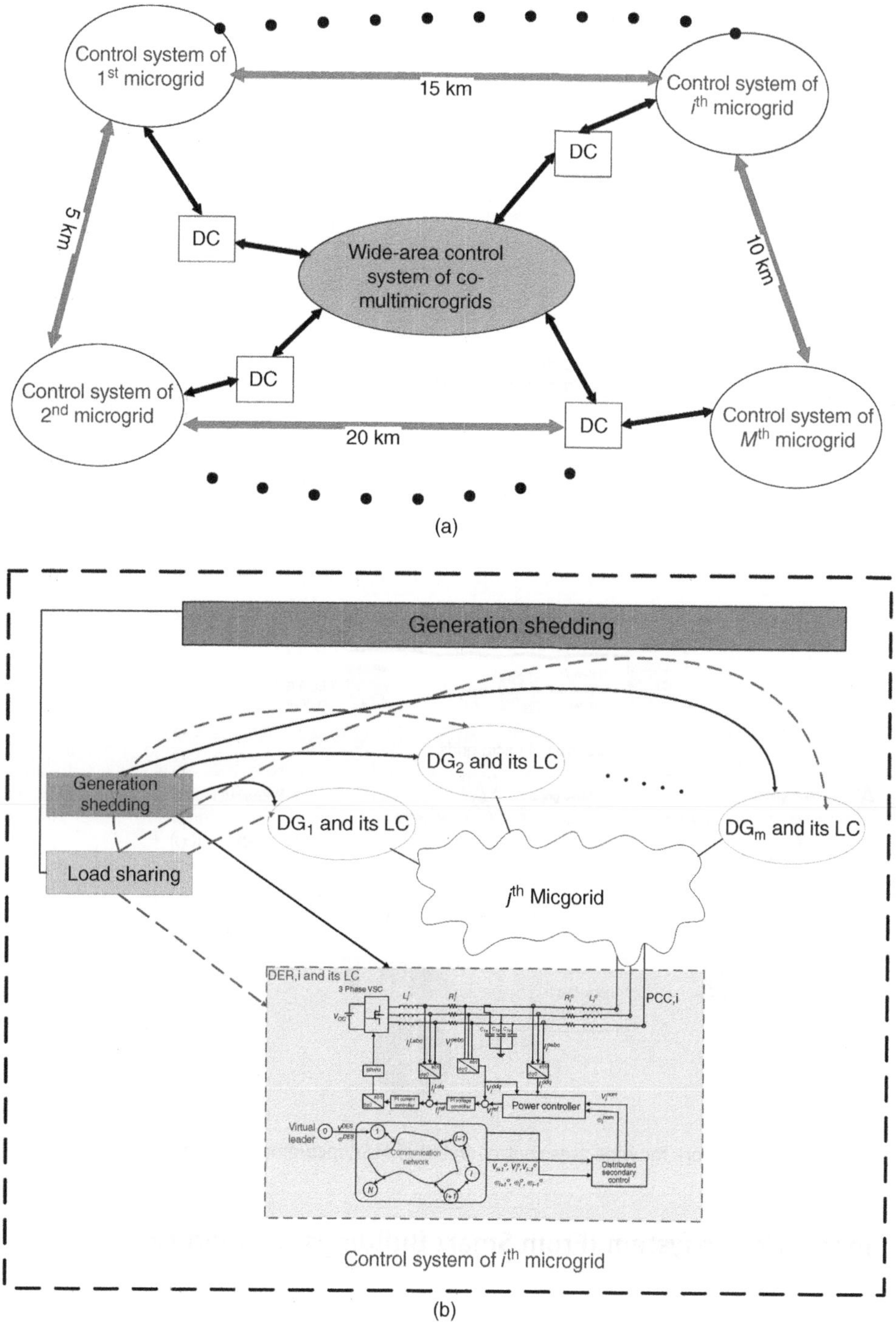

Figure 1.19 (a) WACS communication network for sample smart MG (distances are typical) and (b) control system of ith MG.

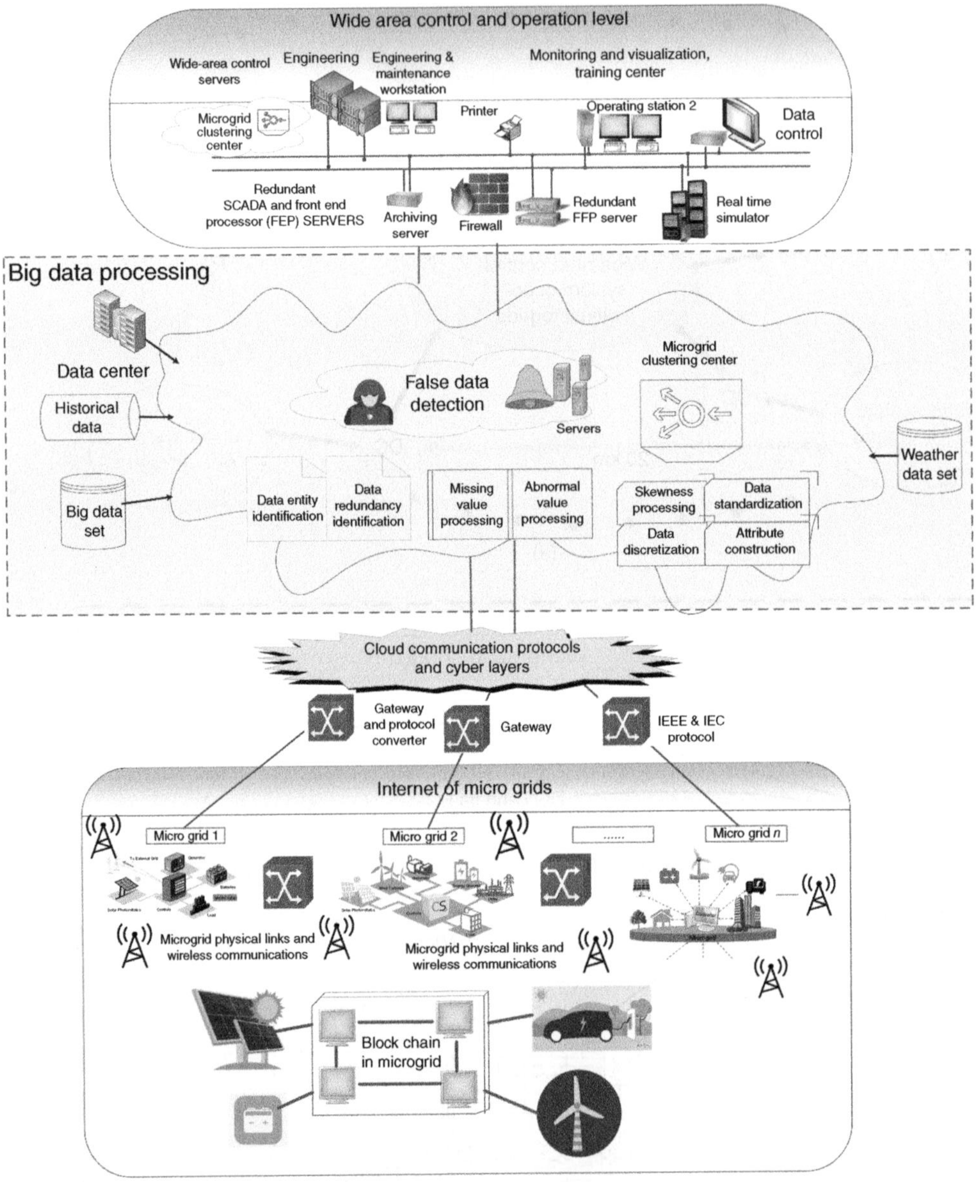

Figure 1.20 The big framework for wide-area control and EI of MGs, including big data processing.

1.6 Smart Grid Ecosystem (From Smart Buildings to Smart Grid)

Let's see the current situation as follows:

- Development of telecommunication networks (with more economical solutions)
- Advancement of IT
- Integration of DERs
- Distribution companies are under great pressure to meet new electrical energy needs

Table 1.1 Assimilation of smart human with smart grid.

Smart human characteristics	The similar characteristics in smart grid	Objective	How it can be realized?
Five Senses	The sensing and measurement equipment	Measurement of the parameters	CT, PT, PMU, μ-PMU, and smart measurement devices
The nervous system of the human body	The communication network	Data transmission	PLC (Power Line Carrier), AMI (Advanced Metering Infrastructure), HAN (Home Area network), fiber optic, ...
The brain	The control system	The data analytics and sending the commands	Demand response, OPF, Volt/VAR optimization, AGC, Oscillation damping, ...
The Muscles	Automation system/equipment and actuators	The execution of the commands to affect the system	DGs, HVDC lines, FACTS/Custom power devices, ESSs, DERs, Automatic switches, Reclosers, ...

- Less than 75% of substations in North America are equipped with communication and information exchange systems, and about 15–20% of distribution feeders have automation equipment
- The origin of 90% of outages is in the DN.

DER can be DG, ESS, EV, or controllable load. Therefore, we need the intelligent behavior of the networks like a human being (increasing the intelligence quotient [IQ] of the electricity network to the extent that the network's response is intelligent from our point of view).

By looking at the current situation, we found high energy demand, restructuring and creating competition, and low financial resources that will result in exploitation of the power system within the limits of exploitation, loss of service quality, and reduced security. These conditions allow us to have a more efficient network, namely, the emergence of a smart grid.

By the way, if we assimilate the smart grid to a smart human, we can present the summaries in Table 1.1. Another definition for a smart grid is "a network with two-way power and information/data exchange, based on communication with the following capabilities: monitoring, data acquisition, data processing, and control." It refers to a smart network whose elements can exchange two-way information (visibility), make decisions, and implement them on the spot (controllability).

The main components of smart grids are:

- **Advanced metering infrastructure (AMI):** A smart grid cannot be implemented without a smart meter and data management system.
- **Distribution and outage management (DMS):** The possibility of real-time management under critical and unstable conditions. Software programs must be able to identify and act spontaneously to have the least impact on subscribers.
- **Distribution/substation automation systems (DAS/SAS):** Automation in the DN lags far behind the transmission system. Automatic switches are practically not used in DNs.

There are several objectives followed by smart grids, including:

- **Self-healing:** A smart grid should be self-healing to heal itself
- A smart grid should motivate consumers to participate actively in the operations of the grid

- Enough level of robustness
- **Cyber-resiliency:** It should be resilient to energy or information theft and/or cyberattacks
- **Power quality:** It should provide higher power quality that will save money wasted from outages
- **Integration of DERs:** It should accommodate all (possible) generation and storage options
- **Enabling power market:** It should have an active electricity market
- **Asset management and efficient operation:** It should optimize assets and operate efficiently
- Reducing the environmental pollution.

Self-healing refers to the engineering design in which the elements of the network that are in trouble are isolated automatically and restored to their normal function with little or no human intervention. As a result, following self-repair operations, providing services, and delivering electric energy to subscribers will be possible without or with minimum interruptions. The benefits of self-healing include:

- Network operation with sufficient security margin and minimal risk,
- Activating corrective and remedial operations,
- Limiting the destructive effects of network errors to the smallest possible area, improving power quality and reliability.

The self-healing necessitate the following items:

- Simultaneous access to required data and parameters
- A fast computational and analytical tool that can estimate the state of the network and damage conditions within a few seconds and estimate definite or random problems in the system
- Continuous and simultaneous evaluation of system status
- Anticipate possible problems
- Diagnosing network errors
- Fast-switching equipment that can reduce the corresponding destructive effects.
- Advanced protective relays with simultaneous and adaptive adjustment
- Redundancy (high availability).

For the self-healing implementation, we should consider control and monitoring layers so that they operate hierarchically. The closer the layers are to the power grid equipment, the shorter the response time, and the farther the layers are from the endpoints and monitor a larger area, the longer the response time. Response time from 10 ms to several hours. The self-healing concept can also be developed to protect power networks. This concept requires a protection philosophy that automatically adjusts network protection relays according to different system conditions and satisfies the coordination of relays in these conditions.

Smart grid projects and technologies are complex and involve many different stakeholders across multiple organizations.

Many utilities have aggressively sought to implement the benefits of smart grid technologies, only to discover they started too early or did not fully understand the risks involved. Smart ecosystem analysis can help utilities and vendors understand how their initiative will fit within the industry and how to manage their project through the complex environment. The SGE is a complex network of interconnected elements.

The SGE shown in Figure 1.21 consists of four key elements associated with some keywords as follows:

- **Smart grid:** Bi-directional energy flows, remote control/automation of power, and integrated distributed energy
- **Smart city:** Complex interconnected infrastructures and services

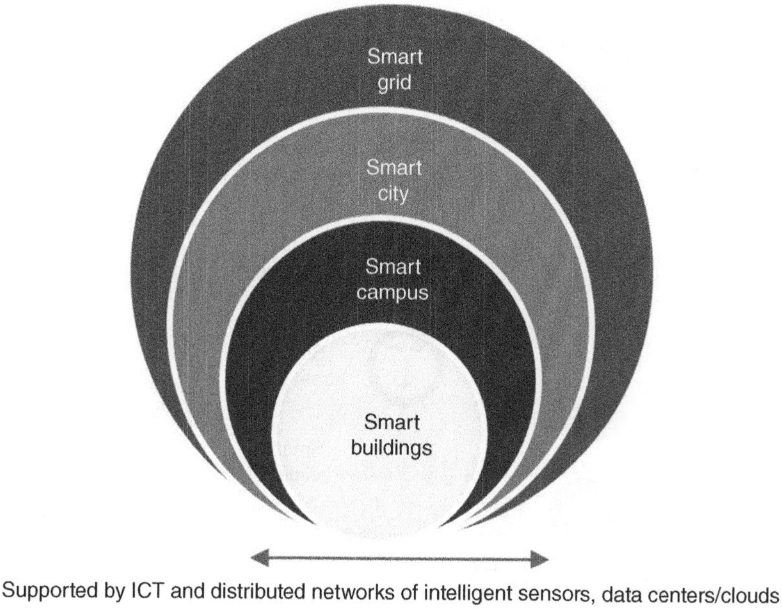

Figure 1.21 The smart grid ecosystem.

- **Smart campus:** A collection of buildings managed by the same facility manager
- **Smart buildings:** Intelligent building automation systems, smart devices, productive users, and grid integration.

In Chapter 11, we will discuss the smart building concept and the developed *BEMOSS* platform. If we install the *BEMOSS* platform in different buildings such that a facility manager can manage this collection of buildings, a smart campus will be created. A complex system of interconnected infrastructure and services will create a smart city in the next step. These three blocks will emerge in smart city, where we have (i) bi-directional flows of energy, (ii) remote control/automation of power, (iii) integrated distributed energy, etc.

It should be noted that the smart building concept mentioned in this ecosystem is not just a load. It is a grid-interactive efficient building (GEB), an asset for the smart grid. It can come to its service when needed to provide security and resilience. To this end, buildings can operate in combination with DER such as PV and BESS in the presence of EV charging stations. To clarify more on the SGE, smart city and smart grid are briefly discussed.

1.6.1 Smart City

1.6.1.1 Smart City Definition and Elements

There is no single consensus definition of a smart city. Still, there is some agreement that a smart city is one in which ICT facilitates improved insight into and control over the various systems that affect the lives of residents. A smart city is an urban development vision to integrate ICT and IoT technology in a secure fashion to manage a city's assets (Figure 1.22). To be fully "smart," a city must be "connected."

Smart city building blocks include:

- **Smart care:** Hospital and healthcare
- **Smart grid/energy:** Reliable and low-price electricity and availability when needed

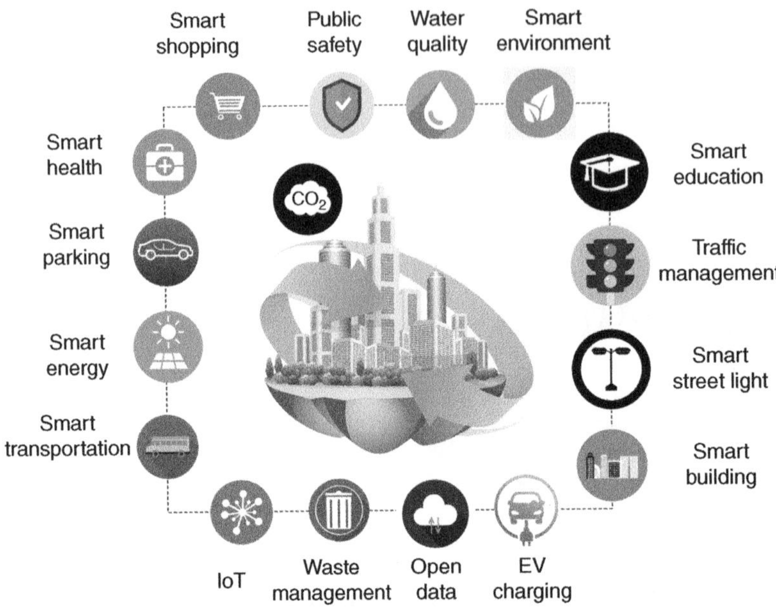

Figure 1.22 Smart city building blocks.

- **Smart society:** People living, working, and enjoying the environment
- **Smart office/buildings:** Smart home/buildings and smart government
- **Smart mobility:** Smart and connected transportation
- **Smart space:** Park and entertainment places and smart agriculture.

As illustrated in Figure 1.22, a smart connected city is a system of interconnected systems, including health care, employment, retail/entertainment, transportation, energy distribution, residences, public services, education, etc. The SoS is tied together by ICTs that transmit and process data about all sorts of activities within the city.

1.6.1.2 Range of Deployments in a Smart City

Cities worldwide are deploying technology to gather data and trying to become cleaner, reduce traffic, and improve urban life. Starting with energy management (reliable and low-cost energy), disaster preparedness (flood, …), public safety (police, Ambulance, …), parking spot assistance (finding open parking lots with an app), paying bills online, facilitating emergency vehicle movement (giving priority to ambulance and fire truck), and much more.

1.6.1.3 Smart Traffic Control

Three main elements of smart traffic control are as follows:

- A smart traffic crossing sensitive to traffic volume
- Synchronized traffic lights for smooth flow
- Emergency vehicle priority access.

Figure 1.23 shows an optical-based traffic signal preemption system for emergency and transit vehicles. In this case, the detector can sense the signal sent by the emitter installed in ambulances, transit buses, etc. As a result, the signal turns green to allow the ambulance to pass quickly in an emergency. To make the process more reliable, the confirmation device should send an acknowledgment to the ambulance and confirm that the signal has been received and the ambulance can pass safely.

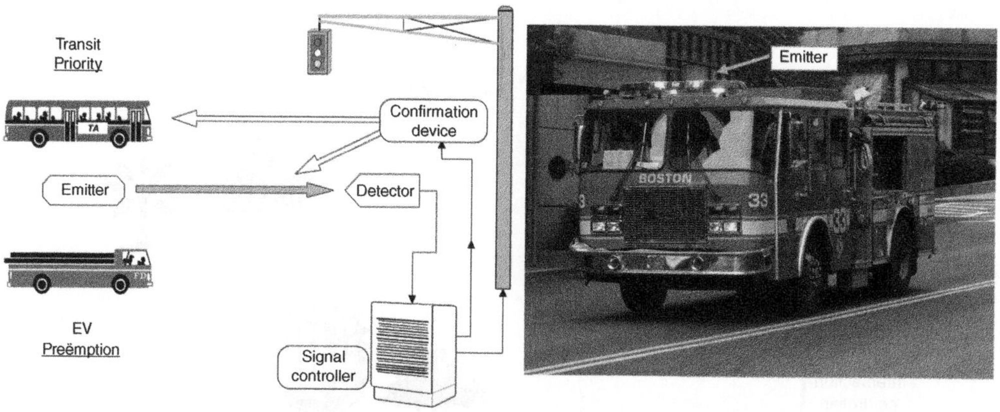

Figure 1.23 Optical-based traffic signal preemption system for emergency and transit vehicles.

1.6.1.4 Connected Transportation

In a smart city, connected transportation will allow vehicles and travelers to share data with equipment and procure mobility as a service, whenever and wherever. As a case in point, one may drive to a metro substation, park the car there, get to a place close to the destination, rent a bicycle, and go to the final destination. As another example, the connected transportation system allows buses more than a minute behind schedule to receive priority at traffic lights automatically.

1.6.1.5 Hierarchical Framework for Intelligent Traffic Management in Smart Cities

Intelligent transportation systems and smart cities are closely coupled, where various technologies are integrated to improve the quality and efficiency of urban services using advanced technologies and data analysis. For intelligent traffic management purposes, AI, ML, and real-time data analytics are employed to manage traffic flow. As shown in Figure 1.24, we may have intersection controllers to identify bottlenecks, suggest alternative routes to drivers, and change traffic light patterns based on current traffic conditions to reduce wait times and improve flow.

1.6.1.6 Smart Lamppost Embedded with Camera and Sensor

As shown in Figure 1.25, a camera on Lamppost scans the space around it. Therefore, it can locate empty parking spaces and broadcast the information to an app, and one can find what street and location parking is available. Other lamp post capabilities embedded with camera and sensor are as follows:

- **Crowd analytics**: Lamppost can detect and analyze crowd congestion and patterns to detect different situations, such as traffic congestion, bus breakdowns, unruly crowds, etc.
- **Face detection**: Artificial-based cameras embedded in Lamppost can analyze race, gender, age, face detection compared to the cloud database, etc.
- **Environmental sensor**: Weather data such as temperature, humidity, pressure, rainfall, air quality, etc., can be recorded by Lamppost and sent to autonomous cars to improve road situational awareness.
- **Personal mobility device**: Detect the speed of mobility devices and bicycles and alert the related agency if the speed is higher than the limits.

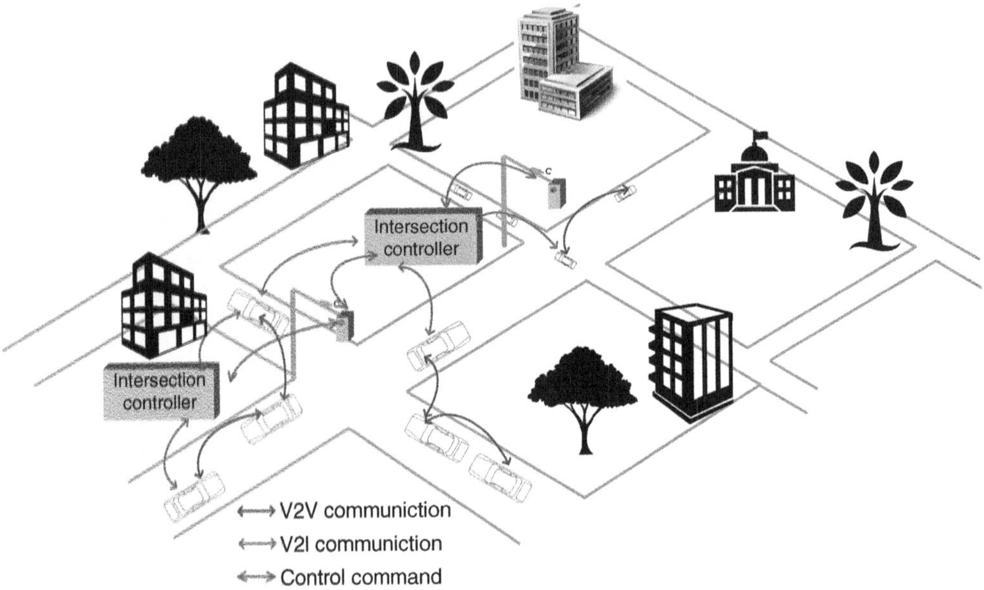

Figure 1.24 Intelligent traffic management in smart cities.

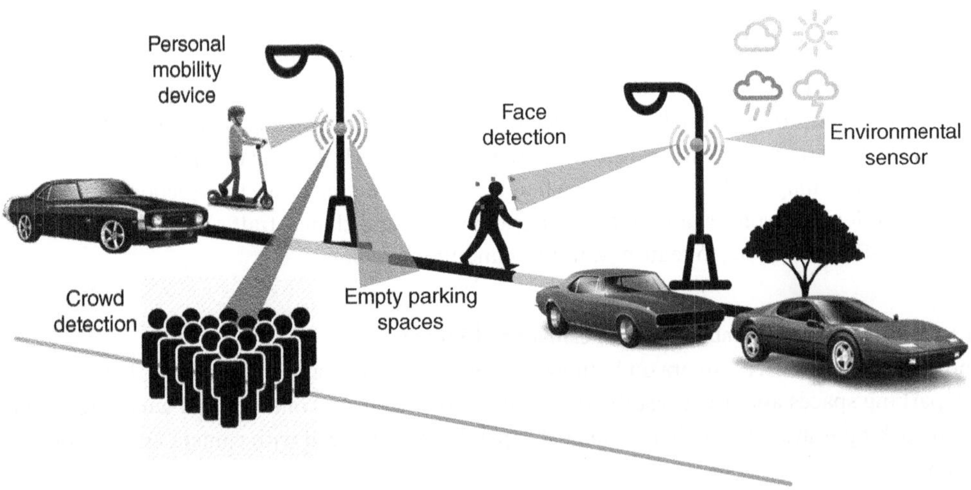

Figure 1.25 Smart lamppost with camera and sensor.

1.6.1.7 Smart Garbage Bin

Installing a sensor at a trash can helps municipality people see the trash level and send a signal when it is full. Therefore, they don't have to come when the trash isn't full and will save money and time. Moreover, a trash compactor will push down plastic and paper to increase the space. Typically, the trash bin should be emptied one to three times per day, depending on traffic. On the contrary, smart ones only need to be emptied four times a week by employing a trash compactor and sending a signal to a city office/or truck whenever it is full (Figure 1.26).

Figure 1.26 Smart garbage bin concept.

1.6.2 Smart Grid

1.6.2.1 Smart Grid Definition

A smart grid is a concept with many elements where monitoring and controlling each component of the chain of generation, transmission, distribution, and end-use allows electricity delivery and use to be more efficient (Figure 1.27). The goal is to make the grid smarter, safer, reliable, and more cost-effective using advanced sensors, communication technologies, and distributed computing. We have two following definitions that distinguish between traditional and modern power systems.

- **Historically: Demand-Driven Supply**

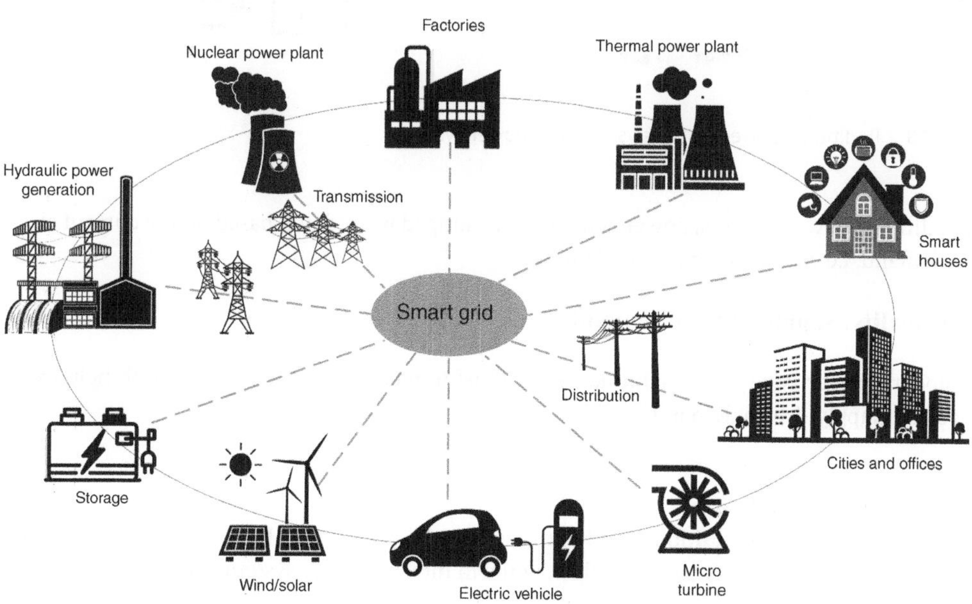

Figure 1.27 Smart grid elements.

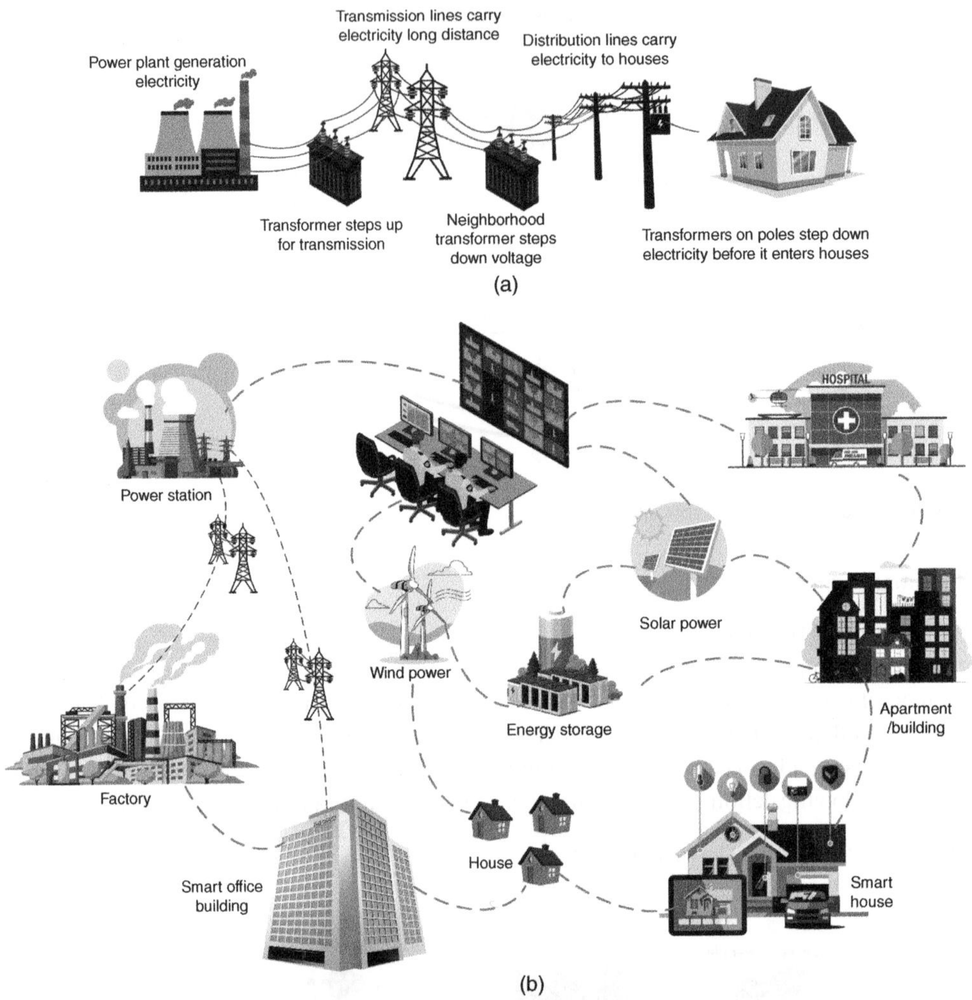

Figure 1.28 Traditional power systems vs. modern power systems.

In traditional power systems, power generation is ramped up or down based on the current electricity demand, i.e., supply responds to demand.

- **New Reality: Supply-Driven Demand**

In modern power systems, demand needs to adjust to meet fluctuating supply with help from storage, i.e., supply-driven demand.

1.6.2.2 Evolution of the Grid

The difference between a traditional grid and a smart grid is shown in Figure 1.28. In traditional power systems, we have one-way power flow, while in modern power systems, the two-way power flow concept is introduced.

1.6.2.3 One-Way Power Flow in Traditional Power Systems

1) **Linear flow**: In traditional power systems, power flows in a single direction—from generation to consumers. Power plants generate electricity, which is transmitted over long distances through high-voltage transmission lines and distributed to customers through local DNs.
2) **Centralized generation**: Electricity is typically generated at large, centralized facilities, often powered by fossil fuels, nuclear energy, or large-scale hydroelectric plants. These facilities are located away from consumption centers due to their size and operational requirements.
3) **Limited interaction with consumers**: Consumers are passive recipients of electricity. The utility's interaction with the consumer is generally limited to billing and basic services. In this case, there is minimal consumer engagement in the electricity generation/distribution process.

1.6.2.4 Two-Way Power Flow in Modern Smart Grids

1) **Interactive flow**: Moder power systems and smart grids enable a two-way flow of electricity and information. Besides receiving power from the grid, consumers can generate electricity (often through renewable sources like solar panels or EVs) and feed it back into the grid. They can make money in this case, and we have a power market concept.
2) **Distributed generation and MGs**: Modern power systems facilitate DG, where electricity is generated from multiple sources close to the point of consumption. MGs are small-scale power grids that can operate independently or in conjunction with the main grid, contributing to the resilience and flexibility of the overall power system.
3) **Active consumer participation**: Consumers transform into "prosumers"—producers and consumers of electricity. They can manage their energy usage more actively, participate in demand response programs, and sell excess energy to the grid.
4) **Integration of renewable energy**: In smart grids, renewable resources are integrated. The variability of renewables like solar and wind is managed more effectively, with energy storage technologies and smart management systems helping to balance supply and demand.

Therefore, after the revolution in power systems (two-way *power flow*)

- Relay capabilities and their coordination should be studied.
- Utilities must update their equipment with reverse-power-flow logic.
- Employing demand response and PV can help Shave Peak Load.
- The impact that PV/Wind might have on the grid side should be investigated.
- The intermittent behavior of PV/wind should be considered in two-way power flow calculations.
- PV tends to bring the voltage up at the point of interconnection. Also, when clouds decrease PV output voltage drops, it results in voltage flickers and fluctuations on the customer's side and should be considered.
- **PV/wind/EV positive impact:** Decrease CO_2 emission (decarbonization)
- In grid to vehicle (G2V) and V2G modes, the stochastic behavior of both vehicle location and its consumption/generation should be taken into account for two-way power flow calculation.

This transition from one-way to two-way power flow marks a significant evolution in how electricity is generated, distributed, and consumed, leading to the transformation of the power industry toward greater efficiency and sustainability. However, transitioning to a two-way power flow system introduces new challenges that need to be addressed, such as ensuring grid stability in case of fluctuating renewable energy sources, enhancing cybersecurity to protect against potential threats, and developing sophisticated energy management systems to effectively balance supply and demand in real-time, which are investigated in this book.

1.6.2.5 Smart Grid Building Blocks

The key building blocks of a smart grid are as follows:

1) **Advanced metering infrastructure (AMI)**: AMI includes smart meters that record real-time electricity usage and send it to a control center. These meters allow for two-way communication between consumers and utilities, enabling dynamic pricing, demand response, and enhanced energy consumption management.
2) **Communication networks**: In smart grids, we have power flow and data flow concepts together (Figure 1.29). Robust and secure communication infrastructure is crucial for real-time data exchange between different smart grid equipment. It can be wireless and wired technologies, facilitating the flow of information for monitoring, control, and management of the grid.
3) **Sensors and IoT devices**: Different sensors and IoT devices are employed throughout the grid to collect data on electrical loads, equipment health, grid stability, and other critical parameters. 3. Sensors and IoT devices provide the required data for real-time analytics and decision-making.
4) **Distributed energy resources (DERs)**: DERs include small-scale electricity generation or storage technologies like wind turbines, solar panels, and battery storage systems. They are often located near electricity consumption points. They can operate independently or interact with traditional power.
5) **Energy management systems (EMS)**: EMS are software tools that analyze the data collected from various components in the power grid. Then, EMS uses all gathered information (e.g., breaker status [open, close], Governer setpoints as an analog signal, transformer tap changer, ...) to optimize energy production, distribution, and consumption, ensuring stability and security in power system operation.

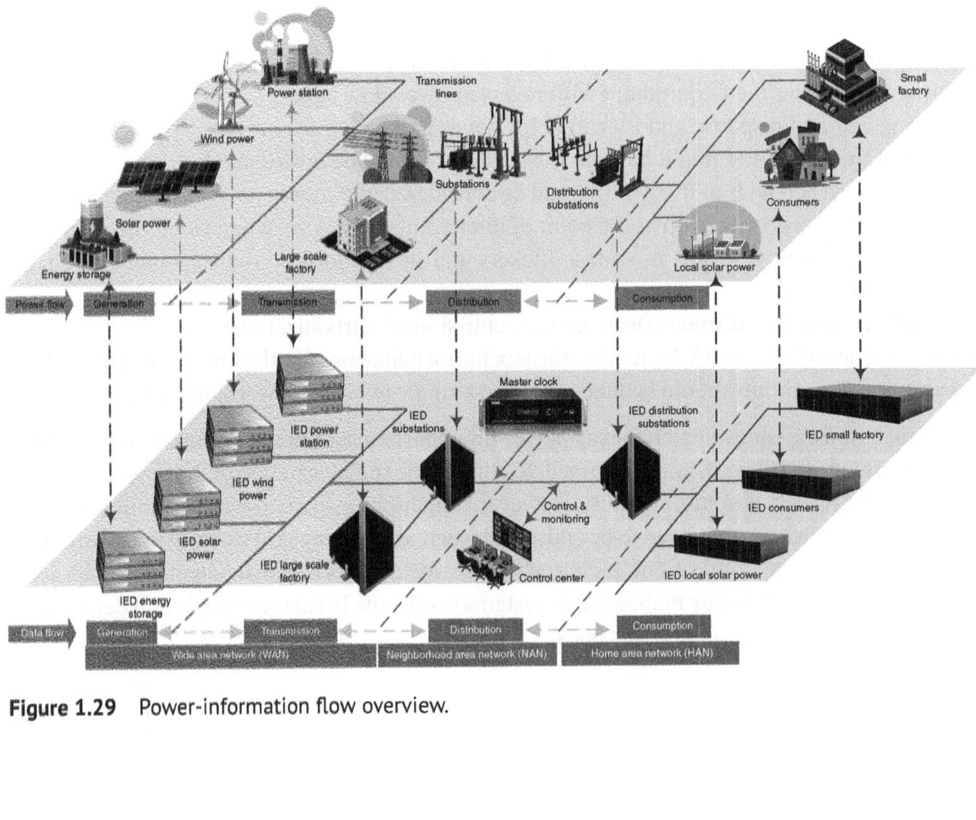

Figure 1.29 Power-information flow overview.

6) **Grid automation technologies**: This includes technologies for automating various grid operations, like switching, fault detection, and self-healing capabilities. Automation enhances the reliability and resilience of the grid, especially in response to outages or disturbances.
7) **Cybersecurity measures**: Cybersecurity is a critical component as smart grids rely heavily on data and networked communications. This involves deploying advanced security protocols and systems to protect the grid from cyber threats and ensure data integrity and privacy.
8) **Demand response (DR) systems**: DR systems enable consumers to play an active role in grid management by reducing or shifting their electricity usage during peak periods in response to signals or incentives from utilities.
9) **Big data analytics and AI**: The use of big data analytics and AI algorithms is crucial for processing and interpreting the vast amounts of data generated by the grid. This facilitates predictive maintenance, load forecasting, anomaly detection, and other intelligent functionalities. The big data concept (Figure 1.30) and related challenges and solutions in modern power systems are discussed in Chapters 37 and 38.
10) **Regulatory and policy frameworks**: Effective regulatory policies and frameworks are necessary to facilitate the integration of these technologies into the existing power systems, ensuring smooth operation, fair pricing, and the protection of consumer rights.
11) **Electric vehicles (EVs):** As an increasingly significant component of the SGE, EVs represent a shift toward sustainable transportation and contribute to grid stability. With storage capabilities, EVs can act as mobile energy resources, offering services like V2G, where the battery power can be fed back into the grid during peak demand times. This integration of EVs into the grid requires advanced management systems to ensure efficient energy distribution and charging infrastructure planning

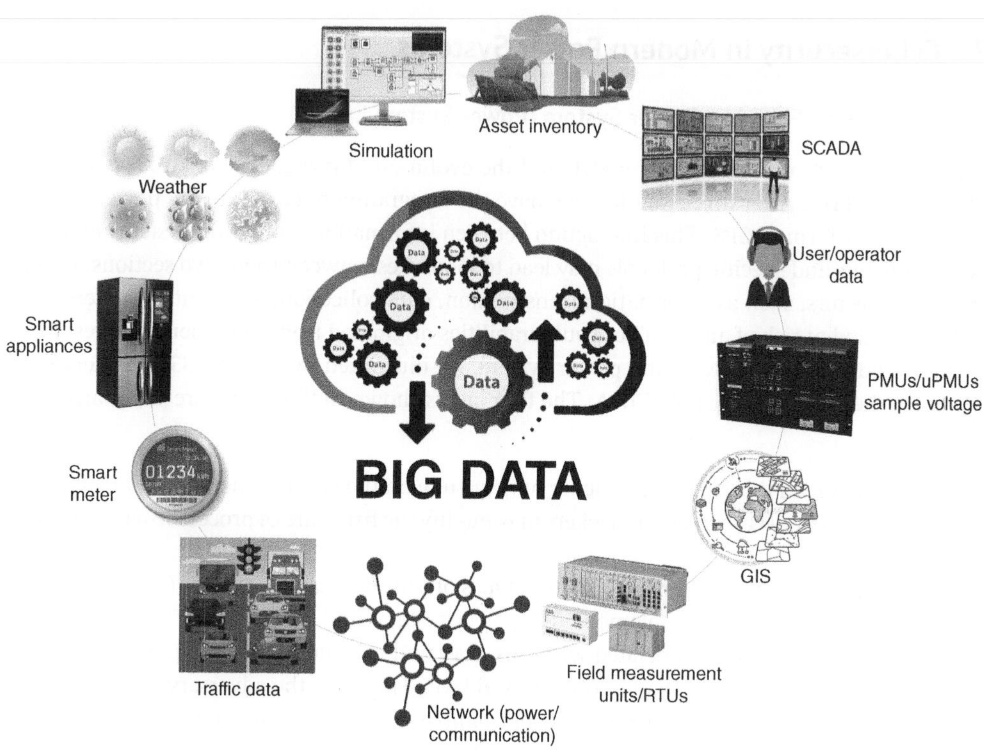

Figure 1.30 Big data emergence in moden power systems.

1.6.2.6 Intelligent Interconnected Microgrids

Figure 1.31 represents a part of an advanced CPPS, illustrating the concept of intelligent interconnected MGs. Central to this figure is a WACS, the hub for coordinating and managing multiple MGs. These MGs are interconnected and independently operable, showcasing a multi-cluster configuration. This design highlights the intricate interplay between individual MG control and centralized management. Each MG, as a localized grouping of energy resources and loads, operates autonomously and in concert with others, reflecting a sophisticated blend of self-sufficiency and communal energy sharing. The WACS, with its advanced communication and control capabilities (e.g., load shedding, load sharing, ...), ensures optimal operation, stability, and efficiency across the network. This setup enhances the resilience and reliability of the power system and exemplifies the potential for scalable and flexible energy distribution in modern CPPSs.

Figure 1.32 expands on interconnected MGs by introducing a detailed view of the cyber-physical co-multi-MG system. The red dotted lines represent the cyber network, which is crucial for the exchange of information. These lines connect various MGs at the tertiary level, facilitating communication and coordination for efficient energy management. Additionally, they link DERs at the secondary level, ensuring that smaller, decentralized energy-producing units are integrated seamlessly into the broader grid system. This representation underscores the importance of cyber networks in maintaining the harmony and efficiency of complex multi-MG structures.

Integrating advanced technologies, communication networks, sensors, and IoT devices into power systems while offering numerous benefits in terms of efficiency, sustainability, and flexibility also inevitably increases the vulnerability of power systems to cybersecurity threats and attacks. Cybersecurity is a crucial aspect of modern power systems and is discussed in the next section.

1.7 Cybersecurity in Modern Power Systems

1.7.1 Information Flow in a Power System (Power System Layers)

In the previous part, we discussed the SGE and the evolution of power systems from traditional to modern networks. After this evolution, we have the integration of ICTs into the power system, which makes it a form of CPS. This interaction between information and power systems in different environments and specific protocols may lead to loopholes between these two sections. In this case, various sectors, such as information transmission, data collection, and control centers, will be vulnerable and at risk of attack. These vulnerabilities originated from the cyber layer, resulting in security and stability issues in the power system. As illustrated in Figure 1.33, attackers may penetrate to these layers and apply FDIA. The four layers shown in this figure are summarized as follows:

i) **Physical layer:** In this case, the attacker may intrude on the control, measurement, and protection devices. As a case in point, hackers may modify the firmware of processor-based devices (e.g., RTU) [51].
ii) **Communication layer:** There are different types of communication systems (e.g., fiber optic, satellite, PLC, WLAN, etc.) implemented in power systems and smart grids. The attacker may exploit the vulnerabilities associated with this layer and execute FDIA. The probability of an attack in this layer is higher than in the physical layer. However, the adversary should be near physical devices or the communication layer to implement FDIA. This causes a limitation for hackers and decreases the chances of attacks [52].

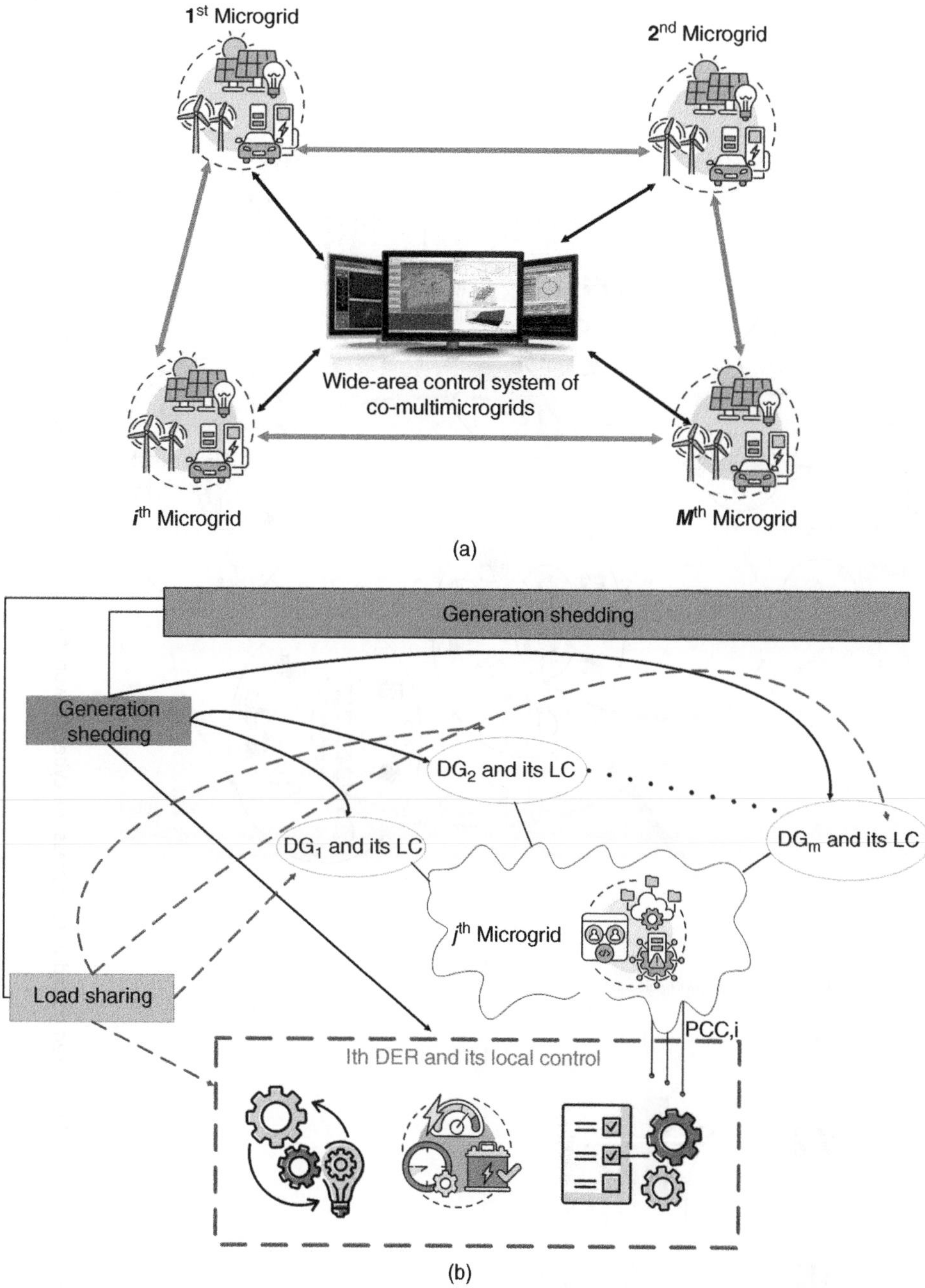

Figure 1.31 Big data emergence in moden power systems (a) WACS communication network for MG and (b) MG control system.

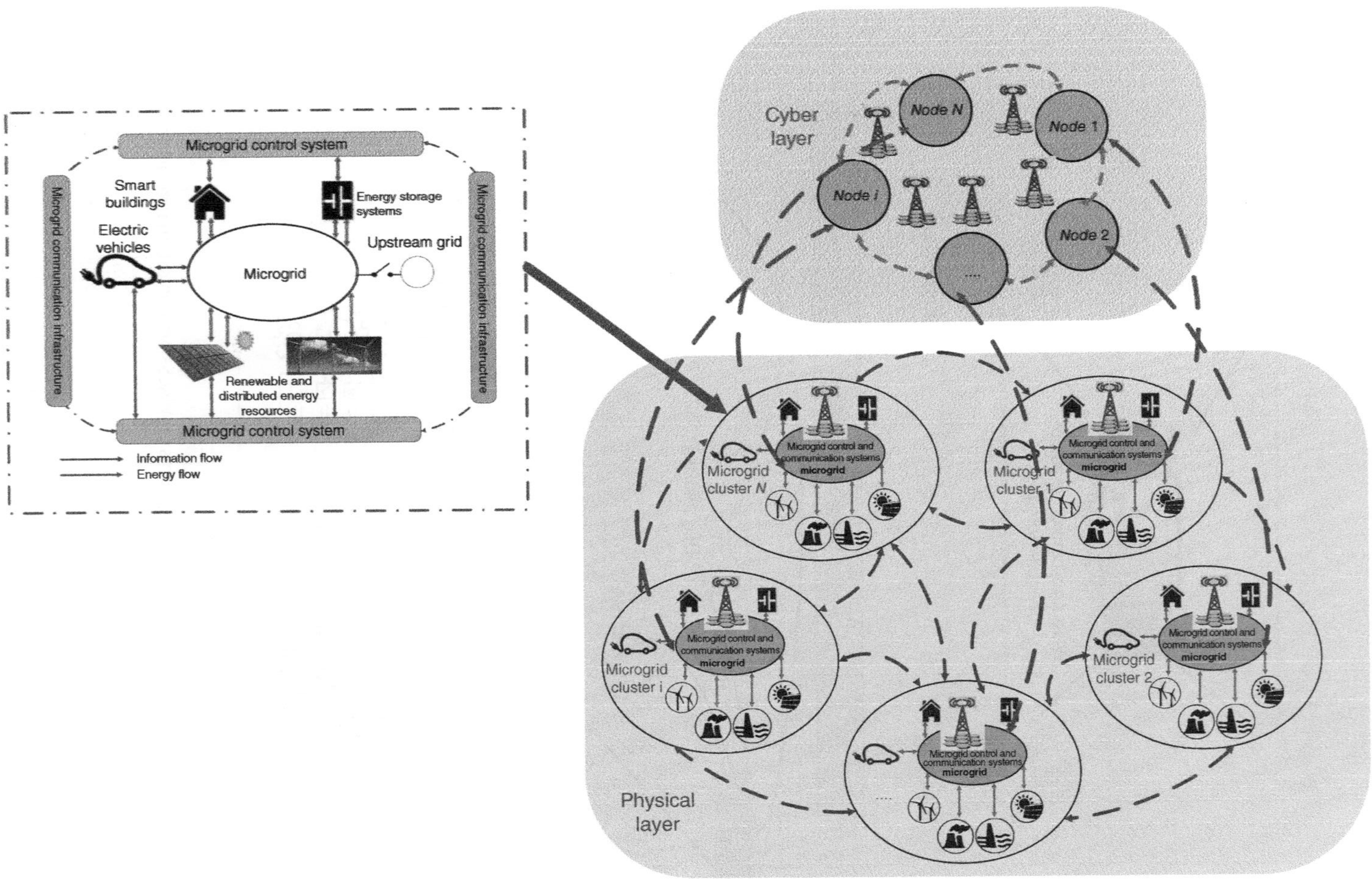

Figure 1.32 In cyber-physical co-multi-MGs, the red dotted lines express the cyber network for information exchange among MGs at the tertiary level and DERs at the secondary level.

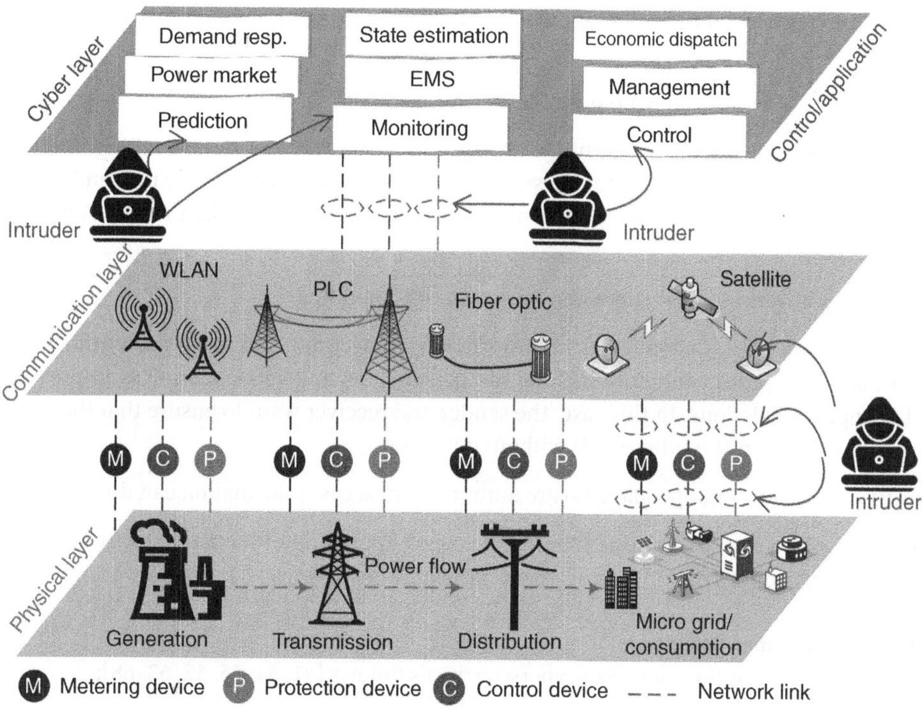

Figure 1.33 Power system layers and their vulnerabilities.

iii) **Network layer:** The network layer transfers variable-length network packets from a source to a destination. In this case, adversaries in different locations may intrude remotely on any node in the network. As a case in point, attackers may use ethernet-based communication protocol (e.g., IEC61850), penetrate the power system, and manipulate a node in the network through TCP/IP modbus or other protocols [53].

iv) **Cyber layer:** In this type of attack, a hacker penetrates the control and application parts of the power system to manipulate the process. The application layer is like a power system's brain; all of the analyses (e.g., state estimation, optimal power flow, economic dispatch, contingency analysis, power market mechanism, etc.) are calculated in this part. This layer is a crucial one and has a direct effect on the power system application. As a case in point, the intruder may manipulate the calculation of power exchange and cause imbalanced demand and supply, increase the cost of energy distribution, and disrupt the energy distribution.

1.7.2 Cyber-Security Definition

Therefore, cybersecurity is one of the crucial challenges in modern power systems and should be addressed. The key definitions of cybersecurity are as follows:

According to NIST, security protects information systems from unauthorized access, use, disclosure, disruption, modification, and destruction to provide confidentiality, integrity, and availability [46, 54–62].

Attacks can be classified into passive and active categories. In the passive attack, the intruder tries to extract the message from the communications channel. On the contrary, the data is modified in an active attack, and false information is delivered to the receiver.

Confidentiality, integrity, and availability are known as CIA triad and are defined by NIST as follows:

- **Confidentiality**: Rules limiting who has access to information.

Confidentiality is synonymous with privacy. It measures the prevention of data falling into the hands of people without authorization to access this information. In this case, only the sender and receiver should understand the message contents (i.e., the sender encrypts the message, and the receiver decrypts the message).

- **Integrity**: Rules governing how and when information is modified.

In the IT world, integrity is about ensuring information is accurate and always stays that way. Common measures to protect integrity include file permissions and version controls to prevent accidental changes or deletions. In this case, the sender and receiver want to ensure that the message does not alter (in transit or afterward) without detection.

- **Availability**: Assurance that people who are authorized to access information can do so.

Ensuring availability requires routine maintenance and upgrading of hardware, software, and operating system environments.

1.7.2.1 Key Terms in Cyber-Security

There are four key terms in the context of cybersecurity as follows [41, 54, 55, 57, 63–68]:

- **Vulnerability:**

Vulnerability is a flaw, loophole, oversight, weakness, or error that an attacker can exploit to perform unauthorized actions within a network and violate the system security policy. An attacker must have at least one useful tool or technique to connect to a system's weakness to exploit a vulnerability.

- **Threat**:

Natural or human-made events can potentially harm computer systems and organizations. The cause could be physical, such as someone stealing a computer that contains vital data. Also, it could be non-physical, such as a virus attack.

- **Exploit:**

An exploit is a specific way (e.g., code) that takes advantage of a vulnerability and breaches the security of an IT system.

- **Risk:**

The probability of exposure or loss resulting from a cyber-attack or a situation involving exposure to danger.

1.7.2.2 Types of Actors in Cyber-Attack and Their Motivations

The primary actors in cyber-attacks can be divided into four groups as follows [54, 63]:

(i) **Internal users:** They are the most likely to initiate security problems, whether intentionally or not. The most common case is that they sell critical information or install malware.
(ii) **Hackers**: They may be attackers that private or public organizations pay. They may also be someone with enough knowledge and time to manipulate the system and bother (not paid).

(iii) **Governments**: They have enough power and money to cause substantial financial and operational losses. They may attack infrastructure, spy, and monitor critical information.
(iv) **Hacktivism**: They are similar to governments but lack financial power. They try to reveal sensitive and confidential information to the public and carry out distributed denial of service (DDoS) campaigns.

These groups may have different motivations:

- **Play**: They may want to demonstrate their capability to other hacker communities.
- **Money**: Some are working with a criminal organization, and large companies (or the government) can make money by threatening.
- **Political actions and movements**: They may fight for specific reasons. As a case in point, they may change people's minds for an upcoming mayoral election in a state by hacking power system utilities and causing blackouts.
- **Hiring**: They may want to demonstrate their capabilities to other competitors so they can hire them or buy their services.

1.7.2.3 Attacks Classification

Attacks can be classified into passive and active categories. In the passive attack (Figure1.34a), which can be an eavesdropping style of attack or traffic analysis, an intruder tries to extract the

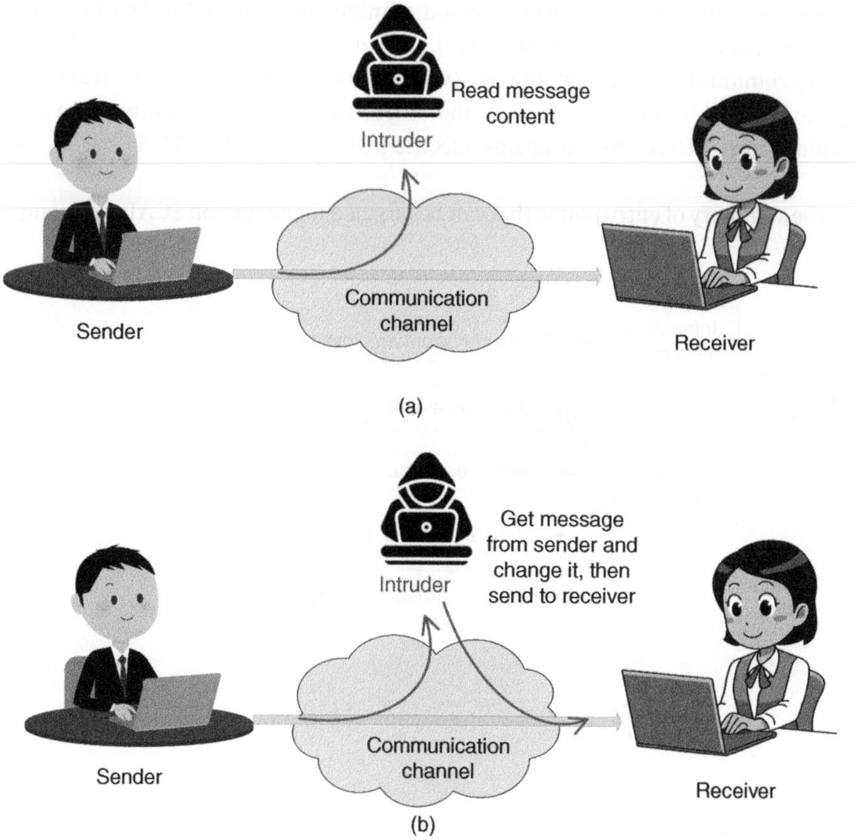

Figure 1.34 Attack classification, (a) passive attack and (b) active attack. In a real network, the sender and receiver may be clients/servers, routers, smart meters, etc.

message from the communications channel. The sender and receiver cannot detect this attack for a long time. In this case, the intruder collects the data, and messages from the sender to the receiver are authenticated and can pass the integrity check. Although confidentiality is violated, the sender and receiver do not have any evidence that an intruder is copying the message [46, 53, 55, 58, 59, 64, 69–76].

In an active attack (Figure 1.34b), the data is modified, and false data is delivered to the receiver. An active attack can be categorized into four types: (i) masquerade (the intruder would pretend to be the sender), (ii) replay (the adversary observes and records a communication sequence to replay it later), (iii) modification of the messages, (iv) denial of service (impede network availability by bombarding a network system with large volumes of meaningless network traffic) [54, 63].

Based on the issues mentioned above, passive attacks are difficult to detect. Conversely, we can define some measures to identify the anomaly caused by active attacks. It is challenging to prevent a network from an active attack, but the goal is to diagnose the attack as soon as possible and prevent the system from any disruption. In a real network, the sender and receiver may be clients/servers, routers, smart meters, etc.

1.7.2.4 Different Types of Attacks

Figure 1.35 generally categorizes security threats. Different types of attacks are: (i) code manipulation (changing the software/firmware), (ii) command manipulation, (iii) false data injection attack (manipulating the data without affecting the code), (iv) network-based attack (physical communication link), (v) communication-based attacks, (vi) database manipulation, (vii) password cracking, etc. Some are related to computer science, and others can be in the communication category. Figure 1.36 illustrates different attacks against electric power systems [46, 53–55, 58, 59, 63, 64, 69–76].

Table 1.2. shows the summary of entry points that can be targeted for attacks on SCADA systems.

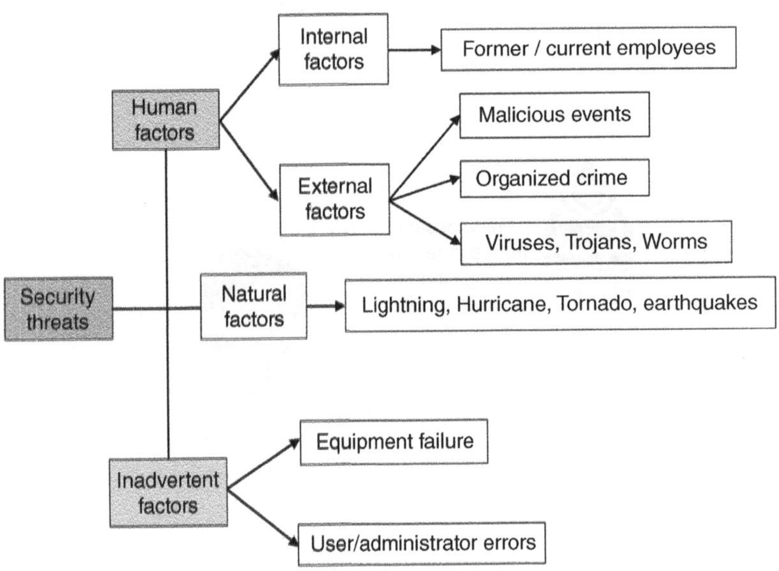

Figure 1.35 General classification of threats in a typical system.

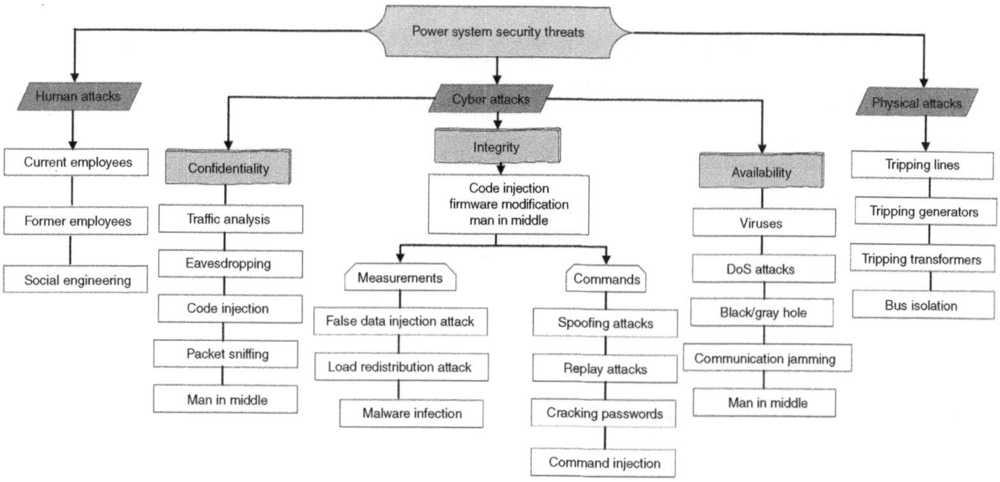

Figure 1.36 Different attacks in power systems.

Table 1.2 Attacks on SCADA systems [54, 63].

Entry point	Description	Likely attacks
1	• Radio communication between RTUs/PLCs and sensors/actuators	• Communication jamming, command injection, and false data injection
2	• RTUs/PLCs and communication with SCADA servers	• Code injection/firmware modification, malware infection, denial of service, jamming, replay attack, command injection, false data injection, black hole/gray hole, network isolation, and rogue node
3	• Control network, including SCADA servers and engineers' workstations	• Malware infection, denial of service, and man-in-the-middle.
4	• Communication gateway/link between control and corporate network (e.g., connection between primary and secondary historian)	• Denial of service and false data injection (database-based)
5	• Corporate network	• Malware infection and social engineering
6	• Internet and networks of partners	• Web-based attacks (malware, SQL injection, etc.) and social engineering

1.7.2.5 Most Frequently Targeted Industries

Figure 1.37 depicts the top ten targeted industries in 2018 to provide a complete picture of the threat landscape. Based on the 2020 report issued by IBM, the top targeted industries in 2019 and their standing compared to 2018 are shown in Table 1.3. It can be observed that the energy sector is the ninth-most targeted industry in the ranking of 2019 and includes 6% of all attacks and incidents. The energy sector is crucial in the economic, national security, and daily activity of industries and people. Therefore, energy companies are the backbone of the country's critical structure and are a target for cyber-attacks [54, 63].

The objective of attacks may be financial material, customer data, and trade secrets similar to other attacks in different industries (e.g., financial services). However, the possibility of physical

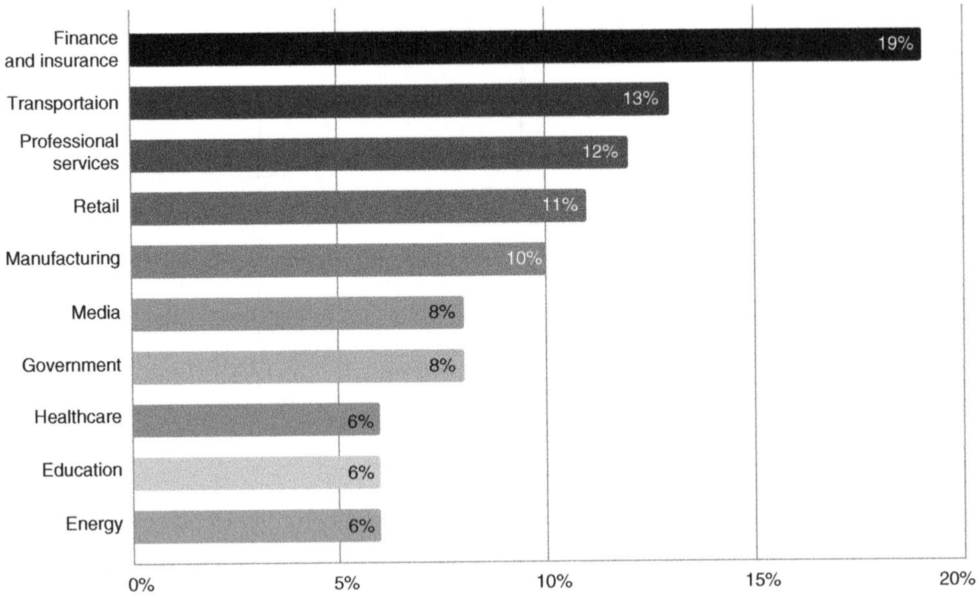

Figure 1.37 Most frequently targeted industries in 2018 [54, 63].

Table 1.3 Top 10 targeted industries ranked by attack volume, 2019 vs. 2018.

Sector	2019 Rank	2018 Rank	Change
Financial services	1	1	—
Retail	2	4	2
Transportation	3	2	−1
Media	4	6	2
Professional services	5	3	2
Government	6	7	1
Education	7	9	2
Manufacturing	8	5	−3
Energy	9	10	1
Healthcare	10	8	−2

disruption and destruction of industrial control systems (ICSs) and the SCADA systems can be the most important goals for attackers. The successful attack on the ICS and SCADA systems may lead to devastating effects on customers utilizing power, gas, oil, or any other resource coming from the energy sector. As a case in point, the adversaries may control power system operation, send the wrong command, change setpoints, and perform other destructive actions. Consider a hacker who manipulates the active power signal and forces the relay or operator in the control room to open the related lines to prevent overloading wrongly. This wrong action affects the financial side of the energy company and, more importantly, causes national security issues. Therefore, false data detection and prevention are the most critical issues in recent years that should be addressed comprehensively [54, 63].

1.7.3 Consequences of FDIA Attack on the Power System

FDIA attacks have different physical, economic, and human life consequences in the power system. In this part, different types of FDIA on the power system are described, along with their effects.

1.7.3.1 Load Redistribution Attack

The authors in [73] presented one of the particular types of FDIAs called load redistribution (LR). This attack may affect system operation by targeting the security-constrained economic dispatch (SCED). LR attack misleads the state estimation process, and consequently, SCED re-dispatches the generation outputs wrongly. In this case, the incorrect solutions may result in infeasible operating states and, in the worst case, cause load-shedding events, which is initially unnecessary.

1.7.3.2 Energy Deceiving Attack

The authors in [74] have investigated another type of attack named "Energy Deceiving Attack," in which the adversary tries to penetrate the routing process of energy distribution. In this case, attackers may manipulate the power system and inject false data in the following ways:

(i) the false quantity of energy that demand nodes request, (ii) the false amount of energy that supply nodes could generate, and (iii) the erroneous states of the energy links. The forged data injected by these attacks will cause imbalanced demand and supply, increase energy distribution costs, and disrupt the energy distribution.

1.7.3.3 Power Market Operation Attack

In the deregulated power systems, the electricity market is operated by independent system operators (ISOs), e.g., PJM. These regulators are third-party entities and should be independent of power suppliers and users. One of the significant responsibilities of ISOs is to determine the market-clearing electricity price. ISOs employ the locational marginal price (LMP) method to calculate day-ahead/real-time prices and manage transmission congestions. Therefore, the power market can be one of the main targets for attackers. The economic impact of FDIA on electric power market operations has been investigated in [59]. The actual SCADA measurements are employed to calculate real-time market prices to find the final settlement prices in the power market. If the adversaries manipulate system measurements and SE results, the electric energy price may be affected by fake SCED. This manipulation of nodal prices may bring financial profits to attackers.

1.7.3.4 Energy Theft

The adversaries can manipulate the data in the grid or their meters to reduce the electrical bill and make unauthorized profit [77].

1.8 Conclusions

As we conclude this foundational chapter, we have laid the groundwork for understanding the complex and multifaceted world of smart CPPSs. We have traversed through the structural overview and roadmap of the book, delved into the general concepts of CPSs, and explored the multi-layered nature of CPPSs. This chapter has contextualized the emerging technologies and the organization of different concepts within the CPPS framework, setting the stage for in-depth discussions in the upcoming chapters.

We examined the transition from conventional power DNs to advanced smart grids, highlighting the evolution of ADNs, the intricacies of MGs and VPPs, and the innovative concepts of the IoE and MGs. The SGE section brought into focus the seamless integration of smart cities and grids, addressing everything from smart traffic control to intelligent, interconnected MGs.

A critical aspect we touched upon is cybersecurity in modern power systems. Understanding the flow of information in these systems and recognizing the various cybersecurity threats and their potential consequences is crucial for safeguarding our future energy infrastructure.

As readers proceed to the subsequent chapters, they will find various contributions from various authors, each bringing their expertise to bear on different aspects of CPPSs. These chapters will delve deeper into specific topics, offering advanced insights and solutions to the challenges we face in smart CPPSs.

This chapter has set the tone for a journey through the dynamic and rapidly evolving field of smart CPPSs. With the foundation now firmly in place, readers are well-equipped to navigate the detailed and specialized discussions that follow, gaining a deeper understanding of the challenges and innovative solutions in this vital sector of modern infrastructure.

References

1 "Gartner hype cycle." [Online]. Available: https://en.wikipedia.org/wiki/Gartner_hype_cycle

2 "Hype Gartner webpage." [Online]. Available: https://www.gartner.com/en%0A

3 "Gartner Identifies Five Emerging Technology Trends That Will Blur the Lines Between Human and Machine," 2018.

4 "What's New in the 2023 Gartner Hype Cycle for Emerging Technologies".

5 Maier, M.W. (1998). Architecting principles for systems-of-systems. *Systems Engineering: The Journal of the International Council on Systems Engineering* 1 (4): 267–284.

6 Jamshidi, M. (2008). *System of Systems Engineering: Innovations for the 21st Century*. John Wiley & Sons, Inc.

7 Bruinenberg, J. et al. 2012. *CEN - CENELEC - ETSI: Smart Grid Coordination Group* - Smart Grid Reference Architecture Report 2*.0*.

8 Liu, L.W.W., Gong, Q., Han, H., and Wang, Z. (2018). Reliability modeling and evaluation of active cyber physical distribution system. *IEEE Transactions on Power Systems* 33 (6): 7096–7108. https://doi.org/10.1109/TPWRS.2018.2854642.

9 Dharmawardena, H. and Venayagamoorthy, G.K. (2022). Distributed volt-var curve optimization using a cellular computational network representation of an electric power distribution system. *Energies (Basel)* 15 (12): https://doi.org/10.3390/en15124438.

10 Skeen, K.A. and Venayagamoorthy, G.K. (2023). Network of microgrids: opportunities and challenges. *2023 IEEE PES Grid Edge Technologies Conference & Exposition (Grid Edge)*, pp. 1–5. https://doi.org/10.1109/GridEdge54130.2023.10102727.

11 Zhong, X., Jayawardene, I., Venayagamoorthy, G.K., and Brooks, R. (2017). Denial of service attack on tie-line bias control in a power system with PV plant. *IEEE Transactions on Emerging Topics in Computational Intelligence* 1 (5): 375–390. https://doi.org/10.1109/TETCI.2017.2739838.

12 Beasley, C., Zhong, X., Deng, J. et al. (2014). A survey of electric power synchrophasor network cyber security. *IEEE PES Innovative Smart Grid Technologies,* Europe, pp. 1–5. https://doi.org/10.1109/ISGTEurope.2014.7028738.

13 Venayagamoorthy, G.K., Sharma, R.K., Gautam, P.K., and Ahmadi, A. (2016). Dynamic Energy Management System for a Smart Microgrid. *IEEE Transactions on Neural Networks and Learning Systems* 27 (8): 1643–1656. https://doi.org/10.1109/TNNLS.2016.2514358.

14 Chowdhury, S., Chowdhury, S.P., and Crossley, P. (2009). *Microgrids and Active Distribution Networks,* 1ste. London, UK: The Institution of Engineering and Technology https://doi.org/10.1049/PBRN006E.

15 Baghaee, H.R., Mirsalim, M., Gharehpetian, G.B., and Ali, T.H. (2016). Reliability/cost-based multi-objective Pareto optimal design of stand-alone wind/PV/FC generation microgrid system. *Energy* 115 (1): 1022–1041. https://doi.org/10.1016/j.energy.2016.09.007.

16 Gharehpetian, G.B., Baghaee, H.R., and Shabestary, M.M. (2021). *Microgrids and methods of analysis.* NY, USA: Elsevier Academic Press.

17 Baghaee, H.R., Mirsalim, M., Gharehpetian, G.B., and Talebi, H.A. (2017). A generalized descriptor-system robust H∞ control of autonomous microgrids to improve small and large signal stability considering communication delays and load nonlinearities. *International Journal of Electrical Power & Energy Systems* 92: 63–82. https://doi.org/10.1016/j.ijepes.2017.04.007.

18 Baghaee, H.R., Mirsalim, M., Gharehpetian, G.B., and Talebi, H.A. (2017). Three-phase AC/DC power-flow for balanced/unbalanced microgrids including wind/solar, droop-controlled and electronically-coupled distributed energy resources using radial basis function neural networks. *IET Power Electronics* 10 (3): 313–328. https://doi.org/10.1049/iet-pel.2016.0010.

19 Danley, D.R. (2019). Defining a Microgrid Using IEEE 2030.7.

20 Baghaee, H.R., Mirsalim, M., Gharehpetian, G.B., and Talebi, H.A. (2018). A Decentralized Robust Mixed H2/H∞ Voltage control scheme to improve small/large-signal stability and FRT capability of islanded multi-der microgrid considering load disturbances. *IEEE Systems Journal* 12 (3): 2610–2621. https://doi.org/10.1109/JSYST.2017.2716351.

21 Firouzi, M. and Gharehpetian, G.B. (2013). Improving fault ride-through capability of fixed-speed wind turbine by using bridge-type fault current limiter. *IEEE Transactions on Energy Conversion* 28 (2): 361–369. https://doi.org/10.1109/TEC.2013.2248366.

22 Baghaee, H.R., Mirsalim, M., Gharehpetian, G.B., and Talebi, H.A. (2017). A new current limiting strategy and fault model to improve fault ride- through capability of inverter interfaced DERs in autonomous microgrids. *Sustainable Energy Technologies and Assessments* 24: 71–81. https://doi.org/10.1016/j.seta.2017.02.004.

23 Wang, Y. and Ren, B. (2018). Fault ride-through enhancement for grid-tied PV systems with robust control. *IEEE Transactions on Industrial Electronics* 65 (3): 2302–2312. https://doi.org/10.1109/TIE.2017.2740858.

24 Eskandari, M. and Savkin, A.V. (2021). On the impact of fault ride-through on transient stability of autonomous microgrids: nonlinear analysis and solution. *IEEE Transactions on Smart Grid* 12 (2): 999–1010. https://doi.org/10.1109/TSG.2020.3030015.

25 Baghaee, H.R., Mirsalim, M., Gharehpetan, G.B., and Talebi, H.A. (2018). Nonlinear load sharing and voltage compensation of microgrids based on harmonic power-flow calculations using radial basis function neural networks. *IEEE Systems Journal* 12 (3): 2749–2759. https://doi.org/10.1109/JSYST.2016.2645165.

26 Guerrero, J.M., Vasquez, J.C., Matas, J. et al. (2011). Hierarchical control of droop-controlled AC and DC microgrids - a general approach toward standardization. *IEEE Transactions on Industrial Electronics* 58 (1): 158–172.

27 Baghaee, H.R., Mirsalim, M., and Gharehpetian, G.B. (2016). Real-time verification of new controller to improve small/large-signal stability and fault ride-through capability of multi-DER microgrids. *IET Generation, Transmission and Distribution* 10 (12): 3068–3084. https://doi.org/10.1049/iet-gtd.2016.0315.

28 Raeispour, M., Atrianfar, H., Baghaee, H.R., and Gharehpetian, G.B. (2020). Robust sliding mode and mixed H2/H1 output feedback primary control of AC microgrids. *IEEE Systems Journal* 1–12. https://doi.org/10.1109/JSYST.2020.2999553.

29 Biglarahmadi, M., Ketabi, A., Reza Baghaee, H., and Guerrero, J.M. (2022). Integrated Nonlinear Hierarchical Control and Management of Hybrid AC/DC Microgrids. *IEEE Systems Journal* 16 (1): 902–913. https://doi.org/10.1109/JSYST.2021.3050334.

30 Afshari, A., Karrari, M., Baghaee, H.R. et al. (2020). Cooperative fault-tolerant control of microgrids under switching communication topology. *IEEE Transactions on Smart Grid* 11 (3): 1866–1879. https://doi.org/10.1109/TSG.2019.2944768.

31 Rocabert, J., Luna, A., Blaabjerg, F., and Rodríguez, P. (2018). Control of power converters in AC microgrids. *IEEE Transactions on Power Electronics* 27 (11): 4734–4749. https://doi.org/10.1109/TPEL.2012.2199334.

32 Baghaee, H.R., Mirsalim, M., Gharehpetian, G.B. et al. (2021). OC/OL protection of droop-controlled and directly voltage-controlled microgrids using TMF/ANN-based fault detection and discrimination. *IEEE Journal of Emerging and Selected Topics in Power Electronics* 9 (3): 3254–3265.

33 Rouzbahani, H.M., Karimipour, H., and Lei, L. (2021). A review on virtual power plant for energy management. *Sustainable Energy Technologies and Assessments* 47: 101370. https://doi.org/10.1016/j.seta.2021.101370.

34 Panahazari, M., Koscak, M., Zhang, J. et al. (2023). A hybrid optimization and deep learning algorithm for cyber-resilient DER control. 2023 *IEEE Power and Energy Society Innovative Smart Grid Technologies Conference,* ISGT 2023. https://doi.org/10.1109/ISGT51731.2023.10066345.

35 Li, X., Zhao, D., and Guo, B. (2018). Decentralized and collaborative scheduling approach for active distribution network with multiple virtual power plants. *Energies* 11 (11): https://doi.org/10.3390/en11113208.

36 Seven, S., Yao, G., Soran, A. et al. (2020). Peer-to-peer energy trading in virtual power plant based on blockchain smart contracts. *IEEE Access* 8: 175713–175726. https://doi.org/10.1109/ACCESS.2020.3026180.

37 Wang, X., Liu, Z., Zhang, H. et al. (2019). A review on virtual power plant concept, application and challenges. *2019 IEEE Innovative Smart Grid Technologies - Asia (ISGT Asia)*, pp. 4328–4333. https://doi.org/10.1109/ISGT-Asia.2019.8881433.

38 Schlund, J. (2018). Blockchain-based orchestration of distributed assets in electrical power systems. *Energy Informatics* 1 (1): 39. https://doi.org/10.1186/s42162-018-0054-y.

39 Zhang, J., Xu, Z., Xu, W. et al. (2019). Bi-objective dispatch of multi-energy virtual power plant: deep-learning-based prediction and particle swarm optimization. *Applied Sciences* 9 (2): https://doi.org/10.3390/app9020292.

40 Ravi, M.A. and Sudharshan, S. (2022). Utilization of electric vehicles for vehicle-to-grid services: progress and perspectives. *Energies (Basel)* 15 (2): 589. https://doi.org/10.3390/en15020589.

41 Dwivedi, A.D., Srivastava, G., Dhar, S., and Singh, R. (2019). A decentralized privacy-preserving healthcare blockchain for IoT. *Sensors* 19 (2): https://doi.org/10.3390/s19020326.

42 Van Mierlo, J. et al. (2021). Beyond the state of the art of electric vehicles: a fact-based paper of the current and prospective electric vehicle technologies. *World Electric Vehicle Journal* 12 (1): https://doi.org/10.3390/wevj12010020.

43 Banerjee, A., Dutta, B., Mandal, T. et al. (2022). Blockchain in IoT and beyond: case studies on interoperability and privacy. In: *Blockchain Based Internet of Things*, 113–138. Springer.

44 Dedeoglu, V., Jurdak, R., Putra, G.D. et al. (2019). A trust architecture for blockchain in IoT. *Proceedings of the 16th EAI International Conference on Mobile and Ubiquitous Systems: Computing, Networking and Services*, pp. 190–199.

45 Majeed, U., Khan, L.U., Yaqoob, I. et al. (2021). Blockchain for IoT-based smart cities: recent advances, requirements, and future challenges. *Journal of Network and Computer Applications* 181: 103007.

46 Attkan, A. and Ranga, V. (2022). Cyber-physical security for IoT networks: a comprehensive review on traditional, blockchain and artificial intelligence based key-security. *Complex & Intelligent Systems* 8 (4): 3559–3591. https://doi.org/10.1007/s40747-022-00667-z.

47 Aminifar, F., Abedini, M., Amraee, T. et al. (2022). A review of power system protection and asset management with machine learning techniques. *Energy Systems* 13 (4): 855–892. https://doi.org/10.1007/s12667-021-00448-6.

48 Bavaresco, M.V., D'Oca, S., Ghisi, E., and Lamberts, R. (2019). Technological innovations to assess and include the human dimension in the building-performance loop: a review. *Energy and Buildings* 202: 109365. https://doi.org/10.1016/j.enbuild.2019.109365.

49 Chen, F., Huang, C., Wang, L. et al. (2017). Flexibility evaluation of distribution network with high penetration of variable generations. *2017 IEEE Conference on Energy Internet and Energy System Integration, EI2 2017 - Proceedings*, vol. 2018-Janua, pp. 1–6. https://doi.org/10.1109/EI2.2017.8245479.

50 Hua, H., Wei, Z., Qin, Y. et al. (2021). Review of distributed control and optimization in energy internet: from traditional methods to artificial intelligence-based methods. *IET Cyber-Physical Systems: Theory & Applications* 6 (2): 63–79. https://doi.org/10.1049/cps2.12007.

51 Konstantinou, M.M. and Charalambos (2019). Hardware-layer intelligence collection for smart grid embedded systems. *Journal of Hardware and Systems Security* 3 (2): 132–146.

52 Bou-Harb, E., Fachkha, C., Pourzandi, M. et al. (2013). Communication security for smart grid distribution networks. *IEEE Communications Magazine* 51 (1): 42–49. https://doi.org/10.1109/MCOM.2013.6400437.

53 Macwan, R., Drew, C., Panumpabi, P. et al. (2016). Collaborative defense against data injection attack in IEC61850 based smart substations. *IEEE Power and Energy Society General Meeting (PESGM)*, Boston, MA, USA: IEEE. https://doi.org/10.1109/PESGM.2016.7741376.

54 "National Institute of Standards and Technology." [Online]. Available: https://www.nist.gov/

55 Gunduz, M.Z. and Das, R. (2020). Cyber-security on smart grid: threats and potential solutions. *Computer Networks* 169: 107094. https://doi.org/10.1016/j.comnet.2019.107094.

56 Hellani, H., Sliman, L., Samhat, A.E., and Exposito, E. (2021). On blockchain integration with supply chain: overview on data transparency. *Logistics* 5 (3): https://doi.org/10.3390/logistics5030046.

57 Andrew, J., Isravel, D.P., Sagayam, K.M. et al. (2023). Blockchain for healthcare systems: architecture, security challenges, trends and future directions. *Journal of Network and Computer Applications* 215: 103633. https://doi.org/10.1016/j.jnca.2023.103633.

58 Wang, P. and Govindarasu, M. (2020). Multi-agent based attack-resilient system integrity protection for smart grid. *IEEE Transactions on Smart Grid* 11 (4): 3447–3456. https://doi.org/10.1109/TSG.2020.2970755.

59 Xie, L., Mo, Y., and Sinopoli, B. (2011). Integrity data attacks in power market operations. *IEEE Transactions on Smart Grid* 2 (4): 659–666. https://doi.org/10.1109/TSG.2011.2161892.

60 Sun, K., Chen, J., and Viboud, C. (2020). Early epidemiological analysis of the coronavirus disease 2019 outbreak based on crowdsourced data: a population-level observational study. *Lancet Digit Health* 2 (4): e201–e208:https://doi.org/10.1016/S2589-7500(20)30026-1.

61 Mehedi, S.K.T., Shamim, A.A.M., and Miah, M.B.A. (2019). Blockchain-based security management of IoT infrastructure with Ethereum transactions. *Iran Journal of Computer Science* 2 (3): 189–195. https://doi.org/10.1007/s42044-019-00044-z.

62 Rajasekaran, A.S., Azees, M., and Al-Turjman, F. (2022). A comprehensive survey on blockchain technology. *Sustainable Energy Technologies and Assessments* 52: 102039. https://doi.org/10.1016/j.seta.2022.102039.

63 "International Business Machines Corporation (IBM)." [Online]. Available: https://www.ibm.com

64 Tian, J., Wang, B., Li, J., and Wang, Z. (2022). Adversarial attacks and defense for CNN based power quality recognition in smart grid. *IEEE Transactions on Network Science and Engineering* 9 (2): 807–819. https://doi.org/10.1109/TNSE.2021.3135565.

65 Upendra Vishwanath, Y.S., Esakkirajan, S., Keerthiveena, B., and Pachori, R.B. (2023). A generalized classification framework for power quality disturbances based on synchrosqueezed wavelet transform and convolutional neural networks. *IEEE Transactions on Instrumentation and Measurement* 72: https://doi.org/10.1109/TIM.2023.3308235.

66 Panarello, A., Tapas, N., Merlino, G. et al. (2018). Blockchain and IoT integration: a systematic survey. *Sensors* 18 (8): https://doi.org/10.3390/s18082575.

67 Yadav, A.K., Singh, K., Amin, A.H. et al. (2023). A comparative study on consensus mechanism with security threats and future scopes: blockchain. *Computer Communications* 201: 102–115. https://doi.org/10.1016/j.comcom.2023.01.018.

68 Li, Q., Meng, S., Zhang, S. et al. (2019). Safety risk monitoring of cyber-physical power systems based on ensemble learning algorithm. *IEEE Access* 7: 24788–24805. https://doi.org/10.1109/ACCESS.2019.2896129.

69 Otokwala, U., Petrovski, A. and Kalutarage, H. (2021). Effective detection of cyber attack in a cyber-physical power grid system. *Advances in Information and Communication: Proceedings of the 2021 Future of Information and Communication Conference (FICC), Volume 1*, Springer, pp. 812–829.

70 Jamali, M., Baghaee, H.R., Gharehpetian, G.B., and Anvari-Moghaddam, A. (2023). Distributed cooperative event-triggered control of cyber-physical AC microgrids subject to denial-of-service attacks. *IEEE Transactions on Smart Grid* 14 (6): 4467–4478. https://doi.org/10.1109/TSG.2023.3259545.

71 Liang, G., Zhao, J., Luo, F. et al. (2017). A review of false data injection attacks against modern power systems. *IEEE Transactions on Smart Grid* 8 (4): 1630–1638. https://doi.org/10.1109/TSG.2015.2495133.

72 Shrestha, R. and Nam, S.Y. (2019). Regional blockchain for vehicular networks to prevent 51% attacks. *IEEE Access* 7: 95033–95045. https://doi.org/10.1109/ACCESS.2019.2928753.

73 Yuan, K.R. and Yanling, Z.L. (2011). Modeling load redistribution attacks in power systems. *IEEE Transactions on Smart Grid* 2 (2): 382–390. https://doi.org/10.1109/TSG.2011.2123925.

74 Lin, J., Yu, W., Yang, X. et al. (2012). On false data injection attacks against distributed energy routing in smart grid. *IEEE/ACM Third International Conference on Cyber-Physical Systems.* Beijing, China.

75 Habibi, M.R., Baghaee, H.R., Dragicevic, T., and Blaabjerg, F. (2022). Secure MPC/ANN-based false data injection cyber-attack detection and mitigation in DC microgrids. *IEEE Systems Journal* 16 (1): 1487–1498.

76 Kumar, S.A.P., Bhargava, B., Macêdo, R. et al. (2017). Securing iot-based cyber-physical human systems against collaborative attacks. 2017 *IEEE International Congress on Internet of Things (ICIOT)*, IEEE, pp. 9–16.

77 Jokar, P., Arianpoo, N., and Leung, V.C.M. (2016). Electricity theft detection in AMI using customers' consumption patterns. *IEEE Transactions on Smart Grid* 7 (1): 216–226. https://doi.org/10.1109/TSG.2015.2425222.

2

Global Demand Response Status: Potentials, Barriers, and Solutions

Sanchari Deb[1], Elahe Doroudchi[2], Sergio Motta[3], Matti Aro[4], and Amir Safdarian[5]

[1] *School of Engineering, Newcastle University, Newcastle, UK*
[2] *Department of Electrical Engineering, School of Technology and Innovations, University of Vaasa, Vaasa, Finland*
[3] *Department of Electrical Engineering, Aalto University, Espoo, Finland*
[4] *Smart Energy and Built Environment, VTT Technical Research Centre of Finland, Espoo, Finland*
[5] *Volue Oy, Helsinki, Finland*

2.1 Background

A key challenge for future electric energy systems is their decarbonization considering more effective ways of electricity management. After the United Nations (UN) Paris Agreement [1], there is more pressure on governments to alter the conventional electric energy systems by employing more renewable sources of energy. When the penetration of new types of energies increases, new challenges arise. The renewable sources of energy are intermittent, and forecasting their electricity production is difficult. Thus, new technologies should be applied to overcome the coming challenges caused by the high penetration of the new methods of electricity production.

From the customers' point of view, the electricity cost is the main issue. Customers do not usually care how the electricity is produced, but they are more concerned about the electricity price and their electricity bills. At the same time, electricity providers are often concerned about their profit, which is affected by the way they manage their income and costs. To this end, electricity providers need to consider their customers' satisfaction and also governments' desire about how energy should be produced. To overcome this challenge, the demand response (DR) concept has been introduced to help producers and consumers better manage electricity use as well as to better optimize energy costs. DR programs suggest solutions for bilateral benefits to customers and electricity providers [2]. This way, the customer benefits increase while the electricity provider needs are also met. These programs offer solutions by reducing or shifting electricity usage during peak periods in response to time-based rates or other forms of financial incentives.

From stakeholders' point of view, DR programs have significant benefits too. When electricity use is shifted or reduced during peak hours, electricity providers would have more flexibility in controlling the possible electricity shortage and sudden unexpected changes that require immediate decisions and actions from them. Furthermore, DR program features and capabilities should be well-introduced to the stakeholders and the customers to fully take advantage of this solution. In addition, the technology that is necessary to enable DR programs at the customer side should be developed and operated functionally. Managing between the stakeholder decision on possible load shifting or reducing electricity usage and the real implementation requires fast, proper, and reliable technologies. This could be provided using artificial intelligence (AI) systems.

Smart Cyber-Physical Power Systems: Fundamental Concepts, Challenges, and Solutions, Volume 1, First Edition.
Edited by Ali Parizad, Hamid Reza Baghaee, and Saifur Rahman.

AI, machine learning (ML), and neural networks (NNs) are the terms that have been heard a lot, but do not have a long history in the management and control of electric energy systems. Emerging new concepts including but not limited to demand-side management and DR concept have forced electric energy systems to come up with intelligent and fast ways to securely facilitate the communication and operation of the decisions and their implementation on the customer side. For the effective implementation of DR programs, there are numerous issues that need to be considered, from load and electricity price forecasting to identifying the right consumers to participate in DR schemes and creating automated systems that manage demand-side resources [3]. AI methods have been applied across the spectrum of DR by providing the tools for prediction, real-time efficient control of distributed systems, and decision-making, while adapting to an ever-changing environment and learning from human behavior. Furthermore, AI solutions have been extensively used by researchers in order to find solutions where traditional approaches could not provide results that are sufficiently efficient or reliable [4].

To this end, this chapter provides a comprehensive review of AI- and ML-based methods for solving planning and operation of smart electric energy systems incorporating DR programs followed by two case studies.

2.2 Global Status of DR Programs

Countries all around the world are moving away from fossil fuel–based electric energy production and toward renewables. In the Net Zero Emissions (NZEs) by 2050 Scenario by the International Energy Agency (IEA), the share of renewables would increase from 29% in total output to over 60% by 2030. Hand in hand with that comes the need for the realization of DR programs since consumption and generation need to be always equal in electric energy systems. Traditionally, generation has followed consumption, which has been relatively easy with fossil fuel–based generation. However, flexibility from the generation side is set to decrease as many markets are decommissioning conventional power plants capable of providing flexibility [5].

Demand-side management refers to any sort of action taken to modify the level of electricity consumption in electric energy systems. The actions under the demand-side management concept can be categorized into two groups, namely energy efficiency programs and DR programs. Energy efficiency programs target the amount of electric energy that is consumed for a service. The programs aim to reduce electricity consumption without compromising the quality of the service. Replacing incandescent and fluorescent lamps with light-emitting diode (LED) lamps can be considered as an activity in line with the energy efficiency concept. DR programs target the way electricity is consumed. In these programs, consumers are motivated to alter the style of their electricity consumption during peak hours when the electricity price is much higher, or supply reliability is jeopardized. These actions mainly aim to optimally manage electricity demand to maintain power balance and avoid grid congestion. This is possible either by shifting electricity consumption to times of excess production of variable renewable energy sources (VRESs) or by reserving flexible consumption capacity for later use if disturbances occur or forecasts fail. In the NZE scenario, 500 GW of DR potential is forecasted to be required by 2030 to support the integration of additional renewables. This means a tenfold increase in the deployment level of DR potentials compared to 2020.

It is worthwhile to mention that mainly since flexibility has been traditionally activated only from large generation units, the national regulation does not necessarily acknowledge DR as a viable source of flexibility. In recent years, this has changed in many countries as the need for new sources of flexibility has increased.

DR programs can offer flexibility in both directions effectively by either increasing or decreasing electricity consumption. A variety of electricity consumers including but not limited to the following can participate in DR programs:

- Process industry (using intermediate storages, heating, ventilation, and air conditioning (HVAC), data centers, chemical and metallurgy, bioforestry)
- Households (HVAC, hot water tanks)
- Electric vehicles (EVs) (public, private)
- Agriculture (irrigation, lighting)
- Public infrastructure (water treatment, district heating, schools, libraries)
- Commercial buildings (food markets, shopping centers, banks)
- Sector coupling (hydrogen, district heating, transportation)

It is worthwhile to mention that integration of large masses of EVs may pose a threat to stability of electric energy systems, but with smart charging programs, EVs could provide flexibility to help replace conventional power generation with renewables. To achieve flexibility, several jurisdictions such as the European Union, Japan, and Canada have lately announced targets and strategies on EVs and their charging infrastructure.

DR programs can be divided into two categories, namely explicit programs and implicit programs. Explicit DR programs are incentive-based aiming to provide services for other stakeholders in the energy domain. These programs include frequency regulation, congestion management, voltage control, and balance responsible party (BRP) services. In contrast, implicit DR programs are price-based with the goal of serving the end user itself. These include programs such as electricity bill reduction by spot price optimization and local load balancing.

A recent study [5, 6] by authors conducted a survey to map the status of DR in different parts of the world. In total, 12 responses were received from eight different countries, including Australia, Canada, Denmark, Finland, Germany, India, Netherlands, and Norway. From the responses, realized market-based DR is mainly provided from industrial processes, while smaller and distributed loads such as EVs and household appliances are mostly still in the piloting stage. According to the responses, the most untapped potential was also still in the industrial processes, but also buildings and EVs were identified to hold a lot of potential.

In the study, the main barriers for utilizing untapped DR potential are identified and depicted in Figure 2.1. In the figure, each bar indicates the percentage of surveys where the respective reason was mentioned as a barrier for utilizing untapped DR potential. At the time of the survey, most challenges related to the deployment of DR were seen to be on the incentive side, linked to the lack of any established marketplaces for consumption units to participate. Lack of technical solutions and lack of regulation allowing aggregation of smaller units together were also seen as a major barrier for utilizing the untapped potential. The AI and ML technologies, which have recently received significant attention can help to mitigate the concern on technical solutions to some extent (Figure 2.2). The next section provides a review on the existing literature on the application of AI and ML technologies in the DR domain. Finally, it is worthwhile to mention that the last bar in the figure (i.e., other) stands for issues with vendor products compliant with DR standards, no readily available end-to-end DR solutions, and lack of resources at utility to reach out, monitor, and incentivize DR.

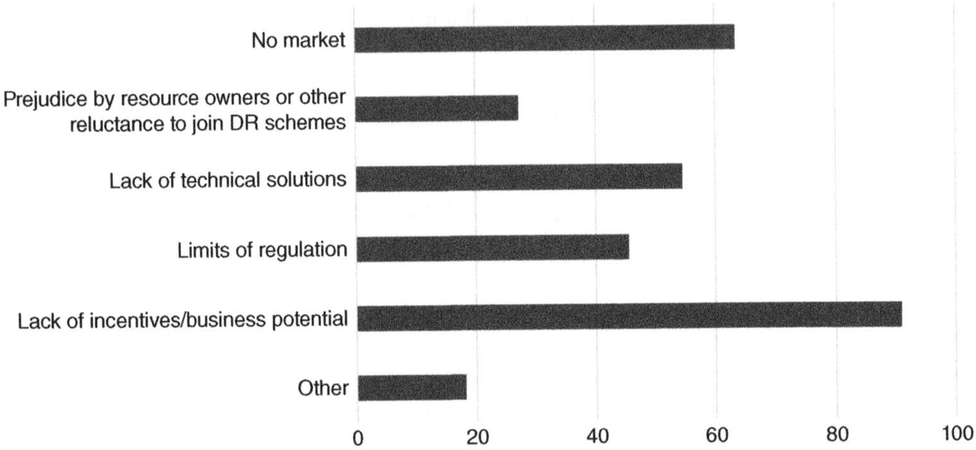

Figure 2.1 Barriers for utilizing untapped DR potential.

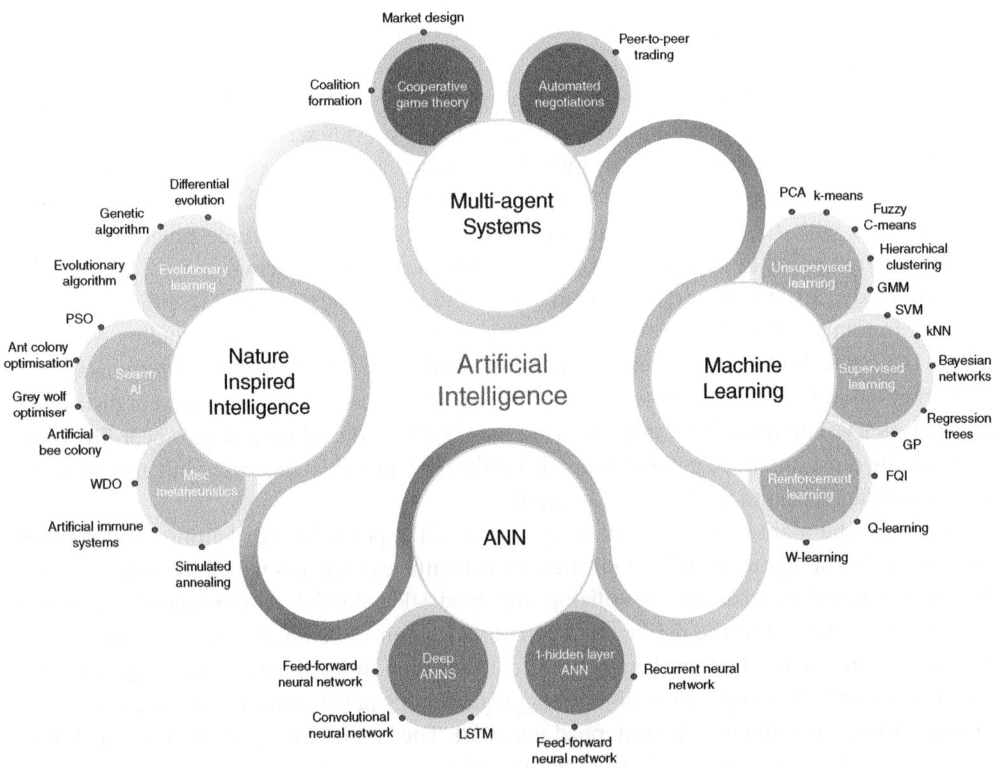

Figure 2.2 The multiple fields of AI [2].

2.3 AI and ML Applications in DR

AI is a very broad topic that has presented massive growth in recent years, both in its research and in applications to different fields such as medicine and e-commerce. With the improvement in computational capacity and data availability, it has become viable to use AI as a tool for solving a multitude of challenges.

Within AI, the field of ML has been receiving a lot of attention as a toolbox to solve pressing challenges and opportunities. The ability of a computer to learn, either based on historical data and pattern detection or through feedback and reinforcement, has the potential to be a game changer in applications that require fast and repetitive actions. In electric energy system operations, DR is a topic that could greatly benefit from the advantages presented by ML methods [2].

DR is also a very broad field with many potential use cases. The addition of flexibility in the consumption of electricity may come from a multitude of vectors. Energy resources in residential buildings, such as HVAC systems [7] or electrical appliances [8], can offer a high degree of flexibility and thus present themselves as promising channels for DR. Moreover, in industry, DR also finds promising applications either in manufacturing [9] or in the service sector [10].

The implementation of AI technologies in DR may take different formats: through the development of clustering, pattern, and anomaly detection methods for electricity consumption and appliance load monitoring; forecasting methods for electric load, energy pricing, or weather behavior; or AI-based agents for managing DR activities automatically, with no requirements for active user inputs. Other alternatives are also possible, and this field is constantly growing and identifying new applications of AI and specifically ML in electric energy system operations.

Clustering and pattern detection methods can be based on both supervised and unsupervised learning agents that learn from identifying patterns in the data. The former is typically used for classification problems, where patterns are matched to existing information, while the latter can be used to build clusters of behaviors and patterns without a clear classification. In such a case, input data are associated within clusters that share similar characteristics. Clustering can be extensively used in DR applications for the detection of patterns in load consumption, thus identifying flexibility potential and actions to explore this flexibility. A relevant example of such applications is in DR actions involving EV charging [11] and for pattern extraction of electricity consumption data [12, 13]. Supervised learning methods are also widely used in this application, for the classification of appliance loads for their flexible control (nonintrusive load monitoring [NILM]). Reference [14] utilized deep NNs (DNNs) for NILM to estimate the electricity consumption of individual appliances via aggregated load data. This approach enables the identification of appliances' flexibility potential and acts as an enabler for their further use in DR initiatives. Reference [15] presented a comprehensive review of NILM techniques and their use in residential applications. This paper showcases a number of alternatives for NILM techniques to be used in households, focusing on their use in energy management systems (EMSs). The detection of events and the classification of loads are some of the most advanced and interesting applications of AI for DR. Reference [16] explored an event-based NILM approach through a combined support vector machine (SVM), genetic algorithm (GA), and grid search cross-validation (GridSearchCV) model and validates this model specifically in DR applications.

Forecasting is another key application of AI technologies for DR, as it provides the basis for decision-making and control of flexible resources. Reference [17] presented a review on probabilistic load forecasting, giving multiple applications and methods for predicting electricity demand for industrial and residential settings and through multiple forecasting horizons. Reference [18] used a supervised learning method, least squares SVM (LSSVM), as the forecasting tool for load and electricity prices in multiple different markets. The study presented an interesting alternative for forecasting critical aspects of DR. Reference [19] utilized supervised learning methods (MP5 predictive models based on regression trees) to forecast weather, electricity pricing, and building temperature. These parameters are integrated into a building EMS to reduce energy costs in a residential environment. A co-simulation model of a typical house in the context of the Republic of Ireland was developed, and DR assets considered were a heat pump and a thermal storage system. Two control algorithms were tested comparing a rule-based approach with a predictive-based approach. The predictive algorithm presented a higher reduction in costs when compared with the rule-based approach.

Reference [20] proposed a supervised learning–based strategy for realizing DR potential of HVAC loads in a large office building. An optimization problem restrained by the building thermal conditions and electricity price was solved, defining an optimal DR schedule through a DNN decision-making agent.

DR also offers an exciting realm of possibilities for applications with reinforcement learning (RL), with the creation of AI-based agents for the monitoring, operation, and control of DR actions. With this approach, the AI agent learns not by identifying patterns or tracing input–output pair functions but through a feedback loop that indicates whether its behavior was productive or nonproductive. Instead of having a clear answer in the shape of an input–output pair, RL agents face consequences in terms of success or failure. Reference [21] explored the trend of developing AI algorithms targeting the built environment, exploring the challenges for building controls, and including energy use and DR applications. RL was the main strategy analyzed in the comprehensive review.

Applications such as automated EMSs are dependent on RL agents for decision-making and optimization. Reference [22] combined deep learning with RL methods for optimizing the scheduling of a building EMS for DR and energy efficiency. In the study, the optimization was not only limited to electricity loads but also integrated local solar energy generation in the building energy optimization. The application of AI-based agents in EMSs was explored in [23] with a deep RL-based agent for the operation of DR activities. The proposed algorithm presents a better performance than other established methods such as model predictive control (MPC) for multiple buildings and facilities, demonstrated on a real experimental setting.

A rather important segment in the field of ML for DR is in multi-agent systems and swarm intelligence applications. Swarm intelligence relates to the field of AI that takes a nature-based inspiration to design algorithms to solve tasks that involve multiple agents behaving as a biological swarm. This is the case of small controllable appliances in residential or industrial environments. The aggregation of the flexibility provided by a large fleet of flexible appliances and loads, able to provide DR services, can be a significant resource for maintaining grid stability through participation in frequency markets. Moreover, the swarm intelligence applications, together with the identification of value-sharing strategies in buildings and energy communities, can pave the path toward a fully flexible grid with the automated participation of consumers through DR actions at the local level. Load flexibility applied at a local level through swarm intelligence applications may greatly contribute to the mitigation of grid congestion and power quality issues by enabling end consumers to actively participate in energy markets through the offering of their flexibility resources.

This granular approach for DR actions, focusing on smaller-scale appliances through the implementation of swarm intelligence algorithms, was explored in [24]. The study proposed a framework for the management of such devices in a smart home and demonstrated an application of an artificial bee colony algorithm for the performance optimization of the EMS. Such an approach expands on the concept of residential DR, enabling smaller appliances to be considered as DR assets through pushing intelligence into edge devices. This in turn requires more robust aggregation techniques for the exploitation of the home flexibility potential.

2.4 Case Study

In this section, the application of AI technologies in enabling DR potentials is better demonstrated via two case studies. In the first case study, an optimization model for optimally calculating time-varying electricity prices is developed and solved via the Q-learning approach. To capture demand-side behavior toward electricity price changes, dynamic price elasticity of demand has been applied. According to the results, Q-learning approach showcased significant performance in devising time-varying prices. In the second case study, RL is used to solve the optimization problem for a home energy manager. The performance of the method is compared with two well-known metaheuristic approaches. It was shown that the RL-based method outperforms the metaheuristic approaches in two different cases with and without energy storage.

2.4.1 Case Study I: Time-Varying Pricing

This case study aims to examine a time-based pricing program. Under real-time pricing (RTP) or dynamic pricing scheme, consumers are charged with prices that are variable in nature. Typically, the price varies over a short time duration. The prices are generally quoted hourly or daily in advance to reflect marginal supply cost. The required data are taken from the day-ahead market of New England [25]. The demand model used for the pricing is represented by a composite demand function (CDF) considering different clusters of consumers. CDF is the case when a service has multiple uses or clusters. Dynamic price elasticities are also considered, and a comprehensive DR (CDR) model is proposed considering all practical aspects like consumer clusters and price elasticities. In the model, the total procurement cost of the system demand is minimized. There are a few technical and financial constraints considered in the model. The most important constraint is the cap over quoted prices. The cap is to represent regulations limiting the maximum price that can be offered to consumers. The above model helps a retail energy provider (REP) agent in an agent-based retail environment to offer day-ahead real-time prices to its customers. The optimal real-time prices are determined through an economically optimized manner based on the principles of Q-learning. Here, Q-learning is used for solving this problem because of the following reasons:

- Great performance
- Less sensitive to changes in input parameters
- Direct learning of the optimal policy

Numerical studies are conducted based on New England day-ahead market data to investigate the performance of the proposed model. Figure 2.3 shows the historical winter and summer load of New England [25]. As can be observed in the figure, the summer demand is much higher than the winter demand for certain hours of the day. The agent of REP produces the best optimal prices to its active customers through Q-learning-based approach.

Figure 2.4 represents the RTP prices for the winter and summer periods. The prices offered for a summer day are much higher than the winter prices. The reason is that system peak hours are during the summer season when energy procurement cost is the highest. According to the figure, the proposed prices are so that the system demand is shifted to hours with lower demand.

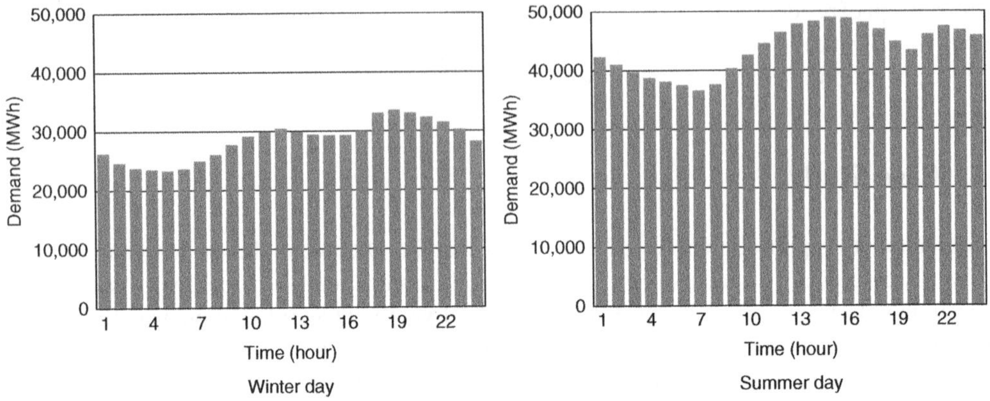

Figure 2.3 New England hourly demand during winter and summer [25].

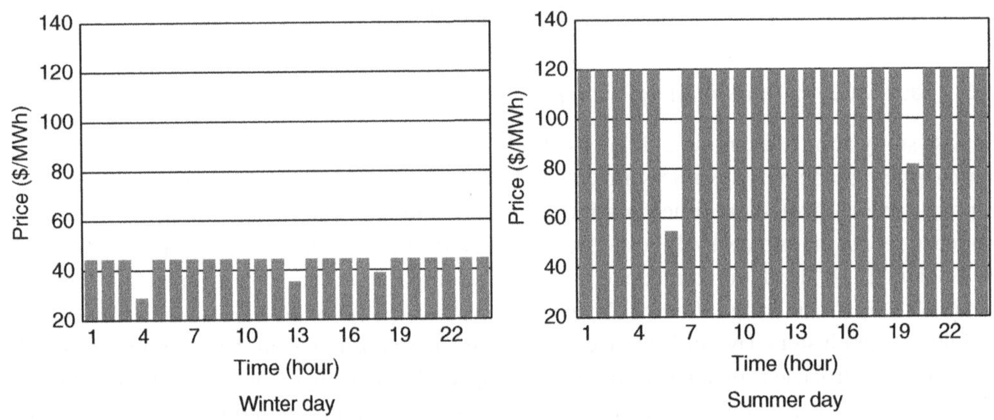

Figure 2.4 Optimized RTP prices during winter and summer days.

During the winter day, the cheapest hour is 4 am when the market price has the least value. During the day, minimum system load occurs from 3 am to 6 am. The cheapest electricity price during the summer day is at 6 am when the minimum market price is experienced. During the day, minimum system load occurs from 6 am to 8 am. It is worthwhile to note that RTP prices are equal to the considered cap in many hours since reducing the price directly leads to lower revenue. The prices are lower than the cap in local minimums to motivate loads from periods with higher market prices to periods with lower market prices. An interesting observation here is that if the correlation between system demand and wholesale market price is not unity, then implementation of RTP prices does not necessarily result in valley filling. Here, the least market price during winter occurs at 4 am, while the minimum demand is observed at 5 am.

Figure 2.5 presents time-of-use (TOU) prices for the winter and summer periods. TOU pricing, if well designed, can help and empower the public to lower their monthly utility bills and reduce pollution, thereby promoting to take advantage of cleaner, less-expensive renewable energy by using nonessential appliances during off-peak times. The TOU pricing scheme implemented here has three price levels. As can be seen, TOU prices are higher during periods with higher demand and lower during lower demand. As an example, TOU prices are lower during early morning hours

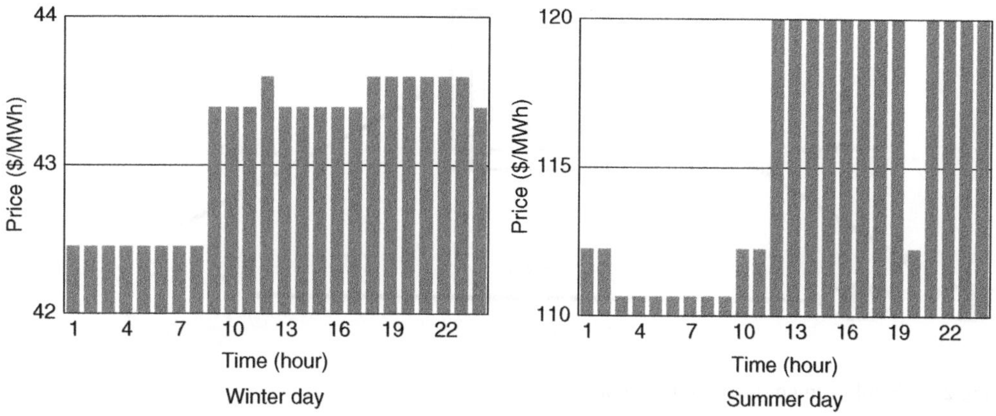

Figure 2.5 Optimized TOU prices during winter and summer.

Table 2.1 Monetary benefit achieved by RTP and TOU pricing schemes.

Season	Profit gained by RTP prices [k$]	Profit gained by TOU prices [k$]
Winter	22,421	13,282
Summer	88,089	11,296

when the system experiences lower demand. As can be seen, electricity prices are much lower during winter when electricity demand is lower.

Using the RTP and TOU prices, the total system cost for procuring demand from the market can be decreased. The decrease is considered as monetary profit here. The following Table 2.1 presents the monetary profit gained by applying the RTP and TOU prices during winter and summer. It is observed that the profits are quite high for RTP compared to those achieved by the TOU pricing scheme. According to the results, the profit gained by the RTP prices is about 69% higher than that achieved by the TOU prices during winter. The value is about 680% during summer. According to these results, it can be concluded that using the DR potential is more beneficial during periods with higher and more volatile market prices (i.e., summer season here). As another observation, it can be concluded that TOU prices can lead to acceptable performance in systems with less volatile market prices or during periods when market prices are less volatile (i.e., winter season here). The TOU prices are unable to perform significantly during the summer season when market prices are very volatile.

2.4.2 Case Study II: Home Energy Management

Energy management in residences is important for utilities. Here, a case study for home energy management having EVs and uninterrupted power supply (UPS) as energy storage is presented. HVAC, water heater, and electric water pump are the loads of the house. The objective function is the minimization of net cost of electricity for a typical day [26]. The system parameters are the same as in [26]. The optimization problem is solved by RL [27]. RL is an area of ML concerned with how intelligent agents ought to take actions in an environment in order to maximize the notion of cumulative reward. Figure 2.6 shows the working of RL for solving optimization problems. In RL,

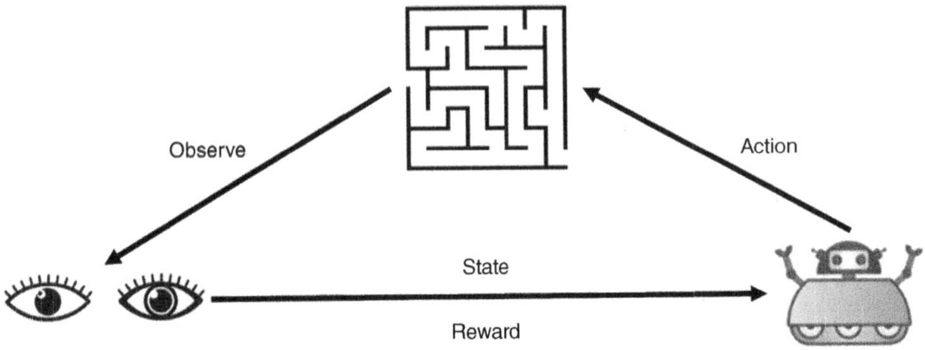

Figure 2.6 Reinforcement learning flowchart.

a method to reward desired behaviors and punish negative behavior is devised by the developers. Thus, this approach assigns positive values to desired actions for encouraging the agent and negative values to undesired behaviors. In the case of optimization problems, RL finds the best optimal solution by this methodology. Here, the performance of the RL-based approach is compared with metaheuristics such as GA and particle swarm optimization (PSO).

The optimization is performed for two scenarios. In the first scenario, there is no energy storage, and in the second scenario, EV and UPS are present as loads as well as energy storage medium. The optimization results with and without energy storage are reported in Tables 2.2 and 2.3, respectively. It is observed that the total cost obtained by RL without energy storage is 312.27 INR that is less than GA and PSO, thereby exhibiting the superior performance of RL. The energy consumption without incorporating energy storage obtained by RL is 314.22 kWh that is better than GA as well as PSO. RL performs 4.54% better than PSO for scenario 1 where there is no energy storage. RL performs 22% better than GA for scenario 1. The total cost obtained by RL incorporating energy storage is 440.15 INR that is less than the results obtained by GA and PSO. Similarly, the energy consumption obtained by RL with energy storage is better compared to that obtained by GA and PSO.

Table 2.2 Optimization results without energy storage.

Algorithm	Electricity consumption [kWh]	Total electricity cost [Rs]
PSO	328.44	568.33
GA	384.25	700.12
RL	314.22	312.27

Table 2.3 Optimization results with energy storage.

Algorithm	Electricity consumption [kWh]	Total electricity cost [Rs]
PSO	365.78	630.74
GA	400.15	700.30
RL	340.19	440.15

It is worthwhile to mention that the total electricity cost associated with the simulation with energy storage is higher than that in the simulation without energy storage. This is because the presence of EVs and UPSs, although they provide energy storage capability, leads to higher electricity consumption.

2.5 Discussion

This section summarizes the whole chapter and elaborates on the future trends in the relevant areas. The chapter delves into AI and ML applications for DR. DR is an effective means for energy management and supporting renewable penetration. This chapter starts by providing a background of DR, thereby explaining its definition, components, global status, and AI applications to DR. Globally, it is observed that market-based DR is provided primarily from industrial processes. The smaller loads like EVs are still in the pilot stage. The applications of AI and ML for DR are also reviewed lucidly in this chapter. It is concluded that the use of AI approaches in DR can be in different ways like clustering creation, anomaly detection, appliance load monitoring, load forecasting, and energy pricing. It is observed that RL and metaheuristics have tremendous applications in energy management incorporating DR.

Two case studies on DR are also presented. The first case study investigated a time-based pricing program by using Q-learning for the day-ahead market of New England. The optimal RTP prices were determined through an economically optimized manner based on the principles of Q-learning. The efficacy of the Q-learning approach was visible from the reported results. The second case study focused on home energy management having EVs and UPS as energy storage. An RL-based approach was used for solving the optimization problem with and without energy storage. The performance of RL was compared with GA and PSO. It was observed that RL performed better than GA as well as PSO.

References

1 United Nations Paris Agreement (2015). https://unfccc.int/sites/default/files/english_paris_agreement.pdf.

2 Antonopoulos, I., Robu, V., Couraud, B. et al. (2020). Artificial intelligence and machine learning approaches to energy demand-side response: a systematic review. *Renewable and Sustainable Energy Reviews* 130: 109899. https://10.1016/j.rser.2020.109899.

3 Huang, W., Zhang, N., Kang, C. et al. (2019). From demand response to integrated demand response: review and prospect of research and application. *Protection and Control of Modern Power Systems* 4: 12. https://doi.org/10.1186/s41601-019-0126-4.

4 Simões, M.G. (2021). *Artificial Intelligence for Smarter Power Systems: Fuzzy Logic and Neural Networks*. London, United Kingdom: The Institution of Engineering and technology (IET). ISBN: 9781839530005.

5 IEA (2021). *Demand Response*. Paris: IEA https://www.iea.org/reports/demand-response.

6 K. Maki, Aro, M. and Bindner, H. (2021). Global analysis on demand response status and further needs for joint research. *CIRED 2021—The 26th International Conference and Exhibition on Electricity Distribution*. pp. 3122–3125. https://doi.org/10.1049/icp.2021.2187.

7 Rotger-Griful, S., Jacobsen, R.H., Nguyen, D., and Sørensen, G. (2016). Demand response potential of ventilation systems in residential buildings. *Energy and Buildings* 121: 1–10. https://doi.org/10.1016/j.enbuild.2016.03.061.

8 D'hulst, R., Labeeuw, W., Beusen, B. et al. (2015). Demand response flexibility and flexibility potential of residential smart appliances: experiences from large pilot test in Belgium. *Applied Energy* 155: 79–90. https://doi.org/10.1016/j.apenergy.2015.05.101.

9 Dehghan-Dehnavi, S., Fotuhi-Firuzabad, M., Moeini-Aghtaie, M. et al. (2021). Decision-making tree analysis for industrial load classification in demand response programs. *IEEE Transactions on Industry Applications* 57 (1): 26–35. https://doi.org/10.1109/TIA.2020.3032932.

10 Wohlfarth, K., Klobasa, M., and Gutknecht, R. (2020). Demand response in the service sector—theoretical, technical and practical potentials. *Applied Energy* 258: 114089. https://doi.org/10.1016/j.apenergy.2019.114089.

11 Sun, C., Li, T., Low, S.H., and Li, V.O.K. (2020). Classification of electric vehicle charging time series with selective clustering. *Electric Power Systems Research* 189 (August): 106695. https://doi.org/10.1016/j.epsr.2020.106695.

12 Wang, Y., Jia, M., Gao, N. et al. (2022). Federated clustering for electricity consumption pattern extraction. *IEEE Transactions on Smart Grid* 3053 (c): 1–16. https://doi.org/10.1109/TSG.2022.3146489.

13 Liu, Q., Kamoto, K.M., Liu, X. et al. (2019). Low-complexity non-intrusive load monitoring using unsupervised learning and generalized appliance models. *IEEE Transactions on Consumer Electronics* 65 (1): 28–37. https://doi.org/10.1109/TCE.2019.2891160.

14 Cui, G., Liu, B., Luan, W., and Yu, Y. (2019). Estimation of target appliance electricity consumption using background filtering. *IEEE Transactions on Smart Grid* 10 (6): 5920–5929. https://doi.org/10.1109/TSG.2019.2892841.

15 Ruano, A., Hernandez, A., Ureña, J. et al. (2019). NILM techniques for intelligent home energy management and ambient assisted living: A review. *Energies (Basel)* 12 (11): 1–29. https://doi.org/10.3390/en12112203.

16 Rehman, A.U., Lie, T.T., Vallès, B., and Tito, S.R. (2021). Non-invasive load-shed authentication model for demand response applications assisted by event-based non-intrusive load monitoring. *Energy and AI* 3: 100055. https://doi.org/10.1016/j.egyai.2021.100055.

17 Hong, T. and Fan, S. (2016). Probabilistic electric load forecasting: a tutorial review. *International Journal of Forecasting* 32 (3): 914–938. https://doi.org/10.1016/j.ijforecast.2015.11.011.

18 Zhang, Y., Deng, C., Zhao, R., and Leto, S. (2020). A novel integrated price and load forecasting method in smart grid environment based on multi-level structure. *Engineering Applications of Artificial Intelligence* 95, no. November 2019: 103852. https://doi.org/10.1016/j.engappai.2020.103852.

19 Pallonetto, F., de Rosa, M., Milano, F., and Finn, D.P. (2019). Demand response algorithms for smart-grid ready residential buildings using machine learning models. *Applied Energy* 239 (October 2018): 1265–1282. https://doi.org/10.1016/j.apenergy.2019.02.020.

20 Kim, Y.J. (2020). A supervised-learning-based strategy for optimal demand response of an HVAC system in a multi-zone office building. *IEEE Transactions on Smart Grid* 11 (5): 4212–4226. https://doi.org/10.1109/TSG.2020.2986539.

21 Wang, Z. and Hong, T. (2020). Reinforcement learning for building controls: the problem, opportunities and challenges. *Applied Energy* 269 (1): 300. https://doi.org/10.1016/j.apenergy.2020.115036.

22 Mocanu, E., Mocanu, D.C., Nguyen, P.H. et al. (2019). On-line building energy optimization using deep reinforcement learning. *IEEE Transactions on Smart Grid* 10 (4): 3698–3708. https://doi.org/10.1109/TSG.2018.2834219.

23 Lu, R., Bai, R., Luo, Z. et al. (2022). Deep reinforcement learning-based demand response for smart facilities energy management. *IEEE Transactions on Industrial Electronics* 69 (8): 8554–8565. https://doi.org/10.1109/TIE.2021.3104596.

24 Bui, K.H.N., Agbehadji, I.E., Millham, R. et al. (2020). Distributed artificial bee colony approach for connected appliances in smart home energy management system. *Expert Systems* 37 (6): 1–14. https://doi.org/10.1111/exsy.12521.

25 Market data | Nord Pool. nordpoolgroup.com (accessed 9th January 2022).

26 Sisodiya, S., Kumbhar, G.B., and Alam, M.N. (2018, March). A home energy management incorporating energy storage systems with utility under demand response using PSO. In: *2018 IEEMA Engineer Infinite Conference (eTechNxT)*, 1–6. New Delhi, India: IEEE.

27 Sutton, R.S. and Barto, A.G. (2018). *Reinforcement Learning: An Introduction*. Cambridge, MA: MIT Press.

3

Smart Power/Energy Management and Optimization in Microgrids

Talal Saleh[1], Omar Mohamed[1], Seyed Farhad Zandrazavi[1], and Miadreza Shafie-khah[2]

[1]*School of Technology and Innovations, University of Vaasa, Vaasa, Finland*
[2]*Research and Innovation Division, Nowocert, Dublin, Ireland*

3.1 Introduction

The global consumption of energy is expected to rise by 28% in 2040, meaning an increase from 575 to 736 quadrillion Btu between 2015 and 2040 [1]. Growing energy demand has led to an increasing need for the penetration of renewable energy sources into the grid. Consequently, the operators of power systems may encounter many challenges in order to meet the energy demand while trying to integrate new renewable energy sources into the existing grids.

The generated energy from renewable energy sources such as photovoltaic (PV) systems is not always matched with the peak demand for electricity usage. As a result, the benefits of integrating the PV units into DNs are still limited [2]. So as to cope with this limitation, energy storage is presented as one of the most effective methods to increase the benefits obtained by the usage of PVs [3]. Furthermore, the use of batteries is regarded as one of the most commonly used energy storage devices, which has a direct effect on the increased usage of PV and its integration into the grid [4]. Different energy storage techniques are available such as pumped hydro-storage, compressed air, battery, capacitors, and thermal systems [5]. The energy storage systems (ESSs) are capable of storing energy during the excess generation from renewable energy sources. The stored energy can be converted back to electricity during peak times or when it is needed. This solution is not only capable of increasing the exploitation and penetration of renewable energy sources, but it is also capable of providing more ancillary services and better control of supply and demand. However, there are many other applications for ESS in power systems such as energy arbitrage, load leveling, renewable energy integration, spinning reserve, and customer-side peak shaving [6–10].

With a decrease in PV panel prices, an increase in grid electricity prices, and governmental support, rooftop PV systems are becoming prevalent in modern societies [11]. Due to the reduction in prices of electricity exported to the grid, local PV self-consumption is gaining attention [12, 13]. The energy generated from rooftop PV can be stored in individual household batteries. Nevertheless, instead of the utilization of individual energy storage at the end-user property, an ESS at a community level can be adopted, which is also known as community energy storage (CES) [14]. Recent research has shown that in an area of high rooftop PV adoption, CES systems are more techno-economically viable compared to individual energy storage for distributed PV integration, since CES can be installed close to the load center in connection with the end user's renewable energy sources [11].

Smart Cyber-Physical Power Systems: Fundamental Concepts, Challenges, and Solutions, Volume 1, First Edition.
Edited by Ali Parizad, Hamid Reza Baghaee, and Saifur Rahman.

There are many other economic and social advantages to the usage of CES such as enabling the user to meet its peak demand by storing the peak generated capacity and consuming it when needed. Moreover, a CES can play a vital role in maintaining the stability of the grid by fulfilling the needs of the community first and later supplying power to the main grid. In addition, CES deployment may incentivize the community members to be involved in the socio-economic enhancement of their community [15].

To propose optimal CES with economic benefits, this research has been conducted by forming communities with a certain number of households and real PV generation data. One such study has proposed a method that has formed a community along the roadside by performing a 15-minute simulation of PV generation and household demand profiles [11]. However, one of the issues during such simulation is the lack of location data associated with the available open-source electricity meter data, which limit the actuality of load profiles of the community's households and the actual rooftop PV generation profiles. In this study, a method is proposed by forecasting actual solar PV generation and load consumption of a certain number of households selected to form a community in Konstanz, a city in Germany.

For forecasting the PV generation and load consumption of the selected community, a machine learning technique is performed by training a gradient-boosting regressor (GBR) model. GBR uses decision trees to solve the problem, and it works by modifying the sample and setting them to a negative gradient, which keeps the distribution constant [16]. The regressor calculates the difference between the prediction and the known values in which the difference remaining is known as residual.

Solar and load forecasting are carried out in this study to achieve approximate load and PV profiles for the community's load consumption and solar energy generation. This method eliminates the need for securing an individual's electricity consumption and/or solar PV generation data, which may not be easily obtained due to privacy and security concerns. Moreover, with the approximate forecasting data, the optimal CES capacity can be determined via calculation or simulation.

3.2 Materials and Methods

3.2.1 Datasets

Forecasting electricity generation or load consumption by supervised machine learning requires datasets to train the models. Datasets used for forecasting in this study are at one-hour time intervals. The actual measurement of a single PV module dataset (2017–2018) is obtained from IEEE DataPort, which is measured at SolarTech Lab, Politecnico di Milano, Italy [17]. The dataset contains one-minute time intervals of PV module power generation and weather data of Politecnico di Milano, Italy, which is resampled at one-hour time intervals to obtain hourly PV power generation. Initially, by utilizing this dataset, a gradient-boosting regression model is trained for January 2017. Based on monthly weather datasets from January to December 2017, PV module output power is forecasted for the entire year (2017). It is noteworthy that PV module datasets from Italy were used as no actual measurement data were found for power generation by PV modules in Konstanz, Germany. In addition, since the PV power generation is forecasted based on weather data, any weather data input of the respective place will predict the PV module power output of that particular place.

By performing household load forecasting, the actual energy consumption of the aggregated houses in the selected community can be forecasted. A detailed dataset of hourly load consumption of aggregated households is used to train a GBR model against weather data. The load consumption dataset is gathered from an open-source power system with data from 11 households measured in Southern Germany [18]. It is worth mentioning that the weather data of Konstanz City are not open-source data and they are obtained using student benefits from a weather website [19].

3.2.2 Community Selection

To practically adopt a CES and propose a battery sizing, the initial step is to form an energy community to which CES will be supplying stored PV generation. In Konstanz, the rooftop PV adoption rate is high with a large number of rooftop PV arrays already installed on buildings. Using satellite images, a community with several households of different sizes and installed rooftop PV modules is taken into consideration at Mondrauteweg, Konstanz, as shown in Figure 3.1.

Large buildings with more area have a greater probability of high load consumption compared to smaller buildings. Based on this assumption, the buildings are categorized into four types concerning load consumption, presented in Table 3.1. It is assumed that all the buildings are occupied by dwellers with no empty households.

The number of PV modules already installed on the rooftop is calculated by using satellite images. Large amounts of PV arrays can be seen installed on rooftop buildings, which can be counted straightforwardly (Figure 3.2). The excess power from these PV arrays will be stored in the CES during off-peak hours, while during peak hours stored energy from PV units can be injected into the community on a priority basis, instead of indiscriminate exporting of the energy to the main grid.

An analysis of the power consumption of buildings in a proposed community is conducted using the load consumption dataset [19]. The measured load consumption of four different buildings in Southern Germany is analyzed, and the average hourly consumption of these four buildings is calculated for 2016. Based on these average values, the buildings are categorized into large, medium-large, medium-small, and small buildings. It is assumed that the buildings with an equal amount of area consume approximately the same power on average.

Figure 3.1 Selected community.

Table 3.1 Buildings with rooftop PVs and load consumption in 2016.

Building type	Total buildings	Buildings with rooftop PV	Rooftop PV modules	Total consumption (kW)	Average hourly consumption (kWh)
Large buildings	6	4	186	5151.6	0.5864
Medium-large buildings	6	0	0	4148.2	0.4722
Medium-small buildings	3	0	0	2676.7	0.3047
Small buildings	7	2	96	1769.4	0.2014
Total buildings	**22**	**6**	**282**	**13,745.9**	**1.5647**

Figure 3.2 Photovoltaic panels on building rooftop.

3.2.3 Solar PV Forecasting

Solar modules, which can consist of either 36 solar cells (12 V) or 72 solar cells (24 V), are constructed by combining multiple solar cells. Cells in individual modules can be connected in series to increase the output voltage of the module, or the cells can be connected in parallel to increase the output current of the module with the same voltage of each cell [20]. Similarly, a number of modules can be connected in series or parallel to maximize the output power with the same current (series connection) or same voltage (parallel connection), respectively. In this study, a dataset of a single module consisting of 72 cells obtained from IEEE DataPort is used to train a machine learning model [17].

PV power generation from a single module is forecasted using a trained GBR machine learning model and weather data from Konstanz City for 2017. In this study, for the sake of simplification, it is assumed that all the PV modules would approximately generate the same amount of power, so the forecasted PV power generation would be the same for all the rooftop PVs. Based on this assumption, PV generation for the entire community can be calculated. To this end, forecasted PV power generation is multiplied by the total number of PV modules in the community to obtain the respective total power generation, as presented in (3.1):

$$E_{To}^{PV} = \Omega_{pv} E_{Mo}^{PV} \tag{3.1}$$

where Ω_{pv}, E_{Mo}^{PV}, and E_{To}^{PV} denote the number of PVs, PV module generation forecast, and total PV generation, respectively. It is noteworthy that the total number of PV modules is estimated using a satellite image as the respective numbers are presented in Table 3.1.

3.2.4 Household Load Forecasting

By deploying load forecasting, the hourly, yearly, and total load consumption of aggregated households is calculated. The GBR model is trained with weather data to target the aggregated load of the community in 2016. In the next step, weather data for 2017 are used by the model to predict the total hourly energy consumption of the community for that year. Similarly, using weather data from 2021, the machine learning model can predict the energy consumption of aggregated households in the selected community for 2021.

3.2.5 CES Battery Capacity Calculation

To minimize costs and optimize the operation of the CES, it is necessary to calculate the size of batteries that will store the excess energy generated by rooftop PV arrays. The energy stored in these batteries can be supplied to the community as a priority, and any additional energy may also be exported to the grid. In other words, by controlling the respective power electronic interfaces (bidirectional converters), the CES can supply energy to the community during peak hours or store energy during off-peak hours.

Based on the forecasted load consumption of the community, the battery size is calculated. The forecasted load consumption in 2017 is averaged into 24 hours, which is summed to obtain the total daily energy consumption (in kWh) by the community. In (3.2) and (3.3), equations linked to the capacity calculation of battery energy storage for the community are presented in which C_{Ah}^{CES} and C_{kWh}^{CES} denote the capacity of CES in ampere-hour and kilowatt-hour. Furthermore, E_{avg}^{CMG} and τ^{aut} represent the total energy consumption per day of the community (in kW) and the number of autonomy days, while P_{bs}^{loss}, ϕ_{bs}^{dis}, and V_{bs}^{nom} are power loss, depth of discharge, and the nominal voltage of batteries, respectively. It is noteworthy that, in this study, the values of P_{bs}^{loss}, ϕ_{bs}^{dis}, and α are considered 0.85, 0.6, and 1000, respectively.

$$C_{Ah}^{CES} = \frac{E_{avg}^{CMG} \tau^{aut}}{P_{bs}^{loss} \phi_{bs}^{dis} V_{bs}^{nom}} \tag{3.2}$$

$$C_{kWh}^{CES} = \frac{C_{Ah}^{CES} V_{bs}^{nom}}{\alpha} \tag{3.3}$$

The capacity of batteries can be calculated for different levels of the battery's nominal voltages. For this calculation, three nominal voltages are considered that are available in the market (Table 3.2). It should be noted that the size of batteries in this section is based on the aforementioned calculations and is not the optimal one yet. The optimal size of batteries is determined in the next section.

Table 3.2 Battery energy storage for the community.

No.	Nominal voltage (V)	Battery capacity (kAh)	Battery energy (kWh)
1	24	16.97	407.28
2	48	8.48	407.04
3	576	0.707	407.23

3.2.6 CES Battery Sizing Optimization

Hybrid Optimization Model for Multiple Energy Resources (HOMER) software is used in order to optimize the battery sizing and analysis of the overall system cost. The daily average load is 307.19 kWh, the load average is 12.8 kW, the peak load is about 21.1 kW, and the load factor is 0.51. The respective average monthly load profile is presented in Figure 3.3. As illustrated, the maximum demand for energy occurs between May and October. Figure 3.4 shows the load box-and-whisker chart, and the total predicted PV power generation for each month is given in Table 3.3. These data are used as an input to the system; the system schematic diagram is presented in Figure 3.5. As depicted in Figure 3.5, the PV arrays and batteries are connected to the direct current (DC) bus and the alternating current (AC) bus via a converter, while the load is directly connected to the AC bus in the presence of the main grid.

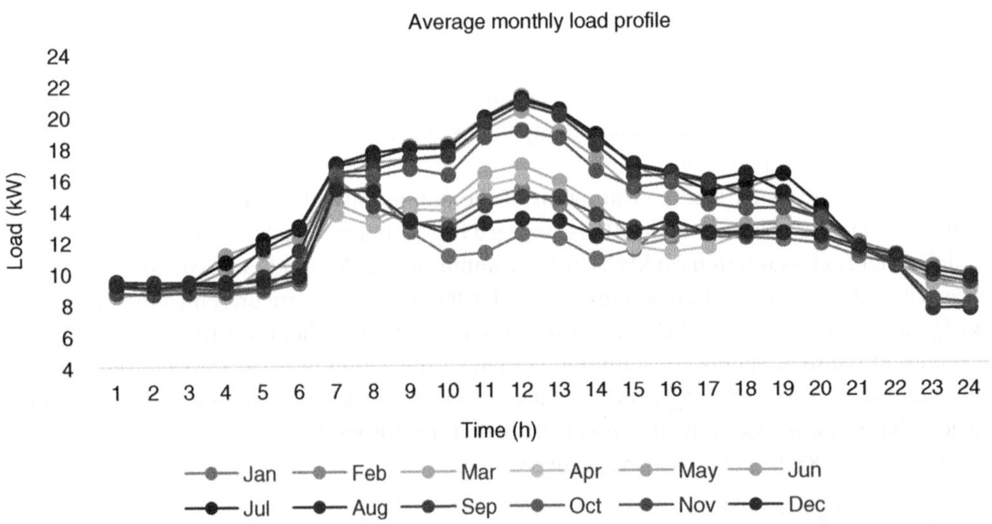

Figure 3.3 Load profile for each month.

Table 3.3 Total predicted energy generation (kWh) for each month.

Jan.	Feb.	Mar.	Apr.	May	June	July	Aug.	Sep.	Oct.	Nov.	Dec.
919.12	3753	9116	9790.7	13,045.8	13,888.5	13,230.6	12,070.4	9003.03	7172.16	2539.2	1562.1

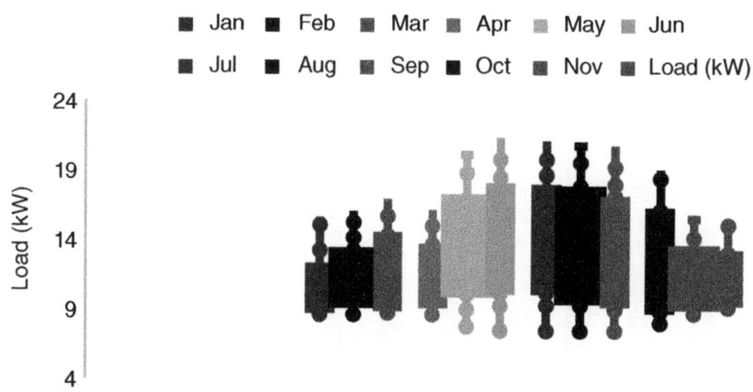

Figure 3.4 Load box-and-whisker chart.

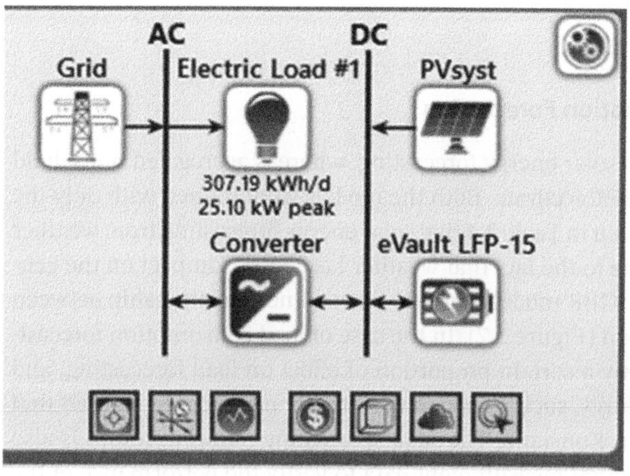

Figure 3.5 System schematic diagram.

The monthly electricity generation from the main grid and the PV arrays can be seen in Figure 3.6. The grid shared 64,023 kWh/yr (39.8%), and the PV arrays shared 97,013 kWh/yr (60.2%). The total renewable penetration is found to be around 60.2%, and the maximum peak load is about 25 kW. Two 14.4 kWh lithium-ion batteries were used so as to provide sufficient power to the system. During the summertime, the generated power is more than the load requirement, and the excess energy is sold to the grid. After adopting energy storage, CES can be configured to supply electricity to the community without selling it to the grid.

3.3 Simulation and Results

In this section, the results of the solar energy forecasting model, load forecasting model, and HOMER software simulation are presented. Solar energy forecasting and load forecasting were conducted using a GBR machine learning model. Predicted energy and load consumption datasets were used to simulate the aggregated household load consumption and study the battery storage system for the community.

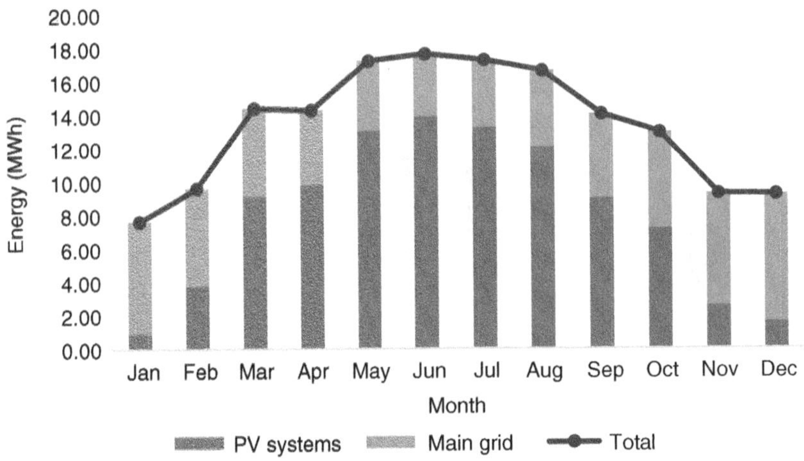

Figure 3.6 Monthly electric production.

3.3.1 Solar Energy and Load Consumption Forecasting

Accuracy up to 93% was achieved for the solar energy forecasting, whereas aggregated household load consumption forecasting was only 35% accurate. Both the models were trained with only the weather data of Konstanz City. It can be seen in Table 3.4 that solar energy forecasting from weather data achieved high accuracy, which is due to the fact that weather has a direct impact on the generation of solar power, and therefore, the GBR model is able to form a linear relationship between weather features and PV power generation (Figure 3.7). In the case of load consumption forecasting, weather data were found to have only a certain proportion of effect on load forecasting, and for more accurate forecasting, other features, such as data on electricity price, could be used that were not available as open-source data in Konstanz. The solar forecasting model accuracy is also rechecked with the upcoming month's actual measurement data to verify the performance of the model (Table 3.4).

To perform simulation on HOMER, hourly PV generation and load consumption of the community were required as an input to perform battery optimization and cost analysis for the proposed community. Both the machine learning models were given input of 365-day weather data (from 2017) of Konstanz City. Peak power produced by the community's PV arrays is predicted to be around 50 kW, whereas on average PV power is 30 kW for the year 2017 (Figure 3.8). The load consumption model was capable of predicting the pattern of load consumption in the selected community, but the aggregated household peak consumption was not predicted by the model.

Table 3.4 Accuracies of forecasting models.

Forecasting models	Training score	Test score	Cross-validation score	Verified accuracy score
Solar PV forecasting	0.9314	0.9121	0.9294	0.88
Aggregated load consumption forecasting	0.384	0.342	0.20	—

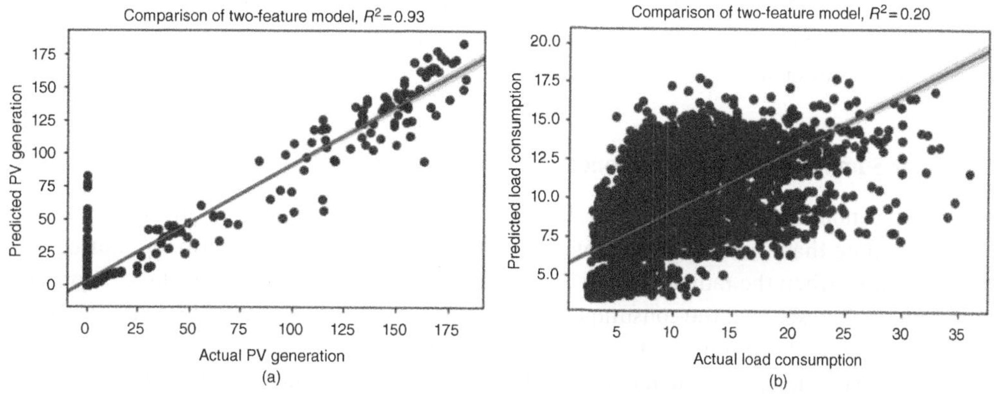

Figure 3.7 GBR model performance. (a) Solar power (Wh) forecasting model liner performance. (b) Aggregated household load consumption (kWh) forecasting model liner performance.

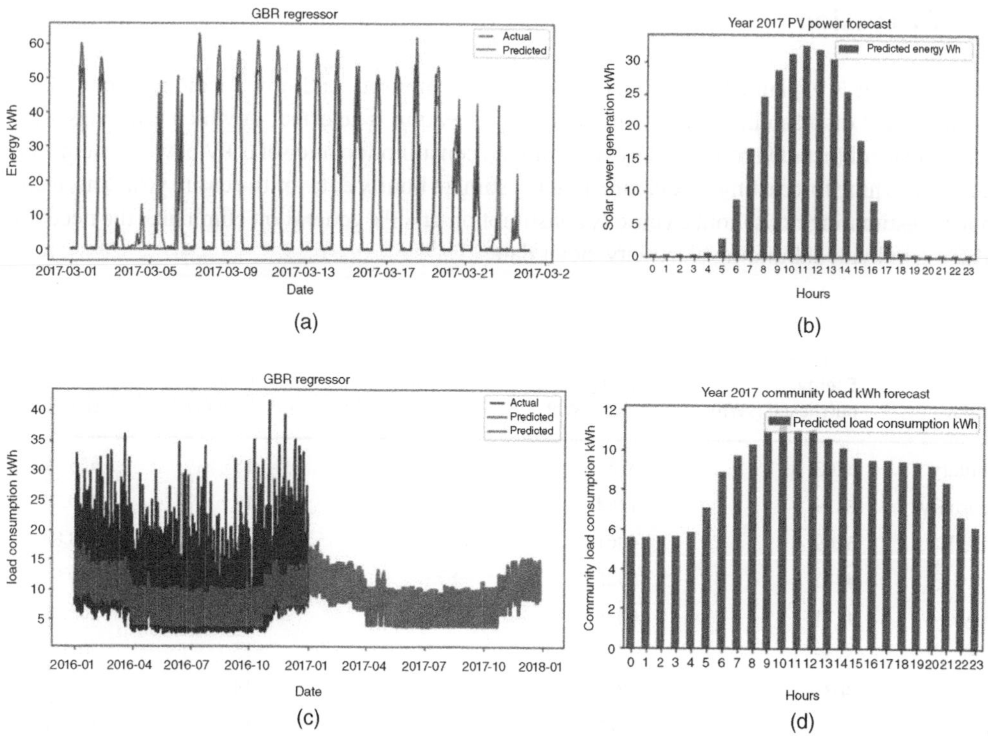

Figure 3.8 Forecasted PV generation and load consumption. (a) PV generation prediction against actual measurement. (b) Average hourly PV generation pattern by community PV arrays. (c) Aggregated household load consumption prediction against actual measurement. (e) Average hourly load consumption pattern by proposed community.

According to Figure 3.8, it can be seen that the low energy consumption by households was predicted by the model, but the peak consumption prediction using only weather data is not considered to be sufficient.

3.3.2 Cost Summary and System Economics

The system economics are presented in Table 3.5; the total energy purchased during the winter months was more than the energy generated from the PV arrays; on the other hand, during the summer months, when the radiation from the sun is greater than in winter, the total energy generated from the PV exceeded the consumption. As a result, the excess energy can be either sold or stored in the proposed community storage battery. In order to be utilized during the peak period, overall, the total cost for energy charge was around €4233.69. The overall cost summary for the proposed system is given in Table 3.6. PV arrays are already installed, so the installation cost is zero. It is noticeable that the major cost is for purchasing the community storage battery, followed by the system converter cost.

3.4 Discussion

In this study, it is explained how to select a proper community and how to adopt suitable CES systems. Based on a satellite image, an appropriate location is selected, in which buildings with high intensity of installed rooftop PVs exist. For load consumption and solar PV power generation, machine learning forecasting models are used to estimate hourly data for the entire year. This technique for estimating household electricity consumption and PV generation eliminates the need for gathering individual building electricity meter data.

Table 3.5 System economics.

Month	Energy purchased (kWh)	Energy sold (kWh)	Net energy purchased (kWh)	Peak load (kW)	Energy charge (€)
January	6703	326	6377	18	653.98
February	5839	1484	4355	18	509.72
March	5269	4861	408	17	283.84
April	4472	5325	−853	17	180.95
May	4171	6316	−2145	15	101.29
June	3752	6497	−2745	15	50.32
July	4031	5848	−1817	14	110.67
August	4571	5387	−816	14	187.72
September	5050	3675	1375	16	321.29
October	5785	2754	3031	16	440.77
November	6724	716	6008	17	636.59
December	7658	184	7473	18	756.55
Annual	**64,023**	**43,373**	**20,650**	**18**	**4233.69**

Table 3.6 Cost summary.

Component	Capital (€)	Replacement (€)	O&M (€)	Fuel (€)	Salvage (€)	Total (€)
Battery 1	10,500.00	9276.10	0.00	0.00	−1257.68	18,518.42
Battery 2	10,500.00	9276.10	0.00	0.00	−1257.68	18,518.42
Grid	0.00	0.00	54,731.06	0.00	0.00	54,731.06
PV system	0.00	0.00	12,927.52	0.00	0.00	12,927.52
System converter	14,083.75	5975.37	0.00	0.00	−1124.62	18,934.49
System	**35,083.75**	**24,527**	**67,658.57**	**0.00**	**−3639.98**	**290,529.91**

Solar PV generation is forecasted based on a single PV module. In this approach, an assumption is made that all the PV modules installed on rooftops can generate the same amount of energy. A machine learning model is trained with solar generation measurement data of a single PV module. The forecasted data are multiplied by total PV panels in a selected community to obtain total PV generation. The results indicate that the PV power generation forecasting model reached 93% accuracy.

Load consumption is also forecasted by using weather data and is categorized based on the area of buildings. It is assumed that all the buildings are occupied by dwellers and buildings with more area have more consumption as the number of residents and electric devices is greater in larger buildings. Actual measured data for different buildings, obtained by an open-source platform, are assumed to be the data for the buildings of the proposed community in Konstanz. However, to model the community with more accuracy, electrical devices and their energy consumption by the buildings can be estimated, which can be used to forecast load consumption by the community.

The accuracies of both the machine learning models can be improved by adopting other machine learning or deep learning techniques. Regression models such as K-nearest neighbor regression and Gaussian process regression can improve the accuracy of load forecasting [21]. Moreover, the physics-constrained long short-term memory (LSTM) deep learning model has also been shown to produce high accuracy scores for PV day-ahead hourly forecasting [22]. Furthermore, for load consumption prediction, prototypical building modeling is also considered a key factor in load prediction, which directly influences the prediction accuracy [23]. Hence, it can be used to build and calibrate community buildings to predict the energy consumption of the buildings.

With the forecasted data obtained for one year, the battery size for the CES has been calculated and simulated. The results show that the proposed community battery requires 407 kWh of battery size to provide energy to the community with an autonomous CES supply of electricity for up to one day. However, the optimized result from the HOMER simulation shows that two batteries, each with 14.4 kWh capacity, are sufficient to store and deliver the extra power to the community. Moreover, a system economics and cost summary are also presented, which highlight the CES investment cost and the system costs that might be incurred with the adoption of CES.

3.5 Conclusions

In conclusion, the study presents a machine learning–based technique to predict community load consumption and solar energy generation by rooftop PVs using weather parameters. The proposed method eliminates the need for obtaining household owner's energy consumption or solar

PV generation, data which might be difficult to obtain due to privacy concerns. However, it should be noted that the datasets used in this study for community load consumption forecasting give an estimate prediction. For more realistic community load consumption forecasting, prototypical building modeling can be considered in further studies. Furthermore, the results show that the machine learning technique can successfully contribute to the selection of communities and the determination of the optimal size of batteries for the communities while considering the proliferation of renewable energy-based distributed generation units in the communities. The results demonstrate that the optimal battery sizing can lead to a substantial increase in PV penetration in the community. As a result, thanks to machine learning techniques and analyses as well as the proper integration of distributed energy resources including battery energy storage, the communities of the future may become more efficient, green, and self-sufficient.

References

1 Shelagh, W., van der Burg, L. Fossil fuel subsidy reform: from rhetoric to reality. https://odi.org/en/publications/fossil-fuel-subsidy-reform-from-rhetoric-to-reality/ (accessed 29 June 2021).

2 Denholm, P. and Margolis, R.M. (2007). Evaluating the limits of solar photovoltaics (PV) in traditional electric power systems. *Energy Policy* 35: 2852–2861. https://doi.org/10.1016/j.enpol.2006.10.014.

3 Friedman, B., Ardani, K., Feldman, D. et al. (2013). *Benchmarking Non-Hardware Balance-of-System (Soft) Costs for U.S. Photovoltaic Systems, Using a Bottom-Up Approach and Installer Survey*, 2ee. Golden, CO (United States): National Renewable Energy Lab. (NREL) p. NREL/TP-6A20-60412, 1107461.

4 Heras-Saizarbitoria, I., Cilleruelo, E., and Zamanillo, I. (2011). Public acceptance of renewables and the media: an analysis of the Spanish PV solar experience. *Renewable and Sustainable Energy Reviews* 15: 4685–4696. https://doi.org/10.1016/j.rser.2011.07.083.

5 Díaz-González, F., Sumper, A., Gomis-Bellmunt, O., and Villafáfila-Robles, R. (2012). A review of energy storage technologies for wind power applications. *Renewable and Sustainable Energy Reviews* 16: 2154–2171. https://doi.org/10.1016/j.rser.2012.01.029.

6 Vazquez, S., Lukic, S.M., Galvan, E. et al. (2010). Energy storage systems for transport and grid applications. *IEEE Transactions on Industrial Electronics* 57: 3881–3895. https://doi.org/10.1109/TIE.2010.2076414.

7 Kazempour, S.J., Moghaddam, M.P., Haghifam, M.R., and Yousefi, G.R. (2009). Electric energy storage systems in a market-based economy: comparison of emerging and traditional technologies. *Renewable Energy* 34: 2630–2639. https://doi.org/10.1016/j.renene.2009.04.027.

8 Alam, M.J.E., Muttaqi, K.M., and Sutanto, D. (2014). A novel approach for ramp-rate control of solar PV using energy storage to mitigate output fluctuations caused by cloud passing. *IEEE Transactions on Energy Conversion* 29: 507–518. https://doi.org/10.1109/TEC.2014.2304951.

9 Wang, G., Ciobotaru, M., and Agelidis, V.G. (2014). Power smoothing of large solar PV plant using hybrid energy storage. *IEEE Transactions on Sustainable Energy* 5: 834–842. https://doi.org/10.1109/TSTE.2014.2305433.

10 Awad, A.S.A., Fuller, J.D., El-Fouly, T.H.M., and Salama, M.M.A. (2014). Impact of energy storage systems on electricity market equilibrium. IEEE Trans. Sustain. *Energy* 5: 875–885. https://doi.org/10.1109/TSTE.2014.2309661.

11 Barbour, E., Parra, D., Awwad, Z., and González, M.C. (2018). Community energy storage: a smart choice for the smart grid? *Applied Energy* 212: 489–497. https://doi.org/10.1016/j.apenergy.2017.12.056.

12 Luthander, R., Widén, J., Nilsson, D., and Palm, J. (2015). Photovoltaic self-consumption in buildings: a review. *Applied Energy* 142: 80–94. https://doi.org/10.1016/j.apenergy.2014.12.028.

13 Weniger, J., Bergner, J., Tjaden, T., et al. (2014). Economics of residential pv battery systems in the self-consumption age. *29th European Photovoltaic Solar Energy Conference and Exhibition*, 3871–3877, 7 pages, 11368 kb, https://doi.org/10.4229/EUPVSEC20142014-7DO.14.3.

14 Roberts, B.P. and Sandberg, C. (2011). The Role of Energy storage in development of smart grids. *Proceedings of the IEEE* 99: 1139–1144. https://doi.org/10.1109/JPROC.2011.2116752.

15 Parra, D., Swierczynski, M., Stroe, D.I. et al. (2017). An interdisciplinary review of energy storage for communities: challenges and perspectives. *Renewable and Sustainable Energy Reviews* 79: 730–749. https://doi.org/10.1016/j.rser.2017.05.003.

16 Duffy, N. and Helmbold, D. (2002). Boosting methods for regression. *Machine Learning* 47: 153–200. https://doi.org/10.1023/A:1013685603443.

17 Leva, S., Nespoli, A., Pretto, S. et al. (2020). *Photovoltaic Power and Weather Parameters.* IEEE Dataport https://dx.doi.org/10.21227/42v0-jz14.

18 Data Platform—Open Power System Data. https://data.open-power-system-data.org/household_data/latest/ (accessed 29 June 2021).

19 Solcast—Solar Forecasting & Solar Irradiance Data. https://solcast.com/ (accessed 29 June 2021).

20 Ahmad, M. (2017). *Operation and Control of Renewable Energy Systems*, Chapter 6, 125–152. Newark: John Wiley & Sons, Incorporated. Accessed June 29, 2021. ProQuest Ebook Central. ISBN. ISBN: 9781119281702.

21 Linton, T. (2015). Forecasting hourly electricity consumption for sets of households using machine learning algorithms. Independent Thesis Advanced Level. Sweden: KTH.

22 Luo, X., Zhang, D., and Zhu, X. (2021). Deep learning based forecasting of photovoltaic power generation by incorporating domain knowledge. *Energy* 225: 120240. https://doi.org/10.1016/j.energy.2021.120240.

23 Xu, L., Pan, Y., Lin, M., and Huang, Z. (2017). Community load prediction: methodology and a case study. *Procedia Engineering* 205: 511–518. https://doi.org/10.1016/j.proeng.2017.10.405.

4

Smart City Energy Infrastructure as a Cyber-Physical System of Systems: Planning, Operation, and Control Processes

Mahdi Nozarian[1], *Alireza Fereidunian*[2,*], *and Masoud Barati*[3]

[1]*Faculty of Electrical Engineering, K.N. Toosi University of Technology, Tehran, Iran*
[2]*Faculty of Electrical Engineering, K.N. Toosi University of Technology, Tehran, Iran*
[3]*Electrical and Computer Engineering Department, University of Pittsburgh, Pittsburgh, PA, USA*

4.1 Introduction

In the twenty-first century, human-made systems have experienced significant increasing growth in scale, complexity, and integration [1, 2]. The networked connection of smart things is one of the main factors in this complexity and integration, while by smart things we mean cyber-physical systems (CPSs), including sensors, actuators, and recorders that monitor and directly influence the physical environment. Specifically, the integration of stand-alone CPSs that offer services beyond the services of any isolated CPSs creates a new category of systems called a CPS of systems (CPSoS). The interaction between various autonomous components makes a CPSoS a highly complex entity [3–5].

Compared to the conventional system engineering processes, the system of systems (SoS) framework expresses fundamental differences, as it integrates a set of heterogeneous and autonomous systems that may have different ages and technologies, which are sometimes even not necessarily well-adapted to collaborate together [6]. Accordingly, the traditional knowledge of system engineering may not be able to manage the process in these complex systems, and new scientific knowledge is required to analyze and develop these CPSoSs. In this regard, SoS engineering can be helpful as an opportunity for the system engineering community to study the complex behavior of integrated twenty-first-century systems [4, 6].

Compared to the conventional system engineering process, the SoS framework shows fundamental differences, as it integrates a set of heterogeneous and autonomous systems that are sometimes not designed to work together and may have different ages and technology. Accordingly, the traditional knowledge in system engineering does not show not so much ability to manage the process in these complex systems; thus, a new knowledge base is required to analyze and manage CPSoSs. In this regard, SoS engineering can be regarded as an opportunity for system engineering communities to study the complex behavior of integrated twenty-first-century systems [4, 6].

Modeling of critical infrastructures is one of the application fields of SoS engineering, as they can be introduced as CPSs [7–9]. In the smart city, several distributed autonomous CPSs integrate and interact through complex relationships toward providing new functionalities. This is performed by interconnecting the different physical systems based on information and communication technologies (ICTs) and Internet of Things (IoT) networks. This integration results in a smart city CPSoS

Smart Cyber-Physical Power Systems: Fundamental Concepts, Challenges, and Solutions, Volume 1, First Edition.
Edited by Ali Parizad, Hamid Reza Baghaee, and Saifur Rahman.

framework that consists of heterogeneous systems developed by different technologies and data formats, which distinct organizations own. The main processes that can benefit from this integration include planning, operation, and control/management of the smart city [10, 11]. Accordingly, the smart city can be considered a system that can integrate different public/private heterogeneous, autonomous CPSs across other domains: environment, health care, energy and water, transportation, government service, and crisis management [12, 13].

Each of the autonomous CPSs of the smart city, across various domains, can be identified as a CPSoS likewise. The transportation system can exemplify this [14], and the smart city energy infrastructure is introduced and discussed as another CPSoS in this chapter. As in a city energy system, real facilities can be considered EHs; a city can be seen as a set of EHs [15]. The EHs in smart city energy infrastructure are the same constituent CPSs that pool their capabilities to create a new, more functional, and reliable smart city energy CPSoS. While introducing and describing the related concepts, this chapter's authors aim to discuss the planning, operation, and control/management process in smart city energy CPSoS.

The chapter is organized as follows: Section 2 introduces general concepts, definitions, and types of CPSoS. A literature review on applications of CPSoS is accomplished in Section 3. In Section 4, it has been shown that a smart city can be introduced as a CPSoS, and its functional CPSs can consist of heterogeneous and autonomous public and private systems. Section 4.5 describes smart city energy infrastructure as a CPSoS and reviews the adaptation of its Maier key characteristic. The proposed framework for planning, operation, and control of the smart city energy CPSoS is provided in Section 4.5. A brief discussion on the complexity and emergence is included in Section 4.6. Finally, this chapter is concluded and summarized in Section 4.7.

4.2 Cyber-Physical System of Systems

4.2.1 Definitions

There are numerous definitions of SoS, as a unified standard definition for a system cannot be found among researchers and philosophers. In [14], In [16], Boardman et al. have collected different definitions from the literature, journal and conference papers, presentations, and published papers by industry, government, and academia. Jamshidi defines an SoS as "a super system or integration of complex systems coordinated together in such a way to achieve a wider goal with possible higher significance" [2]. Most of the definitions of the SoS focus on introducing the two terms of "system (the first one)" and "systems (the second one), or constituent/component systems (CSs)" as two principal terms in the SoS description [1, 2].

- System: The system facilitates the CS interactions.
- CSs: CSs are a part of one or more systems. Each CS of an SoS is a functional autonomous system with its development, management goals, and resources but interacts within the SoS to provide unique capabilities.

Specifically, the networked connection and integration of smart things, i.e., CPSs, including sensors, actuators, and recorders that monitor and directly influence the physical environment, can provide novel services [5]. This integration of stand-alone CPSs that offer services beyond the services of any isolated CPSs creates a new category of systems called CPSoS. CPSoS is a highly complex entity as it integrates a wide range of heterogeneous autonomous CPSs that are the same CSs [17–19]. Some key differences between CPSs and CPSoS are briefly described in Figure 4.1 [20–23].

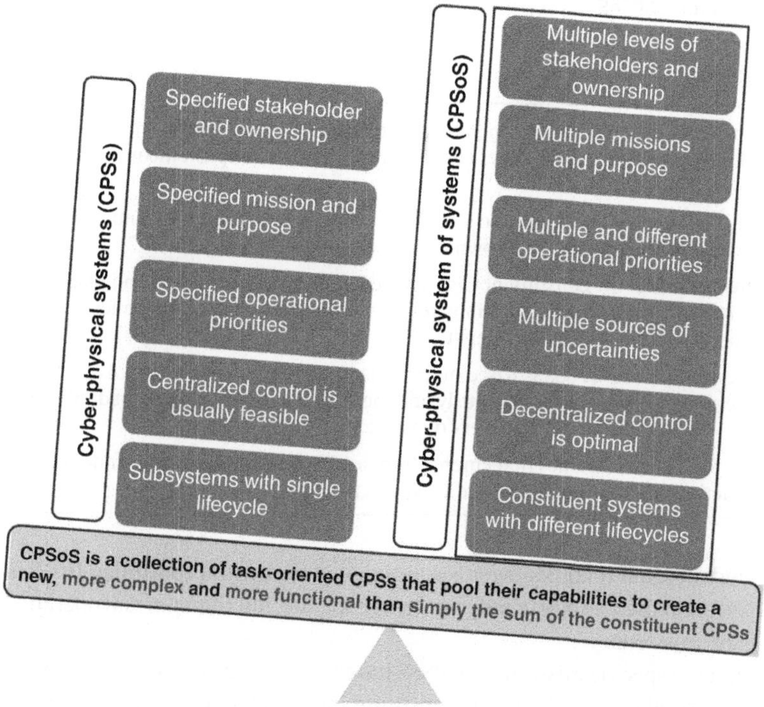

Figure 4.1 Key differences of CPSs and CPSoS.

A smart grid CPSoS employs CPSs to ensure a smooth flow of electric energy from power plants to consumers, as is the case with automated teller machine (ATM) CPSoS, in which the CPS cooperates in withdrawing money from their home bank, displaying the transaction in the currency of the home bank, in the remote country denoted in the currency of the host country [5, 21].

4.2.2 Characteristics

CPSoS is mainly used to describe an integrated set of CPSs that are autonomously operable yet are interacted for a common goal. System scientists postulated various characteristics of an SoS. Meanwhile, Maier (1998) defined five key attributes of an SoS as follows: managerial independence, operational independence, evolutionary development processes, geographical distribution, and emergent behavior. Maier believes that operational and managerial independence distinguishes an SoS from traditional systems; thus, the systems that do not exhibit these principal characteristics cannot be introduced as an SoS regardless of other characteristics. Table 4.1 describes these characteristics, known as Maier [21, 24, 25].

4.2.3 Types

CPSoS are usually classified based on the degree of CPS autonomy to mission accomplishing or their authority over CS behavior. Accordingly, CPSoSs can be classified into the four following categories [6, 24, 25], summarized in Figure 4.2.

1) Directed: These CPSoS are controlled by a central manager, controller, or stakeholder and are designed and implemented to achieve specific missions. Their CSs maintain their operational

Table 4.1 Maier characteristics for a CPSoS.

Characteristics	Description
Operational independence	Each constituent CPS, while operating independently to achieve its individual goals, also cooperates with other constituent CPSs to accomplish the missions of the whole CPSoS
Managerial independence	Each constituent CPS belongs to a specific company or organization and can be managed independently by its owner
Geographic distribution	In many CPSoS, the constituent CPSs are geographically distributed and placed in different areas geographically. Many researchers consider this feature a less important or secondary characteristic
Evolutionary development	The evolution of a CPSoS is based on its constituent CPS evolution processes. These processes may occur asynchronously and mean that a CPSoS evolves gradually. In other words, this structure is a very dynamic
Emergent behavior	Behavioral characteristics that emerge from a CPSoS result from independent CPS interactions that cannot be created locally in any CSs

and managerial independence, but their behavior is a function of the central controller. In other words, constituent CPSs retain their ability to independently operate, but their normal operation is a function of the CPSoS missions.

2) Acknowledged: Although the specific missions of CPSoS have been recognized and endorsed by the constituent CPSs, nevertheless, the CPSs that constitute independent ownership maintain their goals, budget, and approach to development and sustainability.
3) Collaborative: In this structure, the constituent CPSs interact voluntarily and with different degrees of cooperation to realize the specific purposes and missions shared by the central management.
4) Virtual: There are no central controllers and no comprehensive missions and purposes, and even such purposes are often neither designed nor expected. In this framework, the constituent CPSs can cooperate in a scattered and uncoordinated environment where a CPSoS structure may not be recognizable.

These four types of CPSoS are important, because they state that the operational and managerial independence of CPSs may differ in various types. This indicates that the realization of Maier characteristics for a CPSoS is not absolute, but relative, as it can vary from obviously identifiable to blurry and unrecognizable in different structures.

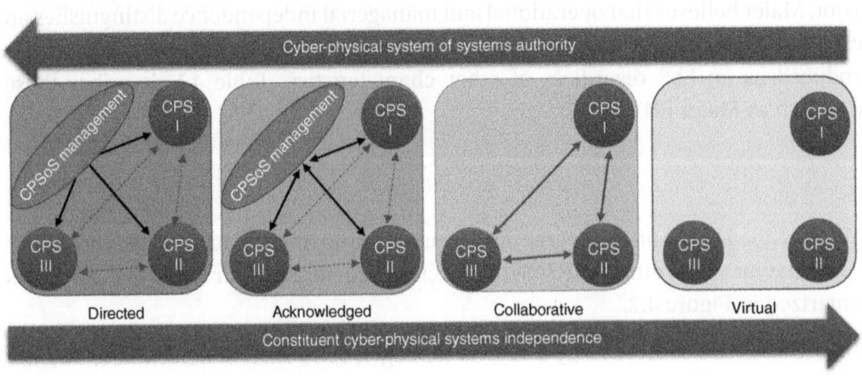

Figure 4.2 CPSoS classification.

4.3 Cyber-Physical System of System Application Domains

The increasing complexity and interconnectedness of developed systems in the twenty-first century have attracted interest from researchers in what has been known as SoS engineering. In this regard, researchers in most engineering research domains are using the SoS engineering concept as optimal and efficient planning, operation, and management framework for their specialized systems [26–29]. In this context, some researchers framed the national infrastructure as a CPSoS [30–33]. Especially, the constituent sectors in critical infrastructure CPSoS can be viewed as a CPSoS. Transportation [34–41], health care [42–46], information and communication [47–56], environment and disaster management [57–61], food supply and industrial facilities [62–70], financial and governmental management [71–75], and energy [76–83] are from these sectors as described in Figure 4.3.

4.3.1 Transportation

In the transportation sector, Reference [34] proposes a risk-based SoS framework to control the road transport flows of dangerous goods (DGs) through a real-time flow assignment problem description. The air transportation design optimization problem is modeled as an SoS-based design problem in [35]. Reference [36] presents a method based on SoS to detect, track, identify, decide, act, and assess the actions of customs and border protection (CBP) forces. Reference [37] characterizes the stakeholders in urban freight transport projects in the urban logistics ecosystem based on an SoS framework. Reference [38] describes the great British railway as it developed from over 300 individual railways to its eventual integration into a single railway through an SoS-based approach. Similarly, the California high-speed rail system (CHSRS) and the future hydrogen-based transportation economy in the United States can be mentioned as other examples [39, 40]. Reference [41] formulates a novel approach to effectively govern the maritime transportation SoS (MTSoS).

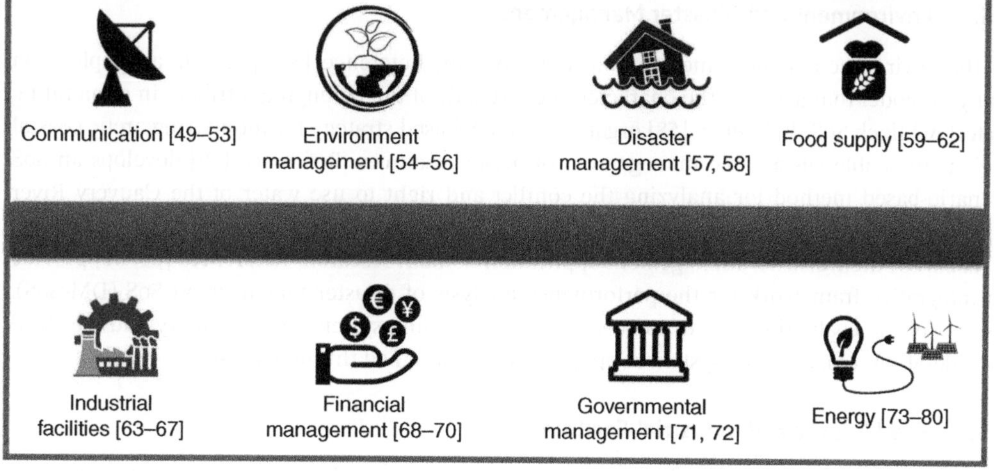

Figure 4.3 Constituent CPSs in critical infrastructure CPSoS.

4.3.2 Health Care

Specifically in the healthcare application domain, Reference [42] proposes a health information SoS management framework that facilitates informed decision-making. Reference [43] describes a human health management SoS scheme and its practical applications that focus on health management, medical diagnosis, and surgical support. Reference [44] devises a methodological framework for modeling healthcare SoS that improves the health level of a population through. As mental illness in cities results from dysfunctional coordination between different city systems and structures, Reference [45] explores the origins of the complexity of the SoS of mental health in cities. Reference [46] depicts the information technology (IT) influence at three hierarchical levels, a micro-level, a meso-level, and a macro-level, in the environment of the U.S. Medicaid SoS program.

4.3.3 Information and Communications

The information technology application domain [47] discusses the efficient fusing of data from multiple sources and effectively controlling the data distribution in a dynamic service-oriented architecture-based command and control SoS. Reference [48] introduces the software mediators as first-class entities of SoS software architectures that aim to offer communication, conversion, coordination, and control capabilities. Reference [49] proposes a model-based framework that provides a novel Unified Modeling Language (UML) profile to model and analyze information quality requirements for CPSoS named information quality (IQCPSoS). Reference [50] proposes designing a generic, programmable position tracking platform, namely CQtracker, constructed and established on SoS concept service architecture to deliver data system as a service. Reference [51] discusses an SoS architecture for the Internet of Things (IoT) systems organized by independent components.

In communication-based research, references [52, 53] present an SoS-based method to model the complex telecom investment problem and the reliability analysis of telecommunication networks, respectively. Reference [54] characterizes the challenges and benefits of adopting an SoS approach in designing, developing, and maintaining mobile safety-critical applications. Reference [55] presents an approach based on the SoS concept for wavelength-division multiplexing (WDM) and time-division multiplexing (TDM) fiber-to-the-home (FTTH) telecommunication networks. Reference [56] develops a model-based optimization framework for satellite SoS architectures.

4.3.4 Environment and Disaster Management

In the environment management application domain, Reference [57] presents an application analysis model that assesses the character and strength of the belonging attribute in a migrating waterfowl flock SoS. Reference [58] organizes an SoS-based strategy for monitoring arctic coastal regions to enable sustainable management of ocean resources. Reference [59] develops an SoS scenario-based method for analyzing the conflict and right to use water of the Cauvery River dispute.

As part of their disaster management application domain research, References [60, 61] provide an integrative framework for the performance analysis of disaster management SoS (DM-SoS). As part of the evaluation of a deep-ocean tsunami warning system, the reliability and logistical considerations of a complex system of operations are combined through stochastic Petri nets.

4.3.5 Food Supply and Industrial Facilities

In the food supply application domain, Reference [62] proposed an SoS approach for seamless information exchange in the agriculture application domain. In [63], an SoS framework is described in the complex interactions modeling among food, energy, and water systems. In [64], many European Union (EU) projects propose an overview of the synergy of the SoS, ontology, IoT, and space technologies to support accurate farming. In particular, this research considers the European project Internet of Food & Farm 2020 (IoF2020), which investigates the potential of Internet technologies in the European food and agricultural industry. In [65], an SoS approach is developed in a formal policy development framework to systematically address, in an integrative and adaptive fashion, significant global challenges, such as the current food crises and their interactions with other essential natural, societal, and technological systems.

In the industrial facilities application domain, Reference [66] provides a framework under the prism of the SoS approach along three principal collaboration axes (lifecycle, value chain, and enterprise) for the assessment of collaborative industrial automation. Reference [67] examines the SoS architecture concepts of automated material handling systems (AMHSs), their attributes, and new capabilities that must be deployed to ensure that the SoS can continue to deliver on the objectives. Reference [68] provides an understanding of SoS testing in large-scale industry settings concerning challenges and how to address them. Reference [69] proposed a framework with the revised rough Decision-Making Trial and Evaluation Laboratory (DEMATEL) method to capture and evaluate smart industrial product-service SoS (SiP-S3) requirements, a new extension of industrial product-service system (PSS) via smart technology and SoS. Reference [70] analyzes the complexity and fundamental features of an industrial case study SoS to understand and overcome a particular subset of challenges.

4.3.6 Financial and Governmental Management

In the financial management application domain, Reference [71] develops an SoS dynamic approach to model the U.S. financial obligations of the collateralized debt obligation market. Reference [72] empirically investigates the structural evolution of global economic systems in eleven countries with relative autonomous economic systems in an SoS-based framework. Reference [73] evaluates the complexity of financial SoS that interact in complicated ways.

In the governmental management application domain, Reference [74] develops a framework based on SoS engineering concepts as a mechanism for more effective governance in the critical infrastructure systems of the United States, and Reference [75] provides a framework to analyze the pillars of governance in SoS.

4.3.7 Energy

The oil and gas infrastructure can be investigated in SoS content in the energy sector. From different aspects, an electrical power grid infrastructure also can be planned, operated, and controlled by the Sequential Oral Sensory (SOS) approach. In a general attitude, Reference [76] presents an SoS approach to understanding, analyzing, and designing modern energy smart grids using model-based system engineering (MBSE) methodologies with system modeling language (SysML). Reference [78] proposed a chance-constrained SoS-based decision-making approach for stochastic scheduling of power systems that comprise the independent system operator (ISO) and distribution companies (DISCOs) as autonomy systems. Reference [79] presents an SoS-based decentralized decision-making framework to determine an economical and secure hourly generation schedule for a transmission system with active distribution grids.

Taking a closer look at the distributed generation system domain, Reference [77] addresses the problem of proactive planning and optimizing the operation of SoS over a time horizon while considering the characteristics of each constituent system and complex interactions among them. In [79], the incorporation of reconfiguration into the expansion planning of smart distribution networks is addressed. Reference [81] develops an SoS framework for the reliability analysis of distributed generation systems to analyze the impact of degraded communication networks. Reference [82] presents an SoS framework that defines both DISCO and microgrids (MGs) as independent systems and determines the process of information exchange among them.

From a more microperspective, Reference [83] studies resilience and energy management in multi-MG systems based on SoS concepts. In this research, the MG is formed by four sub-MGs that pool their resources and capacities to provide more functionality.

4.4 Smart City Cyber-Physical System of Systems

4.4.1 Smart City

The whole population of the cities will reach 70% by 2050 due to the world's urbanization continuing to grow [13]. This growth has increased the need for sustainable smart environments that offer a high-quality life for citizens more than ever. The European Commission defines a smart city as a place that provides more efficient management of traditional networks and services through digital solutions for the benefit of its inhabitants. In this definition, digital technologies are helpful to have upgraded water supply and waste disposal systems, smarter urban transport networks, efficient light and heat management in buildings, more interactive and responsive city administration, and safer public spaces [13, 84]. IEEE Smart Cities emphasizes the smart city indicators, as the integration of technology, government, and society [85]:

- Smart economy
- Smart mobility
- Smart environment
- Smart people
- Smart living
- Smart governance

4.4.2 Smart City as a Cyber-Physical System of Systems

"Growth is inevitable and desirable, but destruction of community character is not. The question is not whether your part of the world is going to change. The question is how."—Edward T. McMahon. This quote acknowledges today's rapid urbanization. With swift growth in technology and associated services, communities are moving into a future based on digitalization, connectivity, and integration [86].

Modern cities must move from an island thinking of working toward an integrated system thinking. The realization of this thinking in the smart city context is provided by integrating many city functional systems. Constructing a smart city is like playing a symphony and requires a lot of orchestration or instrumentation. This means that an integrated framework for smart cities requires the ability to operationally integrate functional systems to provide faster, better, more stable, and cost-effective services to all stakeholders.

Meanwhile, the smart city sectors are often developed and deployed sparsely, each of which has been created and designed to solve specific problems in particular city areas. Accordingly, the

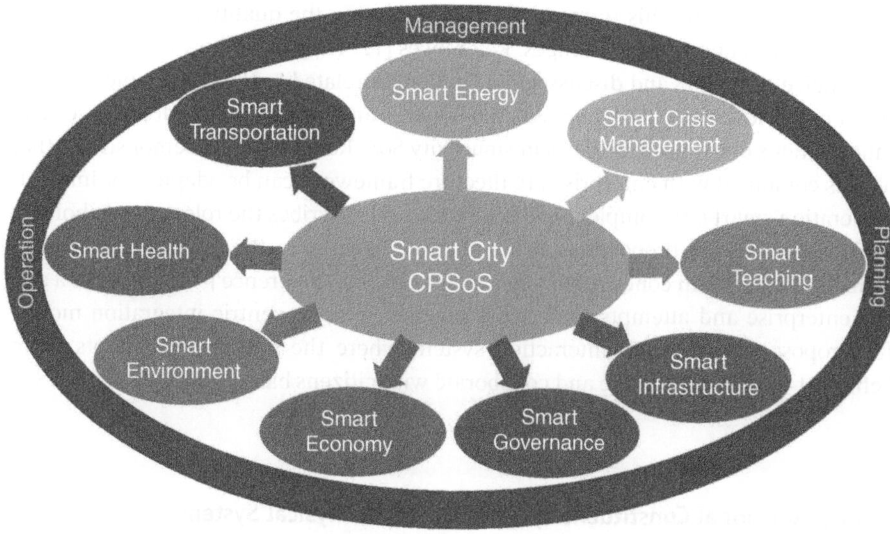

Figure 4.4 Constituent CPSs of smart city CPSoS.

implementation of the smart city concept is formed with bottom-up or part-holistic approaches in most of the world's city densities, such as the European cities of Amsterdam, Manchester, Stockholm, and Helsinki. Based on this, cities have become smarter through decentralized innovations and the gradual implementation of projects, each focusing on a specific goal.

Although these programs and systems may be relatively mature in some specific areas, it is easy to see that there is no interaction between them. This situation may lead to unmanageable and unstable different systems, blocking the way for more efficient, flexible responses and limiting the smart concept in that city. Additionally, there are some essential issues to consider when adopting the decentralized approach. How can these different projects be integrated to achieve a comprehensive vision in the smart city? How can they be encouraged to work together to provide new services to citizens in various fields? The answer is that seeing a smart city as an SoS can be helpful in solving this challenge, even if it poses challenges.

Based on reports and research, the main point is that adopting such a view in the city context can provide more effective capabilities, functions, and management than a traditional system. Improving the capacity of different city organizations and coordinating their activities is one of the significant advantages of this integrated perspective. For example, during a natural disaster, all managers will have a standard view of the situation and can make better decisions about where and how to use available resources [10, 12, 87, 88].

Specifically, a smart city can be defined as a CPSoS, which integrates CPSs, relying on exchanging information to achieve a specific higher goal. Smart city CPSoS facilitates interaction between the CPSs like smart homes and buildings, smart energy, smart transportation, smart communication, smart health, smart waste management, and smart education. As shown in Figure 4.4, smart city CPSoS utilizes a data-driven information network to tie all other infrastructure systems together to ensure a predictive approach to making day-to-day life much better, safer, and more comfortable. In order to achieve the benefits of constituent CPSs being integrated into the smart city based on the CPSoS framework, three main procedures can be benefited [49, 85, 86, 89–91].

Today, some research based on this framework seeks to discuss the quality of this integration and accurately identify its benefits or challenges. References [12, 13] present how a smart city can be holistically modeled as an SoS and discuss some challenges related to the development of SoS thinking in smart cities. Reference [86] discusses the benefits and challenges of adopting vertical farming and autonomous driving as two CPSs in smart city SoS. Reference [92] demonstrates that SoS methodologies combined with enterprise architecture framework can be adapted for innovating the next-generation smart city complex SoS. Reference [93] describes the roles of stakeholders in smart city planning that are proposed to bring tangible benefits to the urban economy and society based on SoS concepts. In conceptualizing a smart city SoS, Reference [94] considers a city as a large-scale enterprise and attempts to design a business process-centric integration model. Reference [13] proposes an idealized interaction system where the government adopts more effective and efficient ways to coordinate and collaborate with citizens based on the smart city SoS framework.

4.4.3 Smart City Functional Constituent Systems as Cyber-Physical System of Systems

As denoted before, functional constituent systems in smart cities can be viewed as a CPSoS. The city transportation systems can exemplify this [14, 88, 95, 96]. For example, in IoT-based transportation development, Barcelona's smart city has a network of traffic lights that integrate with other CPSs named "green light." In Linz, for another example, several tram sets are integrated into the control center and other systems via machine-to-machine communication. This integrated transportation system increases energy efficiency and allows the waiting passenger to see whether a vehicle is empty, half full, or complete. Furthermore, another good example is Budapest's public transportation system that uses a modular traffic control and passenger information system from T-Systems Hungary, FUTAR. About 2300 vehicles are integrated into monitoring and control systems to help the Budapest Transport Center optimize routes, implement integrated control policies, and inform passengers in real time [88]. Similarly, the transportation system in a smart city area in Singapore is discussed as a CPSoS consisting of autonomous constituents with one common objective to transport passengers safely and efficiently [14].

From a more general perspective, the smart city can be viewed as an interactive system of functional CPSs. These CPSs must effectively manage the processes, people, things, and data elements to meet the objectives and planned expectations. Accordingly, CPSs also can be driven by four systems, as indicated in Figure 4.5 [97]:

- System of records (SOR)
- System of engagement (SOE)
- System of things (SOT)
- System of insight (SOI)

Although many studies have determined the smart city platform as a CPSoS, some researchers believe that other systems can drive each functional constituent CPSs. Besides, a smart city from a functionality point of view can be considered as a CPSoS. Due to the specific geographical distribution of its constituents, it may be modeled as a CPSoS with geographically distributed constituent CPSs, similarly. In the next section, the energy infrastructure, another functional CPSs in the smart city SPSoS with geographically distributed constituent systems, is discussed in an SoS engineering framework.

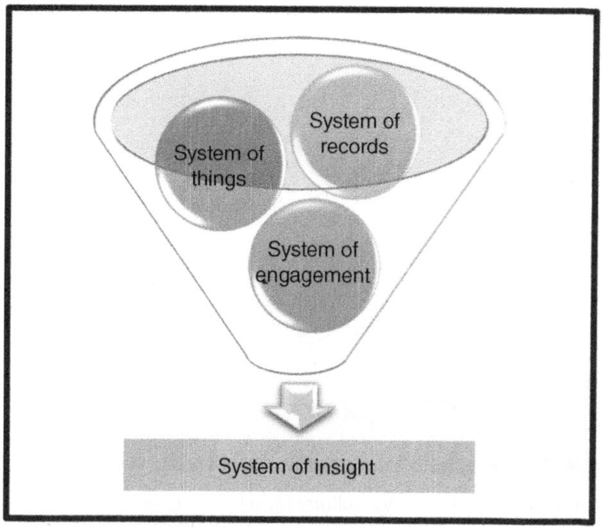

Figure 4.5 Functional CPSs in the smart city will be driven by four systems.

4.5 Smart City Energy Cyber-Physical System of Systems

4.5.1 Energy Hub

Power system operation and management encounter significant challenges as intermittent renewable energy resources are increasingly integrated into power grids. Meanwhile, as an integrated modeling and management framework, the EH can benefit effectively from combined cooling, heat, and power (CCHP), energy storage, and demand-side management applications to secure and reliable operation of power grids. EHs, as an efficient framework for the generation, consumption, conversion, and storage of different energy carriers, have been considered by many researchers as the prospect of future reliable energy systems.

Extensive research on the EH concept proves that the different energy carriers' integration capability within the EHs can lead to more satisfactory performance than independently programmed and controlled systems. One of the essential features of EHs is establishing additional connections between the input and the output. For example, the electrical demands can be supplied directly from the upstream electricity grid or through the CHPs, local renewable resources, or energy storage. This redundancy in energy supply leads to essential advantages in the consumers' reliable, economical, and low carbon supply.

Generally, the structure considered for the input and output connection of the EHs is according to Figure 4.6. In Figure 4.6, L is the vector of the demand of the EHs, and P has introduced the input energy carrier vector. These two vectors (demand and input energy carriers) are connected through the C matrix. For example, an EH can include a transformer, boiler, CHP unit, and a heat storage device that receives electricity and natural gas as input and supplies electricity and heat in output [98–101].

4.5.2 Micro-Energy Hub and Macro-Energy Hub

The EHs have no limitations in modeling, which means that a home energy system or the entire city energy system can be modeled as an EHs. The EHs, based on the consumption sectors, can be

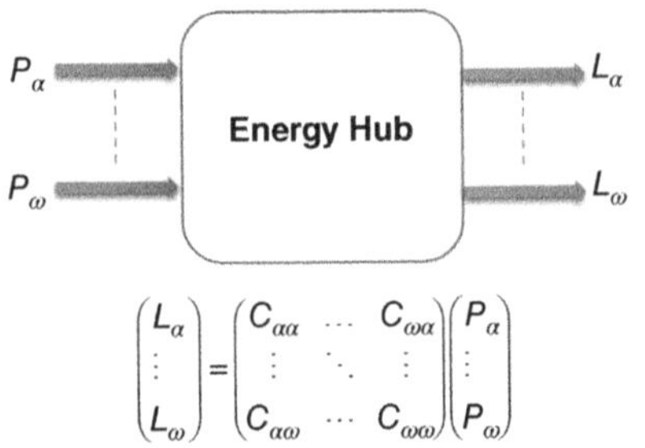

Figure 4.6 General model of EHs.

Figure 4.7 Existing real facilities in cities that can be considered as a MIEH.

divided into types named MIEHs. As the existing real facilities in cities can be considered as EHs, residential, commercial, industrial, agricultural, and critical area energy systems can be considered as MIEHs, as shown in Figure 4.7.

One of the most popular research topics in recent years is the integration of EHs that can be controlled by a central manager, which creates a concept called MAEH. MAEH is a set of MIEHs that can be controlled and programmed together. Integrated control and management of MIEHs can bring benefits to the whole system. For example, when residential and commercial buildings with diverse consumption patterns are controlled together as a MAEH, the generation resources available in each can be used to cover capacity shortages in another. Explicitly, the energy resources and storage facilities of one MIEH can supply part of the peak demand in another MIEH since the peak demand in these sectors usually does not coincide. Additionally, low-grade heat generated by an industrial MIEH can be utilized in the heat network of a MAEH to supply the heat demand of buildings, as well as organic and agricultural waste in an agriculture MIEH can be used to produce biofuels.

The MAEH model can cover many things, including a residential complex, an urban area, commercial buildings, an industrial town, a village, or even an entire city. However, although from the consumer's perspective, such as MIEHs, optimization of the energy consumption costs can be followed as an objective function, and from a controller perspective, such as MAEH, more significant objectives can be considered [101–103].

4.5.3 Smart City Energy Infrastructure as a Cyber-Physical System of Systems

As mentioned earlier, one of the functional constituent CPSs of the smart city that can be identified and introduced as an SoS is the smart city energy infrastructure. In a city area, real facilities such as industrial plants, buildings (airports, hospitals, and shopping malls), rural areas, and island energy systems (trains, ships, and airports) can be identified as EH [15]. Based on the idea of this research, a city consisting of different residential, industrial, commercial, official, educational, and hospital areas can be viewed as a MAEH system of MIEH CPSs designed for each of the areas.

One of the essential advantages of the EHs is the utilization of advanced, efficient resources and storage technologies that effectively manage intermittent renewable energy resources and facilitate their integration into the distribution networks. Accordingly, the MAEH-based management of energy infrastructure enables cities to meet their sustainable development goals with respect to energy. We can expect the realization of the concept of smart for that city and to name it smart city "A smart city (...) is a healthy, energy-efficient city that uses renewable energy sources as much as possible and is a pioneer in the deployment of advanced smart technologies" [104, 105].

In smart city energy infrastructure, constituent MIEH CPSs pool their capabilities to create a new, more functional, and reliable smart city energy CPSoS. In the other world, a smart city energy infrastructure is a multicarrier energy system that provides various types of energy for autonomous consumption from the autonomous distributed generation controlled by local CPSs. Collaboration in smart city energy CPSoS can be possible through existing information, electricity, natural gas, and other infrastructures, as shown in Figure 4.8.

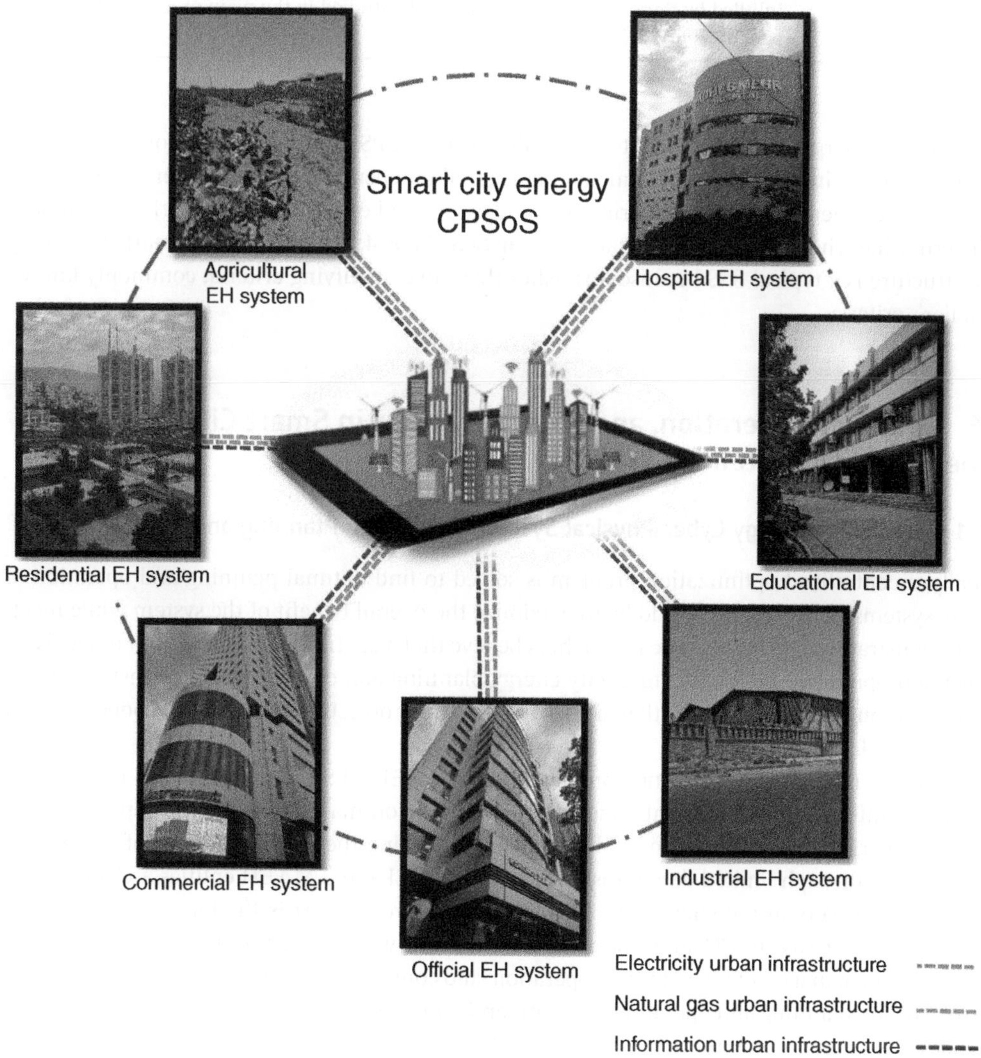

Figure 4.8 Smart city energy CPSoS.

Table 4.2 The qualifying Maier's criteria for smart city energy CPSoS.

Maier's criteria	Criteria qualification in smart city energy CPSoS
Operational independence	Each MIEH can be operated in independent mode as an isolated unit
Managerial independence	Each MIEH is self-contained that can manage itself and run in an independent isolated mode
Geographic distribution	Each MIEH according to its owner's location has been developed in different or distributed areas in city
Evolutionary development	Each MIEH can be built or developed evolutionary and dynamically due to the equipment lifetime, demand growth over time, urbanization, and urban land expansion
Emergent behavior	A smart city energy CPSoS achieves the functions not attainable by its individual MIEHs separately. As the principal purposes of the smart city energy, CPSoS are fulfilled by these behaviors; this should be studied in the main processes that benefit from MIEH CS integration

Smart city energy infrastructure, to be qualified as a CPSoS, must satisfy some qualifying standard criteria, including operational and managerial independence of constituent CPSs, evolutionary development, geographic distribution, and emergent behavior. These criteria are standard in determining whether a system is qualified as an SoS. Table 4.2 shows that the smart city energy infrastructure is a typical SoS because it satisfies these five qualifying criteria, commonly known as Maier's criteria.

4.6 Planning, Operation, and Control Process in Smart City Energy Cyber-Physical System of Systems

4.6.1 Smart City Energy Cyber-Physical System of Systems: Planning and Operation

Usually, a centralized optimization problem is solved to find optimal planning and operation of energy systems. The objective could be to maximize the overall benefit of the system while meeting the constraints. However, some researchers believe that a centralized optimization model is no longer appropriate approach for smart city energy planning and operation, as its constituent systems can be autonomous systems that aim to optimize their objective function and cooperate with others [79, 80].

To solve the planning and operation problems of each MIEH CPS and the entire SoS-based smart city energy infrastructure, a decentralized SoS-based decision-making framework is proposed. In this framework, the final solution of the SoS will determine the operating point of all its constituent CPSs through limited data transferring, as the MIEHs and MAEH entities are considered autonomous players in the smart city energy infrastructure. MAEH is the representative of the electricity DISCO, gas DISCO, and smart city municipality, which is responsible for the smart city energy system management in planning, operation, and control. The MAEH solves distribution system expansion planning and operation problems and optimizes economic and reliability indicators in smart city energy infrastructure. However, MIEHs are thinking of maximizing their profits or having a highly reliable energy supply from their perspective. These MIEHs can be interconnected

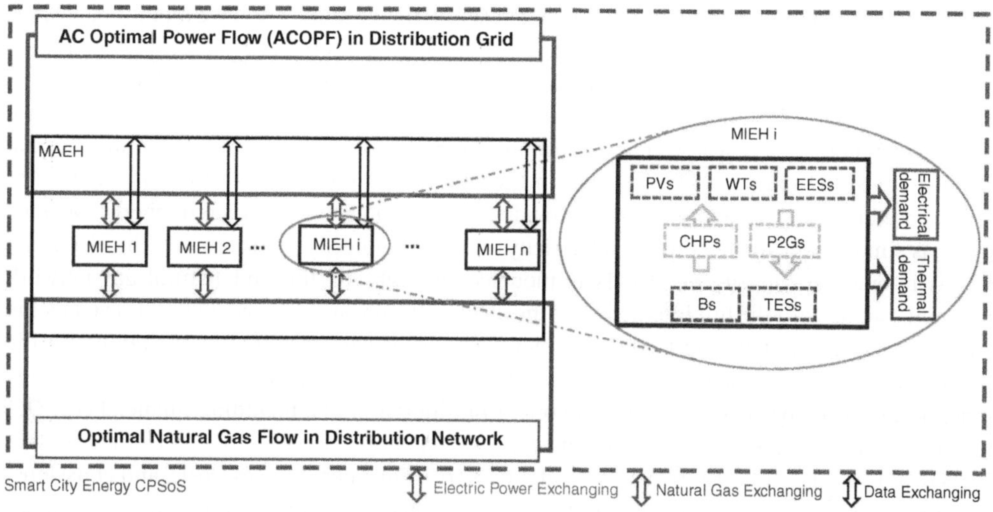

Figure 4.9 MIEHs and MAEH cooperation in planning and operation process of smart city energy CPSoS.

through the existing urban power and natural gas distribution grids and data-sharing infrastructures, as shown in Figure 4.9.

In this SoS-based framework, some shared variables and adaptive parameters are needed in addition to the constant parameters and decision variables that are generally required for constituent CPS modeling. Shared variables are common among two independent entities, and they reflect the effects of different conditions of autonomous entities on each other. Adaptive parameters are constant, which would be determined for one constituent from other constituents' CPSs [78, 79, 82, 106]. In addition, according to the parameter and variable data exchange, each entity could be defined as a "client" or "origin" [82]. An entity that has sent a signal to receive some data about its adaptive parameters or shared variables from another entity so-called client, and an entity that has received the signals to share data named the origin. Hereof, if the MIEH sends signals to MAEH to know the energy price data, at each bus, MIEH is a client and MAEH is an origin.

General Framework for Smart City Energy Cyber-physical System of System Planning Problem

Set: $MIEH_i \rightarrow i = \{1, 2, 3, \ldots, n\}$

Horizon: 1 – 5 years

Objective Function:

Minimize :

$((IC) + \tau\,(OC + MC)), (CENS + CIC), (EL), (Em)$

Constraints:

subject to:

Natural gas and electrical network constraints,

Power and gas exchanged with the upstream grid limitations,

Financial limitations in the distribution system upgrading and MIEHs development,

Invested units constraints,

Total penetration level of DERs in the distribution grid,

DETs investments restriction

Display: (*Optimal Investment Decisions of DERs in MIEHs*), (IC, OC, MC), $(CENS, CIC)$, (EL, Em)

The proposed planning framework seeks to determine the multiple energy generation, conversion, and storage system capacity and placement in smart city energy infrastructure constrained by electricity and natural gas network and MIEH component constraints. In addition to electrical energy storage (EES) and combined heat and power (CHP), wind turbine and photovoltaic resources can provide electric power as the primary source of power, and CHP, boilers, and thermal energy storage (TES) may also be invested in thermal demand–supply, as shown in Figure 4.9. The proposed multicarrier energy system provides cooperation between different energy carriers, as well as simultaneously meeting the electricity and heating needs. Furthermore, the electricity and gas interaction capacity of MIEHs is modeled through ACOPF and natural gas network constraints, along with the full interaction of the electricity and gas network that is allowed by the power-to-gas (P2G) system and CHP investment. The objective function and constraints in the proposed framework can include various terms in the planning process of entire smart city energy infrastructure. The objective function can consist of different parts, including financial (*IC*, *OC*, *MC*: investment, operation, and maintenance costs), reliability (*CENS*, *CIC*: cost of energy not supplied and customer interruption cost), and others (*EL*, *Em*: energy losses, costs, and emissions) [107, 108]. Furthermore, various constraints can be considered, as summarized in the following.

General Framework for Smart City Energy Cyber-physical System of System Operation Problem

Set: $MIEH_i \rightarrow i = \{1, 2, 3, \ldots, n\}$

Horizon: 24 hours

Objective Function:

Minimize :

$(OC + MC), (CENS + CIC), (EL), (Em)$

Constraints:

Subject to :

Natural gas and *electrical network constraints,*

Power and gas exchanged with the upstream grid limitations,

MIEHs units constraints

Display: (*Optimal Operation Conditions of DERs in MIEHs*), (*OC*), (*CENS, CIC*), (*EL, Em*)

However, in the smart city energy infrastructure, when the MIEHs were designed based on the described SoS-based decision-making framework, a similar framework is also required in the optimal operation schedule [109]. In this framework, the final solution of the CPSoS will determine the operating point of all its constituent MIEH CPSs through limited data transferring between the MIEHs and MAEH. In the operation process, the above-mentioned objective function is deformed by ignoring the investment and expansion-related part and more accurate modeling of operation-related terms. Minimum and maximum voltage levels and the maximum capacity of feeders that can be modeled as ACOPF formulation, natural gas distribution limitation, and power and gas exchange with the upstream grid limitations are some constraints that MAEH should consider. Similarly, more detailed limitations in distributed energy resource (DER) operation and load balance in each MIEH are essential constraints that MIEHs should consider as follows:

4.6.2 Smart City Energy Cyber-Physical System of Systems: Control

In the control process of the smart city energy CPSoS, centralized approaches are prone to infeasibility due to the extensive computation and communication requirements since it is an interconnected energy system with Maier's criteria, including operational and managerial independence, as well as geographic distribution. However, a fully decentralized approach is likewise impossible as the MIEH CPSs lack a degree of coordination. Therefore, in the control process of the smart city energy CPSoS, centralized and decentralized control schemes are needed that are realized through a hierarchical control scheme consisting of three control levels as described in [110]. Subsequently, while considering MIEHs as MGs, the three-level control hierarchy can be used accordingly [110, 111].

- As the first level of the hierarchical control scheme, primary or internal control features the fastest response and is focused on local measurements for DERs.
- At the second level, the secondary control or the MIEH energy management system aims at the reliable and economical operation of the MIEH, setting the DER optimal set points.
- At the highest level of control, tertiary control coordinates the MIEHs, considering the missions of the MAEH, in interconnected operational mode.

4.7 Emergence in Smart City Energy Cyber-Physical System of Systems

System theory uses specific features to describe the behavioral characteristics of a CPSoS, called "emergence." Emergence appears when the constituent CPSs work as a united CPSoS, which CPSs do not have this specific behavior or characteristic themselves. Information theory defines emergence as the difference between the output and input CPSoSs [112]. It means that information must emerge within the framework of the CPSoS, which probably emerges from constituent CPS interaction. Therefore, emergence could be seen as the difference in accessible information at a certain level, which is more applicable in chemistry. For example, the color of the chemical materials is not observable on the atomic scale of the same substance; however, the interaction of atoms creates this characteristic color for the substance [113, 114].

The emergent behavior, as one of the other Maier's criteria of smart city energy CPSoS, relates to awareness of the deterministic or stochastic performance of this CPSoS. One of the essential characteristics of any SoS is emergent behavior. The word emergence means to become known or come to light. This implicitly acknowledges that something is added to the existing knowledge that was not known before emergence happens. Emergent behavior means it provides new functions resulting from the cooperation between individual constituent systems that are not performed in isolation by autonomous CPSs [115–117].

The concept of emergence has both ontological and phenomenological meanings:

- The concept of emergent systems as an ontological concept corresponds to the irreducibility of the characteristics, knowledge, methods, causes, or explanations of the CPSoS to its CPs.
- Emergence as a phenomenological concept is the unexpected observable output of the CPSoS that emerged from interactions between the CPSs.

Although emergent is often assumed to be negative, it may bring on positive results or have no effect on the SoS, or, in other words, it is neutral [115–117].

- Emergent behavior is often assumed to have negative results because it is not a designed function. This behavior is considered negative when it fulfills SoS purposes.

- Emergent behavior may also be advantageous and deliver unexpected positive consequences. It is considered positive when it keeps the constituent CPSs operational and healthy in their optimum performance ranges and fulfills the CPSoS purpose.
- Emergent behavior can move to either positive or negative territories with a proper control process when considered neutral, which is neither positive nor negative.

4.8 Conclusions

Nowadays, traditional approaches cannot be efficient in describing the planning, operation, and control process in twenty-first-century systems as they have increased in complexity, more and more. Additionally, the growing integration of these complex systems drives the researchers to focus on implementing other frameworks such as SoS to describe their behavior. Moreover, the networked connection of smart things, i.e., CPSs including sensors, actuators, and recorders that monitor and directly influence the physical environment, can be mentioned as one of the main factors in this complexity and integration.

One of these complex integrated systems is smart city energy. This chapter addresses the planning, operation, and control process in smart city energy infrastructure as a CPSoS. After expressing the general concept, characteristics, and different types of CPSoS, an overview of the SoS application domains, including transportation, health care, information and communication, industrial facilities, financial, food supply, and energy as the national infrastructure sectors, is provided. Then, a smart city as an interacted system of functional CPSs is described as a CPSoS. Additionally, it is emphasized that these functional CPSs can consist of heterogeneous and independent public and private systems and be modeled as a CPSoS.

One of the functional constituent CPSs of the smart city that can be identified and introduced as an SoS is the smart city energy system. As the city energy systems can be defined as a set of MIEHs, a city consisting of different residential, industrial, commercial, educational, and critical areas, etc., can be mentioned as a MAEH consisting of these MIEHs. MAEH represents the electricity DISCO, gas DISCO, and smart city municipality responsible for the smart city energy system management in planning, operation, and control.

The MAEH performance in the planning area is to solve a distribution system expansion planning problem to optimize economic and reliability indicators in smart city energy infrastructure. At the same time, MIEHs can aim to maximize their profits or to have a highly reliable electricity supply from their perspective. The MAEH in the operation area should schedule the integrated smart city energy infrastructure to realize its reliability and financial objective in interaction with the upstream grid and MIEHs, as well as the MIEH objectives.

To manage and control the smart city energy CPSoS behavior, it must be possible to apply the control theory. Therefore, to control the behavior of a CPSoS, one must first be aware of the deterministic or stochastic performance of any SoS that is effective in determining the nature of one of the essential characteristics of any CPSoS that so-called emergent behavior. The concept of emergence has both ontological and phenomenological meanings. Emergence as an ontological concept is the same as irreducibility in system characteristics, knowledge, methods, causations, or explanations about the CPSoS to its CPSs. Emergence as a phenomenological concept is the unexpected observable output of the CPSoS that emerged from interactions between the CPSs.

According to different scientific theories, emergence is not necessarily an absolute concept, and there are arguments that an emergence exists from “simple” to “spooky.” It is an issue of how to control theories that can execute over these four categories. Simple and weak emergent are placed in the deterministic region, and all the engineering approaches available through the control engineering discipline are applicable.

References

1 Nanayakkara, T., Sahin, F., and Jamshidi, M. (2018). *Intelligent Control Systems with an Introduction to System of Systems Engineering*. United States: CRC Press.

2 Jamshidi, M. (ed.) (2017). *Systems of Systems Engineering: Principles and Applications*. United States: CRC Press.

3 Bondavalli, A., Bouchenak, S., and Kopetz, H. (2016). *Cyber-Physical Systems of Systems: Foundations—A Conceptual Model and Some Derivations: The AMADEOS Legacy*, vol. 10099. United States: Springer.

4 van Lier, B. (2018). Cyber-Physical systems of systems and complexity science: the whole is more than the sum of individual and autonomous cyber-physical systems. *Cybernetics and Systems* 49 (7–8): 539–566.

5 Engell, S., 2014. Cyber-physical systems of systems—definition and core research and innovation areas. *European Roadmap on Research and Innovation in Engineering and Management of Cyber-Physical Systems of Systems*, 111.

6 Fereidunian, A., Lesani, H., Zamani, M.A. et al. (2015). A complex adaptive system of systems approach to human–automation interaction in smart grid. In: *Contemporary Issues in Systems Science and Engineering*, 425–500. United States: Wiley.

7 Otto, A., Hall, J.W., Hickford, A.J. et al. (2014). A quantified system-of-systems modeling framework for robust national infrastructure planning. *IEEE Systems Journal* 10 (2): 385–396.

8 Hall, J.W., Henriques, J.J., Hickford, A.J., et al. (2013, October). Systems-of-systems analysis of national infrastructure. *Proceedings of the Institution of Civil Engineers-Engineering Sustainability* Vol. 166, No. 5, pp. 249–257. Thomas Telford Ltd.

9 Darbandsari, A. Nozarian, M. and Fereidunian, A. (2022). Data-driven maintenance of smart cities critical infrastructure. *IEEE Smart Cities Newsletter*. https://smartcities.ieee.org/

10 Naphade, M., Banavar, G., Harrison, C. et al. (2011). Smarter cities and their innovation challenges. *Computer* 44 (6): 32–39.

11 Kojury-Naftchali, M. and Fereidunian, A. (2021). Smart energy-aware cities: customer characterization by energy data analytics to improve demand response performance. In: *Flexible Resources for Smart Cities*, 21–43. Cham: Springer.

12 Cavalcante, E., Cacho, N. and Lopes, F. et al. (2016, December). Thinking smart cities as systems-of-systems: a perspective study. *Proceedings of the 2nd International Workshop on Smart*, pp. 1–4.

13 Cavalcante, E., Cacho, N., Lopes, F., et al. (2017, September). Challenges to the development of smart city systems: a system-of-systems view. *Proceedings of the 31st Brazilian Symposium on Software Engineering*, pp. 244–249.

14 Lee, O.L., Im Tay, R., and Too, S.T. 2019, May). A smart city transportation system of systems governance framework: a case study of Singapore. *2019 14th Annual Conference System of Systems Engineering (SoSE)*, pp. 37–42. IEEE.

15 La Scala, M., Bruno, S., Nucci, C.A. et al. (2021). *From Smart Grids to Smart Cities: New Challenges in Optimizing Energy Grids*, vol. 2. United States: John Wiley & Sons.

16 Boardman, J. and Sauser, B. 2006, April). System of Systems-the meaning of of. *2006 IEEE/SMC International Conference on System of Systems Engineering*, p. 6. IEEE.

17 de C Henshaw, M.J. (2016). Systems of systems, cyber-physical systems, the internet-of-things... whatever next? *Insight* 19 (3): 51–54.

18 Klein, J. and Van Vliet, H. (2013, June). A systematic review of system-of-systems architecture research. *Proceedings of the 9th International ACM Sigsoft Conference on Quality of Software Architectures*, pp. 13–22.

19 Vargas, I.G., Gottardi, T. and Braga, R.T.V. 2016, May). Approaches for integration in system of systems: a systematic review. *2016 IEEE/ACM 4th International Workshop on Software Engineering for Systems-of-Systems (SESoS)*, pp. 32–38. IEEE.

20 Choi, T.M. (2018). A system of systems approach for global supply chain management in the big data era. *IEEE Engineering Management Review* 46 (1): 91–97.

21 Henshaw, M., Dahmann, J. and Lawson, B., 2019. *Systems of systems (SoS)*. https://www.sebokwiki.org/wiki/Systems_of_Systems_(SoS).

22 INCOSE, 2018. *INCOSE Systems of Systems Primer*. https://www.incose.org/publications/technical-product-catalog/sos-primer.

23 Popper, S.W., Bankes, S.C., and Callaway, R. (2004). *System of Systems Symposium: Report on a Summer Conversation, 320*. Arlington, VA: Potomac Institute for Policy Studies.

24 Maier, M.W. (1998). Architecting principles for systems-of-systems. *Systems Engineering: The Journal of the International Council on Systems Engineering* 1 (4): 267–284.

25 de Barros Paes, C.E., Neto, V.V.G., Moreira, T., and Nakagawa, E.Y. (2018). Conceptualization of a system-of-systems in the defense domain: an experience report in the Brazilian scenario. *IEEE Systems Journal* 13 (3): 2098–2107.

26 Jamshidi, M.O. (2008). System of systems engineering-new challenges for the 21st century. *IEEE Aerospace and Electronic Systems Magazine* 23 (5): 4–19.

27 Hipel, K.W., Jamshidi, M.M., Tien, J.M., and White, C.C. III, (2007). The future of systems, man, and cybernetics: Application domains and research methods. *IEEE Transactions on Systems, Man, and Cybernetics, Part C (Applications and Reviews)* 37 (5): 726–743.

28 Zeigler, B.P. and Sarjoughian, H.S. (2017). Modeling and simulation of systems of systems. In: *Guide to Modeling and Simulation of Systems of Systems*, 3–11. Cham: Springer.

29 Bianchi, T., Santos, D.S. and Felizardo, K.R. 2015, May). Quality attributes of systems-of-systems: a systematic literature review. *2015 IEEE/ACM 3rd International Workshop on Software Engineering for Systems-of-Systems*, pp. 23–30. IEEE.

30 Tolone, W.J., Johnson, E.W., Lee, S.W., Xiang, W.N., Marsh, L., Yeager, C. and Blackwell, J. 2008, October). Enabling system of systems analysis of critical infrastructure behaviors. *International Workshop on Critical Information Infrastructures Security*, pp. 24–35. Berlin, Heidelberg: Springer.

31 Katina, P.F. and Keating, C.B. (2015). Critical infrastructures: a perspective from systems of systems. *International Journal of Critical Infrastructures* 11 (4): 316–344.

32 Eusgeld, I., Nan, C., and Dietz, S. (2011). "System-of-systems" approach for interdependent critical infrastructures. *Reliability Engineering & System Safety* 96 (6): 679–686.

33 Rinaldi, S.M., Peerenboom, J.P., and Kelly, T.K. (2001). Identifying, understanding, and analyzing critical infrastructure interdependencies. *IEEE Control Systems Magazine* 21 (6): 11–25.

34 Roncoli, C., Bersani, C., and Sacile, R. (2012). A risk-based system of systems approach to control the transport flows of dangerous goods by road. *IEEE Systems Journal* 7 (4): 561–570.

35 Marwaha, G. and Kokkolaras, M. (2015). System-of-systems approach to air transportation design using nested optimization and direct search. *Structural and Multidisciplinary Optimization* 51 (4): 885–901.

36 Flanigan, D. and Brouse, P. (2013). Evaluating the allocation of border security system of systems requirements. *Procedia Computer Science* 16: 631–638.

37 Lagorio, A., Pinto, R., and Golini, R. (2017). Urban Logistics Ecosystem: a system of system framework for stakeholders in urban freight transport projects. *IFAC-PapersOnLine* 50 (1): 7284–7289.

38 Kemp, D. and Evans, R. (2016). Steampunk System of Systems Engineering: a case study of successful system of systems engineering in 19th century Britain. *INSIGHT* 19 (3): 27–29.

39 Hoehne, O. (2016). Rail systems viewed from a system of systems perspective. *INSIGHT* 19 (3): 36–38.

40 Duffy, M. and Sandor, D. (2008, June). 4.4.1 A system-of-systems framework for the future hydrogen-based transportation economy. *INCOSE International Symposium*, Vol. 18, No. 1, pp. 507–521.

41 Mansouri, M., Gorod, A., Wakeman, T.H., and Sauser, B. (2009). Maritime transportation system of systems management framework: a system of systems engineering approach. *International Journal of Ocean Systems Management* 1 (2): 200–226.

42 Okami, S. and Kohtake, N. (2017). Transitional complexity of health information system of systems: managing by the engineering systems multiple-domain modeling approach. *IEEE Systems Journal* 13 (1): 952–963.

43 Hata, Y., Kobashi, S., and Nakajima, H. (2009). Human health care system of systems. *IEEE Systems Journal* 3 (2): 231–238.

44 Grigoroudis, E. and Phillis, Y.A. (2013). Modeling healthcare system-of-systems: a mathematical programming approach. *IEEE Systems Journal* 7 (4): 571–580.

45 Moustaid, E., Kornevs, M., Lindencrona, F., and Meijer, S. (2020). A system of systems of mental health in cities, digging deep into the origins of complexity. *Administration and Policy in Mental Health and Mental Health Services Research* 47 (6): 961–971.

46 Randall, R.G. and Heffner, M. 2019, May). Medicaid IT as a system of systems. *2019 14th Annual Conference System of Systems Engineering (SoSE)*, pp. 254–259. IEEE.

47 Rothenhaus, K.J., Michael, J.B., and Shing, M.T. (2009). Architectural patterns and auto-fusion process for automated multisensor fusion in soa system-of-systems. *IEEE Systems Journal* 3 (3): 304–316.

48 Garcés, L., Oquendo, F., and Nakagawa, E.Y. (2019). Software mediators as first-class entities of systems-of-systems software architectures. *Journal of the Brazilian Computer Society* 25 (1): 1–23.

49 Gharib, M., Lollini, P., and Bondavalli, A. (2021). IQCPSoS: a model-based approach for modeling and analyzing information quality requirements for cyber-physical system-of-systems. *Journal on Data Semantics* 10 (3): 267–289.

50 Chu, V.W., Wong, R.K., Chi, C.H. et al. (2017). The design of a cloud-based tracker platform based on system-of-systems service architecture. *Information Systems Frontiers* 19 (6): 1283–1299.

51 Nikolopoulos, B., Dimopoulos, A.C., Nikolaidou, M. et al. (2019). A system of systems architecture for the internet of things exploiting autonomous components. *International Journal of System of Systems Engineering* 9 (2): 167–199.

52 Tsilipanos, K., Neokosmidis, I., and Varoutas, D. (2015). Modeling complex telecom investments: a system of systems approach. *IEEE Transactions on Engineering Management* 62 (4): 631–642.

53 Tsilipanos, K., Neokosmidis, I., and Varoutas, D. (2012). A system of systems framework for the reliability assessment of telecommunications networks. *IEEE Systems Journal* 7 (1): 114–124.

54 Bondavalli, A., Ceccarelli, A., Lollini, P. et al. (2016). System-of-systems to support mobile safety critical applications: open challenges and viable solutions. *IEEE Systems Journal* 12 (1): 250–261.

55 Rokkas, T., Neokosmidis, I., Katsianis, D., and Varoutas, D. (2012). Cost analysis of WDM and TDM fiber-to-the-home (FTTH) networks: A system-of-systems approach. *IEEE Transactions on Systems, Man, and Cybernetics, Part C (Applications and Reviews)* 42 (6): 1842–1853.

56 LaSorda, M., Borky, J.M., and Sega, R.M. (2018). Model-based architecture and programmatic optimization for satellite system-of-systems architectures. *Systems Engineering* 21 (4): 372–387.

57 Felder, W.N. and Baldwin, W.C. (2019). Estimation of the belonging metric in a hypothetical system-of-systems. *IEEE Systems Journal* 13 (2): 1936–1944.

58 Honoré-Livermore, E., Birkeland, R., and Haskins, C. (2020, July). Addressing the sustainable development goals with a system-of-systems for monitoring arctic coastal regions. *INCOSE International Symposium* 30 (1): 604–619.

59 Sharma, A., Schweizer, V. and Hipel, K.W. 2020, October). Analyzing cauvery river dispute using a system of systems approach. *2020 IEEE International Conference on Systems, Man, and Cybernetics (SMC)*, pp. 3969–3975. IEEE.

60 Fan, C. and Mostafavi, A. (2019). Metanetwork framework for performance analysis of disaster management system-of-systems. *IEEE Systems Journal* 14 (1): 1265–1276.

61 Volovoi, V. and Peterson, D.K. (2011, December). Coupling reliability and logistical considerations for complex system of systems using stochastic Petri nets. *Proceedings of the 2011 Winter Simulation Conference (WSC)*, pp. 1746–1757. IEEE.

62 Weinert, B. and Uslar, M. 2020, June). Challenges for system of systems in the agriculture application domain. *2020 IEEE 15th International Conference of System of Systems Engineering (SoSE)*, pp. 000355–000360. IEEE.

63 Xiao, Y., Hipel, K.W. and Fang, L. 2019, October). A system of systems framework for the water-energy-food nexus. *2019 IEEE International Conference on Systems, Man and Cybernetics (SMC)*, pp. 994–999. IEEE.

64 Kupriyanovsky, V., Lipuntsov, Y., Grinko, O., and Namiot, D. (2018). Agriculture 4.0: synergy of the system of systems, ontology, the internet of things, and space technologies. *International Journal of Open Information Technologies* 6 (10): 46–67.

65 Hipel, K.W., Fang, L., and Heng, M. (2010). System of systems approach to policy development for global food security. *Journal of Systems Science and Systems Engineering* 19 (1): 1–21.

66 Colombo, A.W., Bangemann, T. and Karnouskos, S. 2013, February). A system of systems view on collaborative industrial automation. *2013 IEEE International Conference on Industrial Technology (ICIT)*, pp. 1968–1975. IEEE.

67 Haddadin, M. and Moreno, W. (2021, May). Automated material handling systems: system of systems architecture examination semiconductor manufacturing perspective. *2021 32nd Annual SEMI Advanced Semiconductor Manufacturing Conference (ASMC)*, pp. 1–6. IEEE.

68 Ali, N.B., Petersen, K. and Mäntylä, M.V. (2012, September). Testing highly complex system of systems: an industrial case study. *Proceedings of the 2012 ACM-IEEE International Symposium on Empirical Software Engineering and Measurement*, pp. 211–220. IEEE.

69 Liu, Z. and Ming, X. (2019). A framework with revised rough-DEMATEL to capture and evaluate requirements for smart industrial product-service system of systems. *International Journal of Production Research* 57 (22): 7104–7122.

70 Potts, M., Sartor, P., Johnson, A. and Bullock, S. 2019, October). Deriving key features of a system-of-systems complexity evaluation framework from an industrial case study analysis. *2019 International Symposium on Systems Engineering (ISSE)*, pp. 1–8. IEEE.

71 Osmundson, J.S., Langford, G.O. and Huynh, T.V. (2009, July). 6.3.3 Emergent behavior in an unregulated financial system of systems: economic meltdown. *INCOSE International Symposium*, Vol. 19, No. 1, pp. 1014–1029.

72 Khashanah, K. and Li, Y. (2016). Dynamic structure of the global financial system of systems. *Modern Economy* 7 (11): 1303.

73 Freedman, R.S. (2014). Understanding the complexity of financial systems of systems. *NYU Tandon Research Paper* (2512307).

74 Mansouri, M., Gorod, A. and Wakeman, T.H. et al. (2009, May). A systems approach to governance in maritime transportation system of systems. *2009 IEEE International Conference on System of Systems Engineering (SoSE)*, pp. 1–6.

75 Darabi, H.R., Gorod, A. and Mansouri, M. 2012, July). Governance mechanism pillars for systems of systems. *2012 7th International Conference on System of Systems Engineering (SoSE)*, pp. 374–379. IEEE.

76 Lopes, A.J., Lezama, R., and Pineda, R. (2011). Model based systems engineering for smart grids as systems of systems. *Procedia Computer Science* 6: 441–450.

77 Antal, M., Pop, C., Cioara, T. et al. (2020). A system of systems approach for data centers optimization and integration into smart energy grids. *Future Generation Computer Systems* 105: 948–963.

78 Kargarian, A., Fu, Y., and Wu, H. (2015). Chance-constrained system of systems based operation of power systems. *IEEE Transactions on Power Systems* 31 (5): 3404–3413.

79 Kargarian, A. and Fu, Y. (2014). System of systems based security-constrained unit commitment incorporating active distribution grids. *IEEE Transactions on Power Systems* 29 (5): 2489–2498.

80 Kargarian, A., Mohammadi, J., Guo, J. et al. (2016). Toward distributed/decentralized DC optimal power flow implementation in future electric power systems. *IEEE Transactions on Smart Grid* 9 (4): 2574–2594.

81 Mo, H.D., Li, Y.F., and Zio, E. (2016). A system-of-systems framework for the reliability analysis of distributed generation systems accounting for the impact of degraded communication networks. *Applied Energy* 183: 805–822.

82 Marvasti, A.K., Fu, Y., DorMohammadi, S., and Rais-Rohani, M. (2014). Optimal operation of active distribution grids: a system of systems framework. *IEEE Transactions on Smart Grid* 5 (3): 1228–1237.

83 Mehrjerdi, H. (2020). Resilience improvement with zero load curtailment by multi-microgrid based on system of systems. *IEEE Access* 8: 198494–198502.

84 European Commission, Smart Cities. (n.d.). Brussels: EC. https://ec.europa.eu/info/eu-regionaland-urban-development /topics/cities-and-urban-development/city-initiatives/smart-cities_en (accessed 28 April 2022).

85 IEEE, Smart Cities Community. (n.d.). Washington, DC: IEEE. https://smartcities.ieee.org (accessed 28 April 2022).

86 Kulkarni, K. (2019, December). Smart city as system of systems: subject of study-vertical farming and autonomous driving in smart city. *INCOSE International Symposium*, Vol. 29, pp. 505–517.

87 Harrison, C. and Donnelly, I.A. (2011, September). A theory of smart cities. *Proceedings of the 55th Annual Meeting of the ISSS-2011*, Hull, UK.

88 Makhdum, F. and Mian, K., 2012. *Smarter city: A system to systems.* https://www.diva-portal.org/smash/record.jsf?pid=diva2:831525.

89 Boroomand, F., Fereidunian, A., and Zamani, M.A. 2010, October). Cyber security for smart grid: a human-automation interaction framework. *2010 IEEE PES Innovative Smart Grid Technologies Conference Europe (ISGT Europe)*, pp. 1–6. IEEE.

90 Khosravi, M. and Fereidunian, A. (2021). Fuzzy realizations of adaptive autonomy in smart grid. In: *Applications of Fuzzy Logic in Planning and Operation of Smart Grids*, 105–151. Cham: Springer.

91 Verma, R. (2021). Smart city healthcare cyber physical system: characteristics, technologies and challenges. *Wireless Personal Communications*, pp. 1–21.

92 Mohammed, H. (2019). Smart cities as complex system of systems: challenges and open research problems. *ISPIM Conference Proceedings*, pp. 1–17. The International Society for Professional Innovation Management (ISPIM).

93 Lu, H., Guo, F., Huang, F. et al. (2013). The construction of smart city based on SoS. *International Conference on Advanced Computer Science and Electronics Information ICACSIS*, pp. 34–37.

94 Javidroozi, V., Shah, H., Cole, A. et al. 2015, December). Towards a city's systems integration model for smart city development: a conceptualization. *2015 International Conference on Computational Science and Computational Intelligence (CSCI)*, pp. 312–317. IEEE.

95 Assaad, R., Dagli, C., and El-Adaway, I.H. (2020). A system-of-systems model to simulate the complex emergent behavior of vehicle traffic on an urban transportation infrastructure network. *Procedia Computer Science* 168: 139–146.

96 Alaguvelu, R., Curry, D.M. and Dagli, C.H. 2016, June). Fuzzy—Genetic algorithm approach to generate an optimal meta-architecture for a smart, safe & efficient city transportation system of systems. *2016 11th System of Systems Engineering Conference (SoSE)*, pp. 1–6. IEEE.

97 Srinivasa Rao, V. (2019). Smart city is a complex system of systems. http://www.businessworld.in/article/Smart-City-Is-A-Complex-System-Of-Systems/28-04-2019-169768/

98 Mohammadi, M., Noorollahi, Y., Mohammadi-Ivatloo, B., and Yousefi, H. (2017). Energy hub: from a model to a concept—a review. *Renewable and Sustainable Energy Reviews* 80: 1512–1527.

99 Geidl, M., Koeppel, G., Favre-Perrod, P. et al. (2006). Energy hubs for the future. *IEEE Power and Energy Magazine* 5 (1): 24–30.

100 Sadeghi, H., Rashidinejad, M., Moeini-Aghtaie, M., and Abdollahi, A. (2019). The energy hub: An extensive survey on the state-of-the-art. *Applied Thermal Engineering* 161: 114071.

101 Nozarian, M. and Fereidunian, A. (2020). Smart city as an smart energy hub: a bibliographic, analytic and structural review. *Iranian Electric Industry Journal of Quality and Productivity* 9 (4): 62–82.

102 Mohammadi, M., Noorollahi, Y., Mohammadi-ivatloo, B. et al. (2018). Optimal management of energy hubs and smart energy hubs—a review. *Renewable and Sustainable Energy Reviews* 89: 33–50.

103 Baboli, P.T., Damavandi, M.Y., Moghaddam, M.P. et al. (2015, July). A mixed integer modeling of micro energy-hub system. In: *2015 IEEE Power & Energy Society General Meeting*, 1–5. IEEE.

104 Van Beuzekom, I., Mazairac, L.A.J., Gibescu, M., et al. 2016, April). Optimal design and operation of an integrated multi-energy system for smart cities. *2016 IEEE International Energy Conference (ENERGYCON)*, pp. 1–7. IEEE.

105 EUROCITIES. (May 2011). Response to public consultation on smart cities & communities initiative. www.eurocities.eu (November 2014).

106 Arasteh, H., Bahramara, S., Kaheh, Z. et al. (2021). A system-of-systems planning platform for enabling flexibility provision at distribution level. In: *Flexibility in Electric Power Distribution Networks*, 41–65. United States: CRC Press.

107 Fereidunian, A., Lesani, H., Lucas, C. et al. (2006). A systems approach to information technology (IT) infrastructure design for utility management automation systems. *Iranian Journal of Electrical and Electronic Engineering* 2 (3): 91–104.

108 Fereidunian, A., Hosseini, M.M., and Abbasi Talabari, M. (2017). Toward self-financed distribution automation development: time allocation of automatic switches installation in electricity distribution systems. *IET Generation, Transmission & Distribution* 11 (13): 3350–3358.

109 Mohammadi H.N.S.M., Heydari, S., Mirsaeedi, H., et al. (2015). Optimally operating microgrids in the presence of electric vehicles and renewable energy resources. *2015 Smart Grid Conference (SGC)*, pp. 66–72.

110 Olivares, D.E., Mehrizi-Sani, A., and Etemadi, A.H. (2014). Trends in microgrid control. *IEEE Transactions on Smart Grid* 5 (4): 1905–1919.

111 Guerrero, J.M., Vasquez, J.C., Matas, J. et al. (2010). Hierarchical control of droop-controlled AC and DC microgrids—A general approach toward standardization. *IEEE Transactions on Industrial Electronics* 58 (1): 158–172.

112 Pourafzal, A. and Fereidunian, A. 2020, December). A complex systems approach to feature extraction for chaotic behavior recognition. *2020 6th Iranian Conference on Signal Processing and Intelligent Systems (ICSPIS)*. IEEE.

113 Safarihamid, K., Pourafzal, A. and Fereidunian, A. 2021, December). A joint-entropy approach to time-series classification. *2021 7th International Conference on Signal Processing and Intelligent Systems (ICSPIS)*. IEEE.

114 Pourafzal, A., Fereidunian, A., and Safarihamid, K. (2023). Chaotic time series recognition: a deep learning model inspired by complex systems characteristics. *International Journal of Engineering*. 36 (1): 1–9.

115 Nozarian, M. and Fereidunian, A. (2021). Analysis of emergent behavior of reliability in the system of systems including energy hubs. *Journal of Modeling in Engineering* 19 (66): 1–21.

116 Johnson, J.J. IV, Tolk, A., and Sousa-Poza, A. (2013). A Theory of emergence and entropy in systems of systems. *Procedia Computer Science* 20: 283–289.

117 Mittal, S. and Rainey, L. (2015, July). Harnessing emergence: the control and design of emergent behavior in system of systems engineering. *Proceedings of the Conference on Summer Computer Simulation*.

5

Metaverse Local Energy Market in Smart City: A Descriptive Model and Strategic Development Analysis

Mohammad Ghafourian Nasiri[1], *Zahra Iranpour Mobarakeh*[2], *Mahdi Nozarian*[2], *Alireza Fereidunian*[2], *Sabrieh Choobkar*[3], *and Hossein Jobran*[2]

[1] *Faculty of Industrial and Systems Engineering, Tarbiat Modares University, Tehran, Iran*
[2] *Faculty of Electrical Engineering, K.N. Toosi University of Technology, Tehran, Iran*
[3] *Faculty of Information and Communication Technologies (ICT) Research Group, Niroo Research Institute (NRI), Tehran, Iran*

> Ah, Love! could you and I with Him conspire
> To grasp this sorry Scheme of Things entire
> Would not we shatter it to bits–and then
> Re-mould it nearer to the Heart's Desire
>
> Omar Khayyam Neyshabouri, (1048–1131)

5.1 Introduction

This book chapter is commenced with a verse by Omar Khayyam Neyshabouri (Persian polymath and poet) that reflects upon the nature of our world. In these poetic lines, Khayyam muses on the idea of transforming the existing world into a new one, one that offers greater ease and pleasure of living. Interestingly, as we venture into the twenty-first century, we find ourselves discussing the concept of the current term "metaverse," which remarkably aligns with the alternative realm envisioned by Khayyam.

Thinking of a new world, digital transformation has lightened long-term wishes for building and entering a virtual sphere as a metaverse. This virtual society not only unlocks a novel virtual environment, but also provides direct effects on the real world and enhances the quality of lives as well. Specifically, this chapter is going to discover "how metaverse can be the hope of better conditions to achieve a desirable level of urban management in energy transferring?". Then, it is going to explore the local energy market (LEM) in the metaverse. Studying the capabilities of the metaverse in this field can indicate part of the reason why cities are progressing toward "metacities."

Determining the role of each technology in the future of cities is the main step in the development and utilization of that technology and its expansion in various activities and situations of everyday life. Various activities can be speeded up by using metaverse or an activity that is not possible in the real world can be done.

Smart Cyber-Physical Power Systems: Fundamental Concepts, Challenges, and Solutions, Volume 1, First Edition.
Edited by Ali Parizad, Hamid Reza Baghaee, and Saifur Rahman.

5.2 Background

The term metaverse was first used by Neal Stevenson, the author of the science fiction book "Snowfall" in 1992 [1]. Then, it was introduced to the world by Mark Zuckerberg, under the cover of Instagram and Facebook and the title "Meta."

Today, modern energy management systems (EMSs) require high-quality connections for the ever-increasing number of equipment and devices in the physical environment. EMSs employ high-speed communication networks to transfer a variety of data between their modules. The emergence of digital twins plays an irreplaceable role in modern energy systems in order to achieve energy efficiency [2, 3].

In the realm of smart cities, energy management and planning are crucial aspects that demand a comprehensive approach. Addressing this need, [4] emphasize the importance of considering all intervention areas, stakeholders, and technologies when designing energy systems. The selection of appropriate system parameters, energy constraints, geographical information, and optimization algorithms is highlighted as key factors in developing efficient and cost-effective solutions. In the context of smart cities, concepts of energy prosumption and utilization of data generated by metering devices and energy sensors hold great significance.

[5] delve into the importance of sharing and utilizing these data through an integrated layered architecture. The proposed architecture aims to provide real-time monitoring, community forecasting, and increased renewable energy trading. Ongoing work involves validating each layer of the architecture with qualitative case data from energy companies in Norway.

The paradigm shift toward distributed generation in urban energy systems is explored in [6]. The research presents a methodology that synthesizes district-scale planning and design procedures within an optimization framework. This computational framework, serving as an automated EMS, prioritizes efficiency and environmental constraints.

The utilization of digital twins in energy services, recommendation systems, and demand-side management is explored in [3]. The study validates data-driven twin technologies as effective tools for improving energy consumer behavior and achieving efficiency. Barriers to adoption are also identified, and policy recommendations are provided in this study for their widespread implementation.

[7] delve into the history, properties, and application of digital twins in the energy management sector. The review highlights challenges associated with their implementation, such as the need for transdisciplinary models and managing model heterogeneity. Co-simulation and co-modeling techniques are suggested as potential solutions.

Lastly, [8] provide an overview of digital twin applications within smart grids. While progress is being made in their implementation and integration, there is still a lack of understanding regarding their full potential and seamless integration with energy management processes.

By achieving a precise integration between digital and physical realms through the concept of a digital twin, it becomes feasible to forecast energy consumption levels across various time periods. This capability enables the prediction of consumer behavior, facilitating the implementation of essential measures and the identification of correct and incorrect habits. Thus, a digital replica of the real world serves as a crucial foundation for the metaverse. Positioned at the core of convergence between physical and virtual domains, digital twins present numerous prospects across diverse fields.

By considering insights from these studies, we can see the interconnectedness of different research areas within the realm of smart cities, energy management, digital twins, and the green metaverse network. Integration of digital twin technologies, intelligent recommendation systems, and optimization frameworks can contribute to the development of efficient and sustainable LEMs. These advancements pave the way for a greener future, where energy systems are optimized, consumer-centric, and environmentally friendly.

Considering the topic of this chapter, it is necessary to provide several key definitions. These definitions will help clarify important concepts and establish a common understanding within the research context.

5.3 Concepts

5.3.1 Smart City

A smart city is a platform where existing traditional networks, services, and resources, with the help of information, digital, and telecommunication technologies, become more flexible, efficient, and sustainable. The data-driven decision-making processes improve the performance of the city for the benefit of its residents [9–12]. A smart city is an urban development vision that integrates multiple information and communication technology (ICT) solutions in a secure fashion to manage a city's assets, which include the city's physical infrastructure, people, community services, and information systems [2, 13–16].

5.3.2 Metaverse

There is no widely accepted or standard definition of the metaverse provided in scientific sources due to its evolving and interdisciplinary nature. Metaverse is a word comprised of two words, meta and world, which means extra-world (extra-universe) and is a developing three-dimensional digital world in which a kind of daily life and economic activities are possible. The concept of metaverse has gained popularity in recent years, particularly in the realms of virtual reality, augmented reality, and online gaming [1, 17].

5.3.3 Digital Twin

A digital twin is a virtual model of a process, product, or service. This coupling of virtual and physical worlds allows data analysis and monitoring systems to fix problems before they happen, prevent failures, create new opportunities, and even plan for the future using simulations. Also, these digital twins help to realize that the metaverse is a mirror of the real world. A digital twin is a virtual representation or digital replica of a physical object, system, or process that is continuously updated and synchronized in real time, incorporating data from various sources such as sensors, simulations, and historical records. It enables real-time monitoring, analysis, and optimization of the physical counterpart's performance, behavior, and maintenance throughout its lifecycle. The digital twin provides a digital platform for modeling, simulation, and decision-making, offering insights into the physical entity's operation, facilitating predictive maintenance, and supporting data-driven decision-making for improved efficiency, sustainability, and performance.

5.3.4 Local Energy Market

Today, energy demand and production are increasing and the centralized system of energy trade is facing a challenge in terms of fair distribution of available energy [18]. With increasing integration of distributed energy resources, a distributed power generation is taking shape where a large number of small-scale production units with different capacities are connected to the distribution network, leading to two-way power flows [19]. A LEM is defined as a localized trading platform that allows for the exchange of energy resources within a specific geographical region. This market enables energy producers, consumers, and prosumers (individuals who both produce and consume energy) to partake in the trading process, which is based on the dynamics of supply and demand. LEMs commonly utilize advanced metering infrastructure and digital technologies to facilitate transactions that are transparent and efficient. Moreover, these marketplaces aim to promote the integration of renewable energy sources and aid in the optimization of energy resources at the local level [20, 21].

The current chapter focuses on the advantages and disadvantages of metaverse in the context of LEMs. It is important to note that this particular study contributes to the existing body of knowledge by proposing a novel approach to modeling LEMs using business process model and notation (BPMN). By leveraging this method, the chapter aims to enhance the understanding of dynamics and complexities involved in LEMs within the metaverse environment. The exploration of advantages and disadvantages of such markets provides insights for policymakers, researchers, and industry professionals in designing and implementing efficient and sustainable energy systems. These insights can inform future research endeavors and policy discussions in the realm of energy management and sustainable development. Our previous research can be considered as one of the first researches in the field of behavioral modeling. Metaverse deals in smart city management in four areas of energy, health, transportation, and a case study of the LEM. In that research, an attempt has been made to find an answer to the question of what benefits the entry of smart cities into the metaverse space and finally becoming a metaverse smart city brings for the residents and operators of the city. In this way, by defining scenarios of daily life, in three important operational areas of the smart city, the role of metaverse in solving these situations is investigated in comparison with the current conditions.

5.3.5 Metacity

Digital technologies have started to change urban structures and will penetrate more in future municipal forms with no doubt. A metacity is a smart city that integrates information and digital technologies with traditional social services to improve their efficiency and performance.

Metacity approach provides a lot of potential in several fields in the urban redesign and structural renewal of systems. The following include some main ecosystems that face change in a metacity:

Management Policies: Metacity offers decision strategies since it utilizes various data aggregation and analysis, hence providing several analytical information for managers, decision-makers, and directors of state or private sections in any large or small scales.

Buildings: Civil architecture of a metacity should follow a special framework to make its commercial buildings and residential houses connected to intelligent structures.

Transportation: A metacity model consists of efficient planning for the accessibility of resources and services. It also provides civilians with the best mobility and daily journey experiences.

Energy: In the metacity context, interaction of energy ecosystems is fundamental. Digital advancements in the cooperated structure of energy platforms offer the best supply to urban areas and allow for consumption optimization and waste reduction [17].

In the next section, a LEM is described as a case study, which is a trading platform for energy exchange.

5.4 Case Study: Local Energy Market in Metaverse

5.4.1 Descriptive Model

With the introduction of Bitcoin, trust in blockchain and its use in energy distribution were noticed. A blockchain system based on energy trading has been introduced in which there is no need for a third party to supply energy. The operator of the distribution system is considered as a node to perform the double auction between the seller and the consumer. In fact, each node confirms the transaction and distributes the copy of the transaction to all other nodes [18].

Energy distribution in the LEM can be completely done by blockchain. With the elimination of the third person in the buying and selling of energy, many exchanges and transactions are still carried out by the people themselves, which requires spending a lot of time and money. Now, by adding a digital twin system and simulating the system used by buyers in the metaverse, the role of the buyer can be reduced because the simulated system in metaverse alone can detect the need for energy and its amount, transfer this request to the user in the blockchain, and purchase energy at the lowest possible price. Also, on the side of the energy seller or producer, by correctly simulating the production system in the metaverse and having the amount of energy produced at any time, information on the amount of energy that can be distributed from the metaverse is given to blockchain and reduces the role of producer in doing this business. Finally, the producer only produces and the consumer only consumes, and in the meantime, all the transactions, calculations, and details of the energy business are done by simulations done in the metaverse and with the help of blockchain.

Comparing LEM processes with the metaverse model can be achieved using BPMN. A BPMN visualizes and optimizes the various processes involved in the LEM in the metaverse, from energy generation and storage to energy distribution, trading, consumption, payment, and regulatory compliance. The visualization presented in Figure 5.1 shows the BPMN model of LEM value chain, which only presents key processes.

An alternative way of expressing the proposed model for LEM into the metaverse realm is outlined below at Figure 5.2.

Studying these two models provides us with valuable insights into how the metaverse can alter the manner in which we generate, disseminate, and consume energy. As the metaverse BPMN model shows, a LEM in the metaverse could provide a decentralized, transparent, and efficient way for energy producers and consumers to trade energy. However, like any new transformation, there are also challenges and potential drawbacks to consider.

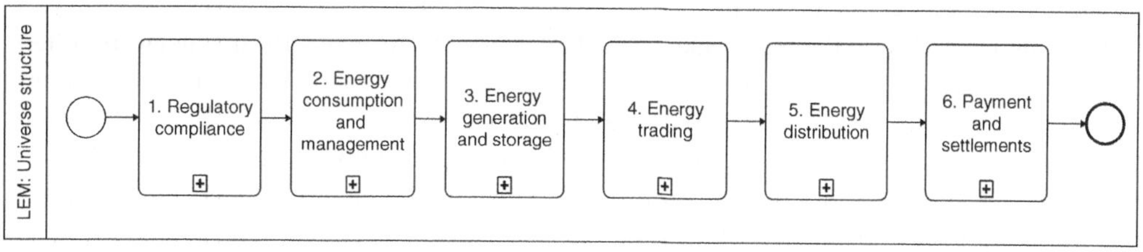

Figure 5.1 Business process model and notation (BPMN) model of local energy market (LEM) value chain.

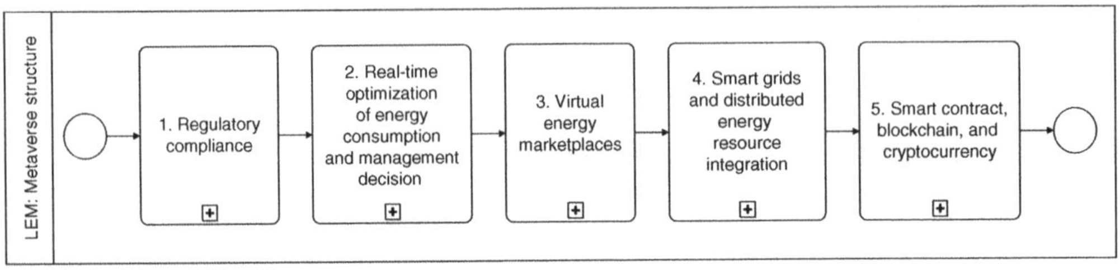

Figure 5.2 BPMN model of an alternative way for LEM value chain in the metaverse.

5.4.2 SWOT Analysis

5.4.2.1 SWOT Methodology

The SWOT methodology is a strategic planning tool used to evaluate the strengths, weaknesses, opportunities, and threats (SWOT) of a business, project, or organization. It involves identifying internal strengths and weaknesses, as well as external opportunities and threats, to facilitate strategic decision-making and improve performance. To identify the strengths (S) and weaknesses (W) in a SWOT analysis, it is more appropriate to focus on specific internal factors such as resources, capabilities, processes, and organizational culture. These can be assessed through various approaches, including internal audits, performance evaluations, feedback from stakeholders, and 7S McKinsey framework.

The 7S McKinsey framework is a management model that examines seven aspects (strategy, structure, systems, shared values, skills, staff, and style) of an organization to assess its effectiveness and identify areas for improvement. In the SWOT analysis, opportunities (O) refer to favorable external factors that an organization can leverage to its advantage. These include emerging markets, technological advancements, changing consumer preferences, or industry trends. Threats (T) in the SWOT analysis refer to external factors that can potentially harm the organization or pose challenges. These may include competitive pressures, regulatory changes, economic downturns, or technological disruptions. The Political, Economic, Sociological, Technological, Legal, and Environmental (PESTLE) analyses identify external opportunities and identify threats by examining the sociocultural, technological, economic, and political factors.

Combining the SWOT analysis, PESTLE analysis, and 7S McKinsey framework makes the LEM study in the metaverse more robust, insightful, and strategic. It enables a thorough examination of internal and external factors, allowing for better strategic planning, risk management, and overall performance improvement in the evolving energy landscape of metaverse.

5.4.2.2 Strength and Weakness

Here, we present a SWOT analysis of LEM in the metaverse, defining its SWOT. By understanding the SWOT analysis, we gain insights into how LEM in the metaverse can be developed and optimized to achieve its full potential.

The strengths of LEM in the metaverse are evident in several areas. Firstly, the metaverse offers a unique environment for thorough testing of innovative energy solutions through simulation, enabling the exploration of various scenarios and fostering the development of agile and flexible business models. Secondly, improved monitoring of energy consumption patterns and behaviors within the metaverse provides better data insights, allowing for enhanced energy efficiency and accurate energy consumption forecasting [22].

The decentralized platform of the metaverse enables direct peer-to-peer transactions, bypassing the need for intermediaries like conventional utilities. This characteristic makes the metaverse an ideal solution for LEM. Additionally, improved integration among energy equipment and stakeholders, including utility providers, consumers, and regulators, facilitates instantaneous insights for informed decision-making and more effective energy consumption management.

On the other hand, the weaknesses of LEM in the metaverse include significant investment required to establish infrastructure and technological support necessary for energy market digital twins. Additionally, the complexity of integrating diverse stakeholders with differing interests and priorities poses a challenge. Technical expertise requirements may limit participation, and the relatively new nature of technology could hinder widespread adoption and its application in LEM. Furthermore, a small customer base within the metaverse can make it challenging for energy companies to achieve economies of scale and operate profitably.

The strengths of LEM in the metaverse can be extracted as follows:

Innovative energy solutions can be thoroughly tested through the simulation of various scenarios. This capability enables more **agile and flexible** business models.

Improved monitoring of energy consumption patterns and behaviors results in better data insights for improving energy efficiency and forecasting energy consumption.

Metaverse's decentralized platform enables direct **peer-to-peer transactions**, by passing the need for intermediaries such as conventional utilities, making it an ideal solution for the LEM.

Improved **integration** among energy equipment and interested parties, comprising utility providers, consumers, and regulators, furnishing instantaneous insights for informed judgment and more effective management of energy consumption.

Better automation of energy systems and processes, resulting in higher efficiency and **lower operational costs**, while also reducing the carbon footprint.

The utilization of the platform leads to a decrease in energy expenditure by limiting the dependency on a centralized power source, thus enhancing the **feasibility and affordability of energy** for all.

Metaverse provides a **user-friendly interface**, making it easy for all parties to participate in the energy market.

The use of metaverse technology guarantees the **authenticity and reliability** of transactions thanks to its foolproof security measures and unparalleled transparency levels.

The weaknesses of LEM in the metaverse can be extracted as follows:

The need for **significant investment** to create the infrastructure and technology needs to support energy market digital twins.

Complexity of integrating various stakeholders with various/diverse interests and priorities.

Technical expertise requirements could limit participation.

The **technology is still relatively new** and is not widely adopted, which limits its application in the LEM.

The **small customer base** on metaverse makes it difficult for energy companies to achieve economies of scale and operate profitably.

5.4.2.3 Opportunity and Threats

The metaverse holds immense potential to revolutionize the energy landscape by enhancing reliability, accessibility, and cost-effectiveness. By reducing intermediaries in traditional power systems, the metaverse can make energy more reliable, accessible, and affordable for all. It also enables the creation of new business models that foster agility and flexibility in energy markets.

Furthermore, the metaverse presents an opportunity to address critical issues such as energy access and poverty, particularly in developing countries. Improved data sharing among stakeholders within the metaverse facilitates better communication and collaboration among market players, leading to enhanced efficiency and informed decision-making.

However, several threats must be considered. Digital twin infrastructure, crucial for the functioning of the metaverse, may suffer from underinvestment due to a short-term focus on energy investments, impacting its long-term sustainability. Concerns regarding data privacy and cybersecurity challenges are on the rise among the general public, which may affect the trust and widespread adoption of metaverse-based energy markets.

Implementation of blockchain technology in energy markets within the metaverse could face restrictions or ethical issues due to unclear regulatory guidelines. Additionally, the dynamic nature of the energy industry, with constant technological advancements, may pose limitations and compatibility issues within the LEM in the metaverse.

The opportunity of the LEM in the metaverse can be extracted as follows:

Metaverse has the potential to revolutionize the LEM, making energy more **reliable and accessible** for all and reducing the cost of energy with the reduction in intermediaries in traditional power generation and distribution systems.

Creation of **new business models** for agile and flexible energy markets.

Metaverse provides an opportunity to **address issues** around energy access and energy poverty in developing countries.

Improved data sharing for stakeholders results in **better communication** and collaboration among market players.

The platform has a lot of potential to **promote the adoption of sustainable energy** sources such as solar and wind, providing a platform for renewable energy projects that will bring significant benefits in energy security, carbon reduction, and financial savings.

As consumers become more aware of the benefits of blockchain technology and decentralized energy markets, **demand** for such platforms is likely to **increase**.

The threats of the LEM in the metaverse can be extracted as follows:

Digital twin infrastructure is suffering from **underinvestment in the long run** because of a narrow focus on short-term energy investments.

Increasing concerns in the general public about data **privacy and cybersecurity** challenges.

The implementation of blockchain in energy markets might be restricted or provoke ethical issues due to the **blurred regulatory guidelines**.

The energy industry is constantly evolving, and advancements in technology may create limitations in the LEM on metaverse, such as **compatibility issues**.

5.5 Discussions and Conclusions

This chapter explores the potential role of the metaverse in smart city energy management. Inspired by the idea of transforming the world into a better place, similar to the vision of the metaverse, the chapter examines how this concept can contribute to achieving desirable urban management in energy transfer. By studying the capabilities of the metaverse in this field, the article sheds light on why cities are progressing toward becoming metacities. Understanding the role of metaverse technologies in different domains can expedite processes, save time, and reduce costs.

Later on, this chapter presents a case study on LEM in the metaverse, exploring its potential benefits and drawbacks. By leveraging blockchain technology, energy distribution in the LEM can be decentralized and eliminate the need for intermediaries. Incorporating digital twin systems and simulations in the metaverse, the buyer's role will be reduced as the system autonomously detects energy needs and purchases it at the lowest price. Similarly, producers can optimize energy distribution by simulating the production systems in the metaverse, streamlining the process, and reducing their involvement.

The chapter utilizes BPMN to visualize and optimize processes involved in the LEM in the metaverse. The descriptive model highlights the potential transformation of energy generation, distribution, trading, consumption, payment, and regulatory compliance in the metaverse. Additionally, a SWOT analysis is conducted to assess the SWOT associated with LEM in the metaverse. This analysis provides insights into the development and optimization of the market within the evolving energy landscape.

Integration of the metaverse into smart city energy management holds promise for efficient and sustainable LEMs. By leveraging digital twin technologies, intelligent recommendation systems, and optimization frameworks, cities can strive toward greener futures that prioritize efficiency, consumer-centric approaches, and environmental sustainability. The chapter emphasizes the need for further research, policy discussions, and industry collaboration to fully realize the potential of the metaverse in energy management and sustainable development.

References

1 Safarianpour, S., Osanloo, M., and Mousavy, A.H. (2022). *Metaverse Podcast*. Castbox https://castbox.fm/va/4731369.

2 O'Dwyer, E., Pan, I., Charlesworth, R. et al. (2020). Integration of an energy management tool and digital twin for coordination and control of multi-vector smart energy systems. *Sustainable Cities and Society* 62: 102412. https://doi.org/10.1016/j.scs.2020.102412.

3 Onile, A.E., Machlev, R., Petlenkov, E. et al. (2021). Uses of the digital twins concept for energy services, intelligent recommendation systems, and demand side management: a review. *Energy Reports* 7: 997–1015. https://doi.org/10.1016/j.egyr.2021.01.090.

4 Calvillo, C.F., Sánchez-Miralles, A., and Villar, J. (2016). Energy management and planning in smart cities. *Renewable and Sustainable Energy Reviews* 55: 273–287. ISSN 1364-0321, https://doi.org/10.1016/j.rser.2015.10.133.

5 Anthony Jnr, B., Abbas Petersen, S., Ahlers, D., and Krogstie, J. (2019). API deployment for big data management towards sustainable energy prosumption in smart cities-a layered architecture perspective. *International Journal of Sustainable Energy* 39 (3): 263–289. doi: 10.1080/14786451.2019.1684287.

6 Manfren, M., Caputo, P., and Costa, G. (2011). Paradigm shift in urban energy systems through distributed generation: methods and models. *Applied Energy* 88 (4): 1032–1048. https://doi.org/10.1016/j.apenergy.2010.10.018.

7 Lamagna, M., Groppi, D., Nezhad, M.M. et al. (2021). A comprehensive review on digital twins for smart energy management system. *International Journal of Energy Production and Management* 6 (4): 323–334. https://doi.org/10.2495/EQ-V6-N4-323-334.

8 Cioara, T., Anghel, I., Antal, M. et al. (2022). An overview of digital twins application in smart energy grids. *2022 IEEE 18th International Conference on Intelligent Computer Communication and Processing (ICCP)* (pp. 25–30). https://doi.org/10.1109/ICCP56966.2022.10053945.

9 Akhavan-Rezai, E., Haghifam, M.R., and Fereidunian, A. (2009). Data-driven reliability modeling, based on data mining in distribution network fault statistics. In: *2009 IEEE Bucharest PowerTech*, 1–6. IEEE https://doi.org/10.1109/PTC.2009.5281796.

10 Caragliu, A., Del Bo, C., and Nijkamp, P. (2011). Smart cities in Europe. *Journal of Urban Technology* 18 (2): 65–82. https://doi.org/10.1080/10630732.2011.601117.

11 Nozarian, M. and Fereidunian, A. (2020). Smart city as an smart energy hub: a bibliographic, analytic and structural review. *Iranian Electric Industry Journal of Quality and Productivity (IEIJQP)* 9 (4): 62–82. https://doi.org/10.29252/ieijqp.9.4.62, http://ieijqp.ir/article-1-717-fa.html.

12 Nozarian, M., Fereidunian, A., and Barati, M. (2023). Reliability-oriented planning framework for smart cities: from interconnected micro energy hubs to macro energy hub scale. *IEEE Systems Journal* 17 (3): 3798–3809.

13 Etemadi, K. and Fereidunian, A. (2022). A comparative study of two people-centric smart cities evaluation frameworks, IEEE Smart Cities Newsletter. https://smartcities.ieee.org/newsletter/june-2022/a-comparative-study-of-two-people-centric-smart-cities-evaluation-frameworks (accessed June 2022).

14 Fereidunian, A. (2020). Smart city: people-centric development, based on energy-awareness, by advanced metering infrastructure (AMI) data analytics. *Invited Talk PowerPoint*, Tehran, Iran (December 15, 2020). K. N.Toosi University of Technology (KNTU). https://doi.org/10.13140/RG.2.2.19448.21764.

15 Kojury-Naftchali, M. and Fereidunian, A. (2021). Smart energy-aware cities: customer characterization by energy data analytics to improve demand response performance. In: *Flexible Resources for Smart Cities* (ed. M. Shafie-khah and M.H. Amini), 21–43. Springer.

16 Nozarian, M., Fereidunian, A., Hajizadeh, A., and Shahinzadeh, H. (2023). Exploring social capital in situation-aware and energy hub-based smart cities: towards a pandemic-resilient city. *Energies* 16 (18): 6479.

17 Iranpour Mobarakeh, Z., Nozarian, M., Ghafourian Nasiri, M. et al. (2022). Evaluating the role of metaverse in smart city management by behavioral modeling, in the field of energy, health, transportation and a case study on local energy market. *8th International Conference on Industrial and Systems Engineering*, Mashhad, Iran (September 2022).

18 Gurmani, M. U., Sultana, T., Ghaffar, A. et al. (2020). Energy trading between prosumer and consumer in P2P network using blockchain. *14th International Conference on P2P, Parallel, Grid, Cloud and Internet Computing (3PGCIC-2019)*, Antwerp, Belgium: University of Antwerp (pp. 875–886). https://doi.org/10.1007/978-3-030-33509-0_82.

19 Khorasany, M., Mishra, Y., and Ledwich, G. (2018). Market framework for local energy trading: a review of potential designs and market clearing approaches. *IET Generation, Transmission & Distribution* 12 (22): 5899–5908. https://doi.org/10.1049/iet-gtd.2018.5309.

20 Noori, A., Tavassoli, B., and Fereidunian, A. (2021). Incentivizing peer-to-peer energy trading in microgrids. *29th Iranian Conference on Electrical Engineering (ICEE)*, Tehran, Iran (2021), pp. 323–328. https://doi.org/10.1109/ICEE52715.2021.9544467.

21 Saghafi, M., Fereidunian, A., and Lesani, H. M. (2019). Peer-to-peer energy trading by smart contracts using blockchain technology. *The 5th International Conference on Technology and Energy Management (IEAC)*, Tehran. https://civilica.com/doc/855197.

22 Kojury-Naftchali, M., Fereidunian, A., and Lesani, H. (2016). Identifying susceptible consumers for demand response and energy efficiency policies by time-series analysis and supplementary approaches. In: *2016 24th Iranian Conference on Electrical Engineering (ICEE)*, 1130–1135. IEEE.

6

Cooperative and Distributed Control Strategies of Microgrids

Mahmood Jamali[1] *and Mahdieh S. Sadabadi*[2]

[1] *School of Electrical and Electronic Engineering, The University of Sheffield, Sheffield, United Kingdom*
[2] *Department of Electrical and Electronic Engineering, The University of Manchester, Manchester, United Kingdom*

6.1 Introduction

The concept of microgrids is an appealing alternative for overcoming the challenges of the integration of renewable distributed energy resources (DERs) and modern loads into power grids [1]. Microgrids are small-scale electrical networks heterogeneously composed of DERs, energy storage systems, and loads that operate locally as a single controllable entity. Microgrids are dominated by power electronic converters interfacing with renewable-based DERs, energy storage systems, and loads. Microgrids can be either connected to the main grid and operate in a grid-tied mode or seamlessly disconnected during power outages and operate in an islanded mode [2], this is recognized as the most significant feature of microgrids in enhancing the resilience of power systems. The islanded operation of microgrids brings several challenges from the control perspective that should be properly addressed [3].

The advanced functionalities and reliable operation of islanded inverter-interfaced microgrids are achieved by means of a hierarchical control system consisting of primary, secondary, and tertiary control [4–6]. The primary-level control that is usually based on a droop control mechanism has a decentralized structure [5]. The secondary control is mostly responsible for frequency synchronization, voltage regulation, and proportional power sharing among DERs and compensates for any steady-state error in the voltage and frequency of microgrids caused by implementing the primary controller [5]. The tertiary control satisfies global criteria for an economic operation in microgrids [5]. Conventionally, a centralized controller was employed for the secondary control of microgrids. However, due to the drawbacks of centralized control systems in terms of high bandwidth, high cost, and having a single point of failure, distributed control systems are preferred for secondary control in islanded microgrids. A distributed control framework necessitates information flow among DERs' local controllers through a communication network [7, 8]. There exists intensive literature on distributed secondary control in inverter-rich microgrids, e.g., see [7] and the references therein.

The conventional secondary control in islanded microgrids is based on consensus-based control algorithms that regulate voltages and synchronize frequencies of DERs. Although conventional secondary control algorithms ensure voltage regulation and frequency synchronization as well as power sharing among DER units, they assume ideal and reliable conditions for sensing, actuating, and communicating information. However, such an assumption might not be necessarily valid for

Smart Cyber-Physical Power Systems: Fundamental Concepts, Challenges, and Solutions, Volume 1, First Edition.
Edited by Ali Parizad, Hamid Reza Baghaee, and Saifur Rahman.

cyber-physical microgrids due to cyberattacks and faults that might compromise the reliability of sensors, actuators, and communication links. In this chapter, the main focus is to consider the impact of faults on the secondary control of inverter-dominated microgrids.

With the increasing integration of power electronic devices in microgrids, there has been increasing concern about the reliability of power electronic converters. Generally, semiconductor switches are vulnerable to a diverse range of non-ideal conditions, such as faults originating from environmental condition changes like temperature, contact problems, metallization failures, electrical overstress, and packing failures [9, 10]. Any faults and failures in the power electronic switches appear in the output signals of converters, affecting the control-loop system [11]. Such faults challenge the reliability and safety of critical loads within the microgrid, as the fault might propagate into the microgrid and lead to serious performance degradation at the microgrid level and unscheduled costly maintenance or result in overall microgrid failure and physical damages.

From a control perspective, faults can be categorized according to purposes, limitations of the system, phenomenological reasons, and stage of formation [12]. Note that several types of software/hardware damages that might influence control input channels could also be considered faults in the control loop.

The other remarkable unreliability in the control-loop structure is the saturation, which represents a common kind of nonlinearity in the actuators caused by restrictive constraints on control gains and signals [13]. Due to the duty cycle limitation, nonidealities, and software/hardware limits, the non-perfect conditions of the control paradigm are observed in the power electronic components in practice. Such control input constraints might severely impact normal control actions and even threaten the stability of microgrids. It is common to employ high-gain feedback methods to cope with any input saturation [14]. However, when the control system encounters faults, this approach cannot take the state trajectories to a desirable set.

The main questions that are of central relevance in the context of reliability in islanded inverter-interfaced microgrids and that will be discussed in detail in this chapter are as follows:

- What happens if the reliability of such information in microgrids' secondary control systems is compromised?
- How to design a control scheme that can ensure normal operation in microgrids where it might be challenging to guarantee the reliability of actuating information, e.g., due to faults or saturation?

Although the conventional cooperative and consensus-based secondary control mechanisms for microgrids consider robustness to modeling uncertainties (e.g., load variation and plug-and-play operation of DERs) into their design, they are not intrinsically robust to cyber stresses and faults in microgrids. Hence, it is crucial to enhance microgrids' reliability and their resilience against the unreliability of used data in the secondary control systems of microgrids. This chapter aims to provide an answer to the above-mentioned questions. In particular, the main focus of this chapter is on unreliable actuating information in the secondary control layer of inverter-interfaced microgrids. It is assumed that the control command dictated by the secondary controller that is sent to the primary level for specifying the nominal values for the voltage and frequency might confront restrictive conditions and unreliable circumstances. Various ranges of non-ideal conditions, e.g., noise, fault, and saturation, can significantly affect the stationary performance of the secondary controller, leading the microgrid to undesired operational points. More specifically, the actuator fault in the control configuration could worsen the performance of microgrids since large control commands are necessary to alleviate the fault impacts. Moreover, transient performance and convergence time might be negatively impressed in the presence of the actuator fault. The goal of this chapter is to design a finite-time distributed secondary controller heeding *"bounded fault"* and *"saturation"* in the secondary control of islanded inverter-interfaced microgrids.

6.1.1 Notation, Preliminaries, and Chapter Structure

6.1.1.1 Notation

Throughout this chapter, $||.||$ stands for Euclidian norm of vectors or matrices. I_N is an $N \times N$ square identity matrix. $col\{d_1, d_2, \ldots, d_N\}$ represents an $N \times 1$ vector and $diag\{d_1, d_2, \ldots, d_N\}$ denotes a diagonal matrix, where $\{d_1, d_2, \ldots, d_N\}$ are on its diagonal. For any $Q \in \mathbb{R}^{N\times N}$, the symbols $\lambda_i(Q)$, $\lambda_{min}(Q)$, and $\lambda_{max}(Q)$ are the i-th eigenvalue, minimum, and maximum eigenvalues of Q, respectively. For a real symmetric matrix Q, the positive definite operator is expressed by $Q > 0$.

6.1.1.2 Graph Theory

For a microgrid with N-DER units, the graph topology for communication is represented by an undirected graph $\mathcal{G} = (\vartheta, \mathcal{E})$, where $\vartheta = \{v_i : i \in \mathbb{N}\}$ is the set of nodes and $\mathcal{E} \subseteq \{\vartheta \times \vartheta\}$ is the set of edges. $\mathcal{A}$ is the adjacency matrix with elements $a_{ij} > 0$ if the i-th DER unit receives (transmits) data from (to) the j-th DER unit. The Laplacian matrix of the graph associated with the adjacency matrix ($\mathcal{A}$) is defined as $\mathcal{L} = [l_{ij}] \in \mathbb{R}^{N\times N}$, where $l_{ii} = \sum_{i\neq j} a_{ij}$ for $i = j$. In the communication topology, only the leader node has access to the reference values of voltage and frequency. The diagonal pinning gain matrix indicating the leader node is defined by $\overline{\mathcal{G}} = diag\{a_{i0}\}$, where $a_{i0} > 0$ if the data is accessible for the i-th DER unit, otherwise, $a_{i0} = 0$.

6.1.1.3 Chapter Structure

The rest of this chapter is organized as follows. Required preliminaries for modeling, controller design, and stability analysis in islanded inverter-interfaced microgrids are presented in Section 6.2. The fault-tolerant control mechanism for voltage restoration is provided in Section 6.3. The effectiveness of the presented distributed secondary control mechanism, along with the comparison with several relevant studies, is presented in Section 6.4. Finally, Section 6.5 concludes the chapter by providing some concluding remarks about the resilience and reliability of distributed secondary controllers in islanded inverter-interfaced microgrids.

6.2 Fault-Tolerant Secondary Control Schemes in Islanded AC Microgrids

Faults and nonlinearity conduce to integral winding up in the control loop, where poor transient performance such as long convergence time could be caused. It is important to be highlighted that in classic linear time-invariant systems, the state trajectories asymptotically converge to set points at the stationary, rendering the asymptotic stability of the closed-loop system. However, in contrary to finite-time control approaches, convergence cannot be guaranteed in a limited time. The finite-time control methods show more robustness against uncertainty and measurement errors. Furthermore, the convergence rate—which is an important performance index in control systems—of such methods is swifter than the conventional distributed control [15]. From a practical view, sensitive loads in microgrids are required to be supplied at the nominal frequency and voltage. Thus, it is essential to accelerate the voltage regulation and frequency synchronization process in a finite time [16]. In [17], it is also proved that the finite-time control approach can even preserve a finite-time convergence rate under graph topology alteration. For example, some types of finite-time control methods have been presented to accelerate the restoration in the secondary control layer (see, e.g., [18, 19]).

Inspired by the aforementioned issues, a finite-time secondary control scheme is designed for voltage regulation and frequency synchronization in islanded AC microgrids under actuator fault and saturation. To do so, distributed auxiliary dynamics based on the conventional cooperative control algorithm are proposed. The proposed control scheme relies on nonlinear auxiliary dynamics that can assure finite-time stability of microgrids subject to input limitations and faults. The proposed control scheme is capable to mitigate unknown multiplicative faults in the control input channels of each DER unit. Different from the existing studies on fault-tolerable secondary control problems, the multiplicative fault term is also time-varying. Furthermore, the proposed controller does not take any assumptions regarding the maximum number of faulty units. Thus, it can be guaranteed the semi-global stability of the microgrid even if all DER units controlling input channels are under faults. Except for the "*boundedness*" of the fault, there is no need to have any numeric data about the fault signal, such as frequency and amplitude. By employing the proposed control scheme, the frequency and voltage restoration are achieved in a finite time. Microgrid control systems can take advantage of the finite-time control approach as it also provides robust performance and disturbance rejection.

6.2.1 Dynamics of Inverter-Interfaced Microgrids

The secondary level is responsible for voltage and frequency restoration due to the steady-state error originating from the primary controller. To return the set of trajectories to the desired set points, it is required to design a compensator that prescribes command signals from the secondary control scheme to the primary level. The droop control method is commonly derived to attain the control purposes in microgrids expressed by:

$$\omega_i = \omega_i^n - m_i^P P_i \tag{6.1}$$

$$\begin{cases} v_{odi} = v_i^n - n_i^Q Q_i^v \\ v_{oqi} = 0 \end{cases} \tag{6.2}$$

where v_{odi} and v_{oqi} are the direct and quadrature components of terminal voltage, ω_i^n and v_i^n are the reference values provided by the secondary control layer, P_i and Q_i are the active and reactive powers of the i-th DER unit, and m_i^P and n_i^Q are the droop coefficients, respectively.

The physical dynamics of microgrids, including a voltage resource, filters, and output connectors, can be formulated in the state-space form as follows:

$$\begin{cases} \dot{\delta}_i = \omega_i^n - m_i^P P_i - \omega_{com} \\ \dot{P}_i = \omega_{ci}(v_{odi} i_{odi} + v_{oqi} i_{oqi} - P_i) \\ \dot{Q}_i = \omega_{ci}(v_{odi} i_{oqi} - v_{oqi} i_{odi} - Q_i) \\ \dot{i}_{ldi} = \dfrac{-R_{f_i}}{L_{f_i}} i_{ldi} + \omega_i i_{lqi} + \dfrac{v_i^n - n_i^Q Q_i - v_{odi}}{L_{f_i}} \\ \dot{i}_{lqi} = \dfrac{-R_{f_i}}{L_{f_i}} i_{lqi} - \omega_i i_{ldi} - \dfrac{v_{oqi}}{L_{f_i}} \\ \dot{v}_{odi} = \omega_i v_{oqi} + \dfrac{i_{ldi} - i_{odi}}{C_{f_i}} \\ \dot{v}_{oqi} = -\omega_i v_{odi} + \dfrac{i_{lqi} - i_{oqi}}{C_{f_i}} \\ \dot{i}_{odi} = \dfrac{-R_{f_i}}{L_{ci}} i_{odi} + \omega_{com} i_{oqi} + \dfrac{v_{odi} - v_{bdi}}{L_{ci}} \\ \dot{i}_{oqi} = \dfrac{-R_{f_i}}{L_{ci}} i_{oqi} - \omega_{com} i_{odi} + \dfrac{v_{oqi} - v_{bqi}}{L_{ci}} \end{cases} \tag{6.3}$$

where δ_i is the angle of the i-th DER unit reference, ω_{ci} is the cut-off frequency of the filter, i_{odi} and i_{oqi} are the direct and quadrature components of the converter current, i_{ldi} and i_{lqi} are the direct and quadrature components of the output current of the LCL filter, ω_{com} is the common rotating frequency, R_{f_i}, L_{f_i}, C_{f_i}, and L_{ci} are the LCL filter elements, and v_{bdi} and v_{bqi} are the terminal voltage of the output connector.

The equations in Eq. (6.3) can be written in a compact form as follows:

$$\begin{cases} \dot{x}_i = f_i(x_i) + W(x_i)H_i + r_{i1}(x_i)u_{i1} + r_{i2}(x_i)u_{i2} \\ y_{i1} = g_{i1}(x_i) \\ y_{i2} = g_{i2}(x_i) + h_i u_{i2} \end{cases} \tag{6.4}$$

where $H_i = [\omega_{com}\ v_{bdi}\ v_{bqi}]^T$ is the disturbance, the state vector is $x_i = [\delta_i\ P_i\ Q_i\ i_{Ldi}\ i_{Lqi}\ v_{odi}\ v_{oqi}\ i_{odi}\ i_{oqi}]^T$, $u_i = \left[u_{i1}\ u_{i2}\right]^T$ is the input vector, and $y_i = \left[y_{i1}\ y_{i2}\right]^T$ is the output vector, respectively. The full details of the functions $f_i(x_i)$, $W(x_i)$, $r_{i1}(x_i)$, $r_{i2}(x_i)$, $g_{i1}(x_i)$, $g_{i2}(x_i)$, and h_i can be acquired from Eq. (6.3).

Let us define $F_i(x_i) = f_i(x_i) + W_i(x_i)H_i$. Then, using input-output feedback linearization, the dynamics of voltage for each DER unit can be rewritten as follows:

$$\begin{cases} \dot{y}_{i1} = \dot{v}_{odi} \\ \ddot{y}_{i2} = \ddot{v}_{odi} = L^2_{F_i g_{i1}} + L_{r_{i1}} L_{F_{i1} g_{i1} u_{i1}} \end{cases} \tag{6.5}$$

where $L_{F_{i1} g_{i1}} = \frac{\partial g_i}{\partial x_i} F_i(x_i)$ and $L^2_{F_i g_{i1}} = \frac{\partial L_{F_{i1} g_{i1}}}{\partial x_i} F_i(x_i)$ indicate Lie Derivatives of g_{i1} with respect to F_i [20]. The dynamics in Eq. (6.5) can be written in the following formation.

$$\dot{y}_i = Ay_i + Bv_i^v(t) + D^v d_i^v \tag{6.6}$$

where $v_i^v(t) = L^2_{F_i g_{i1}} + L_{r_{i1}} L_{F_{i1} g_{i1} u_{i1}}$ is the auxiliary input, $y_i = \left[v_{odi}\ \dot{v}_{odi}\right]^T = \left[y_{i1}\ y_{i2}\right]^T$, $A = \begin{bmatrix} 0 & 1 \\ 0 & 0 \end{bmatrix}$, $B = \begin{bmatrix} 0 & 1 \end{bmatrix}^T$, $D^v = \begin{bmatrix} 1 & 0 \end{bmatrix}^T$, and d_i^v represents the external disturbance. Given the dynamic in Eq. (6.6), the goal of the secondary controller is to generate a control command in order for voltage regulation while the microgrid is under the constraints in the control channels.

It is worth to be mentioned that the feedback linearization technique is implemented to pave the way for the secondary controller design procedure with no need to linearize the dynamic in Eq. (6.4). The feedback linearization is just used for modeling purposes and not for devising the controller. As discussed in [21, 22], and the references therein, the accuracy of the feedback linearization method could be enhanced by overcoming the impact of noise in measurements and parameter incorrectness.

6.2.2 Fault Model and Saturation

Actuator fault signals in control-loop configurations can be modeled in the form of additive and multiplicative terms. In this section, the focus is on the multiplicative fault, which is assumed to be a time-varying but "*bounded*" signal. This assumption is common in the literature and practical applications (see [23, 24]). Taking into consideration the saturation in the actuators, the control input for the voltage dynamics in Eq. (6.6) is written as follows:

$$v_i^v = \theta_i^v(t) sat\left(v_i^v(t)\right) + \varphi_i^v(t) v_i^{vs}(t) \tag{6.7}$$

where

$$sat\left(v_i^v(t)\right) = \begin{cases} v_M^v & v_i^v \geq v_M^v \\ v_i^v & -v_M^v < v_i^v < v_M^v \\ -v_M^v & v_i^v \leq -v_M^v \end{cases} \tag{6.8}$$

with $\theta_i^v(t)$ as the multiplicative fault, $0 \leq \theta^v_{-i}(t) \leq \theta_i^v(t) \leq \overline{\theta}_i^v(t) \leq 1$, $\varphi_i^v \in \{0, 1\}$ is to indicate the happening of the stuck fault, $v_i^{vs}(t)$ is a constant value when there exists the stuck fault in the actuators, and v_M^v is the maximum allowable gain passing through the actuator. It is clear that whenever $\theta_i^v(t) = 1$ and $\varphi_i^v = 0$, there are no faults and when $\varphi_i^v = 1$, the i-the DER unit is under stuck faults.

6.3 Finite-Time Fault-Tolerant Voltage Control

The goal of this part is to design a cooperative voltage controller for the secondary layer in the presence of the fault and saturation, as presented in Eqs. (6.7 and 6.8). Rigorous stability analysis along with analytic forms of stationary error and the convergence time is also provided. Unlike the low-and-high gain controllers commonly used for additive disturbance, the proposed control mechanism can mitigate the fault impacts as well.

The consensus law at the secondary control level for each DER unit in islanded microgrids is stated as follows:

$$e_i^v(t) = \sum_{j=1}^{N} a_{ij}(y_i(t) - y_j(t)) + a_{i0}(y_i(t) - y^{ref}) \tag{6.9}$$

where $a_{ij}(y_i(t) - y_j(t))$ stands for the exchanged data among i-th and j-th DER unit. Adopting Kronecker product properties, Eq. (6.9) can be compactly declared as follows:

$$e^v(t) = (\beta \otimes I_N)y(t) \tag{6.10}$$

where $e^v = col\left\{e_1^v, e_2^v, \ldots, e_N^v\right\}$ and $\beta = \overline{\mathcal{G}} + \mathcal{L}$.

Consider the following parametric Riccati equation for any $\gamma > 0$.

$$A^T P(\gamma) + P(\gamma)A - P(\gamma)BB^T P(\gamma) + Q(\gamma) = 0 \tag{6.11}$$

where $P(\gamma)$ is an $N \times N$ positive definite matrix and the unique solution of Eq. (6.11) and $Q(\gamma) > 0$ is an $N \times N$ feedback matrix, respectively. The distributed finite-time fault-tolerant voltage control for the i-the DER unit is proposed as follows.

$$v_i^v(t) = -\alpha_{i1}^v(t) q_i^v(t) - b_{ij}^v(\gamma, t) \tag{6.12}$$

and,

$$b_{ij}^v(\gamma, t) = \frac{\alpha_{i2}^{v^2}(t)\, q_{ij}^v(\gamma, t)}{\alpha_{i2}^v(t)\left\|q_{ij}^v(\gamma, t)\right\| + \varepsilon^v(\gamma)\Upsilon(t)} \tag{6.13}$$

where $q_i^v = B^T P(\gamma) e_i^v(t)$, $\Upsilon(t)$ is a bounded time-varying function, $\varepsilon^v(\gamma) = \frac{\lambda_{min} Q(\gamma)}{\lambda_{max}(P(\gamma))}$, and $q_{ij}^v(\gamma, t)$ and $b_{ij}^v(\gamma, t)$ are the j-th element of $q_i^v(\gamma, t)$ and $b_i^v(\gamma, t)$, respectively. The adaptive law for α_{i1}^v and α_{i2}^v are expressed as follows:

$$\dot{\alpha}_{i1}^v = v_M^v\left(-\sigma_{i1}^v \alpha_{i1}^v(t) + \|q_i^v(t)\|^2\right) \tag{6.14}$$

$$\dot{\alpha}_{i2}^v = v_M^v\left(-\sigma_{i2}^v \alpha_{i2}^v(t) + \|q_i^v(t)\|\right) \tag{6.15}$$

where σ_{i1}^v and σ_{i2}^v are positive constants.

Note that the first part presented in Eq. (6.12) is allocated as the low-gain term of the controller, with v_M^v in the adaptive law Eq. (6.14) adjusted according to the saturation level. On the other hand, the second term of Eq. (6.12) is a high-gain part to mitigate the fault impacts.

Assumption 6.1 The considered disturbance and fault is "*bounded*" satisfying $\theta_i^v \leq \frac{\lambda_{max}(B_2 B_2^T)\bar{\delta}_i}{\lambda_{min}(B_2 \theta_i B_2^T)}\left(v_M^v - s_t\right)$, where $\bar{\delta}_i$ is the upper bound of $\delta_i^v(t) = \varphi_i^v(t)v_i^{vs} + d_i^v(t)$, $0 < s_t < v_M^v$, and B_2 is the obtained matrix of full-rank decomposition $B = B_1\ B_2$.

Lemma 6.1 [14] Consider a nonlinear dynamic system $\dot{y} = f(y)$ with an equilibrium point at the origin and a Lyapunov candidate $V(y)$. The system is globally stable in a finite-time if there exist positive scalars $z_a > 0$, $c_z > 0$, and $0 < \eta < \infty$ such that

$$\dot{V}(y) \leq -z_a V(y) + \eta. \tag{6.16}$$

The state trajectory of the system is restricted to $\Lambda = \{y \,|\, V(y) \leq \frac{\eta}{z_a(1-c_a)}\}$ and the upper bound for the settling time is $t_s \leq \frac{1}{z_a c_a} \ln \frac{(1-c_a)V(y(0))}{\eta}$.

The following theorem illustrates that the proposed control scheme can ensure the global stability of the microgrid and voltage regulation of DER units while the control input channels are subjected to the fault and saturation in Eqs. (6.7 and 6.8).

Theorem 6.1 Consider an undirected and connected graph topology for an N-DER unit microgrid and control input constraints presented in Eqs. (6.7 and 6.8). By applying the proposed controller in Eqs. (6.12 and 6.13), the voltage regulation at the steady-state can be achieved in a finite time. Also, the stationary error and the convergence time are restricted to the following sets, respectively.

$$\left\{\bar{y} \,\middle|\, \|\bar{y}\| \leq \sqrt{\frac{\varrho^v}{\lambda_1 \lambda_{min}(P(\gamma))(1-\xi)}}\right\} \tag{6.17}$$

and,

$$t_s \leq \frac{1}{\varepsilon^v(\gamma)\xi} \ln \frac{(1-\xi)V(\bar{y}(0))}{\varrho^v} \tag{6.18}$$

where $\bar{\mathbf{y}}$ is the tracking error, $\mathbf{0 < \xi < 1}$, $\varrho^v = \sum_{i=1}^{N}\sum_{j=1}^{2}\kappa_i \sigma_{ij}^v \alpha_{ij}^v + 2\sum_{i=1}^{N}\kappa_i Y(t)$ with a positive constant κ_i, and λ_1 is the minimum eigenvalue of $\boldsymbol{\beta}$.

Proof: Adopting Eqs. (6.7 and 6.12), the dynamic of error considering control input constraints are described as follows.

$$\dot{\bar{y}}(t) = (I_N \otimes A)\bar{y}(t) + (I_N \otimes B)\theta(t)sat(v^v(t)) + (I_N \otimes B)\delta^v(t) \tag{6.19}$$

where $sat(v^v(t)) = \left[sat\left(v_1^v(t), \ldots, v_N^v(t)\right)\right]^T$ and $\delta^v(t) = \left[\delta_1^v(t), \ldots, \delta_N^v(t)\right]^T$.

Given the Riccati equation in Eq. (6.11), the Lyapunov candidate is chosen as follows:

$$V = \bar{y}^T(\beta \otimes P(\gamma))\bar{y}. \tag{6.20}$$

Taking time derivative of Eq. (6.20) along with the dynamic in Eq. (6.6), one can yield

$$\dot{V}(\bar{y}) = 2\bar{y}^T(\beta \otimes P(\gamma))\dot{\bar{y}}. \tag{6.21}$$

Invoking Eq. (6.10) into the above equation, it is obtained that

$$\dot{V}(\bar{y}) = 2\bar{y}^T(\beta \otimes P(\gamma)A)\dot{y} + 2\bar{y}^T(\beta \otimes P(\gamma)B)\,\theta\, sat(v^v) + 2\bar{y}^T(\beta \otimes P(\gamma)B)\delta = V_1 + V_2 \tag{6.22}$$

where V_1 and V_2 are defined as follows.

$$V_1 = 2\bar{y}^T(\beta \otimes P(\gamma)A)\bar{y} - 2\sum_{i=1}^{N} \alpha_{i1}^v q_i^{v^T}(\gamma,t)\theta_i q_i^v(\gamma,t) - 2\sum_{i=1}^{N} \frac{\bar{\delta}}{\kappa}\frac{q_i^{v^T}(\gamma,t)\theta_i q_i^v(\gamma,t)}{q_i^v(\gamma,t)}$$
$$+\, 2\bar{y}^T(\beta \otimes P(\gamma)B)\delta^v, \tag{6.23}$$

$$V_2 = 2\bar{y}^T(\beta \otimes P(\gamma)B)sat(v^v) + 2\sum_{i=1}^{N} \alpha_{i1}^v q_i^{v^T}(\gamma,t)\theta_i q_i^v(\gamma,t) + 2\sum_{i=1}^{N} \frac{\bar{\delta}}{\kappa}\frac{q_i^{v^T}(\gamma,t)\theta_i q_i^v(\gamma,t)}{q_i^v(\gamma,t)}. \tag{6.24}$$

Based on Lemma 6.1, one can obtain that

$$V_1 \le 2\bar{y}^T(\beta \otimes P(\gamma)A)\bar{y} - 2\sum_{i=1}^{N} \kappa_i \alpha_{i1}^v q_i^{v^T}(\gamma,t)\theta_i\, q_i^v(\gamma,t) - 2\sum_{i=1}^{N} \bar{\delta}_i \left\| q_i^{v(\gamma,t)} \right\| + 2\sum_{i=1}^{N} q_i^v(\gamma,t)\delta_i \tag{6.25}$$

where $2\kappa_i\alpha_{i1}^v \ge \left(\frac{1}{\lambda_1}\right), i \in \vartheta$, and $\|\delta_i\| \le \bar{\delta}_i$. Considering Eqs. (6.14 and 6.15), the above inequality can be expressed as follows.

$$V_1 \le \bar{y}^T\left(\beta \otimes P(\gamma)A + \beta \otimes A^T P(\gamma) - \frac{1}{\lambda_1}\beta^2 \otimes P(\gamma)BB^T P(\gamma)\right)\bar{y}. \tag{6.26}$$

As it is assumed that the graph topology is connected and undirected, there can be found an orthogonal matrix ψ such that $\psi^T\beta\psi = \Xi$, where $\Xi = diag\{\lambda_1, \ldots, \lambda_N\}$ with $0 < \lambda_1 \le \lambda_2 \le \ldots \le \lambda_N$. Let us define $\tilde{y} = (\psi^T \otimes I_N)\bar{y}$, then, Eq. (6.26) can be written as follows:

$$V_1 \le \tilde{y}^T\left(\Xi \otimes P(\gamma)A + \Xi \otimes A^T P(\gamma) - \frac{1}{\lambda_1}\Xi^2 \otimes P(\gamma)BB^T P(\gamma)\right)$$
$$\le \tilde{y}^T(\Xi \otimes Q(\gamma))\tilde{y} \tag{6.27}$$

and,

$$V_1 \le -\frac{\lambda_{min}(Q(\gamma))}{\lambda_{max}(P(\gamma))}\tilde{y}^T(\beta \otimes P(\gamma))\tilde{y} = -\varepsilon^v(\gamma)V_1. \tag{6.28}$$

Taking similar steps for V_2, it yields that

$$V_2 \le \sum_{i=1}^{N}\sum_{j=1}^{2} \chi_{ij}(\gamma) \le \sum_{i=1}^{N}\sum_{j=1}^{2} 2\bar{\theta}_{ij}^v \varepsilon(\gamma)\Upsilon(0) \tag{6.29}$$

where

$$\chi_{ij}(\gamma) = -2q_{ij}(\gamma,t)\theta_{ij}\left(sat\left(-\alpha_{ij}^v q_{ij}^v(\gamma,t) - b_{ij}(\gamma)\right) - \alpha_{i1}^v q_{ij}^v(\gamma,t) - r_{ij}^v(\gamma,t)\right) \tag{6.30}$$

and $r_i^v(\gamma,t) = \left(\frac{\bar{\delta}_i}{\kappa_i}\right)\left(\frac{q_i^v(\gamma,t)}{\|q_i^v(\gamma,t)\|}\right)$, $q_{ij}^v(\gamma,t)$, and $r_{ij}^v(\gamma,t)$ stand for the j-th elements of $q_i^v(\gamma,t)$ and $r_i^v(\gamma,t)$, respectively.

Combining Eqs. (6.27 and 6.29),

$$\dot{V}(\bar{y}) \le \varepsilon(\gamma)V(\bar{y}) + \sum_{i=1}^{N}\sum_{j=1}^{2} 2\bar{\theta}_{ij}^v \varepsilon(\gamma)\Upsilon(0). \tag{6.31}$$

According to Lemma 6.1, the error system in Eq. (6.19) is globally stable, and the trajectory $\bar{y}$ is restricted to the set in Eq. (6.17). Also, the settling time limit could be obtained as in Eq. (6.18). This completes the proof.

Note that by some modifications, the proposed secondary controller in Eq. (6.12) could also be applied for frequency synchronization and active power management. To do so, recalling the droop characteristic in Eq. (6.1), the control command enforced by the secondary layer for the frequency is implemented as follows.

$$\omega_i^n = \int \left(\tilde{u}_i^{\omega} + \tilde{u}_i^{P}\right) d\tau \tag{6.32}$$

where $\tilde{u}_i^{\omega}$ and $\tilde{u}_i^{P}$ are the auxiliary frequency and active power control inputs under input control constraints Eqs. (6.7 and 6.8), respectively. By taking a similar procedure and some modifications, global stability and stationary error for the frequency and active power can also be achieved.

6.4 Case Studies

In this section, the performance and effectiveness of the proposed fault-tolerant secondary control scheme are evaluated for an islanded inverter-based microgrid in MATLAB/Simulink. The test microgrid is shown in Figure 6.1 and its parameters besides primary control gains are presented in Table 6.1. The DER units communicate through an "*undirected and connected*" communication graph depicted in Figure 6.2. It is important to note that in this topology, only DER unit 1 can have access to the reference values. The simulations are carried out in several scenarios in which the fault and saturation are reflected on. In addition, the performance of the proposed control scheme is compared with some related studies, illustrating the insufficiency of existing approaches in the literature. In order to implement the proposed secondary control mechanism, it is only required to solve the Ricatti equation in Eq. (6.11) once before the simulation run.

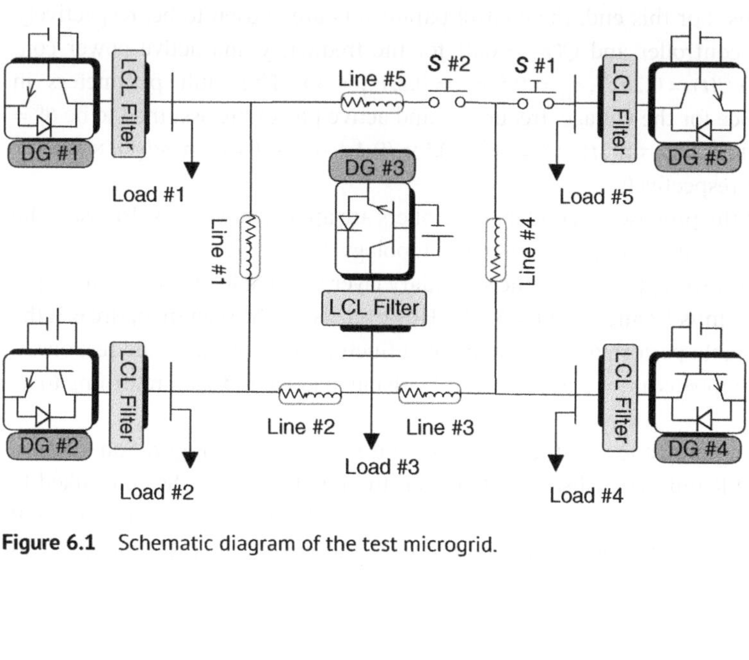

Figure 6.1 Schematic diagram of the test microgrid.

Table 6.1 Test microgrid parameters.

Descriptions	DG#1&#2&#3	DG#4&#5
m_P	12.5×10^{-5}	9.4×10^{-5}
n_Q	1.5×10^{-3}	1.3×10^{-3}
K_{PV}	0.05	0.1
K_{IV}	390	420
K_{PC}	10.5	15
K_{IC}	16,000	20,000
LC Filters	$R_f = 0.3\ \Omega, L_f = 1.5\ \text{mH}, C_f = 47\ \mu\text{F}$	
Output Connectors	$R_c = 0.05\ \Omega, L_c = 0.35\ \text{mH}$	
Converters	$f_{sw} = 8\ \text{kHz}$	$T_s = 0.00002s$
Lines	Lines # 2 & # 3 & # 4	Lines # 1 & # 5
	$0.12 + 0.1\ \Omega$, $0.175 + 0.58\ \Omega$	

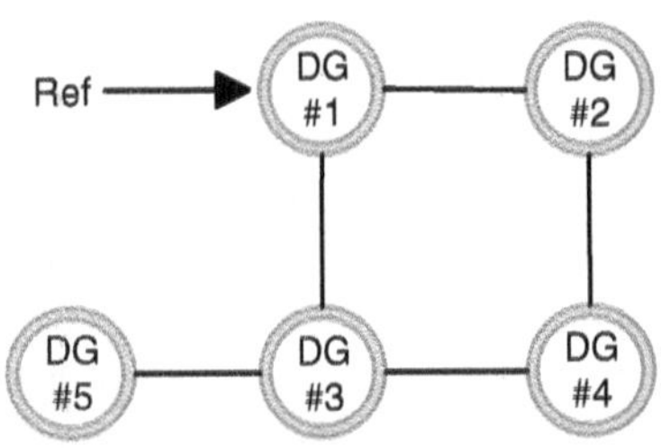

Figure 6.2 Communication graph topology among DER units.

6.4.1 Simulation Results

The simulation evaluates the performance of the proposed control scheme in Eq. (6.12) under faults and input saturations. For this end, the control parameters are chosen to be, respectively, $Q(\gamma) = I_2$ for the voltage controller and $Q(\gamma) = 0.81$ for the frequency and active power controllers, $\Upsilon^{v}(t) = 0.05e^{-2t}, \Upsilon^{\omega}(t) = 0.2e^{-3t}, \sigma^{v}_{i1} = 0.5, \sigma^{v}_{i2} = 0.2, \upsilon^{v}_{M} = 30$. The fault parameters in Eq. (6.7) and the disturbance for the voltage, frequency, and active power are assumed to be $\theta^{v}_{i} = 0.6 + (0.5 + (0.7 - 0.5) \times rand(1)) \times \sin(3t)$, $\theta^{\omega}_{i} = \theta^{P}_{i} = 0.5 + (0.4 + (0.9 - 0.4) \times rand(1)) \times \sin(4t)$, and $d^{v}_{i} = d^{\omega}_{i} = d^{P}_{i} = \cos(t)$, respectively.

It is worth notifying that the proposed secondary controllers are applied at $t = 0.7s$. To assess the performance of the system to any changes in microgrid topology, Switch #1 is closed at $t = 1.5s$, and is opened at $t = 5s$. It is also supposed that the secondary layer is not able to command proper control signals for the DER units #2 and #3 at $t = 5 - 7s$. However, as can be seen in Figure 6.3, the microgrid shows an acceptable performance, though some fluctuations appear in the responses. The load change scenario happens at $t = 2.5s$ to $t = 4.5s$ by rising load #3 and then returning to its initial value.

In order to investigate the robustness of the microgrid and the control configuration under a plug-and-play scenario, DER unit #5 is disconnected from the topology at $t = 8s$ and linked to the microgrid at $t = 10s$. From Figure 6.3, it can be seen that some fluctuations in the active and reactive power trajectories during the plug-and-play; however, the microgrid is still stable. It can

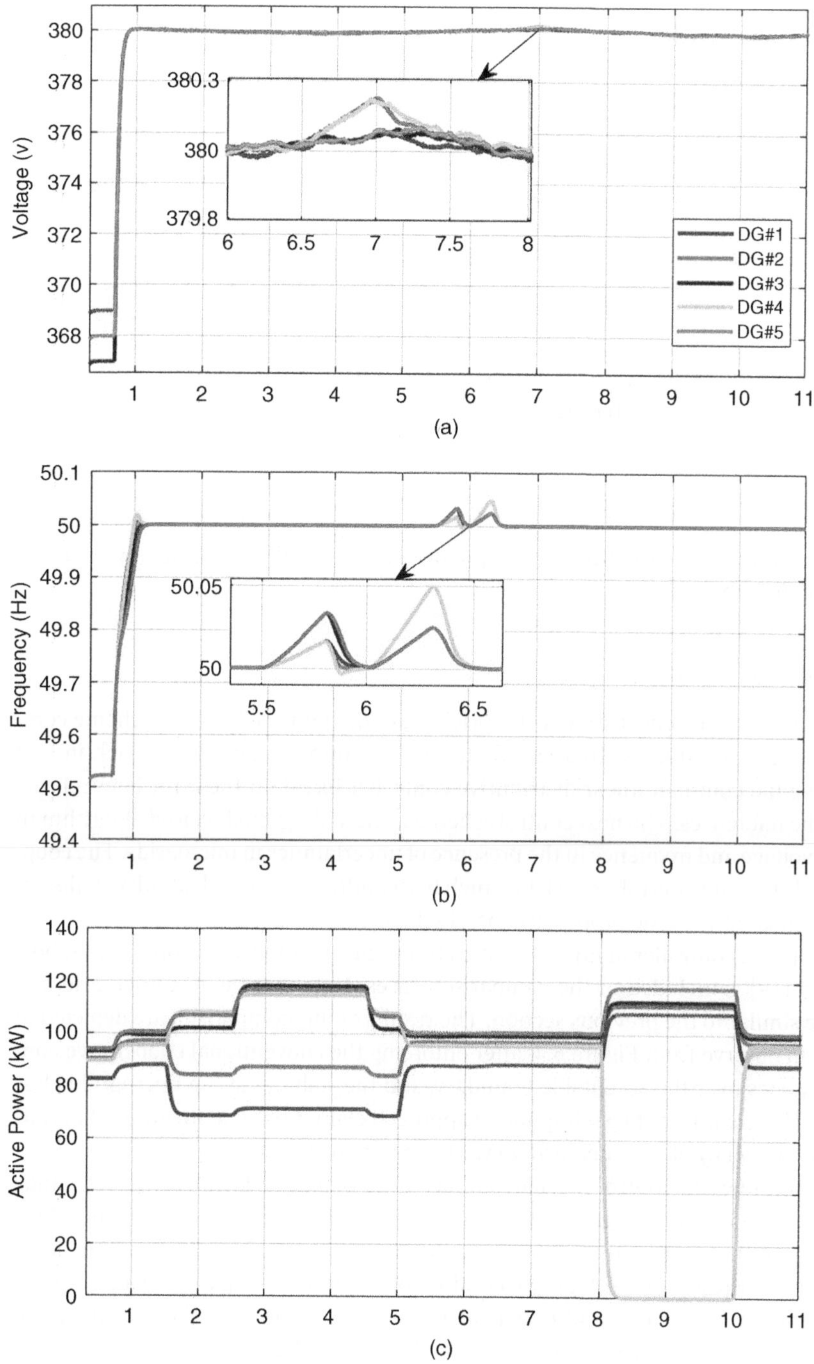

Figure 6.3 Performance of the proposed secondary fault-tolerant control mechanism for DG units: (a) Voltage, (b) Frequency, (c) Active power, and (d) Reactive powers.

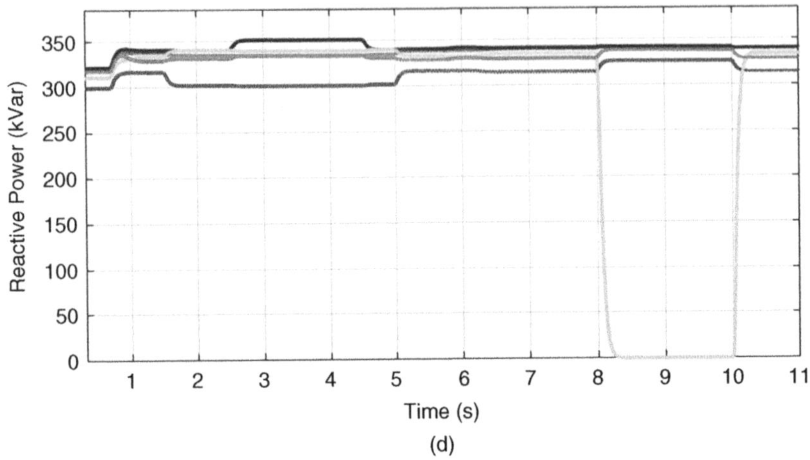

Figure 6.3 *(Continued)*

be concluded that the proposed control scheme can guarantee the stability of the microgrid under input constraints and disturbances.

6.4.2 Comparative Case Study

This part is allocated to assay the effectiveness and privilege of the proposed control scheme compared to the previous research studies. To do so, first, the conventional control approach in [25] is selected, employing the common linear distributive controller based on the error between the current values and the references. The next control scheme is the sliding-mode-based algorithm in [26], which restores voltage and frequency in the presence of uncertainties in microgrids. The cooperative secondary fault-tolerant controller [23], in which both multiplicative and additive faults are considered, is also compared with the proposed control scheme.

To compare the proposed controller in Eq. (6.8) with a finite-time method, the cooperative robust finite-time control in [19] is included in the comparison cases. Note that the test microgrid and input constraints are similar to the previous section. The results of this comparison are depicted in Figure 6.4. As one can observe from Figure 6.4, after enforcing the conventional cooperative controller, the frequency goes over the acceptable boundary, and the stability of the microgrid is lost under the fault. It is also clear that the sliding-mode approach cannot provide the desired performance and takes the frequency of the microgrid to values less than 49.5 Hz.

As expected, the fault-tolerant control mechanism has more robust performance under faults; however, due to the input saturation, it cannot show a fast and proper transient performance. Furthermore, the voltage of the microgrid controlled by the finite-time method does not converge to the reference value as a result of the fault. Generally, such control mechanisms cannot properly react under the input constraints and either take a longer time to be settled or make the microgrid unstable. The control efforts (control commands) of the selected secondary controllers are also presented in Figure 6.5, demonstrating that the proposed method requires less energy to perform.

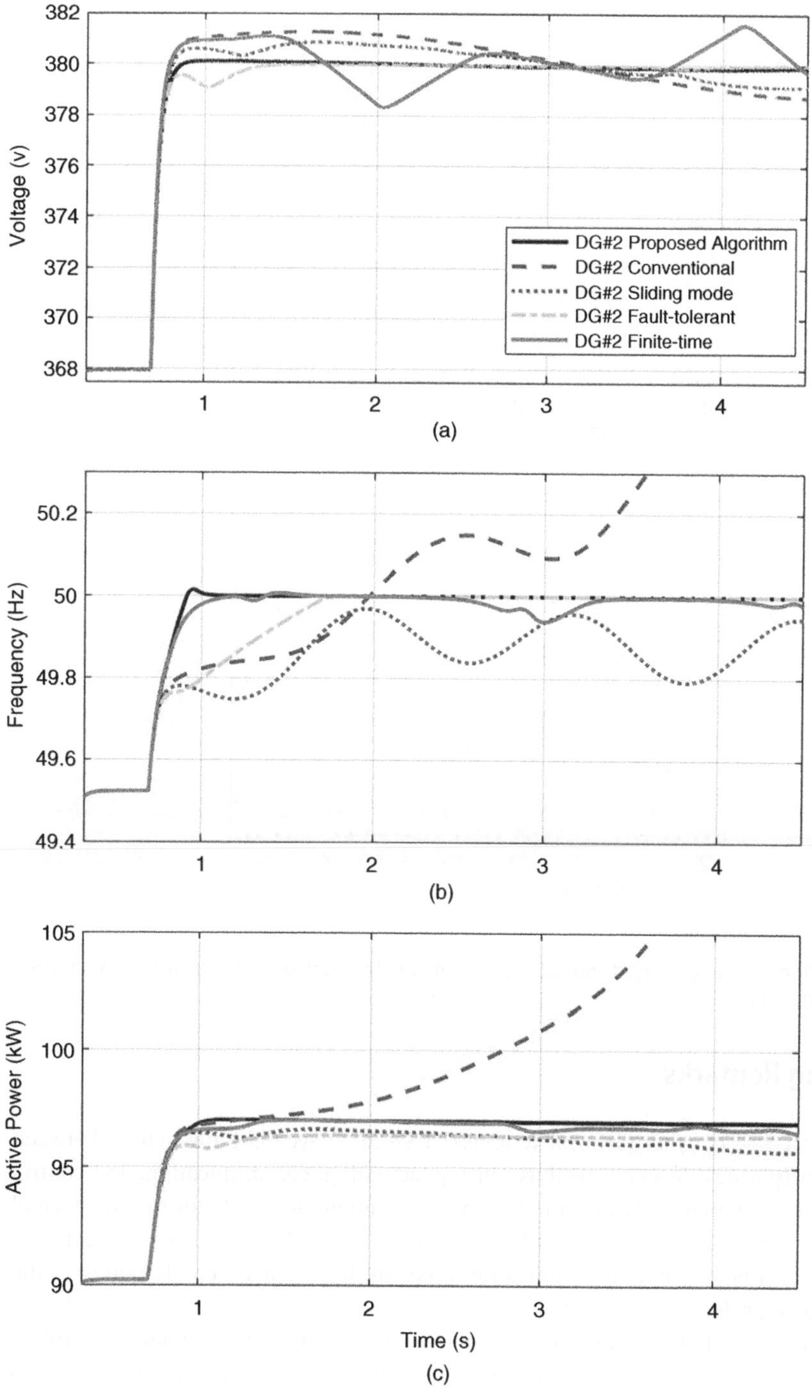

Figure 6.4 Comparison of the proposed control mechanism with methods in *[19, 23, 25, 26]*. (a) Voltage, (b) Frequency, and (c) Active power of DER unit #2.

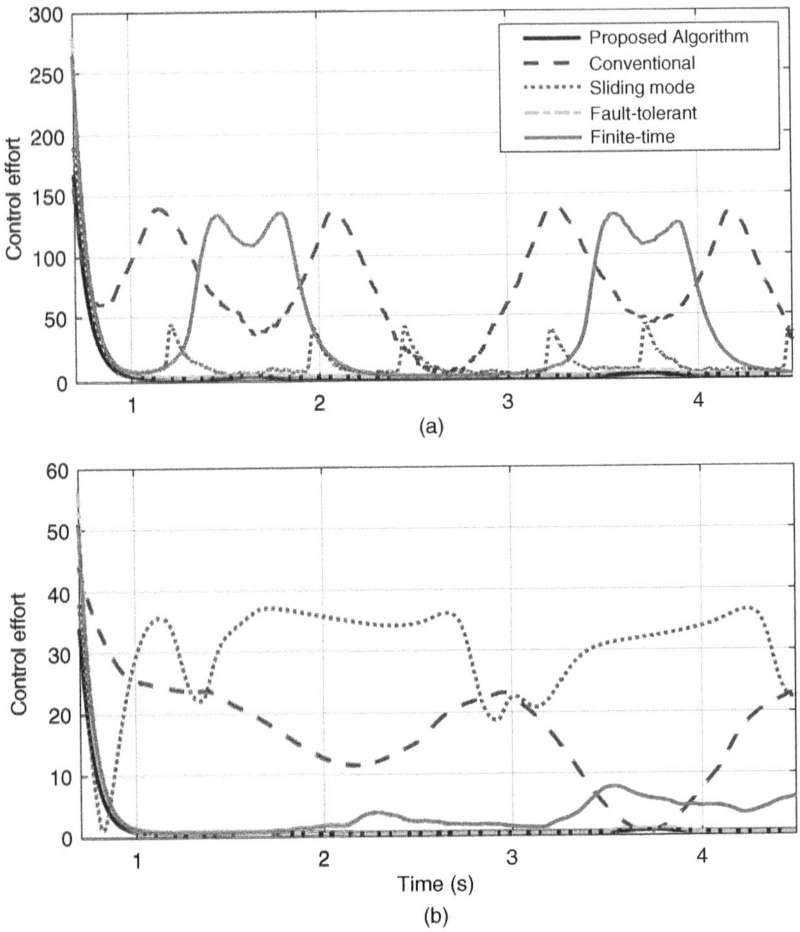

Figure 6.5 Control efforts of the proposed control mechanism and the control methods in *[19, 23, 25, 26]*. (a) Voltage and (b) Frequency.

6.5 Concluding Remarks

Microgrids represent a class of cyber-physical systems (CPS)s that have tight interactions between physical components (physical layer) as well as sensing, actuating, communication, and control elements (cyber layer). Any manipulations in the cyber layer might adversely impact the normal operation and stability of microgrids and lead to physical damage to loads and power electronic devices. The sensors, actuators, and communication links are the most venerable points in the control systems of microgrids.

The main focus of this chapter was on unreliable actuators in the secondary control of inverter-interfaced microgrids. It was assumed that the control input channels are subject to faults and control input saturations. A finite-time cooperative control scheme was presented for the secondary control layer of islanded inverter-interfaced microgrids by considering faults and input saturations. The proposed control approach relies on auxiliary dynamics to mitigate the effect of input constraints. It was shown that voltage regulation and frequency synchronization can be achieved in a finite time, and the stationary error along with the settling time can be numerically obtained. Simulation results and a comparative case study were presented to show the effectiveness of the proposed secondary controller.

In addition to faults, cyberattacks might influence the integrity, availability, and confidentiality of sensing and actuating information, as well as communication infrastructure in cyber-physical microgrids. Hence, it is essential to deploy a cybersecurity mechanism in microgrids' control systems, combined with different detection and mitigation schemes and operating at different control levels, to decrease the vulnerabilities of cyber-physical microgrids and enhance their resilience with respect to sophisticated and unknown cyberattacks, referred to as stealthy attacks. Although the main focus of this chapter was on actuators' faults, we refer interested readers to relevant references (e.g., [27–29], and references therein) for general information on cybersecurity issues, resilient control, and attack detection/mitigation methods in islanded inverter-interfaced AC microgrids.

References

1 Olivares, D.E., Mehrizi-Sani, A., Etemadi, A.H. et al. (2014). Trends in microgrid control. *IEEE Trasnactions on Smart Grid* 5 (4): 1905–1919.

2 Hatziargyriou, N., Asano, H., Iravani, R., and Marnay, C. (2007). Microgrids. *IEEE Power Energy Magazine* 5: 78–94.

3 Sadabadi, M.S., Shafiee, Q., and Karimi, A. (2017). Plug-and-play voltage stabilization in inverter-interfaced microgrids via a robust control strategy. *IEEE Transactions on Control Systems Technology* 25 (3): 781–791.

4 Bidram, A. and Davoudi, A. (2012). Hierarchical structure of microgrids control system. *IEEE Transactions on Smart Grid* 3 (4): 1963–1976.

5 Guerrero, J.M. and J. C. V. J. M. L. G. d. V. a. M. C. (2014). Hierarchical control of droop-controlled AC and DC microgrids—a general approach toward standardization. *IEEE Transactions on Industrial Electronics* 58 (1): 158–172.

6 Shafiee, Q., Dragičević, T., Vasquez, J.C., and Guerrero, J.M. (2014). Hierarchical control for multiple DC-microgrids clusters. *IEEE Transactions on Energy Conversion* 29 (4): 922–933.

7 Espina, E., Llanos, J., Burgos-Mellado, C. et al. (2020). *Distributed control strategies for microgrids: an overview. IEEE Access* 8: 193412–193448.

8 Dragičević, T., Lu, X., Vasquez, J.C., and Guerrero, J.M. (2016). DC microgrids—part I: a review of control strategies and stabilization techniques. *IEEE Transactions on Power Electronics* 31 (7): 4876–4891.

9 Isermann, R. (2006). *An Introduction from Fault Detection to Fault Tolerance*. s.l. Springer.

10 Yang, S., Bryant, A., Mawby, P. et al. (2011). An industry-based survey of reliability in power electronic converters. *IEEE Transactions on Industry Applications* 47 (3): 1441–1451.

11 Mirafzal, B. (2014). Survey of fault-tolerance techniques for three-phase voltage source inverters. *IEEE Transactions on Industrial Electronics* 61 (10): 5192–5202.

12 Avizienis, A., Laprie, J.-C., Randell, B., and Landwehr, C. (2004). Basic concepts and taxonomy of dependable and secure computing. *IEEE Transactions on Dependable and Secure Computing* 1: 11–33.

13 Wang, B., Chen, W., and Zhang, B. (2019). Semi-global robust tracking consensus for multi-agent uncertain systems with input saturation via metamorphic low-gain feedback. *Automatica* 103: 363–373.

14 Yin, Y., Wang, F., Liu, Z., and Chen, Z. (2021). Finite-time leader-following consensus of multiagent systems with actuator faults and input saturation. *IEEE Transactions on Systems, Man, and Cybernetics: Systems* 52 (5): 3314–3325.

15 Meng, D. and Jia, Y. (2011). Finite-time consensus for multi-agent systems via terminal feedback iterative learning. *IET Control Theory & Applications* 5 (18): 2098–2110.

16 Jamali, M., Baghaee, H.R., Sadabadi, M.S. et al. (2022). Distributed finite-time fault-tolerant control of isolated AC microgrids considering input constraints. *IEEE Transactions on Smart Grid* 13 (6): 4525–4537.

17 Deng, Z., Xu, Y., Sun, H., and Shen, X. (2018). Distributed, bounded and finite-time convergence secondary frequency control in an autonomous microgrid. *IEEE Transactions on Smart Grid* 10 (3): 2776–2788.

18 Bidram A., Davoudi A., and Lewis F. L. (2014). Finite-time frequency synchronization in microgrids. *IEEE Energy Conversion Congress and Exposition (ECCE)*, pp. 2648–2654.

19 Dehkordi, N.M., Sadati, N., and Hamzeh, M. (2016). Distributed robust finite-time secondary voltage and frequency control of islanded microgrids. *IEEE Transactions on Power Systems* 32 (5): 3648–3659.

20 Khalil, H. (2002). *Nonlinear Systems*. s.l. Prentice hall Upper.

21 Afshari, A., Karrari, M., Baghaee, H.R. et al. (2021). Robust cooperative control of isolated AC microgrids subject to unreliable communications: a low-gain feedback approach. *IEEE Systems Journal* 16 (1): 55–66.

22 Ge, P., Zhu, Y., Green, T.C., and Teng, F. (2020). Resilient secondary voltage control of islanded microgrids: an ESKBF-based distributed fast terminal sliding mode control approach. *IEEE Transactions on Power Systems* 36 (2): 1059–1070.

23 Afshari, A., Karrari, M., Baghaee, H.R. et al. (2019). Cooperative fault-tolerant control of microgrids under switching communication topology. *IEEE Transactions on Smart Grid* 11 (3): 1866–1879.

24 Li, X., Wen, C., Chen, C., and Xu, Q. (2021). Adaptive resilient secondary control for microgrids with communication faults. *IEEE Transactions on Cybernetics* 52 (8): 8493–8503.

25 Bidram, A., Davoudi, A., Lewis, F.L., and Qu, Z. (2013). Secondary control of microgrids based on distributed cooperative control of multi-agent systems. *IET Generation, Transmission & Distribution* 7 (8): 822–831.

26 Pilloni, A., Pisano, A., and Usai, E. (2017). Robust finite-time frequency and voltage restoration of inverter-based microgrids via sliding-mode cooperative control. *IEEE Transactions on Industrial Electronics* 65 (1): 907–917.

27 Bidram, A., Poudel, B., Damodaran, L. et al. (2020). Resilient and cybersecurity distributed control of inverter-based islanded microgrids. *IEEE Transactions on Industrial Informatics* 16 (6): 3881–3894.

28 Sadabadi, M.S., Sahoo, S., and Blaabjerg, F. (2021). A fully resilient cyber-secure synchronization strategy for AC microgrids. *IEEE Transactions on Power Electronics* 36 (12): 13372–13378.

29 Sahoo, S., Yang, Y., and Blaabjerg, F. (2021). Resilient synchronization strategy for AC microgrids under cyber attacks. *IEEE Transactions on Power Electronics* 36 (1): 73–77.

7

Interconnected Microgrid Systems: Architecture, Hierarchical Control, and Implementation

Tung Lam Nguyen[1], Yu Wang[2], Ha Thi Nguyen[3], and Tran The Hoang[4]

[1] *New York Power Authority, Albany, NY, USA*
[2] *Chongqing University, Chongqing, China*
[3] *University of Connecticut, Storrs, CT, USA*
[4] *University of Auckland, Auckland, New Zealand*

7.1 Introduction

Microgrids (MGs) are essential subsystems in future power systems, consisting of distributed generators (DGs), controllable and noncontrollable loads, energy storage systems, and advanced control and communication systems [1, 2]. However, a single MG may not provide reliable operation during extreme events like natural disasters or grid failures. To improve the overall reliability and resilience of the system, a group of MGs within a given area can create an interconnected MG system, as shown in Figure 7.1. The electrical structures of such systems can be classified based on voltage levels, current types, and connections. For example, low-voltage (LV) MGs can be linked through LV tielines or boosting transformers to form a medium-voltage (MV) network based on the distances and boundaries between MGs. Furthermore, interconnected MGs can be classified on the basis of their current type, which can be AC, DC, or hybrid AC/DC. Additionally, MGs can be interconnected in serial or parallel configurations.

Controlling and operating interconnected MG systems can be more challenging than single MG systems due to the coupling of more electrical components and communication infrastructures. To ensure stable and flexible operation, several studies aim to apply existing single MG control frameworks to interconnected MG systems [3–6]. For example, Wu et al. [5] propose a two-layer distributed hierarchical control scheme for AC interconnected MG systems in power distribution networks, while [3] extend the hierarchical distributed control framework for DC MGs to MG clusters. Zhao et al. [6] investigate the droop-based control and small signal stability of PV-based multiple MG clusters, and [4] propose a hierarchical consensus control framework to manage multiple MG clusters in the Energy Internet with multi-site experimental validation. However, the coordination with upper-level optimized dispatch is not studied in the above research. Additionally, due to ownership and privacy concerns, access to data and communication of each MG is often limited, making a distributed control structure with minimal data exchange with MG-level control highly preferred.

This chapter discusses the structure, hierarchically distributed control, and implementation of the interconnected MGs, including loads and different distributed energy resources (DERs) integration in MGs. To ensure proper operation, the control system is designed to achieve multiple

Smart Cyber-Physical Power Systems: Fundamental Concepts, Challenges, and Solutions, Volume 1, First Edition.
Edited by Ali Parizad, Hamid Reza Baghaee, and Saifur Rahman.

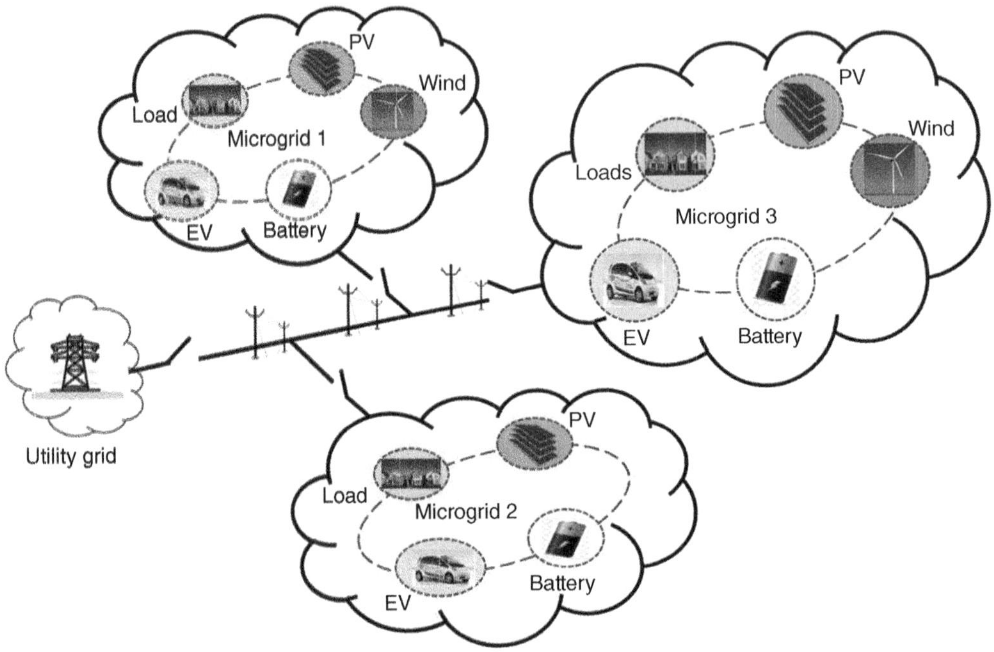

Figure 7.1 An interconnected microgrid system.

functionalities in different layers and time scales [7, 8]. Primary control functions bridge the MG-level control of DERs and the networked MG-level control of MGs. Secondary control maintains frequency/voltage at reference values and achieves arbitrary power-sharing, while tertiary control aims for optimal operation. As a test case in point, a multiagent system (MAS) for implementing distributed control under peer-to-peer communication network is employed to manage the components within an individual MG as well as coordination between MGs and their neighbors. In real-world applications, the MAS is a cluster of entities located at distinctive places. Messages are transferred between neighborhood agents through a communication network. An agent is a program running on a processor (e.g., PLC and microprocessor) for specific purposes. Challenges in implementing MAS in cyber-physical networked MG systems need further intensive investigation for practical deployment. A Cyber Hardware-in-the-loop setup is presented as a real-time cyber-physical platform to simulate the complex multi-domain system. The main parts of the platform are a network emulation, a real-time power system simulation, and the agent system. All components of the platform operate asynchronously, reflecting the practical implementation of the system. The container concept for the agent operation makes the platform scalable and flexible, and the implementation is manageable and reusable.

7.2 Architecture

An interconnected MGs system is described as a complex cyber-physical system that includes connected electrical and communication systems. The system can operate in either islanded or grid-connected modes. This study focuses on an interconnected island MG system that uses a static transfer switch (STS) to isolate it from the main grid. Each MG in the interconnected MG system is equipped with a group of dispatchable DGs and local loads connected to a point of

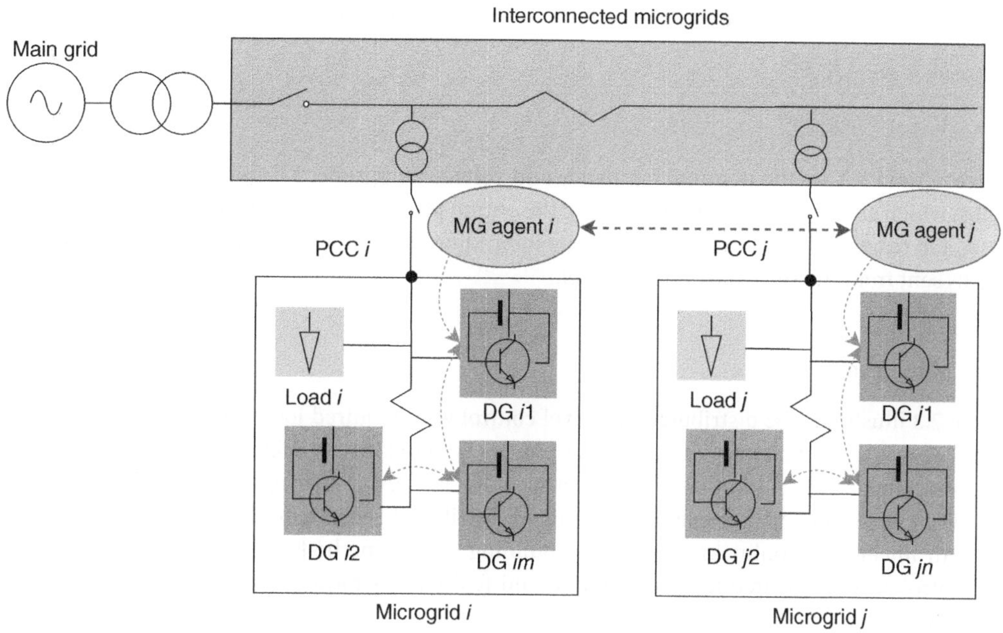

Figure 7.2 Architecture of the interconnected microgrid systems with network communication.

common coupling (PCC). LV/MV transformers connect each MG to a specific bus of the MV grid in the interconnected MG system. Renewable energy resources (RES) and constant power loads (CPLs) on the same bus are treated as a lumped load, and the RESs operate in the maximum power point tracking mode [4]. The DG units in this study are considered as power converters with DC voltage sources, with a focus on AC interconnected MGs.

The communication system in an interconnected MG system consists of an upper-level communication network between MGs and a lower-level communication network between DGs within each MG, as shown in Figure 7.2. In the interconnected MG system communication network, each MG agent measures the tie line information of each MG and exchanges it with neighboring agents to generate reference frequency and voltage signals. All DGs track these reference signals on the basis of the existing distributed secondary control for each MG.

7.3 Hierarchical Control of Interconnected MGs

7.3.1 DG Level

The primary objective of droop control is to replicate the behavior of a synchronous generator, which is typically achieved through the coordination of the turbine governor, the voltage regulator, and the inertia of the generator. When working with islanded AC MGs, the inverter-interfaced DGs are controlled as voltage source inverters. As such, in addition to the droop control, an inner control loop is necessary that comprises both a current control loop and a voltage control loop. Moreover, to achieve precise power sharing among each unit, a virtual impedance loop can also be employed. Droop control is widely used to control the magnitude of voltage and frequency in the case of inverter-interfaced DGs in islanded MGs.

The dynamic droop characteristic for ith DG is shown as follows:

$$\omega_i = \omega^* - K_i^P P_i^m \tag{7.1}$$

$$V_i = V^* - K_i^Q Q_i^m \tag{7.2}$$

where ω^* and V^* are the nominal frequency and voltage amplitude. K_i^P and K_i^Q are droop coefficients, which are commonly chosen based on the output power rating. P_i^m and Q_i^m are the measured active and reactive power output. ω_i and V_i are then adjusted to return to nominal values by control signals sent from the secondary control level.

7.3.2 MG Level

Figure 7.3 illustrates the distributed MG-level control with required local and neighbor information. The primary function of the secondary control is to follow the reference values provided by the interconnected MG-level control and ensure precise power-sharing of the DG within each MG. The state-of-the-art has extensively studied distributed secondary control methods, including both linear and nonlinear approaches. Building on the methods proposed in [9–11], a linear control scheme for frequency restoration and real power-sharing is designed for the ith DG as follows:

$$\omega_{DG_i} = \omega_{MG_k}^* - K_{DG_i}^P P_{DG_i} + \Omega_{DG_i} \tag{7.3a}$$

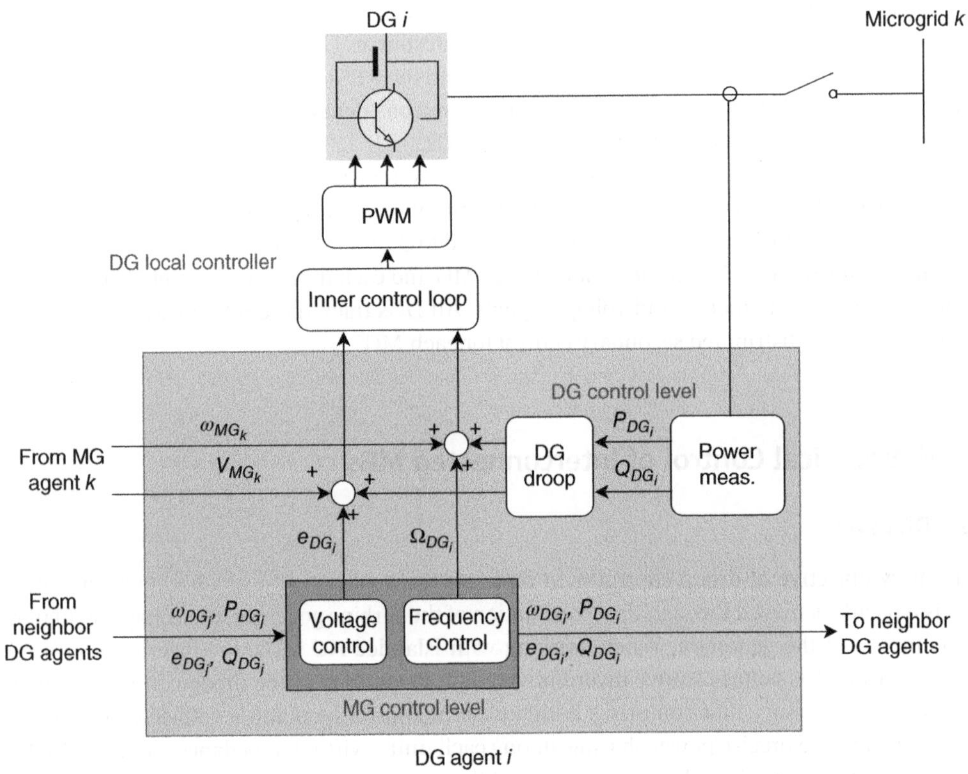

Figure 7.3 Distributed MG control level.

$$\dot{\Omega}_{DG_i} = \sum_{j=1}^{N_{MG_k}} a_{ij}^k \left(\omega_{DG_j} - \omega_{DG_i}\right) + g_i^k \left(\omega_{MG_k} - \omega_{DG_i}\right)$$
$$+ \sum_{j=1}^{N_{MG_k}} a_{ij}^k \left(K_{DG_j}^P P_{DG_j} - K_{DG_i}^P P_{DG_i}\right) \tag{7.3b}$$

Similarly, a linear control for voltage regulation and reactive power-sharing is designed as follows:

$$V_{DG_i} = V_{MG_k}^* - K_{DG_i}^Q Q_{DG_i} + e_{DG_i} \tag{7.4a}$$

$$\dot{e}_{DG_i} = \sum_{j=1}^{N_{MG_k}} a_{ij}^k \left(e_{DG_j} - e_{DG_i}\right) + g_i^k \left(V_{MG_k} - V_{DG_i}\right) \tag{7.4b}$$
$$+ \sum_{j=1}^{N_{MG_k}} a_{ij}^k \left(K_{DG_j}^Q Q_{DG_j} - K_{DG_i}^Q Q_{DG_i}\right)$$

where Ω_{DG_i} and e_{DG_i} are control signals from secondary control. a_{ij}^k is the communication coefficient between DGs i and j in MG k, $a_{ij}^k > 0$ if there is a link, otherwise, $a_{ij}^k = 0$. g_i^k is the pinning gain of the DG i in kth MG, where $g_i^k > 0$ if the DG can directly receive ω_{MG_k} and V_{MG_k}, and $g_i^k = 0$ otherwise. N_{MG_k} is the total number of DGs in MG k. ω_{MG_k} and V_{MG_k} are the frequency and voltage reference values calculated by interconnected MG level control.

7.3.3 Interconnected MG Level

The control hierarchies within interconnected MGs are differentiated based on response speed and infrastructure requirements, such as communication requirements. During a minor disturbance, the primary control provides a rapid response to mitigate any frequency or voltage variations within the MG. The secondary control then activates to regulate frequency, voltage, and power-sharing. Finally, the tertiary control, as the slowest response and highest control level, operates to minimize network losses. The entire peer-to-peer control framework for interconnected MG systems can be achieved through neighboring communications.

The two control layers with a fully distributed structure are presented in Figure 7.4 and introduced in the following subsections.

7.3.3.1 Primary Control Layer in Interconnected MGs

Similar to the droop control employed for DGs in a single MG, a droop control scheme is proposed for each MG to operate autonomously using only local measurements. The interconnected MGs consist of N buses, with the sets of buses, MGs, and lines denoted as $\mathcal{N}$, $\mathcal{M}$, and $\mathcal{V}$, respectively. The droop control for MG k can be defined as follows:

$$\omega_{MG_k} = \omega^* - K_k^P P_k, k \in \mathcal{M} \tag{7.5}$$

$$V_{MG_k} = V^* - K_k^Q Q_k, k \in \mathcal{M} \tag{7.6}$$

where ω^* and V^* are the nominal frequency and voltage amplitude at the interconnected MG level. P_k and Q_k are the power exchange among MG k and the interconnected MGs. K_k^P and K_k^Q are droop coefficients of each MG.

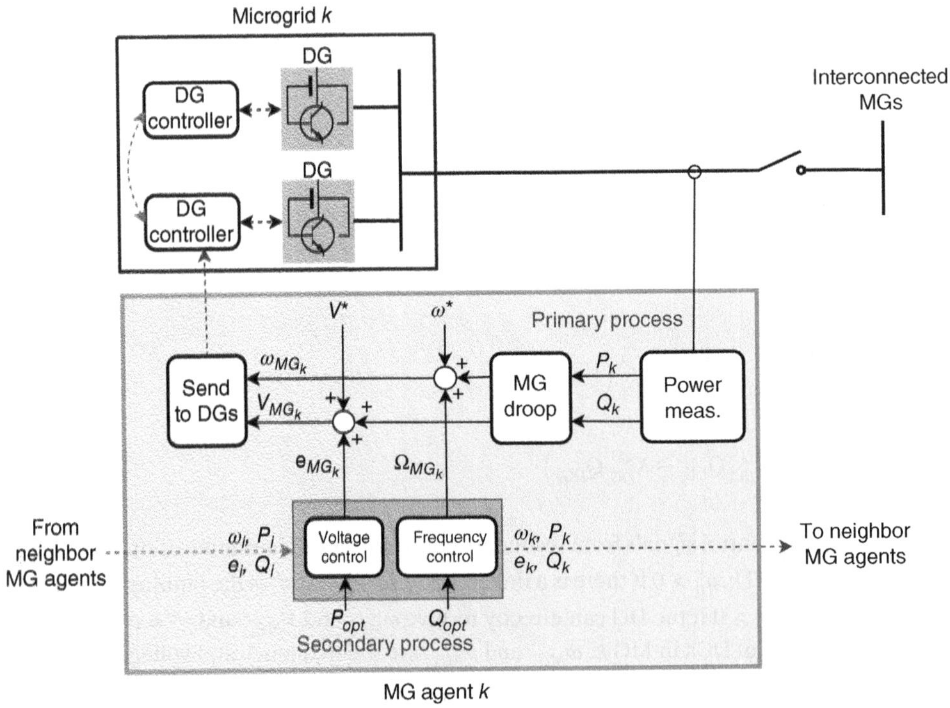

Figure 7.4 Distributed interconnected MG control level.

7.3.3.2 Secondary Control Layer in Interconnected MGs

The distributed secondary control of interconnected MGs aims to achieve three objectives: (i) restoring frequency, (ii) restoring the voltage at the PCC bus, and (iii) achieving arbitrary power-sharing among MGs.

Based on consensus algorithms, the distributed secondary control laws for each MG can be designed as follows:

$$\omega_{MG_k} = \omega^* - K_k^P P_k + \Omega_{MG_k} \tag{7.7a}$$

$$\dot{\Omega}_{MG_k} = \sum_{h=1}^{M} a_{kh}\left(\omega_h - \omega_k\right) + g_k\left(\omega^* - \omega_k\right) + \sum_{h=1}^{M} a_{kh}[K_h^P\left(P_h - P_h^{opt}\right) - K_k^P\left(P_k - P_k^{opt}\right)] \tag{7.7b}$$

$$V_{MG_k} = V^* - K_k^Q Q_k + e_{MG_k} \tag{7.8a}$$

$$\dot{e}_{MG_k} = \sum_{h=1}^{M} a_{kh}\left(e_h - e_k\right) + \left(V^* - V^{PCC}\right) + \sum_{h=1}^{M} a_{kh}\left(K_h^Q\left(Q_h - Q_h^{opt}\right) - K_k^Q\left(Q_k - Q_k^{opt}\right)\right) \tag{7.8b}$$

$$k, h \in \mathcal{M}$$

where Ω_{MG_k} and e_{MG_k} are control signals from secondary control of the interconnected MG system. a_{kh} is the communication coefficient between MGs k and h. g_k is the pinning gain of the MG k. M is the total number of MGs in the interconnected MGs.

The interconnected MG system with proposed control laws in (7.7) and (7.8) will converge to:

$$\lim_{t\to\infty} |\omega^* - \omega_k(t)| = 0 \tag{7.9}$$

$$\lim_{t\to\infty} |V^* - V^{PCC}(t)| = 0 \tag{7.10}$$

$$\lim_{t\to\infty} |K_h^P \left(P_h(t) - P_h^{opt}\right) - K_k^P \left(P_k(t) - P_k^{opt}\right)| = 0 \tag{7.11}$$

$$\lim_{t\to\infty} |K_h^Q \left(Q_h(t) - Q_h^{opt}\right) - K_k^Q \left(Q_k(t) - Q_k^{opt}\right)| = 0 \tag{7.12}$$

The distributed secondary control in this level ensures that the system frequency of each bus in the interconnected MG system is restored to its reference value. Additionally, the PCC voltage of the system is also restored to its reference value. The real and reactive power is regulated to follow the dispatch signals P_k^{opt} and Q_k^{opt}, respectively, received from an optimal upper control process. Any mismatch due to load variations are shared among the MGs based on the droop coefficients K_k^P and K_k^Q, which are defined in (7.5) and (7.6) for each MG.

7.4 The Multi-Agent System

The proposed distributed hierarchical control architecture is implemented using a MAS with peer-to-peer communication. An agent is an entity that possesses the ability to receive local measurements, communicate with other agents, perform calculations, and provide appropriate signals to the DG level controllers. Instead of gathering all data to a central controller, each agent needs only local and adjacent information but can return system-level signals to accomplish global objectives. The neighbor agents are defined on the basis of the electrical connection of an interconnected MG system.

7.4.1 Agent

Agents are designed as independent entities that rely on limited system knowledge. Each agent updates the state of the power network, performs calculations, and makes control decisions.

In an MG agent, the local controller implements the droop control with only local measurements and serves as the primary control. To implement the proposed fully distributed multi-layer control, we design agents for practical system implementation. To achieve all control objectives simultaneously, the agent contains two separate processes running in parallel: the primary process and the secondary and tertiary processes.

7.4.2 Primary Process

In the primary process, each MG-agent utilizes its virtual droop control laws in (7.5) and (7.6) to compute tracking reference values ω_{MG_k} and V_{MG_k} for DGs within the MG. This control process only requires local measurements and is always in operation to ensure system stability in the absence of secondary control processes.

Algorithm 7.1 The primary process in agent k.

1: P_k, Q_k ▷ obtain local measurements at node k
2: Calculate control signals ω_{MG_k}, e_{MG_k} based on droop law and Ω_k, e_k signals from the secondary process ▷ equations (7.5), (7.6)
3: Send ω_{MG_k}, e_{MG_k} to local controller of MG k
4: **redo** from step 1

7.4.3 Secondary Process

All MGs, along with their DGs, contribute to the secondary control process to restore the voltage and frequency of the system. In the distributed control scheme, MG agents iteratively exchange locally sensed information to calculate consensus laws (7.7) and (7.8).

Algorithm 7.2 outlines the iterative step for the secondary process. Initially, agents collect local measurements ω_k, e_k, P_k, and Q_k from devices and exchange messages $\omega_k, e_k, K_k^P\left(P_k - P_k^{Ter}\right), K_k^Q$, and $\left(Q_k - Q_k^{Ter}\right)$ among neighbors. Control signals are computed and sent to controllers. The frequency reference value in the local controller is adjusted by the signal Ω_k, while the voltage reference value is adjusted by the signal e_k.

Algorithm 7.2 The secondary process in agent k.

1: $\mathcal{N}^k$ ▷ list of neighborhood agents
2: ω_k, e_k, P_k, Q_k ▷ obtain local measurements at node k
3: Calculate control signals Ω_{MG_k}, e_{MG_k} based on local measurements, and neighborhood information ▷ equations (7.7), (7.8)
4: Send Ω_k, e_k to the primary process
5: **redo** from step 1

7.5 The Implementation on a Real-Time Cyber-Physical Testbed

7.5.1 Experimental Setup

In this section, the validation of the agent design using the proposed control framework is presented. An interconnected MG system with six buses operated in islanded mode is considered. The interconnected MG test system, as shown in Figure 7.5, comprises three parallel MGs, with 3, 3, and 4 DGs in MG-1, MG-2, and MG-6, respectively. The remaining buses contain the loads. The physical and cyber topology of the test interconnected MG system is demonstrated in Figure 7.6, where the communication topology among agents is designed to be the same as the electrical network in the interconnected MG level. The experimental setup in the laboratory consists of two main parts, which are further illustrated as follows.

7.5.1.1 Physical System

The electrical components in the interconnected MGs and local controllers of DGs in MGs are encompassed by the physical system, which is simulated in real-time using the OPAL-RT simulator. Each control layer requires measurements of the interconnected MGs, which are sent to each agent. The agent calculates the control signals ω_{MG_k} and V_{MG_k} and returns them to the MG-level control

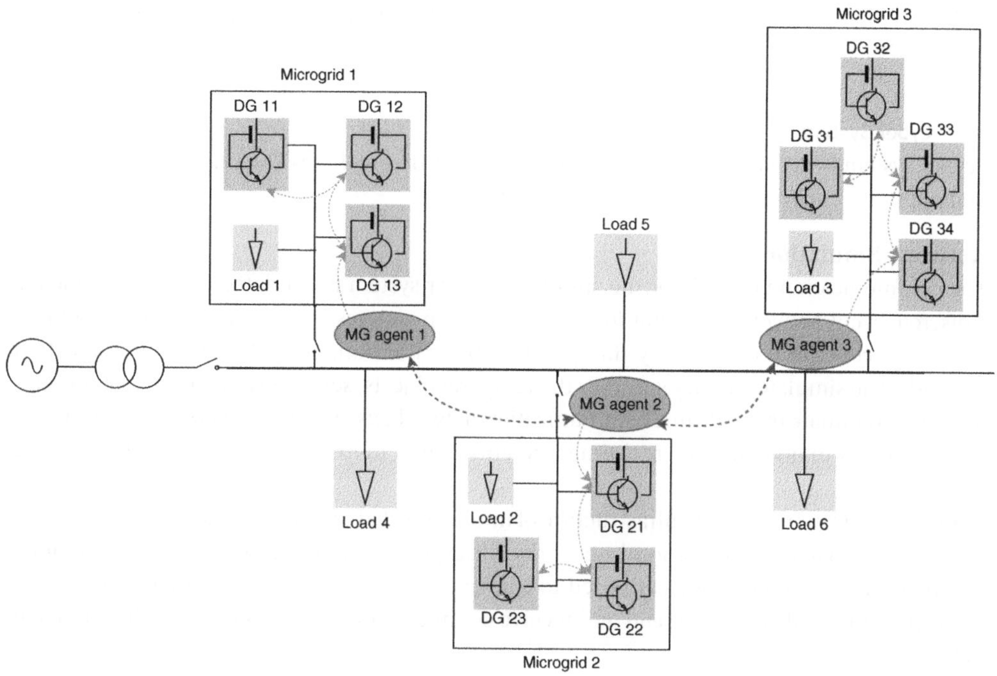

Figure 7.5 The test case of interconnected microgrid system.

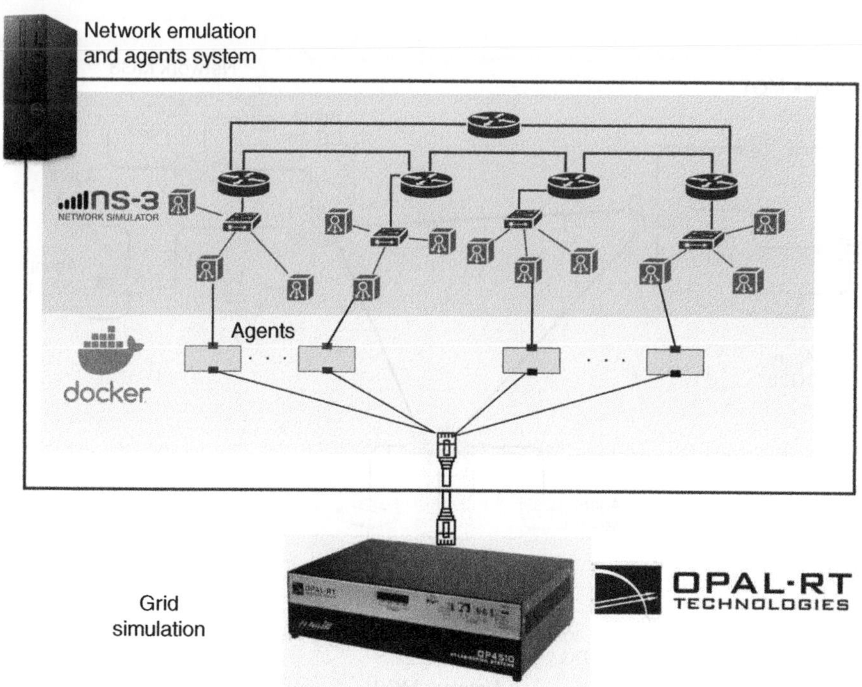

Figure 7.6 Real-time cyber-physical test bed.

in OPAL-RT. Signal exchange between the agent and OPAL-RT occurs through the user datagram protocol (UDP).

7.5.1.2 Cyber System

The cyber system is executed on a Linux-based computer that includes Docker containers used to implement the agent system and a network simulator *ns*3.

a) **Network Emulation in ns3**
   The communication network for the simulated power system is emulated using *ns*3, which is a discrete event network simulator for communication systems. *ns*3 has a modular implementation and contains a core library that provides the basic framework for the communication network. The simulator library specifies the temporal objects, schedules, and events of the simulation. To emulate the real-time communication network, *ns*3 is run on a dedicated server, and it is important to note that the real-time implementation of *ns*3 uses system time to schedule events.
   The communication network infrastructure of interconnected MGs is emulated in *ns*3, as shown in Figure 7.7. There are three local networks corresponding to the three MGs in the system, and point-to-point connections are used to connect MG agents in different areas. Each agent is attached to a node in *ns*3 to send and receive communication packets through the emulated network.
b) **Multi-Agent System in Docker Containers**
   The Docker container is a lightweight and portable software package designed to encapsulate an application along with its dependencies. It simplifies the deployment and management of complex software systems by providing an isolated environment that ensures consistent operation

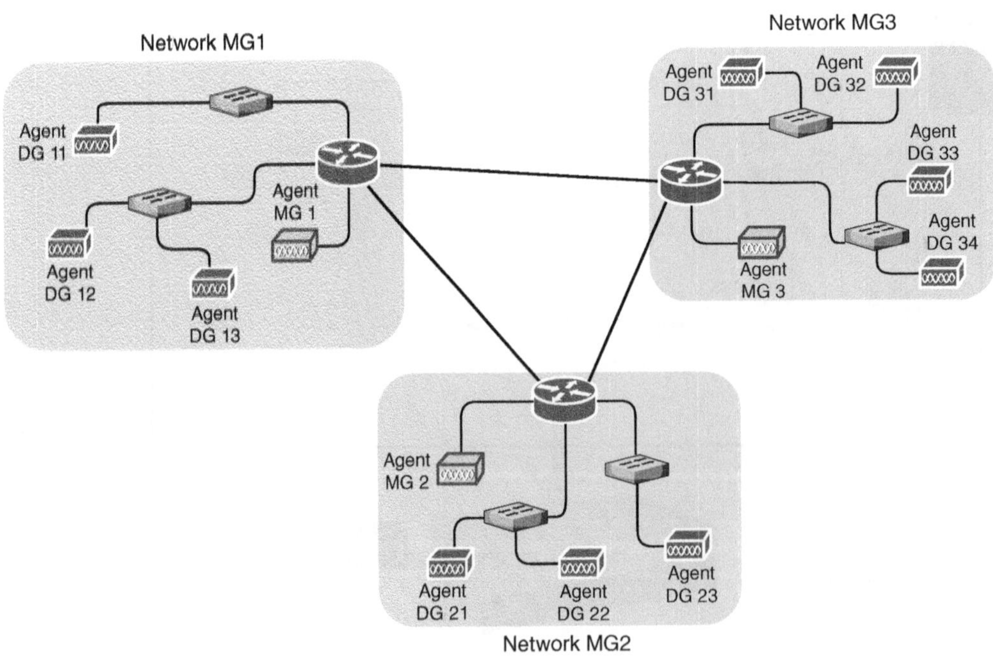

Figure 7.7 Communication network.

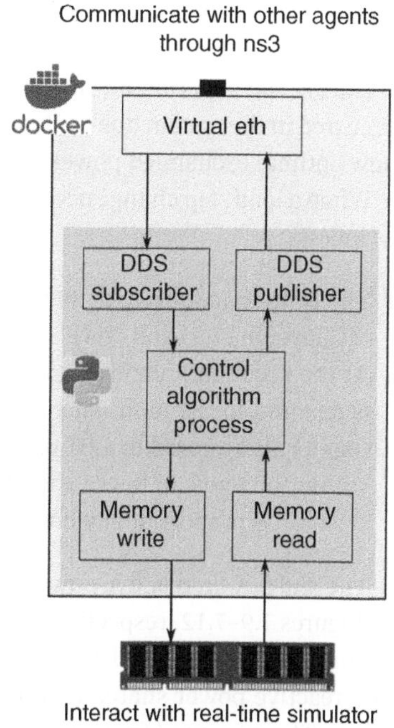

Figure 7.8 Agent design in a Docker container.

across diverse environments. In the present setup as shown in Figure 7.8, the Docker containers are located on the same host computer as the *ns*3 network emulation system. The container acts as an intermediary between the real-time simulation in *Opal – RT* and *ns*3 and is equipped with a virtual network interface and the ability to access data stored within the shared memory of the host computer.

To interface with the network emulation in *ns*3, the virtual network interface serves as a gateway for exchanging data among containers. This interface is linked to Linux bridges that establish connectivity with the host operating system, and tapping devices are attached to these bridges to intercept packets that cross them. The packets are then delivered to user space, where *ns*3 can access them through a specialized NetDevice that transmits them to an *ns*3 ghost node. To communicate with other agents, the DDS middleware with the Real-time Publisher/Subscriber (RTPS) protocol is used.

The Linux operating system provides a shared memory mechanism that facilitates data sharing and interprocess communication, allowing multiple processes to concurrently access the same memory region and improving system efficiency and performance. In the present setup, the combination of Docker containers and shared memory establishes a data buffer on shared memory. This data buffer serves dual purposes of (i) collecting measurement signals from *Opal – RT* and transferring them to a container, and (ii) collecting control signals from the container and transferring them to *Opal – RT*.

The computation part involves implementing the distributed control algorithm using data collected from measurements in the power system simulation and information exchanged via a communication network.

7.5.2 Experimental Results

To validate the performance of the proposed method, an HIL experiment test lasting 480 seconds was conducted. The experiment investigated data collected and recorded from two sources: the logging files of the agents were analyzed to check the calculations in each iteration, and the measurement data saved in the simulator were observed to analyze the system's operation.

During the experiment, five milestones were taken into account: t_2 and t_4 when disturbances occurred in the system due to changes in load power, and t_1, t_3, and t_5 when the MG agents received new optimal exchanged power P^{opt} and Q^{opt} from the upper control processes.

When a load step change occurs in the interconnected MG system, the objectives are summarized as follows:

- The primary control calculates the control inputs Ω_{MG_k} and e_{MG_k} for the DG controllers of the MG level in (7.3) and (7.4).
- At the secondary control level, which has a response speed of seconds, the system frequency is restored to the nominal value of 50 Hz. The PCC voltages of the interconnected MG system (bus 1) are restored to 1.00 p.u., while the voltages of the remaining buses are guaranteed to be within the range of lower and upper thresholds. The real and reactive power outputs of the DGs are shared proportionally based on their rated capacity.

The real power, reactive power, frequency, and bus voltage profiles during the HIL test are shown in Figures 7.9–7.12, respectively. The results are presented in a time sequence as follows:

For 0 s $\leq t < t_2$, the OPAL-RT starts at $t = 0$ s. As illustrated in Figures 7.9b and 7.10b, the real and reactive power shared among DGs in each MG are the same, based on the DG droop coefficients. The measured frequencies are gradually restored to nominal values, as shown in Figure 7.11. The PCC voltage at bus-1 of the interconnected MGs is restored to 1 p.u. At t_1 = 68 s, the optimal

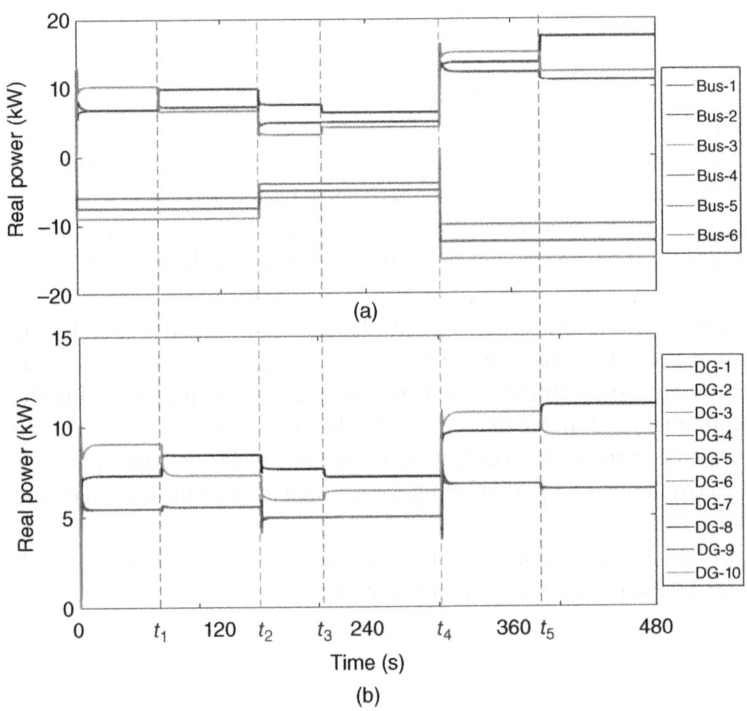

Figure 7.9 Real power injection from each DG and bus in the interconnected MG system. (a) Real power of each NMG bus and (b) real power of each DG.

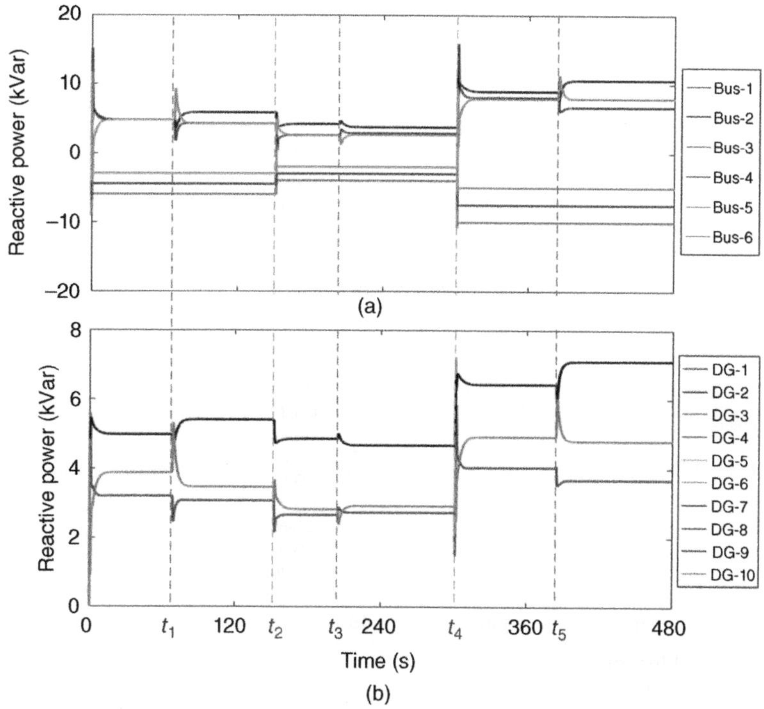

Figure 7.10 Reactive power injection from each DG and bus in the interconnected MG system. (a) Reactive power of each NMG bus and (b) reactive power of each DG.

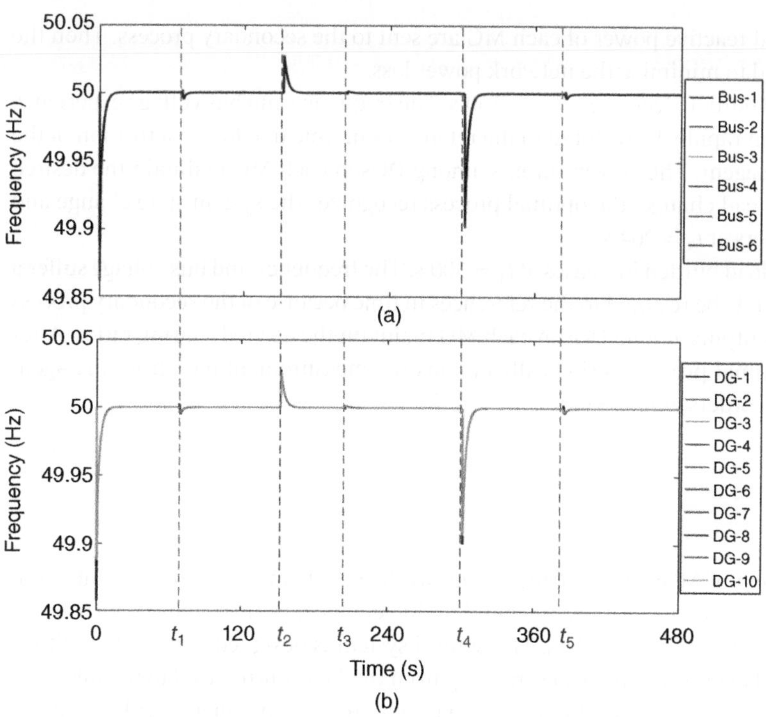

Figure 7.11 The frequency profiles of each DG and bus in interconnected MGs. (a) Frequency of each NMG bus and (b) frequency of each DG.

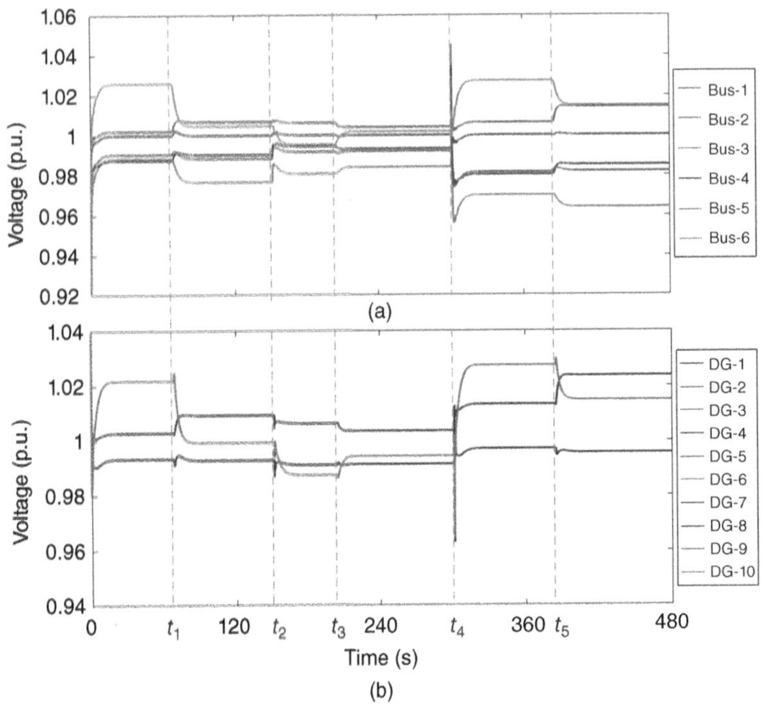

Figure 7.12 The voltage profiles of each DG and bus in interconnected MGs. (a) Voltage of each NMG bus and (b) Voltage of each DG.

set-points for the real and reactive power of each MG are sent to the secondary process. Then the dispatch order is executed to minimize the network power loss.

For $t_2 \leq t < t_4$, the load burden decreases at $t_2 = 150$ s. The frequency and bus voltage experience a sudden increase, but can rapidly be restored to the references in time due to the activation of the secondary process in the agent. The power outputs among DGs in each MG maintain the desired sharing ratios. After the load change, the optimal process recognizes the system state change and updates the dispatch orders at $t_3 = 204$ s.

For $t_4 \leq t < 480$ s, the load burden increases at $t_4 = 300$ s. The frequency and bus voltage suffer a sudden drop, but can rapidly be restored to the references in time because of the secondary process in the agent. The power outputs among DGs in each MG maintain the desired sharing ratios. After the load change, the optimal process periodically obtains the measurement data from the agent and updates the dispatch orders at $t_5 = 384$ s.

7.6 Conclusions

The growing incorporation of MGs into distribution networks results in the creation of interconnected MG systems. This chapter outlines a general architecture and a hierarchical distributed control system for these interconnected MGs. The control system is designed to provide multiple functions across various layers and time scales, ranging from the MG control level to the interconnected MG control level. A MAS is introduced to manage controller components within a single MG and to coordinate with neighboring MGs.

A Cyber Hardware-in-the-loop setup, incorporating the container concept, is presented as a real-time cyber-physical platform for simulating the complex multi-domain system. The successful implementation of the MAS and control system demonstrates the feasibility and effectiveness of the proposed approach. The suggested control system can be expanded to accommodate other MG types, including DC and hybrid AC/DC MGs. Moreover, the MAS can be further enhanced to incorporate additional functions, such as fault detection and diagnosis, and can be applied to other cyber-physical systems.

References

1 Lasseter, R.H., Eto, J.H., Schenkman, B. et al. (2011). Certs microgrid laboratory test bed. *IEEE Transactions on Power Delivery* 26 (1): 325–332.

2 Olivares, D.E., Mehrizi-Sani, A., Etemadi, A.H. et al. (2014). Trends in microgrid control. *IEEE Transactions on Smart Grid* 5 (4): 1905–1919.

3 Shafiee, Q., Dragičević, T., Vasquez, J.C., and Guerrero, J.M. (2014). Hierarchical control for multiple DC-microgrids clusters. *IEEE Transactions on Energy Conversion* 29 (4): 922–933.

4 Wang, Y., Nguyen, T.L., Syed, M.H. et al. (2020). A distributed control scheme of microgrids in energy internet paradigm and its multisite implementation. *IEEE Transactions on Industrial Informatics* 17 (2): 1141–1153.

5 Wu, X., Wu, X., Xu, Y., and He, J. (2018). A hierarchical control framework for islanded multi-microgrid systems. *2018 IEEE Power Energy Society General Meeting (PESGM)*, 1–5.

6 Zhao, Z., Yang, P., Wang, Y. et al. (2019). Dynamic characteristics analysis and stabilization of PV-based multiple microgrid clusters. *IEEE Transactions on Smart Grid* 10 (1): 805–818.

7 Guerrero, J.M., Chandorkar, M., Lee, T., and Loh, P.C. (2013). Advanced control architectures for intelligent microgrids; Part I: Decentralized and hierarchical control. *IEEE Transactions on Industrial Electronics* 60 (4): 1254–1262.

8 Wang, Y., Nguyen, T.-L., Xu, Y. et al. (2020). Peer-to-peer control for networked microgrids: multi-layer and multi-agent architecture design. *IEEE Transactions on Smart Grid* 11 (6): 4688–4699.

9 Bidram, A., Lewis, F.L., and Davoudi, A. (2014). Distributed control systems for small-scale power networks: using multiagent cooperative control theory. *IEEE Control Systems* 34 (6): 56–77.

10 Guo, F., Wen, C., Mao, J., and Song, Y.D. (2015). Distributed secondary voltage and frequency restoration control of droop-controlled inverter-based microgrids. *IEEE Transactions on Industrial Electronics* 62 (7): 4355–4364.

11 Wang, Y., Nguyen, T.L., Xu, Y. et al. (2019). Cyber-physical design and implementation of distributed event-triggered secondary control in islanded microgrids. *IEEE Transactions on Industry Applications* 55 (6): 5631–5642.

8

Internet of Energy, and Internet of Microgrids (IoE, IoM)*

Jonatas Boas Leite[1] *and Mladen Kezunovic*[2]

[1]*Electrical Engineering Department, Sao Paulo State University – UNESP, Sãao Paulo, Ilha Solteira, Brazil*
[2]*Department of Electrical and Computer Engineering, Texas A&M University – TAMU, College Station, TX, USA*

8.1 Introduction

The Internet of Things (IoT) concept comes from the ubiquitous computing initially envisioned by Weiser, 1991 that realized the intense integration among computers and the natural human environment, allowing the computers to be an indispensable background [1]. Around 30 years later, the mobile phone in the hands of a great portion of Earth's population provides increasing amounts of computing, sensing, and communication capabilities, representing the most successful example of a ubiquitous system. The cellular communication technology has also contributed to the implementation of smart grid applications [2], which eventually led to the use of IoT in the smart grids [3].

The IoT technology has been applied in several industrial sectors, such as retail, healthcare, transportation, etc. Saleem et al. [4]. In the energy management sector, for instance, home energy management system (HEMS) is optimizing the residential energy consumption and production by combining software tools, home-area network, and energy sensors [5]. The energy management system (EMS) is integrating advanced metering, information technology (IT) and operational technology (OT) to allow the real-time, or near real-time data feeds to a wide range of equipment to reduce operating costs and increase reliability and productivity [6]. Moving the supervisory control and data acquisition (SCADA) to the cloud [7] may lead to potential savings due to reduced setup time and technical staff to manage software and hardware [8].

The centralization through cloud computing provides consolidated resources and management but, at the same time, has weakness because limited communication capabilities with the devices near the process environment. Multi-tier computing networks can overcome these challenges [9]. The integration of cloud, fog, edge, and sea computing leads to a multi-tier computing network architecture with resource allocation likewise organizational structure of a large company: cloud is equivalent to the CEO with highest decision-making authority and most information sources while the devices, or things, form a sea of computing technologies similar to customers with demands for different services and applications. Between the highest and lowest hierarchy levels, there are managers (fog computing) and front-line staff (edge computing) [10].

*Integrating IoT Devices with Distribution Energy Management System by Harmonizing Their Logical Models Using IEC Standards 61970/61968 and 61850.

Smart Cyber-Physical Power Systems: Fundamental Concepts, Challenges, and Solutions, Volume 1, First Edition.
Edited by Ali Parizad, Hamid Reza Baghaee, and Saifur Rahman.

In the power industry, HEMS has contributed to the expansion of demand response concepts into the customer energy environment, and advanced distribution management system (ADMS) have promoted the IT and OT integration. In [11], the distribution automation is achieved by an IoT-based SCADA integrated with fog computing as a bridge between IoT node devices and the cloud. The distribution automation is divided in three sensing areas: smart meters; feeder sensors; and intelligent electronic devices (IEDs), using 3G/4G/5G and 6LowPAN communications.

Traditionally, microprocessor-based protective relaying integrates a variety of protective and control functions within a rack-mounted case located in the substation control rooms. These protection devices have dominated the market and have become the norm for a large majority of utilities [12]. Restricting factors, such as complicated instruction manuals and excessive input data required for setting the logic, make it difficult for the protective relays to become widely adopted as the distribution network SCADA's remote terminal unit (RTU).

Recently, distribution system is undergoing deep transformation stimulated by smart grid concepts that employ self-healing strategies for automatic restoration of electrical networks with a high level of decentralized preventive control methodologies, such as described in [13]. The use of plug-and-play equipment over cloud-based SCADA should lead to the implementation of self-healing schemes as ubiquitous system. Some challenges are yet to be overcome as the communication protocol running in the control room is different from the ones used in the substation and/or feeder automation (FA) [14].

The IEC standard 61850 covers the general communication to and from IEDs inside and outside of substations [15]. The IEC common information model (CIM), which is described in a set of three standards, namely IEC [16–18], provides data model for the EMS. Both 61850 and CIM are essential for smart grid deployment where the connections between the CIM concept utilized in EMS and IEC 61850 concept primarily utilized in substations are established [19]. This benefits control center applications by facilitating access to IEC 61850 model and data items through the CIM. The models from CIM and IEC 61850 can be integrated with the domain-specific proprietary models, and by deploying microgrid control system solution the abstract models can be communicated between nodes to support the plug-and-play ability of the controllers [20].

This work includes the harmonization solution for the interaction between CIM and IEC 61850 to enable IoT capabilities in smart grid applications. The modeling concept shows how the IoT-based distribution network operation utilizing logical models over more semantic languages makes autonomous algorithms feasible. The concept illustrates how the system architecture with IoT node devices dispersed along the distribution network must be implemented to take advantage of timing measurements on the fully automated procedures of the fault location, isolation, and service restoration (FLISR) operation, or self-healing.

The chapter is organized as follows: Section 8.2 depicts how the IoT node device is modeled and interfaced to the utility datacenter through the integration of the distribution network operation with the multi-tier computational model. Section 8.3 evaluates the implemented logic model at an IoT node. Section 8.4 is dedicated for conclusions. References are given at the end.

8.2 Interfacing of the IoT Node for Self-Healing Strategies

The ubiquitous system and open hardware platform provide more flexibility when implementing self-healing strategies. The cybersecurity and interoperability arise as an imminent challenge, [21, 22]. The use of IoT system standards with multi-tier computing model is a way to overcome these challenges.

8.2.1 Requirements for Multi-Tier Computation Implementation

A smart grid interoperability becomes a fundamental requirement for a robust, reliable, and secure electrical grid. The way to achieve the grid interoperability is through system requirement specification, use of standards, and extensive testing. Interoperability in the smart grid domain is made even easier by using the smart grid architecture model (SGAM) framework for modern electric power systems [20]. The SGAM framework and methodology present the design of smart grid use cases in an architecturally and technologically transparent solution. This framework has five layers covering the smart grid plane, i.e., electrical domains and information management zones.

We illustrate how the FLISR systems supports both manual and automatic network operation to restore the failed distribution feeder using the IoT concepts. Today, three legacy systems are used: FA system; ADMS; and asset and maintenance management system (AMMS). Figures 8.1–8.3 map these three systems with their components and connections, in the SGAM framework, where the communication layer requirements are represented by squares while the information layer requirements are in ovals.

The FA system, in Figure 8.1, plays an important role in meeting power quality expectations, like the decentralized IED-based multi-agent system in [23]. Many FA components, mainly IEDs, are in the field zone (equipment to protect, control, and monitor the process). The architecture in Figure 8.2 of ADMS allows operators to reduce client outage duration using applications such as fault location and restoration switching analysis. The geographic information system (GIS) in the enterprise zone provides data to various systems, such as the outage management system (OMS) in the zone of operation. It facilitates the deployment of visualization tools for improving the resiliency of distribution networks, as in [24]. Condition monitoring and management of field crews are part of the AMMS, Figure 8.3 shows the interaction related to field devices.

Since distributions network may cover a whole city and some parts of surrounding rural areas, the amount of field devices near the energy process zone, or grid, comprises a "sea" of computing technologies, which have to communicate with management services on operation in enterprise zones, or datacenter, to make possible the fully automatic FLISR. Multi-tier computing resources are, thus, required from the datacenter to the grid through enterprise, operation, station, and field zones. In

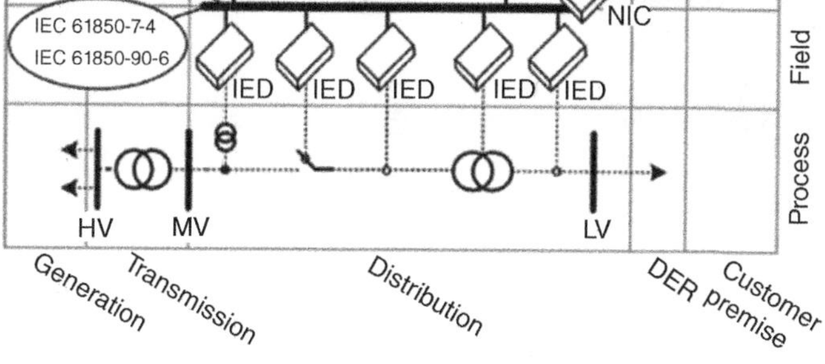

Figure 8.1 Diagram of FA based on SGAM.

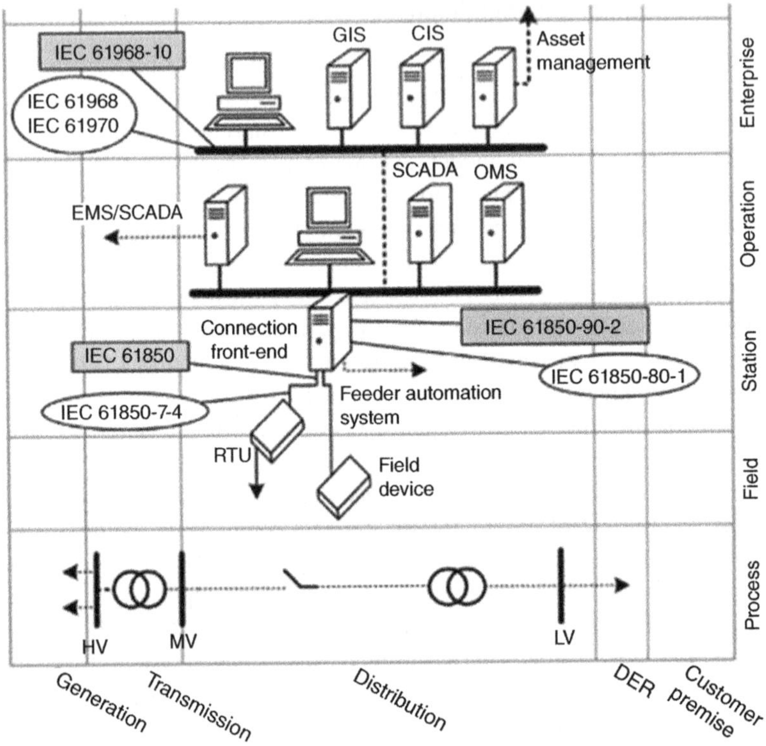

Figure 8.2 Diagram of ADMS based on SGAM.

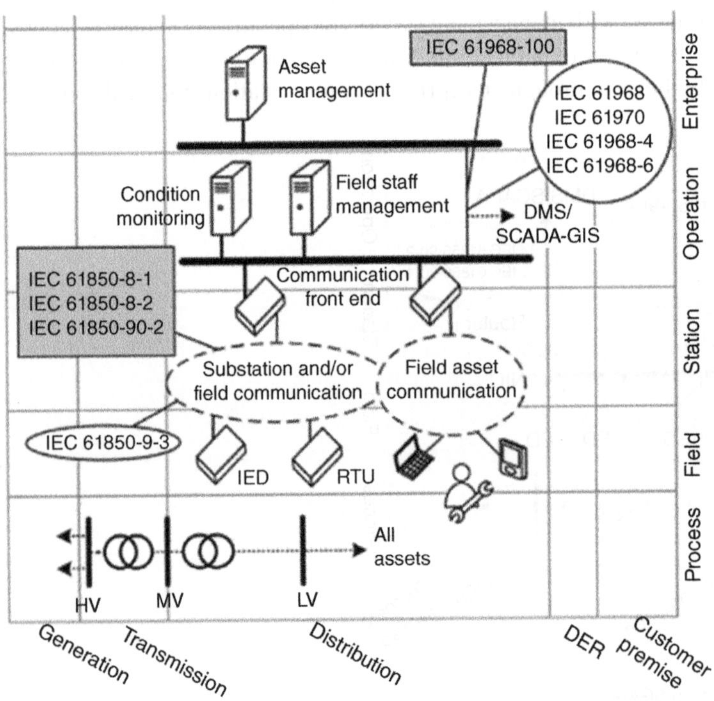

Figure 8.3 Diagram of AMMS based on SGAM.

the deployment procedure, first, network operations are integrated through an IoT platform based on electrical power industry standards, and then a multi-tier computing model is proposed to overcome the cloud computing shortcomings.

8.2.2 Integrated Environment of Network Operations

The integrated environment is supported by IEC 61850 and IEC 61970/61968 standards that are recognized as essential resources in the integration of distribution network operations and smart grid realization, [25]. In terms of cloud-based SCADA, [26] promote an integrated environment of distribution network operations using concepts of the SGAM framework. It assumes the integration of distribution management system (DMS) and OMS with SCADA to enable fully- and semi-automatic procedures of FLISR operation. In this work, a step further is made to advance the interoperability between the utility datacenter services with near-to-energy-process devices by deploying the IoT node device model in the integrated environment of network operation, as depicted in Figure 8.4.

The environment with SCADA, OMS, and DMS integrated as a network operation service, i.e., a web service that exchanges information using the enterprise service bus (ESB), depends on the harmonization solution between CIM and IEC 61850. The direct data exchange with ESB involves

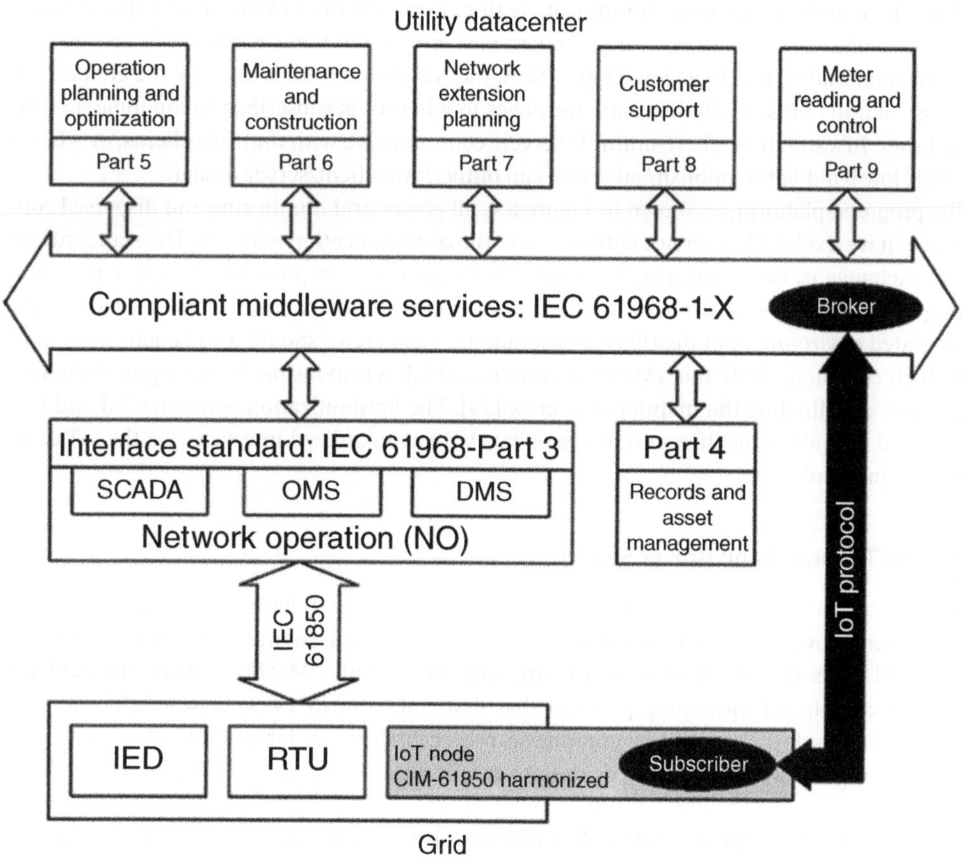

Figure 8.4 Harmonized IoT node device in the distribution system.

standardized interfaces and shaping the message structure in order to be CIM compliant. ESB messages are constructed using the XML scheme (XSD) through defined information exchange by the standardized interfaces [27]. The use of standardized messages provides a set of exchanged data needed for distribution network procedures, such as those found in the automatic FLISR.

8.2.3 IoT Platform

The IoT concepts should seem familiar to power system specialists since they see the IoT technologies as just another highly distributed version of the Information and Communication Technology (ICT) already in use. The cost reduction of solar systems, electric vehicles, and energy storage becomes an indispensable integration requirement of distributed energy resources (DER) throughout the distribution grid. However, IoT's technologies need a new modeling and architecture solutions for their application in power systems, as described in [28]. The proposed integrated architecture of distribution operation is implemented using the IoT platform with IEC 61968-100 defined messaging. In this way, all capabilities from the publishing/subscribing message system of the IoT platform are included in the management of distribution networks.

With client/server, a client directly sends a request to the server, which then directly responds to the client. In an ADMS, it results in the need to know the endpoint of every IoT node device. This approach works well when traffic on a network is low volume, but it is not suitable for a situation where multiple servers communicate with multiple clients. On the other hand, with the publish/subscribe approach, the server and client have no need to know each other's address, but instead each one subscribes to a global topic. Now, the ADMS only needs to know the endpoint of the broker, and the broker will distribute messages to whoever is subscribed to the topic used by the publisher. In a situation where multiple servers communicate with multiple clients, or when a network is low bandwidth, publish/subscribe can outperform client/server model.

In the proposed platform, as shown in Figure 8.4, all power grid monitoring and dispersed control devices have a subscriber service with access to the communication network. The standardized message exchange is, then, performed through the communication protocol of the IoT platform, allowing the intensive aggregation of IoT node devices close to the energy exchange process with the integrated environment of distribution operations. Field crews should also benefit from integrating their cell phones with the OMS infrastructure, which is responsible for managing switching requests and coordinating the maintenance crew [29]. The harmonization between CIM and IEC 61850 should thus allow the direct exchange of data with standardized interfaces via the IoT communication protocol.

8.2.4 MQTT Protocol and Broker Service

When different IoT protocols are evaluated, Message Queuing Telemetry Transport (MQTT) takes preference over the Data Distribution Service (DDS) and Neural Autonomic Transport System (NATS), as it is more open source-friendly and mature. MQTT is ISO's standard for publish/subscribe-based messaging protocol that first came out in 1999. Lightweight and fast, MQTT is ideal for devices with low computing power, offering variable levels of security that ensure a message is sent and received, despite it is not as robust as the DDS protocol in this category.

MQTT is a broker-based protocol. The MQTT broker, also commonly referred to as a server, can be hosted locally or through a cloud-based service. In the broker system, a publisher posts a message to the broker which then passes that message on to any client subscribed to the topic specified

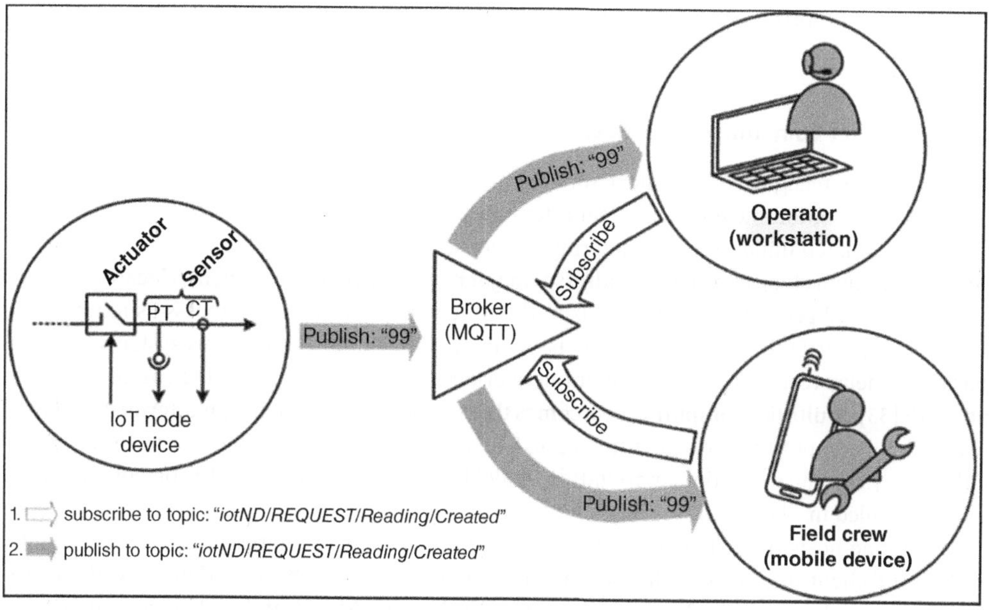

Figure 8.5 Publishing/subscribing message system over MQTT protocol.

by the publisher. These topics can be narrowed down in scope so that a user can be subscribed to a specific subtopic. Figure 8.5 shows a simple illustration of this publish/subscribe model. In a real-world scenario with IoT node devices scattered across heterogeneous networks, the ADMS should at least know the master resource identifier (`mRID`) or name of the IoT node devices to which it is sending a message. When IoT node devices are organized in hierarchical groupings, this messaging design might be more useful.

Topic management with MQTT is relatively simple at its core but allows for greater customization if desired. If a publisher sends a message for a topic that has no current subscribers, the broker will simply discard the information received unless the publisher has specified that the topic should be retained. For example, in the simplified diagram in Figure 8.5, if the publisher sent out its first message for `iotND/REQUEST/Reading/Created` before there were any subscribers present, then this message would only ever be seen if the publisher had notified the broker that the message is to be retained. If subscribers to the topic `iotND/REQUEST/Reading/Created` connect to the broker after this message has been sent by the publisher, and if the message is set to be retained, then each subscriber should receive this message upon subscribing to the topic. This feature can be used in a situation where subscribers need the most recent value for created readings without having to wait for the publisher to push a new value.

MQTT is one of the most practical options for utilities to use when developing the distributed computing environments needed for distribution network operation integration. Reasons why it stands out are the following [30]:

- MQTT is an open-source IoT protocol with a wealth of resources available online;
- The electric power research institute (EPRI) team evaluated MQTT performance, support resources, and scalability and highly recommends its use;
- MQTT could be a solid choice for venturing beyond client/server architecture into a publish/subscribe model due to the protocol's speed, support, and scalability.

Once the MQTT protocol envelops the IEC 61968-100-based messages, the IoT node device should be CIM-IEC 61850 harmonized, i.e., fully understandable by the utility datacenter.

8.2.5 Multi-Tier Computational Model

Indeed, MQTT is recognized as a most suitable protocol for IoT platforms with high reliability, secure multicast communications, and persistent messages [31]. In [32], five MQTT broker applications are evaluated under a smart metering system scenario. MQTT broker resides, typically, in the cloud. Although cloud computing addresses major needs of IoT systems, such as location awareness, mobility, and geo-distribution, it fails in the timely decision-making process as required by the power industry. Only cloud computing is not enough to support ubiquitous systems because of hard time-delay restrictions, intermittent network connectivity, and restricted communication bandwidth [33]. Multi-tier computing overcomes these issues by introducing an intermediate layer between cloud computing infrastructure and IoT node devices, bridging applications in the cloud and the edge [34]. The multi-tier computation model in [35] deploys the MQTT protocol into three layers, as detailed in Figure 8.6.

The layer one hosts the IoT node devices near the power grid, or energy process. They enable to run MQTT client instances. In the second layer, there are remote brokers that assist the main broker in the cloud computing. The layer three, thus, is the main MQTT broker. This computational model guarantees the messages exchanging among IoT node devices and utility data centers as the proposed IoT platform in Figure 8.4.

8.2.6 Harmonized IoT Node

IEC 61850 and the CIM standards specify structures and mechanisms for accessing power utility data. The IEC 61850 models are concentrated on the functions that can be seen and controlled, but not on how they work (what the algorithms are). The CIM standards also include a set of application-neutral services, a set of XML messages (generic service payloads) that are used to exchange CIM data. Comparing the two, it can see a similar document structure [36]. The three common parts of these standards are:

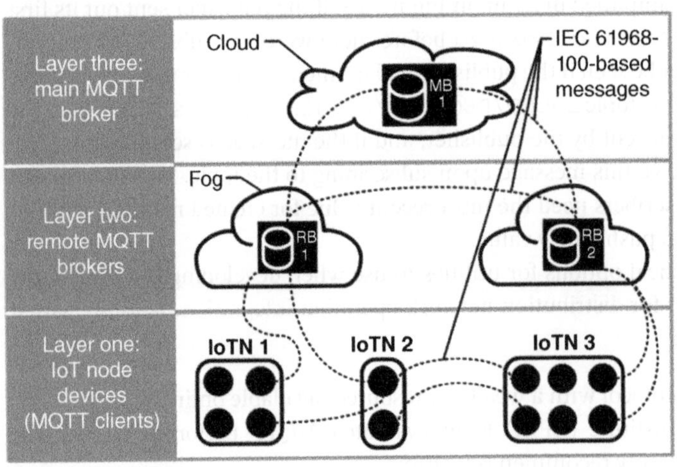

Figure 8.6 Connection of IoT node devices into multi-tier computational model with IEC 61968-100-based messages.

Information Model provides definitions and unique names for each object type into a description of the schema, in terms of classes, attributes, and relationships, for a collection of related real-world objects;

Services provide access to utility devices and applications and allow components to be treated as black boxes;

Technology Profiles describe the lower-level technologies that actually implement the services and carry messages between devices or applications.

These similarities support the use of a harmonized model that extends IEC 61970 CIM with new 61850 packages, where technological profiles are the most important standard structure to implement an IoT node device. Typically, a CIM profile is a subset of the more general model. CIM objects, to be interchanged, must be available and have the same interpretation at both sides of the communication link. As CIM has plenty of optional features, both sides must have an agreement about the options to be used. Thus, an object containing the proposed class `iotNodeDevice` (see Figure 8.7) could have all the attributes that appear in the class definition, while the other side object has none of them. Both objects comply with the definition of `iotNodeDevice` because the multiplicity of the attributes is zero or one `[0 or 1]`. In other words, the attributes are optional.

In Figure 8.7, the IoT node device profile is built through the Enterprise Architect (EA®) software together with a harmonized model package, TC57CIM, using the unified model language (UML). The proposed IoT node device profile has classes from the `IEC61850::` package such as `LNClass`, `LNinst`, `LNode`, and `LogicalDevice`. The package `Collections::`, from IEC 61850 scope, has the class `61850GroupTypeWithData` that is associated with the class `Measurement Value` of the `Meas::` package from the IEC61970 scope, for example. The `StringMeasurementValue`, `AnalgueValue`, `BooleanValue`, and `DisceteValue` provide the abstract modeling of the functional points for monitoring and controlling the IoT

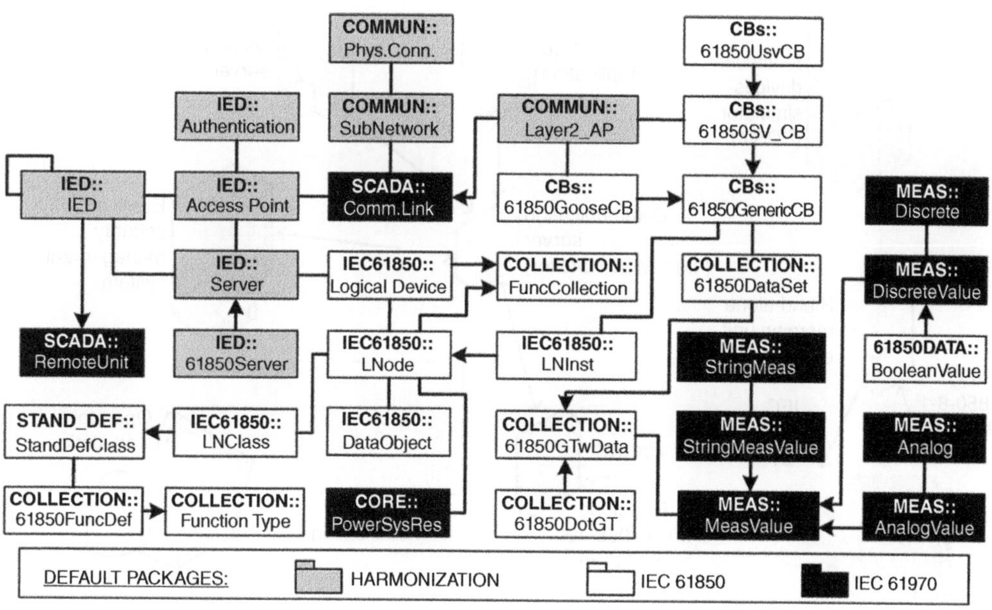

Figure 8.7 Abstract modeling of the IoT node device in UML.

Node device. In the harmonization scope, the `IED::` and `Communication::` packages provide classes comprising the models of the access points and communication interfaces to exchange information with the utility data center and/or other IoT Node devices.

In terms of implementation, a class is in fact a model, or data type used for creating objects that can be understood as a small software module. In this sense, an object can be executed like any software module, can receive input data and can produce results. Thus, the use of a computer board support packages, which contain a basic set of software, drivers, and boot configurations working as the main kernel of a single board computer (SBC), makes possible the implementation of the proposed IoT node device with the abstract modeling from standard classes using any generic programming language.

8.3 Performance Assessment Results

The IoT node device embedded in an SBC operates and interacts with real-time data, hence, its functionality requires validation in a real-time hardware-in-the-loop simulation environment [37].

8.3.1 Hardware-in-the-Loop Test Setup

In Figure 8.8, the stand-alone merging unit emulator collects current and voltage measurements from the power distribution network simulator (EDSIM) as described in [38] and sends this information to the IoT node device using sample values (SV) standardized by IEC 61850-9-2 because the IoT node device is not connected to the real-world electricity grid. After processing these measurements, the IoT node device can act under the simulated distribution system by sending GOOSE

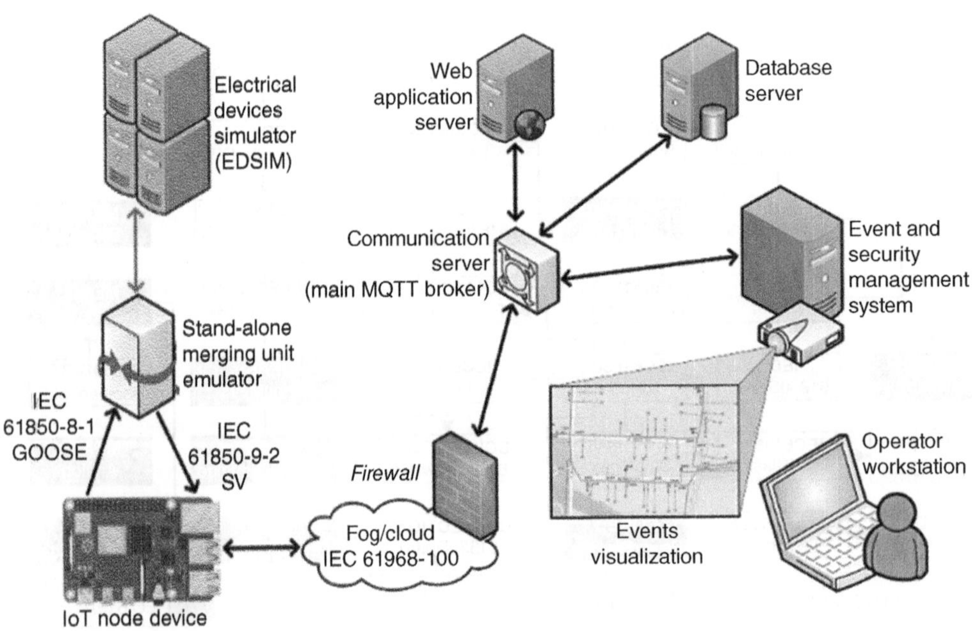

Figure 8.8 Hardware-in-the-loop setup for validation experiments.

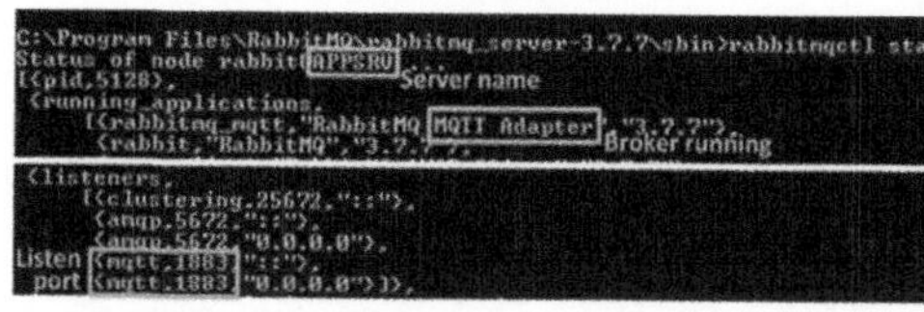

Figure 8.9 Status information about the RabbitMQ®broker.

commands, instead of its actuation to the circuit breaker as in real-world power grid. In addition to the work described in [30], the operator can now act on the IoT node device using standardized messages through the communication server that supports IoT platform subscribers by hosting the MQTT broker.

The communication server has two network adapters: one to private network connections, including utility data center services, and another for the "public" network with IoT node devices near the electricity network energy exchange process. A firewall to safeguard the private network can then be hosted in the communication server, as well. The communication server hosts the main MQTT broker. In Figure 8.9, the command line outputs display the status of the assembled node with the name rabbit@APPSRV that are running the MQTT adapter and listening the TCP port number 1883 as default.

From a real-world perspective, this test setup environment should assist the operator by providing the event and security management system that must enable viewing of events in geographical information system (GIS) applications. Although the web application and database server also receive IEC 61968-100 messages, they are out the scope of the performed experiment where the functionalities of the IoT node device are checked as well as the standardized messages involved in an automatic FLISR procedure.

8.3.2 SV Traffic and IoT Node Device GUI

The sensors of the IoT node device are current and voltage transformers connected in strategic points into the electricity grid. As the hardware-in-the-loop test environment employs a power system simulator, the sensors' measurements arrive at the IoT node device via an SV data payloads that are standardized by the IEC 61850-9-2 protocol. Figure 8.10 shows an example of SV payload received from the electrical simulator. The structure of the SV payload has a series of application service data units (ASDU) encapsulated inside an application protocol data unit (APDU). The sequence SV ASDU brings a header, with items from (1) to (5), before all data that are instantaneous sampled values. The implemented IoT node device abstract model deals each SV as a `DiscreteValue`, which is associated at least to two `AnalogValues`. In this example, every SV is a `Discrete` with name `instMag` that is associated to two Analogs named `scaleFactor` and `offset`.

After the calculation of the real measured magnitude, a phasor estimator algorithm as described in [39] can also be executed in order to update the current and voltage phasors as new SV packages that are acquired by the IoT node device. In this way, a graphical user interface (GUI) is able to display the set of measurements and other derivative values, as is shown in Figure 8.11, where calculated root mean square (RMS) values of voltage and current are displayed with their estimated phasors and frequency. In the proposed IoT platform, the existence of a GUI is not a condition to the operation of the IoT node device, but it is an advance able to aid the field crew technicians.

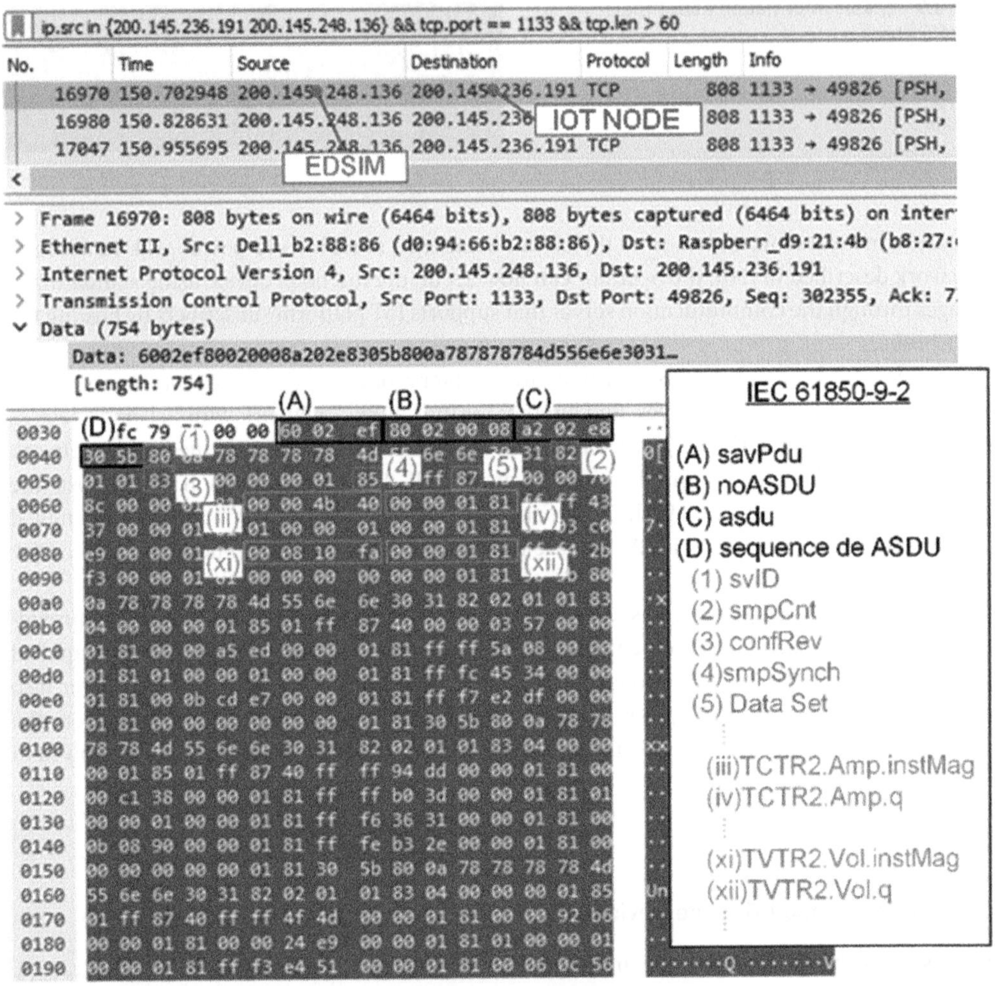

Figure 8.10 Monitoring capture of the SV payload from EDSIM to the IoT node device.

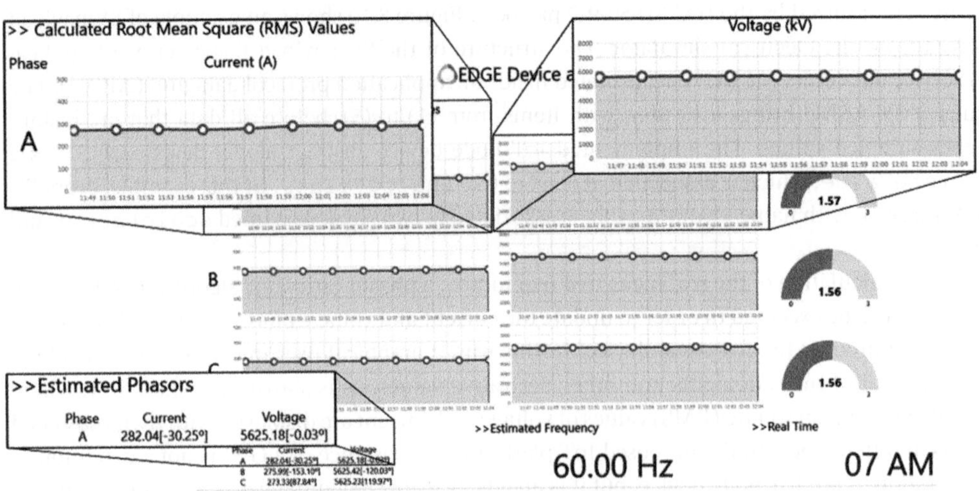

Figure 8.11 Harmonized IoT node device in the distribution system.

8.3.3 CIM-IEC 61850 Harmonized Messages

The evaluation of the message exchange through the MQTT protocol is done considering the automatic FLISR, i.e., the message sequence of the FLISR procedure for SCADA-detected outage and SCADA switching, as defined in [40]. This flow of information from FLISR considers a self-healing network, with the monitoring and control of the system provided by SCADA and automatic outage response directed by the fault management function of a DMS. The sequence diagram in Figure 8.12 illustrates the message flow. Network control (NO-CTL), or SCADA, initiates this procedure when it detects an unexpected change in the state of a protection device and informs network operations (NO-NMON) of the unexpected event. Switching plans to isolate the fault and restore power are subsequently requested upon notification of the incident. NO-NMON must then direct NO-CLT to execute each step of the selected switching plan. The NO-CLT must inform the point of operation through standardized messages to an IoT Node device that performs control of switching equipment, such as a circuit breaker.

To perform the switching operation, first, the NO-CLT sends a request message to the IoT node device asking the creation of the `iotNodeDeviceControl` in order to change a `BooleanValue`. Figure 8.13 presents the Wireshark®capture analysis [41] in the request message of the NO-CLT service at the ADMS application. In Figure 8.12, the MQTT brokers are transparent but, as it is established in the multi-tier computational model in Figure 8.6, every exchanged message out the network operation is sent to the ESB with the broker service. The broker receives the request message in the private network interface and, after 0.9 ms, sends it again to the public network to all IoT node devices subscribing to the topic. The request message has three parts: `<Header>`, `<Request>` and `<Payload>`. The last part informs the controlled point to be operated through the attribute `pathName` of the class `BooleanValue`.

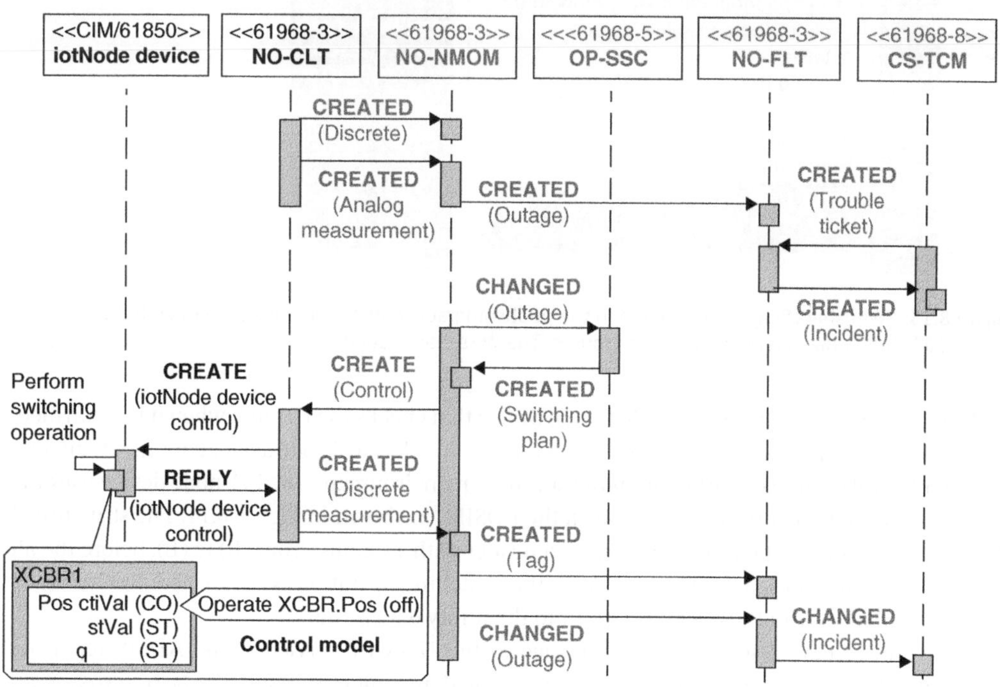

Figure 8.12 Message flow for automatic FLISR with SCADA detection and switching.

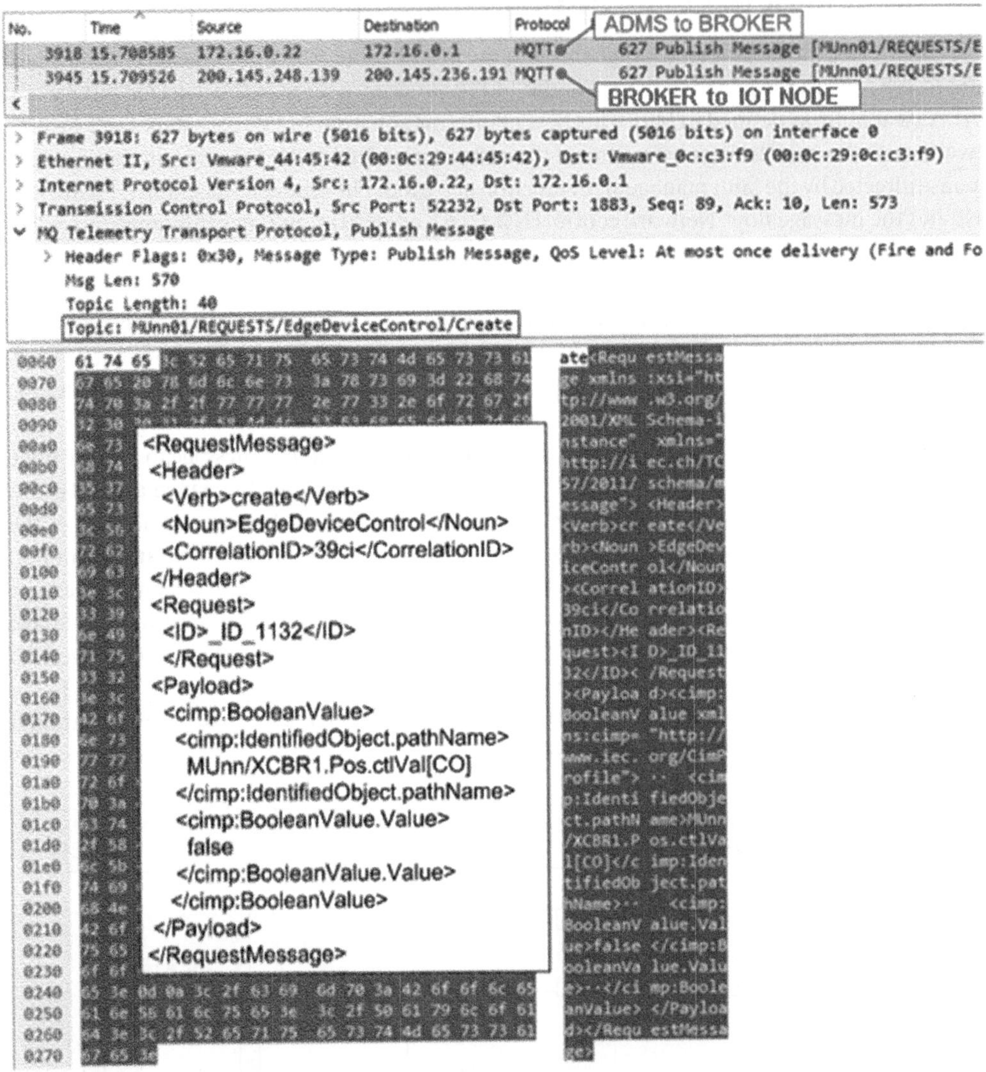

Figure 8.13 Wireshark®captures analysis in the publishing/subscribing message system with IEC 61968-100 standardized request message (source: The Wireshark Foundation).

The request message arrives at the IoT node device that checks the ID, and, when it is confirmed, the Control Model is executed to operate the Boolean value. The IoT node device commands directly the circuit breaker in real-world but, in the simulation environment, the IoT node device uses the GOOSE message to report the new status in the EDSIM, as shown in Figure 8.14. The structure of the GOOSE protocol data unit (PDU) brings a header with 11 items, from (B) to (L), before the all data that are three Boolean data types. After the switching operation is performed, the IoT node device then replies with the operation result to the utility data center.

The IoT node device sends a reply message to the NO-CLT service at the ADMS reporting the result of the `iotNodeDeviceControl` created in order to change a `BooleanValue`.

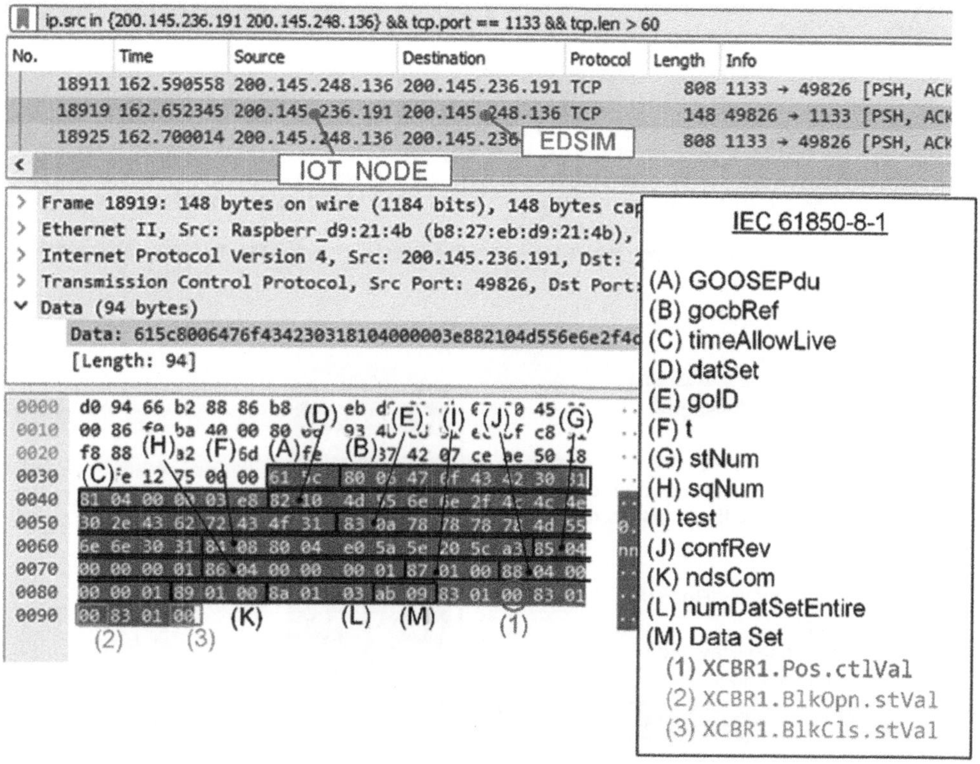

Figure 8.14 Monitoring capture of the GOOSE payload from IoT node device to the EDSIM.

Figure 8.15 displays the Wireshark®capture analysis in the reply message of the IoT node device. Before reaching the ADMS, the IoT node device message passes through the MQTT broker that sends this message to all subscribers of the topic `ADMS/REPLY/Queue`. Like the request message, the reply message is comprised by `<Header>`, `<Reply>` and `<Payload>`. In the header, the `<CorrelationID>` provides a mean to associate this arriving message with those one sent initially. The payload with `<Result> OK` indicates that the requested operation was successfully performed.

The standardized message exchange between the two sides, the IoT node device and ADMS service, just happens because both sides are connected to the MQTT broker. Figure 8.16 shows a logging script of the ADMS service side. First, the service tries connecting to the APPSRV server through the port 1883, see Figure 8.9. After the connection with the server is established, the ADMS service, or client sends a Connect message to the broker containing, for example, the client username and password. The broker responds with `ConnAck` message informing the acceptance of the connection. Subsequently, the client subscribes to the topic named `ADMS/REPLY/Queue` in order to receive standardized response message from remote devices at the ADMS integrated architecture. In addition, it is possible to notice the publication of the messages that creates the IoT node device control, as is shown previously in Figure 8.13.

In summary, the IoT node device abstract model was evaluated in hardware-in-the-loop test simulation environment by sniffing the standardized messages. The employed SBC was the Raspiberry

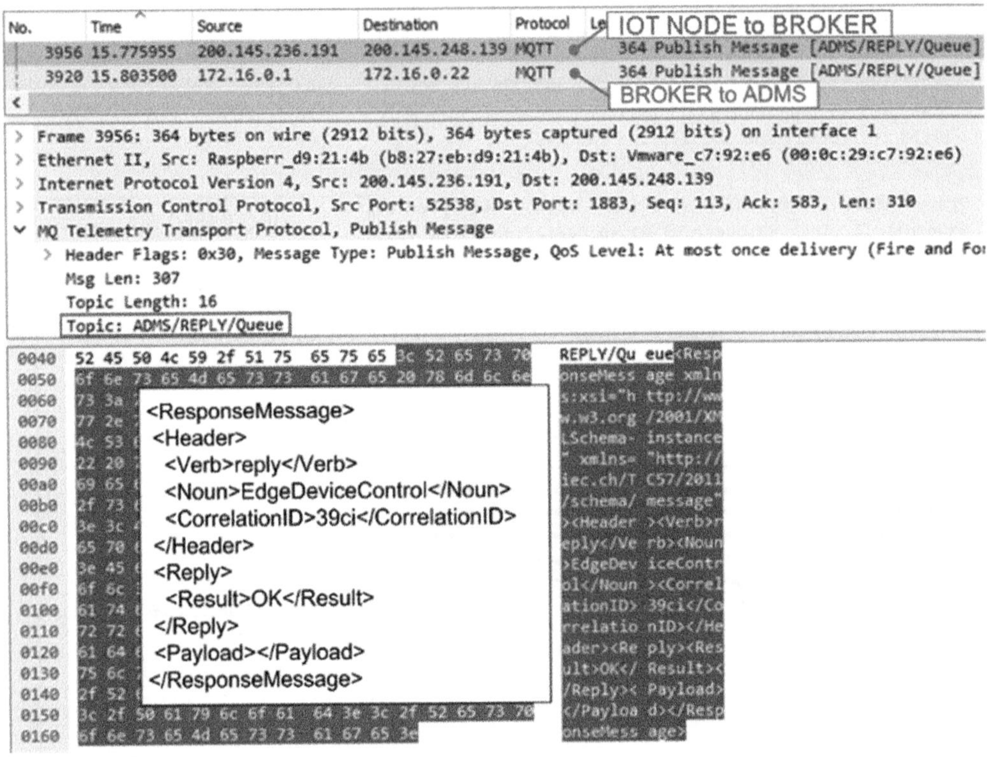

Figure 8.15 Wireshark captures analysis in publishing/subscribing message system with IEC 61968-100 standardized request message.

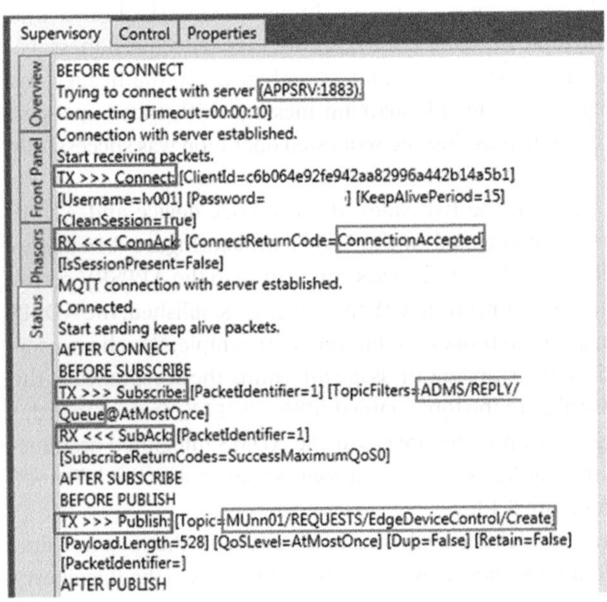

Figure 8.16 Screenshot from the supervisory with the logging script of an ADMS service.

Pi 3®under Windows 10 IoT Core. In this way, the abstract model was implemented using the C-Sharp (C#) programming language. Due to the complexity and importance of the cybersecurity consideration, details of this topic are left for future work.

8.4 Concluding Remarks

The following are the paper findings:

- The integrated distribution network operation using the SGAM framework, where the IoT platform allows the implementation of the IoT node device abstract model into SBC is implemented;
- The IoT node device abstract model is designed as a CIM profile by using Enterprise Architect®software together with the harmonized model package;
- The IoT node device abstract model use is evaluated in a hardware-in-the-loop test simulation environment by sniffing the sequence of messages that are exchanged between the grid and utility data center using standardized messages of IEC 61968-100 encapsulated by the MQTT protocol;
- The implementation of the CIM-IEC 61850 harmonized IoT node device facilitated by the MQTT protocol demonstrates the feasibility of decentralized distribution automation based on IoT devices.

References

1 Oliveira, L.B., Pereira, F.M.Q., Misoczki, R. et al. (2018). The computer for the 21st century: present security & privacy challenges. *Journal of Internet Services and Applications* 9: 1–25.

2 Zheng, W., Sun, K., Zhang, X. et al. (2020). Cellular communication for ubiquitous Internet of Things in smart grids: present and outlook. *2020 Chinese Control and Decision Conference (CCDC)*, 5592–5596.

3 Saleem, Y., Crespi, N., Rehmani, M.H., and Copeland, R. (2019). Internet of Things-aided smart grid: technologies, architectures, applications, prototypes, and future research directions. *IEEE Access* 7: 62962–63003.

4 Saleem, Y., Crespi, N., Rehmani, M.H., and Copeland, R. (2022). Industrial Internet of Things: applying IoT in the industrial context. https://www.ifm.eng.cam.ac.uk/uploads/DIAL/industrial-internet-of-things-report.pdf (accessed 09 May 2022).

5 Li, S., Song, W., Yang, J., and Chen, A. (2019). A real-time electricity scheduling for residential home energy management. *IEEE Internet of Things Journal* 6: 2602–2611.

6 Hossein Motlagh, N., Mohammadrezaei, M., Zakeri, B., and Hunt, J. (2020). Internet of Things (IoT) and the energy sector. *Energies* 13: 1–27.

7 Church, P., Mueller, H., Ryan, C. et al. (2015). Moving SCADA systems to IaaS clouds. *2015 IEEE International Conference on Smart City/SocialCom/SustainCom (SmartCity)*, 908–914.

8 Osman, F.A., Hashem, M.Y.M., and Eltokhy, M.A.R. (2022). Secured cloud SCADA system implementation for industrial applications. *Multimedia Tools and Applications* 81: 9989–10005.

9 Yang, Y., Luo, X., Chu, X., and Zhou, M.T. (2020). Fog computing architecture and technologies. In: *Springer Nature: Fog-Enabled Intelligent IoT Systems*, 39–60. Springer International Publishing.

10 Yang, Y., Chen, X., Tan, R., and Xiao, Y. (2021). *Computing and Service Architecture for Intelligent IoT*. Hoboken, NJ: Wiley.

11 Tom, R.J. and Sankaranarayanan, S. (2017). IoT based SCADA integrated with fog for power distribution automation. *2017 12th Iberian Conference on Information Systems and Technologies (CISTI)*, 1–4.

12 Gurevich, V. (2018). *Microprocessor-based Relays: Prospects and Challenges*. Ukraine: CRC Press.

13 Leite, J.B. and Mantovani, J.R.S. (2017). Development of a self-healing strategy with multiagent systems for distribution networks. *IEEE Transactions on Smart Grid* 8: 2198–2206.

14 Northcote-Green, J. and Wilson, R.G. (2017). *Control and Automation of Electrical Power Distribution Systems*. Boca Raton, FL: CRC Press.

15 IEC 61850-1-2 (2020). *Communication Networks and Systems for Power Utility Automation - Part 1-2: Guideline on Extending IEC 61850*. Edition 1.0. International Electrotechnical Commission.

16 IEC 61970-301 (2020). *Energy Management System Application Program Interface (EMS-API) - Part 301: Common Information Model (CIM) Base*. International Electrotechnical Commission.

17 IEC 61968-13 (2021). *Application Integration at Electric Utilities - System Interfaces for Distribution Management - Part 13: CIM RDF Model Exchange Format for Distribution*. International Electrotechnical Commission.

18 IEC 62325-301 (2021). *Framework for Energy Market Communications - Part 301: Common Information Model (CIM) Extensions for Markets*. International Electrotechnical Commission.

19 Schumilin, A., Duepmeier, C., Stucky, K., and Hagenmeyer, V. (2018). A consistent view of the smart grid: bridging the gap between IEC CIM and IEC 61850. *2018 44th Euromicro Conference on Software Engineering and Advanced Applications (SEAA)*, 321–325.

20 Belu, R. (2022). *Smart Grid Fundamentals: Energy Generation, Transmission and Distribution*. CRC Press.

21 Schütz, J., Uslar, M., and Meister, J. (2021). A case study research on interoperability improvement in smart grids: state-of-the-art and further opportunities. *Open Res Europe* 1 (33): 33.

22 Gopstein, A., Nguyen, C., O'Fallon, C. et al. (2022). *NIST Framework and Roadmap for Smart Grid Interoperability Standards, Release 4.0*. Special Publication (NIST SP). National Institute of Standards and Technology. https://doi.org/10.6028/NIST.SP.1108r4.

23 Leite, J.B., Mantovani, J.R.S., and Kezunovic, M. (2018). Distribution system self-healing implementation using decentralized IED-based multi-agent system. *2018 IEEE PES Transmission & Distribution Conference and Exhibition - Latin America (T&D-LA)*, 1–5.

24 Leite, J.B., Mantovani, J.R.S., Dokic, T. et al. (2019). Resiliency assessment in distribution networks using GIS-based predictive risk analytics. *IEEE Transactions on Power Systems* 34: 4249–4257.

25 Kim, J.S., So, S.M., Kim, J.T. et al. (2019). Microgrids platform: a design and implementation of common platform for seamless microgrids operation. *Electric Power Systems Research* 167: 21–38.

26 Leite, J.B., Peralta, R.A.V., and Mantovani, J.R.S. (2021). Restoration switching analysis in the integrated architecture for distribution network operation. *Electric Power Systems Research* 194: 107069.

27 Marcadet, D. and Lambert, É. (2016). RiseClipse: Why working at the model level is better for validating data conforming to IEC standards. *2016 IEEE Power Systems Computation Conference (PSCC)*, 1–5.

28 EPRI (2018). Program on Technology Innovation: Understanding Internet of Things Architectures. *2018 Technical Report: 3002014370*.

29 Gellings, P.E. (2020). *Smart Grid Planning and Implementation*. Denmark: River Publishers.

30 EPRI (2018). Program on Technology Innovation: Evaluating IoT Messaging Protocols for DER Management. *2018 Technical Report: 3002014370.*

31 Venanzi, R., Kantarci, B., Foschini, L., and Bellavista, P. (2018). MQTT-driven node discovery for integrated IoT-fog settings revisited: the impact of advertiser dynamicity. *2018 IEEE Symposium on Service-Oriented System Engineering (SOSE)*, 31–39.

32 MQTT.ORG (2015). Benchmark of MQTT servers. http://www.scalagent.com/IMG/pdf/Benchmark_MQTT_servers-v1-1.pdf (accessed 12 March 2023).

33 Yang, Y. (2019). Multi-tier computing networks for intelligent IoT. *Nature Electronics* 2: 4–5.

34 Venanzi, R., Kantarci, B., Foschini, L., and Bellavista, P. (2018). MQTT-driven sustainable node discovery for Internet of Things-fog environments. *2018 IEEE International Conference on Communications (ICC)*, 1–6.

35 Veeramanikandan, M. and Sankaranarayanan, S. (2019). Publish/subscribe based multi-tier edge computational model in Internet of Things for latency reduction. *Journal of Parallel and Distributed Computing* 127: 18–27.

36 Berry, T. (2019). Introduction to IEC 62361-102 CIM-61850 harmonization. *25th International Conference on Electricity Distribution*, 1–5.

37 Reiz, C. and Leite, J.B. (2021). Hardware-in-the-loop simulation to test advanced automation devices in power distribution networks. *IEEE Transactions on Power Delivery* 36: 2194–2203.

38 Leite, J.B. and Mantovani, J.R.S. (2015). Development of a smart grid simulation environment, Part I: Project of the electrical devices simulator. *Journal of Control, Automation and Electrical Systems* 26: 80.

39 Bedse, P.R. and Jangle, N.N. (2018). Review on PMU using recursive DFT algorithm. *2018 International Conference on Computing, Power and Communication Technologies (GUCON)*, 375–377.

40 IEC 61968-3 (2021). *Application Integration at Electric Utilities - System Interfaces for Distribution Management - Part 3: Interface for Network Operations*. International Electrotechnical Commission.

41 Chappell, L. (2017). *Wireshark 101: Essential Skills for Network Analysis*, 2e. Chappell University.

9

Voltage Regulation and Reactive Power Optimization for Integration of Distributed Energy Resources into Smart Grids

Firdous Ul Nazir[1], *Bikash C. Pal*[2], *and Rabih A. Jabr*[3]

[1]*Glasgow Caledonian University, Glasgow, United Kingdom*
[2]*Imperial College London, London, United Kingdom*
[3]*American University of Lebanon, Beirut, Lebanon*

9.1 Introduction

The most significant change that the distribution systems are undergoing is the rapid integration of distributed energy resources (DERs) like solar panels, wind farms, electricity storage, and electric vehicles. This paradigm shift toward DER-rich distribution networks is driven by the ambitious climate targets of global economies [1]. DERs exhibit characteristics that are vastly different from the traditional synchronous generators. DERs, like wind and solar power plants, are weather-dependent, which makes their power output highly variable, besides being non-dispatchable [2]. On the other hand, the traditional synchronous generators can be dispatched at a deterministic set point. These different characteristics of DERs lead to improper behavior of traditional voltage control strategies of the distribution network [3]. Such improper behavior includes infeasible operations, such as the excursion of node voltage magnitudes outside the permissible limits. These infeasible operations are highly undesirable as they are associated with huge penalties and inhibit further integration of DERs with the grid. In fact, maintaining the voltage profile of the network within acceptable limits is one of the main factors in deciding the DER hosting capacity of the network. Moreover, the voltage variations in distribution systems are more susceptible to the nodal real power injections because of the higher resistance to reactance ratio of distribution lines. This further complicates the voltage regulation of the distribution networks in the presence of uncertain DERs connected to the network. Thus, advanced voltage control and reactive power optimization schemes, commonly known as Volt/Var control (VVC), are required to increase the integration of DERs with the power network. Stochastic optimization is the primary tool for this purpose. This type of decision-making allows random variables to be used in the dataset of the problem. Thus, the solutions obtained are more robust than their deterministic counterparts. Apart from the basic stochastic optimization, more advanced versions, namely, chance-constrained and robust optimization, also find huge applications.

Motivated by the above concerns, researchers have recently focused on modeling the VVC problem through these advanced stochastic optimization routines. A stochastic VVC program was set up by considering the uncertainties present in the hourly solar irradiance, wind power, and load forecasts by the authors in [4]. The optimization problem was subsequently solved by employing an evolutionary technique known as the modified teaching-learning algorithm. Another study considered a robust reactive power optimization framework for modeling the stochasticity present in the

Smart Cyber-Physical Power Systems: Fundamental Concepts, Challenges, and Solutions, Volume 1, First Edition.
Edited by Ali Parizad, Hamid Reza Baghaee, and Saifur Rahman.

DER outputs, which was solved through the column and constraint generation algorithm [5]. In yet another study, a chance-constrained approach to solving the VVC problem under data uncertainty was presented, whose solution approach is powered by the gradient descent algorithm [6].

This chapter primarily explores the chance-constrained version in the context of the VVC for radial distribution systems. The chance-constrained programming studies the stochastic behavior of the various constraints pertaining to the deterministic optimization model through their probability spaces. These probability spaces of the constraints directly or indirectly depend on the probability functions of the uncertain data available to the programming model. The chance-constrained optimization finally aims at satisfying the probabilistic constraints with a predefined probability level. In the context of the VVC, the chance-constrained programming will ensure that the magnitudes of the node voltages and the branch currents are maintained within the acceptable limits with a predefined probability level, despite the uncertain nodal injections due to DER and load powers.

The chance-constrained problems are traditionally solved by reformulating the non-convex probabilistic constraints to obtain their deterministic counterparts by using the information from probability density functions [7, 8]. However, these deterministic counterparts, obtained through reformulation, turn out to be more complex than the original primary constraints. For example, a linear constraint under probabilistic guarantees results in a second-order conic constraint. In this way, the complexity of the reformulated chance-constrained version is elevated compared to the original deterministic problem. Moreover, in the context of VVC, carrying out the necessary reformulations for chance-constrained programming is very difficult in practice due to the non-linear nature of the power flow model of the network. This chapter discusses a new methodology for solving the chance-constrained problems by using the stochastic information present in the scenarios (possible values of the uncertain parameters). This scenario approach preserves the basic structure of the constraints used in the original problem, thereby not elevating the complexity of the problem. Furthermore, the hard to obtain probability distribution functions of the chance constraints are not needed to ensure the solution.

Many of the advanced VVC routines employ convexification and relaxation techniques, like second-order cone programming (SOCP) and semi-definite programming, to deal with the non-linear power flow model of the power network. However, these convexification techniques introduce auxiliary variables in the formulation. For example, the SOCP technique introduces the square of voltage magnitudes as a variable. In the process, the actual voltage magnitude variable is not explicitly available to the optimizer. This makes it impossible to accurately capture the voltage dependence of the network loads in the VVC routines. Two approximate models are presented in this chapter to deal with this issue. Finally, the use of these load models for energy conservation functionalities of the VVC is briefly discussed.

9.2 Traditional Volt/Var Control

VVC is a primary function of the smart grid distribution management system, which aims at satisfying all the operational constraints of the distribution networks while also increasing their efficiency. The main aim of VVC is to ensure that the system voltage profile is always acceptable by restricting the voltage magnitude of each node within its upper and lower limits. Some VVC routines also allow secondary objectives like minimizing net active power loss of the network or minimizing the power demand on the substation transformer [9]. VVC routines achieve these objectives by determining a periodical schedule of the network voltage control devices (VCDs), viz., switched shunt capacitors, load tap changing transformers, and reactive power injections from the DERs.

The VVC techniques are classified into two broad categories based on the communication infrastructure between the VCDs: centralized VVC and distributed VVC [10]. As the name suggests, the centralized VVC techniques employ a centralized controller that decides the optimum control settings of the VCDs based on the current network topology, generation pattern, and demand profile. Since the centralized controller requires the present state of each node of the system to come up with its decisions, thus an extensive communication infrastructure is needed. In contrast, the distributed VVC techniques need lesser communication infrastructure [11]. The distributed VVC techniques divide the entire network into various zones equipped with their local controllers. The voltage control is performed separately in each zone, with little or no communication between adjacent sections. Some distributed schemes allow a small overlap of the adjacent zones, and by forcing the control variables in these overlapping parts to be equal, the final solutions are assured to be the same as those of the centralized VVC schemes [12].

The most common and practical solution methodology of the VVC has been realized by employing the discrete coordinated descent algorithm [13], which is commensurate with the real-time computing requirements and scales well for large practical networks. This algorithm employs a systematic search methodology in which the current solution is investigated in each possible direction for a better solution. The algorithm terminates when no further improvement in the solution is possible. This algorithm was further improved by guiding the search in the right direction through sensitivity coefficients computed at the current solution. This made the algorithm approximately twice as fast [14]. In addition, the well-known interior point methods have been employed to solve the VVC problem by relaxing the discrete variables to continuous spaces and successively using quadratic discretization penalties to skew outputs toward the available discrete levels of the variables [15, 16]. Some of the other techniques, which have been successfully used to solve the VVC problem by the researchers include mixed-integer linear programming [17], dynamic programming [18], genetic algorithms [19], and particle swarm optimization [20].

9.3 Network Model

Consider a radial power distribution network with n buses; its topology can be described by a directed tree, $\Gamma = (\mathcal{N}, \mathcal{B})$, where $\mathcal{N} = 1, 2, \ldots, n$ represents the node set and $\mathcal{B} = 1, 2, \ldots, n-1$ represents the directed edges of the tree. The tree is further assumed to be rooted at node 1, representing the transmission system feed-in point, with the edges directed away from the root. Each node $j \in \mathcal{N}$ is characterized by a particular voltage, denoted by $\tilde{V}_j$, whose magnitude and square of magnitude are represented by $|V_j|$ and u_j, respectively, and a known complex power injection $\tilde{S}_j^l = P_j^l + iQ_j^l$ from its shunt-connected load. The loads are modeled through the *ZIP* modeling concept, *Z*, *I*, and *P* refer to the constant impedance, constant current, and constant power, respectively, such that $P_j^l = \alpha_j^P + \alpha_j^I|V_j| + \alpha_j^Z|V_j|^2$ and $Q_j^l = \beta_j^P + \beta_j^I|V_j| + \beta_j^Z|V_j|^2$. Additionally, some nodes host shunt capacitors and/or DERs. The reactive power injected by the shunt capacitor connected at node j is denoted by $Q_j^c = b_j^c u_j$, where b_j^c is the susceptance of the capacitor, which varies in discrete steps. In contrast, the complex power injected by DER at this node is $\tilde{S}_j^d = P_j^d + iQ_j^d$. Furthermore, Q_j^d is allowed to depend on P_j^d as per a predefined quadratic function, which is $A_j + B_j P_j^d + C_j\left(P_j^d\right)^2$. This type of reactive power control is in spirit with the two-stage stochastic programming concept, where the second stage provides a recourse opportunity through optimal adjustment of Q_j^d to correct any bad effect once the uncertainty in P_j^d reveals itself. Let the operating power factor of a DER, which is determined by various coefficients appearing in its reactive power control law, be denoted by pf^d. Every edge $(ij) \in \mathcal{B}$ of the tree represents a distribution line and is distinguished by its series impedance $r_{ij} + \mathrm{i}x_{ij}$. The distribution lines are modeled using the π-equivalent concept, and thus let us represent the resultant shunt admittance at a node j by $\tilde{y}_j = g_j + ib_j$. Let the complex power flowing from node i to node j on (ij) be $\tilde{S}_{ij} = P_{ij} + \mathrm{i}Q_{ij}$ with the square of the associated current magnitude equal to l_{ij}. The corresponding branch current itself is denoted by $\tilde{I}_{ij}$. This branch current is associated with a real power loss, given by $r_{ij}l_{ij}$ because of the resistance of the branch, and let the total power loss in the network be denoted by $\mathcal{P}$. The nominal tap setting of the transformer on edge (ij) is assumed to be t_{ij}. Let the upper and lower limits of a quantity X be represented by $\overline{X}$ and $\underline{X}$, respectively. We also assume that the set of nodes hosting DERs and capacitors is represented by $\mathcal{D} \subset \mathcal{N}$ with $\mathbf{card}(\mathcal{D}) = n_d$ and $\mathcal{C} \subset \mathcal{N}$ with $\mathbf{card}(\mathcal{C}) = n_c$, respectively. Similarly, the set of edges hosting the transformers is denoted by $\mathcal{T} \subset \mathcal{B}$ with $\mathbf{card}(\mathcal{T}) = n_t$.

Any form of OPF for this network encounters errors associated with the forecasted values of P_j^l, Q_j^l, and P_j^d. These errors are assumed to be Gaussian in this chapter. This dataset is relaxed to a highly accurate and countably finite set, denoted by $\mathcal{S}$ whose cardinality is n_s, by selecting enough scenarios from its probability space. Let the probability of occurrence of a particular scenario be denoted by π, and the scenario index for a specific data point be indicated by a superscript s on its corresponding quantity. The probability of any random event is represented by $\mathbb{P}$ and the expectation of any random variable by $\mathbb{E}$.

9.4 Chance-Constrained Volt/Var Control

The chance-constrained VVC ensures that the node voltage magnitudes are maintained within the acceptable limits with a predefined probability/robustness level, despite the uncertain nodal injections due to DER active power fluctuations. Mathematically, the general chance-constrained VVC framework employs probabilistic operational constraints and aims to minimize the expected value of the network active power loss as follows:

$$\begin{aligned}
\min_{u} \quad & \mathbb{E}(\mathcal{P}) \\
\text{s.t.} \quad & f(\tilde{V}, \tilde{I}, \tilde{S}, \zeta) = 0 \\
& \mathbb{P}(\underline{|V_j|} \leq |V_j| \leq \overline{|V_j|}) \geq 1 - \varepsilon, \quad \forall j \in \mathcal{N} \\
& \mathbb{P}(|I_{ij}| \leq \overline{|I_{ij}|}) \geq 1 - \varepsilon, \qquad\quad \forall (ij) \in \mathcal{B} \\
& \underline{pf_j^d} \leq pf_j^d, \qquad\qquad\qquad\quad\;\; \forall j \in \mathcal{D}
\end{aligned} \tag{9.1}$$

The equality constraints of the above framework represent the power flow model of the network. This framework considers the randomness of the uncertain parameter ζ, and subsequently forces the power system operational constraints up to the desired probability level of $1 - \varepsilon$. The solutions are said to be ε –level robustly feasible. However, computing the probability functions for the operational constraints of bus voltages and branch currents is a big challenge for the framework [21].

9.4.1 Computationally Feasible Approach for Probabilistic Constraints

To deal with the computational challenge associated with the probabilistic constraints of the framework (1), a scenario-based solution approach is used. According to this approach, the uncertain parameter is randomly sampled with respect to its probability space to collect a minimum number, N, of samples. Then the deterministic constraints of the original chance-constrained problem are forced to be feasible for all these samples. This procedure ensures that the resulting solutions are feasible to the original chance-constrained problem with a predefined confidence level [22]. However, the confidence level can only be guaranteed if the deterministic optimization problem is either convex or a mixed-integer program whose continuous relaxation is convex. The work of Esfahani *et al.* [22] further proves that if such a mixed-integer program has θ continuous and Θ binary variables, then the number of samples to be generated is given by the following relation:

$$2^{\Theta}\sum_{i=0}^{\theta-1}\binom{N}{i}\varepsilon^{i}(1-\varepsilon)^{N-i}\leq\gamma \tag{9.2}$$

Where γ is the confidence parameter implying that with probability no smaller than $1-\gamma$, the sample-based solution will be ε-level robustly feasible to the original chance-constrained problem. γ is normally chosen small enough to be negligible.

This scenario approach to solving the chance-constrained problems not only preserves the basic structure of the optimization but also works equally well with any probability distribution, even when the underlying probability distribution is not known, but the samples are available from measurements, as is the case in most practical situations. Thus, the scenario approach is distributionally robust since the value of N in Eq. (9.2) is probability independent.

9.4.2 A Two-Stage Scenario-Based Optimization Framework for Chance-Constrained Volt/Var Control

The chance-constrained VVC framework of Eq. (9.1) can be efficiently solved by the scenario-based approach discussed in Section 9.4.1 by setting up a two-stage control problem. The two-stage control problem allows the network VCDs to operate at different time scales based on their speed of operation. The slow mechanical VCDs, viz., capacitors, OLTCs, and VRs, are kept fixed throughout the optimization horizon – an hourly time window. These controls are referred to as "here and now" decisions because their decisions are made before the uncertainty is revealed. On the other hand, fast VCDs like the DER inverters are allowed to adjust themselves in real-time in response to the fluctuations of the uncertain parameter. These controls are therefore referred to as "wait and see" decisions. Although the VCDs are adjusted in two separate stages, an hourly slow stage and a real-time fast stage, these decisions are made by solving a single optimization problem. This is possible by incorporating a pre-defined rule, which relates the second-stage inverter reactive power (control action) to its active power (the uncertain parameter). The pre-defined rule is chosen to be a univariate quadratic equation whose coefficients are optimized by the two-stage scenario-based optimizer. The pre-defined rule has the following general form:

$$Q_j^{d,s} = A_j + B_j P_j^{d,s} + C_j\left(P_j^{d,s}\right)^2 \tag{9.3}$$

The quadratic rule Eq. (9.3) is a good choice as it closely conforms with the standard $Q(P)$ characteristic piecewise linear curves suggested by the German grid code [23]. However, it is evident that the optimization framework allows any polynomial decision rule, in which case the corresponding constraint is always linear in the decision variables (coefficients of the decision rule) from the optimization point of view. Thus, the two-stage scenario-based VVC optimization problem has the following structure:

$$\min_{o}\sum_{s=1}^{n_s}\pi^s\sum_{(ij)=1}^{n-1}r_{ij}l_{ij}^s \tag{9.4}$$

$$\text{s.t. } f^s(\widetilde{V^s},\widetilde{I^s},\widetilde{S^s},\zeta^s)=0\ \forall s\in\mathcal{S} \tag{9.5}$$

$$\underline{|V_j|}\leq|V_j^s|\leq\overline{|V_j|}\forall j\in\mathcal{N},\quad\forall s\in\mathcal{S} \tag{9.6}$$

$$|I_{ij}^s| \leq \overline{|I_{ij}|} \forall (ij) \in \mathcal{B}, \quad \forall s \in \mathcal{S} \tag{9.7}$$

$$-P_j^{d,s}\left(\frac{\sqrt{1-(\underline{pf^d})^2}}{\underline{pf^d}}\right) \leq \underbrace{A_j + B_j P_j^{d,s} + C_j\left(P_j^{d,s}\right)^2}_{Q_j^{d,s}} \leq P_j^{d,s}\left(\frac{\sqrt{1-(\underline{pf^d})^2}}{\underline{pf^d}}\right)$$

$$\forall j \in \mathcal{D}, \quad \forall s \in \mathcal{S} \tag{9.8}$$

The optimization framework set up in Eqs. (9.4–9.8) is free of probabilistic constraints and thus is much easier to solve than the original chance-constrained version of Eq. (9.1). However, the size of the problem has significantly increased as all the generated scenarios dictated by Eq. (9.2) are considered in this single program. The objective function Eq. (9.4) seeks to minimize the expected active power loss in the network across all the scenarios. The feasible region for this optimization program is demarcated by the constraints Eqs. (9.5–9.8) and is defined by the vector of decision variables o. Equation (9.5) describes the load flow model of the network written for each considered scenario. Equations (9.6) and (9.7), respectively, ensure that the magnitudes of the node voltages and branch currents are strictly restricted in their allowable limits for all the scenarios. The amount of reactive power available from each DER is limited by Eq. (9.8), which defines a power-factor control that ensures that the DER inverter's power factor does not deteriorate beyond $\underline{pf^d}$.

The modeling of the VCDs is incorporated in the load flow model of the network through Eq. (9.5), which needs to be a convex (or mixed-integer whose continuous relaxation is convex) structure for the computationally feasible approach of Section 9.4.1 to be applicable. The branch flow model based on second-order cone relaxations (for more details on this relaxation, interested readers are referred to the Farivar-Low models [24]) is ideal for this purpose. The voltage control parameters like transformer tap position for the transformer hosted on branch (ij), denoted by t_{ij}, capacitor susceptance for the capacitor connected at node j, denoted by b_j^c, and DER reactive power for the DER hosted by the bus j, denoted by Q_j^d are appropriately included in the Farivar-Low models, as given below:

$$P_j^l + P_j^d = \sum_{k:j\to k} P_{jk} - \sum_{i:i\to j}(P_{ij} - r_{ij}l_{ij}) + g_j u_j \forall j \in \mathcal{N} \tag{9.9}$$

$$Q_j^l + u_j b_{cj} + Q_j^d = \sum_{k:j\to k} Q_{jk} - \sum_{i:i\to j}(Q_{ij} - x_{ij}l_{ij}) + b_j u_j \forall j \in \mathcal{N} \tag{9.10}$$

$$\frac{u_j}{t_{ij}^2} = u_i - 2(r_{ij}P_{ij} + x_{ij}Q_{ij}) + \left(r_{ij}^2 + x_{ij}^2\right) l_{ij} \forall (ij) \in \mathcal{B} \tag{9.11}$$

$$P_{ij}^2 + Q_{ij}^2 \leq l_{ij}u_i \forall (ij) \in \mathcal{B} \tag{9.12}$$

The superscript s has been dropped from the above model to simplify the presentation. It is clear that Eqs. (9.9–9.11) are linear in the variables P_{ij}, Q_{ij}, u_i, l_{ij}, b_j^c, and t_{ij} except for the bilinear terms $u_j b_j^c$ and $\frac{u_j}{t_{ij}^2}$ in second and third equality, respectively. These bilinear terms are linearized through the Big-M methodology [25]. If no capacitor is connected to a particular bus, then the term $u_j b_{cj}$ will be absent from the equation pertaining to that bus. Similar logic holds for other terms featuring any of these variables. Besides, (12) represents a second-order convex cone.

9.5 Solution Algorithm

The two-stage scenario-based VVC problem, presented in Section 9.4.2, is set up to ensure that the probability of node voltage violations during the optimization horizon is kept under desirable robustness limits $(1 - \varepsilon)$ in the face of uncertain nodal power injections. However, the optimization framework needs to consider all the generated scenarios simultaneously, as dictated by Eq. (9.2), leading to a huge size problem. Therefore, to keep the program tractable, only a small but appropriate subset of the generated scenarios is considered through a systematic procedure known as a scenario enforcement algorithm [26]. This algorithm guarantees that the solutions obtained are feasible to the original set of scenarios, thereby preserving the a priori robustness and confidence bounds Eq. (9.2). The scenario enforcement algorithm is summarized by the following steps:

(I) Load the power system line data, the predicted value of the load, and DER data. Initialize the solution scenario set by the predicted value of load and DER data. The solution-scenario set contains all the scenarios that the optimization problem must consider in the current iteration. Set the iteration counter to 1.

(II) Solve the base-case VVC problem, which consists of only the predicted values of the uncertain quantities, free of any errors. The base-case problem is the special case of the VVC model presented in Eqs. (9.4–9.8), which is obtained by considering just the sole scenario given by predicted load and DER data. Compute the schedule of the slow VCDs and the coefficients of the pre-defined rule.

(III) Generate N number of samples from the Gaussian forecast errors as dictated by Eq. (9.2). Compute N scenarios by adding the generated forecast error samples with the predicted values of the load and DER data [27].

(IV) Form the bus admittance matrix, Y_{bus}, that accounts for the most recent schedule of the slow VCDs, which in the first iteration corresponds to the schedule computed in step II. Carry out a proper ordering of the columns of Y_{bus} for factorization sparsity and perform its LU factorization.

(V) Compute N power flow solutions using the current injection algorithm [28] for the N scenarios generated in step III. Note that the Y_{bus} remains constant for all these load flows, ensuring that all the N load flow solutions are available quickly enough for the whole algorithm to converge within practical times. The DERs are allowed to inject reactive powers according to the computed coefficients of the pre-defined decision rule. Check the load flow solution of each scenario for any constraint violation and hence collect the set of violated scenarios. Let the cardinality of the violated scenario set be N_V.

(VI) If $N_V = 0$, stop and dispatch the schedule of the slow classical VCDs and the coefficients of the pre-defined decision rule, else go to the next step. Note that when N_V becomes zero, the current schedule is sufficient to drive the operational constraints of all the N scenarios to their feasible region.

(VII) Choose the most dominant subset of cardinality N_{VR} from the violated scenario set by running the submodular scenario reduction algorithm [29]. The scenario reduction algorithm not only trims down the size of the violated scenario set but also retains most of the stochastic information. The probabilities of the original scenario set are redistributed according to the rule presented by Growe-Kuska et al. [30], according to which the probability of a preserved scenario is given by the sum of its own probability and the probability of all the deleted scenarios, which are closest to it with respect to their Euclidean distances. Add these N_{VR} scenarios to the solution-scenario set.

(VIII) Run the two-stage scenario-based chance-constrained VVC optimization program as set up in Section 9.4.2, considering the scenarios present in the solution-scenario set. Update the schedules of the slow VCDs and the coefficients of the pre-defined decision rule.

(IX) Increment the iteration counter by one and go to step IV.

It is evident that the two-stage scenario-based optimization program set up in Eqs. (9.4–9.8) is solved in step VIII of the scenario enforcement algorithm. The preceding steps of this algorithm are required to prepare the solution-scenario set, which comprises a subset of violated scenarios that contain most of the stochastic information while at the same time being considerably smaller in size. Ensuring that there are no violated scenarios before the algorithm converges in step VI guarantees that the schedule obtained in the final iteration is robust enough to respect the operational constraints of all the scenarios generated in step III of the algorithm. The corresponding flowchart of the scenario enforcement algorithm is shown in Figure 9.1.

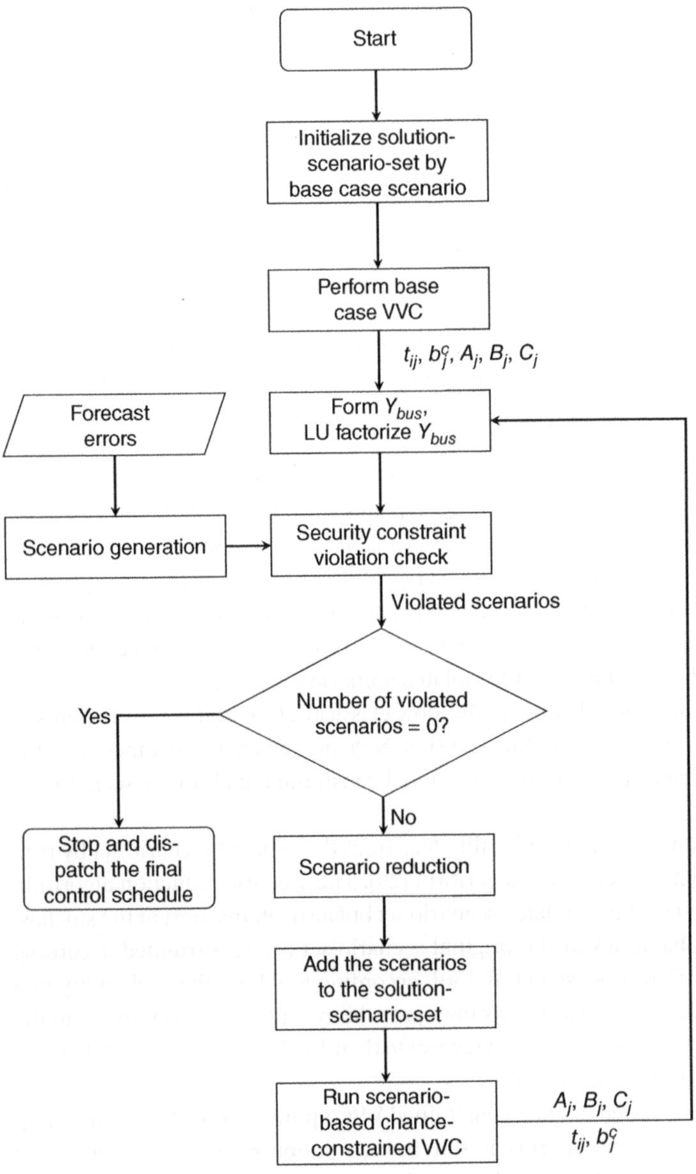

Figure 9.1 The scenario enforcement algorithm.

9.6 Results

The scenario enforcement algorithm was programmed in MATLAB, and the two-stage scenario-based VVC optimization framework was solved using the CPLEX 12.7 optimization studio. The computational tasks were performed on a 3.5 GHz Intel Xeon E5 processor with 64 GB of RAM. The tests were carried out on the 95 bus UK generic distribution system (UKGDS-95). The complete data of this system is available online at [21]. The UKGDS-95 bus network is fed by a 33/11 kV substation transformer, as shown in Figure 9.2. Apart from the substation on-load tap changing

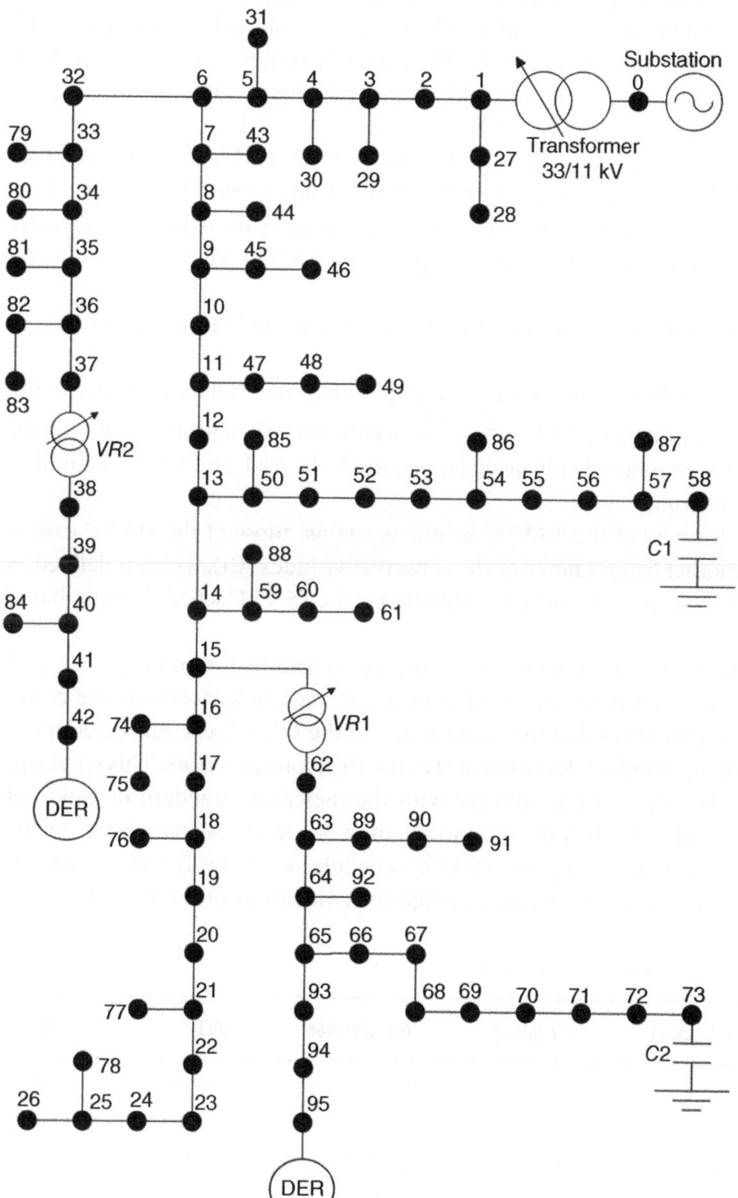

Figure 9.2 95 bus UK Generic Distribution System.

transformer (OLTC), two other voltage regulating (VR) transformers are also present, all of which have an operating range of 0.9–1.1 spanned by 33 equal steps. The shunt capacitor banks, denoted by C1 and C2, can supply a maximum of 300 kvar in steps of 20 kvar. Further, the DERs connected at buses 42 and 95 have an installed capacity of 1.1 and 1.2 MW, respectively. The capacity of the interfacing DER inverters in each case is overrated so that they can operate at 0.95 lead/lag pf during extreme conditions. The standard deviation of the load forecast errors is set equal to 5% of the nominal value [31].

The sample size N is estimated from Eq. (9.2) by choosing a value of $\varepsilon = 0.02$, implying a 98% robustness level, while the confidence parameter γ is set to 10^{-10}. To guarantee these thresholds, Eq. (9.2) dictates that a minimum of 16,272 independent scenarios should be generated. The number of scenarios considered in this study is 30,000, which is sufficient to guarantee that, almost with complete confidence, the solution would be 98% robustly feasible for the original chance-constrained problem.

The schedules obtained from solving the deterministic VVC (D-VVC) and two-stage scenario-based VVC (TS-VVC) are validated against 50,000 Monte-Carlo (MC) trials. Furthermore, the effectiveness of the TS-VVC over the D-VVC is quantified by introducing parameters like the scenario failure rate (SFR) and voltage violation probability (VVP). These parameters are defined as follows:

SFR: SFR is defined as the fraction of scenarios from the set of generated MC trials that have at least one voltage violation.

VVP: VVP is defined as the fraction of the total node voltage magnitude violations across all the MC trials and is mathematically given by $VVP = \frac{\sum_{i=1}^{nMC} nVV_i}{nMC * nB}$, where nMC denotes the total number of MC trials, nVV_i denotes the number of voltage violations in i^{th} MC trial, and nB is the number of buses in the network under study.

The level of the conservativeness of the TS-VVC solutions against those of the D-VVC case is quantified by introducing another factor known as the conservative index (CI), which is defined as the increase in the expected active power loss of the network for the TS-VVC schedule from that of the D-VVC schedule.

Tables 9.1 and 9.2 give the values of the schedules of the VCDs obtained from the D-VVC and TS-VVC approaches, while Table 9.3 shows the resulting voltage magnitude violations and corresponding CIs. Table 9.3 clearly indicates that the performance of the D-VVC schedule deteriorates significantly with the increasing standard deviation (σ^d) in the DER forecast errors. This is evident because the forecast errors become more prominent with the increasing standard deviation of the probability distribution. Table 9.3 further shows that the voltage violations present in the D-VVC schedule are effectively removed by the TS-VVC schedule. With the D-VVC schedules, more than 40% of the scenarios suffer from voltage magnitude violations on at least one node.

Table 9.1 Control set points for the slow classical VCDs.

	σ^d	C1 (kvar)	C2 (kvar)	Transformer	*VR*1	*VR*2
D-VVC	—	60	60	1.04375	0.98125	1.00000
TS-VVC	0.10	60	60	1.03125	0.98125	1.00625
	0.15	80	60	1.01875	0.97500	1.01250
	0.20	60	60	1.02500	0.96875	1.00000
	0.25	60	60	1.03125	0.987500	0.98750

Table 9.2 Coefficients of the pre-defined rule for DER reactive power dispatch.

	σ^d	DER at Bus 42			DER at Bus 95		
		A	*B*	*C*	*A*	*B*	*C*
D-VVC	—	0.0362	0	0	0.0092	0	0
TS-VVC	0.10	−0.0207	0.7830	−2.4847	−0.0426	0.9553	−4.2604
	0.15	−0.0033	0.4537	−1.0692	−0.0053	0.3055	−1.2849
	0.20	−0.0017	0.4176	−1.0874	−0.0223	0.5956	−2.4770
	0.25	−0.0013	0.4785	−1.2543	−0.0005	0.2327	−1.2480

Table 9.3 Robustness performance of the TS-VVC.

σ^d	SFR%		VVP%		
	D-VVC	TS-VVC	D-VVC	TS-VVC	CI%
0.10	42.18	0.67	6.50	0.01	2.29
0.15	48.86	0.28	10.92	0	5.86
0.20	53.22	0.11	14.75	0	6.34
0.25	55.85	5.67	17.91	0.25	3.18

The corresponding schedules for the TS-VVC version are more robust and eliminate the voltage violations in almost all the scenarios. Moreover, the highest value attained by VVP corresponding to the TS-VVC schedule is 0.25%, hence conforming with the a priori probability bounds set for Eq. (9.2). The price of the robustness of the TS-VVC solutions is a slight increase in the expected power loss of the network, as is given by the *CI* in Table 9.3. This slight conservativeness can be traded-off for the robustness of the solutions because robustness is a desirable feature in power system control where infeasible operations may result in a huge penalty.

9.7 Approximate Load Models for Advanced VVC Functions

The load flow model given in Eqs. (9.9–9.12) is essential for the proper working of the scenario-based algorithm, thanks to the second-order conic convex relaxation of Eq. (9.12). However, this convexification is not directly amenable to the famous ZIP (constant impedance, constant current, and constant power) and exponential load models. This leads to the adoption of simplified constant power (Const-P) load models in most cases, but the voltage dependence of the loads is crucial in determining the net nodal injections. The distribution automation project at B.C. Hydro reported a response of 1.6% and 3.1%, respectively, in the active and reactive power demand per 1% change in voltage magnitude in one of their spring days trials [32]. The accurate load models are more important for studies relating to VVC.

It is noteworthy that a load model having both the constant impedance and constant power terms, referred to as the ZP model, can be easily handled by the conic convexified load flow model. There are two methods, based on binomial approximation and linear regression analysis, which seek to find an equivalent ZP model for the original load [33]. For easy presentation, only the expressions considering the real power demand are given below. However, both the equivalent models are easily extendable to the reactive power loads using a similar procedure.

9.8 Binomial Approximation Method

The binomial approximation method (BAM) makes use of the binomial series, which is written in terms of generalized binomial coefficients as follows:

$$(1+x)^{\delta} = \sum_{k=0}^{\infty} \binom{\delta}{k} x^k \tag{9.13}$$

The infinite series in Eq. (9.13) can be approximated to the first two terms, with high accuracy, provided $|x| \ll 1$ and $|\delta x| \ll 1$. The BAM uses this assumption and the fact that under normal operating conditions, the voltage magnitudes at all the nodes in the network are close to 1 pu. Thus, the monomial exponent of the node voltage magnitude can be written as follows:

$$|V|^{\eta} = (1+\Delta|V|)^{\eta} \approx 1 + \eta\Delta|V| \tag{9.14}$$

Letting $u = 1 + \Delta u$ and setting $\eta = 2$ in Eq. (9.14), we obtain the following key relation:

$$\Delta u \approx 2\Delta|V| \tag{9.15}$$

9.8.1 ZIP Loads

The real power demand is modeled in the ZIP framework by a second-order polynomial as given below:

$$P^l = \alpha^P + \alpha^I|V| + \alpha^Z|V|^2 \tag{9.16}$$

The BAM for Eq. (9.16) is derived using the relation Eq. (9.15) as follows:

$$P^l = \alpha^P + \alpha^I(1+\Delta|V|) + \alpha^Z u \tag{9.17}$$

$$P^l \approx \alpha^P + \alpha^I\left(1+\frac{\Delta u}{2}\right) + \alpha^Z u \tag{9.18}$$

$$P^l \approx \alpha^P + \alpha^I\left(1+\frac{u-1}{2}\right) + \alpha^Z u \tag{9.19}$$

$$P^l \approx \underbrace{\left(\alpha^P + \frac{\alpha^I}{2}\right)}_{\alpha_1^P} + \underbrace{\left(\alpha^Z + \frac{\alpha^I}{2}\right)}_{\alpha_1^Z} u \tag{9.20}$$

The relation Eq. (9.20) gives the BAM model for the ZIP loads, characterized by two parameters, α_1^P and α_1^Z, and is thus an equivalent ZP model. It is evident from this relation that the constant current parameter of the original ZIP model gets split equally between the constant power and constant impedance parameters of the equivalent ZP model.

9.8.2 Exponential Loads

The exponential load model expresses the power demand at a node as an exponential function of the voltage magnitude at that node, as given below:

$$P^l = P_0|V|^{\lambda} \tag{9.21}$$

The BAM model for the exponential loads, defined by Eq. (9.21), is also an equivalent ZP model, which can be derived by employing Eq. (9.14) and the same procedure as used for the ZIP loads in Eqs. (9.17–9.20). The eqauivalent model is written as:

$$P^l \approx \underbrace{P_0\left(1-\frac{\lambda}{2}\right)}_{\alpha_2^P} + \underbrace{\frac{P_0\lambda}{2}}_{\alpha_2^Z} u \tag{9.22}$$

The equivalent ZP model for the exponential loads, as given by Eq. (9.22), is defined by two parameters α_2^P and α_2^Z.

9.9 Linear Regression Method

The linear regression method (LRM) also seeks to find an equivalent ZP model for the ZIP and exponential loads of the network. LRM is powered by an estimation procedure based on minimizing the summation of squared residuals between the estimated model and the actual load values. Let us suppose that the parameters of the estimated model are $\widehat{\alpha^P}$ and $\widehat{\alpha^Z}$. Also, let $\mathbf{v}$ be the real-valued vector obtained after sampling the voltage magnitude in a specified range of interest to collect n data points. The Hadamard product of $\mathbf{v}$ by itself is represented by $\mathbf{v}_2$, and similarly $\mathbf{v}_\mathbf{m}$ represents the vector obtained by $\mathbf{m}$ times multiplying $\mathbf{v}$ by itself in an element-wise fashion. Then the following relation holds:

$$\mathbf{V}\begin{pmatrix}\widehat{\alpha^P}\\ \widehat{\alpha^Z}\end{pmatrix} = \mathbf{p}+\mathbf{r} \tag{9.23}$$

Here, $\mathbf{V}\in\mathbb{R}^{n\times 2}$ is defined as the matrix $[\mathbf{1}\ \mathbf{v}_2]$, $\mathbf{p}$ is the vector of the real power values corresponding to $\mathbf{v}$, and $\mathbf{r}$ denotes the residual vector. Note that a boldface lowercase letter represents a vector and a boldface uppercase letter represents a matrix. The LRM aims to minimize the sum of squares of the residual vector to have the best possible equivalent ZP model. The solution of such an LRM estimator is given as:

$$\begin{pmatrix}\widehat{\alpha^P}\\ \widehat{\alpha^Z}\end{pmatrix} \approx (\mathbf{V}^T\mathbf{V})^{-1}\mathbf{V}^T\mathbf{p} \tag{9.24}$$

9.9.1 ZIP Loads

For a ZIP load, each voltage magnitude sample collected in the vector $\mathbf{v}$ corresponds to a particular load sample of the vector $\mathbf{p}$ as per the following relation:

$$\mathbf{p} = \underbrace{[\mathbf{1}\ \mathbf{v}_2\ \mathbf{v}]}_{W}\begin{pmatrix}\alpha^P\\ \alpha^Z\\ \alpha^I\end{pmatrix} \tag{9.25}$$

The matrix $\mathbf{W}\in\mathbb{R}^{n\times 3}$. Using (25) in (24), the LRM estimate for the ZIP load is given as:

$$\begin{pmatrix}\widehat{\alpha^P}\\ \widehat{\alpha^Z}\end{pmatrix} \approx \underbrace{(\mathbf{V}^T\mathbf{V})^{-1}\mathbf{V}^T\mathbf{W}}_{\mathbf{C}}\begin{pmatrix}\alpha^P\\ \alpha^Z\\ \alpha^I\end{pmatrix} \tag{9.26}$$

The transformation matrix $\mathbf{C}$ relates the equivalent LRM with the actual ZIP model, and it is given by:

$$\mathbf{C} = \begin{pmatrix} 1 & 0 & c^p(\mathbf{v}) \\ 0 & 1 & c^z(\mathbf{v}) \end{pmatrix} \tag{9.27}$$

Where $c^p(\mathbf{v}) = \frac{\mathbf{v}\,\mathbf{v}_4 - \mathbf{v}_2\mathbf{v}_3}{\mathbf{v}_4 - (\mathbf{v}_2)^2}$ and $c^z(\mathbf{v}) = \frac{\mathbf{v}_3 - \mathbf{v}\,\mathbf{v}_2}{\mathbf{v}_4 - (\mathbf{v}_2)^2}$. The overline denotes the mean value operator for a vector, thus $\mathbf{v}$ is the mean value of the vector $\mathbf{v}$. It is evident that these coefficients, $c^p(\mathbf{v})$ and $c^z(\mathbf{v})$, are adjustable and their value depends on the vector of the collected voltage magnitude samples only.

9.9.2 Exponential Loads

The full static model for the exponential loads, as shown in Eq. (9.21), is non-linear in nature. This makes it impossible to find closed-form solutions for the LRM estimator for the exponential loads. Therefore, the following procedure should be followed to get the equivalent ZP parameters:

- Sample the voltage magnitude in the desired range and obtain the voltage vector $\mathbf{v}$. This vector is used to calculate the real-valued matrix $\mathbf{V} = [\mathbf{1}\ \mathbf{v}_2]$.
- For each sampled value in $\mathbf{v}$, obtain a corresponding load value using Eq. (9.21). The resultant vector is designated as $\mathbf{p}$.
- Finally, use Eq. (9.24) to estimate the parameters of the exponential model.

9.10 Results

The BAM and LRM approaches for modeling static loads are tested on both the nodal and the network levels to validate their effectiveness. On the nodal level, the accuracy of node power injections is observed, while their effect on the voltage profile of the network is observed by performing network-level load flow studies.

9.10.1 Nodal Level Accuracy

The real power demand at a node is studied for a typical voltage range considering the different load models. Figures 9.3 and 9.4 show the comparison of these different approaches. The parameters of the ZIP model are taken from [34], where $\alpha^p = 0.466$, $\alpha^z = 0.025$, and $\alpha^i = 0.51$. While the parameters for the exponential model are taken from [35], in which the most prevalent value of β throughout the world is estimated to be 0.7. The value of P_0 is 1. For the LRM approach, the vector $\mathbf{v}$ is computed by taking linearly spaced samples with a step length of 0.01 in the range of 0.7–1.3. Thus, the values of $c^p(\mathbf{v})$ and $c^z(\mathbf{v})$ are, respectively, 0.4877 and 0.4969. It is obvious from Figures 9.3 and 9.4 that the traditional method of approximating the static loads with the Const-P-type results in appreciable discrepancy in the net injection, which is efficiently reduced by the BAM and LRM approaches. However, the BAM approach works very well for the normal operating range, while the LRM approach should be employed when the network is under voltage stress. This is because the LRM method is more accurate when the voltage is outside the normal operating range.

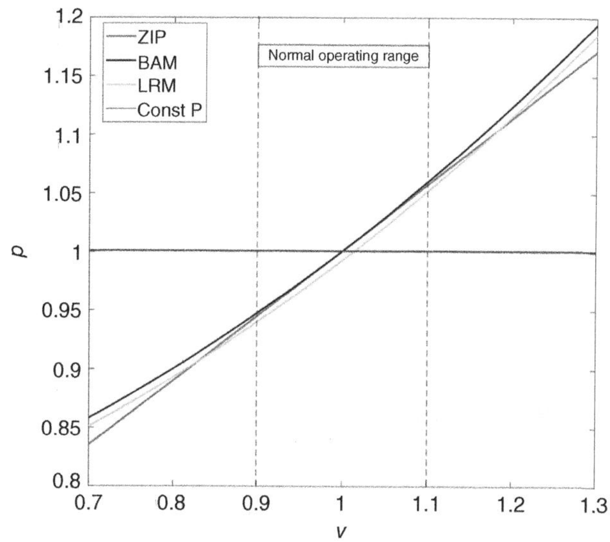

Figure 9.3 Accuracy of the BAM and LRM for the ZIP loads.

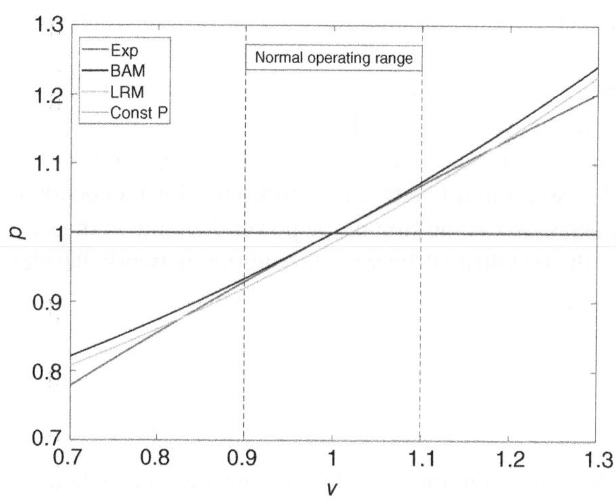

Figure 9.4 Accuracy of the BAM and LRM for the exponential loads.

9.10.2 Effect on the Network Voltage Profile

To understand the effect of the BAM and LRM approaches on the network voltages, the load flow studies are carried out on the UKGDS-95 bus system, whose active and reactive power loads are slightly modified by splitting them equally between the constant power, constant current, and constant impedance parts at the unity voltage magnitude. The benchmark load flow solution is obtained by employing the current injection algorithm [28]. The Const-P, BAM, and LRM approaches are tested separately through second-order conic optimization-based load flow studies [36]. Figure 9.5 shows the percentage normalized error of these load flow studies against the benchmark. It is clear that the BAM and the LRM approaches can effectively reduce the voltage errors present in the traditional constant power modeling approach.

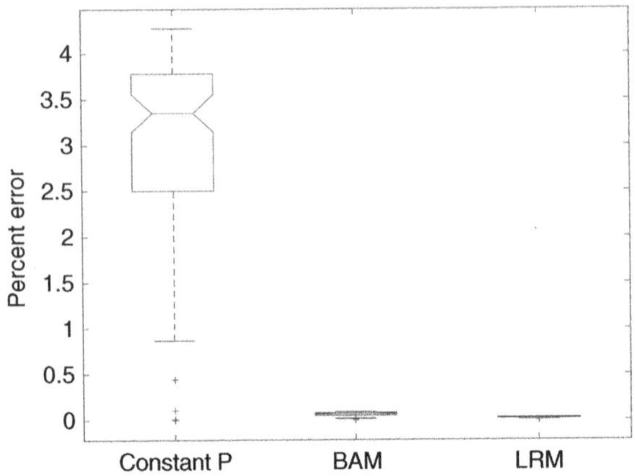

Figure 9.5 Percentage normalized error in the network voltage magnitudes.

9.11 Conservation Voltage Reduction

Conservation voltage reduction (CVR) is the intentional operation of the network toward the lower end of the acceptable voltage profile, aiming to reduce the overall energy demand on the utility. If the customer loads are mostly voltage-dependent, following either the ZIP or the exponential models, lowering the voltage magnitude at their connection point will decrease the net demand on the network. CVR can be conveniently integrated with the VVC routines by properly modeling the voltage dependence of the network loads through the BAM and LRM approaches discussed in the previous section. Moreover, the objective function of CVR is minimizing the net demand of the network, which consists of the network losses and the network-connected voltage-dependent loads. This objective of CVR is more easily expressed as the sum of the power demand on the substation transformer and power demands on the installed DERs for a distribution network. Besides energy conservation, CVR can also help with peak shaving and reducing greenhouse emissions.

9.12 Conclusions

This chapter presented an overview of the VVC problem for a distribution network and a detailed discussion on the two-stage scenario-based VVC approach to achieve highly robust solutions when the network load demand and DER power production are uncertain. The probabilistic constraints of the chance-constrained VVC problem are replaced by enough scenarios from the corresponding probability spaces to guarantee the a priori robustness levels with high confidence. This scenario-based approach is powered by the convexification of the power flow model, and the scenario-enforcement algorithm is used to solve the final problem tractably. The algorithm dispatches the optimum schedule of the classical VCDs and optimizes the coefficients of the pre-defined rule, which relates the DER reactive and active powers. Numerical tests on the UKGDS-95 bus network reveal that the scenario-based algorithm outperforms the traditional deterministic VVC in terms of robust solutions. However, the price of robustness is paid by a slight increase in the expected active power loss of the network.

The conic power flow convexification techniques used by the VVC optimizers cannot properly handle the voltage dependence of the network loads. Two different methods for approximating the voltage dependence of the network loads with high accuracy are the BAM and LRM. Under normal operating conditions, BAM is more accurate, while LRM performs better when the network is under voltage stress. The appropriate load modeling through BAM and LRM further enables the VVC routines to be equipped with the CVR functionalities, through which the net energy demand on the generators is reduced. This is achieved by optimally keeping the voltage-dependent loads at the lower end of the acceptable voltage profile.

References

1 Collinson A., Dai F., Beddoes A. et al. (2003). Solutions for the connection and operation of distributed generation. Technical Report. *EA Technology Ltd.*

2 Miller L., Cibulka L., Brown M. et al. (2013). Distribution system voltage management and optimization for integration of renewables and electric vehicles: status and state of the art. Report. *California Institute for Energy and Environment.*

3 Farag, H.E.Z., El-Saadany, E., and Seethapathy, R. (2012). The evolution for voltage and reactive power control in smart distribution systems. *International Journal of Emerging Electric Power Systems* 13: art. 6.

4 Niknam, T., Zare, M., and Aghaei, J. (2012). Scenario-based multiobjective volt/var control in distribution networks including renewable energy sources. *IEEE Transactions on Power Delivery* 27 (4): 220–2019.

5 Ding, T., Liu, S., Yuan, W. et al. (2016). A two-stage robust reactive power optimization considering uncertain wind power integration in active distribution networks. *IEEE Transactions on Sustainable Energy* 7 (1): 301–311.

6 Agalgaonkar, Y.P., Pal, B.C., and Jabr, R.A. (2015). Stochastic distribution system operation considering voltage regulation risks in the presence of PV generation. *IEEE Transactions on Sustainable Energy* 6 (4): 1315–1324.

7 Zhang, H. and Li, P. (2011). Chance constrained programming for optimal power flow under uncertainty. *IEEE Transactions on Power Apparatus and Systems* 26 (4).

8 López J.C., Muñoz J.I., Contreras J. et al. (2012). Optimal reactive power dispatch using stochastic chance-constrained programming. *2012 Sixth IEEE-PES Transmission and Distribution Latin America Conference and Exposition (TD-LA),* pp. 1–7. Montevideo.

9 Roytelman, I., Wee, B.K., Lugtu, R.L. et al. (1998). Pilot project to estimate the centralized volt/VAr control effectiveness. *IEEE Transactions on Power Apparatus and Systems* 13 (3): 864–869.

10 Tahir M., Nassar M.E., El-Shatshat R. et al. (2016). A review of Volt/Var control techniques in passive and active power distribution networks. *2016 IEEE Smart Energy Grid Engineering (SEGE),* Oshawa, Canada, pp. 57-63.

11 Calderaro, V., Conio, G., Galdi, V. et al. (2014). Optimal decentralized voltage control for distribution systems with inverter-based distributed generators. *IEEE Transactions on Power Apparatus and Systems* 29 (1): 230–241.

12 Nazir, F.U., Pal, B.C., and Jabr, R.A. (2020). Distributed solution of stochastic Volt/VAr control in radial networks. *IEEE Transactions on Smart Grid* 11 (6): 5314–5324.

13 Roytelman, I., Wee, B.K., and Lugtu, R.L. (1995). Volt/var control algorithm for modern distribution management system. *IEEE Transactions on Power Apparatus and Systems* 10 (3): 1454–1460.

14 Jabr, R.A. and Dzafic, I. (2016). Sensitivity-based discrete coordinate-descent for Volt/VAr control in distribution networks. *IEEE Transactions on Power Apparatus and Systems* 31 (6): 4670–4678.

15 Liu, M.B., Canizares, C.A., and Huang, W. (2009). Reactive power and voltage control in distribution systems with limited switching operations. *IEEE Transactions on Power Apparatus and Systems* 24 (2): 889–899.

16 Liu, M., Tso, S.K., and Cheng, Y. (2002). An extended nonlinear primal-dual interior-point algorithm for reactive-power optimization of large-scale power systems with discrete control variables. *IEEE Transactions on Power Apparatus and Systems* 17 (4): 982–991.

17 Borghetti, A. (2013). Using mixed integer programming for the volt/VAr optimization in distribution feeders. *Electric Power Systems Research* 98: 39–50.

18 Lu, F.C. and Hsu, Y. (1995). Reactive power/voltage control in a distribution system using dynamic programming. *IEE proceedings—Generation Transmission and Distribution* 142 (6): 639–645.

19 Augugliaro, A., Dusonchet, L., Favuzza, S., and Sanseverino, E.R. (2004). Voltage regulation and power losses minimization in automated distribution networks by an evolutionary mutiobjective approach. *IEEE Transactions on Power Apparatus and Systems* 19 (3): 1516–1527.

20 Auchariyamet S. and Sirisumrannukul S. (2010). Optimal daily coordination of volt/VAr control devices in distribution systems with distributed generators. *45th International Universities Power Engineering Conference UPEC2010*, Cardiff, Wales, pp. 1–6.

21 Agalgaonkar Y. P. (2014). Control and operation of power distribution system for optimal accommodation of PV generation. Dissertations. Imperial College London.

22 Esfahani, P.M., Sutter, T., and Lygeros, J. (2015). Performance bounds for the scenario approach and an extension to a class of non-convex programs. *IEEE Transactions on Automatic Control* 60 (1): 46–58.

23 Samadi, A., Eriksson, R., Söder, L. et al. (2014). Coordinated active power-dependent voltage regulation in distribution grids with PV systems. *IEEE Transactions on Power Delivery* 29 (3): 1454–1464.

24 Farivar, M. and Low, S.H. (2013). Branch flow model: relaxations and convexifications—part I and II. *IEEE Transactions on Power Apparatus and Systems* 28 (3): 2554–2572.

25 Ahmadi, H., Martí, J.R., and Dommel, H.W. (2015). A framework for Volt-VAR optimization in distribution systems. *IEEE Transactions on Smart Grid* 6 (3): 1473–1483.

26 Nazir, F.U., Pal, B.C., and Jabr, R.A. (2019). A two-stage chance constrained Volt/Var control scheme for active distribution networks with nodal power uncertainties. *IEEE Transactions on Power Systems* 34 (1): 314–325.

27 Aghaei, J., Niknam, T., Azizipanah-Abarghooee, R., and Arroyo, J.M. (2013). Scenario-based dynamic economic dispatch considering load and wind power uncertainties. *International Journal of Electrical Power & Energy Systems* 47: 351–367.

28 Dzafic, I., Jabr, R.A., Halilovic, E., and Pal, B.C. (2014). A sensitivity approach to model local voltage controllers in distribution networks. *IEEE Transactions on Power Apparatus and Systems* 29 (3): 1419–1428.

29 Wang, Y., Liu, Y., and Kirschen, D.S. (2017). Scenario reduction with submodular optimization. *IEEE Transactions on Power Apparatus and Systems* 32 (3): 2479–2480.

30 Growe-Kuska N., Heitsch H., and Romisch W. (2003). Scenario reduction and scenario tree construction for power management problems. *Proceedings IEEE Bologna PowerTech Conference*, pp. 1–7.

31 Jabr, R.A. (2013). Adjustable robust OPF with renewable energy sources. *IEEE Transactions on Power Apparatus and Systems* 28 (4): 4742–4751.

32 Dwyer, A., Nielsen, R.E., Stangl, J., and Markushevich, N.S. (1995). Load to voltage dependency tests at B.C. hydro. *IEEE Transactions on Power Apparatus and Systems* 10 (2): 709–715.

33 Nazir, F.U., Pal, B.C., and Jabr, R.A. (2021). Approximate load models for conic OPF solvers. *IEEE Transactions on Power Systems* 36 (1): 549–552.

34 Martí, J.R., Ahmadi, H., and Bashualdo, L. (2013). Linear power-flow formulation based on a voltage-dependent load model. *IEEE Transactions on Power Delivery* 28 (3): 1682–1690.

35 Milanovic, J.V., Yamashita, K., Villanueva, S.M. et al. (2013). International industry practice on power system load' modelling. *IEEE Transactions on Power Apparatus and Systems* 28 (3): 3038–3046.

36 Jabr, R.A. (2006). Radial distribution load flow using conic programming. *IEEE Transactions on Power Apparatus and Systems* 21 (3): 1458–1459.

10

The Role of Data Analysis in Hosting Capacities of Distribution Power Systems for Electric Vehicles

Alireza Ghadertootoonchi, Mehdi Davoudi, Mohaddeseh Koochaki, and Moein Moeini-Aghatie

Department of Energy Engineering, Sharif University of Technology, Tehran, Iran

Nomenclature

P_t^{E2G}	Energy sent to the grid by the EV (kWh)
P_t^{E2H}	Energy sent to the house by the EV (kWh)
P_t^{dis}	Discharged energy from the EV (kWh)
η_{EV}^{dis}	EV discharge efficiency
Min_{EV}^{dis}	Minimum allowable energy output of the EV (kWh)
σ_t	Charge and discharge control (0, charge; 1, discharge)
P_t^{ch}	Received energy by the EV (charge, kWh)
Max_{EV}^{ch}	Maximum allowable energy input of the EV (kWh)
Max_{EV}^{dis}	Maximum allowable energy output of the EV (kWh)
Min_{EV}^{ch}	Minimum allowable energy input of the EV (kWh)
SOE_t^{EV}	State of the charge of the EV
η_{EV}^{ch}	EV charge efficiency
Δt	Time interval (hour)
$SOE^{EV,max}$	Maximum allowable state of the charge
$SOE^{EV,min}$	Minimum allowable state of the charge
$SOE^{EV,dep}$	State of the charge at the departure time
t_{arr}	Arrival time
t_{dep}	Departure time
RR	Rolling resistance
P_{RR}	Power required to overcome the RR (W)

Smart Cyber-Physical Power Systems: Fundamental Concepts, Challenges, and Solutions, Volume 1, First Edition.
Edited by Ali Parizad, Hamid Reza Baghaee, and Saifur Rahman.

P_{Drag}	Power required to overcome the air resistance (W)
C_{RR}	RR coefficient (–)
C_d	Drag coefficient (–)
m	Vehicle mass (including passengers, kg)
g	Gravitational acceleration (~9.8 m/s^2)
V_{EV}	EV velocity (m/s)
A	Vehicle frontal area (m^2)
V_{wind}	Wind speed (m/s)
D_{wind}	Wind direction (degree)
D_{EV}	EV motion direction (degree)
ρ	Air density (~1.225 kg/m^3)
α	Road slope (degree)

10.1 EVs' Power Demand Forecast Methods

Recently, the adoption of renewable generations such as photovoltaic arrays, wind turbines, etc. has seen a dramatic increase in the power distribution section [1, 2]. Due to the volatile output of these energy sources, some new uncertainties have flourished in the power systems. Moreover, the recent trend toward utilizing electric vehicles (EVs), which one of their main features is their uncertain demand profiles, has aggravated this situation. The growth in penetration of EVs in the electricity grid will increase the burden on the grid as these new loads are highly stochastic. There is a concern that severe consequences will occur if no prediction is available for EVs' power demand, including unwanted power outages in the power grid [3, 4]. Therefore, one subject that has recently gained considerable attention is accurately predicting the electricity demand of EVs in different timeframes and locations; having such a prediction allows the grid operator to manage the network and maintain its stability more effectively. In addition, EVs can also be utilized to preserve the supply-demand balance, flatten the load profile, and reduce the effect of uncertainty caused by renewable energy sources.

In this regard, different forecasting methods can be categorized mainly based on two distinct viewpoints. Some studies have considered the aggregated approach, while others have used the vehicle-centered technique [3]. In the first method, the projected load is a total load of all charging stations, and there is no information about the drivers, their residence addresses, the number of their trips, and the time of each trip. This approach predicts the charging stations' load, considering their historical data, and is less accurate than the vehicle-centered method [5–7]. In the vehicle-centered approach, the share of each car in electricity demand is examined separately [8, 9]. For example, in study [5], to accurately estimate demand, the demand for EVs is modeled separately, and then by summing these demands, the total demand is estimated. Although this type of demand estimation is more accurate than the aggregated approach [10], access to the required information is problematic. Since, firstly, it needs more information, and secondly, the required data are usually related to the car trips, and due to privacy issues, access to this information is difficult [3]. In this regard, the models and methods for predicting EV charging energy consumption can be categorized into three major groups: Statistical, stochastic, and machine learning (ML). The following describes each of these methods.

10.1.1 Statistical Models

In this approach, a statistical distribution is obtained based on available data such as the number and time of trips. The distribution can then be used to analyze the electricity consumption of EVs. For example, authors in [11] use surveys of EV owners to examine their preferences for choosing the place and time of charging during the day. The gathered information is then used to analyze the electricity demand of charging stations.

In another study [12], the driving behavior of EV owners is analyzed, and the results indicate that the normal distribution is a proper probability distribution function (PDF) for EV power consumption. Another research [13] used the Weibull distribution to model the EV's charging time in a vehicle-centered approach. Despite the acceptable results obtained using the Weibull distribution in the mentioned study, such research cannot take into account and analyze the complex behavior of the drivers. In another study [6], historical data showed two peak points in electricity demand during the day; thus, several probability distributions were used to represent the electric load.

Examination of the available data shows that the three available features in most cases are

- The charging time,
- The amount of power received by the vehicles from the charging stations, and
- The time the vehicle arrives at the station.

Hence, many studies have used these three features to model power demand. For example, [14, 15] utilized charging time and energy consumption to develop a Gaussian mixture for load forecasting. Also, the arrival time at the charging station can be modeled with the exponential distribution.

Another statistical method is density estimation, which estimates the density function of a random variable using the observed samples. One related method is the kernel density estimator (KDE), which has been used in several studies to estimate EV loads [16–18].

10.1.2 Stochastic Models

Many real-world phenomena, such as earthquakes and sea waves, are modeled on random processes. Numerous studies have been conducted on predicting EV loads using stochastic processes.

A group of random processes is known as the temporal random process. This method is usually suitable for modeling the load of an EV or a charging station. In cases where it wants to be used for multiple charging stations, it loses accuracy due to not considering the dependence between charging stations. For instance, in [19], the Markov chain models the EV's power demand at a charging station. The Markov chain is a random model for sequential decision-making in which each event depends only on its previous event. It is easier to model events in the real world with this method, which has been used in many studies [19, 20].

Another method referred to as a temporal random process is the auto-regressive integrated moving averages (ARIMA). ARIMA is one of the models used in time series to predict the future, having a sufficient number of initial observations. The method can be used to predict the load of EVs [21]. In this regard, the authors in [22] used historical data on parking demand in charging stations. First, the parking demand in EVs parking lots is predicted, and then using the ARIMA method predicts power demand over the medium term in these stations. In another study [18], daily driving patterns and distances are considered modeling input, and charge demand is predicted. In addition, this study has optimized the ARIMA parameters to improve the results and reduce the mean squared error (MSE).

Another category of stochastic processes is known as spatiotemporal. Since this method considers time and location simultaneously, it can be used to model different stations' loads accurately. It should be noted that the use of such a method on a large scale is problematic due to the presence of spatial information.

Different studies have been conducted concerning the spatiotemporal approach. In [23], a uniform distribution is used to model the time dimension, the Poisson process is used for the location, and a model is developed to predict the power consumption of EVs. In another study [24], Monte Carlo simulation is employed, and charging demand is modeled, considering the location of charging stations for urban and rural areas. The spatiotemporal approach is generally not widely used in urban areas due to its complexity. However, [25] presents a model for predicting spatiotemporal charge demand at fast-charging stations located in urban areas. The researchers gathered data from real-time closed-circuit televisions in Seoul, South Korea, to model the problem. Initially, a Markov chain traffic model determined the entry rate of EVs to charging stations, and then, the charging demand was estimated.

Another study presents a method for predicting EV energy consumption using spatiotemporal models considering weather and traffic [26]. This way, the effect of temperature on battery capacity and air conditioning power is modeled, and the traffic problem is addressed. The daily charge demand by EVs in different regions is then generated by Monte Carlo using model inputs and used in the probabilistic model used in this study to forecast demand.

Authors in [27] used a stochastic simulation method to estimate the charge demand of EVs. This method uses three correlated variables: arrival time at the charging station, departure time from the charging station, and distance traveled during this period. Because these variables have nonstandard distribution functions, the copula distribution function is used to form a joint distribution function. The stochastic simulation method is used to generate artificial data. The reason for using this method is that, firstly, the generated artificial data can consider all the uncertainties of EVs' behavior. Secondly, these synthetic datasets are very useful when shipping data are unavailable.

Electric car battery replacement stations are used to increase driving distance. Due to the randomness of the battery replacement pattern, the article [28] uses stochastic modeling to predict this. Features important in this modeling are as follows: (i) the number of hours after which the EV needs to replace the battery, (ii) charging start time, (iii) charging time, and (iv) travel distance. Then, another simulation model using Monte Carlo to estimate energy consumption is presented using these features.

10.1.3 Machine Learning

Artificial intelligence (AI) and ML methods are other methods used to predict load. Due to the high uncertainty of drivers' behavior and many factors affecting it, there is a need for more accurate methods than other methods [29].

One of the methods based on ML is the linear model (LM); in large networks where the implementation of other models complicates the problem, linear modeling is used for simplicity. For instance, in [30], by modeling the voltage level of each EV with LM, it was revealed that this parameter significantly impacts the level of EV demand. However, due to its incapability to consider complexities in modeling which made it far from reality, the LM cannot understand the irregular patterns in the electricity demand of charging stations and its impact on future electricity demand.

Big Data-based methods such as clustering are also helpful for load prediction. In study [31], the demand for EV power is estimated using clustering data on driving patterns in Denmark. Another example of Big Data techniques is the random forest (RF) method. RF is a learning-based algorithm [31] used for classification and regression based on the decision tree (DT). In studies [32, 33], RF was used to predict the load demand of EVs, and it was observed that this algorithm is also suitable for predicting the load demand of several stations. Another study [10] proposed a Big Data-based method for estimating EV charging demand. This article uses data from South Korean traffic and weather. Cluster analysis was used to classify traffic patterns, and DT was used to create classification criteria. It should be noted that the higher the dimensions of the system or the forecast accuracy, the more the computational load of these models will be.

To reduce the computational load of the system, deep learning techniques such as artificial neural network (ANN) or recurrent neural network (RNN) can be used [7, 34]. ANN-based techniques are commonly used for networks whose datasets are not time-dependent [34], and RNNs are commonly utilized for time-dependent networks. ANN-based methods are highly efficient and accurate in most complex problems [35, 36].

In the article [37], several ANN-based methods were compared, and the result showed that the long short-term memory (LSTM) method is highly efficient and offers better results than other methods. Authors in [38] used a back-propagation (BP) neural network model to estimate demand. This study uses climatic information to improve the model results.

Another example of a ML technique is the Q-learning reinforcement learning (RL) algorithm. In [39], this technique was used to predict charging stations' charging demand, and it was observed that it also improves the predictions made by RNN and ANN techniques. In a nutshell, the methods above can be categorized based on Figure 10.1.

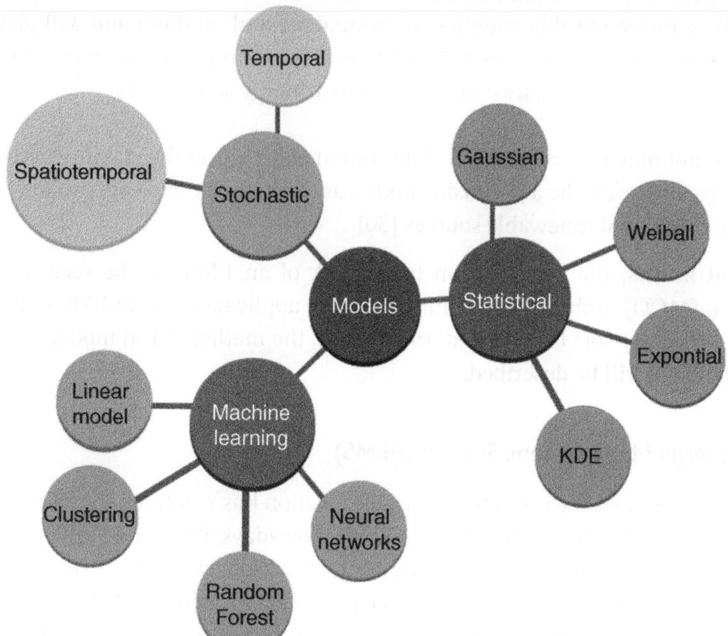

Figure 10.1 Summary of the models used in the EVs' load forecasting.

10.2 Review of EVs' Energy Management Strategies

The growing trend of using EVs remarkably impacts the electricity grid and the load profile. It is because the electricity demand for the EVs is highly uncertain, and they are pure consumers if one does not consider vehicle-to-grid integration [40, 41]. In this regard, EVs can be utilized to increase the grid's stability by providing previously stored electricity to the grid (i.e., vehicle-to-grid) or the house (i.e., vehicle-to-house). An efficient energy management system (EMS) must be able to schedule the interaction between EV, home, and grid to smooth the load profile [42] and minimize the related costs [43].

A proper EMS has many different aspects. These objectives must be taken into account when developing an EMS. Some of these factors will be described in the following:

1) Economic
   The developed EMS must be able to increase the economic profit or decrease the operational energy cost by controlling the charge and discharge of the EV [44]. Paying attention to the economic aspect is crucial as it motivates the car owner to participate in energy management programs.
2) User satisfaction
   An appropriate EMS must be able to take into account the user preferences and convenience [45, 46]. It means that the strategy must be flexible so that the users can modify it based on their preferences, even if those priorities are against the other criteria. Therefore, the EMS does not force the user to follow a specific strategy (e.g., demand response). Also, EVs' EMS must be in a way that does not affect the required energy services (e.g., required passenger-kilometers.)
3) Environmental friendliness
   Considering the current concerns about climate change and global warming, the EMS should be able to manage the energy flow by taking into consideration the environmental limitations. Coordination between EVs and renewable energies is essential for both of them and will play an important role in reducing carbon emissions [47, 48]. However, one must be aware that EVs cause less pollution only when the electricity comes from clean energy sources [49].
4) Utility-related aspects
   The EMS must be able to not only increase the user's profit but also respect the stability of the electricity grid. In this regard, EVs can be a promising option as their storage capacity can lower costs and upsurge the facilitation of renewable sources [50].

Regarding the mentioned factors, the optimization framework of an EMS can be seen as a multi-objective optimization (MOO) problem. In the following, the applications of an EMS in the houses, charging stations, or parking lots is described. In addition, the mathematical model of an EV which can be used in the EMS will be described.

10.2.1 EVs and House Energy Management System (HEMS)

The growing trend of adopting EVs as a means of private transportation has redefined the role of a private vehicle in the house energy management system (HEMS). Nowadays, EVs can be integrated with the house's energy system to lower the daily electricity costs. They can also eliminate the need for an ESS and reduce investment costs by satisfying electricity storage requirements [51].

The general structure of an EV-integrated HEMS is presented in Figure 10.2. As is clear, the framework contains renewable energy sources, different types of loads (i.e., home appliances), battery or ESS, and EV. There is also the possibility of transacting electricity with the electricity grid.

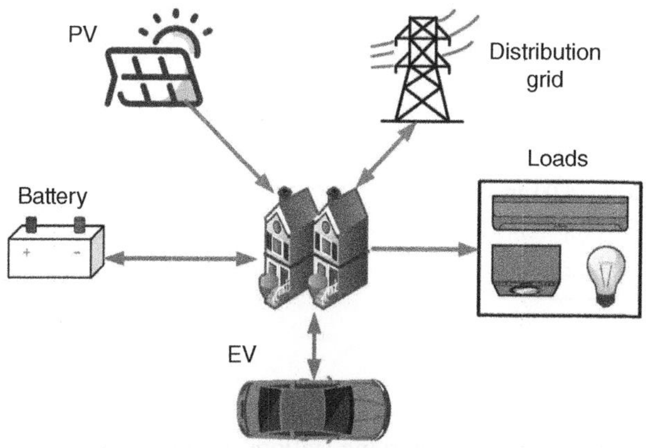

Figure 10.2 HEMS with EVs (Source: [52]).

The presence of EVs in HEMS will increase the electricity demand. Yet, it provides a significant source of flexibility that will benefit the homeowners and utilities [53]. For instance, the HEMS can use EVs to efficiently utilize renewable energies. These sources of energy are highly stochastic. Therefore, EVs' storage capacity can help homeowners flexibly manage renewable energy generation, become greener, and reduce their dependency on the electricity grid [54].

However, it is worth mentioning that when applying the optimization rules to schedule the energy performance of an EV in a HEMS, user comfort must not be sacrificed to reduce the operational costs. Even though the car owners' participation in demand response programs proved profitable [55], the EMS must not force them to do so.

Researchers commonly model EVs as an energy storage system that assists the battery [56]. For instance, an EV-integrated HEMS is defined and analyzed in [57]. The study includes an energy hub consists of home appliances, an electricity storage system (ESS), solar energy, and a plug-in electric vehicle (PEV). In the proposed method, PEV operational time is divided into two sections. First is the interval in which the PEV performs as an ESS and assists the main ESS or battery, and the second is the interval in which the PEV performs as a pure load so that it can store enough energy to operate properly during the day. It is also possible for the user to define a satisfactory level of stored energy that must be reached by the time the vehicle is leaving the house (departure time). The result of the study proved the effectiveness of integrating EV and ESS as the approach reduced the operational cost by 28% compared to utilizing ESS alone.

Authors in [58] optimized an energy hub including rooftop PV, EV, home appliances, and ESS with a mixed-integer linear programming (MILP) model. The considered EV features are illustrated in Table 10.1. Despite the simplicity of the presented method, its main drawback is that the uncertainty of different parameters, like arrival time and initial state of energy (SOE), is ignored. The optimization result demonstrated that during the night, when PV is unavailable, most of the electricity demand could be supplied by EV power, which will decrease the dependency of smart homes on the electricity grid.

In another study, a smart house with renewable energies, ESS, and EVs was modeled and optimized using the MILP method, a well-known and popular approach for modeling and optimizing energy hubs [59]. The study also emphasized the important effect of the EV load stochasticity on the optimum energy flow and daily cost of the smart house. The daily operational cost of the house in the deterministic approach was 11.9 cents, while in the stochastic approach, it was −23.22 cents (the negative sign indicates the profit).

Table 10.1 Typical EV characteristics.

Battery capacity [kWh]	25
Maximum charging/discharging rate [kW]	3.3
Charging/discharging efficiency [%]	95
Arrival time	6.00 pm
Initial SOE [kWh]	8
Minimum SOE [kWh]	4.8

Source: [58].

10.2.2 EMS in EV Charging Stations

In addition to residential energy systems such as HEMS, charging stations are other users that require effective and efficient EMS. The more the EV penetration, the more the demand for parking lots equipped with charging systems and therefore the importance of EMS in this area [60]. In addition, adding the storage capacity of EVs to commercial buildings or workplaces can benefit the buildings too. For instance, the optimal scheduling of EVs in a workplace building with renewable energies can reduce the operational cost by about 7% [61].

The energy management strategy in these stations (Figure 10.3) must be in a way that minimizes the electricity cost and reduce the electricity consumption during the peak load. One option to achieve this goal is to utilize energy storage systems that can benefit from the wholesale electricity market's time of use (TOU) pricing. Such a method is implemented in [63] considering the TOU pricing in the Singapore electricity market and proved effective since it reduced the electricity cost of the charging station by at least 19% and at utmost 56% through an iterative scheduling algorithm.

The main uncertainty faced when the goal is to optimize a parking lot is the model of EVs entering and leaving the parking. To tackle this problem, a study of the presented EVs in that particular area (where the parking lot is located) is needed as done in [64], which analyzed the EVs in the German market and categorized the EVs into ten different groups based on Table 10.2.

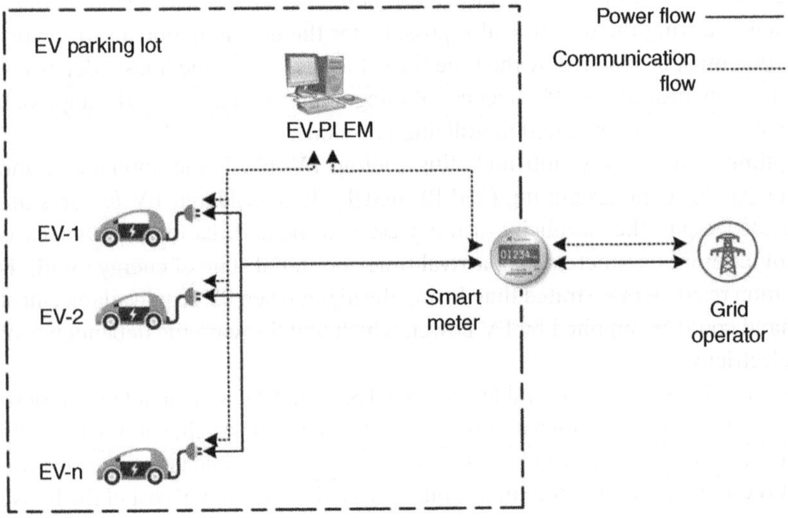

Figure 10.3 Schematic of a PLEMS (Source: [62]).

Table 10.2 The main EV types in the German market and their battery capacity.

Vehicle brand	Volkswagen Grp	Renault	Hyundai Grp	Tesla	Daimler Grp.	BMW Grp	Groupe PSA	Nissan	Jaguar	Honda	All others
Market share (%)	38.3	16.2	12.0	8.7	8.7	8.0	3.3	2.4	0.6	0.2	1.4
Battery capacity (kWh)	40	24	32	8.5	17	12.3	21	20.4	23.4	27	25

Source: [64]/IEEE/CC BY 4.0.

Having the proposed classification, it is now possible to optimally schedule the parking lot or the energy hub since now there is a probability associated whit each type of EV.

Along with energy transactions with the electricity grid, the parking lots in an area can transact required energy between each other in order to effectively supply electricity to the users [65]. Doing so can reduce the electricity cost by at least 2.41% and utmost 12.09%, based on the attained results in [60].

Along with financial profitability, the charging stations' operational reliability must also be considered. According to [66], the charging station's reliability has two main aspects: first, the service life of each device inside the charging station [67] and, second, the impact of the station on the electricity grid [68]. The load variance of the charging station can be utilized as an index of the effect of the charging station on the grid. Such an index can be used as an objective in the EMS of the charging stations to minimize their impact on the distribution grid [66]. Now that the role of EVs in HEMS and parking lot energy management system (PLEMS) is explained, it is time to introduce the mathematical equations which can be employed for modeling EVs in these EMSs.

10.2.3 The Mathematical Model of an EV in an EMS

The general LM, the central part of EV modeling in reviewed studies, is described in (Eqs. 10.1–10.6). Although batteries' charge/discharge process is nonlinear, the linear equations are commonly used for this purpose. The primary motivation for adopting such a scheme is that the solution is guaranteed to be globally optimal [69].

$$P_t^{E2G} + P_t^{E2H} = P_t^{dis} \times \eta_{EV}^{dis} \tag{10.1}$$

$$Min_{EV}^{dis} \times \sigma_t \leq P_t^{dis} \leq Max_{EV}^{dis} \times \sigma_t \tag{10.2}$$

$$Min_{EV}^{ch} \times (1 - \sigma_t) \leq P_t^{ch} \leq Max_{EV}^{ch} \times (1 - \sigma_t) \tag{10.3}$$

$$SOE_t^{EV} = SOE_{t-1}^{EV} + \eta_{EV}^{ch} \times P_t^{ch} \times \Delta t - P_t^{dis} \times \Delta t \tag{10.4}$$

$$SOE^{EV,min} \leq SOE_t^{EV} \leq SOE^{EV,max} \tag{10.5}$$

$$SOE_t^{EV} \geq SOE^{EV,dep} \; t = t_{dep} \tag{10.6}$$

Equation (10.1) states that the discharged power from the EV can be sent to either the grid or the house. Equation (10.2) defines the upper and lower bounds of the EV discharge rate; the binary variable is to prevent charging and discharging simultaneously. The charging limitation constraint is brought in Eq. (10.3). The state of energy of the EV battery is represented in Eq. (10.4). The battery state of the charge must be in a particular range based on Eq. (10.5). Finally, Eq. (10.6)

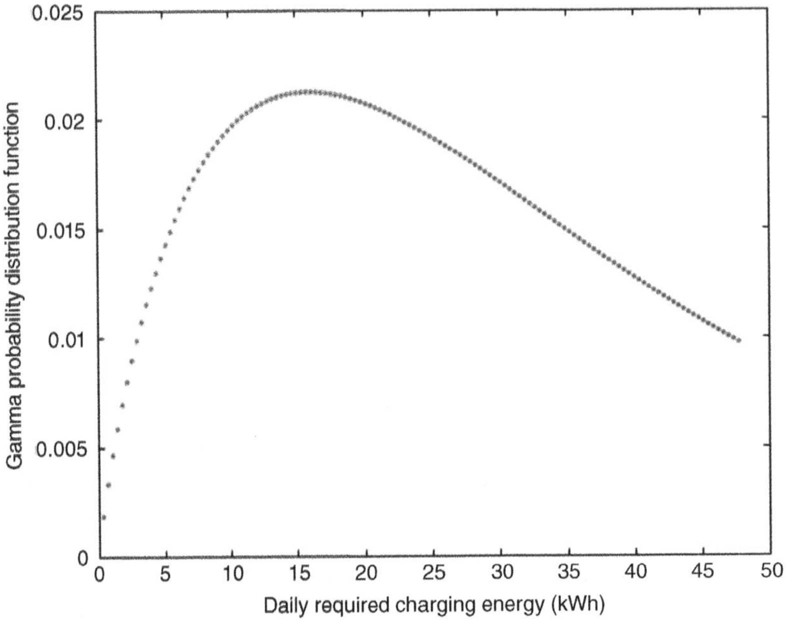

Figure 10.4 PDF of daily required charging energy (Source: [70]).

shows the amount of energy that must be stored in the EV battery at the departure time. It should be mentioned that in all the above equations, except the last one, $t \in [t_{arr}, t_{dep}]$.

The initial amount of energy stored in the EV battery depends on two factors: first, the SOE at the departure time and, second, daily traveling time. The latter is an uncertain parameter that brought stochasticity into the energy management problem. Different probability functions can be used to model the stochasticity of the daily time travel. For instance, in [70], the gamma PDF is considered, and based on that, the daily required charging energy is calculated. The PDF of the daily required charging energy is shown in Figure 10.4.

The normal probability function is another alternative that can be used for modeling the uncertainty of the initial state of the charge, departure state of the charge, arrival time, and departure time as proposed by [60] in Table 10.3.

Another two uncertain parameters are t_{arr}, the time the EV arrives at the charging station [71] and arrival rate. For instance, EVs' arrival and departure rates can be estimated based on historical data [72]. In [73], real data of a university campus are gathered to estimate the arrival rate of EVs. The results are depicted in Figure 10.5.

As is clear from Figure 3, there are two peaks in the arrival rate of EVs at the charging station. Such a pattern can cause unsuitable power peaks in the local distribution grid. Therefore, the EMS

Table 10.3 Normal probability distribution parameters for EV characteristics.

	Initial SoC	Departure SoC	Arrival time	Departure time
Mean	0.5	0.6	9 am	5 pm
St. deviation	0.25	0.25	3	3

Source: [60].

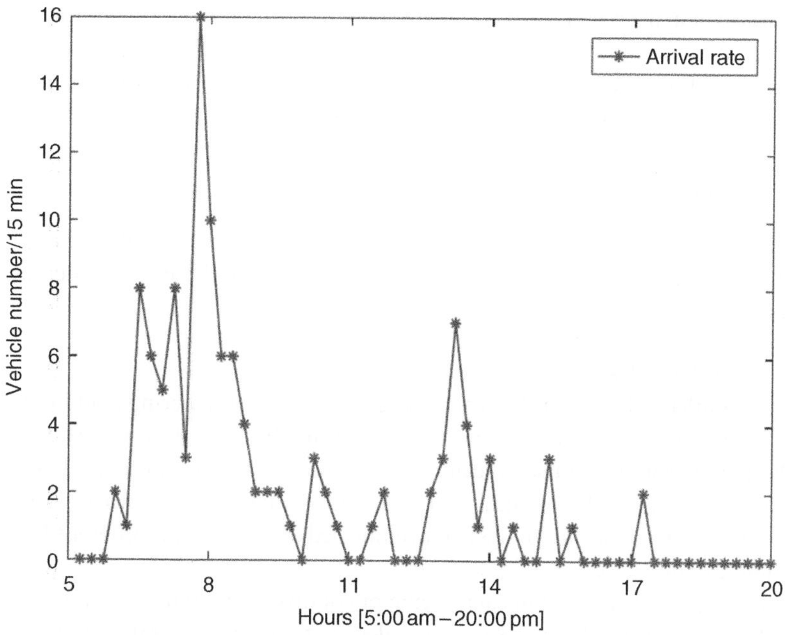

Figure 10.5 The arrival rate of EVs (Source: [73]).

must be able to efficiently manage these peaks and recede them as much as possible. One option to mitigate the peak consumption is to use ESS and renewable energies such as solar systems. Using ESS and on-site renewable energies can result in a 53% power reduction during the peak [73].

10.2.4 Conclusion

Overall, the energy management strategy of EVs can be regarded as an optimization model that tries to optimally allocate the electricity between EVs, homes, charging stations or parking lots, and the electricity grid. The EMS must ensure the profitability of the strategy from different perspectives, namely, the car and or homeowner, the charging station of the parking owner, and the grid operator. It is also worth mentioning that the profit here is not just the economic profit. For instance, from the distribution system operator's (DSO's) point of view, the reliability of the grid as well as its stability must be respected. Also, from the owners' point of view, the strategy must be flexible and take into account their convenience. In addition, to increase the sustainability of the transportation system, the EMS should operate in a way that reduces the environmental effects.

10.3 Uncertainties Regarding EVs and Their Impact on the Power Networks

Increasing the use of EVs has various effects on the electricity grid that must be taken into account. EVs are very different from other electrical appliances used by households. The most important feature of EVs that will be examined in this section is the various uncertainties associated with them. These uncertainties can be related to distance traveled, driving patterns, weather conditions, battery status, and arrival and departure times.

Additionally, if the goal is to plan the operation of a parking lot or charging station for EVs, uncertainty related to the vehicle model should also be considered. Accordingly, vehicle type, network condition, and electricity price should be taken into consideration when optimizing a parking lot or a charging station. In the following, the mentioned uncertainties will be explained in more detail.

10.3.1 Uncertainties Related to EVs

As mentioned earlier, an EMS will face different uncertainties when trying to optimally manage the energy flow in the power grid in the presence of the EVs. In this section, these stochasticities are discussed and demonstrated.

10.3.1.1 Uncertainty in the Distance Traveled

The distance traveled by EVs affects the amount of charge they require when entering a building, parking lot, or charging station. However, the distance traveled cannot be accurately modeled for each vehicle on a daily basis and should be considered an uncertain parameter. Authors in [71] used the 2m PEM method, a statistical method for handling uncertainty, to manage the distance traveled stochasticity. The upper and lower bounds for daily distance traveled are derived for different EVs entering a charging station (Figure 10.6).

It should be noted that the distance traveled in each region or country is different from another. For example, a review of EV data in the United States revealed that the average annual mileage (Figure 10.7) is 17,200 km/yr, equivalent to 47 km/d. Nevertheless, the average daily mileage is 45 km in Denmark and 34 km in the UK [74].

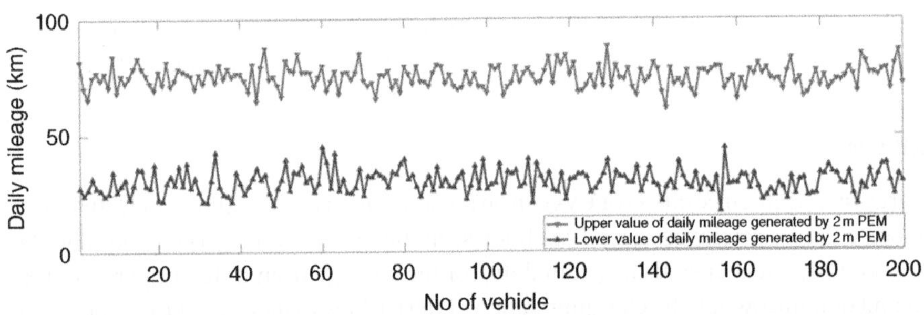

Figure 10.6 Upper and lower limit of distance traveled by EVs (Source: [71]/with permission of ELSEVIER).

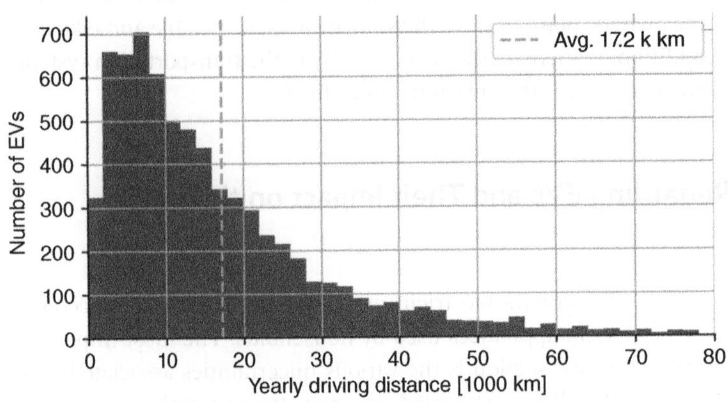

Figure 10.7 Distance traveled in a year in the USA (Source: [74]).

Table 10.4 Commuting probability for different reasons.

Reason		PDF (%)	Average distance (km)	Average duration (min)
A	Work	56.69	26.60	37.70
B	Shopping	14.54	16.80	31.10
C	Study	10.30	19.00	33.40
D	Health care	9.55	22.80	38.70
E	Others	8.92	16.80	31.10

Source: [75].

One of the factors that can help estimate the distance traveled is the reason for the trip. Moving from one place to another can have various reasons, such as going to work, shopping, studying, and health care. Each of these reasons requires a certain distance traveled. For instance, a study in Brazil has shown that the most common reason for traveling in this country is to get to work, which requires an average of 26 km of travel [75]. More information is provided in Table 10.4.

10.3.1.2 Uncertainty due to Driving Patterns

Various factors affect the driving pattern of a vehicle (not necessarily an EV). These factors can be classified into six categories, as shown in Figure 10.8 [76].

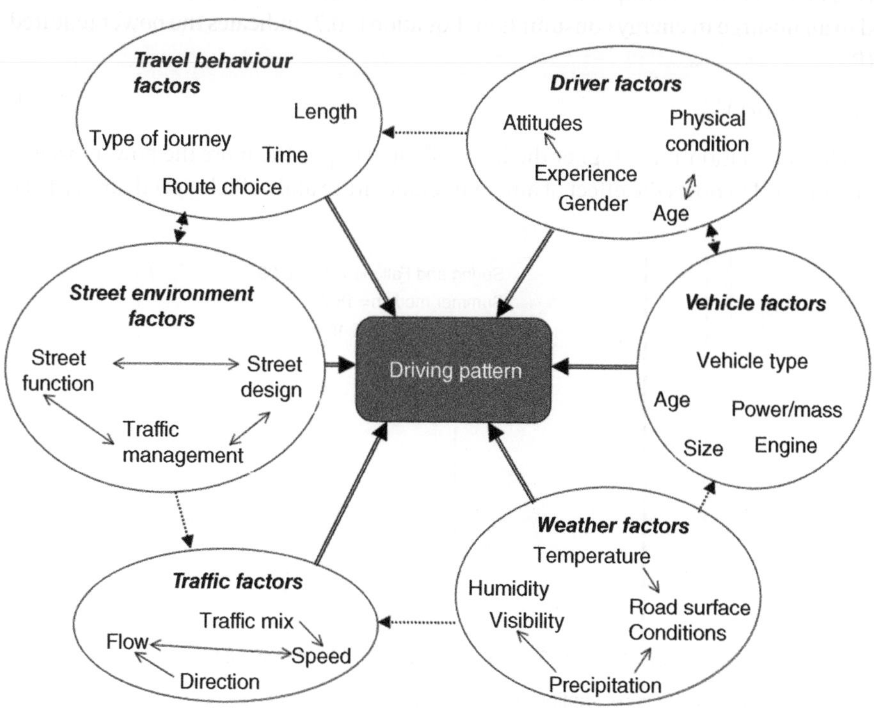

Figure 10.8 Factors that affect the driving pattern (Source: [76]/with permission of ELSEVIER).

For example, traffic affects the energy consumed per kilometer (or mile) traveled. Energy consumption per mile will increase by 7–10.5% in a situation with more traffic [77]. Another influential factor is the weather, which will be discussed in more detail in the following subsection.

Also, as shown in Figure 3, vehicle type is one of the subsets of vehicle factors that affect the driving pattern [78]. For instance, interviews with 11 electric car owners in Lisbon show that three-quarters of them drove slower and more cautiously than when using a nonelectric car [79].

10.3.1.3 Uncertainty due to the Weather Condition

Weather conditions are a factor that can be considered to model the energy consumption of EVs. The primary problem with this parameter, like other EV-related parameters, is its uncertainty. This section will examine the effect of air temperature, precipitation, wind speed, and direction on the energy consumption of EVs.

Air Temperature Air temperature affects electricity consumption in two ways: It (i) affects battery performance, and (ii) increases vehicle energy demand to adjust the inside temperature. Therefore, the effect of temperature on energy consumption can be studied seasonally. For example, a study of the energy consumption of personal EVs in Beijing showed that the average energy consumption per 100 km in winter is 39% higher than in autumn and spring. The effect of season on energy consumption is shown in Figure 10.9 [80].

Figure 10.10 shows the effect of air temperature on the energy required to travel 100 km. As can be seen, the best operating temperature range is between 12 °C and 28 °C. Lowering the temperature below 12° can increase consumption by up to 60% [80].

Another factor affected by the air temperature is the rolling resistance (RR) between the tire and the road surface. The lower the air temperature, the more the friction between the tire and the road, which will lead to an upsurge in energy consumption. Equation (10.7) indicates the power required to overcome RR.

$$P_{RR} = C_{RR} mg \cos(\alpha) . V_{EV} \tag{10.7}$$

As can be seen from Equation 1, the higher the RR coefficient (C_{RR}), the more the power lost due to the friction. Figure 10.11 shows the effect of ambient temperature and vehicle speed on C_{RR} [81].

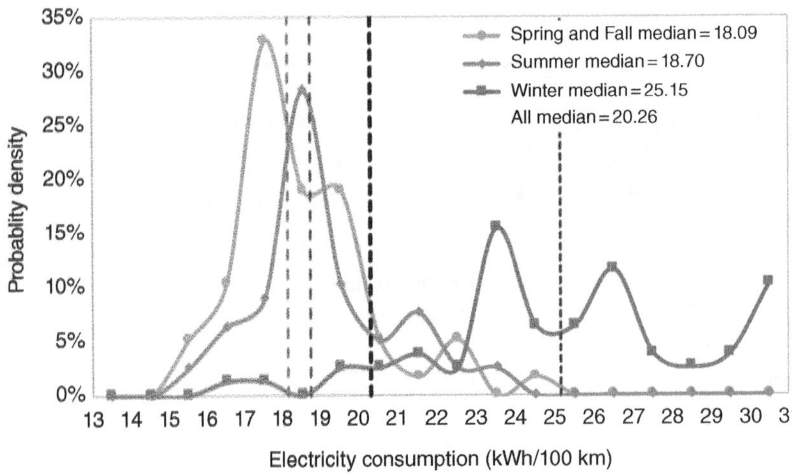

Figure 10.9 Effect of season on EV electricity consumption (Source: [80]/with permission of ELSEVIER).

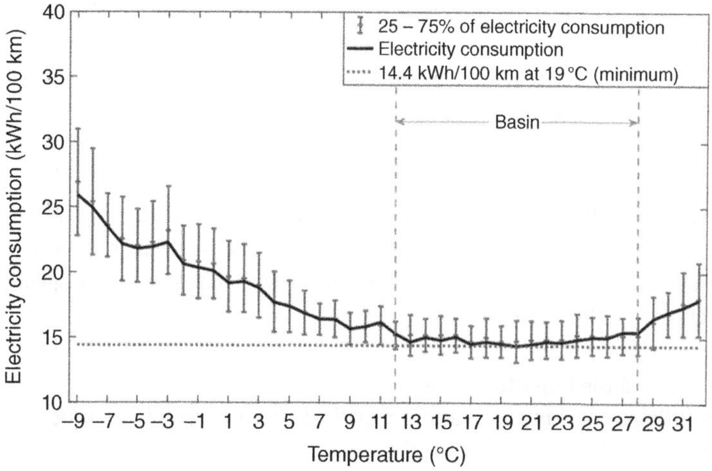

Figure 10.10 The effect of temperature on EV electricity consumption (Source: [80]/with permission of ELSEVIER).

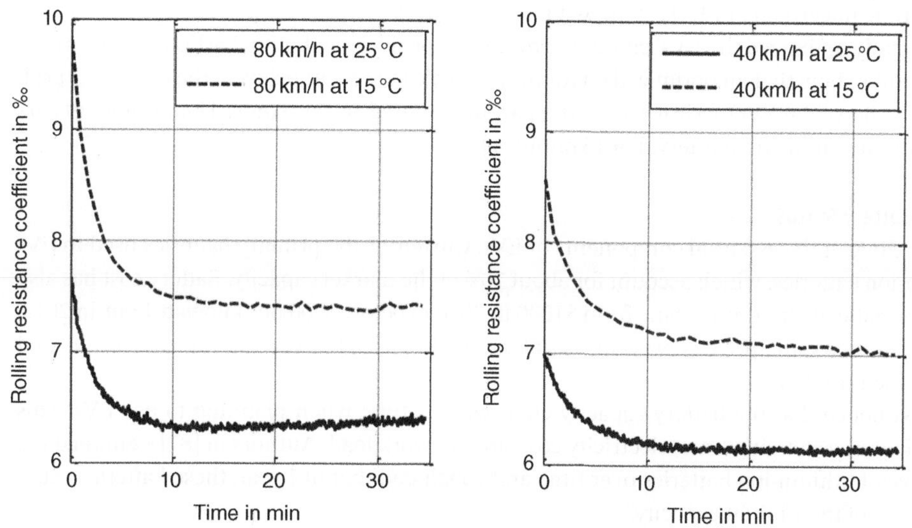

Figure 10.11 The effect of ambient temperature and vehicle speed on rolling resistance coefficient (Source: [81]/with permission of Springer Nature).

The RR coefficient has decreased with increasing temperature, and the effect of temperature is greater than the effect of speed. Nonetheless, one should be aware that the tire model also considerably affects the results. For instance, in some tires, the RR reduction with temperature growth is more intense than the others [82].

Wind Speed and Direction Since one of the main forces that must be overcome to move the vehicle is the drag force, any change in wind speed and direction will change the power required to overcome the air resistance. Equation (10.8) indicates the effect of wind speed and direction on the energy needed to overcome the drag force [83].

$$P_{Drag} = \frac{1}{2} C_d \rho A [V_{EV} - V_{wind} \cos(D_{wind} - D_{EV})]^2 . V_{EV} \tag{10.8}$$

According to Eq. (10.8), if the difference between the wind angle (D_{wind}) and the angle of motion of the vehicle (D_{EV}) is equal to 180°, that is, the wind blows from the front of the vehicle, the power required to overcome the drag force (P_{Drag}) will be maximum. Also, with increasing vehicle speed (V_{EV}), the required power increases. Another important point is that air density also affects the required power.

Precipitation Rainfall can increase the RR between the car tire and the road by affecting the road surface. The RR equation was introduced in Eq. (10.7). It has been shown that the thicker the water layer collected on the road surface, the higher the RR coefficient. The growth depends on the speed and type of tire and has been reported to be 7–10.5% [84].

10.3.1.4 Uncertainty in the Arrival and the Departure Time

EV entry and exit time to the building or parking lot is another critical parameter with uncertainty. Arrival and departure times are crucial because they directly impact the electricity cost needed to charge the car. If many vehicles are charged during peak hours, the cost of charging will be higher and more burden will be on the electricity grid. Also, a proper estimation of arrival and departure hours can help optimize network management. In [71], the 2m PEM method was used to consider the arrival time uncertainty of EVs, as shown in Figure 10.12.

Another study optimized the EVs' charging procedure considering the stochasticity of the arrival rate. The results show that an optimized charging strategy can prevent peaks in the electricity grid. Figure 10.13 shows the EVs' arrival rate in the parking lot used in this study. The number of cars has been counted in the time interval of 15 minutes [85].

10.3.1.5 Battery Status

Batteries are one of the essential components of EVs. Currently, the primary batteries used in EVs are lithium-ion batteries, which account for about 70% of the market capacity. Battery cost has also dropped dramatically in recent years, from $1000 in 2010 to below $200 per kilowatt-hour in 2019. In the coming years, the price is expected to fall below $100/kWh, lowering the cost of EVs and increasing their use [86].

One must notice that the battery capacity changes over time when planning to use EVs. This effect will increase charging time, electricity cost, and network load. Authors in [87] examined the performance of lithium-ion batteries over time and concluded that in 1 year, these batteries could lose up to 20% of their initial capacity.

As mentioned, a change in the battery status due to degradation also affects the operational cost. Authors in [88] examined this effect and concluded that the higher the battery capacity, the higher

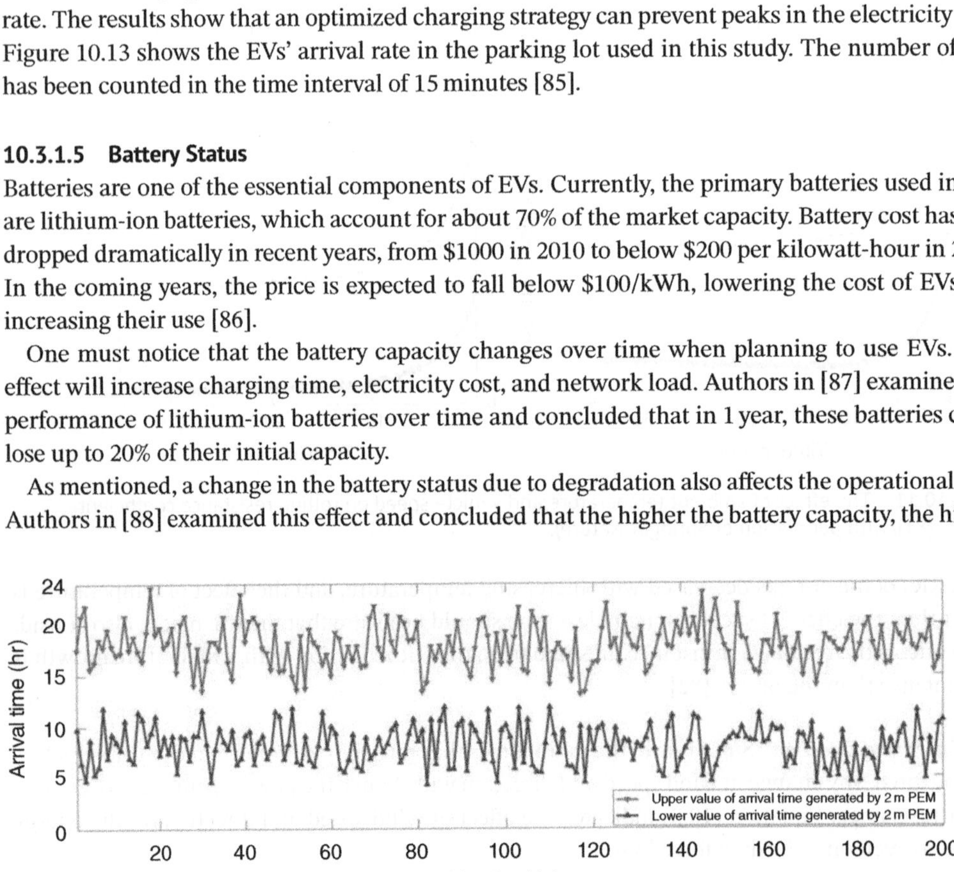

Figure 10.12 The lower and upper bounds of EV arrival time (Source: [71]/with permission of ELSEVIER).

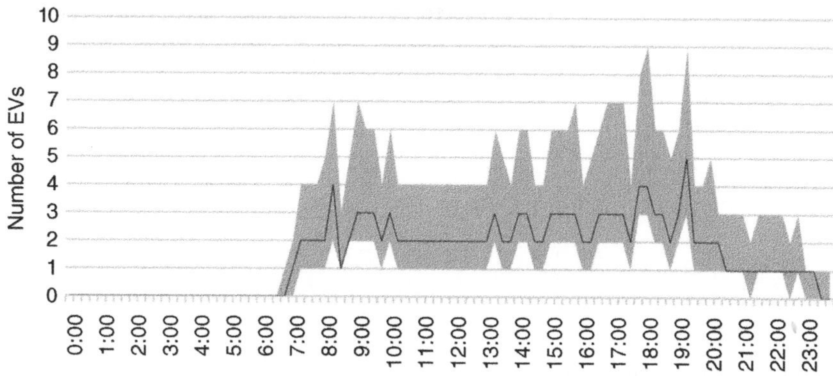

Figure 10.13 Arrival rate of EVs counted every 15 minutes (the gray area indicates 25% to 75% quantiles) (Source: [85]/with permission of ELSEVIER).

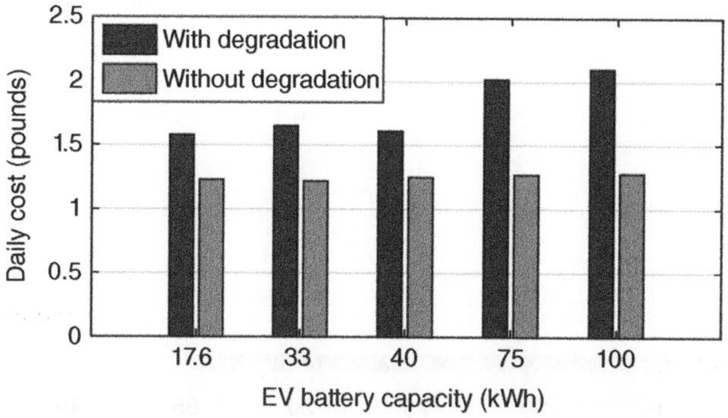

Figure 10.14 The effect of battery degradation on the daily cost (Source: [88]).

the cost of degradation. Figure 10.14 shows the battery degradation effect on the cost, considering the TOU tariff for batteries with different capacities. The degradation cost can be calculated based on battery investment cost, the nominal number of cycles, battery capacity, and nominal depth of discharge (DOD). It should be noted that the number of cycles depends on the operational temperature and DOD [89].

As discussed above, another factor determining the number of battery operating cycles is the depth of charge and discharge. The greater the depth of charge and discharge, the lower the number of operating cycles of the battery, which should be taken into account when planning and optimizing the network of EVs. Figure 10.15 shows the results of various studies investigating the effect of discharge depth on the number of cycles [90].

10.3.1.6 Uncertainty in Vehicle Types

The EV types are another uncertainty with which charging stations and EV parking lots are faced. It is possible to calculate the probability of the presence of each EV model at the charging station by estimating the number of active vehicles in each region or country. The number of EVs in each region or country can be obtained from the statistics of that country. For instance, Figure 10.16 shows in what quantities different EVs were sold in the United Kingdom in 2021 [91].

Considering the number of EVs sold in 2021, the probability of an EV being from one of the above models can be calculated. For instance, the probability of an EV being a Tesla model 3 is 33%. However, one must be aware that such probabilities are imprecise as only the number of EVs sold in 2021 is considered.

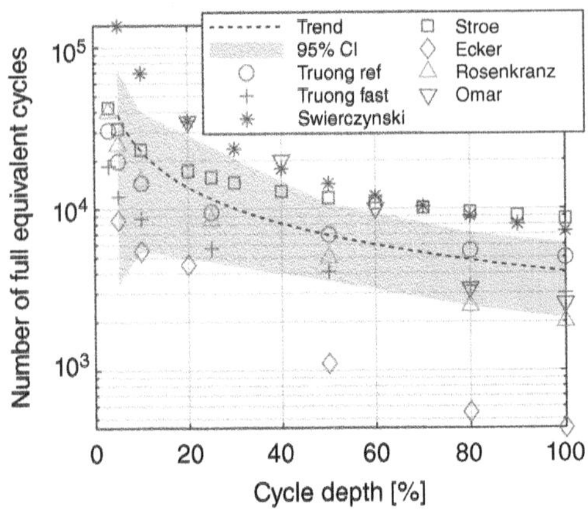

Figure 10.15 Effect of charge and discharge depth on the number of cycles (Source: [90]/with permission of ELSEVIER).

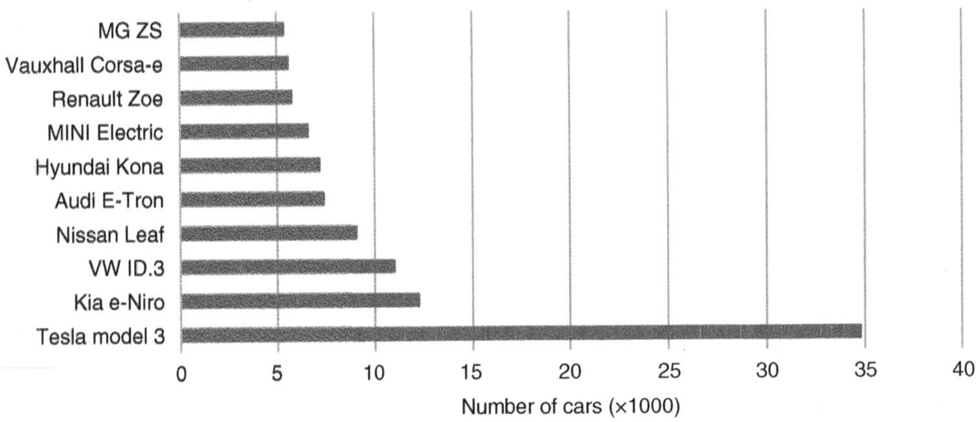

Figure 10.16 Number of EVs sold in the UK by the model (Source: [91]/with permission of ELSEVIER).

The mentioned stochasticities will affect the power grid by increasing the voltage variation due to EVs' charging and discharging process. Therefore, the presence of an EMS that can manage the stability of the grid as well as the uncertainties is necessary. The next section discussed the effect of the EVs and the electricity grid.

10.3.2 Effect of EVs on the Electricity Grid

How EVs affect the power grid is another matter that should be considered when modeling and optimizing their use [92]. In [93], the behavior of EVs was modeled using a Markov chain, and the effect of increasing their penetration on the electricity grid was investigated. According to the results, increasing EV penetration in the power grid leads to sudden voltage and power consumption changes if no proper energy management strategy is employed. Figure 10.17 shows the voltage change in different EV penetration percentages.

When EVs connect to the distribution network, the electricity flow and power generation cost will change [94]. Nevertheless, the change is not in a negative direction necessarily. The power grid can also benefit from integrating EVs into the grid. Such a phenomenon can lead to a reduction in power supply costs and an improvement in the grid load factor. It is a result of the controlled charging (applied by an EMS), which can shift the demand from the peak to the off-peak. A study

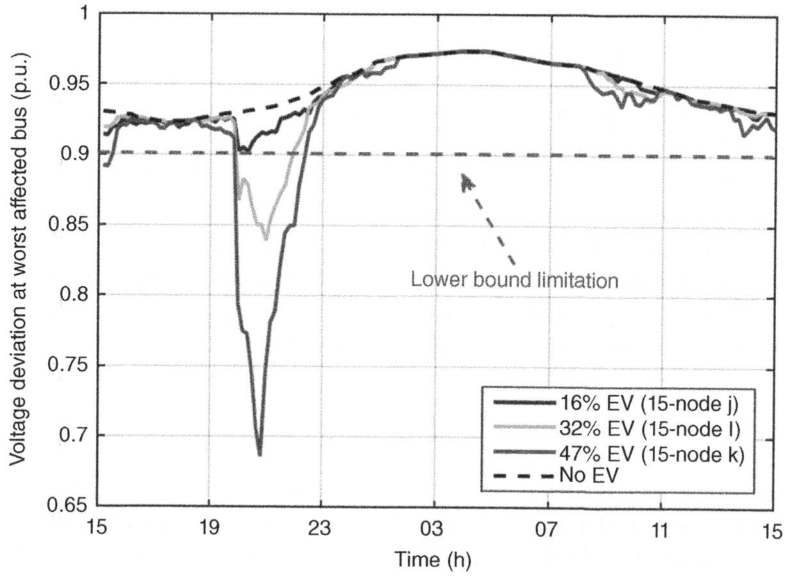

Figure 10.17 Change in the grid voltage under uncontrolled charging (Source: [93]/with permission of ELSEVIER).

in China shows that 50% EV penetration in the power grid can reduce power supply costs by 17.1% [95]. However, it should be noted that the penetration level itself is subjected to the capacity of the distribution network and transformers [96].

The mentioned controlled charging procedure can shift the charging process to the middle of the night when the electricity price is low and prevent feeder overload and voltage drop during peak. It can also increase the diversity of charging load profiles [97].

10.3.3 Conclusion

In general, increasing the use of EVs increases the electrical load of the power grid. Various solutions such as using batteries in charging stations, using renewable energies such as solar energy, and optimizing energy management strategies have been proposed to tackle such a problem. Each of these reduces the pressure on the power grid and helps its stability. Even so, one must always take into account the effects of the EV-related uncertainties when trying to overcome the stated problems. In this regard, an appropriate optimization strategy can efficiently use the available data to estimate the behavior of the EVs taking into account their uncertain behavior. These uncertainties can be related to distance traveled, driving patterns, weather conditions, battery status, and arrival and departure times. In addition, the optimization strategy must not ignore the EVs' effect on the power grid. The strategy should work in a way that increases the solidity and stability of the grid.

10.4 Data Analyses Application in Technical Issues of EVs

Recently, attention to data collection and knowledge extraction from data has increased dramatically. Various algorithms have been developed that can effectively handle a large amount of data and extract proper knowledge from datasets.

These algorithms receive a large number of samples and try to develop a black-box model capable of finding meaningful relationships between the input and output data (i.e., the training stage). The trained model can be used in various applications in the next step. For instance, in the case of EVs, the historical data can be used to develop a model for predicting the charging demand of EVs.

Besides, the use of heuristic and metaheuristic optimization algorithms in this area has experienced remarkable growth. Many studies have optimized the performance of EVs with the help of algorithms such as genetic algorithm (GA) and ant colony optimization (ACO). These algorithms can be used to find the optimal route for EVs to minimize the distance and traveling time.

In order to summarize the mentioned methods, the present section discusses the applications of different learning-based and heuristic approaches in the realm of EVs. The application of ML and heuristic and metaheuristic optimization algorithms are examined in this regard.

10.4.1 Machine Learning

ML algorithms can be categorized into different groups, each of which has its merits and is chosen according to the type of data collected and the purpose of model development. These algorithms are usually divided into supervised, unsupervised, and RL (Figure 10.18). Different studies have used these methods to model EVs' performance. The key findings of these researches are examined in this section.

In the following sections, each of these approaches will be presented and discussed.

10.4.1.1 Supervised Learning

The overall procedure for developing a supervised ML algorithm is illustrated in Figure 10.19. Based on the process, the first stage is data gathering; the next is preparing the gathered data in a way that can be used in the supervised ML algorithm. After that, the data will be divided into two parts: the training dataset and the test dataset. The first set will be used to train the model, and the latter is for evaluating its performance.

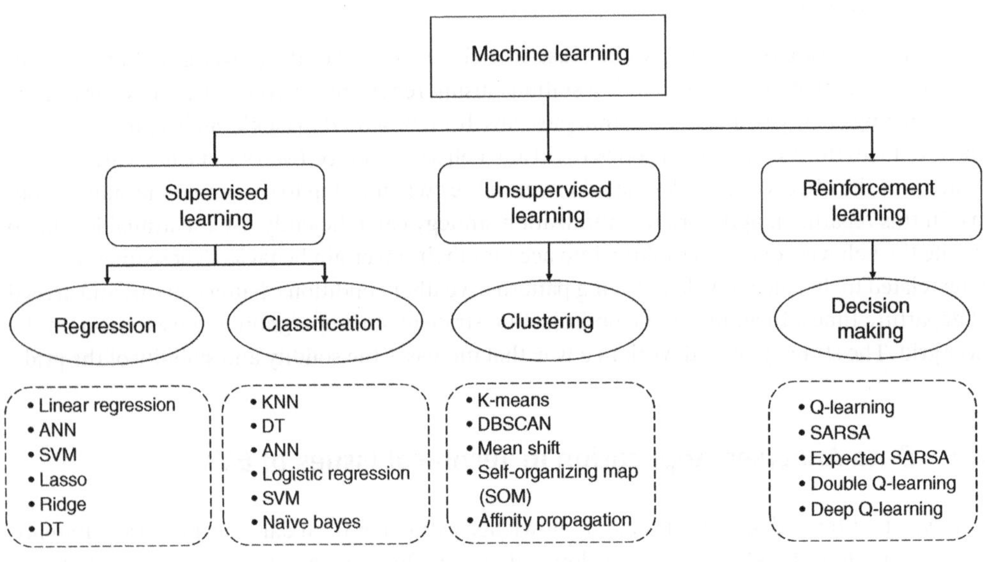

Figure 10.18 Main categories of ML algorithms and widespread examples of them.

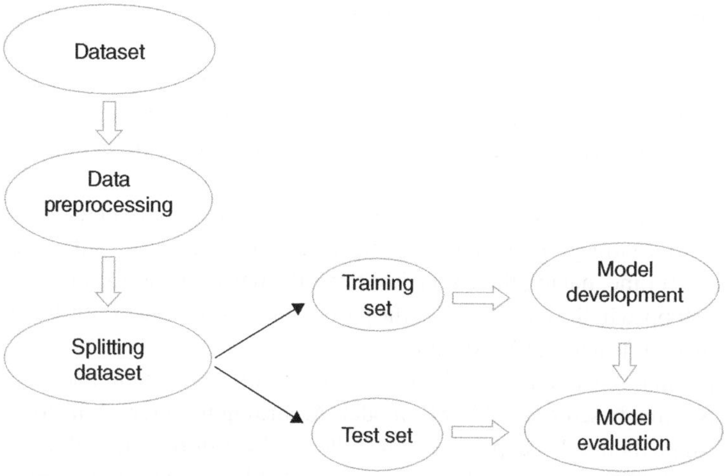

Figure 10.19 Supervised ML model development procedure (Source: [98]/MDPI/CC BY 4.0).

As illustrated in Figure 1, supervised models are further divided into two sections: regression and classification models. In the following, the application of each of these methods is discussed.

Regression Methods Regression algorithms are a subset of supervised learning models which are useful when the goal is to develop a model for predicting continuous variables such as charging energy consumption, battery state of the charge, and remaining range. The following discussed the applications of regression methods in EVs' EMS.

Charging Energy Consumption Estimation ML algorithms can estimate the energy required to charge a car which is helpful for managing the hosting capacity of the network. For instance, various algorithms such as support vector regressor (SVR), DT, RF, and K-nearest neighbor (KNN) were used to estimate the battery energy requirement [99]. Historical data on EV charging were collected from the University of California Los Angeles (UCLA) smart grid energy research center charging station on the UCLA campus and the EA technology website to achieve this goal. The above models were used to estimate EVs' stay duration and energy consumption. The study results showed that the SVR algorithm is more appropriate for EV stay duration estimation. On the other hand, the RF algorithm is more suited for energy consumption estimation. The study clearly shows that it may be necessary to use a separate algorithm to analyze and predict distinct EV-related indicators.

Another study introduced a publicly available dataset to estimate EV stay duration, arrival time, and energy consumption using the Gaussian mixture model. The model was developed based on the normal probability distribution. The study showed that the reliability of estimating the two mentioned parameters with the help of collected historical data is higher than asking users to predict these parameters directly through surveys or questionnaires [14].

Having a reasonable estimate of the arrival and departure time and energy requirements of each EV can help reduce costs and improve the smart charging process. The availability of an estimate of the EV departing time from the charging station is critical because it helps the EMS bring the battery charge to the desired level before the EV leaves the parking lot.

In addition, considering the effect of EVs' charging patterns on the distribution network reliability, the arrival and departure time uncertainties make the optimum grid scheduling difficult, especially in the presence of public charging stations. Therefore, accurately predicting these two parameters can help the DSO to organize the grid properly. Time series algorithms can be utilized to tackle this problem. For instance, in [100], RNNs, LSTM, and gated recurrent units (GRU) are used for arrival and departure time prediction. The LSTM outperformed the other two algorithms in predicting arrival time. However, the performance of GRU was more promising in predicting the departure time.

In [101], a linear regression model has been used to estimate the mentioned parameter. It is shown that the higher the model accuracy, the better the battery performance at the departure time. The model estimates the exit time, having the EV arrival time, the day of the week, and the average time when that EV has stopped in the parking lot in the past. Figure 10.20 shows the critical features considered in the study for estimating EV exit time.

These features and their importance are obtained with the help of the extreme gradient boosting (XGBoost) algorithm, which is a tree-based ensemble ML model. According to Figure 2, historical departure time is essential in predicting EV departure time, followed by historical arrival time. The reason there is a focus on predicting arrival and departure time is that having a proper estimation of these two parameters helps the EMS estimate the EV energy requirement and schedule the charging process to minimize operational costs.

The linear relationship between the EV charging energy demand and its stay duration at the charging station has been analyzed in [102]. In this study, the start and end time of charging is estimated by collecting data related to EV users. Then, the stay duration is calculated by taking the difference between these two. Finally, the relationship between energy consumption and stay duration is investigated. The steps of this analysis are further explored below. First, historical data on the arrival and departure of EVs at the charging point are obtained. An example of this data is given in Figure 10.21.

As can be seen from Figure 4, there is a clear pattern for EVs' entry and exit times. Arrival time is before 10 am in most cases, and departure time is between 3 pm and 8 pm. The relationship

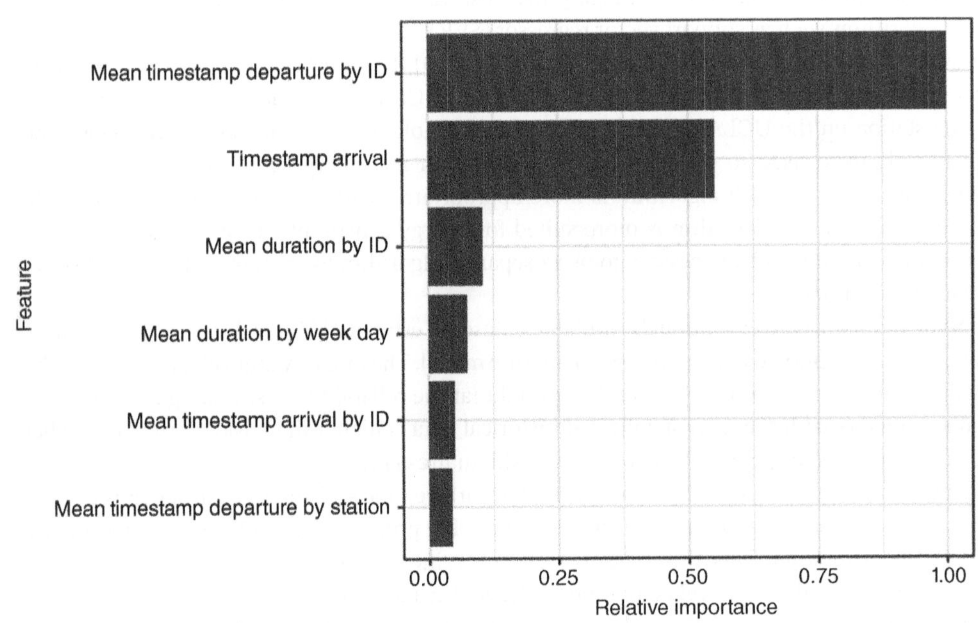

Figure 10.20 Significance of each feature in estimating the EV departure time (Source: [101]).

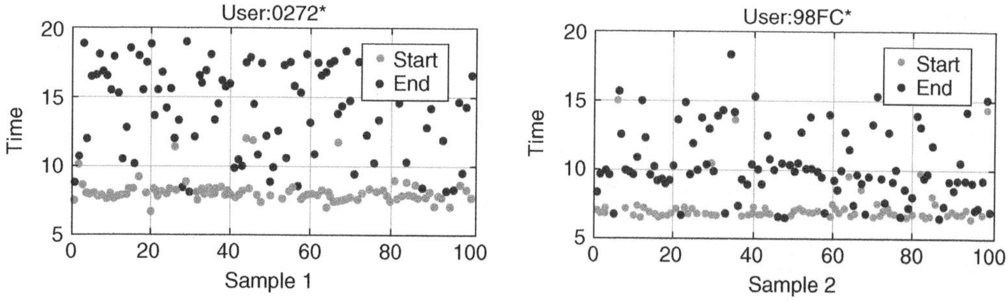

Figure 10.21 EV entry and exit times to and from the charging point (Source: [102]).

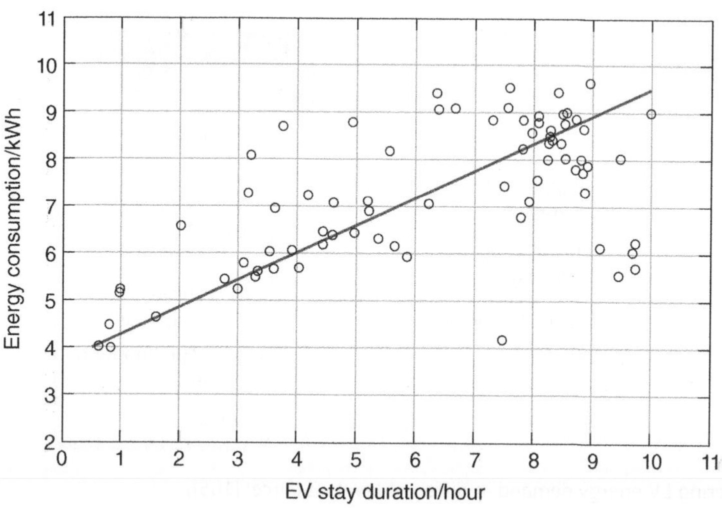

Figure 10.22 Linear relationship between stay duration and energy consumption (Source: [102]).

between stay duration and the energy consumption is investigated in the next step. Figure 10.22 shows that the relationship is linear.

Due to the linearity of the expression, the linear regression algorithm can be used to estimate the amount of energy required for each EV.

Travel Energy Consumption and Range Estimation In spite of estimating EV energy demand in parking lots (i.e., charging demand), ML models can also be used in estimating EV energy demand during a trip. The overall structure of such a model is presented in Figure 10.23. As is vivid, the ML model gets the data on different operational parameters and finds the relationship between them and energy consumption. It is worth mentioning that the alternative to ML models is the white-box models, which are based on mathematical equations and a comprehensive understanding of the physical structure of the vehicle [104].

Different parameters which are affecting EV energy consumption and can be regarded as the ML model input are introduced in Figure 10.24. The parameters are in different levels; however, in most studies, meso- and micro-level parameters are considered [105].

Various data about the trip must be gathered if one wants to predict EV energy demand during a trip. For instance, in [106], features such as total distance traveled, the average speed during the trip, ambient temperature (which affects the need for heating, ventilation, and air conditioning

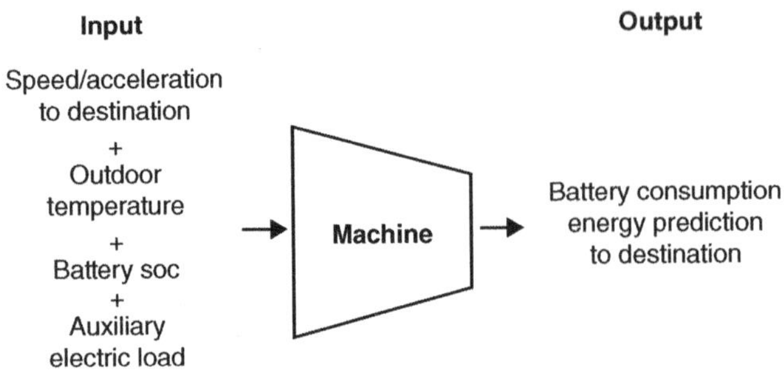

Figure 10.23 Input-output linkage by an ML model (Source: [103]).

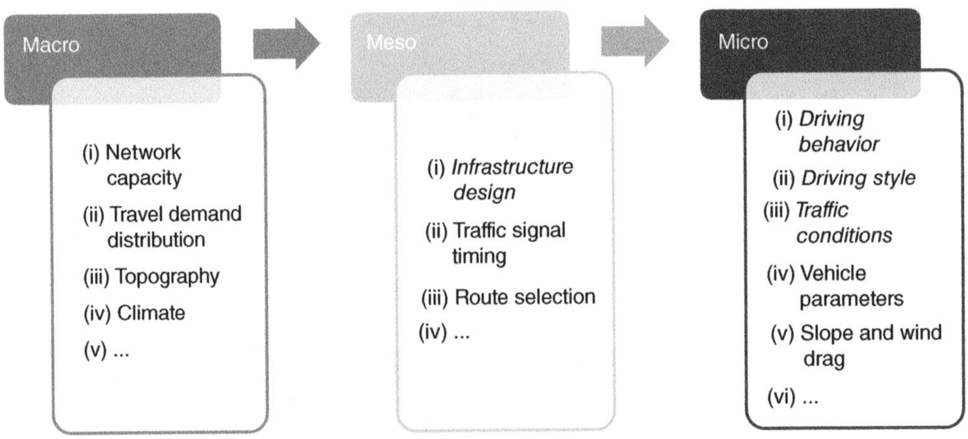

Figure 10.24 Parameters affecting EV energy demand in different levels (Source: [105]).

(HVAC) inside the cabin), and positive and negative elevations of the path are considered for model development. The study results show that the most prominent effect belongs to the distance, and after that, positive elevation plays an integral role in vehicle energy consumption. On the other hand, negative elevation reduces energy consumption. Another study used the battery state of the charge (limited to 20–100%) and minimum cell temperature (MCT, in °C) to predict the driving range (km) [107]. The introduced model is represented in Eq. (10.9).

$$Range = a_0 + a_1 \times SOC + a_2 \times MCT \tag{10.9}$$

In addition to linear regression methods, which are straightforward, more sophisticated supervised learning algorithms can also be used for range prediction. For instance, in [98], three well-known supervised learning algorithms were investigated to predict EV energy demand in a trip. The algorithms were linear regression, SVR, and XGBoost. The results show that the highest accuracy is attained using XGBoost, followed by SVR and linear regression. Also, the importance of traveled distance is emphasized. Based on the developed model, it can be concluded that distance traveled and ambient temperature are two significant factors that affect the energy demand of an EV [108].

Another study has also proved the XGBoost algorithm's effectiveness [109]. In this study, different ML algorithms for range estimation were compared, and the best algorithm was XGBoost, followed by light gradient-boosting machine (LightGBM) and gradient-boosting regression tree

(GBRT). While these three algorithms had a mean absolute percentage error (MAPE) of less than 4%, the others (Lasso regression, elastic net regression, RF, bagging, and NN) had a MAPE of more than 8.5%.

In addition, deep learning regression methods have also proved to be effective in range estimation [110]. For instance, in [103], an LSTM-deep neural network (DNN)-based model was developed for range estimation based on gathered data from a Hyundai Kona. The vehicle driving data were gathered during about 1000 hours or 160,000 km. The data were used to train the LSTM-DNN model. The LSTM is a time series prediction model that learns the EV's long-term dynamics, and the DNN builds a nonlinear network model to concretize the input-output energy balance relation. The model can predict the remaining range with more than 90% accuracy.

Another approach that one should takes into consideration is to use blended ML models. Such a method can be considered to predict the remaining driving range of EVs based on historical data. For instance, in [111], a combination of two algorithms, XGBoost and LightGBM, was analyzed for range estimation. Different features were used as input in the study, namely, the battery minimum and maximum temperature, the cumulative output energy of the motor, and cumulative output energy of the battery. The additional features are whether the vehicle is in braking, accelerating, or stop condition; driving time; and driving pattern (attained from clustering different driving patterns with the K-means clustering algorithm.) The result of the study declared that the accuracy of the blended model is more than that of separate models. Therefore, other studies can also consider utilizing blended models. A similar approach used RF, naïve Bayes, Adaboost, and GBoost to predict a household EV charging demand. The study results once again emphasized the effectiveness of such ensemble models [112].

Even though, energy consumption and range estimation are two major categories which are studied using regression methods. The application of the method is not limited to this area. For instance, the authors in [30] developed a linear regression model to analyze the effect of the time in which charging is done (day or night and whether it is peak time or not) as well as hourly temperature and whether the EV is 230 V or 400 V. The result shows that charging process is faster during the peak time. Also, the night is a better time for charging since during the day the process is slower. In addition, it is concluded that the charging speed is lower when the temperature decreases, and the charging speed of 400 V EVs is approximately three times higher than that of 230 V EVs.

Classification Methods The above studies utilized regression methods to estimate and predict EV-related factors such as arrival and departure time, distance traveled, stay duration (in the charging station), and charging energy consumption. In addition, EV energy demand during a trip and remaining range estimation were also studied using regression algorithms. However, it does not mean that classification algorithms (other supervised learning subsections) are not useful for these applications.

These algorithms can classify users based on their preferences, driving patterns, and arrival or departure time behavior and predict their charging behavior, distance traveled, and charging station selection [113]. For instance, in [114], logistic regression is used to predict whether an EV driver will use a fast-charging station or not. The authors gathered data from 130 private EVs in Beijing for seven months. The results showed a significant relationship between the EV initial SOC and fast-charging usage: the less the initial SOC, the more the probability of using fast-charging stations. Also, an increase in the travel time or duration will increase the probability of using a fast-charging station. Another interesting finding was that the drivers who used fast-charging stations were likelier to use them on the next trip.

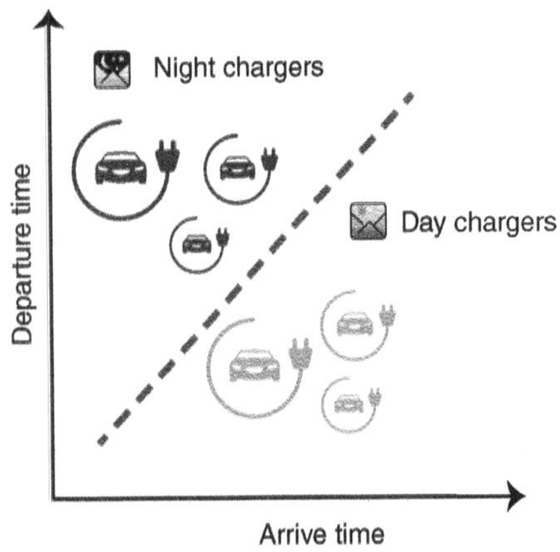

Figure 10.25 EV owner's classification based on the time of charge (Source: [116]).

In another study, the EVs were classified based on their beginning and final location. The authors used PyCaret ML library to tests different classification algorithms and concluded that LightGBM is an appropriate method that can predict the start location with more than 85% accuracy and the end location with more than 75% accuracy. After classifying EVs based on their start and end location, their traveled distance and energy consumption can be predicted [115].

Classification algorithms can also be used to make different groups of EV drivers based on their behavior. For instance, multiple drivers charging data can be gathered, and a classification algorithm can be developed that categorizes the drivers into two groups: those who charge their EVs during the day and those who do so during the night (Figure 10.25). The model then will be used to predict the behavior of new drivers [116].

10.4.1.2 Unsupervised Learning

In classification algorithms, the class of the data is known. For instance, in the previous study [21], the logistic regression algorithm was aware of the possible classes, which were (i) the driver will use fast-charging stations and (ii) the driver will not utilize them. However, there are cases with unknown classes. In such a situation, the data are available, but the algorithm does not know the possible classes. Therefore, instead of classification, a clustering algorithm must be used. For instance, the authors in [117] used the DBScan clustering algorithm to categorize EVs based on their arrival and departure times. A similar study was conducted to categorize EVs based on their charging behavior [118].

One of the things that can happen when using EVs is range anxiety, which means that the driver is worried about whether the battery has enough energy to travel a certain distance. One way to develop a more accurate estimate of how far an EV can travel is to collect large amounts of data and analyze it. In [119], data from a large number of EVs were collected. Then, different energy consumption patterns were clustered using the SOM method which is an unsupervised learning method. An accurate prediction of the amount of energy required to travel a distance is provided by analyzing each cluster. Using this method, an estimate of the battery's state of health is also obtained.

10.4.1.3 Reinforcement Learning

In addition to the mentioned supervised and unsupervised ML algorithms, RL methods are also used to reduce charging time and cost, maximize charging station profit, maximize distribution system profit, satisfy charging demand, and manage the hosting capacity of the network [120]. The schematic of an RL algorithm is presented in Figure 10.26.

Figure 10.26 The schematic of a RL algorithm (Source: [120]).

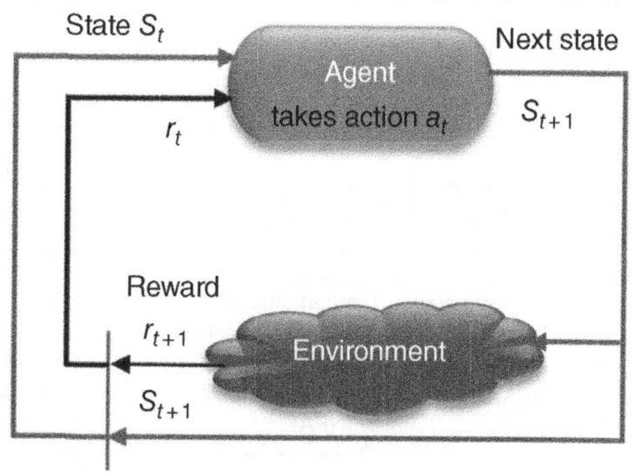

As mentioned earlier, RL is a method to solve a dynamic programming problem in an iterative way effectively. The agent starts in a state s and performs an action a. Based on the chosen action, the environment moved to state s' from state s, and a reward is given to the agent. Since the agent aims to maximize its long-term reward, it will learn which action should be taken in each state so that the attained reward is more. For instance, in [121], the authors considered current battery SOC, electricity price, and remaining time until departure as state features and maximized the EV driver benefit using the Q-learning algorithm. The general structure of the Q-learning (QL) algorithm which is one of the commonly used RL algorithms is presented in Figure 10.27.

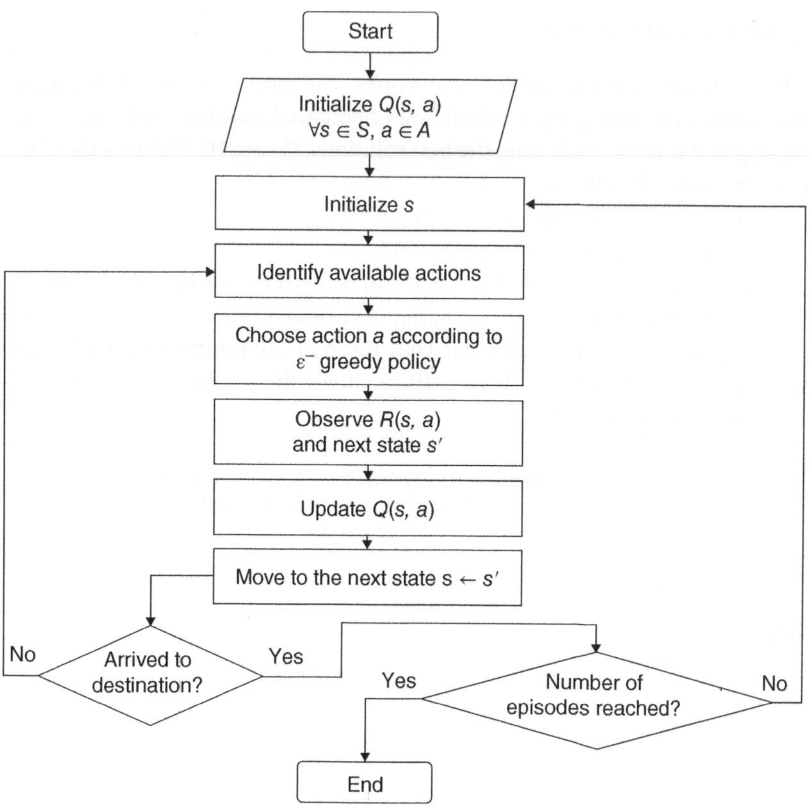

Figure 10.27 The general structure of Q-learning algorithm (Source: [122]).

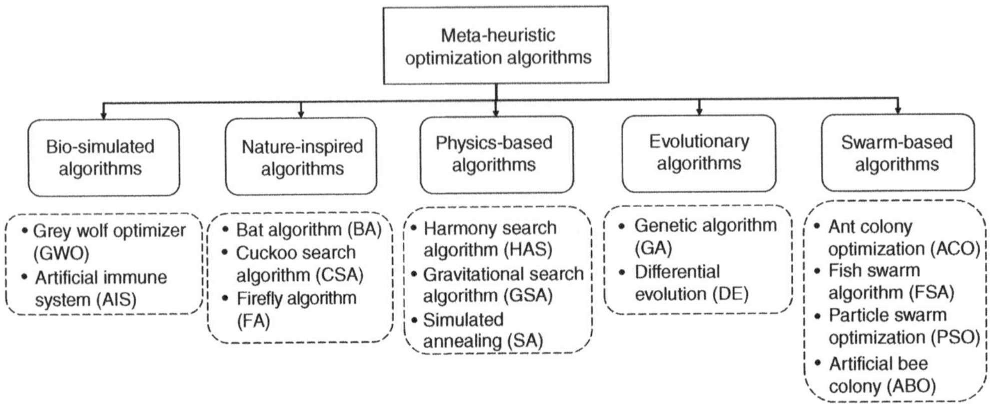

Figure 10.28 Different categories of heuristic optimization algorithms (Source: inspired from: [123, 124]).

Q-learning algorithm is used in [122] in order to find and learn the optimal route for an EV. To do so, the QL agent moves from one state to another and, at each state, it performs different actions based on a predefined policy. Based on the outcome of action a in state s, it updates $Q(s, a)$ which represents the long-term value of action a in state s. It then moves to the next state, if the new state is the terminal state, the algorithm will terminate. Otherwise, the procedure will be repeated.

10.4.2 Heuristic Optimization Algorithms

Similar to ML models, these algorithms have different categories. The most famous heuristic algorithm is the GA, which is based on generating a population of potential solutions and improving them through an iterative procedure by combining the best solutions. Figure 10.28 shows the classification of heuristic optimization algorithms.

Heuristic algorithms can be used for the optimal routing of EVs [125]. For example, a GA was used in the study [126] to find the path with the least cost and time.

ACO is another well-known optimization algorithm that can be used for EV routing problems. Naturally, ants have the ability to explore the surrounding environment and find the best way between their nest and the place which contains food. They do so by communicating with each other, which will result in swarm intelligence. Figure 10.29 shows the performance of ants in overcoming a barrier by finding the best route.

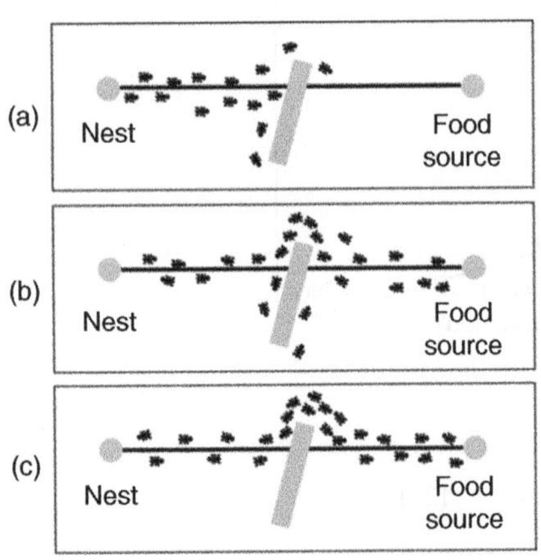

Figure 10.29 (a) Ants facing a barrier in their path, (b) they explore available options to overcome the barrier, and (c) the best available option is found (Source: [127]).

Such a method can be used for EV routing problems as well. For instance, the authors in [128] used the ACO algorithm to find the best routes regarding energy efficiency and charging stations' availability. The results show that the algorithm can successfully find the optimum EV fleet route. The effect of driving patterns and traffic flow on the optimal route is analyzed in [129], and the ACO algorithm solved the problem. The results indicate that the ACO method can find the best route with better computational time compared to traditional methods like DP.

10.4.3 Conclusion

In conclusion, ML-based models and heuristic optimization algorithms can be used to overcome EV-related issues such as the uncertainty in charging energy demand. The methods can also be used to effectively manage the hosting capacity of the network as they help the DSO to manage the power flow in the grid more effectively. In this regard, the commonly used ML models and heuristic optimization algorithms are demonstrated in this section. Supervised learning regression models are used in different studies to develop a model for range estimation, vehicle energy consumption prediction, charging speed prediction, and arrival and exit time forecast. Also, they can be used to predict the charging energy demand of EVs. On the other hand, supervised learning classification models categorize EV users based on their driving and charging behaviors. In addition, unsupervised algorithms (i.e., clustering methods) can divide gathered data into different groups when the information about the classes is unavailable. In the next step, specified EMS can be developed for each group. In addition, RL algorithms can be used to maximize the profit of EV owners, charging station managers, and the DSO. They also can be used for optimal routing problems. For instance, the ACO algorithm proved to find the best route in a graph that minimizes the energy consumption and travel time.

10.5 Concluding Remarks

With regard to the current trends, it is predicted that the market size of EVs will reach around $212 billion in 2030, which means that there probably will be a 348% growth in the market size [130]. With this in mind, it is critical to develop more efficient EMS so that they can effectively handle the growing number of EVs.

The future EMSs must meet several criteria to be capable of reducing the adverse impacts of EV penetration growth on the power grid. They also will need a considerable amount of data to predict the EVs' energy demand. The presented sections tried to identify required data and how one can utilize them to develop an EMS.

The first section considers the importance of demand prediction of EVs. There is a concern that severe consequences will occur if no forecast is available for EVs' power demand, including unwanted blackouts in the power grid. Therefore, having such a prediction allows the grid operator to manage the network and maintain its solidity. There are two general methods for EV energy demand prediction, aggregated and vehicle-centered. In the first, the total energy demand of charging stations is forecasted based on the available historical data; in the latter, each car's behavior is investigated separately. Even though the second approach is more accurate and sophisticated, it has not been utilized widely due to the remarkable amount of data that must be gathered and analyzed. Overall, various predictive models and algorithms can be used for both methods. The three primary options could be statistical methods (using statistical distributions), stochastic methods (using temporal and spatiotemporal random processes), and ML models. One should be aware that the ML models require large datasets.

Section 2 discussed the structure of the currently studied EMSs and their characteristics. It was mentioned that the EMS must be able to schedule the interaction between EV, home, and grid to smooth the load profile [42] and minimize the related costs [43]. The new HEMS should be able to adopt EVs and utilize their storage capacity to lower operational costs. EVs can also eliminate the need for an ESS and reduce investment costs by playing the role of an ESS [51]. Another domain in which the EMS can play an integral role is charging stations. The energy management strategy in these stations must be in a way that minimizes the electricity cost and reduce the electricity consumption during the peak load. The EMS must ensure the profitability of the strategy from different perspectives, namely, the car or homeowner, the parking owner's charging station, and the grid operator. It is also worth mentioning that the profit here is not just the economic profit. For instance, the grid's reliability and stability must be respected from the DSO's point of view. Also, from the owners' point of view, the strategy must be flexible and consider their convenience.

The EMS will face different uncertainties in optimizing the grid energy flow. It is essential to take into account these stochasticities since the optimal solution may vary significantly using deterministic approaches. The main uncertainties are distance traveled, driving pattern, weather conditions, arrival and departure time, battery status, and vehicle type. Driving patterns include driver age and experience, street environment, vehicle, and traffic factors. In addition, important weather parameters are ambient temperature, wind speed and direction, and precipitation. Degradation and DOD are the main uncertainties concerning batteries. Finally, the uncertainty of EV type is of great importance in charging stations and parking lots.

Data analysis approaches can be employed to predict and reduce uncertainty in EMSs. In this regard, Section 4 presented a comprehensive discussion on applying ML algorithms and heuristic optimization methods in EMSs. ML models are categorized into three main groups: supervised learning, unsupervised learning, and RL. Supervised methods are further divided into two groups, regression and classification methods. Regression has been widely used for charging energy demand prediction, travel energy consumption, and range estimation. Classification algorithms can classify users based on their preferences, driving patterns, and arrival or departure time behavior and predict their charging behavior, distance traveled, and charging station selection [113]. In classification algorithms, the class of the data is known. However, there are cases with unknown classes. In such a situation, the data are available, but the algorithm does not know the possible classes. Therefore, an unsupervised clustering algorithm must be used instead of supervised classification. The method categorizes EVs based on their arrival and departure times or charging behavior. In addition to the mentioned supervised and unsupervised ML algorithms, RL methods are also used to reduce charging time and cost, maximize charging station profit, maximize distribution system profit, and satisfy charging demand [120]. In addition, heuristic optimization algorithms can be used for the optimal routing of EVs or optimal energy management of a charging station.

In conclusion, the discussed studies and methods in the presented sections can help decision-makers identify which data must be gathered and how one can employ the gathered data to develop a proper EMS for increasing the hosting capacity of the electricity grid and maintaining its stability. The better the data quality, the better the uncertainties can be identified. Also, high-quality data help the ML methods to be able to predict future outcomes more accurately.

References

1 Davoudi M., Moeini-Aghtaie M., and Mosaddegh H.-R. (2019). Introducing a novel method for improving the design of off-grid photovoltaic systems. *2019 Smart Grid Conference (SGC)*, pp. 1–5. https://doi.org/10.1109/SGC49328.2019.9056592.

2 Davoudi M., Sadeh J., and Davoudi M. (2019). Analysis of DFIG during unsymmetrical grid fault by using crowbar circuit. *2019 Iranian Conference on Renewable Energy & Distributed Generation (ICREDG)*, pp. 1–6. https://doi.org/10.1109/ICREDG47187.2019.194198.

3 Amara-Ouali, Y., Goude, Y., Massart, P. et al. (2021). A review of electric vehicle load open data and models. *Energies* 14 (8): https://doi.org/10.3390/en14082233.

4 Riki, M., Koochaki, M., Moeini-Aghtaie, M. et al. (2022). A novel privacy-preserved voluntary outage management model on a transactive energy scheme. *IET Generation Transmission and Distribution* n/a, no. n/a: https://doi.org/10.1049/gtd2.12535.

5 Galus, M.D., Vayá, M.G., Krause, T., and Andersson, G. (2013). The role of electric vehicles in smart grids. *WIREs Energy and Environment* 2 (4): 384–400. https://doi.org/10.1002/wene.56.

6 Flammini, M.G., Prettico, G., Julea, A. et al. (2019). Statistical characterisation of the real transaction data gathered from electric vehicle charging stations. *Electric Power Systems Research* 166: 136–150. https://doi.org/10.1016/j.epsr.2018.09.022.

7 Zhu, J., Yang, Z., Guo, Y. et al. (2019). Short-term load forecasting for electric vehicle charging stations based on deep learning approaches. *Applied Sciences* 9 (9): https://doi.org/10.3390/app9091723.

8 Garcia-Valle, R. and Vlachogiannis, J.G. (2009). Letter to the editor: electric vehicle demand model for load flow studies. *Electric Power Components & Systems* 37 (5): 577–582. https://doi.org/10.1080/15325000802599411.

9 Olivella-Rosell, P., Villafafila-Robles, R., Sumper, A., and Bergas-Jané, J. (2015). Probabilistic agent-based model of electric vehicle charging demand to analyse the impact on distribution networks. *Energies* 8 (5): https://doi.org/10.3390/en8054160.

10 Kim, Y. and Kim, S. (2021). Forecasting charging demand of electric vehicles using time-series models. *Energies* 14 (5): https://doi.org/10.3390/en14051487.

11 Moon, H., Park, S.Y., Jeong, C., and Lee, J. (2018). Forecasting electricity demand of electric vehicles by analyzing consumers' charging patterns. *Transportation Research Part D Transport and Environment* 62: 64–79. https://doi.org/10.1016/j.trd.2018.02.009.

12 Sun K., Sarker M. R., and Ortega-Vazquez M. A. (2015). Statistical characterization of electric vehicle charging in different locations of the grid. *2015 IEEE Power & Energy Society General Meeting*, pp. 1–5. https://doi.org/10.1109/PESGM.2015.7285794.

13 Khoo, Y.B., Wang, C.-H., Paevere, P., and Higgins, A. (2014). Statistical modeling of electric vehicle electricity consumption in the Victorian EV Trial, Australia. *Transportation Research Part D Transport and Environment* 32: 263–277. https://doi.org/10.1016/j.trd.2014.08.017.

14 Lee, Z.J. and Low, S.H. (2019). *ACN-Data: Analysis and Applications of an Open EV Charging Dataset*. Association for Computing Machinery.

15 Lahariya M., Benoit D., and Develder C. (2020). Defining a synthetic data generator for realistic electric vehicle charging sessions. *Proceedings of the Eleventh ACM International Conference on Future Energy Systems* (pp. 406–407). https://doi.org/10.1145/3396851.3403509.

16 Liang M., Li W., Yu J., and Shi L. (2015). Kernel-based electric vehicle charging load modeling with improved latin hypercube sampling, *2015 IEEE Power & Energy Society General Meeting*, pp. 1–5. https://doi.org/10.1109/PESGM.2015.7285758.

17 Chen, L., Huang, X., and Zhang, H. (2020). Modeling the charging behaviors for electric vehicles based on ternary symmetric kernel density estimation. *Energies* 13 (7): https://doi.org/10.3390/en13071551.

18 Chung Y., Khaki B., Chu C., and Gadh R. (2018). Electric vehicle user behavior prediction using hybrid kernel density estimator. *2018 IEEE International Conference on Probabilistic Methods Applied to Power Systems (PMAPS)*, pp. 1–6. https://doi.org/10.1109/PMAPS.2018.8440360.

19 Sokorai, P., Fleischhacker, A., Lettner, G., and Auer, H. (2018). Stochastic modeling of the charging behavior of electromobility. *World Electric Vehicle Journal* 9 (3): https://doi.org/10.3390/wevj9030044.

20 Häggström, O. (2002). *Finite Markov Chains and Algorithmic Applications*. Cambridge: Cambridge University Press.

21 Amini, M.H., Kargarian, A., and Karabasoglu, O. (2016). ARIMA-based decoupled time series forecasting of electric vehicle charging demand for stochastic power system operation. *Electric Power Systems Research* 140: 378–390. https://doi.org/10.1016/j.epsr.2016.06.003.

22 Amini M. H., Karabasoglu O., Ilić M. D., Boroojeni K. G., and Iyengar S. S. (2015). ARIMA-based demand forecasting method considering probabilistic model of electric vehicles' parking lots. *2015 IEEE Power & Energy Society General Meeting*, pp. 1–5. https://doi.org/10.1109/PESGM.2015.7286050.

23 Jiang H., Ren H., Sun C., and Watts D. (2017). The temporal-spatial stochastic model of plug-in hybrid electric vehicles. *2017 IEEE PES Innovative Smart Grid Technologies Conference Europe (ISGT-Europe)*, pp. 1–6. https://doi.org/10.1109/ISGTEurope.2017.8260233.

24 Neaimeh, M., Wardle, R., Jenkins, A.M. et al. (2015). A probabilistic approach to combining smart meter and electric vehicle charging data to investigate distribution network impacts. *Applied Energy* 157: 688–698. https://doi.org/10.1016/j.apenergy.2015.01.144.

25 Arias, M.B., Kim, M., and Bae, S. (2017). Prediction of electric vehicle charging-power demand in realistic urban traffic networks. *Applied Energy* 195: 738–753. https://doi.org/10.1016/j.apenergy.2017.02.021.

26 Yan, J., Zhang, J., Liu, Y. et al. (2020). EV charging load simulation and forecasting considering traffic jam and weather to support the integration of renewables and EVs. *Renewable Energy* 159: 623–641. https://doi.org/10.1016/j.renene.2020.03.175.

27 Lojowska, A., Kurowicka, D., Papaefthymiou, G., and van der Sluis, L. (2012). Stochastic modeling of power demand due to EVs using copula. *IEEE Transactions on Power Apparatus and Systems* 27 (4): 1960–1968. https://doi.org/10.1109/TPWRS.2012.2192139.

28 Dai, Q., Cai, T., Duan, S., and Zhao, F. (2014). Stochastic modeling and forecasting of load demand for electric bus battery-swap station. *IEEE Transactions on Power Delivery* 29 (4): 1909–1917. https://doi.org/10.1109/TPWRD.2014.2308990.

29 Jahangir, H., Tayarani, H., Ahmadian, A. et al. (2019). Charging demand of plug-in electric vehicles: forecasting travel behavior based on a novel rough artificial neural network approach. *Journal of Cleaner Production* 229: 1029–1044. https://doi.org/10.1016/j.jclepro.2019.04.345.

30 Mies, J.J., Helmus, J.R., and Van den Hoed, R. (2018). Estimating the charging profile of individual charge sessions of electric vehicles in The Netherlands. *World Electric Vehicle Journal* 9 (2): https://doi.org/10.3390/wevj9020017.

31 Kristoffersen, T.K., Capion, K., and Meibom, P. (2011). Optimal charging of electric drive vehicles in a market environment. *Applied Energy* 88 (5): 1940–1948. https://doi.org/10.1016/j.apenergy.2010.12.015.

32 Gerossier, A., Girard, R., and Kariniotakis, G. (2019). Modeling and forecasting electric vehicle consumption profiles. *Energies* 12 (7): https://doi.org/10.3390/en12071341.

33 Lu, Y., Li, Y., Xie, D. et al. (2018). The Application of Improved Random Forest Algorithm on the Prediction of Electric Vehicle Charging Load. *Energies* 11 (11): https://doi.org/10.3390/en11113207.

34 Schmidhuber, J. (2015). Deep learning in neural networks: An overview. *Neural Networks* 61: 85–117. https://doi.org/10.1016/j.neunet.2014.09.003.

35 Chakraborty S., Tomsett R., Raghavendra R., et al. (2017). Interpretability of deep learning models: a survey of results. *2017 IEEE SmartWorld, Ubiquitous Intelligence & Computing, Advanced & Trusted Computed, Scalable Computing & Communications, Cloud & Big Data Computing, Internet of People and Smart City Innovation (SmartWorld/SCALCOM/UIC/ATC/CBDCom/IOP/SCI)*, pp. 1–6. https://doi.org/10.1109/UIC-ATC.2017.8397411.

36 Zhang, Q. and Zhu, S. (2018). Visual interpretability for deep learning: a survey. *Frontiers of Information Technology & Electronic Engineering* 19 (1): 27–39. https://doi.org/10.1631/FITEE.1700808.

37 Zhu, J., Yang, Z., Mourshed, M. et al. (2019). Electric vehicle charging load forecasting: a comparative study of deep learning approaches. *Energies* 12 (14): https://doi.org/10.3390/en12142692.

38 Zhao, W., Dai, T., Wang, L. et al. (2018). Short-term load forecasting considering meteorological factors and electric vehicles. *IOP Conference Series: Materials Science and Engineering* 439: 32114. https://doi.org/10.1088/1757-899x/439/3/032114.

39 Dabbaghjamanesh, M., Moeini, A., and Kavousi-Fard, A. (2021). Reinforcement learning-based load forecasting of electric vehicle charging station using Q-learning technique. *IEEE Transactions on Industrial Informatics* 17 (6): 4229–4237. https://doi.org/10.1109/TII.2020.2990397.

40 Davoudi, M., Moeini-Aghtaie, M., and Ghorani, R. (2021). Developing a new framework for transactive peer-to-peer thermal energy market. *IET Generation Transmission and Distribution* 15: https://doi.org/10.1049/gtd2.12150.

41 Davoudi, M. and Moeini-Aghtaie, M. (2021). Local energy markets design for integrated distribution energy systems based on the concept of transactive peer-to-peer market. *IET Generation Transmission and Distribution* n/a, no. n/a: https://doi.org/10.1049/gtd2.12274.

42 Shi C. and Luo J. (2019). EV energy management strategies considering renewable energy consumption and load smoothing. *iSPEC 2019—2019 IEEE Sustainable Power Energy Conference Grid Modernization Energy Revolution, Proceedings*, no. 1, pp. 1897–1902. https://doi.org/10.1109/iSPEC48194.2019.8975314.

43 Lakshminarayanan, V., Chemudupati, V.G.S., Pramanick, S.K., and Rajashekara, K. (2019). Real-time optimal energy management controller for electric vehicle integration in workplace microgrid. *IEEE Transactions on Transportation Electrification* 5 (1): 174–185. https://doi.org/10.1109/TTE.2018.2869469.

44 Arras P., Tabunshchyk G., Korotunov S., et al. (2020). Cost optimization simulation for electric vehicle charging infrastructure. *2020 IEEE European Technology and Engineering Management Summit (E-TEMS)*, pp. 1–4. https://doi.org/10.1109/E-TEMS46250.2020.9111715.

45 Aswantara I. K. A., Ko K. S., and Sung D. K. (2013). A centralized EV charging scheme based on user satisfaction fairness and cost. *2013 IEEE Innovative Smart Grid Technologies-Asia (ISGT Asia)*, pp. 1–4. https://doi.org/10.1109/ISGT-Asia.2013.6698730.

46 Jebril, I., Hammad, M., Aboushi, A. et al. (2021). User satisfaction of electric-vehicles about charging stations (home, outdoor, and workplace). *Turkish Journal of Computer and Mathematics Education* 12: 3589–3593.

47 Guo, Q., Sun, H., Wang, Y. et al. (2012). Current energy management technologies research in China considering EVs integration. *IEEE Power Energy Society General Meeting* 51025725: 1–6. https://doi.org/10.1109/PESGM.2012.6344876.

48 Kikusato, H., Mori, K., Yoshizawa, S. et al. (2019). Electric vehicle charge-discharge management for utilization of photovoltaic by coordination between home and grid energy management systems. *IEEE Transactions on Smart Grid* 10 (3): 3186–3197. https://doi.org/10.1109/TSG.2018.2820026.

49 Buekers, J., Van Holderbeke, M., Bierkens, J., and Int Panis, L. (2014). Health and environmental benefits related to electric vehicle introduction in EU countries. *Transportation Research Part D Transport and Environment* 33: 26–38. https://doi.org/10.1016/j.trd.2014.09.002.

50 Kempton, W. and Letendre, S.E. (1997). Electric vehicles as a new power source for electric utilities. *Transportation Research Part D Transport and Environment* 2 (3): 157–175. https://doi.org/10.1016/S1361-9209(97)00001-1.

51 García J. J., Enrich R., and Torrent-Moreno M. (2013). A greening energy positive tool for energy management in infrastructures and buildings of public use: Technical and economic assessment framework for the integrated management of PV and EV systems. *2013 World Congress Sustainable Technologies WCST*, pp. 42–46, https://doi.org/10.1109/WCST.2013.6750402.

52 Yan D., Li T., Ma C. et al. (2020). Cost effective energy management of home energy system with photovoltaic-battery and electric vehicle. *IECON Proceedings (Industrial Electronics Conference* (October 2020)), pp. 3611–3616.https://doi.org/10.1109/IECON43393.2020.9255317.

53 Blonsky M., Munankarmi P., and Balamurugan S. P. (2021). Incorporating residential smart electric vehicle charging in home energy management systems. *IEEE Green Technology Conference*, vol. 2021-April, pp. 187–194. https://doi.org/10.1109/GreenTech48523.2021.00039.

54 Satoya D., Yamashita D., and Yokoyama R. (2015). Community energy management with electric vehicles for effective use of solar energy. *Proceedings—2014 4th International Conference Artificial Intelligence with Application Engineering Technology ICAIET 2014*, pp. 241–246. https://doi.org/10.1109/ICAIET.2014.47.

55 Park H., Bae S., Chang M. et al. (2019). A community-scale energy management system for demand response participation of households with DERs and EVs. *2019 IEEE 4th International Future Energy Electronic Conference IFEEC*, pp. 1–5. https://doi.org/10.1109/IFEEC47410.2019.9015047.

56 Davoudi, M., Jooshaki, M., Moeini-Aghtaie, M. et al. (2022). Developing a multi-objective multi-layer model for optimal design of residential complex energy systems. *International Journal of Electrical Power & Energy Systems* 138: 107889. https://doi.org/10.1016/j.ijepes.2021.107889.

57 Hou, X., Wang, J., Huang, T. et al. (2019). Smart home energy management optimization method considering energy storage and electric vehicle. *IEEE Access* 7: 144010–144020. https://doi.org/10.1109/ACCESS.2019.2944878.

58 Chandra L. and Chanana S. (2018). Energy management of smart homes with energy storage, rooftop PV and electric vehicle. *2018 IEEE International Students' Conference Electrical Electronics and Computer Science SCEECS 2018*, pp. 1–6. https://doi.org/10.1109/SCEECS.2018.8546857.

59 Thomas D., Deblecker O., Genikomsakis K. et al. (2017). Smart house operation under PV and load demand uncertainty considering EV and storage utilization. *IECON 2017—43rd Annual Conference of the IEEE Industrial Electronics Society*, pp. 3644–3649. https://doi.org/10.1109/IECON.2017.8216618.

60 Mohammad, A., Zamora, R., and Lie, T.T. (2021). Transactive energy management of PV-based EV integrated parking lots. *IEEE Systems Journal* 15 (4): 5674–5682. https://doi.org/10.1109/JSYST.2020.3043327.

61 Wu, D., Zeng, H., Lu, C., and Boulet, B. (2017). Two-stage energy management for office buildings with workplace EV charging and renewable energy. *IEEE Transactions on Transportation Electrification* 3 (1): 225–237. https://doi.org/10.1109/TTE.2017.2659626.

62 Sengor, I., Erdinc, O., Yener, B. et al. (2019). Optimal energy management of EV parking lots under peak load reduction based DR programs considering uncertainty. *IEEE Transactions on Sustainable Energy* 10 (3): 1034–1043. https://doi.org/10.1109/TSTE.2018.2859186.

63 Chaudhari K. and Ukil A. (2016). TOU pricing based energy management of public EV charging stations using energy storage system. *Proceedings IEEE International Conference Industrial Technology*, vol. 2016-May, pp. 460–465. https://doi.org/10.1109/ICIT.2016.7474795.

64 Mokaramian, E., Shayeghi, H., Sedaghati, F. et al. (2022). An optimal energy hub management integrated EVs and RES based on three-stage model considering various uncertainties. *IEEE Access* 10: 17349–17365. https://doi.org/10.1109/ACCESS.2022.3146447.

65 Zhang, R., Cheng, X., and Yang, L. (2019). Flexible energy management protocol for cooperative EV-to-EV Charging. *IEEE Transactions on Intelligent Transportation Systems* 20 (1): 172–184. https://doi.org/10.1109/TITS.2018.2807184.

66 Shi S., Zhang Y., Ni L., et al. (2021). Energy management method for energy storage system in PV-integrated EV charging station. *Proceedings 2021 IEEE International Conference Power Electronics Computer Application ICPECA 2021*, pp. 427–431. https://doi.org/10.1109/ICPECA51329.2021.9362623.

67 Ghavami M., Essakiappan S., and Singh C. (2016). A framework for reliability evaluation of electric vehicle charging stations. *2016 IEEE Power and Energy Society General Meeting (PESGM)*, pp. 1–5. https://doi.org/10.1109/PESGM.2016.7741872.

68 Deb S., Kalita K., and Mahanta P. (2017). Impact of electric vehicle charging stations on reliability of distribution network. *2017 International Conference on Technological Advancements in Power and Energy (TAP Energy)*, pp. 1–6. https://doi.org/10.1109/TAPENERGY.2017.8397272.

69 Van Der Meer, D., Mouli, G.R.C., Mouli, G.M.E. et al. (2018). Energy management system with PV power forecast to optimally charge EVs at the workplace. *IEEE Transactions on Industrial Informatics* 14 (1): 311–320. https://doi.org/10.1109/TII.2016.2634624.

70 Akil M., Dokur E., and Bayindir R. (2020). Energy management for EV charging based on solar energy in an industrial microgrid. *9th International Conference Renewable Energy Research Application ICRERA*, pp. 489–493. https://dopi.org/10.1109/ICRERA49962.2020.9242663.

71 Pal, A., Bhattacharya, A., and Chakraborty, A.K. (2021). Placement of public fast-charging station and solar distributed generation with battery energy storage in distribution network considering uncertainties and traffic congestion. *Journal of Energy Storage* 41, no. April: 102939. https://doi.org/10.1016/j.est.2021.102939.

72 Liu, Z., Wu, Q., Shahidehpour, M. et al. (2019). Transactive real-time electric vehicle charging management for commercial buildings with PV on-site generation. *IEEE Transactions on Smart Grid* 10 (5): 4939–4950. https://doi.org/10.1109/TSG.2018.2871171.

73 Y. Wu, A. Ravey, D. Chrenko et al. (2018). A real time energy management for EV charging station integrated with local generations and energy storage system. *2018 IEEE Transportation Electrification Conference Expo, ITEC 2018*, pp. 977–984. https://doi.org/10.1109/ITEC.2018.8450235.

74 A. Thingvad, Calearo, L., Andersen, P. B. et al. (2019). Value of V2G frequency regulation in great britain considering real driving data. *Proceedings 2019 IEEE PES Innovation Smart Grid Technology Europe ISGT-Europe 2019*, pp. 1–5. https://doi.org/10.1109/ISGTEurope.2019.8905679.

75 Sausen J. P., Abaide A. R., Adeyanju O. M. et al. (2019). EV demand forecasting model based on travel survey: a Brazilian case study. *2019 IEEE PES Conference Innovation Smart Grid Technology ISGT Latin America 2019*. https://doi.org/10.1109/ISGT-LA.2019.8894955.

76 Ericsson, E. (2000). Variability in urban driving patterns. *Transportation Research Part D Transport and Environment* 5 (5): 337–354. https://doi.org/10.1016/S1361-9209(00)00003-1.

77 Jonas T. and Macht G. A. (2020). Quantifying the impact of traffic on the energy consumption of electric vehicles. *Proceedings 2020 IISE Annual Conference*, pp. 985–990.

78 Langbroek, J.H.M., Franklin, J.P., and Susilo, Y.O. (2018). How would you change your travel patterns if you used an electric vehicle? A stated adaptation approach. *Travel Behaviour and Society* 13 (April): 144–154. https://doi.org/10.1016/j.tbs.2018.08.001.

79 Rolim, C.C., Gonçalves, G.N., Farias, T.L., and Rodrigues, Ó. (2012). Impacts of electric vehicle adoption on driver behavior and environmental performance. *Procedia—Social and Behavioral Sciences* 54: 706–715. https://doi.org/10.1016/j.sbspro.2012.09.788.

80 Hao, X., Wang, H., Lin, Z., and Ouyang, M. (2020). Seasonal effects on electric vehicle energy consumption and driving range: A case study on personal, taxi, and ridesharing vehicles. *Journal of Cleaner Production* 249: 119403. https://doi.org/10.1016/j.jclepro.2019.119403.

81 Ficht A. and Lienkamp M. (2015). Rolling resistance modeling for electric vehicle consumption BT. *6th International Munich Chassis Symposium 2015*, pp. 775–798.

82 Ejsmont, J., Taryma, S., Ronowski, G., and Swieczko-Zurek, B. (2018). Influence of temperature on the tyre rolling resistance. *International Journal of Automotive Technology* 19 (1): 45–54. https://doi.org/10.1007/s12239-018-0005-4.

83 I. Tal, A. Olaru, and G. M. Muntean 2013. EWARPE—energy-efficient weather-aware route planner for electric bicycles. *Proceedings—International Conference Network Protocol ICNP*. https://doi.org/10.1109/ICNP.2013.6733680.

84 Ejsmont, J., Sjögren, L., Świeczko-Żurek, B., and Ronowski, G. (2015). Influence of road wetness on tire-pavement rolling resistance. *Journal of Civil Engineering and Architecture Engineering* 9 (11): 1302–1310. https://doi.org/10.17265/1934-7359/2015.11.004.

85 Wang, Z., Jochem, P., and Fichtner, W. (2020). A scenario-based stochastic optimization model for charging scheduling of electric vehicles under uncertainties of vehicle availability and charging demand. *Journal of Cleaner Production* 254: https://doi.org/10.1016/j.jclepro.2019.119886.

86 Fortune Business Insight (2021). Electric vehicle battery market size, share and Convid-19 impact analysis by battery type, vehicle type and regional forecats, 2021–2028, *Fortune Business Insight*. https://www.fortunebusinessinsights.com/industry-reports/electric-vehicle-battery-market-101700 (accessed 27 May 2022).

87 S. Grolleau, A. Delaille, and H. Gualous (2014). Predicting lithium-ion battery degradation for efficient design and management. *2013 World Electric Vehicle Symposium Exhibition EVS 2014*, pp. 1–6. https://doi.org/10.1109/EVS.2013.6914799.

88 Sun, Y., Yue, H., Zhang, J., and Booth, C. (2019). Minimization of residential energy cost considering energy storage system and EV with driving usage probabilities. *IEEE Transactions on Sustainable Energy* 10 (4): 1752–1763. https://doi.org/10.1109/TSTE.2018.2870561.

89 Zhou, B., Liu, X., Cao, Y. et al. (2016). Optimal scheduling of virtual power plant with battery degradation cost. *IET Generation Transmission and Distribution* 10 (3): 712–725. https://doi.org/10.1049/iet-gtd.2015.0103.

90 Brinkel, N.B.G., Schram, W.L., AlSkaif, T.A. et al. (2020). Should we reinforce the grid? Cost and emission optimization of electric vehicle charging under different transformer limits. *Applied Energy* 276, no. May: 115285. https://doi.org/10.1016/j.apenergy.2020.115285.

91 D. Powell (2022). Electric car statistics—data and projections, *Heycar*. https://heycar.co.uk/blog/electric-cars-statistics-and-projections (accessed 27 May 2022).

92 Golla, N.K. and Sudabattula, S.K. (2021). Impact of plug-in electric vehicles on grid integration with distributed energy resources: a comprehensive review on methodology of power interaction and scheduling. *Materials Today Proceedings* https://doi.org/10.1016/j.matpr.2021.03.306.

93 Tahmasebi, M., Ghadiri, A., Haghifam, M.R., and Miri-Larimi, S.M. (2021). MPC-based approach for online coordination of EVs considering EV usage uncertainty. *International Journal of Electrical Power & Energy Systems* 130, no. March: 106931. https://doi.org/10.1016/j.ijepes.2021.106931.

94 Chen, Z., Li, C., Chen, X., and Yang, Q. (2020). Towards optimal planning of EV charging stations under grid constraints. *IFAC-PapersOnLine* 53 (2): 14103–14108. https://doi.org/10.1016/j.ifacol.2020.12.1005.

95 Wu, W. and Lin, B. (2021). Benefits of electric vehicles integrating into power grid. *Energy* 224: 120108. https://doi.org/10.1016/j.energy.2021.120108.

96 Stiasny, J., Zufferey, T., Pareschi, G. et al. (2021). Sensitivity analysis of electric vehicle impact on low-voltage distribution grids. *Electric Power Systems Research* 191 (October 2019): 106696. https://doi.org/10.1016/j.epsr.2020.106696.

97 F. G. Dias, D. Scoffield, M. Mohanpurkar et al. (2018). Impact of controlled and uncontrolled charging of electrical vehicles on a residential distribution grid. *International Conference Probabilistic Methods Application to Power System PMAPS 2018—Proceedings*, pp. 1–5. https://doi.org/10.1109/PMAPS.2018.8440511.

98 Pokharel, S., Sah, P., and Ganta, D. (2021). Improved prediction of total energy consumption and feature analysis in electric vehicles using machine learning and shapley additive explanations method. *World Electric Vehicle Journal* 12 (3): https://doi.org/10.3390/wevj12030094.

99 Chung, Y.-W., Khaki, B., Li, T. et al. (2019). Ensemble machine learning-based algorithm for electric vehicle user behavior prediction. *Applied Energy* 254: 113732. https://doi.org/10.1016/j.apenergy.2019.113732.

100 Mouaad B., Farag M., Benabdelaziz K. et al. (2022). Electric vehicles arrival and departure time prediction based on deep learning: the case of Morocco. *2022 2nd International Conference on Innovative Research in Applied Science, Engineering and Technology (IRASET)*, Meknes, Morocco, pp. 1–8. https://doi.org/10.1109/IRASET52964.2022.9738115.

101 Frendo, O., Gaertner, N., and Stuckenschmidt, H. (2021). Improving smart charging prioritization by predicting electric vehicle departure time. *IEEE Transactions on Intelligent Transportation Systems* 22 (10): 6646–6653. https://doi.org/10.1109/TITS.2020.2988648.

102 Y. Xiong, C. Chu, R. Gadh et al. (2017) Distributed optimal vehicle grid integration strategy with user behavior prediction. *2017 IEEE Power & Energy Society General Meeting*, pp. 1–5. https://doi.org/10.1109/PESGM.2017.8274327.

103 Kim, D., Shim, H., and Eo, J. (2021). *A machine learning method for EV range prediction with updates on route information and traffic conditions*. In: *Proceedings of the AAAI Conference on Artificial Intelligence*, 12545–12551. https://doi.org/10.1609/aaai.v36i11.21525.

104 Miri, I., Fotouhi, A., and Ewin, N. (2021). Electric vehicle energy consumption modelling and estimation—A case study. *International Journal of Energy Research* 45 (1): 501–520. https://doi.org/10.1002/er.5700.

105 Hu, K., Wu, J., and Schwanen, T. (2017). Differences in energy consumption in electric vehicles: an exploratory real-world study in Beijing. *Journal of Advanced Transportation* 2017: 4695975. https://doi.org/10.1155/2017/4695975.

106 De Cauwer, C., Van Mierlo, J., and Coosemans, T. (2015). Energy consumption prediction for electric vehicles based on real-world data. *Energies* 8 (8): https://doi.org/10.3390/en8088573.

107 Sun, S., Zhang, J., Bi, J., and Wang, Y. (2019). A machine learning method for predicting driving range of battery electric vehicles. *Journal of Advanced Transportation* 2019: 4109148. https://doi.org/10.1155/2019/4109148.

108 Al-Wreikat, Y., Serrano, C., and Sodré, J.R. (2022). Effects of ambient temperature and trip characteristics on the energy consumption of an electric vehicle. *Energy* 238: 122028. https://doi.org/10.1016/j.energy.2021.122028.

109 Wang, Y.U. (2020). Machine learning-based method for remaining range prediction of electric vehicles. *IEE ACCESS* 8: https://doi.org/10.1109/ACCESS.2020.3039815.

110 George D. and Sivraj P. (2021). Driving range estimation of electric vehicles using deep learning. *2021 Second International Conference on Electronics and Sustainable Communication Systems (ICESC)*, pp. 358–365. https://doi.org/10.1109/ICESC51422.2021.9532912.

111 Zhao, L., Yao, W., Wang, Y., and Hu, J. (2020). Machine learning-based method for remaining range prediction of electric vehicles. *IEEE Access* 8: 212423–212441. https://doi.org/10.1109/ACCESS.2020.3039815.

112 Ai S., Chakravorty A., and Rong C. (2018). Household EV charging demand prediction using machine and ensemble learning. *2018 IEEE International Conference on Energy Internet (ICEI)*, pp. 163–168. https://doi.org/10.1109/ICEI.2018.00037.

113 Sun, C., Li, T., Low, S.H., and Li, V.O.K. (2020). Classification of electric vehicle charging time series with selective clustering. *Electric Power Systems Research* 189: 106695. https://doi.org/10.1016/j.epsr.2020.106695.

114 Yang, Y., Tan, Z., and Ren, Y. (2020). Research on factors that influence the fast charging behavior of private battery electric vehicles. *Sustainability* 12 (8): https://doi.org/10.3390/su12083439.

115 Aguilar-Dominguez, D., Ejeh, J., Dunbar, A.D.F., and Brown, S.F. (2021). Machine learning approach for electric vehicle availability forecast to provide vehicle-to-home services. *Energy Reports* 7: 71–80. https://doi.org/10.1016/j.egyr.2021.02.053.

116 Shahriar, S., Osman, A., Dhou, S., and Nijim, M. (2020). Machine learning approaches for EV charging behavior: a review. *IEEE Access* 8: 168980–168993. https://doi.org/10.1109/ACCESS.2020.3023388.

117 Sadeghianpourhamami, N., Refa, N., Strobbe, M., and Develder, C. (2018). Quantitive analysis of electric vehicle flexibility: a data-driven approach. *International Journal of Electrical Power & Energy Systems* 95: https://doi.org/10.1016/j.ijepes.2017.09.007.

118 Xiong Y., Wang B., Chu C. et al. Electric vehicle driver clustering using statistical model and machine learning. *2018 IEEE Power & Energy Society General Meeting (PESGM)*, pp. 1–5. https://doi.org/10.1109/PESGM.2018.8586132.

119 Lee, C.-H. and Wu, C.-H. (2015). A novel big data modeling method for improving driving range estimation of EVs. *IEEE Access* 3: 1980–1993. https://doi.org/10.1109/ACCESS.2015.2492923.

120 Abdullah, H.M., Gastli, A., and Ben-Brahim, L. (2021). Reinforcement learning based EV charging management systems–a review. *IEEE Access* 9: 41506–41531. https://doi.org/10.1109/ACCESS.2021.3064354.

121 Shi W. and Wong V. W. S. (2011). Real-time vehicle-to-grid control algorithm under price uncertainty. *2011 IEEE International Conference on Smart Grid Communications (SmartGridComm)*, pp. 261–266. https://doi.org/10.1109/SmartGridComm.2011.6102330.

122 Dorokhova, M., Ballif, C., and Wyrsch, N. (2021). Routing of electric vehicles with intermediary charging stations: a reinforcement learning approach. *Frontiers in Big Data* 4: https://doi.org/10.3389/fdata.2021.586481.

123 Kochenderfer, M.J. and Wheeler, T.A. (2019). *Algorithms for Optimization*. The MIT Press.

124 Kumar, A. and Bawa, S. (2020). A comparative review of meta-heuristic approaches to optimize the SLA violation costs for dynamic execution of cloud services. *Soft Computing* 24: https://doi.org/10.1007/s00500-019-04155-4.

125 G. Zhenfeng, L. Yang, J. Xiaodan et al. (2017). The electric vehicle routing problem with time windows using genetic algorithm. *2017 IEEE 2nd Advanced Information Technology, Electronic and Automation Control Conference (IAEAC)*, pp. 635–639. https://doi.org/10.1109/IAEAC.2017.8054093.

126 Zhu, Y., Lee, K.Y., and Wang, Y. (2021). Adaptive elitist genetic algorithm with improved neighbor routing initialization for electric vehicle routing problems. *IEEE Access* 9: 16661–16671. https://doi.org/10.1109/ACCESS.2021.3053285.

127 Tang, Z., Sonntag, M., and Gross, H. (2019). Ant colony optimization in lens design. *Applied Optics* 58 (23): 6357–6364. https://doi.org/10.1364/AO.58.006357.

128 M. Mavrovouniotis, G. Ellinas, and M. Polycarpou (2018). Ant colony optimization for the electric vehicle routing problem. *2018 IEEE Symposium Series on Computational Intelligence (SSCI)*, pp. 1234–1241. https://doi.org/10.1109/SSCI.2018.8628831.

129 K. Tangrand and B. A. Bremdal (2016). Using ant colony optimization to determine influx of EVs and charging station capacities. *2016 IEEE International Energy Conference (ENERGYCON)*, pp. 1–6. https://doi.org/10.1109/ENERGYCON.2016.7514018.

130 Statista (2022)Projections for the global battery electric vehicle market size between 2019 and 2030, *Statista*. https://www.statista.com/statistics/1254513/battery-electric-vehicle-market-size-forecast/.

11

Energy Efficiency in Smart Buildings Through IoT Sensor Integration

Implementation of BEMOSS™ Platform

Saifur Rahman and Ali Parizad

Advanced Research Institute (ARI), Virginia Tech, National Capital Region, Arlington, VA, USA

11.1 Introduction

As illustrated in Figures 11.1 and 11.2, and according to the revised Energy Information Administration (EIA) report on September 2022, from the first Commercial Buildings Energy Consumption Survey (CBECS) in 1979 to the 2018 CBECS, the number of buildings has increased from 3.8 to 5.9 million (56%), and the amount of commercial floorspace has increased from 51 billion square feet to ~97 billion square feet (89%) [1]. This growing demand for electricity in the world has to be addressed from both supply side and demand side complementarily. On the demand side, building energy efficiency programs combined with demand response (DR) can decrease energy and peak power consumption. Most of the commercially available Building Energy Management System (BEMS) solutions cater to larger buildings [2] (100,000 ft^2 or more) and are independently controlled with proprietary solutions, which are geared to new buildings with built-in supervisory control and data acquisition (SCADA) systems [3] (e.g., Honeywell heating, ventilation, and air-conditioning (HVAC) control and Phillips lighting control).

There are millions of commercial buildings in the United States that are smaller and older and do not have built-in SCADA systems required for commercially available BEMS. Based on 2018 reports (Figure 11.3) [1], more than half of US commercial buildings were built between 1960 and 1999 and it reveals the following information about the old building:

- Buildings built between 1960 and 1999 accounted for more than 50% of both total number of buildings and floorspace.
- One-quarter of buildings (25%) were built after 2000, accounting for 29% of total floorspace.
- Buildings built before 1960 represented 21% of buildings but only 17% of total floorspace.
- The median year of construction is 1982.

A building size category is shown in Figure 11.4. Also, different categories of commercial buildings (by floorspace) for 2018 are depicted in Figure 11.5.

Figure 11.6 shows that four end uses are common among more than three-fourths of commercial buildings. It can be observed that lighting, heating, and cooling are in more than three-fourths of commercial buildings, accounting for at least 90% of the total floor space.

In the United States, buildings consume over 40% of the total energy consumption, while over 90% of these buildings are either small-sized (<5000 ft^2) or medium-sized (between 5000 and 50,000 ft^2). Commercially available BEMSs to monitor and control such building are not

Smart Cyber-Physical Power Systems: Fundamental Concepts, Challenges, and Solutions, Volume 1, First Edition.
Edited by Ali Parizad, Hamid Reza Baghaee, and Saifur Rahman.

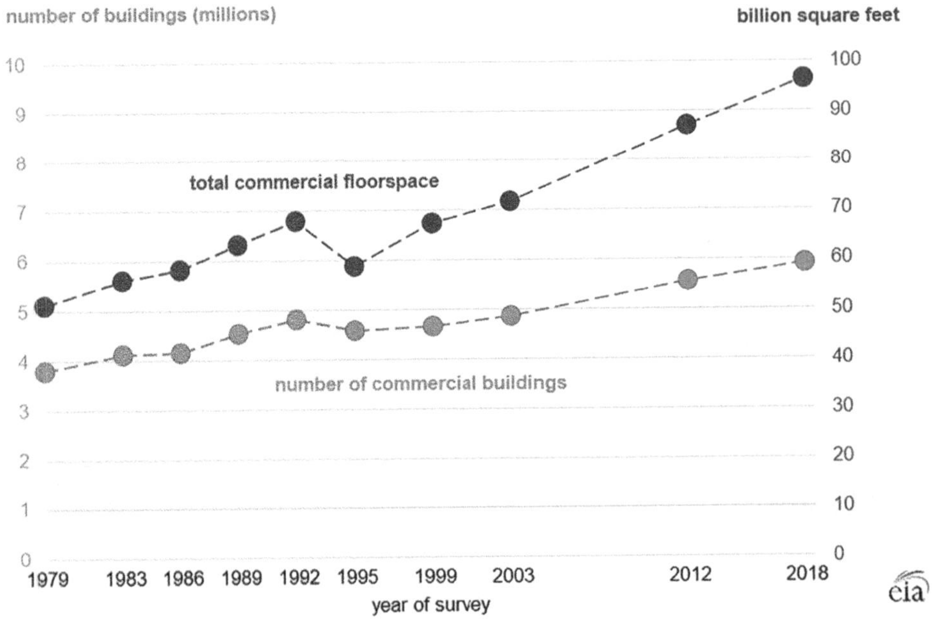

Figure 11.1 The number of commercial buildings and floorspace, 1979–2018 [1].

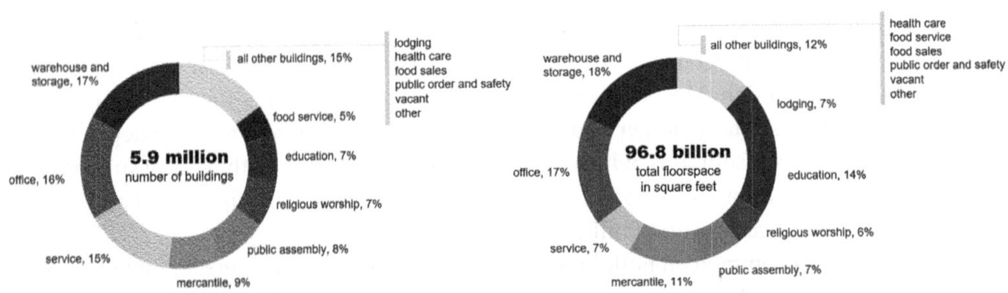

Figure 11.2 Percentage of commercial buildings and floorspace by principal building activity, 2018 [1].

cost-effective. Electric utilities offer incentives to building operators to participate in DR programs [4], which are limited to larger (100,000 ft^2) buildings due to the availability of central BEMSs in such buildings. Small- and medium-sized building (SMB) owner/operators can benefit from such programs only if they can use customizable BEMS platforms, which can interface with plug-and-play devices like smart thermostats and dimmable lights. It is not cost-effective for SMB owner/operators to deploy BEMSs costing hundreds of thousands of dollars, and they do not. Without such systems, operations of HVACs, lighting, and plug loads are not optimized, which results in excessive energy usage. IoT-enabled smart buildings offer in-building device mobility, occupant comfort, and indoor activity automation, but the deployment and management of a large number of IoT devices (Figure 11.7) require smart wireless networks. To address this issue, a low-cost platform is needed to improve energy efficiency in commercial buildings with scalable building automation systems (BASs).

To this end, and based on the lessons learned from the Android revolution in the cell phone industry [5], it is expected that open-source platforms can quickly and effectively penetrate

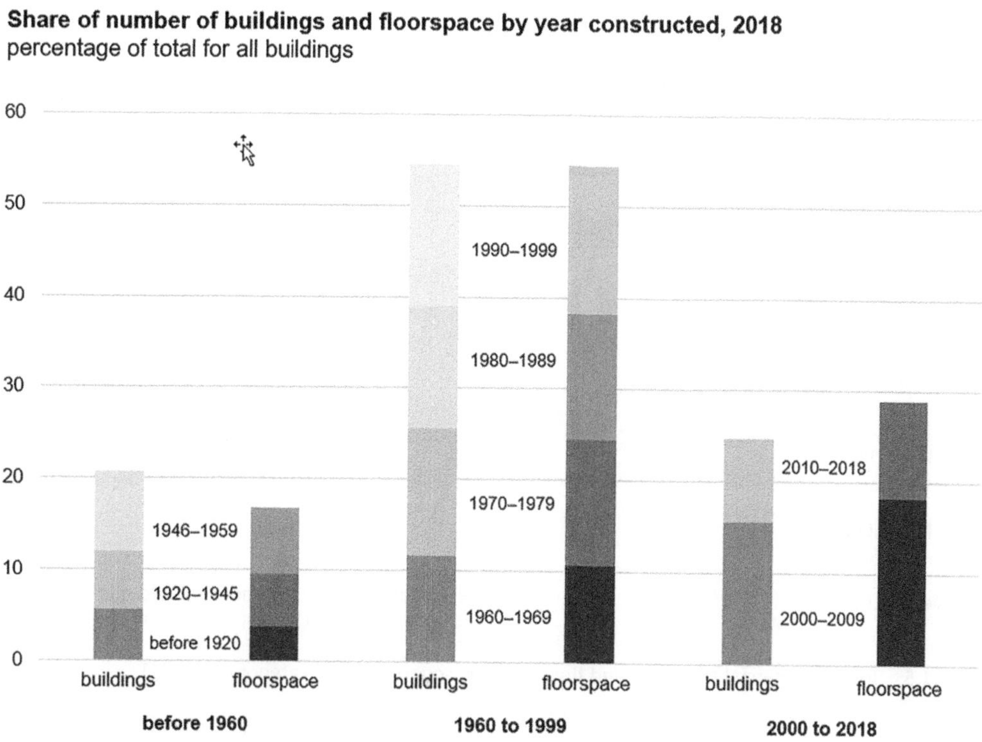

Figure 11.3 Share of number of buildings and floorspace by year constructed, 2018 [1].

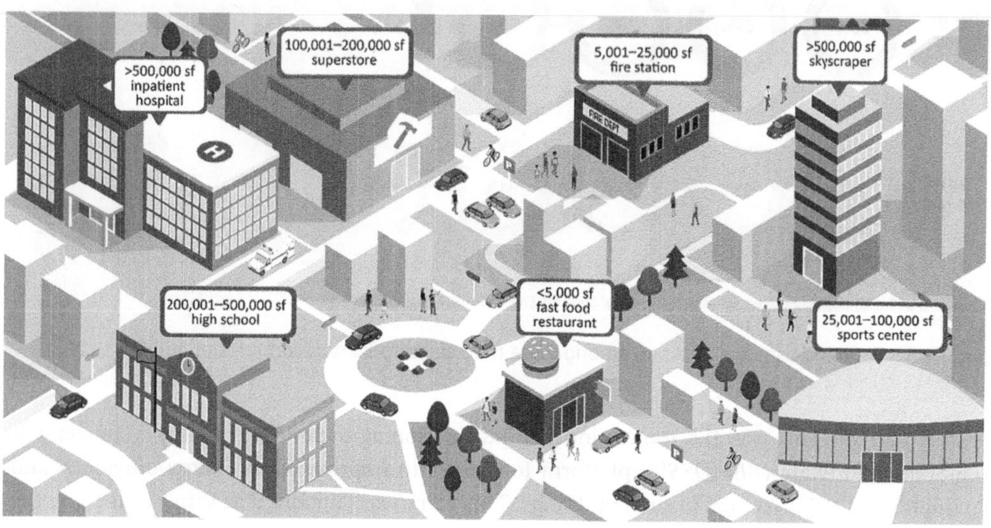

Figure 11.4 Illustration of typical building size category [1].

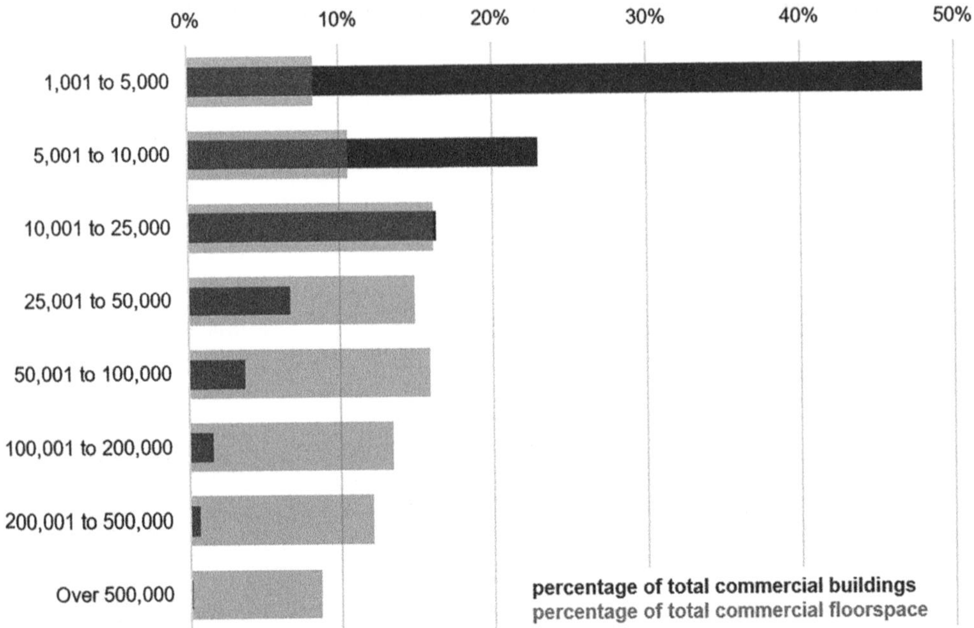

Figure 11.5 Total commercial buildings and floorspace by square footage category, 2018 [1].

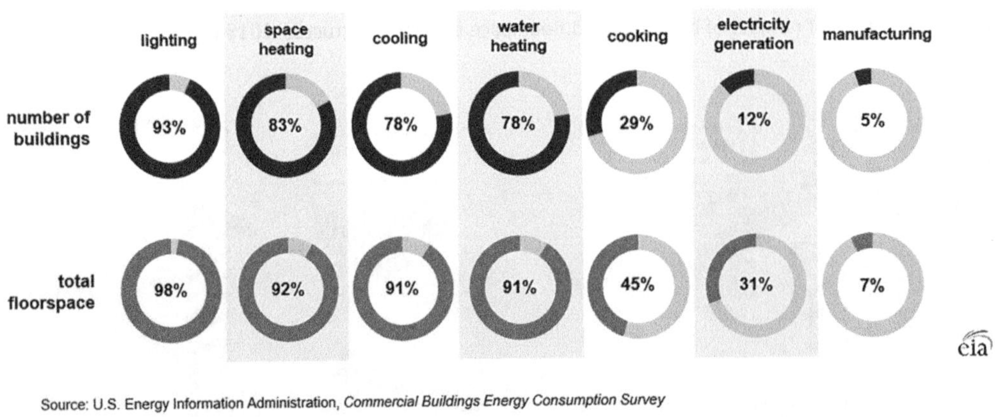

Figure 11.6 Four end uses are common among more than three-fourths of commercial buildings [1].

the SMB market. This was the motivation behind developing a Building Energy Management Open-Source Software (*BEMOSS™*) platform for BEMS to improve SMB energy efficiency and help implement DR programs in such buildings. This work was supported at Virginia Tech by the Building Technology Office of the US Department of Energy. The *BEMOSS™* platform opens up the demand-side ancillary service market [6] and creates opportunities for building owner/operators. Also, it can help accelerate the development of market-ready products like embedded BEMS and communication device controllers for HVAC, lighting, and plug loads. Moreover, it enables utilities and independent system operators (ISOs) to actively leverage DR [7] as a partial substitute for generation reserve or transmission upgrade.

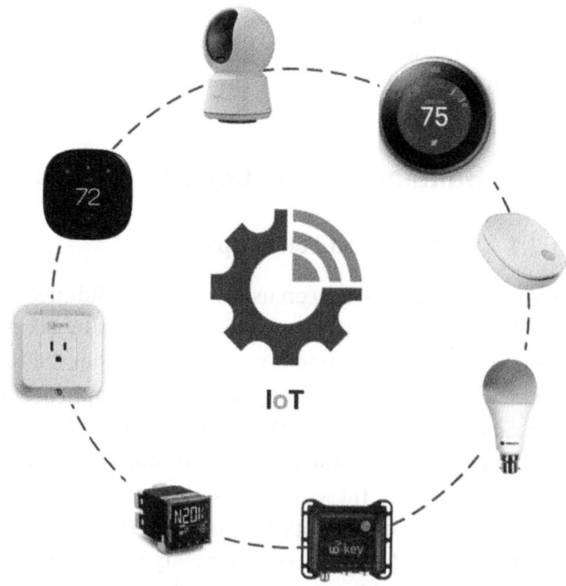

Figure 11.7 IoT-enabled smart building devices.

Detailed architecture and operating principles behind the *BEMOSS™* platform are described in this chapter. This shows how it is possible to effectively manage electrical energy demands in SMBs, thus making commercial buildings smart and energy-efficient. The developed *BEMOSS™* system has the following attributes:

- **Application:** Open source provides wide accessibility to BEMOSS™, encouraging hardware manufacturers to accelerate the development of energy management tools. The application can be made widely usable with inputs from associations like Air-conditioning, Heating and Refrigeration Institute (AHRI) and Association of Home Appliance Manufacturers (AHAM) and manufacturers like Danfoss, Philips, and GE Appliances.
- **Usability:** Various application protocols like Smart Energy (SE) Profile, Open Automated DR (openADR), and Building Automation and Control Networks (BACnet) and communication interfaces like Ethernet, Wi-Fi, ZigBee, and HomePlug (programmable logic controller [PLC]) are being used by equipment/device manufacturers. The proposed BEMOSS™ software has application gateway features to provide interoperability between various standards. Development of plug-and-play-compatible controllers and software tools includes gateway features to provide flexibility to integrate different equipment and device controllers. User interface (UI) has network polling features to locate devices in its vicinity so that users can configure and integrate with the BEMOSS™. It provides increased flexibility through the plug-and-play ability to integrate and incorporate many device controllers and sensors.
- **Advanced monitoring:** BEMOSS™ provides real-time monitoring of the building energy consumption and devices' status through a web interface. Building owners can monitor system performance and control various devices remotely. This can also avoid on-site physical monitoring, which is expensive and can be considered intrusive.
- **Cost-effectiveness:** BEMOSS™ provides advanced algorithms for energy efficiency improvement and DR. DR is achieved by receiving time-of-use pricing signals from the utility power meter and can defer usage of appliances or change HVAC set points for DR. Real-time cost-saving calculation algorithm can perform system tracking and provide possible performance enhancements that can be monitored on the UI and offer different grades of comfort and savings.

Many smart appliances are capable of generating forecasted load profiles based on their settings that can be used to calculate expected building loads. These features can be leveraged to limit a demand restrike.

11.2 Building Automation Solution Landscape

11.2.1 BEMS Product Landscaping

BEMSs have commonly been used in larger buildings or campuses with large energy consumption. Each building can generate significant savings by taking energy efficiency and automation initiatives to justify the investment necessary. This has led to many commercial BEMS products, which provide enterprise solutions to individual equipment or a group of equipment. Most commercial solutions are proprietary with a wide variety of features like device customization, UIs, protocol interoperability, wired and wireless communication accessibility, advanced monitoring, and energy optimization. Since these solutions have been developed based on existing usage for larger buildings, they mostly cater to applications like HVAC, lighting, security, fire safety, and surveillance in such buildings. Most of these proprietary solutions have been developed to provide user monitoring and control and improve energy efficiency which shows direct savings which is significant for larger buildings. Since large buildings have a wide variety of equipment, most BEMS solutions are not plug-and-play-capable. Multiple hardware devices need to be integrated to achieve complete monitoring and control of the system. Some of these solutions are provided in Table 11.1.

SMBs do not generally fit into the specifications of most commercial BEMS solutions for the fact that the nature of applications is much simpler, like modular HVAC units, limited lighting and refrigeration, and plug loads. Most of the control of these systems is taken care of by stand-alone systems like thermostats, occupancy sensors, and photocells. Some of the devices, like smart appliances, are capable of remote monitoring and control with physical communication protocols like ZigBee, Wi-Fi, Ethernet, and HomePlug. On the application protocol level, most of the devices fall into the category of ZigBee, SE Profile, BACnet, OpenADR, etc.

The solutions for SMBs should have both customization and visualization potential offered by large building solutions as well as plug-and-play capability to integrate with a wide variety of devices. Our developed *BEMOSS*™ provides a bridge between the solutions offered for large buildings and those for SMBs.

11.2.2 Customizable BEMS Concept

Products like Echelon SmartServer [8] offer third-party customizability on their BEMS platforms. Their integrated development environment supports the concept of a freely programmable module (FPM), which allows user-developed logic to be deployable on a hardware controller. The hardware integration to devices, communication gateways, and sensor equipment can be handled on the BEM device like Echelon SmartServer and can be accessed through web interfaces. This scalable technology offers an open environment for creating software modules for BEMS, whereas the hardware integration is still proprietary.

11.2.3 Green Button

This initiative [9] provides electricity customers an option to securely download their own energy usage information from their utility or electricity supplier. Green Button was developed as a

Table 11.1 Building automation solution landscape.

Organization	Product	Application	Protocols supported
Johnson Controls	Metasys	HVAC, fire safety, lighting, security	BACnet, LonWorks, N2, IEEE 802.15.4 Wireless
Honeywell	Enterprise Buildings Integrator	HVAC, security, fire safety, surveillance, energy optimization	BACnet, Modbus, LonWorks, OPC
SIEMENS	APOGEE	HVAC, fire and security, lighting control, industrial control	BACnet, Modbus, LonWorks, OPC
novar	Opus	HVAC, refrigeration, lighting	LonWorks, BACnet, Modbus
ECHELON	i.LON SmartServer 2.0	HVAC, lighting, security, fire safety	LonWorks, Modbus, M-Bus, Digital I/O, SOAP/XML, PulseCount Input, Custom Driver Support
PHILIPS	Philips Dynalite	Lighting	Proprietary
ALERTON	BACtalk	HVAC	BACnet
Carrier turn to the experts	i-Vu	HVAC	BACnet (ARCNET, MS/TP, and PTP), Modbus (RTU and ASCII), N2, and LonWorks

public–private partnership initiative and supported by the National Institute of Standards and Technology (NIST). This industry-led effort provides electricity customers with easy-to-access and computer-friendly data from the "Green Button" portal on electric utilities' websites. This information enables consumers to use a large variety of freeware web and smartphone tools to make more informed energy decisions, optimize the size and cost-effectiveness of distributed generation (DG) and storage resources for their home, or verify that energy efficiency retrofit investments are performing as promised. Consumers can also integrate this information with social networking websites like Facebook and compete with friends to save energy and lower their carbon emissions.

11.3 BEMOSS™ Features

BEMOSS™ is designed to make it easy for hardware manufacturers to interface their devices with *BEMOSS*™ seamlessly. Some of the *BEMOSS*™ features are as follows:

11.3.1 Open Source

BEMOSS™ is an open-source operating system and is built upon VOLTTRON™. VOLTTRON™ is developed by the Pacific Northwest National Laboratory (PNNL).

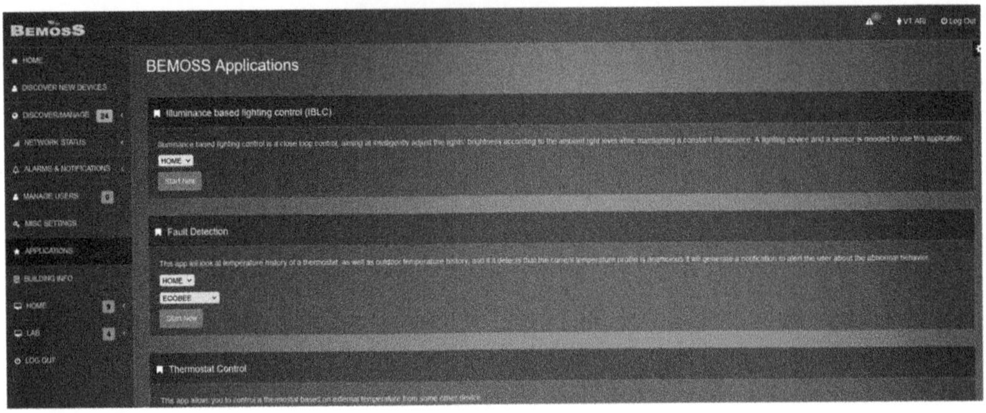

Figure 11.8 BEMOSS™ applications.

11.3.2 Interoperability

BEMOSS™ is designed to work with load control devices from different manufacturers that operate on different communication technologies and data exchange protocols. These include both new commercially available products that operate on Ethernet and Wi-Fi, as well as legacy devices that operate on serial communications using Modbus remote terminal unit (RTU) and BACnet master–slave/token-passing (MS/TP) protocols. Figure 11.8 shows *BEMOSS™* applications and how users can start new applications such as illuminance-based lighting control (IBLC), fault detection, and thermostat control.

- **IBLC**: IBLC is a closed-loop control, aiming at intelligently adjusting the lights' brightness according to ambient light level while maintaining a constant illuminance. A lighting device and a sensor are needed to use this application.
- **Fault detection**: This application will look at the temperature history of the thermostat and outdoor temperature history and detect that the current temperature profile is anomalous. It will generate a notification to alert the user about abnormal behavior.
- **Thermostat control**: This application allows users to control a thermostat based on external temperature from some other devices.

11.3.3 Plug and Play

BEMOSS™ can automatically discover supported devices in a building (Figure 11.9). Discovered devices will be connected to the *BEMOSS™* platform only after a manual or automated authentication protocol is exercised. Figure 11.10 shows that after discovery IoT devices enter the pending state until authenticated. Figure 11.11 shows the list of authenticated devices. Upon authentication, the IoT device can start monitoring and controlling the desired function.

11.3.4 Alarm and Notifications

BEMOSS™ can send the users alarms (Figure 11.12) via email/text to notify them about different situations. As a case in point, it can send an alarm about the lighting loads' status, brightness, and other settings.

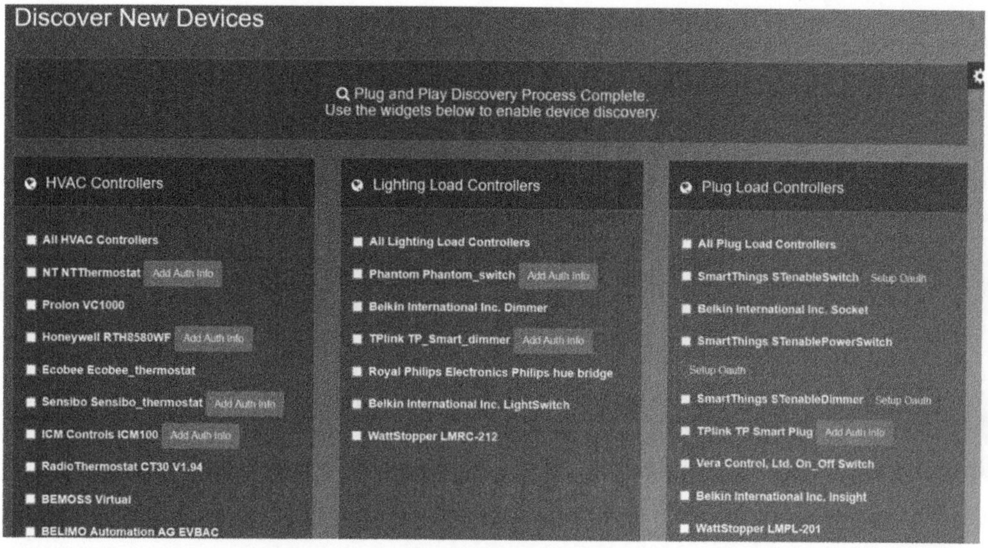

Figure 11.9 Discover new devices by BEMOSS™ platform.

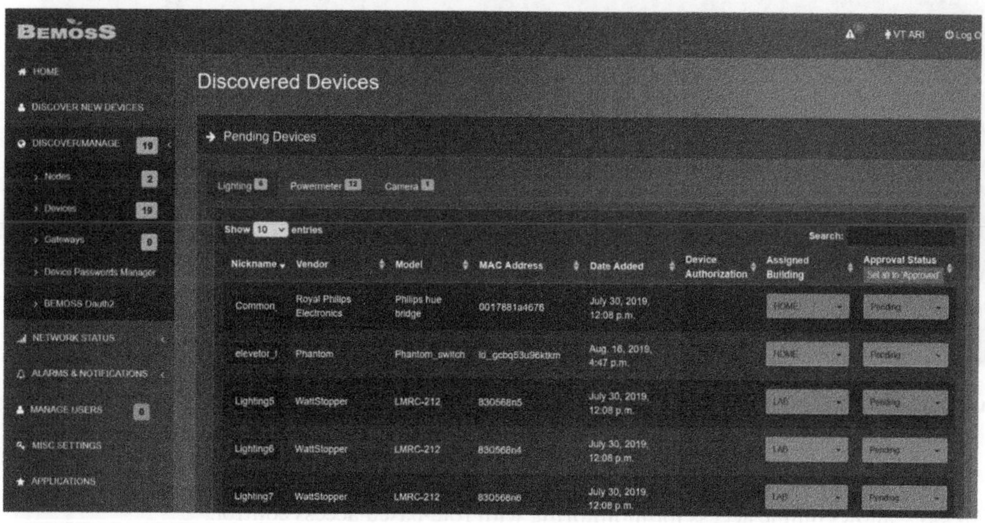

Figure 11.10 Discover new devices by BEMOSS™ platform (pending).

11.3.5 Cost-Effectiveness

BEMOSS™ is built upon a robust open-source platform that can operate on a low-cost single-board computer, such as the ODROID. This feature can contribute to its rapid deployment in small- or medium-sized commercial buildings.

11.3.6 Scalability and Ease of Deployment

BEMOSS™ has a multilayer architecture and can be initially deployed in one zone, easily expanding to the entire building.

Figure 11.11 Discover new devices by BEMOSS™ platform (approved).

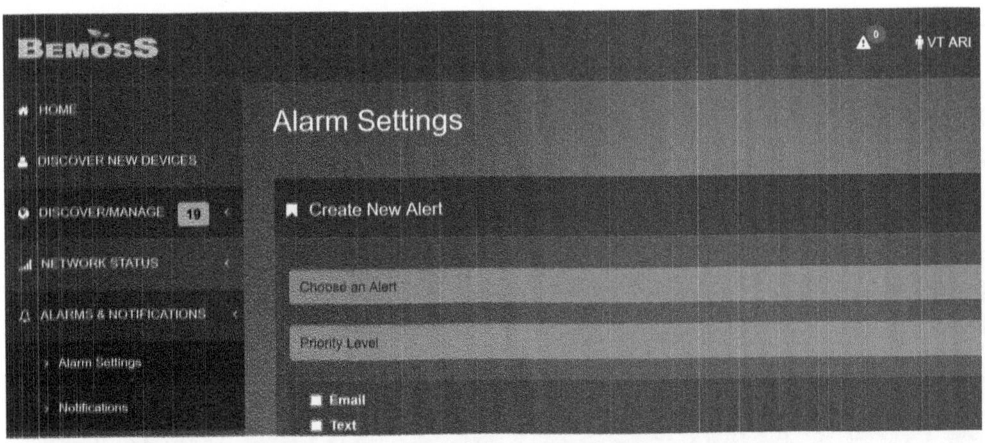

Figure 11.12 Alarm and notification.

11.3.7 Ability to Provide Online Access

BEMOSS™ allows online access for monitoring with role-based access control.

11.3.8 Security

BEMOSS™ utilizes built-in security features provided by VOLTTRON™ and provides enhanced security features for encrypted communications (Figure 11.13).

11.4 Targeted Buildings and Loads

The CBECS estimates that 5.9 million buildings in the United States with 96 billion square feet of total commercial floorspace used 6787 trillion British thermal units (TBtu) of energy in 2018 (Figure 11.14) [10]. According to the CBECS, commercial buildings are categorized based on

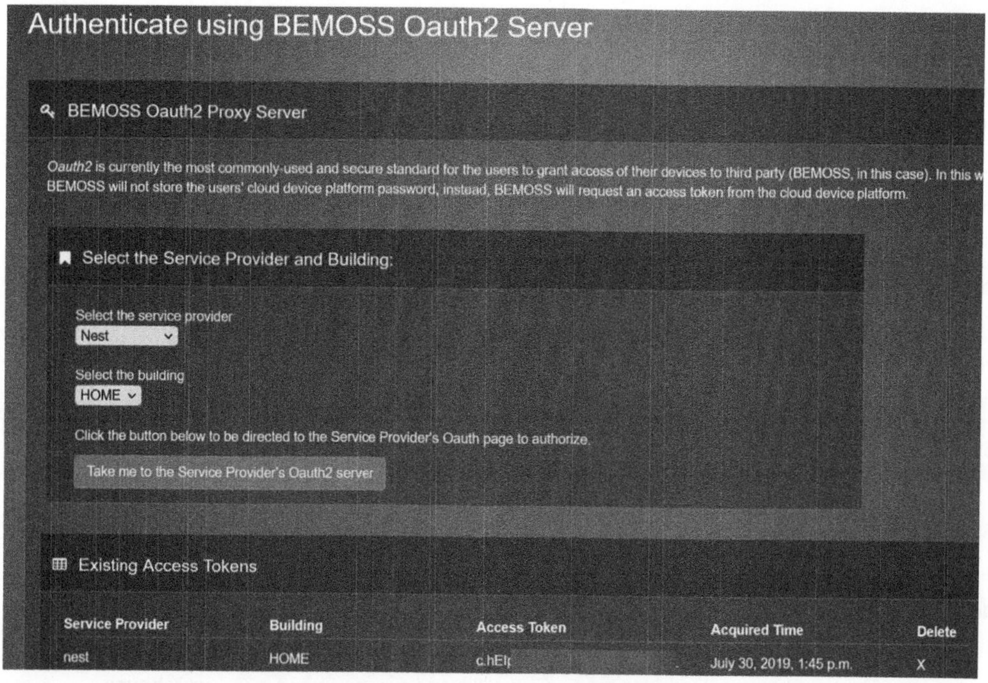

Figure 11.13 Authenticate using BEMOSS™ OAuth 2.

principal activity, which is the primary business, commerce, or function carried on within each building [11, 12].

According to the survey in 2018, we conclude [10]:

- Warehouse and storage, office, and service buildings together accounted for almost one-half (48%) of all commercial buildings.
- Warehouse and storage, office, and education buildings accounted for one-half of total commercial building floorspace.
- Office, mercantile, and education buildings accounted for 43% of energy consumption.

Commercial buildings in the United States, on average, have a size of 16,300 ft^2. Notably, lodging, education, and healthcare facilities tend to be the largest, with average sizes of 33,700, 31,100, and 29,300 ft^2, respectively. Contrastingly, food service buildings are typically the smallest, averaging 4800 ft^2. The healthcare category's average size is largely influenced by the substantial size of inpatient healthcare buildings like hospitals, which average 264,800 ft^2. Outpatient healthcare buildings are much smaller, averaging around 13,700 ft^2 (Figure 11.15).

As shown in Figure 11.16, buildings associated with food services, food sales, and inpatient healthcare were found to have the highest energy intensity. In contrast, buildings that were vacant showed the lowest energy intensity.

According to the 2018 report (Figure 11.17), office buildings are the largest consumers of electricity, using 775 TBtu, which is about three times the amount of natural gas they used (250 TBtu). Education buildings, on the other hand, consumed the most natural gas at 328 TBtu. The categories of service, religious worship, public order and safety, and vacant buildings are unique in that their electricity and natural gas usage did not show a significant difference.

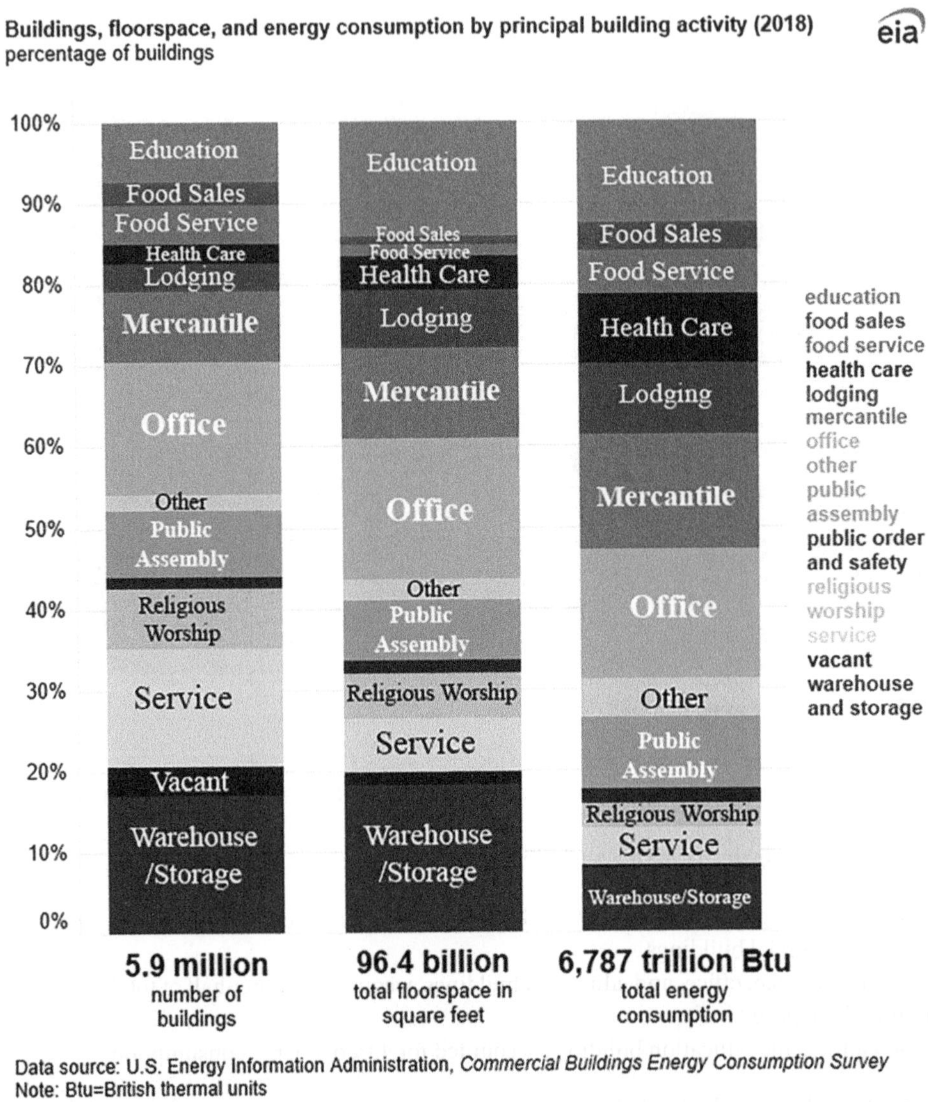

Figure 11.14 Energy use by type of US commercial building, total: 6787 trillion British thermal units (TBtu) (source: US Energy Information Administration/Public domain) [10].

Inpatient healthcare buildings exhibited the highest energy intensity for space heating, at 62.6 thousand British thermal units (MBtu) per square foot in the 2018 report (Figure 11.18). They also led in ventilation energy intensity, registering 28.6 MBtu per square foot. Meanwhile, the greatest energy intensity for cooling was observed in food service buildings, where it reached 19.1 MBtu per square foot. These figures highlight the varying energy demands across different building types and their specific uses.

Figure 11.19 illustrates electricity use in commercial buildings by load type. There are three major loads in commercial buildings with respect to load types: HVAC, lighting, and plug loads. According to the data from EIA published in 2018 [13], energy consumption by HVAC equipment, i.e., space heating (32%), cooling (9%), and ventilation (11%), accounts totally for 52% of the total consumption

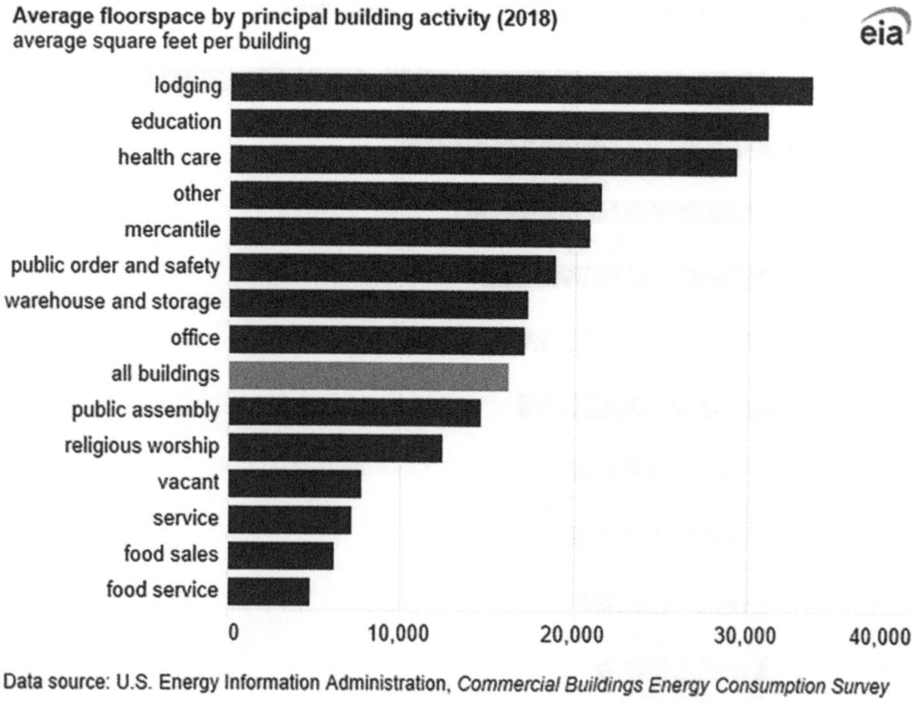

Figure 11.15 Average floorspace by principal building activity (2018) (source: US Energy Information Administration/Public domain) [10].

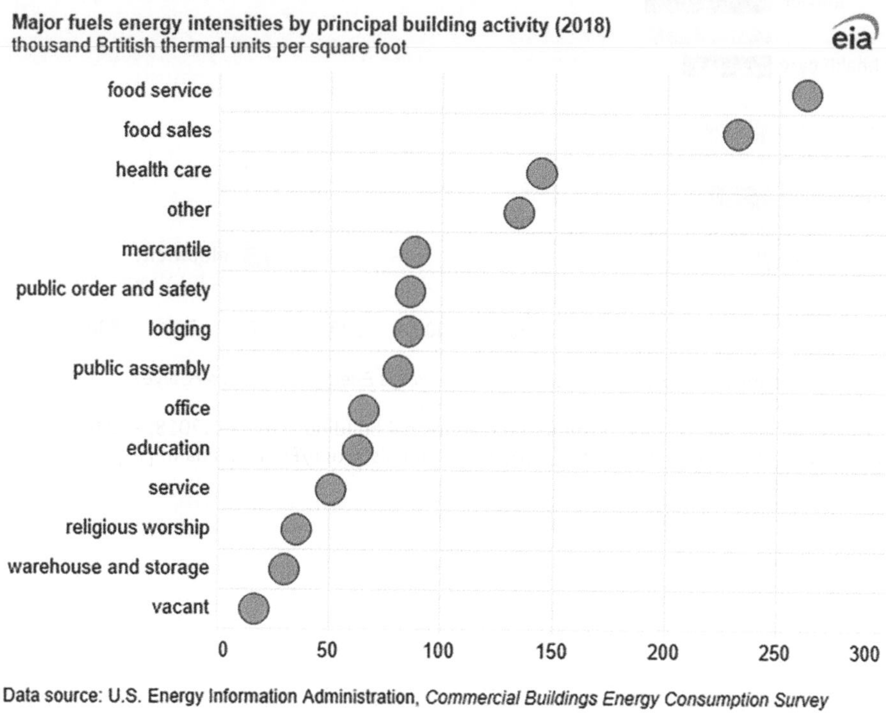

Figure 11.16 Major fuel energy intensities by principal building activity (2018) (source: US Energy Information Administration/Public domain) [10].

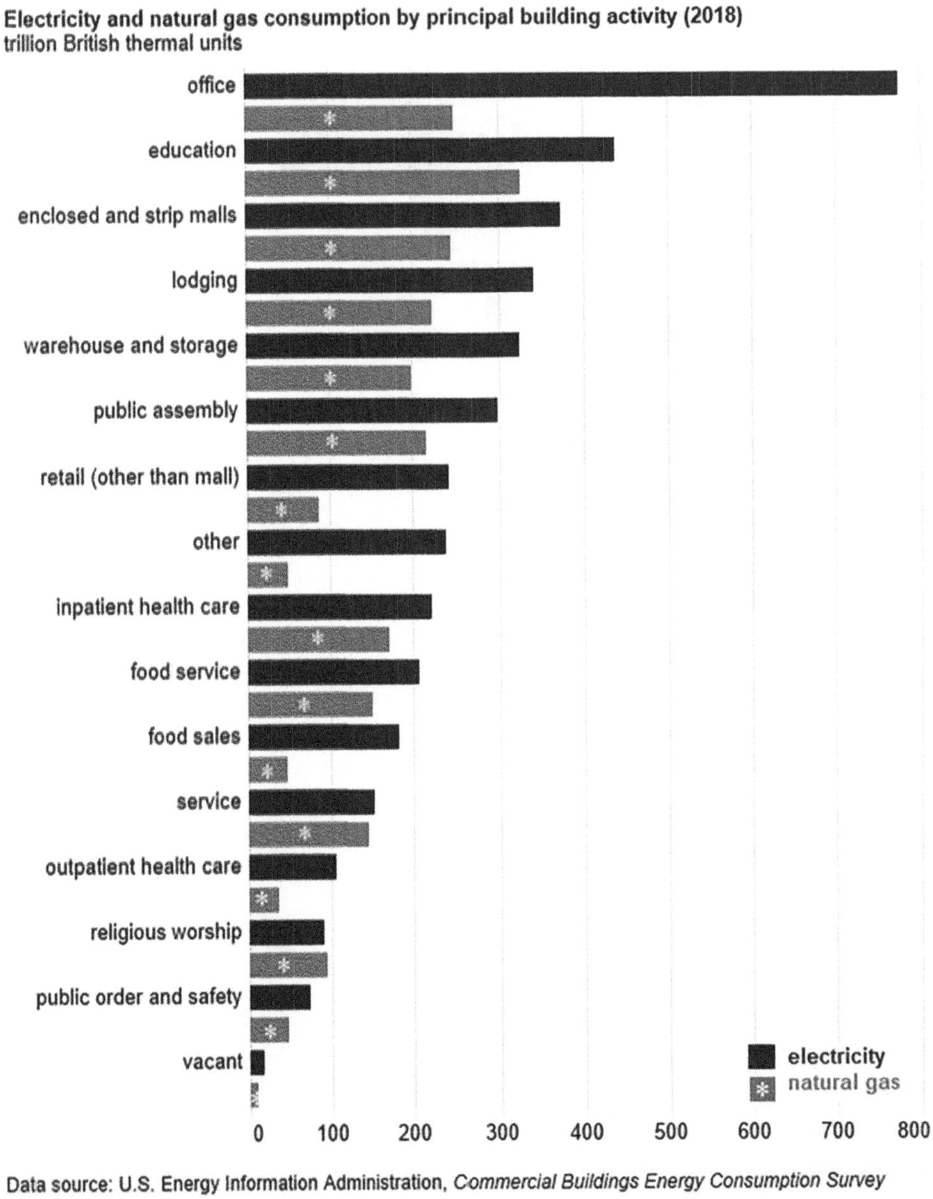

Figure 11.17 Electricity and natural gas consumption by principal building activities (2018)—edited by authors for black/white print (source: US Energy Information Administration/Public domain) [10].

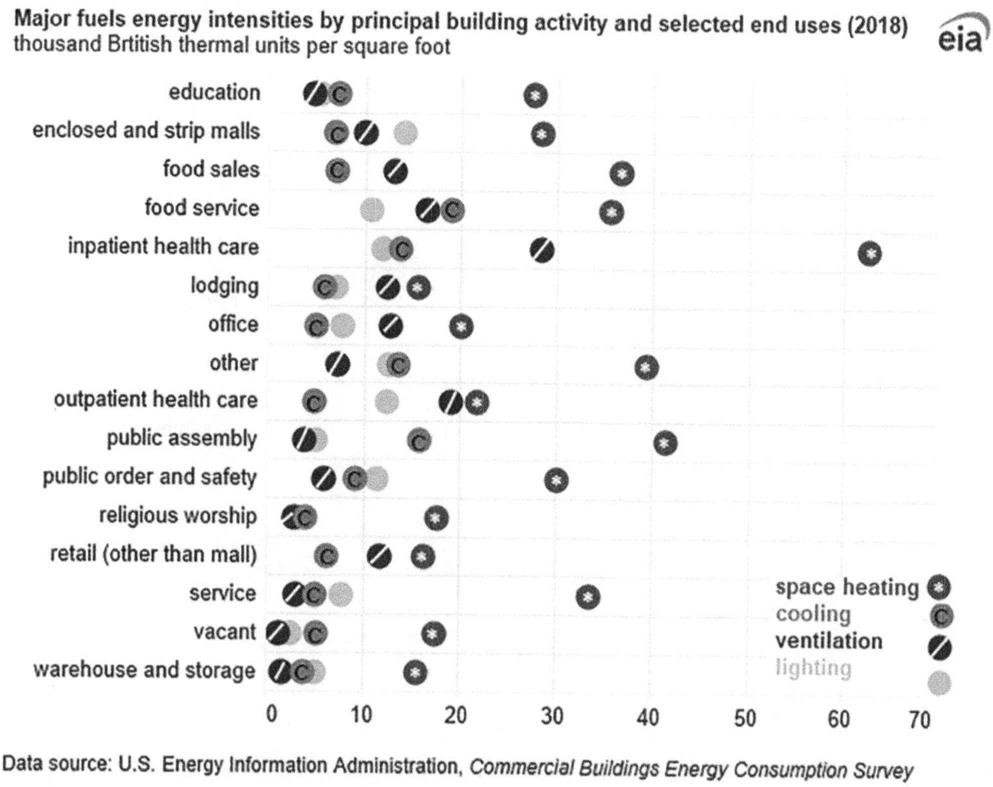

Figure 11.18 Major fuel energy intensities by principal building activities and selected end uses (2018) (source: US Energy Information Administration/Public domain) [10].

in buildings. Moreover, lighting loads constitute the majority share of electricity use at 10%. Other loads include plug loads, computers, refrigeration, and water heating. Many buildings target HVAC, lighting, and plug loads for energy savings through advanced technologies.

In March 2020, EIA projects air-conditioning energy consumption to grow faster than any other use in both residential and commercial buildings [14]. As shown in Figure 11.20, the square footage of commercial buildings is expected to increase by 34%. Moreover, office buildings consume more energy for air-conditioning than any other building type, accounting for 25% of the energy consumed for air-conditioning in the US commercial sector in 2050.

11.5 BEMOSS™ Architecture

11.5.1 BEMOSS™ Architecture for Small Commercial Building (One-Floor Building)

Figure 11.21 shows the *BEMOSS*™ architecture for a small commercial building with a few load controllers of each type. In this architecture, only one single-board computer (e.g., ODROID) embedded with the *BEMOSS*™ software platform is used to enable monitoring and control features of all load controllers in the building. This embedded system can communicate with different types of load controllers, i.e., thermostats, lighting load controllers and plug load controllers, and

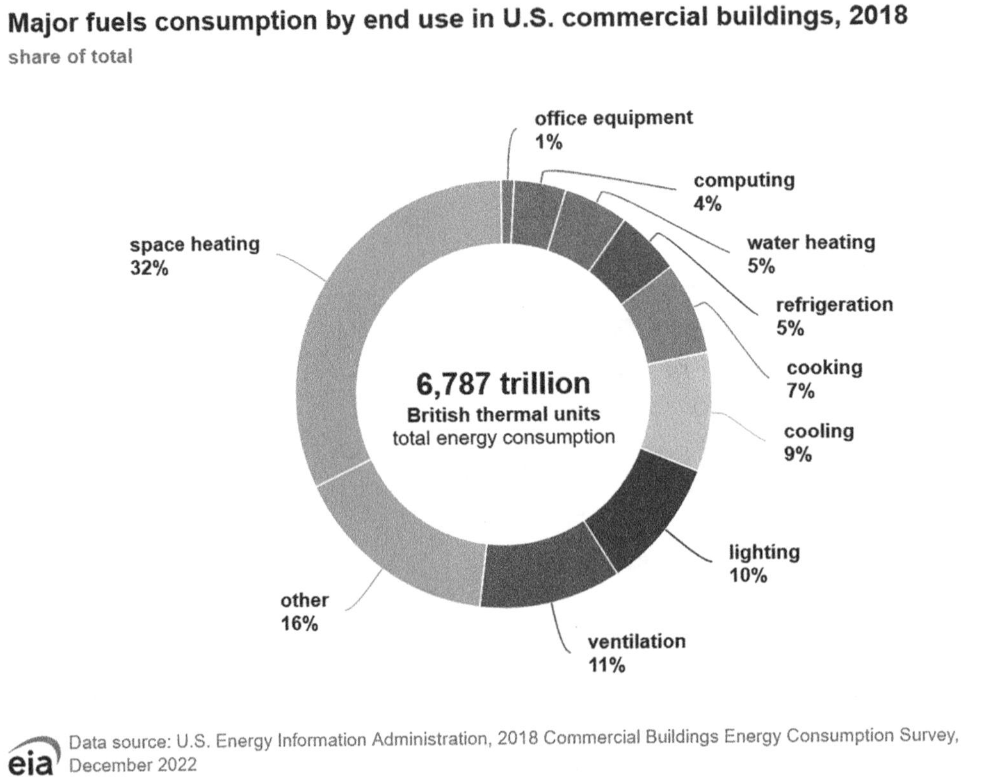

Figure 11.19 Consumption in US commercial buildings by end uses, 2018 (source: US Energy Information Administration/Public domain) [13].

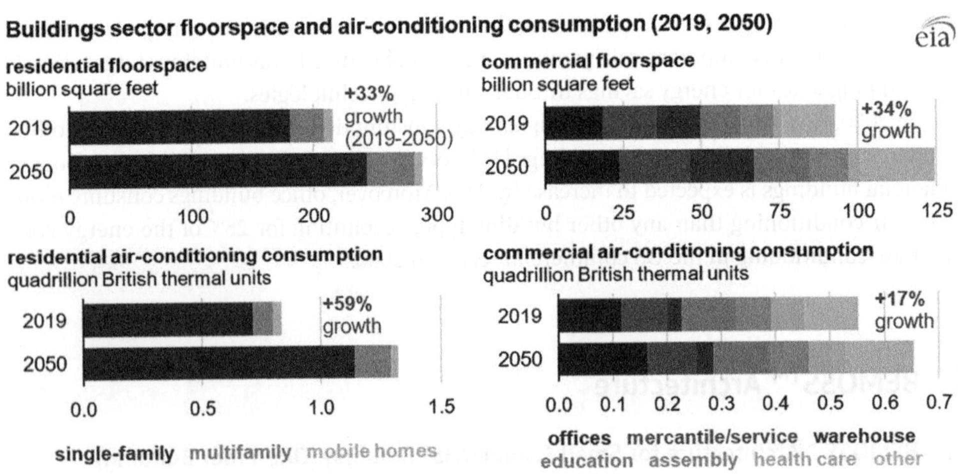

Figure 11.20 Building sector floorspace and air-conditioning consumption (2019, 2050) (source: US Energy Information Administration/Public domain) [14].

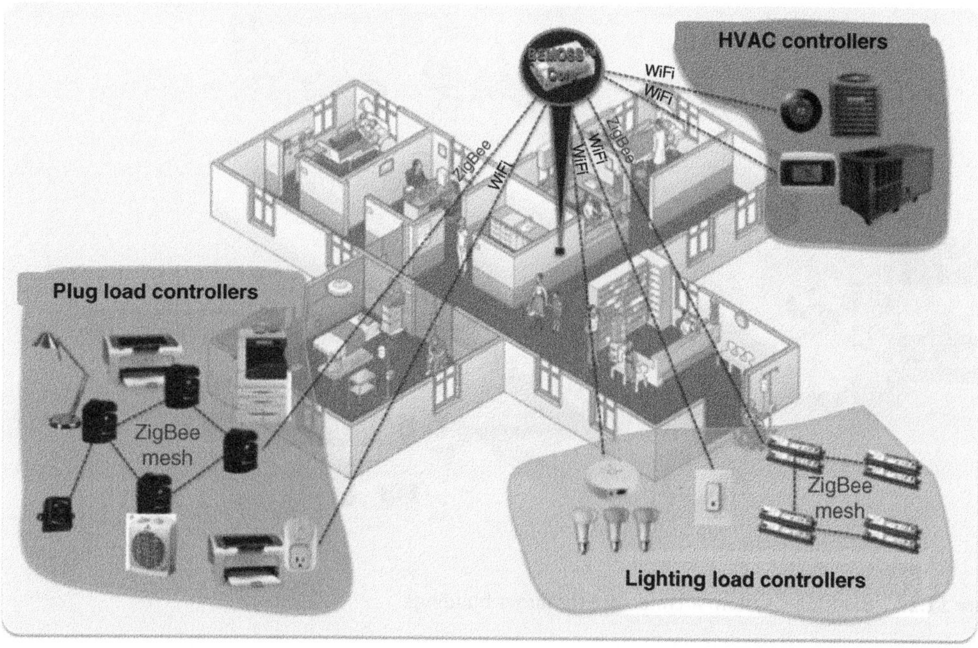

Figure 11.21 BEMOSS™ system architecture for small buildings with a few load controllers.

sensors/power meters via wireless signals (either Wi-Fi or ZigBee). It should be noted that local and remote monitoring and control via a smartphone or a tablet are also enabled.

11.5.2 BEMOSS™ Architecture for Large Commercial Buildings (Multi-Floor Buildings)

BEMOSS™ can also be set up to deploy its multilayer architecture feature for multi-floor buildings with a larger number of devices. In this architecture, a *BEMOSS™* node is responsible for monitoring and controlling devices on one floor. Each *BEMOSS™* node communicates with each other and also communicates with the *BEMOSS™* Core. On the other side, the *BEMOSS™* Core is responsible for supervising the overall system operation, managing multiple *BEMOSS™* nodes, and allowing local and remote access for monitoring and controlling all devices in buildings (Figure 11.22).

11.5.3 Software Architecture

BEMOSS™ has a hierarchical control architecture. The central controller, which hosts *BEMOSS™* software, can be accessed through a web interface. The supervisory control structure helps system optimization be performed at the central controller while the slave controllers provide real-time control to the end-use equipment. It also helps the successful operation of the controllers in case of communication failures.

As illustrated in Figure 11.23, the *BEMOSS™* system comprises four process layers:

1) **UI (Monitoring and Control) Layer** The *BEMOSS™* UI layer provides cloud-based UI apps that can be accessed through a web browser with 128-bit password encryption. This layer has two components: UI and user management. *BEMOSS™* web UI is a dashboard-type interface

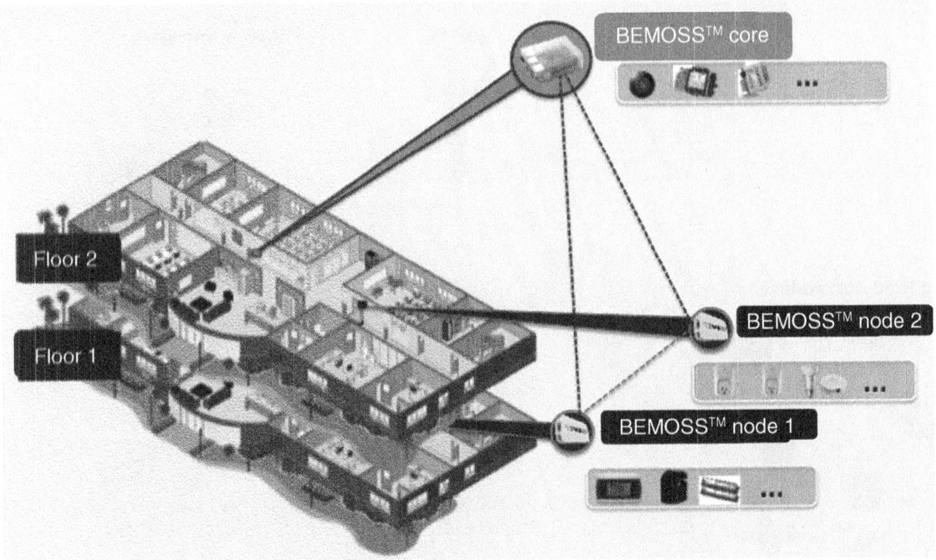

Figure 11.22 BEMOSS™ system architecture for larger buildings.

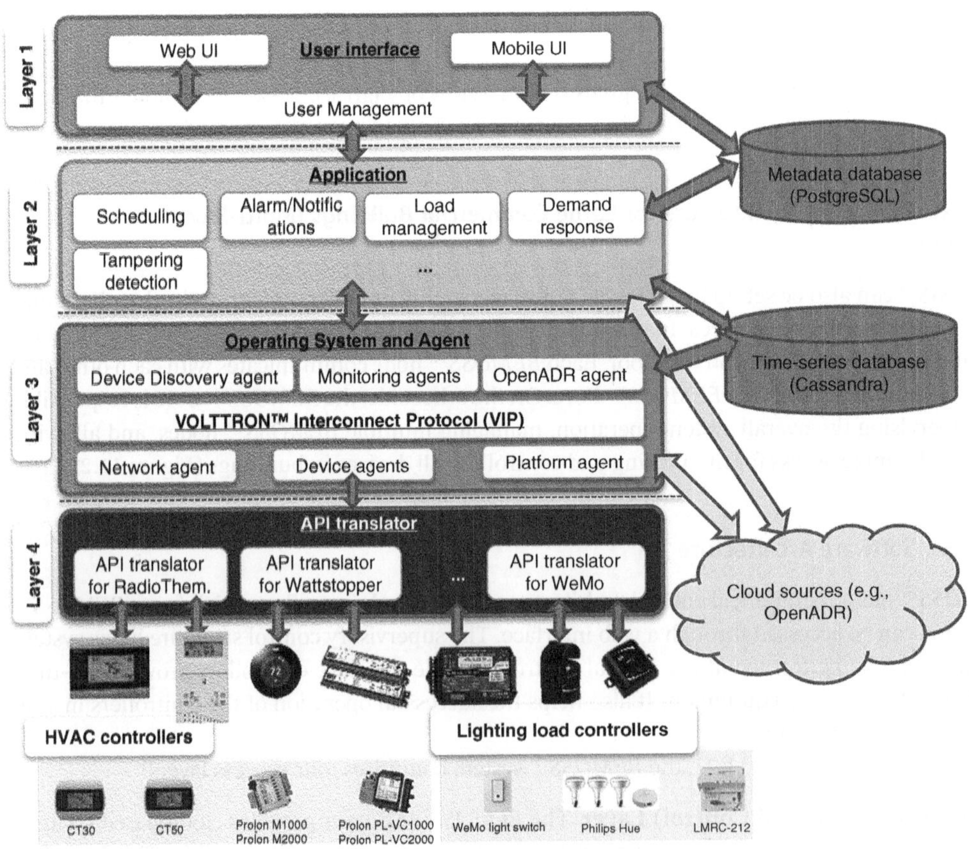

Figure 11.23 BEMOSS™ architecture.

with visuals and graphs to show the current settings of devices in each zone. This layer provides building energy usage monitoring, offering optimal strategies for improving energy efficiency based on system conditions. Authenticated users can also control these devices through an on-site interface.

Regarding user management in *BEMOSS*™, role-based access control is implemented to allow different levels of access to different individuals. For example, building engineers have full authority to adjust set points and schedules of loads in buildings. In contrast, tenants will have limited access to view the current status and historical load data, or control selected loads in specific zones. In *BEMOSS*™, this role-based access control is achieved using access control lists.

2) **Application and Data Management Layer** This layer embeds algorithms to allow monitoring and control of hardware devices interfaced with *BEMOSS*™. Examples of possible applications include DR, price-based management, planning and scheduling, behavior pattern analysis, load management, and alarm/notifications. Apache Cassandra is selected for storing *BEMOSS*™ time-series data in a distributed fashion. A relational database management system (PostgreSQL) is used to satisfy the need to store the metadata for identifying users, devices, and process controls.
3) **Operating System and Framework Layer** In this layer, VOLTTRON™, a distributed agent platform developed by PNNL, is chosen as the software platform for *BEMOSS*™. As shown in Figure 11.23, several *BEMOSS*™ agents have been developed, including a device discovery agent, device agents (e.g., thermostat agents, lighting load agents, and plug load agents), and other system agents. All agents communicate over VOLTTRON™ VIP. The entire *BEMOSS*™ system is also designed to allow email/Short Message Service (SMS) notifications through its alarm/notification app.
   - *Device discovery agent* is responsible for detecting the presence of devices in a building, querying their model numbers, identifying their application programming interfaces (APIs), and launching a control agent to monitor/control the discovered device. With this approach, there is no need to manually identify each device beforehand using a technique like a bar code or Quick Response (QR) code.
   - *Device agent* includes device agents for thermostat, lighting, and plug load. These agents are instantiated to monitor, communicate, and control hardware devices after being discovered by the device discovery agent. Once a device agent is initiated, it is assigned particularly to one hardware device.
   - *Other system agents,* such as the network agent, approval helper agent, app launcher agent, and platform monitor agent, are responsible for the functioning of the *BEMOSS*™ system by facilitating the packaging, installation, starting, stopping, monitoring, and managing the agent execution.
4) **Connectivity Layer** This layer takes care of the communication between the operating system and framework layer and all physical hardware devices. To allow *BEMOSS*™ to communicate with hardware devices that use different communication technologies, data exchange protocols, and device functionalities (different device APIs), the *BEMOSS*™ team created several API interfaces. Each API interface allows *BEMOSS*™ agents to communicate with a group of devices based on their unique APIs. Basically, API interfaces provide a translation service for *BEMOSS*™ agents so that agents can get readings and send control commands to devices (without knowing their APIs) using simple function calls: *getDeviceStatus* and *setDeviceStatus*. Also, an auto-API translator code is developed to allow the automated generation of APIs for different devices that follow supported protocols.

In addition to these, there are also parallel *BEMOSS*™ databases that store all information about *BEMOSS*™ and help smooth its functioning.

11.6 BEMOSS™ Auxiliary Functions

Auxiliary functions of *BEMOSS*™, such as building information, managing users/gateway, and password manager, are shown in Figures 11.24–11.27.

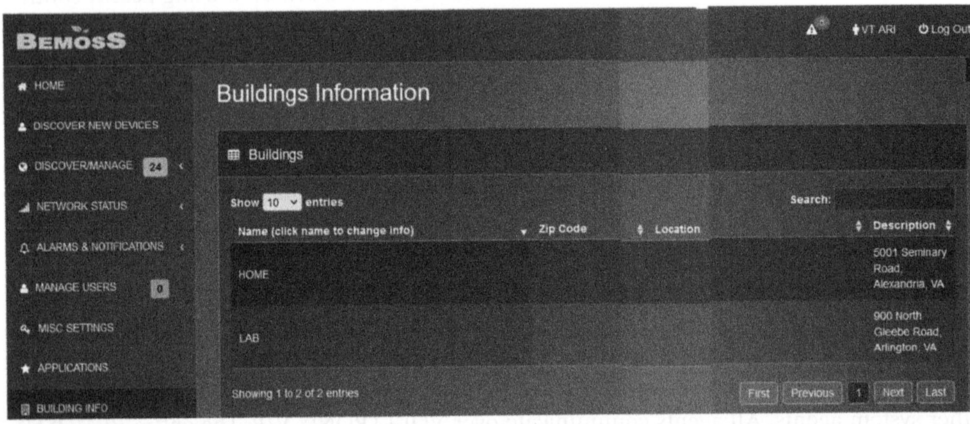

Figure 11.24 Building information.

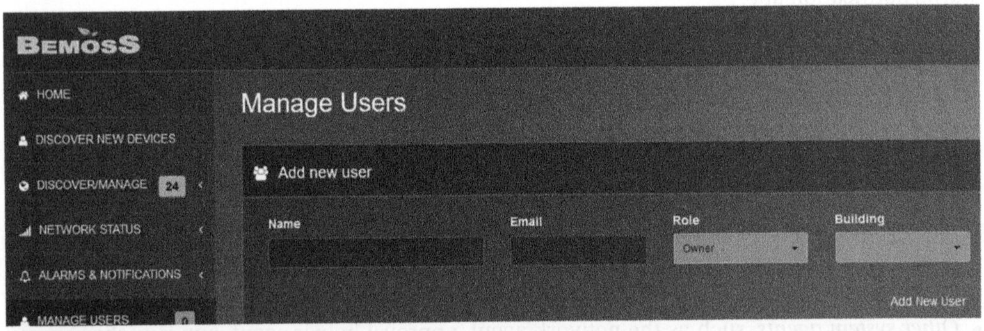

Figure 11.25 Manage users.

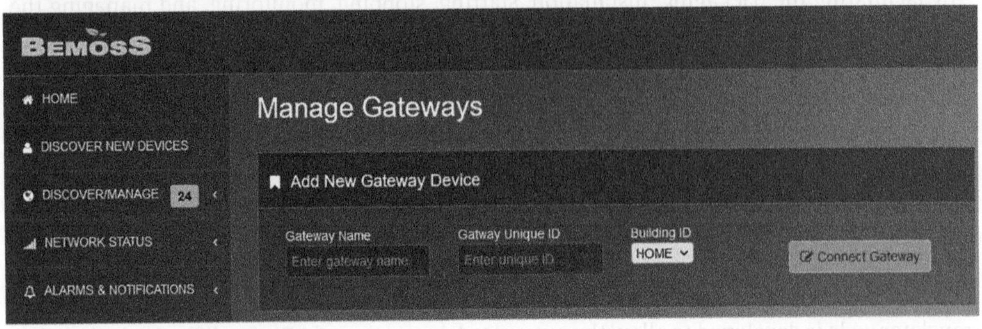

Figure 11.26 Manage gateways.

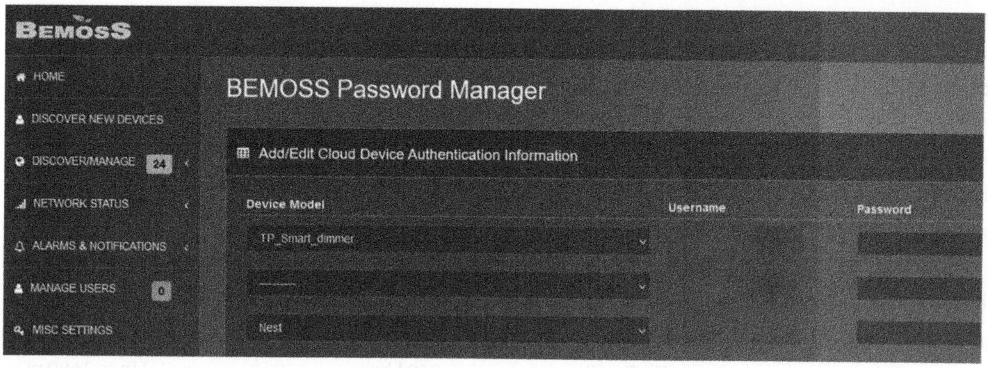

Figure 11.27 Password manager.

11.7 Multiple-protocol Interoperability

BEMOSS™ can communicate with different devices from different manufacturers, while they operate on different communication technologies and data exchange protocols (Figure 11.28). The supported communication technologies and protocols are as follows:

Communication Technologies

- Ethernet (IEEE 802.3)
- Serial Interface (RS-485)
- ZigBee (IEEE 802.15.4)
- Wi-Fi (IEEE 802.11)

Data Exchange Protocols

- BACnet (internet protocol [IP] and MS/TP)
- Modbus (RTU and transmission control protocol [TCP])
- Web (e.g., Extensible Markup Language (XML), JavaScript Object Notation (JSON), and Really Simple Syndication (RSS)/Atom)
- SE
- ZigBee API
- OpenADR

Figure 11.28 Multiple-protocol interoperability.

Table 11.2 Communication technologies.

Technology	Standard/protocol	Max. theoretical data rate	Coverage range
Wired communication technologies			
Ethernet	IEEE 802.3	10 Mbps–1 Gbps	Up to 100 m
Serial	RS-485	100 kbps–35 Mbps	Up to 1200 m
Wireless communication technologies			
ZigBee	ZigBee	250 kbps	Up to 100 m
	ZigBee Pro	250 kbps	Up to 1600 m
Wi-Fi	802.11x	2–600 Mbps	Up to 100 m

Table 11.3 Data exchange protocol.

		Allow communications over			
Data exchange protocol	Application	Ethernet	Serial	Wi-Fi	ZigBee
BACnet (IP)	Building automation	✓	—	✓	—
BACnet (MS/TP)		—	✓	—	—
Modbus (RTU)	Legacy device communication	—	✓	-	-
Modbus (TCP)		✓	—	✓	—
Web (e.g., XML, JSON, and RSS/Atom)	Numerous applications	✓	—	✓	—
ZigBee API	Home/building automation	—	—	—	✓
OpenADR	Demand response	✓	—	✓	—
Smart Energy (SE)	Smart grid	—	—	✓	✓

Tables 11.2 and 11.3 show detailed information about communication technologies and data exchange protocols.

BEMOSS™ release supports selected HVAC, lighting and plug load controllers (Table 11.4).

Therefore, *BEMOSS*™ can support multiple IoT devices through industry-standard protocols and communication technologies. Figure 11.29 shows different devices with different communication technologies implemented in the Virginia Tech Advanced Research Institute (VT-ARI) building.

11.8 Test Results

11.8.1 VT-ARI Lab

To evaluate the functionality of *BEMOSS*™, a laboratory setup (Figure 11.30) is implemented at VT-ARI. The goal is to control all devices through the *BEMOSS*™ UI.

Table 11.4 List of supported hardware.

Device model	Vendor	Protocol
HVAC controller		
CT30 w/Wi-Fi USNAP Module	Radio thermostat	Wi-Fi
CT50 w/Wi-Fi USNAP Module	Radio thermostat	Wi-Fi
PL-M1000RTU/M2000RTU	Prolon	Modbus RTU
VC1000/VC2000	Prolon	Modbus RTU
Lighting load controller		
WeMo Light Switch	Belkin	Wi-Fi
Philips Hue	Philips	Wi-Fi/Ethernet
LMRC-212-U	Wattstopper	BACnet MS/TP
Plug load controller		
WeMo Switch	Belkin	Wi-Fi
WeMo Insight Switch	Belkin	Wi-Fi
LMPL-201	Wattstopper	BACnet MS/TP
Sensor		
LMLS-400	Wattstopper	BACnet MS/TP

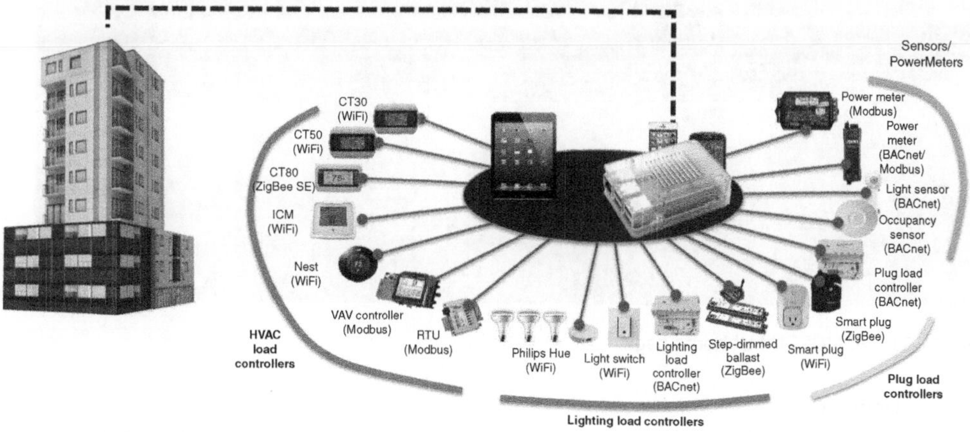

Figure 11.29 Supporting multiple IoT devices by BEMOSS™.

It can be observed from Figure 11.31 that different devices, including HVAC, lighting, plug load, sensor, power meter, and distributed energy resources (DERs), are programmed and integrated into *BEMOSS*™.

To demonstrate the *BEMOSS*™ performance, the functionality of the following devices is investigated.

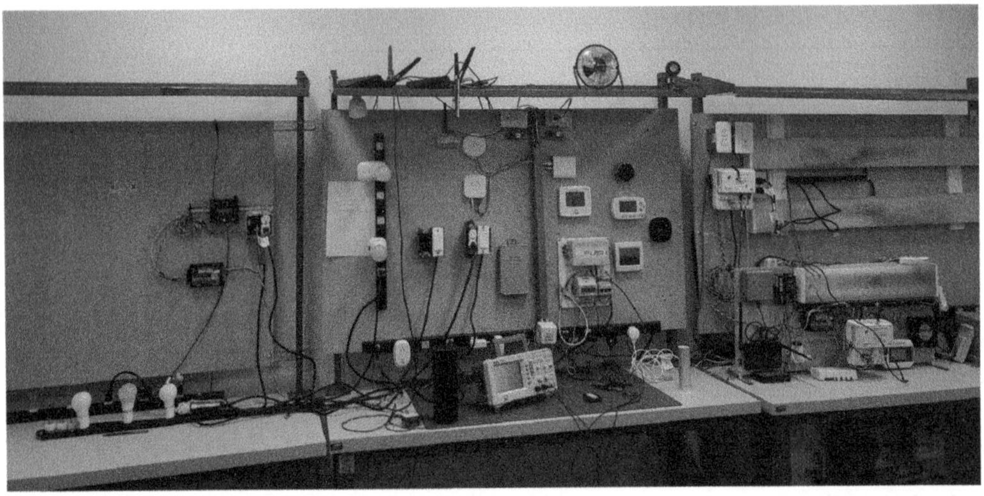

Figure 11.30 VT-ARI lab.

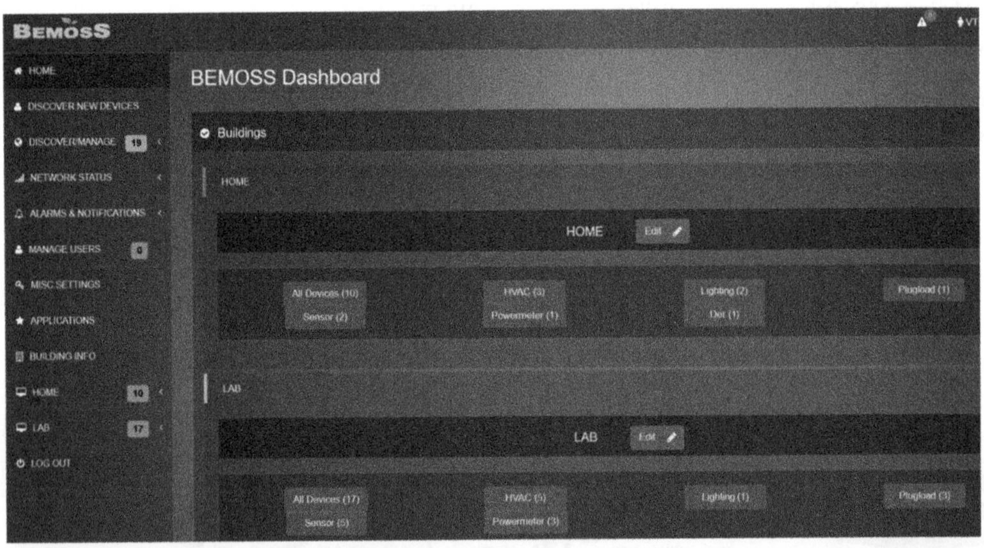

Figure 11.31 BEMOSSTM UI home page.

11.8.1.1 Lighting

As shown in Figure 11.32, the device status for the office light is off and it is in "Online" mode (Figure 11.32a), meaning that the device (Figure 11.32b) is communicating with the *BEMOSS*™. In case, we change the status of the office-light switch to "on," the light will turn on (Figure 11.32d). Also, the brightness can be controlled through *BEMOSS*™ UI. As a case in point, we may change the brightness to 38% (Figure 11.32c) and save energy while maintaining working visibility in the room.

11.8.1.2 Thermostat

In this part, an Ecobee Inc. (*ECOBEE*) thermostat (Figure 11.33a, b) is controlled by the *BEMOSS*™ UI. As shown in Figure 11.33a), this thermostat is connected to a fan. The fan represents the AC, and we control the thermostat operation via *BEMOSS*™ UI. Figure 11.33c, d shows the status page

(a) (b)

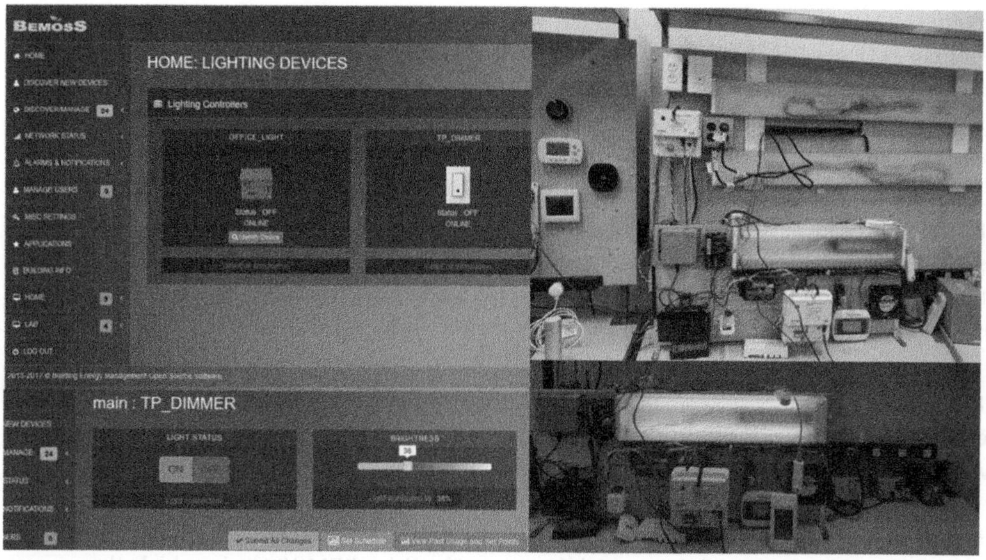

(c) (d)

Figure 11.32 Lighting control.

(a) (b)

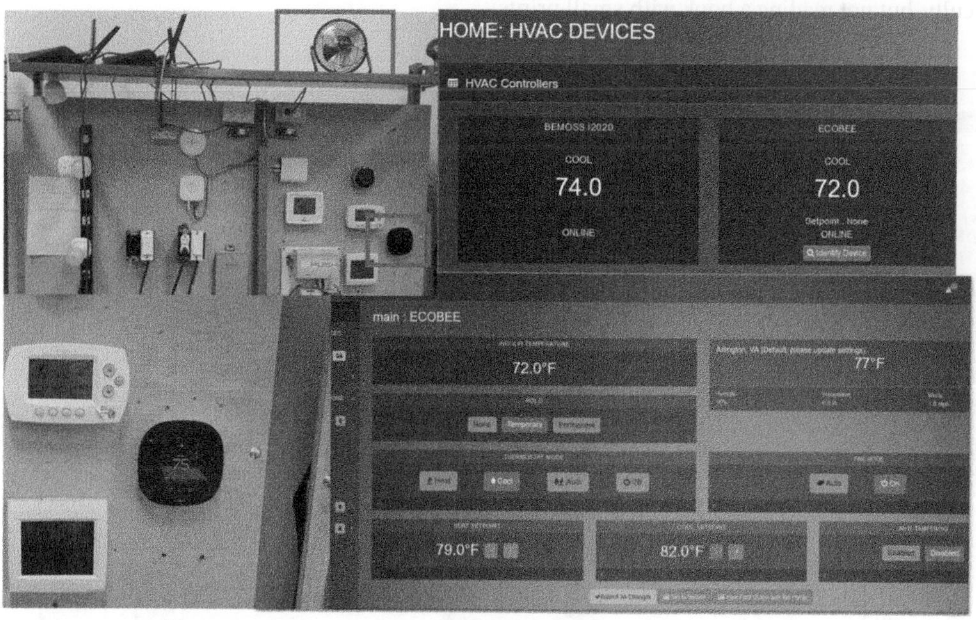

(c) (d)

Figure 11.33 Thermostat control.

related to the *ECOBEE* thermostat where the indoor air temperature is 74 °F when the cooling set point is 82 °F; therefore, the fan does not work. If we change the cooling set point to a value lower than 74 °F (e.g., 72 °F), the fan will start working. Therefore, we could control the thermostat through *BEMOSS*™ UI. Moreover, we are able to see the historical changes as well as set schedules with different HVAC set points for different periods, e.g., morning, evening, and midnight (more details are provided in the next section).

11.8.1.3 Plug Load

The *BEMOSS*™ can also control different plug loads, as shown in Figure 11.34a,c,d. Plug loads can be turned on through *BEMOSS*™ UI by changing the status switch (Figure 11.34b). Moreover, the plug load's power, voltage, and current can be monitored in the *BEMOSS*™.

11.8.1.4 Power Meter

The *BEMOSS*™ allows operators to monitor different measurements such as current, voltage, and power. As shown in Figure 11.35, the power consumption is almost 87 W when the office light is set to 100% brightness.

In case the brightness is changed to 25%, the power consumption is reduced to 32 W (Figure 11.36). This is one of the crucial features offered by *BEMOSS*™, in which the energy can be saved by adjusting the brightness. To this end, one can set different schedules for different days per hours, applying different brightness to save energy in a commercial building. For a test case, researchers have gauged the reaction of occupants in an office by reducing the brightness to 50% during the lunch hour. At that lighting level, most office functions can continue without difficulty, but not reading a book with small prints.

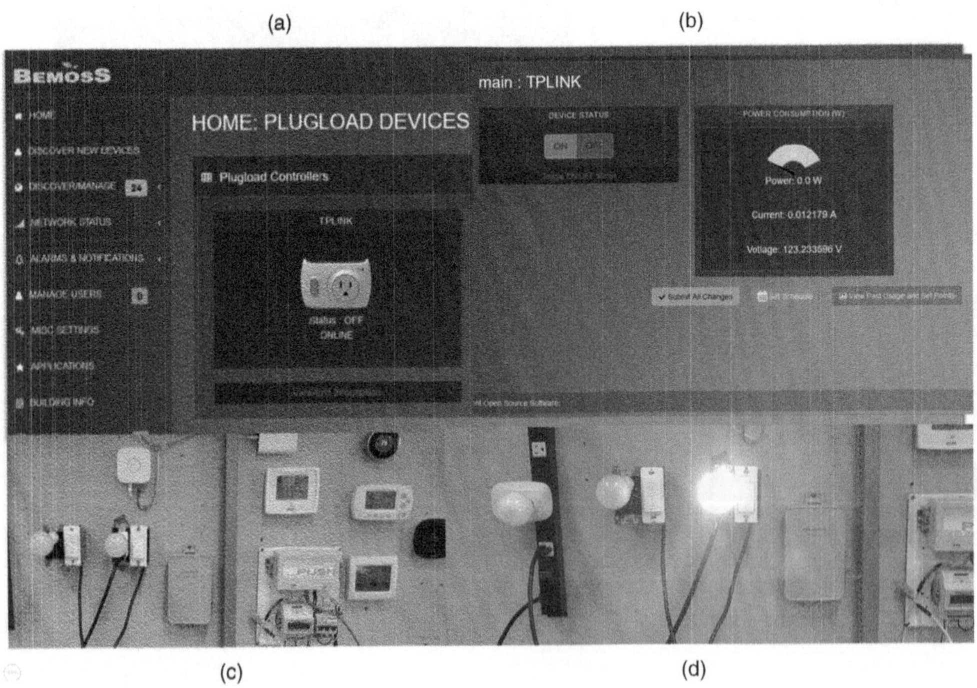

Figure 11.34 Plug load control.

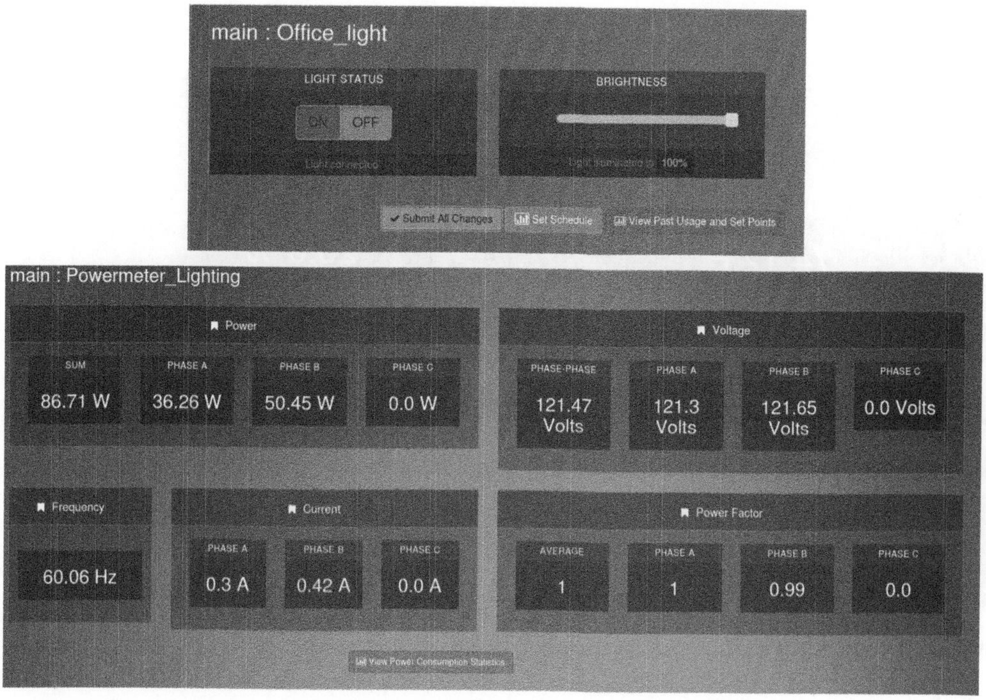

Figure 11.35 Power meter data (brightness is 100%).

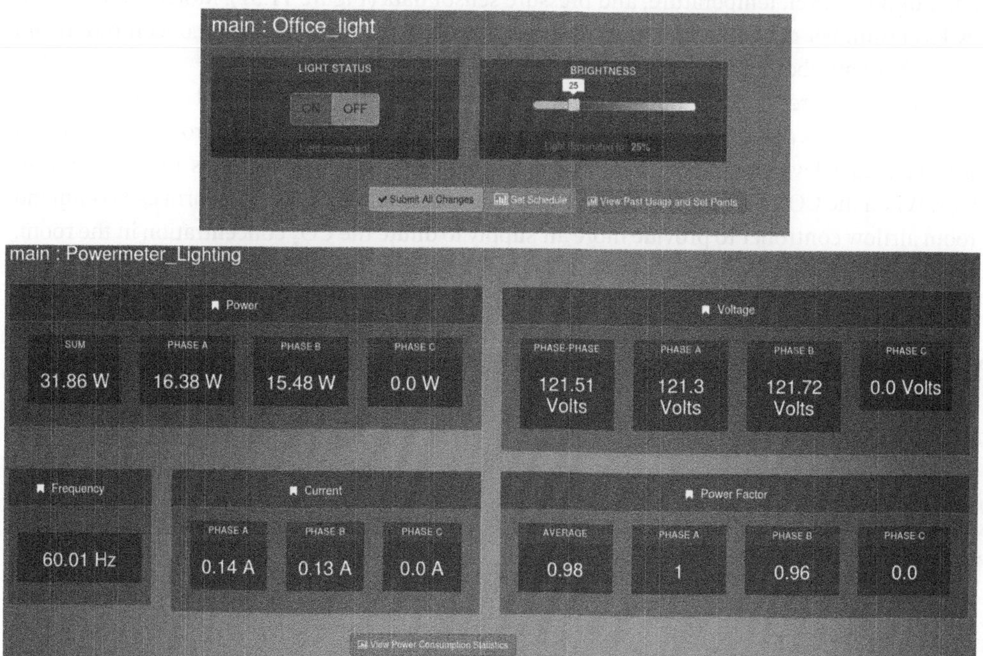

Figure 11.36 Power meter data (brightness is 25%).

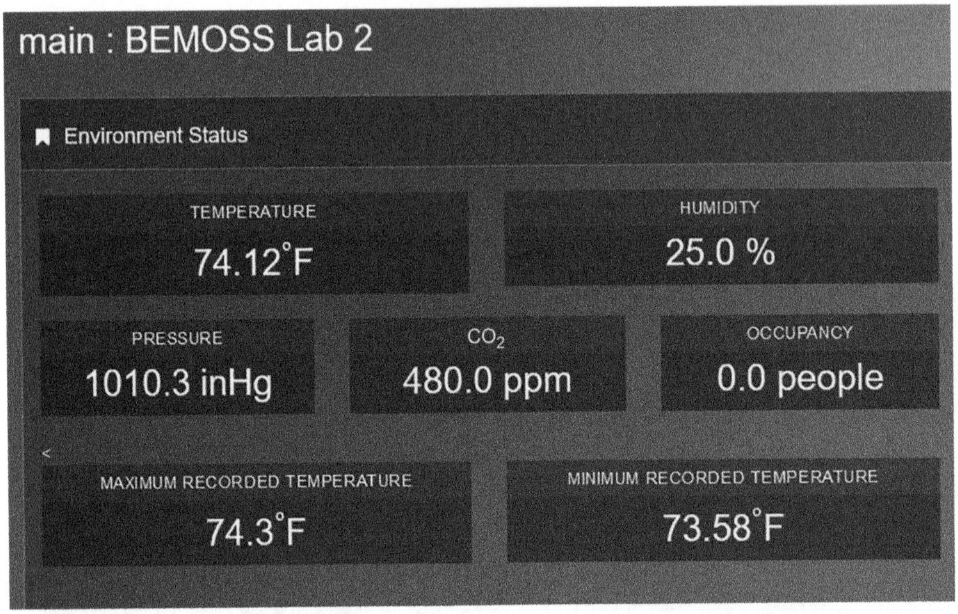

Figure 11.37 CO_2 sensor data.

11.8.1.5 CO_2 Sensor

BEMOSS™ also has the capability to read various sensor data like carbon dioxide (CO_2) concentration, humidity level, temperature, and pressure sensor data (Figure 11.37). Moreover, the occupancy level (number of people in a room) can be estimated by employing CO_2 concentration and defining a base number for CO_2 concentration.

For example, Figure 11.38 shows that the number of people in a room around 08:00 pm (March 1, 2022) is 5. This was determined based on the CO_2 concentration data in a classroom and using an empirical formula developed from a historical dataset of various occupancy levels and CO_2 concentrations. When the CO_2 level exceeds a predefined number, the *BEMOSS*™ platform can command the room airflow controller to provide more air supply to dilute the CO_2 concentration in the room. It should be noted that a high level of CO_2 may cause dizziness and headache.

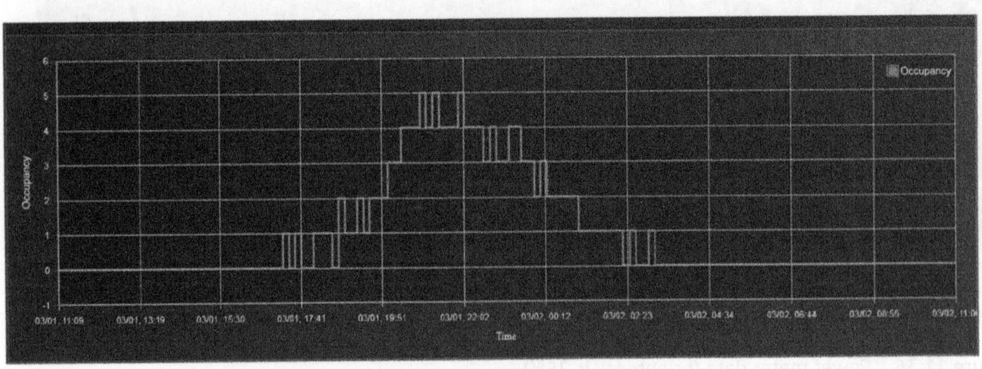

Figure 11.38 Occupancy estimation using a CO_2 sensor.

Figure 11.39 Illuminant sensor.

11.8.1.6 Illuminant Sensor

An illuminant sensor is another integrated part of *BEMOSS*™ such that one can monitor the lux level in a room. Figure 11.39 shows the lux level in the ARI lab, which is 264 Lux. One can use this information to control the room's brightness during the day when there is sunlight. This can save energy and help occupants to be in a preferred comfort zone.

11.8.1.7 Distributed Energy Resources (DERs)

A rooftop solar panel is installed on the top of the ARI building whose status can be monitored through *BEMOSS*™ UI. The real-time data show that the solar panel is generating 1389 W of AC power. Other information, such as efficiency, current, voltage, irradiance, wind velocity, and power factor, can be found on the rooftop solar page (Figure 11.40).

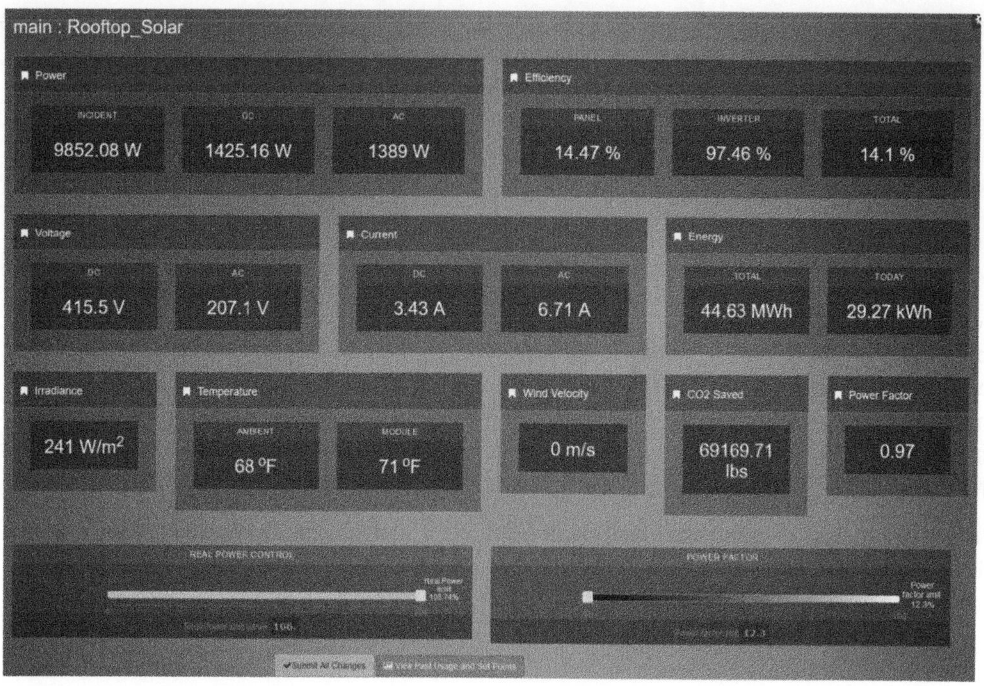

Figure 11.40 Distributed energy resources.

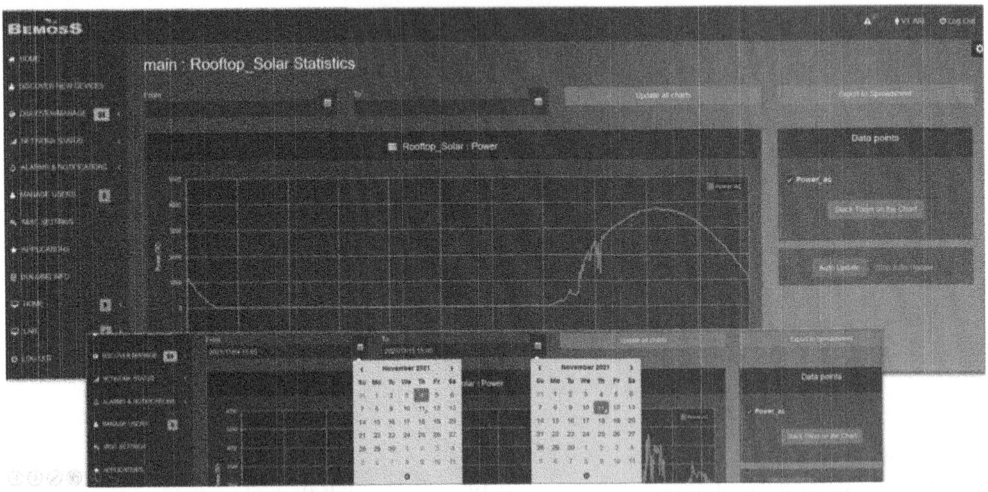

Figure 11.41 Historical data (rooftop solar power).

11.8.2 BEMOSS™ Historical Data and Schedule Capabilities

In all scenarios, *BEMOSS*™ can store data in the database and be shown to the user in a graphical environment. Figure 11.41 shows the rooftop solar historical data.

As an example for analysis, some fluctuations can be seen in Figures 11.42 and 11.43, which refer to cloudy and sunny situations.

Also, historical data related to temperature, humidity, pressure, CO_2, noise, and occupancy are shown in Figure 11.44–11.48.

Moreover, *BEMOSS*™ can set schedules in multiple periods and define different brightness at different hours of a day. This is an essential feature that can save energy. As shown in Figure 11.47, the user can set brightness to 100% from 7:30 to 10:30 am and then set 41% from 10:30 to 19:30 pm (when more automated work is done in the warehouse) and finally turn it off after 19:30 pm when the operation is suspended for the day. Then, *BEMOSS*™ will take care of everything and will do this loop (or a different one, as needed) automatically every day.

11.8.3 Practical Tests and Energy-Saving Results

To evaluate the performance of our proposed energy efficiency software, *BEMOSS*™ software and related hardware are installed and deployed in different buildings. The results demonstrate a 20% energy saving for HVAC and a 25% saving for lighting (Figure 11.48). Moreover, we can have other benefits from our proposed *BEMOSS*™ platform. Since *BEMOSS*™ is monitoring DER, plug loads, lighting, HVAC, etc., in real-time mode, it can generate an alarm whenever there is a fault in any of these devices. Thus, the following auxiliary benefits can be offered by the *BEMOSS*™ platform.

- **Improved operations and maintenance:** *BEMOSS*™ analytical platform enables operators to detect faults when devices operate outside standard thresholds enabling building operators to investigate prior to device failure.

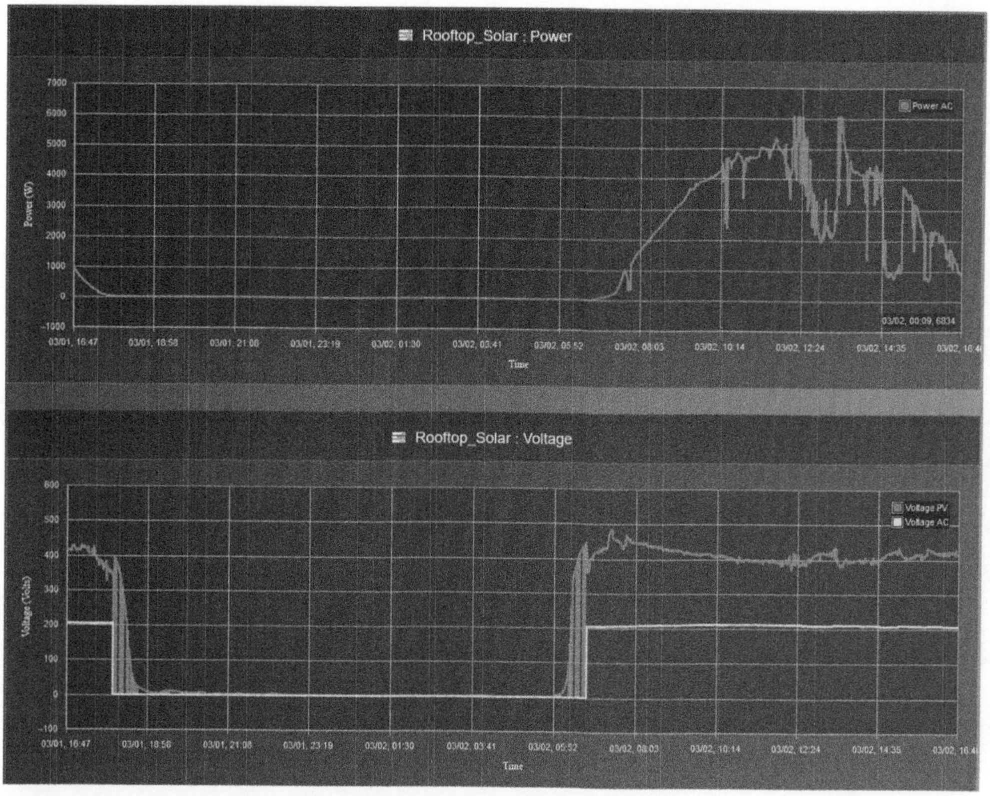

Figure 11.42 Historical data (rooftop solar power and voltage).

- **Occupant satisfaction:** Spaces controlled by *BEMOSS*™ have been more comfortable due to more consistent temperature profiles and healthier air quality through consistent monitoring of environmental factors (CO_2 levels, particulate matter (PM) 2.5, etc.).

The *BEMOSS*™ platform is deployed in four buildings in Alexandria, Arlington, and Blacksburg as shown in Figure 11.49.

11.8.3.1 Real-Time Monitoring of Classroom (Building 1, Virginia Tech Academic Building, Alexandria)

We have deployed our proposed solution (*BEMOSS*™) to the VT classroom building. Different wireless sensors including motion sensor, plug load controller, CO_2, and thermostat have been installed in this classroom (Figure 11.50). All of the data from these sensors (e.g., CO_2, noise, temperature, and relative humidity) are sent to the *BEMOSS*™ core (Raspberry Pi). The thermostat can control the operation of the unit on the rooftop. Moreover, the smart plug load is used to control the printer, computer, liquid crystal display (LCD), etc. To this end, we can program smart plug load and turn off these devices after 8:00 pm to 8:00 am.

Figure 11.51 shows the real-time monitoring of temperature, humidity, pressure, noise, etc., in the classroom. It can be seen that the CO_2 concentration increases when students come to the class

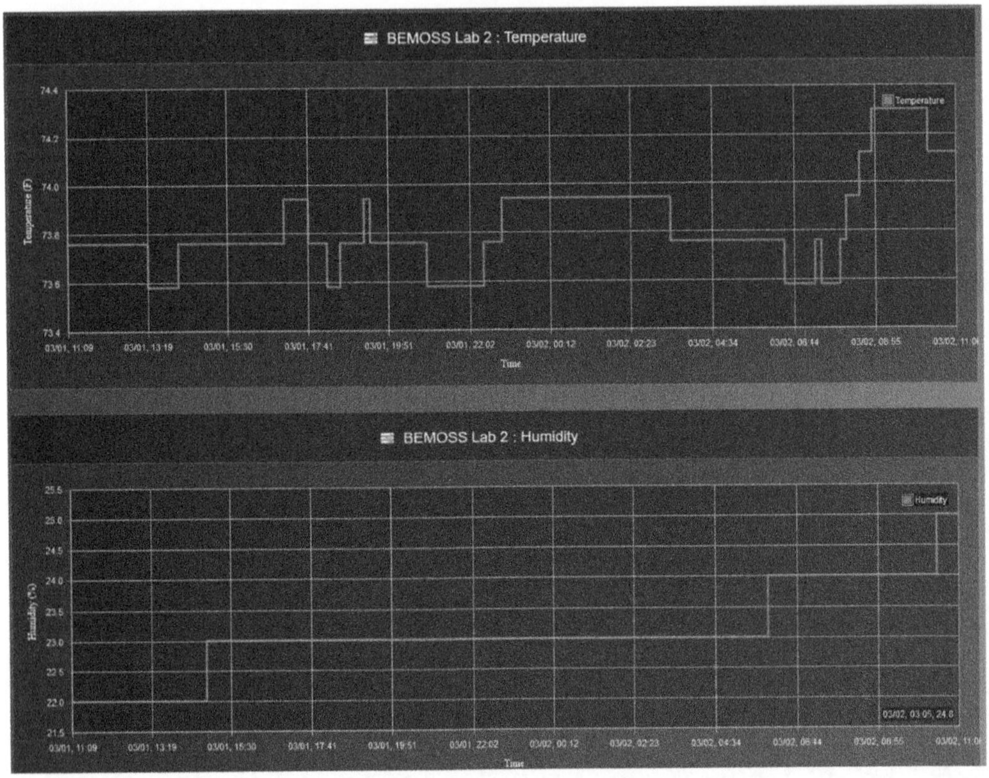

Figure 11.43 Historical data (rooftop solar current and temperature).

and decreases after finishing the class. The maximum CO_2 is 1100 PPM, which is more than the normal value (the normal CO_2 level is 750 for a classroom) and may lead to dizziness and headache. The *BEMOSS*™ platform can address this issue by automatically sensing the CO_2 level and flushing fresh air whenever it is needed.

Figure 11.52 shows the process of energy savings in the Alexandria building after increasing set points by 2 °F in a classroom. It can be observed that the energy consumption decreases from 5.29 kWh on Day 1 to 4.75 kWh on Day 2. Therefore, increasing 2 °F set point could result in 10% energy saving.

11.8.3.2 Energy and Peak Savings from HVAC Control (Building 1, Virginia Tech Academic Building, Alexandria)

The following devices are deployed in the building to investigate the effect of controlling HVAC in energy and peak saving:

- 6 thermostats: two per floor
- 6 power meters: for six rooftop air conditioners
- 1 Li-ion battery: to see the peak load reduction
- 1 environmental sensor

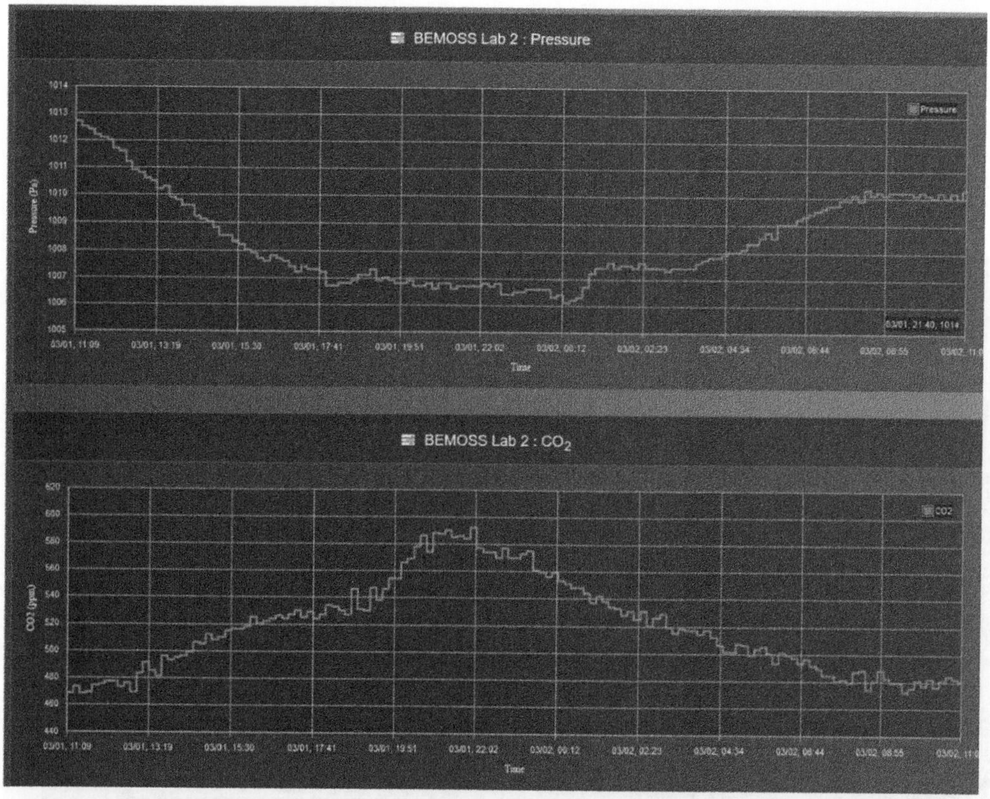

Figure 11.44 Historical data (temperature and humidity).

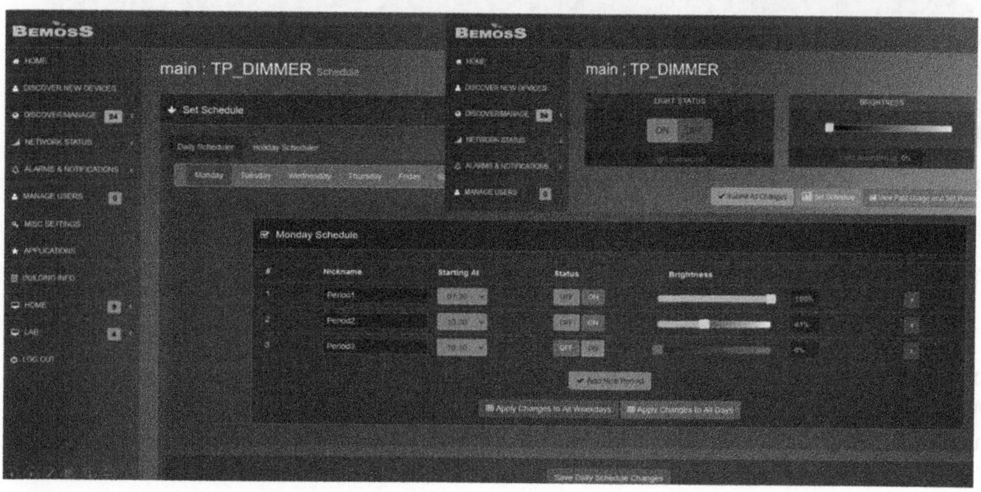

Figure 11.45 Historical data (pressure and CO_2).

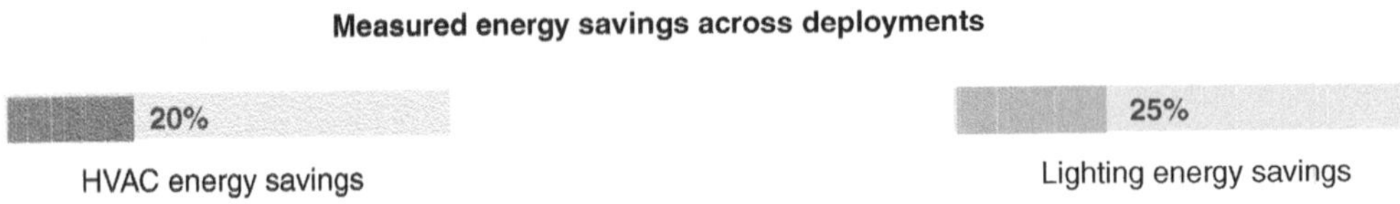

Figure 11.46 Historical data (noise and occupancy).

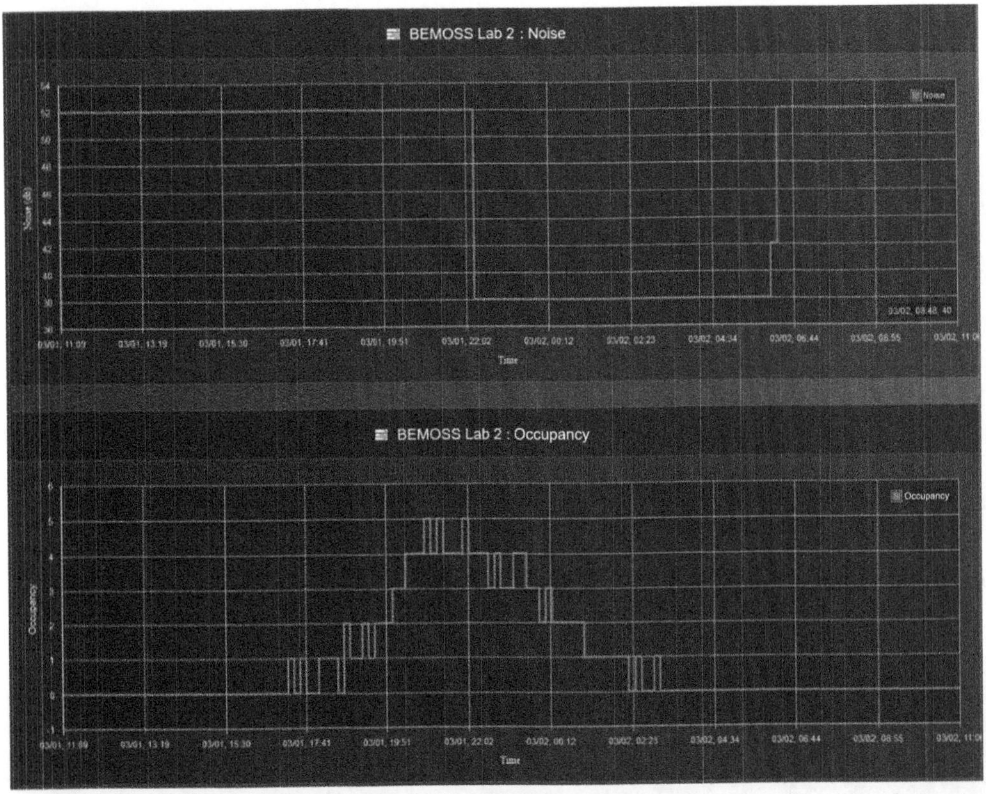

Figure 11.47 Set schedule.

Table 11.5 shows the comparison of compressor consumption before and after installing the *BEMOSS*™ platform. It should be noted that the data were only measured for one floor for three months, i.e., June, July, and August. As can be observed, the compressor consumption before installing *BEMOSS*™ in 2014 was around 8340 kWh. After the installation of the *BEMOSS*™ platform, the consumption was reduced to 6071 kWh. Therefore, energy saving is almost 27%.

Figure 11.53 shows the real-time trend before and after demand reduction. Green, blue, and red lines represent the temperature, air conditioner consumption, and thermostat set point, respectively. As can be seen in Figure 11.53a, the thermostat set point is set to74 °F. In normal operation, when the temperature exceeds the HVAC set point, the AC turns on, and therefore, the power consumption increases up to 3.2 kW. After some time, the temperature goes down below the HVAC set point, and consequently, AC turns off. This process repeats frequently. Since from 1:00 to 5:00 pm

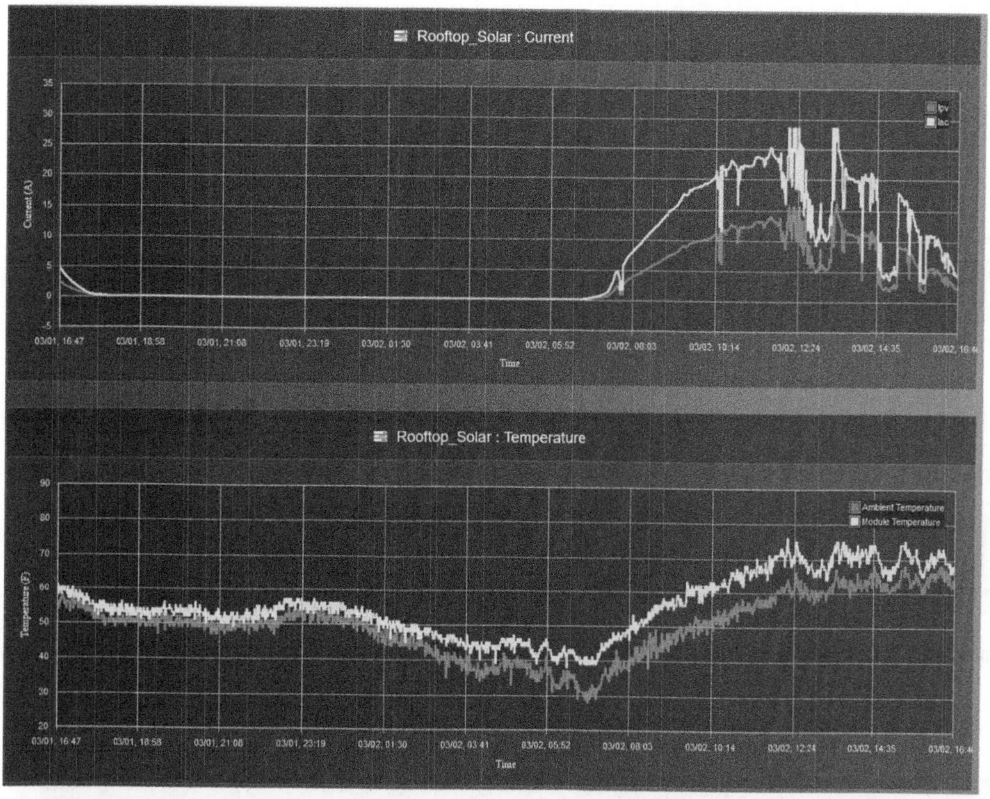

Figure 11.48 BEMOSS™ measured energy saving across deployments.

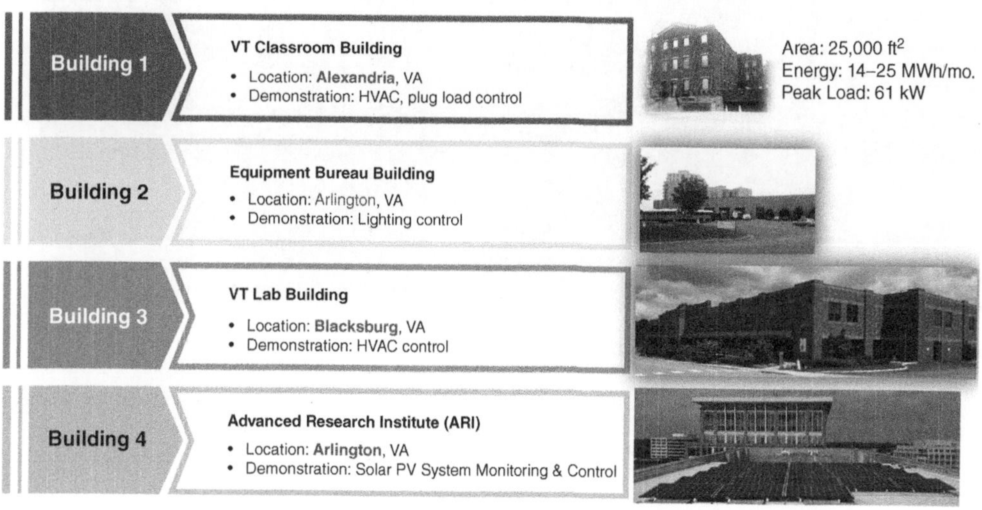

Figure 11.49 BEMOSS™ deployment in four buildings.

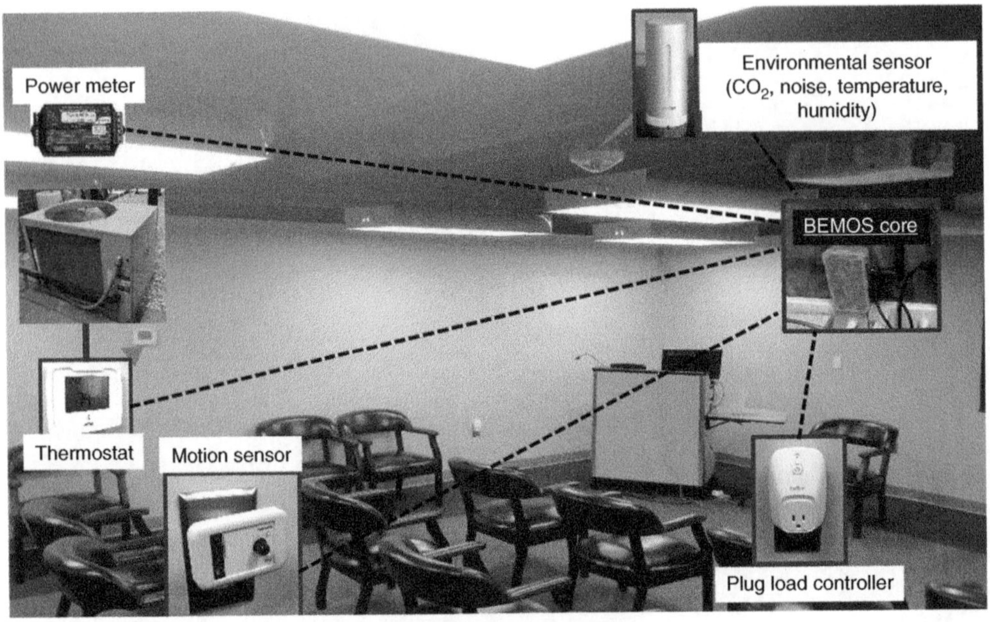

Figure 11.50 Deployment of wireless devices in VT classroom building.

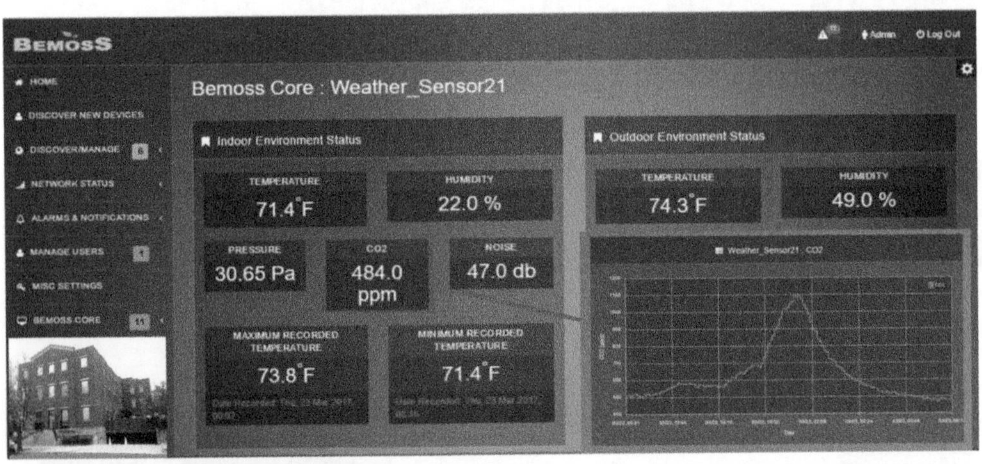

Figure 11.51 Indoor/outdoor environmental monitoring.

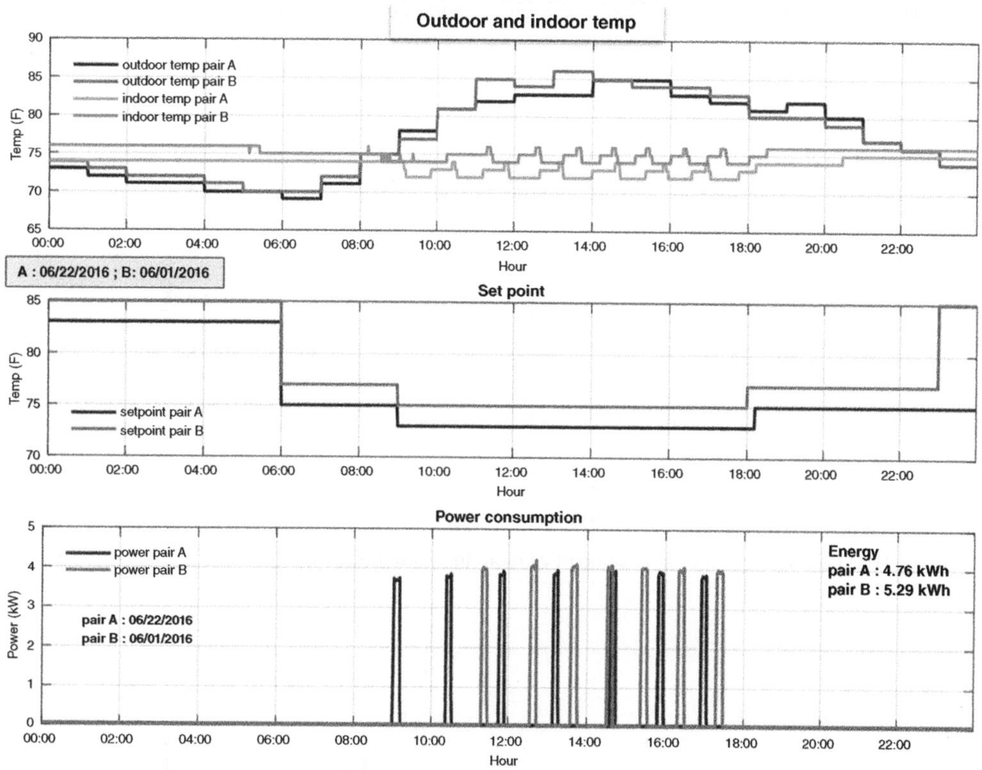

Figure 11.52 Energy savings—after increasing set points by 2 °F in a classroom.

Table 11.5 Comparison of compressor consumption before and after installing the *BEMOSS*™ platform.

Summer months (June–July–August)	
Compressor consumption 2014 (before *BEMOSS*™)	8340 kWh
Compressor consumption 2016 (after *BEMOSS*™)	6071 kWh
Average savings: 26.8%	

electricity price is too high, we should change our strategy and turn off the AC for these peak hours to save money. To this end, we can start pre-cooling before 1:00 pm, e.g., from 12:00 pm when the electricity price is low, and then turn the AC off. It can be seen from Figure 11.53b that the temperature reaches almost 77 °F at 5:00 pm. At this time, our control ends and the AC can be turned on.

Table 11.6 compares these two cases. It is seen that energy usage reduces from 2.72 to 1.42 kWh. Moreover, max demand decreases from 3.98 to 0.5 kW, which is a great reduction.

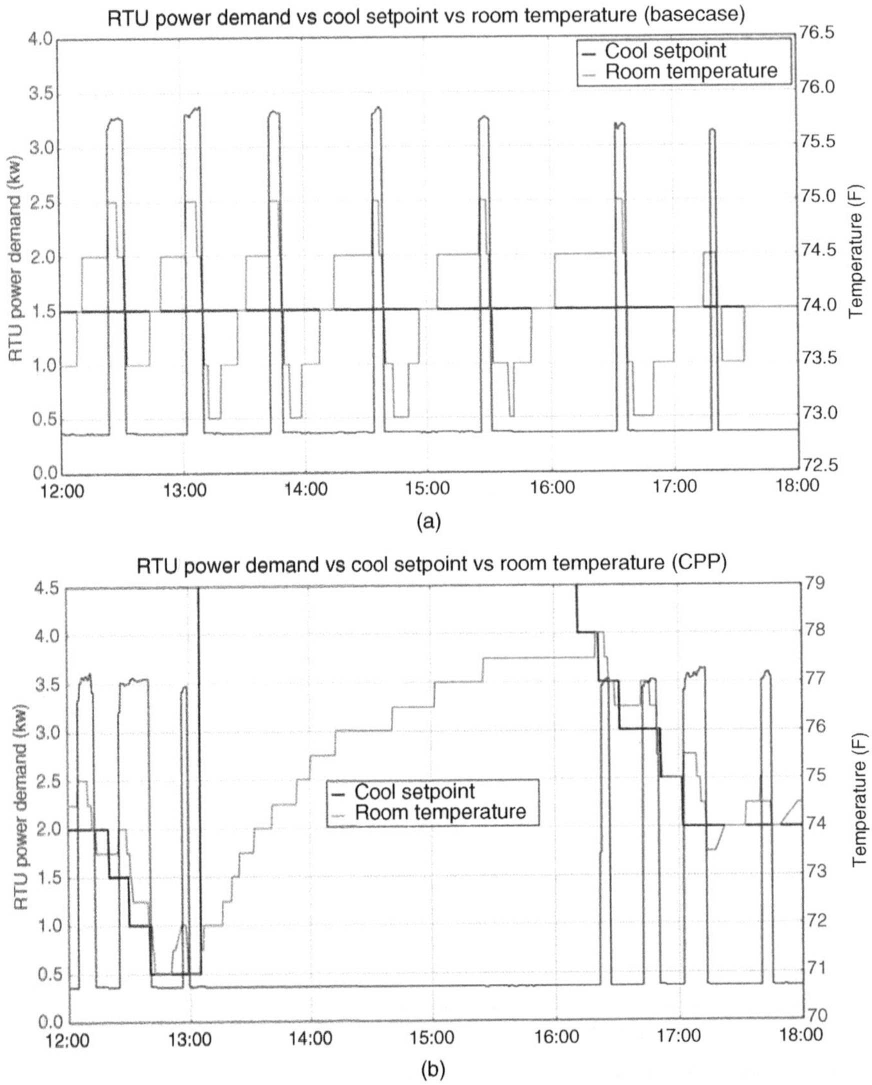

Figure 11.53 (a) Temperature profile before BEMOSS™ demand reduction and (b) temperature profile after BEMOSS™ demand reduction.

Table 11.6 Comparison of energy usage before/after HVAC control.

	Base case (without *BEMOSS*™)	**With *BEMOSS*™**
Set point	74 °*F*	77 °*F*
Energy usage	2.72 kWh	1.42 kWh
Max demand	3.98 kW	0.5 kW

11.8.3.3 Energy Savings by Controlling Light Intensity (Building 2, Equipment Bureau Building, Arlington)

To investigate the impact of controlling light intensity on energy saving, the *BEMOSS*™ and related devices, i.e., three lighting controllers and one power meter, have been deployed in an office building in Arlington, Virginia. In the office, we have a working area with skylight and ceiling lights (Figure 11.54a). In this area, a lot of light comes from the sky while the ceiling lights are working at full capacity. A solution for energy saving for this area is to dim the light when the solar light intensity increases during the day. To this end, a light sensor is installed to control the lighting intensity and save energy for the working area.

Another part of this office is the staff working area shown in Figure 11.54b, where the staff does not turn off the lights when they go for lunch. The same situation happens in the conference room shown in Figure 11.54c. To save energy in these two areas, two switches with the capability of scheduling and dimming are installed. Therefore, we can dim the light during lunchtime (e.g., 12:00–1:00 pm) by scheduling.

Based on occupant requirements, the light intensity level was reduced during October–December 2016, resulting in the average kWh energy savings. Table 11.7 shows that we could save almost 35% of energy with the proposed dimming control method.

Table 11.8 compares the total energy consumption before/after lighting control showing almost 35% energy saving.

Figure 11.54 (a) Office working area with skylight and ceiling lights, (b) staff working area, and (c) conference room.

Table 11.7 An average energy savings of 35% was achieved through dimming control.

Oct 2016	Nov 2016	Dec 2016	Jan 2017	Feb 2017	Mar 2017	Apr 2017	May 2017	Jun 2017	Average
33.7%	33.9%	34.4%	33.4%	35.9%	36.2%	35.0%	36.0%	36.3%	34.5%

Table 11.8 Comparison of consumption before/after lighting control.

Month	Total measured energy consumption (kWh)	Total calculated energy consumption without dimming (kWh)	Energy savings by dimming (%)
October 2016	264.37	399.90	33.89%
November 2016	278.13	423.78	34.37%
December 2016	280.76	426.40	34.16%
Total (October–December)	823.26	1250.08	34.14%

Table 11.9 Scheduled dimming level from 6:30 am to 9:00 pm.

Area	Open office area A	Open office area B	Chief office's desk area	Chief office's meeting area	Conference room A	Conference room B	Lights
Percentage change	50%	45%	60%	50%	50%	45%	Off after 9:00 pm

Table 11.9 shows the scheduled dimming level from 6:30 am to 9:00 pm. We can use machine learning (ML) and monitor the people's activities in a building to optimize the process and do dimming smartly.

11.8.3.4 Energy Savings by Increasing Set Point (Building 3, Retail Office Building, Blacksburg, VA)

In another case study, the cooling set point is increased by 5 °F and the energy saving is investigated. To have a fair comparison, we have selected two weekdays with similar weather condition, i.e., an average ambient temperature of 71 °F (Figure 11.55).

Figure 11.56a,b shows the power consumption in two different cases, i.e., June 6, 2016: day-time cool set point 70 °F and May 27, 2016: day-time cool set point 75 °F.

Figure 11.55 Blacksburg retail office building.

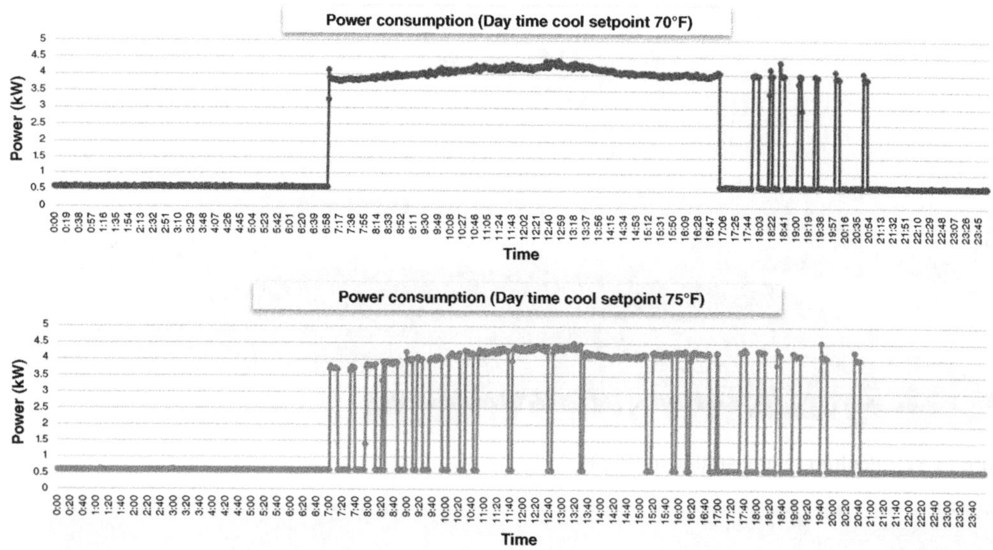

Figure 11.56 (a) June 6, 2016: day-time cool set point 70 °F and (b) May 27, 2016: day-time cool set point 75 °F.

Table 11.10 Energy savings by increasing set point by 5 °F in one suite.

Case	Day-time cool set point	Total daily energy usage	Energy saving
June 6, 2016	70 °F	52.1	7.4 kWh (14.2%)
May 27, 2016	75 °F	44.7	

It can be seen from Table 11.10 that increasing the set point from 70 to 75 °F results in around 14% energy saving.

11.8.3.5 Solar PV System Monitoring and Control (Building 4, Advanced Research Institute, Arlington, VA)

Since *BEMOSS*™ is an open-source platform, it can control other devices besides HVAC, plug load, and lighting. To this end, the operation and functionality of rooftop solar installed at VT-ARI (Figure 11.57) can be monitored, as shown in Figure 11.58 (e.g., alternating current (AC)/direct current (DC) output, temperature, solar radiation, and photovoltaic (PV) output). Also, we have a smart inverter control capability that is accessible through the *BEMOSS*™ platform.

11.8.3.6 Peak Load Reduction by Battery Energy Storage System (BESS)

Moreover, we have installed a *BEMOSS*™ in the building (Figure 11.59) to reduce peak consumption during peak hours by discharging the battery. This battery is charged when the energy price is low and will be controlled to be discharged during peak load, when the energy price is high.

As shown in Figure 11.60, the battery status, state of charge (SOC), and output power can be monitored through the *BEMOSS*™ platform.

Figure 11.57 Rooftop solar installed at VT-ARI.

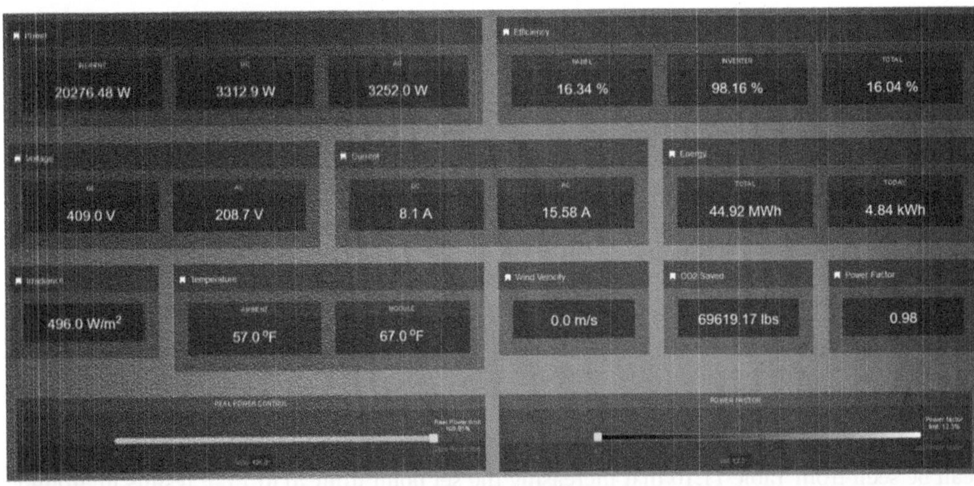

Figure 11.58 Solar PV system monitoring and control.

Figure 11.59 Battery energy storage system installed in the building.

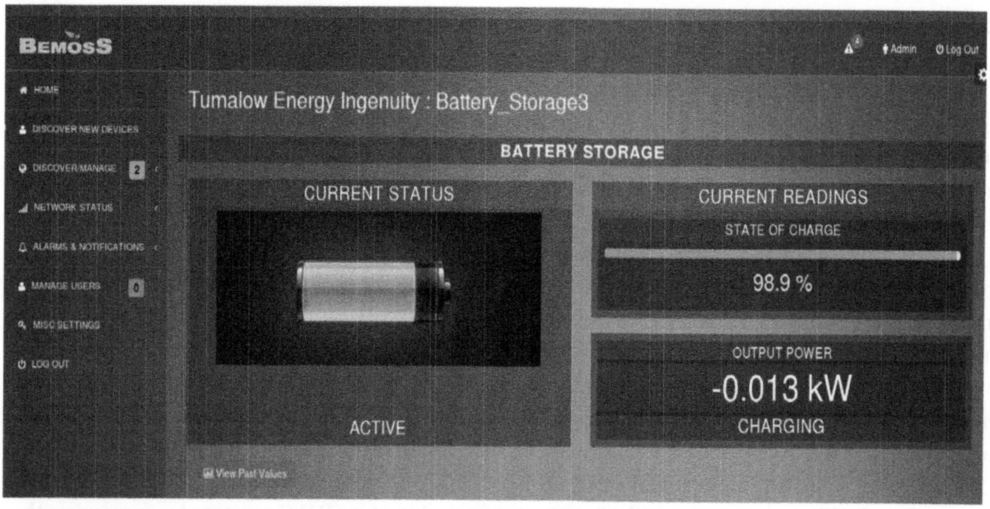

Figure 11.60 Battery energy storage system monitoring and control.

11.9 BEMOSS™ Platform for Campus Applications

Consider that we have many buildings in which each building has HVAC, lighting load, plug load, sensors, PV, battery, and security camera. In this case, the *BEMOSS™* software can be installed in a cloud, e.g., Amazon Web Services (AWS), and for each building, a dedicated account is created. On the other side and on a cloud, we have a facility manager who has full access to all of the buildings and can monitor and control them to reduce energy consumption. Moreover, the power company can be connected to the cloud. Therefore, we have access to the power company's database and can get the real-time price of energy (Figure 11.61).

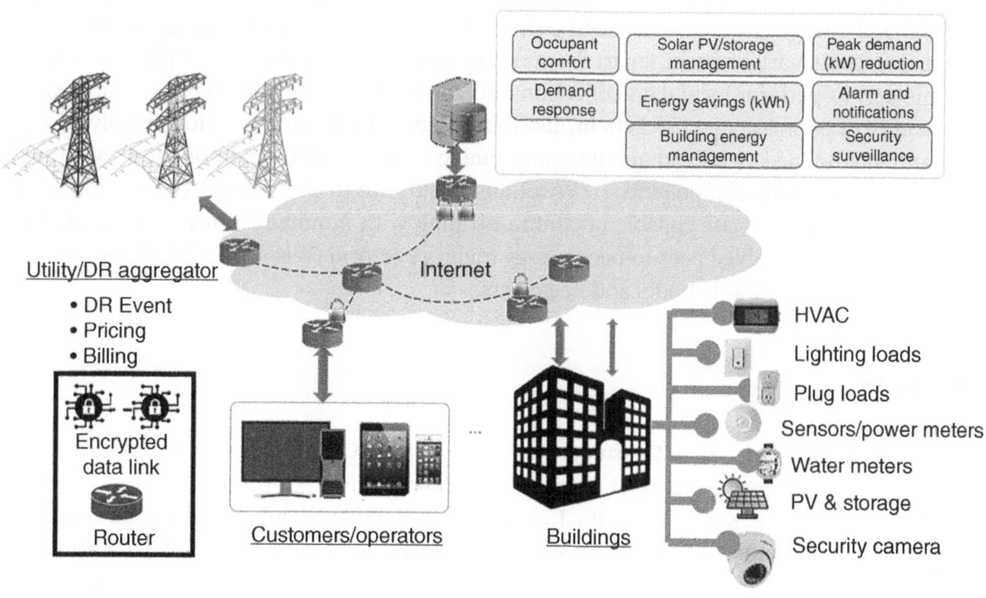

Figure 11.61 Transition from smart building to smart campus.

11.10 Conclusion

BEMOSS™ is a *B*uilding *E*nergy *M*anagement *O*pen-*S*ource *S*oftware solution that is engineered to improve sensing and control of all IoT-enabled equipment in commercial buildings. It can monitor and control three major loads in buildings, i.e., (i) HVAC; (ii) lighting loads; and (iii) plug loads along with other devices and sensors (e.g., air quality and CO_2). It has been developed in consultation with industry.

In this chapter, we demonstrated that the proposed IoT sensor platform, i.e., *BEMOSS*™, has the capability to make a building smart by integrating disparate sources and providers in one unified communication platform. *BEMOSS*™ is a robust open-source operating system for building energy management that is built completely using open-source software tools. As a result, *BEMOSS*™ can be used as an affordable and low-cost energy-saving platform for commercial buildings and monitor and control HVAC, lighting loads, plug loads, solar/PV, security cameras, air quality, etc. The practical application of this platform has been evaluated in four buildings in Alexandria, Arlington, and Blacksburg. The results demonstrated that BEMOSS™ could save energy by adjusting thermostat set points and light intensity as (i) 10–15% energy savings for HVAC systems by increasing the temperature set point and (ii) 30–35% energy savings for lighting by reducing light intensity level.

To achieve intelligent control, ML algorithms have been implemented in the proposed platform to build an accurate ML model based on historical data and occupant preferences to save energy (kWh) while increasing occupant comfort. To this end, we can reduce peak demand using DR protocols. As a result, *BEMOSS*™ could improve energy efficiency and facilitate peak load savings in commercial buildings. *BEMOSS*™ will make your life easier!

11.11 Exploring Other Capabilities of the BEMOSS™ Platform

The *BEMOSS*™ platform has been commercialized by BEM Controls LLC with a new name—WiseBldg©. This is based on an energy Internet platform that provides higher operational efficiency, grid resiliency, and cost reduction for both utilities and end customers, as well as invaluable behavioral and energy insights by leveraging (i) existing IoT-enabled smart devices in homes and buildings (such as smart inverters, smart thermostats, smart light switches, smart plugs, and smart power meters) to perform local data collection and control with no additional hardware investments; and (ii) a permissioned blockchain/distributed ledger technology to automatically execute smart contracts/defined logics and maintain immutable records of all transactions and operational data. There are two main components: (i) WiseBldg© platform—a cloud-based, open-architecture energy management software capable of communicating with hundreds of IoT-enabled devices; and (ii) a blockchain-enabled peer-to-peer energy trading platform (WiseMrkt©) that manages the complex exchange of kilowatt-hours and negawatts.

References

1 Commercial Buildings Energy Consumption Survey Results [Online]. http://www.eia.gov/consumption/commercial. (accessed October 2024)

2 Katipamula, S., Piette, K.A., Kuruganti, T., et al. (2012, Oct). Small and medium sized commercial building monitoring and controls needs: a scoping study. PNNL-22169, [Online]. http://www.pnnl.gov/main/publications/external/technical_reports/PNNL-22169.pdf (accessed October 2024).

3 Bloom, E., and Gohn, B. (2012). Smart buildings: ten trends to watch in 2012 and beyond. Pikes research, [Online]. chrome-extension://efaidnbmnnnibpcajpcglclefindmkaj/https://www.infrastructureusa.org/wp-content/uploads/2012/05/SB10T-12-Pike-Research.pdf ((accessed October 2024)

4 California Public Utility Commission. Demand response monthly reports [Online]. https://www.cpuc.ca.gov/industries-and-topics/electrical-energy/electric-costs/demand-response-dr/dr-provided-by-utilities/demand-response-monthly-reports. (accessed October 2024).

5 Android open-source repository. source.android.com/ (accessed October 2024).

6 MacDonald, J., Cappers, P., Callaway, D., et al..Demand response providing ancillary services: a comparison of opportunities and challenges in the US wholesale markets. Grid Interoperability Forum 2012 [Online]. http://www.gridwiseac.org/pdfs/forum_papers12/macdonald_paper_gi12.pdf (accessed October 2024).

7 Eto, J. H., Nelson-Hoffman, J., Torres, C., et al. (2007, May). Demand response spinning reserve demonstration. LBNL-62761, [Online]. https://www.osti.gov/servlets/purl/925589 (accessed October 2024).

8 Echelon SmartServer 2.0 [Online]. https://www.echelon.com/products/controllers/smartserver/ (accessed June 2013).

9 DOE Green Button Initiative [Online]. http://www.greenbuttondata.org/ (accessed October 2024).

10 Principal Building Activities Overview [Online]. https://www.eia.gov/consumption/commercial/pba/overview.php. (accessed October 2024)

11 Building Type Definition [Online]. https://www.eia.gov/consumption/commercial/building-type-definitions.php. (accessed October 2024)

12 Commercial buildings in depth [Online]. https://www.eia.gov/energyexplained/use-of-energy/commercial-buildings-in-depth.php. (accessed October 2024)

13 Commercial buildings [Online]. https://www.eia.gov/energyexplained/use-of-energy/commercial-buildings.php. (accessed October 2024)

14 EIA projects air-conditioning energy use to grow faster than any other use in buildings [Online]. https://www.eia.gov/todayinenergy/detail.php?id=43155(accessed October 2024)

12

Optimal Dispatch of Smart Energy System Based on Cyber–Physical–Social Integration

Jizhong Zhu, Ziyu Chen, Wanli Wu, and Chenke He

South China University of Technology, School of Electric Power Engineering, Guangzhou, China

12.1 Introduction

The increasing importance of social behaviors includes transaction, management, and user selection in energy system research. The cyber–physical system (CPS) is extended to cyber–physical–social system (CPSS), which comprises cyber network, physical system, and social factors. The application of CPSS in the energy system can be called the CPSS in energy [1]. Due to the development of the distributed energy trading market, users with distributed power sources participate in power transactions, and the producer and consumer with the dual role of electricity sale/utilization have emerged. Users who actively manage their energy consumption, production, and storage are formed, which also better reflects human interests and behaviors [2]. In the optimization process, energy management is essential for the renewable energy (RE) consumption and economic operations of the power system. Electric vehicle (EV) is one of the most common controllable loads, whose demand is agile. Thousands of EVs are aggregated to become a flexible storage model. In addition, the incentive analysis should be premeditated to encourage the flexible part to actively partake in the demand response [3], which relates to the social factors.

The bidirectional energy interaction characteristics of EV empower them to play a proactive role in the operation of power systems, especially at the distribution grid level. Firstly, EV aggregators can enhance grid stability through the intelligent scheduling of EV charging/discharging. To be specific, EV aggregators can utilize big data and artificial intelligence technologies to predict power demand and supply, thereby achieving more precise schedules. For instance, they can project electricity demand within specific time horizons based on historical data and arrange the charging/discharging of EVs according to these forecasts. Additionally, they can utilize real-time (RT) electricity pricing information to adjust schedule strategies, such as charging at low-price periods and discharging during high-price periods. This approach not only helps balance the power grid to prevent overloads or shortages but also saves electricity costs for EV users. Moreover, EV aggregators can leverage their scale advantage to engage in various electricity markets (e.g., energy markets [4, 5], reserve markets [6, 7], and frequency regulation markets [8]) to generate more benefits. This not only brings economic benefits to the aggregator but also assists the power market by better balancing demand and supply, enhancing the grid's stability and reliability.

However, realizing the above objectives is a complex process that calls for collaborative efforts from multiple stakeholders—EV users, EV aggregators, and power system operators. EV users play a crucial role in this process. Their charging habits can significantly impact the balance of the grid.

Smart Cyber-Physical Power Systems: Fundamental Concepts, Challenges, and Solutions, Volume 1, First Edition.
Edited by Ali Parizad, Hamid Reza Baghaee, and Saifur Rahman.

Therefore, they need to adaptively adjust their charging behaviors based on the grid's operation. EV aggregators have the responsibility to guide and incentivize these adaptive behaviors. They can provide RT signals to charge or discharge based on electricity demand. Furthermore, they can develop incentive mechanisms, such as economic discounts or rewards for flexible resources, to encourage users to charge or discharge at optimal periods. Power system operators are responsible for maintaining the overall stability and safety of the power grid. They can provide a reasonable and supportive policy environment that fosters adaptive power consumption behaviors. This could involve policies that promote grid-interactive regulations that ensure the safety of bidirectional energy transfer or initiatives that reward grid-supportive behaviors. By cultivating a conducive policy environment, power system operators can facilitate the integration of EVs into the power grid, promoting a more sustainable and resilient energy system. A more efficient and sustainable power system through the integration of EVs requires concerted efforts from EV users, EV aggregators, and power system operators. Each stakeholder has a unique role to play, and their interactions and cooperation are vital for the overall success of this venture.

In addition, the large-scale access of demand response resources such as EV load needs to achieve the balance of multiparty interests under the game of all parties and also seek the balance of the overall interests and the maximum acceptance of flexible load. The investment subject of EV charging stations and their power supply network is no longer limited to distribution network operators, but gradually to a multi-subject transformation including power generators, users, and load aggregators [9]. The planning and optimization of EV charging stations and their power supply network are the most important links in distribution network planning [10]. Generally, the objective function is selected for the planning object, the constraint conditions are considered, the corresponding planning model is established, and finally, some optimization algorithm is selected for solving [11, 12]. Traditional EV charging station planning and optimization are relatively simple in terms of planning objects, planning objectives, planning factors considered, and corresponding optimization solutions. With the introduction of a large number of new planning factors such as EV load, as well as the application of fast charging technology, power change technology, and vehicle network interaction technology, new planning models and planning methods are bound to emerge [13, 14].

The planning scheme is the basis of planning decision for EV charging station and its power supply network [15]. The evaluation of EV charging station and its power supply network planning mainly include the establishment of evaluation system and selection of evaluation method. The evaluation system includes economic evaluation, EV charging convenience evaluation, reliability evaluation, and safety evaluation [16]. Compared with reliability or economy as a single index, comprehensive evaluation can reflect the energy efficiency level of the current distribution network more comprehensively. By determining the comprehensive evaluation index and efficient decision-making method, the energy efficiency improvement point can be accurately found, and then, the EV charging station and its power supply network can be effectively managed and planned.

12.2 CPSS Model

Compared with CPS, the main improvement of CPSS is the existence of human social space. In each optimization task, firstly, each energy supplier or energy demander obtains the current operating parameters of the corresponding distributed equipment from the physical system. Then, each of them makes respective dispatch decision by interacting with others in the social system based on the computation in the cyber system and the intercourse in a communication network. Finally, the optimal dispatch scheme is sent to each distributed device for achieving optimal control in physical system [17].

12.2.1 The Structure of the CPSS

Figure 12.1 shows an overall CPSS framework with the wind turbine–photovoltaic–energy storage (WT–PV–ES) combined generation (CG) system. The framework includes an essential communication network and three core systems, which are cyber system, physical system, and social system [18]. Cyber system provides basic data computing and storage services, as a platform for running various CPSS applications, including dispatch, control, planning, and deployment. The network of secondary control and protection information flow is reflected by cyber system. For the physical system, it belongs to the physical entity of the CPSS, which embodies the primary power network topology. In the study, this part mainly presents the smart energy system (SES) containing WT–PV–ES hybrid generation. Moreover, there are distribution networks, smart meters, and loads. As to social system, power users, electricity sale companies, power generation companies, and power grid enterprises constitute complex social relationships and have their own goals and benefits. The social system considers a cooperative or competitive game network formed by each power dispatch decision-making terminal in the case of open energy management (EM) transactions [19]. In addition, the communication network enables every part of the CPSS to participate effectively in the interaction, which makes the whole system more closely connected and the resources can be configured more convenient. It includes fourth-/fifth-generation wireless communication network, optical network, wired communication network, firewall, and Internet. In general, the four parts are connected to each other by signal flow and energy flow.

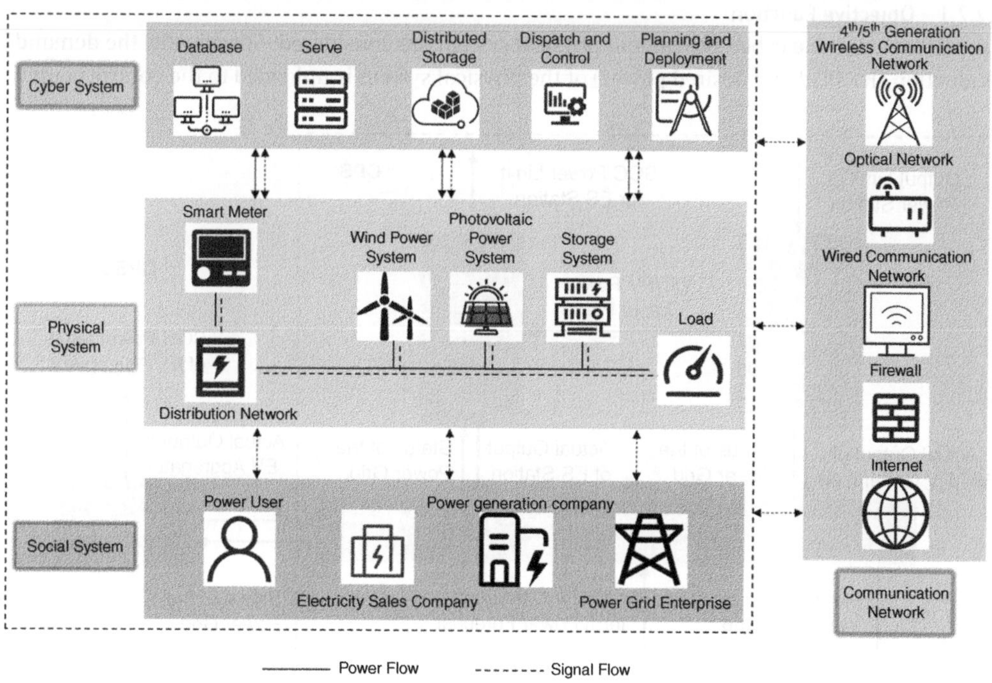

Figure 12.1 Framework of a CPSS with WT–PV–ES CG.

12.2.2 The Optimal Dispatch Based on CPSS

The uncertainty and intermittence of WT/PV power generation increase the fluctuation range of power generation capacity, and its large-scale access to the power grid will affect the safe and stable operations of the system [20]. The WT–PV–ES CG system can integrate WT–PV complementary power generation system with the ES station, which makes full use of the ES station to reduce the volatility of WT–PV output. In order to absorb RE to the greatest extent, this section adopts two strategies to optimize the CG, which is based on CPS and CPSS, respectively. The hybrid dispatch model is illustrated in Figure 12.2.

The CPSS model is constructed based on the CPS model, and apart from social factors, the cyber system and physical system are consistent. By using the method of controlling variables, the impact of considering social factors on the consumption of RE can be reflected. In the CPSS model, the control messages (price and policy signals) are transmitted from the control center to the aggregators and then transferred to the coalitions. The EV aggregator is a centralized control infrastructure for a large number of EVs, which can respond to the dispatch signal and feedback energy to the power grid during peak load period. In addition, according to market price signals, under the premise of meeting the demand for EV travels, reasonable charging and discharging plans can be used to obtain profits. The model considers the cyber, physical, and social systems. The cyber system contains the data collection or forecast of WT/PV/TM generation power, controllable load, uncontrollable load, electricity prices, issuance of dispatch, and control instructions. The physical system includes the constraints on the state of charge (SOC) of ES/EV, WT/PV/TM power, and power grid security. The social system embodies the influence of electricity prices on the charging and discharging behaviors of various types of EVs at different time intervals, so that controllable loads are encouraged to actively participate in optimization dispatch.

12.2.2.1 Objective Function

In the CPS model, the cyber system and physical system are considered. Specifically, the demand or induction (RT load and desired power) of the physical system is uploaded to the control center.

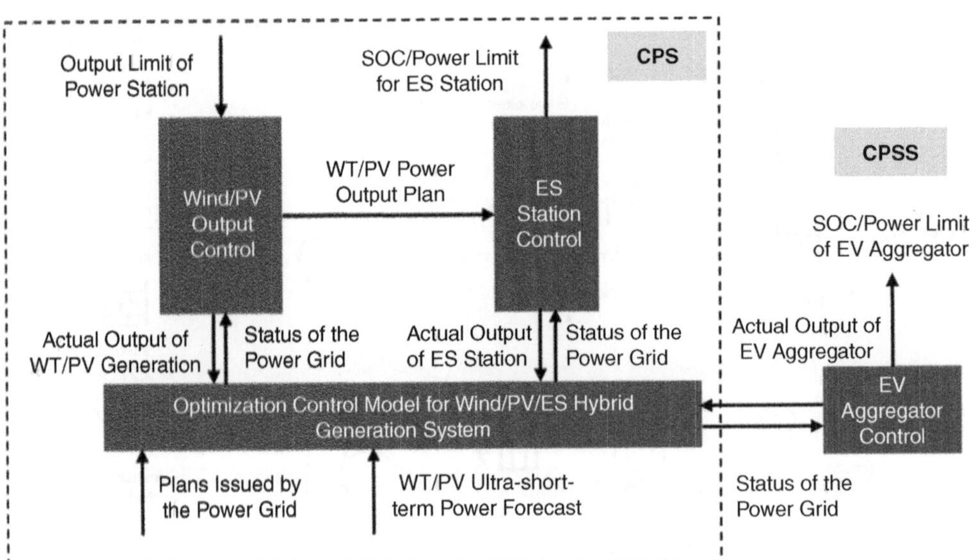

Figure 12.2 The optimal dispatch based on CPS and CPSS.

The cyber system mainly includes the data collection or forecast of WT/PV/thermal (TM) power and load and the issuance of dispatch and control instructions. The physical system involves the constraints on battery SOC, WT/PV/TM power, and power grid security. The tracking plan is the available RE that follows the dispatch plan of the CG system. In a certain period of time, when the tracking error ΔP^{f}_{CG} between forecast RE and WT–PV–ES CG system is the smallest, the RE consumption can be maximized. The objective function is the maximum consumption of RE and the lowest cost of generation system operation and maintenance (OM) at the same time. The two objectives are linearly combined by the above two considerations to obtain a comprehensive goal; the objective function of model A is to minimize the comprehensive goal $O_{CPS}(t)$:

$$\min O_{CPS}(t) = \omega_1 \cdot \sum_{t=1}^{N} \Delta P^{f}_{CG}(t) + \omega_2 \cdot \sum_{t=1}^{N} C_G(t) \tag{12.1}$$

$$\Delta P^{f}_{CG}(t) = P^{f}_{RE}(t) - P^{d}_{RE}(t) - P^{ch}_{ES}(t) + P^{dis}_{ES}(t) \geq 0 \tag{12.2}$$

$$P^{f}_{RE}(t) = P^{f}_{WT}(t) + P^{f}_{PV}(t) \tag{12.3}$$

$$P^{d}_{RE}(\mathrm{t}) = P^{d}_{WT}(\mathrm{t}) + P^{d}_{PV}(\mathrm{t}) \tag{12.4}$$

where $t = 1, 2, \dots, N$ means the corresponding time interval, $N = 96$. ω_1 and ω_2 mean the weights of two objectives. P^{f}_{RE} and P^{d}_{RE} are the forecast and dispatch values of RE outputs. P^{f}_{WT} is the forecast value of WT power, P^{f}_{PV} indicates the forecast value of PV power, P^{d}_{WT} shows the dispatch value of WT power, and P^{d}_{PV} denotes the dispatch value of PV power. P^{ch}_{ES} and P^{dis}_{ES} represent the charging and discharging power of ES station, respectively.

The cost of power generation OM $C_G(t)$ involves WT, PV, TM, and ES, which can be expressed as

$$C_G(t) = C^{OM}_{WT}(t) + C^{OM}_{PV}(t) + C^{OM}_{TM}(t) + C^{OM}_{ES}(t) \tag{12.5}$$

$$C^{OM}_{WT}(t) = m^{OM}_{WT} \cdot P^{d}_{WT}(t) \cdot \Delta t \tag{12.6}$$

$$C^{OM}_{PV}(t) = m^{OM}_{PV} \cdot P^{d}_{PV}(t) \cdot \Delta t \tag{12.7}$$

$$C^{OM}_{TM}(t) = m^{OM}_{TM} \cdot P_{TM}(t) \cdot \Delta t \tag{12.8}$$

$$C^{OM}_{ES}(t) = m^{OM}_{ES} \cdot \left(P^{ch}_{ES}(t) + P^{dis}_{ES}(t)\right) \cdot \Delta t \tag{12.9}$$

where m^{OM}_{WT}, m^{OM}_{PV}, m^{OM}_{TM}, and m^{OM}_{ES} are the OM cost coefficients of WT, PV, TM, and ES, respectively. C^{OM}_{WT}, C^{OM}_{PV}, C^{OM}_{TM}, and C^{OM}_{ES} indicate the OM cost values of WT, PV, TM, and ES, respectively. However, the investment cost of each generator, the depreciation cost of ES, and the penalty cost of abandoning WT/PV are not considered in this part, which are planning issues and another means of consuming RE.

In the CPSS model, the control messages (price and policy signals) are transmitted from the control center [21] to the aggregators and then transferred to the coalitions. The EV aggregator can respond to the dispatch signal and feedback energy to the power grid during peak load period. In addition, according to market price signals, under the premise of meeting the demand for EV travels, reasonable charging and discharging plans can be used to obtain profits. The model considers the cyber, physical, and social systems. The cyber system contains the data collection or forecast of $WT/PV/TM$ generation power, controllable load, uncontrollable load, electricity prices, and issuance of dispatch and control instructions. The physical system includes the constraints on SOC of *ES/EV*, *WT/PV/TM* power, and power grid security. The social system embodies the influence of electricity prices on the charging and discharging behaviors of various types of EVs at different time intervals, so that controllable loads are encouraged to actively participate in optimization dispatch. The objective function is the maximum consumption of *RE* and the lowest cost of controllable load and generation system *OM* at the same time. The three objectives are linearly combined by the above three considerations to obtain a comprehensive goal; the objective function of the CPSS model is to minimize the comprehensive goal $O_{CPSS}(t)$:

$$\min O_{CPSS}(t) = \omega_1 \cdot \sum_{t=1}^{N} \Delta P^{f}_{CG}(t) + \omega_2 \cdot \sum_{t=1}^{N} C_G(t) + \omega_3 \cdot \sum_{t=1}^{N} C_L(t) \tag{12.10}$$

where ω_3 means the weight of the third objective. In general, the weights are not fixed values, which depend on the proportions of actual demands. The controllable load cost $CL(t)$ is the sum of the charging payment $C^{ch}_{AG}(t)$ and discharging revenue $C^{dis}_{AG}(t)$ of the aggregator.

$$C_L(t) = C^{ch}_{AG}(t) - C^{dis}_{AG}(t) \tag{12.11}$$

$$C^{ch}_{AG}(t) = \sum_{k}^{\mu} P^{ch}_{EV_k}(t) \cdot m^{ch}_{AG}(t) \cdot \Delta t \tag{12.12}$$

$$C^{dis}_{AG}(t) = \sum_{k}^{\mu} P^{dis}_{EV_k}(t) \cdot m^{dis}_{AG}(t) \cdot \Delta t \tag{12.13}$$

where $\sum_{k}^{\mu} P^{ch}_{EV_k}(t)$ and $\sum_{k}^{\mu} P^{dis}_{EV_k}(t)$ indicate the sum of the charging and discharging power of the kth *EV* in the *EV* set μ at time t. $m^{ch}_{AG}(t)$ and $m^{dis}_{AG}(t)$ represent the RT electricity prices of the aggregator charging and discharging at time t.

12.2.2.2 Constraint Function

In the network with M buses, a brunch is denoted as ij, $(i, j \in M)$, if it points from bus i to bus j; the DistFlow equations [22] used to model the power flows for all buses at time t are displayed as follows:

$$P_j(t) = P_{ij}(t) - r_{ij} \frac{P^2_{ij}(t) + Q^2_{ij}(t)}{V^2_i(t)} \tag{12.14}$$

$$Q_j(t) = Q_{ij}(t) - x_{ij} \frac{P^2_{ij}(t) + Q^2_{ij}(t)}{V^2_i(t)} \tag{12.15}$$

$$V^2_j(t) = V^2_i(t) - 2(r_{ij}P_{ij}(t) + x_{ij}Q_{ij}(t)) + \left(r^2_{ij} + x^2_{ij}\right) \frac{P^2_{ij}(t) + Q^2_{ij}(t)}{V^2_i(t)} \tag{12.16}$$

Sending end of each branch ij, P_j, and Q_j indicates the injected active and reactive power at bus j. V_i and V_j mean the voltage magnitude at buses i and j; r_{ij} and x_{ij} are the resistance and reactance of branch ij. Note that the *PV/WT* is usually designed to operate at a unity power factor; thus, it provides only active power, and the reactive power is not considered in the study. The transmission power flow capacity of the branch ij and the voltage magnitude of bus i are constrained as

$$-P^{\max}_{ij} \le P_{ij}(t) \le P^{\max}_{ij} \tag{12.17}$$

$$V^{\min}_i \le V_i(t) \le V^{\max}_i \tag{12.18}$$

where $P_{ij}^{\max}$ expresses the upper limit of the branch ij transmission power flow. $V_i^{\min}$ and $V_i^{\max}$ are the lower and upper bounds of the voltage magnitude at bus i, respectively. The line loss of the branch ij can be denoted as

$$P_{ij}^{loss}(t) = r_{ij}\frac{P_{ij}^2(t)}{V_i^2(t)} \tag{12.19}$$

The constraint function of WT power system is

$$0 \le P_{WT}^d(t) \le P_{WT}^f(t) \tag{12.20}$$

The constraint function of PV power system is

$$0 \le P_{PV}^d(t) \le P_{PV}^f(t) \tag{12.21}$$

The constraint function of TM power generation set output is

$$P_{TM}^{\min}(t) \le P_{TM}(t) \le P_{TM}^{\max}(t)$$

where $P_{TM}^{\min}$ and $P_{TM}^{\max}$ denote the minimum and maximum output values of TM power generation set.

Due to the discharge rate of the storage battery and the power limit of the converter, the ES station has a maximum charging/discharging power constraint. The charging and discharging power constraints are as follows:

$$0 \le P_{ES}^{ch}(t) \le U_{ES}(t) \cdot P_{ES}^{ch,\max} \tag{12.22}$$

$$0 \le P_{ES}^{dis}(t) \le (1 - U_{ES}(t)) \cdot P_{ES}^{dis,\max} \tag{12.23}$$

where $P_{ES}^{ch,\max}$ and $P_{ES}^{dis,\max}$ mean the maximum charging and discharging power of the ES station, respectively. The binary variable $U_{ES}(t)$ indicates the state of ES station (1 for charging state and 0 for other states). In addition, the SOC_{ES} is defined as the ratio of the remaining capacity of the ES battery to the rated capacity.

$$SOC_{ES}(t) = \frac{E_{ES}(t)}{E_{ES}^u(t)} \tag{12.24}$$

$$SOC_{ES}^{\min} \le SOC_{ES}(t) \le SOC_{ES}^{\max} \tag{12.25}$$

where $E_{ES}(t)$ means the remaining capacity of ES station at time t, E_{ES}^u is the rated capacity of ES station, and $SOC_{ES}(t)$ denotes the SOC of ES station at time t. $SOC_{ES}^{\max}$ and $SOC_{ES}^{\min}$ indicate the upper and lower limits of SOC of ES station.

$$SOC_{ES}(t) = SOC_{ES}(t-1) + \Delta SOC_{ES}(t) \tag{12.26}$$

$$\Delta SOC_{ES}(t) = \frac{P_{ES}^{ch}(t) \cdot \eta_{ES}^{ch} - \frac{P_{ES}^{\text{dis}}(t)}{\eta_{ES}^{\text{dis}}}}{E_{ES}^u} \cdot \Delta t \tag{12.27}$$

where $\Delta SOC_{ES}(t)$ represents the electricity change at time t; specifically, a positive number means charging and a negative number indicates discharging. η_{ES}^{ch} and η_{ES}^{dis} ES indicate the charging and discharge efficiency, respectively.

The constraint function of the electric power balance is defined as

$$P_{RE}^d(t) - P_{ES}^{ch}(t) + P_{ES}^{\text{dis}}(t) + P_{TM}(t) = P_L(t) + P_{AG}(t) + P_{loss}(t) \tag{12.28}$$

In the CPS model, the charging and discharging behaviors of aggregator do not consider the influence of social factors. Specifically, the uncontrollable load of EV aggregator P_{AG}^{un} of time t is equal to the sum of the EV charging power at this time; the EV can be fully charged with the maximum charging power $P_{EV_k}^{ch,\max}$:

$$P_{AG}^{un}(t) = \sum_{k}^{\mu} P_{EV_k}^{ch,\max}(t) \tag{12.29}$$

where $P_{EV_k}^{ch,\max}$ is the maximum charging power of kth EV.

In the CPSS model, price demand response (PDR) load of the EV aggregator is the controllable. For a benefit coalition, the power of EV aggregator to participate in the optimal dispatch of the power system is

$$P_{AG}^{co}(t) = \sum_{k}^{\mu} P_{EV_k}^{ch}(t) - \sum_{k}^{\mu} P_{EV_k}^{\text{dis}}(t) \tag{12.30}$$

The EV aggregator-rated capacity can be displayed as

$$E_{AG}^{u}(t) = \sum_{k}^{u} E_{EV_x}^{u}(t) \tag{12.31}$$

which varies with the number and characteristics of EVs arriving at the EV aggregator at different time intervals. The EV aggregator remaining capacity at time t can be displayed as

$$E_{AG}(t) = \sum_{k}^{u} [E_{EV_k}^{u}(t) \cdot SOC_{EV_k}(t) \tag{12.32}$$

The charging and discharging power constraints of the kth EV are as follows

$$0 \leq P_{EV_k}^{ch}(t) \leq U_{EV_k}(t) \cdot P_{EV_k}^{ch,\max} \tag{12.33}$$

$$0 \leq P_{\text{EV}_k}^{\text{dis}}(t) \leq (1 - U_{\text{EV}_k}(t)) \cdot P_{\text{EV}_k}^{\text{dis,max}} \tag{12.34}$$

where $P_{EV_k}^{ch,\max}$ and $P_{EV_k}^{\text{dis,max}}$ mean the maximum charging and discharging power of the kth EV, respectively. The binary variable $U_{EV_k}(t)$ is the state of kth EV (1 represents charging status and 0 for other states). The SOC of the kth EV $SOC_{EV_k}(t)$ is the remaining capacity at time t $E_{EV_k}(t)$ to the rated capacity $E_{EV_k}^{u}$, which can be displayed as

$$SOC_{EV_k}(t) = \frac{E_{EV_k}(t)}{E_{EV_k}^{u}} \tag{12.35}$$

$$SOC_{EV_k}^{\min} \leq SOC_{EV_k}(t) \leq SOC_{EV_k}^{\max} = SOC_{EV_k}^{lea} \tag{12.36}$$

where $SOC_{EV_k}^{\max}$ and $SOC_{EV_k}^{\min}$ denote the upper and lower limits of the kth EV SOC. $SOC_{EV_k}^{lea}$ means that the SOC of the kth EV must reach the upper limit when it leaves:

$$SOC_{EV_k}(t) = SOC_{EV_k}(t-1) + \Delta SOC_{EV_k}(t) \tag{12.37}$$

$$\Delta SOC_{EV_k}(t) = \frac{P_{EV_k}^{ch}(t) \cdot \eta_{EV_k}^{ch} - \frac{P_{EV_k}^{\text{dis}}(t)}{\eta_{EV_k}^{\text{dis}}}}{E_{EV_x}^{u}} \cdot \Delta t \tag{12.38}$$

where $\Delta SOC_{EV_k}(t)$ represents the electricity change at time t; specifically, a positive number means charging and a negative number means discharging. $\eta_{EV_k}^{ch}$ and $\eta_{EV_k}^{\text{dis}}$ mean the charging and discharging efficiency of kth EV, respectively.

12.2.2.3 Study Case

The CPS model and the CPSS model are mixed-integer programming (MIP) problems. The model A and model B are tested on the IEEE 33 bus system. *TM* power generation set, *WT–PV–ES CG* system, *EV* aggregator, and uncontrollable load are located at bus {0, 1, 18, 22}, respectively. The *WT*/*PV* power and load data are provided by the Elia Group. The RT electricity prices of *EV* aggregator charging and discharging are demonstrated in Figure 12.3; the parameters of the CPSS system are indicated in Tables 12.1 and 12.2.

The consumption level of *RE* is reflected by the value of *RE* curtailment, and the curtailment value of new energy P_{cur} is the minimum absolute value of the deviation between the actual value of *WT–PV* power generation and the dispatch value of the *WT–PV–ES CG* power generation system, where the portion that is not sufficient for scheduling is compensated by the timely output of the power system:

$$P_{cur}(t) = \min \Delta P^{a}_{CG}(t) = |P^{a}_{RE}(t) - P^{d}_{RE}(t) - P^{ch}_{ES}(t) + P^{dis}_{ES}(t)| \tag{12.39}$$

where $P^{a}_{RE}(t)$ is the actual value of *RE* generation. $\Delta P^{a}_{CG}(t)$ is the actual dispatch error of the *WT–PV–ES CG* system during time period *t*.

Table 12.3 shows the performance comparison of each subobjective based on CPS and CPSS optimization dispatch models. After considering social factors, not only the consumption of *RE* and the charging cost of *EV* aggregators have been significantly optimized, but also the *OM* costs of the power generation system have also been improved. This is mutually beneficial for power systems, distributed power stations, and *EV* users.

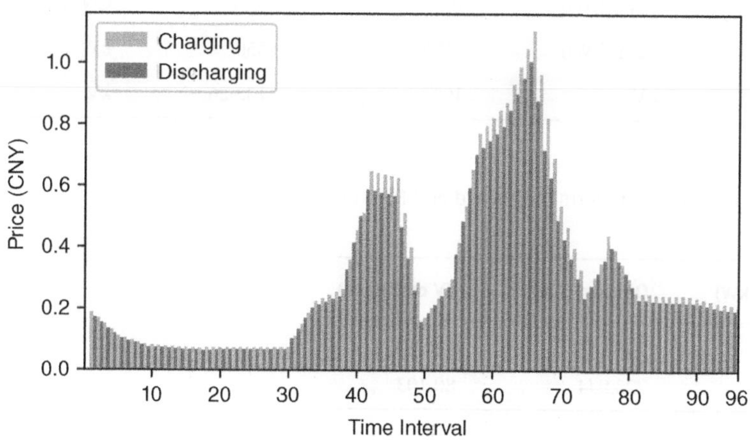

Figure 12.3 The real-time electricity prices of EV aggregator charging and discharging.

Table 12.1 Types, number, and parking periods of different EVs.

Label	Types	Number	Parking period
EVA	Taxi	EV_1–EV_{10}	00:00–08:00
EVB	Lorry	EV_{11}–EV_{20}	08:00–20:00
EVC	Private EV	EV_{21}–EV_{30}	20:00–08:00 (next day)

Table 12.2 Parameters of the CPSS.

Parameter	Value	Unit	Parameter	Value	Unit
$P_{ES}^{ch,\max}$	145	kW	$P_{ES}^{dis,\max}$	165	kW
$\eta_{ES}^{ch/dis}$	0.95	—	E_{ES}^{u}	825	kWh
$SOC_{ES}^{\min}$	0.15	—	$SOC_{ES}^{\max}$	0.9	—
$P_{TM}^{\min}$	40	kW	$P_{TM}^{\max}$	85	kW
$SOC_{ES}(1)$	0.5	—	Δt	15	min
E_{EVA}^{u}	26	kWh	$SOC_{EVA}(1)$	0.5	—
$P_{EVA}^{ch,\max}$	6.8	kW	$P_{EVA}^{dis,\max}$	7.8	kW
$\eta_{EVA}^{ch/dis}$	0.95	—	E_{EVB}^{u}	38	kWh
$SOC_{EVB}(1)$	0.4	—	$P_{EVB}^{ch,\max}$	9.5	kW
$P_{EVB}^{dis,\max}$	11.2	kW	$\eta_{EVB}^{ch/dis}$	0.9	—
E_{EVC}^{u}	32	kWh	$SOC_{EVC}(1)$	0.5	—
$P_{EVC}^{ch,\max}$	8.6	kW	$P_{EVC}^{dis,\max}$	10.4	kW
$\eta_{EVC}^{ch/dis}$	0.95	—	$SOC_{EV}^{\min}$	0.3	—
$SOC_{EV}^{\max}$	0.9	—	m_{ES}^{OM}	0.05	CNY/kWh
m_{WT}^{OM}	0.01	CNY/kWh	m_{PV}^{OM}	0.01	CNY/kWh
m_{TM}^{OM}	0.1	CNY/kWh	$P_{ij}^{\max}$	350	kW
$V_{i}^{\min}$	11.394	kV	$V_{j}^{\min}$	13.926	kV

Table 12.3 Comparison of optimization performance based on CPS and CPSS models.

Model	RE curtailment(kW)	OM cost (CNY)	EV cost (CNY)
CPS	1,126.675	796.233	120.170
CPSS	610.262	726.944	80.103

12.3 The Cooperative Operation in V2G

12.3.1 Collaboration Potential of Multiple EV Aggregators

In the day-ahead (DA) market, an EV aggregator places bids based on projected EV behaviors and historical electricity prices provided by the power company. However, in the RT market, the aggregator must adjust these bids according to updated forecast data. If there is a substantial deviation in electricity due to inaccuracies in EV behavior forecasting, the aggregator could face penalties for putting strain on the power system. Given that different EV aggregators can experience diverse RT operational states, it is feasible to alleviate this issue by integrating assistance from other

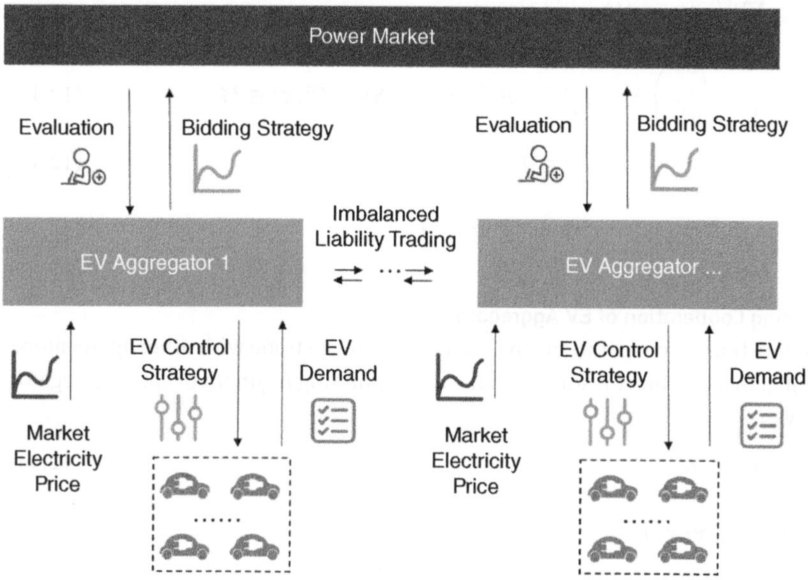

Figure 12.4 The coordinated framework of multiple EV aggregators participating in RT market.

aggregators. As such, a combined bidding optimization model for multiple EV aggregators in the RT market can help mitigate the adverse effects of EV uncertainty on both the aggregator and the power system. The structure of the suggested model is depicted in Figure 12.4. Each aggregator within the coalition is tasked with making internal power bidding decisions and trading imbalance liabilities with other aggregators externally.

12.3.1.1 Internal Bidding of EV Aggregator

Numerous existing studies have delved into the power market bidding model of a singular EV aggregator [23]. Prior to the closure of RT market bidding, the EV aggregator has the capability to update and submit bid quantities based on the most recent EV forecasts. The aggregate model is commonly used in devising optimal bidding strategies for EV aggregators [24, 25]. This model succinctly represents the collective characteristics of a group of EV owners with only four parameters (energy and power limits) at each time step. In this scenario, the aggregator is not required to predict detailed individual EV information, thereby streamlining the computation of bidding decisions. The model must take into account the equivalent battery power constraints Eqs. (12.39–12.44) and energy constraints Eqs. (12.45–12.47) as outlined below:

$$0 \leq P_i^{c,t} \leq P_i^{+,t}, \qquad \forall t \in T^{rt}, \forall i \in M \tag{12.40}$$

$$P_i^{-,t} \leq P_i^{d,t} \leq 0, \qquad \forall t \in T^{rt}, \forall i \in M \tag{12.41}$$

$$P_i^{+,t} = \sum_{\forall h \in H} P_{i,h}^{+,t}, \qquad \forall t \in T^{rt}, \forall i \in M \tag{12.42}$$

$$P_i^{-,t} = \sum_{\forall h \in H} P_{i,h}^{-,t}, \qquad \forall t \in T^{rt}, \forall i \in M \tag{12.43}$$

$$P_i^{c,t}\left(1 - s_i^t\right) - P_i^{d,t} s_i^t = 0, \qquad \forall t \in T^{rt}, \forall i \in M \tag{12.44}$$

$$P_i^t = P_i^{c,t} + P_i^{d,t} = \sum_{\forall h \in H} \left(P_{i,h}^{c,t} + P_{i,h}^{d,t} \right), \quad \forall t \in T^{rt}, \forall i \in M \tag{12.45}$$

$$E_i^{-,t} \leq E_i^{t-1} + \left(\eta^c P_i^{c,t} + \frac{P_i^{d,t}}{\eta^d} \right) \Delta t + E_i^{arr,t} - E_i^{dep,t} \leq E_i^{+,t}, \forall t \in T^{rt}, \forall i \in M \tag{12.46}$$

$$E_i^{+,t} = \sum_{\forall h \in H} E_{i,h}^{+,t}, \quad \forall t \in T^{rt}, \forall i \in M \tag{12.47}$$

$$E_i^{-,t} = \sum_{\forall h \in H} E_{i,h}^{-,t}, \quad \forall t \in T^{rt}, \forall i \in M \tag{12.48}$$

12.3.1.2 External Trading Cooperation of EV Aggregator

In the imbalanced liability trading cooperation, any EV aggregator can trade with other aggregators. The quantity of net imbalanced liability trading and the corresponding payment of each aggregator are described as follows:

$$e_i^t = \sum_{j \in M \backslash i} e_{ij}^t, \quad \forall i \in M, \forall t \in T^{rt} \tag{12.49}$$

$$\pi_i^t = \sum_{j \in M \backslash i} \pi_{ij}^t, \quad \forall i \in M, \forall t \in T^{rt} \tag{12.50}$$

$$e_{ij}^t + e_{ji}^t = 0, \quad \forall i \in M, \forall j \in M \backslash i, \forall t \in T^{rt} \tag{12.51}$$

$$\pi_{ij}^t + \pi_{ji}^t = 0, \quad \forall i \in M, \forall j \in M \backslash i, \forall t \in T^{rt} \tag{12.52}$$

12.3.1.3 Coordinated Optimization Model

For every EV aggregator, its optimization objective is to minimize its total cost consisting of cost C_i^t in RT market and payment π_i^t in imbalanced liability trading. The details are formulated as follows:

$$\min \ C_i + \pi_i = \sum_{\forall t \in T^{rt}} \left(C_i^t + \pi_i^t \right), \quad \forall i \in M \tag{12.53}$$

$$C_i^t = C_i^{e,t} + C_i^{p,t} + C_i^{bd,t} = pr^t \left(P_i^t - P_i^{da,t} \right) \Delta t + \left(pe_i^{+,t} + pe_i^{-,t} \right) \Delta t + \frac{C^b k_{bc} P_i^{d,t} \Delta t}{100 B \eta^d}, \forall i \in M, \forall t \in T^{rt} \tag{12.54}$$

$$\Delta P_i^t + e_i^t = P_i^t - P_i^{da,t}, \quad \forall i \in M, \forall t \in T^{rt} \tag{12.55}$$

$$\Delta P_i^t = \Delta P_i^{+,t} - \Delta P_i^{-,t}, \quad \forall i \in M, \forall t \in T^{rt} \tag{12.56}$$

$$pe_i^{+,t} \geq k_{pe} \left(\Delta P_i^{+,t} - sign_i^{da,t} k_{th} P_i^{da,t} \right), \quad \forall i \in M, \forall t \in T^{rt} \tag{12.57}$$

$$pe_i^{-,t} \geq k_{pe} \left(\Delta P_i^{-,t} - sign_i^{da,t} k_{th} P_i^{da,t} \right), \quad \forall i \in M, \forall t \in T^{rt} \tag{12.58}$$

$$\Delta P_i^{+,t}, \Delta P_i^{-,t}, pe_i^{+,t}, pe_i^{-,t} \geq 0, \quad \forall i \in M, \forall t \in T^{rt} \tag{12.59}$$

$$C_i + \pi_i \leq C_i^{Non}, \quad \forall i \in M \tag{12.60}$$

The first item in Eq. (12.41) indicates the cost of electricity purchased in RT energy market. The electricity deviation between RT and DA schedules is settled according to the electricity price in RT market. The second term penalizes excessive electricity deviations under the cases of positive deviation or negative deviation. The last item denotes the compensation cost for battery degradation caused by discharging. Constraint Eq. (12.42) indicates that the aggregator electricity deviation between DA and RT markets can be undertaken by itself and other trading aggregators. Constraint Eq. (12.43) further analyzes the electricity deviation's up and down components. Constraints Eqs. (12.44 and 12.457) define the penalties for electricity deviations when the deviation exceeds thresholds. Otherwise, these constraints are out of work because their right-hand sides are not positive. The nonnegative constraints in Eq. (12.468) guarantee that the variables are not negative. Individual rationality in Eq. (12.47) is a necessary condition. The cost of EV aggregator participating in cooperation cannot exceed the lowest cost in independent operation.

12.3.1.4 Optimization Based on GNB Theory

As a classical negotiation approach in the cooperative game model, generalized Nash bargaining (GNB) theory shows many advantages, which has been widely used to construct the cooperative model of multiple agents. The proposed cooperation among multiple EV aggregators based on GNB theory can be formulated as follows:

$$
\begin{aligned}
&\max \prod_{\forall i \in M'} \left[C_i^{Non} - (C_i + \pi_i)\right]^{\alpha_i} \\
&\text{subject to } (12.39) - (12.51), (12.53) - (12.59) \\
&\text{variables}: \left\{P_i^{c,t}, P_i^{d,t}, \Delta P_i^t, e_i^t, \pi_i^t, \Delta P_i^{+,t}, \Delta P_i^{-,t}, pe_i^{+,t}, pe_i^{-,t}\right\} \\
&\forall i \in M', \forall t \in T^{rt}
\end{aligned}
\tag{12.61}
$$

where C_i^{Non} is the lowest cost when *EV* aggregator *i* rejects the cooperation and α_i is the bargaining power of *EV* aggregator.

12.3.2 Case Study

12.3.2.1 Case Settings

Detailed simulations for the interaction with three EV aggregators have been conducted to evaluate the performance of the proposed model. EV arrival time, departure time, and arrival state of energy (SOE) are modeled in truncated Gaussian distributions. EV departure SOE is set to be uniformly distributed between 80% and 100% [26]. Assume the main uncertainties come from the differences in EV parameters: arrival/departure time and arrival state. Each EV aggregator forecasts 1000 EVs can be scheduled, while almost 750, 800, and 1200 in fact.

12.3.2.2 Cooperation Impacts

Take the cooperation at 19:00 (electricity price peak time) as an example to show the scheduling variation by cooperation. According to DA forecasting EV information, EV aggregators 1 and 2 are likely to overestimate available EVs in this slot, while EV aggregator 3 did the opposite. Figure 12.5(a) shows the impacts of cooperation on EV aggregator individual and coalition power consumption. The latest RT forecast shows that EV aggregators 1 and 2 probably do not have enough subordinate EVs to support that, resulting in the power deviation as 1384 and 1105 kW. Since EV aggregator 3 underestimates available EV capacity in this slot, EV aggregator 3 provides electricity around −1960 kW, which is within the deviation allowed range but not the most ideal. The coalition power of these three EV aggregators is less than DA bidding power. Compared with the case without EV aggregators' cooperation, there are some variations in EV aggregators' power. As we have just analyzed that EV aggregators 1 and 2 have reached their discharging upper bounds with limited EV capacity, their discharging power cannot be higher even after cooperation. By

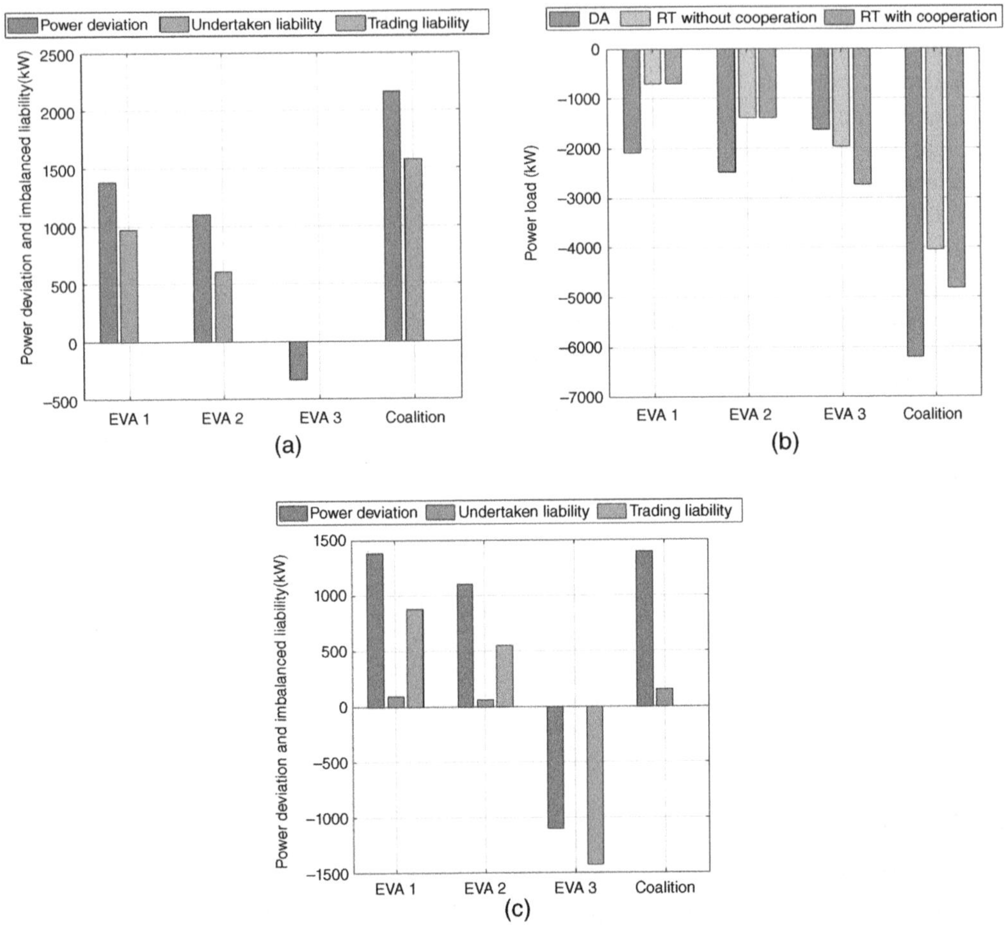

Figure 12.5 The impacts of cooperation: (a) power load, (b) power deviation and imbalanced liability before cooperation, and (c) power deviation and imbalanced liability after cooperation.

contrast, EV aggregator 3 increases its discharging power to −2730 kW. This discharging behavior improves the power output of the aggregator coalition, and the coalition deviation is reduced simultaneously. This figure certifies the positive effect of the proposed cooperation model on power balance. Figure 12.5(b) and (c) interpret the effect of the proposed cooperation on every aggregator's imbalanced liability. We can find that only EV aggregator 3 does not need to undertake imbalanced liability before the cooperation. Since there is no imbalanced liability trading service before the cooperation, the liabilities of every aggregator are dependent on their power deviations fully. After the cooperation, the imbalanced liabilities of EV aggregators 1 and 2 have been reduced obviously, which are trading to EV aggregator 3. Although EV aggregator 3 cannot eliminate all the imbalanced liability by adjusting its power consumption, the coalition discharging power shortage has been improved greatly.

This trading is derived from the coupled constraint Eq. (12.48) in the model, which breaks the bounds of different EV aggregators. Any aggregator in the coalition can adjust its power consumption to affect others' imbalanced liabilities. They are willing to do contribution to the coalition through changed behavior, which may increase their electricity cost or battery compensation cost,

Table 12.4 Cost components of all EV aggregators.

Metric		EVA 1	EVA 2	EVA 3	Coalition
Cost without cooperation ($)	Energy	2.55	−43.93	86.10	44.72
	Penal	316.79	303.45	85.08	705.33
	Compensation	2.23	6.86	5.19	14.27
	Payment	0	0	0	0
	Final	321.58	266.38	176.36	764.32
Cost with cooperation ($)	Energy	5.26	−44.69	77.57	38.14
	Penal	85.66	81.15	61.53	228.34
	Compensation	2.16	6.77	6.52	15.46
	Payment	118.73	95.38	−214.11	0
	Final	211.81	138.62	−68.48	281.95

but they can gain more benefit from others' payment. Thus, cooperation surplus allocation is vital, which decides the stability of the coalition.

Table 12.4 shows the cost components of EV aggregators 1, 2, and 3 before and after cooperation. After solving P1, RT energy market cost, penal cost, and battery compensation cost are varied with the new power consumption schedule. The energy cost of EV aggregator 1 and the battery compensation cost of EV aggregator 3 increase slightly because of more control demand. More importantly, the coalition cost was reduced from \$764.32 to \$281.95. From the mathematic perspective, the global optimal solution for the coupled optimization problem is always superior to the sum of locally optimal solutions for subproblems because of a larger variable feasible space. After the solving process, the payment and final cost of all EV aggregators are determined. From the individual perspective, the costs of all EV aggregators are all reduced to varied degrees through cooperation. The above numerical results prove the effectiveness of the proposed cooperation in cost saving.

12.4 Framework of a Charging Station with Battery Swapping Mode

The charging station with battery swapping mode consists of a battery centralized charging station (BCCS), battery distribution station (BDS), and battery logistics system, which is shown in Figure 12.6. BCCS is a centralized charging and control center facility for EV batteries. Through BCCS, EV batteries can be centrally controlled and optimized for charging, and the grid can be flexibly charged and discharged, improving the operational flexibility and economy of the grid.

12.4.1 Planning Model

12.4.1.1 Objective

The annual planning comprehensive cost C is as follows:

$$C = \sum_{i=1}^{N_C} \left(A_i^B + A_i^S + A_i^O + A_i^T\right) \tag{12.62}$$

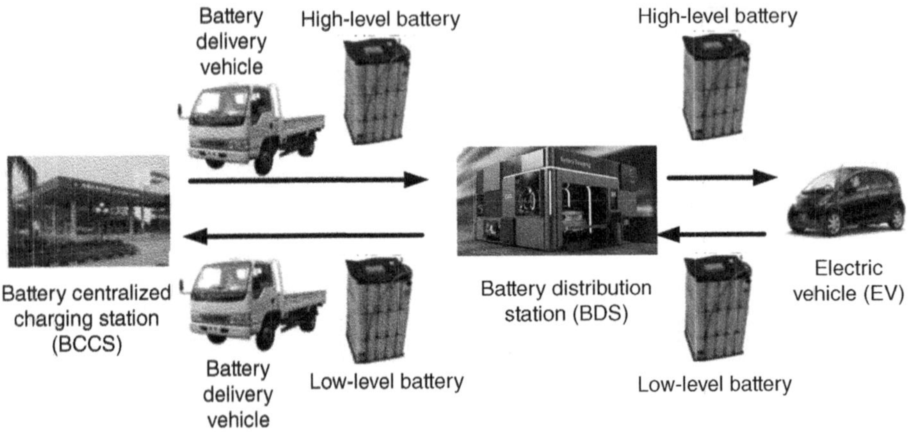

Figure 12.6 The charging station with battery swapping mode.

where A_i^B, A_i^S, A_i^O, and A_i^T are the energy storage cost, infrastructure cost, operation–maintenance cost, and battery transportation cost, respectively, which can be calculated as Eqs. (12.50–12.65). N_C is the planning quantity of BCCS.

1) Energy Storage Cost

$$A_i^B = \left[\frac{a_B^p r(1+r)^{y_B}}{(1+r)^{y_B}-1} + \frac{a_B^d r}{(1+r)^{y_B}-1}\right] S_i^B \tag{12.63}$$

where r is the discount rate. a_B^p and a_B^d are the costs of purchase and disposing of unit capacity of energy storage, respectively. y_B is the useful life of energy storage. S_i^B is the capacity of energy storage BCCS i.

2) Infrastructure Cost

$$A_i^S = \frac{r(1+r)^{y_B}}{(1+r)^{y_B}-1}\left[a_{Cha}N_i^{Cha} + a_{Tra}S_i^{Tra} + a_{Lin}L_i^{Lin} + a_{Are}S_i^{Are} + a_i^{Dis} + a_i^{Fix}\right] \tag{12.64}$$

where y_B is the useful life of BCCS. N_i^{Cha}, S_i^{Tra}, L_i^{Lin}, and S_i^{Are} are the number of chargers, the capacity of transformer, the length of power line, and the land area of BCCS i, respectively. a_{Cha}, a_{Tra}, a_{Lin}, and a_{Are} are the unit costs of charger, the capacity of transformer, the length of power line, and the land area, respectively. a_i^{Dis} and a_i^{Fix} are costs of distribution equipment and fixed investment of BCCS i, respectively.

3) Operation–Maintenance Cost

$$A_i^O = N_d o_{eo} e_i^{ex} + \frac{r(1+r)^{y_B}}{(1+r)^{y_B}-1}\left(o_{Tra}a_{Tra}N_i^{Tra} + o_{Dis}a_i^{Dis}\right) \tag{12.65}$$

where N_d is the days per year. o_{eo} is the unit operation–maintenance cost of energy storage. o_{Tra} and o_{Dis} are the ratio of operation–maintenance cost to purchase cost of transformers and distribution equipment, respectively. e_i^{ex} is the charging of electricity of energy storage.

4) Battery Transportation Cost

$$A_i^T = n_T N_d a_T \sum_{t=1}^{T} d_{i,t}^T \tag{12.66}$$

where T is hours per day. n_T is the number of round trips of battery delivery vehicle in a complete battery logistics. a_T is the unit driving cost of battery delivery vehicle. $d^T_{i,t}$ is the driving distance of battery delivery vehicle at time t.

The decision variables in Eqs. (12.62–12.65) are calculated using Eqs. (12.–12.).

The driving distance from BDS $j1$ to target BCCS of a battery delivery vehicle is given by Eq. (12.):

$$\left[i(j), d^T_{i,t}\right] = r_{Roa} \min \left\{ \left\| \left(x^C_i, y^C_i\right), \left(x^B_j, y^B_j\right) \right\|_2 \mid \forall \left(x^C_i, y^C_i\right) \right\}, \forall t \tag{12.67}$$

where $i(j)$ is the number of the target BCCS of the BDS j. (x^C_i, y^C_i) and $\left(x^B_j, y^B_j\right)$ are the coordinates of BCCS i and BDS j, respectively. r_{Roa} is the bending coefficient of roads in the planning area. $\|a, b\|_2$ is the two-dimensional Euclidean distance between a and b.

The capacity of energy storage of a BCCS can be calculated by Eq. (12.):

$$S^B_i = (1 + r_B) S_{EV} \max \left(N^{ex}_{i,t}\right), \forall t \tag{12.68}$$

where $N^{ex}_{i,t}$ is the number of batteries that are transported by battery delivery vehicles to the BCCS i at time t. S_{EV} is the capacity of energy storage of a EV. r_B is the margin coefficient of the energy storage in BCCS.

The load of EVs in a BCCS can be given by using Eq. (12.):

$$\begin{cases} N^{ex}_{i,t} = \sum_{j=1}^{n^{BDS}_i} n^B_{j,t} \\ E_{EX,i} = S_{EV} \sum_{t=1}^{T} N^{ex}_{i,t} \end{cases} \tag{12.69}$$

where n^{BDS}_i is the number of BDSs that batteries supplied through BCCS i. $n^B_{j,t}$ is the demand number of battery of EVs of BDS j at time t.

After that, Eq. (12.) gives the number of chargers in BCCS i.

$$N^{Cha}_i = \left\lceil \frac{(1 + r_{Cha}) S^B_i}{S_{EV} n^{Cha}_n} \right\rceil_1 \tag{12.70}$$

where n^{Cha}_n is the number of batteries that can be charged simultaneously by a single charger. $\lceil a \rceil_1$ is the upward rounding operator, and a is rounded up in units of 1.

Then, Eq. (12.) provides the calculation method of capacity of transformer in a BCCS:

$$S^{Tra}_i = \left\lceil \frac{P^{Cha}_n N^{Cha}_i}{\cos \phi_T} \right\rceil_T \tag{12.71}$$

where P^{Cha}_n is the rated power of a charger. $\cos \phi_T$ is the rated power factor of the load. $\lceil a \rceil_T$ means a is round up by the standard capacities of transformer.

Due to the large capacity of BCCS, it is necessary to build BCCS near the substation. A dedicated power supply line needs to be installed from the substation to BCCS, and the length of the power supply line to be built is calculated using Eq. (12.):

$$L^{Lin}_i = r_{Lin} \min \left\{ \left\| \left(x^B_i, y^B_i\right), \left(x^T_k, y^T_k\right) \right\|_2 \mid \forall \left(x^T_k, y^T_k\right) \right\} \tag{12.72}$$

where r_{Lin} is the margin coefficient of the power line. $\left(x^T_k, y^T_k\right)$ is the coordinate of substation k.

Afterward, Eq. (12.) calculates the area of a BCCS.

$$S_i^{\mathrm{Are}} = (1 + r_{\mathrm{Are}})\left(m_{\mathrm{Cha}} N_i^{\mathrm{Cha}} + m_{\mathrm{B}} S_i^{\mathrm{B}} + m_{\mathrm{Tra}}\right) \tag{12.73}$$

where r_{Are} is the margin coefficient of the area of a BCCS. m_{Cha} is the area of a charger. m_B is the area of unit capacity of energy storage. m_{Tra} is the area of a transformer.

12.4.1.2 Constraints

The system needed to meet the grid flow constraint:

$$\begin{cases} P_T - P_C = U_i \sum\limits_{j\in\Pi(i)} U_j(G_{i,j}\cos\theta_{i,j} + B_{i,j}\sin\theta_{i,j}) \\ Q_T - Q_C = U_i \sum\limits_{j\in\Pi(i)} U_j(G_{i,j}\sin\theta_{i,j} - B_{i,j}\cos\theta_{i,j}) \end{cases} \tag{12.74}$$

where P_T and Q_T are active and reactive power injected into by substation. P_C and Q_C are the active and reactive load power of BCCS. U_i is voltage amplitude of node *i*. $G_{i,j}$ and $B_{i,j}$ are the real and imaginary parts of power line *i,j* admittance. $\theta_{i,j}$ is the phase angle difference in power line *i,j*.

12.4.2 Case Study

The planning area is shown in Figure 12.7. The time frame and research period of the planning project is 20 years, and the interest rate is 6%. Figures 12.8–12.11 give the planning results.

It can be seen that the change trend of annual comprehensive cost *F* is basically consistent with the change trend of energy storage capacity. As the number of BCCS (NC) increases, F and energy storage capacity gradually increase. When $NC = 3, 4$, and 5, the annual comprehensive cost is lower;

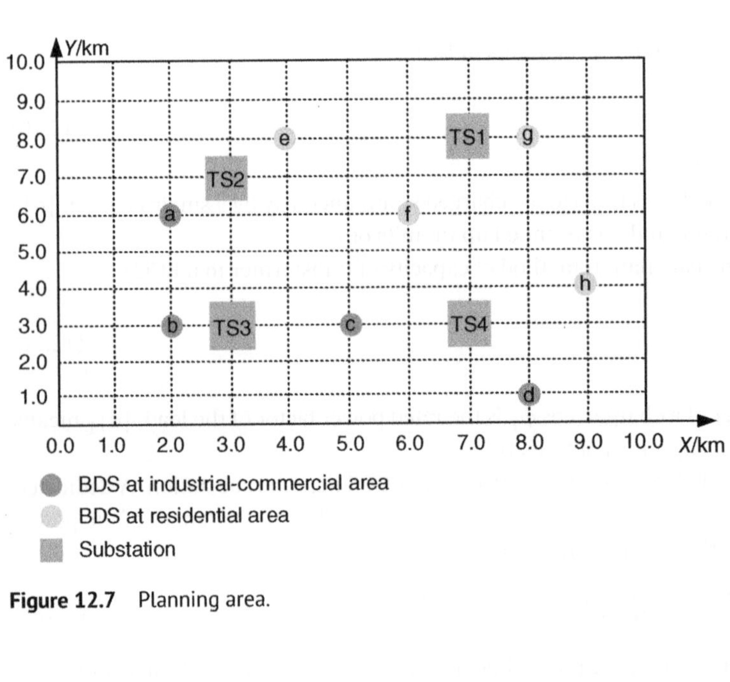

Figure 12.7 Planning area.

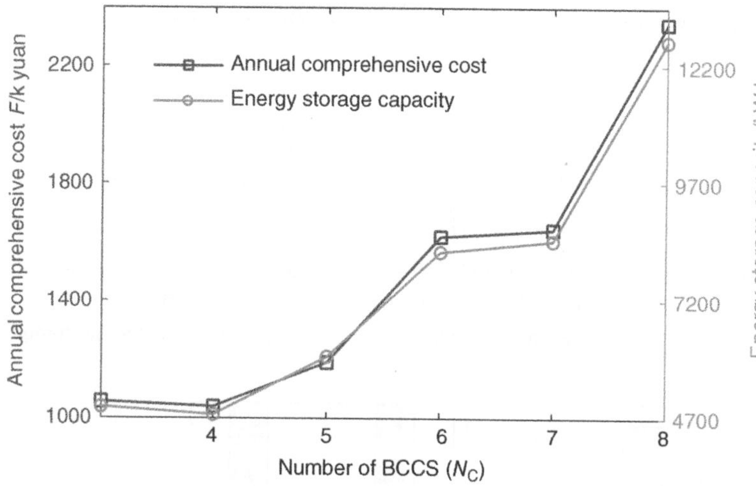

Figure 12.8 Annual comprehensive cost.

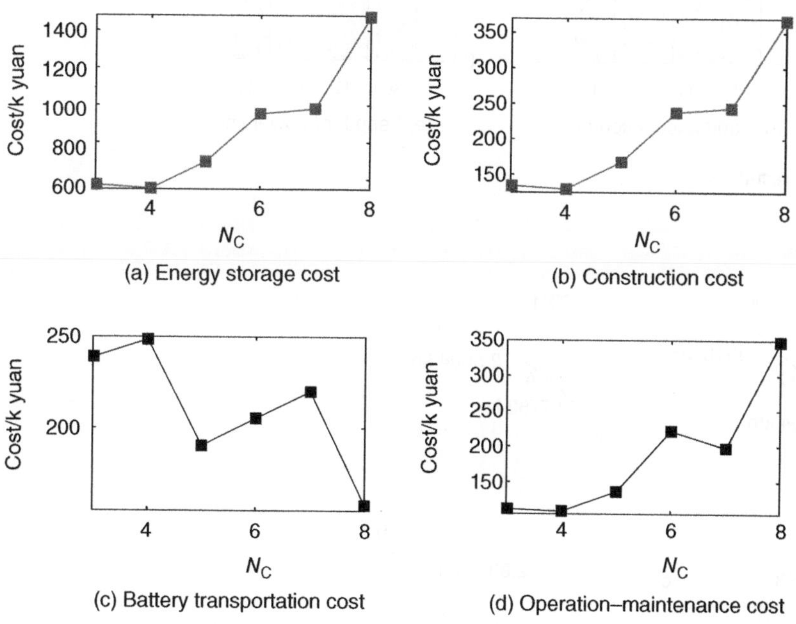

Figure 12.9 Each cost of planning.

when $NC = 4$, the annual comprehensive cost is lower than 17.1679 k yuan; and when $NC = 3$ and 5, the annual comprehensive cost is 150.8579 k yuan. The energy storage capacity of $NC = 4$ is 165 kWh lower than that of $NC = 3$, the energy storage cost is 19.1025 k yuan lower, the construction cost is 45,413 k yuan lower, and the operation–maintenance cost is 340.41 k yuan lower. To sum up, the planning scheme of BCCS Building 4 is the best.

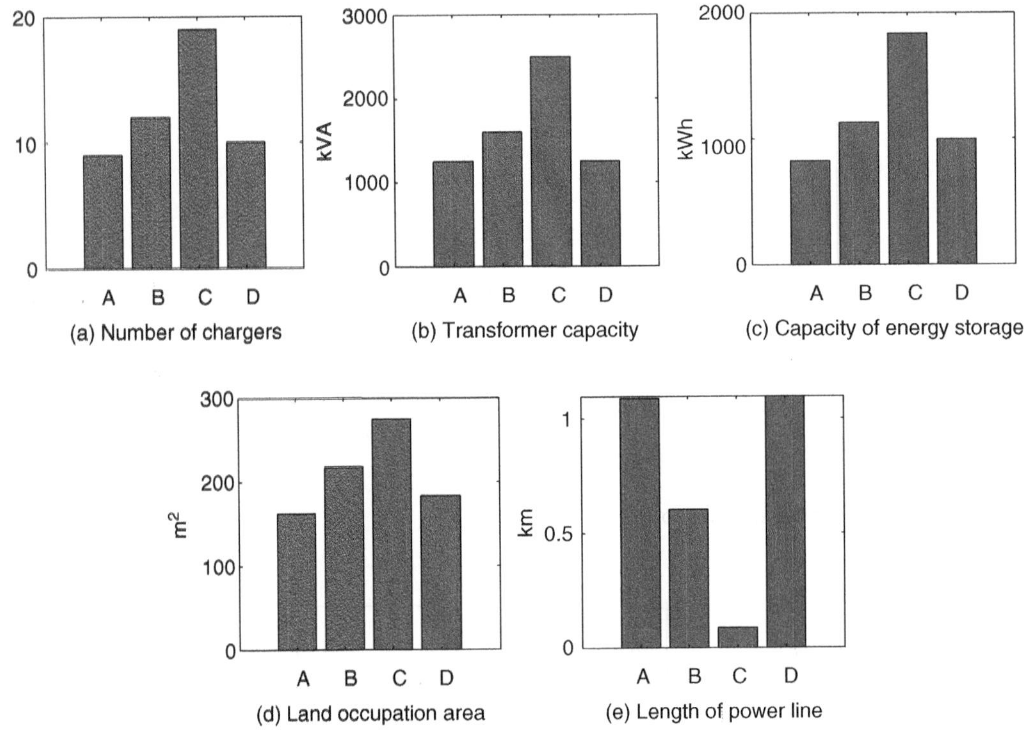

Figure 12.10 Planning result.

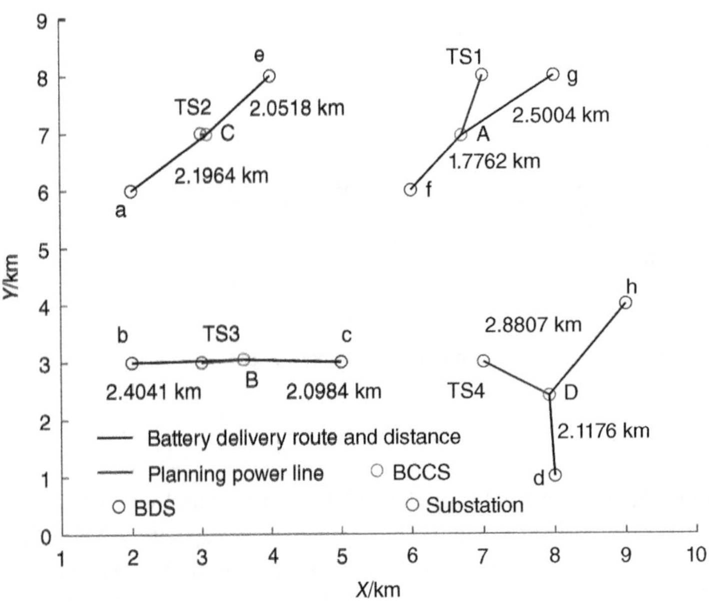

Figure 12.11 Locations of BCCSs.

12.5 Conclusion

Firstly, this chapter proposes a cyber–physical–social model and establishes an optimal dispatch model based on CPS and CPSS, in which the EV aggregator, as a flexible load, is incentivized to actively participate in the optimization scheduling of the power system through PDR. The optimization results obtained by the CPSS-based model are superior to those based on the CPS model. The application of CPSS in energy transformation has achieved mutual benefit and win-win situation among various entities in the new power system.

Secondly, in the interaction between EV and grid, this chapter indicates the significant importance of studying the coordinated optimization of different EV aggregators to reduce the adverse effect of EV uncertainty. In addition, EV aggregator's cooperative operation model in RT market with imbalanced liability trading can save EV aggregator coalition cost and promote grid power balance.

Thirdly, in this chapter, the planning model of BCCS considers the economic performance of the grid in terms of investment and operation of charging and switching facilities. The optimization results meet the requirements of different types of switching loads in the planned area. The correctness of the BCCS planning method proposed is verified.

References

1 Xue, Y. and Yu, X. (2017). Beyond smart grid—cyber–physical–social system in energy future [point of view]. *Proceedings of the IEEE* 105 (12): 2290–2292.

2 Morstyn, T., Farrell, N., Darby, S.J., and McCulloch, M.D. (2018). Using peer-to-peer energy-trading platforms to incentivize prosumers to form federated power plants. *Nature Energy* 3 (2): 94–101.

3 Zheng, W. and Hill, D.J. (2021). Incentive-based coordination mechanism for distributed operation of integrated electricity and heat systems. *Applied Energy* 285: Art. no. 116373.

4 Vagropoulos, S.I. and Bakirtzis, A.G. (2013). Optimal bidding strategy for electric vehicle aggregators in electricity markets. *IEEE Transactions on Power Systems* 28 (4): 4031–4041.

5 Wu, W., Zhu, J., Liu, Y. et al. (2013). A coordinated model for multiple electric vehicle aggregators to grid considering imbalanced liability trading. *IEEE Transactions on Smart Grid* in press.

6 Duan, X., Hu, Z., and Song, Y. (2021). Bidding strategies in energy and reserve markets for an aggregator of multiple EV fast charging stations with battery storage. *IEEE Transactions on Intelligent Transportation Systems* 22 (1): 471–482.

7 Zhang, H., Hu, Z., Xu, Z., and Song, Y. (2017). Evaluation of achievable vehicle to-grid capacity using aggregate PEV model. *IEEE Transactions on Power Systems* 32 (1): 784–794.

8 Tan, J. and Wang, L. (2017). A game-theoretic framework for vehicle-to-grid frequency regulation considering smart charging mechanism. *IEEE Transactions on Smart Grid* 8 (5): 2358–2369.

9 He, C., Zhu, J., Li, S. et al. (2022). Sizing and locating planning of EV centralized-battery-charging-station considering battery logistics system. *IEEE Transactions on Industry Applications* 58 (4): 5184–5197.

10 C. He, J. Zhu, S. Li et al. (2021). Sizing and locating planning of EV centralized-battery-charging-station considering battery logistics system. *2021 IEEE 4th International*

Electrical and Energy Conference (CIEEC), Wuhan, China, pp. 1–6, https://doi.org/10.1109/CIEEC50170.2021.9510699, http://ieeexplore.ieee.org/document/9510699.

11 C. He, J. Zhu, J. Lan et al. (2021). Optimal planning of electric vehicle battery centralized charging station based on EV load forecasting. *2021 6th International Conference on Power and Renewable Energy (ICPRE)*, Shanghai, China, pp. 1192–1197, https://doi.org/10.1109/ICPRE52634.2021.9635414, http://ieeexplore.ieee.org/document/9635414.

12 He, C., Zhu, J., Lan, J. et al. (2022). Optimal planning of electric vehicle battery centralized charging station based on EV load forecasting. *IEEE Transactions on Industry Applications* 58 (5): 6557–6575.

13 Zhu, J., He, C., Cheung, K. et al. Coordination planning of integrated energy system and electric vehicle charging station considering carbon emission reduction. *IEEE Transactions on Industry Applications* https://doi.org/10.1109/TIA.2023.3298330.

14 C. He, J. Zhu, Y. Liu et al. (2022). Coordination planning of integrated energy system and electric vehicle charging station considering carbon emission reduction. *2022 IEEE/IAS Industrial and Commercial Power System Asia (I&CPS Asia)*, Shanghai, China.

15 C. He, J. Zhu, H. Zhu et al. (2023). Coordinated planning of charging swapping stations and active distribution network. *2023 IEEE 6th International Electrical and Energy Conference (CIEEC)*, Hefei, China, pp. 4129–4134.

16 C. He, J. Zhu, Y. Liu et al. (2022). Joint planning of electric vehicle charging swapping station and energy hub. *2022 5th International Conference on Renewable Energy and Power Engineering (REPE)*, Beijing, China, pp. 448–454.

17 Zhang, X., Yu, T., Xu, Z. et al. (2018). A cyber-physical-social system with parallel learning for distributed energy management of a microgrid. *Energy* 165: 205–221.

18 Xu, L., Guo, Q., Yang, T., and Sun, H. (2019). Robust routing optimization for smart grids considering cyber-physical interdependence. *IEEE Transactions on Smart Grid* 10 (5): 5620–5629.

19 Cheng, L. and Yu, T. (2019). Game-theoretic approaches applied to transactions in the open and ever-growing electricity markets from the perspective of power demand response: an overview. *IEEE Access* 7: 25727–25762.

20 Lin, Z., Chen, H., Wu, Q. et al. (2020). Mean-tracking model based stochastic economic dispatch for power systems with high penetration of wind power. *Energy* 193: 116826.

21 Chen, Z., Zhu, J., Li, S. et al. (2022). Detection of false data injection attacks on load frequency control system with renewable energy based on fuzzy logic and neural networks. *Journal of Modern Power Systems and Clean Energy* 10 (6): 1576–1587.

22 Chen, Z., Zhu, J., Dong, H. et al. (2022). Optimal dispatch of WT/PV/ES combined generation system based on cyber-physical-social integration. *IEEE Transactions on Smart Grid* 13 (1): 342–354.

23 Zhao, T., Li, Y., Pan, X. et al. (2018). Real-time optimal energy and reserve management of electric vehicle fast charging station: hierarchical game approach. *IEEE Transactions on Smart Grid* 9 (5): 5357–5370.

24 Xu, Z., Hu, Z., Song, Y., and Wang, J. (2017). Risk-averse optimal bidding strategy for demand-side resource aggregators in day-ahead electricity markets under uncertainty. *IEEE Transactions on Smart Grid* 8 (1): 96–105.

25 Xu, Z., Callaway, D.S., Hu, Z., and Song, Y. (2016). Hierarchical coordination of heterogeneous flexible loads. *IEEE Transactions on Power Systems* 31 (6): 4206–4216.

26 Sarker, M.R., Dvorkin, Y., and Ortega-Vazquez, M.A. (2016). Optimal participation of an electric vehicle aggregator in day-ahead energy and reserve markets. *IEEE Transactions on Power Systems* 31 (5): 3506–3515.

13

Power Distribution Systems Self-Healing

Konrad Schmitt[1], Manohar Chamana[2], Meisam Mahdavi[3], Stephen Bayne[1], and Luciane Neves[4]

[1]*Department of Electrical and Computer Engineering, Texas Tech University, Lubbock, TX, USA*
[2]*National Wind Institute, Texas Tech University, Lubbock, USA*
[3]*Department of Electrical Engineering, University of Brasilia, Brasilia, Brazil*
[4]*Department of Electromechanical and Power Systems, Federal University of Santa Maria, Santa Maria, Brazil*

13.1 Introduction

Electric power delivery to end users is one of the primary responsibilities of distribution systems (DSs). Built with a meshed topology, DSs are traditionally operated in a radial configuration, as their intrinsic asymmetry and unbalancing make the meshed operation extremely challenging. Customers are connected through single-, two-, and three-phase arrangements throughout the network and phases. To electrically connect these customers, the DSs have single-, two-, and three-phase branches, creating asymmetrical and, consequently, unbalanced properties. With the predominant radial operation, compounded by several pieces of equipment and presenting a higher resistance to reactance (R/X) ratios compared to meshed transmission systems, DSs have substantially higher losses and unique challenges in their modeling and operation [1].

With the intention of monitoring and measuring the quality of power delivery service, reliability indices are factors used to represent the utilities' performance. These metrics are computed based on different parameters and conditions defined by the IEEE 1366–2012 Guide for Electric Power Distribution Reliability Indices [2]. DSs' reliability metrics aim to measure the ability of the system to deliver uninterrupted service to customers. Even though there can be planned and unplanned interruptions, the unplanned ones are the most challenging events. Planned interruptions are events where the utility needs to perform maintenance or another type of service that requires part of the system to be de-energized for safety purposes. In these scenarios, customers are notified in advance, and a detailed plan for reconfiguration and de-energization is developed. An unplanned interruption is an unpredictable event where the utility must act fast enough to minimize its impact, independent of the condition. The standard defines two types of unplanned interruptions: temporary and sustained. Temporary interruptions last less than five minutes, while sustained last more than that. This time threshold is crucial for restoration analysis, as only sustained outages are counted in the reliability indices, meaning that if the utility is able to restore as many customers as possible under a permanent fault event through SH actions, it will significantly reduce the event's impact on the reliability indices.

The need to meet reliability indices levels has encouraged utilities to improve traditional distribution management systems' monitoring and control capabilities, expanding it to Advanced

Smart Cyber-Physical Power Systems: Fundamental Concepts, Challenges, and Solutions, Volume 1, First Edition.
Edited by Ali Parizad, Hamid Reza Baghaee, and Saifur Rahman.

Distribution Management Systems (ADMS). The ADMS concept embraces several online and offline monitoring and control features to ensure power delivery with reliability, continuity, and quality. This solution is usually centralized at the Distribution Operations Center (DOC) but may also have distributed elements. By having data about the network topology, such as electrical and physical information of connections, line segments, equipment, and customers, ADMS can obtain real-time field measurements to perform further analysis on the network to automatically respond to events or support operators' decisions with more detailed information.

Distribution network reconfiguration (DNR) is one of the primary ADMS features. Given a network topology, its operational configuration can be changed through the maneuvering of switches. The reconfiguration brings flexibility to the complex distribution networks which are mounted with different passive and active elements, such as switches, step-voltage regulators, capacitor banks, and meters. Regarding switches, DSs can have a large variety of types, such as circuit breakers, reclosers, load breakers, motor-operated switches, and sectionalizers, which differ from each other in operational aspects. Even though circuit breakers and reclosers are able to extinguish short-circuit current levels, load breakers are designed only to interrupt load current levels. On the other hand, motor-operated switches and sectionalizers do not have the capability to extinguish the arc current and, by this, should not be operated under load. During DNR, it is crucial to understand the switches' control capabilities, which can be divided into tele-controlled switches (TCS) and manually controlled switches (MCS). Nowadays, most circuit breakers, reclosers, load breakers, and motor-operated switches are TCSs, while sectionalizers are MCSs and need linemen's intervention to be maneuvered.

With newer and more accessible ADMS technologies, the grid modernization movement has brought a broader necessity to improve energy delivery by incorporating intelligent electronic devices (IED). IEDs are devices allocated over the network that can provide electrical and status measurements through communication venues and protocols. Even with an investment cost related to it, increasing the number of TCSs intends to reduce costs related to power delivery by enabling a more efficient DNR [3]. The DNR concept is also used for outage events. Instead of maintaining the system de-energized according to the post-event scenario until the damage is repaired, switching maneuvers can isolate faults and partially restore de-energized customers. With this approach, only the faulted region would be kept de-energized until linemen repair the structural damage. However, when TCSs are operated in an automated manner based on a predefined algorithm, the restoration action is commonly called Self-Healing (SH) or, most recently, Fault Location Isolation and Service Restoration (FLISR) or Fault Detection, Isolation and Restoration (FDIR) [4]. SH is able to utilize available field measurements to identify an event, locate the fault, and then compute a sequence of switching commands that can isolate the damaged area and restore as many customers as possible. According to the level of autonomy, the SH actions can be automatic or guided by operators' supervision. The main goal of SH is to provide a fast response to an outage event that can minimize the number of impacted customers and facilitate the linemen's efforts [5–7].

13.2 Historical Notes

The power outage restoration of DSs has been advancing by incorporating automation, communication, and computational tools to improve fastness and accuracy with SH schemes. The development of SH solutions has been made through matching the need for faster responses, availability of TCSs, and computational processing capabilities. Although SH applications and methods have just been reaching maturity, many utilities worldwide have already started the transition from traditional restoration solutions to SH schemes.

13.2.1 Background

The concern of integrating TCSs to increase network flexibility and resilience started back in 1971 [8]. Consequently, the DNR problem was first introduced in 1975 [9], where classic optimization methods were used to minimize network losses through switching maneuvers. Since then, DNR has been mostly used for power loss minimization. However, with the advantage of changing the network configuration over time to improve power delivery, the DNR application was also expanded to optimize other operational aspects such as power quality, voltage profile, stability, reliability, resiliency, loading, maintenance, unbalancing, repair, and restoration.

However, DNR analyses are complex due to the large number of nodes, unbalanced characteristics, and lack of observability in DSs, which were facilitated by the use of computational platforms. Since the 1960s, computers have been used for online analysis of DS [10], but computational platforms have only gained enough strength to enable ADMS solutions during the 1990s [11, 12]. ADMSs are designed as a centralized platform compounded by different synchronous and asynchronous tasks that can improve the quality, reliability, assertiveness, and fastness of DS operation and planning decisions [13]. Techniques such as distribution power flow, configuration processing, topology processing, state estimation, load estimation, conservative voltage reduction, Volt/VAr control, DNR, hosting capacity analysis, maintenance and outage planning, and outage restoration are part of modern ADMS solutions [1].

To bring flexibility to the operators, ADMS has also been broken down into different management platforms, such as Outage Management Systems (OMS), which focus on maintenance, outage planning, and restoration, and Distributed Energy Resources Management Systems (DERMS), responsible for ensuring proper coordination between DERs over the network through hosting capacity and Volt/VAr controls. The trend to have specific ADMS tools implies that each feature is becoming more sophisticated and complex, requiring a dedicated platform to perform its analysis and decisions as expected. Moreover, studies have shown that operators' decisions can be deceived and significantly impact the system's reliability levels [14]. Based on that, even though operators' experience and broad view of the system are unique and difficult to replace, ADMS features tend to become more independent and rely less on human intervention. This trend is based on the need for accuracy and speed in a highly dynamic modern DSs, where actions delayed by seconds can imply poor power quality delivery or even cascade events. The DNR aiming for an outage restoration problem is unique in its need for fast response. As soon as a fault is extinguished through protection schemes, the DNR solution must be able to isolate the damaged region and restore potential customers. The faster decisions are made and actions are taken through the ADMS, the less impacted the utility's reliability metrics will be.

Power systems faults are categorized into temporary and permanent faults. Temporary faults are short circuits that last a couple of seconds, usually happening in non-insulated overhead branches whenever the vegetation touches the conductor or when cables that lack proper spacers touch each other [15]. Through reclosing coordination, protective devices are programmed to protect the system from these temporary faults and attempt to reclose, usually three or four times based on the open interval timeout for each shot. Approximately 80% of the DSs' faults are temporary and are eliminated by themselves after the reclosing attempts [16, 17]. The remaining 20% are permanent faults caused mainly by damage or failure of power delivery structures, such as distribution towers and poles, which usually originate from traffic accidents, weather events, vegetation, and equipment failure [18]. Due to the radial configuration, a permanent fault event implies a power outage to all customers connected downstream to the tripped relay or blown fuse [19, 20], and at this moment, the utility must work as fast and precisely as possible to restore the impacted customers. Figure 13.1 presents a conceptual outage resiliency curve, highlighting the actions and impacts of an outage event restoration.

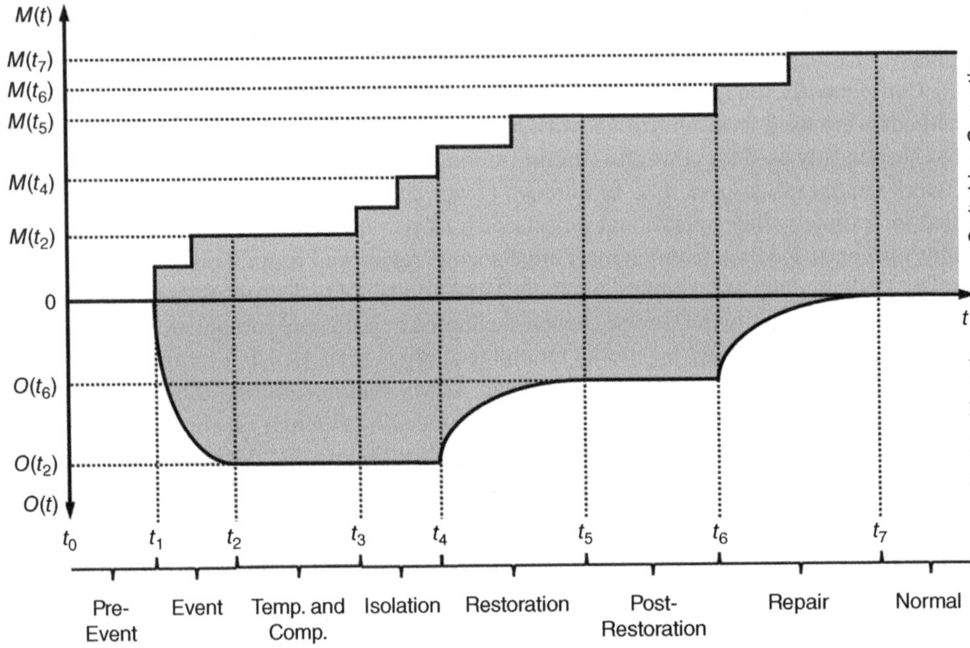

Figure 13.1 Outage resiliency curve.

The outage resilience curve can be defined by factors such as energized elements and loads or unserved power demand based on actions to recover the system. Figure 13.1 considers the resilience curve based on the unserved demand, $O(t)$, and switching operations $M(t)$. During an outage event, the system's pre-event state is defined by the network configuration and demand levels. As soon as an event happens at t_1, there is a short time window where protective devices operate according to their predefined protection pickups between t_1 and t_2. The protection scheme's performance depends on the predefined settings and other factors such as fault type and location. The scheme's actions will define the amount of de-energized loads. To allow proper operation of the protective elements, there is a temporization from the first protective device lockout to allow the protection scheme operation completion before achieving the post-event network configuration, $O(t_2)$, as represented between t_2 and t_3. As an illustration, between t_3 and t_4 the system performs fault isolation, where the damage is electrically isolated, and some loads may be restored. The restoration maneuvers start at t_4 and $O(t_6)$ is reached when all restoration actions are exhausted. The amount of unserved power in $O(t_6)$ will then only be restored by the intervention of linemen to repair the structural damage(s) between t_6 and t_7. With the damage fixed, additional maneuvers are taken to reconfigure the system back to its normal state at t_7.

13.2.2 Traditional Outage Restoration

Traditionally, the restoration problem is based on improving restored loads with a minimum number of switch maneuvers to achieve a new network configuration [21]. Initially, in non-automated and passive DS, customers would call the utility to notify of a power outage, which would be passed over to the operators. The DOC would dispatch crews to locate the damaged equipment or infrastructure to then make decisions on possible ways to operate MCSs and restore part of the customers while the physical damage is repaired. With the integration of automated devices, the DOC started to dispatch crews to maneuver MCSs while operators were remotely maneuvering TCSs with the intention of accelerating the restoration time and reducing its impact [22]. This approach has been used for many years by utilities due to the proper integration of remote and local maneuver actions.

The initial automated restoration solutions were primarily designed to support operators' decisions when dispatching crews for manual network maneuvers and repairs. One of the first studies on service restoration proposed an integrated restoration with power loss reduction [23]. In the sequence, other studies aimed to improve the operators' actions by having a guided step-by-step restoration process in the DOC [24, 25]. These early studies were mostly based on a set of logical conditions to locate the fault and restore customers based on the network topology and pre-fault configuration. In [26, 27], heuristic techniques have been used to improve the performance of DSs service restoration. Aiming to optimize different operational aspects, an integrated service restoration and capacitor control problem was formulated and proposed in [28].

Having in mind that the repair is constrained by the crews' location at the service dispatch moment, skills, equipment, available information to locate the damage, and the severity of the damage, it is imperative that the restoration process is as efficient and fast as possible because the repair may take hours to be completed. With advancements in communication capabilities, hardware processing, and techniques to restore customers while optimizing other operational aspects, the SH concept was introduced to DS. SH is the capability of DS to act quickly, automatically, and independently from operators to identify, locate, isolate, and restore outages [4]. Ideally, most of the customers would be re-energized within five minutes from the fault event to minimize the impact on reliability indices. With that, SH schemes use traditional restoration methods based on DNR but are designed and modeled aiming for fast response in online computations.

13.3 Self-Healing Concept

The SH concept integrates field measurement and network information into a predefined algorithm that automatically takes decisions during an outage event. The solution must incorporate techniques to identify an event, locate the fault, isolate the damage, and then restore as many customers as possible before any operator or linemen crew intervention. Moreover, some SH may also be able to automatically bring the network back to the pre-fault configuration as soon as the damage is fixed. All actions are based on switching maneuvers to precisely reconfigure the network and de-energize or re-energize specific areas.

To perform all these actions quickly, all network maneuvers must be based on TCSs. Otherwise, linemen crews' intervention would be required, which would considerably slow down the total time and would be out of the SH scope. Besides, it is important to highlight that DSs are not mounted with one TCS per segment, as it would be technically and financially infeasible. On the contrary, DSs are formed by cells, which are sets of interconnected nodes and branches bounded by a set of TCSs. Moreover, TCSs may or may not be protective devices. TCSs that are protective devices will also form protective zones, defined based on the utility's protective scheme. Based on these concepts, Figure 13.2 illustrates an outage scenario and the SH process.

Switches S1, S2, S4, and S5 are protective devices, while switch S3 is a motor-operated TCS, and the green color represents an open and red a closed position. In this example, the five switches define four cells (S1–S2, S2–S3, S3–S4, and S4–S5). In the first step, the network is in its pre-fault state and configuration, where all customers are being supplied. With a permanent fault in a lateral branch between S3 and S4, the short-circuit fault is supplied by the substation connected to the left, and the current passes through S1, S2, and S3. From the instantaneous and temporized

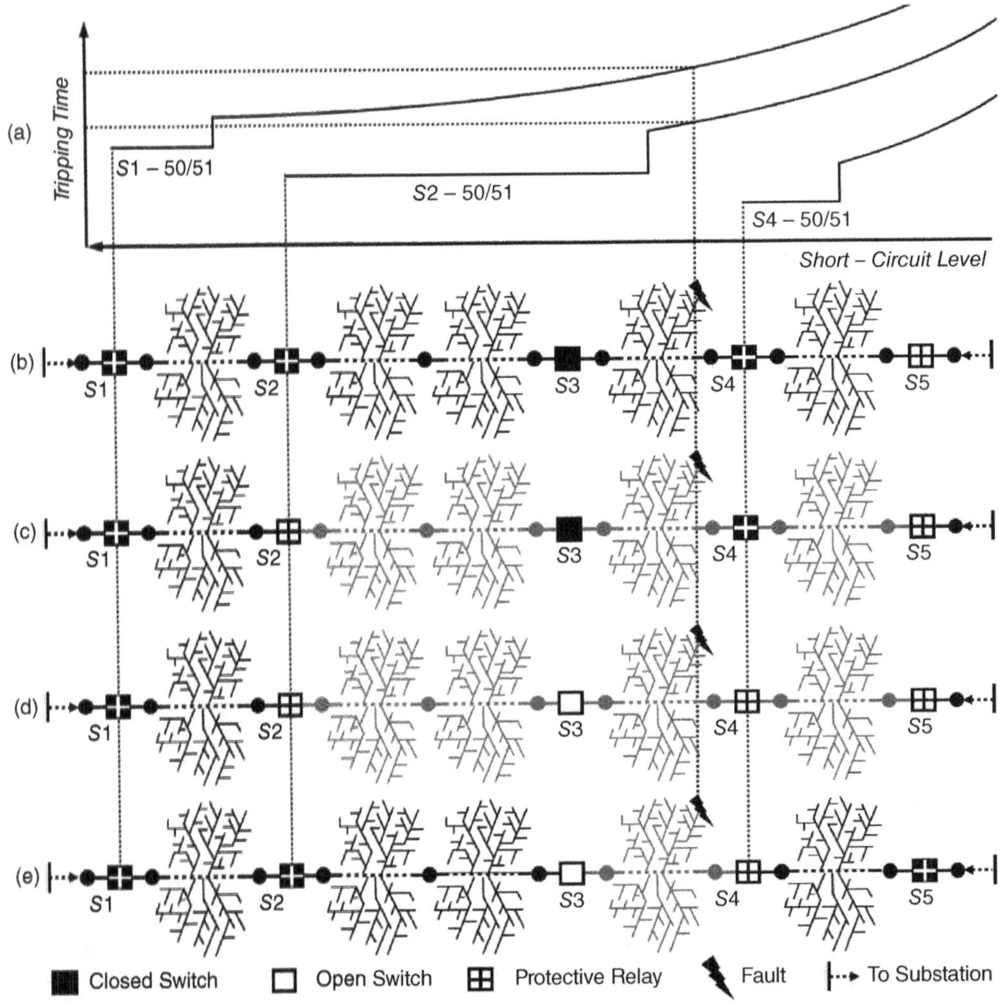

Figure 13.2 Conceptual restoration process: (a) over-current protection scheme, (b) step 1, (c) step 2, (d) step 3, and (e) step 4.

over-current protection coordination, American National Standards Institute (ANSI) elements 50 and 51, respectively, S2 should trip for the proposed fault. Assuming a proper protective scheme operation, S2 trips and locks out, as presented in step 2, de-energizing cells S2–S3, S3–S4, and S4–S5. Considering a proper SH solution, by understanding that S1, S2, and S3 measured fault currents and that S3 is not a protective device but is the most downstream TCS of the feeder, the isolation plan identifies that S3 and S4 can isolate the fault and that S2 could be reclosed. With the fault isolated at step 3, S2 can be reclosed to re-energize cell S2–S3. At the restoration analysis, S5 can be closed to re-energize the cell S4–S5 through the adjacent feeder at step 4. For instance, before transferring cell S4–S5 to the adjacent feeder, it is crucial to ensure that the load transfer will not originate unacceptable levels of loading and voltage to the system's backbone and laterals.

Depending on the communication structure, SH solutions can follow a distributed, centralized, or hierarchical topology [29]. The distributed SH concept is based on a set of logic according to each device's location and potential communication with neighboring devices. Envisioning the integration of fuzzy methods, [30] presented a distributed SH model that accounts for DERs through a multiagent topology. In [31], a fully distributed optimization problem for the SH solution is proposed. As a continuation, Shen et al. [29] formulated a distributed SH model to restore unbalanced distribution networks. However, distributed control schemes can only achieve suboptimal restoration solutions. Even though distributed SH solutions are faster, only centralized control techniques are able to provide global optimal solutions that respect operational conditions. The current trend of distribution utilities worldwide to deploy fast communication links between DOC and field equipment has encouraged the movement to centralized and robust SH solutions. Besides, centralized ADMS systems are not limited to hardware and processing capabilities as they are usually hosted on servers, and by hosting other ADMS tools together, they can use and exchange information with each other.

13.3.1 Protection Scheme

A protection scheme aims to ensure that the electrical system is maintained stable and reliable by isolating components that are under fault while leaving as much of the network as possible in operation. Power outages primarily originate from equipment or infrastructure damage, which creates short-circuits and high current levels. Even though many devices may operate during such events, only digital relays and IEDs are able to report measurements and status through communication protocols. Based on that, this chapter only focuses on digital relays as protective devices, but there are other elements responsible for protecting power systems, such as fuses.

Traditionally, SH schemes are not related to protection schemes. The protection scheme must properly identify and extinguish the fault quickly and precisely while de-energizing as few elements as possible. Independent of the coordination settings, fault type, or conditions, the SH scheme will take place after the protection schemes are finalized to locate and isolate the fault efficiently and then perform an automated restoration.

13.3.2 Fault Location and Isolation

It is common to have recloser devices as part of the DS protection scheme mounted with an Automatic-Reclosing protection function (ANIS 79). As soon as a fault happens in the system, reclosers may attempt to reclose to reestablish service in case of temporary faults. In this scenario, the relay will have a successful reclosing, and no further action is needed on the SH's behalf. However, in case of a permanent fault, relays may cycle as many times as they have been set to until one or more of them reach a lockout state, remaining in an open position. In an SH scheme, an outage is first identified through the lockout of one or multiple protective devices [32]. As it may take a couple of seconds for the information to reach the ADMS, it is a common practice to have a time delay from receiving the first lockout before starting any analysis.

With the lockout flag received and a temporization completed, the SH solution should first locate the fault, which can follow different approaches but is commonly designed based on field IEDs' measurements and statuses. TCSs can provide enough information to properly distinguish which one has measured the short-circuit current levels, while other devices, such as fault indicators (FI), can also provide auxiliary information to locate the fault more accurately. Assuming that new switches won't be installed in the network during the SH analysis period, sets of nodes and branches forming each cell and zone can be computed beforehand with graph theory techniques. Once the faulted node or branch is located, the faulted cell and, consequently, the TCSs bounding it can be identified. These TCSs must be opened and remain open until proper inspection is made and the physical damage is repaired.

13.3.3 Outage Restoration

Once the fault is properly located and isolated, the SH solution can continue to a restoration analysis. The restoration solution must ensure an operational condition that respects predefined constraints, such as nodal voltage and branch current levels. By reconfiguring the network, the power flow changes, and voltage drops or branch loading may increase to unacceptable ranges. Most of the distributed SH solutions are not able to account for these constraints, but centralized optimization models can be modeled in such a way that the final solution will ensure reliable electrical conditions.

DNR is a Mixed-Integer Programming (MIP) problem with unbounded constraints that forms a nonconvex Mixed-Integer Non-linear Programming (MINLP). The solution must satisfy multiple sets of linear and non-linear constraints based on power flow and operational conditions [33]. As a key difference between traditional DNR and SH problems, DNR operates in normal conditions, while SH must maintain de-energized faulted cell. Otherwise, the solution may try to close one or multiple of the isolation switches and, by this, re-energize the fault. To achieve this type of solution, additional constraints must be added to the DNR problem to enforce the isolation TCSs to an open state.

13.3.4 Switching Control Sequence

The switching control sequence (SCS) is a technique used to coordinate a sequential order of switching operations to achieve a new configuration from an initial state that respects constraints, such as not looping the network or closing it into the fault [34]. Due to the fastness of SH solutions, the SCS method must be efficiently designed to accurately coordinate maneuvers.

The SCS can be split into isolation control sequence (ICS) and restoration control sequence (RCS). The ICS is a set of maneuvers to open and close TCSs to isolate the fault, while the RCS is responsible for restoring the system. The benefit of differentiating the sets of actions is that the ICS happens before the RCS. Even more, the ICS actions can be properly set to happen in parallel to the restoration computation. In this scenario, as soon as the switches responsible for the isolation are identified, the optimization computation uses this information to optimize the post-isolation network reconfiguration while the ICS commands are sent and take place in parallel with the restoration computation. However, before performing the RCS, ensuring that the ICS actions have been completed is crucial.

13.3.5 Ongoing Challenges

Even with the extensive literature already developed by academia and expertise obtained by utilities' deployed solutions on the topic of DNR, there are still many challenges in designing efficient SH models that can be easily formulated and quickly computed without compromising power systems stability and reliability conditions. Some of the DSs' characteristics that bring challenge to the field of SH are:

- A large number of nodes.
- A limited number of real-time measurements.
- An asymmetrical network design with unbalanced loads.
- A large R/X ratio, particularly for underground cables.

Besides the intrinsic characteristics of DSs, there are other more complex challenges in performing SH. For instance, protection is one of the biggest challenges. Ideally, the closest protection device to the fault would be operating before any other device operates. However, due to the high loading currents and small short-circuit levels, the coordination between recloser-recloser and recloser-fuse becomes challenging in DSs, and the chances for miscoordinations increase. Besides that, depending on the levels of distributed energy resources (DERs) capacity and generation in the system, parts of the DS can be subjected to reverse power flow [35]. Considering that traditional DSs were designed for radial operation, a reverse power flow can significantly compromise control and protection schemes, increasing the chances of protection misbehaving. Whenever a miscoordination happens, a protective device that is not the closest one to the fault operates and de-energizes customers that are not part of the fault zone. With that, recent studies on SH have also been integrating miscoordination analysis into the isolation process so the protection misbehaving can be corrected before the restoration process starts [36]. Otherwise, customers that could be re-energized will be kept de-energized until the damage is repaired, or the fault may be wrongly located. Most importantly, the ongoing SH studies must account for how the DERs' integration may impact the outage identification, location, and isolation and how the restoration will account for potential DERs' reconnection after their point-of-interconnection is re-energized.

Besides, due to the uncertain time crews may take to repair the damage and the possibility of infeasible solutions, some other optimization models have proposed sectionalizing the distribution networks into islanded microgrids by managing DERs and loads [37–39]. Even though using distributed resources to form microgrids over the DS can improve reliability indices, its coordination and the definition of agents' roles are still big challenges that have slowed down this search topic. The asymmetrical nature of DS with single- and two-phase load and generation connections makes the microgrid formation and operation extremely difficult when, most of the time, a DNR-based SH approach would be able to address the outage.

13.4 Mathematical Formulation

To provide a perspective of time in reference to Figure 13.2, the variables and parameters are denoted based on "t_0" for a pre-fault, "t_1" for a post-fault, "t_2" for a post-isolation, and "t_3" for a post-restoration network configuration and conditions, or generically as "t." The SH problem starts with a post-fault network configuration and conditions. At this stage, all the protection scheme's elements and functions have already finalized their actions to extinguish the fault. From the final state of each protective device and with the assistance of other IEDs' information, the fault can be accurately located.

The restoration part of the SH problem can be approached on different fronts, but the presented SH model follows an optimal DNR approach and is modeled as a deterministic Mixed-Integer Conic Programming (MICP) Multi-Objective Optimization (MOO). Continuous parameters, such as power demand, are fixed and assumed to be known, but in real-world applications, they may be obtained from field IEDs and state estimation techniques. The MOO solution will define the current flow and nodal voltage, along with the status of each switch, which will form the optimal network operational condition. Having the nodal power demand as the main input parameter, the solution is constrained by equality and inequality relations that represent the network topology, power flow, and operation limitations according to each parameter and variable. The exact equations are linearized by techniques presented in the literature to provide a conic model that is fast and easily solvable by commercial solvers without compromising the solution's accuracy. However, before presenting the SH model, it is equally important to model the network elements properly and reliably. For the sake of simplicity, all variables, parameters, and formulations below are based on per-unit (p.u.) values.

13.4.1 Network Model

13.4.1.1 Line Model

DSs' line branches are commonly modeled as short or medium-length transmission lines. The admittance is usually pure capacitive, i.e., only compounded by susceptance. Even though the consideration of mutual capacitance doesn't significantly impact the DSs' power flow results, the shunt capacitance presents a significant contribution as well as series and mutual impedances. Based on that, the medium-length transmission line model does not consider mutual capacitance. If the total shunt admittance of the line, assumed as purely capacitive, is divided into two equal parts and placed at the sending and receiving end nodes, the circuit is called nominal-π [40]. Figure 13.3 illustrates a nominal-π network branch model and its electrical parameters.

As the shunt admittance of branch ij, $Y^{sh}_{ij,\phi}$, is assumed as purely capacitive in the nominal-π model, then $Y^{sh}_{ij,\phi} = jB^{sh}_{ij,\phi}$ and the shunt power is also strictly reactive, where $B^{sh}_{ij,\phi}$ is the total shunt susceptance. Being Γ_i the set of branches connected to node i, the total shunt and reactive power for a given node i, phase ϕ and time t, $Q^{sh}_{i,\phi,t}$, in a nominal-π transmission line model can be obtained by Eq. (13.1).

$$jQ^{sh}_{i,\phi,t} = |V_{i,\phi,t}|^2 \cdot \sum_{jl\in\Gamma_i} \left(\frac{Y^{sh}_{jl,\phi}}{2}\right)^* = -j|V_{i,\phi,t}|^2 \cdot \sum_{jl\in\Gamma_i} \frac{B^{sh}_{jl,\phi}}{2}$$

$$\forall i \in \Omega_N, \phi \in \Omega_\Phi \tag{13.1}$$

Where the superscript "*" indicates the conjugate operator, $V_{i,\phi,t}$ is the voltage phasor, Ω_N is the set of nodes and Ω_Φ is the set of phases $\{a,b,c\}$.

13.4.1.2 Load Model

Each node can be compounded by complex power consumption, $P^c_{i,\phi,t} + jQ^c_{i,\phi,t}$, and generation, $P^g_{i,\phi,t} + jQ^g_{i,\phi,t}$. In a polynomial load model, also known as ZIP, the load is modeled as a composition of three components: constant impedance, constant current, and constant power. The constant impedance term varies its demand proportionally to the square of the voltage magnitude, and the

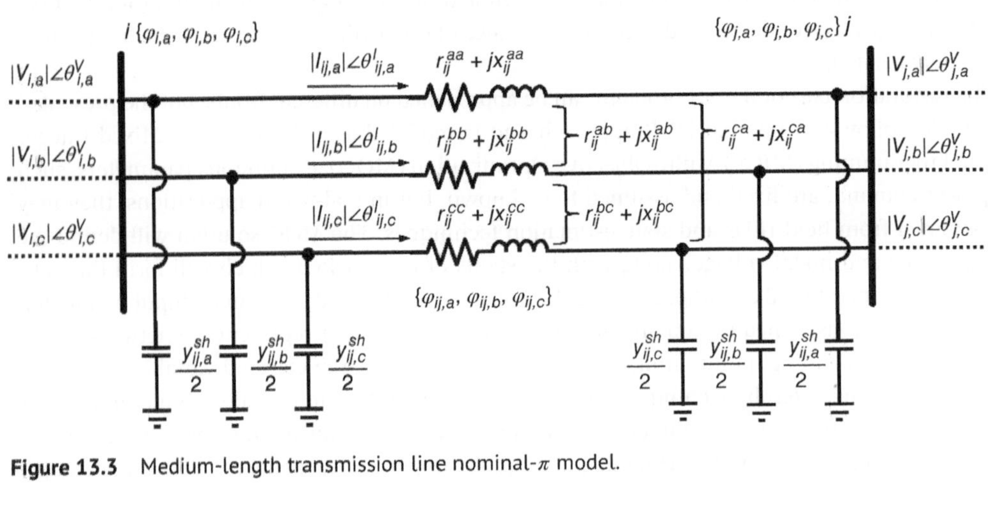

Figure 13.3 Medium-length transmission line nominal-π model.

constant current varies the power demand directly proportional to the voltage magnitude, while a constant power term doesn't vary its demand based on the voltage levels. Following a ZIP load model, the node's active and reactive power demand, $P^d_{i,\phi,t}$ and $Q^d_{i,\phi,t}$, respectively, is then obtained by Eq. (13.2).

$$\begin{cases} P^d_{i,\phi,t} = P^c_{i,\phi,t} \cdot \left[\alpha_i \cdot |V_{i,\phi,t}|^2 + \beta_i \cdot |V_{i,\phi,t}| + \gamma_i\right] - P^g_{i,\phi,t} \\ Q^d_{i,\phi,t} = Q^c_{i,\phi,t} \cdot \left[\alpha_i \cdot |V_{i,\phi,t}|^2 + \beta_i \cdot |V_{i,\phi,t}| + \gamma_i\right] - Q^g_{i,\phi,t} \\ \qquad \forall\, i \in \Omega_N, \phi \in \Omega_\Phi \end{cases} \tag{13.2}$$

The ZIP factors for constant impedance, current, and power load models are defined by α_i, β_i, and γ_i, respectively, which respect the condition of Eq. (13.3). For the present formulation, the factors are considered constant over time and the same across phases. It is also important to highlight that generation is considered to be constant power, and by this, it doesn't have ZIP factors.

$$\alpha_i + \beta_i + \gamma_i = 1 \;\forall\, i \in \Omega_N \tag{13.3}$$

13.4.2 Isolation Control Sequence

There are many techniques to perform fault location in DSs. However, the present model follows a centralized logic-based algorithm. This type of solution relies on field devices' flags to properly identify the faulted cell. Considering FIs, $FI_{ij,t}$, are available and can be obtained from all TCSs present in the network, this information can be used to locate the fault once it is identified and temporized.

Being, Π^s_j the set of branches connecting the substation node s to node j based on the pre-fault network configuration and Ω_S the set of branches with a TCS, Algorithm 13.1 presents the computation sequence to locate the closest TCS to the fault per feeder. The main outputs are two sets of switches, one to be open and another to be closed, $\Omega^{ICS}_{S,O}$, and $\Omega^{ICS}_{S,C}$, respectively, as part of the ICS. From lines 5 to 9, the algorithm identifies the furthest downstream TCS that reported a fault indication in reference to the pre-fault network configuration. The reminder of the algorithm locates and appends the furthest downstream switch to the $\Omega^{ICS}_{S,O}$. The other upstream switches in that feeder that felt the fault should be appended to closed. The SCS will then be responsible for checking the switch status and only sending the command in case the command is to change its current status.

By identifying the TCS responsible for the fault isolation and using the pre-fault network configuration, it is possible to locate the outage cell downstream of this device. The cells' outage status variable, $C^x_{o,t}$, is then updated to 0 in case of an outage; otherwise, it is updated to 1. The outage cell's boundary switches are appended to $\Omega^{ICS}_{S,O}$, so they can be opened to isolate the fault. With this information, the SCS computes the required steps to isolate the faulted cell. This process happens based on $\Omega^{ICS}_{S,O}$ and $\Omega^{ICS}_{S,C}$. The SCS initially opens all switches in the $\Omega^{ICS}_{S,O}$, and monitors their state until all operations are completed. As soon as these TCSs are in an open state, the system sends closing commands to TCSs in $\Omega^{ICS}_{S,C}$.

13.4.3 Optimal Restoration Model

Exploring feasible solutions and finding the optimal one in typically large and nonconvex network models is challenging and may require substantial time and processing capabilities [41]. Even though the power flow constraints to define the network topology in a DNR problem are non-linear, it is possible to simplify and linearize the model without compromising its solution

Algorithm 13.1 Fault Location Analysis

Input $FI_{ij,t1}$
Output $\Omega^{ICS}_{S,O}$, $\Omega^{ICS}_{S,C}$
Procedure
$\Omega^{ICS}_{S,O} = []$
$\Omega^{ICS}_{S,C} = []$
$var\_length = 0$
$var\_branch = 0$
for ij **in** Ω_S **do**
  if $FI_{ij,t1}$ **then**
    if $|\Pi^s_i| > var\_length$ **then**
      $var\_length = |\Pi^s_i|$
      $var\_branch = ij$
for ij **in** Ω_S **do**
  if $FI_{ij,t1}$ **then**
    if $ij = var\_branch$ **then**
      append ij to $\Omega^{ICS}_{S,O}$
    else
      append ij to $\Omega^{ICS}_{S,C}$

quality. Besides requiring extensive computational time and capabilities, non-linear optimization models are complex and not easily solvable by commercial software. Linearization must trade the model accuracy through simplifications to obtain fastness without impacting the solution's feasibility and quality.

The restoration problem aims to obtain the optimum switches' statuses to define a new network configuration that can minimize the total unserved kW and can be achieved with the minimum amount of switching maneuvers. The solution must ensure radiality along with post-reconfiguration current flow and nodal voltage levels within the operational limits. The following formulation presents a MICP MOO that is extended from traditional DNR models into an SH model [36]. Linearization tools and techniques are used to efficiently improve the optimization time response without compromising the quality of the decisions.

13.4.3.1 Objective

The optimization objective is responsible for defining the solution's goal and which variables will be minimized. As presented, the restoration problem traditionally aims to minimize the amount of de-energized customers with the minimum quantity of switching maneuvers, which forms a MOO. Being $y^b_{ij,t}$ the branch ij and $y^n_{i,t}$ the node i energization statuses, and $P^d_{i,\phi,t}$ active power demand, Eq. (13.4) presents the optimization objective.

$$\min \left[\sum_{i \in \Omega_N} \sum_{\phi \in \Omega_\Phi} w_{1,i} \cdot P^d_{i,\phi,t0} \cdot \left(1 - y^n_{i,t3}\right) + w_2 \cdot \sum_{ij \in \Omega_{B,S}} \left| y^b_{ij,t3} - y^b_{ij,t2} \right| \right] \tag{13.4}$$

To minimize both the amount of unserved power and the quantity of switching maneuvers, $w_{1,i}$ and w_2 weights are used to normalize the two objectives into costs. $w_{1,i}$ is the value of lost load (VOLL) of node i, which depends on several social-economical aspects of the region and varies according to the type of customers connected to node i. In [42], the VOLL was presented for different commercial and industrial (C&I) and residential customers during the winter of 2023, shown below:

- Medium and large C&I customers: \$17.2/kWh
- Small C&I customers: \$132.3/kWh
- Residential customers: \$0.9/kWh

w_2 is the cost per switching operation, which is usually not provided by the literature but can be calculated based on the device's initial cost, planning horizon, and endurance. A typical TCS has an average endurance of 10,000 operations, an initial cost of approximately \$15,000, and an annual maintenance cost of 2% of the annualized investment. Considering 15 years as a planning horizon and an inflation rate of 3%, the cost per maneuver is around \$1.86.

Even though the cost per switching operation is significantly smaller than the VOLL, during a restoration, the main goal is to reestablish power as fast as possible. Hence, this term of the objective function is crucial to ensure an optimal solution close to the pre-fault network configuration, and by this, that does not require many maneuvers to be achieved.

13.4.3.2 Cell Constraints

Distribution networks are organized in cells, which are groups of nodes and branches bounded by TCS. This concept defines that the energization or de-energization of a cell implies the energization or de-energization of all nodes and branches within the cell. Considering that outage detection and isolation happen before to the restoration computation, it is crucial to constrain the restoration model to de-energize branches and nodes within the fault cell. Being $C^{e}_{o,t}$ the binary variable representing the energization status of cell o and $C^{\alpha}_{o,t}$ is the outage status parameter of cell o, the outage constraint is defined by Eq. (13.5).

$$0 \leq C^{e}_{o,t3} \leq C^{\alpha}_{o,t2} \ \forall \, o \in \Omega_C \tag{13.5}$$

In case an outage is identified in cell o, $C^{\alpha}_{o,t}$ will be 0, and then the cell status $C^{e}_{o,t}$ will be forced to become 0. In case cell o doesn't contain an outage, its status will be defined by the optimization algorithm as energized ($C^{e}_{o,t} = 1$), or de-energized ($C^{e}_{o,t} = 0$). Being $\Omega^{o}_{C,N}$ the set of nodes within cell o, and $y^{n}_{i,t}$ the binary energization status of node i, the nodes' statuses based on the cell energization are defined by constraint in Eq. (13.6) [43]. Similarly, being $\Omega^{o}_{C,NS}$ the set of non-switchable branches within cell o and $y^{b}_{ij,t}$ the energization status of branch ij, the branches' energization statuses are defined by Eq. (13.7).

$$y^{n}_{i,t3} = C^{e}_{o,t3} \ \forall \, i \in \Omega^{o}_{C,N}, o \in \Omega_C \tag{13.6}$$

$$y^{b}_{ij,t3} = C^{e}_{o,t3} \ \forall \, ij \in \Omega^{o}_{C,NS}, o \in \Omega_C \tag{13.7}$$

Being $\Omega^{o}_{C,S}$ the set of switchable branches bounding cell o, the status of these switches depends on the status of the cell's energization status, as presented by Eq. (13.8).

$$0 \leq y^{b}_{ij,t3} \leq C^{e}_{o,t3} \ \forall \, ij \in \Omega^{o}_{C,S}, o \in \Omega_C \tag{13.8}$$

With this relation, the TCSs-bounding cell o will be forced to zero (de-energized) whenever there is an outage within the cell, as per Eq. (13.5). Otherwise, the TCSs' statuses will be defined by the optimization solution as closed or open.

13.4.3.3 Operational Constraints

Whenever a node is energized, its voltage level must be maintained within acceptable limits. Similarly, branch currents must be maintained under overcurrent protection pickups and cabling thermal limits. As the nodal voltage magnitude is a variable, its squared value becomes a non-linearity in the system. Based on that, a new variable, $U_{i,\phi,t}$, can be used to linearize the model. Being $U_{i,\phi,t} = |V_{i,\phi,t}|^2$, the results for voltage magnitude can be obtained by simply getting the squared root of $U_{i,\phi,t}$.

Being V^m and V^M the minimum and maximum operational voltage limits in p.u., respectively, Eq. (13.9) ensures that voltage levels of energized nodes are within these boundaries. According to ANSI, V^m and V^M are defined as 0.95 and 1.05 p.u., respectively.

$$y^n_{i,t3} \cdot (V^m)^2 \leq U_{i,\phi,t3} \leq y^n_{i,t3} \cdot (V^M)^2 \ \forall\, i \in \Omega_N, \phi \in \Omega_\Phi \tag{13.9}$$

Being $J_{ij,\phi,t} = |I_{ij,\phi,t}|^2$, then the maximum branch current is limited by Eq. (13.10), where I^M_{ij} is the maximum allowed current flow in branch ij and $\varphi_{ij,\phi}$ is the phase indicator of branch ij and phase ϕ.

$$0 \leq J_{ij,\phi,t3} \leq \left(I^M_{ij}\right)^2 \cdot y^b_{ij,t3} \cdot \varphi_{ij,\phi} \ \forall\, ij \in \Omega_B, \phi \in \Omega_\Phi \tag{13.10}$$

The apparent power, $S_{ij,\phi,t}$, equation compounds a circle equation, $(S_{ij,\phi,t})^2 = U_{i,\phi,t} \cdot J_{ij,\phi,t} = \left(P^b_{ij,\phi,t}\right)^2 + \left(Q^b_{ij,\phi,t}\right)^2$, where $P^b_{ij,\phi,t}$ and $Q^b_{ij,\phi,t}$ are the branch ij active and reactive power flows, respectively, which is a non-linear constraint. A linearization method uses a conic approach, defined by Eq. (13.11), which relates the voltage and current with active and reactive power flows [44].

$$U_{i,\phi,t3} \cdot J_{ij,\phi,t3} \geq \left(P^b_{ij,\phi,t3}\right)^2 + \left(Q^b_{ij,\phi,t3}\right)^2 \ \forall\, ij \in \Omega_B, \phi \in \Omega_\Phi \tag{13.11}$$

13.4.3.4 Power Flow Constraints

Kirchhoff's current law must be considered to ensure nodal power balance. The law defines that the total power leaving the node must equal the total power arriving at the node, accounting for branch power flows and nodal power demand. As the branch power flows are variables resulting from the optimization, the nodal power demand is a constant parameter defining the optimization solution.

By decoupling active and reactive power, Eqs. (13.12) and (13.13) present the nodal power balance equations for active and reactive power, respectively, where power demand is defined as negative for generation and positive for consumption. Also, being $\boldsymbol{J}_{ij,t} = [J_{ij,a,t}\ J_{ij,b,t}\ J_{ij,c,t}]^T$, where the superscript "T" indicates the transpose operator, the power losses between each adjacent branch are accounted.

$$\sum_{ij \in \Omega_B} P^b_{ij,\phi,t3} = \sum_{jl \in \Omega_B} \left[P^b_{jl,\phi,t3} + (\boldsymbol{R}_{jl} * \boldsymbol{J}_{jl,t3})_\phi\right] + P^d_{j,\phi,t0} \cdot y^n_{j,t3}$$
$$\forall\, j \in \Omega_N, ij, jl \in \Omega_B, \phi \in \Omega_\Phi \tag{13.12}$$

$$\sum_{ij \in \Omega_B} Q^b_{ij,\phi,t3} = \sum_{jl \in \Omega_B} \left[Q^b_{jl,\phi,t3} + (\boldsymbol{X}_{jl} * \boldsymbol{J}_{jl,t3})_\phi\right] + Q^d_{j,\phi,t0} \cdot y^n_{j,t3} - U_{j,\phi,t3} \cdot \sum_{jl \in \Gamma_j} \frac{B^{sh}_{jl,\phi}}{2}$$
$$\forall\, j \in \Omega_N, ij, jl \in \Omega_B, \phi \in \Omega_\Phi \tag{13.13}$$

In this proposed model, the load's demand dependency on the voltage level is neglected, and the demand is considered constant. As presented in Eq. (13.2), the constant current and constant impedance ZIP load models depend on the voltage magnitude and its squared value, respectively.

By using variable substitution for the voltage, $U_{i,\phi,t} = |V_{i,\phi,t}|^2$, the square root of $U_{i,\phi,t}$ would be a non-linearity in the system. The literature has proposed DNR optimization models to account for ZIP load models [45].

Kirchhoff's voltage law is another constraint that must be considered in power systems optimization. Being, $\boldsymbol{P}_{ij,t} = \left[P^b_{ij,a,t} \; P^b_{ij,b,t} \; P^b_{ij,c,t}\right]^T$ and $\boldsymbol{Q}_{ij,t} = \left[Q^b_{ij,a,t} \; Q^b_{ij,b,t} \; Q^b_{ij,c,t}\right]^T$, Eqs. (13.14) and (13.15) define the voltage drop between two nodes based on the branch characteristics and the active and reactive power flow.

$$U_{i,\phi,t3} - U_{j,\phi,t3} \geq 2 \cdot (\widehat{\boldsymbol{R}}_{ij} * \boldsymbol{P}_{ij,t3} + \widehat{\boldsymbol{X}}_{ij} * \boldsymbol{Q}_{ij,t3})_\phi - \left(2 - y^b_{ij,t3} - \varphi_{ij,\phi}\right) \cdot M$$
$$\forall\, i \neq j \in \Omega_N, ij \in \Omega_B, \phi \in \Omega_\Phi \tag{13.14}$$

$$U_{i,\phi,t3} - U_{j,\phi,t3} \leq 2 \cdot (\widehat{\boldsymbol{R}}_{ij} * \boldsymbol{P}_{ij,t3} + \widehat{\boldsymbol{X}}_{ij} * \boldsymbol{Q}_{ij,t3})_\phi + \left(2 - y^b_{ij,t3} - \varphi_{ij,\phi}\right) \cdot M$$
$$\forall\, i \neq j \in \Omega_N, ij \in \Omega_B, \phi \in \Omega_\Phi \tag{13.15}$$

To identify a satisfactory solution, the Big-M concept is utilized to relax the constraints (13.14) and (13.15). If branch ij is energized ($y^b_{ij,t} = 1$), and there is an actual conductor for phase ϕ ($\varphi_{ij,\phi} = 1$), then the M term will be disregarded, where that term becomes zero. On the other hand, when the branch is de-energized ($y^b_{ij,t} = 0$) or there is no phase conductor ($\varphi_{ij,\phi} = 0$), $P^b_{ij,\phi,t}$ and $Q^b_{ij,\phi,t}$ will be zero, according to (13.10) and (13.11), and then Kirchhoff's law is not applicable to the branch. Besides that, in [46], a Kirchhoff voltage law formulation for a three-phase unbalanced DSs optimization problem was presented. The study used equivalent branch matrices of resistance, $\widehat{\boldsymbol{R}}_{ij}$, and reactance, $\widehat{\boldsymbol{X}}_{ij}$, to efficiently decouple voltage phase and magnitude. In this approach, it is assumed that voltage magnitudes are similar between phases, and the phase unbalancing is not too severe, so the voltages are nearly balanced across the phases. Based on these conditions, $\widehat{\boldsymbol{R}}_{ij}$ and $\widehat{\boldsymbol{X}}_{ij}$ can be obtained by Eqs. (13.16) and (13.17), respectively.

$$\widehat{\boldsymbol{R}}_{ij} = \mathbb{R}\{\boldsymbol{\Lambda} * \boldsymbol{\Lambda}^H\} \odot \boldsymbol{R}_{ij} + \mathbb{I}\{\boldsymbol{\Lambda} * \boldsymbol{\Lambda}^H\} \odot \boldsymbol{X}_{ij} \tag{13.16}$$

$$\widehat{\boldsymbol{X}}_{ij} = \mathbb{R}\{\boldsymbol{\Lambda} * \boldsymbol{\Lambda}^H\} \odot \boldsymbol{X}_{ij} - \mathbb{I}\{\boldsymbol{\Lambda} * \boldsymbol{\Lambda}^H\} \odot \boldsymbol{R}_{ij} \tag{13.17}$$

Where the superscript "H" is the Hamiltonian operator, "$\odot$" denotes the Hadamard product (dot product) operator, and $\boldsymbol{\Lambda}$ is the vector of relative phase unbalance per phase, defined as $\boldsymbol{\Lambda} = \left[1 \; e^{-j \cdot 2 \cdot \pi/3} \; e^{j \cdot 2 \cdot \pi/3}\right]^T$.

13.4.3.5 Radiality Constraints

Formulating radiality constraints in DSs can be challenging, as the network can have a highly meshed topology and connections to different substations. By defining two binary variables for each branch indicating forward, $\beta^+_{ij,t}$, and reverse, $\beta^-_{ij,t}$, power flow directions, it is possible to ensure a network that corresponds to a spanning tree regardless of the power flow direction [47]. Equation (13.18) defines that each couple of nodes defining a branch connection should respect the parent-child relation if they are energized.

$$\beta^+_{ij,t3} + \beta^-_{ij,t3} = y^b_{ij,t3} \; \forall\, ij \in \Omega_B \tag{13.18}$$

In case of node i is the parent of node j ($\beta^+_{ij,t} = 1$ and $\beta^-_{ij,t} = 0$) or node j is the parent of node i ($\beta^+_{ij,t} = 0$ and $\beta^-_{ij,t} = 1$), then the branch should be considered as energized ($y^b_{i,t} = 1$), otherwise is de-energized ($y^b_{ij,t} = 0 \therefore \beta^+_{ij,t} = 0$ and $\beta^-_{ij,t} = 0$).

Algorithm 13.2 Restoration Control Sequence

```
Input y^b_{ij,t3}, y^b_{ij,t2}
Output Ω^{RCS}_{S,O}, Ω^{RCS}_{S,C}
Procedure
Ω^{RCS}_{S,O} = []
Ω^{RCS}_{S,C} = []
for ij in Ω_S do
    if y^b_{ij,t3} < y^b_{ij,t2} then
        ij appended to Ω^{RCS}_{S,O}
    else if y^b_{ij,t3} > y^b_{ij,t2} then
        ij appended to Ω^{RCS}_{S,C}
```

Equation (13.19) imposes that every energized node has only one parent node from which it receives power. At the same time, each node may have none, one, or multiple child nodes. Analogically, de-energized nodes do not have any parent or child nodes.

$$\sum_{ij,ji\in\Gamma_i} \left(\beta^{+}_{ji,t3} + \beta^{-}_{ij,t3}\right) = y^{n}_{i,t3} \; \forall\, i \in \Omega_N \tag{13.19}$$

It is important to highlight that Eqs. (13.18) and (13.19) ensure radiality in small and large networks that may have connectivity to one or several substations while respecting the de-energized nodes by the solution. However, nodes' connectivity is imposed by the power flow equations, Eqs. (13.12)–(13.15) [48].

13.4.4 Restoration Control Sequence

The restoration process provides the optimal state for each branch, $y^b_{ij,t3}$, which defines the optimal restoration network configuration. Being $y^b_{ij,t2}$ the branches' statuses after the isolation, a set of switches that should be open and closed to perform the restoration, $\Omega^{RCS}_{S,O}$ and $\Omega^{RCS}_{S,C}$, respectively, RCS is computed by Algorithm 13.2.

The algorithm logic will compare the current and previous status of each branch that has a TCS. In case the previous state is closed, and the current one is open, an open command is appended to $\Omega^{RCS}_{S,O}$. Similarly, in case the previous state was open, and the current one is closed, then a close command is emitted by appending the switch to $\Omega^{RCS}_{S,C}$. With the RCS computed, devices in $\Omega^{RCS}_{S,O}$ will first be opened, and then in $\Omega^{RCS}_{S,C}$ will be closed. This stage-by-stage SCS approach is capable of avoiding operation scenarios that may cause system instability and/or cascade tripping, such as network looping, closing into the fault, and re-energizing customers that are not part of the optimal restoration solution.

13.5 Case Studies

With the lack of available solutions flexible enough for unbalanced distribution networks and the need to develop fast and accurate ADMS tools, the presented SH solution was developed in Python. The entire method is hardcoded, and the restoration optimization uses Pyomo as the optimization modeler and CPLEX as the solver. The code is hosted in an Intel NUC i7-1165G7 4-Core, 16GB RAM, 512GB PCIe SSD, with Ubuntu Linux as OS.

In the sequence, the proposed test system, assumptions, and considerations are shown, along with the results obtained from offline simulations for each test case.

13.5.1 Considerations

The 123 nodes test system was first introduced by [49]. This circuit is characterized by overhead and underground lines, asymmetrical branches, unbalanced loads, and 12 protective relays. The base power is 5 MVA, and the nominal voltage is 4.16 kV. Even though the system is based on one feeder, there is one main substation and four open connections. These characteristics have made this test system a reference network for DNR and power outage restoration methods validation.

A modified version of the 123 nodes test system is adopted for the present analysis. As the number of switches and their location in the original system don't allow a suitable SH process, and the primary goal of using this system is to validate the proposed solution in an asymmetrical and unbalanced network, other load breakers have been inserted into the initial network. Figure 13.4 shows the modified 123 nodes DS with the additional breakers and simplifications, and the nodes' renumbering.

With 14 switches, five normally open and nine normally closed, the system increases from 123 to 130 nodes due to the addition of pseudo nodes between the actual nodes and line branches to fit the additional switches. Of the switches, 12 are protective relays (based on original switches), and 4 are load breakers (additional switches). Table 13.1 shows the switches' location and type, where "NO" stands for normally open and "NC" for normally closed states.

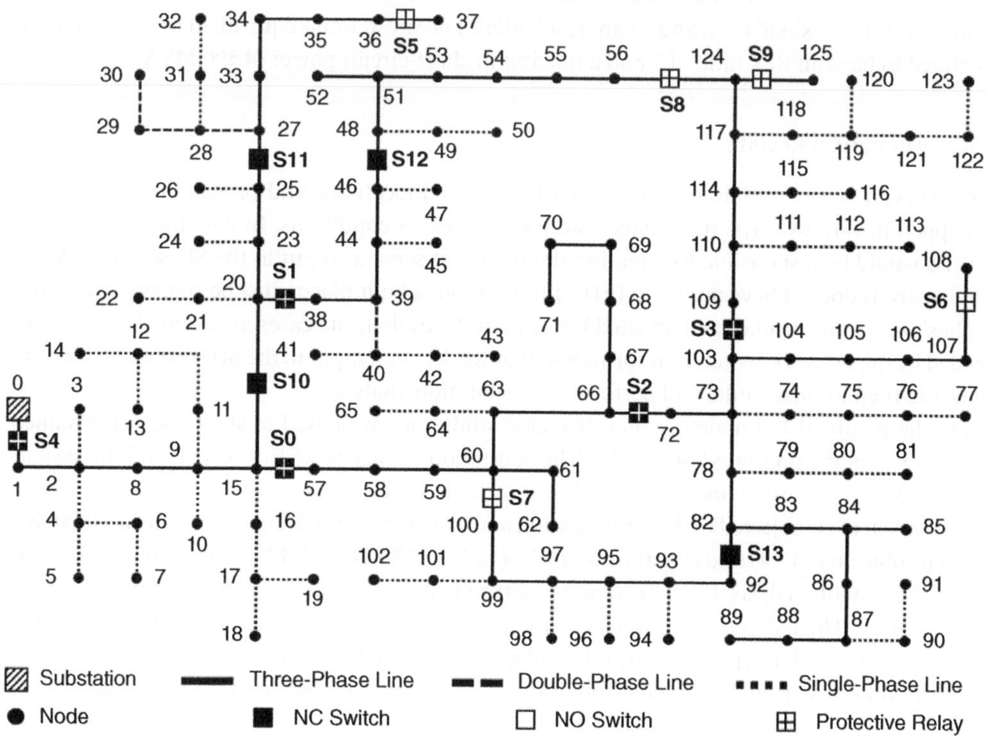

Figure 13.4 Modified 123 nodes distribution system schematic.

Table 13.1 Modified 123 nodes switches' information.

Switch	From node	To node	Type	State
S0	15	57	Recloser	NC
S1	20	38	Recloser	NC
S2	66	72	Recloser	NC
S3	103	109	Recloser	NC
S4	0	1	Recloser	NC
S5	36	37	Recloser	NO
S6	107	108	Recloser	NO
S7	60	100	Recloser	NO
S8	56	124	Recloser	NO
S9	124	125	Recloser	NO
S10	15	126	Breaker	NC
S11	25	127	Breaker	NC
S12	46	128	Breaker	NC
S13	82	129	Breaker	NC

Of the 130 nodes, 85 have load connections. As the type of customer per node is not defined by the literature, it can be assumed based on [42]. As presented in Table 13.2, five nodes are large and medium C&I, 47 are small C&I, and 33 are residential. The substation's equivalent upstream circuit is assumed to have an R/X ratio of 7 and a maximum short-circuit power of 500 MVA.

13.5.2 Numerical Results

As any DS can be susceptible to different fault types and locations, and protection performance can happen in various ways, the outage can have numerous conditions. In a real-world scenario, the IEDs would be responsible for creating the flags that serve as input to the SH solution. As an offline analysis doesn't have access to IEDs' information, a fault place must be suggested, and the switches' flags must be manually created based on it. With that, the cases are defined by the flags provided by the network switches, information that works as an input to the SH solution to identify events and then localize faults and perform the restoration analysis.

From the modified 123 nodes system, two case studies are proposed to show the performance and efficiency of the presented SH model. The substation voltage is set to 1.05 p.u. and the system loading to 50%. The cases' conditions are shown in Table 13.3.

The test system is subjected to the fault location and isolation and SCS logics, and the MICP MOO, formed by objective (13.4) and constraints (13.5)–(13.15), (13.18), and (13.19), presented above. To support the benefit analysis, the percentage of served load (PSL) and cost of unserved load (CUL) factors are used. The PSL represents the percentage of the system's load that is energized over time. On the other hand, CUL represents the total outage cost related to the de-energized loads and their VOLL over time. The results obtained from offline simulations are shown in the sequence.

Table 13.2 Modified 123 nodes customers' information.

Node	Customer type	VOLL [$/kWh]	Node	Customer type	VOLL [$/kWh]
2	Small C&I	132.3	58	Residential	0.9
3	Small C&I	132.3	59	Residential	0.9
5	Small C&I	132.3	61	Residential	0.9
6	Residential	0.9	62	Small C&I	132.3
7	Small C&I	132.3	64	Residential	0.9
8	Residential	0.9	65	Small C&I	132.3
10	Small C&I	132.3	66	Small C&I	132.3
11	Small C&I	132.3	67	Residential	0.9
13	Small C&I	132.3	68	Small C&I	132.3
14	Small C&I	132.3	69	Medium and large C&I	17.2
16	Small C&I	132.3	70	Medium and large C&I	17.2
18	Small C&I	132.3	71	Small C&I	132.3
19	Small C&I	132.3	74	Residential	0.9
21	Residential	0.9	75	Residential	0.9
22	Small C&I	132.3	76	Residential	0.9
24	Small C&I	132.3	77	Small C&I	132.3
26	Small C&I	132.3	79	Residential	0.9
30	Small C&I	132.3	80	Residential	0.9
31	Residential	0.9	81	Small C&I	132.3
32	Small C&I	132.3	82	Medium and large C&I	17.2
33	Residential	0.9	83	Residential	0.9
34	Residential	0.9	85	Small C&I	132.3
35	Residential	0.9	86	Residential	0.9
39	Small C&I	132.3	88	Residential	0.9
41	Small C&I	132.3	89	Small C&I	132.3
42	Residential	0.9	90	Residential	0.9
43	Small C&I	132.3	91	Small C&I	132.3
45	Small C&I	132.3	92	Residential	0.9
46	Small C&I	132.3	93	Small C&I	132.3
47	Small C&I	132.3	94	Small C&I	132.3
49	Small C&I	132.3	96	Small C&I	132.3
50	Small C&I	132.3	98	Small C&I	132.3
51	Medium and large C&I	17.2	100	Small C&I	132.3
52	Small C&I	132.3	101	Residential	0.9
53	Medium and large C&I	17.2	102	Small C&I	132.3
54	Residential	0.9	104	Residential	0.9
55	Residential	0.9	105	Residential	0.9

(continued)

Table 13.2 (Continued)

Node	Customer type	VOLL [$/kWh]	Node	Customer type	VOLL [$/kWh]
106	Residential	0.9	118	Residential	0.9
111	Residential	0.9	120	Small C&I	132.3
112	Small C&I	132.3	121	Residential	0.9
113	Small C&I	132.3	122	Small C&I	132.3
115	Residential	0.9	123	Small C&I	132.3
116	Small C&I	132.3	—	—	—

Table 13.3 Test cases conditions.

	Flags		
Case	Trip	Lockout	Fault Indicator
I	S2	S2	S0, S2, S4
II	S4	S4	S4, S10

13.5.2.1 Case I

Figure 13.5 shows the TCSs' statuses over time for this case scenario in "ss.ms" format. The solution first identifies an outage event through the lockout of S2. By receiving this flag, the SH algorithm performs a ten-second temporization to allow enough time for all other potential protection functions to operate before initiating any computation. After the temporization, the FI flags from S0, S2, and S4, as well as the trip of S2, are identified. From this information, it is understood that only S2 operated for the event, and the fault can be located by analyzing the FIs. The solution efficiently locates the fault downstream S2. This cell is isolated by TCSs S2, S3, S6, and S13. As S2 and S6 are already in an open position, the algorithm issues an opening command to S3 and S13 as part of the ICS.

Knowing that the fault is within S2, S3, S6, and S13 and that these switches must be maintained open, the algorithm performs the restoration optimization analysis constrained by keeping this cell

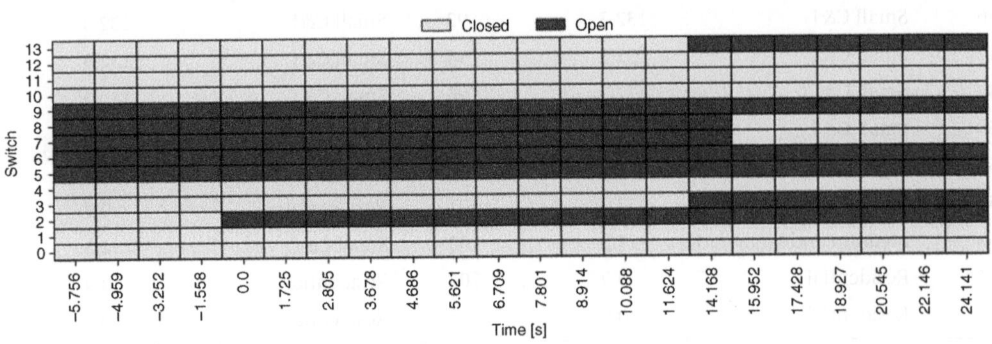

Figure 13.5 Case I switching status over time.

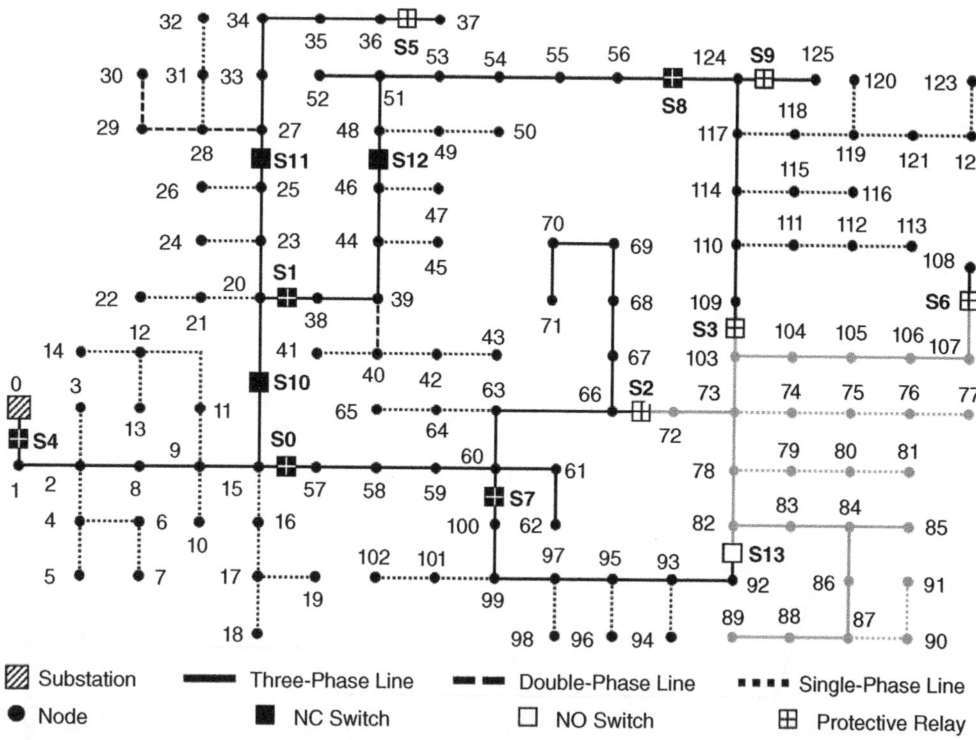

Figure 13.6 Case I network configuration after restoration.

de-energized and finding a solution that maintains these switches open. The result decides to close S7 and S8 to re-energize customers under outage that are outside the faulted cell. Figure 13.6 shows the network configuration after the SH actions were completed.

The final network configuration was achieved after 15.952 seconds from the lockout identification, and the system presented a minimum voltage level of 1.01 p.u. and maximum branch current loading of 49.92%.

From an impact perspective, Figure 13.7 shows the outage breakdown. The event de-energized 40.43% of the system's load, and the restoration was able to restore 40.47% of the impacted customers, leaving 1.29 MW of load de-energized.

On the economic side, Figure 13.8 shows the CUL against PSL over time in "ss.ms" format. The outage de-energized 2.16 MW of load, which would have implied a total event cost of $126,174.00 if the entire outage had been sustained. However, as the solution was able to restore 0.87 MW of customers's load within a couple of seconds, the outage cost was reduced to $42,609.00, where 0.87 MW faced a temporary outage, while 1.29 MW faced a sustained outage.

13.5.2.2 Case II

Figure 13.9 shows the TCSs' statuses over time for this case scenario in "ss.ms" format. The solution first identifies an outage event through the lockout of S4, the substation breaker, which de-energizes the entire feeder. After performing a 10-second temporization, the algorithm identifies the FI flags from S4 and S10, as well as the trip of S4. The solution efficiently locates the fault downstream S10. Even though the fault is downstream S10, S4 operated properly to protect the system from the fault, as S10 doesn't have protective capabilities. However, being S10 a TCS, it can be commanded and

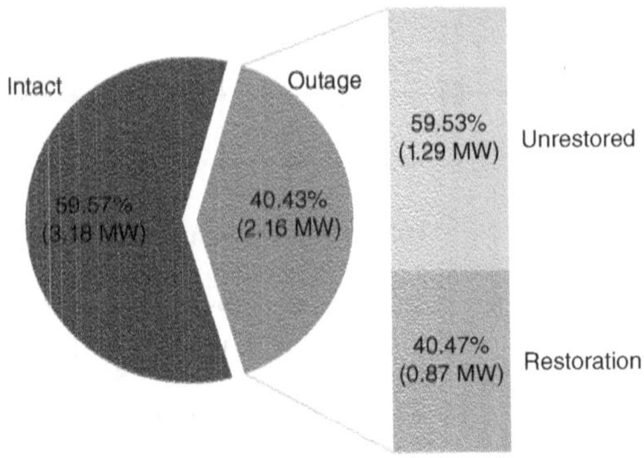

Figure 13.7 Case I outage restoration breakdown.

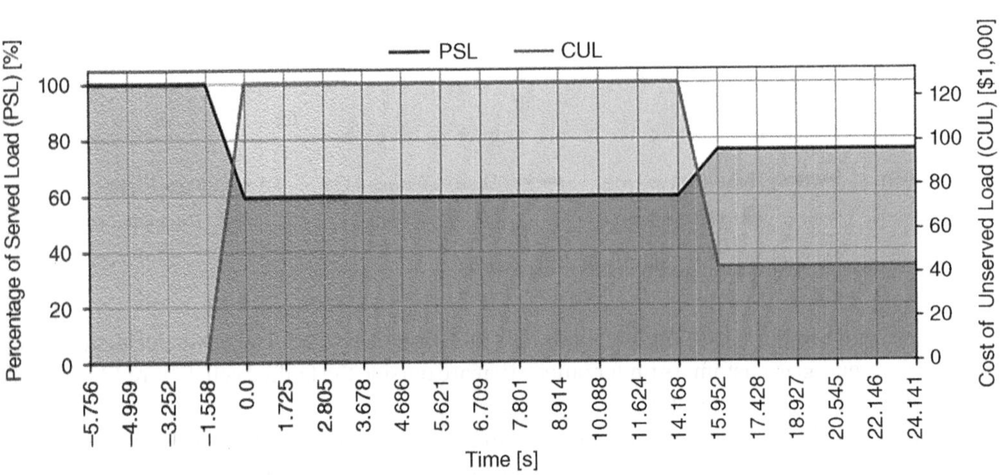

Figure 13.8 Case I restoration benefit over time.

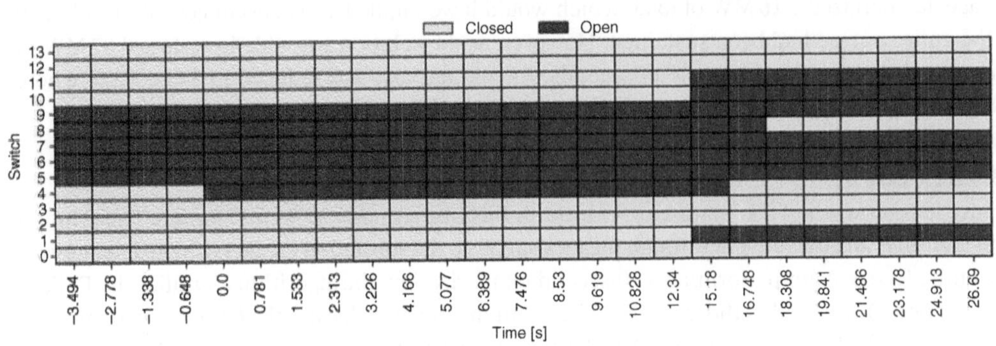

Figure 13.9 Case II switching status over time.

used for the SH's isolation process. With that, the solution identifies that the faulted cell is isolated by TCSs S1, S10, and S11. As all three switches are in closed position, the solution issues an opening command to each of them as part of the ICS. Besides, the solution issues a closing command to S4 as soon as S1, S10, and S11 are in an open position. This action allows the re-energization of the majority of the network's customers. However, S4 can only be reclosed after the faulted cell is successfully isolated. Otherwise, cascade events may be created.

Knowing that the fault is within S1, S10, and S11 and that these switches must be maintained open and that S4 was closed back, the solution performs the restoration optimization analysis constrained by keeping this cell de-energized. The solution decides to close S8 to re-energize customers under outage that are outside the faulted cell. Figure 13.10 shows the network configuration after the SH actions were completed.

The final network configuration was achieved after 18.308 seconds from the lockout identification, and the system presented a minimum voltage level of 1.00 p.u. and maximum branch current loading of 55.04%.

From an impact perspective, Figure 13.11 shows the outage breakdown. The event de-energized 100% of the system's load with the substation breaker trip. The proper isolation actions re-energized 67.81% of the customers, while the restoration re-energized an additional 21.85%, totalizing a restoration of 89.65% of the system.

On the economic side, Figure 13.12 shows the CUL against PSL over time in "ss.ms" format. The outage de-energized 5.34 MW of load, which would have implied a total event cost of $365,888.25 if the outage had been entirely sustained. However, as the solution restored 4.79 MW of customers's load within a couple of seconds, the outage cost was reduced to $35,964.00, where 4.79 MW faced a temporary outage, while 0.55 MW faced a sustained outage.

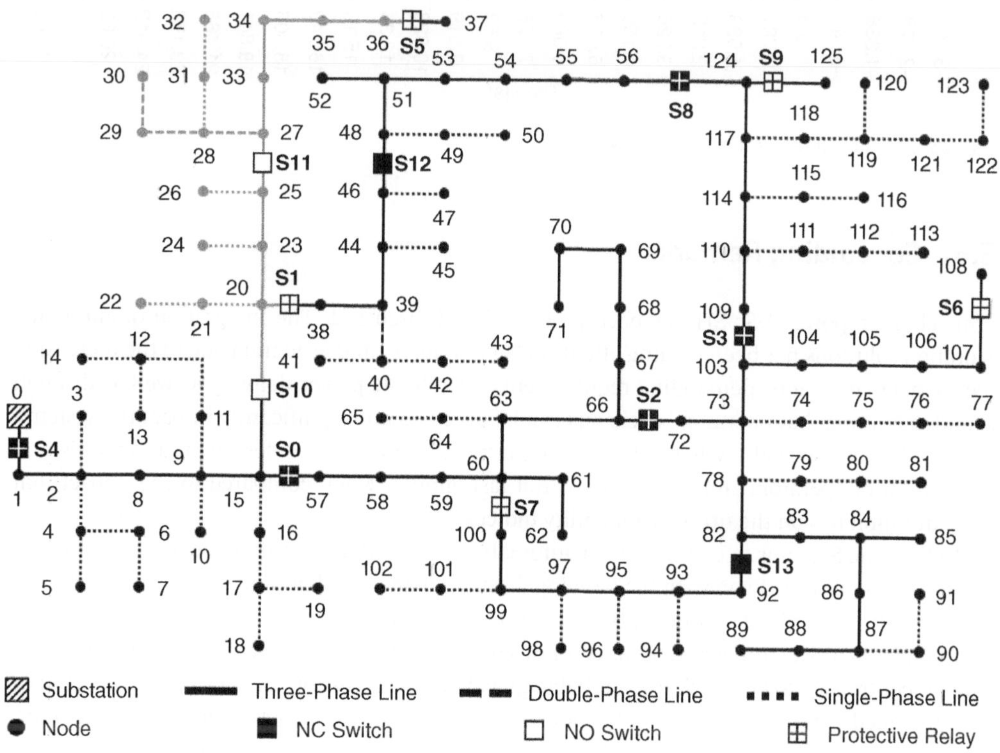

Figure 13.10 Case II network configuration after restoration.

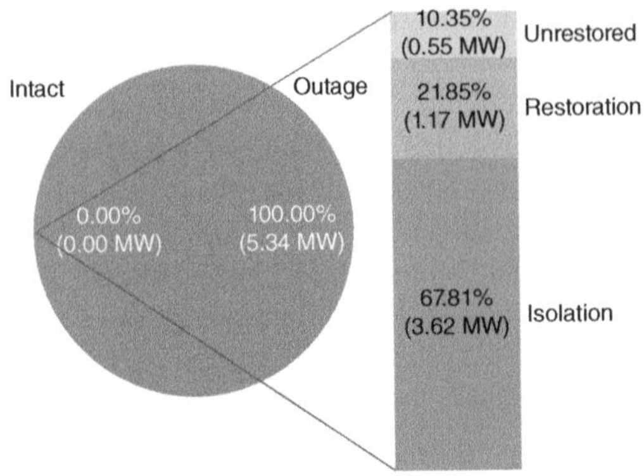

Figure 13.11 Case II outage restoration breakdown.

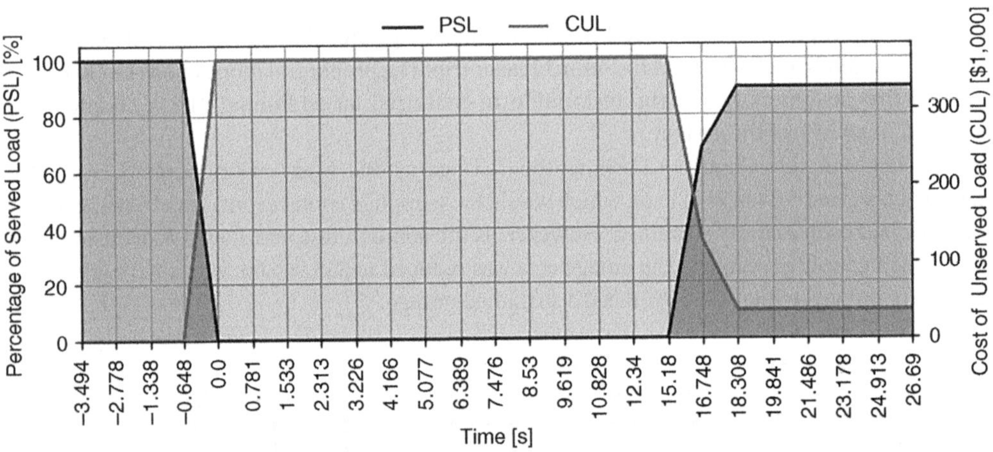

Figure 13.12 Case II restoration benefit over time.

13.6 Concluding Remarks

In this chapter, power DSs SH has been presented and discussed. The integration of automatic restoration solutions has been gaining attention from industry and academia due to the increasing need to meet customers' demanding requirements for reliable power delivery as well as through the understanding that operators' decisions can be deceived and significantly impact the system's reliability levels, especially when taken during critical moments. Besides, manual analysis and coordination of operators and field crews is not fast enough when compared to the impact that outage duration has on the utilities' reliability indices.

With that, the SH aims to compound an automatic solution that identifies, locates, isolates, and restores power outages. Even with the ongoing discussion and development of centralized, hierarchical, and distributed SH solutions, the constant growth and advancements of ADMS solutions have been showing a stronger trend of centralized control schemes. Different from a distributed topology, centralized SH can achieve a reliable global optimal solution, besides having the chance to be integrated with other ADMS tools, such as network parameters, power flow, metering information, and state estimation, among others.

Table 13.4 Self-healing benefits summary.

Case	Min. voltage [p.u.]	Max. current [%]	Comp. time[a] [s]	Outage cost[b] Initial [$]	Final [$]	Reduction [%]	System load Impacted [%]	Restored [%]
I	1.0160	49.92	15.952	126,174.00	42,609.00	66.23	40.43	16.29
II	1.0015	55.04	18.308	365,888.25	35,964.00	90.17	100	89.65

a) Computational time in an offline environment.
b) Cost related to the customers' de-energization.

A centralized SH model has been formulated and presented in this chapter. The fault identification, location, and isolation method is developed based on commonly available field IEDs' flags. Through the trip, lockout, and FI flags, a set of logic based on the pre-fault network configuration can successfully identify and locate the cell containing the fault. From this, it can be isolated by locating the TCSs bounding it. As soon as the fault is properly isolated, a MICP MOO model is proposed to restore the system‘s customer optimally. The MICP model is constrained by the outage cell, which must be maintained de-energized by the solution, as well as operational constraints, such as nodal voltage levels, loading, and radial configuration. With the proper modeling of the constraints and normalization of the optimization objective into cost, the proposed model can obtain a restoration solution that maximizes the number of energized customers with minimum switching maneuvers while respecting the operational conditions. The model was tested with a standard distribution test feeder for two different outage conditions, from which it has shown the successful performance of an SH solution for all cases. A summary of the SH benefits is presented in Table 13.4.

From the perspective of SH solutions development, proposed models must be developed and tested under different simulation environments before successfully confirming their deployability. Offline and Software-In-the-Loop simulations are important for initial validation and modeling. Still, such SH solutions must be tested, accounting for the performance of actual IEDs, in special digital relays. Hardware-in-the-loop testbeds can push the solution's testing to a controllable environment that brings crucial aspects of a real-world application, such as physical and processing time delays, communication protocols, and data exchange challenges. On the other hand, SH fault location and restoration solutions must account for the increasing integration of DERs. DERs contribute to the fault and create flags that may mislead the fault location. Besides, the restoration problem must consider the DERs' generation levels along with the customers' consumption, as after a fault, the DERs may take minutes to be reconnected, and the new configuration must ensure operational stability before and after their reconnection.

References

1 Thomas, M.S. and McDonald, J.D. (2015). *Power System SCADA and Smart Grids*. Boca Raton, FL: CRC Press.

2 IEEE Guide for Electric Power Distribution Reliability Indices. *IEEE Std 1366-2012 (Revision of IEEE Std 1366-2003)*, pp. 1–43, May 2012, https://doi.org/10.1109/IEEESTD.2012.6209381.

3 Yuan, Y., Dehghanpour, K., Bu, F., and Wang, Z. (2020). Outage detection in partially observable distribution systems using smart meters and generative adversarial networks. *IEEE Transactions on Smart Grid* 11 (6): 5418–5430. https://doi.org/10.1109/TSG.2020.3008770.

4 Shen, F., Wu, Q., and Xue, Y. (2020). Review of service restoration for distribution networks. *Journal of Modern Power Systems and Clean Energy* 8 (1): 1–14. https://doi.org/10.35833/MPCE.2018.000782.

5 Arif, A., Wang, Z., Chen, C., and Wang, J. (2020). Repair and resource scheduling in unbalanced distribution systems using neighborhood search. *IEEE Transactions on Smart Grid* 11 (1): 673–685. https://doi.org/10.1109/TSG.2019.2927739.

6 Arif, A., Wang, Z., Wang, J., and Chen, C. (2018). Power distribution system outage management with co-optimization of repairs, reconfiguration, and DG dispatch. *IEEE Transactions on Smart Grid* 9 (5): 4109–4118. https://doi.org/10.1109/TSG.2017.2650917.

7 Arif, A., Ma, S., Wang, Z. et al. (2018). Optimizing service restoration in distribution systems with uncertain repair time and demand. *IEEE Transactions on Power Systems* 33 (6): 6828–6838. https://doi.org/10.1109/TPWRS.2018.2855102.

8 Task Group State of the Art Distribution System Design (1984). Bibliography on distribution automation 1969–1982. *IEEE Transactions on Power Apparatus and Systems* PAS-103, no. 6: 1176–1182. https://doi.org/10.1109/TPAS.1984.318446.

9 Merlin, A. and Back, H. (1975). Search for a minimal-loss operating spanning tree configuration in an urban power distribution system. *5th Power Systems and Computer Conference*, London.

10 Berg, R., Hawkins, E.S., and Pleines, W.W. (1967). Mechanized Calculation of Unbalanced Load Flow on Radial Distribution Circuits. *IEEE Transactions on Power Apparatus and Systems* 89 (4).

11 Cheng, C.S. and Shlrmohammadi, D. (1995). A three-phase power flow method for real-time distribution systems analysis. *IEEE Transactions on Power Systems* 10 (2).

12 Shirmohammadi, D., Liu, W.E., Lau, K.C., and Hong, H.W. (1996). Distribution automation system writh real-time analysis tools real-time applications for DA systems. *IEEE Computer Applications in Power* 9 (2).

13 Sakis Meliopoulos, A.P., Polymeneas, E., Tan, Z. et al. (2013). Advanced Distribution Management System. *IEEE Transactions on Smart Grid* 4 (4): 2109–2117. https://doi.org/10.1109/TSG.2013.2261564.

14 Bessani, M., Fanucchi, R.Z., Delbem, A.C.C., and Maciel, C.D. (2016). Impact of operators' performance in the reliability of cyber-physical power distribution systems. *IET Generation, Transmission and Distribution* 10 (11): 2640–2646. https://doi.org/10.1049/iet-gtd.2015.1062.

15 Graziano, R.P., Kruse, V.J., and Rankin, G.L. (1992). Systems analysis of protection system coordination: a strategic problem for transmission and distribution reliability. *IEEE Transactions on Power Delivery* 7 (2): 720–726. https://doi.org/10.1109/61.127073.

16 Lee, J.H., Jeon, S.G., Kim, D.K. et al. (2020). Temporary fault ride-through method in power distribution systems with distributed generations based on PCS. *Energies (Basel)* 13 (5): https://doi.org/10.3390/en13051123.

17 Oh, J.-H., Yun, S.-Y., Kim, J.-C., et al. (2000). Particular Characteristics associated with temporary and permanent fault on the multi-shot reclosing scheme. *Power Engineering Society Summer Meeting.*

18 Salman, A.M., Li, Y., and Stewart, M.G. (2015). Evaluating system reliability and targeted hardening strategies of power distribution systems subjected to hurricanes. *Reliability Engineering and System Safety* 144: 319–333. https://doi.org/10.1016/j.ress.2015.07.028.

19 Chen, C., Wang, J., and Ton, D. (2017). Modernizing distribution system restoration to achieve grid resiliency against extreme weather events: an integrated solution. *Proceedings of the IEEE* 105 (7): 1267–1288. https://doi.org/10.1109/JPROC.2017.2684780.

20 Singh, M. (2017). Protection coordination in distribution systems with and without distributed energy resources—a review. *Protection and Control of Modern Power Systems* 2 (1): https://doi.org/10.1186/s41601-017-0061-1.

21 Dimitrijevic, S. and Rajakovic, N. (2015). Service Restoration of Distribution Networks Considering Switching Operation Costs and Actual Status of the Switching Equipment. *IEEE Transactions on Smart Grid* 6 (3): 1227–1232. https://doi.org/10.1109/TSG.2014.2385309.

22 Zhou, Q., Shirmohammadi, D., Liu, E. et al. (1997). Distribution feeder reconfiguration for service restoration an. *IEEE Transactions on Power Systems* 12 (2).

23 Liu, C.-C., Lee, J.S., and Venkata, S.S. (1988). An expert system operation aid for restoration and loss reductions of distribution systems. *IEEE Transactions on Power Systems* 3 (2).

24 Hotta, K., Nomura, H., Takemoto, H. et al. (1990). Implementation of a real-time expert system for a restoration guide in a dispatching center. *IEEE Transactions on Power Systems* 5 (3).

25 Shirmohammadi, D. (1992). Service restoration in distribution networks via network reconfiguration. *IEEE Transactions on Power Delivery* 7 (2).

26 Hsu, Y.-Y. et al. (1992). Distribution system service restoration using a heuristic search approach. *IEEE Transactions on Power Delivery* 7 (2).

27 Morelate, A.L. and Monticelli, A. (1989). Heuristic search approach to distribution system restoration. *IEEE Transactions on Power Delievery* 4 (4).

28 Miu, K.N., Chiang, H.-D., and Mcnulty, R.J. (2000). Multi-tier service restoration through network reconfiguration and capacitor control for large-scale radial distribution networks. *IEEE Transactions on Power Systems* 15 (3).

29 Shen, F., Lopez, J.C., Wu, Q. et al. (2020). Distributed self-healing scheme for unbalanced electrical distribution systems based on alternating direction method of multipliers. *IEEE Transactions on Power Systems* 35 (3): 2190–2199. https://doi.org/10.1109/TPWRS.2019.2958090.

30 Elmitwally, A., Elsaid, M., Elgamal, M., and Chen, Z. (2015). A fuzzy-multiagent self-healing scheme for a distribution system with distributed generations. *IEEE Transactions on Power Systems* 30 (5): 2612–2622. https://doi.org/10.1109/TPWRS.2014.2366072.

31 Nejad, R.R. and Sun, W. (2022). Enhancing active distribution systems resilience by fully distributed self-healing strategy. *IEEE Transactions on Smart Grid* 13 (2): 1023–1034. https://doi.org/10.1109/TSG.2021.3127518.

32 Jiang, Y., Liu, C.C., Diedesch, M. et al. (2016). Outage management of distribution systems incorporating information from smart meters. *IEEE Transactions on Power Systems* 31 (5): 4144–4154. https://doi.org/10.1109/TPWRS.2015.2503341.

33 Mahdavi, M., Alhelou, H.H., Bagheri, A. et al. (2021). A comprehensive review of metaheuristic methods for the reconfiguration of electric power distribution systems and comparison with a novel approach based on efficient genetic algorithm. *IEEE Access* 9: 122872–122906. https://doi.org/10.1109/ACCESS.2021.3109247.

34 Lopez, J.C., Franco, J.F., Rider, M.J., and Romero, R. (2018). Optimal restoration/maintenance switching sequence of unbalanced three-phase distribution systems. *IEEE Transactions on Smart Grid* 9 (6): 6058–6068. https://doi.org/10.1109/TSG.2017.2703152.

35 Schmitt, K., Negri, C.A., Daneshvardehnavi, S. et al. (2021). Short circuit and arc flash study on a microgrid facility. *Asian Basic and Applied Research Journal* 3 (1): 54–63.

36 Schmitt, K., Chamana, M., Mahdavi, M. et al. (2024). Power distribution systems optimal outage restoration with miscoordination detection. *IEEE Transactions on Power Delivery* 39 (3): 1723–1735.

37 Sun, W., Ma, S., Alvarez-Fernand, I. et al. (2018). Optimal self-healing strategy for microgrid islanding. *IET Smart Grid* 1 (4): 143–150.

38 Wang, Z. and Wang, J. (2015). Self-healing resilient distribution systems based on sectionalization into microgrids. *IEEE Transactions on Power Systems* 30 (6): 3139–3149. https://doi.org/10.1109/TPWRS.2015.2389753.

39 Wang, Z., Chen, B., Wang, J., and Chen, C. (2016). Networked microgrids for self-healing power systems. *IEEE Transactions on Smart Grid* 7 (1): 310–319. https://doi.org/10.1109/TSG.2015.2427513.

40 Kersting, W.H. (2002). *Distribution System Modeling and Analysis*. Las Cruces: CRC Press.

41 Macedo, L.H., Franco, J.F., Mahdavi, M., and Romero, R. (2018). A contribution to the optimization of the reconfiguration problem in radial distribution systems. *Journal of Control, Automation and Electrical Systems* 29 (6): 756–768. https://doi.org/10.1007/s40313-018-0415-6.

42 M. Sullivan, M. Collins, J. Schellenberg, and P. Larsen, "*Estimating Power System Interruption Costs: A Guidebook for Electric Utilities*," 2018.

43 Zhang, Q., Ma, Z., Zhu, Y., and Wang, Z. (2021). A two-level simulation-assisted sequential distribution system restoration model with frequency dynamics constraints. *IEEE Transactions on Smart Grid* 12 (5): 3835–3846. https://doi.org/10.1109/TSG.2021.3088006.

44 Mahdavi, M., Alhelou, H.H., Hatziargyriou, N.D., and Al-Hinai, A. (2021). An efficient mathematical model for distribution system reconfiguration using AMPL. *IEEE Access* 9: 79961–79993. https://doi.org/10.1109/ACCESS.2021.3083688.

45 Mahdavi, M., Schmitt, K., Bayne, S., et al. (2023). Reconfiguration of power distribution systems in the presence of voltage-dependent loads. *IEEE Texas Power and Energy Conference (TPEC)*. Institute of Electrical and Electronics Engineers Inc.https://doi.org/10.1109/TPEC56611.2023.10078635.

46 Robbins, B.A. and Domínguez-García, A.D. (2016). Optimal reactive power dispatch for voltage regulation in unbalanced distribution systems. *IEEE Transactions on Power Systems* 31 (4): 2903–2913. https://doi.org/10.1109/TPWRS.2015.2451519.

47 Jabr, R.A., Singh, R., and Pal, B.C. (2012). Minimum loss network reconfiguration using mixed-integer convex programming. *IEEE Transactions on Power Systems* 27 (2): 1106–1115. https://doi.org/10.1109/TPWRS.2011.2180406.

48 Ahmadi, H. and Martí, J.R. (2015). Mathematical Representation of Radiality Constraint in Distribution System Reconfiguration Problem. *International Journal of Electrical Power & Energy Systems* 64: 293–299. https://doi.org/10.1016/j.ijepes.2014.06.076.

49 Kerting, W.H. (1991). Radial Distribution Test Feeders. *IEEE Transactions on Power Systems* 6 (3): 975–985. https://doi.org/10.1109/59.119237.

14

Resiliency, Reliability, and Security of Cyber-Physical Power System

Mohsen Chegnizadeh[1], *Mahmoud Fotuhi-Firuzabad*[1], *and Sajjad Fatahian dehkordi*[2]

[1] *Department of Electrical Engineering, Sharif University of Technology, Tehran, Iran*
[2] *Department of Electrical Engineering and Automation, Aalto University, Espoo, Finland*

Abbreviations

AI	Artificial Intelligence
ANN	Artificial Neural Network
BART	Bayesian Additive Regression Trees
BP	Back Propagation
CNN	Convolutional Neural Networks
DBN	Deep Belief Networks
DG	Distributed Generation
DL	Deep Learning
DRL	Deep Reinforcement Learning
DSO	Distribution System Operator
DTI	Decision Tree Induction
ELM	Extreme Learning Machines
ELM-NN	Extreme Learning Machine Neural Networks
ESS	Energy Storage System
FEMA	Federal Emergency Management Agency
HILF	High Impact Low Frequency
HILP	High Impact Low Probability
ICT	Information and Communications Technology
KNN	k-Nearest Neighbors algorithm
KR	Knowledge Representation and Reasoning
LSTM	Long Short-term Memory
MARS	Multivariate Additive Regression Splines
MCTS	Monte Carlo Tree Search
MINLP	Mix Integer Linear Programming
ML	Machine Learning
MLP	Multilayer Perceptions
MP	Multilayer Perceptron

Smart Cyber-Physical Power Systems: Fundamental Concepts, Challenges, and Solutions, Volume 1, First Edition.
Edited by Ali Parizad, Hamid Reza Baghaee, and Saifur Rahman.

MPS	Mobile power Sources
NB	Naive Bayes
NERC	North American Electric Reliability Corporation
NIAC	National Infrastructure Advisory Council
PDF	Probability Distribution Function
PHR	Predictable, High-impact, and Rare
RF	Random Forest
RL	Reinforcement Learning
RNN	Recurrent Neural Networks
SAE	Sparse Automatic Encoder
SVM	Support Vector Machine
UHR	Unpredictable, High Impact, and Rare
V2G	Vehicle to Grid
WAMS	Wide Area Management Systems
WHCEA	White House Council of Economic Advisors

14.1 Introduction and Motivation

The resiliency, reliability, and security of critical infrastructures, which are paramount to national prosperity and societal well-being, have gained significant attention in recent years. Recognizing this critical necessity, a surge of innovative approaches with the aim of managing the risks of these systems has emerged [1].

Power systems are unquestionably the bedrock and lifeblood of modern society, serving as its most crucial critical infrastructure. As the US National Academy of Sciences stated, "the world runs on electricity", and its proper functioning forms the backbone of nearly all critical infrastructure and human lives, both social and economic [2]. Consequently, any capacity deficit within this system can ripple through, crippling operations across most other critical sectors. From financial transactions and water supply to banking systems and communication technology, all rely on electricity and suffer severe disruptions without it [3].

Based on the above facts and the integral role that electricity plays, future power systems should possess eight key characteristics: affordable, safe, accessible, clean, reliable, secure, flexible, and resilient [3, 4]. To accomplish such clear-sighted vision, this system has traditionally been designed according to concepts such as reliability, so as to be able to respond appropriately in the occurrence of predictable events with limited impacts [2]. Despite the efforts that have been made, widespread blackouts have recently occurred around the world, which demonstrates the inadequacy of existing concepts and tools for achieving these goals. As compared to previous cases, these blackouts, which could result in significant direct and indirect costs on governments and electricity institutions, have three distinct characteristics:

1) In general, these widespread blackouts are primarily caused by natural events and human errors, whether intentional or unintentional.
2) The physical damages caused by these events could be widespread throughout the power system, spanning a broad geographical area and affecting various levels of generation, transmission, and distribution.
3) Ultimately, the impacts of these events could extend beyond the power system, and therefore disrupt many other critical infrastructures. This is due to two main factors: (i) these events not only disrupt the power system but also cause damage to many other critical infrastructures,

and (ii) critical national infrastructures are interdependent, thus making them susceptible to cascading failures [5].

The significant number of widespread blackouts and cascading outages indicate the vulnerability and weakness of the power system in dealing with such events. U.S. power system outages are largely caused by severe weather events, according to the White House Council of Economic Advisors (WHCEA) report. A $300 billion estimate has been made for the costs associated with power outages caused by natural disasters in the country between 2003 and 2012. [6]. Additionally, the power system faces other emerging challenges, fuelled by rapid technological advancements, increasing reliance on cyberspace, and other factors. Needless to mention, such emerging challenges may pose new risks to electric power systems, hence accentuating the need to scrape the clichéd methods intended to ensure routine operation and achieve high levels of reliability.

Since the Boston storm of 1930 spurred a renewed focus on power system risk management research, it has become increasingly apparent that a paradigm shift beyond classical reliability is crucial to safeguard the effective operation of power systems in the face of severe and large-scale events. Just as widespread blackouts in the past resulted in efforts to improve the electricity infrastructure, leading to the Electric Power Reliability Act of 1967 and the establishment of North American Electric Reliability Corporation (NERC), similarly, new risks arising from the emerging challenges demand an innovative approach to address them effectively. In essence, a power system in the twenty-first century cannot solely rely on the reliability concept. To effectively manage these emerging challenges, particularly high impact low frequency (HILF) events, a system with exceptional flexibility and adaptability is indispensable, transcending the limitations of traditional power systems to cope with such events. This realization has propelled engineers to formulate and implement the "resiliency" paradigm. Under this concept, even though stressors and stresses cannot be perfectly predicted, the system's proactive and adaptive capabilities can effectively mitigate the associated risks.

Despite envisioning eight key qualities, including resiliency, attaining them for future power systems becomes ever more intricate. Not only emerging challenges but also complex systems, interdependent infrastructures, and data deluges from smart grids complicate this pursuit. Recently, to overcome the deficiencies of traditional methods to assure the accomplishment of such qualities, machine learning (ML) techniques have been successfully and broadly used. Specifically, such techniques are applied in the reliability, resiliency, and security evaluation of large-scale cyber-physical power systems.

ML, a subfield of artificial intelligence (AI), excels at handling problems where devising the closed form of a problem is intricate or impossible. These techniques can learn from known data to make forecasts without using mathematical formulations [7]. For instance, the risk evaluation methods for cyber-physical power systems become increasingly complex as a result of the interdependencies between power systems and other infrastructures [8]. In addition, in some cases, quick analysis is needed to make timely decisions, or there is a compromise between the solution accuracy and computational burden, which makes its adoption challenging [9].

Motivated by the above challenges, this chapter focuses on the application of ML in the resiliency, reliability, and security of modern power systems. So, as a part of this chapter, we examine the crucial concepts of resiliency, reliability, and security and explain why these concepts are crucial for the modern power system. Furthermore, we introduce existing ML applications like forecasting and risk analysis before highlighting successful real-world implementations that enhance electricity grid resilience. Equipped with this theoretical foundation, we can take the next step: a compelling case study. Here, we delve into the tangible potential of ML, illustrating how it empowers us to evaluate power system resiliency to bolster grid preparedness.

14.2 Conceptual and Definitional Studies

14.2.1 Introduction

The aim of this section is to address the definitions of resiliency, reliability, and risk as well as illustrate the differences between such paradigms. Moreover, the significance of these concepts in power system studies is discussed in this section. Finally, some of the existing works on assessing and enhancing the resiliency of cyber-physical power systems are introduced.

14.2.2 Rethinking Security of Power Systems in the Age of HILF Events

Due to climate changes, there has been a dramatic increase in the frequency and intensity of large-scale events. Hurricane names like Irma, Harvey, and Maria are sadly familiar to the electric power industry, as these devastating storms have not only caused the loss of countless lives but also highlighted the urgent need for new strategies to safeguard the power grid against these increasingly emerging challenges [2]. Severe weather is a leading cause of widespread power outages in the United States. Considering this: In 2012, Hurricane Sandy resulted in power outages for 8.5 million homes and businesses, affecting tens of millions of individuals, with some areas remaining without electricity for weeks. [10]. Between 2003 and 2012, nearly 679 power outages impacting at least 50,000 customers were attributed to weather events in the United States, with 80–90% stemming from distribution system failures [10]. Looking ahead, climate changes and aging infrastructure are expected to further exacerbate these risks, increasing the system's vulnerability and the potential consequences of outages.

Beyond severe weather events, recent experiences have exposed the stark vulnerability of current power systems to cyber and physical attacks. This susceptibility stems, in part, from the growing reliance on information and communications technology (ICT) within power grids. On the one hand, ICT has fostered the development of smart grids, enhancing efficiency and control. On the other hand, it has inadvertently opened the door to malicious attacks seeking to critically damage the system [4].

The modern power grid faces threats—from climate change to malicious cyberattacks and physical sabotage—all of which underscore the need to identify their defining characteristics. Here, a fundamental question looms: What are the prominent characteristics of extreme weather events and cyberattacks? Based on their features, these threats can be categorized as follows [11]:

1) **Unforeseen high-impact rare events (UHR):** This category, known as "unidentified", encompasses rare, catastrophic events with unforeseen or unobserved consequences. The 2004 Indian Ocean tsunami and earthquakes exemplify UHR events, defined by their severe impact and unpredictability.
2) **Partially predictable, high-impact rare events (PHR):** Namely "unknown", PHRs might seem apt, but can be partially modeled. These rare, high-impact events often include weather phenomena like tornadoes and floods, some of which can be predicted to varying degrees [12].

The extent to which a conceptual model can describe a phenomenon determines whether it is considered rare, unknown, or unidentified. In terms of probability distributions, probability distribution functions (PDFs) can be defined as mathematical models for stochastic phenomena [13]. In this case, three states are possible:

1) **Fully specified PDF:** grasping all consequences and probabilities
   - In this state, both the consequences and their associated probabilities are fully understood.
   - An example is generating unit unavailability, also known as the forced outage rate (FOR).
2) **Partially known PDF:** consequences known, probabilities hidden
   - Here, the potential consequences are known, but their associated probabilities remain elusive.
   - This state aptly describes PHR events. While we can envision such events and their impact on the power grid, assigning exact probabilities has been proved to be challenging.
3) **Unknowable PDF:** unveiling consequences only in hindsight
   - In this state, even identifying possible consequences beforehand is impossible.
   - UHR events fall into this category, as their nature is often only revealed after their occurrences.

Considering the preceding discussion, PHR and UHR events, also known as HILF or high impact low probability (HILP), reside within the realm of resilience studies. In this regard, HILF events are of three categories (Figure 14.1): intentional attacks, stochastic events, and natural disasters (technical, human, deliberate sabotage, cyber). HILF events are a type of threat whose characteristics change over time: namely the type, time, intensity, extent, and location of the event.

Here lies the critical difference between the concepts of reliability and resilience. While reliability concerns events with known consequences and probabilities, HILF events—or high-impact, low-frequency events—fall squarely within the realm of resilience studies.

14.2.3 From Risk and Reliability to Power Grid Resilience

In this section, we will explore resilience definitions by first defining two concepts, risk and reliability. Risk analysis is well understood by researchers and engineers. In infrastructure like the power system, risk hinges on two factors: the likelihood of disruptions and their potential consequences. It boils down to the probability of something bad happening and the severity of it.

Reliability, as defined by NERC, measures how effectively the electrical system elements deliver power to customers within acceptable limits. According to Institute of Electrical and Electronics

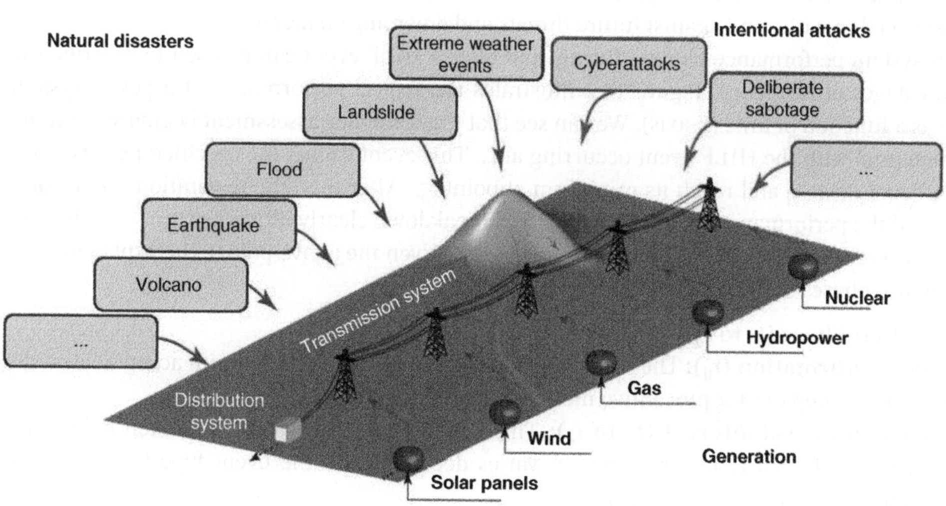

Figure 14.1 Threats facing power systems.

Engineers (IEEE), a system's reliability shows the likelihood that it fulfills its function under specific conditions in a given period of time.

As the definitions indicate, risk management encompasses a broader concept than reliability. To be more precise, reliability aims to manage the risks of the power system against certain events, with known consequences and probabilities. The idea of resilience, on the other hand, refers to managing the risk of the power system against different categories of events, including those that have high impacts and low frequencies. In conclusion, risk management is a more encompassing concept than either of the two paradigms. With these points in mind, the following section will discuss the details of the definition of resilience.

As a concept, resilience describes how a system can respond to HILF events. Taking its Latin root "resilire" (meaning "to bounce back") [14], the term suggests a system's ability to bounce back after unexpected shocks and challenges [15]. However, defining resilience comprehensively remains a challenge due to its adaptability across different disciplines. The paradigm has been interpreted differently in fields such as economics, social sciences, and complex systems research since Holling's 1973 definition [16–20]. In modern power systems, this paradigm is essential to preparing for withstanding, surviving, and recovering from disruptions.

In 2010, the National Infrastructure Advisory Council (NIAC) provided a comprehensive definition of infrastructure resilience. Subsequently, the NIAC-provided resilience definition has also been endorsed by NERC, which consequently is now referenced in power systems [21]. This definition is as follows:

> "*The effectiveness of a resilient infrastructure depends on its ability to predict, absorb, adapt to, and/or rapidly recover from a wide-scale disruptive event* [20]."

Resilience in power systems isn't one dimensional; it unfolds over time. Associated studies fall into two distinct zones: the immediate response in the moments after a shock, and the gradual evolution in the face of emerging challenges. In the short term, a power system needs to be robust (resisting disruption), redundant (having backup options), flexible, and fast healing (restoring quickly). However, long-term resilience demands adaptability. Therefore, the grid must be transformed to withstand against future threats and emerging challenges.

Power system performance before, during, and after a HILF event can be used to describe the temporal stages of resilience. Figure 14.2 illustrates the typical performance of a power system (y-axis) as a function of time (x-axis). We can see that the resilience assessment begins at t_0 on the horizontal axis, with the HILF event occurring at t_e. This event causes the specified performance index to decrease at t_d and reach its minimum at point t_m. After that, the restoration phase commences and the performance index increases. This breakdown clearly illustrates the three distinct stages of resilience: prevention, survival, and recovery. Given the above points, the explanation of the resilience phases is as follows:

1) **Prevention Phase (t_0 to t_d):**
   - **Event confirmation (t_0):** The system confirms the impending event (with acceptable confidence) and prepares for preventive measures.
   - **Preparedness subinterval (t_0 to t_e):** This subinterval involves actions taken before the event's actual occurrence. Its duration varies depending on the event type (e.g., zero for earthquakes, hours for tornadoes).
   - **Robustness Subinterval (t_e to t_d):** This subinterval focuses on maintaining system robustness during the event itself. Its effectiveness depends on the system's structure, control and protection programs, and the event's nature (e.g., tornadoes threaten overhead grids more than floods, while floods impact underground grids more).

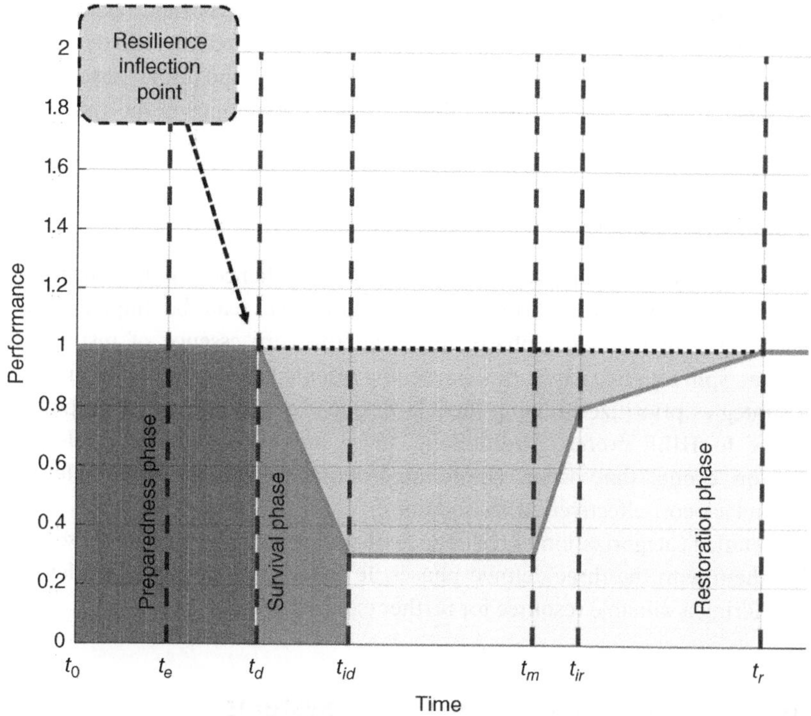

Figure 14.2 Performance of the power system after a HILF event.

- This time span $[t_0, t_d]$ enables system operators to foresee possible harm and initiate proactive measures to lessen the impact of events on the system. For instance, altering the operating point of the system (such as optimal load balancing with resilience-focused constraints) is a crucial measure that can be implemented during the prevention phase.

2) **Survival Phase (t_d to t_m):**
   - This phase starts at t_d, the time at which the performance of the system begins to decrease.
   - The performance of the system reaches its minimum at t_m.
   - Control, automation, and protection schemes are primarily responsible for this phase. Automation plays a crucial role in swift reaction, minimizing damage, and initiating recovery processes.
   - The objective of this stage is to maximize system functionality despite the shock.
3) **Recovery phase (t_m to t_r):**
   - **System restoration (t_m to t_{ir}):** In this subinterval, network loads are rapidly reconnected, which prioritizes critical areas and minimizes their downtime. Also, temporary solutions and bypasses may be employed to keep the system in operational mode, while permanent repairs are underway. By means of these measures, the system gradually regains functionality, approaching acceptable operating levels.
   - **Infrastructure Recovery (t_{ir} to t_r):** This subsection involves repairing or replacing damaged components and structures, enabling the grid to return to its pre-event functionality. Simultaneously, long-term measures are implemented to enhance the grid's resilience and prevent similar vulnerabilities from recurring in the future. Upon completion of this stage, the network recaptures its full strength and resilience, positioned to confront future challenges.

- During this phase (t_m to t_r), the emphasis shifts to re-establishing the normal operational mode of the system. Network performance must revert to acceptable standards, and damaged infrastructure necessitates comprehensive repair. By implementing expeditious restoration strategies and investing in durable infrastructure enhancements, we empower the system to rebound rapidly and emerge from adversity more resilient.

14.2.4 Enhancing Power Grid Resilience

Building on previous discussions, we recognize that power system resilience manifests in three distinct phases: prevention, survival, and recovery. Diverse strategies can be implemented to enhance this resilience throughout each phase, reflecting the very essence of resilience. Broadly, these strategies are split into two categories: hardening oriented and operation oriented.

Operation-oriented strategies prioritize utilizing flexible responsive resources to make power systems more responsive to HILF events. Additionally, these measures empower real-time decision-making based on events that occur. Hardening-oriented measures, while offering inherent robustness, often lack cost-effectiveness in isolation.

Table 14.1 provides a detailed categorization of resilience-enhancing actions based on these two perspectives and aligns them with the three defined phases. It reviews and categorizes relevant references and papers, offering a valuable resource for further exploration.

14.3 Application of Machine Learning in Power Systems

14.3.1 Introduction

AI algorithms and ML methods have been widely applied in numerous research initiatives within the power system studies; e.g. forecasting, security assessment, risk assessment, the fault identification of distribution system, and the power interruption length forecast. Therefore, this section intends to explore the general applications of AI and ML in power system studies and, specifically, how these data-driven techniques bolster the resilience of this sector.

14.3.2 Data Analysis and AI Algorithms for Enhancing the Resilience of the Power System

Over the past few years, significant research works have focused on modernizing the power system to improve its resilience. While traditional approaches rely heavily on mathematical models, the complexity of the current and future power systems, as well as their associated uncertainties, pose challenges in effectively implementing such models. Unlike traditional methods, artificial intelligence (AI) can overcome many of these limitations. As a result, AI contributes primarily to the resilience of power systems in two ways:

1) The possibilities of increasing the power system resilience have been improved as a result of the growing amount of data coming from wide area management systems (WAMS) systems and smart devices in energy networks. These systems and devices provide network parameters and geographical information to operators, which enable them to have a timely and precise view of the power system status and situation. They also contain valuable insights that can be used to improve the system resilience during weather-related disturbances, power outages, and transient events. In addition to the required huge amount of data, historical operational data

Table 14.1 Categorization of resilience-enhancing actions.

	Remedial action	Prevention phase	Survival phase	Recovery phase
Hardening-oriented strategies	Undergrounding	[22, 23]		
	Upgrading transmission and distribution line equipment	[23, 24]		
	Evolving transmission and distribution substations	[23]		
	Siting power system equipment	[23–25]		
	Creating redundancy in the system	[23, 24]		
Operation-oriented strategies	Using flexible resources such as distributed generation (fixed and mobile) and energy storage	[23–26]	[27–35]	[22, 25, 31, 36–40]
	Demand side management		[30, 37]	
	Decentralized control	[23, 41]	[28–32]	[22, 40, 42]
	Forecasting	[26, 41, 43]	[44]	[36]
	Preventive control	[23–26]		
	Microgrids	[24–26, 41]	[23, 27–33, 36, 44, 45]	[22, 25, 31, 36–40, 42, 46, 47]
	Situational awareness		[37]	
	Self-healing		[23, 27–29, 37, 44]	[40]
	Switching	[24, 41]	[28–31, 48, 49]	[25, 31, 36–40, 42]

also hold a wealth of valuable information that can help in improving the system resilience. Nevertheless, such a great deal of data must be analyzed, which shows the necessity of developing and applying AI-based algorithms to analyze the system resilience.

2) AI algorithms have become more accurate and faster because of progress in this field. This matter is very important, especially for studies in the field of resilience, in which speed of decision-making is very important.

14.3.3 A Review of AI-Driven Power System Studies and Machine-Learning-Empowered Resilience Strategies

Driven by the benefits discussed above, AI techniques are finding increasing applications in power systems, particularly in the area of resilience. Three main groups of AI techniques used in this field of research are ML, deep learning (DL), and reinforcement learning (RL).

The first group, traditional ML, enables knowledge acquisition from data without requiring explicit programming with mathematical equations. This is particularly advantageous in scenarios where extracting a closed-form formula proves challenging. These algorithms excel in classifying observed data (supervised learning), clustering similar patterns (unsupervised learning), and predicting system outputs based on historical behavior and data (regression modeling). This group encompasses five primary models:

1) **Neural network**: These networks were in use until the advent of BP algorithms. This set of algorithms encompasses multilayer perceptions (MLP), artificial neural network (ANN), and extreme learning machines (ELM).
2) **Kernel methods**: These algorithms are utilized for classification tasks. For instance, support vector machine (SVM) is a part of this group. This machine employs a kernel function to project each point from the original space to a higher-dimensional space and compute the distance between them in the transformed space.
3) **Tree-based methods**: This approach typically involves classification or partitioning the predictor space into several regions to classify data.
4) **Probabilistic modeling**: This model is grounded in probability theory; it uses statistical methods to construct models to address classification or regression problems. Some of these methods encompass NB and logistic regression.
5) **Ensemble learning**: This approach merges multiple ML models to achieve superior outcomes, such as adaptive boosting [7] and random forest (RF).

ML-based approaches have been implemented in different spheres of power system studies. These areas include the following [26–28]:

- Reliability,
- Control and stability assessment of the system,
- Frequency control and assessment,
- Output evaluation,
- Economic load dispatch,
- Short-term state evaluation and forecasting of the system,
- Power system planning,
- Transient stability,
- Voltage stability assessment,
- Prediction of the system's inertia,
- Sizing and siting planning of distributed generation units, and
- Improving system resilience and robustness.

The second group is DL. DL is a subfield of AI and a type of deep architecture [8] that seeks to learn from functions that exhibit a high level of abstraction. DL models this process by using a deep graph with multiple processing layers (consisting of multiple layers of linear and nonlinear transformations). In other words, its basis is on learning knowledge representation and reasoning (KR) in the model layers. The main DL models are deep belief networks (DBN), convolutional neural network (CNN), recurrent neural networks (RNN), and long short-term memory (LSTM).

The third category, referred to as RL, obtains information about its environment from an agent and learns to select actions that maximize its reward. So, it could learn to self-improve by evaluating the feedback generated from its experiences.

Several other studies were conducted in the electric and energy sectors concerning the use of ML applications. These include forecasting [using extreme learning machine neural networks (ELM-NN)] [50], security assessment [using decision tree induction (DTI), MLP, and k-nearest neighbors algorithm (KNN)] [51], risk assessment (using semi-parametric regression models, parametric regression models, non-parametric regression models, ANN, and SVM) [51], the fault identification of distribution system (using ANN and SVM) [52], and power interruption length forecast [(using regression, regression trees, Bayesian additive regression trees (BART), and multivariate additive regression splines (MARS)] [53]. Considering this chapter's focus on applying ML to power system resilience, two categories of aforementioned studies stand out: (i) forecasting and (ii) grid restoration applications of AI.

As mentioned, resilience improvement strategies can be applied at three levels (pre-event, during event, and post event). In addition, in current study, these actions are classified into two categories: hardening (group A) and operational (group B) strategies. In the field of resilience improvement actions at the post-event level, studies conducted with AI-based approaches are very limited. These studies have been largely limited to the recovery period to address challenges associated with the time-consuming nature of simulations or solving complex optimization algorithms.

To tackle the issue of an online generator startup following a power outage, reference [54] has combined the Monte Carlo tree search (MCTS) algorithm with the sparse autoencoder (SAE) for making real-time decisions. In this research, a substantial number of samples could be produced through the offline procedure, after which the SAE can rapidly approximate the maximum power that a unit could produce in a specific state. This approximation could be employed to enhance the efficiency of the MCTS search procedure in the simulation.

Leveraging smart meter data, [55] developed a data-driven framework to predict cold-wave-induced demand surges in customers. An SVM model, trained on historical outage data, drives the predictions.

In reference [56], decision-making on recovery actions has been made through RL, including the routing and scheduling of electric vehicles. On the other hand, switch reconfiguration [57] and distributed energy resources (DER) control [58] are among other actions that have been taken in past studies to improve the system resilience based on machine-learning-based approaches. In addition, coordination of controllable distributed generation units and scheduling of mobile energy storages, through DRL, has been proposed in [59]. In this reference, coordination of operational actions between flexible resources (including distributed generation units and energy storages) has been able to improve the possibility of recovering critical loads in microgrids. In addition to distribution networks, learning-based approaches have also been used in studies to recover other levels of power systems. Reference [60] has used neural networks to generate worst-case load and renewable energy generation (wind power) scenarios for security assessment. These studies are summarized in Table 14.2.

Table 14.2 Some of the studies on the applications of learning-based methods in the resilience of power systems.

#	ML type		Level	Remedial action type	References
1	Multi-agent RL	RL	Distribution system	Electric vehicle scheduling for resilience improvement	[56]
2	Deep RL	RL	Distribution system	Decision-making on operating parameters including switching for reconfiguration	[57]
3	Q-learning	RL	Distribution system	Tap changers of transformers, connecting micro-grids to the upstream network	[58]
4	Q-learning	RL	Distribution system	Operation of distributed generation (fixed and mobile)	[59]
5	Convolutional neural networks	Neural networks	Transmission system	Load recovery under worst-case scenario	[60]

14.3.4 Data Augmentation and Synthesis Approaches

In the context of the applications of ML in power systems, it is very important to review the available tools for addressing data scarcity. Limited or inaccessible data can significantly hinder model training. Therefore, this section delves into the available tools and techniques for overcoming this critical obstacle.

ML models require abundant, high-quality data to perform well. However, data scarcity can occur for a variety of reasons. Two common methods for addressing data scarcity are data augmentation and synthetic data generation.

Data augmentation stands as a prevalent technique in enhancing the performance and robustness of ML models, particularly in domains like ML and deep learning (DL) [61]. This method entails artificially increasing the size, variety, and quantity of training datasets. These transformations are implemented on existing data, generating new synthetic samples. Data augmentation is particularly valuable in AI applications that rely on pattern recognition and prediction from extracted patterns. Specific data augmentation methods include jittering, time warping, window slicing, seasonal decomposition, data scaling, time series interpolation, feature engineering, generative models, day-night transformation, sequential sampling, etc. Additionally, DL methods can be employed for this purpose. DL models, such as generative models and RNNs, possess the ability to comprehend data divergences accurately and generate realistic artificial samples. Some data generation techniques that utilize DL include generative adversarial networks (GAN), variational autoencoders (VAE), sequence to sequence (Seq2Seq), temporal convolutional network (TCN), and temporal GANs.

The advantages of data augmentation extend to:

1) **Enhanced model robustness:** By introducing variations to the data, data augmentation enables AI models to learn from a broader spectrum of patterns. This enhances their accuracy and resilience to unforeseen data.
2) **Addressing data scarcity**: Data augmentation plays a pivotal role in scenarios with limited data availability. It generates new data points from the existing pool, effectively increasing the training dataset's size and quality for ML and DL models.

3) **Preventing overfitting**: By providing a wider range of data, data augmentation helps prevent overfitting, ultimately improving overall performance.
4) **Improving robustness to noise and variations**: Data augmentation techniques train models to handle noise and variations in data. This enhances their compatibility with real-world scenarios, enhancing robustness.
5) **Enhancing generalization ability**: Data augmentation promotes generalization ability by exposing models to a broader range of data patterns. This enables them to generalize better to unforeseen data and improve overall performance.

Synthetic data generation is a process that creates new or simulated data that closely mimic the features and patterns found in real datasets. Unlike real data, which are collected through observations or measurements of the real world, synthetic data are produced entirely by algorithms, models, or simulations. This can be useful in situations where real data are scarce or difficult to obtain [62]. Both data augmentation and synthetic data generation can be effective in addressing data scarcity. The best approach depends on the specific situation. Key characteristics of synthetic data generation include:

1) **Artificial creation**: Synthetic data do not rely on direct measurements or observations from the real world. Instead, it is generated through computational processes.
2) **Imitation of real data**: The objective of generating synthetic data is to produce data that accurately mirrors the statistical characteristics, distribution, and trends found in actual datasets.
3) **Enhanced variety and complexity**: Synthetic data generation can introduce diverse and complex samples that do not exist in the original dataset. This is particularly valuable in situations where the available real data are limited.
4) **Addressing data scarcity**: Synthetic data generation are often used to address the problem of limited real data, which can hinder the effective training of ML and DL models.

Various techniques are used for synthetic data generation, including:

1) **Generative adversarial networks (GANs):** GANs create realistic and diverse synthetic data through a competitive process between two models.
2) **Variational autoencoders (VAEs):** VAEs produce synthetic data that are similar to existing data by learning the underlying structure of the data.
3) **Simulations and modeling:** Simulations and modeling can be used to generate synthetic data that represent real-world scenarios, such as weather patterns or financial markets.
4) **Rule-based generation:** Rule-based generation uses a set of rules to create synthetic data.
5) **Data interpolation and extrapolation:** Data interpolation and extrapolation techniques can be used to fill in missing data or to generate data that extend beyond the range of the existing data.

14.4 Case Study

14.4.1 Introduction

Envision a power network with certain elements that are susceptible to harm from an imminent hurricane. The trajectory and severity of the hurricane can be forecasted using meteorological data from weather agencies. Each component's condition can be categorized as either "damaged" (outage) or "functional" (in service) depending on two variables: the hurricane's wind velocity and the component's proximity to the hurricane's center. Historical hurricane data show that damaged components are more likely with higher wind speeds and smaller distances from the hurricane's center. In this section, our objective is to effectively predict power system equipment outages in response to upcoming hurricanes using advanced ML techniques.

In this section, we will utilize the naive Bayes, logistic regression, SVM, KNN, decision tree, ANN, and random forest techniques to conduct the study. Recognizing the potentially hindering effects of data scarcity in training these models, this section initially investigates a data synthesis approach employed to address this concern.

Following the discussion of data synthesis, the confusion matrix and its associated evaluation criteria, including accuracy, precision, sensitivity, and specificity, will be introduced and meticulously explained. These metrics could be applied to assess the performance of the various classification models employed in the study.

Finally, the case study and its corresponding results will be presented and dissected in detail. This comprehensive analysis will provide valuable insights into the effectiveness of the employed classification models in addressing the research question.

14.4.2 Data Synthesis Approach

To synthesize the data, the initial step involves ascertaining the probability of failure for network elements from fragility curves, considering each of the potential wind paths, individual network elements, and the associated wind speeds and distances. Since our objective is a classification problem, the data must be discrete and comprise two values: 0 for intact elements and 1 for failed elements. To transform the probabilities into values of 0 and 1, random numbers with a Gaussian distribution between 0 and 1 are initially generated. Subsequently, the probabilities are compared to these random numbers: if the probability of failure for an element exceeds the corresponding random number, it is assigned the value 1; otherwise, it is assigned the value 0. These data are subsequently used to train the ML algorithms. Notably, the data are two dimensional, entailing that the input features for each data point encompass wind speed and distance from the storm center, while the output represents the equipment status (failure or intact).

14.4.3 Confusion Matrix

A confusion matrix is utilized to assess the effectiveness of a classification method. The confusion matrix is a tabular representation that encapsulates the classification outcomes of a model. It shows the number of samples that were correctly and incorrectly classified.

The confusion matrix has four cells, each of which represents a different type of classification outcome. The cells are labeled as follows:

- **True positive (TP):** The sample was actually positive and was correctly classified as positive.
- **False negative (FN):** The sample was actually positive but was incorrectly classified as negative.
- **True negative (TN):** The sample was actually negative and was correctly classified as negative.
- **False positive (FP):** The sample was actually negative but was incorrectly classified as positive.

The confusion matrix can be employed to compute an array of performance indicators, including accuracy, precision, and recall.

Accuracy is the proportion of all samples that have been accurately categorized. It is computed in the following manner:

$$\text{Accuracy} = \frac{TP + TN}{TP + FN + FP + TN} \tag{14.1}$$

Additionally, precision is the proportion of samples that were identified as positive and were indeed positive. It is computed as follows:

$$\text{Precision for Positive Class} = \frac{TP}{TP + FP} \tag{14.2}$$

$$\text{Precision for Negative Class} = \frac{TN}{TN + FN} \tag{14.3}$$

Recall, also known as sensitivity, is a measure that represents the proportion of TP instances that were accurately identified as such. Here's how it's computed:

$$\text{Recall} = \frac{TP}{TP + FN} \tag{14.4}$$

Other performance metrics that can be calculated from the confusion matrix include:

- **Specificity**: The percentage of samples that were classified as negative that were actually negative.
- **F1 score**: A weighted average of precision and recall.
- **Error rate**: The percentage of all samples that were incorrectly classified.

The confusion matrix is a valuable tool for understanding the performance of a classification algorithm. It provides a clear and concise summary of the classification results, and it can be used to calculate a variety of performance metrics.

14.4.4 Case Study and Numerical Results

Assuming a hurricane of category 3 on the Saffir–Simpson scale traverses the test system, it commences at the coordinates [$x = 30\,\text{km}$, $y = -\ 100\,\text{km}$] on the ocean surface. The air density and Coriolis frequency of earth's rotation are $\rho = 1.225\,\text{kg/m}^3$ and $f = 7.2392 \times 10\text{–}5\,\text{Hz}$, respectively. The hurricane commences advancing with approach angles of $3\pi/4$, $5\pi/8$, and $\pi/2$. The hurricane makes landfall at [$y = -50\,\text{km}$], and its translation velocity, background wind speed, and decay parameters are extracted from reference [63]. Considering three storm paths, we produce 900 failure scenarios for the network components using the synthesis approach described.

This study utilizes the modified distribution feeder of Roy Billinton test system (RBTS) bus 4 shown in Figure 14.3, employing data from [64]. The distribution system experiences the effects of a hurricane for a period of 24 hours, commencing at hour 0. As the hurricane transects the system, distinct line segments are exposed to varying wind speeds, contingent upon their proximity to the hurricane's eye. For ML algorithms, the data are divided into two parts: training and testing, with a ratio of 75% to 25%. Tables 14.3 to 14.8 show the performance metrics of six ML algorithms: naive Bayes, logistic regression, support vector machine (SVK), KNN, decision trees, and artificial neural networks with a logistic activation function. It is worth noting that the best performance of the nearest neighbor algorithm in terms of accuracy occurs at $K = 44$. Furthermore, for

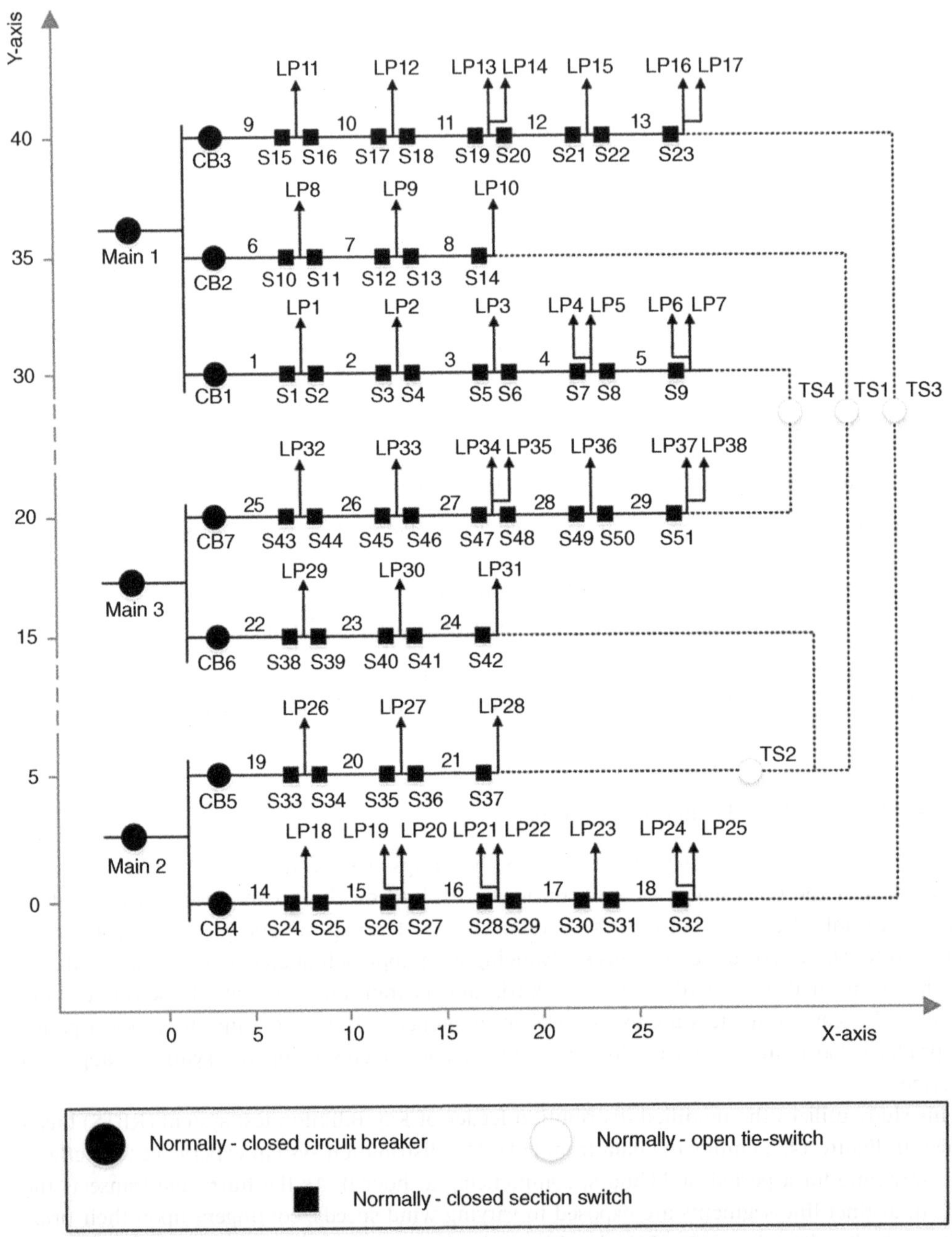

Figure 14.3 Modified distribution feeder of RBTS bus 4 (Source: [64]/John Wiley & Sons/CC BY 4.0).

the decision tree algorithm, the highest accuracy is obtained for a depth of 6. For single-layer neural networks, it is observed that the best accuracy corresponds to the logistic activation function with seven neurons.

Based on the above results, the SVM algorithm with a quadratic kernel and with the parameter $C = 10$ is the best algorithm according to the accuracy criterion.

Table 14.3 Performance evaluation parameters for the algorithm naive Bayes.

	Precision	Recall	F1-Score	Accuracy
Operational (0)	0.74	0.76	0.75	0.725
Damaged (1)	0.71	0.68	0.69	

Table 14.4 Performance evaluation parameters for the algorithm logistic regression.

	Precision	Recall	F1-Score	Accuracy
Operational (0)	0.75	0.75	0.75	0.724
Damaged (1)	0.70	0.69	0.70	

Table 14.5 Performance evaluation parameters for the algorithm SVMs.

Kernel		*C* = 1	*C* = 10	*C* = 100	*C* = 1000
Linear	F1-Score (operational/damaged)	0.74/0.69	0.74/0.69	0.74/0.69	0.74/0.70
	Accuracy	0.721	0.721	0.721	0.722
Quadratic	F1-Score (operational/damaged)	0.76/0.69	0.76/0.69	0.76/0.69	0.75/0.68
	Accuracy	0.726	0.728	0.727	0.725
Cubic	F1-Score (operational/damaged)	0.76/0.69	0.76/0.68	0.76/0.68	0.75/0.67
	Accuracy	0.718	0.724	0.723	0.724
Gaussion	F1-Score (operational/damaged)	0.76/0.67	0.76/0.67	0.76/0.67	0.75/0.67
	Accuracy	0.717	0.718	0.721	0.723

Table 14.6 Performance evaluation parameters for the algorithm K-nearest neighbors.

	Precision	Recall	F1-Score	Accuracy
Operational (0)	0.72	0.80	0.76	0.723
Damaged (1)	0.72	0.63	0.68	

Table 14.7 Performance evaluation parameters for the algorithm decision trees.

	Precision	Recall	F1-Score	Accuracy
Operational (0)	0.70	0.81	0.75	0.705
Damaged (1)	0.72	0.59	0.65	

Table 14.8 Performance evaluation parameters for the algorithm artificial neural networks with a logistic activation function.

	Precision	Recall	F1-Score	Total number	Accuracy
Operational (0)	0.74	0.76	0.75	1367	0.727
Damaged (1)	0.70	0.68	0.69	1136	

14.5 Conclusion

Our growing dependence on electricity necessitates strengthening our power systems, now facing the perils of cyberattacks, extreme weather events, and technological disruptions. The "resilience" paradigm, emphasizing adaptability and proactive measures, emerges as a beacon of hope in this challenging environment. ML, empowered by the vastness of data, drives this transformative revolution. This chapter delves into the concepts of resiliency, reliability, and security, illuminating their critical roles in the modern power grid. We explore existing ML applications in forecasting and risk analysis, followed by a case study showcasing ML's effectiveness in assessing power system vulnerabilities. This case exemplifies how ML facilitates proactive interventions, bolstering the resilience of the power grid. Ultimately, this chapter shows the pivotal role of ML in safeguarding our invaluable power infrastructure, paving the way for a secure and dependable energy future.

Acknowledgments

M. Chegnizadeh and M. Fotuhi-Firuzabad appreciate the support from INSF.

References

1 Treverton, G. (2017). Global Trends: Paradox of Progress. *NICDNI Wash. Pp Vi Ix.*

2 National Academies of Sciences and Medicine (2017). *Enhancing the Resilience of the Nation's Electricity System.* National Academies Press.

3 Johansson, T.B., Patwardhan, A.P., Nakićenović, N., and Gomez-Echeverri, L. (2012). *Global Energy Assessment: Toward a Sustainable Future.* Cambridge University Press.

4 National Conference of State Legislatures, Anderson, G., Cleveland, M., and Shea, D. (2019). *Modernizing the Electric Grid: State Role and Policy Options.* National Conference of State Legislatures.

5 Li, Y., Li, Z., Wen, F., and Shahidehpour, M. (2018). Minimax-regret robust co-optimization for enhancing the resilience of integrated power distribution and natural gas systems. *IEEE Transactions on Sustainable Energy* 11 (1): 61–71.

6 Preston, B.L. et al. (2016). Resilience of the US electricity system: a multi-hazard perspective. *DOE Rep. August*, p. 52.

7 Alimi, O.A., Ouahada, K., and Abu-Mahfouz, A.M. (2020). A review of machine learning approaches to power system security and stability. *IEEE Access* 8: 113512–113531.

8 Duchesne, L., Karangelos, E., and Wehenkel, L. (2020). Recent developments in machine learning for energy systems reliability management. *Proceedings of the IEEE* 108 (9): 1656–1676.

9 Guo, Y., Yang, Z., Feng, S., and Hu, J. (2018). Complex power system status monitoring and evaluation using big data platform and machine learning algorithms: a review and a case study. *Complexity* 2018.

10 National Research Council (2013). *The resilience of the electric power delivery system in response to terrorism and natural disasters: summary of a workshop*. National Academies Press.

11 Braun, M., Hachmann, C., and Haack, J. (2020). Blackouts, restoration, and islanding: a system resilience perspective. *IEEE Power and Energy Magazine* 18 (4): 54–63.

12 Science and Technology Select Committee (2014). *The Resilience of the Electricity System*. House Lords UK.

13 Moreno, R. et al. (2020). From reliability to resilience: Planning the grid against the extremes. *IEEE Power and Energy Magazine* 18 (4): 41–53.

14 Panteli, M. and Mancarella, P. (2015). The grid: Stronger, bigger, smarter?: Presenting a conceptual framework of power system resilience. *IEEE Power and Energy Magazine* 13 (3): 58–66.

15 Suri, N. and Cabri, G. (2014). *Adaptive, Dynamic, and Resilient Systems*. New York: Auerbach Publications.

16 Pimm, S.L. (1984). The complexity and stability of ecosystems. *Nature* 307 (5949): 321–326.

17 Perrings, C. (2006). Resilience and sustainable development. *Environment and Development Economics* 11 (4): 417–427.

18 Pisano, U. (2012). Resilience and sustainable development: theory of resilience, systems thinking. *European Sustainable Development Network (ESDN)* 26: 50.

19 Guttromson, R. and Watson, J. (2016). Defining, measuring, and improving resilience of electric power systems. In: *Smart Grid Handbook*, 1–21.

20 Hollnagel, E., Woods, D.D., and Leveson, N. (2006). *Resilience Engineering: Concepts and Precepts*. Ashgate Publishing, Ltd.

21 Righi, A.W., Saurin, T.A., and Wachs, P. (2015). A systematic literature review of resilience engineering: Research areas and a research agenda proposal. *Reliability Engineering and System Safety* 141: 142–152.

22 Lei, S., Wang, J., Chen, C., and Hou, Y. (2016). Mobile emergency generator pre-positioning and real-time allocation for resilient response to natural disasters. *IEEE Transactions on Smart Grid* 9 (3): 2030–2041.

23 Ma, S., Chen, B., and Wang, Z. (2016). Resilience enhancement strategy for distribution systems under extreme weather events. *IEEE Transactions on Smart Grid* 9 (2): 1442–1451.

24 Ma, S., Li, S., Wang, Z., and Qiu, F. (2019). Resilience-oriented design of distribution systems. *IEEE Transactions on Power Apparatus and Systems* 34 (4): 2880–2891.

25 Yang, L.-J., Zhao, Y., Wang, C. et al. (2019). Resilience-oriented hierarchical service restoration in distribution system considering microgrids. *IEEE Access* 7: 152729–152743.

26 Ding, Y., Morstyn, T., and McCulloch, M.D. (2022). Distributionally robust joint chance-constrained optimization for networked microgrids considering contingencies and renewable uncertainty. *IEEE Transactions on Smart Grid* 13 (3): 2467–2478.

27 Roche, R., Celik, B., Bouquain, D. et al. (2015). A framework for grid-edge resilience improvement using homes and microgrids coordination. *Presented at the 2015 IEEE Eindhoven PowerTech*, IEEE, pp. 1–6.

28 Pashajavid, E., Shahnia, F., and Ghosh, A. (2015). Development of a self-healing strategy to enhance the overloading resilience of islanded microgrids. *IEEE Transactions on Smart Grid* 8 (2): 868–880.

29 Saleh, M.S., Althaibani, A., Esa, Y. et al. (2015). Impact of clustering microgrids on their stability and resilience during blackouts. *Presented at the 2015 International Conference on Smart Grid and Clean Energy Technologies (ICSGCE)*, IEEE, pp. 195–200.

30 Hussain, A., Bui, V.-H., and Kim, H.-M. (2017). Resilience-oriented optimal operation of networked hybrid microgrids. *IEEE Transactions on Smart Grid* 10 (1): 204–215.
31 Farzin, H., Fotuhi-Firuzabad, M., and Moeini-Aghtaie, M. (2016). Enhancing power system resilience through hierarchical outage management in multi-microgrids. *IEEE Transactions on Smart Grid* 7 (6): 2869–2879.
32 Che, L. and Shahidehpour, M. (2014). DC microgrids: economic operation and enhancement of resilience by hierarchical control. *IEEE Transactions on Smart Grid* 5 (5): 2517–2526.
33 Liang, L., Hou, Y., Hill, D.J., and Hui, S.Y.R. (2016). Enhancing resilience of microgrids with electric springs. *IEEE Transactions on Smart Grid* 9 (3): 2235–2247.
34 Allicance, G. (2013). Improving electric grid reliability and resilience: lessons learned from superstorm sandy and other extremes events. *GridWise Alliance, Tech. Rep.*
35 Chen, C., Wang, J., Qiu, F., and Zhao, D. (2015). Resilient distribution system by microgrids formation after natural disasters. *IEEE Transactions on Smart Grid* 7 (2): 958–966.
36 Panteli, M., Trakas, D.N., Mancarella, P., and Hatziargyriou, N.D. (2016). Boosting the power grid resilience to extreme weather events using defensive islanding. *IEEE Transactions on Smart Grid* 7 (6): 2913–2922.
37 Ding, T., Lin, Y., Li, G., and Bie, Z. (2017). A new model for resilient distribution systems by microgrids formation. *IEEE Transactions on Power Apparatus and Systems* 32 (5): 4145–4147.
38 Chanda, S. and Srivastava, A.K. (2016). Defining and enabling resiliency of electric distribution systems with multiple microgrids. *IEEE Transactions on Smart Grid* 7 (6): 2859–2868.
39 Liu, X., Shahidehpour, M., Li, Z. et al. (2016). Microgrids for enhancing the power grid resilience in extreme conditions. *IEEE Transactions on Smart Grid* 8 (2): 589–597.
40 Wang, Z. and Wang, J. (2015). Self-healing resilient distribution systems based on sectionalization into microgrids. *IEEE Transactions on Power Apparatus and Systems* 30 (6): 3139–3149.
41 Li, G. et al. (2013). Risk analysis for distribution systems in the northeast US under wind storms. *IEEE Transactions on Power Apparatus and Systems* 29 (2): 889–898.
42 Arjomandi-Nezhad, A., Fotuhi-Firuzabad, M., Moeini-Aghtaie, M. et al. (2020). Modeling and optimizing recovery strategies for power distribution system resilience. *IEEE Systems Journal* 15 (4): 4725–4734.
43 Gan, W. et al. (2020). Coordinated planning of transportation and electric power networks with the proliferation of electric vehicles. *IEEE Transactions on Smart Grid* 11 (5): 4005–4016.
44 Ma, S., Arif, A. and Wang, Z. (2019). Resilience assessment of self-healing distribution systems under extreme weather events. *Presented at the 2019 IEEE Power & Energy Society General Meeting (PESGM)*, IEEE, pp. 1–5.
45 Montoya, M., Sherick, R., Haralson, P. et al. (2013). Islands in the storm: integrating microgrids into the larger grid. *IEEE Power and Energy Magazine* 11 (4): 33–39.
46 Schneider, K.P., Tuffner, F.K., Elizondo, M.A. et al. (2016). Evaluating the feasibility to use microgrids as a resiliency resource. *IEEE Transactions on Smart Grid* 8 (2): 687–696.
47 Balasubramaniam, K., Saraf, P., Hadidi, R., and Makram, E.B. (2016). Energy management system for enhanced resiliency of microgrids during islanded operation. *Electric Power Systems Research* 137: 133–141.
48 Dorostkar-Ghamsari, M.R., Fotuhi-Firuzabad, M., Lehtonen, M., and Safdarian, A. (2015). Value of distribution network reconfiguration in presence of renewable energy resources. *IEEE Transactions on Power Apparatus and Systems* 31 (3): 1879–1888.
49 Yang, Z., Dehghanian, P., and Nazemi, M. (2020). Seismic-resilient electric power distribution systems: harnessing the mobility of power sources. *IEEE Transactions on Industry Applications* 56 (3): 2304–2313.

50 Teo, T.T., Logenthiran, T. and Woo, W.L. (2015). Forecasting of photovoltaic power using extreme learning machine. *Presented at the 2015 IEEE Innovative Smart Grid Technologies-Asia (ISGT ASIA)*, IEEE, pp. 1–6.

51 Li, S., Ding, T., Jia, W. et al. (2021). A machine learning-based vulnerability analysis for cascading failures of integrated power-gas systems. *IEEE Transactions on Power Apparatus and Systems* 37 (3): 2259–2270.

52 Thukaram, D., Khincha, H., and Vijaynarasimha, H. (2005). Artificial neural network and support vector machine approach for locating faults in radial distribution systems. *IEEE Transactions on Power Delivery* 20 (2): 710–721.

53 Nateghi, R., Guikema, S.D., and Quiring, S.M. (2011). Comparison and validation of statistical methods for predicting power outage durations in the event of hurricanes. *Risk Analysis: An International Journal* 31 (12): 1897–1906.

54 Sun, R., Liu, Y., and Wang, L. (2018). An online generator start-up algorithm for transmission system self-healing based on MCTS and sparse autoencoder. *IEEE Transactions on Power Apparatus and Systems* 34 (3): 2061–2070.

55 Bu, F., Dehghanpour, K., Wang, Z., and Yuan, Y. (2019). A data-driven framework for assessing cold load pick-up demand in service restoration. *IEEE Transactions on Power Apparatus and Systems* 34 (6): 4739–4750.

56 Qiu, D., Wang, Y., Zhang, T. et al. (2022). Hybrid multiagent reinforcement learning for electric vehicle resilience control towards a low-carbon transition. *IEEE Transactions on Industrial Informatics* 18 (11): 8258–8269. https://doi.org/10.1109/TII.2022.3166215.

57 Bedoya, J.C., Wang, Y., and Liu, C.-C. (2021). Distribution system resilience under asynchronous information using deep reinforcement learning. *IEEE Transactions on Power Apparatus and Systems* 36 (5): 4235–4245.

58 Li, Y., Xu, Z., Bowes, K.B. et al. (2021). Reinforcement learning-enabled seamless microgrids interconnection. *Presented at the 2021 IEEE Power & Energy Society General Meeting (PESGM)*, IEEE, pp. 1–5.

59 Yao, S., Gu, J., Zhang, H. et al. (2020). Resilient load restoration in microgrids considering mobile energy storage fleets: a deep reinforcement learning approach. *Presented at the 2020 IEEE Power & Energy Society General Meeting (PESGM)*, IEEE, pp. 1–5.

60 Zhao, J., Li, F., Chen, X., and Wu, Q. (2021). Deep learning based model-free robust load restoration to enhance bulk system resilience with wind power penetration. *IEEE Transactions on Power Apparatus and Systems* 37 (3): 1969–1978.

61 Shorten, C. and Khoshgoftaar, T.M. (2019). A survey on image data augmentation for deep learning. *Journal of Big Data* 6 (1): 1–48.

62 Nikolenko, S.I. (2019). Synthetic data for deep learning. *ArXiv Prepr.* ArXiv190911512

63 Kaplan, J. and DeMaria, M. (1995). A simple empirical model for predicting the decay of tropical cyclone winds after landfall. *Journal of Applied Meteorology and Climatology* 34 (11): 2499–2512.

64 Hosseini, M.M. and Parvania, M. (2020). Quantifying impacts of automation on resilience of distribution systems. *IET Smart Grid* 3 (2): 144–152.

15

Cyberattacks on Power Systems

Alfan Presekal[1], Vetrivel Subramaniam Rajkumar[1], Alexandru Ştefanov[1], Kaikai Pan[2], and Peter Palensky[1]

[1] *Intelligent Electrical Power Grids, Department of Electrical Sustainable Energy, Faculty of Electrical Engineering, Mathematics and Computer Science, Delft University of Technology, Delft, South Holland, The Netherlands*
[2] *College of Electrical Engineering, Zhejiang University, Hangzhou, Zhejiang, China*

15.1 Introduction

Power grids are undergoing a fast-paced process of digitalization for enhanced monitoring and control capabilities and grid intelligence. Infrastructure and participants are enhanced and supported by information and communication technologies (ICTs). Further integration of digital technologies is vital for the development of the future power grid, for example, next-generation operational technologies (OTs), Internet of Things (IoT), digital substations, artificial intelligence, and Big Data analytics. All this is expected to increase sustainability, affordability, and resilience of the power system. The latter, however, is also challenged by all these new elements. Opening up the energy system to everyone by means of ICTs requires careful considerations with regard to data privacy and information security in general. This combined with the trend toward distributed renewable generation, electrification of virtually all aspects of our lives, and easy market participation for all energy system participants; the cybersecurity and resilience requirements of the power grid become even more critical. The increased digitalization raises questions, especially with regard to vulnerabilities, threats, and cybersecure operation of the power system. It is well recognized that information technology (IT)-OT systems are vulnerable to cyberattacks. Furthermore, the combination of heterogeneous, co-existing smart and legacy technologies generates significant vulnerabilities and security challenges. With respect to security of supply and reliability of the future energy system provision, special attention is needed for new vulnerabilities and threats that come with digitalization. Accordingly, cyber resilience aspects are critical for a further power grid digitalization.

Examples of cybersecurity incidents related to power grids already exist around the world. On December 23, 2015, cyberattacks were conducted on the power grid in Ukraine that resulted in power outages, which affected 225,000 customers. More sophisticated cyberattacks on the Ukrainian power grid followed on December 17, 2016, resulting in a power outage in the distribution network where 200 MW of load was unsupplied. On March 9, 2020, it was reported that the IT network of the European Network of Transmission System Operators for Electricity (ENTSO-E) had been compromised in a cyber intrusion. Fortunately, the compromised IT network

Smart Cyber-Physical Power Systems: Fundamental Concepts, Challenges, and Solutions, Volume 1, First Edition.
Edited by Ali Parizad, Hamid Reza Baghaee, and Saifur Rahman.

was not connected to any operational electric transmission system. However, this indicates that interconnected power grids may become targets. Such laborious cyberattacks conducted by powerful adversaries are a real threat to the security of the modern society. Cyberattacks on power systems can initiate cascading failures and result in a catastrophic blackout, ending up in a doomsday scenario especially if it is considered that the world may experience a global crisis such as a pandemic. The power outage can disrupt the entire energy chain including water supply, heating, and gas networks. Without power, hospitals and other critical services are severely affected. A disruption of service may lead to financial loss, damages, chaos, or even a loss of lives.

The complexity of cyberattacks on power systems is likely to increase. Cybersecurity and resilience to cyberattacks are emergent challenges for future power grids. The grid operational resilience ensures security of supply and a stable system operation following high-impact, low-frequency disturbances. Cyber resilience is defined as the capability of the power system to anticipate, absorb the shock, adapt, and rapidly recover from cyberattacks. A cyber kill chain contains the series of stages and steps used to trace the typical phases of a cyberattack from early reconnaissance to attack execution, which can result in a physical disruption of power system operation, instability, or even a blackout. The kill chain of cyberattacks on grid operators may start by exploiting vulnerabilities in the utility IT system through phishing emails and similar methods. Malware is installed to open gateways and facilitate remote access for system reconnaissance, weaponization, and OT targeting. Attackers can latterly move from the IT system into the OT system, which is used for power system operation, by stealing login credentials, escalating access privileges, and discovering networked IT-OT systems and hosts. In the OT system, they can tamper with the Supervisory Control and Data Acquisition (SCADA) system, disconnect power plants and entire substations, and cause physical damage to equipment by interfering with their control systems. To improve the cyber resilience of power grids, it is necessary to identify potential threats and IT-OT system vulnerabilities, classify and review major types of cyberattacks on power grids, analyze their impact on system operation and stability, and develop mitigation techniques to improve the four stages of system resilience.

This chapter provides state-of-the-art and essential knowledge of threats and cyberattacks on power systems. It reviews major cyberattacks on power grids and industrial control systems (ICSs) and provides a detailed taxonomy of cyberattacks. In this chapter, we classify the types of attacks into six categories, that is, phishing, malware, network-based attacks, man-in-the-middle (MITM) attacks, host-based attacks, and denial of service (DoS). The impact of cyberattacks on grid operation is analyzed in terms of loss of load, cascading effects, and equipment damage. A case study of a cyberattack scenario and simulation results are provided.

15.2 Cyber Kill Chain

The cyber kill chain is a framework for cybersecurity investigation and intelligence-driven defense. It is derived from a military model, originally established to identify, prepare to attack, engage, and destroy a target. Kill chains are used to understand, anticipate, recognize, and combat advanced persistent threats (APTs), social engineering attacks, ransomware, security breaches, and advanced attacks [1]. A cyber kill chain usually consists of seven stages, which represent the typical phases of a cyberattack, that is, reconnaissance, weaponization, delivery, exploitation, installation, command and control (C2), and actions and objectives. However, the order of these stages is not fixed. Based on the cyberattack scenario, they can vary and even some can run in parallel. Figure 15.1 depicts the cyber kill chain stages.

1) *Reconnaissance* is the stage where adversaries gather information about the targeted system such as network topology, communication protocols, Internet Protocol (IP) addresses, and running applications. Reconnaissance can be either active or passive. Active reconnaissance is an intrusive activity to test the target and extract more information through direct interaction. Hence, although it can uncover sensitive and critical information, it has a higher chance of detection by a defense system, for example, intrusion detection system (IDS). On the other hand, passive reconnaissance does not rely on direct interaction with the targeted system and therefore is stealthier. Passive reconnaissance is also known as passive information gathering. Passive reconnaissance was conducted in the early stages of the cyberattack in Ukraine 2015 to gather information about the distribution system operators (DSOs) and consequently launch a spear phishing campaign. Additionally, active reconnaissance was also conducted to extract information about network topology and active hosts in the targeted IT-OT system.
2) *Weaponization* is the process of preparing the attack vector. Weapons used in cyberattacks can be represented by malware or tools to achieve remote code execution. For an effective weaponization, adversaries first need to understand the targeted system and decide on appropriate tools to conduct the attack. Examples of weaponization include the Stuxnet malware used in the cyberattack on the Iranian nuclear reactor and BlackEnergy3 malware used in the cyberattack on the Ukrainian power grid.
3) *Delivery* is a mechanism to transfer the weaponized bundle to the victim or targeted system. Some of the most predominant delivery methods include email attachments, websites, software applications, and Universal Serial Bus (USB) removable media. For example, in the Ukraine 2015 and 2016 cyberattacks, delivery was achieved through targeted spear phishing emails containing malicious attachments. On the other hand, in the Stuxnet attack, delivery was achieved through USB removable media.
4) *Exploitation* is a stage where attackers exploit vulnerabilities to execute code on the targeted system, for example, vulnerabilities of operating systems, service applications, and communication protocols. For example, in the Ukraine 2015 cyberattack, attackers exploited vulnerabilities present in the Windows active directory (AD) server to steal login credentials.
5) *Installation* is a process to deploy malware on assets, such as backdoors, Trojans, and botnets. These malicious applications allow adversaries to gain control of the targeted system and maintain their persistence. For example, in the Ukraine 2015 cyberattack, the installation process was done through a malicious macro-script hidden within a Microsoft Excel file.
6) *C2* allow adversaries to remotely control the targeted system by exploiting system vulnerabilities using malicious applications. C2 typically can be found in botnets where the adversaries can remotely gather information and deliver commands.
7) *Actions and Objectives* represent the stage where the intruders accomplish their goals. In most cases, adversaries remain in a stealthy mode until they reach their final goal and reveal their true objective. In the Ukraine 2015, cyberattack for example, the attackers' objective was to cause a power outage. This was achieved taking over control of the SCADA system and opening multiple circuit breakers to disconnected circuits and cause a direct power outage in the distribution network.

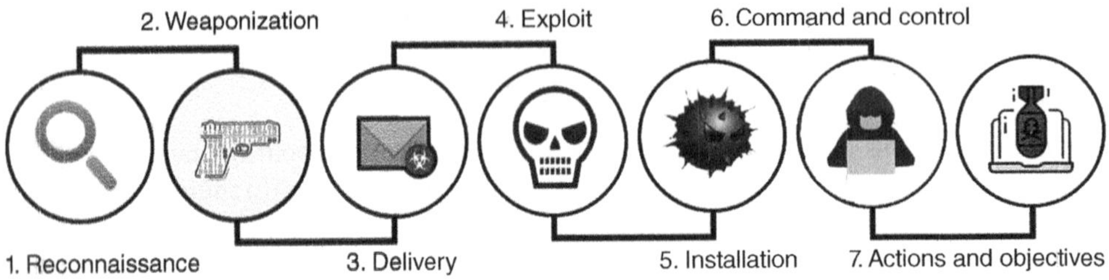

Figure 15.1 Cyber kill chain.

The cyber kill chain is suitable to investigate various cyberattack strategies and APTs. Most recent trends show that there are emerging APTs targeting power grids as demonstrated repeatedly in Ukraine. These are reviewed in Section 15.3, based on the aforementioned cyber kill chain framework.

15.3 Review of Major Cyberattacks

15.3.1 Cyberattacks on Industrial Control Systems

ICS is a broad term, typically used to describe the integration of physical processes with sensors, actuators, communication networks, and controllers in order to support various industries and critical infrastructures. Large-scale ICSs deploy SCADA for monitoring and control of the physical processes. Hence, an ICS is a cyber-physical system, integrating both cyber and physical aspects of the industrial process. Such systems are commonly found in power plants, manufacturing, and processing facilities. Due to their cyber-physical nature, ICSs are prone to cyberattacks. Compared to cyberattacks on IT systems, cyberattacks on ICS of critical infrastructures can have a devastating impact on the modern society with safety and financial implications. Cybersecurity threats targeting ICS have existed for several decades now. A historical record of cyberattacks on ICSs between 1982 and 2017 is reviewed in [2–4], and a summary is given in Table 15.1. Based on historical trends, it can be concluded that cyberattacks targeting ICS are on the rise.

Most of the cyberattacks on ICS have not resulted in direct physical damages. However, nearly all of them have resulted in data breaches and disruptions to system operations. Nevertheless, there are some incidents that have caused a direct physical impact. One such recorded cyberattack was a Trojan software attack on the SCADA system of a Siberian pipeline in 1982. The attack triggered an explosion in the pipeline equivalent to three kilotons of TNT [5]. This cyberattack put human lives at risk and caused immense physical damage.

The most notable and well-known cyberattack targeting ICSs is Stuxnet, which was reported in 2010. Stuxnet was the first, widely known, cyberwarfare weapon. Reports about the Stuxnet malware have been presented in [6–8]. Being a weaponized malware, its creators had strong knowledge of the SCADA system operation. According to the investigation carried out in [7], it triggered a shift in rotational frequency of the motor's programmable logic controller (PLC), damaging to the uranium enrichment process. In the future, ICSs of critical infrastructures may become obvious targets for state-sponsored cyberattacks. Hence, cybersecurity of ICS is an important concern for industries and governments, alike.

Table 15.1 Summary of cyberattacks targeting ICSs.

Cyberattack	Year	Description
Siberian pipeline explosion	1982	A Trojan attack was conducted on the SCADA system of a Siberian pipeline. This attack caused an explosion equivalent to 3 kilotons of TNT.
Chevron emergency alert system hacking	1992	A former Chevron employee hacked and disabled the emergency alert system. The emergency alert system was down for more than 10 hours putting at risk people from more than 22 states in Canada.
Salt River project	1994	Hackers gained unauthorized access and took control of the SCADA system of a 131-miles canal for five hours.
Worcester Massachusetts airport	1997	Hackers disabled for six hours the system which controlled the telephone lines in the airport.
Gazprom	1999	Hackers conducted a Trojan attack on Gazprom (a Russian gas company) and gained control of the gas flow pipelines.
Maroochy Water	2000	Hackers gained control to the water facil ity's SCADA system and released 265,000 gallons of untreated sewage.
California system operator	2001	Hackers infiltrated into a process control system in California.
Davis-Besse nuclear power plant	2003	A Structured Query Language (SQL) Slammer worm infected the control system of the nuclear power plant. Safety parameters and process display computers were disabled for several hours.
CSX corporation	2003	A computer virus named Sobig infected and disabled the train signaling system in Florida, USA. The virus reportedly spread via email attachments.
Tahoma Colusa canal	2007	A former employee installed an unauthorized application to the Canal SCADA system.
Turkey pipeline explosion	2008	Attackers exploited vulnerabilities in the security camera software to gain physical access to the control center. Subsequently, they triggered a pipeline explosion.
Stuxnet	2010	A weaponized malware attack targeted the uranium enrichment processes at Iranian nuclear facilities. The malware caused centrifuges to spin abnormally, while blindsiding operators.

(Continued)

Table 15.1 (Continued)

Cyberattack	Year	Description
Night Dragon	2010	Various malware attacks targeted oil, petrochemical, and energy companies.
Duqu	2011	A malware attack targeted specific organizations including the industrial control systems of various manufacturers. It served as an ICS reconnaissance tool.
Flame	2012	A malware targeted oil companies in the Middle East and North Africa. The malware's main function is to steal data from targets.
Malware infection of the SCADA system in a power plant	2012	The SCADA system of a power plant in USA was infected with malware via an USB drive during maintenance. The infection caused a three-week restart delay of the power plant.
New York dam	2013	Iranian hackers reported the launch of a cyberattack on industrial control system of Bowman Dam in New York.
Havex	2013	A malware campaign was conducted, mainly targeting industrial control systems.
German steel mill	2014	Cyberattacks using malware targeted the SCADA system of a German steel mill causing significant damages.
Kemuri water company	2016	Hackers gained access to the SCADA system and manipulated the control applications.
TRITON	2017	A malware attack was conducted on a Saudi Arabian petrochemical plant targeting the industrial safety systems.

15.3.2 Cyberattacks on Power Grids

Increased power grid digitalization, driven by the energy transition, has introduced cyberattacks on power grids as a real modern-day threat. Such advanced cyberattacks come with worrying ramifications. They can be classified as high-impact, low-frequency events with a wide range of effects. It is worth noting, however, that there are only a handful of known real cases of cyberattacks specifically targeting power grids. Nevertheless, such attacks have established the means and provided a glimpse of the possible disastrous consequences. The most well-known examples of cyberattacks targeting power grids, that is, Ukraine attacks in 2015 and 2016, are extensively dealt with in the subsequent subsections.

15.3.2.1 Ukraine 2015

On December 23, 2015, at 15:30 local time, a cyberattack was conducted on the power grid in Ukraine. This attack is the first publicly known cyberattack targeting power systems, which resulted into a power outage. The attackers compromised the SCADA systems of three DSOs and disconnected seven 110 kV and twenty-three 35 kV substations from the distribution network. The attack was conducted successfully. The power outage affected 225,000 customers [9, 10].

Figure 15.2 presents the typical IT-OT systems of utilities. The IT network is used for the non-operational side of the business, for example, asset and resource management, geographic information system, legal, finance, human resources, and payroll. The IT and OT systems in the control center are interconnected. However, security controls such as firewalls are in place to keep the IT-OT networks segmented, and OT network separates from a direct connection to the Internet. Segmentation of the control center from the corporate IT network is implemented for cybersecurity considerations. The OT systems gather data from substation bays and station control systems, for example, voltage and current magnitudes, active and reactive powers, circuit breaker status, and transformer tap positions, and send them to the SCADA servers in the control center. The real-time data are used by grid operators for power system operation. However, the cyberattack in Ukraine 2015 proved that the utility IT-OT systems with their current security controls are not impenetrable. The IT network segment served as the entry point for the attackers. Thereon, the attackers continued to the control center and substations with the objective to cause a blackout. Based on the results of forensic investigation, the cyber kill chain is used to divide the

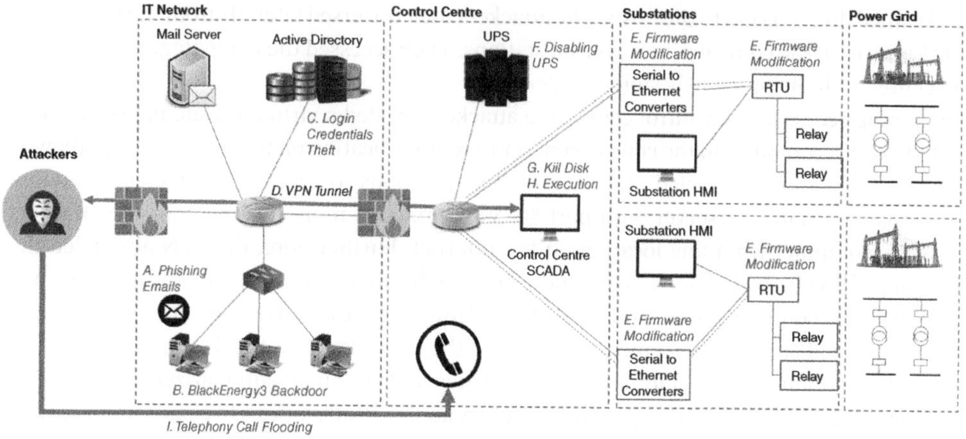

Figure 15.2 Cyberattack on the power grid in Ukraine in 2015.

cyberattack into nine stages in a chronological sequence. This indicates that the attackers adopted sophisticated attack techniques at every stage.

Spear phishing served as the entry point of the attack in the corporate IT network. The attackers specifically targeted three DSOs, pretending to be officials from the Ukraine Ministry of Energy. The malicious emails, seemingly originating from trustworthy sources, contained a weaponized Microsoft Excel file attachment. The attackers exploited Microsoft Excel's macro vulnerabilities to install their malware in the DSO IT network. A macro is a program created to automate tasks in Microsoft Excel using Visual Basic Application (VBA) scripts. When the recipients enabled the macro, the VBA script was launched, installing BlackEnergy3 malware on the computer. After a successful installation, the malware connected to attackers' remote C2 server using IP address 5.149.254.114 and port number 80. The remote IP address and port number were specifically coded in the malware. The connection via port 80 seemed innocuous. This is because port 80 is a common port for website traffic via hypertext transfer protocol (HTTP). With the connection to the C2 server enabled, the attackers could remotely control the BlackEnergy3 malware.

BlackEnergy3 provides various functionalities such as network scanner, file stealer, password stealer, keylogger, screenshot capturer, and network discovery. BlackEnergy3 malware played a very important role during the early stages of active reconnaissance and C2. Using the network scanner and network discovery modules, the malware found information about the IT network configuration, for example, network segments, network topology, and hosts connected. With BlackEnergy3, the attackers could also steal passwords using a keylogger, steal files, and capture screenshots of the targeted computers. This preliminary information was important to plan the following attack stages. The attackers transferred all the collected information directly to the remote C2 server.

From the active reconnaissance stage, the attackers found a vulnerable AD server which became a breach point of the IT-OT system. The AD is a Windows-based database service for IT networked system operations. An AD manages access permissions for a user by acting as a central authentication and authorization authority for managed accounts, hosts, and services. AD makes the IT system operation easier and more flexible with centralized authentication and authorization instead of using segregated services. AD databases also store usernames, passwords, and host and services information. Hence, the AD serves as a critical point in the entire DSO IT network. In the Ukraine 2015 cyberattack, the attackers compromised the AD server to gain access to a majority of hosts in the IT network. From the compromised AD server, attackers obtained user login credentials and gained direct access to the hosts. The attackers then traversed laterally from the IT network to the OT system in the control center and substations. They accessed the computers located in the control center with the previously stolen credentials.

After gaining access to the control center, the attackers created a virtual private network (VPN) connection from a computer in the control center to a remote location on the Internet. VPN allowed the attackers to perform tunneled and encrypted connections to access the targeted computer. Instead of using a static C2 server with port 80, VPN provided more flexibility for the attackers to access the computer from any location on the Internet. Furthermore, the VPN also aided the attackers in evading firewall detection and hide their real locations. At this point, the attackers had already gained full control to launch the final attack. However, the attackers remained stealthy, carrying out additional actions to magnify the impact of the cyberattack on the distribution network.

The attackers conducted four additional stages, that is, E, F, G, and I, summarized in Table 15.2, to increase the attack severity. In stage E, the attackers accessed substations to compromise remote terminal units (RTUs) and the serial-to-Ethernet converters. A serial-to-Ethernet converter connects Ethernet communications in the substation, for example, IEC 104, to serial communications

Table 15.2 Ukraine 2015 cyber kill chain.

Stage	Process	Cyber kill chain
A	Attackers send phishing emails containing Microsoft Excel file attachments. The file attachment is weaponized with BlackEnergy3 malware. When the recipient opens the Excel file, a macro function is launched and BlackEnergy3 is installed onto the computer.	Weaponization, delivery, installation
B	The attackers use the BlackEnergy3 and compromised computers to serve as backdoors for remote C2 actions. Further investigation of BlackEnergy3 has been presented in [11, 12]. Using BlackEnergy3, attackers also performed active reconnaissance and lateral movement in the IT network.	C2 and reconnaissance
C	The attackers successfully locate and compromise the AD server on the IT network. From the AD database, the attackers steal user login credentials. Using the stolen credentials, the attackers access devices located in the control center.	Exploitation
D	After gaining access to the OT system, attackers create a VPN tunnel from the Internet directly into the control center. This VPN tunnel allows direct access and remote execution. According to the investigation in [9], attackers compromised the IT network and control center OT system six months before the final attack execution on December 23, 2015. However, during this period, their malicious activities went undetected and the attack became an APT. The attackers could have launched their attack much earlier. However, they maintained stealth and created a more devastating impact. To increase the attack severity, stages E, F, G, and I were conducted.	Installation, exploitation, and C2
E	The attackers modify the firmware of the serial-to-Ethernet converters and RTUs in substations.	Exploitation and installation
F	The attackers disable the UPS backup power supply in the control center.	Exploitation and installation
G	The attackers install KillDisk in the control center and erase hard disk data, leaving no traces of activity.	Exploitation and installation
H	On December 23, 2015, at 15:30 local time, attackers start the cyberattack execution on distribution network operation by remotely controlling the SCADA system and opening circuit breakers in substations. This step successfully disconnected seven 110 kV and twenty-three 35 kV substations directly causing power outages. Under normal conditions, power system restoration is done remotely from the control center or manually from substations. However, the remote control was unsuccessful as the RTU and serial-to-Ethernet converter firmware were compromised. The compromised UPS in the control center could not provide backup power to SCADA servers, exacerbating the situation. Furthermore, the SCADA system failed to operate because KillDisk execution wiped out the server's hard disks.	C2 and actions and objectives
I	The attackers also conducted a telephone call flooding attack on the DSO call center from foreign numbers so that customers could not report the power outages. Due to this complicated situation, restoration efforts took around six hours.	Actions and objectives

with the control center, for example, IEC 101. These devices are typically embedded without an operating system. The processes are controlled by firmware. Hence, the attackers also created malicious firmware and replaced the legitimate firmware in the RTUs and serial-to-Ethernet converters to make them dysfunctional upon reboot. The malicious firmware ensured that grid operators could not remotely control the substations to perform emergency restorative actions. In stage F, the attackers accessed the uninterruptible power supply (UPS) units in the control center. Subsequently, they disabled the UPS backup power supply to cause an outage in the control center as well during power outage in the distribution network. In stage G, the attackers used KillDisk to erase hard drives and delete data in the control center servers and computers and make them unbootable. KillDisk is a module of BlackEnergy3 malware for deleting data, registry, and system configuration. In stage I, immediately after the final attack execution on December 23, attackers also conducted a telephone call flooding attack on the DSO call center from foreign numbers so that customers could not report the power outages. Due to these attack stages, the power outage could only be restored after six hours. Thus, the attackers successfully executed one of the two most significant cyberattacks targeting power systems to date. Table 15.2 provides a summary of the attack stages and mapping to the cyber kill chain.

15.3.2.2 Ukraine 2016

On December 17, 2016, at 23:53 local time, a second cyberattack was conducted on the power grid in Ukraine. This is the first publicly acknowledged cyberattack, involving malware that targeted power systems and resulted into a power outage. It affected the SCADA system at the transmission level targeting a single 330 kV substation. The cyberattack resulted into a power outage in the distribution network where the total unsupplied load was 200 MW. Compared to the 2015 attack, the 2016 attack was more advanced in terms of the attack technique. Thankfully, this attack resulted in a much lower impact, compared to the previous one. The 2016 attack employed sophisticated malware named CRASHOVERRIDE or Industroyer. Further studies related to this attack are presented and discussed in [13–17]. Based on these investigations and the cyber kill chain framework, we classify the Ukraine 2016 attack into seven stages in chronological sequence as represented in Figure 15.3 and summarized in Table 15.3.

The overall lower impact of the cyberattack in 2016 can be attributed to several causes. The primary reason was the limited success of the malicious payload injection. This was probably caused by a rigid attack technique, predefined within the code by the attackers. Most likely, they prepared the protocol payload module based on their test systems, which was later deployed on the targeted system. Hence, the attackers did not have an opportunity to fully test their attack methods on a real-world ICS. Such an attack scenario while being sophisticated comes with the cost of easy detection by system operators. Nonetheless, the attackers successfully demonstrated advanced attack strategies with a deeper understanding and insight of the targeted power system. In the future, such types of attack may potentially become more common, leading to catastrophic damages to the power grid infrastructure. The overall comparison between Ukraine 2015 and 2016 cyberattacks is provided in Table 15.4.

15.4 Taxonomy of Cyberattacks on Power Grids

There are many attack techniques that can potentially be deployed to specifically target power grids. In this section, we classify such types of attacks into six categories, that is, phishing, malware, network-based attacks, MITM attacks, host-based attacks, and DoS. Figure 15.4 shows the taxonomy of cyberattacks on power systems and ICS. We delve in depth into each cyberattack category targeting ICSs and power grids in the following subsections.

Table 15.3 Ukraine 2016 cyber kill chain.

Stage	Process	Cyber kill chain
A	In January 2016, phishing emails targeting the power grid operator in Ukraine were reported. The attackers may have used a different strategy for phishing compared to the previous year. Unfortunately, detailed information in this regard is not available. Nonetheless, phishing emails are assumed to be the entry point of the attack.	Delivery
B	Slowik et al. identified that malware was clearly involved in the early stages of the cyberattack [17]. The malware was used to conduct reconnaissance, initial lateral movement, and credential theft within the IT system. Unlike the 2015 attack, there is no substantial information on how the attackers actually stole credentials.	Weaponization and installation, reconnaissance
C	The attackers established a VPN tunnel to bypass the firewall and perform remote C2 via the compromised server.	Installation and C2
D	The attackers used a MySQL server as the central point of the attack, executing commands from within the IT network to the control center OT systems. At this stage, the attackers also gathered information on type of protocols being used in the control center OT network. This knowledge and information were vital for the attackers to prepare for the subsequent attack stages.	Exploitation, C2, and reconnaissance
E	Next, the attackers sent a malicious text file to the SCADA server in the OT network segment. Upon arriving at the designated host, the file changed into an executable file containing CRASHOVERRIDE malware. CRASHOVERRIDE is a unique malware, designed and created based on the knowledge of ICS protocols such as IEC 101, IEC 104, IEC 61850, and open platform communication (OPC). As a result, the malware could deliver crafted traffic based on those specific protocols.	Weaponization, delivery, and installation
F	In the final stages of the attack, the attackers took control of the compromised hosts in the control center to send malicious payloads to substations and open circuit breakers. These attacks were launched on December 17, 2016, at 23:53 local time. The attacks affected the SCADA system at the transmission level focusing on a single 330 kV/110 kV/10 kV substation, resulting in a distribution-level outage. Soon after, operators swiftly responded to the attack and transferred controls into manual mode.	C2 and actions and objectives
G	After the final attack, the wiper module which was also a part of the CRASHOVERRIDE overwrote system service registry entries to null values to render the system unbootable. The wiper module removed files relating to ICS operations to prevent swift IT-OT recovery and power system restoration.	Actions and objectives

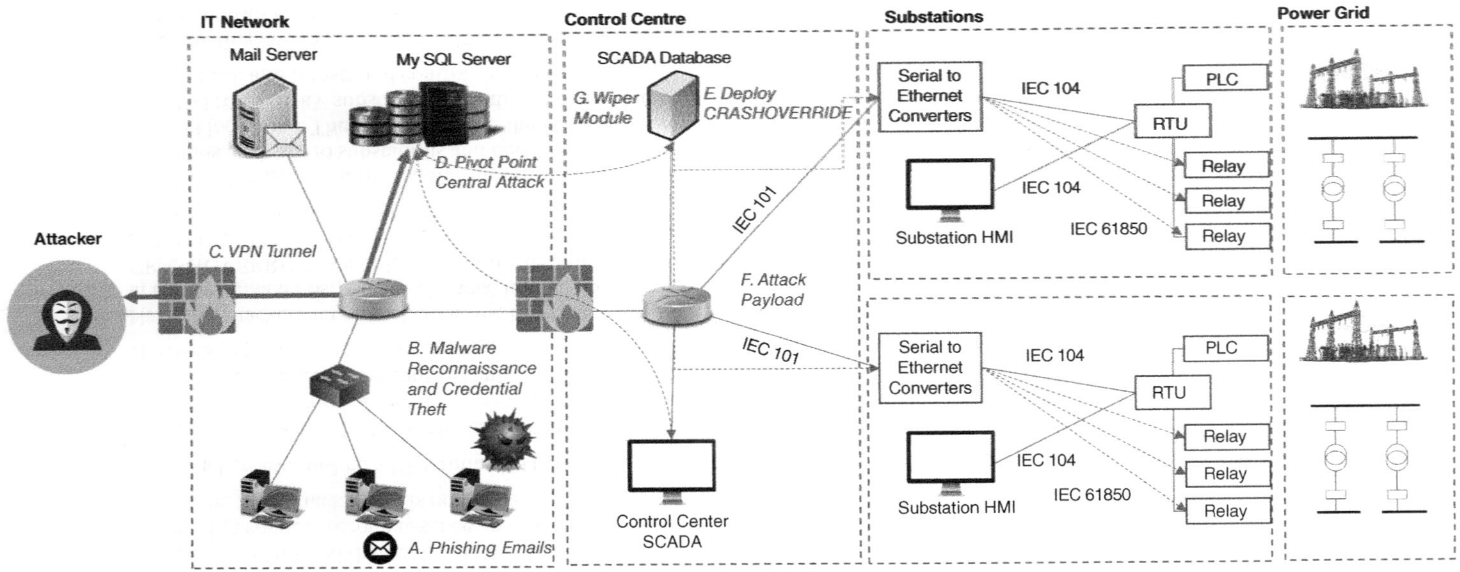

Figure 15.3 Cyberattack on the power grid in Ukraine in 2016.

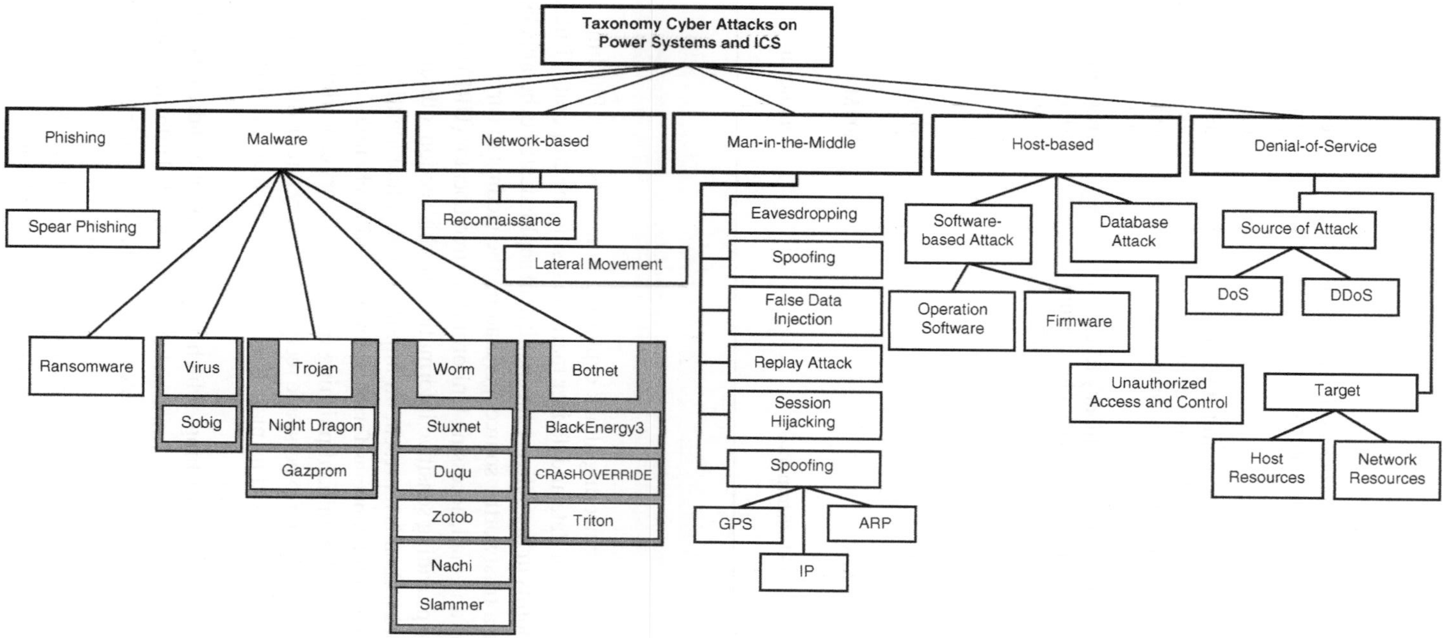

Figure 15.4 Taxonomy of cyberattacks on power systems and ICS.

Table 15.4 Comparison of Ukraine 2015 and 2016 cyberattacks.

Metric	Ukraine 2015 attack	Ukraine 2016 attack
Malware	BlackEnergy3	CRASHOVERRIDE
Role of the malware	Remote C2, reconnaissance, credential theft, delete system files, and corrupt OT systems	Remote C2, reconnaissance, credential theft, delete system files, corrupt OT systems, and launch scheduled attack
Final attack stage	Attackers took control of SCADA via remote desktop and opened multiple circuit breakers in real time	Automated malware was launched with malicious payload instructions to open circuit breakers
Increase attack severity	Compromise UPS, modify RTU firmware, erase system files, corrupt OT systems using KillDisk, and conduct telephone call flooding	Erase system files and corrupt OT systems using wiper module
Affected substations	30	1
Power outage duration	Six hours	30 minutes
Load unsupplied	135 MW	200 MW
Affected customers	225,000 customers	Unknown

15.4.1 Phishing

Phishing is a type of social engineering attack exploiting human factors. Phishing attacks aim to obtain sensitive information or data and deliver malware inside the corporate IT network, thereby often serving as entry points in the delivery mechanism of sophisticated cyberattacks. In phishing, attackers pretend to be a trustworthy source, persuading the victims to carry out certain actions. These could be opening an email, accessing Uniform Resource Locator (URL) links, downloading files, and unknowingly providing sensitive information. In this section, our discussion is focused on email phishing attacks. Emails represent an important individual identity on the Internet and also carry organizational significance as formal communication channels. Due to these reasons, emails have become the most dominant phishing media. Attackers can arbitrarily send phishing emails to any email address, aiming to persuade recipients to carry out further actions, as described above. These actions may be followed by malicious outcomes such as information breaches, malware installation, and financial losses.

Spear phishing is a variant of phishing, which targets a specific individual or organization. Before launching a spear phishing attack, an attacker gathers detailed information about the target. Information gathering can be done via the Internet or social engineering. Based on preliminary information, the attacker crafts the phishing email content to be relevant and strongly related to the target, prompting a higher success rate compared to arbitrary phishing. In general, state-sponsored attackers or cyber criminals are the ones behind spear phishing attacks. Consequently, spear phishing serves as an important entry point for APTs. For example, in Ukraine 2015, the attackers pretended to be from the Ukrainian Ministry of Energy, which was strongly related to the targeted

DSOs. Hence, the spear phishing campaign was successful in deceiving the DSO employees to access the malicious email attachments, resulting in the installation of malware.

Protection against phishing attacks can be achieved by strong organizational policies and corporate IT security. This can include verifying and screening email address sources, URL links, and file attachments in emails. Through such measures, malicious emails can then be flagged and filtered, serving as the initial line of defense against phishing attacks. However, the most critical defense method is raising user awareness. With proper awareness and training, users can recognize phishing emails.

15.4.2 Malware

Malware is a piece of malicious code that exploits software vulnerabilities in a targeted system. The occurrence of malware-related attacks has been on the rise, especially over the last decade [18]. There are various types of malware, for example, ransomware, virus, Trojan, worm, and botnet, examples of which are presented in Figure 15.4. Advanced types of malware such as worms and botnets, owing to their intelligence and controllability, are gaining popularity as modern cyberattack vectors. Furthermore, given their ability to propagate and spread independently, they have become the cyber weapons of choice targeting ICSs. Therefore, in this section, we focus on the most well-known examples of worms and botnets involved in cyberattacks on ICS and power grids, that is, Stuxnet, BlackEnergy, CRASHOVERRIDE, and Triton. Table 15.5 shows an overview of the comparison of the capabilities between most commonly found types of malware in ICS-related cyber incidents.

Table 15.6 shows a brief history of malware involved in ICS cyber-related incidents from 1982 to 2017. It can be seen that worms were popular in the early 2000s due to their capabilities to spread across the network. However, an attacker cannot fully control a worm, leading to an uncontrollable spread. This may result in loss of stealth and unexpected exposure and discovery. For example, the Stuxnet worm was originally created only to target Iranian nuclear reactors. Unfortunately, it ended up spreading across the world [7], leading to its discovery in 2010. As a result, botnets have become the malware of choice in recent years. A botnet can be controlled remotely and acts as a vector for APTs [9, 16].

15.4.2.1 Stuxnet

Stuxnet is a worm discovered in June 2010, designed specifically to target PLCs in ICSs. It exploited unprecedented zero-day vulnerabilities present in the Microsoft Windows operating system, Siemens STEP7 PLC software, and remote procedure call (RPC) server mechanism. Stuxnet also employed the rootkit technique to hide from commercial antivirus software. It was later identified as a state-sponsored and developed cyber weapon to target the uranium enrichment facilities in Iran. Detailed technical studies about Stuxnet are presented in [7, 24].

Table 15.5 Comparison of malware capabilities.

No.	Capabilities	Viruses	Trojans	Worms	Botnets
1	Self-replication	✓		✓	
2	Backdoor		✓		✓
3	Remote control		✓		✓
4	Network spread			✓	✓

Table 15.6 History of malware involved in major ICS cyber-related incidents.

No.	Target	Type of malware (name)	Year
1	SCADA gas pipeline, Siberia [2]	Trojan horse	1982
2	SCADA gas pipeline, Russia [2]	Trojan horse (Gazprom)	1999
3	Web service RTU, PLC [19]	Worm	2002–2003
4	SCADA database [19]	Worm (Slammer)	2003
5	SCADA petrochemical plant [19]	Worm (Nachi)	2003
6	Train signaling system, USA [2]	Virus (Sobig)	2003
7	SCADA automotive manufacturing plants [20]	Worm (Zotob)	2005
8	Energy companies [21]	Trojan horse (Night Dragon)	2009
9	SCADA nuclear centrifuge [8]	Worm (Stuxnet)	2010
10	SCADA reconnaissance [2]	Worm (Duqu)	2011
11	SCADA steel mill, Germany [22]	Botnet	2014
12	SCADA power grid, Ukraine [11]	Botnet (BlackEnergy3)	2015
13	SCADA power grid, Ukraine [17]	Botnet (CRASHOVERRIDE)	2016
14	SCADA petrochemical plant, Saudi Arabia [23]	Botnet (Triton)	2017

Stuxnet spread through three main mechanisms. The initial entry point was via USB flash drives in computers located within the SCADA system. When a USB drive is plugged into a computer, the Windows operating system will execute autorun.inf or Windows .lnk as an autorun mechanism. Exploiting this vulnerability, Stuxnet copied itself onto the computer's hard drive. The second spreading mechanism was through Windows network shares, wherein Stuxnet had the capability to duplicate itself into a shared folder on the same network. Through this mechanism, Stuxnet exploited the Server Message Block (SMB), popularly known as Samba, which is a protocol for file and folder sharing. The third spreading mechanism exploited vulnerabilities present in Siemens' software, that is, Siemens WinCC and Siemens STEP7. Stuxnet identified and compromised access to computers that ran Siemens WinCC. Furthermore, it also replicated itself into WinCC computers via an SQL injection command. Siemens STEP7 is an application used to program Siemens PLCs. Hence, through STEP7, Stuxnet spread into PLC devices as well. If a computer was running Siemens STEP7, Stuxnet modified the Windows Dynamic Link Library (DLL) and associated executable files. With this infection, Stuxnet added malicious code into the PLC devices, allowing complete control over them. In the reported incidents, it caused centrifuges to spin abnormally, while blindsiding operators.

Taking a closer look at its structure, Stuxnet mainly consists of two modules, that is, user mode and kernel mode. The user mode module has four functions: (i) searching function for specific targets, (ii) privilege escalation, (iii) malicious code injection into PLC, and (iv) installation kernel mode. It is interesting to note that the malware also had a time limit date set to June 24, 2012. Hence, it was functional only until that specific date. The kernel mode made Stuxnet work on the lower system levels, well below the application level of the user mode. By implementing the kernel mode, Stuxnet was maliciously launched during every Windows bootup process. This system-level execution made it more persistent and impervious to antivirus protection. The main functionality of Stuxnet was defined in its code. However, it was also designed to remotely communicate to the C2 server. This mechanism allowed adversaries to remotely update the malware. However, this remote update was never performed, in order to limit the uncontrollable spread of Stuxnet outside of its designated target.

15.4.2.2 BlackEnergy

BlackEnergy is a malware, initially identified in 2007 as an HTTP-based botnet for distributed denial of service (DDoS) attacks. It identified and targeted multiple file extensions, including Microsoft Office, Java, and executable files. More specifically, BlackEnergy injected malicious files into the Windows System32 folder on a target Windows machine. However, its most infamous use was as a weaponized malware during the cyberattack on the power grid in Ukraine, in December 2015.

In this attack, BlackEnergy3 exploited the SMB protocol to propagate across the IT-OT networks. SMB was used as the attack vector due to its capabilities to bypass typical firewalls. Through the infected SMB, BlackEnergy3 replicated across hosts in the IT network, delivering malicious payloads. In addition, it was accompanied by an RPC to serve as a backdoor module. This was done to establish a connection between the infected hosts and the attacker's C2 server. RPC allowed the malware to receive commands from the remote C2 server, and relay critical information back. Such advanced remote capabilities allowed adversaries to perform early reconnaissance using its many plugins during the attack. These plugins were programmed with many functionalities, such as executing file operations, that is, enumerate, execute, download, and overwrite; stealing credentials; discovering networks; and self-destructing. The self-destruction function of BlackEnergy3 was executed via KillDisk. Furthermore, BlackEnergy3 also contained information regarding the current time and location. This allowed it to run malicious activities during non-peak hours, for example, at midnight. Overall, with all these capabilities, BlackEnergy3 has been proven to be a vicious tool for cyberattacks on power grids. Further detailed investigation of BlackEnergy3 is presented in [12].

15.4.2.3 CRASHOVERRIDE

CRASHOVERRIDE or Industroyer was the root cause of the cyberattack on the power grid in Ukraine in December 2016. Its many features such as backdoors, intrusion, and reconnaissance strategies are quite similar to BlackEnergy3. However, according to its code level investigations, there is no strong connection between the two malware. It is very likely that CRASHOVERRIDE was a new type of malware, specifically created for the Ukraine 2016 attack [13]. It mainly comprised of three components, that is, backdoor, payload, and launcher.

The backdoor of CRASHOVERRIDE can be further subclassified into the main backdoor and the additional backdoor. The main backdoor served as the main controller, connecting to a remote C2 server via hypertext transfer protocol secure (HTTPS). Remote commands were encapsulated as HTTPS traffic, while source and destination addresses for establishing connectivity to C2 server were hardcoded. This clearly shows that CRASHOVERRIDE was created on purpose to specifically target the Ukrainian power grid. Using the backdoor, attackers could define a specific time to activate the malware allowing them to perform a multitude of actions. This includes remote control for process execution, switch execution into a specific user account, download file from C2 server, copy files, start and stop services, change registry values, and execute shell commands. The additional backdoor was set up as contingency, in case of a failure of the main backdoor. It was deployed in the form of a Trojan file disguised as a Notepad executable file. This additional backdoor had a different configuration and connected to a different C2 server.

The payload component described the payload for specific protocols such as IEC 101, IEC 104, IEC 61850, and OPC. Hence, in order to craft malicious packets, an adversary must possess proper knowledge about power grid communication and automation standards. These payloads are typically stored using the DLL file extension on the Windows operating system. Exploiting such mechanisms, adversaries saved critical operational data related to network configurations such as IP addresses and running protocols inside .ini extension files.

In order to carry out the attack, adversaries executed the launcher module that triggered the malicious payload execution and packet delivery into targeted substations, based on predefined protocols. These malicious packets were expected to open circuit breakers and cause a blackout. Finally, to remove all traces of the attack, the launcher module also executed the data wiper function. Data wipers changed the registry value on the Windows operating system to make it unbootable and also deleted files on the infected computers. Subsequently, the data wiper triggered process termination causing the operating system to crash. Besides these three main components, CRASHOVERRIDE possessed additional capabilities such as port scanner and DoS. The port scanner identified open ports on the target's IP address. On the other hand, the DoS tool sent malicious packets to Siemens SIPROTEC devices to make them unresponsive. Such capabilities were expected to increase the severity of the cyberattack. However, the attack did not work as expected. One of the probable reasons was the hardcoded nature of CRASHOVERRIDE that made it less flexible.

15.4.2.4 Triton

Triton is a botnet that targeted the Triconex safety instrument system (SIS) from Schneider Electric at a Saudi Arabian petrochemical processing plant in 2017 [23]. SIS is an automated mechanism in ICS to prevent operational failures and protect from hazards such as fires and explosions. Hence, Triton's objective was to disrupt SIS functioning to allow the potential occurrence of catastrophic incidents. This malware was an APT as almost all of its operations were carried out in a stealthy manner. Investigations show that Triton is arguably the stealthiest malware targeting ICS to date. Hence, it is very fortunate that it was uncovered. The Triton attack was exposed because the adversaries made a mistake in triggering the safety system mechanism, thereby shutting down the entire ICS. Otherwise, it is estimated that Triton would have probably remained undetected with potentially catastrophic consequences.

Triton employed social engineering techniques to gain access to the ICS network. The plant operators received or downloaded a trilog.exe file. This file pretended to be a legitimate Schneider Triconex SIS application. The executable file served as a vector to initiate the cyberattack. The malicious file then injected the Triton payload into the memory of the Triconex SIS controller. In addition, it also injected two files inject.bin and imain.bin to the SIS devices. Inject.bin contained payload data for the attack, and imain.bin became a backdoor for allowing remote execution. To carry out such a major attack, adversaries had good knowledge of how the SIS system worked. By exploiting the payload of the SIS protocols and executing remote control, attackers conducted a coordinated attack on the SIS protection systems. There were three possible attacks performed by Triton. The first was to shutdown the SIS process itself. The second and third were to reprogram and persistently maintain the SIS in an unsafe state [25]. It is worth mentioning that there is no detailed technical information available regarding Triton's attack process. Nonetheless, Pinto et al. investigated the technical mechanism of the Triton attack process using a replicated attack environment [23] where some of the attack stages were simulated based on assumptions.

15.4.3 Network-Based Attacks

The communication network is the backbone for data exchange between connected hosts in IT-OT systems. Thereby, a successful cyberattack can potentially target multiple aspects of the communication network, including physical connections, device information, and protocols in use. Active network reconnaissance and lateral movement are the two most common network-based attacks.

15.4.3.1 Network Reconnaissance

Network reconnaissance is the process of discovering information related to the computer network such as connected hosts, network topology, protocols, applications, and services running on the IT-OT network. It can also be used to discover vulnerabilities. Attackers typically employ the Internet Control Message Protocol (ICMP) to identify active hosts connected to the communication network. Based on a prediction of the range of active IP addresses in the network, an attacker launches a ping ICMP scan using tools such as tcpdump or nmap. The scan provides a list of active host IP addresses that responded to the ping message. This can be mitigated by filtering ICMP packets and discarding them. Another variant of the attack involves attackers using a Transmission Control Protocol (TCP) scan by launching TCP sync packets to the list of IP addresses. Active hosts respond with an acknowledge packet. However, for this attack one also needs to consider the number of active ports, which is time and resource intensive. Hence, TCP scanning is more challenging. In any case, as a result of network reconnaissance, attackers can obtain a list of active host IP addresses connected to the network.

From the list of active hosts, attackers can obtain further details such as the running services by scanning for active ports. For example, if port 80 is open, it may imply that the host is running an HTTP service/webserver. If port 25 is open, then the host is probably running a Simple Mail Transfer Protocol (SMTP) mail server. Taking this further, attackers can also find a detailed version of the running applications through host fingerprinting and obtain potential vulnerabilities. Therefore, in a cyberattack scenario on power grids, network reconnaissance plays an important role in discovering the target hosts and protocols on the IT-OT systems. For example, in Ukraine 2015, attackers successfully identified vulnerabilities in the Microsoft AD server, paving the way for subsequent access to the OT systems in the control centre through login credential theft. Similarly, in Ukraine 2016, attackers successfully detected active protocols such as IEC 101, IEC 104, and IEC 61850 used for communication within substations.

15.4.3.2 Lateral Movement

Lateral movement is the attack process of progressively propagating throughout the targeted communication network. It starts from the most vulnerable host, serving as the entry point, moving through multiple hosts to reach the final OT target. This attack typically exploits user login credentials to access various hosts and move laterally within the IT-OT network. This technique is a common mechanism, often found in APTs. Consequently, lateral movement was heavily employed in the cyberattacks targeting the power grid in Ukraine in 2015 and 2016. In 2015, the attackers' final objective through lateral movement was to access the SCADA system in the control centre and cause a blackout. IT-OT network segmentation is one solution to mitigate the lateral movement threat. However, as seen in Ukraine repeatedly, despite network segmentation, the attacks were still successful. This is because network segmentation alone cannot guarantee a complete protection against APTs. Hence, power grid operators must complement network segmentation with additional security controls such as next-generation firewalls (NGFs) and intrusion detection and prevention systems (IDPSs) to minimize the threat of lateral movement.

15.4.4 Man-in-the-Middle Attacks

MITM attack is a type of cyberattack classified based on the location of the adversaries. In this attack, adversaries are located between two or more hosts, allowing them to maliciously observe and be involved in their communication traffic. In this section, we focus on potential MITM attacks targeting power grids, which are classified into five categories, that is, eavesdropping, spoofing, false data injection (FDI), replay attacks, and session hijacking.

15.4.4.1 Eavesdropping

Eavesdropping, also known as sniffing or snooping, is an attack where adversaries intercept information transmitted over the network. The main objective of eavesdropping is to intercept and gather information about the contents of the transmitted data. In comparison with other attacks, adversaries seek to only observe and not change legitimate communication. To mitigate the threat of eavesdropping, encrypted protocols can be used for communication. Due to encryption, adversaries or third parties cannot easily decipher the contents of the transmitted data. However, in a real SCADA system, most of the dataflows are unencrypted. Moreover, communications between SCADA end devices such as station control systems, RTUs, protection relays, and merging units mainly work based on a broadcast communication mechanism. For example, broadcast communication through the Distributed Network Protocol (DNP3) is widely adopted in SCADA communications [26]. Such broadcast mechanisms are susceptible to eavesdropping and sniffing attacks. Adversaries can easily intercept communications by gaining unauthorized access to the OT system and exploiting the vulnerabilities of the broadcast communication mechanism. Therefore, eavesdropping plays an important role in an APT, especially during the reconnaissance stage. Hence, it can be used as a stealthy mechanism to gather network intelligence through passive means [27]. Valli et al. present a study about eavesdropping attacks targeting smart grids in [28]. The eavesdropping is mainly focused on the advanced metering infrastructure (AMI). AMI establishes communications through wireless channels between a smart grid operator and smart metering devices. Adversaries may exploit the vulnerabilities present in wireless networks to intercept communications and capture transmitted data. Research also shows that eavesdropping can lead to privacy concerns for smart grid users, as seen in [29, 30].

15.4.4.2 Spoofing

Spoofing is an active attack where adversaries pretend to be legitimate entities and disrupt normal communications. Such attacks can be realized through many forms of spoofing such as emails, website URLs, text messages, Global Positioning System (GPS), and IP addresses. The most widely researched spoofing attack targeting power grids is based on GPS spoofing of phasor measurement unit (PMU) data. There are multiple studies in this direction, as discussed in [31–33]. PMUs provide magnitudes and phase angles of fundamental power system parameters such as voltages and currents, using a common time source for synchronization [34]. Hence, GPS spoofing attacks, mainly targeting the timing signals used for synchronization, may lead to distortion of observed PMU data, which includes phase angle errors [35]. There are two types of GPS signals widely in use for civilian and military applications. GPS signals for military purposes are encrypted, while civilian ones are not. Typical PMUs for power grids function based on civilian GPS signals. This may allow adversaries to spoof GPS signals by exploiting the lack of encryption and using a portable device without a direct access to the power grid communication network. Currently, there are two approaches to mitigate GPS spoofing attacks. The first is through GPS spoofing signal detection using parameter such as signal-to-noise ratio [36], and the second is via anomaly detection in power system measurements.

Another commonly reported type of spoofing attack on power systems is IEC 61850 spoofing. The IEC 61850 standard is a modern power system communications standard used for substation automation and protection in digital substations. It enables information exchange through different communication protocols, of which two are of utmost importance. The generic object-oriented substation event (GOOSE) and sampled values (SV) protocols are used to communicate critical substation events and measurements within a substation, respectively. Although it provides increased benefits, IEC 61850 is not cybersecure. Due to strict operational constraints and timing requirements for power system protection schemes, the standard does not implement any encryption. This makes it particularly vulnerable to packet sniffing and spoofing type of attacks. Such types of spoofing attacks are well reported and have been investigated extensively in literature [37, 38]. Multiple vulnerabilities and exploits, specifically targeting GOOSE and SV protocols are widely discussed in [39–41].

The premise of all these discussions is similar. Due to the lack of encryption in IEC 61850, an attacker with access to the substation communication infrastructure can wreak havoc. By carefully monitoring IEC 61850 traffic via the process and station buses, it is possible to craft spoofed GOOSE packets that can maliciously open circuit breakers. When such spoofed packets are sent to a target relay, it is tricked into opening the circuit breaker. This is successful as the packet is made to appear to be originating from the station or a bay controller, within the same substation. It is also possible to inhibit protection functionality of relays due to spoofing of SV measurement data, causing protection equipment to not operate during a critical fault conditions [42]. The spoofing attack causes a relay to get blocked from further operations due to multiple concurrent input SV streams. With the target protection device blocked, other relays in the system may trip during fault conditions. Such types of spoofing attacks can have disastrous consequences for power system operation and stability. A well-targeted spoofing attack can not only compromise but also disable equipment and components within a digital substation. Subsequently, this can instigate major system instabilities, and may even induce cascading failures, due to the sudden loss of multiple components. In a doomsday scenario, attackers may trigger a system-wide collapse, that is, a blackout, by compromising critical digital substations, leading to catastrophic damages. Nonetheless, such types of spoofing attacks can be mitigated by adopting proper cybersecurity measures, as discussed in [43].

IEC 62351-6 is a standard that specifically addresses cybersecurity of IEC 61850. It recommends an additional field to the GOOSE and SV data payloads for security-related information. This field contains a Rivest-Shamir-Adleman (RSA)-based digital signature to ensure payload integrity. Through this mechanism, sending and receiving intelligent electronic devices (IEDs) are clearly identified and it becomes impossible to manipulate the payload. Similarly, the standard also recommends usage of hash-based message authentication codes (HMACs) using cryptographic algorithms such as SHA-256 to ensure data integrity of GOOSE and SV frames. Such techniques can prevent spoofing and sniffing attacks. However, the suggested use of the digital signatures and security based on RSA and HMAC algorithms comes with associated costs. For protection applications where a 4 ms or lower response time is strictly required, such measures are unsuitable. This is because encryption and decryption are computationally demanding [43, 44]. Furthermore, the usage of RSA and HMAC-based authentication keys for IEDs and equipment necessitates a dedicated key management infrastructure within the digital substation. Hence, such security mechanisms have not gained widespread use, yet.

15.4.4.3 False Data Injection

The most extensively researched type of cyberattack on power systems is the FDI attack. An FDI attack operates under the assumption that attackers have access to the station control systems and RTUs in substations and/or the SCADA master in the control center. Consequently, they can inject falsified SCADA measurements, maliciously introducing correlated and consistent power flow measurements into state estimation (SE), aiming to mislead system operators. Nowadays, SE is an integral tool in the energy management system for contingency analysis, security-constrained optimal power flow, and pricing calculation algorithms. The critical nature of SE highlights the importance of making it accurate and secure for power system operation. However, as discussed above, the SCADA system is vulnerable to FDI attacks. In [45], Liu et al. introduced a class of

FDI attacks that can perturb the estimated states without being detected by the safeguard scheme within the SE process. The interesting part of such attack is that the adversary is assumed to have the knowledge of the targeted power system including the power network topology and parameters and thus can exploit such knowledge to systematically generate multiple FDIs on power flow measurements [46]. It has been illustrated that such FDI attacks can bring potential economic damages by manipulating the nodal price of market operations [47] or even physical impact such as a line overload [48]. The FDI attack may seem difficult to conduct as the adversary needs to be equipped with enough knowledge of the target power system and vast attack resources to manipulate multiple measurement data channels. However, the complexity and functionalities of malware in recent cyber incidents on ICSs provide credible means to realize the FDI attack [49]. Notably, in addition to FDI on SE, recent research also considers attack scenarios where other critical applications, for example, automatic generation control, are targeted [50]. In addition, studies on power system vulnerability analysis have been carried out to explore how FDI attacks can achieve the desired targets with incomplete system knowledge or very few attack resources, using both static and dynamic (time-variant) FDI strategies [51, 52]. Detection and mitigation techniques at the physical layer of the power system are proposed in [53, 54] based on both model-based and data-driven detectors from control-theoretic domains or machine learning areas.

15.4.4.4 Replay Attack

A replay attack is a variant of the MITM attack where attackers record communication traffic and replay it to mimic legitimate entities. Pidikiti et al. investigated replay attacks on the SCADA system exploiting IEC 101 and IEC 104 protocols in [55]. These SCADA protocols were originally created without cybersecurity considerations. Nevertheless, these protocols do implement a packet checksum mechanism to prevent replay attacks to a certain extent. However, the size of the checksum is small and limited by the packet frame size and bandwidth. This condition leads to unreliable checksums to ensure data integrity. Therefore, this vulnerability can be exploited to launch replay attacks on SCADA systems. On the other hand, replay attacks in IT systems are more common and usually prevented by using authentication and secure session mechanisms. For example, a countermeasure to replay attacks was proposed using Kerberos authentication protocol [56]. This protocol would force the network hosts to authenticate themselves. After a successful authentication, a secure session is established between hosts. Such a session is typically valid only for a limited period of time preventing the reuse of session information. However, adoption of such prevention mechanisms in OT systems is challenging. For example, in IEC 101 and IEC 104, the limited packet frame size makes it difficult to add more data to improve protocol security. In addition to the aforementioned authentication mechanisms at the cyber layer, research efforts have also been undertaken to study the replay attack from the perspective of the physical power system layer. Such research is focused on detection methods to secure the control process of the SCADA system in power grids [57]. However, it is to be mentioned, there could still exist sufficient conditions under which plausible replay attacks may remain stealthy irrespective of the detection mechanism used. This is applicable even to a control-theoretic approach wherein the attacker has access to all the necessary data channels and executes the replay attack at a suitable time [58].

15.4.4.5 Session Hijacking

Communication sessions are interactive information exchanges between two or more networked devices for a limited time duration. Typical session establishment is initiated through authentication between hosts via secure protocols. Therefore, a session hijacking attack aims to bypass these protocols, allowing adversaries to circumvent authentication mechanisms and gain unauthorized access to legitimate communications. Kleinmann et al. presented a study on session hijacking in SCADA systems by exploiting Modbus protocol [59]. Modbus was originally designed only for serial communications between field devices in substations. To improve its flexibility, it was later upgraded to implement TCP. This modification allowed Modbus to work on Ethernet connections using IP addresses providing more data faster. The session establishment in Transmission Control Protocol/Internet Protocol (TCP/IP) works based on a three-way handshake mechanism. However, TCP/IP is widely known to be susceptible to cyberattacks including session hijacking [60] and thereby compromising Modbus as well. Besides session hijacking at the protocol level, hijacking can also be conducted through web applications or human-machine interfaces (HMIs) of SCADA systems. A successful session hijacking attack allows adversaries to assume the identity of the compromised devices/users and provides unauthorized access and control of the OT system. Burgers et al. presented a session hijacking case study and mitigation techniques for SCADA [61]. This research focuses on session hijacking via web-based applications, which use login authentication for session establishment.

15.4.5 Denial-of-Service Attacks

DoS is a cyberattack with the objective of preventing legitimate access for users/networked devices to specific system resources such as network connections, computing capabilities, and application services. The term "DDoS" refers to a coordinated DoS attack originating from multiple, distributed sources to increase attack severity and prevent tracking and identification of attackers' origin. A single DoS attack can be mitigated by blocking the sole attack source. Conversely, for a DDoS attack, blocking all attack sources is challenging, making its mitigation difficult. DoS attacks can further be classified into bandwidth depletion and resource depletion attacks [62]. The bandwidth depletion DoS attack aims to overload the bandwidth capacity of a target communication network. This can be achieved by either directly flooding the communication channel with bogus traffic or via third parties, which send multiple legitimate requests at the same time in an amplification attack. As a consequence, in either instance, legitimate communication traffic is affected, which significantly reduces the overall network performance. The resource depletion attack aims to overwhelm the target's resource usage, for example, computing resources of a targeted host, by exploiting protocols and known response mechanisms. For example, as previously mentioned, TCP/IP implements a three-way handshake mechanism, allowing two hosts to initiate communication with a preliminary request and response mechanism. Adversaries may exploit this mechanism by sending a multitude of malicious requests to the targeted host. Consequently, the targeted host is kept busy responding to all malicious requests, leading to the disruption of a proper response to legitimate ones.

DoS attacks can target SCADA systems of power grids. Studies about SCADA susceptibility to DoS attacks are reported in [63, 64]. Petrovic et al. demonstrated DoS attacks on SCADA systems using OPNET communication network simulator [63]. The attacks significantly reduced SCADA network throughput and processing capabilities, directly affecting power system monitoring and control. Similarly, Kalluri et al. demonstrated a DoS attack exploiting IEC 104 protocol used in substations, affecting the processing and communication capabilities of RTUs [64]. Carcano et al. demonstrated a resource exhaustion attack targeting IEC 62351 [65], highlighting its cybersecurity shortcomings. In summary, the DoS attack is a potential threat against data availability in power grids, as it prevents successful communication of measurements and controls. Attackers can either jam the SCADA communication channels or compromise field devices and prevent them from communicating data. They may also attack the routing protocols or flood the network with bogus traffic [66]. DoS attacks on power systems may be modeled to analytically study the impact of data absence on power system monitoring and control. By properly designing DoS attack sequences, attackers can corrupt the normal operation of controllers and consequently impact power system stability [67, 68]. Mitigation techniques are discussed in [69].

15.4.6 Host-Based Attacks

A host-based attack as the name suggests is an attack targeting various hosts in IT-OT systems, such as SCADA servers and HMIs, databases, application servers, station control systems, RTUs, protection relays, and merging units. In this section, we classify host-based attacks into three categories, that is, software-based, database, and unauthorized access and control attacks.

15.4.6.1 Software-Based Attacks

Software-based attacks on power grids exploit vulnerabilities present in software used in IT-OT systems such as SCADA and energy management systems. Usually, the software applications and security controls in OT systems inherit the same vulnerabilities present in regular IT systems. The main issue is that software and security controls of such IT systems may be patched, and their vulnerabilities may be mitigated more often than in OT systems. It is more difficult to update the OT systems of critical infrastructures as this process can affect the normal operation of physical facilities. A disruption of service such as electricity supply to customers may result in regulatory penalties and financial loss. Furthermore, extensive commissioning is needed after each update process that prolongs the voluntary outage for maintenance.

Most SCADA system solutions provided by vendors were developed before the emergence of cybersecurity concerns [70]. Software vulnerabilities in SCADA systems can be classified into three categories, that is, improper input validation, software or source code, and resources control vulnerabilities [71]. As a result of input validation vulnerabilities, SCADA software is susceptible to modification attacks such as data injection and buffer overflow. SCADA source codes have also been found to contain improper security mechanisms and vulnerabilities such as the null pointer dereference vulnerability [72]. Resources control vulnerabilities are strongly related to software updates and patches. Corporate IT security typically pushes software updates and operating system patches over the IT network. However, SCADA software updates and patching in control centers and substations are more difficult to implement. This is due to the blend of state-of-the-art and legacy end devices, in addition to continuous operational requirements of the SCADA system in production.

In August 2020, 19 software vulnerabilities were exposed by JSOF, an Israeli cybersecurity company. These vulnerabilities, dubbed Ripple20, affected ICS devices using the proprietary Treck TCP/IP stack software libraries. Two of the most severe vulnerabilities are related to TCP/IP tunneled packet fragmentation [73] and DNS packet decompression mechanisms [74]. The Treck software library has widely been adopted in IoT networked devices by several vendors across a whole range of industries including manufacturing, healthcare, and power grids. It is a cause for serious concern, as shown in [74], that a specific payload injection could remotely turn off an UPS device. Therefore, we can infer that Ripple20 is a real-world example of challenges pertaining to updates and security of software in ICSs, further complicated by global supply chains.

15.4.6.2 Database Attacks

A database is an essential element of the SCADA system as it stores real-time information from substations along with user access credentials. Zhu et al. categorize a database attack as an important cyberattack vector targeting SCADA systems [75]. Most common databases work based on SQL. Thus, one of the popular attacks targeting databases is SQL injection. This attack exploits input handling of the database system. When a database cannot correctly parse and handle inputs, it may lead to database access violations and illegitimate manipulation. In the worst-case scenario, with the breached confidential database information, adversaries can gain unauthorized control of the SCADA master. Consequently, databases have proven to be an important attack element in real-world cyberattacks on power grids. For example, in Ukraine 2015 attack, adversaries gained access to the control center using stolen credentials from the Windows AD database [74]. AD is one of the most critical applications since early 2000 as it offers flexibility and interoperability of service authentication and authorization. However, a breach in security measures of AD can lead to a breach of the entire system since AD serves as the central authentication and authorization point. There are many publicly available tools to exploit AD security. For example, Mimikatz can be used to exploit AD hashes and Kerberos ticketing mechanism. Nonetheless, there are counter measures to prevent cyberattacks targeting AD. One of the options is the application of Microsoft Credential Guard. Credential Guard is a virtualization-based isolation technology for Local Security Authority Subsystem Service (LSASS) which prevents attackers from stealing credentials and prevents hash attacks. Another option is the implementation of tiered (multi-level) administrator models. The tiered admin model can prevent attackers from gaining top-level privileges in an AD. Another common practice to prevent AD breaches is to implement secure credential policies. For example, a user has to change passwords periodically and use strong combination of characters. Multi-level or two-factor authentication mechanisms through mobile phone messages and emails can also be applied to increase the overall system access security.

15.4.6.3 Unauthorised Access and Control

Access authorization typically uses an authentication mechanism applied to secure hosts, software, and web services. Unauthorized access occurs when an adversary gains access to the system without legitimate credentials. Hence, unauthorized access and control can be achieved if attackers circumvent the authentication mechanisms. There are many techniques to achieve this objective such as credential theft using a keylogger, database breaches, brute force attacks, and buffer overflows. It is also possible to gain unauthorized access using penetration testing tools such as Metasploit. This exploits system vulnerabilities by injecting malicious payloads on the target system. The most basic form of unauthorized access is achieved through the guest (non-administrator) mode. However, in this mode, attackers' options are limited. Thereby, to increase attack severity, attackers can perform privilege escalation and become administrators allowing them complete control over the compromised system. SCADA systems typically employ Windows-based operating systems. However, Windows is vulnerable to unauthorized access attacks. Thus, Windows operating systems in IT-OT systems must be regularly updated and protected with firewalls and antiviruses. Researchers have also proposed solutions to prevent unauthorized access in SCADA systems. Taylor et al. proposed a SCADA authentication technique using a custom key distribution mechanism [76] applicable to DNP3 protocol, to prevent unauthorized access and control. Similarly, Vaidya et al. proposed an authentication and authorization for substation-level communications [77]. This method implements multi-level authentication and uses public key certificates to authenticate and authorize access to the substation automation system. Other approaches to prevent and reduce the risk of unauthorized access and control include measures such as securing the host operating systems and implementing security perimeters and IDPS.

15.5 Impact of Cyberattacks on Power Grids

Cyberattacks on power grids are considered high impact, low frequency disturbances with a wide range of effects. These could include, but are not limited to, equipment damages, loss of load, and power system instability. In the worst case, sophisticated cyberattacks may also cause system-wide cascading failures, leading to a blackout. Hence, this section discusses the various potential

Table 15.7 Summary of known cyberattacks on power grids and their impact.

No.	Attack	Year	Category	Impact
1	Malware infection of SCADA system, Europe [3]	2003	Service disruption	Three-day loss of management functions in distribution substations
2	Aurora experimental cyberattack, USA [78]	2007	Physical damage	Physical damage to 2 MW synchronous generator
3	Power plant malware infection, USA [3]	2012	Service disruption	Three-week restart delay of power plant
4	Cyberattack on power grid, Ukraine [9]	2015	Service disruption	Power outage affecting 225,000 customers for six hours
5	Cyberattack on power grid, Ukraine [17]	2016	Service disruption	Power outage in distribution network, 200 MW unsupplied load

impacts of cyberattacks on power grids, ranging from component to system level. Table 15.7 summarizes the known cyberattacks on power grids and their impact. Four of the attacks shown in Table 15.7 are real, except the Aurora attack. The Aurora project was an experimental cyberattack that led to the physical destruction of a 2 MW synchronous generator. This was mainly done as a demonstration to raise awareness about cybersecurity and associated threats. The significant cyberattack on power grids, so far, is the Ukraine 2015 attack. It has been confirmed that this attack directly led to a power outage, affecting over a quarter-million customers for a duration of over six hours. Besides the real-world examples of cyberattacks on power grids, research has also been carried out to investigate the potential impacts of cyberattacks on power system operation [79–81]. Such empirical studies discuss various cyberattack scenarios and associated effects. A doomsday scenario would entail a cyber-induced cascading failure culminating in a complete blackout. Hence, the subsequent subsection firstly provides an overview of the cascading failure mechanism, followed by various cyberattack scenarios and their impact analysis, as reported in the literature.

15.5.1 Overview of the Cascading Failure Mechanism

Any major power system blackout is preceded by the phenomenon of cascading failures. A cascading failure, as the name suggests, is a successive failure of power system elements that can lead to a complete system collapse, that is, a blackout. Most cascading failures are initiated by one or a set of multiple related events. These can include line flashovers, protection maloperation, and human error. Historically, most of these events tend to be caused by a combination of equipment failures, for example, ageing equipment, environmental conditions, and human factors. Depending on the operating state of the system and severity of the initiating events, the entire power system may enter an emergency state. Without proper control actions or remedial measures, the system is highly vulnerable to further cascading effects. In such a case, various outcomes are possible. One such outcome, commonly observed in historical blackouts such as Italy 2003 and USA 2003 [82] is as follows. Due to the initial set of events, overloading of parallel transmission lines occurs, to account for power redistribution. Eventually, these lines are also overloaded beyond their limits and start tripping, initiating a cascading process of transmission line disconnections. After a certain time, the effect of these outages is felt on system dynamics. Transient instability can occur in a matter of a few seconds due to the large disturbances. Generators may lose synchronism due to sudden loss of transmission lines. This will also affect system voltages, causing major voltage drops. Consequently, in the case of heavy system loading, voltage stability problems may also arise. An inability to meet growing reactive power demands can eventually result in a voltage collapse. If left unchecked, such dynamic phenomena can result in islanding, that is, formation of smaller clusters in the system with a mismatch of supply and demand. Ultimately, the power system reaches a so-called point of no return [83]. From this point onward, the entire cascading process is rapid, involving loss of multiple generator units and loads. This domino effect is uncontrollable, culminating in a blackout. It is worth mentioning here that such a sequence of events is based on historical cascading failures and blackouts, involving physical system events. In case of a well-targeted and coordinated cyberattack, the effects can be severely magnified. As shown in [84, 85], cyberattacks targeting bulk power systems may initiate cascading failures. A sophisticated cyberattack can target multiple substations, disconnecting many lines and tampering with control setpoints. As a result, power system instability and associated phenomena discussed above may be induced much faster. Consequently, in comparison with previous blackouts, the point of no return may be reached much sooner in case of cyberattacks.

15.5.2 Impact Analysis

As previously mentioned, the physical impact of cyberattacks on power grids is wide-ranging. There are many empirical studies reported in literature, covering these impacts. The most reported consequence of a cyberattack on power grid infrastructure is equipment damage. The Aurora experiment is a good real-world example of such possible effects. The attack demonstrated how rapid opening and closing of a generator's circuit breaker cause an out of phase reconnection and permanent equipment damage. Along similar lines [85], extensively discusses switching attacks on generators. This work clearly highlights how sophisticated cyberattacks can not only disconnect generators and cause equipment damages but also initiate cascading failures. By applying a fast, switching attack on a generator's main circuit breaker, transient instability can be induced, destabilizing the entire power grid in a matter of a few seconds. Other possible impacts include damage to equipment such Flexible Alternating Current Transmission System (FACTS) devices and On-Load Tap Changers (OLTCs) through setpoint modification [86, 87]. These devices are critical in ensuring voltage stability, and such equipment damage can trickle down and affect the entire power system. Loss of load is another commonly reported result of cyberattacks on power grids. If a cyberattack affects the system frequency, automatic measures such as load shedding are undertaken to preserve system integrity. Additionally, switching or data modification attacks can directly lead to loss of load [88]. The worst possible outcome, however, is that of cyber-induced cascading failures and a blackout. As discussed in [84], a cyberattack on multiple substations in any power system may lead to a blackout. Cyberattacks targeting specific grid components or equipment can impact power system stability. For example, targeting voltage control mechanisms such as Static VAR Compensators (SVCs) and Static Synchronous Compensators (STATCOMs) can severely affect voltage stability. By carrying out data modification or MITM attacks, as stated in [86, 87], reactive power compensation is severely affected. As a result, voltages throughout the system can be influenced. Sustained under voltages can lead to emergency load shedding and, in the worst case, a voltage collapse. As part of a coordinated effort, such attacks can induce system-wide cascading failures and even a blackout.

15.6 Study Case and Simulation Results

This section discusses a study case, involving an example of a digital substation and spoofing of the IEC 61850 communication traffic. The layout of a digital substation and its communication

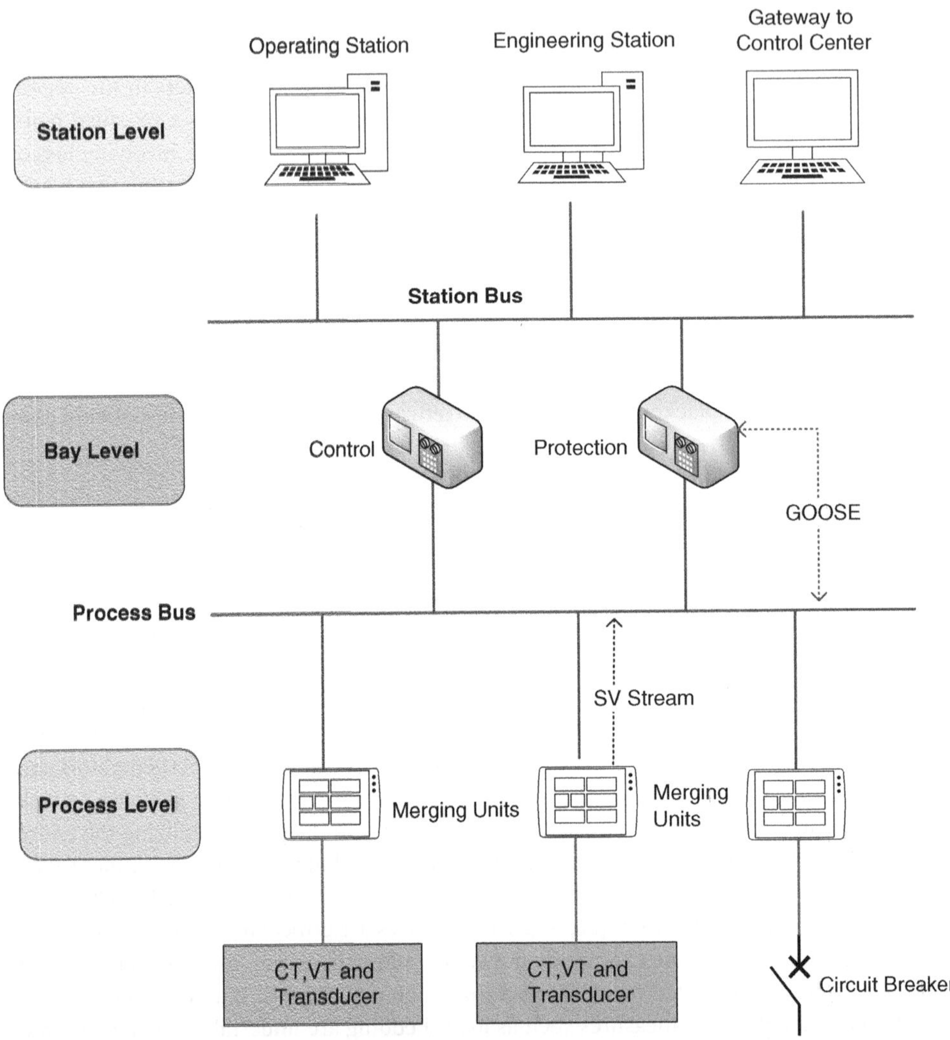

Figure 15.5 Layout of communication architecture in a digital substation.

network based on IEC 61850 is shown in Figure 15.5. It comprises of the station, bay, and process levels. In this work, our focus is on the bay level where the process bus interconnects IEDs and relays, enabling power system automation and protection applications. Typical communication networks in digital substations employ a fiber ring at the bay level. Additionally, IEC 61850 implements a publisher-subscriber communication mechanism for GOOSE and SV messages. In this context, the status and trip signals are communicated as multicast GOOSE messages via the process bus between various IEDs at the bay level, as represented in Figure 15.5. As discussed in Section 15.4.4.2, since IEC 61850 traffic is not encrypted, it is susceptible to spoofing and MITM attacks. This forms the basis for the discussed study case and simulation results.

15.6.1 Attack Scenario

The test system under study is the IEEE-39 bus transmission system. The system comprises of 27 user-defined substations. The digital substation under consideration consists of three IEDs, protecting the bays for three high-voltage lines. In the attack scenario, an adversary has gained access to the substation communication network via malicious means such as phishing and malware, as outlined in Section 15.4. With unauthorized access to the IEC 61850 traffic within the digital substation, the adversary conducts network reconnaissance and sends crafted GOOSE packets to the target relays. Due to the correct nature of the spoofed packets, the relays open their associated circuit breakers, leading to a sudden *N-3* contingency. This occurs at simulation time *t*. Under normal circumstances, the system operator would quickly take cognizance of the situation and corrective actions would be applied, that is, the circuit breakers will be closed. However, in this attack scenario, in addition to the GOOSE spoofing, attackers also launch a DoS attack on the utility's OT system. This hinders timely corrective actions as commands sent from the control center do not reach the substation on time. Consequently, system stability is jeopardized, initiating cascading failures and culminating with an extensive power outage. The exact sequence of events and impact of the cyberattack are discussed in the following subsection.

15.6.2 Simulation Results

The aforementioned cyberattack results in the malicious disconnection of lines 05-06, 04-05, and 05-08, along with a DoS attack that restricts remedial actions by the system operator. Consequently, the prolonged cyberattack affects power system stability. Multiple lines are disconnected by distance relays operating on sustained under voltages and over currents. Such a phenomenon was also observed during the North American cascading failures and blackout in 2003. A critical line tripped due to incorrect operation of zone 3 distance protection aggravating the domino effect and leading to the spread of the cascading phenomenon, ultimately ending in a large-scale blackout [82]. An example of such a trip is shown through Figures 15.6b-d for line 08-09. As a result of multiple line disconnections, generators in the system are extremely stressed and operating close to their limits. Finally, in the absence of remedial actions, generator G3 is tripped by its interface protection due to a high rate of change of frequency (ROCOF) condition, well exceeding the ROCOF setting of 2 Hz/s, as seen in Figure 15.6a. A similar condition leads to the loss of generator G2. Now, due to the loss of generation, system frequency starts plummeting and emergency load shedding is activated to preserve system integrity. Ultimately, the cyberattack leads to a partial blackout with 10 busbars being de-energized and a loss of load amounting to 772 MW. The entire power system after the cyberattack is shown in Figure 15.7. The area that suffers a power outage is indicated along with the two generators lost and loads left unsupplied. The cascading failure sequence is summarized in Table 15.8.

15.7 Conclusion

Power grids are undergoing a fast-paced process of digitalization, opening up the energy system to everyone by means of ICTs. However, future grid digitalization will require careful considerations with regard to data privacy and cybersecurity. It is now well recognized that IT-OT systems are vulnerable to cyberattacks. Hence, cyber resilience requirements of the power grid are more critical than ever before. The complexity of cyberattacks on power systems is likely to increase. To improve the cyber resilience of power grids, it is needed to identify potential threats and IT-OT system vulnerabilities, classify and review major types of cyberattacks on power grids, analyze their impact on system operation and stability, and develop mitigation techniques. Hence, this chapter provided the state-of-the-art and essential knowledge of threats and cyberattacks on power systems. It reviewed

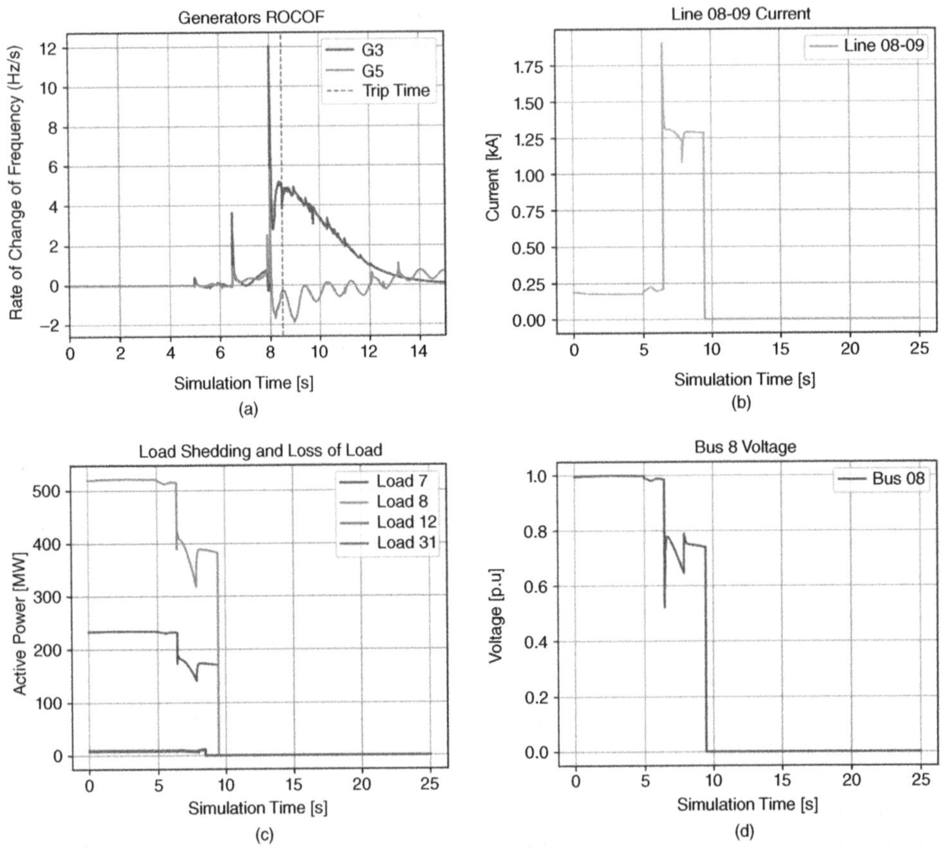

Figure 15.6 Simulation results showing impact of a cyber-induced cascading failures and partial blackout. (a) Generator ROCOFs, (b) Line 08-09 current, (c) Load shedding and loss of load, and (d) Bus 08 voltage.

major cyberattacks on power grids and ICSs and provided a detailed taxonomy of cyberattacks. The most common security controls implemented in power grids include antiviruses, firewalls, network segmentation, and IDSs. Worryingly, even these security control mechanisms may be outdated or insufficient. Consequently, cyberattacks on power grids exploiting various threat vectors can have a catastrophic impact on system operation. This chapter provided indicative simulation results of such a hypothetical cyberattack scenario. Results show that sophisticated attacks may not only cause loss of load but also induce cascading failures resulting in a blackout. Therefore, the urgent need of the hour is to develop comprehensive defense, mitigation, and incident response techniques to enhance power grid cyber resilience.

Acknowledgement

This work was supported by the Designing Systems for Informed Resilience Engineering (DeSIRE) program of the 4TU Center for Resilience Engineering (4TU.RE) and the EU H2020 project, ERI-Grid 2.0 with Grant Agreement Number 870620. DeSIRE is funded by the 4TU-program High Tech for a Sustainable Future (HTSF). 4TU is the federation of the four technical universities in the Netherlands.

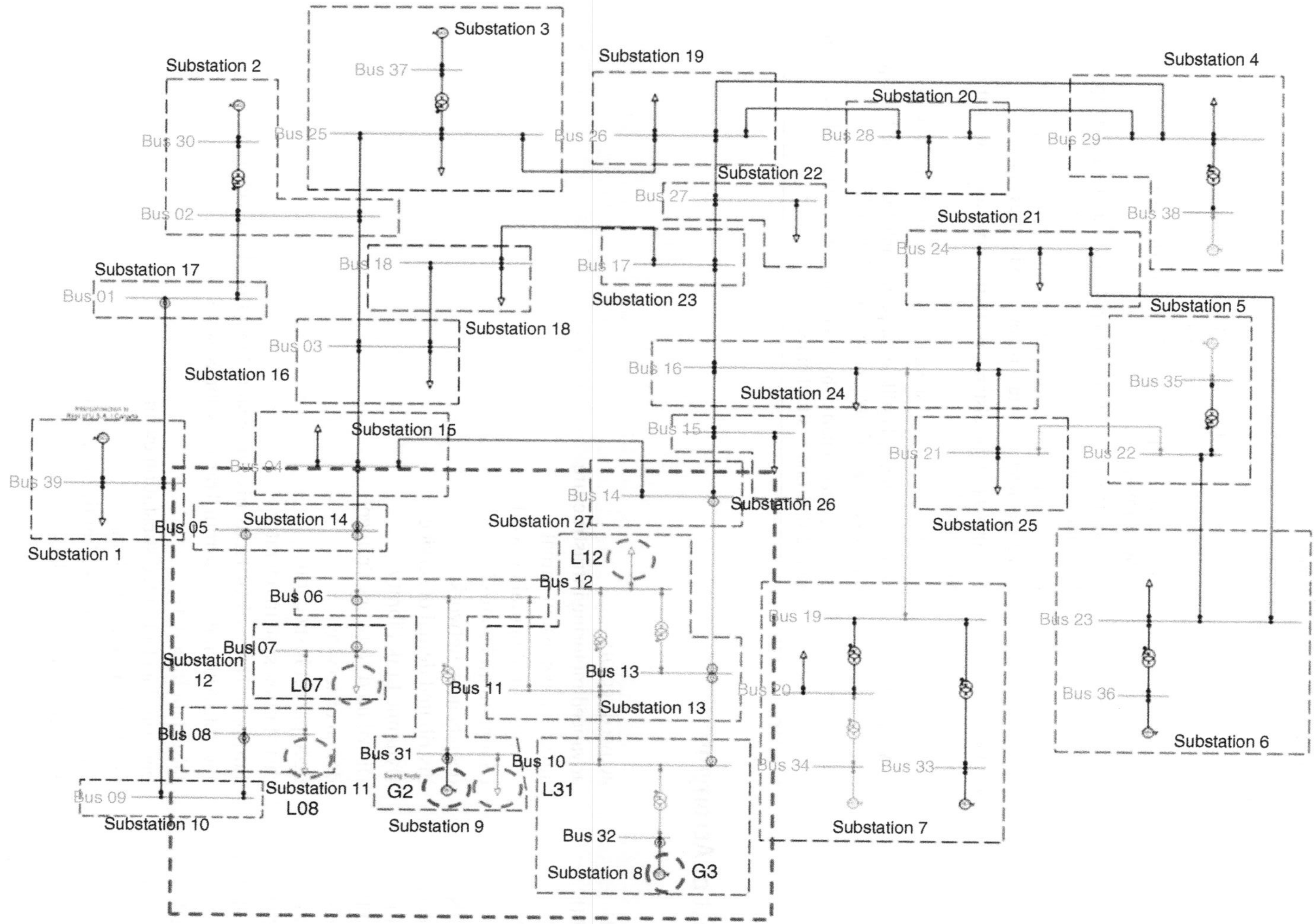

Figure 15.7 Single-line diagram of IEEE-39 bus system after cyberattack.

Table 15.8 Cascading failure sequence.

No	Time [s]	Event
1	0 s	Start of simulation.
2	5 s	Cyberattack on substation 14. Lines 05-06, 04-05, and 05-08 maliciously disconnected.
3	6.477 s	Distance relay trips line 06-07. Power grid is split into two isolated areas.
4	7.886–7.997 s	Lines 10-13 and 13-14 in vicinity of attacked substation tripped by distance protection.
5	8.497 s	Generators G3 and G2 tripped due to ROCOF interface protection and disconnected. System is now heavily stressed.
6	9.474 s	Line 08-09 trips on distance protection. Two areas now left unsupplied.
7	12.073–12.548 s	Underfrequency load shedding activated. All loads shed by 6.5%.
8	13.05–18.609 s	System frequency is still below permissible limits. Underfrequency load shedding activated in steps of 5.9% and 7%.
9	25 s	System suffers a partial blackout with total loss of load amounting to 772 MW. End of simulation.

List of Acronyms

AD	active directory
AMI	advanced metering infrastructure
APT	advanced persistent threat
CNN	convolutional neural network
C2	command and control
DBN	deep believe network
DDoS	distributed denial of service
DLL	Dynamic Link Library
DNP	Distributed Network Protocol
DoS	denial of service
DPI	deep packet inspection
DSO	distribution system operator
ENTSO-E	European Network of Transmission System Operators for Electricity
FACTS	Flexible Alternating Current Transmission System
FDI	false data injection
GOOSE	generic object-oriented substation event
GPS	Global Positioning System
HMAC	hash-based message authentication code
HMI	human-machine interface

HTTP	hypertext transfer protocol
HTTPS	hypertext transfer protocol secure
ICMP	Internet Control Message Protocol
ICS	industrial control system
ICT	information and communication technology
IDPS	intrusion detection and prevention system
IDS	intrusion detection system
IED	intelligent electronic device
IoT	Internet of Things
IP	Internet Protocol
IT	information technology
LSASS	Local Security Authority Subsystem Service
LSTM	long short-term memory
MAC	message authentication code
MITM	man-in-the-middle
NGF	next-generation firewall
OLTC	On-Load Tap Changer
OPC UA	Open Platform Communication Unified Architecture
OSI	Open Systems Interconnection
OT	operational technology
PLC	programmable logic controller
PMU	phasor measurement unit
ROCOF	rate of change of frequency
RPC	remote procedure call
RSA	Rivest-Shamir-Adleman
RTU	remote terminal unit
SCADA	Supervisory Control and Data Acquisition
SE	state estimation
SIS	safety instrument system
SMB	Server Message Block
SMTP	Simple Mail Transfer Protocol
SNTP	Simple Network Time Protocol
SQL	Structured Query Language
STATCOM	Static Synchronous Compensator
SV	sampled values
SVC	Static VAR Compensator
TCP	Transmission Control Protocol
TCP/IP	Transmission Control Protocol/Internet Protocol
TLS	Transport Layer Security Protocol
UPS	uninterruptible power supply
URL	Uniform Resource Locator
USB	Universal Serial Bus
VBA	Visual Basic Application
VPN	virtual private network

References

1 Hutchins, E., Cloppert, M. and Amin, R. (2011). Intelligence-driven computer network defense informed by analysis of adversary campaigns and intrusion kill chains, pp. 1–14, [Online]. https://lockheedmartin.com/ content/dam/lockheed-martin/rms/documents/cyber/LM-White-Paper-Intel-Driven-Defense.pdf. (accessed July 7, 2023).

2 Miller, B. and Rowe, D.C. A survey of SCADA and critical infrastructure incidents. *Proceedings of the International Conference on Research in Info Tech*, Calgary, Canada (October 2012), pp. 51–56.

3 Noguchi, M. and Ueda, H. (2018). An Analysis of the Actual Status of Recent Cyberattacks on Critical Infrastructures. *NEC Tech Journal Spec Issue Cybersec* 2 (2): 9–24.

4 K.E. Hemsley and R. E. Fisher. (2018, December) History of Industrial Control System Cyber Incidents. *Idaho Natl Lab (INL) Tech Report, USA,* [Online]. https://www.osti.gov/biblio/1505628 (accessed: July 7, 2023).

5 T. Daniela. Communication security in SCADA pipeline monitoring systems. *Proceedings of the International Conference on Networking in Education and Research*, Iasi, Romania (August 2011), pp. 1–5.

6 Chen, T.M. and Abu-Nimeh, S. (2011). Lessons from Stuxnet. *Computer* 44 (4): 91–93.

7 Falliere, N., Murchu, L.O., and Chien, E. (2011). W32. Stuxnet Dossier, Symantec Security Response, Version 1.4, February 2011. *Symantec Sec Response* 4: 1–69.

8 Langner, R. (2011). Stuxnet: Dissecting a Cyberwarfare Weapon. *IEEE Security and Privacy* 9 (3): 49–51.

9 R. Lee, M. Assante and T. Conway. (2016, March). Analysis of Cyber Attack on the Ukrainian Power Grid. *Electricity Information Sharing Center (E-ISAC) Tech Report*, pp. 1–26, [Online]. https://media.kasperskycontenthub.com/wp-content/uploads/sites/43/2016/05/20081514/E-ISAC_SANS_Ukraine_DUC_5.pdf (accessed July 20, 2023).

10 D. E. Whitehead, K. Owens, D. Gammel et al. Ukraine cyber-induced power outage: analysis and practical mitigation strategies. *Proceedings of the International Conference for Protective Relay Engineers*, Texas, USA (April 2017), pp. 1–8.

11 A. Cherepanov and R. Lipovsky. Blackenergy-what we really know about the notorious cyber attacks. *Proceedings of the International Conference Virus Bulletin*, Denver, USA (October 2016), pp. 1–8.

12 S. Shrivastava. BlackEnergy - Malware for Cyber-Physical Attacks. (2016, May). *iTrust Cent Res Cyber Sec Analysis Report*, pp. 1–15, [Online]. https://itrust.sutd.edu.sg/wp-content/uploads/2016/10/itrust-analysis-blackenergy.pdf (accessed july10, 2023).

13 A. Cherepanov. (2017, June). Win32/Industroyer: A New Threat for Industrial Control Systems. *ESET Tech Report*, pp. 1-17, [Online]. https://www.welivesecurity.com/wp-content/uploads/2017/06/Win32_Industroyer.pdf (accessed July 10, 2023).

14 J. Henry and C. I. Systems. (2018, April). Sandia ICS Security Capabilities to Investigate CRASHOVERRIDE. *Tech Report, USA*, [Online]. https://www.osti.gov/servlets/purl/1575340 (accessed July 10, 2023).

15 J. Slowik. Anatomy of an attack: detecting and defeating CRASHOVERRIDE. *Proceedings of the International Conference Virus Bulletin*, Montreal, Canada (October 2018), pp. 53–75.

16 Dragos Inc. 2017, March). CRASHOVERRIDE: Analysis of the Threat to Electric Grid Operations. *Tech Report*, pp. 1–35, [Online]. https://www.dragos.com/wp-content/uploads/CrashOverride-01.pdf (accessed July 15, 2023).

17 J. Slowik. 2019, August Crashoverride: Reassessing the 2016 Ukraine Electric Power Event as a Protection-Focused Attack. *Dragos Inc. Tech Report*, pp. 1–16, [Online]. https://www.dragos.com/wp-content/uploads/CRASHOVERRIDE.pdf (accessed July 15, 2023).

18 Panda Security. (2017, November). PandaLabs Annual Report 2017. *Tech Report*, pp. 1–42, [Online]. https://www.pandasecurity.com/en/mediacenter/src/uploads/2017/11/PandaLabs_Annual_Report_2017.pdf (accessed July 13, 2023).

19 R. J. Turk. (2005, October). Cyber Incidents Involving Control Systems. *Idaho Natl Lab (INL) Tech Report, USA*, pp. 1–58, [Online]. https://www.osti.gov/biblio/911775 (accessed July 20, 2023).

20 R. Derbyshire, B. Green, D. Prince et al. An analysis of cyber security attack taxonomies. *Proceedings of IEEE European Symposium on Security and Privacy Workshops*, London, UK, (April 2018), pp. 153–161.

21 McAfee Labs. (2011, February). Global energy cyberattacks: 'Night Dragon' Version 1.4. McAfee Inc. White Paper, [Online]. https://www.mcafee.com/blogs/wp-content/uploads/2011/02/McAfee_NightDragon_wp_draft_to_customersv1-1.pdf (accessed July 13, 2023).

22 R. M. Lee, M. J. Assante, and T. Conway. (2014, December). German steel mill cyber attack. ICS SANS Case Study Paper, pp. 1–15, [Online]. https://assets.contentstack.io/v3/assets/blt36c2e63521272fdc/bltc79a41dbf7d1441e/607f235775873e466bcc539c/ICS-CPPE-case-Study-2-German-Steelworks_Facility.pdf (accessed July 15, 2023).

23 A. Di Pinto, Y. Dragoni, and A. Carcano. TRITON: the first ICS cyber attack on safety instrument systems. Understanding the malware, its communications and its OT payload. *Proceedings Black Hat USA*, Las Vegas, USA (August 2018), pp. 1–26.

24 P. Mueller, and B. Yadegari. (2012, June). The Stuxnet Worm. *University of Arizona Tech Report*, pp. 1–12, [Online]. https://www2.cs.arizona.edu/~collberg/Teaching/466-566/2012/Resources/presentations/topic9-final/report.pdf (accessed July 11, 2023).

25 C.G. Blake Johnson, D. Caban, M. Krotofil et al. (2017, December). Attackers deploy new ICS attack framework 'TRITON' and cause operational disruption to critical infrastructure. *FireEye Threat Research Blog*, [Online]. https://www.fireeye.com/blog/threat-research/2017/12/attackers-deploy-new-ics-attack-framework-triton.html (accessed July 20, 2023).

26 Amoah, R., Camtepe, S., and Foo, E. (2016). Securing DNP3 Broadcast Communications in SCADA Systems. *IEEE Trans Industrial Informatics* 12 (4): 1474–1485.

27 Ibrahim Diyeb, I.A., Saif, A., and Al-Shaibany, N.A. (2018). Ethical Network Surveillance using Packet Sniffing Tools: A Comparative Study. *Int Journal Computer Net Info Sec* 10 (7): 12–22.

28 C. Valli et al. Eavesdropping on the smart grid. *Proceedings of Australian Digital Forensics Conference*, Perth, Australia (December 2013), pp. 54–60.

29 Yuan, H., Xia, Y., Yuan, Y., and Yang, H. (2021). Resilient strategy design for cyber-physical system under active eavesdropping attack. *Journal of the Franklin Institute* 358 (10): 5281–5304.

30 Kumar, P., Lin, Y., Bai, G. et al. (2019). Smart Grid Metering Networks: A Survey on Security, Privacy and Open Research Issues. *IEEE Comm Surveys & Tutorials* 21 (3): 2886–2927.

31 Fan, Y., Zhang, Z., Trinkle, M. et al. (2015). A Cross-Layer Defense Mechanism Against GPS Spoofing Attacks on PMUs in Smart Grids. *IEEE Trans Smart Grid* 6 (6): 2659–2668.

32 Xue, A., Xu, F., Xu, J. et al. (2020). Online Pattern Recognition and Data Correction of PMU Data Under GPS Spoofing Attack. *Journal of Modern Power Sys and Clean Energy* 8 (6): 1240–1249.

33 Risbud, P., Gatsis, N., and Taha, A. (2019). Vulnerability Analysis of Smart Grids to GPS Spoofing. *IEEE Trans Smart Grid* 10 (4): 3535–3548.

34 Sterzbach, B. (1997). GPS-based Clock Synchronization in a Mobile, Distributed Real-Time System. *Real-Time Syst* 12 (1): 63–75.

35 Bi, T., Guo, J., Xu, K. et al. (2017). The Impact of Time Synchronization Deviation on the Performance of Synchrophasor Measurements and Wide Area Damping Control. *IEEE Trans Smart Grid* 8 (4): 1545–1552.

36 F. Zhu, A. Youssef, and W. Hamouda. Detection techniques for data-level spoofing in GPS-based phasor measurement units. *Proceedings of the International Conference on Selected Topics in Mobile and Wireless Networking*, Cairo, Egypt (June 2016), pp. 1–8.

37 M. Kabir-Querrec, S. Mocanu, J. Thiriet et al. A test bed dedicated to the study of vulnerabilities in IEC 61850 power utility automation networks. *Proceedings of the International Conference on Emerging Tech and Factory Automation*, Berlin, Germany (November 2016), pp. 1–4.

38 J. G. Wright and S. D. Wolthusen. Stealthy injection attacks against IEC61850's GOOSE messaging service. *Proceedings of IEEE PES Innovative Smart Grid Tech Conference*. Europe, Sarajevo, Bosnia (July 2018), pp. 1–6.

39 El Hariri, M., Youssef, T.A., and Mohammed, O.A. (2016). On the Implementation of the IEC 61850 Standard: Will Different Manufacturer Devices Behave Similarly Under Identical Conditions? *Electronics* 5 (4).

40 T. A. Youssef, M. El Hariri, N. Bugay et al. IEC 61850: technology standards and cyber-threats. *Proceedings of the IEEE International Conference on Environment and Electrical Engineering*, Florence, Italy (June 2016), pp. 1–6.

41 N. Kush, E. Ahmed, M. Branagan et al. Poisoned GOOSE: exploiting the GOOSE protocol. *Proceedings for the Australasian Information Security Conference*, Auckland, New Zealand (January 2014), pp. 17–22.

42 V. S. Rajkumar, M. Tealane, A. Ştefanov et al. (2020)Cyber Attacks on Power System Automation and Protection and Impact Analysis. *IEEE PES Innovative Smart Grid Technologies Conference, Europe*, The Hague, Netherlands, pp. 247–254.

43 Hong, J., Liu, C., and Govindarasu, M. (2014). Integrated Anomaly Detection for Cyber Security of the Substations. *IEEE Trans Smart Grid* 5 (4): 1643–1653.

44 A. Elgargouri, R. Virrankoski, and M. Elmusrati. IEC 61850 based smart grid security. *Proceedings of the IEEE International Conference on Industrial Technology*, Seville, Spain (June 2015), pp. 2461–2465.

45 Liu, Y., Ning, P., and Reiter, M.K. (2009). False Data Injection Attacks Against State Estimation in Electric Power Grids. *ACM Trans Info Syst Sec* 14 (1): 1–33.

46 Pan, K., Teixeira, A., Cvetkovic, M., and Palensky, P. (2019). Cyber Risk Analysis of Combined Data Attacks Against Power System State Estimation. *IEEE Trans Smart Grid* 10 (3): 3044–3056.

47 Jia, L., Kim, J., Thomas, R.J., and Tong, L. (2014). Impact of Data Quality on Real-Time Locational Marginal Price. *IEEE Trans Power Systems* 29 (2): 627–636.

48 Liang, J., Sankar, L., and Kosut, O. (2016). Vulnerability Analysis and Consequences of False Data Injection Attack on Power System State Estimation. *IEEE Trans Power Systems* 31 (5): 3864–3872.

49 Liang, G., Weller, S.R., Zhao, J. et al. (2017). The 2015 Ukraine Blackout: Implications for False Data Injection Attacks. *IEEE Trans Power Systems* 32 (4): 3317–3318.

50 A. Ashok, P. Wang, M. Brown et al. Experimental evaluation of cyber attacks on automatic generation control using a CPS security testbed. *Proceedings of the* IEEE PES GM, Denver, USA (October 2015), pp. 1–5.

51 Reda, H.T., Anwar, A., and Mahmood, A. (2022). Comprehensive survey and taxonomies of false injection attacks in smart grid: attack models, targets, and impacts. *Renew. Sustain. Energy Rev.* 163 (112423): 1–24.

52 Sridhar, S. and Govindarasu, M. (2014). Model-Based Attack Detection and Mitigation for Automatic Generation Control. *IEEE Trans Smart Grid* 5 (2): 580–591.

53 Pan, K., Palensky, P., and Esfahani, P.M. (2020). From Static to Dynamic Anomaly Detection with Application to Power System Cyber Security. *IEEE Trans Power Systems* 35 (2): 1584–1596.

54 Yu, J.J.Q., Hou, Y., and Li, V.O.K. (2018). Online False Data Injection Attack Detection with Wavelet Transform and Deep Neural Networks. *IEEE Trans Industrial Informatics* 14 (7): 3271–3280.

55 Pidikiti, D.S., Kalluri, R., Kumar, R.K.S., and Bindhumadhava, B.S. (2013). SCADA Communication Protocols: Vulnerabilities, Attacks and Possible Mitigations. *CSI Trans ICT* 1 (2): 135–141.

56 Dua, G., Gautam, N., Sharma, D., and Arora, A. (2013). Replay Attack Prevention in Kerberos Authentication Protocol using Triple Password. *Int Journal of Computer Net & Comm (IJCNC)* 5 (2): 1–12.

57 A. Hoehn and Ping Zhang. Detection of replay attacks in cyber-physical systems. *Proceedings of the American Control Conference*, Boston, USA (July 2016), pp. 290–295.

58 Y. Mo and B. Sinopoli. Secure control against replay attacks. *Proceedings of the International Conference on Communication, Control, and Computing*, Monticello, USA (October 2009), pp. 911–918.

59 Kleinmann, A., Amichay, O., Wool, A. et al. (2017). Stealthy Deception Attacks Against SCADA Systems. *Comp Sec* 93–109.

60 De Vivo, M., De Vivo, G.O., Koeneke, R., and Isern, G. (1999). Internet Vulnerabilities Related to TCP/IP and T/TCP. *ACM SIGCOMM Comp. Comm. Rev.* 29 (1): 81–85.

61 Burgers, W., Verdult, R., and Van Eekelen, M. (2013). Prevent Session Hijacking by Binding the Session to the Cryptographic Network Credentials. In: *Secure IT Systems. NordSec*, Lecture Notes in Comp Sci, vol. 8208 (ed. N.H. Riis and D. Gollmann), 33–50. Berlin, Heidelberg: Springer.

62 S. M. Specht and R. B. Lee. Distributed denial of service: taxonomies of attacks, tools, and countermeasures. *Proceedings of the International Conference Parallel and Distributed Comp Systems*, San Francisco, USA (September 2004), pp. 543–550.

63 J.D. Markovic-Petrovic and M.D. Stojanovic. Analysis of SCADA system vulnerabilities to DDoS attacks. *Proceedings of the International Conference on Telecom in Modern Satellite*, Cable and Broadcasting Services, Nis, Serbia (October 2013), pp. 591–594.

64 R. Kalluri, L. Mahendra, R.K.S. Kumar et al. Simulation and impact analysis of denial-of-service attacks on power SCADA. *Proceedings of the National Power System Conference*, Bhubaneswar, India (September 2016), pp. 1–5.

65 A. Carcano, A. Di Pinto, Y. Dragoni et al. The future of securing intelligent electronic devices using the IEC 62351-7 standard for monitoring. *Proceedings Black Hat USA*, Las Vegas, USA (August 2019), pp. 1–21.

66 Lu, A.Y. and Yang, G.H. (2019). Switched Projected Gradient Descent Algorithms for Secure State Estimation Under Sparse Sensor Attacks. *Automatica* 103 (1): 503–514.

67 Vijayshankar, S., Chang, C.-Y., Utkarsh, K. et al. (2023). Assessing the impact of cybersecurity attacks on energy systems. *Applied Energy* 345 (121297): 1–12.

68 Pan, K., Dong, J., Rakhshani, E., and Palensky, P. (2020). Effects of Cyber Attacks on AC and High-Voltage DC Interconnected Power Systems with Emulated Inertia. *Energies* 13 (21): 5584.

69 Schenato, L. (2009). To Zero or to Hold Control Inputs With Lossy Links? *IEEE Trans Auto Control* 54 (5): 1093–1099.

70 Ranathunga, D., Roughan, M., Nguyen, H. et al. (2016). Case Studies of SCADA Firewall Configurations and the Implications for Best Practices. *IEEE Trans Network and Service Mgmt* 13 (4): 871–884.

71 Upadhyay, D. and Sampalli, S. (2020). SCADA (Supervisory Control and Data Acquisition) Systems: Vulnerability Assessment and Security Recommendations. *Computers & Security* 89 (101666).

72 Department of Homeland Security Office of Cybersecurity and Communication. 2015NCCIC/ICS-CERT FY 2015 annual vulnerability coordination report, pp. 1–14, [Online]. https://us-cert.cisa.gov/sites/default/files/Annual_Reports/NCCIC_ICS-CERT_FY%202015_Annual_Vulnerability_Coordination_Report_S508C.pdf (accessed July 15, 2023).

73 M. Kol and S. Oberman. (2020, June)CVE-2020-11896 RCE and CVE-2020-11898 info leak. JSOF Inc. White Paper, pp. 1–27, [Online]. https://www.jsof-tech.com/wp-content/uploads/2020/06/JSOF_Ripple20_Technical_Whitepaper_June20.pdf

74 M. Kol, A. Schon, and S. Oberman. (2020, August). CVE-2020-11901. JSOF Inc. White Paper, pp. 1–23, [Online]. https://www.jsof-tech.com/wp-content/uploads/2020/06/JSOF_Ripple20_Technical_Whitepaper_June20.pdf (accessed July 5, 2023).

75 B. Zhu, A. Joseph, and S. Sastry. A taxonomy of cyber attacks on SCADA systems. *Proceedings of the International Conference on IoT, Cyber, Physical and Social Comp*, Dalian, China (October 2011), pp. 380–388.

76 C.R. Taylor, C.A. Shue, and N.R. Paul. A deployable SCADA authentication technique for modern power grids. *Proceedings of the IEEE International Energy Conference*, Cavtat, Croatia (May 2014), pp. 696–702.

77 Vaidya, B., Makrakis, D., and Mouftah, H.T. (2013). Authentication and Authorization Mechanisms for Substation Automation in Smart Grid Network. *IEEE Network* 27 (1): 5–11.

78 D. Salmon, M. Zeller, A. Guzman et al. Mitigating the aurora vulnerability with existing technology. *Proceedings of the Annual Western Protective Relay Conference*, Washington, USA (October 2009), pp. 1–7.

79 Liu, R., Vellaithurai, C., Biswas, S.S. et al. (2015). Analyzing the Cyber-Physical Impact of Cyber Events on the Power Grid. *IEEE Trans Smart Grid* 6 (5): 2444–2453.

80 Raman, G., AlShebli, B., Waniek, M. et al. (2020). How Weaponizing Disinformation Can Bring Down A City's Power Grid. *PLoS One* 15 (8): 1–14.

81 Sun, C.C., Hahn, A., and Liu, C.C. (2018). Cyber Security Of A Power Grid: State-of-the-Art. *Int. J. Electr. Power Energy Syst.* 99: 45–56.

82 Andersson, G., Donalek, P., Farmer, R. et al. (2005). Causes of the 2003 Major Grid Blackouts in North America and Europe, and Recommended Means to Improve System Dynamic Performance. *IEEE Trans Power Sys* 20 (4): 1922–1928.

83 Pourbeik, P., Kundur, P.S., and Taylor, C.W. (2006). The Anatomy of a Power Grid Blackout - Root Causes and Dynamics of Recent Major Blackouts. *IEEE Power Energy Mag* 4 (5): 22–29.

84 Ten, C.W., Yamashita, K., Yang, Z. et al. (2018). Impact Assessment of Hypothesized Cyberattacks on Interconnected Bulk Power Systems. *IEEE Trans Smart Grid* 9 (5): 4405–4425.

85 Liu, S., Chen, B., Zourntos, T. et al. (2014). A Coordinated Multi-Switch Attack for Cascading Failures in Smart Grid. *IEEE Trans Smart Grid* 5 (3): 1183–1195.

86 B. Chen, S. Mashayekh, K.L. Butler-Purry et al. Impact of cyber attacks on transient stability of smart grids with voltage support devices. *Proceedings of the IEEE PES GM*, Vancouver, Canada (July 2013), pp. 1–5.

87 B. Chen, K.L. Butler-Purry, S. Nuthalapati et al. Network delay caused by cyber attacks on SVC and Its impact on transient stability of smart grids. *Proceedings of the IEEE PES GM*, National Harbor, USA (July 2014), pp. 1–5.

88 A. Castillo, B. Arguello, G. Cruz et al. Cyber-physical emulation and optimization of worst-case cyber attacks on the power grid. *Proceedings of the Resilience Week (RWS)*, San Antonio, USA (November 2019), pp. 14–18.

16

Vulnerabilities of Machine Learning Algorithms to Adversarial Attacks for Cyber-Physical Power Systems

Tapadhir Das[1], *Raj Mani Shukla*[2], *Mohammed Ben-Idris*[3], *and Shamik Sengupta*[4]

[1]*Department of Computer Science, University of the Pacific, Stockton, California, USA*
[2]*School of Computing and Information Science, Anglia Ruskin University, Cambridge, UK*
[3]*Department of Electrical and Computer Engineering, Michigan State University, East Lansing, Michigan, USA*
[4]*Department of Computer Science and Engineering, University of Nevada, Reno, Reno, Nevada, USA*

16.1 Introduction

With increased interest in optimal throughput to sustain the massive demand for performance, modern-day cyber systems are progressively pressured to achieve faster speeds, increased efficiency, and minimal latencies. This has facilitated improvements in contemporary computing environments that have increased computational capabilities, computing resources, and processing power. This has led to the rise in machine learning (ML) and artificial intelligence (AI) technology across various domains, including cyber-physical power systems (CPPSs). A general architecture of modern-day CPPS is illustrated in Figure 16.1. In CPPS, ML algorithms can be applied in a wide variety of fields, including but not limited to electric vehicle (EV) power predictions [1], energy trading in electric distribution systems [2], optimal scheduling for battery swapping stations [3], creation of high performance and efficient solar cells [4], performance estimation and monitoring of power generation plants [5], and energy consumption in smart home environments [6]. The usage of ML in CPPS has progressively increased making modern-day power systems intelligent, efficient, and optimal with faster speeds, and minimal latencies.

However, increased usage of ML in CPPS has made the security of the algorithms a prominent issue [7, 8]. According to the National Institute of Standards and Technology (NIST), multiple stages in the ML pipeline can be classified as potential targets of attack (TA) with various techniques and in-depth knowledge about the system [9]. This can put CPPS at a tremendous risk as it can result in loss of revenue and reputation for the organizations, leading to the destruction of resources, properties, and public endangerment. Some prominent TAs within ML systems include:

- **Input Domain**: This domain includes the input sensors to the ML pipeline or framework. Potential attacks in this domain can involve malicious tampering with the collected sensor data that are to be fed into the ML pipeline.
- **Data Preprocessing**: This domain includes systems for data preparation before being fed into the ML pipeline. Attackers can manipulate collected datasets that are being preprocessed and maliciously alter them so that tampered data are presented to the ML pipeline for training.
- **Machine Learning Models**: The primary TA within the pipeline is the ML model itself. Attackers can poison data that are being processed or create generative adversarial examples of data that

Smart Cyber-Physical Power Systems: Fundamental Concepts, Challenges, and Solutions, Volume 1, First Edition.
Edited by Ali Parizad, Hamid Reza Baghaee, and Saifur Rahman.

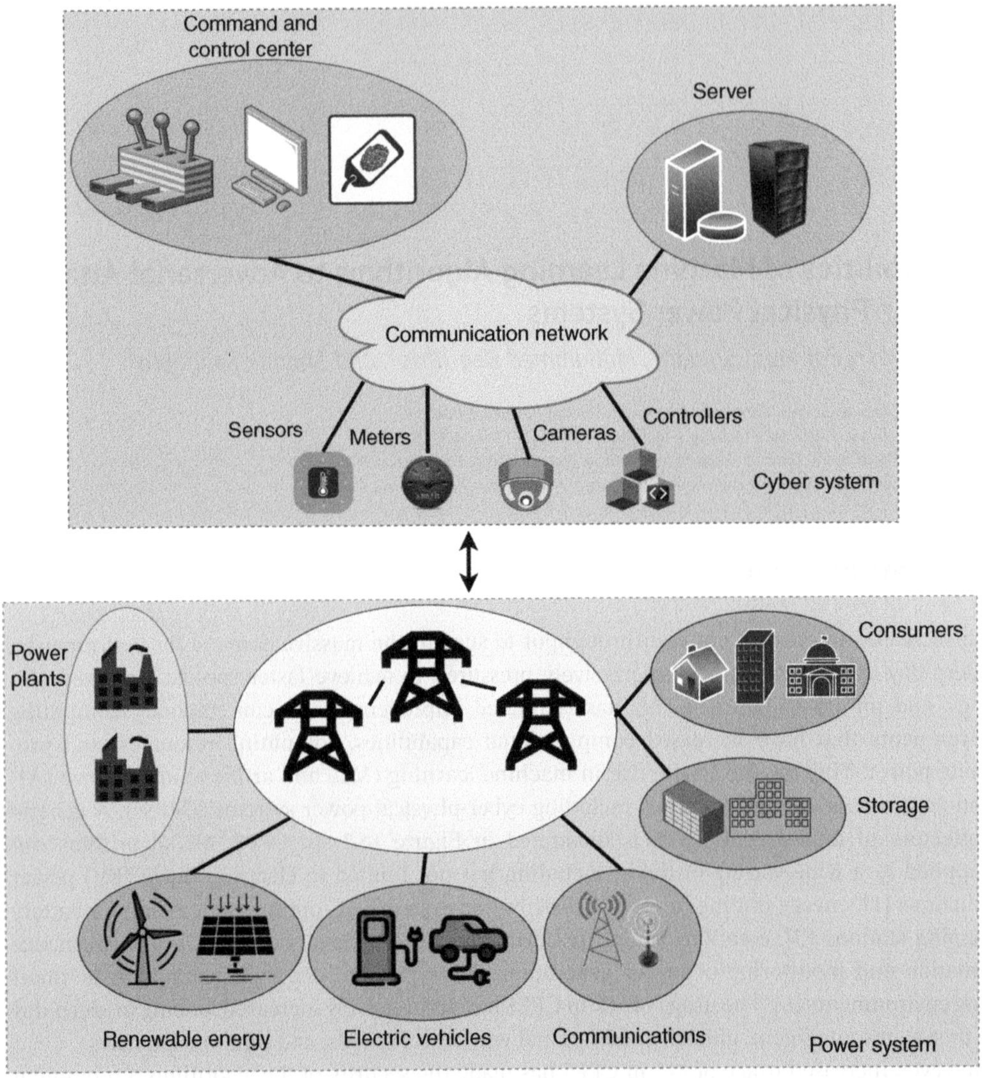

Figure 16.1 Cyber-physical power system.

deceive ML algorithms to make misclassifications and false predictions. These attacks can occur during both the training and testing phases of the model.

- **Output Domain**: This domain includes the output sensors from the ML pipeline or framework. Potential attacks in this domain can involve malicious tampering of the output sensor or display data that are interpretable to the remainder of the system and the system operators.

Adversarial attacks refer to cyber attacks whose aim is to conduct adversarial manipulations on ML algorithms. They consist of adversarial methods that can manipulate training data and exploit model sensitivities to adversely affect the performance of ML classification and regression tasks. Adversarial attacks can be detrimental to critical infrastructures like CPPS as compromised ML pipelines can be catastrophic and put the public and environment at risk. Adversarial techniques against ML systems can be applied to both the training and testing phases of the algorithms.

Attacks on the training try to acquire or influence the training data or the model's performance. Some common adversarial training attacks include:

- **Data Access Attack**: In this attack, the whole or a subset of the training data is accessed to create a malicious substitute model that can be sent for testing.
- **Poisoning Attacks**: In this attack type, data and/or models are directly or indirectly tampered with. Indirect poisoning attacks are done when attackers do not have access to the original data, but instead poison data before preprocessing. Direct poisoning attacks consist of immediate manipulation of the original or preprocessed data or model and can be conducted using data injection, data manipulation, and logic corruption.

Attacks to the testing phase do not tamper with the target model or data, but, instead, generate adversarial inputs that can evade proper classification by the model or infer information about the model or training data. Common adversarial testing attacks include:

- **Evasion Attack**: In this attack, adversaries generate adversarial examples of inputs that can evade proper classification by the model. The adversary solves a constrained optimization problem to find a small input disturbance that causes a massive change in the loss function that can facilitate misclassification.
- **Oracle Attacks**: In this attack, the goal of the adversary is to collect and infer information about the model or training data. The adversary does not know the model architecture itself, but the obtained input output pairings can be used to train a substitute model that will operate like the target model.

Multiple scientists, academic laboratories, and industrial organizations have tried to address the topic of adversarial attacks and have also attempted at generating strategies to protect ML algorithms against adversarial threats in various domains and fields. In the area of CPPS, researchers are actively investigating methods to protect power systems against these threats [10, 11]. The proposed methods and techniques are found to be efficient against certain classes of attacks, but none of these can be used as a primary or one-stop solution for all kinds of prospective adversarial attacks on ML algorithms. Moreover, implementation of many of these defense mechanisms can lead to degradation of performance, thereby leading to inefficiencies for the concerned model [12].

16.2 Vulnerabilities of ML Algorithms to Adversarial Attacks

As previously mentioned, ML algorithms can become vulnerable to adversarial attacks. A visual representation of the TA within an ML pipeline for power systems is illustrated in Figure 16.2. In a typical ML pipeline, adversarial attacks can be targeted in the four main TAs: input domain, data preprocessing, ML models, and output domain. Each of these has its unique vulnerabilities that can be exploited by cybercriminals to corrupt the functionality of the ML pipeline and environment.

16.2.1 Input Domain

The input domain to a CPPS ML pipeline includes the various sensors that can aggregate data. These sensors can measure an assortment of physical parameters in power generation, transmission lines, substation distribution lines, energy storage, and customers. Examples of these sensors are transformers, phasor measurement units, smart meters, temperature sensors, humidity sensors, accelerometers, Internet protocol (IP) network cameras, pyranometers

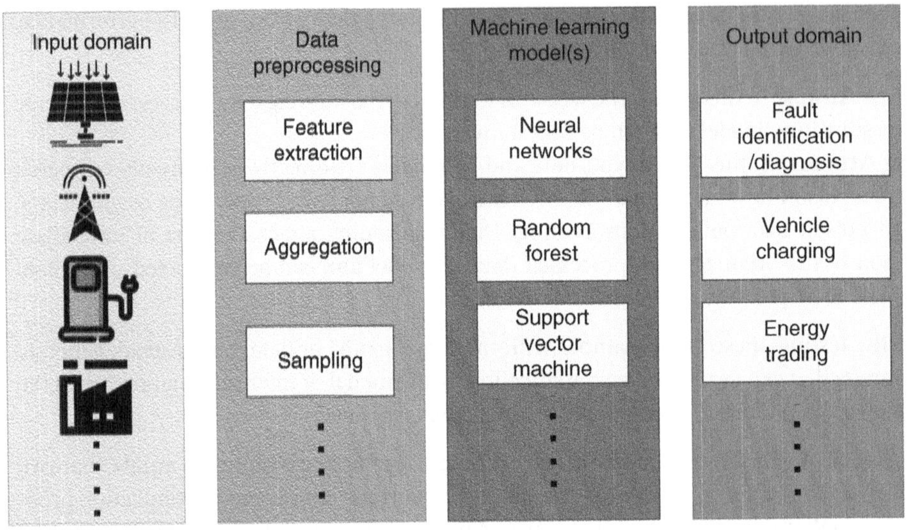

Figure 16.2 Power system ML pipeline.

and pyrheliometers, gas sensors, ultrasound, and ultra-highfrequency sensors, torque sensors, discharge rate sensors, load-leveling sensors, occupancy sensors, and power quality monitors [13].

The widespread usage and variety of these sensors in CPPS make them an attractive TA for adversarial attacks due to their extensive attack space. Also, many of these sensors are available commercial off-the-shelf and inexpensive, which means they do not possess innate security functionality for their protection [14]. This can entice adversaries to maliciously tamper with the input domain sensors. Common methods to tamper with the collected data in this domain include false data injection attacks and data manipulation [15–19]. False data injections can be used to inject malicious, yet believable, data amid uncorrupted data in such a way that it makes the ML algorithm malfunction. Although considerable research has been documented on false data injections, their biggest threat is that they are optimally designed, which makes them a considerable challenge to detect, unaided, because it becomes difficult to differentiate between legitimate and manipulated data. The main objective of the adversarial perturbation is, mainly, to impact the classifier predictions rather than the data. Data manipulation techniques can be used to maliciously alter the labels associated with inputs to create noise in the collected data, which can cause performance degradation.

For example, sensors within power protection devices may be designed to detect anomalies or outlier readings in voltage, current, and frequency measurements. When a sensor reading is observed outside of normal operational conditions, it raises an alarm. In case of adversarial manipulation, the sensors may provide the same normal output in addition to optimal noise which might still raise the alarm if the final reading is outside normal thresholds, but if this sensor reading is further used down the pipeline as an input to another ML algorithm for predictions like electricity demand, electric transmission ampacity, and peak electricity load, the prediction results would be significantly different. This makes adversarial attacks difficult to detect and more distinctive from normal anomaly detection. A visual illustration of adversarial attacks that cause perturbations on input sensor readings is shown in Figure 16.3.

Other methods to disrupt the functionality of input sensors and chips include physically tampering with them to modify their physical characteristics, which can provide correct temperatures but additional noise. Methods to tamper with these chips and sensors include changing their

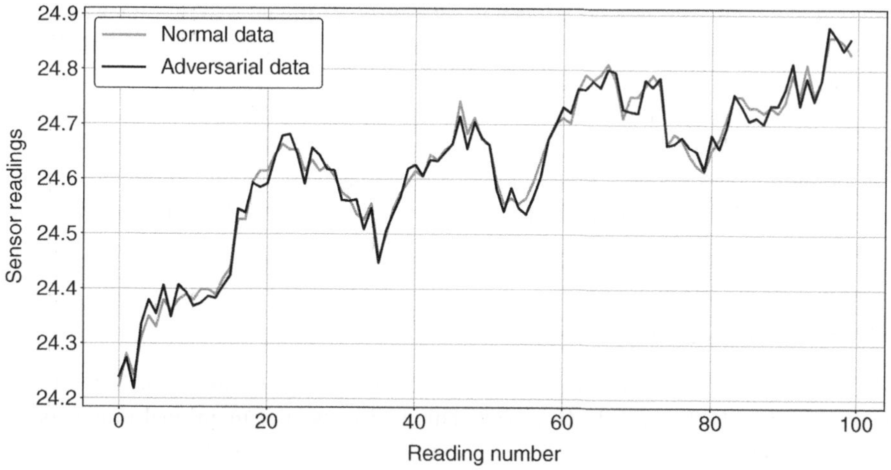

Figure 16.3 Adversarial attacks on input sensors.

position, orientation, and location. They can also include physically damaging these devices using force. Other tampering methods for physical devices can include placing stickers on power system equipment to alter their images/videos that are used to classify the abnormal state of the equipment, causing it to provide significantly different results [20–23]. Additionally, the equipment's design may get modified optimally by physical tampering, which can provide false predictions even if the actual measurements remain the same.

16.2.2 Data Preprocessing

The next major TA where adversarial attacks can occur is the data preprocessing stage. This is where aggregated data from the input domain must be preprocessed before continuing. The main reason for performing data preprocessing is to transform the aggregated data into a uniform, useful, and easily understandable format. Real-world raw sensor data can be messy as they are captured from analog sensors, can have inconsistent formatting, may have been exposed to noisy environments, may have undergone randomization during data collection, may have experienced human errors, and can be incomplete. Data preprocessing allows the data to be presented in a more efficient, convenient, and complete manner so that data analysis and data mining can be easily performed. Examples of data preprocessing techniques include noise removal, frequency and time analysis, data sampling, dimensionality reduction, and feature extraction.

ML methods like autoencoders and their different forms like variational autoencoders are often used to perform the data preprocessing like feature reduction and denoising. Similarly, various clustering mechanisms like hierarchical clustering could be used to find a common subset of sensors/devices to improve accuracy. It has been found that the autoencoders, as well as clustering techniques, are also susceptible to adversarial perturbation.

Using data preprocessing methods makes this stage another attractive TA for adversaries looking to disrupt the ML pipeline in CPPS. Certain methods can be used by adversaries to attack the data during preprocessing. For example, adversaries can maliciously manipulate noise removal algorithms, so that the noise remains in the dataset, causing misclassifications. They can also introduce malicious magnitudes in certain data parameters so that feature extraction algorithms do not perform optimally and lead to higher cardinality in the data. Visual data that are captured from CPPS network cameras can also be tampered with through image scaling attacks where the attacker carefully manipulates images so that they change their appearance when scaled to a specific dimension [24].

These preprocessing data manipulation techniques, as seen above, vary from normal data manipulation because the manipulation and addition of noise are optimally conducted. The perturbations that are conducted on the parameters for data preprocessing may be negligible to the naked eye. However, their impact on the classifier and its predictions can be dramatically different. This can result in negative operational performance with the CPPS.

16.2.3 Machine Learning Models

The primary TA, where adversarial attacks can make a maximal impact and make the ML algorithm vulnerable in the pipeline, is the actual ML model. This is because the main goal of adversarial attacks is to circumvent the predictive capabilities of the ML model and instead, maliciously manipulate it, to evade detection and degrade system performance. Within this TA, attacks can be conducted in both the training and testing phases of the model operation [9]. Attacks during the training phase are geared to retrieve or influence the training data or the ML model itself. Attacks in the testing phase do not tamper with the actual model or the training data. These attack types try to generate adversarial inputs that can evade detection or proper classification from the ML model. Table 16.1 encompasses these attacks and provides a summary of the classification of adversarial attacks against ML models.

16.2.3.1 Training Phase Attacks

Some of the most common attacks in the training phase include poisoning attacks and data access attacks. Poisoning attacks, also known as causative attacks, consist of directly or indirectly modifying the ML or training data. By "poisoning" these entities, the adversary attempts to disrupt the entire learning process ultimately. Poisoning attacks come in two categories: indirect and direct poisoning. Indirect poisoning attacks are done when attackers do not have access to the original data, but instead poison the data before preprocessing. In direct poisoning attacks, the ML

Table 16.1 Adversarial attack types against ML models.

Machine learning attack	Point of compromise	Method of attack
Training phase		
Data injection	Training data	Introducing adversarial inputs in training data
Data manipulation	Training data	Modifying training data directly before use
Logic corruption	Machine learning model	Tampering with model directly
Testing phase		
Evasion	Model loss function	Optimally determining small perturbation in the loss function, leading to misclassifications
Oracle	Model input–output pair	Training substitute model by providing input and observing outputs

or training data are directly maliciously modified. Common methods to conduct direct poisoning attacks include:

- **Data Injection**: In data injection attacks, the adversary can have access to the training data, but not the model. Adversarial inputs are introduced within the original training data. This alters the underlying distribution of the training data but keeps the original features and labels of the data. These injected inputs get optimized by linear programming techniques that can shift the model decision boundary, during unsupervised learning, or by gradient descent/ascent on the testing error to deteriorate ML classification during supervised learning.
- **Data Manipulation**: Similar to data injection attacks, in data manipulation attacks, the adversary can have access to the training data, but not the model. The adversary poisons the training data directly, specifically by modifying the data before it can be used for training. Two primary ways to perform data manipulation include adversarial modification of the labels, also known as label manipulation, and the input data, known as input manipulation.
- **Logic Corruption**: In logic corruption, the adversary can tamper with the ML algorithm directly. This can change the learning process and the model itself. It is very difficult to generate counter-strategies against these adversaries who can alter the learning logic in the ML algorithm.

The other prominent attack category for training phase attacks includes data access attacks. In data access attacks, some or the entirety of the training data is illegally accessed. The goal for this is to generate a substitute model that can be used to evaluate the effectiveness of potential inputs before submitting them to attacks during the testing phase of the ML operation.

16.2.3.2 Testing Phase Attacks

The other end of the spectrum when it comes to adversarial threats against ML models is the testing phase attacks. These attacks are also referred to as exploratory attacks, and they perform no tampering or malicious influence on the ML model or training data. Instead, these attacks generate adversarial inputs that can evade proper classification by the model or can collect and infer information about the model or training data. Two primary threats for testing phase attacks include:

- **Evasion Attacks**: In evasion attacks, the adversary's main goal is to solve a constrained optimization problem to find a small input perturbation that causes a significant change in the loss function, leading to an incorrect prediction or misclassification. These attacks include gradient-based search algorithms, which require knowledge of the model, or substitute model, to compute the gradients in the loss function across input–output pairs.
- **Oracle Attacks**: In Oracle attacks, an application programming interface is used by an adversary to present the model with inputs and to observe the model's outputs. Even when the adversary has no access to the model, the input–output pair obtained can be used to generate a substitute model like the target ML model. This substituted model can be used to generate adversarial examples against the target model.

16.2.4 Output Domain

The output domain from the ML pipeline is another TA for adversarial threats within a CPPS ML pipeline. This domain includes the output sensors or visualizers from the ML pipeline, which denote the forecasted predictions and classifications from the pipeline. These results are pertinent within the CPPS environment as they can be used to conduct procedures like computing or forecasting electricity demand, electric transmission ampacity, and peak electricity load. Examples of output sensors in CPPS can include smart meters, IP network cameras, gauges, and visualization interfaces/dashboards.

Although an important component within the ML pipeline, it is the least influential target for adversaries when conducting attacks against ML models. In this phase, adversaries have no access to the training data or the target model to perform any manipulation, which means that they have no control over the learning and evaluation process. Their main option is to tamper with the output of the ML pipeline. Though this stage has the least amount of influence, it can still be problematic for CPPS. Using data manipulation attacks like label manipulation, adversaries can change the predicted labels at the output, which can make the ML pipeline look tampered with, when, in reality, it is working correctly. Correspondingly, the adversary can manipulate the labels or regressed values to make it look like the CPPS component is working normally when in reality it is malfunctioning or under attack. This can confuse the system operators who are relying on the predictions of these ML pipelines to make necessary decisions for the CPPS. The effects of these attacks can be exacerbated if you have CPPS ML pipelines that are daisy-chained to promote more automation and less human involvement. Manipulated outputs could negatively influence the decisions in ML pipelines further down the line that rely on these outputs. This can cause malfunctions in the CPPS.

For instance, a smart thermostat in a power plant may provide environmental readings like temperature and humidity levels to a system administrator. However, if this smart thermostat is compromised, then adversaries can manipulate the values/labels that get visualized on the device. It might show the temperature at dangerous levels when it is working safely. It can also show safe operational conditions, when, in fact, the temperature is outside normal thresholds. This can cause extensive confusion and operational misuse. Furthermore, if this temperature reading is further used down the pipeline as an input to another ML algorithm for predictions like electricity demand, electric transmission ampacity, and peak electricity load, the prediction results would be significantly different. This can exacerbate the effects of the compromised output domain on a larger portion of the CPPS.

16.3 Theoretical Foundations and Applications of Adversarial Attacks

Adversarial attacks against ML algorithms revolve around the concept of generating an adversarial sample where the adversary injects a small and imperceptible perturbation within the input sample. This leads the ML algorithm to predict an incorrect output. This can be denoted as a non-targeted attack or targeted attack, depending on the results of this attack. Nontargeted and targeted attacks are two primary intentions behind adversarial attacks. A targeted attack refers to making the classifier output specific incorrect results, while a nontargeted attack refers to making the classifier achieve any incorrect output [25]. A visual illustration of an adversarial attack against an ML model is provided in Figure 16.4.

This section provides the fundamental theoretical foundations for adversarial attacks. Let the mathematical model of the trained ML classifier be defined as $f(,;\theta) : X \rightarrow Y$. Here, θ denotes the model parameter, Y is the label of the input, and X represents the input into the model. For any data sample x, the ML classifier predicts a label $\hat{l}(x,\theta)$. The mathematical model for the adversarial sample is generated by finding the minimum perturbation p. After introducing this perturbation into the input data sample, this sample can influence the classifier to predict, classify, or forecast an incorrect output. The mathematical model of the perturbation p is represented as

$$\min \ ||p||_2, \text{s.t.} \hat{l}(x,\theta) \neq \hat{l}(x+p,\theta), \text{ and } x+p \in X \tag{16.1}$$

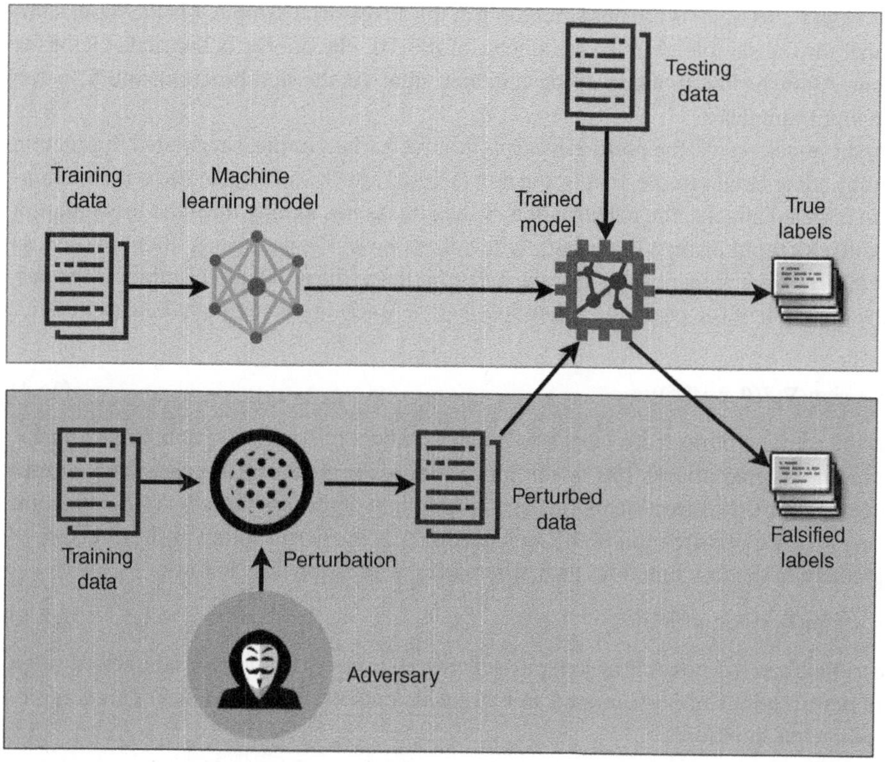

Figure 16.4 Adversarial attack against ML model.

Here, p can be of two distinctive types: one targeted and one nontargeted. When generating the adversarial sample for a specific input sample for a targeted attack, p is the specific perturbation. When generating the adversarial sample for a specific input sample for a nontargeted attack, p is a universal perturbation. Equation (16.1) is a nonconvex function and therefore cannot be solved directly. Therefore, the computation for p is conducted using some continuous approximation techniques like limited-memory Broyden–Fletcher–Goldfarb–Shanno [26] and Jacobian-based saliency map attack [27]. A prominent method for this is the fast gradient sign method (FGSM) [28], which is employed due to its high computational efficiency for generating adversarial perturbations for data samples.

The FGSM algorithm makes use of the gradient of the ML classifier loss function as the perturbation for the data sample, in conjunction with a scale factor that is responsible for controlling the intensity of the perturbation. Additionally, it utilizes a sign function to determine the direction of the perturbation. In these circumstances, the loss function is adjusted to get the optimal perturbation p using (16.2):

$$p = \epsilon * \text{sign}(\nabla_x J(\theta, x, y)) \tag{16.2}$$

The required gradient of the loss function can be computed efficiently using the back propagation method for the neural network. The injection of this perturbation into the input data sample generates the adversarial sample. The mathematical expression for this procedure is given in (16.3):

$$x' = x + \epsilon * \text{sign}(\nabla_x J(\theta, x, y)) \tag{16.3}$$

Here, the $\epsilon * \text{sign}(\nabla_x J(\theta, x, y))$ is the perturbation p of the adversarial sample, x is the input sample, y is the label for the sample, θ is the parameter of the ML classifier, ϵ is the scale factor for the perturbation, $J(\cdot)$ is the loss function of the classifier, sign$(\cdot)$ is the sign function, and ∇_x is the gradient of the input samples x.

According to the attack result, the perturbation of the adversarial sample is regulated to generate the corresponding adversarial sample, that is, the non targeted attack and targeted adversarial samples. In the non targeted attacks, the perturbation of the attacks has to maximize the loss function of the ML classifier's output against the input. As the direction of the positive gradient of the loss function is the direction of the increasing loss, the perturbation of the non targeted attack is the perturbation in the direction of the positive gradient of the loss function. Its mathematical expression is as follows:

$$p = -\epsilon * \text{sign}(\nabla_x J(\theta, x, y^{true})) \tag{16.4}$$

Here, y^{true} is the actual ground-truth label for the data sample. In contrast to non target attacks, the perturbation in targeted attacks has to minimize the loss function of the classifier's output against the target class. As the direction of the negative gradient of the loss function is the direction of the decreasing loss, the perturbation of the targeted attack is the perturbation in the direction of the negative gradient of the loss function. Its mathematical expression is as follows:

$$p = -\epsilon * \text{sign}(\nabla_x J(\theta, x, y^{target})) \tag{16.5}$$

Using these techniques, it is possible to generate adversarial samples from input samples, through the addition of perturbations of non targeted and targeted attacks to those samples. This is represented in the following equations:

$$x_{adversarial} = x + \epsilon * \text{sign}(\nabla_x J(\theta, x, y^{true})) \tag{16.6}$$

$$x_{adversarial'} = x - \epsilon * \text{sign}(\nabla_x J(\theta, x, y^{target})) \tag{16.7}$$

The above equations provide a fundamental explanation of the theoretical foundations for adversarial attacks.

16.4 Attack Models Under Different Scenarios Including Full, Limited, and No Knowledge About the Target Model

Adversarial attacks are geared with multiple potential objectives in mind. However, the main roadmap is the same for all intentions. An adversary attempts to provide an input x to an ML classifier such that it produces an incorrect prediction, output, or forecast [12]. The intended objective of the adversary can be deduced from the inaccuracy of the ML classifier. Depending on the impact of the adversarial attack on the integrity of the output of the classifier, adversarial objectives can be primarily classified as:

- **Confidence Reduction**: The main idea behind this objective is to reduce the confidence of a prediction for a target model. For example, a classifier can alert the CPPS to an ongoing fault. However, this prediction can occur with a lesser confidence or probability than normal due to an adversarial attack.
- **Misclassification**: The main idea behind this objective is to directly alter the output classification of an input data sample to another class. For example, a CPPS classifier misclassifies a line-to-ground fault as a line-to-line fault due to an adversarial attack [29].

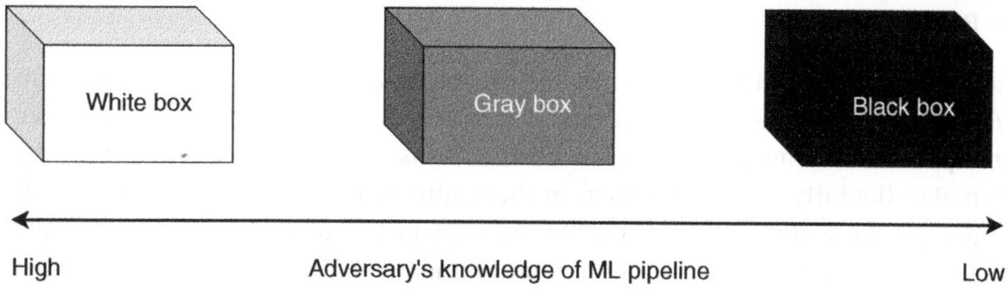

Figure 16.5 Adversarial attack categories.

- **Targeted Attack**: The main idea behind this objective is to manipulate the input data sample in such a way that the ML classifier predicts it as the output of a particular desired target class. For example, a CPPS classifier misclassifies a line-to-ground fault as a three-phase fault due to an adversarial attack, which can alter the appropriate response measures for the CPPS.
- **Nontargeted Attack**: The main idea behind this objective is to manipulate the input data sample in such a way that the ML classifier predicts it as any other class other than the intended class. For example, a CPPS classifier misclassifies a line-to-ground fault to other fault classes due to an adversarial attack, which can alter the appropriate response measures for the CPPS.

In the domain of adversarial attacks, attack models can be applied from three primary scenarios: full, partial, and no knowledge, which are described as follows. Full-knowledge attacks are referred to as white-box attacks; partial-knowledge attacks are referred to as gray-box attacks; and no-knowledge attacks are known as black-box attacks. The relationship between the three adversarial attack categories is illustrated in Figure 16.5. A summary of the adversarial attack types against target models is provided in Table 16.2.

Table 16.2 Adversarial attack types on target model.

Adversarial attack	Knowledge of target model	Point of compromise
White box	Complete	• Algorithm architecture • Training data distribution • Optimization function • Model parameters • Training method • Loss function
Gray box	Partial	• Algorithm architecture • Model parameters • Training method • Loss function • Training data distribution
Black box	Low	• Model settings • Model inputs • Nodel outputs

16.4.1 White-Box Attacks

A white-box attack refers to a situation where an adversary has complete knowledge of an ML classifier f. The adversary has total information about the training algorithm used, like gradient-descent optimization, and can access the input data distribution μ. They also know the model parameters θ of the fully trained classifier. In these situations, the adversary can use this information to analyze the feature space where the model will be vulnerable—i.e., the model can have a high error rate. The complete access to the internal architectures and weights or the classifier makes the white-box attack problematic to defend against.

16.4.2 Gray-Box Attacks

A gray-box attack refers to a situation where an adversary has partial or limited knowledge about the ML classifier f, which can include the model architecture, model parameters θ and their magnitudes, training method, loss functions, and training data distribution μ. Compared to white-box attacks, these attacks are more practical for an adversary.

16.4.3 Black-Box Attacks

The most practical knowledge-based adversarial attack is the black-box attack. This refers to the situation where an adversary does not know the internal functionality and parameters of an ML classifier. It utilizes information about the settings, inputs, and outputs to exploit the model. Black-box attacks can be classified into various categories:

- **None Adaptive Attack**: In this scenario, an adversary can only access the training data distribution μ of a model f. The adversary can, then, train a local model *train'* for a model architecture f' using samples from the original data distribution μ. The goal is to approximate the model learned by the target ML classifier. The adversary can craft adversarial examples on the local model f' using white-box attack strategies and then transfer these crafted inputs to the target model f to trigger misclassifications.
- **Adaptive Attack**: In this scenario, an adversary has no information regarding the training process for a target model f, but can access the model as an unknown trained model. The adversary issues adaptive input samples to this target model and labels a carefully selected dataset. This means that, for any arbitrarily chosen data sample x, the adversary obtains its label y through this unknown target model f. The adversary can then produce *train'* and model architecture f' to train a surrogate model over the input–output pairs of (x, y) that were observed from the target model. Then, this surrogate model can produce adversarial samples by using white-box attack techniques, which can be further used to trigger misclassifications in the target model f.
- **Strict Attack**: In this scenario, an adversary may not possess the data distribution μ but can collect the input–output pairs (x, y) from a target classifier f. However, they cannot change the inputs to observe the changes in output like an adaptive attack. This attack type is comparable to a known-plaintext attack in cryptography and can work on a large set of input–output pairs [30].

The primary objective of black-box attacks is that adversaries try to learn neither the randomness r that is utilized to train a target model f nor the model parameters θ. The primary objective for this attack is to train a local model f' with the data distribution μ or a carefully selected adversarial dataset.

16.5 Real-Life Practical Adversarial Example Generation and Implementation in CPPS

Adversarial example generation on ML algorithms can render them useless. These manipulations can be made to live data that are being processed using ML algorithms in real time. This can cause the ML algorithms to misclassify the inputs as hamper performance. Depending on the application of this ML algorithm, the results could be disastrous. To exacerbate this problem, performing practical adversarial example generation in real life is a trivial task.

A prominent ML domain where real-life practical adversarial examples can be generated is real-time object detection. Real-time object detection is a necessary component in various applications like people counting systems in retail stores, people detection in security systems, and autonomous driving systems like lane, pedestrian, and sign recognition. For example, in autonomous driving, an important function is reliable traffic sign detection. Typically, a cut-out of a localized traffic sign is fed into the ML algorithm from a real-live video stream. An adversary can take a light projector and place a printed object on top of the traffic sign [31]. Other adversaries can use spray paint or paste stickers on top of the traffic sign to create occlusion. This can cause the ML algorithm to not recognize the traffic sign. In this application domain, this kind of adversarial attack can have disastrous consequences. An illustration of this kind of attack is shown in Figure 16.6.

Facial recognition systems have become an integral part of today's computing systems. However, real-life practical adversarial examples can be generated for this technology by placing stickers on faces, wearing color-framed eyewear, and projecting visible light-based perturbations on human

Figure 16.6 Adversarial attack against traffic sign.

faces. These methods can lead to targeted or untargeted misclassification of faces. As facial recognition technologies are relevant in security authorization mechanisms, these adversarial attacks can create security breaches and leaks [32]. Another case for real-life practical adversarial examples can be noticed in live network traffic analysis. Some traffic analysis parties use the deep neural network (DNN)-based classifiers to detect cyber criminals. To circumvent their efforts, adversaries can interfere with the traffic analysis procedure by adversarially perturbing the traffic patterns of the connections that they are intercepting. This perturbation reduces the accuracy of these DNN classifiers. As an example, the traffic analysis "party" can be a malicious Internet service provider (ISP) aiming to deanonymize Tor users by analyzing their Tor connections. Correspondingly, the traffic analysis "adversary" can be benign Tor relays who perturb the traffic patterns of their connections to thwart potential traffic analysis attacks [33].

These practical adversarial examples are also prevalent in CPPS. For instance, a substantial component of modern CPPS is surveillance sensors. These sensors play a significant role in industrial safety, such as within smart grids, nuclear power plants, and other monitoring systems. These sensors are used to obtain the physical or chemical states of key nodes in the system, along with measurements that play a decisive role in the process of state estimation. Examples of these measurements can include temperature, pressure, and other system parameters. This can help monitor the system state in real time. To attack these CPPS surveillance sensors, adversaries can inject perturbations within raw data from the surveillance sensors. This can be done logically through injections, or even by physically tampering with or damaging the sensors. The goal is to introduce these perturbations in such a way that the ML classifier will not treat it as an abnormal state when the recovered system state is different from that of the actual system state [34].

Another practical adversarial example in CPPS can be in the domain of EVs. ML algorithms are prevalent in the EV charging infrastructure. Using historical data or charging load and user behavior, ML algorithms can be used to train to learn the trends and patterns from the data. The predictions from these models can be used to enhance EV charging scheduling strategies along with charging behavior predictions [35]. The EV charging infrastructure relies on the interaction between four main entities communicating and interacting with one another: the EV, the power grid, the central management system, and the EV charging station (EVCS). These components utilize a set of protocols to transfer data, energy, and information. This architectural setup and variation can be inherently vulnerable as each protocol brings its own set of compromises. For instance, most EVCSs contain no physical security or supervision. An adversary can easily damage the EVCS or install malware through the Universal Serial Bus (USB) ports, which can cause theft of data or a denial of service on the EV charging. Adversaries can also abuse the weak links in the system to get access to critical points. For example, an adversary can gain access to vulnerable charging stations in an area and abuse EV charging by disabling the charger. These unauthorized accesses to EVCS can cause a ripple effect on the power grid, exacerbating the issue [36].

16.6 Protection Strategies Against Adversarial Attacks

In this chapter, the main focus has been demonstrating that adversarial attacks can compromise the functionality and performance of ML algorithms. Due to this imminent threat, current research has focused on generating methods, mechanisms, and frameworks to thwart adversarial attacks. Adversarial attacks are difficult to protect against due to the following reasons [12, 37]:

1) **Adversarial example crafting model**: Generation of adversarial samples involves a complex optimization process as it contains nonlinearity and nonconvex properties. There is also a lack

of proper theoretical modeling tools that can provide solutions to these complex optimization problems. This makes it arduous to craft any theoretical argument that can help prospective defense mechanisms rule out adversarial examples.

2) **ML Models Require Outputs for Every Input**: Modification of the ML model to account for dependability against adversarial examples can change the basic objective of the ML model. This will hamper model reliability, performance, and usefulness.

Due to the above reasons, defense against adversarial samples is still an open challenge as there has not been a complete methodology generated that can reliably protect ML algorithms from adversaries. Most of the proposed modern defense strategies are not adaptive to all types of adversarial threats as one method can block one type of attack, but can leave the model exposed and vulnerable to an attacker who may know the underlying mechanism of the model. Additionally, implementation of these defense strategies can incur additional resource and operational overhead, while degrading the performance of the ML algorithm. A summarized table of the protection strategies is illustrated in Figure 16.3 [12]. Current defense mechanisms are explained as follows.

16.6.1 Adversarial Training

The main idea behind adversarial training is to increase model robustness. This is typically achieved by injecting adversarial examples in the training data [38]. This way, the target model is trained with the generated and perturbed samples created by the defender. Augmentation can be done by feeding the model with both legitimate data and the crafted samples, or by learning with a modified objective function. The created model can predict the same classification for both legitimate and perturbed data in the same direction, making the model more robust. The adversarial examples are typically crafted using the methods like those provided in Section 16.3.

The typical method is to train the model on adversarial examples that get crafted on the original model. This can be problematic when trying to defend against black-box attacks, as adversaries can generate malicious samples using a locally trained substitute model that models that emulate the original model. Researchers have also proved that adversarial training can be bypassed using a two-step attack, where random perturbations are applied to an instance first and then any traditional attack is performed on it [39].

16.6.2 Gradient Hiding

To protect against gradient-based adversarial attacks like FGSM, an intuitive countermeasure can be to hide essential information about the model's gradient from the adversary (Table 16.3).

Table 16.3 Protection strategies against adversarial attack categories.

Defensive strategy	Point of protection	Intended attacks
Adversarial training	Training data	Black box
Gradient hiding	Model gradient	Black box
Defensive distillation	Training data	Black box
Feature squeezing	Training data	Black box
Blocking transferability	Machine learning model	Black box
Defense-GAN	Training data	White box, gray box, black box
MagNet	Model outputs	Black box

Gradient-based threats can be ineffective if the ML model is nondifferentiable like a decision tree, nearest neighbor, or a random forest. Despite this benefit, this defense mechanism can be easily circumvented by learning a surrogate black-box model that contains visible gradient information and crafting examples using this.

16.6.3 Defensive Distillation

Distillation can be used as an adversarial training technique to make the model less influenced by adversarial samples [40]. An ML model is trained to classify input samples as either "hard" or "soft" labels, where the final softmax layer generates a probability distribution over all the labels. The "soft" label output of the ML classifier is fed as an input to another identical ML model, trained on the same input samples. This second "distilled" model makes the output smoother and more robust to adversarial attacks. The second model provides a smoother loss function that is more generalized to an unknown dataset and will provide higher performance accuracy even for adversarial perturbed samples. However, the recent advancements in black-box attacks have negatively affected defensive distillation as these methods can bypass defensive distillation methods completely. The main reason behind the success of these new black-box attacks is the strong transferability of adversarial examples across ML models.

16.6.4 Feature Squeezing

Feature squeezing is a model hardening technique, and the main idea behind it is to reduce the complexity of representing the data, which can make the adversarial perturbations disappear due to the low sensitivity. The main heuristics behind this technique include:

- Reduction in color depth on a pixel level making the usage of fewer values to encode colors.
- The usage of a smoothing filter over the images to map multiple inputs to the same value makes the model more resistant and less influenced by noise and adversarial attacks.

The main knock on this technique is that they dramatically reduce the performance and accuracy of the ML models, due to the reduction in dimensionality for representing the data.

16.6.5 Blocking the Transferability

The primary reason for the ineffectiveness of most well-known adversarial defense mechanisms is the strong transferability property of ML algorithms. This means that adversarial examples generated on one classifier are expected to cause similar classifiers to make the same mistake. This transferability property holds, regardless of the underlying architecture of the classifiers or datasets they are trained on. Therefore, a way to protect against black-box attacks would be preventing or blocking the transferability of the adversarial examples [39].

Researchers in [41] proposed a three-step NULL labeling procedure to prevent adversarial examples from being transferred from one ML model to another. The main idea is to augment a new NULL label in the dataset and train the classifier to reject the adversarial examples by classifying them as NULL. The three steps in this technique include:

1) **Initial Training of Target ML Model**: Initial training is conducted on a clean dataset to obtain the classification decision boundaries.

2) **Computing NULL Probabilities**: The probability of belonging to the NULL class is calculated using a probability function for the adversarial examples generated with different amounts of perturbations.
3) **Adversarial Training**: Each clean data sample is then retrained with the original classifier in conjunction with the different perturbed inputs for the sample. The label for the training data is decided based on the NULL probabilities obtained in Step 2.

Transferability blocking is advantageous as the perturbed labels are classified as NULL instead of their original label. To date, this method is regarded as the most effective defense mechanism against adversarial attacks. This technique is also robust enough to reject an adversarial example while not compromising the accuracy of clean data.

16.6.6 Defense-GAN

Another proposed method to protect against adversarial attacks is leveraging the power of a generative adversarial network (GAN) to reduce the efficiency of adversarial perturbations. This can be used to defend against both white-box and black-box attacks. In a normal GAN, a generative model or generator, which emulates the data distribution, and a discriminative model or discriminator, which differentiates between original input and perturbed input, are trained simultaneously. The main idea is to project inputs onto the range of the generator by minimizing the reconstruction error, before feeding the input to the ML classifier. Because of this, legitimate samples will be close to the range of the generator than the adversarial samples, which results in a suboptimal reduction in the potential of adversarial perturbations. This mechanism is effective against adversarial attacks. However, this effectiveness is tied to the expressiveness and generative power of the GAN. Training of GANs can be exhaustive, and improper training can cause the performance of the Defense-GAN to deteriorate significantly.

16.6.7 MagNet

The authors in [42] generated a framework called MagNet, which uses ML classifiers as black boxes to read the output of the classifier's last layer only without modifying the classifier and uses *detectors* to discern between normal and adversarial samples. These detectors check to see whether the distance between the test sample and the manifold exceeds a threshold. The methodology also uses a *reformer* to reform adversarial samples to legitimate samples using autoencoders. MagNet was successful in thwarting attacks against black-box threats. However, its performance deteriorated significantly against white-box attacks. In the work around this situation, researchers started using a variety of autoencoders and randomly picked one to confuse the adversary regarding the selected autoencoder.

16.7 Conclusion and Recommendation

This chapter has illustrated the vulnerabilities that occur within ML pipelines in CPPS and how adversarial attacks can use these vulnerabilities to compromise these algorithms. Due to the ubiquitous nature and usage of ML in CPPS, these threats pose substantial risks to the functionality of the power industry. The chapter started by introducing the usage of ML in CPPS, along with presenting the concept of adversarial attacks that threaten the ML pipelines. Next, it established the current vulnerabilities and TA that exist within ML algorithms and pipelines. These vulnerabilities can be maliciously exploited by adversarial attacks to corrupt their functionality. We also provide the theoretical foundations that occur in crafting an adversarial example for adversarial attacks. We, further, provide these mathematical formulations using the FGSM algorithm, which has been investigated to create adversarial samples. After, attack models under full, limited, and no knowledge of target models are presented to showcase the various situational classifications of adversarial attacks. Real-life practical examples of performing these attacks are also presented, with examples of how these attacks can occur in CPPS. To conclude this chapter, we also present protection strategies that have been attempted to counteract the threat of adversarial attacks on ML algorithms. These methods have shown promising signs, but they can still be conveniently defeated by cyber criminals.

Certain recommendations can be advocated to escalate protection for CPPS ML pipelines against adversarial attacks. These recommendations revolve around developing defense strategies from the data, operational, and architectural viewpoints of the ML pipeline. From the data standpoint, adversarial training can be a useful tool to create more robust models that are capable of distinguishing between normal and perturbed samples. Adversarial training can embed the possibility of adversarial perturbations within the trained ML model, providing prospective protection from adversarial attacks. From the operational standpoint, gradient hiding can be a valuable technique to increase protection against adversarial attacks. Many adversarial attacks are based on gradient-based formulations like FGSM. Hiding or encrypting the essential ML parameters like gradients can help protect the model from gradient-based adversarial threats. From the architectural standpoint, efforts can be made to disrupt the ability of the adversary to conduct adversarial attacks. The utilization of data from various sources and sensors can be used to predict the same output. For instance, predicting the future meter reading of net solar power generation can be conducted using both solar irradiance and meter readings. This method increases the attack surface and adds redundancy to the system, making the adversary's task more difficult as they have to deploy resources to attack both solar and meter readings. Protection of CPPS ML pipelines from adversarial attacks is still an open challenge area. More robust technologies and methods need to be investigated to protect CPPS more vigorously from these attacks.

References

1 Rhode, S., Van Vaerenbergh, S., and Pfriem, M. (2020). Power prediction for electric vehicles using online machine learning. *Engineering Applications of Artificial Intelligence* 87: 103278.

2 Wang, N., Li, J., Ho, S.-S., and Qiu, C. (2021). Distributed machine learning for energy trading in electric distribution system of the future. *The Electricity Journal* 34 (1): 106883.

3 Gao, Y., Yang, J., Yang, M., and Li, Z. (2020). Deep reinforcement learning based optimal schedule for a battery swapping station considering uncertainties. *IEEE Transactions on Industry Applications* 56 (5): 5775–5784.

4 Mahmood, A. and Wang, J.-L. (2021). Machine learning for high performance organic solar cells: current scenario and future prospects. *Energy & Environmental Science* 14 (1): 90–105.

5 Hundi, P. and Shahsavari, R. (2020). Comparative studies among machine learning models for performance estimation and health monitoring of thermal power plants. *Applied Energy* 265: 114775.

6 Schweizer, D., Zehnder, M., Wache, H. et al. (2015). Using consumer behavior data to reduce energy consumption in smart homes: applying machine learning to save energy without

lowering comfort of inhabitants. *2015 IEEE 14th International Conference on Machine Learning and Applications (ICMLA)*, 1123–1129. IEEE.

7 Reznik, L. (2022). *Adversarial Machine Learning*, 315–335. Wiley.

8 Olowononi, F.O., Rawat, D.B., and Liu, C. (2021). Resilient machine learning for networked cyber physical systems: a survey for machine learning security to securing machine learning for CPS. *IEEE Communications Surveys & Tutorials* 23 (1): 524–552.

9 Tabassi, E., Burns, K.J., Hadjimichael, M. et al. (2019, 2019). A taxonomy and terminology of adversarial machine learning. *NIST IR* 1–29.

10 Tian, J., Wang, B., Wang, Z. et al. (2021). Joint adversarial example and false data injection attacks for state estimation in power systems. *IEEE Transactions on Cybernetics* 52 (12): 13699–13713.

11 Paul, S., Ni, Z., and Mu, C. (2019). A learning-based solution for an adversarial repeated game in cyber–physical power systems. *IEEE Transactions on Neural Networks and Learning Systems* 31 (11): 4512–4523.

12 Chakraborty, A., Alam, M., Dey, V. et al. (2018). Adversarial attacks and defences: a survey. *arXiv preprint arXiv:1810.00069*.

13 Song, E.Y., FitzPatrick, G.J., and Lee, K.B. (2017). Smart sensors and standard-based interoperability in smart grids. *IEEE Sensors Journal* 17 (23): 7723–7730.

14 Das, T., ShuklaT, R.M., and Sengupta, S. (2021). Imposters among us: a supervised learning approach to anomaly detection in IoT sensor data. *2021 IEEE 7th World Forum on Internet of Things (WF-IoT)*, 818–823. IEEE.

15 Liu, Y., Ning, P., and Reiter, M.K. (2009). False data injection attacks against state estimation in electric power grids. *Proceedings of the 16th ACM Conference on Computer and Communications Security*, ser. CCS '09, 21–32. New York, NY, USA: Association for Computing Machinery.

16 Sayghe, A., Hu, Y., Zografopoulos, I. et al. (2020). Survey of machine learning methods for detecting false data injection attacks in power systems. *IET Smart Grid* 3 (5): 581–595.

17 Aoufi, S., Derhab, A., and Guerroumi, M. (2020). Survey of false data injection in smart power grid: attacks, countermeasures and challenges. *Journal of Information Security and Applications* 54: 102518.

18 Bhattacharjee, S. and Das, S.K. (2018). Detection and forensics against stealthy data falsification in smart metering infrastructure. *IEEE Transactions on Dependable and Secure Computing* 18 (1): 356–371. https://doi.org/10.1109/TDSC.2018.2889729.

19 Bhusal, N., Gautam, M., and Benidris, M. (2021). Detection of cyber attacks on voltage regulation in distribution systems using machine learning. *IEEE Access* 9: 40402–40416.

20 Sinha, P., Maharana, M.K., Jena, C. et al. (2021). Power system fault detection using image processing and pattern recognition. *2021 IEEE 2nd International Conference on Applied Electromagnetics, Signal Processing, & Communication (AESPC)*, 1–5. IEEE.

21 Shareef, H., Mohamed, A., and Ibrahim, A.A. (2013). An image processing based method for power quality event identification. *International Journal of Electrical Power & Energy Systems* 46: 184–197.

22 Liu, X., Li, Z., Liu, X., and Li, Z. (2016). Masking transmission line outages via false data injection attacks. *IEEE Transactions on Information Forensics and Security* 11 (7): 1592–1602.

23 Anwar, A., Mahmood, A.N., and Ahmed, M. (2015). False data injection attack targeting the LTC transformers to disrupt smart grid operation. In: *International Conference on Security and Privacy in Communication Networks* (ed. J. Tian, J. Jing, and M. Srivatsa), 252–266. Springer International Publishing.

24 Xiao, Q., Chen, Y., Shen, C. et al. (2019). Seeing is not believing: camouflage attacks on image scaling algorithms. *28th USENIX Security Symposium (USENIX Security 19)*, 443–460.

25 Huang, T., Chen, Y., Yao, B. et al. (2020). Adversarial attacks on deep-learning-based radar range profile target recognition. *Information Sciences* 531: 159–176.

26 Szegedy, C., Zaremba, W., Sutskever, I. et al. (2013). Intriguing properties of neural networks. *arXiv preprint arXiv:1312.6199*.

27 Papernot, N., McDaniel, P., Jha, S. et al. (2016). The limitations of deep learning in adversarial settings. *2016 IEEE European Symposium on Security and Privacy (EuroS&P)*, 372–387. IEEE.

28 Goodfellow, I.J., Shlens, J., and Szegedy, C. (2014). Explaining and harnessing adversarial examples. *arXiv preprint arXiv:1412.6572*.

29 Goswami, T. and Roy, U.B. (2019). Predictive model for classification of power system faults using machine learning. *TENCON 2019 - 2019 IEEE Region 10 Conference (TENCON)*, 1881–1885. IEEE.

30 Peng, X., Zhang, P., Wei, H., and Yu, B. (2006). Known-plaintext attack on optical encryption based on double random phase keys. *Optics letters* 31 (8): 1044–1046.

31 Lovisotto, G., Turner, H., Sluganovic, I. et al. (2021). SLAP: Improving physical adversarial examples with Short-Lived adversarial perturbations. *30th USENIX Security Symposium (USENIX Security 21)*, 1865–1882.

32 Shen, M., Liao, Z., Zhu, L. et al. (2019). VLA: A practical visible light-based attack on face recognition systems in physical world. *Proceedings of the ACM on Interactive, Mobile, Wearable and Ubiquitous Technologies* 3 (3): 1–19.

33 Nasr, M., Bahramali, A., and Houmansadr, A. (2021). Defeating DNN-based traffic analysis systems in real-time with blind adversarial perturbations. *30th USENIX Security Symposium (USENIX Security 21)*, 2705–2722.

34 Li, J., Liu, Y., Chen, T. et al. (2020). Adversarial attacks and defenses on cyber–physical systems: a survey. *IEEE Internet of Things Journal* 7 (6): 5103–5115.

35 Shahriar, S., Al-Ali, A., Osman, A.H. et al. (2020). Machine learning approaches for EV charging behavior: a review. *IEEE Access* 8: 168980–168993.

36 ElHussini, H., Assi, C., Moussa, B. et al. (2021). A tale of two entities: contextualizing the security of electric vehicle charging stations on the power grid. *ACM Transactions on Internet of Things* 2 (2): 1–21.

37 OpenAI (2017). Attacking Machine Learning with Adversarial Examples. https://openai.com/blog/adversarial-example-research/ (accessed 15 October 2024).

38 Lyu, C., Huang, K., and Liang, H.-N. (2015). A unified gradient regularization family for adversarial examples. *2015 IEEE International Conference on Data Mining*, 301–309. IEEE.

39 Tramèr, F., Kurakin, A., Papernot, N. et al. (2017). Ensemble adversarial training: attacks and defenses. *arXiv preprint arXiv:1705.07204*.

40 Papernot, N., McDaniel, P., Wu, X. et al. (2016). Distillation as a defense to adversarial perturbations against deep neural networks. *2016 IEEE Symposium on Security and Privacy (SP)*, 582–597. IEEE.

41 Hosseini, H., Chen, Y., Kannan, S. et al. (2017). Blocking transferability of adversarial examples in black-box learning systems. *arXiv preprint arXiv:1703.04318*.

42 Meng, D. and Chen, H. (2017). MagNet: a two-pronged defense against adversarial examples. *Proceedings of the 2017 ACM SIGSAC Conference on Computer and Communications Security*, 135–147.

17

Synchrophasor Data Anomaly Detection for Wide-Area Monitoring and Control in Cyber-Power Systems

A.K. Srivastava[1,2], S. Pandey[3], A. Ahmed[4], S. Basumalik[5], and S.K. Sadanandan[6]

[1]*Smart Grid Resiliency and Analytics Lab, West Virginia University, Morgantown, WV, USA*
[2]*Smart Grid Demonstration and Research Investigation Lab, Washington State University, Pullman, WA, USA*
[3]*Smart Grid and Technology, ComEd, Oakbrook Terrace, IL, USA*
[4]*Intel Corporation, Hillsboro, OR, USA*
[5]*New York Power Authority, Albany, NY, USA*
[6]*DEWA R&D, Dubai, UAE*

17.1 Introduction

This chapter discusses synchrophasor measurement anomaly detection and applications of synchrophasor measurements in wide-area monitoring and control (WAMC) of electric power grids. The synchrophasor technology, widely known as phasor measurement units (PMUs), delivers high-resolution time-stamped voltage and current phasor measurements. The PMU data travels through a hierarchical network of phasor data concentrators (PDCs) at the local level and the control center level before feeding power system applications such as the topology processor, state estimation, electricity markets, fault analysis, energy forecasting, oscillation monitoring, and load modeling, among many others. The complex multi-layered PMU-PDC communication network between the local substation and the main control center may introduce outliers and anomalies due to loss of data during communication failure or data errors. Incorrect and missing measurements have an adverse impact on the operation of power system applications, which may result in erroneous manual or automatic control actions, jeopardizing the reliability of the electric grid. As a result, it is imperative that PMU data quality issues are addressed promptly for improved situational awareness and decision support.

An unsupervised machine learning (ML) algorithm-based *Synchrophasor Anomalies Detection and Classification* (SyADC) tool is developed that enables anomaly identification on PMU data in real-time and provides comprehensive information on the category of anomaly [1]. This tool employs three lightweight, fast, and accurate outlier methods—(i) isolation forest (iforest), (ii) kMeans, and (iii) local outlier probability (LoOP), in a two-level architecture, to detect and classify anomalies with high accuracy. The accurate identification of anomalies helps develop quality-aware applications using synchrophasors for wide area monitoring and control. The developed method is tested on multiple real-world scenarios on the real-time digital simulator (RTDS) considering two application—load modeling and oscillation monitoring. Results show significant improvement in the accuracy of online PMU applications when anomaly data is detected and mitigated using the SyADC tool.

Smart Cyber-Physical Power Systems: Fundamental Concepts, Challenges, and Solutions, Volume 1, First Edition.
Edited by Ali Parizad, Hamid Reza Baghaee, and Saifur Rahman.

17.2 Synchrophasor-Based Wide-Area Monitoring and Control

Wide-area monitoring and control through high-resolution and accurate synchrophasor measurements improve grid situational awareness. Through the advancement in network communication and the ability to process a large volumes of data, the dynamic state, and hence the stability of the power system, can be determined in real-time using PMU measurements. This capability equips power system operators with an early understanding of how far the system is away from the point of collapse and helps design effective corrective action schemes to promptly address any system instability. Additionally, the high-resolution time-stamped PMU measurements aid in the visualization of system events and anomalies in real-time across the entire network. Figure 17.1 shows the various applications of PMU in power systems.

PMUs have widespread use in (i) phase angle monitoring [2], (ii) power oscillation monitoring [3], (iii) voltage stability monitoring [4], (iv) linear state estimation [5], (v) remedial action schemes [6], (vi) line thermal monitoring [7], (vii) power system protection [8], (viii) power damping monitoring [9], and (ix) visualization [10].

For example, PMUs measure the real-time voltage phase angles between two buses or at interfaces giving operators an estimate of the power flow, thereby allowing safe power transfer closer to line limits without violating stability constraints. Power oscillation monitoring is another important application of synchrophasor for WAMC, which helps detect poorly damped low-frequency oscillations between 0.1 and 2 Hz, and distinguish between local or interarea modes of oscillations, thereby serving as an early warning system. Such oscillations, if not damped properly, may lead to widespread blackouts [11]. Real-time PMU-based disturbance analysis has also lead to the development of corrective action schemes such as load shedding, generator rejection, and network switching. Similarly, PMUs measure voltage phasors at each end of the line, leading to voltage stability monitoring in real-time, thereby enabling operators to calculate PV-curve and power margins and safeguarding the system against imminent voltage collapse. With phasor measurements collected up to 240 messages per second, PMUs enable linear state estimation when measurements are expressed in rectangular form. Linear state estimation uses the weighted least square (WLS) method and is iteration free, unlike the traditional supervisory control and data acquisition (SCADA)-based state estimation. Additionally, PMUs can calculate the average transmission line

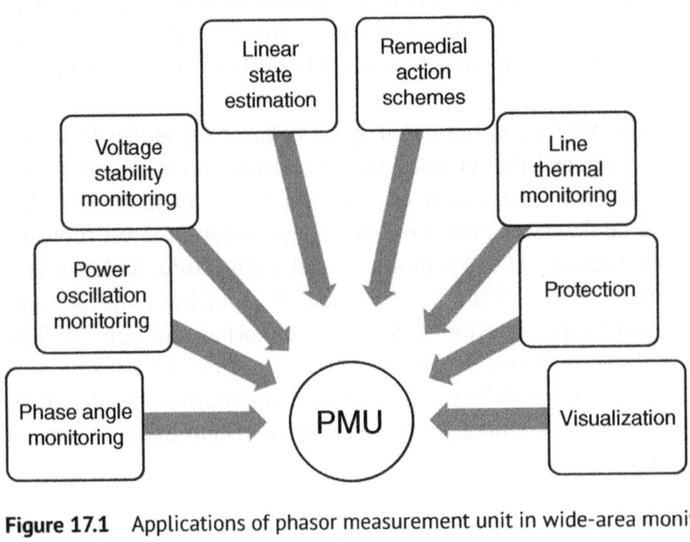

Figure 17.1 Applications of phasor measurement unit in wide-area monitoring and control in electric power systems.

temperature and line losses, which helps operators detect overheated lines. As a result, quick actions such as generation re-dispatch or load shedding can be taken to prevent violation of system operating limits (SOLs). Further, PMUs, through relay functionalities at two ends of the transmission line also provide primary and backup protection during faults. For all of the above WAMC applications, the high-resolution PMU data aids in real-time visualization of power system events, post-event analysis, and creating operator-understandable meaning statistical analytical results.

17.3 Synchrophasor Data Flow, Anomalies, and Impacts

17.3.1 Synchrophasor Data Flow Architecture

Synchrophasors or PMUs provide time-synchronized high-resolution measurements for real-time, wide-area monitoring and control of electric power systems. The high-level data flow architecture of PMUs is shown in Figure 17.2.

Current transformers (CTs) and potential transformers (PTs) measure voltage and current measurements, which are then sampled by synchrophasors before being converted to digital signals using an analog–digital converter. All measurements sampled by the PMU are time-stamped and aligned. Industry-grade PMUs can provide single or 3-phase positive, negative, or zero sequence values in both rectangular and polar formats as 16-bit integer values. The sampling rate can range between 30 and 240 messages per second. Measurements are transported as packet data following the IEEE C37.118.1-2011 Standard for synchrophasors. These packet data flow through the wide area networks (WAN) and concentrate at external C37.118 clients such as PDCs. Industry-graded PDCs have the capability of processing data often from more than 120 concurrent IEEE C37.118-2005 or C37.118-2011 PMUs at 240 messages per second [12]. Data from multiple regional PDCs then concentrates at the central PDC in a control center, from where the data is used in downstream power system-wide area monitoring and control applications or archived [13]. This high-quality time-synchronized voltages, currents, and frequency data can now be utilized for assessing any power system events.

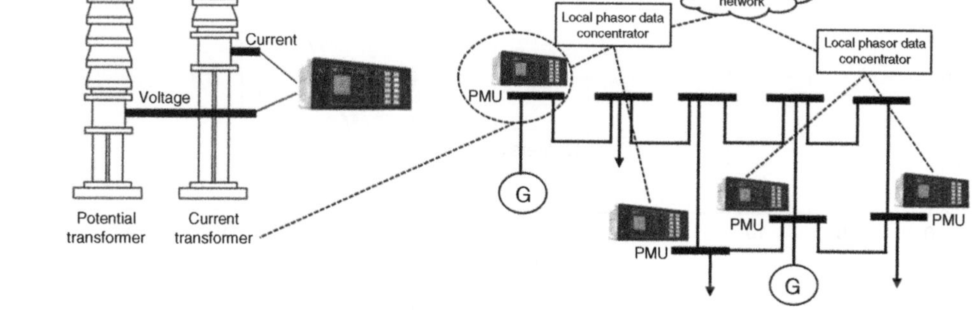

Figure 17.2 PMU-PDC data flow architecture.

17.3.2 Data Anomalies and Impact

As discussed above, PMUs provide high-fidelity voltage and current phasors. This high-resolution synchrophasor data is used for multiple downstream power system applications. The complex measurement and communication infrastructure can introduce bad data, which may be missing measurements or gross outliers. Any such low-quality data might produce incorrect results for power applications, hence may result in incorrect automatic or operator-based actions. Such errors may be broadly classified as follows (Figure 17.3):

1) **Errors from instrument transformers:** Bias in current and potential transformers can cause errors in PMU measurements [14]. Such errors may be introduced due to incorrect calibrations of different classes of instrument transformers or gross errors such as ratio errors and phase errors [15].
2) **Errors in PMUs:** PMUs inherently may have errors within the device itself, for example, anomalies inside the filtering module, missing time-stamps with measurements, gross errors associated with one or more channels, and harmonic interference [16].
3) **Errors in data concentrators:** PDCs aggregate measurements from multiple PMUs. Errors are introduced when measurements are not time-aligned, loss of data due to PMU packet drop, and problems during data compression and streaming. Additionally, multiple data packet loss can also occur at the aggregator end.
4) **Errors in communication network:** Corruption of PMU data inside the network communication layer can occur due to various reasons. Loss of data packet may be due to the loss of network nodes or congestion in the network, which may result in latency and message delivery delay.
5) **Errors in data storage:** Besides concentrating at PDCs, PMU data is also stored for offline usage in databases. Inconsistencies during data archival, errors during storage, loss of data, mixing of PMU IDs, and corruption of data may all ultimately lead to discrepancies.
6) **Errors due to cyber attack:** Targeted cyber intrusions such as false data injection attacks [17] and GPS spoofing [18] may introduce intentional bad data, targeting a specific downstream application.

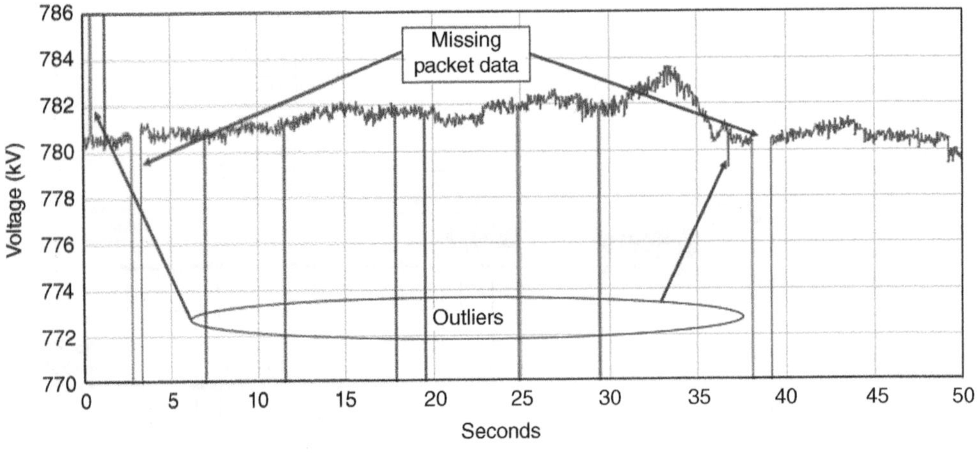

Figure 17.3 Outliers and missing measurements in real-life industrial PMU data.

The above anomalies can occur in a single or in multiple PMUs or at a data concentrator. When downstream PMU applications take as input low-quality measurements, they might produce misleading results, which can lead to incorrect operator actions. Some of the applications that may perform poorly with erroneous PMU data include voltage stability [19], state estimation [20], oscillation monitoring [21], system instability detection [22], load modeling [23], power system restoration [24], islanding detection [25], parameter estimation, event detection [26–28], and protection system failure diagnosis [29–31]. For example, incorrect PMU measurements may result in prediction errors for short-term and long-term voltage stability assessments. Further, bad data results in high residuals in state estimation, leading to incorrect estimates of bus voltages and angles. This would further jeopardize all downstream applications that rely on state estimation such as power flow analysis, contingency analysis, electric markets, ancillary services, controlled islanding, etc. Similarly, noisy PMU data inhibits the identification of different electromechanical modes during oscillation monitoring as well as localization and identification of power system faults. For improving the quality of output of PMU-based applications, it is thus important to accurately detect and classify any suspicious data into bad data or event data.

17.4 Synchrophasor Anomalies Detection and Classification (SyADC)

17.4.1 Background

Multiple approaches have been proposed in the power system literature for the detection, classification, and diagnosis of anomalies in the synchrophasor network. For example, bad data detection using multiple clustering techniques has been proposed in [32–34]. A density-based spatial clustering is used in [34] for online bias detection in PMUs. Three different base detectors—linear regression, Chebyshev, and the DBSCAN method, are employed, together with an unsupervised learning model such as the maximum likelihood (ML) estimator ensemble [32, 33] and expectation maximization (EM) [35] for anomaly detection in PMUs. Considering low-quality PMU data, anomaly detection was performed using an iforest-based online detection in [36]. Outliers due to various power system events were discovered using a moving window-based kernel principal component analysis (PCA) method [37]. For incorrect PMU measurements on raw data streams, Mao et al. [13] applied PCA. A Kalman filter-based algorithm is developed in [38, 39]. Further, event detection and recovery of synchrophasor signals from anomalous data are developed in [36–38]. Other data-driven approaches that have been widely used for the detection and classification of faults include using support vector machines (SVMs), decision trees, and neural networks [40, 41]. Apart from events, various anomaly detection systems to classify cyber-attacks against PMUs have been developed in [17, 42]. A convolutional neural network-based anomaly detector was developed in [17] to identify multiple different faults and cyber-attacks, while a symbolic approximation algorithm was employed in [42] for the same. For the purpose of error classification and detection, anomalies arising from different sources are modeled as outliers in the time-series data. Next, a PMU anomaly data detection and classification tool consisting of an unsupervised stacked ensemble learning algorithm.

17.4.2 The SyADC Tool

This chapter develops a synchrophasor anomaly detection and classification with the following properties: (i) it identifies PMU anomalies in real-time, (ii) it is based on an unsupervised ML

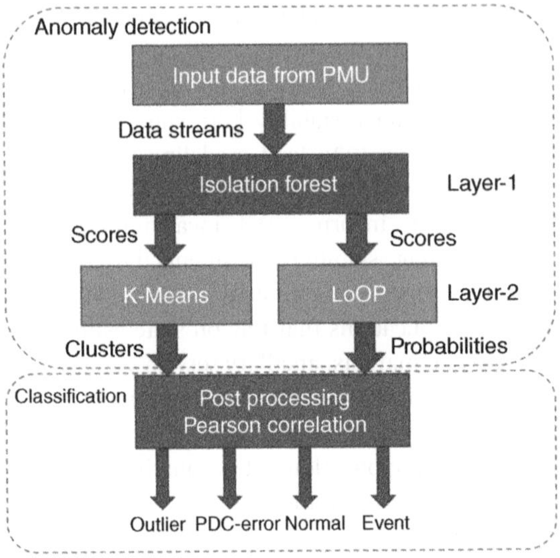

Figure 17.4 SyADC tool for PMU anomaly detection and classification.

algorithm, and (iii) it incorporates an ensemble technique that combines three fast and accurate outlier methods, namely, iforest, kMeans and LoOP, (iv) it serves as a data pre-processing tool that classifies PMU measurements into either bad data or physical events, (v) it gives substantial information on the PMU data quality, and (vi) finally, it is designed to be implemented as a pre-processing data management tool that resolves data quality issues and enhances performance before data is used for power system applications. The SyADC tool is shown in Figure 17.4.

There are two components of the SyADC tool:

1) **Anomaly detection:** This block takes as input the PMU dataset and uses a combination of algorithms, such as iforest [43] with kMeans [44] and LoOP [45], to calculate probabilities of a data being an anomaly or not.
2) **Anomaly classification:** The output of the anomaly detector serves as an input to the next block which classifies PMU data into normal, bad data, event data, or PDC error.

Figure 17.5 shows the classification of PMU anomalies based on the errors associated with them. Chiefly, the classification problem is binary—normal and anomalous data. Anomalies may arise from a single PMU or a set of multiple PMUs. Data anomaly is further divided into bad data and data due to events. Causes of bad PMU data may include CT, PT bias, missing data, outliers, etc. While bad data is mostly restricted to a single PMU, instances of bad data may also occur across multiple PMUs, which may well be an example of a PDC error. On the other hand, event data may arise due to various types of faults, disconnection of generators and loads, switching of capacitor banks, etc. Event signatures are more prominent across a large number of PMUs.

Figure 17.6 illustrates the methodology of the SyADC tool. The working of the tool is described as follows: first, PMU packet data from a central concentrator is set as input to the ensemble anomaly detection method. This detector calculates the probability of each measurement in the data window being an anomaly or not. When the probability score is below a set threshold, there are two possibilities: (i) the data represents steady state (normal) or (ii) instances of missing PMU packet data. When PMU packet data is missing, in case of NaN or "0" measurement, the data concentrator is checked for errors. Otherwise, normal data flow to the downstream power system applications as usual.

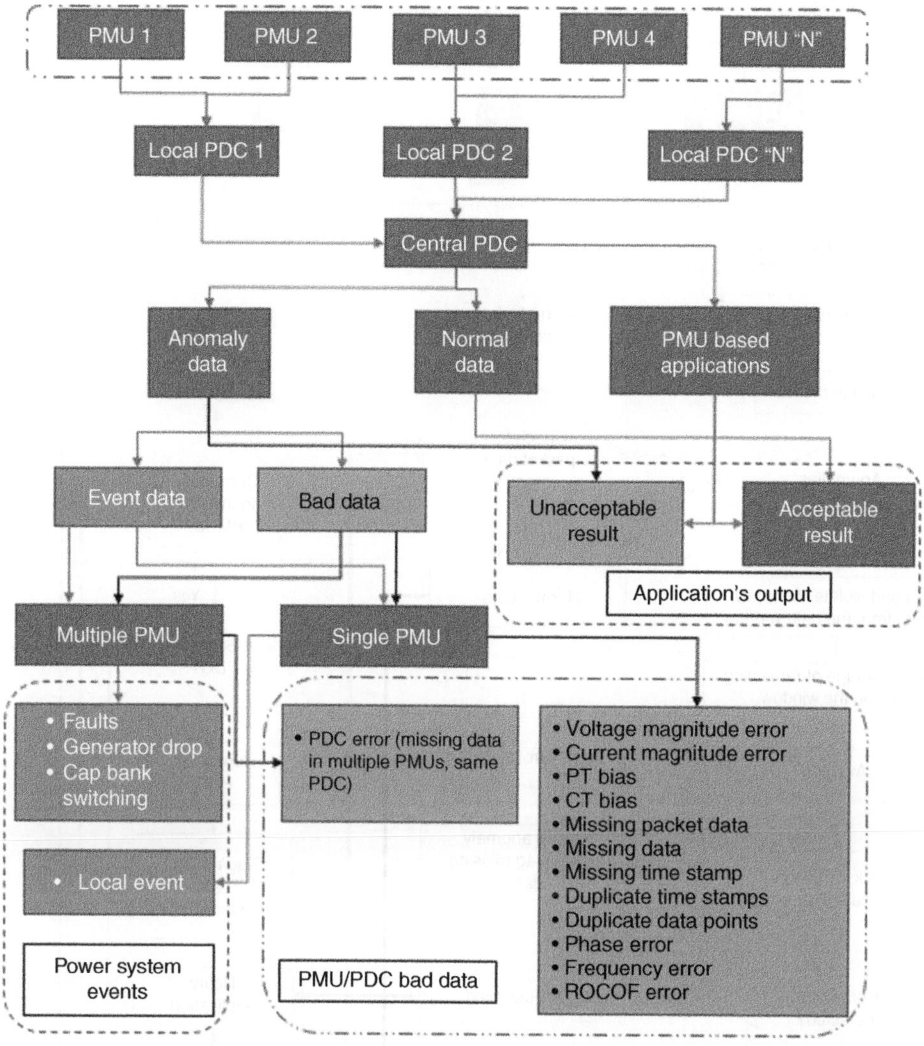

Figure 17.5 Classification of anomalies in PMU.

When the probability score is larger than the set threshold, an anomaly is detected. Such an anomaly may arise due to

1) missing or bad data
2) power system event

The first step is to check for missing data. All detected anomalies are flagged and removed from the data window under consideration. For the rest of the data, the mean is calculated, and missing data is identified when the average is below a set threshold. Next, PDC errors or events are checked for. If the average is larger than the set threshold, the correlation between multiple PMU data stream, as well as between multiple PMUs, is computed. If the correlation is high, then the data window is classified as a power system event, otherwise checked for PDC error.

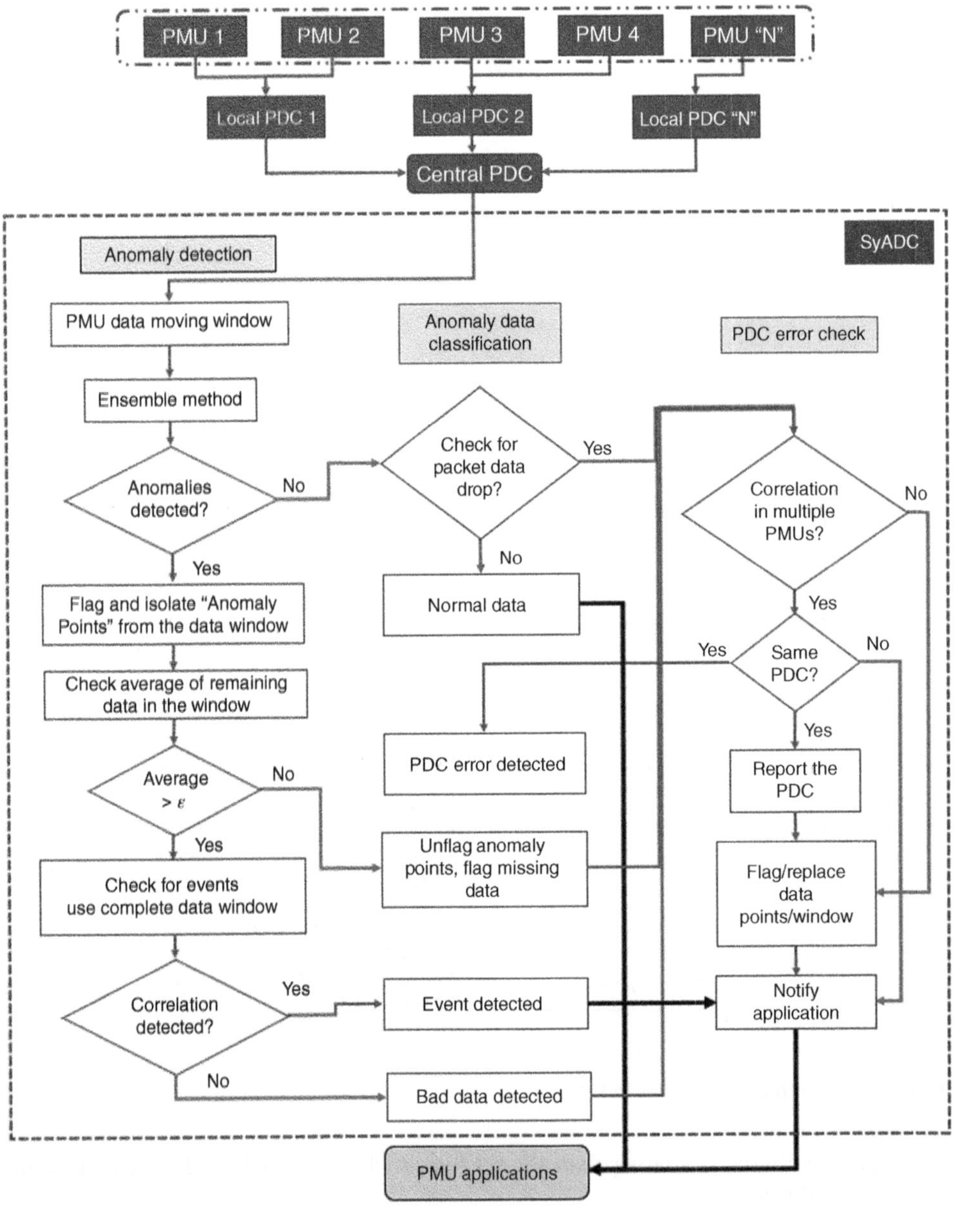

Figure 17.6 Working of the SyADC tool.

Overall, the SyADC tool classifies anomalies into three groups: (i) bad data, (ii) physical event data, and (iii) PDC error. The learning mechanism is as follows—a stack of ensemble-based methods is employed for increasing the accuracy of detection and classification. The first layer consists of an ensemble of base detectors, the output of which serves as the input to the second layer. The results of classification are improved by combining the output of both the layers [46]. The details of the SyADC tool layers are explained next.

17.4.3 Unsupervised Anomaly Detection

In traditional ML approaches, anomalies can be detected with very high accuracy when labels corresponding to anomalous data are available in the training. In such cases, supervised ML models can be deployed. However, the major drawback of this approach is that labeled data sets are not always available, and hence this method cannot be used for real-time online detection and classification. As a result, it is more fitting to use an unsupervised ML algorithm that is able to successfully identify anomalies while requiring no class labels during the initial training process. The philosophy of unsupervised learning is based on the fact that most of the dataset belongs to the normal category, while a minimum fraction of the data is abnormal. Further, the statistical details of the outlier are inherently different from data under normal conditions. Unsupervised algorithms classify infrequent data groups as anomalies. The details of the unsupervised anomaly detection are presented next.

17.4.3.1 Isolation Forest

The iforest [43] is chosen as the unsupervised anomaly detector. The iforest detector has the following advantages:

1) It is capable of detecting anomaly in the shortest possible time as the detector requires only a small sub-sampled dataset during the initial training.
2) The detector consumes low memory and has linear time complexity, hence can be deployed for online purposes, and further, it can be scalable to larger systems.
3) The detector is capable of building a partial model on a smaller size of dataset, hence reduces swamping and masking problems [47–49].
4) The detector is able to consistently perform with sufficient accuracy under different power system test system dynamic situations.

The iforest algorithm identifies anomalies by constructing an ensemble of multiple decision trees for PMU measurement data. These decision trees, also called isolation trees (IT), are similar to binary search trees (BST). Instances of anomalies are classified as those trees with short average path lengths. The iforest algorithm is successfully able to identify anomalies, which are less frequent in number than normal PMU data. First, a partition is created in these IT using a value between the maxima and the minima for a randomly selected feature. Then, each point in the PMU dataset is separated and an anomaly score is generated for decision-making. The entire process of classification is explained as follows.

Consider PMU data window $D = d_i, d_{i+1}, d_{i+2}, \dots, d_{i+w}$ of size w. The set $D \subset \mathcal{R}^N$ is an N-dimensional feature space. Here, each d_i consists of six features for learning—(i) voltage magnitude, (ii) current magnitude, (iii) voltage angle, (iv) current angle, (v) frequency, and (vi) ROCOF. If F_{ij} is the jth feature of the ith data, then

$$d_i = \{F_{i1}, F_{i2}, \dots, F_{in}\} \tag{17.1}$$

With d being an observation in D and w being the sub-sampling size, the anomaly score is calculated as,

$$s(d, w) = 2^{-\frac{E(h(d))}{c(w)}} \tag{17.2}$$

$$c(w) = \begin{cases} 2H(w-1) - \frac{2(w-1)}{x}, & \text{if } w > 2 \\ 1, & \text{if } w = 2 \\ 0, & \text{otherwise} \end{cases} \tag{17.3}$$

$$H(i) \approx \log(i) + e. \tag{17.4}$$

$$E(h(d)) = \frac{1}{Tr}\sum_{i=1}^{Tr} h_i(d) \tag{17.5}$$

Here, the total number of trees is Tr, the path length of observation d is $h(d)$, the average path length of an unsuccessful search is $c(w)$, the number of external nodes is w, and the average $h(d)$ for a collection of ITs is $E[h(d)]$.

The anomaly score, $0 \leq s \leq 1$, is calculated as follows—under normal conditions without an anomaly or event, the average path length $E(h(d))$ is equal to the average path length $c(w)$, and hence $s = 0.5$, regardless of the number of data points [43]. When anomalies are present, a higher score of s closer to 1 is generated. Based on these scores, the next algorithm, LoOP [45], gives the probabilities of the outliers for each PMU measurement data point.

17.4.3.2 Probability Using LoOP

The LoOP is used to ascertain whether a particular data point in the PMU measurement window is an anomaly. The input to the LoOp block is the output of the iforest block. The comparison of a particular data point with its neighbors is done based on the calculated probabilities. The calculated probabilities for LoOP is similar to those using the local outlier factor (LOF) algorithm, which normalizes outlier factors to probabilities [50]. The final scores are calculated using statistical theory, which gives an indication of the observed measurement to be a local anomaly.

Consider ρ as the first layer output, which consists of the set of anomaly scores for each PMU measurement computed over a given window using the iforest algorithm.

$$\rho = \{s_i, s_{i+1}, \dots, s_{i+w}\} \tag{17.6}$$

Next, the LoOP algorithm computes the probability of an observed measurement being an outlier. Let the reference point be R and the observed data point be s. The standard distance, such as the Euclidean or Manhattan distance measure, is used to compute the probability as

$$\sigma(s, R) = \sqrt{\frac{\sum_{r \in R} \text{dist}\,(s, r)^2}{|R|}} \tag{17.7}$$

where dist(s, r) is the distance measure. The probabilistic set distance with $\lambda = 3$, signifying 98% confidence level, is given as,

$$\text{pdist}(\lambda, s, R) = \lambda * \sigma(s, r) \tag{17.8}$$

First, the probabilistic LOF (PLOF) for a PMU measurement point d is defined as,

$$PLOF_{\lambda,k}(s) = \frac{\text{pdist}(\lambda, s, NN_k(s))}{\text{E}_{r \in NN_k(s)}[\text{pdist}(\lambda, r, NN_k(s))]} - 1. \tag{17.9}$$

where k is the neighborhood size and $NN(s)$ is defined as the nearest Euclidean distance between observations under study obtained from the result of the unsupervised learning algorithm iforest. The LoOP is then defined as follows,

$$LoOP_{\lambda,k}(s) = \max\left\{0, \text{erf}\left(\frac{PLOF_{\lambda,k}(s)}{nPLOF \cdot \sqrt{2}}\right)\right\} \tag{17.10}$$

where

$$nPLOF = \lambda.\sqrt{E[PLOF^2]} \tag{17.11}$$

The output of the LoOP block is a set of probabilities L of size w, where $0 \leq l_i \leq 1$, $\forall_{l_i} \in L$. The value of l_i indicates the probability of a PMU measurement data point is an outlier.

17.4.3.3 Clustering with kMeans

The kMeans algorithm is designated to be the second-layer learner. This learner takes as input the output of the iforest block and results a binary classification, labeling the data being either normal or anomaly. Consider the initial set of clusters be designated as $m_1^{(1)}, \dots, m_k^{(1)}$. Each PMU measurement data point s_i is assigned to a cluster, which has the least squared Euclidean distance. This is given by,

$$C_i^{(t)} = \left\{ s_p : \left\| s_p - m_i^{(t)} \right\|^2 \leq \left\| s_p - m_j^{(t)} \right\|^2 \forall_j, 1 \leq j \leq k \right\} \tag{17.12}$$

The clusters are updated in each step as,

$$m_i^{(t+1)} = \frac{1}{\left| C_i^{(t)} \right|} \sum_{x_j \in C_i^{(t)}} s_j \tag{17.13}$$

The objective of the kMeans algorithm can be stated as,

$$argmin_C \sum_{i=1}^{k} \sum_{s \in C_i} \| \rho - \Xi_i \|^2 = argmin_C \sum_{i=1}^{k} |C_i| \; VarC_i \tag{17.14}$$

where Ξ_i denotes the average value of data points in C_i. The formulation in (17.14) is equivalent to minimizing the pairwise squared deviations of data points in each cluster as,

$$argmin_C \sum_{i=1}^{k} \frac{1}{2\left| C_i \right|} \sum_{x,y \in C_i} \| x - y \|^2 \tag{17.15}$$

The output of the kMeans algorithm is a set of cluster labels C of size w, where $c_i = 1$, or $c_i = 0, \forall c_i \in C$. In this case, the value $K = 2$ is set to divide the observations into two groups—normal data is labeled as "0" and outliers as "1."

17.4.3.4 Ensemble of Observations

For a given data point i, the probability of it being an anomaly is obtained by combining the result of the LoOP and kMeans as follows,

$$P = C \cdot L = c_i l_i, c_{i+1} l_{i+1}, \dots, c_{i+w} l_{i+w} \tag{17.16}$$

Here, $c_i = 0$ denotes normal data and $c_i = 1$ denotes outliers. Overall, a value of $p_i \leq 0.1 \; \forall p_i, p_i \in P$ indicates that there are no anomalies. When $p_i > 0.1$, it indicates a possible anomaly in the PMU data stream. The formulation of kMeans classifier PMU data as either normal or anomaly. Since kMeans minimizes the sum-of-squares as its objective, it puts more emphasizes on data points that are far from normal and hence is sensitive to changes. As a result, normal data with small changes may also be erroneously classified as outliers. The output of LoOP shows normal data when l_i is closer to 0. In (17.16), when the results of the kMeans cluster are multiplied by the results from LoOP. The normal data detected by kMeans is masked, preventing incorrect assignment of normal data with small changes as outliers. This combination of the two algorithms, thereby increases the overall accuracy of the ensemble method.

17.4.4 Anomaly Classification

The type of anomaly is decided by computing the correlation between individual PMU data streams. The Pearson correlation is computed as follows:

$$\text{corr}(x, y) = \frac{\text{cov}(X, Y)}{\sigma_x \sigma_y} \tag{17.17}$$

The values of the correlation lie between −1 and +1. A correlation value of $-0.4 \leq \text{corr} \leq 0.4$ suggests no correlation between multiple PMU data measurement streams and is classified as bad data. When the correlation values lie in the range $0.4 < \text{corr} < 0.7$ and $-0.7 < \text{corr} < -0.4$, it indicates a moderate correlation between data streams, which may be caused due to a power system event in a nearby area. Correlation values $0.7 < \text{corr} < 1$ and $-1 < \text{corr} < -0.7$ indicate a strong correlation. In both cases, further steps are required to classify the event. The above correlation values are inferred from previous studies [51–53].

To further classify the event, the SyADC tool computes the correlation between multiple PMUs in the same zone. For a power system event, the event signature is profound across multiple PMUs; hence, if the correlation values for the PMUs communicating to a particular PDC is $-0.4 \leq \text{corr} \leq 0.4$, it is flagged as bad data. Further, when $|\text{corr}| \leq 0.4$, it indicates no correlation between PMU data streams and can be classified as bad data. When $|\text{corr}| > 0.4$, two additional steps are employed to detect the event.

1) The correlation between PMUs reporting to the same PDC is computed, and if $\text{corr} < 0.7$, bad data is flagged. The matrix correlation between two different PMUs is calculated as,

$$\text{corr2}(A, B) = \frac{\sum_{i=1}^{n} \text{corr}(A_{F_i}, B_{F_i})}{n} \tag{17.18}$$

where F_i in (17.18) is the ith feature for the PMU under study.

2) In other cases, the correlation between PMUs reporting to different PDCs is calculated. The correlation between two PDCs is computed as,

$$\text{corr2D}(Z_1, Z_2) = \frac{\sum_{j=1}^{pm_1} \sum_{k=1}^{pm_2} \text{corr2}(PMU_j, PMU_k)}{pm_1 pm_2} \tag{17.19}$$

where PMU_j is the jth PMU in the corresponding PDC. When $\text{corr} < 0.4$ for PMUs in neighboring PDCs, bad data is identified. Otherwise, when $\text{corr} > 0.4$, it is classified as a power system event.

The values of correlation between two PMUs belonging to the same PDC with bad data, PDC errors, and event data are illustrated in Figure 17.7.

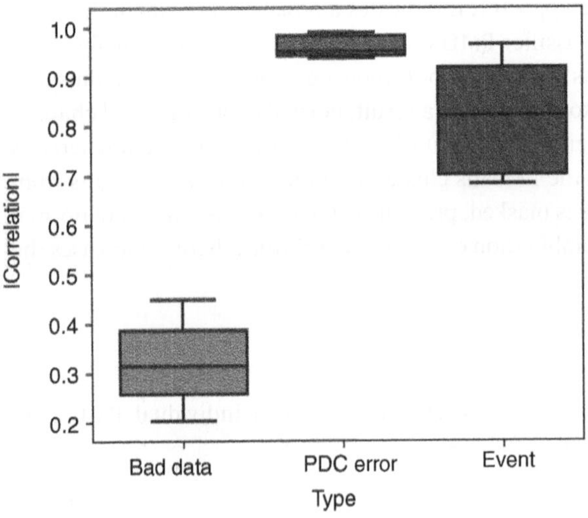

Figure 17.7 Correlation distribution between two PMUs.

A total of 8060 measurement points were analyzed for an IEEE 14 bus simulated in the RTDS. It was observed that the correlation for bad data is around 0.30. For cases of events, the correlation ranged between 0.7 and 0.95. Correlation values between 0.4 and 0.7 indicate an event in a neighboring area. For cases of PDC errors, the correlation was greater than 0.9, and such high values were mostly because for cases of PDC errors, PMU measurements had zero values. The next step is to detect errors in PDC.

$$\mu(D) = \frac{\sum_{j=1}^{n} \sum_{i=1}^{w} F_{ij}}{nw} \tag{17.20}$$

When $\mu(D) \leq 0.2$, it is an indication that there might be a missing PMU packet data in a particular zone. When there is a PDC malfunction, the measurements are reported as "0" or "NAN," depending on data stream settings. In this study, all "NaN" values are removed and substituted with zeros. As a result, when the average value of $\mu(D)$ for all PMU data streams reporting to the same PDC, for a given window size "w", which is zero, it is deemed a PDC error.

17.4.5 Illustrative Example

The proposed method is validated on two test systems modeled in the RTDS. The first example of the IEEE 14 bus system is shown in Figure 17.8.

The data were obtained from actual hardware PMUs (connected to RTDS) at 30 frames per second. A total of 8060 data points were concentrated using the OpenPDC software. The following events are simulated at different buses:

1) faults
2) capacitor bank switching
3) load change
4) generator drop
5) transformer tap change

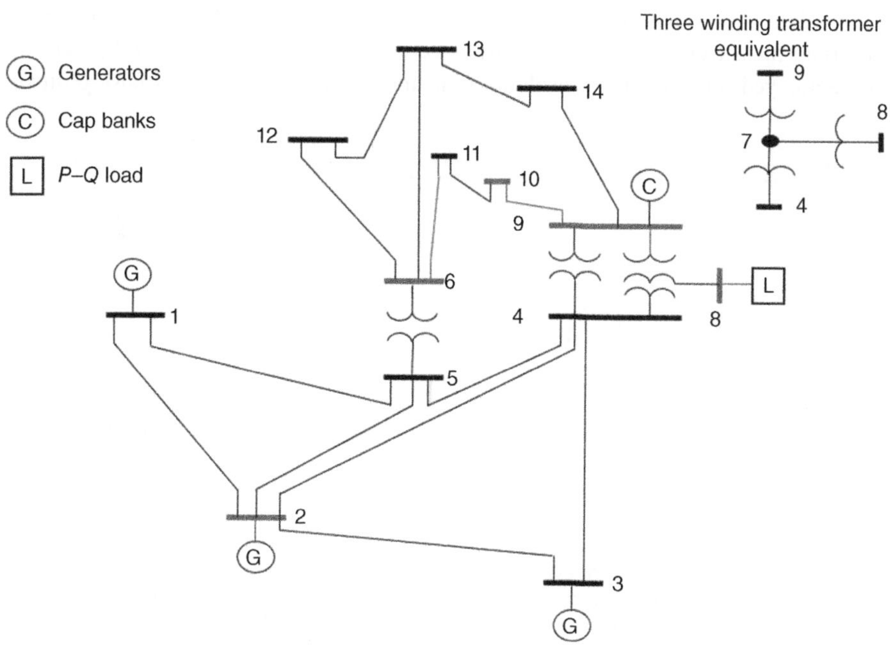

Figure 17.8 The IEEE 14 bus network with PMUs on buses 2, 6, 8, 9, and 10.

Anomalies were synthetically injected in the PMU measurement dataset in five different fractions. The percent (%) of anomalies were kept between 5% and 15%, and the range of bad data was varied between 0.07% and 50%. Figure 17.9 illustrates PMU data for different scenarios at buses 6 and 10 reporting to the same data concentrator.

For a total length of 80 data points, two cases of PDC errors were simulated for PMUs at bus 6 and bus 10. Additionally, an event was simulated on bus 6. Using the SyADC tool, all bad data, events, and PDC errors were identified as anomalous data points. Next, the anomalies were classified using correlation as follows:

1) In scenario A, the majority of the measurement data points were zeros as a result of PDC errors, while only a small number of points are normal. The unsupervised learning algorithm first detected a smaller cluster of anomalies. The average value of $\mu(D)$ using (17.20) was obtained, which was less than the set threshold of 0.2 p.u. As a result, the larger cluster was designated as anomalous, while the smaller cluster as normal PMU measurements.
2) In scenario B, the ensemble algorithm detected all points as normal measurements, as all data points had the same distribution. For further detection, a post-processing step was employed, and $\mu(D)$ was calculated, where $\mu(D) < 0.2$ p.u. Next, the correlation between all PMU measurements reporting to the same PDC was computed, and a strong correlation was obtained, which indicated the scenario of a PDC error.
3) In both scenarios C and D, the unsupervised algorithm detected missing data and outliers. A weak correlation was detected as these were individual anomalies by nature.
4) In scenario E, the unsupervised algorithm detected anomalies. The correlation was calculated in the post-processing step between PMU current and voltage data streams. A strong correlation was obtained, and a correlation with neighboring PMUs was calculated. It was found that an event had occurred at bus 6, and its signature was prominent on bus 10.
5) In scenario F, the unsupervised algorithm correctly detected all data points as normal.

In general, the SyADC tool tackles a classification problem with imbalance class representations, i.e., there are more data points that are normal than which are anomalous. As a result, accuracy is not an appropriate measurement for classification accuracy. As a result, recall and precision metrics [54] are employed for correctness evaluation. in the first step, the metrics false positive

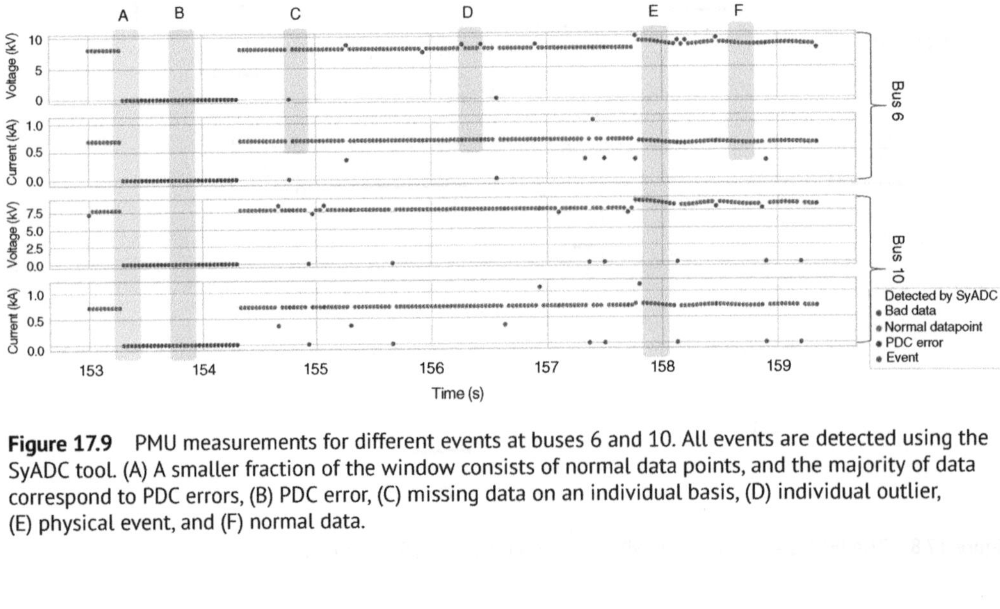

Figure 17.9 PMU measurements for different events at buses 6 and 10. All events are detected using the SyADC tool. (A) A smaller fraction of the window consists of normal data points, and the majority of data correspond to PDC errors, (B) PDC error, (C) missing data on an individual basis, (D) individual outlier, (E) physical event, and (F) normal data.

(FP), false negative (FN), true positive (TP), and true negative (TN) are computed. Then, recall and precision are calculated as,

$$precision = \frac{TP}{TP + FP} \tag{17.21}$$

$$recall = \frac{TP}{TP + FN} \tag{17.22}$$

where,

1) FP is the total number of normal data classified as bad data,
2) FN is the total number of bad data classified as normal data,
3) TP is the total number of bad data classified as bad data, and
4) TN is the total number of normal data classified as normal data.

Table 17.1 shows the information of recall and precision corresponding to events and anomalies at PMU bus 6 and at PMU bus 8, as shown in Figure 17.8.

The processing time for anomaly detection and classification with different data window sizes is shown in Table 17.2. For a window size of 100 data points, the total processing time is <0.13 seconds with a 15% anomaly rate. While a large window size results in a better anomaly detection in SyADC, the total processing time increases, which may not be practical for real-time online applications. From Table 17.2, it is seen that a window size of 10 is able to sufficiently provide the desired classification accuracy, and is suitable for downstream PMU applications.

The evaluation of the SyADC algorithm was further carried out on the IEEE 68 bus system, simulated in RTDS. A total of five PMUs are placed on bus 29, 25, 54, 31, and 41 monitoring lines 29-Load, 25–26, 54–55, 31–10, and 41–40, respectively. Only a small part of the network is visualized in Figure 17.10.

Each PMU measurement is sampled at 30 messages per second, and a total of 12,000 measurements are concentrated. Events simulated include line outage, load change, generator outage, and faults. Anomaly was added at a rate of 15%, and a total of 10 different sets of PMU data were created

Table 17.1 Test results of anomaly detection for buses 6 and 8.

		Anomaly rate				
		5%	7.5%	10%	13%	15%
PMU bus 6	TP	424	608	882	1051	1174
	FP	82	80	67	81	76
	TN	7560	7374	7110	6909	6778
	FN	1	5	8	26	39
	Recall	0.997	0.99	0.99	0.97	0.97
	Precision	0.84	0.89	0.93	0.93	0.94
PMU bus 8	TP	418	609	875	1042	1180
	FP	67	63	43	81	63
	TN	7580	7390	7142	6920	6788
	FN	2	5	7	24	36
	Recall	0.995	0.99	0.99	0.98	0.97
	Precision	0.86	0.90	0.95	0.93	0.95

Table 17.2 Processing time for different window sizes.

Window size (#of data points)	Processing time (s)
10	0.105
20	0.113
50	0.116
100	0.129

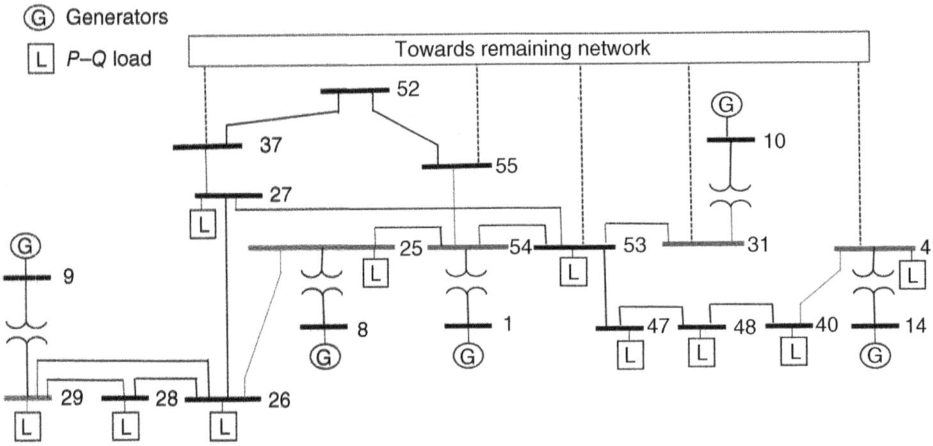

Figure 17.10 IEEE 68 bus system.

for classification and detection. The results of the SyADC tool, in terms of precision and recall, are shown in Table 17.3. It is observed that the algorithm is able to successfully detect anomalies with high precision and recall.

Further, the proposed approach is tested on real-life industrial datasets. Various measurements such as voltage, current, frequency, active power, and reactive power flow are sampled from 60 different PMUs. The SyADC tool was able to successfully detect anomalies, single point missing data, or total packet drops. The results are shown in Figure 17.3, presented at the beginning of the chapter.

Table 17.3 Performance of SyADC on the IEEE 68-bus system.

No.	Precision	Recall	No.	Precision	Recall
1	0.9560	0.9822	6	0.9780	0.9948
2	0.9578	0.9833	7	0.9764	0.9640
3	0.9636	0.9776	8	0.9834	0.9860
4	0.9468	0.9864	9	0.9356	0.9836
5	0.9588	0.9742	10	0.9615	0.9732

Table 17.4 Comparative analysis of SyADC.

Anomaly rate	Method	Precision	Recall
5%	Linear regression [55]	0.78	0.857
	DBSCAN [56]	0.80	0.901
	$PMU_{Ensemble}$ [35]	0.86	0.90
	SyncAD [33]	0.88	0.936
	SyADC	0.86	**0.997**
10%	Linear regression	0.82	0.83
	DBSCAN	0.81	0.93
	$PMU_{Ensemble}$	0.90	0.94
	SyncAD	0.91	0.947
	SyADC	**0.93**	**0.99**
15%	Linear regression	0.86	0.80
	DBSCAN	0.82	0.94
	$PMU_{Ensemble}$	0.92	0.95
	SyncAD	0.93	0.965
	SyADC	**0.95**	**0.97**

Boldface emphasize that the SyADC algorithm achieves better results than state of the art.

17.4.5.1 Comparison with Other Methods

The accuracy and validity of the SyADC tool developed in this chapter are compared with multiple methods available in the literature, such as linear regression [55], DBSCAN [56], $PMU_{Ensemble}$ [35], and SyncAD [33]. Table 17.4 summarizes the results of the comparison. From the area under curve (AUC)-receiver operating characteristic (ROC) shown in Figure 17.11, it is seen that the AUC-ROC for the proposed method is almost about 99%. This demonstrates the superiority of the SyADC tool in detecting and classifying multiple types of anomalies in the synchrophasor system.

17.5 Quality-Aware Synchrophasor-Based Monitoring and Control Applications

Synchrophasor systems and their applications improve the power system's monitoring and control. When PMUs are strategically placed in the power system network, and high-resolution time-stamped data are obtained, the power system operators can capture a real-time snapshot of the electric grid. PMUs have applications in multiple areas, such as state estimation, oscillation monitoring, restoration, islanding detection, load modeling, parameter estimation, and many more. However, PMUs and their related communication network may suffer from data quality issues, data drops, anomalies, etc., which may adversely impact the real-time operation and monitoring. Two examples of quality-aware synchrophasor-based monitoring and control application—load modeling and oscillation monitoring, are presented next.

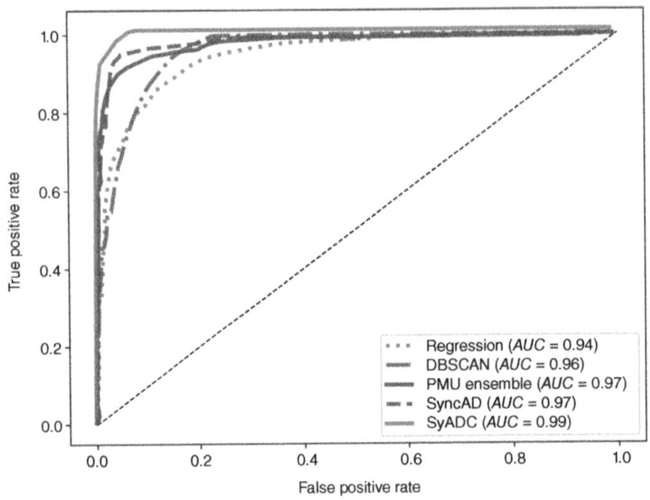

Figure 17.11 The AUC-ROC for 10% anomaly rate in the data.

17.5.1 Load Modeling

An example of a PMU-dependent application is load modeling, which helps power system operators understand the impact of voltage-dependent load behavior [57]. As load behavior directly correlates with voltage stability, capturing the voltage-dependent load behavior aids in efficient control and functional design. The high-resolution time-synced PMU measurements can be used for accurate aggregated load model parameter estimation. We consider realistic PMU data corrupted with 45 dB (SNR) Gaussian noise and data anomalies. Anomalies are inserted as outliers and missing data with zero values. A total of 1000 noisy and bad-data corrupted datasets are used in the study. Simulation is performed on the IEEE 14 bus system using the PSS/E and the RTDS platform.

The first step toward load modeling is data preprocessing. Using an ensemble-based technique [35], similar to the process described in this chapter, all anomalous data points in the PMU measurement stream are identified. Three base detectors are employed for this purpose—linear regression, Chebyshev, and DBSCAN. The aggregated anomaly scores from the three base detectors are set as input to an EM learning algorithm for anomaly identification and mitigation. The anomaly detection and mitigation performance are shown in Table 17.5.

Once anomalies are identified, the next step is to obtain a steady-state measurement value set from the PMU measurement by reducing the impact of noise. This is done by using median and

Table 17.5 Anomaly detection performance.

No.	Precision	Recall	No.	Precision	Recall
1	0.9814	0.99	6	0.9724	1.00
2	0.9719	0.98	7	0.980	0.94
3	0.9716	0.98	8	0.9724	0.98
4	0.9807	0.95	9	0.985	0.99
5	0.9724	1.00			

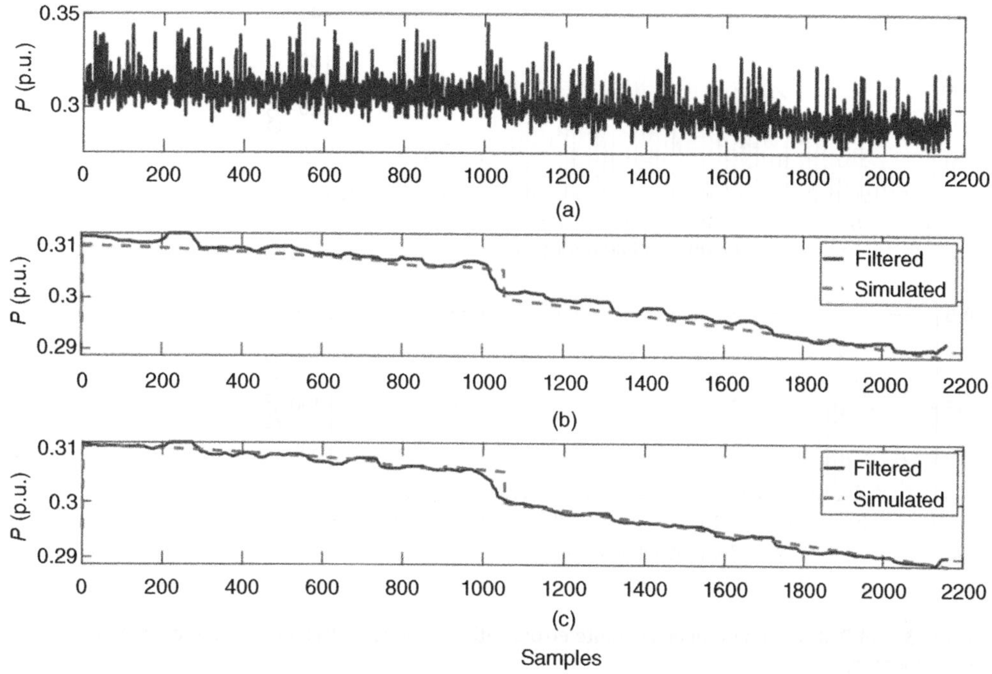

Figure 17.12 (a) Field PMU measurements, (b) filtered PMU measurements without bad-data treatment, and (c) filtered PMU measurements after bad-data treatment.

Kalman filters. Figure 17.12a shows the PMU measurements corrupted with noise and bad data. Figure 17.12b,c show the application of the Kalman filter and median filter without and with bad data removal using the ensemble method.

The third step is to use the error-free synchrophasor data for load parameter estimation. The impedance (Z), current (I), and power (P) (ZIP) load models are used to model a static voltage-dependent load behavior. For active and reactive power, the ZIP model representation is given as,

$$\begin{aligned} W_p &= P_b^{\mathrm{ZIP}}\left(V(t)^2 Z_p + V(t) I_p + P\right) \\ W_q &= Q_b^{\mathrm{ZIP}}\left(V(t)^2 Z_q + V(t) I_q + Q\right) \end{aligned} \tag{17.23}$$

where W_p and W_q are the aggregated real and reactive power measurements. The aggregated real power measurements are a function of constant impedance Z_p, constant current I_p, and constant power P. The aggregated reactive power is similarly written. The PMU measurements are combined, and the load parameters are calculated using two algorithms: (i) recursive least squares with adaptive reconfiguration for ZIP parameter estimation (RLS-RC) and (ii) least squares assisted by variable elimination and adaptive selection of estimation windows (A-LSVE), as described in detail in [57].

The impact of bad data on the load parameter estimation is shown in Figure 17.13 using histogram and probability density function of mean absolute error (MAE) values. By fitting normal distribution on the obtained MAE values, a probability density function (pdf) is obtained, which shows that bad data severely impacts the estimation accuracy.

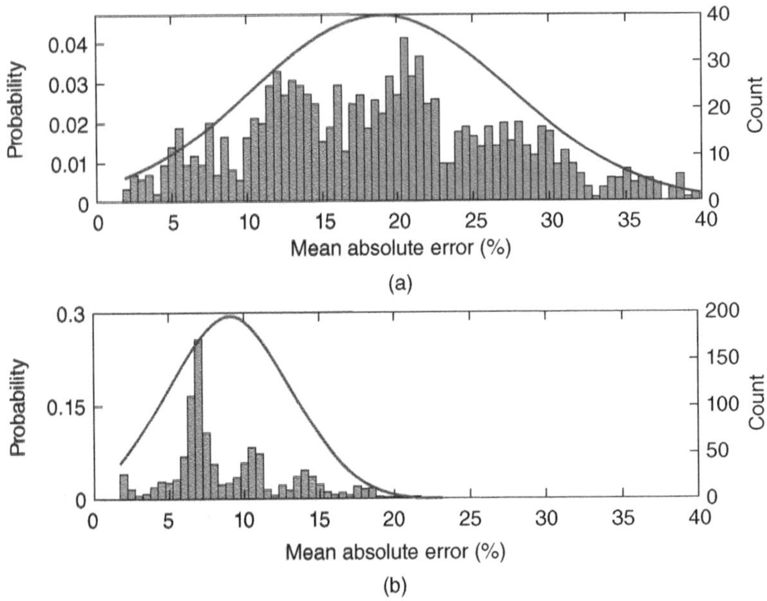

Figure 17.13 (a) Distribution of mean absolute errors without and (b) with bad data detection and correction module.

17.5.2 Oscillation Monitoring

Synchrophasors find widespread use in oscillation monitoring during transient disturbances. Following a power oscillation, time-stamped data obtained from PMUs can help system operators estimate the mode of the system, such as inter-area or local. Consider the disturbance shown in Figure 17.14. Two sets of PMU data are utilized, with a 2-cycle and a 6-cycle estimation window. The first set is from a P-type PMU with damping of 17.74% and a damping frequency of 5.91 Hz. The second data set is from an M-type PMU with a damping of 1.11% and a damping frequency of 4.13 Hz.

The PMU at bus 2 measures the voltage and current phasor at bus 2 and line 2 – 1, respectively, at 60 frames per second. The real power is calculated from these measurements and is shown in Figure 17.15.

A matrix pencil analysis is performed [58], and damping ratio and frequency are estimated from the PMU data. The results are compared with the actual damping and frequency and shown in Table 17.6. Additionally, Table 17.6 illustrates the parameter estimation when anomalies and bad data are injected into the PMU dataset. With anomalies, there is a significant deviation of the estimated parameters of damping and frequency from the actual values. Once measurement errors

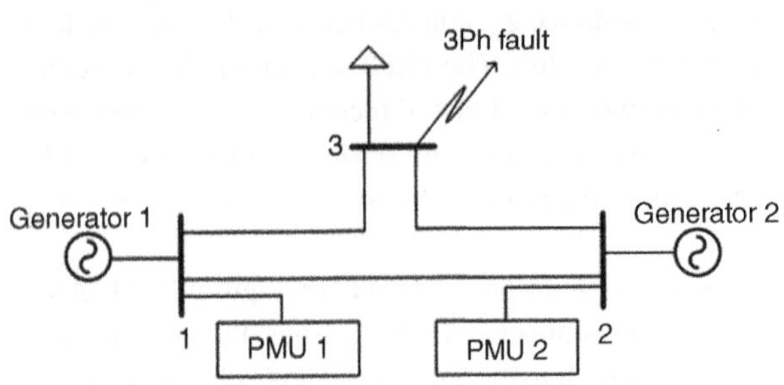

Figure 17.14 Three bus system with fault at bus 3.

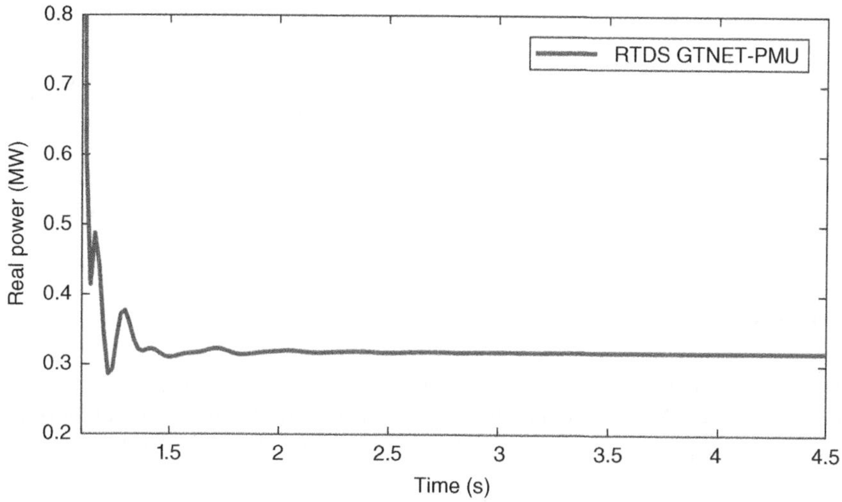

Figure 17.15 Magnitude of real power in line 2 measured from bus 2.

Table 17.6 Oscillation monitoring.

Type of anomaly	PMU setting	True parameters		With anomalies		Without anomalies	
		% Damping	Frequency	% Damping	Frequency	% Damping	Frequency
Voltage magnitude error	2 cycle, P type	17.75	5.92	90.06	5.09	17.68	5.86
	6 cycle, M type	1.11	4.13	Inf	0	1.13	4.2
PT bias	2 cycle, P type	17.75	5.92	17.99	5.94	17.99	5.94
	6 cycle, M type	1.11	4.13	0.48	4.17	0.489	4.17
CT bias	2 cycle, P type	17.75	5.92	12.07	5.24	12.07	5.24
	6 cycle, M type	1.11	4.13	13.83	3.36	13.83	3.36
Missing packet data	2 cycle, P type	17.75	5.92	Inf	0	—	—
	6 cycle, M type	1.11	4.13	Inf	0	—	—
Missing data	2 cycle, P type	17.75	5.92	Inf	0	17.69	5.89
	6 cycle, M type	1.11	4.13	81.47	11.69	1.123	4.19
Duplicate data point	2 cycle, P type	17.75	5.92	17.30	5.86	17.30	5.86
	6 cycle, M type	1.11	4.13	5.75	4.43	5.75	4.43

and missing data are detected and removed using the ensemble learning method [35], it is seen that the estimation error improves. For the cases of CT and PT bias errors, the particular type of anomaly remained undetected. Therefore, the error remained high.

17.6 Summary

This chapter gives an overview of synchrophasor applications in wide area monitoring and control in cyberpower electric grids and illustrates the application of anomaly detection in PMU data measurements using a novel real-time anomaly detection tool called SyADC. This tool is based on an unsupervised anomaly detection algorithm using three fast base detectors to differentiate between PMU errors, physical events, and PDC errors, among other anomalies. The developed method is applied on multiple electric grid test systems simulated on real-time digital simulator and real-life industrial data. It was observed that anomalies were detected with high accuracy with 99.6% precision and 85% recall, compared to other techniques. Further, the SyADC tool was applied in areas of load modeling and oscillation monitoring, where both applications showed significant improvement in performance after anomalies were successfully detected and mitigated.

Acknowledgements

Authors would like to thank financial support from the US Department of Energy and National Science Foundation for this work. We also acknowledge technical support from past researchers at the Washington State University, including Dr. E. Khaledian, Y. Wu, M. Zhou, and Dr. P. Kundu.

References

1 Khaledian, E., Pandey, S., Kundu, P., and Srivastava, A.K. (2020). Real-time synchrophasor data anomaly detection and classification using *Isolation Forest, KMeans*, and *LoOP*. *IEEE Transactions on Smart Grid* 12 (3): 2378–2388.

2 Hiebert, J., Subakti, D., Vinnakota, V.R., and Alam, A. (2019). Operational use of synchrophasor technology for wide-area power system phase angle monitoring at California ISO. In: *Power System Grid Operation Using Synchrophasor Technology* (ed. S.N.D.R. Nuthalapati), 297–305. Springer.

3 Nabavi, S., Zhang, J., and Chakrabortty, A. (2015). Distributed optimization algorithms for wide-area oscillation monitoring in power systems using interregional PMU-PDC architectures. *IEEE Transactions on Smart Grid* 6 (5): 2529–2538.

4 Vournas, C.D., Lambrou, C., and Mandoulidis, P. (2016). Voltage stability monitoring from a transmission bus PMU. *IEEE Transactions on Power Systems* 32 (4): 3266–3274.

5 Liu, R., Srivastava, A.K., Bakken, D.E. et al. (2017). Decentralized state estimation and remedial control action for minimum wind curtailment using distributed computing platform. *IEEE Transactions on Industry Applications* 53 (6): 5915–5926.

6 Wang, Y.-J., Liu, C.-W., and Liu, Y.-H. (2005). A PMU based special protection scheme: a case study of Taiwan power system. *International Journal of Electrical Power & Energy Systems* 27 (3): 215–223. https://doi.org/10.1016/j.ijepes.2004.09.008.

7 Mousavi-seyedi, S.S., Aminifar, F., Azimi, S., and Garoosi, Z. (2014). On-line assessment of transmission line thermal rating using PMU Data. *2014 Smart Grid Conference (SGC)*. https://doi.org/10.1109/sgc.2014.7090880.

8 Neyestanaki, M.K. and Ranjbar, A.M. (2015). An adaptive PMU-based Wide Area backup protection scheme for power transmission lines. *IEEE Transactions on Smart Grid* 6 (3): 1550–1559. https://doi.org/10.1109/tsg.2014.2387392.

9 Haugdal, H. (2022). Application of phasor measurements for online monitoring and adaptive damping control of electromechanical oscillations.

10 Arunkumar, A., Gupta, N., Pinceti, A. et al. (2022). PMUVis: A large scale platform to assist power system operators in a smart grid. *IEEE Computer Graphics and Applications* 42 (6): 84–95.

11 Prasertwong, K., Mithulananthan, N., and Thakur, D. (2010). Understanding low-frequency oscillation in power systems. *International Journal of Electrical Engineering Education* 47 (3): 248–262.

12 Schweitzer, E.O., Whitehead, D., Guzmán, A. et al. (2010). Advanced real-time synchrophasor applications. *Journal of Reliable Power* 2 (2): 1–14.

13 Mao, Z., Xu, T., and Overbye, T.J. (2017). Real-time detection of malicious PMU data. *2017 19th International Conference on Intelligent System Application to Power Systems (ISAP)*, 1–6. IEEE. ISBN: 978-1-5090-4000-1. https://doi.org/10.1109/ISAP.2017.8071368.

14 Komarnicki, P., Dzienis, C., Styczynski, Z.A. et al. (2008). Practical experience with PMU system testing and calibration requirements. *2008 IEEE Power and Energy Society General Meeting-Conversion and Delivery of Electrical Energy in the 21st Century*, 1–5. IEEE.

15 GD Energy. Instrument transformer basic technical information and application.

16 PMU NASPI (2017). PMU data quality: a framework for the attributes of PMU data quality and quality impacts to synchrophasor applications.

17 Basumallik, S., Ma, R., and Eftekharnejad, S. (2019). Packet-data anomaly detection in PMU-based state estimator using convolutional neural network. *International Journal of Electrical Power & Energy Systems* 107: 690–702.

18 Fan, X., Du, L., and Duan, D. (2017). Synchrophasor data correction under GPS spoofing attack: a state estimation-based approach. *IEEE Transactions on Smart Grid* 9 (5): 4538–4546.

19 Momoh, J.A., Xia, Y., and Boswell, G. (2008). Voltage stability enhancement using Phasor Measurement Unit (PMU) technology. *2008 40th North American Power Symposium*, 1–6. IEEE.

20 Göl, M. and Abur, A. (2015). A hybrid state estimator for systems with limited number of PMUs. *IEEE Transactions on Power Systems* 30 (3): 1511–1517.

21 Liu, G., Quintero, J., and Venkatasubramanian, V.M. (2007). Oscillation monitoring system based on wide area synchrophasors in power systems. *2007 iREP Symposium-bulk Power System Dynamics and Control-VII. Revitalizing Operational Reliability*, 1–13. IEEE.

22 Srivastava, A., Pandey, S., Zhou, M. et al. (2017). Ensemble based technique for synchrophasor data quality and analyzing its impact on applications. *North American Synchrophasor Initiative (NASPI)*, Gaithersburg, MD, 1–24. https://doi.org/10.1109/NASPI.2017.7914501.

23 Pandey, S., Srivastava, A.K., Markham, P. et al. (2018). Online estimation of steady-state load models considering data anomalies. *IEEE Transactions on Industry Applications* 54 (1): 712–721.

24 Sarmadi, S.A.N., Dobakhshari, A.S., Azizi, S., and Ranjbar, A.M. (2011). A sectionalizing method in power system restoration based on WAMS. *IEEE Transactions on Smart Grid* 2 (1): 190–197.

25 Guo, Y., Li, K., Laverty, D.M., and Xue, Y. (2015). Synchrophasor-based islanding detection for distributed generation systems using systematic principal component analysis approaches. *IEEE Transactions on Power Delivery* 30 (6): 2544–2552.

26 Dahal, O.P., Brahma, S.M., and Cao, H. (2014). Comprehensive clustering of disturbance events recorded by phasor measurement units. *IEEE Transactions on Power Delivery* 29 (3): 1390–1397.

27 Kundu, P. and Pradhan, A.K. (2018). Real-time event identification using synchrophasor data from selected buses. *IET Generation, Transmission Distribution* 12 (7): 1664–1671.

28 Sun, C., Wang, X., Zheng, Y. et al. (2019). Early warning system for spatiotemporal prediction of fault events in a power transmission system. *IET Generation, Transmission & Distribution* 13 (21): 4888–4899.

29 Dubey, R., Samantaray, S.R., and Panigrahi, B.K. (2017). An spatiotemporal information system based wide-area protection fault identification scheme. *International Journal of Electrical Power & Energy Systems* 89: 136–145.

30 Gholami, A., Srivastava, A.K., and Pandey, S. (2019). Data-driven failure diagnosis in transmission protection system with multiple events and data anomalies. *Journal of Modern Power Systems and Clean Energy* 7 (4): 767–778.

31 Khaledian, P., Johnson, B.K., and Hemati, S. (2018). Power grid resiliency improvement through remedial action schemes. *IECON 2018-44th Annual Conference of the IEEE Industrial Electronics Society*, 774–779. IEEE.

32 Pandey, S., Chanda, S., Srivastava, A., and Hovsapian, R. (2020). Resiliency-driven proactive distribution system reconfiguration with synchrophasor data. *IEEE Transactions on Power Systems* 35 (4): 2748–2758.

33 Pandey, S., Srivastava, A., and Amidan, B. (2020). A real time event detection, classification and localization using synchrophasor data. *IEEE Transactions on Power Systems* 35 (6): 4421–4431.

34 Wang, X., Shi, D., Wang, Z. et al. (2018). Online calibration of phasor measurement unit using density-based spatial clustering. *IEEE Transactions on Power Delivery* 33 (3): 1081–1090.

35 Zhou, M., Wang, Y., Srivastava, A.K. et al. (2018). Ensemble-based algorithm for synchrophasor data anomaly detection. *IEEE Transactions on Smart Grid* 10 (3): 2979–2988. https://doi.org/10.1109/TSG.2018.2816027.

36 Wu, T., Zhang, Y.J., and Tang, X. (2020). Online detection of events with low-quality synchrophasor measurements based on iForest. *IEEE Transactions on Industrial Informatics* 17 (1): 168–178.

37 Chatterjee, K. and Chaudhuri, N.R. (2019). Corruption-resilient detection of event-induced outliers in PMU data: a kernel PCA approach. *2019 IEEE Power & Energy Society General Meeting (PESGM)*, 1–5. IEEE.

38 Chatterjee, K., Mahapatra, K., and Chaudhuri, N.R. (2019). Robust recovery of PMU signals with outlier characterization and stochastic subspace selection. *IEEE Transactions on Smart Grid* 11 (4): 3346–3358.

39 Jones, K.D., Pal, A., and Thorp, J.S. (2015). Methodology for performing synchrophasor data conditioning and validation. *IEEE Transactions on Power Systems* 30 (3): 1121–1130.

40 Al Karim, M., Chenine, M., Zhu, K. et al. (2012). Synchrophasor-based data mining for power system fault analysis. *2012 3rd IEEE PES Innovative Smart Grid Technologies Europe (ISGT Europe)*, 1–8. IEEE. ISBN: 978-1-4673-2597-4. https://doi.org/10.1109/ISGTEurope.2012.6465843.

41 Morais, J., Pires, Y., Cardoso, C., and Klautau, A. (2009). An overview of data mining techniques applied to power systems. *Intechopen.com*.

42 Ma, R., Basumallik, S., and Eftekharnejad, S. (2020). A PMU-based data-driven approach for classifying power system events considering cyberattacks. *IEEE Systems Journal* 14 (3): 3558–3569.

43 Liu, F.T., Ting, K.M., and Zhou, Z.-H. (2008). Isolation forest. *2008 Eighth IEEE International Conference on Data Mining*, 413–422. IEEE.

44 Grira, N. and Crucianu, M. (2004). Unsupervised and semi-supervised clustering: a brief survey. https://api.semanticscholar.org/CorpusID:7238091

45 Kriegel, H.-P., Kröger, P., Schubert, E., and Zimek, A. (2009). LoOP: Local outlier probabilities. *Proceedings of the 18th ACM Conference on Information and Knowledge Management*, 1649–1652. ACM.

46 Chowdhury, A.S., Khaledian, E., and Broschat, S.L. (2019). Capreomycin resistance prediction in two species of mycobacterium using a stacked ensemble method. *Journal of Applied Microbiology* 127 (6): 1656–1664.

47 Aminanto, M.E., Zhu, L., Ban, T. et al. (2019). Automated threat-alert screening for battling alert fatigue with temporal isolation forest. *2019 17th International Conference on Privacy, Security and Trust (PST)*, 1–3. IEEE.

48 Zhang, T., Wang, E., and Zhang, D. (2019). Predicting failures in hard drivers based on isolation forest algorithm using sliding window. *Journal of Physics: Conference Series* 1187 (4): 042084.

49 Zou, Z., Xie, Y., Huang, K. et al. (2019). A docker container anomaly monitoring system based on optimized isolation forest. *IEEE Transactions on Cloud Computing* 10 (1): 134–145.

50 Breunig, M.M., Kriegel, H.-P., Ng, R.T., and Sander, J. (2000). LOF: Identifying density-based local outliers. *ACM SIGMOD Record* 29 (2): 93–104.

51 Akoglu, H. (2018). User's guide to correlation coefficients. *Turkish Journal of Emergency Medicine* 18 (3): 91–93.

52 Artusi, R., Verderio, P., and Marubini, E. (2002). Bravais-Pearson and Spearman correlation coefficients: meaning, test of hypothesis and confidence interval. *The International Journal of Biological Markers* 17 (2): 148–151.

53 Gao, Z., Kong, D., and Gao, C. (2012). Modeling and control of complex dynamic systems: applied mathematical aspects. *Journal of Applied Mathematics* 2012: 869792.

54 Powers, D.M. (2011). Evaluation: from precision, recall and F-measure to ROC, informedness, markedness and correlation. *arXiv preprint arXiv:2010.16061.*

55 Weisberg, S. (2005). *Applied Linear Regression*, vol. 528. Wiley.

56 Ester, M., Kriegel, H.-P., Sander, J., and Xu, X. (1996). A density-based algorithm for discovering clusters in large spatial databases with noise. *Kdd* 96 (34): 226–231.

57 Rizvi, S.M.H., Sadanandan, S.K., and Srivastava, A.K. (2021). Real-time ZIP load parameter tracking using sensitivity-based adaptive window and variable elimination with realistic synchrophasor data. *IEEE Transactions on Industry Applications* 57 (6): 6525–6536.

58 Hua, Y. and Sarkar, T.K. (1990). Matrix pencil method for estimating parameters of exponentially damped/undamped sinusoids in noise. *IEEE Transactions on Acoustics, Speech, and Signal Processing* 38 (5): 814–824. https://doi.org/10.1109/29.56027.

18

Application of State Observers and Filters in Protection and Cyber-Security of Power Grids

Mohammadmahdi Asghari[1], *Amir Ameli*[1], *Mohsen Ghafouri*[2], *and Mohammad N. Uddin*[1]

[1] *Department of Electrical and Computer Engineering, Lakehead University, Thunder Bay, Ontario, Canada*
[2] *Concordia Institute for Information Systems Engineering (CIISE), Montreal, Quebec, Canada*

18.1 Introduction

The main aim of power networks, which are among the most sophisticated interconnected energy systems, is to deliver electricity to consumers in an efficient, reliable, secure, and cost-effective manner. The provided energy from this system is necessary for the functionality, safety, and well-being of society and its critical infrastructures. The operation of this energy delivery system is dependent on the accuracy and authenticity of the data used for control and protection applications. Any inaccuracy in the measured data can significantly affect the reliability and security of power grids. As a result, fault and cyber-attack diagnosis in the sensory networks and measurement signals of power grids, as well as the accuracy of measured values, are of paramount importance. Thus, power grids should be equipped with the required tools and algorithms to detect and mitigate such problems in a timely manner to preserve the reliability of the grid. Generally, the accuracy and authenticity of measurements can be affected by (i) faults in the operation of a component, e.g., internal faults on voltage transformers (VTs); (ii) the presence of noise and the limitations of power grid equipment, e.g., the limited bandwidth of current transformers (CTs); and (iii) intentional manipulation of data with the aim of performance degradation, such as through cyber threats. Generally, the accuracy and authenticity of measurements can be affected by (i) malfunction of measuring devices, e.g., limited bandwidth of CTs; (ii) the presence of noise; and (iii) intentional manipulation of data with the aim of performance degradation, such as through cyber threats.

A fault is a system condition in which a component fails to operate correctly. In a power system, a fault is often associated with an abnormal electric current, e.g., when a short circuit connection occurs between two points of the grid, often resulting in a much greater current level compared to the normal operating range. To protect power grids against faults, protection relays are employed to detect and isolate faults before grid components are damaged. Detecting faults, however, is sometimes challenging. For instance, if an external fault saturates one or both of the CTs of the transformer differential protection, there is a high likelihood of relay malfunction if an internal fault happens simultaneously with the external fault, a phenomenon called cross-country faults [1].

On the other hand, the performance of relays depends on the accuracy of the signals that are fed to them. Although instrumentation equipment used nowadays in power systems is sufficiently

Smart Cyber-Physical Power Systems: Fundamental Concepts, Challenges, and Solutions, Volume 1, First Edition.
Edited by Ali Parizad, Hamid Reza Baghaee, and Saifur Rahman.

precise, some sources of error are unavoidable. Such errors can be, for instance, due to the presence of noise, saturation of CTs, the transient response of capacitor VTs (CVTs), the limited bandwidth of CTs and CVTs, or the geomagnetically induced currents. For instance, it has been proven that the transient response of capacitive VTs distorts the measured voltage by them, thus might result in over-reaching of distance relays [2].

Another incident that might affect the performance of control and protection schemes in power networks is a cyber-attack. Recently, there has been a trend in power networks toward harnessing information technology (IT) through extensive deployment of advanced computers, communication technologies, and intelligent electronic devices (IEDs). As a result, the traditional power system has evolved into a smart cyber-physical system with increased efficiency and reliability. Integration of IT systems, however, has made energy infrastructure prone to a variety of cyber-security problems, which have become particularly important in recent years due to the proliferation of successful cyber-attacks with devastating impacts. Through cyber-attacks, adversaries can disrupt the normal operation of the power grids by misleading the system operator about the current condition of the system. For instance, by manipulating the measurement of sensory networks, attackers can degrade the performance of the control system, e.g., the automatic generation control (AGC) system, or even push the grid toward an unstable condition [3].

Based on the above discussions, it is crucial to develop frameworks that constantly monitor the measured signals to (i) detect and identify faults and cyber-attacks and (ii) enhance the accuracy of measurements in order to improve the reliability of control and protection schemes in power grids. Such frameworks can be developed using observers and filters, which accurately model the system and estimate its states to achieve the above mentioned goals. On this basis, the rest of this chapter first elaborates on the state–space modeling of systems and the most important types of observers and filters. It then discusses how observers and filters can be designed to achieve the specified goals. Then, three case studies are presented to demonstrate the effectiveness of observers and filters in enhancing the authenticity and accuracy of the measured data in power grids. Finally, the concluding remarks are presented.

18.2 State–Space Model of Systems

A state–space model refers to a mathematical depiction of a physical system, characterized by a group of inputs, outputs, and state variables that are interconnected by first-order differential or difference equations. This can be expressed in a universal form as [4]

$$\begin{cases} \dot{x}(t) = f(t, x(t), u_n(t)) \\ y(t) = h(t, x(t), u_n(t)) \end{cases} \tag{18.1}$$

where $x(t) \in R^n$ indicates the state vector of the system; $u_n \in R^m$ is the vector of known and available inputs; and $y(t) \in R^p$ demonstrates the output vector. In this representation, n, m, and p, respectively, represent the number of states, inputs, and outputs.

In linear or linearized nonlinear systems, (18.1) can be expressed in a matrix form as follows [5]:

$$\begin{cases} \dot{x}(t) = Ax(t) + B_n u_n(t) \\ y(t) = Cx(t) + D_n u_n(t) \end{cases} \tag{18.2}$$

where $A \in R^{n\times n}$ represents the state matrix; $B_n \in R^{n\times m}$ is the input matrix for known inputs; $C \in R^{p\times m}$ is the output matrix; and $D_n \in R^{p\times m}$ shows the feed through matrix for known inputs. The block diagram of this state–space model is shown in Figure 18.1a.

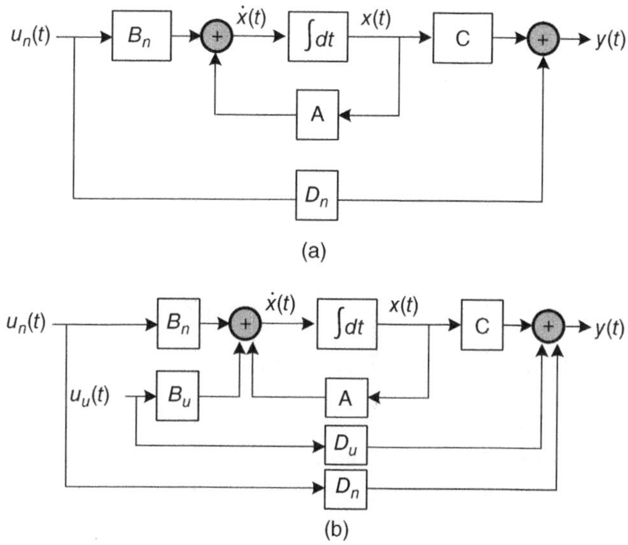

Figure 18.1 Block diagrams of the state-space model for systems (a) without and (b) with unknown inputs.

Since some systems may include inputs whose values are not available, e.g., the power grid load for the AGC system, the state-space model of such systems can be represented as shown in (18.2):

$$\begin{cases} \dot{x}(t) = Ax(t) + B_n u_n(t) + B_u u_u(t) \\ y(t) = Cx(t) + D_n u_n(t) + D_u u_u(t) \end{cases} \tag{18.3}$$

in which $B_u \in R^{n\times N}$, $D_u \in R^{p\times N}$, and $u_u \in R^N$ represent, respectively, the unknown input matrix, the feed through matrix for unknown inputs, and the vector of unknown inputs. To differentiate between the known and unknown inputs, as well as their associated matrices, subscripts n and u are used for known and unknown parameters, respectively. The block diagram of the state-space model of Eq. (18.3) is shown in Figure 18.1b.

In order to prepare (18.2) and (18.3) for numerical analysis and implementation, matrices A, B_u, and B_n should be discretized using the following equations:

$$\mathbb{A} = e^{A\times T_s} \tag{18.4a}$$

$$\mathbb{B}_u = \int_{\tau=0}^{T_s} e^{A\times\tau} B_u d\tau \tag{18.4b}$$

$$\mathbb{B}_n = \int_{\tau=0}^{T_s} e^{A\times\tau} B_n d\tau \tag{18.4c}$$

in which T_s is the discretization time step. Thus, for instance, the discrete state-space model of (18.3) can be represented as follows:

$$\begin{cases} \mathbb{X}[k+1] = \mathbb{A}\mathbb{X}[k] + \mathbb{B}_n \mathbb{U}_n[k] + \mathbb{B}_u \mathbb{U}_u[k] \\ \mathbb{Y}[k] = \mathbb{C}\mathbb{X}[k] + \mathbb{D}_n \mathbb{U}_n[k] + \mathbb{D}_u \mathbb{U}_u[k] \end{cases} \tag{18.5}$$

In this equation, $\mathbb{X}[k] \in \mathbb{R}^n$, $\mathbb{U}_u[k] \in \mathbb{R}^N$, $\mathbb{U}_n[k] \in \mathbb{R}^m$, and $\mathbb{Y}[k] \in \mathbb{R}^p$ are, respectively, the vectors of states, unknown inputs, known input, and outputs at time step k. Additionally, Figure 18.2 represents the block diagram of this model in the discrete domain, in which Z denotes the Z transform. The parameters of continuous and discrete state-space models are summarized in Table 18.1.

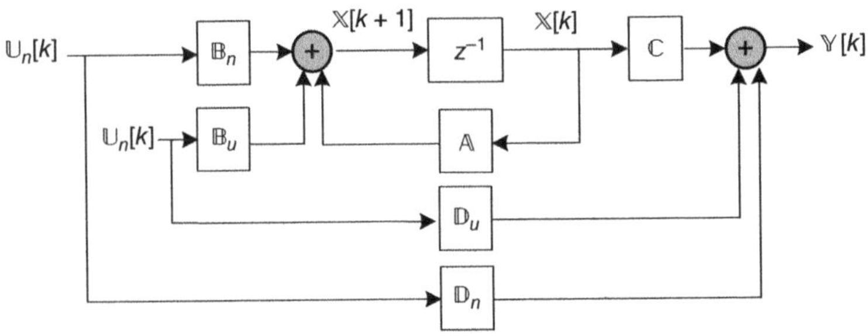

Figure 18.2 The block diagram of the state–space model of systems in the discrete domain.

Table 18.1 The parameters of continuous and discrete state–space models in this chapter.

	Continuous-time	Discrete-time
State vector	$x(t) \in R^n$	$\mathbb{X}[k] \in \mathbb{R}^n$
Known input vector	$u_n \in R^m$	$\mathbb{U}_n[k] \in \mathbb{R}^m$
Unknown input vector	$u_m \in R^N$	$\mathbb{U}_u[k] \in \mathbb{R}^N$
Output vector	$y(t) \in R^p$	$\mathbb{Y}[k] \in \mathbb{R}^p$
State matrix	$A \in R^{n\times n}$	$\mathbb{A} \in \mathbb{R}^{n\times n}$
Input matrix for known input	$B_n \in R^{n\times m}$	$\mathbb{B}_n \in \mathbb{R}^{n\times m}$
Input matrix for unknown input	$B_u \in R^{n\times N}$	$\mathbb{B}_u \in \mathbb{R}^{n\times N}$
Output matrix	$C \in R^{p\times m}$	$\mathbb{C} \in \mathbb{R}^{p\times n}$
Feed through matrix for known input	$D_n \in R^{p\times m}$	$\mathbb{D}_n \in \mathbb{R}^{p\times m}$
Feed through matrix for unknown input	$D_u \in R^{p\times N}$	$\mathbb{D}_u \in \mathbb{R}^{p\times N}$

Since in most systems, especially those that are dealt with in this chapter, the inputs do not directly impact the outputs, matrices $\mathbb{D}_n$ and $\mathbb{D}_u$ are considered to be zero in the rest of this chapter.

18.3 Properties of State–Space Models

This section studies the main properties of linear state–space models for the purpose of designing state observers.

18.3.1 Stability

Consider a linear system as follows:

$$\mathbb{X}[k+1] = \mathbb{A}\mathbb{X}[k] \tag{18.6}$$

The system defined in (18.6) is stable if for any initial values $\mathbb{X}[0]$,

$$\lim_{k\to\infty} \mathbb{X}[k] = 0 \tag{18.7}$$

It is proven in [4] that a linear system is stable if and only if all the eigenvalues of $\mathbb{A}$ are located in the unit circle in the complex domain.

18.3.2 Observability

A system is observable if, for any possible sequence of states and inputs, the current state can be determined from observation of outputs over a finite time interval. The observability matrix of a system in the form of (18.5) for an arbitrary integer L is defined as follows:

$$O_L = \begin{bmatrix} \mathbb{C} \\ \mathbb{C}\mathbb{A} \\ \vdots \\ \mathbb{C}\mathbb{A}^L \end{bmatrix} \tag{18.8}$$

As proven in [6], the system is observable if and only if rank $\left(O_{n-rank(\mathbb{C})}\right) = n$.

18.3.3 Invertibility

The observability matrix shown in (18.8) was constructed under the presumption that all system inputs were fully known. Nevertheless, as explained in Section 18.2, the inputs of some systems might be unknown. To rebuild some or all of the unknown inputs, it is needed to "invert" the system. Assume the initial state $\mathbb{X}[0]$ is known for the system shown in (18.5). The system is said to have an α-delay inverse for a nonnegative integer α, if an input $\mathbb{U}_n[k]$ can be recovered only from the system's outputs up to a time step $\mathbb{Y}[k+\alpha]$. An invertible system is defined as one that possesses an α-delay inverse for a certain finite α-value. The system's inherent delay is the smallest integer α for which an α-delay inverse exists.

The invertibility matrix for a positive integer α can be defined as follows:

$$j_\alpha = \begin{bmatrix} 0 & 0 & \cdots & 0 \\ \mathbb{C}\mathbb{B}_u & 0 & \cdots & 0 \\ \vdots & \vdots & \ddots & \vdots \\ \mathbb{C}\mathbb{A}^{\alpha-1}\mathbb{B}_u & \mathbb{C}\mathbb{A}^{\alpha-2}\mathbb{B}_u & \cdots & 0 \end{bmatrix} \tag{18.9}$$

It is proven in [7] that the system is α-delay invertible for an α that is smaller than or equal to n, if and only if

$$\text{rank}\left(j_\alpha\right) - \text{rank}\left(j_{\alpha-1}\right) = m \tag{18.10}$$

18.4 State Observers and Filters

There are numerous applications where having knowledge of a system's states over a time interval is desired, since these states may include important information about the system. It is often possible to measure some of the states directly by employing appropriate sensors. Using sensors for measuring all states, however, may be either inefficient or impractical. Therefore, in some applications, it becomes necessary to estimate the values of a system's states from the measurements that are already available. State estimation is a technique to estimate the immeasurable states of a system using the inputs and measured outputs of that system. Filters and observers are two categories of state estimation techniques, where observers are often used for estimating the states of deterministic systems, while filters are employed for stochastic ones. The state of a system can be estimated if and only if the system is observable [4].

Numerous state observers and filters have been developed so far, which can be classified differently, e.g., based on their features and capabilities. For instance, Figure 18.3 classifies observers and

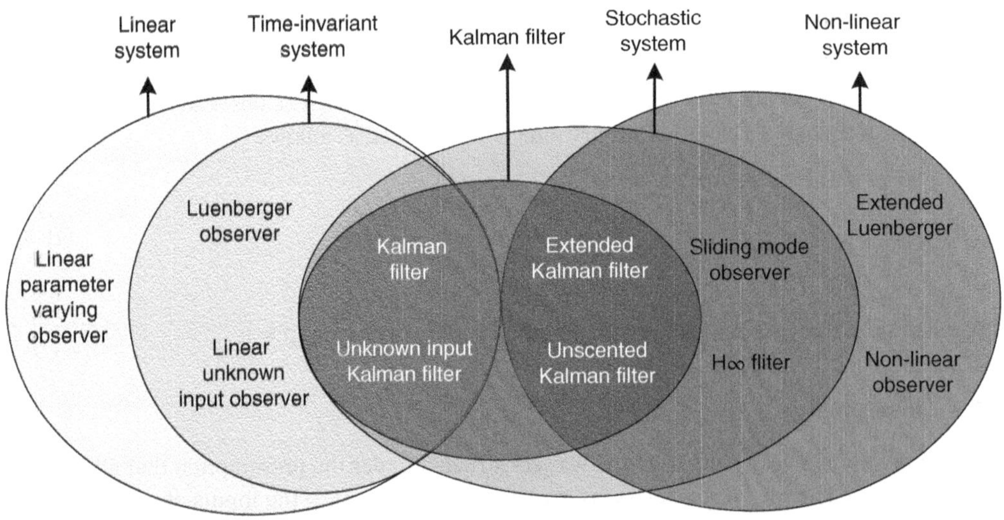

Figure 18.3 Classification of state estimation.

filters based on the type of the system. The observers and filters shown in this figure are explained in further detail in Sections 18.4.1–18.4.7.

18.4.1 Luenberger Observers

The Luenberger observer shown in Figure 18.4 is one of the earliest observers developed for estimating the states of linear systems without unknown inputs. In this observer, the difference between the output of a system and the observer is fed back linearly into the observer to eliminate the estimation error. The observer estimates the states of the system using the following equation:

$$\hat{\mathbb{X}}[k+1] = \hat{\mathbb{X}}[k] + \mathbb{B}_n\mathbb{U}_n[k] + \mathbb{L}(\mathbb{Y}[k] - C\hat{\mathbb{X}}[k]) \tag{18.11}$$

where $\hat{\mathbb{X}}[k]$ is the estimated states. Since $\mathbb{Y}[k] = C\mathbb{X}[k]$, the term $\mathbb{Y}[k] - C\hat{\mathbb{X}}[k]$ presents the error between the measured and estimated states, which approaches zero if the observer tracks the system states perfectly. The weighting matrix $\mathbb{L}$, which is called the observer gain, is designed such

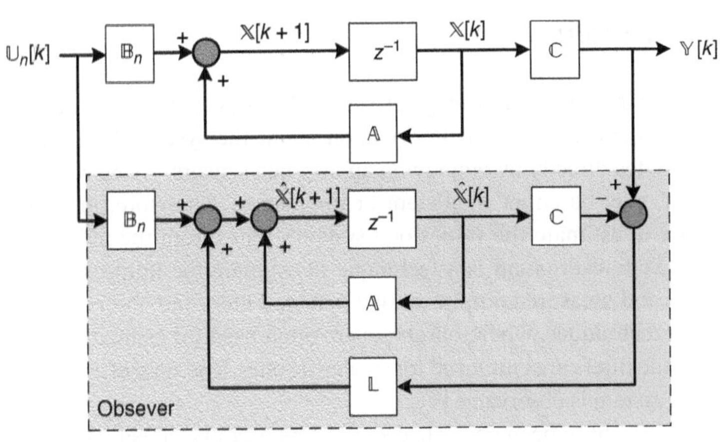

Figure 18.4 The block diagram of the Luenberger observer.

that this term approaches zero. In fact, $\mathbb{L}$ corrects the model output and improves the performance of the observer in tracking the system states.

To design $\mathbb{L}$, the estimation error should be obtained, which is defined as follows:

$$\mathbb{E}[k+1] = \mathbb{X}[k+1] - \hat{\mathbb{X}}[k+1] \tag{18.12}$$

By substituting $\mathbb{X}[k+1]$ and $\hat{\mathbb{X}}[k+1]$ from (18.5) and (18.11), (18.12) is simplified to

$$\mathbb{E}[k+1] = (\mathbb{A} - \mathbb{L}\mathbb{C})\mathbb{E}[k] \tag{18.13}$$

Therefore, if $\mathbb{A}$–$\mathbb{L}\mathbb{C}$ is stable, i.e., all its eigenvalues are located in the unit circle, the estimation error approaches zero regardless of the initial estimation error. Therefore, $\mathbb{L}$ must be designed to stabilize the poles of $\mathbb{A} - \mathbb{L}\mathbb{C}$. This objective is attainable via the implementation of a pole-placement technique, such as the one described in [8]. This technique involves the selection of n stable eigenvalues and the design of $\mathbb{L}$ so that the eigenvalues of $\mathbb{A} - \mathbb{L}\mathbb{C}$ coincide with them. In order to achieve this aim, the method allocates n linearly independent eigenvectors to the chosen eigenvalues. This allocation is done in a manner that the eigenvector matrix is as well-conditioned as possible, as detailed in [9]. Subsequently, $\mathbb{L}$ is determined using these chosen eigenvalues and eigenvectors. Once $\mathbb{L}$ is fully designed, the observer can estimate the states of the system by starting from an initial condition.

The Luenberger observer has been widely used in power system applications for monitoring of power grid components and subsystems, as well as estimating their states. For instance, a Luenberger observer is used in [10] to obtain the states of a wind farm and its associated power grid to be used in subsynchronous control schemes. This type of observer along with a transient frequency drift estimator is used in [11] to obtain the frequency of a weak alternating current (AC) microgrid and improve the system stability. In [12], a Luenberger observer is utilized to control the grid-side current of inductor–capacitor–inductor (LCL) filters using a set of power grid and converter measurements. The observer output is then leveraged to estimate the grid impedance and control the pulsed signals set to the converter. In [13], attacks against the load frequency control (LFC) system of a power grid have been identified using a framework that consists of a Luenberger observer and an artificial neural network (ANN).

18.4.2 Linear Unknown Input Observers

As mentioned before, some systems may include unknown inputs. The problem of estimating the states of a system with unknown inputs has been the subject of a large number of studies in the past three decades, since most faults or cyber intrusions can be generally modeled as unknown inputs to the system [14]. Therefore, unknown input observers (UIOs) and filters are developed for estimating the states of a system by using the system outputs and known inputs. Linear UIOs [3, 15], unknown input Kalman filter (KF) (UIKF) [16, 17], nonlinear UIO [18, 19], and unknown input sliding mode observer (SMO) [20] have been developed so far for estimating states of systems with unknown inputs. This section elaborates on linear UIOs, and UIKFs are discussed in Section 18.4.4. More information about other types of observers and filters with unknown inputs can be found in the relevant references mentioned above.

Consider the state–space model of (18.5), where matrices $\mathbb{D}_n$ and $\mathbb{D}_u$ are zero. The research in [21] establishes that state estimation can be achieved in the presence of unknown inputs through a linear UIO, given that a particular delay α is incorporated into the linear UIO. The necessary quantity of α is contingent on the parameters of the system, as will be further elucidated in this subsection. For the estimation of the system states at time step k, a window of $\alpha + 1$ sampling instants, spanning from time step k to time step $k + \alpha$, is taken into consideration. This window, which is $(\alpha + 1)$ samples in length, progresses over time, and the most recent moment of sampling is denoted as $(k + \alpha)$. Therefore, to estimate the states of the system at time step k, i.e., $\mathbb{X}[k]$, the outputs of the system during this window and the initial states are required. To develop a linear UIO for Eq. (18.5), the system's outputs derived from Eq. (18.5) during the period of the aforementioned window (i.e., from k to $k + \alpha$) need to be arranged in a matrix form as follows:

$$\mathbb{Y}\left[k : k+\alpha\right] = \Theta_\alpha \mathbb{X}\left[k\right] + M_{u,\alpha}\mathbb{U}_u\left[k : k+\alpha\right] + M_{n,\alpha}\mathbb{U}_n\left[k : k+\alpha\right] \tag{18.14}$$

where

$$\mathbb{Y}\left[k : k+\alpha\right] = \left[\mathbb{Y}[k]^T\ \mathbb{Y}[k+1]^T\ \cdots\ \mathbb{Y}[k+\alpha]^T\right]^T \tag{18.15a}$$

$$\Theta_\alpha = \left[\mathbb{C}^T\ (\mathbb{C}\mathbb{A})^T\ \cdots\ (\mathbb{C}\mathbb{A}^\alpha)^T\right]^T \tag{18.15b}$$

and the unknown and known input vectors over the length of the window are equal to

$$\mathbb{U}_u\left[k : k+\alpha\right] = \left[\mathbb{U}_u[k]^T\ \mathbb{U}_u[k+1]^T\ \cdots\ \mathbb{U}_u[k+\alpha]^T\right]^T \tag{18.15c}$$

$$\mathbb{U}_n\left[k : k+\alpha\right] = \left[\mathbb{U}_n[k]^T\ \mathbb{U}_n[k+1]^T\ \cdots\ \mathbb{U}_n[k+\alpha]^T\right]^T \tag{18.15d}$$

and matrices $M_{u,\alpha}$ and $M_{n,\alpha}$ can be expressed by

$$M_{u,\alpha} = \begin{bmatrix} 0 & 0 & \cdots 0 \\ \mathbb{C}\mathbb{B}_u & 0 & \cdots 0 \\ \vdots & \vdots & \ddots \vdots \\ \mathbb{C}\mathbb{A}^{\alpha-1}\mathbb{B}_u & \mathbb{C}\mathbb{A}^{\alpha-2}\mathbb{B}_u & \cdots 0 \end{bmatrix} \tag{18.15e}$$

$$M_{n,\alpha} = \begin{bmatrix} 0 & 0 & \cdots 0 \\ \mathbb{C}\mathbb{B}_n & 0 & \cdots 0 \\ \vdots & \vdots & \ddots \vdots \\ \mathbb{C}\mathbb{A}^{\alpha-1}\mathbb{B}_n & \mathbb{C}\mathbb{A}^{\alpha-2}\mathbb{B}_n & \cdots 0 \end{bmatrix} \tag{18.15f}$$

Using the output vector obtained in (18.14), a linear UIO can estimate the states of the system based on the following equation:

$$\hat{\mathbb{X}}\left[k+1\right] = \mathbb{A}\hat{\mathbb{X}}\left[k\right] + \mathbb{B}_n\mathbb{U}_n\left[k\right] + \mathbb{L}\left(\mathbb{Y}[k : k+\alpha] - \Theta_\alpha\hat{\mathbb{X}}\left[k\right] - M_{n,\alpha}\mathbb{U}_n\left[k : k+\alpha\right]\right) \tag{18.16}$$

where $\hat{\mathbb{X}}\left[k\right]$ is the estimation of $\mathbb{X}\left[k\right]$ and $\mathbb{L}$ is the UIO's gain. The gain matrix should be configured to ensure the accuracy and stability of the UIO. Once the matrix $\mathbb{L}$ has been defined, the development of the UIO is complete. The accuracy and stability of UIO are discussed in the following subsections.

18.4.2.1 Accuracy of Linear UIOs

The linear UIO described by Eq. (18.16) is considered accurate when the error—i.e., the difference between the estimated and actual states—tends toward zero over time. To meet this requirement without the necessity of the unknown input vector $\mathbb{U}_u\left[k\right]$, the error of the UIO, i.e., $\hat{\mathbb{X}}\left[k+1\right] - \mathbb{X}\left[k+1\right]$, is calculated using (18.5) and (18.16):

$$\mathbb{E}\left[k+1\right] = \mathbb{L}\mathbb{Y}\left[k : k+\alpha\right] + \mathbb{L}M_{n,\alpha}\mathbb{U}_n\left[k : k+\alpha\right] + \underbrace{(\mathbb{A} - \mathbb{L}\Theta_\alpha)}_{\mathbb{A}'}\hat{\mathbb{X}}\left[k\right] - \mathbb{A}\mathbb{X}\left[k\right] - \mathbb{B}_u u_u\left[k\right] \tag{18.17}$$

By substituting $\mathbb{Y}[k : k+\alpha]$ from (18.14), (18.17) can be simplified to

$$\mathbb{E}\left[k+1\right] = \mathbb{A}'\mathbb{E}\left[k\right] + \mathbb{L}M_{u,\alpha}\mathbb{U}_u\left[k : k+\alpha\right] - \mathbb{B}_u\mathbb{U}_u\left[k\right] \tag{18.18}$$

It is evident from Eq. (18.18) that the accuracy condition is met, and $\mathbb{E}\left[k+1\right]$ tends toward zero when the final two terms on the right side of Eq. (18.18) nullify each other, that is,

$$\mathbb{L}M_{u,\alpha} = \left[\,B_u\; 0 \cdots 0\,\right] \tag{18.19}$$

As proven in [22], a gain matrix $\mathbb{L}$ satisfies (18.19) if

$$\text{rank}\left(M_{u,\alpha}\right) - \text{rank}\left(M_{u,\alpha-1}\right) = N \tag{18.20}$$

where $M_{u,\alpha-1}$ can be found using $M_{u,\alpha}$ in (18.15e) as follows:

$$M_{u,\alpha-1} = \begin{bmatrix} 0 & 0 & \cdots & 0 \\ \mathbb{C}\mathbb{B}_u & 0 & \cdots & 0 \\ \vdots & \vdots & \ddots & \vdots \\ \mathbb{C}\mathbb{A}^{\alpha-2}\mathbb{B}_u & \mathbb{C}\mathbb{A}^{\alpha-3}\mathbb{B}_u & \cdots & 0 \end{bmatrix} \tag{18.21}$$

In fact, (18.20) is the prerequisite for developing a linear UIO for a system. In other words, it is impossible to estimate a system's states without requiring unknown inputs if this condition is not met.

It can be proven that the following matrix $\mathbb{L}$ satisfies (18.19):

$$\mathbb{L} = [\mathbb{L}_1\; \mathbb{L}_2] \times Q \tag{18.22}$$

where $\mathbb{L}_2 = \mathbb{B}_u$, matrix $\mathbb{L}_1$ is a free $n \times (\alpha-1)N$ matrix from the perspective of the UIO's error, and Q is an $\alpha N \times (\alpha+1)p$ matrix that satisfies

$$QM_{u,\alpha} = \begin{bmatrix} 0 & 0 \\ I_N & 0 \end{bmatrix} \tag{18.23}$$

As a result, if $\mathbb{L}$ is selected according to (18.22) and (18.23), multiplying L by $M_{u,\alpha}$ for any L_1 meets the condition of (18.19). Thus, as $k \to \infty$, the estimation error in (18.18) decreases and approaches zero, and so the accuracy condition is met.

18.4.2.2 Stability of Linear UIOs

A UIO is stable if all the eigenvalues of $\mathbb{A}'$ in (18.17) are located inside the unit circle of the complex plane. Furthermore, the locations of these eigenvalues can influence the speed at which the UIO's error $\mathbb{E}\left[k\right]$ converges to zero. For example, all elements of $\mathbb{E}\left[k\right]$ converge to zero at speeds exceeding σ^k, where σ represents the maximum absolute value of all the eigenvalues of $\mathbb{A}'$. Consequently, even if a large error occurs between $\mathbb{X}_1$ and $\hat{\mathbb{X}}_1$ when the linear UIO starts to operate, by selecting sufficiently small eigenvalues for $\mathbb{A}'$, the estimated states swiftly approach the actual states, maintaining the estimation error at zero thereafter.

As explained in Section 18.4.2.1, $\mathbb{L}_1$ is a free matrix from the UIO's error perspective. As a result, it can be utilized to fulfill the stability requirement. By replacing $\mathbb{L}$ from (18.22) in $\mathbb{A}'$ from (18.17) and dividing $Q\Theta_\alpha$ into two submatrices, S_1 and S_2, having $(\alpha-1)N$ and N rows, respectively, a reformed version of $\mathbb{A}'$ can be represented as

$$\mathbb{A}' = \left(\mathbb{A} - \mathbb{B}_u S_2\right) - \mathbb{L}_1 S_1 \tag{18.24}$$

where

$$\left[\,S_1^T\; S_2^T\,\right]^T = Q\Theta_\alpha \tag{18.25}$$

As demonstrated in [22], an $\mathbb{L}_1$ exists that fulfills this condition and stabilizes the eigenvalues of Eq. (18.24) provided that

$$\text{rank}\begin{pmatrix} \mathbb{A} - zI_n & \mathbb{B}_u \\ \mathbb{C} & 0 \end{pmatrix} = n + N, \quad \forall z \in \mathbb{C}, |z| \geq 1 \tag{18.26}$$

where $\mathbb{C}$ represents the set encompassing all complex numbers. The linear UIO would be unstable if condition (18.26) is not met. Thus, both (18.26) and (18.19) are the necessary conditions for developing a linear UIO. To design matrix $\mathbb{L}_1$, a pole-placement approach—such as the one suggested in [8]—can be employed to stabilize the system poles.

To sum up, the matrix $\mathbb{L}$ is designed to ensure both the accuracy and stability of the linear UIO as defined in Eq. (18.16). Submatrix $\mathbb{L}_2$ and matrix Q are designed to assure accuracy, whereas $\mathbb{L}_1$ is designed to stabilize the UIO. The linear UIO's development procedure is as follows:

1) Determine the least value of α that fulfills the condition set by (18.20).
2) Follow the steps below to design the matrix $\mathbb{L}$ for the estimator defined in Eq. (18.16):
   a) Determine the value of Q that meets the conditions of Eq. (18.23).
   b) Equate $\mathbb{L}_2$ with $\mathbb{B}_u$.
   c) Determine the value of $\mathbb{L}_1$ that ensures the eigenvalues of (18.24) remain stable.
   d) Insert the values of $\mathbb{L}_1$, $\mathbb{L}_2$, and Q into Eq. (18.22) to construct $\mathbb{L}$.
3) At each time step, use (18.16) to estimate the system states.

Linear UIOs have been utilized in different applications in power grids. In [2], a framework based on UIOs is proposed to mitigate measurement inaccuracies caused by capacitive VT. Moreover, in [15], linear UIOs are utilized for detecting false data injection attacks (FDIAs) against line current differential relays (LCDRs) and differentiating them from real internal faults. The authors of [3] have used linear UIOs to detect FDIAs against AGC systems. The problem of fault detection in direct current (DC) microgrids, which include several power-electronic-based loads, has been addressed using UIOs in [23]. Additionally, fault detection and isolation for sensors and actuators, which are used in the LFC scheme of interconnected power systems, are achieved using linear UIOs in [24].

18.4.3 Kalman Filters

As explained in the previous subsection, the error of a linear UIO approaches zero for deterministic systems if the eigenvalues of $\mathbb{A} - \mathbb{L}\mathbb{C}$ are located inside the unit circle. Such observers, however, are not ideal when stochastic disturbances, such as noise, are present. The KF follows true state variables and simultaneously eliminates the effects of noise to improve the estimation performance. To illustrate the concept of KF, process and measurement noises are added to the state–space model presented in (18.5), and the unknown inputs are removed, resulting in the following state–space model:

$$\begin{cases} \mathbb{X}[k+1] = \mathbb{A}\mathbb{X}[k] + \mathbb{B}_n\mathbb{U}_n[k] + w[k] \\ \mathbb{Y}[k] = \mathbb{C}\mathbb{X}[k] + v[k] \end{cases} \tag{18.27}$$

The vectors $w[k]$ and $v[k]$ represent the process and measurement noises, respectively. Both are characterized by having zero mean, being white, and maintaining uncorrelation with each other as well as with the initial states. The covariance matrices corresponding to $w[k]$ and $v[k]$ are denoted by Q and R, respectively.

To minimize the estimation error in the presence of stochastic variables, the 2-norm of the estimation error is defined as follows:

$$\min \left\| \mathbb{X}[k] - \hat{\mathbb{X}}[k \mid j] \right\|^2 \tag{18.28}$$

where k is the time step for which the states are going to be estimated and j denotes the time step at which the measurements are captured. If $k > j$, the problem is called prediction, while it is called filtering and smoothing if $k = j$ and $k < j$, respectively. In this section, the focus is filtering and one-step prediction problems. The measurement vector required by KFs is defined as follows:

$$\mathbb{Y}_j = \{\mathbb{Y}[0], \mathbb{Y}[1], \dots, \mathbb{Y}[j]\} \tag{18.29}$$

Additionally, the optimal state vector and its associated estimation error are defined as follows:

$$\hat{\mathbb{X}}[k \mid j] = E\left\{\mathbb{X}[k] \mid \mathbb{Y}_j\right\} \tag{18.30}$$

$$\tilde{\mathbb{X}}[k \mid j] = \mathbb{X}[k] - \hat{\mathbb{X}}[k \mid j] \tag{18.31}$$

Accordingly, the covariance matrices of the estimation error are as follows:

$$P^-[k+1] = E\left\{\tilde{\mathbb{X}}[k+1 \mid k]\, \tilde{\mathbb{X}}^T[k+1 \mid k]\right\} \tag{18.32}$$

$$P[k+1] = E\left\{\tilde{\mathbb{X}}[k+1 \mid k+1]\, \tilde{\mathbb{X}}^T[k+1 \mid k+1]\right\} \tag{18.33}$$

For time step $k+1$, the state vector $\mathbb{X}[k+1]$ can be predicted by using the state–space model of (18.27) and the measurements at time step k using the following equation:

$$\hat{\mathbb{X}}[k+1 \mid k] = A\hat{\mathbb{X}}[k \mid k] + \mathbb{B}_n\mathbb{U}_n[k] \tag{18.34}$$

Using (18.34), the filtering stage of a KF is formulated as follows:

$$\hat{\mathbb{X}}[k+1 \mid k+1] = \hat{\mathbb{X}}\,[k+1 \mid k] + \acute{\mathbb{K}}[k+1]\ \left(\mathbb{X}[k+1] - \hat{\mathbb{X}}[k+1 \mid k]\right) \tag{18.35}$$

where $\acute{\mathbb{K}}[k+1]$ is an $(n \times n)$ weighting matrix for time step $k+1$, which is chosen such that the covariance of the estimation error, i.e., $P[k+1]$, is minimized. Assuming $\acute{\mathbb{K}}[k+1] = \mathbb{K}[k+1]\mathbb{C}$,

$$\hat{\mathbb{X}}[k+1 \mid k+1] = \hat{\mathbb{X}}[k+1 \mid k] + \mathbb{K}[k+1]\left(\mathbb{Y}[k+1] - C\hat{\mathbb{X}}[k+1 \mid k]\right) \tag{18.36}$$

where $\hat{\mathbb{X}}[k+1 \mid k]$ is the model prediction of $\mathbb{X}[k+1 \mid k]$, which was obtained in (18.34). By substituting (18.34) in (18.36), this equation is modified to

$$\hat{\mathbb{X}}[k+1 \mid k+1] = A\hat{\mathbb{X}}[k] + \mathbb{B}_n\mathbb{U}_n(k) + \mathbb{K}[k+1]\ \left(\mathbb{Y}[k+1] - \mathbb{C}(\mathbb{A}\hat{\mathbb{X}}[k] + \mathbb{B}_n\mathbb{U}_n[k])\right) \tag{18.37}$$

which is known as the general form of the KF. In fact, in this equation, the predicted measurements based on old estimations, i.e., $\mathbb{C}(\mathbb{A}\hat{\mathbb{X}}[k] + \mathbb{B}_n\mathbb{U}_n[k])$, are compared with new measurements, i.e., $\mathbb{Y}[k+1]$, and the difference is multiplied by the correction matrix $\mathbb{K}[k+1]$ to minimize the effects of noise. Thus, the selection of the correction matrix $\mathbb{K}[k+1]$ should be made in such a way as to minimize the covariance matrix associated with the estimation error. As proven mathematically in [25], $\mathbb{K}[k+1]$ can be obtained using

$$\mathbb{K}\left[k+1\right] = P^-\left[k+1\right]\ \mathbb{C}^T\left(\mathbb{C}P^-\left[k+1\right]\ \mathbb{C}^T + R\right)^{-1} \tag{18.38}$$

where

$$P^-\left[k+1\right] = \mathbb{A}P\left[k\right]\mathbb{A}^T + Q \tag{18.39}$$

The block diagram of KFs is shown in Figure 18.5.

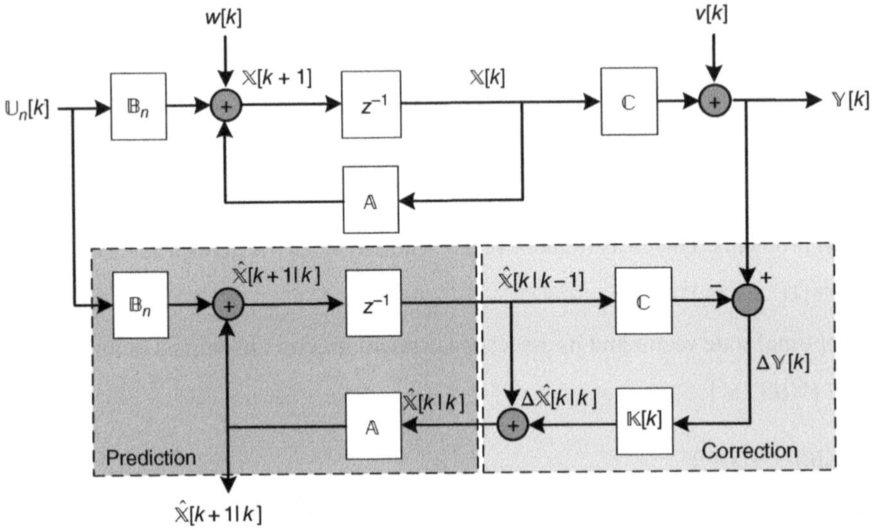

Figure 18.5 The block diagram of KFs.

Although the KF was initially developed for linear systems, it was then modified to deal with nonlinear systems as well. For example, extended KFs (EKFs) [26], unscented KFs (UKFs) [27], square-root UKFs (SR-UKFs) [28], extended particle filters (EPFs) [29], and ensemble KFs (EKFs) [30] are various types of KFs, which are developed for nonlinear systems.

Due to their numerous advantages, KFs have been used in various applications in power grids. In [31], KFs are used to estimate the states of the grids and reconstruct the measurements to mimic a fault-free grid. The comparison between the outputs of the KF and the grid measurements is then used to identify sensor faults. A KF and a recurrent neural network are used to identify FDIAs against the sensory network of a transmission grid in [32]. Aiming at identifying the islanding condition in a power grid with multiple distributed generators, a KF is used in [33] to detect the energy mismatch between the measured and estimated values of the 3rd and 5th harmonics in the system. In [34], measurements of generator terminals are utilized as inputs of a KF to accurately predict the rotor positions and quantify the angular stability of machines. Additionally, in [35], extended KFs are used to ensure the maximum power point tracking in permanent magnet synchronous generator (PMSG)-based wind farms. In [36], two-stage KFs have been used to increase the tolerance of doubly fed induction generator (DFIG)-based wind turbines against faults.

18.4.4 Unknown Input Kalman Filters

Since most practical systems are stochastic and include inputs with unknown values, developing KFs with unknown inputs is crucial for such systems. UIKFs are mainly divided into two categories: (i) unknown inputs are only present in the state equations (not in the output equations) [37, 38] and (ii) unknown inputs are present in both state equations and output equations as direct feedthrough [39, 40]. This chapter focuses only on the first category. More information about the second category can be found in [41].

By adding unknown inputs to the state–space equation of (18.27), the state–space model of linear systems in the presence of noise and unknown input is obtained as follows:

$$\begin{cases} \mathbb{X}[k+1] = \mathbb{A}\mathbb{X}[k] + \mathbb{B}_n\mathbb{U}_n[k] + \mathbb{B}_u\mathbb{U}_u[k] + w[k] \\ \mathbb{Y}[k] = \mathbb{C}x[k] + v[k] \end{cases} \tag{18.40}$$

In this subsection, the goal is to estimate the state vector $\mathbb{X}[k]$ and unknown vector $\mathbb{U}_u[k]$ using the concept of KFs. To this aim, according to [42], the following conditions must be met:

$$\text{rank}(\mathbb{C}) = p \tag{18.41a}$$

$$\text{rank}(\mathbb{B}_u) = N \tag{18.41b}$$

$$N \le p \tag{18.41c}$$

$$\text{rank}(\mathbb{C}\mathbb{B}_u) = N \tag{18.41d}$$

These conditions are generally common for observers or filters with unknown inputs. Another condition that must be met to develop a UIKF is the stability condition, which was previously presented in (18.26) [42]. By rearranging the state equations of (18.40) in the singular system form, the following equation is obtained:

$$\begin{cases} \mathscr{E}\mathscr{X}[k+1] = \mathscr{A}\mathscr{X}[k] + \mathscr{B}_n\mathbb{U}_n[k] + w[k] \\ \mathbb{Y}[k] = \mathscr{C}\mathscr{X}[k] + v[k] \end{cases} \tag{18.42}$$

where

$$\mathscr{X}[k] = \left[\mathbb{X}[k]^T,\ \mathbb{U}_u[k]^T\right]^T \tag{18.43a}$$

$$\mathscr{E} = \left[I,\ \mathbb{B}_u^T\right]^T \mathscr{A} = \left[\mathbb{A}^T,\ 0\right]^T \mathscr{C} = \left[\mathbb{C}^T,\ 0\right]^T \tag{18.43b}$$

Thus, the problem is reduced to semi-state estimation for singular systems. As shown in [42], the KF can be applied to the system of (18.42), and the new state vector, i.e., $\mathscr{X}$, can be estimated. The results are then rearranged to obtain the general equations for estimating the states and unknown inputs of the original system, which are as follows:

$$\hat{\mathbb{X}}[k+1] = \mathbb{A}\hat{\mathbb{X}}[k] + \mathbb{B}_u\hat{\mathbb{U}}_u[k] + \mathbb{L}_x[k+1]\left(\mathbb{Y}[k+1] - \mathbb{C}\left(\mathbb{A}\hat{\mathbb{X}}[k] + \mathbb{B}_u\hat{\mathbb{U}}_u[k]\right)\right) \tag{18.44}$$

$$\hat{\mathbb{U}}_u[k] = \mathbb{L}_d[k+1]\left(\mathbb{Y}[k+1] - \mathbb{C}A\hat{\mathbb{X}}[k]\right) \tag{18.45}$$

where $\hat{\mathbb{U}}_u[k]$ is the optimal estimation of unknown inputs at time step k. Moreover, $\mathbb{L}_x[k+1]$ and $\mathbb{L}_d[k+1]$ are

$$\mathbb{L}_x[k+1] = \left(\underbrace{\left(\mathbb{A}P_x[k]\mathbb{A}^T + Q\right)^{-1}}_{F[k]} + \mathbb{C}^TR^{-1}\mathbb{C}\right)^{-1}\mathbb{C}^TR^{-1} \tag{18.46}$$

$$\mathbb{L}_d[k+1] = P_{dx}[k+1]\mathbb{C}^TR^{-1} \tag{18.47}$$

in which

$$P_x[k] = \left(F[k-1] + \mathbb{C}^TR^{-1}\mathbb{C} - F[k-1]\mathbb{B}_u\left(\mathbb{B}_u{}^TF[k-1]\mathbb{B}_u\right)^{-1}\mathbb{B}_u{}^TF[k-1]\right)^{-1} \tag{18.48}$$

$$P_{dx}[k+1] = P_d[k]\mathbb{B}_u{}^TF[k]\left(F[k] + \mathbb{C}^TR^{-1}\mathbb{C}\right)^{-1} \tag{18.49}$$

In (18.49), $P_d[k]$ is obtained using

$$P_d[k] = \left(\mathbb{B}_u{}^T\mathbb{C}^T\left(R + \mathbb{C}F^{-1}[k]\mathbb{C}^T\right)^{-1}\mathbb{C}\mathbb{B}_u\right)^{-1} \tag{18.50}$$

Indeed, a UIKF does more than just minimize the influence of noise on the estimation of states; it also curtails the effects of noise on unknown inputs. In this pursuit, a UIKF predicts the states and unknown inputs and subsequently refines these predictions in light of incoming feedback. During the prediction stage, the UIKF extends the states $\mathbb{X}$ and the error covariances (namely P_x, Pd, and Pdx) from the preceding time step, aiming to predict the states and error covariances for the current step. When it comes to the correction stage, gain matrices, labeled as $\mathbb{L}_x$ and $\mathbb{L}_d$, are derived from the covariance of the previous step by capitalizing on the mean-square error to minimize the noise's impact on the states and unknown input(s) in the current time step. Finally, the predicted states and covariances are updated using the computed gains in conjunction with the measurements pertaining to the current time step.

UIKFs have been used for various applications in power grids. For instance, UIKFs are utilized in [43] to develop an attack-resilient adaptive controller for Z-source inverters and mitigate FDIAs that target them. UIKFs are also used in [16] to estimate the parameters of synchronous machines by utilizing the quantities obtained from phasor measurement units (PMUs). Using UIKFs, the authors of [44] have developed a technique to detect cyber-attacks against DC microgrids and differentiate them from uncertainties and other disturbances.

18.4.5 Observer for Linear Parameter-varying Systems

So far, all state observers explained in this chapter were for time-invariant systems. In this section, a type of observer is explained for time-variant systems. A linear parameter-varying (LPV) system can be represented by a state–space model whose dynamics change as a function of certain time-varying parameters. In fact, LPV systems nonlinearly depend on time-varying parameters [45]. Examples of LPV systems include (i) systems that smoothly switch between distinct operating conditions, and each condition can be represented by a linear time-invariant model; and (ii) nonlinear systems that are linearized, e.g., by rewriting them as piecewise functions or by parameterized Jacobian matrices [46].

The LPV structure in this chapter is expressed by

$$\begin{cases} \mathbb{X}\left[k+1\right] = \mathbb{A}(\rho[k])\mathbb{X}\left[k\right] + \mathbb{B}_n(\rho[k])\mathbb{U}_n[k] \\ \mathbb{Y}\left[k\right] = \mathbb{C}\mathbb{X}\left[k\right] \end{cases} \tag{18.51}$$

where matrices $\mathbb{A}$ and $\mathbb{B}$ are dependent on $\rho[k]$. The output matrix $\mathbb{C}$ is assumed to be constant. However, the approach in this section can be extended for a system with a parameter-varying output matrix as well.

Analyzing LPV systems is often done by using the polytopic decomposition of the parameter-dependent dynamical matrices. Such a decomposition is possible when the time-varying parameter(s) is (are) bound [46]. Assuming $\rho[k] = (\rho_1 \ \cdots \ \rho_{n_p})$ represents the vectors of n_p time-varying parameters, which are bound as follows:

$$\rho[k] \in \Theta \tag{18.52}$$

in which Θ is a hyper-rectangle and is equal to

$$\Theta = \left\{\rho[k] \in R^{n_p} \mid \rho_1 \in [\rho_1^{min}, \rho_1^{max}], \ldots, \rho_{n_p} \in [\rho_{n_p}^{min}, \rho_{n_p}^{max}]\right\} \tag{18.53}$$

where ρ_i^{min} and ρ_i^{max} $i = 1, \ldots, n_p$ are the minimum and maximum values of $\rho_i(t)$, respectively. The LPV system of (18.51) can be precisely described in a polytopic form within the compact set Θ as follows:

$$\begin{cases} \mathbb{X}\left[k+1\right] = \sum_{i=1}^{\mathcal{N}} \mu_i(\rho[k])(\mathbb{A}_i\mathbb{X}[k] + \mathbb{B}_{n,i}\mathbb{U}_{n,i}[k]) \\ \mathbb{Y}\left[k\right] = \mathbb{C}\mathbb{X}\left[k\right] \end{cases} \tag{18.54}$$

where $\mathcal{N} = 2^{n_p}$ is the number of vertices of the polytope (i.e., the number of linear submodels).

To estimate the states of (18.54), an observer must be developed such that it is stable and its estimation error approaches zero as $k \to \infty$. As proven in [47], to estimate the states of the system in (18.54), an observer in the following form can be employed:

$$\begin{cases} \mathbb{Z}\left[k+1\right] = \sum_{i=1}^{\mathcal{N}} \mu_i(\rho[k]) \left(\mathbb{N}_i\mathbb{Z}\left[k\right] + \mathbb{G}_i\mathbb{U}_n\left[k\right] + \mathbb{L}_i\mathbb{Y}\left[k\right]\right) \\ \hat{\mathbb{X}}\left[k+1\right] = \mathbb{Z}\left[k+1\right] - \mathbb{H}\mathbb{Y}\left[k+1\right] \end{cases} \tag{18.55}$$

The accuracy and stability constraints should be met when designing the matrices $\mathbb{N}$, $\mathbb{G}$, $\mathbb{L}$, and $\mathbb{H}$.

The observer outlined in Eq. (18.55) is considered accurate when $\hat{\mathbb{X}}\left[k\right]$ approaches $\mathbb{X}\left[k\right]$ as k tends toward infinity. To fulfill this requirement, the observer's error, represented by $\mathbb{X}\left[k+1\right] - \hat{\mathbb{X}}\left[k+1\right]$, is derived using Eqs. (18.51) and (18.55):

$$e\left[k+1\right] = \mathbb{X}\left[k+1\right] - \mathbb{Z}\left[k+1\right] + \mathbb{H}\mathbb{C}\mathbb{X}\left[k+1\right] = \underbrace{(I+\mathbb{H}\mathbb{C})}_{\mathbb{P}}\mathbb{X}\left[k+1\right] - \mathbb{Z}\left[k+1\right] \tag{18.56}$$

where I is an $n \times n$ identity matrix. Using (18.51), (18.55), and (18.56), the dynamics of the state estimation error is

$$e\left[k+1\right] = \sum_{i=1}^{\mathcal{N}} \mu_i(\rho[k]) \left[\mathbb{N}_i e[k] + \left(\mathbb{P}\mathbb{A}_i - \mathbb{L}_i\mathbb{C} - \mathbb{N}_i\mathbb{P}\right)\mathbb{X}[k] + \left(\mathbb{P}\mathbb{B}_{n,i} - \mathbb{G}_i\right)\mathbb{U}_n\left[k\right]\right] \tag{18.57}$$

Thus, if the following conditions are satisfied for $\forall i = \{1 \ \ldots \ \mathcal{N}\}$

$$\mathbb{P}\mathbb{A}_i - \mathbb{L}_i\mathbb{C} - \mathbb{N}_i\mathbb{P} = 0 \tag{18.58a}$$

$$\mathbb{P}\mathbb{B}_{n,i} - \mathbb{G}_i = 0 \tag{18.58b}$$

then Eq. (18.57) is reduced to

$$e\left[k+1\right] = \sum_{i=1}^{\mathcal{N}} \mu_i(\rho[k])\mathbb{N}_i e[k] \tag{18.59}$$

As a result, the error of the observer approaches asymptotically to zero if matrix $\sum_{i=1}^{\mathcal{N}} \mu_i(\rho[k])\mathbb{N}_i$ is stable. Such $\mathbb{N}_i$ matrices can be designed based on the Lyapunov stability theory. More information about the design procedure of $\mathbb{N}_i$ matrices can be found in [48]. Once $\mathbb{N}_i$ matrices are defined, they must be used to solve (18.58) and find other unknown matrices.

So far, observers for LPV systems have been used for various applications in power networks. In [49], this type of observer has been used to detect faults in DC microgrids with nonlinear loads. In [50], the LPV observers are used to identify and locate faults in wind turbines and to design fault-tolerant control schemes for them. To improve the performance of the energy management system (EMS) of renewable-energy-based microgrids, an LPV fault-tolerant control scheme has been developed in [51] to estimate the production of energy during healthy operation and identify faulty conditions. An LPV observer has been utilized in [52] to estimate the frequency of a microgrid with multiple grid-connected inverters. The estimated value is then used in a resonant control strategy to inject a high-quality power signal into the grid in the presence of distortions and nonlinear/unbalanced loads.

18.4.6 Observers for Linear Parameter-varying Systems with Unknown Inputs

By adding unknown inputs to the state–space equation of (18.51), the state–space model of LPV systems in the presence of unknown input is obtained as follows:

$$\begin{cases} \mathbb{X}\left[k+1\right] = \mathbb{A}(p[k])\mathbb{X}\left[k\right] + \mathbb{B}_n(p[k])\mathbb{U}_n[k] + \mathbb{B}_u(p[k])\mathbb{U}_u[k] \\ \mathbb{Y}\left[k\right] = \mathbb{C}\mathbb{X}\left[k\right] \end{cases} \tag{18.60}$$

As explained in Section 18.4.5, the polytopic form of (18.60) can be obtained as follows:

$$\begin{cases} \mathbb{X}\left[k+1\right] = \sum_{i=1}^{\mathcal{N}} \mu_i(p[k])(\mathbb{A}_i\mathbb{X}[k] + \mathbb{B}_{n,i}\mathbb{U}_{n,i}[k] + \mathbb{B}_{u,i}\mathbb{U}_{u,i}[k]) \\ \mathbb{Y}\left[k\right] = \mathbb{C}\mathbb{X}\left[k\right] \end{cases} \tag{18.61}$$

As shown in [47], the observer defined in (18.55) can also be employed for LPV systems with unknown input after some modifications. Thus, similar to what was performed in Section 18.4.5, the estimation error (i.e., $\mathbb{X}\left[k+1\right] - \hat{\mathbb{X}}\left[k+1\right]$) should be obtained first as follows:

$$e\left[k+1\right] = \mathbb{X}\left[k+1\right] - \mathbb{Z}\left[k+1\right] + \mathbb{HCX}\left[k+1\right] = \mathbb{PX}\left[k+1\right] - \mathbb{Z}\left[k+1\right] \tag{18.62}$$

Using (18.55), (18.60), and (18.61), the dynamics of the state estimation error is

$$e\left[k+1\right] = \sum_{i=1}^{\mathcal{N}} \mu_i(\rho[k])\left(\mathbb{N}_i e[k] + \left(\mathbb{PA}_i - \mathbb{L}_i\mathbb{C} - \mathbb{N}_i\mathbb{P}\right)\mathbb{X}[k] + \left(\mathbb{PB}_{n,i} - \mathbb{G}_i\right)\mathbb{U}_n\left[k\right] + \mathbb{PB}_{u,i}\mathbb{U}_u\left[k\right]\right) \tag{18.63}$$

Thus, the error of the observer approaches zero if the following condition is met for $\forall i = \{1 \ \ldots \ \mathcal{N}\}$ in addition to the ones previously discussed in (18.58)

$$\mathbb{PB}_{u,i} = 0 \tag{18.64}$$

As a result, by satisfying (18.58) and (18.64), the estimation error becomes equal to $\sum_{i=1}^{\mathcal{N}} \mu_i(\rho[k])\mathbb{N}_i$. This error approaches asymptotically to zero if matrix $\mu_i(\rho[k])\mathbb{N}_i$ is stable. As explained in Section 18.4.5, $\mathbb{N}_i$ matrices can be designed using the Lyapunov stability theory according to [48]. Once $\mathbb{N}_i$ matrices are defined, they must be used to solve (18.58) and (18.64) and find other unknown matrices.

18.4.7 Other Types of Observers and Filters

Sections 18.4.1–18.4.6 elaborated on the most commonly used observers and filters. This section enumerates some other types of observers and filters that might be used for protection and cyber-security application in power grids.

In some applications, it is required that observers are robust against uncertainties. For instance, to accurately detect faults and anomaly, the residual function (RF), i.e., defined as the difference between the actual and estimated outputs, should be robust to uncertainties while being sensitive to faults and anomalies. To this aim, robust control techniques are combined with state observers and filters to improve the robustness. Among such approaches, H_∞ filters and SMOs have received a great deal of attention in recent years [53]. H_∞ filters are similar to KFs; however, KFs minimize the mean-square error of estimated parameters, while H_∞ filters minimize the estimation error of the worst case. Thus, H_∞ filters are more robust against uncertainties than the standard KF. More information about H_∞ filters can be found in [54, 55]. SMOs are another type of robust observer, which is widely used for estimating the states of uncertain nonlinear systems. This kind of observer is able to generate a sliding motion on the RF and ensures a high estimation accuracy [56]. The primary attributes of SMOs encompass their resilience to uncertainties and their ability to reconstruct these uncertainties utilizing the concept of equivalent injection input. More information about SMOs with and without unknown inputs can be found in [57] and [58], respectively.

SMOs and H_∞ filters have been used in power grids in various applications. For example, in [59], an SMO and an adaptive control technique have been implemented into the control loops of a DFIG load side converter (LSC) to improve the stability of the grid. In this work, the SMO estimates the state variables of LSC and sends them to the controller for injection of an appropriate amount of reactive power by the wind turbine. In [60], a new super twisting SMO is designed to estimate the lumped disturbance for the dynamics of rotor speed. The states of a multi area power grid have been estimated using an SMO in [61] and used in an integral-type sliding mode control scheme to reduce the impact of sensor faults and disturbances on the stability of the system. In [62], an H_∞ filter is utilized to obtain the relation between the open circuit voltage and the state of charge of electric vehicles by using current and voltage measurements.

Nonlinear observers and filters are also used for estimating the state of nonlinear systems. Such observers and filters can be classified into three major groups: (i) deterministic, e.g., extended Luenberger and asymptotic observers; (ii) Bayesian, such as EKFs, UKFs, and particle filters; and (iii) hybrid, which is a combination of observers and filters in the previous groups [63, 64].

Nonlinear observers have been utilized in different applications in power grids, such as dynamic state estimation [16, 65–70]. For instance, in [65], EKFs are used for estimating power system states when subjected to disturbances. Particle filters have been utilized for the dynamic state estimation of generators in power systems [68]. Additionally, an algorithm based on EKF is proposed in [71] to detect and address the saturation of CTs. On the other hand, since accurate parameters of turbine governor are not often accessible, the authors of [72] have leveraged UKFs to estimate the governor droop and the dead-band width. In [73], an extended Luenberger–SMO is implemented to estimate rotor flux and resistance to accurately control three-phase induction motors.

18.5 Application of Observers and Filters in Improving the Authenticity and Accuracy of Measured Data

So far, different methods for estimating the states of a system were studied. This section elaborates on how state estimation can be used to enhance the authenticity and accuracy of measurements.

18.5.1 Detecting Faults and FDIAs

Generally, methods to detect faults and FDIAs can be divided into two categories: (i) data-driven and (ii) model-based. Data-driven methods are trained based on historical data of a system to distinguish between normal, faulty, or manipulated data. Although data-driven approaches have received significant attention in diverse applications, their successful deployment requires abundant data for various system conditions, which is typically challenging and time-consuming to attain. Moreover, training data are often limited and not necessarily adequate to correctly detect all possible events [25]. Model-based techniques, on the other hand, model a system accurately and detect faults and FDIAs by monitoring the inconsistencies between the modeled and actual systems. Among various model-based frameworks, the ones that use observers/filters and work based on state estimation have received significant attention. In model-based techniques, first the states of a system are estimated using one of the above mentioned observer/filters (depending on the type and specifications of the system), and some/all of its outputs are calculated using the states. The difference between the measured and estimated outputs, which is called the RF, remains close to zero in normal situations since the system model based on which the observer operates accurately represents the system. Nevertheless, the model mismatch during anomalies, e.g., faults or FDIAs, leads to a large difference between the measured and estimated outputs, and so anomalies can be detected by monitoring the RF in real time and comparing it with a predefined detection threshold.

To identify FDIAs, i.e., to determine what parameters are attacked, J identification observers/filters must be designed, where J is the number of all possible attack inputs. For each observer, all attack inputs except one of them, i.e., $J-1$ attack inputs, are modeled as unknown inputs. Therefore, UIOs or filters must be used to estimate the states of the system in the presence of FDIAs. If an attack input of an observer/filter that is already modeled becomes nonzero, the RF of that observer/filter does not increase. However, its RF grows as soon as the non modeled attack input increases. Hence, the attack can be identified by monitoring the RFs of all the J observers/filters.

To differentiate between FDIAs and faults, the system model must be obtained in the presence of all J attack inputs, which are modeled as unknown inputs. In this design, the observer/filter is insensitive to all modeled attacks, and its RF does not grow during the FDIAs. However, any event that is not considered in the model, e.g., faults, increases the RF of the observer/filter. Thus, faults can be differentiated from cyber-attacks by comparing the RF with a predefined threshold.

18.5.2 Enhancing the Accuracy of Measured Data

The integrity and accuracy of measurements are of paramount importance for power system operation, protection, and control applications. Ideally, measurement devices give an exact indication of the signal being measured. However, in practice sometimes the measurements may contain errors due to various reasons, such as malfunction of measuring devices and noise [74, 75]. For instance, malfunction of CTs due to core saturation may cause the mal operation of protective devices. On the other hand, estimating power system states may lead to strongly biased results when the measurements are contaminated with noise [76]. Therefore, enhancing the accuracy of measured data is crucial.

To improve the accuracy of measured data, observers/filters can be used to either minimize the effect of noise, find the best estimation of measured data, or estimate the states of instrumentation devices to remove their impacts on measured data. For instance, KFs can be used to minimize the impacts of noise on the measured data. Another example is about removing the impacts of CT saturation from current measurements. It has been shown in the literature that CT saturation results in a leading phase angle and a reduced magnitude for measured currents [77]. To address this issue, CTs can be modeled by using LPV sets of state–space equations, and LPV observers can be used to estimate the states of CTs. Thus, by accurately estimating the primary currents of CTs using their secondary currents, protective relays can benefit from an accurate estimation of the primary currents of CTs.

In order to provide more details about the application of state observers in power system protection and cyber-security, three case studies are presented in the following. Case study 1 shows how linear UIOs can be used to detect and identify cyber-attacks against AGC systems. Next, this chapter presents case study 2, in which the limited bandwidth of CTs for traveling-wave (TW)-based applications is addressed using UIKFs. Finally, case study 3 is on how LPV observers can address the challenges of transformer differential protection for single-phase transformers or three-phase transformer banks. More specifically, this case study demonstrates how LPV observers enable the differential scheme to (i) work properly if transformers saturate; (ii) detect internal faults; (iii) detect internal faults while energizing transformers; and (iv) detect inrush currents.

18.6 Case Study 1: Attack Detection and Identification for Automatic Generation Control Systems

The interconnected and complex nature of a power grid requires specific schemes for its monitoring, protection, and control. The efficient and accurate implementation of such schemes often needs information and communication technologies (ICTs) to be used for measuring the data and transmitting it over a cyber layer. On this basis, in recent years, power systems have started to shift to smarter grids with massive integration of cyber technologies, which consequently makes the resultant smart grid a cyber-physical system. As a result, as proved by various recent cyber incidents, smart grids are prone to cyber-attacks.

Among different control schemes in a smart grid, LFC and its secondary control layer, i.e., AGC, have received a surge of interest, since AGC is the only automatic closed-loop system between the cyber and physical layers of a smart grid [78]. Considering an economic dispatch objective, AGC uses the frequency signal and the power exchange between the areas of a power grid to regulate the frequency by adjusting the set points of generators. These measurements are often provided to AGC by distributed network protocol version 3.0 (DNP3), which is commonly used in North American utilities. Additionally, similar to other control schemes, AGC needs different measurement and communication devices, often classified as IEDs. Due to the dependence of AGC on the cyber layer, e.g., measurements and IEDs, this controller is prone to a variety of cyber threats, e.g., denial of service (DoS), malware injection, and FDIAs. Among these threats, FDIAs—in which adversaries manipulate the measured data or the commands of the control center—are among the most impactful ones, since they can severely affect the stability or economics of the grid [14].

To address this problem, this section aims to employ linear UIOs to detect cyber-attacks against the AGC system. Additionally, type identification linear UIOs are presented to find the under-attack parameters of the LFC system and consequently to identify the type of FDIAs.

18.6.1 State–Space Modeling of the LFC System

This subsection develops the state–space model of the LFC system, which will be used to detect attacks against the AGC system. This set of linear differential equations models the equivalent mechanical parts of the generator, the turbine, the governor, and the AGC. It is worth noting that since the LFC phenomenon results in low-frequency oscillations in the grid, fast electrical transients do not affect the model's accuracy. Moreover, since the time constant of the LFC system is larger than the one of the automatic voltage regulators (AVRs), they can be modeled separately. In such a case, only the steady–state operation of AVR will be modeled, and the LFC model can be considered separate from voltage dynamics. On this basis, the model presented in Figure 18.6 is accurate enough for studying LFC. In a multi area power system, the dynamics of area i, as demonstrated in Figure 18.6, can be expressed as

$$\Delta\dot{\omega}_i = \frac{1}{2H_i}\left(\Delta P_{m_i} - \Delta P_{tie_i} - \Delta P_{L_i} - D_{e_i}\Delta\omega_i\right) \tag{18.65}$$

where $\Delta\dot{\omega}_i$ is the time derivative of $\Delta\omega_i$; ΔP_{L_i}, $\Delta\omega_i$, H_i, and D_{e_i} are, respectively, load change, frequency deviation, equivalent inertia constant, and equivalent damping coefficient for the under-study area. Moreover, ΔP_{tie_i} and ΔP_{m_i} depict the sum of tie-lines' power and generators' mechanical power deviations for area i, respectively. These parameters are defined as

$$\Delta P_{tie_i} = \sum_{j\in\delta_i} \Delta P_{tie_{i,j}} \tag{18.66}$$

$$\Delta P_{m_i} = \sum_{g=1}^{G_i} \Delta P_{m_{g,i}} \tag{18.67}$$

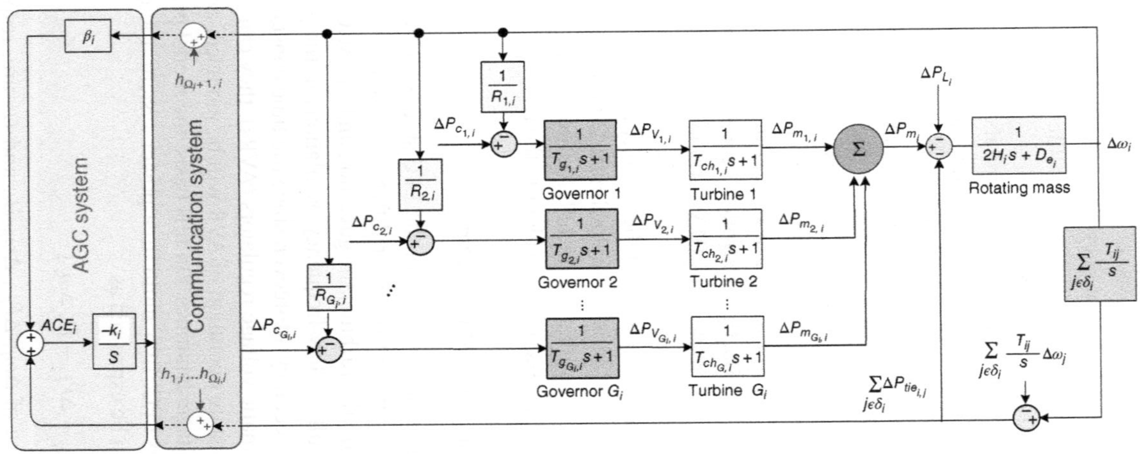

Figure 18.6 The linearized model of LFC system in area "*i*," featuring a single generator operated by AGC.

In these equations, G_i is the number of generators in area i, $\Delta P_{m_{g,i}}$ is the deviation of mechanical power of gth generator in area i, and $\Delta P_{tie_{ij}}$ represents the power deviation of the tie-line connecting areas i and j, and δ_i is the set of areas to which area i is connected. Additionally, the power transfer dynamic of a tie-line can be expressed as

$$\Delta \dot{P}_{tie_{i,j}} = T_{ij}\left(\Delta\omega_i - \Delta\omega_j\right) \tag{18.68}$$

where T_{ij} is the synchronizing power coefficient between areas i and j. It should be mentioned that tie-lines can be modeled in two ways, i.e., separately in the state–space representation or together as the sum of all tie-line powers.

To regulate the frequency and keep it within an acceptable range, the governor of each generator adjusts its turbine's valve position $\Delta P_{v_{g,i}}$ by measuring $\Delta\omega_i$ as well as the generator set points $\Delta P_{c_{g,i}}$ based on

$$\Delta \dot{P}_{v_{g,i}} = -\frac{1}{T_{g_{g,i}}}\left(\frac{1}{R_{g,i}}\Delta\omega_i + \Delta P_{v_{g,i}} - \Delta P_{c_{g,i}}\right) \tag{18.69}$$

where $T_{g_{g,i}}$ and $R_{g,i}$ are, respectively, the droop coefficient and the governor's time constant of the gth generator. Controlling $\Delta P_{v_{g,i}}$ also regulates the mechanical power by adjusting the flow of mechanical energy, i.e., the steam of the turbine. For the gth generator, the turbine dynamic can be formulated as

$$\Delta \dot{P}_{m_{g,i}} = -\frac{1}{T_{ch_{g,i}}}\Delta P_{m_{g,i}} + \frac{1}{T_{ch_{g,i}}}\Delta P_{v_{g,i}} \tag{18.70}$$

In this equation, for the gth generator, the time constant is represented by $T_{ch_{g,i}}$. If generator g is equipped with the AGC, its governor set point, i.e., $\Delta P_{c_{g,i}}$, is determined by the AGC to track load variations ΔP_{L_i}. To calculate this set point for area i, the area control error (ACE) for this area should be obtained first using the following equation:

$$ACE_i = \beta_i \Delta\omega_i + \Delta P_{tie_i} \tag{18.71}$$

where β_i and $\Delta\omega_i$ are the frequency bias and angular frequency deviation of area i, respectively. The ACE signal is calculated every 2–4 seconds. This error signal is then given to a controller—which is often just an integrator—to calculate the set points of governors using the following equation:

$$\Delta \dot{P}_{c_{g,i}} = -k_i \times ACE_i \tag{18.72}$$

where k_i is the gain of the AGC. It is worth noting that when a generator is not equipped with the AGC controller, its governor's set point is determined manually.

To develop the state–space model of the LFC system, the differential equations presented in (18.65)–(18.72) must be written in matrix form. By selecting $\Delta\omega_i$, $\Delta P_{m_{g,i}}$, and $\Delta P_{v_{g,i}}$ of all generators, ΔP_{tie_i}, and $\Delta P_{c_{g,i}}$ of AGC-controlled generators as state variables of area i, the following state–space model can be obtained for area i:

$$\dot{x}_i(t) = A_{ii}x_i(t) + B_{u,i}u_{u,i}(t) + B_{n,i}u_{n,i}(t) + \sum_{j=1:N-\{i\}} A_{ij}x_j(t) \tag{18.73}$$

where x_i and A_{ii} are, respectively, the state vector and matrix of area i; $B_{u,i}$ and $B_{n,i}$ are the input matrices associated with unknown and known inputs, respectively; $u_{u,i}$ is the vector of unknown inputs, which includes ΔP_{L_i}; and $u_{n,i}$ is the vector of known inputs, which includes $\Delta P_{c_{g,i}}$ for non-AGC-controlled generators; additionally, A_{ij} is the state matrix that connects the states of areas i to j. By selecting $\Delta P_{c_{g,i}}$ for AGC-controlled generators, $\Delta\omega_i$, and ΔP_{tie_i}, the output equation of the state–space model for area i is formulated as follows:

$$y_i(t) = C_i x_i(t) \tag{18.74}$$

To find the state–space model of the entire N-area power system for LFC studies, the state–space equations of all areas must be lumped together as explained in [3, 79], resulting in

$$\dot{x}(t) = Ax(t) + B_u u_u(t) + B_n u_n(t) \tag{18.75a}$$

$$y(t) = \mathbb{C}x(t) \tag{18.75b}$$

where

$$x = \left[x_1^T \; x_2^T \; \cdots \; x_N^T\right]^T \tag{18.76a}$$

$$A = \begin{bmatrix} A_{11} & A_{12} & \cdots & A_{1N} \\ A_{21} & A_{22} & \cdots & A_{2N} \\ \vdots & \vdots & \ddots & \vdots \\ A_{N1} & A_{N2} & \cdots & A_{NN} \end{bmatrix} \tag{18.76b}$$

$$u_u = \left[u_{u,1}^T \; u_{u,2}^T \; \cdots \; u_{u,N}^T\right]^T \tag{18.76c}$$

$$u_n = \left[u_{n,1}^T \; u_{n,2}^T \; \cdots \; u_{n,N}^T\right]^T \tag{18.76d}$$

$$B_u = \text{diag}\,\{B_{u,1} B_{u,2} \cdots B_{u,N}\}^T \tag{18.76e}$$

$$B_n = \text{diag}\,\{B_{n,1} B_{n,2} \cdots B_{n,N}\}^T \tag{18.76f}$$

$$\mathbb{C} = \text{diag}\,\{C_1 C_2 \cdots C_N\}^T \tag{18.76g}$$

For instance, in the case of the three-area power system shown in Figure 18.7 (data and specifications can be found in [3]), state equations for Area 1 are

$$\dot{x}_1(t) = A_{11}x_1(t) + B_{u,1}u_{u,1}(t) + B_{n,1}u_{n,1}(t) + \sum_{j=2,3} A_{1j}x_j(t) \tag{18.77}$$

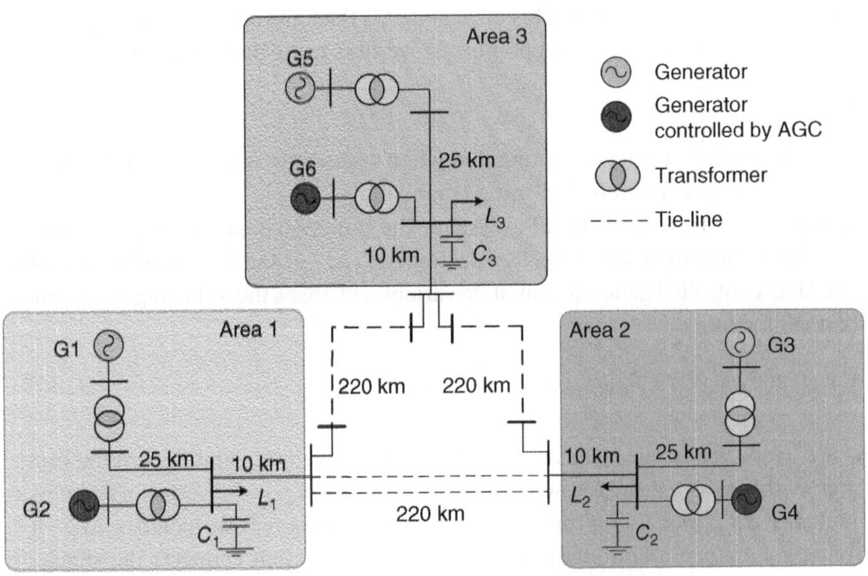

Figure 18.7 Single-line diagram of the three-area test system.

where

$$x_1 = [\Delta P_{tie_1}\ \Delta\omega_1\ \Delta P_{m_{1,1}}\ \Delta P_{m_{2,1}}\ \Delta P_{v_{1,1}}\ \Delta P_{v_{2,1}}\ \Delta P_{c_{2,1}}]^T \tag{18.78a}$$

$$B_{u,1} = \begin{bmatrix} 0 & \frac{-1}{2H_1} & 0\ 0\ 0\ 0\ 0 \end{bmatrix}^T \tag{18.78b}$$

$$B_{n,1} = \begin{bmatrix} 0\ 0\ 0\ 0 & \frac{1}{T_{g_{1,1}}} & 0\ 0 \end{bmatrix}^T \tag{18.78c}$$

$$A_{11} = \begin{bmatrix} 0 & \sum\limits_{j\in\delta_i} T_{ij} & 0 & 0 & 0 & 0 & 0 \\ \frac{-1}{2H_1} & \frac{-D_{e_1}}{2H_1} & \frac{1}{2H_1} & \frac{1}{2H_1} & 0 & 0 & 0 \\ 0 & 0 & \frac{-1}{T_{ch_{1,1}}} & 0 & \frac{1}{T_{ch_{1,1}}} & 0 & 0 \\ 0 & 0 & 0 & \frac{-1}{T_{ch_{2,1}}} & 0 & \frac{1}{T_{ch_{2,1}}} & 0 \\ 0 & \frac{-1}{R_{1,1}T_{g_{1,1}}} & 0 & 0 & \frac{-1}{T_{g_{1,1}}} & 0 & 0 \\ 0 & \frac{-1}{R_{2,1}T_{g_{2,1}}} & 0 & 0 & 0 & \frac{-1}{T_{g_{2,1}}} & \frac{1}{T_{g_{2,1}}} \\ -k_1 & -k_1\beta_1 & 0 & 0 & 0 & 0 & 0 \end{bmatrix} \tag{18.78d}$$

Additionally, $u_{u,1}(t) = \Delta P_{L_1}$ and $u_{n,1}(t) = \Delta P_{c_{1,1}}$. In Eq. (18.77), A_{1j} is a 7×7 matrix, whose elements are zero, except for $-T_{1j}$. This parameter, which is only multiplied by $\Delta\omega_j$, is located in row 1 and column 2. Moreover, the output equation for Area 1 is described by the following equation:

$$y_1(t) = C_1 x_1(t) \tag{18.79}$$

where matrix C_1 is expressed as follows:

$$C_1 = \begin{bmatrix} 1\ 0\ 0\ 0\ 0\ 0\ 0 \\ 0\ 1\ 0\ 0\ 0\ 0\ 0 \\ 0\ 0\ 0\ 0\ 0\ 0\ 1 \end{bmatrix} \tag{18.80}$$

By utilizing (18.75) and (18.4), the discretized attack-free state–space representation of an N-area grid is as follows:

$$\begin{cases} \mathbb{X}[k+1] = \mathbb{A}\mathbb{X}[k] + \mathbb{B}_n\mathbb{U}_n[k] + \mathbb{B}_u\mathbb{U}_u[k] \\ \mathbb{Y}[k] = \mathbb{C}\mathbb{X}[k] \end{cases} \tag{18.81}$$

In this equation, $\mathbb{X}[k] \in \mathbb{R}^n$, $\mathbb{U}_u[k] \in \mathbb{R}^N$, $\mathbb{U}_n[k] \in \mathbb{R}^m$, and $\mathbb{Y}[k] \in \mathbb{R}^p$ are, respectively, the vectors of states, unknown inputs, known input, and outputs at time step k.

18.6.1.1 LFC System State–Space Model in the Presence of FDIAs

In this subsection, the impact of FDIAs on the measurements of frequency and tie-line power measurements, which are the inputs of AGC system, is studied. As shown in Figure 18.6, the attack inputs for the measurements of the tie-lines that are connected to area i are denoted by $h_{1,i}$ to $h_{\Omega_i,i}$, where Ω_i is the cardinality of δ_i, representing the number of areas connected to area i. Moreover, $h_{\Omega_i+1,i}$ is used for the attack input of frequency measurements. In the presence of an FDIA that targets the AGC system, an attacker affects ACE_i by injecting nonzero values into $h_{1,i}$ to $h_{\Omega_i+1,i}$. As a result, (18.72) during FDIAs is modified to

$$\Delta \dot{P}_{C_{g,i}} = -k_i \underbrace{\left(\beta_i \Delta\omega_i + \sum_{j\in\delta_i} \Delta P_{tie_{i,j}} + \beta_i h_{\Omega_{i+1},i} + \sum_{j\in\delta_i} h_{j,i} \right)}_{ACE_i} \tag{18.82}$$

During the attack, only the parameters that are transferred by the communication system are compromised. Consequently, other state equations, presented in (18.65)–(18.70), are not affected. Therefore, using (18.82) and (18.65)–(18.70), the dynamics of area i during the attack can be expressed as

$$\dot{x}_i(t) = A_{ii}x_i(t) + B_{u.i}u_{u,i}(t) + B_{n,i}u_{n,i}(t) + B_{h,i}h_i(t) + \sum_{j\in\delta_i} A_{ij}x_j(t) \tag{18.83}$$

In this equation, $B_{h,i}$ represents the relation between the attack matrix and the system states, and $h_i(t)$ denotes the attack vector, i.e.,

$$h_i(t) = \left[h_{1,i} \; h_{2,i} \; \cdots \; h_{\Omega_i+1,i} \right]^T \tag{18.84}$$

As an example, in a three-area grid, $B_{h,1}$ is

$$B_{h,1} = \begin{bmatrix} 0 & 0 & 0 \\ \vdots & \vdots & \vdots \\ 0 & 0 & 0 \\ -k_1 & -k_1 & -k_1\beta_1 \end{bmatrix} \tag{18.85}$$

It should be noted that the output equation, i.e., (18.79), is not affected by the attack.

Addition of $B_{h,i}h_i(t)$ to (18.83), due to the FDIA, changes (18.75) to

$$\dot{x}(t) = Ax(t) + B_u u_u(t) + B_n u_n(t) + B_h h(t) \tag{18.86}$$

where

$$B_h = diag\left[B_{h,1} \; B_{h,2} \; \cdots \; B_{h,N} \right] \tag{18.87a}$$

$$h(t) = \left[h_1^T(t) \; h_2^T(t) \; \cdots \; h_N^T(t) \right]^T \tag{18.87b}$$

To obtain the discrete state–space representation in the presence of attacks, (18.86) should be transferred to the discrete-time domain. This process has already been discussed in (18.4) for A, B_u, and B_n. Matrix B_h can also be discretized by using an equation similar to (18.4b) and (18.4c), as shown below:

$$\mathbb{B}_h = \int_{\tau=0}^{T_s} e^{A\times\tau} B_h d\tau \tag{18.88}$$

As a result, the discretized state–space representation of an under-attack N-area grid is formulated as

$$\mathbb{X}[k+1] = \mathbb{A}\mathbb{X}[k] + \mathbb{B}_u\mathbb{U}_u[k] + \mathbb{B}_n\mathbb{U}_n[k] + \mathbb{B}_h\mathbb{H}[k] \tag{18.89}$$

18.6.2 Detecting and Identifying FDIAs Using Linear UIOs

In order to estimate the states of the LFC system in normal conditions, a linear UIO, which can be designed based on Eq. (18.81) and according to the procedure discussed in Section 18.4.2, is required. Such a linear UIO can be used to detect the attacks that target the AGC system. Additionally, separate type identification linear UIOs can be developed based on (18.89) to identify the type of attacks. Figure 18.8 illustrates how the above mentioned attack detection and identification schemes can be integrated into the control center of a power grid to diagnose attacks against the AGC system. These schemes are discussed in detail in the following subsections.

18.6.2.1 Attack Detection Scheme

As demonstrated in Section 18.6.1, in the presence of an FDIA, the system model in (18.81) is augmented with a new terms, resulting in (18.89). As detailed in (18.14), this new term also affects the system output vector, i.e., $Y_{k:k+\alpha}$, as follows:

$$\tilde{\mathbb{Y}}\left[k : k+\alpha\right] = \mathbb{Y}\left[k : k+\alpha\right] + \mathbb{V}_{\alpha}\mathbb{H}\left[k : k+\alpha\right] \tag{18.90}$$

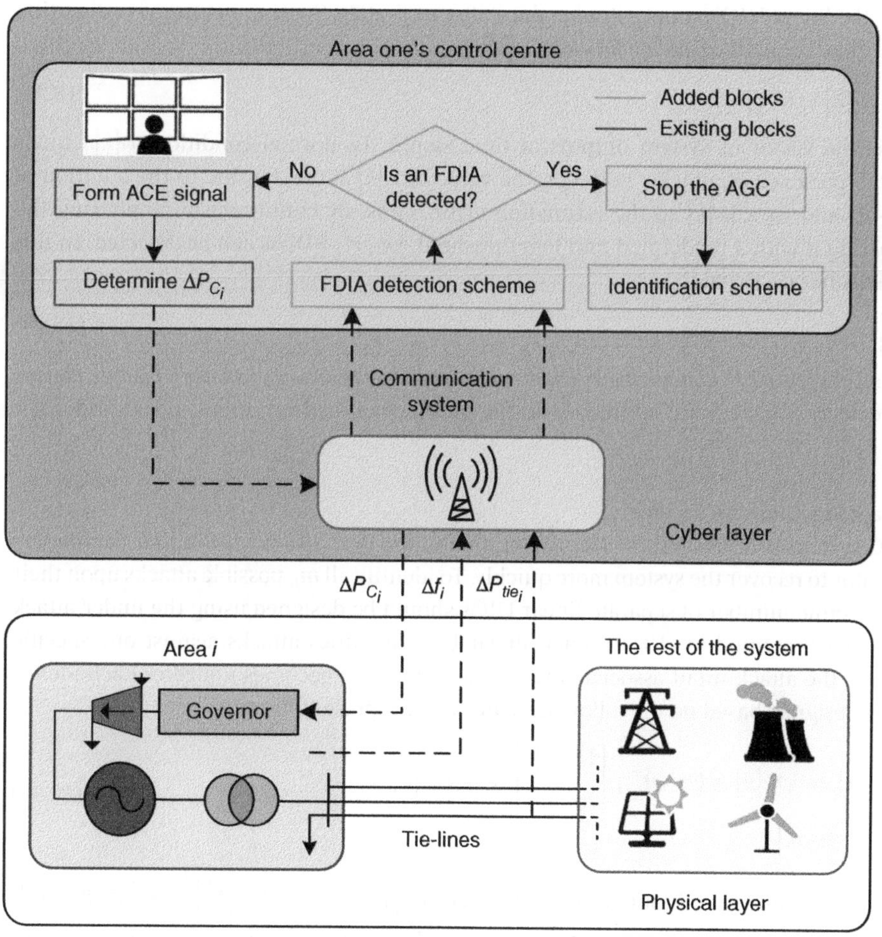

Figure 18.8 Schemes for detecting and identifying attacks in the AGC system.

where $\tilde{\mathbb{Y}}\left[k : k+\alpha\right]$ is the system output in the presence of the attack, and $\mathbb{H}\left[k : k+\alpha\right]$ and $\mathbb{V}_\alpha$ are

$$\mathbb{H}\left[k : k+\alpha\right] = \left[\mathbb{H}\left[k\right]^T \ \mathbb{H}\left[k+1\right]^T \ \cdots \ \mathbb{H}\left[k+\alpha\right]^T\right]^T \tag{18.91a}$$

$$\mathbb{V}_\alpha = \begin{bmatrix} 0 & 0 & \cdots & 0 & 0 \\ \mathbb{C}\mathbb{B}_h & 0 & \cdots & 0 & 0 \\ \vdots & \vdots & \ddots & \vdots & \vdots \\ \mathbb{C}\mathbb{A}^{\alpha-2}\mathbb{B}_h & \mathbb{C}\mathbb{A}^{\alpha-3}\mathbb{B}_h & \cdots & 0 & 0 \\ \mathbb{C}\mathbb{A}^{\alpha-1}\mathbb{B}_h & \mathbb{C}\mathbb{A}^{\alpha-2}\mathbb{B}_h & \cdots & \mathbb{C}\mathbb{B}_h & 0 \end{bmatrix} \tag{18.91b}$$

In an attack-free system, all elements of $\mathbb{H}\left[k : k+\alpha\right]$ are zero. Consequently, $\mathbb{Y}\left[k : k+\alpha\right]$ and $\tilde{\mathbb{Y}}\left[k : k+\alpha\right]$ are equal. In the presence of an FDIA, however, the values of these two will not be equal anymore. Since $\mathbb{L}$ is designed based on (18.19) such that $\mathbb{L}\mathbb{M}_{u,\alpha}\mathbb{U}_u\left[k : k+\alpha\right] - \mathbb{B}_u\mathbb{U}_u\left[k\right] = 0$, the UIO error in the presence of an FDIA increases to

$$e\left[k+1\right] = \mathbb{A}'e\left[k\right] + \mathbb{L}\mathbb{V}_\alpha\mathbb{H}\left[k : k+\alpha\right] - \mathbb{B}_h\mathbb{H}\left[k\right] \tag{18.92}$$

Consequently, the linear UIO cannot estimate the states accurately. Such an inaccurate estimation can be used to detect attacks using the following RF:

$$r\left[k\right] = \mathbb{Y}\left[k\right] - \mathbb{C}\hat{\mathbb{X}}\left[k\right] \tag{18.93}$$

where $\mathbb{Y}\left[k\right]$ is the vector of system outputs at time step k. In normal condition, $r\left[k\right]$ equals $\mathbb{C} \times e\left[k\right]$. In the presence of attacks, however, the value of $r\left[k\right]$ increases due to the addition of $\mathbb{L}\mathbb{V}_\alpha\mathbb{H}\left[k : k+\alpha\right]$ and $\mathbb{B}_h \times \mathbb{H}\left[k\right]$ to the estimation error. Thus, by continuously monitoring the RF and comparing it with a predefined unit less threshold, i.e., r^*, FDIAs can be detected. In this process, an FDIA has occurred if

$$\left\|r\left[k\right]\right\| > r^* \tag{18.94}$$

The detection threshold r^* can be determined such that false-positive and false-negative alarms of the attack detection system are minimized in the presence of load variations, noise, and other disturbances.

18.6.2.2 Attack Identification Scheme

When an LFC system is under attack, identifying the attack type and compromised parameters helps the operator to recover the system more quickly. To identify all m_h possible attacks upon their occurrence, the same number of separate linear UIOs should be designed using the under-attack state–space model presented in (18.89). Each linear UIO identifies attacks against one specific parameter when the attack input associated with that parameter becomes nonzero. Each identification UIO is designed based on the following state–space representation:

$$\begin{cases} \mathbb{X}\left[k+1\right] = \mathbb{A}\mathbb{X}\left[k\right] + \left[\mathbb{B}_u \ \ \mathbb{B}_h^{-j}\right]\begin{bmatrix}\mathbb{U}_u\left[k\right] \\ \mathbb{H}^{-j}\left[k\right]\end{bmatrix} + \mathbb{B}_n\mathbb{U}_n\left[k\right] \\ \mathbb{Y}\left[k\right] = \mathbb{C}\mathbb{X}\left[k\right] \end{cases} \tag{18.95}$$

In this equation, the unknown input vector is denoted by $\left[\mathbb{U}_u\left[k\right]^T \ \left(\mathbb{H}^{-j}\left[k\right]\right)^T\right]^T$; $\mathbb{B}_h^{-j} \in R^{n\times(m_h-1)}$ includes all the columns of $\mathbb{B}_h$ except the jth one; and $\mathbb{H}^{-j}\left[k\right] \in R^{(m_h-1)\times 1}$ is a vector that includes all attack inputs of $\mathbb{H}\left[k\right]$, except the jth one. The main idea behind such a design is that jth identification UIO is insensitive to all attacks that are included in $\mathbb{H}^{-j}[k]$, except the attack that is missing in the model, i.e., the jth one. Therefore, if an attack input of $\mathbb{H}^{-j}[k]$ becomes non-zero, the RFs of the UIO do not increase. However, the RF grows as soon as the jth attack input, which is missing in the model, increases. In this design, the jth attack has occurred when the RF of the jth UIO grows. The system's state–space representation when the jth attack is launched is as follows:

$$\begin{cases} \mathbb{X}[k+1] = \mathbb{A}\mathbb{X}[k] + \begin{bmatrix} \mathbb{B}_u & \mathbb{B}_h^{-j} \end{bmatrix} \begin{bmatrix} \mathbb{U}_u[k] \\ \mathbb{H}^{-j}[k] \end{bmatrix} + \mathbb{B}_h^j \mathbb{H}^j[k] + \mathbb{B}_n \mathbb{U}_n[k] \\ \mathbb{Y}[k] = \mathbb{C}\mathbb{X}[k] \end{cases} \tag{18.96}$$

where $\mathbb{B}_h^j$ signifies the jth column of $\mathbb{B}_h$ and $\mathbb{H}^j[k]$ is the jth element of $\mathbb{H}[k]$. Comparing (18.96) and (18.95) explains why such a design can identify the attacks during which the jth attack input is exploited.

In summary, an attack that targets the jth input of $\mathbb{H}[k]$ is identified by all the linear UIOs except the jth one. Thus, when the main linear UIO detects an attack, and at the same time the RF of the jth identification UIO also surpasses its predefined threshold, it indicates that the adversary has exploited the jth attack input. More specifically, the jth attach has occurred if

$$\left\| r[k] \right\| > r^* \quad \text{and} \quad \left\| r_j[k] \right\| > r_j^* \tag{18.97}$$

in which $r_k[k]$ and r_j^* are, respectively, the RF of the jth linear UIO and its corresponding threshold.

18.6.3 Performance Evaluation

This section evaluates the performance of the above mentioned attack detection and identification schemes for the AGC system. Simulations are performed using (matrix laboratory) MATLAB/Simulink on the three-area test systems. This test system has two generators in each area. As each tie-line is included separately in the state–space equation, the total number of states in each area is 8. There is one known and one unknown input in each area. Moreover, the number of outputs for each area is 4.

To detect and identify cyber-attacks, the thresholds r^* and $r_j^*(j = 1, \ldots, m_h)$ in (18.94) and (18.97) should be determined such that false attack detection during normal operation is avoided. Since load changes are unknown inputs for all linear UIOs, they have no effect on the thresholds. As a result, these thresholds are determined in this case study in the presence of process and measurement noises. In order to consider the effects of noise in the LFC system's state–space model, zero-mean Gaussian process noise—with the covariance matrix of $Cov_1 = 0.03 \times diag\begin{bmatrix} 1 & 1 & 0.03 & 1 & 1 & 1 & 1 & 1 \end{bmatrix}$—is added to each area's state equations in (18.81). Additionally, zero-mean Gaussian measurement noise with the covariance matrix of $Cov_2 = 0.03 \times diag\begin{bmatrix} 1 & 1 & 0.03 & 1 \end{bmatrix}$is also added to the output equations of (18.81). The main UIO's RF with and without noise is shown in Figure 18.9. If the measurements do not include any noise, the main UIO's RF is zero. On the other hand, if measurements are noisy, the maximum of the RF could increase to 0.5. Consequently, 0.5 plus a security margin, e.g., 20%, can be chosen as the FDIA detection threshold.

As mentioned earlier, attack inputs h_{11}, h_{12}, and h_{13} represent FDIAs that target the power and frequency measurements of Area 1 in the test system. For each of the remaining areas, three similar attack inputs are defined. All attack inputs are zero prior to the initiation of attacks, and the ACE signal is computed using (18.71). When the FDIAs begin, however, the attack inputs become nonzero, and as a result, the ACE signal changes to that shown in (18.82). For each area, three

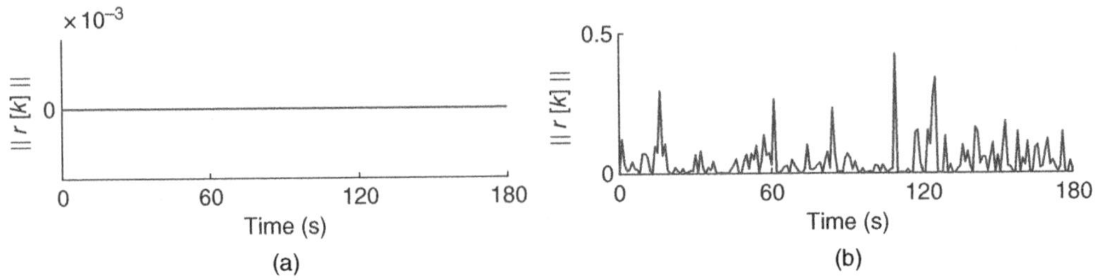

Figure 18.9 Main UIO's RF, (a) without noise and (b) with noise.

identifying UIOs are designed based on the method described in 18.6.2. For example, the identification UIOs for Area 1 are as follows:

- *UIO A* is responsible for detecting FDIAs that focus on the power measurement of the tie-line interconnecting Area 1 and Area 2. The unknown inputs within this linear UIO consist of the vector $\mathbb{U}u\left[k\right]$ and all remaining attack inputs present in $\mathbb{H}\left[k\right]$, with the exception of $h11$.
- *UIO B* is tasked with identifying FDIAs that aim at the power measurement of the tie-line that links Area 1 and Area 3. The unknown inputs of this linear UIO consist of the vector $\mathbb{U}u\left[k\right]$ and all attack inputs within $\mathbb{H}\left[k\right]$, excluding $h12$.
- *UIO C* is designed to recognize attacks that target the frequency measurements of Area 1. The unknown inputs for this linear UIO are represented by the vector $\mathbb{U}u\left[k\right]$ and all attack inputs, except for $h13$.

Using the process that was explained for the main linear UIO, the thresholds of 0.6 are chosen for type identification linear UIOs A, B, and C, as well. To evaluate the performance of the UIOs in detecting and identifying attacks, the following scenarios are studied for Area 1:

- **Scenario 1**: The AGC system is targeted by an attack, in which Δf_1, $\Delta P_{tie_{12}}$, and $\Delta P_{tie_{13}}$ are all multiplied by 1.002. A very small attack multiplier was employed in this scenario to show how slight modifications may have an influence on the RFs of identification and main linear UIOs. The attack starts at $t = 30$ s and lasts until $t = 180$ s. Figure 18.10 illustrates the results of this

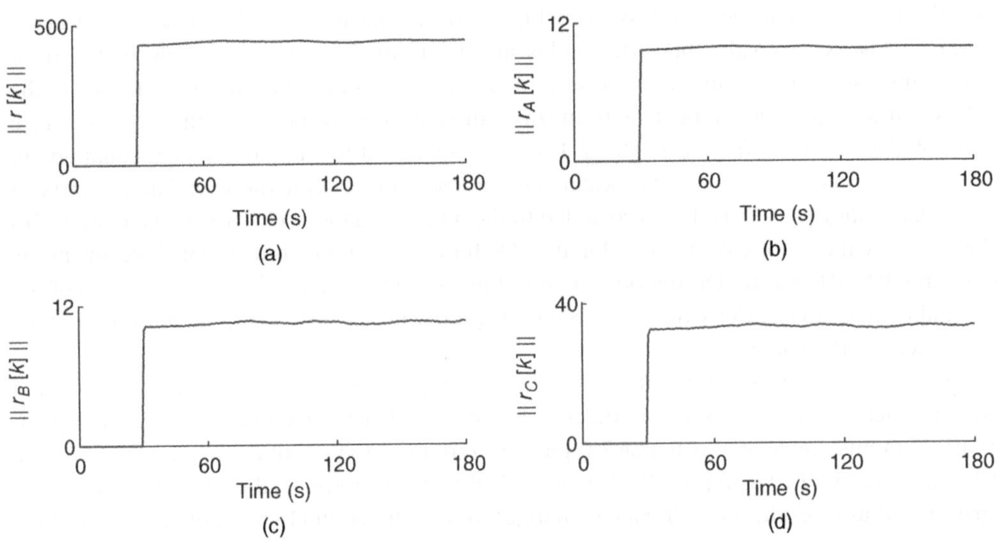

Figure 18.10 RFs of scenario 1, (a) main UIO, (b) UIO A, (c) UIO B, and (d) UIO C.

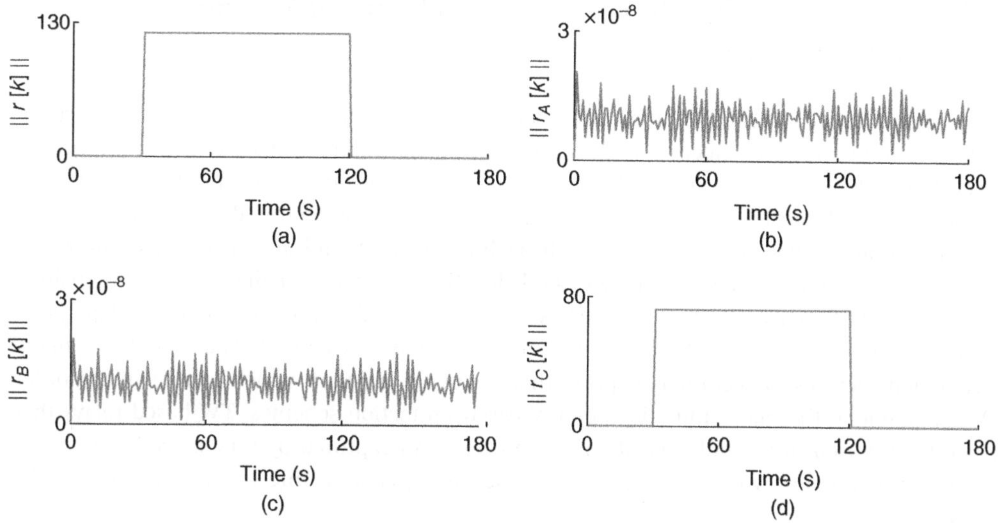

Figure 18.11 RFs of scenario 2, (a) main UIO, (b) UIO A, (c) UIO B, and (d) UIO C.

scenario. All UIOs' RF surpasses the detection threshold, i.e., (0.6), as soon as the attack begins, indicating that all measurements are manipulated.

- **Scenario 2**: In this scenario, frequency measurements are manipulated and dropped by 0.12 Hz between $t = 30$ and $t = 120$ s. The RFs of each linear UIO are presented in Figure 18.11. During the attack, the RFs of the main UIO and UIO C go above their thresholds, meaning that frequency measurements are targeted. Additionally, since the attacks associated with linear UIOs A and B are not in progress, the RFs of these linear UIOs do not rise.
- **Scenario 3**: In this scenario, the ongoing attacks related to UIOs A and B are currently underway, with each tie-line power measurement experiencing a 5% increase between the time periods of 30 to 120 seconds. Figure 18.12 demonstrates that the main UIO's RF, as well as those of UIOs A and B, surpasses 0.6 during the attack, while the RF of UIO C does not grow, since its corresponding attack has not happened (i.e., its attack input is zero). Thus, linear UIOs can detect and identify FDIAs and can determine their start and end times.

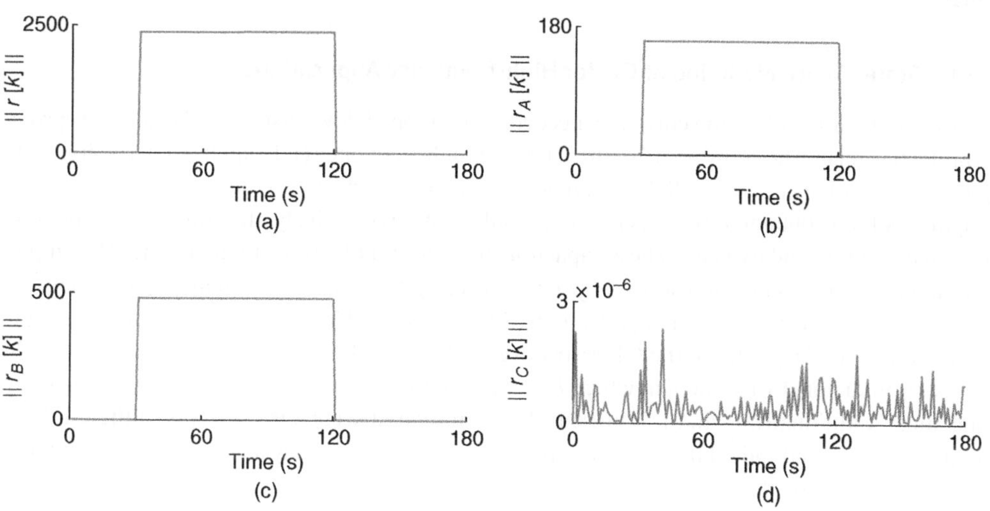

Figure 18.12 RFs of scenario 3, (a) main UIO, (b) UIO A, (c) UIO B, and (d) UIO C.

18.7 Case Study 2: Developing Wide-Band Current Transformers for Traveling-wave-based Protection

The development of TW-based protection schemes provides power systems with opportunities for accurate and fast detection and localization of faults. In TW-based relays, the voltage and current TWs, as well as their reflections from the line terminals, are used to identify the presence of a fault and locate it. These relays can be used for protecting different components, e.g., under-ground cables (UGCs) and over-head lines (OHLs), and also in various power grid domains, e.g., microgrids. The accuracy of fault localization in power lines and distinguishing in-zone and out-of-zone faults in microgrids using the TW-based techniques are highly dependent on the frequency content of TWs captured by measurement device [17]. The inability of many measurement devices to accurately capture this frequency content, which can be in the order of MHz, is among the main limitations of TW-based protection schemes. TW-based relays that are commonly used in the industry often use current TWs captured by CTs, since voltage TWs captured by coupling CVTs (CCVTs) are more severely distorted due to the low bandwidths of CCVTs.

When a fault occurs on a line, a step-like TW is created with an infinite range of frequency components. If the TW is captured by a CT, the CT acts as a low-pass filter and distorts the waveform. As a result, the estimated arrival time of the TW is relatively inaccurate. This problem is more severe if the fault occurs close to the terminals of a line or in short lines. Despite the fact that such an issue may not considerably affect some protection applications, it complicates fault location schemes for UGCs and microgrids. Therefore, there is a need for a compensation technique that addresses the limitation of CT bandwidth for TW-based protection applications.

This section employs UIKF to address the bandwidth limitation of CTs. To address this problem, the UIKF accurately estimates the waveform of TWs using the state–space model of the CT, as well as its output current. A salient feature of UIKF that makes it appropriate for this application is that it minimizes the impacts of process and measurement noises on the estimation error. In the following, Section 18.7.1 describes the state-space representation of CTs. Subsequent elaboration on the compensation algorithm developed for TWs is provided. Finally, Sections 18.7.2 and 18.7.3 demonstrate the effectiveness of UIKFs through simulation studies.

18.7.1 State–Space Modeling of CTs for High-frequency Applications

As shown in Figure 18.13a, the current is needed to be stepped down using a CT before it is given to a protective relay. The equivalent circuit of CTs for high-frequency (HF) applications is shown in Figure 18.13b [80]. By referring all the parameters to the secondary side of the CT, the circuit shown in Figure 18.13c is obtained. In this circuit, C_1 and C_2 are, respectively, the stray capacitances of the primary and secondary sides. These capacitors operate as a filter and attenuate the HF components of the measured current before it is given to the relay; thus, these capacitances are among the major sources of inaccuracy for HF applications. Subscripts 1 and 2 represent the parameters of the primary and secondary sides of the CT, respectively. The core of the CT is modeled using a parallel resistor–inductor (RL) branch, in which L_m and R_c represent the core losses and magnetization inductance. Additionally, I_p and I_s represent the primary and secondary currents of the CT. The burden resistance and inductance, which represent the protective relay and its connecting wires, have been signified by R_b and L_b.

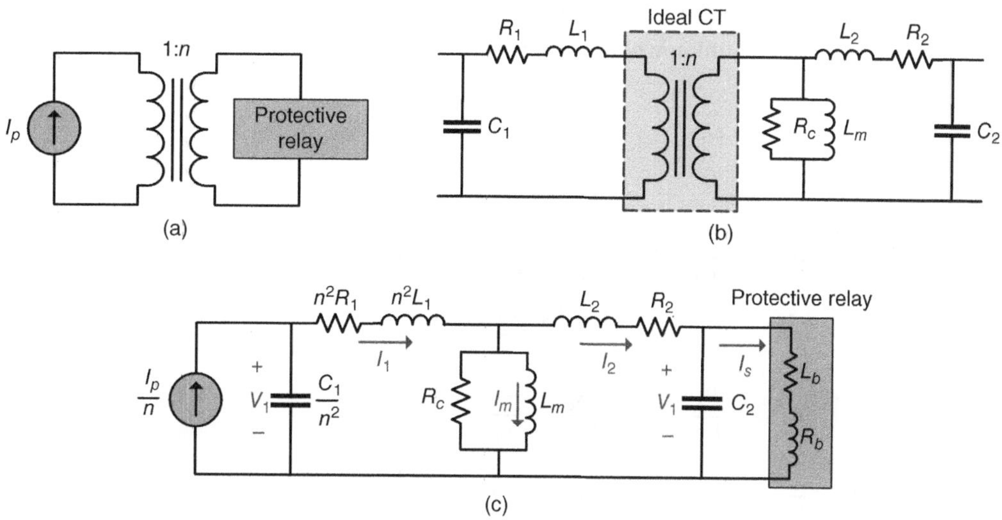

Figure 18.13 Current transformer: (a) physical circuit, (b) HF model of CTs with parasitic capacitors (source: adapted from [80]), and (c) equivalent HF model referred to the secondary side.

To obtain the state–space representation of the CT shown in Figure 18.13c, it is assumed that the CT is operating in its linear region, which means that the core inductance is constant. Such an assumption can be justified since, in TW-based protection applications, the first and second TWs are of interest, and these TWs are captured in a few milliseconds following faults. During this short amount of time, the CT does not saturate if it is appropriately designed based on practical procedures, e.g., in [81]. The differential equations that describe the CT equivalent circuit shown in Figure 18.13c are as follows:

$$\frac{dV_1}{dt} = \frac{n}{C_1}I_p - \frac{n^2}{C_1}I_1 \tag{18.98a}$$

$$\frac{dV_2}{dt} = \frac{1}{C_2}I_2 - \frac{1}{C_2}I_s \tag{18.98b}$$

$$\frac{dI_1}{dt} = -\frac{R_c + n^2R_1}{n^2L_1}I_1 + \frac{R_c}{n^2L_1}I_2 + \frac{R_c}{n^2L_1}I_m + \frac{V_1}{n^2L_1} \tag{18.98c}$$

$$\frac{dI_2}{dt} = -\frac{R_c + R_2}{L_2}I_2 + \frac{R_c}{L_2}I_1 - \frac{R_c}{L_2}I_m - \frac{V_2}{L_2} \tag{18.98d}$$

$$\frac{dI_m}{dt} = \frac{R_c}{L_m}I_1 - \frac{R_c}{L_m}I_2 - \frac{R_c}{L_m}I_m \tag{18.98e}$$

$$\frac{dI_s}{dt} = \frac{1}{L_b}V_2 - \frac{R_b}{L_b}I_s \tag{18.98f}$$

By writing (18.98a)–(18.98f) in a matrix form, the state–space representation of CTs can be written as follows:

$$\dot{x}(t) = Ax(t) + B_uI_p \tag{18.99}$$

in which I_p is an unknown input, x represents the vector of states, and A and B_u denote the state matrix and input matrix for unknown inputs, respectively. These matrices are shown below:

$$A = \begin{bmatrix} 0 & 0 & \frac{-n^2}{C_1} & 0 & 0 & 0 \\ 0 & 0 & 0 & \frac{1}{C_2} & 0 & \frac{-1}{C_2} \\ \frac{1}{n^2 L_1} & 0 & -\frac{R_c + n^2 R_1}{n^2 L_1} & \frac{R_c}{n^2 L_1} & \frac{R_c}{n^2 L_1} & 0 \\ 0 & \frac{-1}{L_2} & \frac{R_c}{L_2} & -\frac{R_c + R_2}{L_2} & \frac{-R_c}{L_2} & 0 \\ 0 & 0 & \frac{R_c}{L_m} & \frac{-R_c}{L_m} & \frac{-R_c}{L_m} & 0 \\ 0 & \frac{1}{L_b} & 0 & 0 & 0 & \frac{-R_b}{L_b} \end{bmatrix} \quad (18.100a)$$

$$x = \left[V_1 \; V_2 \; I_1 \; I_2 \; I_m \; I_s \right]^T \quad (18.100b)$$

$$B_u = \left[\tfrac{n}{C_1} \; 0 \; 0 \; 0 \; 0 \; 0 \right]^T \quad (18.100c)$$

In addition to (18.99), the output equation of the system should be obtained as well. In this subsection, the primary voltage and secondary current have been selected as outputs due to their low-range values, availability for measurement, and being system states. On this basis, the equations that represent the output of the CT can be represented as

$$y(t) = \mathbb{C}x(t) \quad (18.101)$$

where y is the output vector and $\mathbb{C}$ is the output matrix as follows:

$$\mathbb{C} = \begin{bmatrix} 1 & 0 & 0 & 0 & 0 & 0 \\ 0 & 0 & 0 & 0 & 0 & 1 \end{bmatrix} \quad (18.102)$$

Similar to other case studies in this chapter, the state–space equations (18.99) and (18.101) must be discretized. This can be done using (18.4a) and (18.4b). Additionally, in order to have a realistic model, the measurement and process noises should also be considered. As a result, the discretized state–space model representing the equivalent circuit of CTs for HF applications is as follows:

$$\begin{cases} \mathbb{X}[k+1] = \mathbb{A}\mathbb{X}[k] + \mathbb{B}_u \mathbb{I}_p[k] + w[k] \\ \mathbb{Y}[k] = \mathbb{C}\mathbb{X}[k] + v[k] \end{cases} \quad (18.103)$$

In the given equation, $\mathbb{X}[k] \in \mathbb{R}^6$ represents the vector of states at time step k, and $\mathbb{Y}[k] \in \mathbb{R}^2$ represents the vector of outputs at time step k. The variable $\mathbb{I}_p[k] \in \mathbb{R}$ represents the sampled value of the primary current at time step k. The variables $w[k]$ and $v[k]$ correspond to the process and measurement noise vectors, respectively. The noises possess characteristics of being zero mean and white, exhibiting no correlation with one another or the initial states. The covariance matrices for $w[k]$ and $v[k]$ are represented as Q and R, respectively.

18.7.2 Utilizing UIKFs for Estimating the Primary Current of CTs

As previously discussed, in order to accurately estimate the primary current of a CT, the secondary current and primary voltage of the CT and its state–space representation are needed. Given that the primary voltage of the CT has a low range, it can be directly measured without the need for VTs. Additionally, since $\mathbb{I}_p[k]$ in the model presented by (18.103) is an unknown input, UIKF can be used to estimate the system states and accordingly the system output.

In order to develop a stable and accurate UIKF, the procedure presented in Section 18.4.4 should be followed for the system presented by (18.103). As proven in [17], (18.41a)–(18.41d) are met for the state–space model of (18.103), and thus, a UIKF can be designed based on (18.44)–(18.50) to estimate the optimal states and obtain the primary current of the CT.

18.7.3 Performance Evaluation

To evaluate the performance of the UIKFs in estimating the primary current of CTs, two test systems are used. The first test system is the 14-bus Conseil International des Grands Réseaux Electriques (CIGRE) 20 kV benchmark European distribution test grid [82], which is modeled using power system computer-aided design/electromagnetic transient design and control (PSCAD/EMTDC) (Figure 18.14). This section concentrates on the UGC that connects buses 12 and 13. The length of this UGC is 2.5 km, and its relays are named R12 and R13. The CTs of these relays are sized according to the common practice of industry for TW-based applications [81]. The parameters of the CT used for the relays of this UGC, i.e., CT1, are presented in Table 18.2. More information about this test system can be found in [82].

The second test system is the CIGRE North American high-voltage (HV) transmission network benchmark [82], which is modeled using PSCAD/EMTDC (Figure 18.15). The focus of this section is on OHLs 1–2, which connects buses 1 and 2. The voltage level of this OHL is 230 kV, its length is

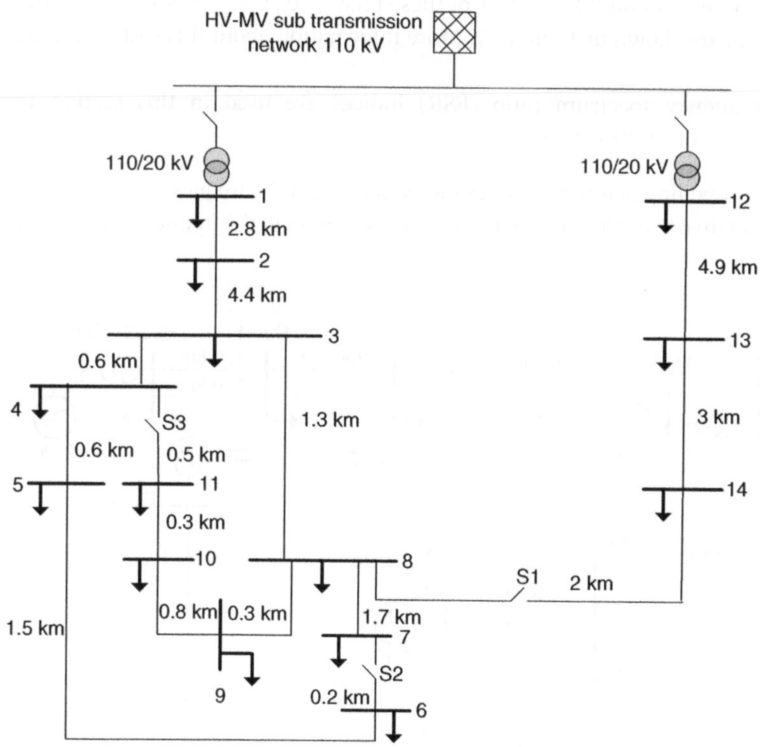

Figure 18.14 CIGRE 20 kV benchmark European distribution test grid (source: adapted from [82]).

Table 18.2 CT parameters.

	CT	
Parameters	**1**	**2**
Accuracy class	C400	C400
Turn ratio	500:5	800:5
Rated power (VA)	25	30
rated frequency (Hz)	60	60
Core mean length (m)	0.448	0.4978
Core area (m^2)	0.00168	0.00192
Burden (Ω)	0.2	0.2
Power factor of the burden	1	1
Secondary resistance (Ω)	0.418	0.668
Secondary leakage inductance (mH)	0.326	0.412
Secondary capacitance (μF)	5	4
Primary resistance (μΩ)	6	8
Primary leakage inductance (μH)	0.25	0.27
Primary capacitance (nF)	15	18

100 km, and its relays are named R1 and R2. The CTs of these relays, i.e., CT2, are sized according to 18.2, and their parameters are shown in Table 18.2. More information about this system can be found in [82].

The following three frequency spectrum ratio (FSR) indices are used in this section to numerically measure the effect of a CT on TWs:

1) $\boldsymbol{FSR_l}$: It measures the dampening of low-frequency elements within the frequency spectrum of the secondary current, relative to the low-frequency elements present in the frequency spectrum of the primary current.

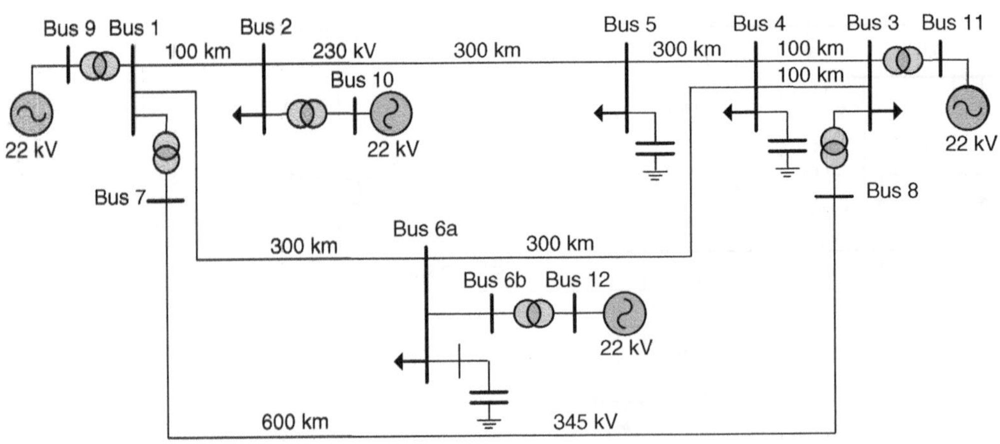

Figure 18.15 CIGRE North American HV transmission network benchmark (source: adapted from [82]).

2) $\boldsymbol{FSR_h}$: This represents the reduction in high-frequency elements in the frequency spectrum of the secondary current, relative to the high-frequency elements within the frequency spectrum of the primary current.
3) $\boldsymbol{FSR_t}$: This indicates the dampening of the entire range of frequency components within the frequency spectrum of the secondary current, in comparison with the entire range of frequency components in the frequency spectrum of the primary current.

These indices are formulated as follows:

$$FSR_l = \sum_{i=1}^{\lceil \frac{K}{2} \rceil} \frac{I_s(i)}{\frac{I_p}{n}(i)} \tag{18.104a}$$

$$FSR_h = \sum_{i=\lceil \frac{K}{2} \rceil + 1}^{K} \frac{I_s(i)}{\frac{I_p}{n}(i)} \tag{18.104b}$$

$$FSR_t = \sum_{i=1}^{K} \frac{I_s(i)}{\frac{I_p}{n}(i)} \tag{18.104c}$$

where i and K are the frequency component identifier and the total number of frequency components, respectively, and $\lceil . \rceil$ is the ceiling function. In fact, these indices compare the first half, the second half, and the entire frequency components of currents at both sides of a CT. If the CT does not attenuate components, FSR indices remain equal to 1. However, the higher the attenuation of the components by the CT, the lower these indices become. The minimum value of an frequency spectrum ratio (FSR) index is 0.

In order to evaluate the performance of UIKFs, the primary currents of the under-study CTs, presented in Table 18.2, are estimated following faults on UGCs 12–13 and OHLs 1–2. Since this method is designed to be implemented in relays, all its required signals are already available for the relay. As a result, no additional measurement or hardware is required for the implementation of the developed method. In order to consider the impact of noise, the process and measurement noises are considered to be zero mean, white, uncorrelated, and with signal-to-noise ratios (SNRs) of 35 dB [83].

The following subsections demonstrate the effectiveness of UIKFs and analyze the impact of different fault parameters, i.e., fault types, resistances, inception angles, and locations, on the performance of the protection scheme. It is worth mentioning that, without loss of generality, only the results for phase A are demonstrated in this section. Moreover, the sampling frequency is assumed to be 50 MHz.

18.7.3.1 Performance Evaluation for UGCs 12–13

The UIKF is utilized to compensate the CT currents associated with R12 and R13. Figure 18.16 demonstrates I_p/n, which signifies the primary current transferred to the secondary side, as well as the actual and compensated secondary currents, respectively, denoted by I_s and $\hat{I}_p/n$. For both relays, the compensated and actual currents completely match with each other, whereas the actual secondaries are entirely different. The results in the frequency domain also verify the time domain comparison, as Figure 18.17 shows. It can be observed from this figure that the primary and compensated secondary currents of the CT have similar fast Fourier transform (FFT) coefficients, and thus, their frequency spectra are almost identical. To quantify this comparison, the FSR indices, presented in (18.104), are calculated based on $\hat{I}_p/n$ and I_p/n. For R12, these indices equal $FSR_l = 0.9986$, $FSR_h = 1.0230$, and $FSR_t = 1.0052$. On the other hand, the FSR indices calculated

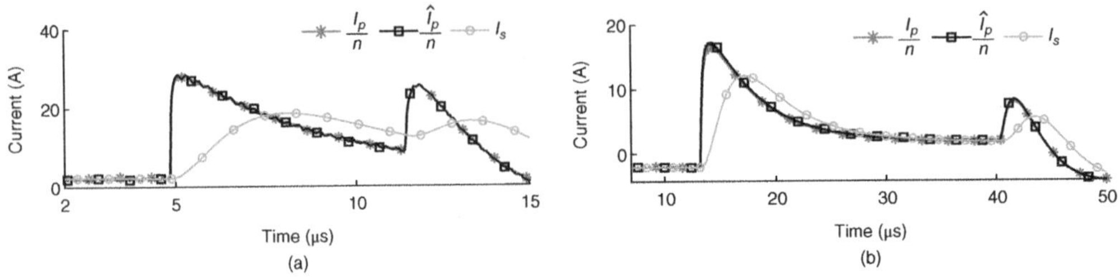

Figure 18.16 Primary, actual secondary, and compensated secondary currents for (a) R12 and (b) R13.

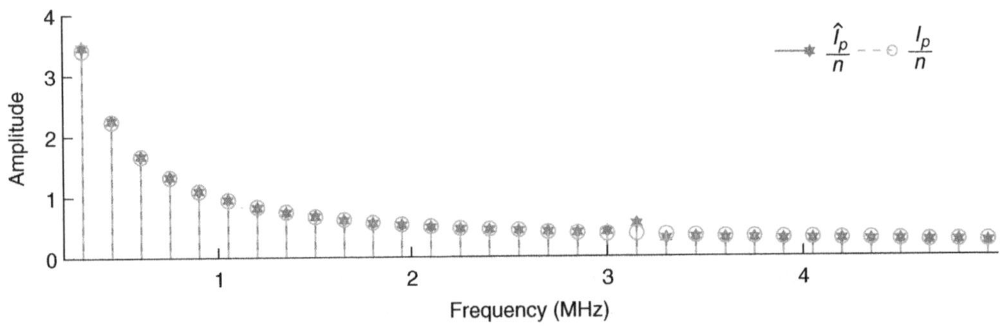

Figure 18.17 Spectral analysis of the first TW within primary and compensated secondary currents of R12.

based on I_s and I_p/n are significantly lower, i.e., $FSR_l = 0.5377$, $FSR_h = 0.5282$, and $FSR_t = 0.5362$. Based on this analysis, it can be confirmed that the compensated secondary currents provide the relays with the same information that the primary currents do. Thus, the relays can accurately identify and locate the faults.

18.7.3.2 Performance Evaluation for OHLs 1–2

The UIKF is utilized for compensating the CT currents associated with R1 and R2. For these relays, the primary, actual secondary, and compensated secondary currents in the time domain are shown in Figure 18.18. Additionally, Figure 18.19 demonstrates the frequency spectra of the first TW waveforms presented in Figure 18.18b. As it can be observed in these figures, the TWs of the primary and compensated secondary currents are identical in both time and frequency domains. This can be also proved by the FSR indices of the first TW for R2. If these indices are calculated based on $\hat{I}_p/n$ and I_p/n, they become $FSR_l = 0.9945$, $FSR_h = 0.9912$, and $FSR_t = 0.9940$. However, for I_s and

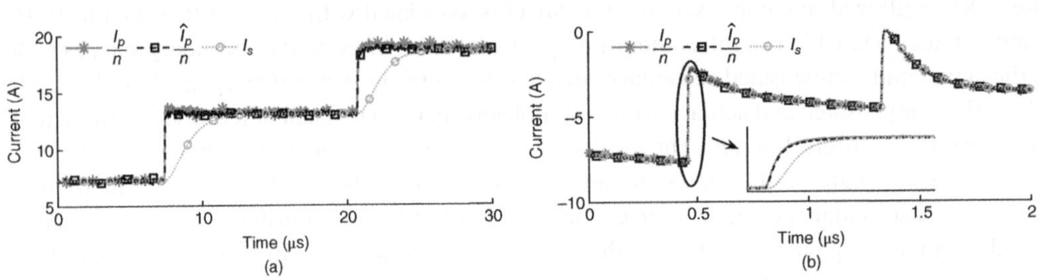

Figure 18.18 Primary, actual secondary, and compensated secondary currents for (a) R1 and (b) R2.

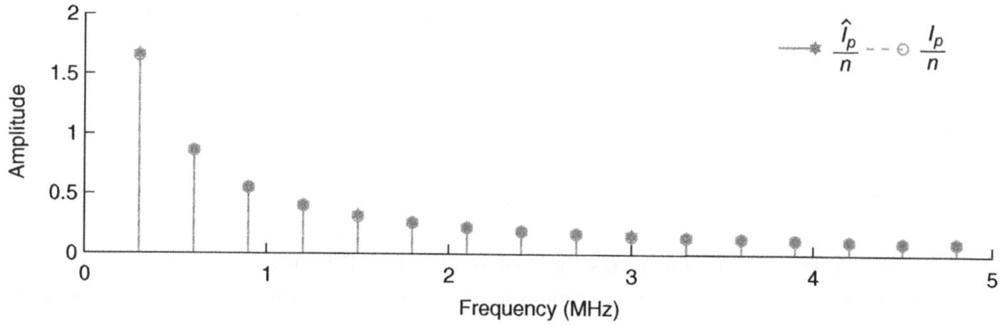

Figure 18.19 Spectral analysis of the first TW within primary and compensated secondary currents of R2.

I_p/n, these indices are obtained as $FSR_l = 0.5377$, $FSR_h = 0.5282$, and $FSR_t = 0.5362$. As a result, UIKF is able to provide the relays with an accurate estimation of the primary current.

18.7.3.3 Impacts of Fault Resistances

To investigate the impacts of fault resistance R_f on the performance of the UIKF-based method, eight fault cases, four on each test system, are simulated with a wide range of fault resistances, from 5 to 250 Ω. The faults are three phase with 90° inception angle at phase A and happen 0.2 km away from R12 on UGCs 12–13 and 1 km away from R1 on OHLs 1–2. The results for actual and compensated secondary currents of CTs are presented in Table 18.3. This table compares FSR indices for the fault cases with and without using the UIKF-based method. The observation reveals that the compensated and actual secondary currents closely align with each other. This is corroborated by the FSR indices, which are near 100%, with a maximum deviation from 100% of just 1.24%. On the other hand, the minimum deviation of these indices for the actual secondary current of CTs is around 36.8%. Thus, as it can be observed, the UIKF-based method successfully compensates the secondary currents of CTs without being affected by the fault resistances.

Given that the traveling distance of TWs is the factor that affects the attenuation of their frequency components, variation of R_f, while the fault location is fixed, does not significantly change the distortion level of TWs. Moreover, due to the fault location being closer to R1 than to R12, the

Table 18.3 Variation of FSR indices across different fault resistances.

		Compensated secondary current			Actual secondary current		
	R_f(Ω)	FSR_l	FSR_h	FSR_t	FSR_l	FSR_h	FSR_t
OHL (R1)	5	0.9928	0.9959	0.9941	0.1714	0.0399	0.1144
	50	0.9928	0.9958	0.9941	0.1718	0.0400	0.1147
	100	0.9929	0.9959	0.9942	0.1708	0.0397	0.1140
	250	0.9927	0.9964	0.9945	0.1711	0.0398	0.1142
UGC (R12)	5	0.9997	0.9909	0.9984	0.6317	0.3646	0.5933
	50	0.9999	0.9879	0.9980	0.6245	0.3508	0.5824
	100	1.0000	0.9878	0.9981	0.6405	0.3717	0.5988
	250	0.9999	0.9876	0.9979	0.6622	0.3990	0.6209

HF components of the TWs received by R1 are less distorted. Consequently, the CT that feeds R1 attenuates the TWs more intensely.

18.7.3.4 Impacts of Fault Types

To investigate the impacts of fault type on the performance of the UIKF-based method, bolted AG, ABG, AB, and ABCG faults are applied to both test systems, 8 km away from R1 on OHLs 1–2 and 750 m away from R12 on UGCs 12–13. For all cases, the voltage angle of phase A is 90° when the faults occur. Table 18.4 demonstrates the FSR indices for all fault cases. It can be observed from this table that the indices for the compensated secondary currents are within the range of (98%, 100%). In particular, for the OHL, 0.3%, 0.19%, and 0.26% are the maximum attenuation for low-, medium-, and high-frequency components, respectively. In contrast, the minimum attenuation values for actual secondary current of CTs for low-, high-, and overall frequency spectra are, respectively, 32.21%, 51.33%, and 34.59% for the UGC and 71.72%, 88.85%, and 77.03% for the OHL. From Table 18.4, it can also be observed that due to greater fault distances in this subsection, the FRS indices are higher compared to those obtained in the previous subsection. Based on these results, it can be observed that the fault type does not affect the accuracy of the UIKF-based method.

18.7.3.5 Impacts of Fault Inception Angles

This subsection discusses the impacts of fault inception angle on the performance of the UIKF-based compensation method. For this purpose, bolted AG faults are simulated 5 km away from R1 on OHLs 1–2 and 1.25 km away from R12 in UGCs 12–13 with 5°, 30°, 45°, and 60° inception angles for phase A. Table 18.5 shows the FSR indices for R1 and R12. From this table, it can be observed that, for both cases, the calculated indices are almost 100% with the maximum deviation of 1.02% from 100%. For instance, for the low-, high-, and overall frequency components of the compensated secondary current of CT1, the maximum errors are, respectively, 0.03%, 0.19%, and 0.25%. On the other hand, for the actual secondary current of CT1, FSR_l, FSR_h, and FSR_t are off by 25.28%, 30.01%, and 25.32%, respectively. Table 18.5 demonstrates similar results for OHLs 1–2 as well. As a result, the UIKF-based method can effectively compensate the secondary current of CTs for different fault inception angles.

Table 18.4 Variation of FSR indices across different fault types.

		Compensated secondary current			Actual secondary current		
	Type	FSR_l	FSR_h	FSR_t	FSR_l	FSR_h	FSR_t
OHL (R1)	AG	0.9970	0.9981	0.9974	0.2688	0.1044	0.2174
	ABG	0.9971	0.9982	0.9974	0.2807	0.1104	0.2278
	AB	0.9972	0.9982	0.9975	0.2827	0.1114	0.2296
	ABCG	0.9971	0.9982	0.9975	0.2828	0.1115	0.2297
UGC (R12)]	AG	1.0000	0.9875	0.9984	0.6779	0.4867	0.6541
	ABG	1.0000	0.9879	0.9985	0.6777	0.4865	0.6540
	AB	0.9999	0.9879	0.9984	0.6767	0.4865	0.6531
	ABCG	0.9999	0.9875	0.9984	0.6771	0.4864	0.6534

Table 18.5 Variation of FSR indices across different fault inception-angles.

	Inception angle	Compensated secondary current			Actual secondary current		
		FSR_l	FSR_h	FSR_t	FSR_l	FSR_h	FSR_t
OHL (R1)	5°	0.9981	0.9946	0.9968	0.2100	0.0523	0.1525
	30°	0.9969	0.9935	0.9956	0.2085	0.0534	0.1476
	45°	0.9969	0.9949	0.9961	0.2067	0.0541	0.1477
	60°	0.9970	0.9951	0.9963	0.2098	0.0549	0.1499
UGC (R12)	5°	1.0002	0.9916	0.9991	0.7373	0.6062	0.7468
	30°	1.0002	0.9906	0.9990	0.7446	0.5900	0.7258
	45°	1.0003	0.9924	0.9993	0.7366	0.5649	0.7152
	60°	1.0002	0.9898	0.9989	0.7472	0.6099	0.7310

Table 18.6 Variation of FSR indices across different fault locations.

	Fault location (km)	Compensated secondary current			Actual secondary current		
		FSR_l	FSR_h	FSR_t	FSR_l	FSR_h	FSR_t
OHL (R1)	10 km	0.9981	0.9963	0.9975	0.3764	0.1733	0.3151
	25 km	0.9994	0.9984	0.9992	0.5297	0.3657	0.4930
	50 km	0.9996	0.9999	0.9997	0.5774	0.5050	0.5670
	90 km	0.9997	0.9995	0.9997	0.6157	0.5360	0.6041

18.7.3.6 Impact of Fault Locations

The impacts of fault location on the accuracy of the UIKF-based method are studied in this subsection. In order to do this, bolted ABCG faults with the inception angle of 90° at phase A are applied to OHLs 1–2 at 10, 25, 50, and 90 km away from R1. The FRS indices for the actual and compensated secondary currents of CT2 are demonstrated in Table 18.6. For all fault locations, the indices are within the range of (99%, 100%). For instance, when the fault occurs at the middle of the OHL, FSR_l, FSR_h, and FSR_t are, respectively, 99.96%, 99.99%, and 99.97%. These values are almost twice the indices obtained for the actual secondary current of CT, which are, respectively, 57.74%, 50.50%, and 56.70%. As a result, the UIKF-based method accurately compensates the CT current. Additionally, from Table 18.6, it can be observed that by decreasing the fault distance from R1, the CT attenuates the HF components more severely. This phenomenon can be attributed to the fact that with decreasing distance, the HF components of the TWs undergo less severe attenuation by the line. Consequently, the primary current retains more HF components as it traverses the CT.

18.8 Case Study 3: Fault Diagnosis in Transformers Using LPV Observers

In a power grid, transformers are often protected using differential relays, which are known to be sensitive, reliable, and selective. The transformer differential scheme, i.e., 87T, uses Kirchhoff's current law to identify a fault based on the input and output currents of the protection zone.

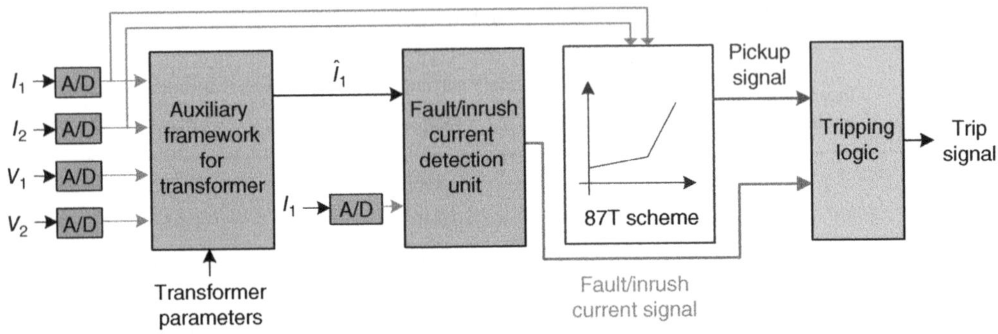

Figure 18.20 Transformer differential scheme after implementing the proposed auxiliary framework.

In such a scheme, the major sources of malfunction are inrush currents and transformer over excitation. To address these issues, harmonic blocking/restraint elements and/or multiple-slope characteristics are used in practice, aiming at minimizing the relay malfunction during these events. However, these approaches negatively affect the speed and sensitivity of the differential scheme or its effectiveness during cross-country faults. Therefore, there is a need for an algorithm that can differentiate between faults and inrush currents or over excitation, without deteriorating the performance of the 87T element.

The existing methodologies in the scholarly literature for differentiating between inrush currents or over excitation and internal faults can be sorted into three categories: (i) application of signal processing techniques such as S-transform [84] and wavelet-based methods [85]; (ii) employment of data-driven techniques like ANNs [86] and support vector machines [87]; and (iii) characterizing specific aspects of transformers during inrush currents or internal faults and distinguishing these two scenarios by comparing real-time features with those previously modeled.

The first category may not be efficacious under certain conditions, such as when shunt capacitors generate significant second harmonics during internal faults. The second category is constrained by the inherent limitations of data-driven techniques, such as the need for large datasets for training, and reliance on the operating point. The third category of studies, despite their effectiveness, often comes with substantial computational demands, which can restrict their practical application.

To address the discussed problem for single-phase transformers or three-phase transformer banks, an auxiliary framework—which utilizes LPV observers along with the voltage and current measurements of the primary and secondary sides of the transformer—is developed for transformer differential relays, as shown in Figure 18.20. The proposed framework (i) can perform well even in the presence of transformer saturation; (ii) is able to identify faults even during energization; (iii) is capable of identifying inrush currents; and (iv) differentiates between internal faults and inrush currents.

In what follows, initially, the state–space representation of transformers is formulated in Section 18.8.1. Then, Section 18.8.2 explains the process of designing an LPV observer for transformers. Afterward, Section 18.8.3 discusses the procedure for designing the framework using the transformer state–space model and LPV observers. Finally, Section 18.8.4 evaluates the performance of the framework.

18.8.1 State–Space Modeling of Transformers

To design the above mentioned framework for single-phase transformers or three-phase transformer banks, the state–space representation of the transformer should be obtained using its equivalent circuit shown in Figure 18.21. The equivalent circuit demonstrated in this figure represents a single-phase transformer or one of the phases in a three-phase transformer bank. The turn ratio is denoted by n; subscripts 1 and 2 are used for the parameters of primary and secondary windings, respectively; R_c is used to model the core power loss; and the magnetization of the core is represented by variable inductance L_m, whose value is dependent on the core flux λ_m and can be obtained using the excitation curve for any flux value [88, 89]. This curve is whether available in the transformer data sheets or can be obtained using an excitation test. To model transformers using the equivalent circuit of Figure 18.21b, the Kirchhoff's voltage law (KVL) equations of the three loops should be written as follows:

$$R_1 I_1 + \frac{d\lambda_1}{dt} = V_1 \tag{18.105a}$$

$$\left(\frac{R_2}{n^2}\right) nI_2 + \frac{d\lambda_2}{dt} = \frac{V_2}{n} \tag{18.105b}$$

$$R_c\left(I_1 + nI_2 - I_m\right) - \frac{d\lambda_m}{dt} = 0 \tag{18.105c}$$

In this equation, λ_1 and λ_2 are the flux linkages of the primary and secondary windings, respectively. These fluxes can be defined as

$$\lambda_1 = \lambda_{l1} + \lambda_m \tag{18.106a}$$

$$\lambda_2 = \lambda_{l_2} + \lambda_m \tag{18.106b}$$

In these equations, λ_l and λ_m are, respectively, the leakage fluxes of the windings and the magnetic flux of the core. Thus, the primary, secondary, and magnetization currents can be expressed as

$$I_1 = \frac{\lambda_1 - \lambda_m}{L_1} \tag{18.107a}$$

$$I_2 = \frac{n\left(\lambda_2 - \lambda_m\right)}{L_2} \tag{18.107b}$$

$$I_m = \frac{\lambda_m}{L_m\left(\lambda_m\right)} \tag{18.107c}$$

Substituting (18.107) in (18.105) results in

$$\frac{d\lambda_1}{dt} = V_1 - \frac{R_1}{L_1}\lambda_1 + \frac{R_1}{L_1}\lambda_m \tag{18.108a}$$

$$\frac{d\lambda_2}{dt} = \frac{V_2}{n} - \frac{R_2}{L_2}\lambda_m + \frac{R_2}{L_2}\lambda_2 \tag{18.108b}$$

$$\frac{d\lambda_m}{dt} = \frac{R_c}{L_1}\lambda_1 + \frac{n^2R_c}{L_2}\lambda_2 - \left(\frac{R_c}{L_1} + \frac{n^2R_c}{L_2} + R_c f\left(\lambda_m\right)\right)\lambda_m \tag{18.108c}$$

where $f(\lambda_m) = 1/L_m(\lambda_m)$ is a function that depends on the value of λ_m. Since for a transformer, the value of L_m is within the range of ($L_m^{min} \neq 0$, L_m^{max}), the value of $f(\lambda_m)$ is also between $\underline{f} = 1/L_m^{max}$ and $\bar{f} = 1/L_m^{min}$ as follows:

$$1/L_m^{max} \leq f\left(\lambda_m\right) \leq 1/L_m^{min} \tag{18.109}$$

Rewriting (18.108) in a matrix form results in the state–space representation of the transformer as follows:

$$\dot{x}(t) = A\left(\lambda_m\right)x(t) + B_n u_n(t) \tag{18.110}$$

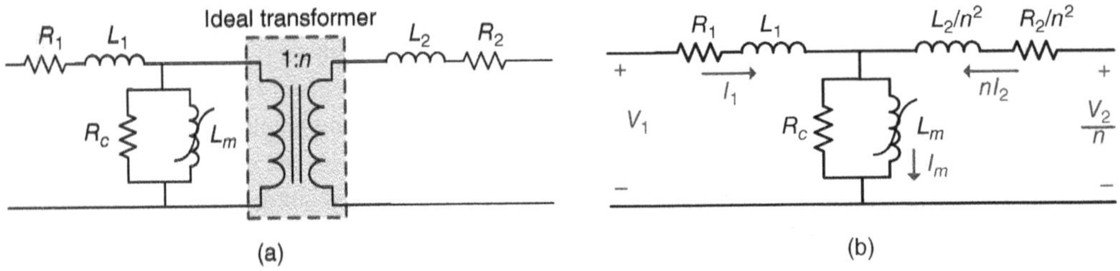

Figure 18.21 (a) Equivalent circuit of a transformer and (b) equivalent circuit referred to the primary side.

In this equation, $A(\lambda_m)$ and B_n are the state and known input matrices, respectively; x and $u_n(T)$ are the vectors of states and known inputs, respectively, all expressed as follows:

$$A(\lambda_m) = \begin{bmatrix} \frac{-R_1}{L_1} & 0 & \frac{R_1}{L_1} \\ 0 & -\frac{R_2}{L_2} & \frac{R_2}{L_2} \\ \frac{R_c}{L_1} & \frac{n^2 R_c}{L_2} & -\frac{R_c}{L_1} - \frac{n^2 R_c}{L_2} - R_c f(\lambda_m) \end{bmatrix} \tag{18.111a}$$

$$x = \left[\, \lambda_1 \; \lambda_2 \; \lambda_m \,\right]^T \tag{18.111b}$$

$$B_n = \begin{bmatrix} 1 & 0 & 0 \\ 0 & \frac{1}{n} & 0 \end{bmatrix}^T \tag{18.111c}$$

$$u_n = \left[\, V_1 \; V_2 \,\right]^T \tag{18.111d}$$

As it can be observed from (18.110) and (18.111), these equations present an LPV state–space model for transformers, where the parameter $f(\lambda_m)$ is bound as shown in (18.109). Thus, an LPV observer can be utilized to obtain an accurate estimation of the transformer states. The design of LPV observers for this problem is discussed in Section 18.8.2.

The output parameter of the model is selected to be the secondary current of the transformer, since (i) it can be measured, (ii) it is a physical parameter, and (iii) it can be expressed as a linear combination of λ_2 and λ_m, as detailed in (18.107b). Thus, the output vector, i.e., y, can be represented as

$$y(t) = \mathbb{C}x(t) \tag{18.112}$$

where $\mathbb{C}$ is the output matrix, which can be formulated as

$$\mathbb{C} = \left[\, 0 \; \frac{n}{L_2} \; -\frac{n}{L_2} \,\right] \tag{18.113}$$

Finally, the set of equations (18.110) and (18.112) is discretized using (18.4a) and (18.4b) to make them suitable for numerical implementation by digital relays. The discretized LPV model of transformers is as follows:

$$\begin{cases} \mathbb{X}\left[k+1\right] = \mathbb{A}(\lambda_m)\mathbb{X}\left[k\right] + \mathbb{B}_n\mathbb{U}_n\left[k\right] \\ \mathbb{Y}\left[k\right] = \mathbb{C}\mathbb{X}\left[k\right] \end{cases} \tag{18.114}$$

In this equation, $\mathbb{X}\left[k\right] \in \mathbb{R}^3$ and $\mathbb{Y}\left[k\right] \in \mathbb{R}$ represent the vector of states and outputs at time step k, respectively, and $\mathbb{U}_n\left[k\right] \in \mathbb{R}^2$ is the sampled value of the primary and secondary voltages at time step k. Equation (18.114) is used in the next subsection to design an LPV observer, which is leveraged in Section 18.8.3 to address the issues that occur in a transformer during saturation and in the presence of inrush currents.

18.8.2 Developing LPV Observers for Transformers

This section explains the process of designing an LPV observer for transformers. As discussed before, the state-space representation of transformers is LPV due to the presence of $f(\lambda_m)$, whose boundaries are expressed in (18.109). Thus, an LPV observer is needed and the transformer model should be rearranged in a polytopic form, as described by (18.54). This polytopic form is shown below:

$$\begin{cases} \mathbb{X}[k+1] = \left(\mu_1 \mathbb{A}\left(\bar{f}\right) + \mu_2 \mathbb{A}\left(\underline{f}\right)\right)\mathbb{X}[k] + \mathbb{B}_n \mathbb{U}_n[k] \\ \mathbb{Y}[k] = \mathbb{C}\mathbb{X}[k] \end{cases} \tag{18.115}$$

where μ_1 and μ_2 are represented as

$$\mu_1 = \frac{f(\lambda_m) - \underline{f}}{\bar{f} - \underline{f}} \tag{18.116a}$$

$$\mu_2 = \frac{\bar{f} - f(\lambda_m)}{\bar{f} - \underline{f}} \tag{18.116b}$$

In this form, matrices $A\left(\bar{f}\right)$ and $A\left(\underline{f}\right)$ have constant elements and μ_1 and μ_2 are variable parameters. By following the procedure explained in Section 18.4.5, the LPV observer can be designed to be accurate (i.e., the estimation error approaches zero as $t \to \infty$) and stable. Thus, the states of transformers can be estimated using (18.55).

18.8.3 Proposed Auxiliary Framework for Transformer Differential Protection

The method presented in Figure 18.20 should be implemented using the state-space model presented in (18.110)–(18.113). This model includes a time-varying inductor L_m, which is a function of core's flux λ_m. Thus, the transformer excitation curve and the value of the flux in each time step can be used to obtain an estimation of this inductor value for the next time step. On this basis, the state-space representation of transformers and LPV observers, as well as values of primary and secondary voltage, are used to calculate the primary current of the transformer, i.e., $\hat{I}_1$, at each time step. In the absence of internal faults, the state-space model provides an accurate representation of the transformer, which results in the estimated primary currents of the transformer matching the actual ones. Nonetheless, in the event of internal faults, the state-space model's accuracy in portraying the transformer is compromised, leading to nonzero deviations in the observer's error during these fault occurrences. As a result, the estimated and actual primary currents deviate from each other. Such an error can be quantified using the following RF:

$$r[k] = \frac{\left|\hat{I}_1[k] - I_1[k]\right|}{\left|I_1[k]\right|} \tag{18.117}$$

Consequently, internal faults can be detected by continuously tracking the RF and comparing it against a predetermined fault detection threshold, denoted as δ. Using this scheme, the 87T element can detect an internal fault when the differential scheme picks up and the following equation is met:

$$r[k] > \delta \tag{18.118}$$

On the other hand, when there are inrush currents or when the transformer saturates, the relay picks up; however, the condition of (18.118) is not satisfied.

18.8.4 Performance Evaluation

This subsection evaluates the performance of the framework developed for addressing the challenges of differential schemes during inrush currents or saturation. To this aim, EMTP-RV software is used to simulate the test system shown in Figure 18.22. A bank of three single-phase transformers with the following parameters is used in this system: 100 MVA, 138/69 V_{LL}, and 60 Hz. Moreover, Figure 18.23b illustrates the magnetization curve of the transformer. The other parameters of the transformer are $R_1 = 0.0027$ p.u., $L_1 = 30.52$ p.u., $R_2 = 0.0003$ p.u., $L_2 = 3.393$ p.u., and $R_c = 500$ p.u. . This transformer is protected using a differential scheme (87T) with a dual-slope characteristic (Figure 18.23a), and it is set based on the default settings of commercial relays [90].

In order to calculate the fault detection threshold, the RF, i.e., $r[k]$, is recorded during a wide range of fault-free conditions; for instance, when the measurements are contaminated with noise, there are uncertainties in the transformer's parameters. The noises are assumed to be independent,

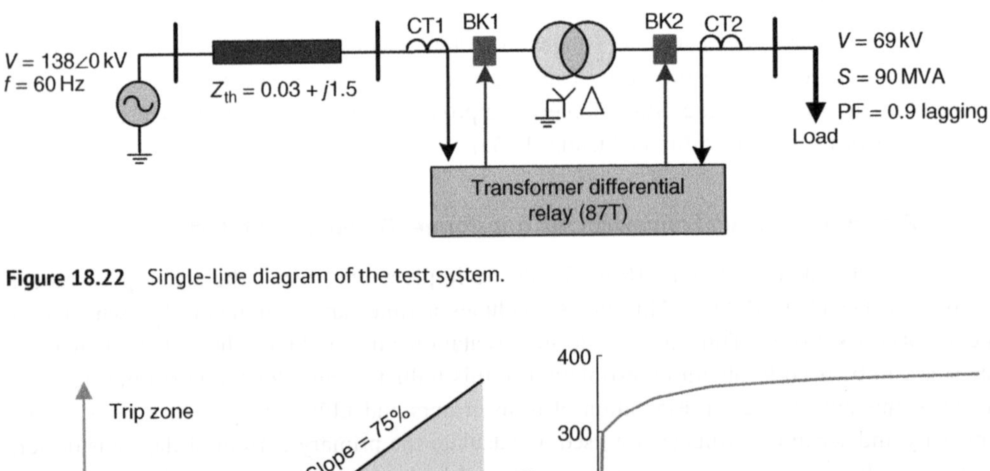

Figure 18.22 Single-line diagram of the test system.

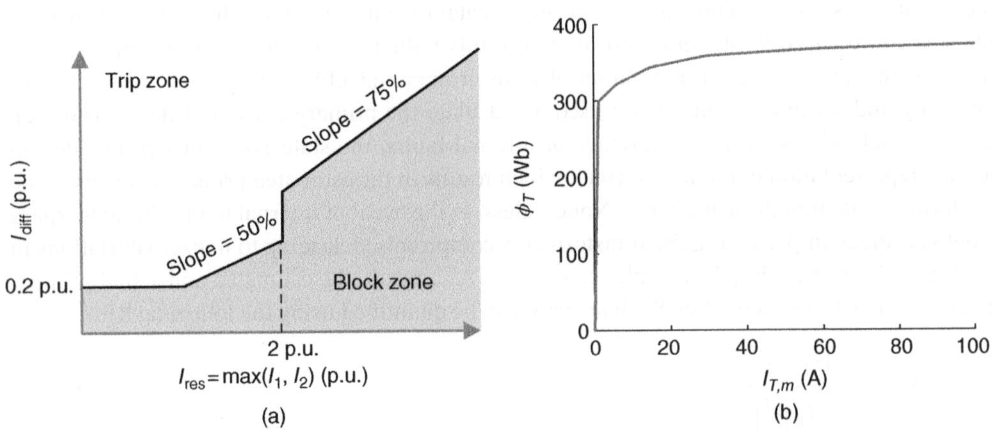

Figure 18.23 (a) The dual-slope characteristic used for the differential relay (source: adapted from [90]) and (b) the magnetization curve of the transformer.

white, and Gaussian, with a SNR of 35 dB. To consider uncertainties, a Gaussian percentage error with a mean value of zero and a standard deviation of 2% has been added to the parameters. Since the proposed framework only depends on the transformer parameters, the uncertainties in the grid, e.g., in operating point and parameters, do not affect the proposed framework. The value of the threshold is selected as the maximum value of the RF plus 10% security margin, which results in $\delta = 4.4\%$. In order to ensure that the obtained threshold results in acceptable performance, it is tested for 500 cases of various internal faults, i.e., turn-to-turn (TT) and turn-to-ground (TG). The results demonstrate that the selected threshold detects 496 cases correctly. The other four faults that were not detected had negligible impact on the current of the transformer. In the Section 18.8.2 evaluate the performance of the proposed method during some challenging situations for the transformer differential scheme.

18.8.4.1 Energizing Transformers in the Presence of Inrush Currents

In this subsection, the under-study transformer is energized at different switching angles (θ), and for each case, $r[k]$ is recorded. The first three cycles of energization waveform for the switching angle of $\theta = 45°$ are shown in Figure 18.24a. As it can be observed from this figure, the actual primary current and the one estimated using the proposed auxiliary framework are the same, showing that the framework can accurately estimate the primary current during energization. Moreover, Figure 18.24b demonstrates the RF associated with Figure 18.24a: The maximum value for the RF is less than the specified threshold. Therefore, the proposed framework specifies that the increase in differential current is due to inrush currents. Additionally, the maximum value of $r[k]$ for various switching angles is presented in Table 18.7. The results of this table show that for all θs, the maximum $r[k]$ remains below the threshold.

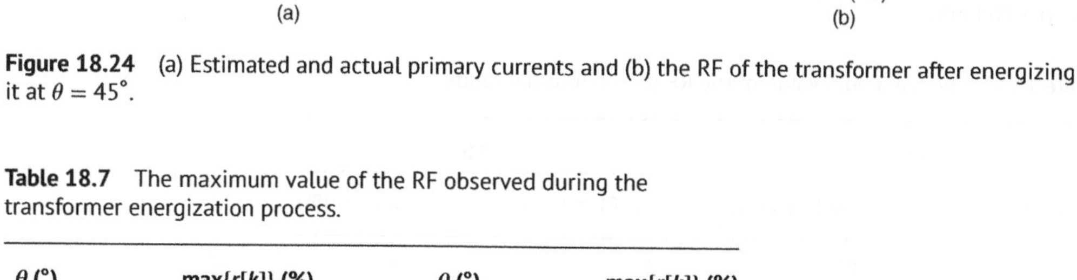

Figure 18.24 (a) Estimated and actual primary currents and (b) the RF of the transformer after energizing it at $\theta = 45°$.

Table 18.7 The maximum value of the RF observed during the transformer energization process.

θ (°)	max{$r[k]$} (%)	θ (°)	max{$r[k]$} (%)
0	1.6	120	3.5
30	1.3	135	3
45	0.9	150	1.8
60	1.7	180	1.5
90	0.2	270	1.2

18.8.4.2 Internal Faults

This section evaluates the performance of the proposed auxiliary framework in detecting TT and TG internal faults. Figures 18.25 and 18.26 illustrate the estimated and measured currents, as well as the RF, for a 20% TT fault that happens at $t = 20$ ms. As these figures demonstrate, before the occurrence of the fault, the estimated and measured currents are overlapping, indicating that the estimation of the current based on the developed model is accurate, and thus, the RF remains low. During the fault, however, $\hat{I}_1$ and I_1 are significantly different, and consequently, the RF surpasses the predefined threshold δ (Figures 18.25b and 18.26b), and thus, the faults are detected. The maximum value of RF during different fault scenarios is presented in Table 18.8. It can be observed that when the severity of faults, i.e., the percentage of short-circuited turns, increases, the value of

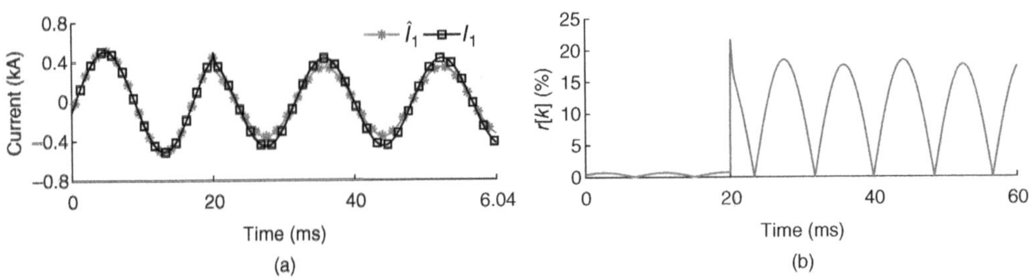

Figure 18.25 (a) The estimated and actual primary currents, and (b) the RF of the transformer for a 20% internal TT fault.

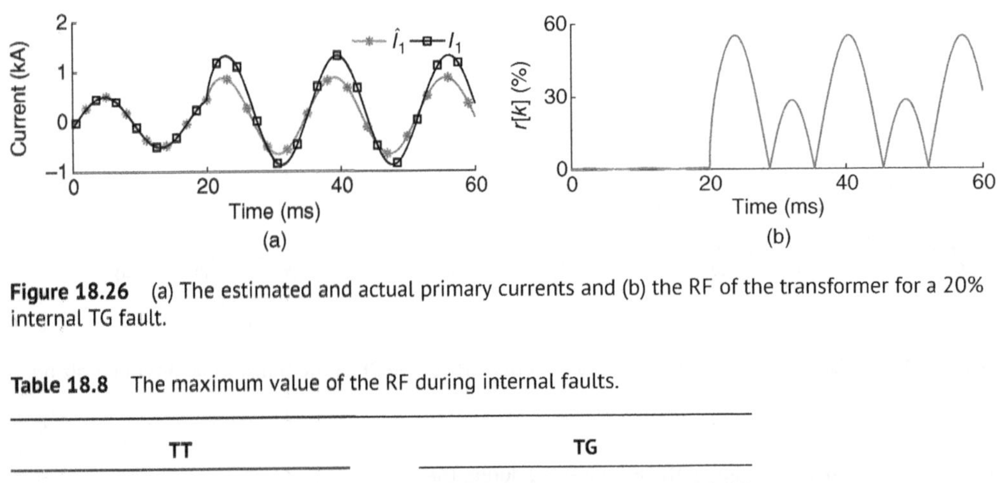

Figure 18.26 (a) The estimated and actual primary currents and (b) the RF of the transformer for a 20% internal TG fault.

Table 18.8 The maximum value of the RF during internal faults.

TT		TG	
% of total turns	**max{*r*[*k*]} (%)**	**% of total turns**	**max{*r*[*k*]} (%)**
5	4.7	5	11.5
10	9.8	10	24.3
20	21.6	20	55
30	34.2	30	93.4
40	60.5	40	130
50	99.8	50	162
75	262.5	75	193

RF also increases. This table demonstrates that the proposed auxiliary framework detects all the faults accurately.

18.8.4.3 Energizing a Faulty Transformer

This subsection analyzes energization of a faulty transformer to verify the effectiveness of the proposed framework. To this aim, at $t = 10$ ms, when the voltage angle is at 30°, the transformer is energized in the presence of 10% TT and TG faults on its wye side. The estimated and actual primary currents, as well as the corresponding RFs, are demonstrated in Figures 18.27 and 18.28. It can be observed from these figures that after energization, estimated and measured primary currents are significantly different. As a result, the RFs exceed δ, and the faults are detected. Moreover, Table 18.9 demonstrates the maximum value of RF for different fault severities. For

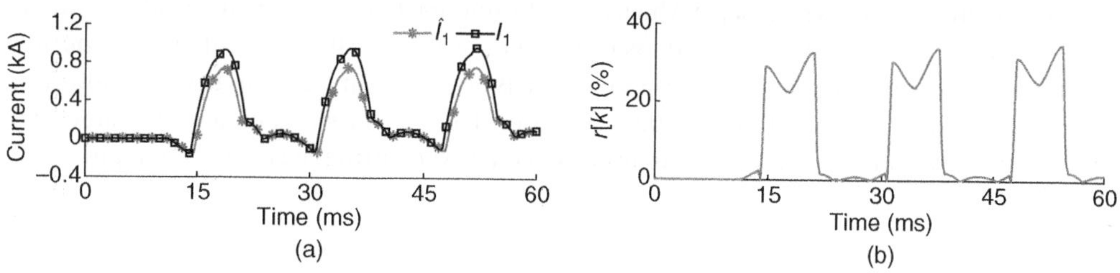

Figure 18.27 (a) The estimated and actual primary currents and (b) the RF during the process of energizing the transformer when a 10% TT fault is present.

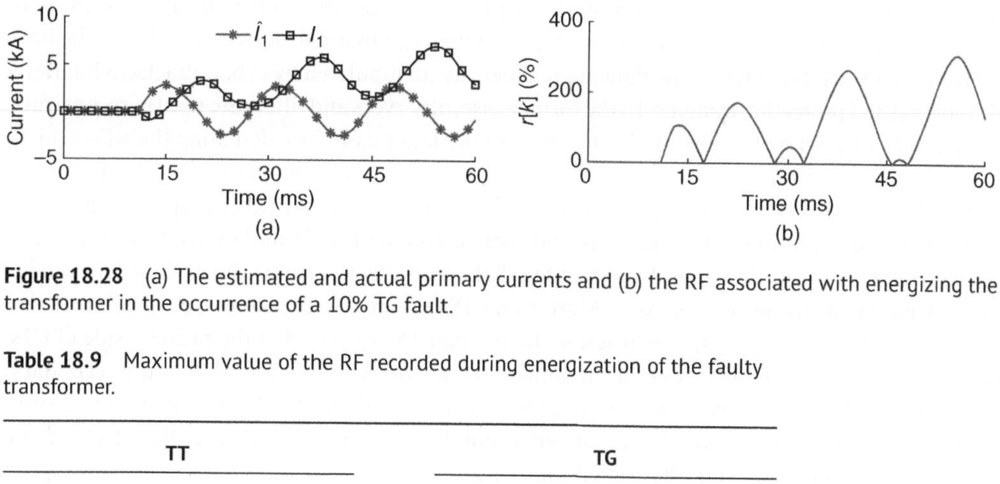

Figure 18.28 (a) The estimated and actual primary currents and (b) the RF associated with energizing the transformer in the occurrence of a 10% TG fault.

Table 18.9 Maximum value of the RF recorded during energization of the faulty transformer.

TT		TG	
% of total turns	**max{$r[k]$} (%)**	**% of total turns**	**max{$r[k]$} (%)**
5	16.49	5	259
10	34.8	10	266
20	75.9	20	282
30	114.3	30	292
40	135	40	285
50	140	50	236
75	212	75	40

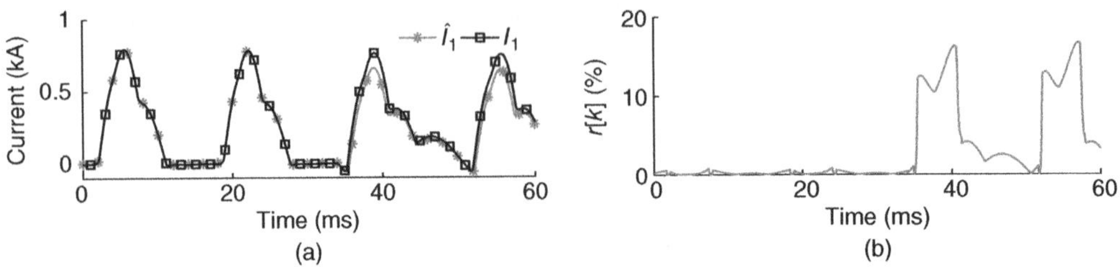

Figure 18.29 (a) The estimated and actual primary currents and (b) the RF corresponding to a 10% TT fault that occurs following the energization of the transformer amid inrush currents.

all cases, the RF surpasses the determined threshold, and all the faults are correctly detected. Moreover, Figure 18.29 shows a case, in which a 10% TT internal fault happens at $t = 30$ ms—when $\theta = 45°$—following energization of the transformer at $t = 10$ ms. As this figure illustrates, during energization and before occurrence of the fault, the RF is negligible and remains below δ. However, as soon as the fault starts, the RF significantly increases. Thus, as the simulation results show, the proposed auxiliary framework can accurately detect faults even during the energization process.

18.9 Conclusion

In this chapter, the problems of data accuracy and authenticity were addressed for some control and protection applications in power systems by properly using observers and filters. Such tools can be used to estimate the data accurately when original signals are distorted, e.g., due to the limited bandwidth of CTs, or when the data are maliciously manipulated by cyber-attacks to interfere with control and protection systems. In the former case, observers and filters are used to reconstruct the original signal using the distorted waveform. In the latter case, by calculating the RF—which is an indicator of data accuracy—cyber-attacks can be detected and differentiated from faults. To investigate the performance of these observer-based solutions, first, different types of observers were categorized, and their specifications and their mathematical formulations were discussed. Then, through a case study, it was demonstrated that cyber-attacks against the AGC system can be detected and identified using observers. Moreover, UIKFs were designed and used to enhance the accuracy of TW-based relays by constructing the original TWs received at the primary side of CTs using the distorted waveform—due to the limited bandwidth of CTs—received at the secondary side. Additionally, LPV observers were utilized for transformer differential relays in order to distinguish between inrush currents and internal faults. Simulation results corroborated the effectiveness of the developed observer- and filter-based solutions.

References

1 Ameli, A., Ghafouri, M., Zeineldin, H.H. et al. (2021). Accurate fault diagnosis in transformers using an auxiliary current-compensation-based framework for differential relays. *IEEE Transactions on Instrumentation and Measurement* 70: 1–14.

2 Ameli, A., Ghafouri, M., Salama, M.M.A., and El-Saadany, E.F. (2022). An auxiliary framework to mitigate measurement inaccuracies caused by capacitive voltage transformers. *IEEE Transactions on Instrumentation and Measurement* 71: 1–11.

3 Ameli, A., Hooshyar, A., El-Saadany, E.F., and Youssef, A.M. (2018). Attack detection and identification for automatic generation control systems. *IEEE Transactions on Power Apparatus and Systems* 33 (5): 4760–4774.

4 Antsaklis, P. and Michel, A.N. (2005). *Linear Systems*. Boston, MA: Birkhauser.

5 Ogata, K. et al. (2010). *Modern Control Engineering*, vol. 5. Upper Saddle River, NJ: Prentice Hall.

6 Ogata, K. (1995). *Discrete-Time Control Systems*, Chapter 5. Prentice-Hall International.

7 Sain, M. and Massey, J. (1969). Invertibility of linear time-invariant dynamical systems. *IEEE Transactions on Automatic Control* 14 (2): 141–149.

8 Kautsky, J., Nichols, N.K., and Van Dooren, P. (1985). Robust pole assignment in linear state feedback. *International Journal of Control* 41 (5): 1129–1155.

9 Wilkinson, J.H. (1988). *The Algebraic Eigenvalue Problem*. Oxford University Press.

10 Ghafouri, M., Karaagac, U., Karimi, H. et al. (2017). An LQR controller for damping of subsynchronous interaction in DFIG-based wind farms. *IEEE Transactions on Power Systems* 32 (6): 4934–4942.

11 Navarro-Rodrıguez, A., Garcıa, P., Georgious, R. et al. (2019). Observer-based transient frequency drift compensation in AC microgrids. *IEEE Transactions on Smart Grid* 10 (2): 2015–2025.

12 Garcıa, P., Sumner, M., Navarro-Rodrıguez, A. et al. (2018). Observer-based pulsed signal injection for grid impedance estimation in three-phase systems. *IEEE Transactions on Industrial Electronics* 65 (10): 7888–7899.

13 Abbaspour, A., Sargolzaei, A., Forouzannezhad, P. et al. (2020). Resilient control design for load frequency control system under false data injection attacks. *IEEE Transactions on Industrial Electronics* 67 (9): 7951–7962.

14 Taha, A.F., Qi, J., Wang, J., and Panchal, J.H. (2016). Risk mitigation for dynamic state estimation against cyber attacks and unknown inputs. *IEEE Transactions on Smart Grid* 9 (2): 886–899.

15 Ameli, A., Hooshyar, A., El-Saadany, E.F., and Youssef, A.M. (2020). An intrusion detection method for line current differential relays. *IEEE Transactions on Information Forensics and Security* 15: 329–344.

16 Ghahremani, E. and Kamwa, I. (2011). Dynamic state estimation in power system by applying the extended Kalman filter with unknown inputs to phasor measurements. *IEEE Transactions on Power Systems* 26 (4): 2556–2566.

17 Ameli, A., Saleh, K.A., El-Saadany, E.F. et al. (2021). Wide-band current transformers for traveling-waves-based protection applications. *IEEE Transactions on Smart Grid* 12 (1): 845–858.

18 Zhu, J.-W., Yang, G.-H., Wang, H., and Wang, F. (2015). Fault estimation for a class of nonlinear systems based on intermediate estimator. *IEEE Transactions on Automatic Control* 61 (9): 2518–2524.

19 Abbasi, A., Poshtan, J., and Moarefianpour, A. (2016). A decentralized approach based on unknown input observers for actuator fault detection and isolation of a class of interconnected nonlinear systems. *Studies in Informatics and Control* 25 (4): 454.

20 Yang, J., Zhu, F., and Zhang, W. (2013). Sliding-mode observers for nonlinear systems with unknown inputs and measurement noise. *International Journal of Control, Automation and Systems* 11 (5): 903–910.

21 Saberi, A., Stoorvogel, A.A., and Sannuti, P. (2000). Exact, almost and optimal input decoupled (delayed) observers. *International Journal of Control* 73 (7): 552–581.

22 Sundaram, S. and Hadjicostis, C.N. (2007). Delayed observers for linear systems with unknown inputs. *IEEE Transactions on Automatic Control* 52 (2): 334–339.

23 Yao, X., Le, V., and Lee, I. (2022). Unknown input observer-based series DC arc fault detection in DC microgrids. *IEEE Transactions on Power Electronics* 37 (4): 4708–4718.

24 Caliskan, F. and Genc, I. (2008). A robust fault detection and isolation method in load frequency control loops. *IEEE Transactions on Power Systems* 23 (4): 1756–1767.

25 Isermann, R. (2005). *Fault-Diagnosis Systems: An Introduction from Fault Detection to Fault Tolerance*. Springer Science & Business Media.

26 Huang, Z., Schneider, K., and Nieplocha, J. (2007). Feasibility studies of applying Kalman filter techniques to power system dynamic state estimation. *2007 International Power Engineering Conference (IPEC 2007)*, 376–382. IEEE.

27 Valverde, G. and Terzija, V. (2010). Unscented Kalman filter for power system dynamic state estimation. *IET Generation Transmission and Distribution* 5 (1): 29–37.

28 Van Der Merwe, R. and Wan, E.A. (2001). The square-root unscented Kalman filter for state and parameter-estimation. *2001 IEEE International Conference on Acoustics, Speech, and Signal Processing. Proceedings (Cat. No. 01CH37221)*, volume 6, 3461–3464. IEEE.

29 Zhou, N., Meng, D., and Lu, S. (2013). Estimation of the dynamic states of synchronous machines using an extended particle filter. *IEEE Transactions on Power Systems* 28 (4): 4152–4161.

30 Zhou, N., Meng, D., Huang, Z., and Welch, G. (2014). Dynamic state estimation of a synchronous machine using PMU data: a comparative study. *IEEE Transactions on Smart Grid* 6 (1): 450–460.

31 Rigatos, G., Serpanos, D., and Zervos, N. (2017). Detection of attacks against power grid sensors using Kalman filter and statistical decision making. *IEEE Sensors Journal* 17 (23): 7641–7648.

32 Wang, Y., Zhang, Z., Ma, J., and Jin, Q. (2022). KFRNN: An effective false data injection attack detection in smart grid based on Kalman filter and recurrent neural network. *IEEE Internet of Things Journal* 9 (9): 6893–6904.

33 Liserre, M., Pigazo, A., Dell'Aquila, A., and Moreno, V. (2006). An anti-islanding method for single-phase inverters based on a grid voltage sensorless control. *IEEE Transactions on Industrial Electronics* 53 (5): 1418–1426.

34 Ashraf, S.M. and Chakrabarti, S. (2021). A single machine equivalent-based approach for online tracking of power system transient stability. *IEEE Transactions on Power Systems* 36 (3): 1688–1696.

35 Bakhtiari, F. and Nazarzadeh, J. (2020). Optimal estimation and tracking control for variable-speed wind turbine with PMSG. *Journal of Modern Power Systems and Clean Energy* 8 (1): 159–167.

36 Xiahou, K., Li, M.S., Liu, Y., and Wu, Q.H. (2018). Sensor fault tolerance enhancement of DFIG-WTs via perturbation observer-based DPC and two-stage Kalman filters. *IEEE Transactions on Energy Conversion* 33 (2): 483–495.

37 Kitanidis, P.K. (1987). Unbiased minimum-variance linear state estimation. *Automatica* 23 (6): 775–778.

38 Darouach, M. and Zasadzinski, M. (1997). Unbiased minimum variance estimation for systems with unknown exogenous inputs. *Automatica* 33 (4): 717–719.

39 Darouach, M., Zasadzinski, M., and Boutayeb, M. (2003). Extension of minimum variance estimation for systems with unknown inputs. *Automatica* 39 (5): 867–876.

40 Gillijns, S. and De Moor, B. (2007). Unbiased minimum-variance input and state estimation for linear discrete-time systems with direct feedthrough. *Automatica* 43 (5): 934–937.

41 Pan, S., Du, P., Li, Y. et al. (2014). The study on an general Kalman filter with unknown inputs. *Proceeding of the 11th World Congress on Intelligent Control and Automation*, 3562–3567. IEEE.

42 Darouach, M., Zasadzinski, M., Onana, A.B., and Nowakowski, S. (1995). Kalman filtering with unknown inputs via optimal state estimation of singular systems. *International Journal of Systems Science* 26 (10): 2015–2028.

43 Ahmadi, A., Asadi, Y., Amani, A.M. et al. (2022). Resilient model predictive adaptive control of networked Z-source inverters using GMDH. *IEEE Transactions on Smart Grid* 13 (5): 3723–3734.

44 Ghafoori, M.S. and Soltani, J. (2022). Designing a robust cyber-attack detection and identification algorithm for DC microgrids based on Kalman filter with unknown input observer. *IET Generation, Transmission & Distribution* 16 (16): 3230–3244.

45 Shamma, J.S. (2012). An overview of LPV systems. In: *Control of Linear Parameter Varying Systems with Applications*, 3–26. Springer.

46 Anstett, F., Millérioux, G., and Bloch, G. (2009). Polytopic observer design for LPV systems based on minimal convex polytope finding. *Journal of Algorithms & Computational Technology* 3 (1): 23–43.

47 Ichalal, D. and Mammar, S. (2015). On unknown input observers for LPV systems. *IEEE Transactions on Industrial Electronics* 62 (9): 5870–5880.

48 de Oliveira, M.S. and Pereira, R.L. (2021). On unknown input observers designs for discrete-time LPV systems with bounded rates of parameter variation. *European Journal of Control* 58: 183–195.

49 Asadi, S., Vafamand, N., Moallem, M., and Dragičević, T. (2021). Fault reconstruction of islanded nonlinear DC microgrids: an LPV-based sliding mode observer approach. *IEEE Journal of Emerging and Selected Topics in Power Electronics* 9 (4): 4606–4614.

50 Casau, P., Rosa, P., Tabatabaeipour, S.M. et al. (2015). A set-valued approach to FDI and FTC of wind turbines. *IEEE Transactions on Control Systems Technology* 23 (1): 245–263.

51 Morato, M.M., Mendes, P.R., Normey-Rico, J.E., and Bordons, C. (2020). LPV-MPC fault-tolerant energy management strategy for renewable microgrids. *International Journal of Electrical Power & Energy Systems* 117: 105644.

52 Ramos, G.A., Soto-Perez, R.A., and Cifuentes, J.A. (2017). A varying frequency LPV-based control strategy for three-phase inverters. *IEEE Transactions on Industrial Electronics* 64 (9): 7599–7608.

53 Safaeipour, H., Forouzanfar, M., and Casavola, A. (2021). A survey and classification of incipient fault diagnosis approaches. *Journal of Process Control* 97: 1–16.

54 Zhong, M.-Y., Shuai, L., and Hui-Hong, Z. (2008). Krein space-based H∞ fault estimation for linear discrete time-varying systems. *Acta Automatica Sinica* 34 (12): 1529–1533.

55 Dong, H., Wang, Z., Bu, X., and Alsaadi, F.E. (2016). Distributed fault estimation with randomly occurring uncertainties over sensor networks. *International Journal of General Systems* 45 (5): 662–674.

56 Zhou, Y., Soh, Y., and Shen, J. (2014). High-gain observer with higher order sliding mode for state and unknown disturbance estimations. *International Journal of Robust and Nonlinear Control* 24 (15): 2136–2151.

57 Spurgeon, S.K. (2008). Sliding mode observers: a survey. *International Journal of Systems Science* 39 (8): 751–764.

58 Xiong, Y. and Saif, M. (2001). Sliding mode observer for nonlinear uncertain systems. *IEEE Transactions on Automatic Control* 46 (12): 2012–2017.

59 Mi, Y., Song, Y., Fu, Y., and Wang, C. (2020). The adaptive sliding mode reactive power control strategy for wind-diesel power system based on sliding mode observer. *IEEE Transactions on Sustainable Energy* 11 (4): 2241–2251.

60 Liu, Y.-C., Laghrouche, S., Depernet, D. et al. (2021). Disturbance-observer-based complementary sliding-mode speed control for PMSM drives: a super-twisting sliding-mode observer-based approach. *IEEE Journal of Emerging and Selected Topics in Power Electronics* 9 (5): 5416–5428.

61 Su, X., Liu, X., and Song, Y.-D. (2018). Fault-tolerant control of multiarea power systems via a sliding-mode observer technique. *IEEE/ASME Transactions on Mechatronics* 23 (1): 38–47.

62 (a) Xiong, R., Yu, Q., Wang, L.Y., and Lin, C. (2017). A novel method to obtain the open circuit voltage for the state of charge of lithium ion batteries in electric vehicles by using H infinity filter. *Applied Energy* 207: 346–353. (b) Yan, J., Sun, F., Choug, S.K. et al. (2017). Transformative innovations for a sustainable future –Part II. *Applied Energy* 207: 1–6.

63 Besançon, G. (2007). *Nonlinear Observers and Applications*, vol. 363. Springer.

64 Wang, X., Li, T., Sun, S., and Corchado, J.M. (2017). A survey of recent advances in particle filters and remaining challenges for multitarget tracking. *Sensors* 17 (12): 2707.

65 Zhao, J., Netto, M., and Mili, L. (2017). A robust iterated extended Kalman filter for power system dynamic state estimation. *IEEE Transactions on Power Systems* 32 (4): 3205–3216.

66 Paul, A., Kamwa, I., and Jóos, G. (2018). Centralized dynamic state estimation using a federation of extended Kalman filters with intermittent PMU data from generator terminals. *IEEE Transactions on Power Systems* 33 (6): 6109–6119.

67 Karimipour, H. and Dinavahi, V. (2015). Extended Kalman filter-based parallel dynamic state estimation. *IEEE Transactions on Smart Grid* 6 (3): 1539–1549.

68 Emami, K., Fernando, T., Iu, H.H.-C. et al. (2015). Particle filter approach to dynamic state estimation of generators in power systems. *IEEE Transactions on Power Systems* 30 (5): 2665–2675.

69 Qi, J., Sun, K., Wang, J., and Liu, H. (2018). Dynamic state estimation for multi-machine power system by unscented Kalman filter with enhanced numerical stability. *IEEE Transactions on Smart Grid* 9 (2): 1184–1196.

70 Zhao, J. and Mili, L. (2019). A decentralized H-infinity unscented Kalman filter for dynamic state estimation against uncertainties. *IEEE Transactions on Smart Grid* 10 (5): 4870–4880.

71 Naseri, F., Kazemi, Z., Farjah, E., and Ghanbari, T. (2019). Fast detection and compensation of current transformer saturation using extended Kalman filter. *IEEE Transactions on Power Delivery* 34 (3): 1087–1097.

72 Bhui, P., Senroy, N., Singh, A.K., and Pal, B.C. (2018). Estimation of inherent governor dead-band and regulation using unscented Kalman filter. *IEEE Transactions on Power Systems* 33 (4): 3546–3558.

73 Kim, J., Ko, J., Lee, J., and Lee, Y. (2017). Rotor flux and rotor resistance estimation using extended Luenberger-sliding mode observer (ELSMO) for three phase induction motor control. *Canadian Journal of Electrical and Computer Engineering* 40 (3): 181–188.

74 Linders, J.R., Barnett, C., Chadwick, J. et al. (1995). Relay performance considerations with low-ratio CTs and high-fault currents. *IEEE Transactions on Industry Applications* 31 (2): 392–404.

75 Zocholl, S.E. and Smaha, D. (1992). Current transformer concepts. *Proceedings of the 46th Annual Georgia Tech Protective Relay Conference*, volume 29, Atlanta, GA.

76 Zhao, J. and Mili, L. (2017). Robust unscented Kalman filter for power system dynamic state estimation with unknown noise statistics. *IEEE Transactions on Smart Grid* 10 (2): 1215–1224.

77 Hargrave, A., Thompson, M.J., and Heilman, B. (2018). Beyond the knee point: a practical guide to CT saturation. *2018 71st Annual Conference for Protective Relay Engineers (CPRE)*, 1–23. IEEE.

78 Law, Y.W., Alpcan, T., and Palaniswami, M. (2014). Security games for risk minimization in automatic generation control. *IEEE Transactions on Power Systems* 30 (1): 223–232.

79 Ameli, A., Hooshyar, A., Yazdavar, A.H. et al. (2018). Attack detection for load frequency control systems using stochastic unknown input estimators. *IEEE Transactions on Information Forensics and Security* 13 (10): 2575–2590.

80 Kondrath, N. and Kazimierczuk, M.K. (2008). Bandwidth of current transformers. *IEEE Transactions on Instrumentation and Measurement* 58 (6): 2008–2016.

81 SEL (2019). *SEL-411L Advanced Line Differential Protection, Automation, and Control System*. Pullman, WA: SEL. https://selinc.com/products/411L/ (accessed 16 October 2024).

82 Strunz, K., Hatziargyriou, N., Andrieu, C. et al. (2014). Benchmark systems for network integration of renewable and distributed energy resources. *CIGRE Task Force C6.04.02*.

83 Hooshyar, A., Sanaye-Pasand, M., Afsharnia, S. et al. (2012). Time-domain analysis of differential power signal to detect magnetizing inrush in power transformers. *IEEE Transactions on Power Delivery* 27 (3): 1394–1404.

84 Ashrafian, A., Rostami, M., and Gharehpetian, G.B. (2012). Hyperbolic S-transform-based method for classification of external faults, incipient faults, inrush currents and internal faults in power transformers. *IET Generation Transmission and Distribution* 6 (10): 940–950.

85 Gaouda, A.M. and Salama, M.M.A. (2010). DSP wavelet-based tool for monitoring transformer inrush currents and internal faults. *IEEE Transactions on Power Delivery* 25 (3): 1258–1267.

86 Pihler, J., Grcar, B., and Dolinar, D. (1997). Improved operation of power transformer protection using artificial neural network. *IEEE Transactions on Power Delivery* 12 (3): 1128–1136.

87 Shah, A.M. and Bhalja, B.R. (2013). Discrimination between internal faults and other disturbances in transformer using the support vector machine-based protection scheme. *IEEE Transactions on Power Delivery* 28 (3): 1508–1515.

88 IEEE PC37.110/D3 (2019). *IEEE Draft Guide for the Application of Current Transformers Used for Protective Relaying Purposes*, 1–89.

89 Sen, P.C. (2007). *Principles of Electric Machines and Power Electronics*. Wiley.

90 GE Vernova (2018). *745 - Transformer Protection System Instruction Manual*. Markham, Ontario, Canada: GE Grid Solutions. https://www.gegridsolutions.com/multilin/catalog/745.htm (accessed 16 October 2024).

19

Anomaly Detection and Mitigation in Cyber-Physical Power Systems Based on Hybrid Deep Learning and Attack Graphs

Alfan Presekal, Alexandru Ştefanov, Vetrivel Subramaniam Rajkumar, and Peter Palensky

Intelligent Electrical Power Grids, Department of Electrical Sustainable Energy, Faculty of Electrical Engineering, Mathematics and Computer Science, Delft University of Technology, Mekelweg 4, 2628 CD, Delft, Zuid Holland, The Netherlands

Abbreviations

AI	Artificial Intelligence
APDU	Application Protocol Data Unit
AUC	Area Under the Curve
CIA	Confidentiality, Integrity, and Availability
DCS	Distributed Control System
DNP3	Distributed Network Protocol 3
DoS	Denial of Service
DPI	Deep Packet Inspection
FDI	False Data Injection
FN	False Negative
FP	False Positive
GC-LSTM	Graph Convolutional Long Short-Term Memory
GCN	Graph Convolutional Network
GNN	Graph Neural Network
GOOSE	Generic Object-Oriented Substation Event
HIL	Hardware-in-the-Loop
ICCP	Inter-Control Center Communications Protocol
ICS	Industrial Control System
IDPS	-Intrusion Detection System and Prevention System
IEC	International Electrotechnical Commission
IED	Intelligent Electronic Device
IoT	Internet of Things
IT	Information Technology
LSTM	Long Short-Term Memory
MITM	Man-in-the-Middle
NGF	Next-Generation Firewall

Smart Cyber-Physical Power Systems: Fundamental Concepts, Challenges, and Solutions, Volume 1, First Edition.
Edited by Ali Parizad, Hamid Reza Baghaee, and Saifur Rahman.

OPC-UA	Open Platform Communication-Unified Architecture
OSI	Open Systems Interconnection
OSINT	Open Source Intelligence
OT	Operational Technology
PPDU	Presentation Protocol Data Unit
RNN	Recurrent Neural Network
RTDS	Real-Time Digital Simulator
SCADA	Supervisory Control and Data Acquisition
SDN	Software-Defined Networking
SIEM	Security Information and Event Management
SPDU	Session Protocol Data Unit
SV	Sample Value
TCP/IP	Transmission Control Protocol/Internet Protocol
TLS	Transport Layer Security
TN	True Negative
TP	True Positive
TPDU	Transport Protocol Data Unit
TSC	Time Series Classification
VM	Virtual Machine

19.1 Power Grid Cyber Resilience

Digitalization is paving the way toward enhanced power grid operational capabilities and intelligence. The adoption of digital technologies is essential for the advancement of the forthcoming power grid. The integration of the Internet of Things (IoT), artificial intelligence (AI), and big data analytics are encompassed within this scope. They enhance power system sustainability, affordability, and resilience. The increased digitalization, however, also implies a greater risk from cyber vulnerabilities and threats. Various power systems facets such as transmission and distribution systems, digital substations, control centers, and wide-area communication networks are vulnerable to cyber-attacks. It is widely acknowledged that the integration of information technology (IT) and operational technology (OT) systems introduces new threats and cybersecurity challenges. When it comes to ensuring the reliability of the future energy system and the security of electricity supply, there is a pressing need to give close attention to the new vulnerabilities and dangers posed by grid digitalization. Therefore, cyber resilience is essential for further digitalization of the power grid.

Cyber attacks on power systems are infrequent yet high-impact disruptions that can result in an extensive range of undesirable outcomes, e.g., load shedding, equipment damage, system instability, and power outages. The ramifications of a cyber attack on electrical power grids transcend the immediate disruptions, including cascading effects on interconnected power systems and other critical infrastructures, e.g., water supply, gas distribution, telecommunication, and transportation systems. The most notable cyber attacks on power grids are the twin attacks on the Ukrainian power grid in 2015 and 2016. These incidents clearly highlighted that cyber attacks on power grids are imminent threats that need to be addressed. Cyber attacks were conducted on the power grid in Ukraine on December 23, 2015, leading to power outages that affected approximately 225,000 customers [1]. On December 17, 2016, more advanced cyber attacks were carried out against the Ukrainian power grid. This attack led to a power outage in the distribution network, where 200 MW of load was unsupplied [2]. The capabilities of the adversaries behind these types of advanced cyber attacks pose an existential threat to the security of modern society. The emergence of cyber attacks on power systems has the potential to trigger cascading failures that can culminate in a catastrophic blackout, ultimately leading to a doomsday scenario. Furthermore, the absence of electricity has a significant impact on all social aspects, which can result in financial losses, damages, chaos, or even a loss of lives.

Extensive research on cyber attacks on power grids was conducted in recent years. We identified three main research directions to address these challenges. The first research direction enhances the security of communication protocols utilized in power grid OT systems, which is essential [3]. The second research direction is toward cyber-physical system (CPS) modeling and co-simulation using testbeds [4–7]. A simulated environment is necessary for power grid cybersecurity due to its nature as critical infrastructure with high availability requirements. Therefore, the implementation of a testbed enables researchers to safely conduct a variety of power system tests and cyber attack simulations. Finally, the third research direction is anomaly detection in power grids due to cyber attacks. It is noteworthy that the predominant focus of anomaly detection in the state of the art pertains to the detection of online attacks on power grids in the context of false data injection (FDI) attack scenarios. This line of research concentrates on analyzing power system measurements to find anomalies in power grids [8–13]. Nevertheless, the cases of cyber attacks on power grids that have been reported in [1, 2, 14] were not associated with the execution of FDI attacks. In the early stages of the cyber kill chain, attackers target IT/OT systems rather than manipulating measurement data. Hence, there is a need for anomaly detection in OT communication traffic.

This chapter provides essential knowledge of cyber attack mitigation for cyber-physical power systems, i.e., (i) secure communication protocols for operational technologies, (ii) cyber-physical co-simulation and penetration testing using cyber ranges, and (iii) network security controls and intrusion detection and prevention systems. Among the wide-scope mitigation, AI is highlighted as an emerging solution. This chapter presents how hybrid deep learning based on graph convolutional long short-term memory is used for anomaly detection in power system OT networks. Unlike traditional signature and supervised learning-based intrusion detection, hybrid deep learning anomaly detection utilizes the OT traffic throughput. It takes the advantage of the OT traffic's deterministic and homogenous characteristics to provide robust and flexible anomaly detection for a wide scope of cyber attacks. The traffic anomalies are incorporated into an attack graph that aids power system operators to identify and localize anomalies of active attacks on power systems in near real time. Cyber attack case studies and cyber-physical co-simulation results are provided to demonstrate the efficiency of hybrid deep learning for anomaly detection.

The chapter is structured as follows. Section 19.1 introduced the overview of power grid cyber resilience. Section 19.2 describes operational technologies' vulnerabilities and secure communication protocols. In Section 19.3, we present cyber-physical co-simulation and penetration testing using cyber ranges. Section 19.4 provides state-of-the-art security controls, and Section 19.5 presents hybrid deep learning for anomaly detection in power system OT networks. Case studies are presented in Section 19.6. The conclusions are discussed in Section 19.7.

19.2 Operational Technologies and Secure Communication Protocols

19.2.1 Cybersecurity of Operational Technology

The term OT pertains to computerized systems that oversee industrial operations, including but not limited to industrial control systems (ICS), supervisory control and data acquisition (SCADA), and distributed control systems (DCS) [15]. SCADA is an OT system architecture designed specifically for managing large and complex processes. It collects data from the field and transmits them to the control center. It includes a control center, local control systems, and local and wide-area communication systems. Meanwhile, the DCS is a comprehensive process control system that comprises a range of components, such as controllers, sensors, actuators, and terminals. DCS systems are typically utilized for on-site control, whereas SCADA systems are commonly employed for remote control purposes. SCADA and DCS are both included under the umbrella term known as ICS.

Typically, OTs have high uptime and availability requirements for mission-critical operations. IT systems, on the other hand, prioritize confidentiality, integrity, and availability (CIA). For OT systems, however, availability and safety have the highest priority [3]. Therefore, cybersecurity controls ensuring confidentiality and integrity may interfere with the high OT availability requirements. As a result, this conflict leads to a tradeoff between the availability and implementation of security controls in OT systems.

In [16], the author demonstrated that OT systems encounter challenges in incorporating cryptography due to the significant computational time required for cryptographic processes. For example, in the International Electrotechnical Commission (IEC) 16850 standard for OT systems, fault isolation and protection of Type 1A/P1 requires a maximum delay of 3 ms. Despite the strength of cryptographic algorithms such as 2048-bit RSA and 1024-bit DSA, the processing time of cryptographic operations, respectively, entailed a total of 61.04 and 14.90 ms. Due to time constraints, this circumstance resulted in the adoption of less secure cryptographic methods that require less computational time, or in most applications, the complete absence of cryptographic measures. Consequently, this situation led to cybersecurity implementation challenges in OT systems compared to IT systems.

According to [17], it has been proposed that the optimal approach for ensuring cybersecurity best practices is to maintain an air gap between the OT and IT systems. However, in recent years, there has been a growing trend toward the IT/OT convergence [16]. Several contemporary IT-based solutions, such as virtualization technology, software-defined networking (SDN), cloud services, and edge computing, are gradually being incorporated into OT systems. These technologies are double-edged swords that offer benefits and introduce potential vulnerabilities at the same time. Therefore, it is crucial to address the potential threats and vulnerabilities that arise in the convergence of IT and OT.

19.2.2 Secure Communication Protocols

In order to successfully mitigate the threat of cyber attacks on power grids, it is important to first understand the relationship between computer networking and cybersecurity. Figure 19.1 presents the mapping between communication network layers and associated cyber threats and countermeasures, based on the well-known open systems interconnection (OSI) seven-layer and transmission control protocol/internet protocol (TCP/IP) four-layer models. The seven-layer OSI abstraction explains the flow of data in computer networks as bits in the physical layer, frames

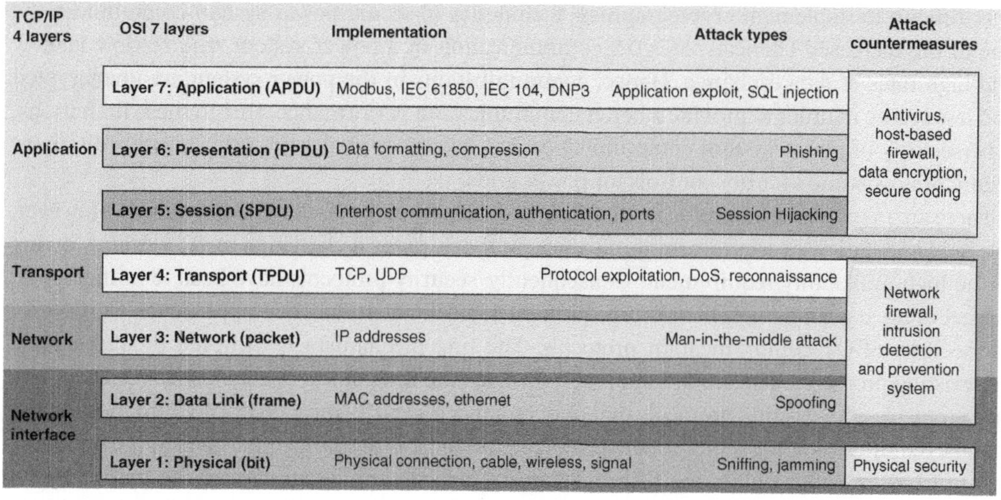

Figure 19.1 Mapping of OSI layers, cyber attacks, and mitigation techniques.

in the data link layer, packets in the network layer, transport protocol data units (TPDUs) in the transport layer, session protocol data units (SPDUs) in the session layer, presentation protocol data units (PPDUs) in the presentation layer, and finally as application protocol data units (APDUs) in the application layer. SCADA communications typically uses APDUs to deliver the payloads, i.e., measurements and controls. Information exchange and delivery is done either through network layer or data link layer. Layer 2 communication is limited to the confines of a substation where the data are exchanged as a frame. Meanwhile, layer 3 communication is used for communication between the substations and control center. Layer 3 communication uses the TCP/IP stack and network routing mechanisms to deliver information.

Figure 19.1 also shows the attack types for each layer of the OSI model and its associated countermeasures. The physical layer is prone to attacks such as sniffing and signal jamming. A suitable solution to protect layer 1 is by using physical security such as physical protection of cable connections.

Information exchange at layer 2 uses physical addresses to identify hosts. This is typically implemented at substations, employing a broadcast mechanism for information delivery. Due to this situation, layer 2 communication is prone to spoofing attacks. Attackers can observe all communication traffic in the network and mimic legitimate traffic to launch a spoofing attack. On the other hand, layer 3 communication works based on internet protocol (IP) addresses. Unlike layer 2, the network layer is a closed-loop communication from source to destination using IP addresses and routing mechanisms. This form of communication is typically used between substations and the control center through a wide area network. However, layer 3 is vulnerable to man-in-the-middle attacks. Attackers can perform IP spoofing to mimic legitimate IP addresses for a successful man-in-the-middle attack. Layer 4 is the transport layer that defines communication protocols. Attacks on this layer mainly exploit protocol operations. For example, the TCP sync mechanism can be exploited to launch a denial of service (DoS) sync flood attack. In order to protect layers 2, 3, and 4, security mechanisms such as network firewalls and intrusion detection and prevention systems can be applied.

For power system communication, typically only layer 7 from the upper layers is used wherein the APDU stores traffic payload. Layers 5 and 6 are typically not used. This is due to the limitation of advanced security implementations in the application layers of power system communications.

It is difficult to implement cryptographical techniques to secure power system communications due to the increased latencies. SCADA communication in a power system requires low latency and high rates of data exchange. Hence, communications in the power system are unencrypted and less secure in order to provide a better communication performance. Due to these limitations, cybersecurity of power system communication has become a vital issue. This chapter discusses secure protocols and security controls for power grids.

There are many standard protocols that have been deployed for power grid operations. However, the implementation of secure communication protocols poses a challenge in OT systems, owing to the high-availability requirement. Consequently, security protocols have been identified to be critical areas requiring significant improvement [3]. We identified five approaches to improve the security of OT communication protocols. The first mechanism is achieved by altering the pre-existing protocols. The second approach involves the integration of established legacy power grid protocols with existing protocols that offer enhanced security measures. The third mechanism is achieved by developing a brand new protocol. The fourth mechanism pertains to the enhancement of key exchange, while the fifth mechanism involves the integration of the protocol with blockchain technology. Figure 19.2 summarizes the secure OT protocol research directions.

The first mechanism proposed an alteration of the existing protocols. The authors in [18] carried out a study utilizing formal methods to examine potential authentication vulnerabilities present in distributed network protocol 3 (DNP3). Upon the identification of vulnerabilities, the authors subsequently suggested the implementation of security enhancements for the DNP3 secure authentication broadcast [19]. The conventional implementation of DNP3 employs a broadcast mechanism for the purpose of verifying the authenticity of communication that is transmitted between the master and remote station. The default broadcast mechanism sends information arbitrarily without a well-defined mechanism. This mechanism may lead to potential vulnerabilities like a man-in-the-middle attack, modification, replay, and injection attacks. The research in [19] proposes a modification of the DNP3 secure authentication broadcast message and checks the validity of the established connection. The proposed solution improves the efficiency and enhances the resiliency of DNP3 broadcast messages against man-in-the-middle attacks. In [20], the authors describe the secure DNP3 protocol with an additional authentication mechanism for enhancing communication integrity. An authentication challenge is issued by the slave when the master station requests a "write" message. The master station sends an authentication response. The slave confirms with acknowledgment and response messages. At this stage, it is inferred that the master station is recognized as a trustworthy and legitimate entity. The authors in [20] also present the security enhancement of the inter-control center communications protocol (ICCP) through the utilization of digital certificates to improve communication integrity.

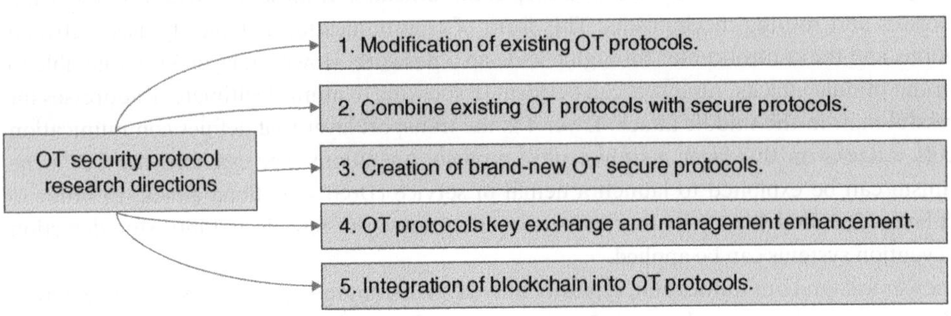

Figure 19.2 Summary of secure protocol research and classification.

In the second direction, there is already research being done with the intention of using a combination of existing protocols to put the approach into practice. Authors in [21] proposed the utilization of Modbus communication via transport layer security (TLS) protocol to create a secure communication channel. The Modbus protocol is a conventional communication standard utilized in power grid systems that lacks security mechanisms. Meanwhile, TLS is considered a broadly adopted mechanism for facilitating secure communication through the use of encrypted data. The proposed mechanism involves the encapsulation and encryption of Modbus information within a TLS packet. The aforementioned mechanism necessitates the process of encapsulating and subsequently de-encapsulating data. Therefore, this approach shows that it is possible to implement power grid communications utilizing pre-existing security protocols.

Instead of modifying existing protocols, the third direction is to create new protocols and standards. An example of a new standard is open platform communication-unified architecture (OPC-UA), which replaces the previous versions of OPC through the integration of cryptographic and authentication mechanisms [20]. Another example is IEC 62351 which aims to mitigate cybersecurity concerns in current protocols via the implementation of cryptographic techniques [22]. Nevertheless, the deployment of cryptographic techniques presents several obstacles. One of the foremost challenges is related to the distribution of keys. Therefore, it comes to the fourth approach using key exchange and management enhancement. Key exchange and management have been identified as a challenge in the SCADA system [23]. Numerous key exchange and management schemes have been suggested to enhance the security of SCADA communication. However, a comprehensive solution to this issue cannot be achieved through a silver bullet solution. The proposed solutions inevitably entail a tradeoff between real-time availability and security. The authors in [24] propose a scheme for the pre-distribution of SCADA network keys. The secret key is transmitted over the untrusted network using a pre-distributed matrix-based key. Each device generates unique keys using an algorithm for key generation based on a preliminary matrix reference. This mechanism prevents a man-in-the-middle attack against the key. Unfortunately, if attackers successfully compromise a device, they may still be able to circumvent the secure communication process.

The fifth proposed solution for enhancing security in power grid communications involves the implementation of blockchain technology. Data in the blockchain are stored in the form of a chain of information to preserve integrity [25]. The authors in [26] present diverse potential applications of blockchain technology in the context of power systems. The primary purpose of blockchain technology is to enhance credibility and safeguard the confidentiality of transactions within the energy sector. The proposal of utilizing blockchain technology to enhance the security of message exchange protocols in ICS was proposed in [27]. It is anticipated that blockchain technology will enhance the mechanisms for protocol identification, methods for authentication, and chain of encrypted information. This type of scenario could be appropriate for limited message transmissions. Nevertheless, the communication traffic of power grids primarily comprises telemetry and measurement data that exhibit a high volume of traffic. Therefore, the implementation of blockchain remains challenging, and there is currently no practical implementation of blockchain to improve the security of power grid communication protocols.

To summarize, the implementation of the first and second mechanisms represents a straightforward approach to promptly enhance the security of power grid communication protocols. These solutions exhibit a high degree of elegance in addressing deficiencies pertaining to data encryption and authentication in legacy power grid protocols. Nevertheless, these mechanisms may lack reliability due to the absence of inherent security within the protocols. The fourth and fifth mechanisms have the potential to serve as alternative solutions for augmenting the key exchange and authentication aspects of the protocol. Nevertheless, similar to the aforementioned alternatives, these approaches are not inherently incorporated within the existent protocols. Therefore, the third mechanism has the potential to emerge as a viable alternative for enhancing protocol security over a longer time frame. New security standards, e.g., IEC 62351, provide guidelines and requirements for implementing security measures to protect the operation and data exchange within OT systems, including protection against cyber threats and unauthorized access. Unfortunately, the implementation of new protocols is a time-intensive process. Moreover, the implementation of new protocols does not always guarantee high reliability and security. For instance, in [28], it was demonstrated that IEC 62351 is still susceptible to resource exhaustion attacks.

19.3 Cyber-Physical System Co-Simulation and Cyber Ranges

A power grid is an example of critical infrastructure that requires a high level of availability. Conducting experiments on actual power grids is a challenging task owing to their stringent operational requirements. Therefore, cyber-physical system (CPS) modeling and simulation are essential components of the research. In this section, we classify the CPS modeling and simulation into two parts. The first part provides an overview of power grid co-simulation testbeds. Meanwhile, the second part elaborates on the integration of cyber ranges in the CPS testbed.

19.3.1 Cyber-Physical Power System Co-Simulation

The utilization of CPS modeling and simulation provides significant importance in the domain of power system resilience research. Many survey papers concerning the current state of the art in smart grid modeling can be found in [4–7, 29]. This section focuses on CPS models with cybersecurity capabilities. The CPS modeling framework comprises two primary components, i.e., the power systems and IT/OT systems. Table 19.1 provides a summary of the CPS model simulators utilized in power systems. There are many power system simulators currently available, including but not limited to real-time digital simulator (RTDS), OPAL-RT, Typhoon HIL,

Table 19.1 Cyber-physical system models for power systems research.

Cyber-physical system	Power system simulator	IT/OT simulator	Protocols
Testbed for analyzing security of SCADA control systems (TASSCS) [30]	Software based	OPNET	DNP3, IEC 61850, OPC UA
SCADASim [31]	Software based	OMNeT++	DNP3, Modbus
Washington State University [32]	RTDS	Mininet, Core	IEC 61850, Modbus, DNP3
DeterLab [33, 34]	Software based	Virtual machine	—
Idaho CPS smart grid cybersecurity testbed (ISAAC) [35]	RTDS	Real hardware	IEC 61850, IEEE C37.118, DNP3
SCEPTRE [36]	PyPower, OpenDSS, PowerWorld	Virtual machine	—

DIgSILENT PowerFactory, GridLab-D, OpenDSS, Siemens PSS/E, Homer, Cymdist, PSAT, and MATPOWER. Numerous communication network simulators are also available, including NS2, NS3, OPNET, OMNeT++, network-based environment for modeling and simulation (NetSim), new sampling/sensor initiative (NeSSi), DeterLab, and Mininet. Therefore, there are numerous potential combinations of power systems and communication network simulators for the purpose of modeling the cyber-physical power system.

According to the state-of-the-art literature review [4–7, 29], RTDS has emerged as the pre-eminent simulator for power systems. RTDS is a computational tool that enables the simulation of power systems in real time, allowing for the accurate representation of the dynamic behavior of these systems in synchronization with the actual system time. This capability is important in the context of testing and validating control systems, protection schemes, and other applications that require timely execution. In the meantime, for IT/OT communication networks, the majority of organizations are moving toward adopting a virtual environment that is based on virtual machines (VMs). Over the past ten years, there has been a rise in alternative communication network simulators for the CPS model of power grids, including OPNET [30, 37, 38], OMNET++ [31, 39], and network simulator 2 (NS2)/network simulator 3 (NS3) [40]. Nevertheless, the fidelity of these simulators is inferior when contrasted with the virtual environment.

In summary, the communication network simulators utilized for CPS modeling of power grids can be classified into four different categories. They are (i) code-/script-based, (ii) software-based, (iii) virtualization-based, and (iv) real hardware implementation. Figure 19.3 displays the clustering and categorization for each respective category. In Figure 19.3, each category is evaluated according to its scalability and level of fidelity. It would be preferable for the CPS model to have higher scalability as well as fidelity. The most realistic and least scalable form of simulation is real hardware. The most scalable simulators, meanwhile, are code-based simulators. Code-based simulators enable the simulation of a network at a large scale. However, the code-based needs to specify what constitutes communication and it requires to manually specify each type of communication functionality in the code. Furthermore, unlike in a real system, the communication process is not

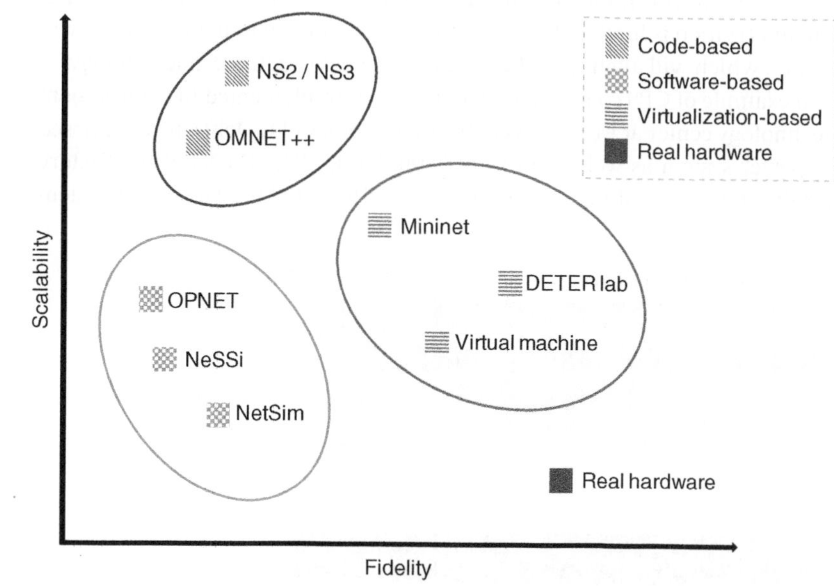

Figure 19.3 Comparison of communication network simulators for CPS modeling.

natural. The subsequent category pertains to simulators that are based on software. The low scalability and low fidelity of these particular simulators leave it a less desirable alternative. Considering the aforementioned factors, it is very likely that the optimal choice for simulation would be based on virtualization.

A VM-based simulator is likely to provide an environment that is nearly identical to that of real hardware. It also can be more scalable than real hardware through hardware virtualization techniques using hypervisor. For instance, DETERLab is classified as a VM because it consists of a cluster of VMs. The other option is Mininet, an operating-system level virtualization, which works based on the Linux namespace over containerization. In contrast to VMs, containers employ virtualization to encapsulate the operating system (OS) and application dependencies, thereby allowing for the sharing of the host OS kernel across multiple containers. In summary, it can be concluded that the most suitable alternatives for communication network simulation are those based on VM and container technologies, as they offer an optimal equilibrium between scalability and high fidelity.

The differences between an application running on actual physical hardware, VM, and containerization are illustrated in Figure 19.4. When compared to actual hardware, VM allows us to run applications in a more isolated manner within the operating system (OS). This feature enables users to simulate a greater number of virtual environments within the IT/OT network. However, as illustrated in Figure 19.4, the VM was required to install the guest OS on top of the host OS. The scenario involving the stacking of OSs is known to significantly consume a substantial amount of resources. To address this challenge, operating-system level virtualization through containerization applications such as Docker and Linux-based namespace have experienced an increase in popularity in the past few years [41]. One of the reasons for this is that they are able to deploy applications directly on top of the host OS by utilizing an isolation mechanism, which optimizes the utilization of available resources. In addition, the utilization of containers enables users to emulate a greater number of hosts and larger networks in comparison to VMs. Due to the aforementioned factors, OS virtualization solutions may become the most suitable network communication simulation tool for modeling power grid CPS. However, the current implementation of power grid CPS models developed through containerization is limited. It is likely that the number of implementations will increase in the near future, which will align with the development of virtualization technology.

Figure 19.5 depicts an example of CPS co-simulation architecture implemented in control room of the future (CRoF) technology center at Delft University of Technology (TU Delft). It is composed of a simulation of the power system as well as an IT/OT simulation. DIgSILENT PowerFactory and RTDS are used for the simulation of the power system, i.e., IEEE 39-bus. The power system

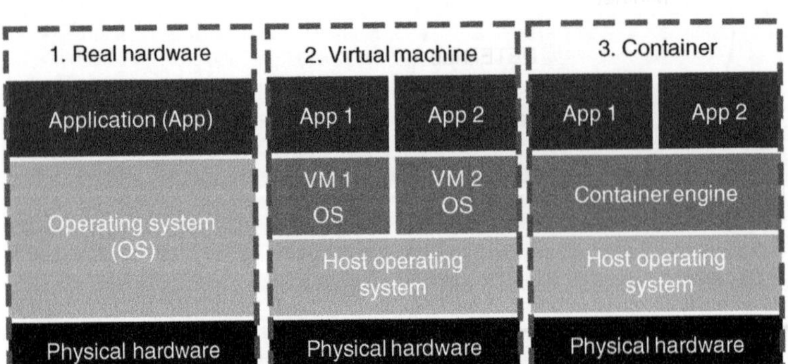

Figure 19.4 Comparison of real hardware, virtual machines, and container-based systems.

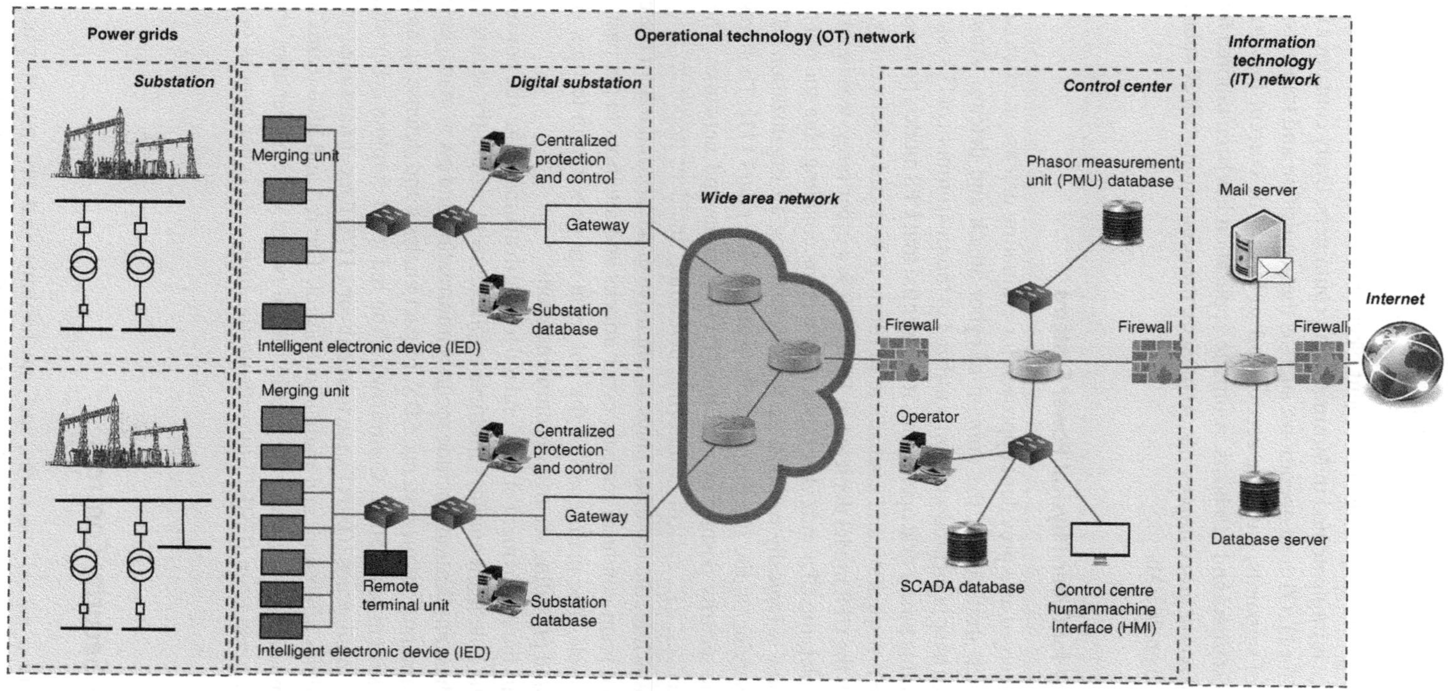

Figure 19.5 CPS architecture in CRoF at TU Delft.

model provides circuit breaker status and measurement data of active and reactive powers; voltages; and currents from busbars, lines, and generators. The implementation of OPC-UA facilitates the interfacing of data exchange between power grids and IT/OT simulation. The implementation of the IT/OT architecture is carried out through the application of Mininet. Each host in the IT/OT network, e.g., merging units, intelligent electronic devices (IEDs), network switches, routers, databases, etc., is implemented in Mininet using containers. Every container incorporates a tailored application for IT/OT host operations, such as the acquisition and transmission of measurement data, control setpoints, database access, and so forth. The current implementation of CPS comprises 27 substations and 210 hosts. A unique application has been tailored for each host to replicate the CPS of power grid components. At present, the simulation of all 27 substations runs on 50,000 lines of code on 26 VMs.

19.3.2 Cyber Range for Cyber-Physical Power Systems

Cyber ranges have emerged as a prevalent approach for evaluating defense mechanisms and simulating potential attack strategies in the domain of cyber attack and defense simulations [42]. Typically, cyber ranges have been predominantly utilized in the environment of IT systems. In order to align with forthcoming power grid operations, it is essential that CPS models possess cyber range capabilities to enable investigation and assessment of future power grid cybersecurity.

In accordance with the CPS model depicted in Figure 19.5, a cyber range was incorporated into CRoF. Figure 19.6 depicts the CPS and cyber range architecture, enabling blue and red teams experiments. The blue team is typically responsible for safeguarding an organization's IT/OT assets and infrastructure, serving as the internal security team or defenders [43]. Their responsibility entails upholding the security posture of both the IT/OT systems and networks. The blue team has several key objectives, e.g., system monitoring, defending, incident response, and cybersecurity assessment. Meanwhile, the red team plays an offensive or adversarial role in the cyber range exercise [43]. The red team conducts realistic cyber attacks and attempts to get past the organization's security controls. The main goals of the red team are penetration testing, vulnerability analysis, reporting, and providing security recommendations.

Figure 19.7 presents the deployment of blue and red team's instruments for the power grid IT/OT systems in CRoF. The blue team employs multiple applications to ensure the secure operation of the power system. These applications include security information and event management (SIEM), intrusion detection and prevention systems, SDN, impact analysis and defense against cascading failures, and power system restoration. Contrariwise, the red team employs cyber attack tools to execute open source intelligence (OSINT), payload delivery, IT/OT reconnaissance, lateral movement, response function inhibition, and malicious control. The red and blue teams are engaged in a cyber range competition to evaluate the capabilities of power system operators and computer security incident response team (CSIRT) to mitigate the impact of cyber attacks on power grid operations.

19.4 Network Security Controls

Security controls are a set of measures and mechanisms that are put in place to ensure the protection of information systems from potential threats, vulnerabilities, and unauthorized access. Security controls have been devised with the purpose of reducing potential hazards

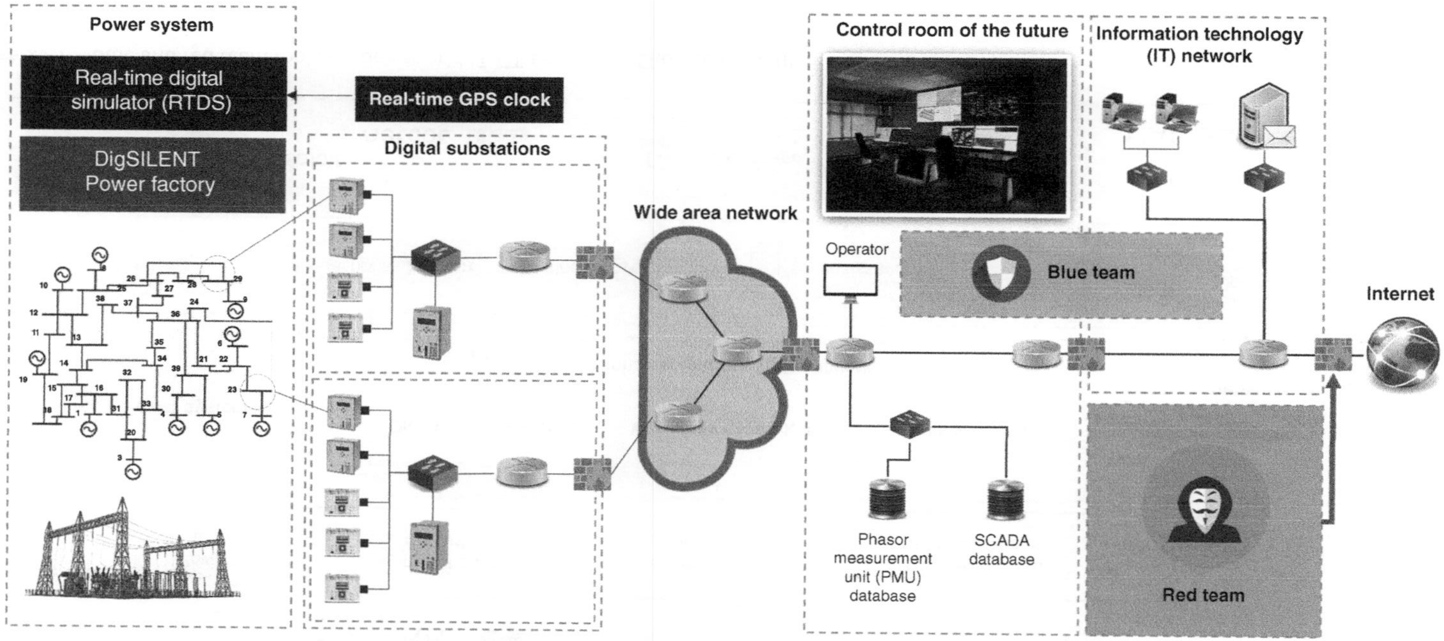

Figure 19.6 CPS and cyber range architecture of CRoF at TU Delft.

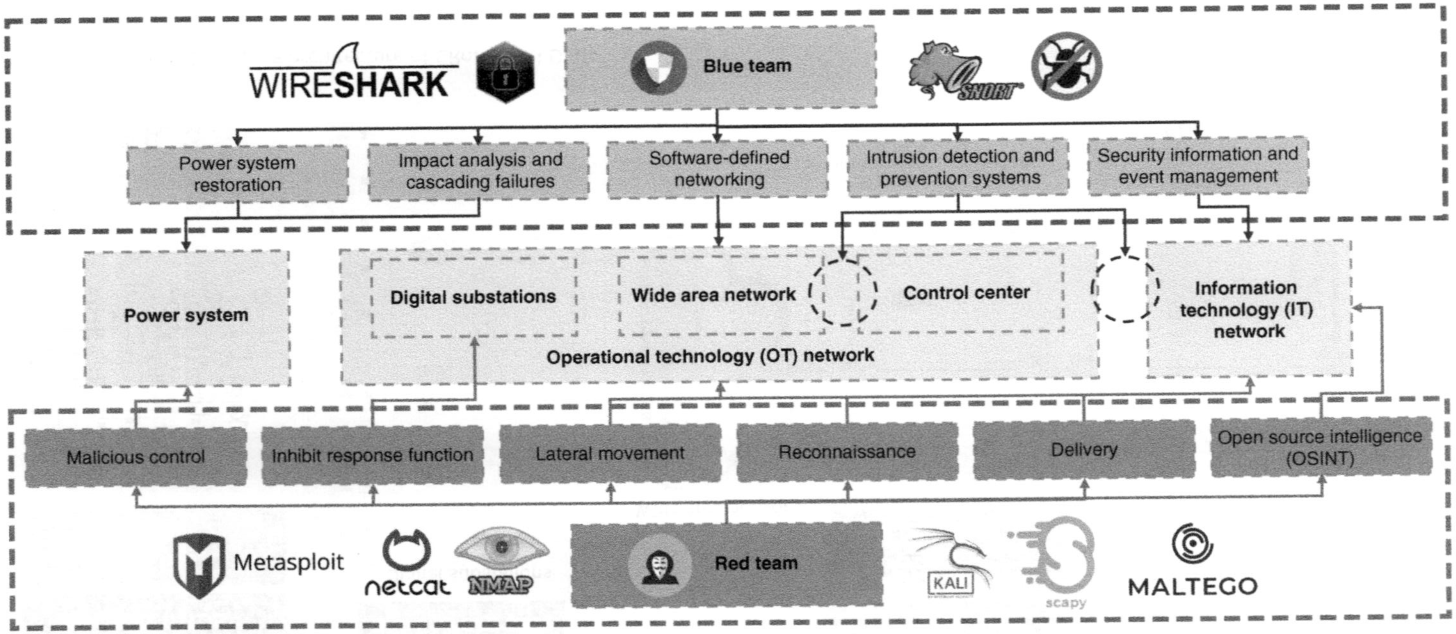

Figure 19.7 Blue and red team tools for power grid IT/OT systems in CRoF at TU Delft.

Table 19.2 Summary of network security control applications.

Security control	Methods	Protocols	References
Firewall	Packet filtering	DNP3	[44]
		Modbus	[45]
	Next-generation firewall/deep packet inspection	Not specified	[46]
		IEC 104	[47–49]
IDPS	Signature based	Not specified	[50, 51]
		IEC 104	[52, 53]
		Modbus	[54–57]
		DNP3	[54, 58]
		Siemens S7	[59]
		IEC 61850	[60–63]
		IEEE C37.118	[64]
	Anomaly based and AI based	Not specified	[65–82]
		IEC 104	[83]
		DNP3	[84–86]

and guaranteeing the confidentiality, integrity, and accessibility of both data and resources. This section discusses the state-of-the-art research conducted on network security controls for power grids, which are divided into two categories, i.e., firewalls and intrusion detection system and prevention systems (IDPS). The summary of network security controls is provided in Table 19.2.

19.4.1 Firewalls

The firewall was initially designed to operate predominantly through conventional IT systems. However, the implementation of a firewall is also a viable measure for enforcing security controls for power grids. In [44], a proposal was made for a Linux-based firewall modification intended for use in power grid applications. The Linux OS features a firewall application that is configured through the implementation of iptables rules. Iptables enables the user to designate IP address origin and destination, port, and packet type for inclusion in either a blacklist or whitelist reference. Furthermore, the study suggests the utilization of an extra 32 bits of header data derived from the DNP3 protocol. The decision to filter is made using 32 bits of information extracted from DNP3 packets. In [45], another variant with a comparable filtering mechanism was proposed for the Modbus protocol. In general, implementing security measures based on firewalls represents a straightforward approach to safeguarding communication networks for power grids. The firewall operates on predetermined rules that are hardcoded and subsequently applies these rules to filter packets accordingly. Unfortunately, a firewall is considered inadequate for dealing with advanced cyber attacks. By utilizing advanced methods of attack, adversaries may circumvent the static firewall rules.

Another type of firewall known as next-generation firewall (NGF) is equipped with the capacity to perform deep packet inspection (DPI). DPI enables NGF to not only inspect the header information of a packet, but also to inspect the contents and contextual information of the packet payload. Several studies have suggested the utilization of DPI applications for enhancing security measures in power grids. For instance, the DPI application for IEC 104 protocol is researched in [47–49] and other OT protocols in [46]. NGF exhibits superior performance when compared to traditional packet-filtering firewalls. Prior knowledge of the traffic is a prerequisite for NGF to effectively execute traffic classification and filtering. Consequently, NGF exhibits limitations in its ability to identify anomalies from new types of cyber attacks.

19.4.2 Intrusion Detection and Prevention Systems

IDPS is a security mechanism that was specifically developed to identify and counteract any malicious actions or unauthorized entry attempts that may occur within an IT/OT system. The operational mechanism involves the monitoring of network traffic, system events, and user activities with the aim of detecting potential security breaches or policy violations. In general, there exist two primary classifications of IDPS, namely signature based and anomaly based.

A signature-based IDPS operates by utilizing a predetermined set of information, i.e., signatures for known cyber attacks, for classifying the network traffic. Numerous studies have been carried out related to the utilization of signature-based IDPS in various power-systems-related communication protocols. These include IEC 104 [52, 53], Modbus [54–57], DNP3 [54, 58], Siemens S7 [59], IEC 61850 [60–63], and IEEE C37.118 [64]. Additionally, certain implementations have been developed for carrying out general OT protocols as described in [50, 51].

An alternative type of IDPS runs through the application of anomaly detection techniques. Rather than depending on pre-defined attack signatures, this approach establishes a standard baseline for typical behavior for the network, systems, and users' activities. The system continuously observes network traffic and system events, seeking out any deviations or anomalies from the normal pattern. An alert is generated if an activity or behavior deviates significantly from what is considered normal. An anomaly-based IDPS is an effective method for detecting previously unseen or zero-day attacks and advanced attack techniques.

Statistical analysis, expert systems, and AI are three techniques that can be employed to identify an anomaly. In recent years, the AI-based technique gained more attention. In general, AI-based methods can be subdivided into machine learning and deep learning. Prior studies have proposed the application of machine learning techniques for IDPS in power grids. The vast majority of the research focuses on IDPS in general and does not address any specific OT protocols [65–82]. Some of them also implement anomaly-based IDPS for specific protocols, e.g., IEC 104 [87], IEC 61850 [83].

Deep learning is a subset of machine learning that involves more complex neural network layers and higher computing demands. Some of the popular deep learning models include convolutional neural network (CNN), recurrent neural network (RNN), long short-term memory (LSTM), and graph neural network (GNN). In [85], the authors proposed IDPS based on deep learning to classify DNP3 traffic. The traffic is classified into four categories, i.e., normal, DoS attack, unsolicited attack, and cold restart attack. Another example, CNN-based attack detection for DNP3 protocol was proposed in [84]. More deep-learning-based IDPS examples are provided in Table 19.2. Although deep learning requires more computational resources, it outperformed traditional machine learning in terms of performance. As a result, the majority of IDPS research in recent years has focused on applications of deep learning.

19.5 Hybrid Deep Learning for Anomaly Detection in Power System OT Networks

This section provides an anomaly-based IDPS solution using hybrid deep learning for power grid OT systems. The vast majority of deep-learning-based IDPS is mainly focused on IT system applications [86, 88–90]. Despite the integration of a utility's IT and OT systems, the traffic patterns exhibit distinctive characteristics. The network traffic in OT systems is generated from automated processes that exhibit deterministic and homogenous behavior, whereas the network traffic in IT systems is composed of user-generated data that exhibit a stochastic behavior [91]. Consequently, the deployment of traffic-based anomaly detection in OT systems differs from IT. In order to solve this challenge, hybrid deep learning techniques are used to develop an IDPS for OT systems.

Deep-learning-based IDPSs are encountering challenges due to their reliance on training datasets for their objectives of anomaly detection and classification. Consequently, it cannot detect new or unknown types of cyber attacks. In order to fill this gap, rather than relying on data that have been specifically labeled for each type of attack, quantitative anomaly is used. The OT communication traffic throughput is utilized in quantitative anomaly detection. The quantification of throughput is represented as a time series, resulting in a distinctive waveform pattern, as demonstrated in [92–94]. Therefore, rather than classifying specific attack types or sequences, the time series traffic flow throughput is classified into two categories, i.e., normal and anomalous. The following subsections provide more detailed explanations of hybrid deep learning for anomaly detection in power system OT networks. They are classified into three parts including wide-area monitoring for OT networks, a hybrid deep learning model for anomaly detection, and attack graph methods for power-system-wide situational awareness in near real time.

19.5.1 Wide-Area Monitoring of OT Networks

The implementation of wide-area monitoring is needed for the purpose of observing traffic behavior within the control center and substation OT networks. Wide area OT traffic monitoring for power grids can be enabled by using SDN. The SDN networking paradigm is founded on the principles of network virtualization and the separation of data and control planes [86]. Figure 19.8 represents the SDN architecture for power grids consisting of three different abstraction layers. These layers are referred to as the data plane, control plane, and management plane. The conventional OT communication networks are represented by the data plane, whereas the control plane provides control capabilities over the data plane. The deployment of various network applications, such as routing algorithms, load balancers, IDPS, attack graph models, and so on, is made possible by the SDN management plane. While SDN is a relatively new concept in computer networking, prior studies have explored its application in cyber-physical power systems, as evidenced by other research [88–92].

Previous studies have utilized SDN to detect anomalies by relying on traffic flow data [93, 94]. However, these works do not aim to identify anomalies caused by cyber attacks in OT networks. Furthermore, an analysis that is critical in nature of the state-of-the-art techniques for detecting anomalies in communication traffic indicates the following. (i) Current SDN applications designed for CPSs lack emphasis on securing OT networks against cyber threats [89–94]. (ii) The rules governing them are exclusively developed on packet flow [94]. (iii) The cyber kill chain is disregarded and stealthy cyber attacks are not taken into account [93, 94].

SDN can be used to perform real-time monitoring of network traffic that originates from the data plane of the power system OT networks. In addition, the primary emphasis of this research is

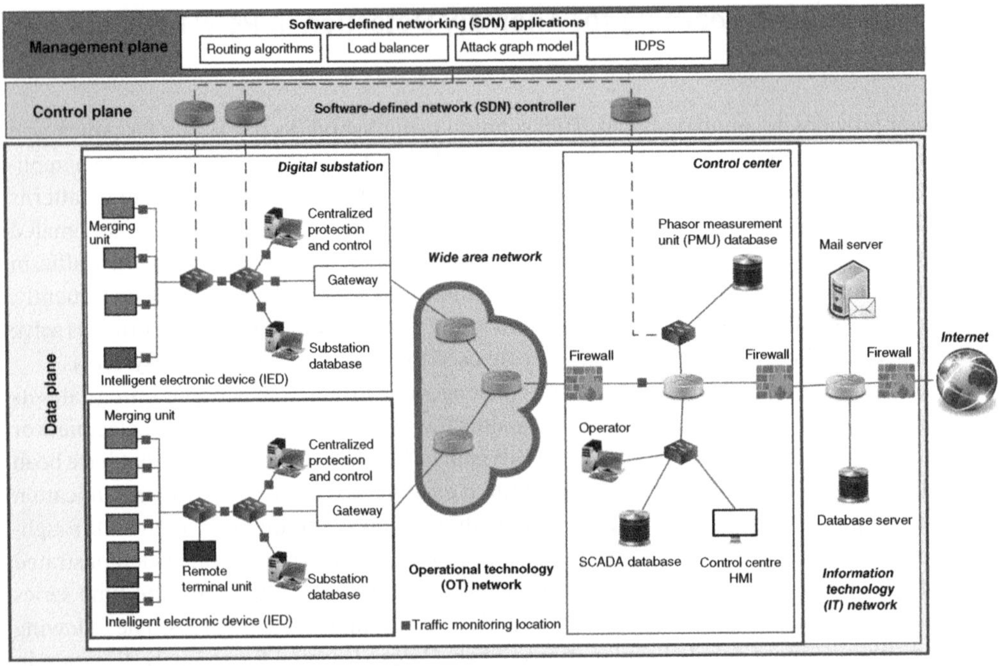

Figure 19.8 SDN architecture for power grid OT networks.

placed on the detection of anomalies during the early stages of the cyber kill chain in order to minimize the severity of the impact during the later stages. Network virtualization enables the SDN controller to monitor and control network traffic as well as implement custom network applications. SDN enhances monitoring and control of OT networks by gathering communication traffic reports in the control center. The traffic observation points are depicted as small red squares that are dispersed throughout the substations and control center. Using these observation points, the real-time OT network traffic is monitored from the control center in order to detect traffic anomalies at each substation and generate an attack graph in near real time. Spatial-temporal data are obtained by collecting OT network traffic throughputs for each observation point. This data are subsequently utilized for hybrid deep-learning techniques.

19.6 Hybrid Deep Learning Model for Anomaly Detection

Previous research has investigated the detection and classification of anomalies using time series data [95–98]. The state-of-the-art time series classification (TSC) techniques have been built on deep learning models [97, 98]. Nevertheless, their efficacy in identifying stealthy attacks is limited due to their inability to detect small changes in network traffic throughput. Furthermore, these techniques exhibit poor performance owing to the presence of imbalanced data, as evidenced by their F1 and geometric mean scores. Therefore, a hybrid deep learning approach can be used to tackle these challenges in detecting anomalies in the traffic of power grid OT networks. The hybrid model employs CNN, graph convolutional network (GCN), and LSTM. The methodology utilizes unsupervised learning techniques to acquire knowledge on the intricate patterns of OT network traffic throughput, and supervised learning techniques to accurately classify the OT traffic.

The proposed method uses graph convolutional long short-term memory (GC-LSTM) to learn the traffic behavior of the OT network. Two machine learning models are applied in GC-LSTM, i.e., GCN and LSTM. The GCN utilizes graph-based representations of the OT network topology and incorporates localized features from neighboring communication nodes in the spatial domain. Subsequently, the LSTM will carry out temporal learning based on the time-series data of the observed OT network traffic. The integration of GCN and LSTM confers the benefit of acquiring knowledge from both the spatial and temporal domains. Several applications utilizing spatial and temporal models based on graphs have been proposed in [99–102].

This chapter presents CyResGrid [103], an innovative approach for predicting nodal features by leveraging the communication network topology and characteristics of neighboring graph nodes based on the OT traffic observation locations. CyResGrid processes depicted in Figure 19.9 consist of four stages, i.e., (i) GC-LSTM training and traffic dispersion graph (TDG), (ii) CNN training for TSC, (iii) near real-time anomaly detection and nodes classification, (iv) attack graph generation and visualization. Initially, TDG and GC-LSTM are implemented for analyzing the power system's OT network traffic as shown in Figure 19.9a. The TDG extract network topology is based on observed traffic in OT network. The GC-LSTM learns the complex behavior of OT traffic data and topology. The prediction output from GC-LSTM subsequently generates traffic for the supervised predictions of CNN as shown in Figure 19.9b. Based on GC-LSTM and CNN training results, a hybrid combination of unsupervised and supervised models is used for OT traffic anomaly detection and nodes classification as shown in Figure 19.9c. Finally, the nodes classification result and network topology information are integrated into an attack graph depicted in Figure 19.9d.

The primary input for the GC-LSTM approach is the graph structure of the OT network topology. TDG is used to derive this particular graph structure. The graph (G) elements are vertices/nodes (V), edges/links (E), and adjacency matrix (A). The adjacency matrix is a representation of elements denoted by $A_{i,j}$, where i and j are node index numbers. $A_{i,j}$ equals 1 when two nodes are connected and 0 when they are not. In Eq. (19.1), the GCN model is predicated on the Hadamard product multiplication ($\odot$) of the weight matrix (W_{gcn}), adjacency matrix (A), and node features derived from the historical traffic data (X_t). The adjacency matrix is a mathematical representation that encapsulates pertinent details concerning the topology of the OT network. The modified adjacency matrix ($\hat{A}$) is obtained by adding the identity matrix (I) to the original adjacency matrix (A). The time series data set (X_t) is modeled by an equation that accounts for a specific time point (t) and the overall number of time observations (T). The node feature matrix (X) contains information about each node (x_i), where n represents the total number of nodes. The equation takes into account the exponent k, which represents the number of hops from a communication node to its neighboring nodes, as described in [101, 104]. Following the acquisition of spatial features through the GCN, the LSTM model is subsequently employed to examine the temporal or time-series characteristics. The functions and processes that occur within an LSTM cell are described in Eqs. (19.2–19.7). The LSTM process comprises six primary sub-equations, namely the forget gate (f_t), input gate (i_t), output gate (o_t), internal cell state (c'_t), transferable cell state (c_t), and hidden state (h_t).

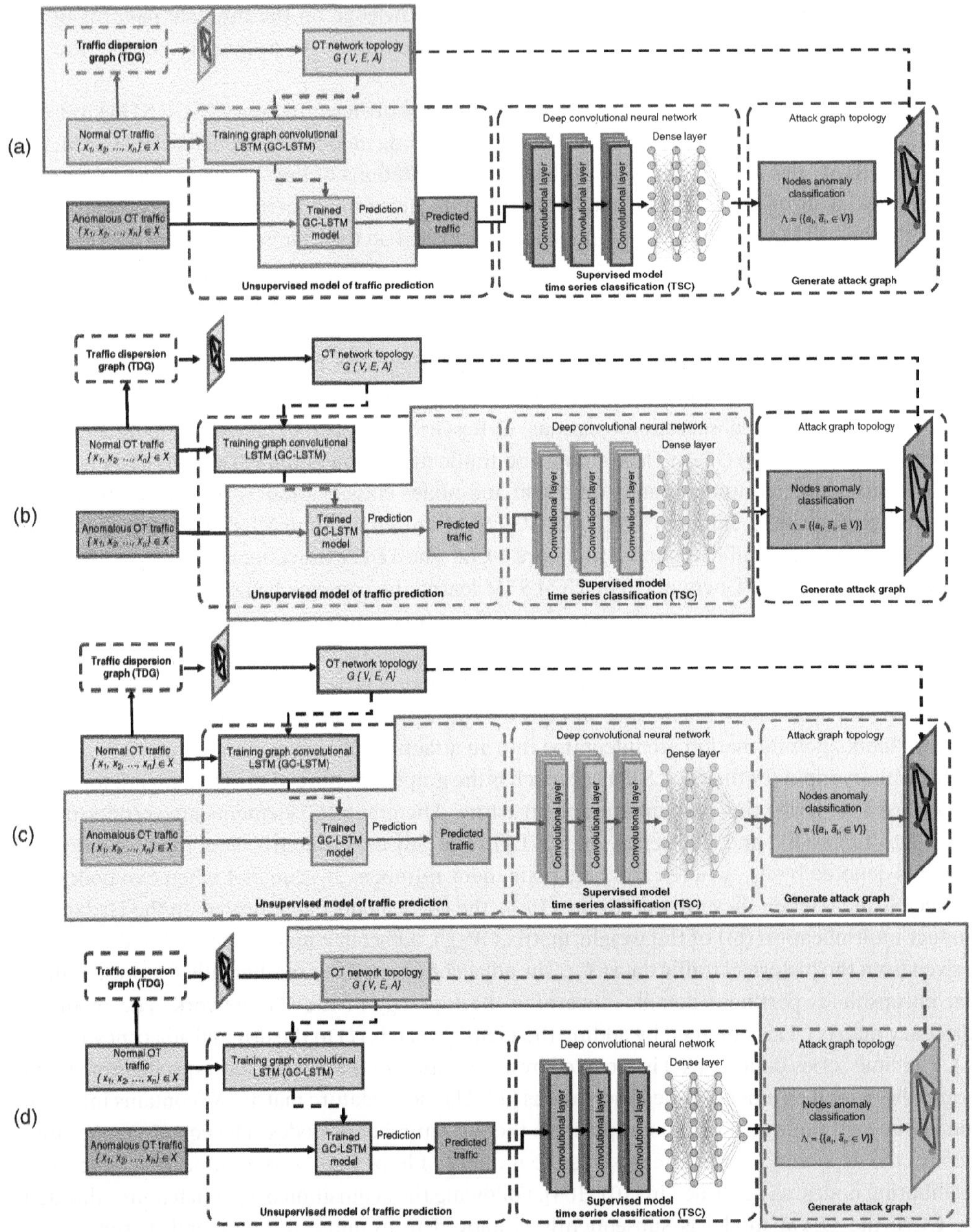

Figure 19.9 CyResGrid attack graph generation processes from (a) GC-LSTM training for traffic prediction and Traffic Dispersion Graph (TDG), (b) CNN training for Time Series Classification (TSC), (c) near real-time anomaly detection and nodes classification based on pretrained models, and (d) attack graph generation and visualization based on nodes classification and known graph topology.

$$GCN_t^k \leftarrow (W_{gcn} \odot \hat{A}^k)X_t \tag{19.1}$$

$$f_t = \sigma\left(\left(W_f GCN_t^k\right) + (U_f h_{t-1}) + b_f\right) \tag{19.2}$$

$$i_t = \sigma\left(\left(W_i GCN_t^k\right) + (U_i h_{t-1}) + b_i\right) \tag{19.3}$$

$$o_t = \sigma\left(\left(W_o GCN_t^k\right) + (U_o h_{t-1}) + b_o\right) \tag{19.4}$$

$$c_t' = \tanh\left(\left(W_c GCN_t^k\right) + (U_c h_{t-1}) + b^{c'}\right) \tag{19.5}$$

$$c_t = (f_t \odot c_{t-1}) + \left(i_t \odot c_t'\right) \tag{19.6}$$

$$h_t = o_t \odot tanh(c_t) \tag{19.7}$$

TSC is implemented using a CNN algorithm with a multilayer convolutional and ReLU activation function, as depicted in Eq. (19.8). The variables under consideration in Eq. (19.8) are the number of layers (l), filter size (m), weight (w), and bias (b). This model is trained to optimize classification performance based on previous GC-LSTM output. We perform hyperparameter tuning based on the number of layers, filters, and kernel size to develop our hybrid deep learning model. The deep learning model is optimized by using the technique of Bayesian optimization [105]. The optimization function seeks to maximize the efficacy of deep learning, as described in Eq. (19.9). The surrogate model and acquisition function are the foundation upon which Bayesian optimization is built. The Gaussian process serves as a surrogate model, enabling the quantification of uncertainty pertaining to regions that are not directly observable. In order to attain the optimal value of the objective function, the expected improvement (EI) is employed as the acquisition function. Iterations are carried out in Bayesian optimization in order to obtain a function that has the best possible performance. Through the iterative process, the CNN with the best performance is obtained that consists of three layers, sixty-four filters, and three kernel sizes. After the optimization, CNN is used to perform binary classification for each node into normal and anomalous. The classification is performed based on TSC from time series throughput data for each node (X). The result from the classification is then used to construct a forensic graph in the following stage.

$$y_i^l = \mathrm{ReLU}\left(\sum^{m-1} w y_{(i)}^{l-1} + b\right) \tag{19.8}$$

$$x^* = \arg\max_x f(x) \tag{19.9}$$

19.7 Attack Graph for Situational Awareness

Attack graphs can be used to model CPS vulnerabilities and exploits. An attack graph is an essential instrument for vulnerability analysis and the development of mitigation strategies. In the context of a communication network, numerous hosts are susceptible to potential vulnerabilities. Consequently, cybersecurity of the entire CPS cannot be relied solely upon the security of an individual host. Hence, it is crucial to detect and classify all susceptible nodes/hosts within a communication network as a group of possible threats in the CPS. Therefore, in this research, the observation and analysis of anomalous OT traffic behavior is used to detect potentially compromised nodes in the control center and substations. The data pertaining to anomalous nodes are subsequently utilized to generate an online attack graph in near real time covering all OT networks of the power grid.

The process of generating an attack graph is described in Algorithm 19.1. The algorithm takes the OT network traffic (X) as its input. The GC-LSTM algorithm is used to predict the OT traffic based on the network traffic data obtained from each substation (X_n). The GC-LSTM architecture generates a series of traffic forecasts (h_t) as its outputs. The corresponding output obtained from the prediction process is subsequently utilized as an input for the CNN-based TSC. Time-series-based anomaly detection is conducted for every node (a) within V. The classifier categorizes individual nodes as either anomalous or normal, utilizing the input OT traffic prediction. Subsequently, the aforementioned data are utilized to formulate the attack graph.

Algorithm 19.1 CyResGrid Attack Graph Generation

Inputs: $S\{s_1, s_2, \dots, s_n\}; X \in s_n$: Substations traffic data
$\{x_1, x_2, \dots, x_n\} \in X$: Nodes traffic data
Output: $\Lambda = \{\{a_i, \bar{a}_i, \in V\}\}$: Nodes classification as attack graph

Iteration for each substation
for s_i in S **do**
  for $t = 1$ to T **do**
    Traffic prediction

$$GCN_t^k \leftarrow (W_{gcn} \odot \hat{A}^k) X\{x_1, x_2, \dots, x_n\}_t$$

$$h_t, c_t = LSTM(X\{x_1, x_2, \dots, x_n\}, GCN_t^k, h_{t-1}, C_{t-1})$$

    Iteration for each node a in V
    for a in V
      Node classification

$$\bar{a}_i = \sum^{m-1} w\, h_{t,(t)}^{l-1} + b$$

    end for
  end for
end for
return: $\Lambda = \{\{a_i, \bar{a}_i, \in V\}\}$

$$\Lambda = \{\{a_i, \overline{a}_i, \in V\}\} \tag{19.10}$$

$$\Lambda = \{\{a_i, \overline{a}_i, \in V\}, \{u_i \notin V\}\} \tag{19.11}$$

There are two different types of attack graphs, which can be comprehended by Eqs. (19.10) and (19.11). Attack graph type I, as described in Eq. (19.10), is generated by applying prior knowledge of the OT network topology and the output of node classification. In the meantime, the attack graph type II presented in Eq. (19.11) takes into consideration unknown nodes based on the TDG. There are two elements of attack graph (Λ) type I as indicated in Eq. (19.10), i.e., normal nodes (a_i) and anomalous nodes ($\overline{a}_i$). Both aforementioned nodes are constituent elements of the set of known nodes (V). On the other hand, the attack graph of type II, which is shown in the Eq. (19.11), consists of one additional element of the unidentified node. Nodes that cannot be identified are regarded as anomalous due to their lack of association with the known nodes (V).

Figure 19.10 depicts an example comparison of attack graph representations of the OT network under normal OT network traffic conditions in Figure 19.10a, and under anomalous traffic conditions in Figure 19.10b and c. The anomalous network traffic conditions are determined based on observed abnormal node behavior shown in red. Subsequently, these nodes are integrated to

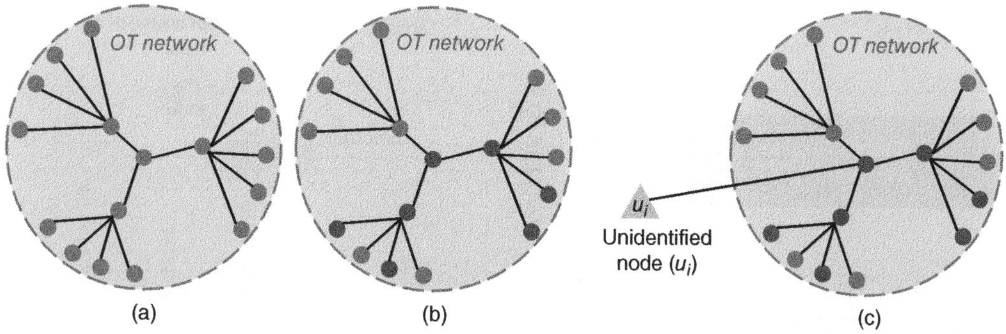

Figure 19.10 Attack graph representation for normal and anomalous traffic: (a) Normal graph: $\Lambda = \{ a_i, \in V\}$, (b) Attack graph type I: $\Lambda = \{ a_i, \bar{a}_i, \in V\}$which contains normal and anomalous nodes, and (c) Attack graph type II: $\Lambda = \{\{a_i, \bar{a}_i, \in V\}, \{u_i \notin V\}\}$which contains normal, anomalous, and unidentified nodes.

construct an attack graph (Λ). There are three elements in the attack graph, i.e., normal nodes (a_i), anomalous nodes ($\bar{a}_i$), and unidentified nodes (u_i). The first attack graph type depicted in Figure 19.10b categorizes nodes as anomalous based on the traffic patterns observed from all identified nodes. In contrast, the attack graph of type II depicted in Figure 19.10c considers all unknown nodes to categorize abnormal behavior. The recognition of unidentified nodes (u_i) is dependent upon acquiring addresses from unknown sources or destinations through the TDG. It is presumed that the presence of the unknown nodes (u_i) indicates an active cyber attack that is being launched from an unlisted host within the known OT network (V).

19.8 Cyber Attack Case Studies

This section presents an analysis of two case studies, which involve instances of cyber attacks on a digital substation and wide area networks. In the first scenario, the digital substation is the target of the cyber attack, whereas in the second scenario, multiple substations are targeted.

19.8.1 Substation Attack Exploiting GOOSE Protocol Vulnerabilities

The primary objective of a cyber attack targeting a digital substation is to alter, disrupt, or incapacitate the functionality of one or more protection, automation, or control devices. Figure 19.11 illustrates a hardware-in-the-loop (HIL) setup employed to execute cyber attacks on a digital substation and implement the anomaly detection using hybrid deep learning and attack graph method. RTDS is used to model the power system in real time. The implementation of data exchange between the RTDS and substation OT communication network is facilitated through the utilization of giga transceiver network (GTNET) cards. The OT network within the substation comprises of IEDs. The IEDs are in compliance with the IEC 61850 standard, including generic object-oriented substation event (GOOSE) and sampled values (SV) messaging. A host was compromised inside the substation from where the cyber attack was conducted.

During simulation, the GTNET cards periodically send IEC 61850 SV packets to IEDs communicating sampled voltage and current measurements. A switch that functions at the data link layer (Layer 2) of the OSI model is connected to the hosts that make up the substation network. As a consequence, in the configuration of the substation network, each packet is sent to all hosts that are connected to the switch. The IEDs can detect a fault simulated in RTDS and issue control commands using IEC 61850 GOOSE to open circuit breakers and clear the fault. The compromised host

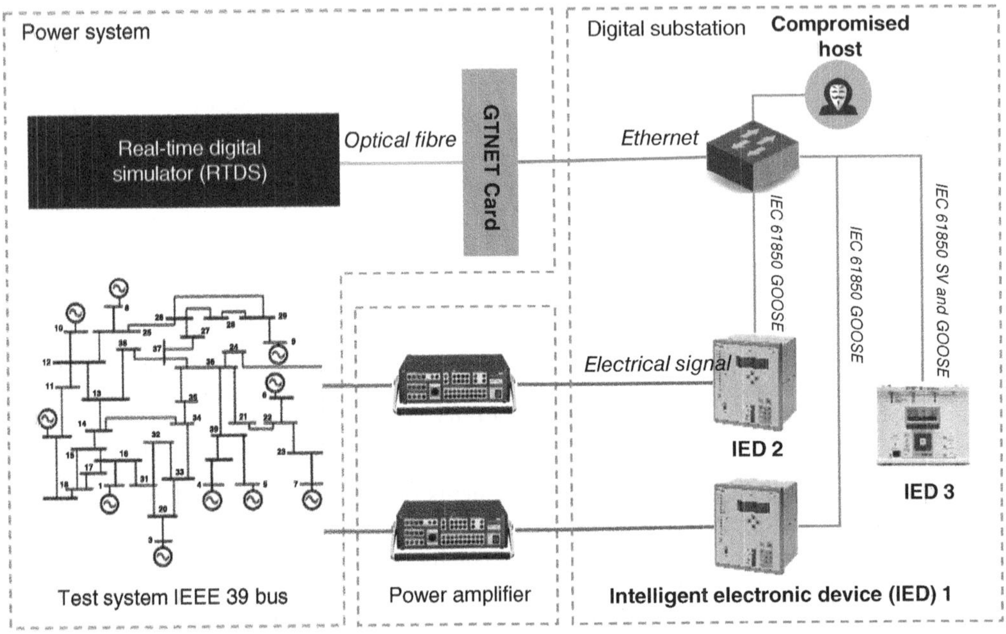

Figure 19.11 Cyber-physical experimental architecture to analyze the impact of cyber attacks on digital substations.

is connected to the Ethernet switch. The attacker uses various tools to perform network reconnaissance and sniff the OT network traffic through the network switch. Following weaponization, the attacker injects spoofed GOOSE packets into the switch to open circuit breakers [106, 107]. Based on the HIL setup and cyber attack scenarios, OT network traffic data are collected from the switch for analysis using CyResGrid. The OT data collection process is carried out through Wireshark, in accordance with the substation network configuration.

Figure 19.12 presents the attack graph results for the cyber attack conducted on the digital substation, i.e., network reconnaissance and GOOSE attacks. There are 85 nodes in total present in each graph (a)–(c). Figure 19.12a depicts the attack graph, while the OT network is operating normally indicated with blue nodes. Meanwhile, Figures 19.12b and c show the attack graphs under GOOSE and network reconnaissance attacks. The anomalous communications are indicated with red nodes. The GOOSE attack is characterized by targeting specific nodes, which are linked to

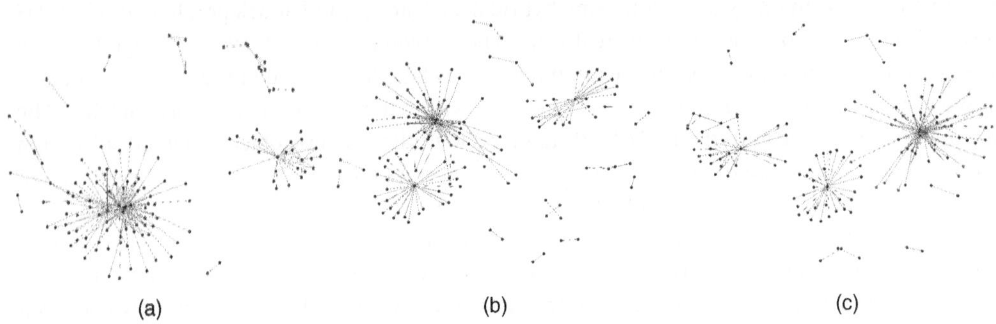

Figure 19.12 Attack graph results for cyber attacks on digital substation under (a) normal traffic, (b) GOOSE attack, and (c) reconnaissance attack.

IEDs and compromised hosts, resulting in anomalous traffic patterns. During a reconnaissance attack, the attackers focus on targeting numerous hosts within the IP address ranges. As a result, a greater number of nodes exhibit anomalous behavior depicted with red, indicating the presence of anomalous OT traffic in the digital substation.

19.8.2 Wide-Area OT Anomaly Detection with Attack Graphs

The monitoring of OT traffic over a wide area is facilitated by SDN using the architecture depicted in Figure 19.8. The traffic data are collected as spatial-temporal dataset in real time. It serves as input for the hybrid deep learning model, which is used to generate attack graphs in near real time. Figure 19.12 depicts the comprehensive attack graph map utilized for the purpose of identifying and visualizing online cyber-attacks on the power grid, i.e., distributed denial of service (DDoS) and network reconnaissance. The attack graph illustrates the OT network deployed in the CPS model of IEEE 39-bus comprising 27 substations and one control center. The control center is depicted by a central node, while the remaining nodes situated at the edges represent the IEDs in substations. Table 19.3 shows nine different levels of cyber-attack intensity with specific time duration. The DDoS attacks were executed with hping3 and network reconnaissance were executed with Nmap. The cyber-attacks last for a total of 345,000 seconds, and data are collected every second to generate the dataset.

Figure 19.13a depicts the attack graph in a normal state, where all nodes are represented in blue. Figures 19.13b and c illustrate the attack graph when subjected to a DDoS attack, both in a single target and multiple targets scenario. In Figure 19.13b, DDoS targets a single node in substation number 7, and in Figure 19.13c, DDoS targets multiple nodes in substations 2–7. The DDoS attacks are initiated from the control center. Consequently, the control center node, substation gateways, and nodes are exhibiting anomalous behavior indicated with red. Based on Figure 19.13b and c, DDoS attacks can be classified with high accuracy using hybrid deep learning. The reason for this is that DDoS attacks generate a more significant traffic increase than normal.

The attack graphs under the reconnaissance attack are depicted in Figure 19.13d, e, and f. The control center is the source of the attacks, which are specifically aimed at substation numbers 7 through 13. During both normal and aggressive scanning, all nodes located within the targeted substations are shown in red. However, under stealthy scanning intensities, some of the targeted nodes do not turn red. This occurrence can be attributed to a false negative generated by the traffic classifier. Notwithstanding the limitations, the CyResGrid methodology has already exhibited superior

Table 19.3 Cyber attack scenarios.

Attack type	Intensity	Tool	Time duration (s)
DDoS	High	*hping3*	30,000
	Medium	*hping3*	30,000
	Low	*hping3*	30,000
Reconnaissance	Paranoid	*Nmap*	75,000
	Sneaky	*Nmap*	50,000
	Polite	*Nmap*	40,000
	Normal	*Nmap*	30,000
	Aggressive	*Nmap*	30,000
	Insane	*Nmap*	30,000

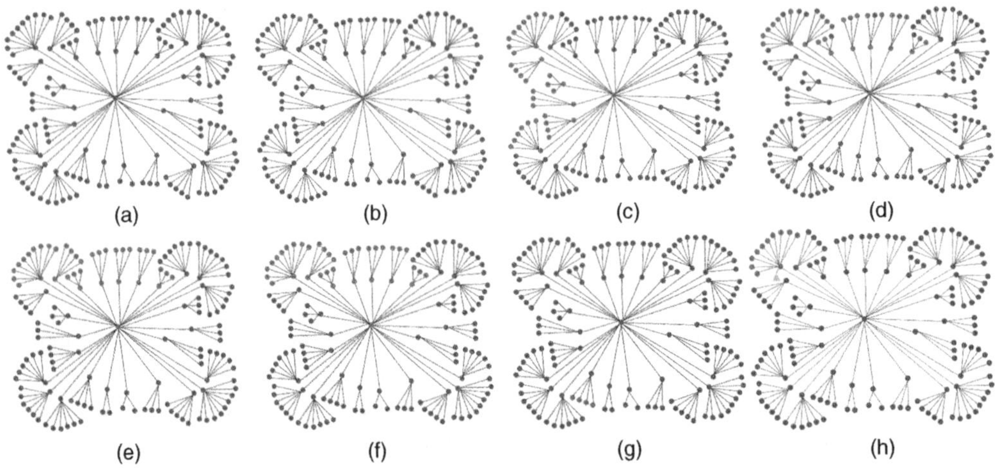

Figure 19.13 Cyber-attack location identification and visualization using attack graph maps including (a) normal, (b) DoS one target, (c) DoS multiple targets, (d) sneaky intensity scanning, (e) normal intensity scanning, (f) aggressive intensity scanning, (g) internal substation attack, and (h) unidentified nodes.

Table 19.4 Performance comparison of anomaly detection methods.

No	Methods	AUC	TN	FP	FN	TP	Accuracy	F1	G mean	Time (s)
Combined attack scenarios										
1	ResNet	0.849	82.27	11.32	3.49	2.92	85.19	28.29	15.50	633
2	Inception	0.961	93.50	0.20	4.10	2.31	95.71	51.76	14.68	976
3	FCN	0.955	88.16	5.43	3.92	2.49	90.65	34.76	14.81	1016
4	MLP	0.758	72.22	21.37	4.86	1.55	73.77	10.55	10.57	113
5	GC-LSTM + Resnet	0.974	93.29	0.31	3.27	3.14	96.42	63.77	17.12	1056
6	GC-LSTM + Inception	0.976	92.10	1.49	3.35	3.06	95.16	55.87	16.79	1409
7	GC-LSTM + FCN	0.972	92.28	1.30	3.68	2.73	95.01	52.26	15.87	1342
8	GC-LSTM + MLP	0.937	93.40	0.19	6.13	0.28	93.68	8.14	5.12	765
9	**CyResGrid**	**0.984**	**93.47**	**0.13**	**3.42**	**2.99**	**96.45**	**65.03**	**17.16**	**714**
Stealthy attack scenarios										
10	ResNet	0.8637	86.94	12.02	0.96	0.08	87.02	1.26	2.69	91
11	Inception	0.9887	98.93	0.02	1.04	0.0004	98.93	0.09	0.22	224
12	FCN	0.9833	87.82	11.13	1.01	0.02	87.85	0.47	1.58	240
13	GC-LSTM + Resnet	0.9524	89.93	9.02	0.95	0.09	90.02	1.87	2.92	226
14	GC-LSTM + Inception	0.9489	89.96	8.99	0.95	0.10	90.05	1.87	2.92	303
15	GC-LSTM + FCN	0.9491	89.96	8.99	0.95	0.10	90.05	1.87	2.92	304
16	**CyResGrid**	**0.9243**	**91.15**	**7.81**	**0.94**	**0.111**	**91.25**	**2.32**	**3.08**	**138**

performance in comparison to state-of-the-art TSC. Table 19.4 presents a comparative analysis of the performance of CyResGrid and other TSC techniques, including ResNets [108], Inception [98], fully convolutional neural network (FCN) [109], and multi layer perceptron (MLP) [110]. As indicated by Table 19.4, the classifiers' performance declines in the event of stealthy attack scenarios. The reason for this phenomenon is that stealthy attacks produce a relatively minor impact on traffic anomalies. As a result, the classifier algorithm is likely to produce higher rates of false negatives (FNs) and false positives (FPs). Additionally, the outcomes generated by the classifier will result in reduced values for various performance metrics, including area under the curve (AUC), true negative (TN), true positive (TP), accuracy, F1, and G mean.

Figure 19.13g depicts a DDoS attack scenario originating from internal substation number 26. The internal substation was identified as the source of the attack, and it is noteworthy that the substation gateway and control center remain unaffected, as denoted by the blue nodes color. This scenario demonstrates that the attack graph has the capability to incorporate a wide-area network monitoring and identify localized anomalies within a substation. The network scanning aimed at substation 7 is depicted in Figure 19.13h, wherein an unidentified node is observed to be the source of the activity, as denoted by an orange triangle. The origin of the attack is categorized as unidentified due to its absence from the lists of recognized nodes within the OT network.

19.9 Conclusions

Given the increasing risk of cyber attacks targeting power grids, strengthening attack detection capabilities in OT systems has become imperative. This chapter provides essential knowledge of cyber-attack mitigation for cyber-physical power systems, i.e., secure communication protocols for operational technologies, penetration testing using cyber ranges and cyber-physical co-simulation, and network security controls including firewalls and intrusion detection and prevention systems. Among the wide-scope mitigation, AI is highlighted as an emerging solution. A hybrid deep learning model is presented that combines GC-LSTM and CNN for detecting anomalies in OT communication networks for power grids. The GC-LSTM algorithm predicts OT traffic based on the spatial and temporal characteristics of the input data. By means of its forecasting capabilities, the data's variability and outliers are mitigated. The utilization of GC-LSTM can enhance the efficacy of TSCs in detecting anomalies. Unlike traditional signature and supervised learning-based intrusion detection, the hybrid deep learning anomaly detection utilizes the OT traffic throughput. It takes advantage of the OT traffic deterministic and homogenous characteristics to provide robust and flexible anomaly detection for a wide scope of cyber attacks at early stages of the cyber kill chain. The traffic anomalies are incorporated into an attack graph that aids power system operators identify and localize anomalies of active attacks on power systems in near real time. Cyber attack case studies and cyber-physical co-simulation results are provided to demonstrate the efficiency of hybrid deep learning for anomaly detection in power grid OT networks.

Acknowledgments

This work was supported in part by the EU Horizon Europe Cooperative Cyber Protection For Modern Power Grids (COCOON) project with Grant Agreement Number 101120221 and ERIGrid 2.0 with Grant Agreement Number 870620.

References

1 Whitehead, D.E., Owens, K., Gammel, D. et al. (Apr. 2017). Ukraine cyber-induced power outage: analysis and practical mitigation strategies. *Proc. Int. Conf. for Prot. Relay Engineers*, Texas, USA, pp. 1–8.

2 Assante, M.J., Lee, R.M. and Conway, T. (2017). ICS defense use case no. 6: modular ICS malware. *Electricity Information Sharing Center (E-ISAC) Tech. Report*, pp. 1–27, 2(1).

3 ENISA (Feb. 2017). Communication network dependencies for ICS/SCADA Systems. [Online]. Available: enisa.europa.eu/publications/ics-scada-dependencies (accessed 27 June 2023).

4 Cintuglu, M.H., Mohammed, O.A., Akkaya, K., and Uluagac, A.S. (2016). A survey on smart grid cyber-physical system testbeds. *IEEE Communication Surveys and Tutorials* 19 (1): 446–464.

5 Sun, C.C., Hahn, A., and Liu, C.C. (2018). Cyber security of a power grid: State-of-the-art. *International Journal of Electrical Power & Energy Systems* 99: 45–56.

6 Gupta, B.B. and Akhtar, T. (2017). A survey on smart power grid: frameworks, tools, security issues, and solutions. *Annals of Telecommunications* 72 (9): 517–549.

7 Montoya, J. et al. (2020). Advanced laboratory testing methods using real-time simulation and hardware-in-the-loop techniques: a survey of smart grid international research facility network activities. *Energies* 13 (12): 3267.

8 Liang, G., Zhao, J., Luo, F. et al. (2017). A review of false data injection attacks against modern power systems. *IEEE Transactions on Smart Grid* 8 (4): 1630–1638.

9 Deng, R., Xiao, G., Lu, R. et al. (2017). False data injection on state estimation in power systems attacks, impacts, and defense: a survey. *IEEE Transactions on Industrial Informatics* 13 (2): 411–423.

10 Musleh, A.S., Chen, G., and Dong, Z.Y. (2020). A survey on the detection algorithms for false data injection attacks in smart grids. *IEEE Transactions on Smart Grid* 11 (3): 2218–2234.

11 Reda, H.T., Anwar, A., and Mahmood, A. (2022). Comprehensive survey and taxonomies of false injection attacks in smart grid: attack models, targets, and impacts. *Renewable and Sustainable Energy Reviews* 163 (112423): 1–24.

12 Sayghe, A. et al. (2020). Survey of machine learning methods for detecting false data injection attacks in power system. *IET Smart Grid* 3 (5): 581–595.

13 Zhang, H., Liu, B., and Wu, H. (2021). Smart grid cyber-physical attack and defense: a review. *IEEE Access* 9: 29641–29659.

14 SANS ICS (Mar. 2016). White Analysis of the Cyber Attack on the Ukrainian Power Grid. *Electricity Information Sharing Center (E-ISAC) Tech. Report*, pp. 1–29, 388(1).

15 Securicon (May 2019). What's the difference between OT, ICS, SCADA and DCS?https://www.securicon.com/whats-the-difference-between-ot-ics-scada-and-dcs/ (accessed 23 June 2023).

16 Hahn, A. (2016). Operational technology and information technology in industrial control systems. In: *Cyber-Security of SCADA and Other Industrial Control Systems*, 51–68. Berlin, Germany: Springer.

17 Ginter, A. (2018). *Secure Operations Technology*. Calgary, Alberta, Canada: Abterra Technologies Incorporated.

18 Amoah, R., Camtepe, S., and Foo, E. (2016). Formal modelling and analysis of DNP3 secure authentication. *Journal of Network and Computer Applications* 59: 345–360.

19 Amoah, R., Camtepe, S., and Foo, E. (2016). Securing DNP3 broadcast communications in SCADA systems. *IEEE Transactions on Industrial Informatics* 12 (4): 1474–1485.

20 Knapp, E.D. and Langill, J.T. (2014). *Industrial Network Security: Securing Critical Infrastructure Networks for Smart Grid, SCADA, and Other Industrial Control Systems*. The Netherlands: Syngress.

21 Ferst, M.K., De Figueiredo, H. F. M., Denardin, G. et al. (2018). Implementation of secure communication with modbus and transport layer security protocols. *Proc. IEEE Int. Conf. Ind. Appl.*, Sao Paulo, Brazil, pp. 155–162.

22 Hussain, S.M.S., Ustun, T.S., and Kalam, A. (2020). A review of IEC 62351 security mechanisms for IEC 61850 message exchanges. *IEEE Transactions on Industrial Informatics* 16 (9): 5643–5654.

23 Rezai, A., Keshavarzi, P., and Moravej, Z. (2017). Key management issue in SCADA networks: a review. *Engineering Science and Technology, an International Journal* 20 (1): 354–363.

24 Pramod, T.C., Boroojeni, K.G., Amini, M.H. et al. (2019). Key pre-distribution scheme with join leave support for SCADA systems. *International Journal of Critical Infrastructure Protection* 24: 111–125.

25 Zheng, Z., Xie, S., Dai, H.-N. et al. (2018). Blockchain challenges and opportunities: a survey. *International Journal of Web and Grid Services* 14 (4): 352–375.

26 Alladi, T., Chamola, V., Rodrigues, J.J.P.C., and Kozlov, S.A. (2019). Blockchain in smart grids: a review on different use cases. *Sensors* 19 (22): 1–25.

27 Brandão, R. (2020). A blockchain-based protocol for message exchange in a ICS network: student research abstract. *Proc. ACM Symp. Appl. Comput.*, Brno, Czech, pp. 357–360.

28 Carcano, A., Di Pinto, A., Dragoni, Y. et al. (2019). The future of securing intelligent electronic devices using the IEC 62351-7 standard for monitoring. *Proc Black Hat USA*, Las Vegas, USA, pp. 1–21.

29 Gunduz, M. Z. and Das, R. (2018). A comparison of cyber-security oriented testbeds for IoT-based smart grids. *Int. Symp. Digit. Forensic Secur.*, Antalya, Turkey, pp. 1–6.

30 Mallouhi, M., Al-Nashif, Y., Cox, D. et al. (2011). A testbed for analyzing security of SCADA control systems (TASSCS). *Proc. IEEE PES Innov. Smart Grid Tech. Conf.*, Anheim, USA, pp. 1–7.

31 Queiroz, C., Mahmood, A., and Tari, Z. (2011). SCADASim a framework for building SCADA simulations. *IEEE Transactions on Smart Grid* 2 (4): 589–597.

32 Sarker, V. et al. (2020). Cyber-physical security and resiliency analysis testbed for critical microgrids with IEEE 2030.5. *Work. on Mode. and Sim. of Cyber-Physical Energy Systems*, Sydney, Australia, pp. 1–6.

33 Mirkovic, J. and Benzel, T. (2012). Teaching cybersecurity with DeterLab. *IEEE Security and Privacy* 10 (1): 73–76.

34 Mirkovic, J., Benzel, T.V., Faber, T. et al. (2010). The DETER project: advancing the science of cyber security experimentation and test. *IEEE Int. Conf. Technol. Homel. Secur.*, Waltham, USA, pp. 1–7.

35 Oyewumi, I.A. et al. (2019). ISAAC: the idaho CPS smart grid cybersecurity testbed. *IEEE Texas Power Energy Conf.*, College Station, USA, pp. 1–6.

36 Johnson, J., Onunkwo, I., Codeiro, P. et al. (2020). Assessing DER network cybersecurity defences in a power-communication co-simulation environment. *IET Cyber-Physical Systems: Theory & Applications* 1–11.

37 Chen, B., Butler-Purry, K. L., Goulart, A. et al. (2014). Implementing a real-time cyber-physical system test bed in RTDS and OPNET. *North Am. Power Symp.*, Pullman, USA, pp. 1–6.

38 Chen, B., Pattanaik, N., Goulart, A. et al. (2015). Implementing attacks for modbus/TCP protocol in a real-time cyber physical system test bed. *Proc. IEEE Int. Work. Tech. Comm. Commun. Qual. Reliab.*, Charleston, USA, pp. 1–6.

39 Allaoua, A., Layadi, T.M., Colak I. et al. (2021). Design and simulation of smart-grids using OMNeT++/matlab-simulink co-simulator. *Int. Conf. on Ren. Energy Research and Appl.*, Istanbul, Turkey, pp. 141–145.

40 Vellaithurai, C.B., Biswas, S.S., and Srivastava, A.K. (2015). Development and application of a real-time test bed for cyber–physical system. *IEEE Systems Journal* 11 (4): 2192–2203.

41 Watada, J., Roy, A., Kadikar, R. et al. (2019). Emerging trends, techniques and open issues of containerization: a review. *IEEE Access* 7: 152443–152472.

42 Yamin, M.M., Katt, B., and Gkioulos, V. (2020). Cyber ranges and security testbeds: scenarios, functions, tools and architecture. *Computers & Security* 88 (101636).

43 Diogenes, Y. and Ozkaya, E. (2018). *Cybersecurity, Attack and Defense Strategies Infrastructure Security with Red Team and Blue Team tactics*. Birmingham, United Kingdom: Packt Publishing.

44 Nivethan, J. and Papa, M. (2016). A Linux-based firewall for the DNP3 protocol. *IEEE Symp. Technol. Homel. Secur.*, Waltham, USA, pp. 1–5.

45 Nivethan, J. and Papa, M. (2016). On the use of open-source firewalls in ICS/SCADA systems. *Information Security Journal* 25 (1–3): 83–93.

46 Li, D., Guo, H., Zhou, J. et al. (2019). SCADAWall: a CPI-enabled firewall model for SCADA security. *Computers & Security* 80: 134–154.

47 Yang, Y., McLaughlin, K., Littler, T. et al. (2013). Intrusion detection system for IEC 60870-5-104 based SCADA networks. *IEEE Power Energy Soc. Gen. Meet.*, Vancouver, Canada, pp. 1–5.

48 Chromik, J., Remke, A., Haverkort, B. R. et al. (2019). A parser for deep packet inspection of IEC-104: a practical solution for industrial applications. *Proc. IEEE/IFIP Int. Conf. Dependable Syst. Networks Ind. Track*, Portland, USA, pp. 5–8.

49 Sommer, R., Amann, J. and Hall, S. (2016). Spicy: a unified deep packet inspection framework for safely dissecting all your data. *ACM Int. Conf. Proc. Ser.*, Los Angles, USA, pp. 558–569.

50 Cruz, T. et al. (2016). A cybersecurity detection framework for supervisory control and data acquisition systems. *IEEE Transactions on Industrial Informatics* 12 (6): 2236–2246.

51 Pan, S., Morris, T., and Adhikari, U. (2015). A specification-based intrusion detection framework for cyber-physical environment in electric power system. *International Journal of Network Security* 17 (2): 174–188.

52 Maynard, P. and McLaughlin, K. (2020). Towards understanding man-on-the-side Attacks (MotS) in SCADA networks. *arXiv* 2004 (14334): 1–9.

53 Yang, Y., McLaughlin, K., Littler, T. et al. (2013). Rule-based intrusion detection system for SCADA networks. *IET Conf. Publ.*, Beijing, China, pp. 8–11.

54 Koutsandria, G., Muthukumar, V., Parvania, M. et al. (2014). A hybrid network IDS for protective digital relays in the power transmission grid. *IEEE Int. Conf. Smart Grid Commun. SmartGridComm*, Venice, Italy, pp. 908–913.

55 Goldenberg, N. and Wool, A. (2013). Accurate modeling of modbus/TCP for intrusion detection in SCADA systems. *International Journal of Critical Infrastructure Protection* 6 (2): 63–75.

56 Almalawi, A., Fahad, A., Tari, Z. et al. (2015). An efficient data-driven clustering technique to detect attacks in SCADA systems. *IEEE Transactions on Information Forensics and Security* 11 (5): 893–906.

57 Erez, N. and Wool, A. (2015). Control variable classification, modeling and anomaly detection in Modbus/TCP SCADA systems. *International Journal of Critical Infrastructure Protection* 10: 59–70.

58 Lin, H., Slagell, A., Kalbarczyk, Z. et al. (2014). Semantic security analysis of scada networks to detect malicious control commands in power grids (poster). *ACM Int. Conf. Proceeding Ser.*, Glasglow, UK, pp. 492–495.

59 Kleinmann, A. and Wool, A. (2017). Automatic construction of statechart-based anomaly detection models for multi-threaded industrial control systems. *ACM Transactions on Intelligent Systems and Technology* 8 (4): 1–21.

60 Koucham, O., Mocanu, S., Hiet, G. et al. (2018). Efficient mining of temporal safety properties for intrusion detection in industrial control systems. *IFAC-PapersOnLine* 51 (24): 1043–1050.

61 Lahza, H., Radke, K., and Foo, E. (2018). Applying domain-specific knowledge to construct features for detecting distributed denial-of-service attacks on the GOOSE and MMS protocols. *International Journal of Critical Infrastructure Protection* 20: 48–67.

62 I. E. C. S. Networks (2016). Multidimensional intrusion detection system for IEC 61850-based SCADA networks. *IEEE Transactions on Power Delivery* 32 (2): 1068–1078.

63 Yoo, H. and Shon, T. (2014). Novel approach for detecting network anomalies for dubstation sutomation based on IEC 61850. *Multimedia Tools and Applications* 74 (1): 303–318.

64 Pan, S., Morris, T., and Adhikari, U. (2015). Developing a hybrid intrusion detection system using data mining for power dystems. *IEEE Transactions on Smart Grid* 6 (6): 3104–3113.

65 Reuter, L., Jung, O. and Magin, J. (2020). Neural network based anomaly detection for SCADA systems. *Conf. Innov. Clouds, Internet Networks Work.*, Paris, France, pp. 194–201.

66 Chaithanya, P.S., Priyanga, S., Pravinraj, S., and Sriram, V.S.S. (2020). SSO-IF: an outlier detection approach for intrusion detection in SCADA systems. In: *Inventive Communication and Computational Technologies*, 921–929. Singapore: Springer.

67 Maglaras, L., Cruz, T., Ferrag, M.A., and Janicke, H. (2020). Teaching the process of building an intrusion detection system using data from a small-scale SCADA testbed. *Internet Technology Letters* 3 (1): e132.

68 Kalech, M. (2019). Cyber-attack detection in SCADA systems using temporal pattern recognition techniques. *Computers & Security* 84: 225–238.

69 Selvarajan, S., Shaik, M., Ameerjohn, S., and Kannan, S. (2020). Mining of intrusion attack in SCADA network using clustering and genetically seeded flora-based optimal classification algorithm. *IET Information Security* 14 (1): 1–11.

70 Derhab, A. et al. (2019). Blockchain and random subspace learning-based IDS for SDN-enabled industrial IoT security. *Sensors (Switzerland)* 19 (14): 1–24.

71 Tamy, S., Belhadaoui, H., Rabbah, M.A. et al. (2019). An evaluation of machine learning algorithms to detect attacks in SCADA network. *Mediterr. Congr. Telecommun.*, Fez, Morocco, pp. 1–5.

72 Suaboot, J. et al. (2020). A taxonomy of supervised learning for IDSs in SCADA environments. *ACM Computing Surveys* 53 (2).

73 Al-Asiri, M. and El-Alfy, E.S.M. (2020). On using physical based intrusion detection in SCADA systems. *Procedia Computer Science* 170: 34–42.

74 Khan, I.A., Pi, D., Khan, Z.U. et al. (2019). Hml-ids: a hybrid-multilevel anomaly prediction approach for intrusion detection in scada systems. *IEEE Access* 7: 89507–89521.

75 Huda, S., Yearwood, J., Hassan, M.M., and Almogren, A. (2018). Securing the operations in SCADA-IoT platform based industrial control system using ensemble of deep belief networks. *Applied Soft Computing Journal* 71: 66–77.

76 Neha, R.S.N., Priyanga, S., Seshan, S., and Sriram, V.S.S. (2020). SCO-RNN: a behavioral-based intrusion detection approach for cyber physical attacks in SCADA systems. In: *Lecture Notes in Networks and Systems*, 911–919. Singapore: Springer.

77 Zizzo, G., Hankin, C., Maffeis, S., and Jones, K. (2019). Intrusion detection for industrial control systems: evaluation analysis and adversarial attacks. *arXiv* 1911 (04278): 1–12.

78 Gao, J. et al. (2019). LSTM for SCADA intrusion detection. *IEEE Pacific Rim Conf. Comm. Comp. Sig. Proc.*, Victoria, Canada, pp. 1–4.

79 Gao, J. et al. (2020). Omni SCADA intrusion detection using deep learning algorithms. *IEEE Internet of Things Journal* 8 (2): 951–961.

80 Ahsan M. and Nygard, K. (2020). Convolutional neural networks with LSTM for intrusion detection. *CATA, San Fransisco, USA*, pp. 69–57.

81 Kim, T.-Y. and Cho, S.-B. (2019). CNN-LSTM neural networks for anomalous database intrusion detection in RBAC-administered model. *Int. Conf. on Neural Information Processing*, Sydney, Australia, pp. 131–139.

82 Praanna, K., Sruthi, S.V.P., Kalyani, K.V., and Tejaswi, A.S. (2020). A CNN-LSTM model for intrusion detection system from high dimensional data. *Journal of Information and Computational Science* 10 (3): 1362–1370.

83 Rrushi, J. L. and Campbell, R. H. (2008). Detecting attacks in power plant interfacing substations through probabilistic validation of attack - effect bindings. *Proc. SCADA Secur. Sci. Symp.*, pp. 1–24.

84 Yang, H., Cheng, L. and Chuah, M.C. (2019). Deep-learning-based network intrusion detection for SCADA systems. *IEEE Conf. Commun. Netw. Secur.*, Washington, USA, pp. 1–7.

85 Altaha, M., Lee, J.-M., Aslam, M., and Hong, S. (2020). Network intrusion detection based on deep neural networks for the SCADA system. *Journal of Physics Conference Series* 1585: 012038.

86 Khraisat, A., Gondal, I., Vamplew, P., and Kamaruzzaman, J. (2019). Survey of intrusion detection systems: techniques, datasets and challenges. *Cybersecurity* 2 (1): 1–22.

87 Lin, C. and Nadjm-tehrani, S. (2019). Timing patterns and correlations in spontaneous SCADA traffic for anomaly detection. *Int. Sym. on Research in Attacks, Intrusions and Defenses (RAID)*, Beijing, China, pp. 73–88.

88 Liu, H. and Lang, B. (2019). Machine learning and deep learning methods for intrusion detection systems: a survey. *Applied Sciences* 9 (20): 1–28.

89 Aldweesh, A., Derham, A., and Emam, A.Z. (2020). Deep learning approaches for anomaly-based intrusion detection systems: a survey, taxonomy, and open issues. *Knowledge-Based Systems* 189 (105124): 1–19.

90 Mishra, P., Varadharajan, V., Tupakula, U., and Pilli, E.S. (2018). A detailed investigation and analysis of using machine learning techniques for intrusion detection. *IEEE Communications Surveys & Tutorials* 21 (1): 686–728.

91 Barbosa, R., Sadre, R. and Pras, A. (Mar. 2012). Difficulties in modeling SCADA traffic: a comparative analysis. *Proc. Passive and Active Measure.*, Berlin, Germany, pp. 126–135.

92 Guan, X., Qin, T., Li, W., and Wang, P. (2010). Dynamic feature analysis and measurement for large-scale network traffic monitoring. *IEEE Transactions on Information Forensics and Security* 5 (4): 905–919.

93 Kind, A., Stoecklin, M.P., and Dimitropoulos, X. (2009). Histogram-based traffic anomaly detection. *IEEE Transactions on Network and Service Management* 6 (2): 110–121.

94 Xu, K., Zhang, Z.L., and Bhattacharyya, S. (2008). Internet traffic behavior profiling for network security monitoring. *IEEE/ACM Transactions on Networking* 16 (6): 1241–1252.

95 Wu, H. (Dec. 2016). A survey of research on anomaly detection for time series. *Proc. 13th Int. Compt. Conf. on Wav. Act. Med. Tech. and Inf. Proc. (ICCWAMTIP)*, Chengdu, China, pp. 426–431

96 Shaukat, K. et al., (Apr. 2021). A review of time-series anomaly detection techniques: a step to future perspectives. *Proc. Future of Information and Communication Conf.*, Vancouver, Canada, pp. 865–877.

97 Fawaz, I., Forestier, G., Weber, J. et al. (2019). Deep learning for time series classification: a review. *Data Mining and Knowledge Discovery* 33 (4): 917–963.

98 Fawaz, I. et al. (2020). Inceptiontime: finding alexnet for time series classification. *Data Mining and Knowledge Discovery* 34 (6): 1936–1962.

99 Lin, W., Wu, D., and Boulet, B. (2021). Spatial-temporal residential short-term load forecasting via graph neural networks. *IEEE Transactions on Smart Grid* 12 (6): 5373–5384.

100 Boyaci, O., Narimani, M.R., Davis, K.R. et al. (2022). Joint detection and localization of stealth false data injection attacks in smart grids using graph neural networks. *IEEE Transactions on Smart Grid* 13 (1): 807–819.

101 Cui, Z., Henrickson, K., Ke, R., and Wang, Y. (2020). Traffic graph convolutional recurrent neural network: a deep learning framework for network-scale traffic learning and forecasting. *IEEE Transactions on Intelligent Transportation Systems* 21 (11): 4883–4894.

102 Deng, L., Lian, D., Huang, Z., and Chen, E. (2022). Graph convolutional adversarial networks for spatiotemporal anomaly detection. *IEEE Transactions on Neural Networks and Learning Systems* 33 (6): 2416–2428.

103 Presekal, A., Ştefanov, A., Rajkumar, V.S. et al. (n.d.) *Attack Graph Model for Cyber-Physical Power Systems using Hybrid Deep Learning. IEEE Transactions on Smart Grid, early access.*

104 Chen, J., Wang, X., and Xu, X. (2021). GC-LSTM: graph convolution embedded LSTM for dynamic link prediction. *Applied Intelligence* 1–16.

105 Snoek, J., Larochelle, H., and Adams, R. (2012). Practical bayesian optimization of machine learning algorithms. *Advances in Neural Information Processing Systems* 25: 1–9.

106 Rajkumar, V. S., Tealane, M., Ştefanov, A. et al. (2020). Cyber attacks on protective relays in digital substations and impact analysis. *Proc. 8th Work. on Mod. and Simu. of Cy.-Phy. En. Sys.*, Sydney, NSW, Australia.

107 Rajkumar, V.S., Tealane, M., Ştefanov, A. et al. (2020). Cyber attacks on power system automation and protection and impact analysis. *Proc. ISGT-Europe*, The Hague, Netherlands, pp. 247–254.

108 He, K., Zhang, X., Ren, S. et al. (Jun. 2016). Deep residual learning for image recognition. *Proc. of the IEEE Conf. on Comp. Vis. and Pat. Recog.*, Las Vegas, USA, pp. 770–778.

109 Long, J., Shelhamer, E. and Darrell, T. (Jun. 2015). Fully convolutional networks for semantic segmentation. *Proc. of the IEEE Conf. on Comp. Vis. and Pat. Recog.*, Boston USA, pp. 3431–3440

110 LeCun, Y., Bengio, Y., and Hinton, G. (2015). Deep learning. *Nature* 521 (7553): 436–444.

20

Attack Detection and Countermeasures at Edge Devices

Fahim Ahmed and Md Tanvir Arafin

Cyber Security Engineering Department, George Mason University, Fairfax, VA, USA

20.1 Introduction

In recent years, the world has experienced a marked explosion in the number of low-power, application-specific computing devices connected via the internet. These devices are mostly connected at the edge of a network for data collection, aggregation, and data-derived computation at a granular level and for providing intelligent system design and control solutions. In current literature, these networked connected components are referred to as *Things* or *edge devices*.

Edge devices solve some of the long-standing problems of intelligent real-time systems design. These devices are ubiquitous in applications—from home appliances to power systems to edge robots. For example, connected sensor nodes (such as phasor measurement units) in a large area power distribution network can provide a detailed picture of the load (i.e., power consumption) within a grid and enable better load-balancing, generation, and control algorithms leading to a *smart* grid. Similarly, edge robots (i.e., small-scale robots and autonomous systems) can be deployed for intelligent rescue operations, autonomous surveillance, and delivery services. Additionally, large-scale deployment of environmental sensors can lead to early warning systems; distributed monitoring of interdependent processes and control parameters in industrial plants ushers intelligent manufacturing solutions; and communication and collaboration between inter- and intra-vehicular sensors can deliver intelligent transportation systems.

As a result, the number of edge devices is predicted to increase significantly over this decade, as shown in Figure 20.1. Interestingly, the problem of designing large-scale smart/intelligent systems has become tractable due to three factors:

a) Increased deployment of low-power data collection/sensing nodes;
b) Improved networking solutions via 4G/5G connectivity;
c) The advent of machine learning (ML) algorithms for ingesting large volumes of data to provide meaningful inference solutions.

Edge devices often serve as the sensors and actuators in an Internet of Things (IoT) ecosystem, whereas a central server provides the control decision. Therefore, increasing the number of connected edge devices results in better ML algorithms and precision control. Over the last decade, this successful scalable and *smart* system design approach (where edge nodes collect data and a central server processes this data using ML models for control decisions) has led to tremendous growth in the number of edge devices. This growth is expected to continue over the next decade, resulting in more than 7 billion edge devices by 2030 [1].

Smart Cyber-Physical Power Systems: Fundamental Concepts, Challenges, and Solutions, Volume 1, First Edition.
Edited by Ali Parizad, Hamid Reza Baghaee, and Saifur Rahman.

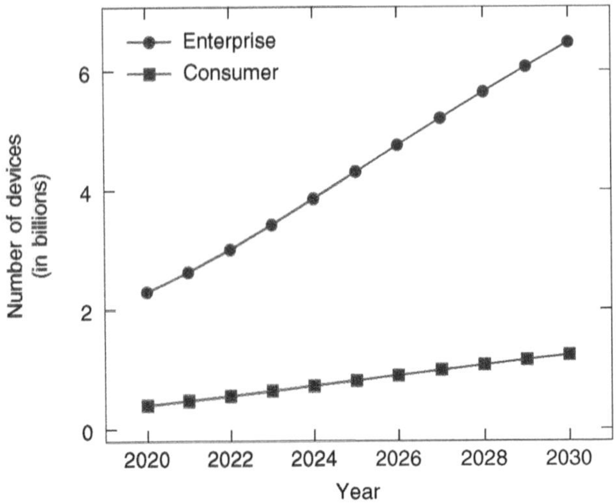

Figure 20.1 Predicted increase in edge devices from 2020 to 2030 in enterprise and consumer sectors (adapted from [1]).

Edge devices are primarily application-specific and designed with budget constraints in terms of power, area, and computation capacity. Hence, the security of these devices has often remained an afterthought in the system designing process. If adopted for critical infrastructure, this security-oblivious *smart* system design has the potential for severe consequences. For example, attackers can target Phasor Measurement Unit (PMUs) in a smart grid via network-based attacks or Global Positioning System (GPS) spoofing attacks to corrupt the phasor measurements and subsequently compromise the entire power system [2].

Unfortunately, weak and vulnerable edge devices are abundant in the current generation of intelligent IoT systems. Given the variety of edge devices and their multitudes of designs and operations, it is challenging to create targeted solutions for attack detection and countermeasure designs. Hence, it is an opportune time to introspect the progress made over the last few decades in attack detection and countermeasure development for the IoT edge and explore the open problems in this field. This chapter will present our exploration, discoveries, and introspection on the advances and opportunities of security research for edge resources.

20.2 Attack Surfaces for Edge Devices

An *attack surface* is defined as a pathway for compromising the integrity of the operation of a network-connected device. This pathway can originate from hardware, software, networking, or other system components connecting the device with the environment.

In edge devices, vulnerabilities originating from multiple surfaces can be exploited using different attack models and attacker goals. Hence, for each distinct attack, the type and intent of the attacker must be understood before designing countermeasures. In general, attack surfaces in edge devices are classified into three categories, i.e., (i) physical/hardware, (ii) software, and (iii) network attack surfaces.

20.2.1 Physical Attack Surface

At the lowest level of system architecture, we have hardware that performs the computation and connects a computation's logic and control decisions to the physical world. Although traditional computer security primarily focuses on software vulnerabilities, increasing hardware exploitation in recent years has exposed the perils of hardware-oblivious security designs. Computation does not occur within a void; instead, it is realized through electronic signals and systems, and thus it leaves physical fingerprints. Therefore, hardware vulnerabilities arise from different attack surfaces, such as side and covert channels, leakage of execution time, fault injection, and the inclusion of malicious hardware components.

Hardware attacks become more prominent for edge devices due to the proximity and availability of the devices to the end user. For example, profiling-based SCA requires data acquisition from similar devices. Thus, mimicking such attacks on large enterprise servers might cost attackers significantly, whereas profiling low-cost edge nodes is cheap and effective. Modern-day cryptography depends on Kerckhoff's principle, where security is inherently built on the secrecy of the cryptographic keys. Thus, using hardware side channels, an attacker can cost-effectively leak keys, and if such keys are used over the network, it becomes easier to infiltrate the network from the edge.

Moreover, attack-oblivious circuit design opens up more straightforward yet effective attack surfaces. For example, firmware used in an edge device is often stored in flash memory. If the flash memory is not protected, an attacker can easily capture the contents using simple flash dump attacks [3]. Thus, attack surfaces on physical hardware in an edge device can have profound implications for the security of the entire IoT ecosystem.

20.2.2 Software Attack Surface

Software security for edge devices is also challenging due to the unique nature and application of these devices. Software attack surfaces on IoT edge devices arise from two distinct sources:

a) Insecure application code, and
b) Vulnerable operating system (OS).

Edge devices are generally budget-constrained regarding computation power and memory. As a result, full-fledged OSs are not the first choice while designing these systems. Therefore, designers use (i) bare-metal software, (ii) custom-made OSes such as Mbed OS [4] and MicroBlaze [5], or (iii) OSes tailored from existing full-fledged lightweight OS distributions such as Yocto [6], BuildRoot [7], and OpenWRT [8]. Interestingly, these design choices lead to different attack surfaces for the system.

Generally, bare-metal programming for embedded systems rarely considers secure coding practices and thus remains vulnerable to simple attacks such as buffer overflows, memory corruption, and code reuse [9]. In addition, security via obfuscation strategy in embedded system development still (falsely) provides acceptable security assumptions to the developers. As a result, embedded systems at the edge remain the most vulnerable to common software attacks.

On the other hand, system-specific OSes such as MbedOS and Microblaze suffer software vulnerabilities due to insecure implementation of applications, unpatched zero days, and relatively uncomplicated reverse engineering efforts. Finally, software in Linux-derived lightweight OSes enjoys better security and management options. However, insecure software writing practices still make the applications vulnerable to standard software attacks ranging from basic buffer overflows to complicated return-oriented programming (ROP) attacks.

20.2.3 Network Attack Surface

The triumph of edge-device-centric intelligent system design has relied on the tremendous development in network connectivity required to maintain communication and control for many

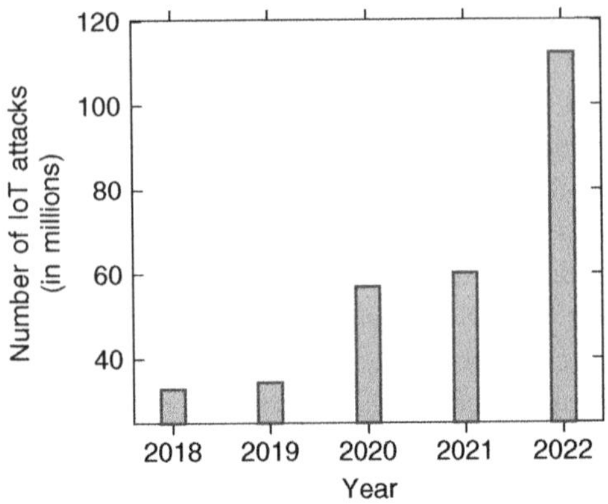

Figure 20.2 Number of IoT cyber attacks worldwide from 2018 to 2022 (source: adapted from [10]).

devices and systems. Less-regulated connectivity to many resource-constrained devices has the potential to create network security nightmares. This is evident in the increasing number of cyber attacks in the IoT ecosystem over the last few years, as shown in Figure 20.2.

Network attack surfaces are fundamentally attributed to weak passwords and default credentials, improper use of cryptography, poorly defined access control, and a lack of understanding of network security fundamentals for the edge. Assess control and resource management for a dynamic large-scale network is a complex problem, which can become highly challenging with resource-constrained devices connected to the network.

20.2.4 Goals of the Attacker

Attacks on edge devices aim at accessing the system to collect secret information, alter the system's runtime behavior, and disrupt or altogether turn off the system. We find that attacks at the edge nodes have some common goals:

G1. **Resource Hijacking**: Resource hijacking can be performed with software-based attacks. The attacker accesses the device first using software vulnerabilities and then, through privilege escalation, gains control over the system. Finally, the attacker manipulates the device to perform malicious tasks while keeping the process discreet from the victim. Common examples of such attacks are utilizing a device's resource for cryptocurrency mining, controlling a device as a botnet to launch a distributed denial of service (DDoS) attack, or using a device to spread spam and malicious software.

G2. **Information Leakage**: The goal of such an attack is to infer secret information (such as cryptographic keys, process algorithms, and user identity) from the device. Standard side channels and eavesdropping are examples of such attacks.

G3. **Subversion**: Subversive attacks usually target the underlying control algorithms in an IoT subsystem. For example, with the rise of ML-large-scale controls, attackers can manipulate the sensor reports or network data from the edge to corrupt the control parameters or decisions.

G4. **Reverse Engineering**: In this scenario, the attacker aims to steal a device's firmware, design architecture, or sensitive intellectual property (IP). It is difficult to cluster reverse engineering (RE) activities as a covert or overt approach since the victim might be aware of such activities due to the reports of stolen devices or counterfeit products. However, the attacker usually aims to minimize the victim's awareness when such activities occur. In addition, RE efforts can also provide the necessary tools and efforts to perform an overt attack.

G5. **Device Function Disruption**: The goal of the attacker, in this case, is that the edge device will not perform as expected after the attack. Fault injection Attacks, such as bit-flip attacks, timing, and power glitches, can corrupt the running processes on a single device. Hence, the output behavior of the device will be unusual. When such attacks are launched over multiple devices, they can lead to service disruptions and system failures.

G6. **Denial of Service**: This attack targets a complete or temporary device shutdown of a single or a group of devices. Network-level attacks, such as DDoS attacks by Botnets and software-level attacks like malware attacks, can serve this purpose.

When building an intelligent system such as a smart grid or an edge robots-based automation solution, the designers should be aware of these common attack goals and surfaces for edge resources and develop their system accordingly with tighter access control and resource management policies.

20.3 Security Issues and Common Attacks in Edge Devices

20.3.1 Security Issues

Based on our discussions in this chapter so far, it is evident that edge devices present unique security challenges. We find that security issues in edge devices depend on factors such as outdated systems and lifecycle management issues, poor automation and control definition, weak privacy measures, and insecure scaling practices.

20.3.1.1 Outdated Systems

In the face of new and exciting products and systems launched each year, the lifecycle management of edge devices has become a challenging problem. Legacy devices coexist with newer ones in a network and can have severe security flaws due to the lack of support for end-of-life (EoL) devices. In addition, many legacy systems at the edge of IoT networks were not designed with security fundamentals at all.

Interestingly, competing financial motives can exist between the consumer and the manufacturer to replace or update existing devices. From an economic point of view, there is very little financial motivation for the manufacturers to continue support such as vulnerability patching, OS updating, and standard software maintenance. These activities require engineering resources, which from a business point of view, would be better utilized if used for new product development.

On the other hand, a user might continue using a functional device even after the manufacturer's support has ended for cost-saving reasons. These outdated devices are easily compromised using known vulnerabilities and zero-day attacks. The issue has become so severe that the FBI recently published an industry notification illustrating the cyber attack opportunities on unpatched and outdated medical devices and systems [11]. Thus, outdated EoL software, devices, and systems at the edge create significant security problems for the IoT ecosystem.

20.3.1.2 Weak Device and Network Management

Managing a network of heterogeneous devices is another complicated design issue for large-scale IoT networks, and this poor management can open up significant attack points to the network. For example, weak default configuration or exploitable side channel on the edge devices can efficiently be utilized to gain access to a system and attain the covert goals (i.e., G1–G3) or the overt ones (i.e., G5–G6).

Moreover, network management issues such as credential handling, access control, security monitoring, patching, updating, and resource management for the IoT have yet to receive extensive evaluation and standardization processes. Therefore, manufacturers and users of edge devices seldom follow network hygiene for these systems.

As a result, common botnets employ simple tactics, such as using the support passwords on edge devices to gain access to the network [12]. Thus, we have observed the success of botnets such as Mirai, which used a small list of (around 60) factory default credentials to hijack more than 300,000 IoT devices in 164 countries within a few months [12, 13]. From a security research point of view, it is clear that such exploitation will continue. As the easier "low-hanging" vulnerabilities (such as support passwords) are resolved, the attackers will deploy more complex attacks on network and device management tools to infiltrate IoT subsystems.

20.3.1.3 Privacy

Privacy is another critical concern in IoT edge devices. Since a significant subset of the devices is deployed in a user-facing environment, protecting user data becomes an imperative security problem to consider. The simplest solution remains to utilize cryptography for protecting user data and confidential information during transmission and storage. However, using cryptography in resource-constrained systems is complicated and sometimes prohibitively expensive. This leads to weaker resource protection and information leakage attacks at the edge. Moreover, end-to-end cryptography measures are difficult and costly to deploy and seldom have any financial incentive for the manufacturer.

20.3.1.4 Economics of Scale

Finally, scaling security fundamentals from the edge to the cloud remains a challenging problem. The unprecedented growth in the sheer volume of edge-based resources and cloud solutions for everyday issues has outpaced these systems' thorough security design and implementation. Therefore, scaling problems are rampant in large IoT ecosystems that support heterogeneous nodes and complex computing and control solutions. As a result, security has often been an afterthought in both the enterprise and consumer space and has only been attended to when major security disasters emerge.

20.3.2 Common Attack Examples

Due to these existing security issues, attacks on edge devices have become increasingly common. These attacks can occur as a single type of attack or a combination of several types, depending on the attacker or the kind of attack. Table 20.1 provides a list of common malware found in IoT edge devices. A few of the well-known attacks are discussed below.

20.3.2.1 Malware for Distributed Denial of Service Attacks

With the increasing number of internet-connected edge nodes, one of the most trivial attacks is to launch DDoS employing compromised edge devices. Some common malware used for such attacks are Mirai, Remaiten, and BASHLITE. Attackers usually scan over the internet for vulnerable edge devices and then infect them with malware that provides basic command and control capabilities, thus creating remotely controlled bots. Once the malware is installed over a large number of devices, the attacker performs a DDoS attack on a victim's website or service using the bots.

Botnet-based DDoS attacks require a command and control capability over many devices. It has been found that the attackers exploit the poor network hygiene of edge devices to build an army of

Table 20.1 Common malware found in IoT edge devices.

Malware	Vulnerability utilized	Common usage
Mirai [12]	Default manufacturer credentials	DDoS
Remaiten [14]	Weak username and password	DoS, malware distribution
Linux.Wifatch [15]	Weak or default telnet credentials	Disconnecting a device
BASHLITE [16]	Common usernames and passwords	DDos, creating C&C network
Linux.Darlloz [17]	CVE-2012-1823	Crypto mining
BrickerBot [15]	Weak telnet credentials	Destroying a device
Fusob [18]	User download	Mobile ransomware
WannaCry [18]	CVE-2017-0144	ransomware
Linux Spike Trojan (MrBlack) [19]	Default credentials	Man-in-the-middle, cookie hijacking
Stuxnet [20]	Zero days in Windows OS	Compromising SCADA systems

botnets [12–14]. In most cases, using factory default credentials or common user IDs and passwords leads to a small but potent dictionary that can compromise many devices. In addition, poor network management, over-generalized access control policies, and non-existing lifecycle maintenance of the devices, as discussed in Section 20.2, contribute to the success of botnet attacks.

20.3.2.2 Ransomware

A ransomware infects a computer system and encrypts critical data and resources with an attacker-provided encryption key. The victim can only retrieve the resource by obtaining the decryption key from the attacker by paying a significant ransom. Ransomware such as Fusob and WannaCry have demonstrated how mobile and cloud computing platforms can be compromised to gain ransom from a victim network [18]. Ransomware gains profitability and practical prominence due to the wide-scale use of cryptocurrencies, which can provide complete anonymity over a financial transaction. Unprotected edge nodes can offer entry points for ransomware, and then poorly managed access control policies in the network lead to the compromise of critical resources.

20.3.2.3 Eavesdropping and Man in the Middle Attacks

Traditional eavesdropping and man-in-the-middle (MITM) attacks are mainly target-specific and part of advanced persistent threats (APTs). These attacks compromise the privacy guarantee of communication in the IoT network. Given the prominence of edge devices in the consumer space, the potential of private information leakage via eavesdropping and MITM attacks is alarming. For example, there have been reports on the hacking of home security cameras, baby monitors, and IP cameras via software vulnerabilities and common/weak credential usage [21].

20.3.2.4 Computer Resource Stealing Attacks

The advent of cryptocurrency mining and command and control (C&C)-based DDoS attacks have made edge nodes attractive targets for resource stealing. Since many edge devices are automatically operated and maintained, resource-stealing malware remains unnoticed as long as there is no significant drop in performance. Thus, the victim ends up unknowingly paying for the computation cost. Since cryptocurrency mining and C&C operations provide lucrative opportunities, malware such as Mirai and Linux.Darlloz compete for device resources and try to retain control by eliminating other malware in a compromised edge device.

Moreover, some malware, such as Linux.Wifatch and Hajime do not steal the resources; instead, they disconnect devices from the network. Although they claim to be white-hat activists who remove vulnerable devices from the network, these attacks cause significant disruption to service.

20.3.2.5 Hardware Attacks

Low-power and resource-constrained hardware in the edge devices is also targeted for hardware-based attacks. Hardware attacks are target-specific and create entry points for other attacks. For example, side channel leakage of cryptographic keys used for encrypting a firmware of a given device leads to bot design or counterfeit manufacturing for that device. On the other hand, fault injection attacks help attackers bypass security measures. Hence, here we discuss these hardware attacks in detail.

Side Channel Analysis (SCA): Physical operation of edge devices can leak security-sensitive information through hardware side channels. In such cases, the attacker observes, extracts, and analyzes physical properties to infer information about cryptographic computation, execution details, and other critical functionalities. SCAs in edge devices are classified into two categories—(i) physical and (ii) non-physical.

- **Physical SCA**: During a physical side-channel attack, the attacker exploits the physical properties of the device, such as electromagnetic leakage, power consumption, and acoustic output. For example, power analysis attacks examine the instantaneous power consumption and use a statistical model (for power analysis during cryptographic computation) to infer information regarding the encryption key [22]. Similarly, acoustic attacks are performed by capturing and analyzing the acoustic waves generated by the chips during encryption using a microphone [23].
- **Non-Physical SCA**: In this attack, the attacker exploits a chip's non-physical parameters to collect the design credentials and the processes. For instance, during the timing attack, the attacker observes the routine runtime to obtain information about the running processes. In addition, cache-based attacks monitor cache hit-and-miss timing differences to reveal system information [24].

Fault-Injection Attack: Unlike SCA, in a fault-injection (FI) attack, the attacker injects external faults into the device to modify the functionality, extract sensitive information, or disable the system. Some common FI attacks are discussed below.

- **Row Hammer Attack**: In this attack, bits in the dynamic random-access memory (DRAM) cells are flipped, thus injecting faults into the device. DRAM cells consist of transistors and capacitors aligned in arrays. Due to the continuous scaling of DRAMs, the cell density is increasing; hence, the electromagnetic coupling effect among the cells is increasing. It has been shown that activating the rows in a DRAM at a very high frequency disturbs the nearby cells. When the disturbance exceeds the threshold, bits in the nearby cells are flipped and corrupted [25].

- **Power Glitch Attack**: During the power glitch attack, the supply voltage is changed aggressively to modify the execution flow, specifically to skip the targeted instruction. This attack usually aims at bypassing security checks, avoiding the number of attempts barrier while launching a brute-force password break attack [26].
- **Clock Glitch Attack**: This attack exploits the clock management capabilities in hardware. A *glitchy clock cycle* is a temporal voltage spike that can be generated by changing the clock source where both have the same clock frequency but slight phase differences. During the glitched clock signal, an invalid signal is received at the register, which corrupts the corresponding routine execution [27].

20.4 Attack Detection Techniques and Countermeasures

Detecting a new threat in an IoT ecosystem requires intelligent monitoring and surveillance services over the entire network. Additionally, hardware attacks can involve physical compromise or cloning of the devices, and therefore, situational and physical awareness of the network is also required to detect such attacks. Moreover, honeypots, zero-day management, and malware research also help discover attacks early. These topics are discussed in detail here.

20.4.1 Common Attack Detection Techniques

20.4.1.1 Real-Time Monitoring and Honeypots

Real-time monitoring and analytic tools are standard in enterprise IoT networks. Different vendor products exist in the commercial domain that provide end-to-end IoT network monitoring solutions. General network attacks, i.e., simple DDoSes, can be detected through these solutions. However, commercial solutions have several drawbacks, like cost, engineering efforts for installation and maintenance, a generalized one-solution-for-all approach, and a lack of compatibility among different solutions. Some of these drawbacks, such as compatibility issues and depth of protection, can be addressed through standardization efforts. On the other hand, other disadvantages, e.g., detecting targeted attacks and protecting against such threats, will require a complete rethinking of network design and architecture for edge devices.

Honeypots provide active threat detection and analysis capabilities to the researchers. Thus, new malware that exploits zero days or performs weak credentials-based attacks can be detected through well-positioned honeypots at the edge [28]. Moreover, a well-engineered honeypot design that redirects or reroutes malware traffic from the network can provide active protection against DDoS attacks [29].

20.4.1.2 Machine Learning Tools

Large-scale attacks on IoT networks and edge devices start with abnormal behavior. Thus, effective anomaly detection techniques are imperative for the early diagnosis of a critical attack. Recent progress in data-oriented large-scale ML is poised to impact anomaly detection in the IoT network significantly. Interestingly, current progress in ML algorithms is mostly in supervised learning scenarios that are excellent in detecting patterns that they have learned from labeled training data. Thus, supervised learning-based models will have good precision and recall performance for attack signatures that the model has already learned. However, such an approach is only partially helpful in detecting novel attacks.

20.4.1.3 Physical Fingerprinting

For hardware-based attacks, real-time malware detection is possible through application signature monitoring [30–32]. Intra-processor SCA through shared memory systems or power lines can also provide real-time process monitoring capabilities to protect processors from running malicious or crypto-mining codes. In addition, the physical operation of edge devices and sensors leaves fingerprints in the collected data that can be leveraged to detect an attack on the node [33, 34].

20.4.2 Countermeasures

Active and passive countermeasures can secure edge resources and devices from future threats. Since vulnerability exploitation and malware propagation depend on common factors such as commonly used credentials and poor network management, as discussed in Section 20.3, a well-informed system design with countermeasures will lead to a robust cyberspace. Endpoint protection, use of lightweight cryptography, proper manufacturer description criteria for secure configuration of edge devices, and security-conscious hardware designs will secure the next generation of consumer and enterprise edge resources.

20.4.2.1 Endpoint Authentication

Secure authentication protocols that do not use factory default credentials and do not allow commonly used passwords provide security against botnet generation. However, designing automated multi-factor authentication protocols for the edge nodes seldom accessed by operators or users is challenging. For these scenarios, carefully designed protocols and hardware-based security primitives such as physical unclonable functions (PUFs), fingerprints, and root of trusts (i.e., trust zones and secure enclaves) can provide better authentication guarantees at the edge [35–37].

20.4.2.2 Lightweight Cryptography

Cryptographic protocols are mandatory on data during transit and at rest for data privacy and resource protection at the edge nodes. Unfortunately, standard cryptographic algorithms such as Advanced Encryption Systems (AES) for private key and Rivest-Shamir-Adleman (RSA) algorithms for public key cryptography demand additional resources from budget-constrained devices. Fortunately, there have been significant advances in lightweight cryptography for small devices in recent years. Recently (February 2023), NIST selected Ascon—a family of cryptographic algorithms for low-power resource-constrained devices [38]. This standardization of lightweight cryptography promises to solve the long-existing standard encryption and key management problems for IoT networks.

20.4.2.3 Manufacturer Usage Descriptions (MUDs)

Unauthorized malicious devices can enter a network by masquerading or cloning harmless edge nodes. Such attacks use ambiguity in the network usage permission of the edge resources. To thwart such exploitation, NIST has recently provided guidelines for manufacturer usage descriptions (MUDs) [39]. MUDs are manufacturer-defined resource usage descriptions that are assigned to individual edge devices. When connected to a new network, a MUD-supported device provides details of the access requirements for its operation. The manufacturer predetermines a given device's network use and resource access. This enables stricter access control policies without compromising the activities of trusted nodes. This strategy also moves part of the security burden to the manufacturer of edge products.

20.4.2.4 Zero Trust Architecture

Recent cyber attacks on networked components leverage the lack of granular access control of resources. For example, attackers first target edge components to access the network and then perform privilege escalation to exploit a critical resource that lacks proper access control mechanisms. This way, data and other resources such as computation, sensor readings, and actuator control can be stolen or exploited via weak and outdated access control policies for shared resources.

To thwart the vulnerabilities arising from traditional enterprise firewalls that use generalized and broad access control policies, zero trust design for network architecture has been proposed [40–43]. Zero trust networks fundamentally differ from traditional static, broad access-controlled designs and employ detailed and dynamic resource monitoring and access policies. The core concepts of zero trust architecture revolve around granular access control, least privilege for resource utilization, dynamic trust validation, and smaller *trust zones*. To provide a well-defined standard framework for zero trust architecture, NIST has recently (August 2020) published a zero trust architecture guideline that provides the necessary details an enterprise network should maintain to design a zero trust solution [41].

The key logical components of a zero-trust architecture are the policy engine, policy administrators, and policy enforcement points, as shown in Figure 20.3. In a zero-trust network, resources are defined with an all-inclusive approach. Hence, not only sensitive data marked for tighter access, relatively standard network components, i.e., computing resources, trusted edge connected devices, and verified low-power sensors and actuators are all tagged as trusted resources in the network. Access to these trusted resources is rigorously maintained using the key components. The policy engine provides the ultimate access decision (i.e., grant, deny, and revoke) for a networked resource to an incoming or existing device/node in the network. The policy engine utilizes standard access policies and real-time threat intelligence data to execute dynamic trust algorithms, ensuring that only trusted entities can access the resource [41].

Zero trust designs enforce that *every* resource access be monitored and controlled by the policy enforcement point and thus move away from the *lazy* and implicit trust solutions that allow open (resource) access to previously trusted entities. This results in a vigilant networking solution that does not inherently trust any given entity even if it was authenticated before (hence the term zero trust). With the broader definition of resource, this zero trust mechanism overseen by a dynamic policy engine and enforced by the enforcement points delivers smaller but effective trust zones at the edge. Hence, zero-trust networks are promising for securing the smart power grid solutions and other critical infrastructures modernized by edge-based technologies.

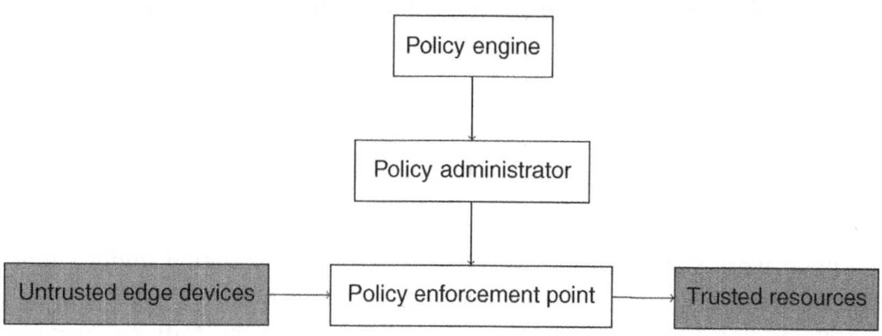

Figure 20.3 Basic components of a zero trust architecture.

20.5 Conclusions and Future Research Directions

Scalable security solutions are imperative for the ever-increasing number of edge resources and devices in the IoT ecosystem. This chapter discusses the common attacks that have exploited edge devices and the existing measures to detect and counter these threats. Over the last decade, simpler exploits such as factory default passwords were practical enough for global-scale cyber-attacks. Fortunately, the situation has improved. However, existing vulnerabilities from the hardware, software, and network layers have the potential to compromise edge devices for the next large-scale cyber attack. Hence, significant research and standardization efforts should be pursued for usable security at the IoT network edge.

Newer solutions like zero trust architecture and MUD-based access control policies offer more granular and application-specific security solutions. Unfortunately, deploying these protocols requires additional resources and support. Therefore, secure transitioning protocols to finer access control should be investigated, which will create an on-ramp for the existing/legacy network.

Early detection of an active threat can significantly reduce the damage. Therefore, novel monitoring and detection techniques should be investigated. Edge nodes in a network usually have predictable behavior, network signature, and hardware fingerprint. These properties need to be carefully fused for robust monitoring and anomaly detection. Research and development initiatives in unsupervised and reinforcement learning algorithms for characterizing and isolating novel threats need to be pursued by the academic community and industry. Translating the revolution in ML for cyber security problems would create a paradigm shift in secure and robust network design.

In addition, proper deployment of cryptography to resolve privacy issues requires carefully engineered effort. Recent developments in lightweight cryptography will usher in a plethora of products and encryption solutions for the edge. However, their integration of the legacy system will be an exciting research avenue. Other advanced primitives, e.g., fully homomorphic encryption algorithms, offer cloud and edge computation on encrypted data. These primitives are enjoying rapid development and are expected to impact the data security of edge devices in the near future.

Finally, hardware security problems in edge devices are under-focused and remain open threats to the safety of future IoT networks. Hardware attacks can be very target-oriented and cause entry points for firmware RE, data breaches, and authentication compromises. Interestingly, hardware-based authentication and security solutions offer more robust entity management and guarantees. In each passing quarter, more security enclaves, trust-zone-supported microprocessors, and microcontrollers enter the edge device market. This hardwired root-of-trust at the device level has the potential to enable robust trust propagation solutions from the edge to the cloud. Therefore, new research initiatives are required to transcend hardware from the weakest to the strongest link in IoT security.

Acknowledgments

This material is based upon work supported by the National Science Foundation under Grant No. 2245156. Any opinions, findings, conclusions, or recommendations expressed in this material are those of the authors and do not necessarily reflect the views of the National Science Foundation.

References

1 Transforma Insights (2022). Number of edge enabled Internet of Things (IoT) devices worldwide from 2020 to 2030, by market. https://www.statista.com/statistics/1259878/edge-enabled-iot-device-market-worldwide/ (accessed 10 August 2023).

2 Arafin, M.T., Anand, D., and Qu, G. (2017). A low-cost GPS spoofing detector design for Internet of Things (IoT) applications. *Proceedings of the on Great Lakes Symposium on VLSI 2017*, 161–166.

3 Tellez, M., El-Tawab, S., and Heydari, M.H. (2016). IoT security attacks using reverse engineering methods on WSN applications. *2016 IEEE 3rd World Forum on Internet of Things (WF-IoT)*, 182–187. IEEE.

4 Alves, J. (2018). Arm Mbed OS 5.10 release: focus on connectivity, firmware management and ease of use.

5 Chu, P.P. (2018). *FPGA Prototyping by SystemVerilog Examples: Xilinx MicroBlaze MCS SoC Edition*. Wiley.

6 Salvador, O. and Angolini, D. (2014). *Embedded Linux Development with Yocto Project*. Packt Publishing Ltd.

7 Petazzoni, T. and Electrons, F. (2012). BuildRoot: a nice, simple and efficient embedded Linux build system. *Embedded Linux System Conference*, volume 2012.

8 Fainelli, F. (2008). The OpenWRT embedded development framework. *Proceedings of the Free and Open Source Software Developers European Meeting*, 106.

9 Clements, A.A., Almakhdhub, N.S., Saab, K.S. et al. (2017). Protecting bare-metal embedded systems with privilege overlays. *2017 IEEE Symposium on Security and Privacy (SP)*, 289–303. IEEE.

10 SonicWall (2023). Annual number of Internet of Things (IoT) malware attacks worldwide from 2018 to 2022 (in millions) [graph], in statista. https://www.statista.com/statistics/1377569/worldwide-annual-internet-of-things-attacks/ (accessed 12 August 2023).

11 FBI Cyber Division. Private industry notification: unpatched and outdated medical devices provide cyber attack opportunities. https://www.ic3.gov/Media/News/2022/220912.pdf (accessed 12 August 2023).

12 Cluley, G. (2016). These 60 dumb passwords can hijack over 500,000 IoT devices into the Mirai botnet. https://grahamcluley.com/mirai-botnet-password/ (accessed 12 August 2023).

13 Devry, J. (2024). Mirai botnet infects devices in 164 countries. https://www.cybersecurity-insiders.com/mirai-botnet-infects-devices-in-164-countries/ (accessed 12 August 2023).

14 Shobana, M. and Rathi, S. (2018). IoT malware: an analysis of IoT device hijacking. *International Journal of Scientific Research in Computer Science, Computer Engineering, and Information Technology* 3 (5): 2456–3307.

15 De Donno, M., Dragoni, N., Giaretta, A., and Mazzara, M. (2016). AntibioTic: Protecting IoT devices against DDoS attacks. *Proceedings of the 5th International Conference in Software Engineering for Defence Applications: SEDA 2016*, 59–72. Springer.

16 Marzano, A., Alexander, D., Fonseca, O. et al. (2018). The evolution of Bashlite and Mirai IoT botnets. *2018 IEEE Symposium on Computers and Communications (ISCC)*, 00813–00818. IEEE.

17 Adat, V. and Gupta, B.B. (2017). A DDoS attack mitigation framework for Internet of Things. *2017 International Conference on Communication and Signal Processing (ICCSP)*, 2036–2041. IEEE.

18 Adams, C. (2018). Learning the lessons of WannaCry. *Computer Fraud & Security* (9): 6–9.

19 Zeifman, I., Gayer, O., and Atias, R. (2015). Lax Security Opens the Door for Mass-Scale Abuse of SOHO Routers. https://www.incapsula.com/blog/ddos-botnet-soho-router.html (accessed 12 August 2023).

20 Langner, R. (2011). Stuxnet: Dissecting a cyberwarfare weapon. *IEEE Security & Privacy* 9 (3): 49–51.

21 Wroclawski, D. (2019). How to keep your home security cameras from being hacked. https://www.consumerreports.org/home-garden/home-security-cameras/keep-home-security-cameras-from-being-hacked-a2927068390// (accessed 12 August 2023).

22 Le, T.-H., Canovas, C., and Clédiere, J. (2008). An overview of side channel analysis attacks. *Proceedings of the 2008 ACM Symposium on Information, Computer and Communications Security*, 33–43.

23 Genkin, D., Pattani, M., Schuster, R., and Tromer, E. (2019). Synesthesia: Detecting screen content via remote acoustic side channels. *2019 IEEE Symposium on Security and Privacy (SP)*, 853–869. IEEE.

24 Doychev, G., Köpf, B., Mauborgne, L., and Reineke, J. (2015). CacheAudit: A tool for the static analysis of cache side channels. *ACM Transactions on Information and System Security (TISSEC)* 18 (1): 1–32.

25 Mutlu, O. and Kim, J.S. (2019). RowHammer: A retrospective. *IEEE Transactions on Computer-Aided Design of Integrated Circuits and Systems* 39 (8): 1555–1571.

26 Gomina, K., Rigaud, J.-B., Gendrier, P. et al. (2014). Power supply glitch attacks: design and evaluation of detection circuits. *2014 IEEE International Symposium on Hardware-Oriented Security and Trust (HOST)*, 136–141. IEEE.

27 Obermaier, J., Specht, R., and Sigl, G. (2017). Fuzzy-glitch: A practical ring oscillator based clock glitch attack. *2017 International Conference on Applied Electronics (AE)*, 1–6. IEEE.

28 Razali, M.F., Razali, M.N., Mansor, F.Z. et al. (2018). IoT honeypot: a review from researcher's perspective. *2018 IEEE Conference on Application, Information and Network Security (AINS)*, 93–98. IEEE.

29 Anirudh, M., Thileeban, S.A., and Nallathambi, D.J. (2017). Use of honeypots for mitigating DoS attacks targeted on IoT networks. *2017 International Conference on Computer, Communication and Signal Processing (ICCCSP)*, 1–4. IEEE.

30 Damshenas, M., Dehghantanha, A., Choo, K.-K.R., and Mahmud, R. (2015). M0droid: An android behavioral-based malware detection model. *Journal of Information Privacy and Security* 11 (3): 141–157.

31 Isohara, T., Takemori, K., and Kubota, A. (2011). Kernel-based behavior analysis for android malware detection. *2011 7th International Conference on Computational Intelligence and Security*, 1011–1015. IEEE.

32 Faruki, P., Bharmal, A., Laxmi, V. et al. (2014). Android security: a survey of issues, malware penetration, and defenses. *IEEE Communications Surveys & Tutorials* 17 (2): 998–1022.

33 Arafin, M.T. and Kornegay, K. (2021). Attack detection and countermeasures for autonomous navigation. *2021 55th Annual Conference on Information Sciences and Systems (CISS)*, 1–6. https://doi.org/10.1109/CISS50987.2021.9400224.

34 Yimer, T., Arafin, M.T., and Kornegay, K. (2020). Securing industrial control systems using physical device fingerprinting. *2020 7th International Conference on Internet of Things: Systems, Management and Security (IOTSMS)*, 1–6. https://doi.org/10.1109/IOTSMS52051.2020.9340160.

35 Arafin, M.T. (2018). Hardware-based authentication for the Internet of Things. PhD thesis. College Park, MD: University of Maryland.

36 Gao, M., Wang, Q., Arafin, M.T. et al. (2017). Approximate computing for low power and security in the Internet of Things. *Computer* 50 (6): 27–34.

37 Zhang, J., Shen, C., Su, H. et al. (2022). Voltage over-scaling-based lightweight authentication for IoT security. *IEEE Transactions on Computers* 71 (2): 323–336. https://doi.org/10.1109/TC.2021.3049543.

38 NIST (2023). NIST Selects 'Lightweight Cryptography' Algorithms to Protect Small Devices. https://www.nist.gov/news-events/news/2023/02/nist-selects-lightweight-cryptography-algorithms-protect-small-devices (accessed 12 August 2023).

39 Lear, E., Droms, R., and Romascanu, D. (2019). Manufacturer Usage Description Specification. *Technical Report Number: RFC 8520.*

40 He, Y., Huang, D., Chen, L. et al. (2022). A survey on zero trust architecture: challenges and future trends. *Wireless Communications and Mobile Computing* 2022 (1): 6476274.

41 Rose, S., Borchert, O., Mitchell, S., and Connelly, S. (2020). *Zero Trust Architecture. NIST Special Publication* 800-207.

42 Bertino, E. (2021). Zero trust architecture: does it help? *IEEE Security & Privacy* 19 (05): 95–96.

43 Syed, N.F., Shah, S.W., Shaghaghi, A. et al. (2022). Zero trust architecture (ZTA): a comprehensive survey. *IEEE Access* 10: 57143–57179.

21

Privacy-Preserving Outage Detection in Modern Distribution Grids: Challenges and Opportunities

Chenhan Xiao, Yizheng Liao, and Yang Weng

Department of Electrical, Computer and Energy Engineering, Arizona State University, Tempe, AZ, USA

21.1 Introduction

In distribution grids, the detection of line outages is essential for system monitoring and control, playing a critical role in the restoration of network stability [1] and the mitigation of customer losses. According to the U.S. Energy Information Administration [2], customers experienced over seven hours of power interruptions in 2021, attributed mainly to severe weather events and power supply shortages. Traditionally, utility companies have installed smart meters with advanced metering infrastructure (AMI) and fault location, isolation, and service restoration (FLISR) systems to report outages in cases of power absence [3]. However, these "last gasp" notifications are limited when customers continue to have power after the line outage from distributed energy resources such as rooftop solar panels, battery storage, and electric vehicles, which are now widely adopted. Additionally, in some urban areas, secondary distribution grids are mesh networks. In this setup, a single line outage induced by circuit faults or human interference may not result in a power outage because of alternative power supply routes. Consequently, smart meters at the customer end also cannot report outages.

To identify these types of line outages, real-time sensor measurements, including voltage magnitudes, phasor angles, and load estimates have been employed and confirmed for their effectiveness [4–9]. However, the utilization of real-time sensor measurements raises privacy concerns, particularly regarding the potential exposure of sensitive information. For example, if a customer's time-series grid data were provided to an untrusted third party, they could deduce appliance usage [10] and unveil details about household occupancy and economic status (as illustrated in the lower half of Figure 21.1) using non-intrusive load monitoring techniques [11–13]. Therefore, it is crucial to safeguard such data against direct disclosure to third parties during the outage detection process.

In pursuing a privacy-aware outage detection procedure, we choose to develop a decentralized randomization scheme based on a probabilistic methodology for encrypting the raw data. Among the methodologies for utilizing sensor measurements in outage detection, both deterministic [4, 8] and probabilistic [14, 15] approaches have been proposed. Deterministic methods typically set a threshold and declare an outage when data changes exceed this threshold. Although these techniques are easy to implement, they do not align with our concept of a randomization scheme for data encryption. In contrast, probabilistic approaches focus on monitoring changes in the probability distribution of sensor measurements, providing a suitable foundation for our approach. The core idea is to alter the absolute values of sensor measurements to protect privacy while

Smart Cyber-Physical Power Systems: Fundamental Concepts, Challenges, and Solutions, Volume 1, First Edition.
Edited by Ali Parizad, Hamid Reza Baghaee, and Saifur Rahman.

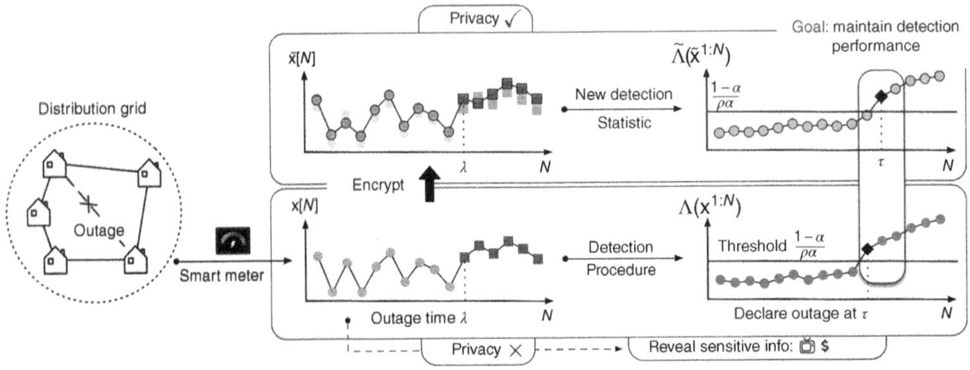

Figure 21.1 An overview of the privacy-aware line outage detection problem in the distribution grid.

preserving the relative changes in data distribution before and after an outage event (see upper half of Figure 21.1).

Specifically, we aim to develop a privacy-aware outage detection procedure based on our prior research [6, 15], which utilizes a probabilistic change point detection (CPD) method known for its guaranteed performance. The CPD approach is adopted for detecting changes in the probabilistic distribution of sensor measurements while adhering to a predefined false alarm tolerance constraint [16]. In our problem, the sensor measurements are modeled as a time-series data stream $\mathbf{x}[n] \in \mathbb{R}^p$, where $n \in \mathbb{N}$ corresponds to the time step. These time-series data are assumed to exhibit distinct probabilistic distributions before and after an outage time $\lambda \in \mathbb{N}$:

$$\mathbf{x}[n] \overset{i.i.d}{\sim} g,\ n < \lambda \quad \text{and} \quad \mathbf{x}[n] \overset{i.i.d}{\sim} f,\ n \geq \lambda \tag{21.1}$$

where g and f represent the distributions before and after the outage, respectively. The CPD framework with sensor data defined in (21.1) has been applied to detect line outages and faults in transmission grids [17, 18] as well as in DC microgrids [19]. These applications benefit from theoretical guarantees regarding optimal detection delay, as studied in [20].

In addition to detecting power line outages using sensor data from electricity customers, many other applications of the CPD framework involve similar privacy concerns related to the use of sensitive data. Such applications include monitoring patient health based on heart rates [21] and evaluating financial conditions using transaction data [22]. Consequently, the development of a privacy-aware CPD that preserves its detection performance has emerged as a substantial area of interest and is the primary focus of this chapter.

To safeguard privacy, recent studies have introduced randomization schemes to encrypt data, effectively concealing sensitive information from potential attackers. In assessing the level of privacy achieved by such randomization schemes, the differential privacy framework [23] is employed, offering a worst-case privacy guarantee. In the context of parametric CPD, where distributions g and f are known in (21.1), [24] utilized noisy approximation algorithms developed by Dwork and Roth [25] to compute a privately approximated change-point maximum likelihood estimation (MLE). In non-parametric CPD scenarios where the distributions g and f are unknown, [26] privately estimated the change points using the Mann–Whitney test [27]. These studies involved encrypting the detection statistic with Laplace noise after a trusted third party collected the raw data $\mathbf{x}[n]$. In cases where a trusted third party is absent, [28] proposed randomizing the raw data with Laplace noise, ensuring that the raw data remains inaccessible to anyone except its original holder. Despite the privacy guarantees offered by existing randomization approaches, there remains an inherent trade-off between privacy and detection performance, often in the form of prolonged detection delays. To the best of our knowledge, safeguarding privacy without compromising detection performance remains out of the reach of existing theory.

In this chapter, we narrow our focus on the parametric setting of CPD for line outage detection. Having knowledge of the distributions g and f allows us to quantify the cost associated with introducing privacy guarantees into the outage detection procedure. Furthermore, it empowers us to design a novel detection statistic aimed at mitigating this cost. To ensure that raw data remains undisclosed, we design a decentralized scheme to directly encrypt the raw data. To demonstrate that this scheme adheres to differential privacy, we employ the concept of Gaussian differential privacy [29], an extension of differential privacy applicable to arbitrary distributions. Despite the privacy guarantee, there is an inevitable compromise in detection performance due to the encryption of data. Specifically, we demonstrate that this compromise results in a prolonged detection delay by examining the Kullback–Leibler (KL) divergence between distributions f and g. These analytical investigations enable us to answer a key question: to what extent will the detection performance be compromised when aiming for a specific level of privacy?

To further resolve the degradation in detection performance, we introduce an innovative detection statistic by considering an unbiased estimation of the optimal statistic with access to raw data. The proposed statistic is shown to closely approximate the optimal case in terms of detection delay while adhering to the false alarm rate (FAR) constraint. Additionally, we discover that this proximity is related to Jensen's inequality, where the Jensen gap can be constrained by controlling the variance of the variable [30]. Building upon this insight, we redesign the proposed statistic to have a reduced variance. Through this process, we demonstrate that the degradation in detection performance can be nearly eliminated.

In summary, our contributions include (i) protecting the raw data through a randomization scheme under the differential privacy framework, (ii) quantifying the trade-off between privacy and detection performance in terms of detection delay and FAR, and (iii) introducing a novel statistic to alleviate, and in some cases eliminate, the impact of encrypted data on detection performance. To validate our contributions, we conduct comprehensive experiments utilizing representative distribution grids and real load profiles, covering 17 distinct outage configurations.

In the following, Section 21.2 introduces the preliminary aspects of our system modeling, the CPD framework, and the differential privacy framework. Section 21.3 presents our privacy-aware detection procedure that does not compromise detection performance. Section 21.4 assesses our method using four distribution grids and real-world load profiles. Section 21.5 concludes of this chapter.

21.2 Preliminaries

21.2.1 System Modeling

To illustrate our probabilistic design for the privacy-aware detection procedure, we define the following variables. The voltage magnitude at each bus $i \in \mathcal{G}$ is modeled as a random variable V_i, where $\mathcal{G} := \{1, 2, \ldots, p\}$ represents the distribution grid as a graph containing $p > 0$ buses. At time step n, we denote the realization of V_i as $v_i[n] \in \mathbb{R}$ in per unit, and we use $\mathbf{v}[n] = \{v_1[n], \ldots, v_p[n]\} \in \mathbb{R}^p$ to represent the collection of voltage magnitudes in the grid $\mathcal{G}$.

Finally, we use the notation $\mathbf{x}[n] = \mathbf{v}[n] - \mathbf{v}[n-1]$ to denote the incremental change in voltage magnitudes.[1]

We utilize voltage increment data because [15] establishes that this data adheres to two multivariate Gaussian distributions, denoted as $g \sim \mathcal{N}(\boldsymbol{\mu}_0, \boldsymbol{\Sigma}_0)$ and $f \sim \mathcal{N}(\boldsymbol{\mu}_1, \boldsymbol{\Sigma}_1)$ before and after a line outage. For the sake of simplicity, we also use the notation $\mathbf{x}^{1:N} = \{\mathbf{x}[1], \dots, \mathbf{x}[N]\}$ to represent all the measurements up to time N.

Based on the modeling, the problem of detecting distribution grid line outages while preserving privacy is formally defined as follows (refer to Figure 21.1 for visualization):

- **Given**: A stream of voltage magnitude increments $\mathbf{x}^{1:N}$ from the smart meters.
- **Find**: The line outage time λ as quickly as possible.
- **Require**: Avoid disclosing the raw data $\mathbf{x}^{1:N}$.

21.2.2 Outage Detection Based on CPD

To detect the outage time λ in (21.1) using voltage magnitude increments $\mathbf{x}^{1:N}$, our previous work [6, 15] follows the Bayesian detection procedure [16, 20]. That is, identifying the outage time is equivalent to performing the hypothesis test:

$$\mathcal{H}_0 : \lambda > N \quad \text{and} \quad \mathcal{H}_1 : \lambda \leq N$$

sequentially given data $\mathbf{x}^{1:N} = \{\mathbf{x}[1], \dots, \mathbf{x}[n], \dots, \mathbf{x}[N]\}$. As data is received in a streaming manner as (N increases), the first time hypothesis $\mathcal{H}_0$ is rejected reveals the value of λ. To determine when to reject $\mathcal{H}_0$, the posterior probability ratio

$$\Lambda(\mathbf{x}^{1:N}) = \frac{\mathbb{P}(\lambda \leq n | \mathbf{x}^{1:N})}{\mathbb{P}(\lambda > n | \mathbf{x}^{1:N})} = \sum_{k=1}^{N} \pi_N^k \prod_{n=k}^{N} \frac{f(\mathbf{x}[n])}{g(\mathbf{x}[n])} \tag{21.2}$$

is calculated at each time step N. $\lambda \in \mathbb{N}$ is assumed to follow a prior distribution π, and we define $\pi_N^k = \frac{\pi(k)}{\sum_{k=N+1}^{\infty} \pi(k)}$ for simplicity. The ratio in (21.2) compares the probabilities of "outage occurred ($\lambda \leq N$)" and "outage did not occur ($\lambda > N$)" given the historical measurements $\mathbf{x}^{1:N}$. A larger ratio indicates that "outage occurred" is more likely than "outage did not occur." Therefore, we declare the outage time λ when the ratio in (21.2) exceeds a predefined threshold. By the Shiryaev–Roberts–Pollaks procedure [16, 20], the following threshold in Theorem 21.1 optimally considers the trade-off between the false alarm and the detection delay.

Theorem 21.1 When λ follows a geometric prior Geo(ρ), we declare the outage time when the posterior probability ratio $\Lambda(\mathbf{x}^{1:N})$ exceeds the threshold $\frac{1-\alpha}{\rho\alpha}$ for the first time as

$$\tau = \inf \left\{ N \in \mathbb{N} : \Lambda(\mathbf{x}^{1:N}) \geq \frac{1-\alpha}{\rho\alpha} \right\} \tag{21.3}$$

The detection procedure (21.3) constrains that the FAR remains below a predefined tolerance level α, i.e., $FAR(\Lambda, f, g) := \mathbb{P}(\tau < \lambda) \leq \alpha$. More importantly, as $\alpha \to 0$, τ is asymptotically optimal for minimizing the average detection delay (ADD) as

$$\mathbb{E}[\tau - \lambda | \tau \geq \lambda] = \inf_{\mathbb{P}(\tau^* \leq \lambda) \leq \alpha} \mathbb{E}[\tau^* - \lambda | \tau^* \geq \lambda]$$

1 For simplicity, we use the notation $\mathbf{x}$ instead of $\Delta\mathbf{v}$.

$$= \frac{|\log \alpha|}{-\log(1-\rho) + D_{KL}(f\|g)} := \underline{ADD}(\Lambda, f, g) \tag{21.4}$$

where $D_{KL}(f\|g)$ denotes the KL divergence between distributions f and g.

21.2.3 Differential Privacy

To assess the level of privacy preservation, we follow the framework of differential privacy [23], which offers worst-case privacy guarantees. Specifically, an algorithm $\mathcal{M} : \mathbb{R}^p \to \mathbb{R}^p$ is (ϵ, δ)-differentially private if, for any neighboring datasets X and X' (differing in at most one element), and for every subset of possible outputs $\mathcal{S}$, the following inequality holds:

$$\mathbb{P}[\mathcal{M}(X) \in \mathcal{S}] \le \exp(\epsilon)\mathbb{P}[\mathcal{M}(X') \in \mathcal{S}] + \delta \tag{21.5}$$

In essence, this property ensures that a potential attacker observing the outcomes of the algorithm $\mathcal{M}$ cannot easily deduce whether a specific individual's information is present in the dataset. While the conventional technique for achieving differential privacy involves the introduction of Laplace noise [24] to raw data, the concept of Gaussian differential privacy [29] extends differential privacy to encompass noises generated from a broader range of distributions.

21.3 Privacy-Aware Line Outage Detection with Boosted Performance

In the aforementioned outage identification procedure (21.3), the increments of voltage magnitude data $\mathbf{x}^{1:N}$ are critical. However, such data may also be used to infer customer's sensitive information [11–13], such as household occupancy (see lower half of Figure 21.1), e.g., when the house owner arrives or leaves home. To protect the raw voltage data of customers, at each time step n when data $\mathbf{x}[n]$ is received, we apply a randomizing scheme to encrypt the raw data directly:

$$\tilde{\mathbf{x}}[n] = \mathbf{x}[n] + \mathbf{e}[n] \tag{21.6}$$

where $\mathbf{e}[n] \in \mathbb{R}^p$ is a random noise vector. The noise $\mathbf{e}[n]$ has to be sufficiently large to hide the characteristics of the raw data while not being too large to impact the detection performance. To establish the suitable level of embedded noise, we evaluate the privacy guarantee within the differential privacy framework in Section 21.3.1 and assess the corresponding degradation in detection performance in Section 21.3.2. We show that, in general, the noise added to the data makes it harder to distinguish whether the data comes from the distribution g or f, leading to a prolonged detection delay. Integrating these analyses, we propose a new statistic in Section 21.3.3 (to replace (21.2)) such that the new detection procedure is both privacy-preserving and has comparable detection performance to the optimal case with access to raw data.

For making the randomizing scheme (21.6) satisfy the differential privacy, we generate noise from the same distribution (Gaussian) as the raw data, i.e., $\mathbf{e}[n] \sim \mathcal{N}(\mathbf{0}, \boldsymbol{D}_e)$. The covariance matrix is designed to be diagonal, i.e., $\boldsymbol{D}_e = \text{diag}(\sigma_e^2, \dots, \sigma_e^2)$, where variance σ_e^2 represents the noise level or amount of noise. A diagonal covariance indicates that each element in the noise vector is independent. In doing so, the scheme (21.6) is equivalent to adding a random noise scalar to each dimension of the data vector, ensuring that each customer's raw data is encrypted before being sent to any third party (see Figure 21.2). Notice that, unlike some works that add noise to the statistics (e.g., $\Lambda(\mathbf{x}^{1:N})$) [24, 26] after raw data is collected, our approach ensures no direct exposure of the raw data.

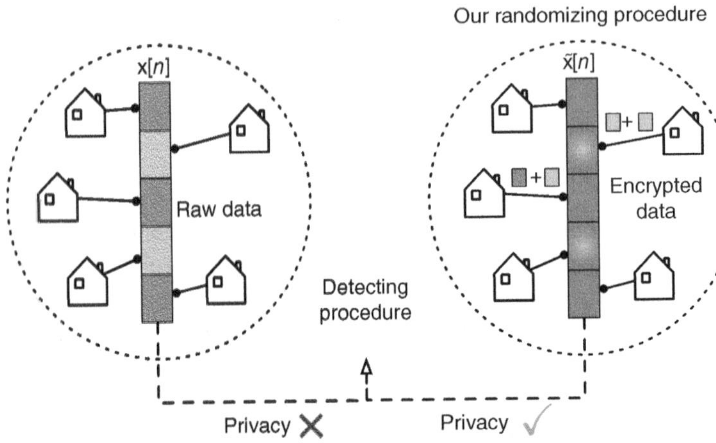

Figure 21.2 The decentralized randomizing scheme (21.6) to protect privacy of each customer i's raw data $\mathbf{x}_i[n]$ in the data vector $\mathbf{x}[n]$.

21.3.1 Differential Privacy Guarantee of the Randomizing Scheme

Applying the randomizing scheme (21.6), the detection procedure will be performed on the encrypted data $\tilde{\mathbf{x}}^{1:N} = \{\tilde{\mathbf{x}}[1], \dots, \tilde{\mathbf{x}}[N]\}$ to find the outage (see Figure 21.2). In this subsection, we quantify how much privacy is preserved w.r.t. the noise level σ_e^2. To achieve this, we prove that (21.6) satisfies the classic (ϵ, δ)-differential privacy mechanism [23].

A differential privacy scheme indicates that by looking at the encrypted data $\tilde{\mathbf{x}}[n]$, an adversary struggles to tell whether any piece of real data $\mathbf{x}_i[n]$ is included. The mathematical definition is given in (21.5). Since the noise $\mathbf{e}[n]$ is independent Gaussian to the raw data $\mathbf{x}[n]$, the encrypted data $\tilde{\mathbf{x}}[n]$ also follows Gaussian. This allows us to detour the proof of classic differential privacy by the tool of Gaussian differential privacy [29]. Specifically, a G_μ-Gaussian differential privacy scheme implies the following: telling whether any piece of real data $\mathbf{x}_i[n]$ is present in the encrypted data $\tilde{\mathbf{x}}[n]$ is more difficult than distinguishing between the distributions $\mathcal{N}(0, 1)$ and $\mathcal{N}(\mu, 1)$. The difficulty is quantified using the trade-off function $T(\mathcal{N}(0, 1), \mathcal{N}(\mu, 1))$, which characterizes the balance between type I and type II errors in distinguishing these distributions [29]. This particular trade-off function is also referred to as G_μ. In Lemma 21.1, we show that our scheme (21.6) is Gaussian differential private.

Lemma 21.1 The randomizing scheme (21.6) is $G_{\frac{\Delta}{edev}}$-Gaussian differential private [29], where $\Delta := \sup_{\mathbf{x}[n],\mathbf{x}'[n]} \|\mathbf{x}[n] - \mathbf{x}'[n]\|$ is the sensitivity of the raw data and $\mathbf{x}[n], \mathbf{x}'[n]$ only differs in exactly one element.

Proof: The encrypted data $\tilde{\mathbf{x}}[n]$ and its neighboring data $\tilde{\mathbf{x}}'[n]$ (i.e., they differ in exactly one element) both follow Gaussian distributions as $\tilde{\mathbf{x}}[n] \sim \mathcal{N}(\mathbf{x}[n], \boldsymbol{D}_e)$ and $\tilde{\mathbf{x}}'[n] \sim \mathcal{N}(\mathbf{x}'[n], \boldsymbol{D}_e)$. Then, we have

$$\begin{aligned} T\left(\tilde{\mathbf{x}}[n], \tilde{\mathbf{x}}'[n]\right) &= T(\mathcal{N}(\mathbf{x}[n], \boldsymbol{D}_e), \mathcal{N}(\mathbf{x}'[n], \boldsymbol{D}_e)) \\ &= G_{\|\mathbf{x}[n]-\mathbf{x}'[n]\|/edev} \geq G_{\frac{\Delta}{edev}} \end{aligned} \tag{21.7}$$

where $T(\tilde{\mathbf{x}}[n], \tilde{\mathbf{x}}'[n])$ is defined as the trade-off function between type I and II errors in differentiating data $\tilde{\mathbf{x}}[n]$ and $\tilde{\mathbf{x}}'[n]$. The inequality is due to the definition of sensitivity, i.e., $\|(\mathbf{x}[n] - \mathbf{x}'[n])/\mathbf{edev}\| \leq \frac{\Delta}{\mathbf{edev}}$.

Given the foundation of Gaussian differential privacy, we are ready to demonstrate that our scheme (21.6) also adheres to the classic (ε, δ)-differential privacy [23].

Corollary 21.1 Provided the $G_{\frac{\Delta}{edev}}$-Gaussian differential privacy, (21.6) satisfies the $(\varepsilon, \delta(\varepsilon))$-differential privacy [29] where

$$\delta(\varepsilon) = \Phi\left(-\frac{\varepsilon edev}{\Delta} + \frac{\Delta}{2edev}\right) - e^{\varepsilon}\Phi\left(-\frac{\varepsilon edev}{\Delta} - \frac{\Delta}{2edev}\right)$$

and Φ is the cumulative distribution function (CDF) of the unit normal distribution.

Satisfying the $(\varepsilon, \delta(\varepsilon))$-differential privacy in Corollary 21.1, our proposed scheme (21.6) ensures that an adversary can not easily determine if the data he observes is real, thus preserving the privacy of raw data. Moreover, we can control the amount of noise to achieve any desired level of privacy guarantee.

In fact, Lemma 21.1 and Corollary 21.1 reveal that the degree of differential privacy is directly related to the noise variance σ_e^2: larger noise results in enhanced privacy protection. The sensitivity Δ is determined by the distribution system and can be approximated using domain expertise. For instance, in power grid analysis, the sensitivity of voltage data can be computed based on its standard operational range (ranging from 0 to 1.1 p.u.).

21.3.2 Quantification of Detection Performance Degradation

While (21.6) enhances privacy protection, it may degrade the ability to detect line outages, potentially leading to increased detection delays and a higher FAR. Therefore, it is crucial to analyze the extent to which detection performance is compromised when utilizing the encrypted data $\tilde{\mathbf{x}}^{1:N}$. Only after completing this analysis can we devise a new solution to mitigate the degradation.

To study the performance degradation, we first note that the encrypted data $\tilde{\mathbf{x}}[n]$ follows a Gaussian distribution due to our choice of independent Gaussian noise for the raw data. Specifically, $\tilde{\mathbf{x}}[n]$ follows $g_e \sim \mathcal{N}(\boldsymbol{\mu}_0, \boldsymbol{\Sigma}_0 + \boldsymbol{D}_e)$ before the outage ($n < \lambda$) and follows $f_e \sim \mathcal{N}(\boldsymbol{\mu}_1, \boldsymbol{\Sigma}_1 + \boldsymbol{D}_e)$ after the outage ($n \geq \lambda$). We use the notation g_e and f_e to denote the "encrypted" distributions, which are the results of introducing independent noise to distributions g and f, respectively.

Having defined g_e and f_e, we can now rigorously measure the performance degradation. In Theorem 21.2, we demonstrate that the "distance" between g_e and f_e is smaller than that between g and f by evaluating their KL divergence. The "closer" the distributions are, the more challenging it is to distinguish them in the outage detection procedure, thus leading to a prolonged detection delay. Intuitively, if the noise term is infinitely large ($\sigma_e^2 \to \infty$), the distributions g_e and f_e will be dominated by the same noise distribution and become impossible to distinguish.

Theorem 21.2 The randomizing scheme (21.6) diminishes the KL divergence between pre- and post-outage distributions:

$$KL_{\Delta} := D_{KL}(f \| g) - D_{KL}(f_e \| g_e) \geq 0 \tag{21.8}$$

$$KL_{\Delta} \leq \mathcal{O}(\sigma_e^2)\left(\|\boldsymbol{\mu}_0 - \boldsymbol{\mu}_1\|_2^2 + \frac{\left(tr(\boldsymbol{\Sigma}_1) - tr(\boldsymbol{\Sigma}_0)\right)^2}{tr(\boldsymbol{\Sigma}_1)}\right) \tag{21.9}$$

Proof: For showing $KL_\Delta \geq 0$, we have

$$2KL_\Delta = \frac{1}{2}(\boldsymbol{\mu}_0 - \boldsymbol{\mu}_1)^T[(\boldsymbol{\Sigma}_0)^{-1} - (\boldsymbol{\Sigma}_0^e)^{-1}](\boldsymbol{\mu}_0 - \boldsymbol{\mu}_1)$$
$$+ \frac{1}{2}\log\frac{|\boldsymbol{\Sigma}_0|\,|\boldsymbol{\Sigma}_1^e|}{|\boldsymbol{\Sigma}_1|\,|\boldsymbol{\Sigma}_0^e|} + \frac{1}{2}tr\{(\boldsymbol{\Sigma}_0)^{-1}(\boldsymbol{\Sigma}_1) - (\boldsymbol{\Sigma}_0^e)^{-1}(\boldsymbol{\Sigma}_1^e)\}$$
$$\geq \frac{1}{2}\sum_{i=1}^{p}\left[(\nu_i - \log \nu_i) - (\mathbf{x}i_i - \log \mathbf{x}i_i)\right]$$

where $\boldsymbol{\Sigma}_i^e = \boldsymbol{\Sigma}_i + \boldsymbol{D}_e$ for $i = 0, 1$. $\nu_1, \ldots, \nu_p$ and $\mathbf{x}i_1, \ldots, \mathbf{x}i_p$ are the eigenvalues of $(\boldsymbol{\Sigma}_0)^{-1}\boldsymbol{\Sigma}_1$ and $(\boldsymbol{\Sigma}_0^e)^{-1}\boldsymbol{\Sigma}_1^e$, respectively. The inequality is due to that matrix $(\boldsymbol{\Sigma}_0)^{-1} - (\boldsymbol{\Sigma}_0^e)^{-1}$ is positive semi-definite. Moreover, since $|\mathbf{x}i_i - 1| \leq |\nu_i - 1|, \forall i = 1, \ldots, p$, we finally obtain $KL_\Delta \geq 0$. Aside from the lower bound as zero, an upper bound of KL_Δ is further derived as

$$KL_\Delta \leq \frac{1}{2}\|\boldsymbol{\mu}_0 - \boldsymbol{\mu}_1\|_2^2\left(\frac{1}{\nu_0^{min}} - \frac{1}{\nu_0^{min} + \sigma_e^2}\right)$$
$$+ \frac{M}{2}\left(\frac{\nu_1^{max}}{\nu_0^{min}} - \log\frac{\nu_1^{max}}{\nu_0^{min}} + \log\frac{\nu_1^{max} + \sigma_e^2}{\nu_0^{min} + \sigma_e^2} - \frac{\nu_1^{max} + \sigma_e^2}{\nu_0^{min} + \sigma_e^2}\right)$$
$$\leq \frac{\sigma_e^2}{2(\nu_0^{min})^2}\left(\|\boldsymbol{\mu}_0 - \boldsymbol{\mu}_1\|_2^2 + M\frac{(\nu_1^{max} - \nu_0^{min})^2}{\nu_1^{max}}\right)$$

where ν_0^{min} is the smallest eigenvalue of $\boldsymbol{\Sigma}_0$ and ν_1^{max} is the largest eigenvalue of $\boldsymbol{\Sigma}_1$.

As a corollary of $D_{KL}(f_e\|g_e) \leq D_{KL}(f\|g)$ in Theorem 21.2, the asymptotic lower bound of ADD in (21.4) is increased when the randomizing scheme is applied:

$$\underline{ADD}(\Lambda, f_e, g_e) \geq \underline{ADD}(\Lambda, f, g)$$

resulting in a prolonged detection delay of finding the outage time given encrypted data $\tilde{\mathbf{x}}^{1:N}$. Theorem 21.2 not only indicates a strict performance degradation but also infers the magnitude of this degradation by deriving the upper bound of KL_Δ. That is, we know approximately how much extra delay is brought w.r.t. the noise variance σ_e^2.

To illustrate the prolongation of detection delay, we present Figure 21.3, comparing two scenarios: the application of the statistic (21.2) to raw data $\Lambda(\mathbf{x}^{1:N})$ and its application to encrypted

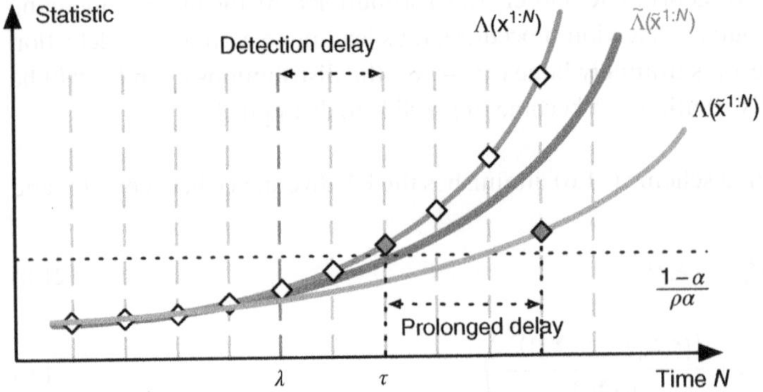

Figure 21.3 Outages are reported when the calculated statistic surpasses the threshold $\frac{1-\alpha}{\rho\alpha}$. See Table 21.1 for a summary of these statistics.

data $\Lambda(\tilde{\mathbf{x}}^{1:N})$. Due to the KL divergence reduction established in Theorem 21.2, $\Lambda(\tilde{\mathbf{x}}^{1:N})$ is typically smaller than $\Lambda(\mathbf{x}^{1:N})$ (we will show this claim later in the chapter), especially after the outage occurrence. This inequality has two intuitive consequences. First, it reduces the likelihood of triggering a false alarm when detecting the outage time using encrypted data, i.e., $FAR(\widetilde{\Lambda}, f_e, g_e) \leq FAR(\Lambda, f, g) \leq \alpha$. Second, encrypted data leads to a prolonged detection delay, i.e., the performance degradation.

To address the performance degradation, as suggested by Figure 21.3, a logical approach is to design a new detection statistic (represented by $\widetilde{\Lambda}$) to process the encrypted data. The new detection procedure is expected to maintain a comparable detection delay to the optimal scenario with access to raw data and still restrict the FAR below α.

21.3.3 A New Statistic to Boost the Detection Performance

In this subsection, we formally introduce a noise-mitigation technique to achieve detection performance comparable to the optimal scenario with access to raw data. We term it "noise-mitigation" since the technique essentially alleviates the performance impact resulting from the privacy-protective noise. To achieve this, we design a new statistic $\widetilde{\Lambda}$ to process the encrypted data $\tilde{\mathbf{x}}^{1:N}$. The new statistic aims to offer an approximation of the optimal statistic $\Lambda(\mathbf{x}^{1:N})$ (Figure 21.3), even when raw data is not available. We refer to the statistic $\Lambda(\mathbf{x}^{1:N})$ as "optimal" due to its demonstrated optimal detection performance when raw data is available (see Theorem 21.1). It's also important to note that this optimal statistic doesn't incorporate privacy protection. To prevent any ambiguity with these statistics, we present Table 21.1 for a comprehensive summary of the detection statistics used in this chapter, along with their relevant attributes.

For designing a new statistic $\widetilde{\Lambda}$ that approximates Λ, we leverage the following insights. While the noise is generated randomly, its pattern, specifically the distribution parameters σ_e^2, are known to utility operators. This insight prompts us to compute the expectation of noise-related terms in the statistic $\Lambda(\tilde{\mathbf{x}}^{1:N})$. By replacing these terms with their respective expectations, we can provide an unbiased estimation of $\Lambda(\tilde{\mathbf{x}}^{1:N})$. Following this rationale, the new design for the statistic $\widetilde{\Lambda}(\tilde{\mathbf{x}}^{1:N})$ is presented in (21.10):

$$\widetilde{\Lambda}(\tilde{\mathbf{x}}^{1:N}) = \sum_{k=1}^{N} \pi_N^k \prod_{n=k}^{N} \frac{\sqrt{|\Sigma_0|} \exp\left(\beta_1[n]\right)}{\sqrt{|\Sigma_1|} \exp\left(\beta_0[n]\right)} \tag{21.10}$$

where $\beta_i[n] := -\frac{1}{2}(\tilde{\mathbf{x}}[n] - \boldsymbol{\mu}_i)^T(\boldsymbol{\Sigma}_i)^{-1}(\tilde{\mathbf{x}}[n] - \boldsymbol{\mu}_i) + \frac{1}{2}\mathbf{e}dev \cdot tr(\boldsymbol{\Sigma}_i^{-1})$ for $i = 0, 1$. We note that $\beta_i[n]$ is an unbiased estimation of the corresponding term in the optimal statistic $\Lambda(\mathbf{x}^{1:N})$, i.e., $\mathbb{E}_{\mathbf{e}\sim\mathcal{N}(\mathbf{0},\mathbf{D}_e)}\beta_i[n] = -\frac{1}{2}(\mathbf{x}[n] - \boldsymbol{\mu}_i)^T(\boldsymbol{\Sigma}_i)^{-1}(\mathbf{x}[n] - \boldsymbol{\mu}_i)$. By the unbiased design, the proposed statistic

Table 21.1 Summarize of detection statistics.

Statistic	Calculation	Privacy	Detection performance
$\Lambda(\mathbf{x}^{1:N})$	(21.2)	×	Optimal
$\Lambda(\tilde{\mathbf{x}}^{1:N})$	(21.2)	✓	Compromised
$\widetilde{\Lambda}(\tilde{\mathbf{x}}^{1:N})$	(21.10)	✓	Sub-optimal
$\widetilde{\Lambda}_\gamma(\tilde{\mathbf{x}}^{1:N})$	(21.13)	✓	Optimal

$\widetilde{\Lambda}(\tilde{\mathbf{x}}^{1:N})$ serves as the desired approximation of the optimal statistic $\Lambda(\mathbf{x}^{1:N})$. This effect is shown in Figure 21.3 and proved in Lemma 21.2.

Lemma 21.2 The proposed statistic $\widetilde{\Lambda}$ in (21.10) satisfies

$$\Lambda(\tilde{\mathbf{x}}^{1:N}) \leq \widetilde{\Lambda}(\tilde{\mathbf{x}}^{1:N}) \leq \Lambda(\mathbf{x}^{1:N}), \quad N \geq \lambda \tag{21.11}$$

Proof: For showing $\Lambda(\tilde{\mathbf{x}}^{1:N}) \leq \widetilde{\Lambda}(\tilde{\mathbf{x}}^{1:N})$, it suffices to show $\frac{f_e(\tilde{\mathbf{x}}[n])}{g_e(\tilde{\mathbf{x}}[n])} \leq |\boldsymbol{\Sigma}_1|^{\frac{1}{2}}/|\boldsymbol{\Sigma}_0|^{\frac{1}{2}} \exp(\beta_1[n] - \beta_0[n])$, where these two terms are denoted as $(\#^e)$ and $(*)$. In fact, we have

$$\log(\#^e) = \frac{p}{2}\log\frac{s_0+\sigma_e^2}{s_1+\sigma_e^2} + \left(\frac{a_0}{s_0+\sigma_e^2} - \frac{a_1}{s_1+\sigma_e^2}\right),$$

$$\log(*) = \frac{p}{2}\log\frac{s_0}{s_1} + \left(\frac{a_0}{s_0} - \frac{a_1}{s_1}\right) - \frac{p\sigma_e^2}{2}\left(\frac{1}{s_0} - \frac{1}{s_1}\right)$$

where $a_0 = \frac{1}{2}\|\tilde{\mathbf{x}}[n] - \boldsymbol{\mu}_0\|^2$ and $a_1 = \frac{1}{2}\|\tilde{\mathbf{x}}[n] - \boldsymbol{\mu}_1\|^2$, and we consider diagonal co-variances $\boldsymbol{\Sigma}_0 = \text{diag}(s_0, \dots, s_0)$ and $\boldsymbol{\Sigma}_1 = \text{diag}(s_1, \dots, s_1)$. When $n \geq \lambda$ (the line outage occurs), we have $a_0 \gg a_1$, which results in $\log(\#^e) \leq \log(*)$.

For showing $\widetilde{\Lambda}(\tilde{\mathbf{x}}^{1:N}) \leq \Lambda(\mathbf{x}^{1:N})$, it suffices to show $\frac{f(\mathbf{x}[n])}{g(\mathbf{x}[n])} \geq |\boldsymbol{\Sigma}_1|^{\frac{1}{2}}/|\boldsymbol{\Sigma}_0|^{\frac{1}{2}} \exp(\beta_1[n] - \beta_0[n])$, where these two terms are denoted as $(\#)$ and $(*)$. In fact, we have

$$2\log\frac{(\#)}{(*)} = p(\sigma_e^2 - \|\mathbf{e}[n]\|^2)\left(\frac{1}{s_0} - \frac{1}{s_1}\right) + 2\frac{b_1}{s_1} - 2\frac{b_0}{s_0}$$

where $b_0 = \langle \mathbf{e}[n], \boldsymbol{\mu}_0\rangle$ and $b_1 = \langle \mathbf{e}[n], \boldsymbol{\mu}_1\rangle$ satisfying $\mathbb{E}_{\mathbf{e}} b_0 = \mathbb{E}_{\mathbf{e}} b_1 = 0$. It indicates that $\mathbb{E}_{\mathbf{e}}[\log\frac{(\#)}{(*)}] = 0$, which further gives us $\mathbb{E}_{\mathbf{e}}\left[\frac{(\#)}{(*)}\right] \geq \exp(0) = 1$ from Jensen's inequality. It implies a higher likelihood that $(\#) > (*)$. Since at every time step $n = k$, we will randomly generate a noise vector $\mathbf{e}[n]$, we can conclude that $\widetilde{\Lambda}(\tilde{\mathbf{x}}^{1:N}) \leq \Lambda(\mathbf{x}^{1:N})$.

From Lemma 21.2, the proposed statistic $\widetilde{\Lambda}(\tilde{\mathbf{x}}^{1:N})$ falls between the $\Lambda(\tilde{\mathbf{x}}^{1:N})$ and $\Lambda(\mathbf{x}^{1:N})$ after the outage event ($n \geq \lambda$), aligning with our expectations in Figure 21.3. Consequently, it will exhibit a reduced FAR compared to the optimal scenario with raw data access, and will help alleviate the prolongation of ADD.

Corollary 21.2 The proposed $\widetilde{\Lambda}$ in (21.10) restricts the FAR below α, and alleviate the prolongation of ADD, i.e.,

$$\begin{aligned} &FAR(\Lambda, f_e, g_e) \leq FAR(\widetilde{\Lambda}, f_e, g_e) \leq FAR(\Lambda, f, g) \leq \alpha \\ &\underline{ADD}(\Lambda, f_e, g_e) \geq \underline{ADD}(\widetilde{\Lambda}, f_e, g_e) \geq \underline{ADD}(\Lambda, f, g) \end{aligned} \tag{21.12}$$

As indicated by the proof of Lemma 21.2, Jensen's inequality hinders the attainment of a "perfect" approximation to the optimal statistic Λ, resulting in a remaining gap between $\widetilde{\Lambda}$ and Λ. To address this matter, a logical approach is to seek specific conditions under which Jensen's inequality converges toward equality. With this in mind, we modify the statistic in (21.10) by introducing a constant term $\gamma \geq 1$ as

$$\widetilde{\Lambda}_\gamma(\tilde{\mathbf{x}}^{1:N}) = \sum_{k=1}^{N} \pi_N^k \prod_{n=k}^{N} \frac{\sqrt{|\boldsymbol{\Sigma}_0|}\exp\left(\beta_1[n]/\gamma\right)}{\sqrt{|\boldsymbol{\Sigma}_1|}\exp\left(\beta_0[n]/\gamma\right)} \tag{21.13}$$

We refer to the constant term γ as the variance scaling factor since it scales the variance of term $\beta_i[n]$ by $1/\gamma^2$ times. When $\gamma = 1$, the statistic in (21.13) degrades to the statistic in (21.10). We employ this variance scaling factor because Jensen's inequality tends to become equality as the variance of the variable approaches zero. To describe the effect of introducing γ to scale β_i, we provide Lemma 21.3. From previous discussing, the term β_i is an unbiased estimation of $\overline{\beta}_i := -\frac{1}{2}(\mathbf{x}[n] - \boldsymbol{\mu}_i)^T(\boldsymbol{\Sigma}_i)^{-1}(\mathbf{x}[n] - \boldsymbol{\mu}_i)$, whose variance is denoted as σ_i^2. Thus, the scaled term β_i/γ used in (21.13) can be modeled in a distribution $\mathcal{P}$ with mean $\overline{\beta}_i/\gamma$ and variance σ_i^2/γ^2. According to the theorem of the Jensen inequality gap, we have the following upper bound w.r.t. to the variance scaling factor γ.

Lemma 21.3 Suppose $|\exp(\beta_i/\gamma) - \exp(\overline{\beta}_i/\gamma)| \leq M|\beta_i/\gamma - \overline{\beta}_i/\gamma|^2$ for some M and any $\beta_i/\gamma \in \mathbb{R}$, for any convex function f, we have an upper bound of Jensen gap as

$$\left[\mathbb{E}\left[f\left(\frac{\beta_i}{\gamma}\right)\right] - f\left(\mathbb{E}\left[\frac{\beta_i}{\gamma}\right]\right)\right] \leq M \int \left|\frac{\beta_i}{\gamma} - \frac{\overline{\beta}_i}{\gamma}\right|^2 d\mathcal{P}\left(\frac{\beta_i}{\gamma}\right) \leq M\frac{\sigma_i^2}{\gamma^2}$$

According to Lemma 21.3, the variance-reduction technique in (21.13) can narrow the gap in the Jensen inequality, consequently achieving a nearly perfect approximation of the optimal statistic. In summary, when implementing the randomization scheme (21.6) to encrypt raw data and utilizing the new statistic (21.13) for outage detection, we outline the privacy-aware line outage detection procedure, referred to as **PLOD**, in Algorithm 21.1. The proposed PLOD offers two key advantages. First, it ensures privacy preservation by using noise for data encryption. Second, the proposed statistic provides an approximation to the optimal statistic when raw data is accessible, thereby achieving a comparable lower bound on detection delay while limiting the FAR to a predefined tolerance level.

Algorithm 21.1 Privacy-aware line outage detection (PLOD) with boosted detection performance.

1: **Input**: New voltage data $\mathbf{x}[n]$
2: **Parameter**: Noise variance σ_e^2, variance scaling factor γ
3: **Output**: Outage time
4: Apply **noise** to encrypt raw data.

$$\tilde{\mathbf{x}}[n] = \mathbf{x}[n] + \mathbf{e}[n],\ \mathbf{e}[n] \sim \mathcal{N}(\mathbf{0}, \text{diag}(\sigma_e^2, \dots, \sigma_e^2))$$

5: Calculate detection **statistic** $\tilde{\Lambda}(\tilde{\mathbf{x}}^{1:n})$ in (21.13).
6: **if** $\tilde{\Lambda}_\gamma(\tilde{\mathbf{x}}^{1:n}) \geq \frac{1-\alpha}{\rho\alpha}$ **then**
7:     **report** outage time n
8: **end if**

21.4 Validation on Extensive Outage Scenarios with Real-World Data

This section evaluates the privacy guarantee, ADD, and FAR of PLOD, comparing it with recent baselines on privacy-aware detection methods.

21.4.1 Dataset Configuration

To assess PLOD across diverse system sizes and environments, we conduct comprehensive experiments using various network configurations. The systems include the IEEE 8-bus and IEEE 123-bus networks [31], along with two representative European distribution systems: a medium voltage (MV) network in an urban area and a low voltage (LV) network in a suburban area [32]. In each of these networks, we select bus 1 as the slack bus.

In recognition of the complexities in real-world distribution grid outage scenarios, we explore situations where alternative power sources come into play following a line outage. In such scenarios, relying solely on the "last gasp" notification becomes less effective, rendering the detection of line outages more challenging. To model this complexity, we conduct simulations for the following two representative scenarios.

- **Mesh Networks**: Mesh networks are often used to model networks in urban areas, where most buses retain non-zero voltages after a line outage as they can receive power from alternative branches. To simulate mesh networks, we introduce loops into the aforementioned systems, ensuring their connectivity remains intact after line outages [6]. As an example, in the IEEE-123 bus network, we introduce loops by adding two branches: one between bus 77 and 120 and another between bus 50 and 56, with admittances matching that of the branch between bus 122 and 123.
- **Radial Networks with DERs**: In such case, some buses continue to receive power from DERs though isolated from the main grid after a line outage. This type of outage scenario is typical in residential areas. To simulate DERs, we select multiple buses to have solar power panels with batteries as energy storage. For solar panels, we use power generation profiles computed using the PVWatts Calculator [33].

To generate more authentic data, we use real residential power profiles from the Duquesne Light Company (DLC) in Pittsburgh, USA. The DLC dataset comprises anonymized and secure hourly (and 15-minute) smart meter readings of active power from over 5000 houses throughout the year 2016.

21.4.2 Implementation Details

The time-series voltage magnitude data are generated using the MATLAB Power System Simulation Package (MATPOWER) in MATLAB R2022b. In each distribution system, we assign active power $p_i[n]$ from the DLC power profile to bus i at time n. The reactive power $q_i[n]$ is determined based on a randomly generated power factor $pf_i[n]$, which follows a uniform distribution Unif(0.9, 1). Using the active and reactive power values, we employ MATPOWER to solve the power flow equations and derive voltage measurements. Additionally, we simulate outage scenarios by setting the admittance of one or multiple branches to zero and solve the power flow equations again.

For more robust evaluation, each experiment is conducted using Monte Carlo simulation with over 1000 replications. In each replication, the voltage sequence in (21.1) is generated by concatenating $\lambda - 1$ records from pre-outage data and 50 records from post-outage data (50 samples are sufficient since the detection delay in our experiments is lower than 50). The outage time λ is randomly generated using a geometric distribution Geo(ρ). This geometric prior is based on our belief that outages can occur independently at any time step, with an equal probability of ρ. We choose $\rho = 0.04$ in our experiments, which is derived from historical outage data, indicating that each time step has a 4% chance of experiencing a line outage.

After obtaining voltage data from MATLAB, the remaining calculations for outage detection in Algorithm 21.1 are implemented using Python 3.8 on a personal computer with a Windows 10 operating system, an Intel Core i7 processor clocked at 2.2 GHz, and 16 GB of RAM.

21.4.3 Baseline Methods

In the following experiments, the optimal Bayesian detection procedure with access to raw data ($\Lambda(\mathbf{x}^{1:N})$) is referred to as the **benchmark**. It should have optimal detection performance but has no privacy guarantee. The same detection statistic applied to encrypted data ($\Lambda(\tilde{\mathbf{x}}^{1:N})$) is referred to as **privacy-only** since it degrades the detection performance. To remove this performance degradation, our proposed method ($\widetilde{\Lambda}_\gamma(\tilde{\mathbf{x}}^{1:N})$) in Algorithm 21.1 is referred to as **PLOD**. The noise level σ_e^2 and the variance scaling factor γ will be further pointed out. We also compare with recent techniques dealing with privacy concerns, including a private approximation of the change-point (**MLE**) [24], and an uncertain likelihood ratio (**ULR**) proposed to replace the original detection statistic [34].

21.4.4 Visualization of Privacy Guarantee

Before evaluating the detection performance of our privacy-aware approach, we first examine how much privacy is preserved when using DLC data in the IEEE 8-bus system. Specifically, as established in Theorem 21.1, applying randomization scheme (21.6) to voltage data $\mathbf{x}^{1:N}$ yields $G_{\frac{\Delta}{edev}}$- Gaussian differential privacy. To visualize this privacy guarantee, we plot the trade-off functions $T(\mathcal{N}(0,1), \mathcal{N}(\frac{\Delta}{\mathbf{edev}}, 1))$ at varying levels of noise variance σ_e^2, and compare them with baseline trade-off functions $G_\mu := T(\mathcal{N}(0,1), \mathcal{N}(\mu, 1))$ when μ takes values of 0.5, 1, and 3.

The results are shown in Figure 21.4. By employing the privacy-preserving scheme (21.6) with a noise variance of $\sigma_e^2 = 5e - 3$, it becomes notably difficult for an adversary to differentiate the encrypted data $\tilde{\mathbf{x}}^{1:N}$ from the raw data $\mathbf{x}^{1:N}$, thereby effectively safeguarding privacy. In quantitative

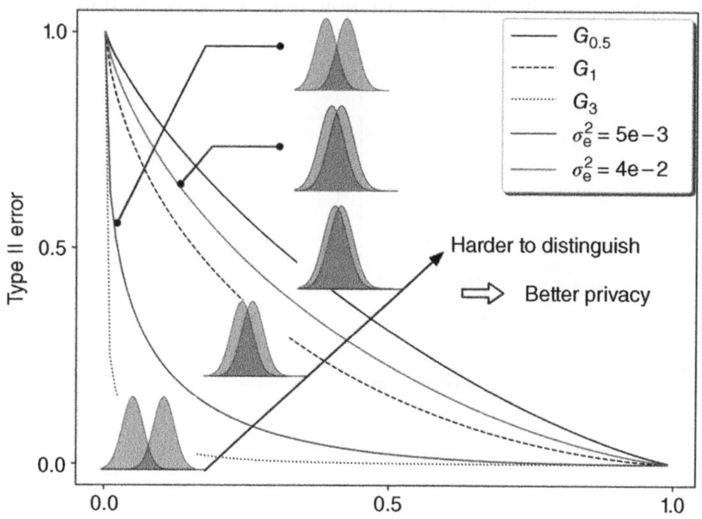

Figure 21.4 The comparison of trade-off functions of distinguishing unit-variance Gaussian distributions using DLC data and IEEE 8-bus system simulation.

terms, distinguishing $\tilde{\mathbf{x}}^{1:N}$ from $\mathbf{x}^{1:N}$ proves to be a more arduous task than distinguishing between the probability distributions $\mathcal{N}(0,1)$ and $\mathcal{N}(3,1)$. When we increase the noise variance to $\sigma_e^2 = 4e-2$, we achieve a heightened degree of privacy protection by making it even more challenging to differentiate encrypted data from genuine data. It is now more difficult than distinguishing between $\mathcal{N}(0,1)$ and $\mathcal{N}(1,1)$.

21.4.5 Evaluation of the Noise-Mitigation Design

Despite the privacy protection shown in Figure 21.4, applying the randomizing scheme (21.6) will compromise detection performance (see Section 21.3.2). Within this subsection, we assess whether our proposed detection procedure PLOD can ameliorate this performance degradation.

Specifically, we compare the old detection statistic in (21.2) to our newly proposed statistic in (21.10), with their logarithms plotted in Figure 21.5. This simulation is performed in the IEEE 123-bus system, chosen to examine the efficacy of PLOD in a large-scale network. As we can see, the optimal statistic $\Lambda(\mathbf{x}^{1:N})$ increases dramatically after the outage time $\lambda = 30$, resulting in a near-zero detection delay. The same statistic applied to encrypted data $\Lambda(\tilde{\mathbf{x}}^{1:N})$ has a privacy guarantee but postpones the detection. Our proposed statistic $\widetilde{\Lambda}(\tilde{\mathbf{x}}^{1:N})$ in (21.10) closely approximates the optimal statistic, thus effectively mitigating the postponing effect while still preserving privacy. It is worth noting that Figure 21.5 also serves to validate the conclusions drawn in Lemma 21.2.

21.4.6 Evaluation of the Variance-Reduction Design

In Figure 21.5, the proposed statistic ($\widetilde{\Lambda}$) still exhibits a gap compared to the optimal statistic (Λ) due to the influence of Jensen's inequality. Our newly designed approach in (21.13) aims to reduce the statistic's variance, ultimately narrowing this gap. To confirm this improvement, we plot the logarithm of the statistic $\widetilde{\Lambda}_\gamma(\tilde{\mathbf{x}}^{1:N})$ as defined in (21.13) for various choices of the variance scaling factor γ and compare them with the optimal statistic $\Lambda(\mathbf{x}^{1:N})$. The results are shown in Figure 21.6. As γ increases from 1 to 3, $\widetilde{\Lambda}_\gamma(\tilde{\mathbf{x}}^{1:N})$ progressively converges to a more precise approximation of the

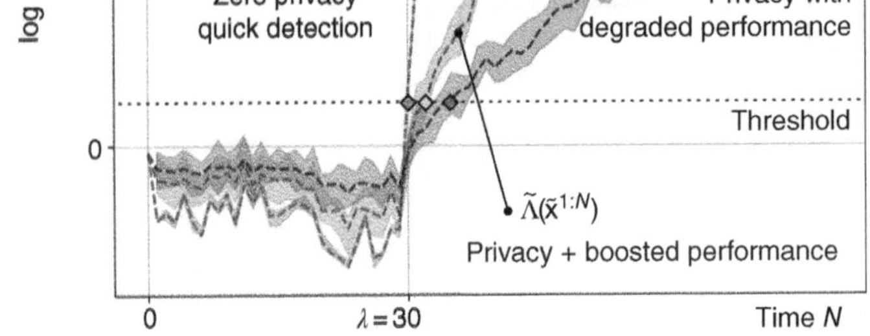

Figure 21.5 The logarithm of various detection statistics in IEEE 123-bus system. $\lambda = 30$, $\sigma_e^2 = 4e-2$, $\gamma = 1$ and $\alpha = 1\%$.

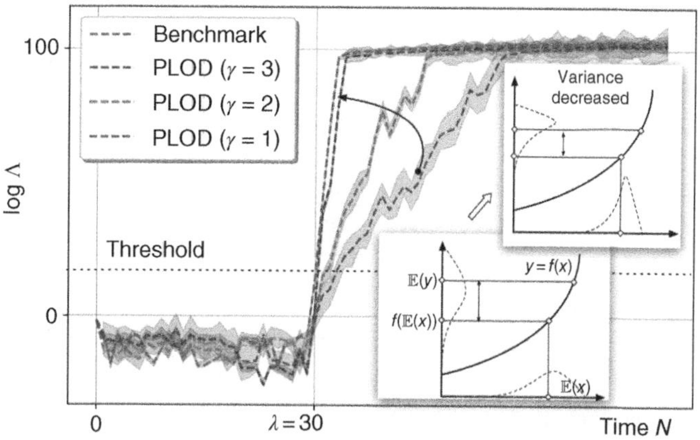

Figure 21.6 The logarithm of detection statistics with different variance scaling factors γ in the IEEE 123-bus system. $\lambda = 30$, $\sigma_e^2 = 4e - 2$, and $\alpha = 1\%$.

optimal statistic $\Lambda(\mathbf{x}^{1:N})$. This convergence is a consequence of the reduced variance in the detection statistic, as shown in two zoomed-in illustrations in Figure 21.6. This reduction in variance effectively constrains the error gap in Jensen's inequality, as elaborated in Lemma 21.3.

In summary, even when only encrypted data $\tilde{\mathbf{x}}^{1:N}$ is accessible, employing PLOD with the statistic from (21.13) can effectively approximate the optimal statistic with access to raw data $\mathbf{x}^{1:N}$. Thus, we can expect PLOD to not only safeguard the privacy of customer data but also maintain the detection performance. We assess the latter aspect in Section 21.4.7.

21.4.7 Evaluation of Detection Performance: Average Detection Delay and False Alarm Rate

After verifying the effects of our various designs within the proposed method PLOD, we finally evaluate its corresponding detection performance when using encrypted data $\tilde{\mathbf{x}}^{1:N}$. The evaluation covers both the ADD and the FAR. To validate the asymptotic optimality of the detection delay in Theorem 21.1, we plot in the upper half of Figure 21.7 the average delay $\mathbb{E}(\tau - \lambda | \tau \geq \lambda)$ divided by $|\log \alpha|$ and the theoretical lower bound $-\log(1-\rho) + D_{KL}(f||g)$. We observe that the detection delay of the benchmark and that of the PLOD both achieve the optimal lower bound asymptotically, while the delay of the privacy-only approach is higher.

The detection rule is also expected to restrict the FAR below α. To verify this, we calculate the empirical FAR $\mathbb{P}(\tau < \lambda)$ and compare it against the upper bound α, as shown in the lower half of Figure 21.7. Our proposed method has a similar performance compared to the benchmark since the empirical false alarm is mainly below the upper bound α (especially when $\alpha \to 0$). This observation demonstrates that our proposed algorithm can quickly detect line outages with a low FAR, even when encrypted data are utilized.

To evaluate PLOD across diverse grid systems under various outage configurations, we present a comprehensive summary of results in Table 21.2. Throughout these experiments, we conduct comparisons not only with the benchmark method but also with two recent relevant techniques that provide privacy-aware approaches for detecting distribution changes. These methods are referred to as **MLE** [24] and **ULR** (Utility Learning-based Rule) [34]. To ensure consistency in the level of differential privacy guarantees across all methods, we apply noise with a variance of $\sigma_e^2 = 4e - 2$ to the raw data.

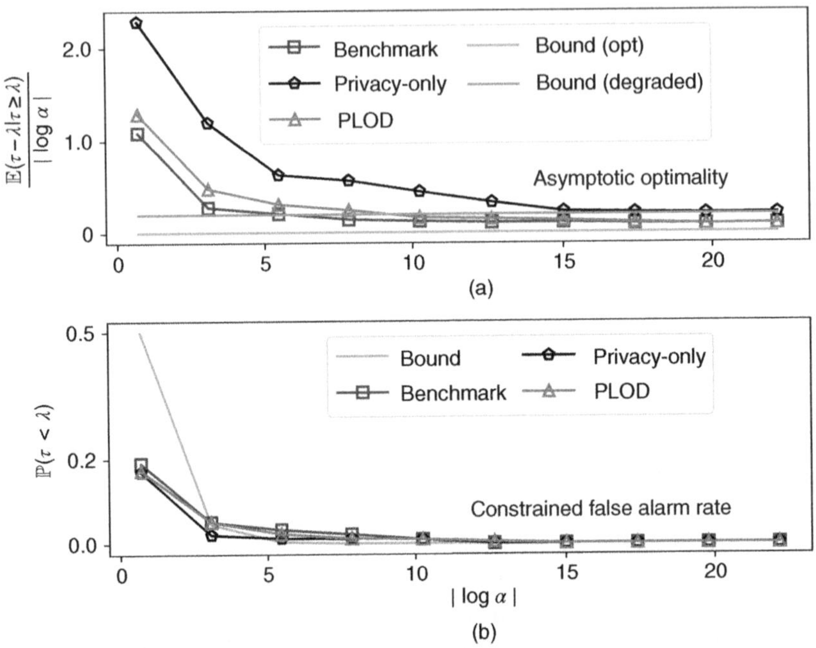

Figure 21.7 The average detection delay (a) and the false alarm rate (b) in various scenarios in the IEEE 123-bus system. $\sigma_e^2 = 4e-2$, $\gamma = 1$ and $\alpha = 1\%$.

Table 21.2 Performance comparison on various systems. $\sigma_e^2 = 4e-2$ and $\alpha = 1\%$.

System	System information			ADD (1 unit) ↓		FAR (%) ↓	
Mesh network	**#Branches**	**#DERs**	**Outage**	**Bench.**	**PLOD**	**Bench.**	**PLOD**
8-bus	9	0	4–7	3.10	$3.35^{+0.25}$	0.9	$1.2^{+0.3}$
8-bus	9	8	4–7	3.82	$4.11^{+0.29}$	1.4	$1.6^{+0.2}$
123-bus	124	0	73–74	0.89	$0.94^{+0.05}$	0.6	$0.8^{+0.2}$
123-bus	124	0	73–74,14–15	0.88	$0.94^{+0.06}$	0.5	$0.7^{+0.2}$
123-bus	124	0	5 branches	0.84	$0.91^{+0.07}$	0.5	$0.7^{+0.2}$
LV suburban	129	0	26–95	3.80	$4.11^{+0.31}$	1.5	$2.0^{+0.5}$
LV suburban	129	30	26–95	3.89	$4.19^{+0.30}$	1.3	$1.7^{+0.4}$
MV urban	48	0	34-35	0.83	$0.89^{+0.06}$	0.4	$0.5^{+0.1}$
MV urban	48	7	34-35	1.45	$1.55^{+0.10}$	0.8	$1.1^{+0.3}$
Radial network	**# Branches**	**# DERs**	**Outage**	**Bench.**	**PLOD**	**Bench.**	**PLOD**
8-bus	7	8	4–7	3.5	$3.8^{+0.3}$	0.8	$1.0^{+0.2}$
8-bus	7	8	2–6	3.58	$3.84^{+0.26}$	0.9	$1.1^{+0.2}$
123-bus	122	12	73–74	1.29	$1.39^{+0.1}$	0.5	$0.6^{+0.1}$
123-bus	122	122	73–74	6.75	$7.31^{+0.56}$	2.8	$3.8^{+1.0}$
LV suburban	114	30	26–95	3.25	$3.53^{+0.28}$	1.4	$1.8^{+0.4}$
LV suburban	114	113	26–95	8.87	$9.58^{+0.71}$	4.0	$5.4^{+1.4}$
MV urban	38	7	34–35	1.45	$1.57^{+0.12}$	0.7	$0.9^{+0.2}$
MV urban	38	7	23–35	1.69	$1.82^{+0.13}$	1.1	$1.5^{+0.4}$

Table 21.2 showcases PLOD's ability to handle a variety of outage scenarios in both mesh (loopy) networks and radial networks with DER penetration. In contrast to the benchmark method, which has access to raw data and therefore carries a privacy risk, PLOD exhibits only marginal degradation in detection delay and FAR. Moreover, Table 21.2 reveals two significant observations. First, in cases where multiple branches undergo simultaneous outages, the ADD tends to be shorter. This can be attributed to the increased KL distance between the distributions g and f when multiple lines are disconnected. Second, in radial networks with a greater number of simulated DERs, the detection of line outages takes more time, primarily due to the smaller KL distance between g and f in such scenarios.

To assess detection performance under varying levels of noise introduced to the raw data, we present Table 21.3, which includes results for both the ADD and FAR. The comparison underscores that our proposed approach consistently outperforms other baseline methods, regardless of the desired level of privacy protection.

21.4.8 Sensitivity Analysis to Data Coverage

In the distribution grid, access to the data of every bus is not guaranteed for several reasons. For instance, rural areas may lack smart meter installations, technical issues can result in data loss, and privacy concerns might lead to data refusal. Thus, an analysis of incomplete smart meter data coverage is necessary to assess PLOD's real-world detection performance.

Based on records from [35], over 107 million smart meters covered 75% of U.S. households by 2021. Therefore, we simulate the scenario where a fraction of buses (ranging from 75% to 100%) is randomly selected to provide voltage measurements for outage detection. The outcomes, illustrated in Figure 21.8, reveal how much additional ADD and FAR is introduced at various coverage ratios in comparison to 100% data coverage. For instance, when applying PLOD with variance scaling factors γ equal to 1, 2, and 3, a 75% data coverage ratio necessitates an additional 2.5, 1.9, and 1.7 data samples, respectively, to detect the outage. Simultaneously, the FAR increases by 9.5%, 11.9%, and 13.1%, respectively. It's noteworthy that when data coverage is not complete, increasing the variance scaling factor γ involves a trade-off: it reduces detection delay at the expense of introducing more false alarms.

Table 21.3 Performance comparison at different noise levels in the IEEE 8-bus system. $\alpha = 1\%$ and $\gamma = 3$.

Noise	Method	ADD (unit)	FAR (%)
	PLOD	**2.71** $\pm$ 0.53	**1.03** $\pm$ 0.02
$\sigma_e^2 = 0.01$	MLE	3.41 $\pm$ 0.72	1.12 $\pm$ 0.15
	ULR	3.80 $\pm$ 0.89	1.57 $\pm$ 0.19
	PLOD	**3.67** $\pm$ 0.65	**0.95** $\pm$ 0.03
$\sigma_e^2 = 0.04$	MLE	4.12 $\pm$ 0.83	1.06 $\pm$ 0.37
	ULR	4.69 $\pm$ 1.01	2.16 $\pm$ 0.74
	PLOD	**4.56** $\pm$ 0.77	**3.82** $\pm$ 7.39
$\sigma_e^2 = 0.09$	MLE	4.85 $\pm$ 0.81	4.45 $\pm$ 0.81
	ULR	4.69 $\pm$ 1.01	3.99 $\pm$ 0.74

a) Bold values highlight that PLOD, compared to the other two baselines, has a smaller Average Detection Delay (ADD) and False Alarm Rate (FAR).

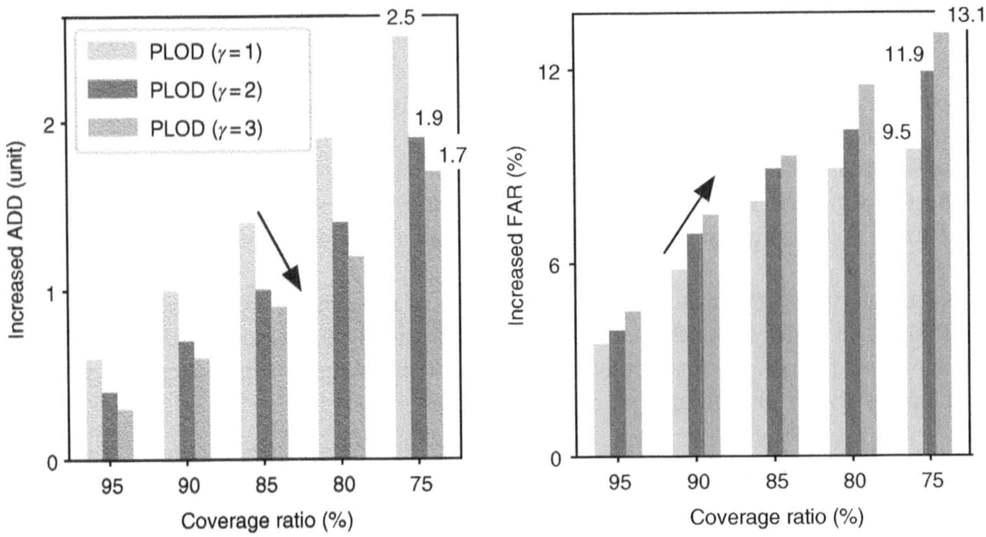

Figure 21.8 Increased ADD (unit) and FAR (%) under different ratios of data coverage compared to 100% coverage in 123-bus loopy system. $\alpha = 1\%$.

We note that our method doesn't rely on the assumption of 100% sensor data coverage across the grid. In reality, power line outages tend to impact a majority of buses in the system, with the extent of impact varying based on their proximity to the outage location. It allows us to identify outages by detecting distribution changes in sensor data from some rather than all buses located near the source of the outage.

21.5 Conclusions

In summary, this chapter presents a robust and privacy-aware method for detecting line outages within distribution grids, effectively achieving a delicate equilibrium between safeguarding privacy and upholding detection performance. Our contributions encompass several pivotal dimensions. First, we establish the direct protection of raw data through a randomization scheme embedded within the differential privacy framework. Second, we quantify the trade-off between privacy preservation and detection performance, taking into account factors such as detection delay and FAR. Finally, we introduce a novel detection statistic that alleviates the adverse impact of encrypted data on detection performance, at times eliminating it entirely.

To validate our contributions, we conduct extensive experiments across a diverse range of network systems, spanning 17 distinct outage configurations. The empirical results underscore the success of our privacy-aware outage detection methodology, achieving a harmonious balance between privacy preservation and detection performance that rivals the optimal case.

References

1 Yuan, J. and Weng, Y. (2022). Physically invertible system identification for monitoring system edges with unobservability. *European Conference on Machine Learning and Principles and Practice of Knowledge Discovery in Databases (ECML PKDD).*

2 Energy Information Administration (2016). Average frequency and duration of electric distribution outages vary by states. https://www.eia.gov/todayinenergy/detail.php?id=35652 (accessed 16 October 2024).

3 U.S. Department of Energy (2014). Fault Location, Isolation, and Service Restoration Technologies Reduce Outage Impact and Duration. *Smart Grid Investment Grant Program.*

4 Babakmehr, M., Harirchi, F., Al-Durra, A. et al. (2019). Compressive system identification for multiple line outage detection in smart grids. *IEEE Transactions on Industry Applications* 55 (5): 4462–4473.

5 He, M. and Zhang, J. (2010). Fault detection and localization in smart grid: a probabilistic dependence graph approach. *2010 1st IEEE International Conference on Smart Grid Communications*, 43–48. IEEE.

6 Liao, Y., Weng, Y., Tan, C.-W., and Rajagopal, R. (2021). Quick line outage identification in urban distribution grids via smart meters. *CSEE Journal of Power and Energy Systems* 8 (4): 1074–1086.

7 Liao, Y., Xiao, C., and Weng, Y. (2022). Quickest line outage detection with low false alarm rate and no prior outage knowledge. *2022 IEEE Power & Energy Society General Meeting (PESGM)*, 1–5. IEEE.

8 Sevlian, R.A., Zhao, Y., Rajagopal, R. et al. (2017). Outage detection using load and line flow measurements in power distribution systems. *IEEE Transactions on Power Systems* 33 (2): 2053–2069.

9 Soleymani, M. and Safdarian, A. (2019). Unsupervised learning for distribution grid line outage and electricity theft identification. *2019 Smart Grid Conference (SGC)*, 1–5. IEEE.

10 Zoha, A., Gluhak, A., Imran, M.A., and Rajasegarar, S. (2012). Non-intrusive load monitoring approaches for disaggregated energy sensing: a survey. *Sensors* 12 (12): 16838–16866.

11 Kalogridis, G., Cepeda, R., Denic, S.Z. et al. (2011). ElecPrivacy: Evaluating the privacy protection of electricity management algorithms. *IEEE Transactions on Smart Grid* 2 (4): 750–758.

12 McDaniel, P. and McLaughlin, S. (2009). Security and privacy challenges in the smart grid. *IEEE Security & Privacy* 7 (3): 75–77.

13 Wood, G. and Newborough, M. (2003). Dynamic energy-consumption indicators for domestic appliances: environment, behaviour and design. *Energy and Buildings* 35 (8): 821–841.

14 Dwivedi, A. and Tajer, A. (2021). Scalable quickest line outage detection and localization via graph spectral analysis. *IEEE Transactions on Power Systems* 37 (1): 590–602.

15 Xiao, C., Liao, Y., and Weng, Y. (2023). Distribution grid line outage identification with unknown pattern and performance guarantee. *IEEE Transactions on Power Systems* 39 (2): 3987–3999.

16 Shiryaev, A.N. (1963). On optimum methods in quickest detection problems. *Theory of Probability and its Applications* 8 (1): 22–46.

17 Chen, Y.C., Banerjee, T., Domínguez-García, A.D., and Veeravalli, V.V. (2016). Quickest line outage detection and identification. *IEEE Transactions on Power Systems* 31 (1): 749–758. https://doi.org/10.1109/TPWRS.2015.2394246.

18 Wei, C., Wiesel, A., and Blum, R.S. (2012). Change detection in smart grids using errors in variables models. *2012 IEEE 7th Sensor Array and Multichannel Signal Processing Workshop (SAM)*, 17–20. https://doi.org/10.1109/SAM.2012.6250460.

19 Gajula, K., Le, V., Yao, X. et al. (2021). Quickest detection of series arc faults on DC microgrids. *2021 IEEE Energy Conversion Congress and Exposition (ECCE)*, 796–801. IEEE.

20 Tartakovsky, A.G. and Veeravalli, V.V. (2005). General asymptotic Bayesian theory of quickest change detection. *Theory of Probability & Its Applications* 49 (3): 458–497.

21 Yang, P., Dumont, G., and Ansermino, J.M. (2006). Adaptive change detection in heart rate trend monitoring in anesthetized children. *IEEE Transactions on Biomedical Engineering* 53 (11): 2211–2219.

22 Hand, D.J. and Blunt, G. (2001). Prospecting for gems in credit card data. *IMA Journal of Management Mathematics* 12 (2): 173–200.

23 Dwork, C., McSherry, F., Nissim, K., and Smith, A. (2006). Calibrating noise to sensitivity in private data analysis. *Theory of Cryptography Conference*, 265–284. Springer.

24 Cummings, R., Krehbiel, S., Mei, Y. et al. (2018). Differentially private change-point detection. *Advances in Neural Information Processing Systems 31.*

25 Dwork, C. and Roth, A. (2014). The algorithmic foundations of differential privacy. *Foundations and Trends®in Theoretical Computer Science* 9 (3–4): 211–407.

26 Cummings, R., Krehbiel, S., Lut, Y., and Zhang, W. (2020). Privately detecting changes in unknown distributions. *International Conference on Machine Learning*, 2227–2237. PMLR.

27 Wilcoxon, F. (1945). Individual comparisons by ranking methods. *Biometrics Bulletin* 1 (6): 80–83.

28 Berrett, T. and Yu, Y. (2021). Locally private online change point detection. *Advances in Neural Information Processing Systems 34*, 3425–3437.

29 Dong, J., Roth, A., and Su, W.J. (2019). Gaussian differential privacy. *arXiv preprint arXiv:1905.02383.*

30 Gao, X., Sitharam, M., and Roitberg, A.E. (2017). Bounds on the Jensen gap, and implications for mean-concentrated distributions. *arXiv preprint arXiv:1712.05267.*

31 Kersting, W.H. (1991). Radial distribution test feeders. *IEEE Transactions on Power Systems* 6 (3): 975–985.

32 Mateo, C., Prettico, G., Gómez, T. et al. (2018). European representative electricity distribution networks. *International Journal of Electrical Power & Energy Systems* 99: 273–280.

33 Dobos, A.P. (2014). PVWatts Version 5 Manual. Golden, CO, USA: National Renewable Energy Lab. (NREL). Technical Report. *NREL/TP-6A20-62641.*

34 Hare, J.Z., Kaplan, L., and Veeravalli, V.V. (2021). Toward uncertainty aware quickest change detection. *2021 IEEE 24th International Conference on Information Fusion (FUSION)*, 1–8. IEEE.

35 Institution for Electronic Innovation (2021). Electric Company Smart Meter Deployments: Foundation for a Smart Grid (2021 Update). https://www.edisonfoundation.net/-/media/Files/IEI/publications/IEI_Smart_Meter_Report_April_2021.ashx (accessed September 2023).

22

Transactive Energy Management and Distribution System Reform Using Market Concepts

Amr A. Mohamed[1], *Bala Venkatesh*[1], *Carlos Sabillon*[2], *and Ali Golriz*[3]

[1]*Electrical and Computer Engineering Department, Toronto Metropolitan University, Toronto, Ontario, Canada*
[2]*Northland Power Inc., Toronto, Ontario, Canada*
[3]*System & Sector Development, Innovation and R&D at Independent Electricity System Operator, Toronto, Ontario, Canada*

Nomenclature

Indices:

t	Hourly time index
i	Bus index
f	Index of phases
a	Aggregator index
p	Prosumer index
e	Energy storage entity index
Z	Index of zone in the distribution system
s	Index of segments in a bid
Γ	Index of feasible system topology

Sets:

Ω_{Pro}	Set of prosumers
Ω_{Agg}	Set of aggregators
Ω_N	Set of all buses
Ω_L	Set of all distribution lines
Ω_F	Set of all phases
Ω_T	Set of all time intervals
Ω_Z	Set of all zones in the distribution system
Ω_{SW}	Set of all switches
Ω_Γ	Set of all feasible topologies, where the protection system has preset values
Ω_γ^Γ	Set of all active switches in feasible topology γ, $\Omega_\gamma^\Gamma \subseteq \Omega_\Gamma$
Ω_N^z	Set of all nodes in zone "z", $\Omega_N^z \subseteq \Omega_N$
Ω_S^i	Set of bid segments for power interaction with TX at node "i"
Ω_S^a	Set of bid segments for power interaction with aggregator "a"
Ω_S^p	Set of bid segments for power interaction with prosumer "p"

Parameters:

NB	Number of buses
NS	Number of substations

Smart Cyber-Physical Power Systems: Fundamental Concepts, Challenges, and Solutions, Volume 1, First Edition.
Edited by Ali Parizad, Hamid Reza Baghaee, and Saifur Rahman.

NL	Number of distribution lines
NZ	Number of zones in the distribution system
$\lambda^{T}_{i,t}$	Marginal price of the transmission market
$\lambda^{D}_{i,t}$	Bid price of the distribution market
$\lambda^{Tdr}_{i,t}$	Marginal price for demand response in the transmission market
$\lambda^{Ddr}_{i,t}$	Bid price for demand response in the distribution market
$\overline{S^{A+}_{i,t,a}}, \overline{S^{P+}_{i,t,a}}$	Injected apparent power limit for aggregator and prosumer
$\overline{P^{E}_{e}}$	Energy storage power limit
η^{+}, η^{-}	Discharging and charging efficiency of ES entity
E^{0}_{e}	Initial energy of ES entity
$y_{(i,f),(k,h)}$	Magnitude admittance element obtained from the bus admittance matrix
$\theta_{(i,f),(k,h)}$	Angle phase of the admittance element obtained from the bus admittance matrix
P^{T}_{0}	Maximum power injected by TS from previous days
K^{DC}_{i}	Demand charge cost

Variables:

P^{T}	Transmitted power from the bulk TS to the DS
P^{A+}, P^{P+}	Injected active power by the aggregator/prosumer
P^{A-}, P^{P-}	Demand active power by the aggregator/prosumer
Q^{A+}, Q^{P+}	Injected reactive power by the aggregator/prosumer
Q^{A-}, Q^{P-}	Demand reactive power by the aggregator/prosumer
P^{Adr}, P^{Pdr}	Demand response power dispatched to the aggregator/prosumer
P^{Ars}, P^{Prs}	Reserve power dispatched to the aggregator/prosumer
P^{sw}, Q^{sw}	Active/reactive power flowing through switches
$P^{inj}_{i,f,t}, Q^{inj}_{i,f,t}$	Total active and reactive power injection
$P^{load}_{i,f,t}, Q^{load}_{i,f,t}$	Total active and reactive power demand
$P^{T}_{i,f,t}, Q^{T}_{i,f,t}$	Transmitted active and reactive power exchange between the bulk TS and the DS
$P^{L}_{i,f,t}, Q^{L}_{i,f,t}$	Remaining active and reactive power demand
$P^{E+}_{f,t,e}, P^{E-}_{f,t,e}$	Discharging and charging power of ES entity
$V_{i,h,t}$	Nodal voltage magnitude
$\emptyset_{i,f,t}$	Angle phase of nodal voltage

Integer variables:

$\mu_{t,e}$	Integer variable of discharging status of ES
$\omega_{m,n}$	Integer variable for state of switches
Γ_{γ}	Integer variable for distribution system topologies

22.1 Introduction

The increasing integration of renewable energy sources, advancements in communication technologies, and the emergence of smart grid systems have created new opportunities and challenges in the electricity sector, specifically in distribution systems. Transactive energy management (TEM) is a promising approach that leverages real-time data, automation, and market mechanisms to efficiently balance energy supply and demand in smart distribution systems. The integration of economic and control mechanisms in TEM provides a pathway toward a more efficient, fair, and transparent energy market.

Recent research has brought considerable attention to transactive energy markets, with comprehensive literature surveys covering TEM architectures and market mechanisms, such as those found in [1, 2]. The primary objective of transactive energy markets is to establish a decentralized market that not only offers advantages in terms of market pricing but also ensures secure and transparent transaction logs for energy transactions [3]. Moreover, the literature underscores the necessity of a versatile marketing concept applicable to transactive energy markets [4]. The importance of committed capacities in energy and ancillary service markets is stressed, emphasizing the need for guaranteed capacities in energy markets [5]. Viewed from the grid's perspective, efficient load management devices, like ice storage units, can enhance energy infrastructure utilization, thereby providing greater reliability and flexibility to grid operators in handling the variability of renewable generation [6]. Additionally, the literature explores optimal trading strategies in multi-energy markets, underscoring the significance of integrated demand response for load-serving entities in such TEM markets [7].

An incentive-compatible decentralized market framework, based on the Vickrey-Clarke-Groves auction model, facilitating energy transactions between district heating and power networks is presented, illustrating the relevance of decentralized market frameworks in transactive energy markets [8]. Research has delved into the trading settlement mechanism of electricity markets, essential for understanding the intricacies of processes involved in transactive energy markets [9]. The importance of virtual power plants in addressing the challenges and risks posed by the intermittent and uncontrollable nature of renewable DERs has been highlighted in [10]. This underscores their role in enhancing overall stability and market competitiveness. Furthermore, the importance of achieving an efficient and economical balance of energy supply and demand in transactive energy markets is emphasized in [11]. Zheng et al. proposed an equilibrium model for a peer-to-peer (P2P) transactive energy market, highlighting the significance of energy sharing in such markets [12]. Peng et al. demonstrated the potential of P2P energy trading in improving energy efficiency and promoting the consumption of renewable energy through a novel deep learning-based transaction method [13]. Further, practical insights into China's electricity market have been provided through a TEM model that reduced market imbalance and promoted renewable DERs' participation in [14]. A fully decentralized P2P market model for local energy communities has been presented and implemented on real projects such as Enerchain and the Energy Collective in [15]. Additionally, handling power transactions in microgrids has been investigated in [16]. This includes the exploration of utilizing blockchain technology and game theory to develop dynamic pricing strategies rooted in leader-follower decision-making processes. In [17], a multi-market-driven approach to energy scheduling in smart microgrids, emphasizing the use of transactive energy to reduce energy costs and enhance system stability, is developed. Moreover, there has been a redefinition of the conceptual architecture of power systems and their agents to adapt to the increased integration of renewable energy, reflecting the changing landscape of electricity markets [18]. Yue et al. showed a case for the development of transactive energy auctions, enabling the exchange of energy among distributed prosumers in campus demonstrations, highlighting the practical applicability of these systems [19].

In terms of the pricing strategies for energy transactions in TEM, a novel transaction mechanism based on dynamic pricing has been introduced to improve energy system reliability [20]. Additionally, Kim et al. discussed the use of dynamic pricing mechanisms to control energy supply and stabilize transaction prices in off-grid systems [21]. A pricing strategy for microgrid power transactions based on leader-follower game theory, focusing on meeting power user requirements and recording transaction volume and economic indicators, is presented in [22]. In [23], a blockchain-based spot market transaction model that incorporates intraday time-of-use pricing mechanisms has been presented. The goal is to mitigate the negative impacts of intraday power price fluctuations.

In summary, recent research has concentrated on overcoming challenges linked to the implementation of transactive energy systems. The emphasis lies in incorporating engineering principles, dynamic pricing strategies, and innovative technologies to facilitate efficient energy transactions and improve the integration of renewable DERs. Nevertheless, a research gap persists, as no comprehensive market model fitted for power networks has been introduced, capturing P2P transactions and the interaction between transmission and distribution networks as well as considering a robust pricing mechanism.

This chapter serves as a comprehensive guide delineating a strategic roadmap for seamlessly integrating TEM with distribution system reform through the application of market structures. The roadmap outlined here aims to navigate the complex terrain of energy sector transformation, shedding light on potential benefits and challenges intrinsic to these innovative approaches. The overarching goal is to underscore the imperative of adapting the energy sector to harness the full potential of advanced technologies and effectively accommodate the dynamic shifts in energy generation and consumption patterns.

Central to this roadmap is the introduction of a TEM market mechanism, strategically designed and implemented as a robust mixed-integer linear programming (MILP) formulation. This meticulous formulation takes into account various critical factors, including three-phase nodal active and reactive power balance equations, voltage and power flow limits, intertemporal constraints, and system reconfiguration. The versatility of this market mechanism extends to its capability to accommodate various technologies, such as privately owned DERs, utility-owned energy storage batteries, and prosumers and aggregator entities.

The roadmap envisions a future where TEM seamlessly intertwines with distribution system reform, fostering an ecosystem that optimizes energy utilization, enhances grid reliability, and maximizes social welfare. By meticulously addressing the intricacies of market structures, the proposed TEM market mechanism becomes a key driver for orchestrating a harmonious integration that aligns with the evolving landscape of energy dynamics. It is an essential step toward creating a resilient and adaptive energy framework that not only embraces technological advancements but also anticipates and responds to the changing needs of a dynamic energy landscape.

22.2 Proposed TEM Market Platform

This section presents the proposed procedure for the TEM market mechanism. Further, it gives details of the mathematical formulation of the TEM market as an extended version of the market model developed in [24].

22.2.1 TEM Market: Settlement Procedure

This section outlines the envisaged process for settling the TEM market. The distribution system operator (DSO) is assumed to assume the role of overseeing the distribution market settlement. Consequently, the DSO will undertake the following responsibilities:

- Predicting demand across distribution feeders.
- Forecasting energy prices at transmission interconnection nodes.
- Gathering bids from participants.

- Administering settlements for the distribution energy market.
- Managing settlements for the service markets.
- Submitting bids to the transmission market for interchange energy.
- Optimizing any additional real-time market settlement activities.

Additionally, the proposed distribution market settlement procedure takes into account various entities, including aggregators and prosumers, permitted to engage in energy and related services trading within the distribution market settlement. An aggregator is presumed to be an entity responsible for consolidating multiple DERs/customers across different meters and transacting energy at a uniform price across all coupling points. Similarly, a prosumer is defined as a DER/demand entity connected behind a single meter, with the ability to trade energy exclusively at this individual node.

Illustrated in Figure 22.1, the settlement process unfolds in three primary stages. The entire market settlement process is outlined as follows:

- In the initial stage, all active participants, including aggregators and prosumers, are required to submit their bids to the DSO.
- In the second stage, the DSO will settle the distribution market. This settlement will be executed based on the selected market model. Several model options are presented later in this study. At this stage, DSO will forecast the nodal demand along the distribution system and the energy price at the transmission interconnection nodes prior to market settlement. Once the distribution market is settled, the DSO will send a bid for the net interchange energy to the independent system operator (ISO) on the transmission side. The ISO will then settle the transmission market and fix the interchange power (PT) and price (λ^T) and make them available to the DSO.
- Finally, in the third stage, the DSO will re-optimize the distribution market based on the fixed values of PT and transmission marginal price (λ^T) to obtain the new distribution local marginal prices (λ^D).

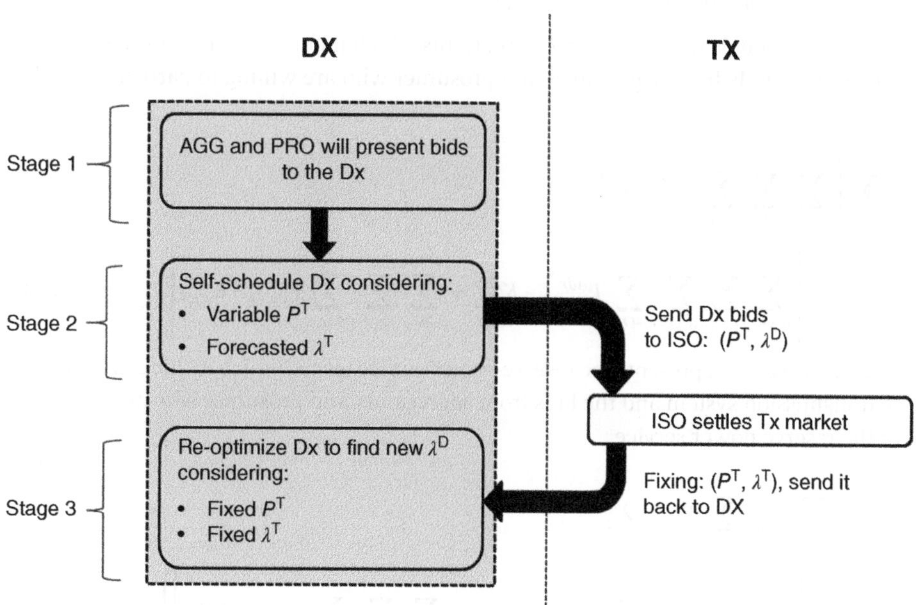

Figure 22.1 Three-stage TEM market settlement procedure [24].

22.2.2 Distribution Market: Mathematical Formulation

The subsequent equations present a comprehensive formulation of the TEM market and its associated services. This formulation is designed to accommodate diverse options for distribution market models, which will be explored in more detail, and it is an expanded model to the one presented in [24]. Equation (22.1) outlines the objective function of the distribution market, aiming to:

- Maximize social welfare, encompassing bids for both demand (SWD) and generation (SWG).
- Maximize social welfare for energy-related services, incorporating demand response (DR), reserve power (RS), and voltage regulation (QS).
- Minimize demand charges (DC).

$$\max \; O.F. = SWD - SWG + DR + RS + QS - DC \tag{22.1}$$

Equation (22.2) represents the demand bids in the social welfare which includes the load from aggregators and prosumers.

$$SWD = \sum_{t\in\Omega_T}\left(\left[\sum_{i\in\Omega_N}\sum_{f\in\Omega_F}\sum_{s\in\Omega_S^a}\sum_{a\in\Omega_{Agg}} P_{i,f,t,s,a}^{A-}\cdot K_{t,s,a}^{D-} + \sum_{f\in\Omega_F}\sum_{s\in\Omega_S^p}\sum_{p\in\Omega_{Pro}} P_{f,t,s,p}^{P-}\cdot K_{t,s,p}^{D-}\right]\right) \tag{22.2}$$

In (22.3), the generation terms of social welfare are represented, including generation bids from transmission system and active agents.

$$SWG = \sum_{t\in\Omega_T}\left(\left[\sum_{i\in\Omega_N}\sum_{f\in\Omega_F}\sum_{s\in\Omega_S^i} P_{i,f,t,s}^{T}\cdot \lambda_{i,t,s}^{T} + \sum_{i\in\Omega_N}\sum_{f\in\Omega_F}\sum_{s\in\Omega_S^a}\sum_{a\in\Omega_{Agg}} P_{i,f,t,s,a}^{A+}\cdot K_{t,s,a}^{D+} + \sum_{f\in\Omega_F}\sum_{s\in\Omega_S^p}\sum_{p\in\Omega_{Pro}} P_{f,t,s,p}^{P+}\cdot K_{t,s,p}^{D+}\right]\right) \tag{22.3}$$

Moreover, Eq. (22.4) models the demand response terms which includes the bids from the transmission system and the bids from aggregators and prosumer who are willing to participate in the DR program.

$$DR = \sum_{t\in\Omega_T}\left(\sum_{i\in\Omega_N}\sum_{f\in\Omega_F}\sum_{s\in\Omega_S^i} P_{i,f,t,s}^{Tdr}\cdot \lambda_{i,t,s}^{Tdr} - \left[\sum_{i\in\Omega_N}\sum_{f\in\Omega_F}\sum_{s\in\Omega_S^a}\sum_{a\in\Omega_{Agg}} P_{i,f,t,s,a}^{Adr}\cdot K_{t,s,a}^{Ddr} + \sum_{f\in\Omega_F}\sum_{s\in\Omega_S^p}\sum_{p\in\Omega_{Pro}} P_{f,t,s,p}^{Pdr}\cdot K_{t,s,p}^{Ddr}\right]\right) \tag{22.4}$$

Similar to (22.4), Eq. (22.5) represents the reserve power service terms, which include the demand bids from the transmission system and the bids from aggregators and prosumer who are willing to participate in the reserve power service.

$$RS = \sum_{t\in\Omega_T}\left(\sum_{i\in\Omega_N}\sum_{f\in\Omega_F}\sum_{s\in\Omega_S^i} P_{i,f,t,s}^{Trs}\cdot \lambda_{i,t,s}^{Trs} - \left[\sum_{i\in\Omega_N}\sum_{f\in\Omega_F}\sum_{s\in\Omega_S^a}\sum_{a\in\Omega_{Agg}} P_{i,f,t,s,a}^{Ars}\cdot K_{t,s,a}^{Drs} + \sum_{f\in\Omega_F}\sum_{s\in\Omega_S^p}\sum_{p\in\Omega_{Pro}} P_{f,t,s,p}^{Prs}\cdot K_{t,s,p}^{Drs}\right]\right) \tag{22.5}$$

The bids of demanded reactive power from the transmission system and the bids from aggregators and prosumers willing to participate in the voltage regulation service are modeled in (22.6).

$$QS = \sum_{t\in\Omega_T}\left(\sum_{i\in\Omega_N}\sum_{f\in\Omega_F}\sum_{s\in\Omega_S^i} Q_{i,f,t,s}^{T}\cdot\lambda_{i,t,s}^{TQ} - \left[\sum_{i\in\Omega_N}\sum_{f\in\Omega_F}\sum_{s\in\Omega_S^a}\sum_{a\in\Omega_{Agg}} Q_{i,f,t,s,a}^{A+}\cdot K_{t,s,a}^{DQ} + \sum_{f\in\Omega_F}\sum_{s\in\Omega_S^p}\sum_{p\in\Omega_{Pro}} Q_{f,t,s,p}^{P+}\cdot K_{t,s,p}^{DQ}\right]\right) \tag{22.6}$$

Finally, Eq. (22.7) represents the demand charges for the distribution system as a load of the transmission system.

$$DC = \sum_{i\in\Omega_N}\left(\left[\max\left(\sum_{f\in\Omega_F}\sum_{s\in\Omega_S^i} P_{i,f,t,s}^{T};\forall_t\varepsilon\Omega_T, P_0^T\right)\right]\cdot K_i^{DC}\right) \tag{22.7}$$

This objective function is subjected to the following constraint:

Active power balance (22.8)–(22.10): It is formulated as a nonlinear, three-phase set of power balance equations, considering the switching based on a zonal approach.

$$\left(\sum_{\substack{(m,n)\varepsilon\Omega_{SW}\\ n=i}} P_{m,n,f,t}^{sw} - \sum_{\substack{(m,n)\varepsilon\Omega_{SW}\\ m=i}} P_{m,n,f,t}^{sw}\right) + P_{i,f,t}^{inj} - P_{i,f,t}^{load} = V_{i,f,t}\sum_{\substack{k\varepsilon\Omega_N^z\\ i\ni\Omega_N^z}}\sum_{h\in\Omega_F} V_{k,h,t}y_{(i,f),(k,h)}cos(\emptyset_{i,f,t}-\emptyset_{k,h,t}-\theta_{(i,f),(k,h)});\forall_i\varepsilon\Omega_N;\forall_f\varepsilon\Omega_F;\forall_t\varepsilon\Omega_T \tag{22.8}$$

$$P_{i,f,t}^{inj} = \left(P_{i,f,t}^{T} - P_{i,f,t}^{Tdr}\right) + \sum_{a\varepsilon\Omega_{AGG}} P_{i,f,t,a}^{A+} + \sum_{p\varepsilon\Omega_{Pro}} P_{f,t,p}^{P+}\cdot f(i,p) + \sum_{e\varepsilon\Omega_{ES}} P_{f,t,e}^{E+}\cdot f(i,e);\forall_i\varepsilon\Omega_N;\forall_f\varepsilon\Omega_F;\forall_t\varepsilon\Omega_T \tag{22.9}$$

$$P_{i,f,t}^{load} = P_{i,f,t}^{L} + \sum_{a\varepsilon\Omega_{AGG}}\left(P_{i,f,t,a}^{A-} - P_{i,f,t,a}^{Adr}\right) + \sum_{p\varepsilon\Omega_{Pro}}\left(P_{f,t,p}^{P-} - P_{f,t,p}^{Pdr}\right)\cdot f(i,p) + \sum_{e\varepsilon\Omega_{ES}} P_{f,t,e}^{E-}\cdot f(i,e);\forall_i\varepsilon\Omega_N;\forall_f\varepsilon\Omega_F;\forall_t\varepsilon\Omega_T \tag{22.10}$$

Reactive power balance (11)–(13): Consistent with the preceding active power balance equations, it is formulated as nonlinear, three-phase set of power balance equations.

$$\left(\sum_{\substack{(m,n)\varepsilon\Omega_{SW}\\ n=i}} Q_{m,n,f,t}^{sw} - \sum_{\substack{(m,n)\varepsilon\Omega_{SW}\\ m=i}} Q_{m,n,f,t}^{sw}\right) + Q_{i,f,t}^{inj} - Q_{i,f,t}^{load} = V_{i,f,t}\sum_{\substack{k\varepsilon\Omega_N^z\\ i\ni\Omega_N^z}}\sum_{h\in\Omega_F} V_{k,h,t}y_{(i,f),(k,h)}sin(\emptyset_{i,f,t}-\emptyset_{k,h,t}-\theta_{(i,f),(k,h)});\forall_i\varepsilon\Omega_N;\forall_f\varepsilon\Omega_F;\forall_t\varepsilon\Omega_T \tag{22.11}$$

$$Q_{i,f,t}^{inj} = \left(Q_{i,f,t}^{T} - Q_{i,f,t}^{Tdr}\right) + \sum_{a\varepsilon\Omega_{AGG}} Q_{i,f,t,a}^{A+} + \sum_{p\varepsilon\Omega_{Pro}} Q_{f,t,p}^{P+}\cdot f(i,p);\forall_i\varepsilon\Omega_N;\forall_f\varepsilon\Omega_F;\forall_t\varepsilon\Omega_T \tag{22.12}$$

$$Q_{i,f,t}^{load} = Q_{i,f,t}^{L} + \sum_{a\varepsilon\Omega_{AGG}} Q_{i,f,t,a}^{A-} + \sum_{p\varepsilon\Omega_{Pro}} Q_{f,t,p}^{P-}\cdot f(i,p);\forall_i\varepsilon\Omega_N;\forall_f\varepsilon\Omega_F;\forall_t\varepsilon\Omega_T \tag{22.13}$$

Reserve power (22.14)–(22.16): It includes limits on the total reserve power of aggregators and prosumers, also taking into account ramping features.

$$P^{Ars}_{i,f,t,a} = min\left(\overline{P^{A+}_{i,f,t,a}} - P^{A+}_{i,f,t,a}, P^{Aramp}_{i,f,t,a}\right); \forall_a \varepsilon \Omega_{Agg}; \forall_i \varepsilon \Omega_N; \forall_f \varepsilon \Omega_F; \forall_t \varepsilon \Omega_T \tag{22.14}$$

$$P^{Prs}_{f,t,p} = min\left(\overline{P^{P+}_{f,t,p}} - P^{P+}_{f,t,p}, P^{Pramp}_{f,t,p}\right); \forall_p \varepsilon \Omega_{Pro}; \forall_f \varepsilon \Omega_F; \forall_t \varepsilon \Omega_T \tag{22.15}$$

$$P^{Trs}_{i,f,t} = \sum_{a\varepsilon\Omega_{AGG}} P^{Ars}_{i,f,t,a} + \sum_{p\varepsilon\Omega_{Pro}} P^{Prs}_{f,t,p} \cdot f(i,p); \forall_i \varepsilon \Omega_N; \forall_f \varepsilon \Omega_F; \forall_t \varepsilon \Omega_T \tag{22.16}$$

Network limits (17) and (18): This set of constraints set limits on the voltage magnitudes at each bus of the grids, as well as on the apparent power flowing through each branch.

$$\underline{V} \leq V_{i,f,t} \leq \overline{V}; \forall_i \varepsilon \Omega_N; \forall_f \varepsilon \Omega_F; \forall_t \varepsilon \Omega_T \tag{22.17}$$

$$\left(P^{sw}_{m,n,f,t}\right)^2 + \left(Q^{sw}_{m,n,f,t}\right)^2 \leq \left(\overline{S^{sw}_{m,n}}\right)^2; \forall_{mn} \varepsilon \Omega_{SW}; \forall_f \varepsilon \Omega_F; \forall_t \varepsilon \Omega_T \tag{22.18}$$

Energy storage (22.19)–(22.22): This set of intertemporal constraints models the operation of utility-owned energy storage devices, determining optimal scheduling and considering limits on maximum capacity, depth of discharge, and number of cycles.

$$0 \leq P^{E+}_{f,t,e} \leq \overline{P^E_e} \cdot \mu_{t,e}; 0 \leq P^{E-}_{f,t,e} \leq \overline{P^E_e} \cdot (1-\mu_{t,e}); \forall_e \varepsilon \Omega_{ES}; \forall_f \varepsilon \Omega_F; \forall_t \varepsilon \Omega_T \tag{22.19}$$

$$\overline{E^E_e} * (1 - DoD) \leq E^0_e - \sum_{h<t\varepsilon\Omega_F} \sum_{f\varepsilon\Omega_F} P^{E+}_{f,h,e} \cdot \eta^+ + \sum_{h<t\varepsilon\Omega_F} \sum_{f\varepsilon\Omega_F} P^{E-}_{f,h,e} \cdot \eta^- \leq \overline{E^E_e}; \forall_e \varepsilon \Omega_{ES}; \forall_t \varepsilon \Omega_T \tag{22.20}$$

$$-\sum_{t\varepsilon\Omega_F} \sum_{f\varepsilon\Omega_F} P^{E+}_{f,t,e} \cdot \eta^+ + \sum_{t\varepsilon\Omega_F} \sum_{f\varepsilon\Omega_F} P^{E-}_{f,t,e} \cdot \eta^- = 0; \forall_e \varepsilon \Omega_{ES} \tag{22.21}$$

$$\sum_{\substack{t\varepsilon\Omega_T \\ t<1}} \mu_{t-1,e} - \mu_{t,e} \leq NC; \forall_e \varepsilon \Omega_{ES} \tag{22.22}$$

Switching (22.23)–(22.26): This set of constraints optimizes the network topology, opening and closing switches according to preset feasible topologies given by the system operator.

$$-\overline{S^{sw}_{m,n}} \cdot \omega_{m,n} \leq P^{sw}_{m,n,f,t} \leq \overline{S^{sw}_{m,n}} \cdot \omega_{m,n}; \forall_{(m,n)} \varepsilon \Omega_{SW}; \forall_f \varepsilon \Omega_F; \forall_t \varepsilon \Omega_T \tag{22.23}$$

$$-\overline{S^{sw}_{m,n}} \cdot \omega_{m,n} \leq Q^{sw}_{m,n,f,t} \leq \overline{S^{sw}_{m,n}} \cdot \omega_{m,n}; \forall_{(m,n)} \varepsilon \Omega_{SW}; \forall_f \varepsilon \Omega_F; \forall_t \varepsilon \Omega_T \tag{22.24}$$

$$\Gamma_\gamma \leq \sum_{(m,n)\varepsilon\Omega^\Gamma_\gamma} \omega_{m,n} / (NZ - NS); \forall_\gamma \varepsilon \Omega_\Gamma \tag{22.25}$$

$$\sum_{\gamma\varepsilon\Omega_\Gamma} \Gamma_\gamma = 1 \tag{22.26}$$

The proposed formulation of the TEM market is implemented as a linearized set of power balance equations is introduced and solved using linear programming (LP). The linearized power flow equations are replacing the set of equations (22.8)–(22.13) in the original formulation. The updated linearized formulation is solved to settle the distribution market with linear optimization.

The linearized set of active power balance is presented as follows in (22.27)–(22.29):

$$\left(\sum_{\substack{(m,n)\varepsilon\Omega_{SW} \\ n=i}} P^{sw}_{m,n,f,t} - \sum_{\substack{(m,n)\varepsilon\Omega_{SW} \\ m=i}} P^{sw}_{m,n,f,t}\right) + P^{inj}_{i,f,t} - P^{load}_{i,f,t}$$
$$= \sum_{(i,j)\varepsilon\Omega_L} \left(P_{i,j,f,t} + P^{loss}_{i,j,f,t}\right) - \sum_{(k,i)\varepsilon\Omega_L} P_{k,i,f,t}; \forall_i \varepsilon \Omega_N; \forall_f \varepsilon \Omega_F; \forall_t \varepsilon \Omega_T \tag{22.27}$$

$$P^{inj}_{i,f,t} = \left(P^{T}_{i,f,t} - P^{Tdr}_{i,f,t}\right) + \sum_{a\varepsilon\Omega_{AGG}} P^{A+}_{i,f,t,a} + \sum_{p\varepsilon\Omega_{Pro}} P^{P+}_{f,t,p} \cdot f(i,p) + \sum_{e\varepsilon\Omega_{ES}} P^{E+}_{f,t,e} \cdot f(i,e); \forall_i \varepsilon\Omega_N; \forall_f \varepsilon\Omega_F; \forall_t \varepsilon\Omega_T \tag{22.28}$$

$$P^{load}_{i,f,t} = P^{L}_{i,f,t} + \sum_{a\varepsilon\Omega_{AGG}} \left(P^{A-}_{i,f,t,a} - P^{Adr}_{i,f,t,a}\right) + \sum_{p\varepsilon\Omega_{Pro}} \left(P^{P-}_{f,t,p} - P^{Pdr}_{f,t,p}\right) \cdot f(i,p) + \sum_{e\varepsilon\Omega_{ES}} P^{E-}_{f,t,e} \cdot f(i,e); \forall_i \varepsilon\Omega_N; \forall_f \varepsilon\Omega_F \forall_t \varepsilon\Omega_T \tag{22.29}$$

The linearized set of reactive power balance is presented as follows in (22.30)–(22.32):

$$\left(\sum_{\substack{(m,n)\varepsilon\Omega_{SW}\\ n=i}} Q^{sw}_{m,n,f,t} - \sum_{\substack{(m,n)\varepsilon\Omega_{SW}\\ m=i}} Q^{sw}_{m,n,f,t}\right) + Q^{inj}_{i,f,t} - Q^{load}_{i,f,t} = \sum_{(i,j)\varepsilon\Omega_L} \left(Q_{i,j,f,t} + Q^{loss}_{i,j,f,t}\right) - \sum_{(k,i)\varepsilon\Omega_L} Q_{k,i,f,t}; \forall_i \varepsilon\Omega_N; \forall_f \varepsilon\Omega_F; \forall_t \varepsilon\Omega_T \tag{22.30}$$

$$Q^{inj}_{i,f,t} = Q^{T}_{i,f,t} + \sum_{a\varepsilon\Omega_{AGG}} Q^{A+}_{i,f,t,a} + \sum_{p\varepsilon\Omega_{Pro}} Q^{P+}_{f,t,p} \cdot f(i,p); \forall_i \varepsilon\Omega_N; \forall_f \varepsilon\Omega_F; \forall_t \varepsilon\Omega_T \tag{22.31}$$

$$Q^{load}_{i,f,t} = Q^{L}_{i,f,t} + \sum_{a\varepsilon\Omega_{AGG}} Q^{A-}_{i,f,t,a} + \sum_{p\varepsilon\Omega_{Pro}} Q^{P-}_{f,t,p} \cdot f(i,p); \forall_i \varepsilon\Omega_N; \forall_f \varepsilon\Omega_F; \forall_t \varepsilon\Omega_T \tag{22.32}$$

The linearized losses and voltage drop equations are presented as follows in (22.33)–(22.36):

$$V^{sqr}_{i,f,t} - V^{sqr}_{j,f,t} = \sum_{h\in F} \left\{2 * \left(\tilde{R}_{i,j,f,h} P_{i,j,f,t} + \tilde{X}_{i,j,f,h} Q_{i,j,f,t}\right) + \tilde{Z}^{2}_{i,j,f,h} I^{sqr}_{i,j,f,t}\right\}; \forall_{(i,j)} \varepsilon\Omega_L; \forall_f \varepsilon\Omega_F; \forall_t \varepsilon\Omega_T \tag{22.33}$$

$$\hat{V}^{sqr}_{j,f,t} I^{sqr}_{i,j,f,t} = P^{2}_{i,j,f,t} + Q^{2}_{i,j,f,t}; \forall_i \varepsilon\Omega_N; \forall_f \varepsilon\Omega_F; \forall_t \varepsilon\Omega_T \tag{22.34}$$

$$P^{loss}_{i,j,f,t} = \sum_{h\in\Omega_F} \tilde{R}_{i,j,f,h} \frac{\left(P_{i,j,f,t} P_{i,j,h,t} + Q_{i,j,f,t} Q_{i,j,h,t}\right)}{V_{j,f,t} V_{j,h,t}} + \tilde{X}_{i,j,f,h} \frac{\left(-Q_{i,j,f,t} P_{i,j,h,t} + P_{i,j,f,t} Q_{i,j,h,t}\right)}{V_{j,f,t} V_{j,f,t}} \tag{22.35}$$

$$Q^{loss}_{i,j,f,t} = \sum_{h\in\Omega_F} \tilde{X}_{i,j,f,h} \frac{\left(P_{i,j,f,t} P_{i,j,h,t} + Q_{i,j,f,t} Q_{i,j,h,t}\right)}{V_{j,f,t} V_{j,h,t}} + \tilde{R}_{i,j,f,h} \frac{\left(Q_{i,j,f,t} P_{i,j,h,t} - P_{i,j,f,t} Q_{i,j,h,t}\right)}{V_{j,f,t} V_{j,f,t}} \tag{22.36}$$

The DSO will predict demand across the distribution system, while all participants, including aggregators and prosumers, have the option to submit bids either as suppliers or loads, or both. Within this TEM framework, the DSO seeks to maximize social welfare, and entities dispatched will receive compensation determined by the marginal price obtained through the distribution market settlement. As previously indicated, the DSO will comprehensively manage all the stages outlined in the market settlement procedure. Some illustrative case studies are developed and presented in the following section.

22.3 Demonstrative Case Studies

To evaluate the effectiveness of the TEM market platform, a modified IEEE 34-bus test system has been utilized. The mathematical framework detailed in previous section is implemented in AMPL software and resolved using the CPLEX commercial solver. Furthermore, the transactive energy management system (TEMS) identifies all optimization variables.

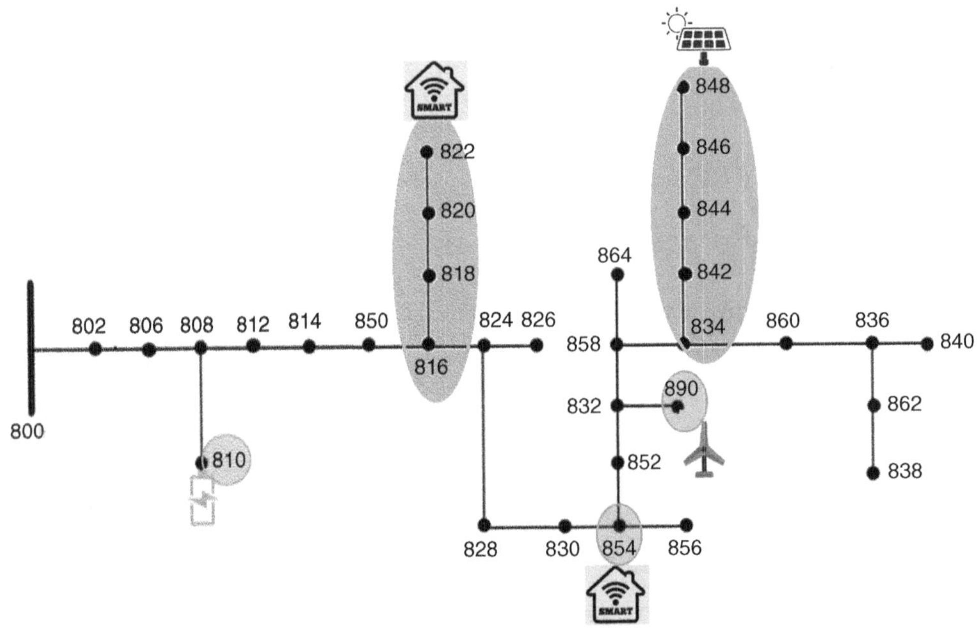

Figure 22.2 Modified 34-bus test system.

The adapted IEEE 34-bus test system is used to show the performance of TEM market as illustrated in Figure 22.2. The 34-bus test system is a modified version derived from the IEEE 34-bus test system, featuring the following specifications. The substation's nominal voltage is established at 24.9 kV. Bus # 800 is equipped with a substation transformer with a capacity of 5 MW. The system has the following entities:

- Aggregator 1: It is an aggregation of generating DERs at buses # 834, 842, 844, 846, and 848.
- Aggregator 2: It is an aggregation of smart loads at buses # 816, 818, 820, and 822.
- Prosumer 1: It is a smart load at bus # 854.
- Prosumer 2: It is a generating DER at bus # 890.
- Utility-owned battery connected at bus # 810.

Table 22.1 shows the power capacity bids for the following participants. It is noteworthy that the bidding processes for participants and the formulation of their aggregation agreements can be structured using various approaches, as detailed in references [25, 26].

The system has a total nominal demand load of 1.78 MW, and Table 22.2 provides a breakdown of the nominal demand load of each bus.

Figure 22.4 provides information on the percentage of nominal load demanded at each hour. Details about distribution line segment data can be found in Appendix 22.A.

22.3.1 Case#1 (TEM Settlement for Day-Ahead Without Utility-Owned Batteries)

In this case, two aggregators (providing energy bids at nine different buses) and two prosumers were connected to the system. Simulations were conducted for a 24-hour day-ahead market settlement. Figure 22.5 shows the percentage of transactions dispatched for each hour within the

Table 22.1 Energy bidding of all participants of TEM market mechanism.

Participant	Connection type	Transaction type	Power capacity (kW)	Price (¢/kWh)
Aggregator 1	3-Phase	P2DSO	600	1.00
Aggregator 2	1-Phase	P2DSO	120	6.00
Prosumer 1	1-Phase	P2DSO	50	8.00
Prosumer 2	3-Phase	P2DSO	150	3.5
Utility-owned battery	3-Phase	T2DSO	100	Hourly as per Figure 22.3
Transmission System	3-Phase	T2DSO	3000	

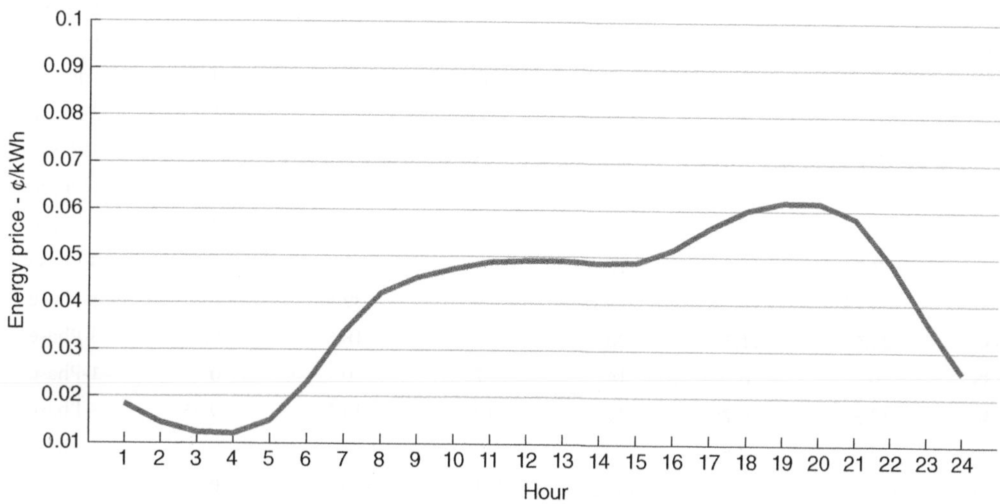

Figure 22.3 Hourly energy price of the transmission system at bus # 800.

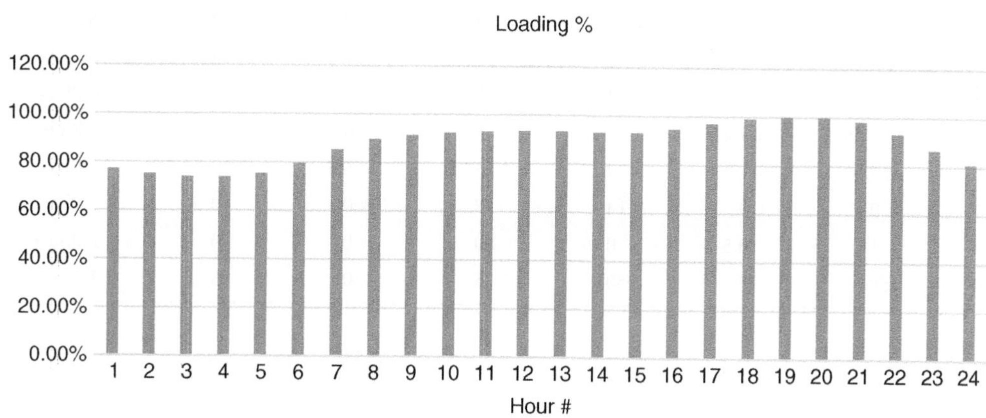

Figure 22.4 Loading percentage per hour.

Table 22.2 Nominal demand summary of the 34-bus test system (kW).

Bus	Phase A		Phase B		Phase C		Load Type
	P	Q	P	Q	P	Q	
800	0	0	0	0	0	0	None
802	0	0	15	7.5	12.5	7	2-Phase
806	0	0	15	7.5	12.5	7	2-Phase
808	0	0	8	4	0	0	1-Phase
810	0	0	8	4	0	0	1-Phase
812	0	0	0	0	0	0	None
814	0	0	0	0	0	0	None
816	0	0	2	1	0	0	1-Phase
818	17	8.5	0	0	0	0	1-Phase
820	84.5	43.5	0	0	0	0	1-Phase
822	67.5	35	0	0	0	0	1-Phase
824	0	0	20	11	2	1	2-Phase
826	0	0	20	10	0	0	1-Phase
828	3.5	1.5	0	0	2	1	2-Phase
830	21	9	10	5	17.5	7.5	3-Phase
832	3.5	1.5	15	0.5	3	1.5	3-Phase
834	35.75	18	13.75	7	39.5	20	3-Phase
836	22.5	11.5	20	10.25	18.5	9.75	3-Phase
838	0	0	14	7	0	0	1-Phase
840	13.5	9.25	19	12	14.5	9.75	3-Phase
842	4.5	2.5	0	0	0	0	1-Phase
844	139.5	107.5	147.5	111	145	110.5	3-Phase
846	0	0	24	11.5	10	5.5	2-Phase
848	20	16	31.5	21.5	20	16	3-Phase
850	0	0	0	0	0	0	None
852	0	0	0	0	0	0	None
854	0	0	2	1	0	0	1-Phase
856	0	0	2	1	0	0	1-Phase
858	8.75	4.25	5.75	3	10	5.25	3-Phase
860	69.5	41	39	25.75	65.5	39.25	3-Phase
862	0	0	14	7	0	0	1-Phase
864	1	0.5	0	0	0	0	1-Phase
890	150	75	150	75	150	75	3-Phase

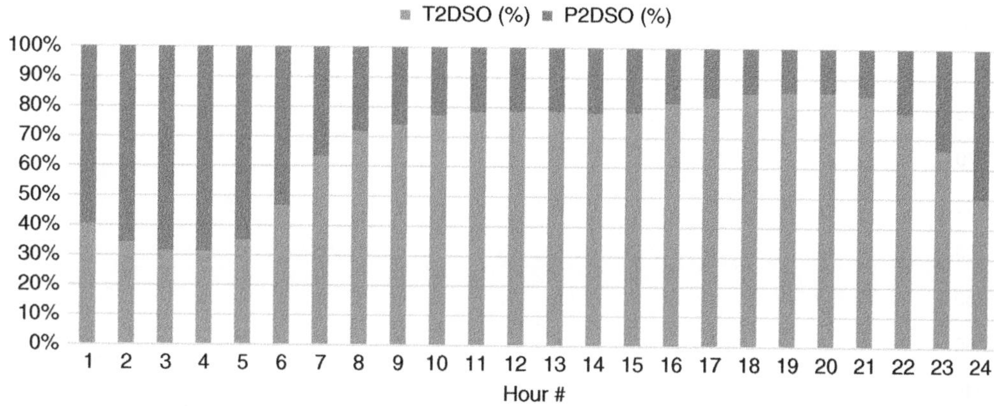

Figure 22.5 Percentage of total transactions transmission versus distribution peer transactions.

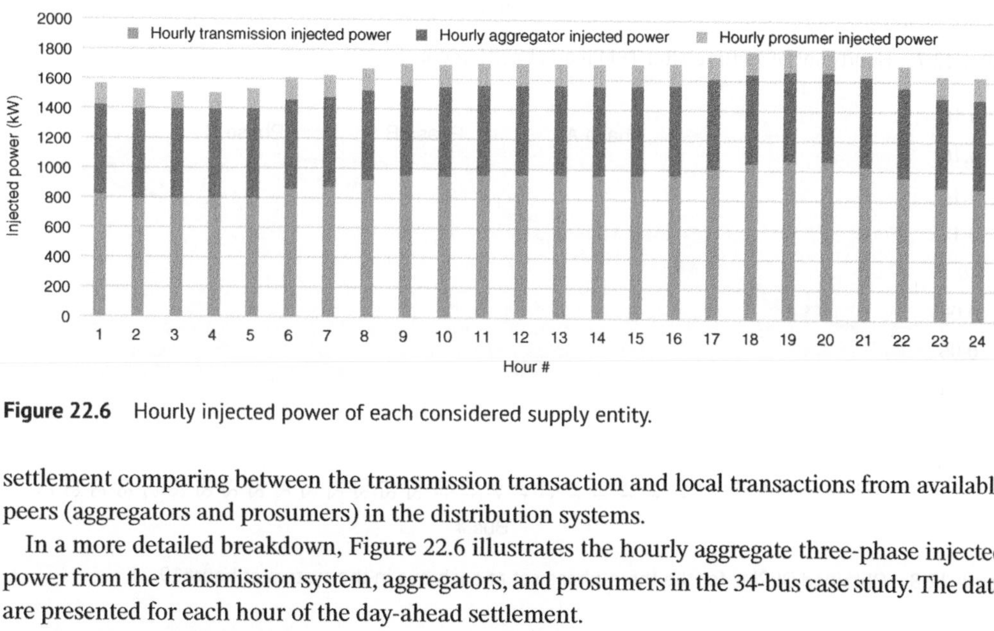

Figure 22.6 Hourly injected power of each considered supply entity.

settlement comparing between the transmission transaction and local transactions from available peers (aggregators and prosumers) in the distribution systems.

In a more detailed breakdown, Figure 22.6 illustrates the hourly aggregate three-phase injected power from the transmission system, aggregators, and prosumers in the 34-bus case study. The data are presented for each hour of the day-ahead settlement.

From the economic perspective, Figure 22.7 provides the hourly social welfare value (objective function) obtained from the TEM market settlement. In this case, the TEM market settlement resulted in a 9.5% enhancement of social welfare during the 24-hour market settlement. Furthermore, it effectively lowered DLMPs across the system by sourcing more cost-effective energy from existing DERs. The obtained hourly DLMPs are presented in Figure 22.8 for each phase.

Furthermore, Figure 22.9 depicts the voltage profile for each bus within the 34-bus system at hour #19, specifically at the maximum load factor (100%) during the day-ahead settlement.

22.3.2 Case#2 (TEM Settlement for Day-Ahead with Utility-Owned Battery)

A utility-owned battery can serve the dual purpose of enhancing the voltage profile within the distribution system and supplying remote communities or rural areas during abnormal operational

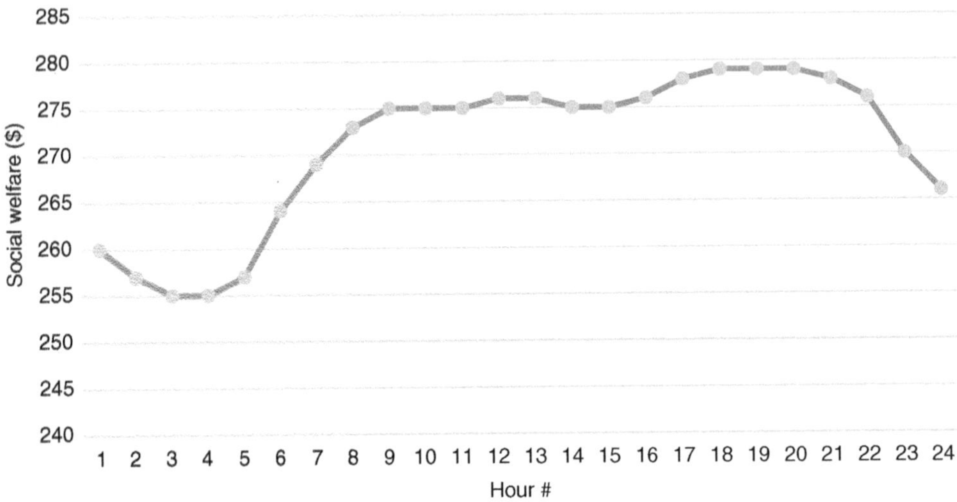

Figure 22.7 Hourly social welfare after TEM market settlement.

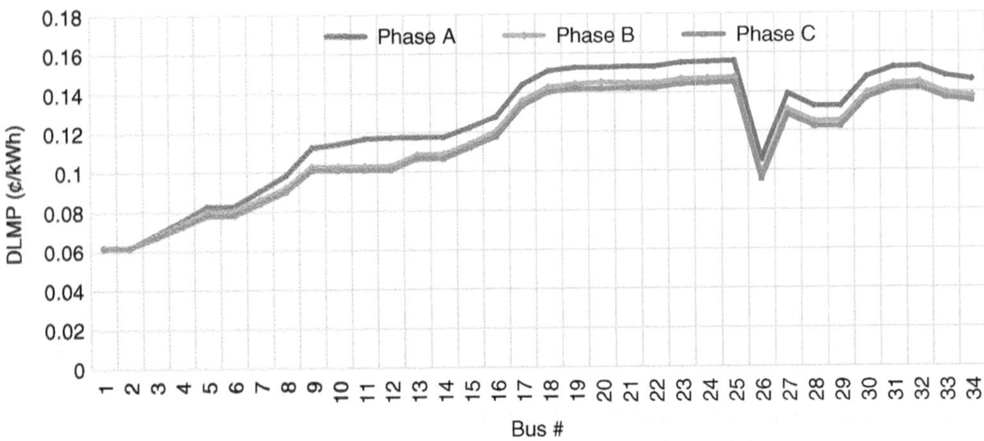

Figure 22.8 Obtained DLPMs for the three-phase set at each bus in the system at hour#19.

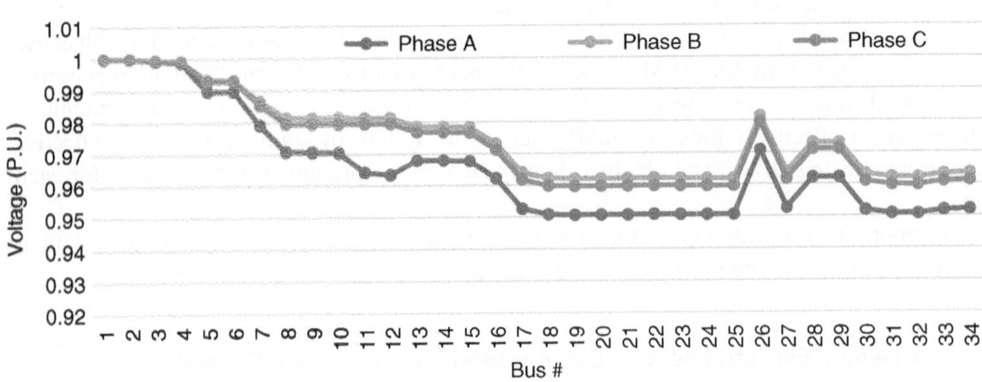

Figure 22.9 The three-phase voltage magnitudes at each bus in the system at hour#19 with maximum load factor.

conditions when utilized by the DSO. In this scenario, in addition to the two aggregators and two prosumers were connected to the system, there is utility-owned battery of 100 kW at bus # 810. The batter storage device has a maximum storage capacity of 500 kWh. The maximum depth of discharge allowed for this device is set to 90%.

The TEM market settlement is conducted for a 24-hour day-ahead market settlement. Figure 22.10 shows the hourly state-of-charge for the battery at bus # 810. It is also worth noticing that the state-of-charge for the battery is kept within limits along the whole time period.

Furthermore, the hourly schedule of utility-owned power to inject power at bus # 810 is demonstrated at Figure 22.11. Results reveal that the battery energy storage injects power during the high-price hours.

The technical advantages resulting from the implementation of the utility-owned battery are illustrated in Figure 22.12, depicting an enhanced voltage profile across the distribution system when compared to the preceding scenario without the battery, as depicted in Figure 22.9.

This scenario exemplifies the TEM market mechanism's capability to settle and generate schedules for all participants, encompassing utility-owned batteries and their corresponding charging and discharging schedules.

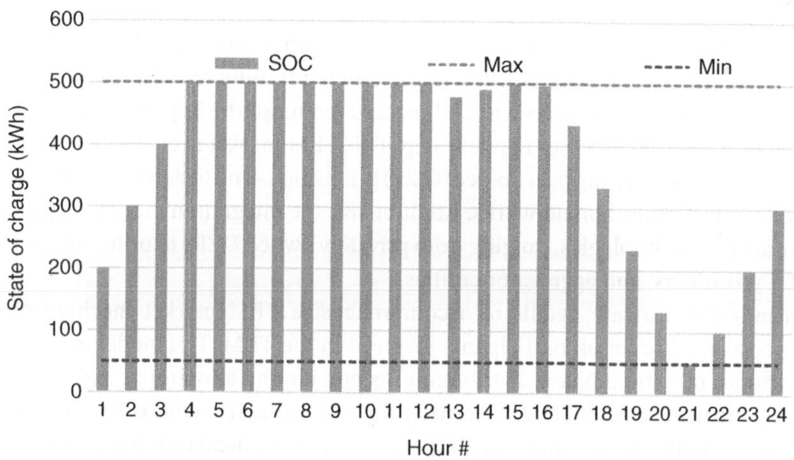

Figure 22.10 Hourly state-of-charge for the battery at bus # 810.

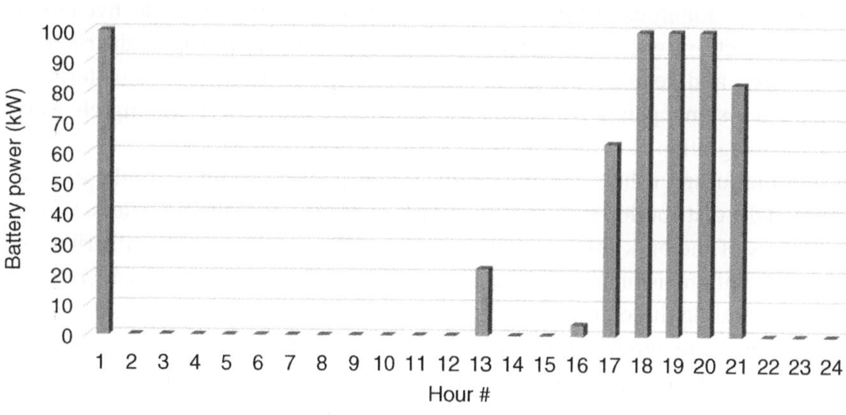

Figure 22.11 Hourly battery power injected at bus # 810.

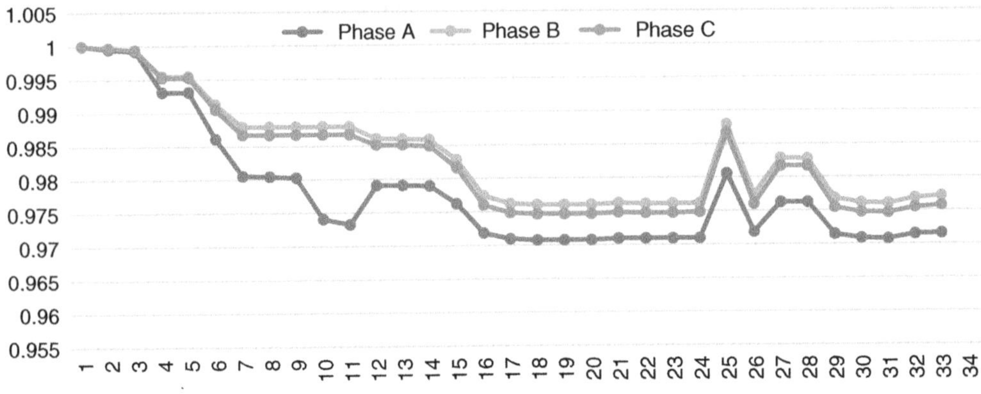

Figure 22.12 The three-phase voltage magnitudes at each bus in the system at hour#19 at maximum load factor after considering the utility-owned battery.

22.4 Conclusion Remarks and Prospects for the Future

The escalating integration of distributed energy resources (DERs) is spurred by a collective drive to reduce costs and enhance the overall efficiency of energy systems. To fully harness the myriad benefits that DERs offer, a holistic approach is essential. First and foremost, the implementation of a TEM market, overseen by a DSO, emerges as a pivotal step. This TEM market acts as a facilitator for various transaction types, encompassing peer-to-peer (P2P) interactions and fostering collaboration between transmission and distribution networks. Additionally, the integration process should be inclusive of a diverse array of technologies, ranging from privately owned DERs to utility-owned energy storage batteries, prosumers, and aggregator entities.

This chapter delves into the intricate details of a comprehensive TEM market mechanism designed to maximize social welfare for all participants involved in the TEM. The model meticulously captures the dynamics of a three-phase distribution system, aiming to optimize economic prospects for DERs while ensuring cost-effective electricity for end-users. It accounts for various transaction types, spanning both energy and ancillary services, and elucidates the nuanced interaction between the bulk electricity market and the TEM under a DSO control model. The TEM mechanism is practically implemented through a MILP formulation, equipped with network reconfiguration capabilities. Rigorous testing on a 34-bus system showcases the effectiveness of the TEM market model in settling energy and ancillary service transactions, providing distribution locational marginal prices, and demonstrating the ability to schedule utility-owned batteries to support local distribution system operations. This multifaceted approach reflects a comprehensive strategy to enhance the integration of DERs within the evolving energy landscape.

Looking ahead into the future, it becomes imperative to grasp and accommodate the distinctive features inherent in the distribution system, particularly concerning ancillary services. Unlike the conventional services of the bulk transmission system, ancillary services in the distribution system may necessitate local provision, presenting a set of unique challenges and considerations. Recognizing and understanding these differences is crucial for developing tailored solutions that effectively cater to the specific requirements of the distribution network.

Furthermore, the evolving landscape of TEM mechanisms introduces a paradigm shift in the operational dynamics of the distribution system. In the context of the operational day-ahead TEM mechanism, it is essential to delve into the intricate interplay between various transactions and their implications for the medium and long-term planning of the distribution system. Understanding how these transactions influence the system's capacity planning, infrastructure development, and overall resilience is paramount for ensuring a seamless integration of DERs and the optimization of the distribution network.

Therefore, as we look toward the future, comprehensive research and analysis are needed to delineate the specific requirements of ancillary services in the distribution system and to establish a nuanced understanding of how the operational intricacies of the TEM mechanism reverberate through the medium and long-term planning phases. This holistic approach will pave the way for resilient, adaptive, and forward-looking strategies that align with the evolving dynamics of the energy landscape and contribute to the sustainable development of distribution systems.

References

1 Capper, T., Gorbatcheva, A., Mustafa, M.A. et al. (2022). Peer-to-peer, community self-consumption, and transactive energy: a systematic literature review of local energy market models. *Renewable and Sustainable Energy Reviews* 162: https://doi.org/10.1016/j.rser.2022.112403.

2 Zia, M.F., Benbouzid, M., Elbouchikhi, E. et al. (2020). Microgrid transactive energy: review, architectures, distributed ledger technologies, and market analysis. *IEEE Access* 8: 19410–19432. https://doi.org/10.1109/ACCESS.2020.2968402.

3 Park, L., Lee, S., and Chang, H. (2018). A sustainable home energy prosumer-chain methodology with energy tags over the blockchain. *Sustainability* 10 (3): 658. https://doi.org/10.3390/su10030658.

4 Grönroos, C. (1991). The marketing strategy continuum: towards a marketing concept for the 1990s. *Management Decision* 29 (1): https://doi.org/10.1108/00251749110139106.

5 He, G., Chen, Q., Chen, K., and Xia, Q. (2016). Optimal offering strategy for concentrating solar power plants in joint energy, reserve and regulation markets. *IEEE Transactions on Sustainable Energy* 7 (3): 1245–1254. https://doi.org/10.1109/tste.2016.2533637.

6 Sehar, F., Pipattanasomporn, M., and Rahman, S. (2016). An energy management model to study energy and peak power savings from pv and storage in demand responsive buildings. *Applied Energy* 173: 406–417. https://doi.org/10.1016/j.apenergy.2016.04.039.

7 Liu, P., Ding, T., Zou, Z., and Yang, Y. (2019). Integrated demand response for a load serving entity in multi-energy market considering network constraints. *Applied Energy* 250: 512–529. https://doi.org/10.1016/j.apenergy.2019.05.003.

8 Davoudi, M., Moeini-Aghtaie, M., and Ghorani, R. (2021). Developing a new framework for transactive peer-to-peer thermal energy market. *IET Generation Transmission & Distribution* 15 (13): 1984–1995. https://doi.org/10.1049/gtd2.12150.

9 L. Zhang, S. Chen, Y. Zhou, H. Zhang, & Y. Li, "*Research on the trading settlement mechanism of electricity market*", 2023. https://doi.org/10.1117/12.2679274

10 Tang, Z., Xiang, J., Duan, Y. et al. (2022). Robust scheduling of virtual power plant with power-to-gas device. *Journal of Physics Conference Series* 2260 (1): 012009. https://doi.org/10.1088/1742-6596/2260/1/012009.

11 Chen, Y., Yong, Y., and Xu, X. (2022). Towards transactive energy: an analysis of information-related practical issues. *Energy Conversion and Economics* 3 (3): 112–121. https://doi.org/10.1049/enc2.12057.

12 B. Zheng, Y. Fan, W. Wei, Y. Xu, S. Huang, & S. Mei, "Distribution optimal power flow with energy sharing via a peer-to-peer transactive market", *Frontiers in Energy Research*, vol. 9, 2021. https://doi.org/10.3389/fenrg.2021.701149

13 Peng, D., Xiao, H., Pei, W. et al. (2022). A novel deep learning-based peer-to-peer transaction method for prosumers under two-stage market environment. *Iet Smart Grid* 5 (6): 430–439. https://doi.org/10.1049/stg2.12078.

14 Li, G., Li, G., and Zhou, M. (2022). Market transaction model design applicable for both plan and market environment of china's renewable energy. *Frontiers in Energy Research* 10: https://doi.org/10.3389/fenrg.2022.862653.

15 Dong, A., Baroche, T., Latimier, R., and Ahmed, H. (2021). Convergence analysis of an asynchronous peer-to-peer market with communication delays. *Sustainable Energy Grids and Networks* 26: 100475. https://doi.org/10.1016/j.segan.2021.100475.

16 He, Y., Tan, W., and Tian, Y. (2023). Micro-grid power transaction management platform based on blockchain technology. Proc. SPIE 12593, Second Guangdong-Hong Kong-Macao Greater Bay Area Artificial Intelligence and Big Data Forum. https://doi.org/10.1117/12.2671681

17 Yue, J., Hu, Z., Anvari-Moghaddam, A., and Guerrero, J. (2019). A multi-market-driven approach to energy scheduling of smart microgrids in distribution networks. *Sustainability* 11 (2): 301. https://doi.org/10.3390/su11020301.

18 Rodríguez-García, J., Ribó-Pérez, D., Alvarez, C., and Peñalvo-López, E. (2019). Novel conceptual architecture for the next-generation electricity markets to enhance a large penetration of renewable energy. *Energies* 12 (13): 2605. https://doi.org/10.3390/en12132605.

19 Yue, J., Hu, Z., Li, C., et al. (2019). Dynamic pricing for microgrids energy transaction in blockchain-based ecosystem. 2019 IEEE Innovative Smart Grid Technologies—Asia (ISGT Asia), Chengdu, China, pp. 1598–1602. https://doi.org/10.1109/isgt-asia.2019.8881671

20 Oh, E. and Son, S. (2019). Transaction mechanism based on two-dimensional energy and reliability pricing for energy prosumers. *Applied Sciences* 9 (7): 1343. https://doi.org/10.3390/app9071343.

21 Kim, J., Lee, J., and Choi, J. (2020). Joint demand response and energy trading for electric vehicles in off-grid system. *IEEE Access* 8: 130576–130587. https://doi.org/10.1109/access.2020.3009739.

22 Wang, P., and Tang, C. (2022). Pricing strategy of micro-grid power transaction based on leader-follower game theory. Fifth International Conference on Mechatronics and Computer Technology Engineering (MCTE 2022). https://doi.org/10.1117/12.2660663

23 Hu, W., Li, H., Hu, Y., and Yao, W. (2019). A blockchain-based spot market transaction model for energy power supply and demand network. *European Journal of Electrical Engineering* 21 (1): 75–83. https://doi.org/10.18280/ejee.210112.

24 Sabillon, C., Mohamed, A.A., Golriz, A. et al. (2021). Comprehensive platform for distribution transactive energy markets. *IET Generation, Transmission & Distribution.* 15 (16): 2344–2355. https://doi.org/10.1049/gtd2.12182.

25 Sabillon, C., Mohamed, A.A., Golriz, A. et al. (2021). Optimal operation of small, numerous, and disparate DERs via aggregation in transactive distribution systems with universal metering. *IET Generation, Transmission & Distribution* 15 (17): 2422–2434. https://doi.org/10.1049/gtd2.12187.

26 Mohamed, A.A., Sabillon, C., Golriz, A. et al. (2021). Value-stack Aggregator Optimal Planning considering Disparate DERs Technologies. *IET Generation, Transmission & Distribution* 15 (18): 2632–2644. https://doi.org/10.1049/gtd2.12205.

Appendix 22.A Line Segment Data of the 34-bus Test System (ohms)

Line FB TB:	Ra → Xa → Rab → Xab → Rac Xac → Rba → Xba → Rb → Xb Rbc→ Xbc → Rca → Xca → Rbc Xbc → Rc → Xc
1 800 802:	0.163302261 → 0.162996854 → 0.025665627 → 0.070595766 → 0.026019887 0.061262784 → 0.025665627 → 0.070595766 → 0.161714207 → 0.165757647 0.025238070 → 0.056083240 → 0.026019887 → 0.061262784 → 0.025238070 0.056083240 → 0.162398294 → 0.164560527
2 802 806:	0.109501129 → 0.109296340 → 0.017209897 → 0.047337471 → 0.017447444 0.041079309 → 0.017209897 → 0.047337471 → 0.108436270 → 0.111147570 0.016923202 → 0.037606204 → 0.017447444 → 0.041079309 → 0.016923202 0.037606204 → 0.108894980 → 0.110344850
3 806 808:	2.040012354 → 2.036197127 → 0.320621381 → 0.881899821 → 0.325046882 0.765309891 → 0.320621381 → 0.881899821 → 2.020173983 → 2.070685645 0.315280226 → 0.700605749 → 0.325046882 → 0.765309891 → 0.315280226 0.700605749 → 2.028719767 → 2.055730925
4 808 810:	0.000000000 → 0.000000000 → 0.000000000 → 0.000000000 → 0.000000000 0.000000000 → 0.000000000 → 0.000000000 → 0.769331808 → 0.408231595 0.000000000 → 0.000000000 → 0.000000000 → 0.000000000 → 0.000000000 0.000000000 → 0.000000000 → 0.000000000
5 808 812:	2.373579375 → 2.369140313 → 0.373046906 → 1.026101250 → 0.378196031 0.890447438 → 0.373046906 → 1.026101250 → 2.350497188 → 2.409268125 0.366832406 → 0.815163375 → 0.378196031 → 0.890447438 → 0.366832406 0.815163375 → 2.360440313 → 2.391868125
6 812 814:	1.881773729 → 1.878254440 → 0.295751587 → 0.813493071 → 0.299833814 0.705946729 → 0.295751587 → 0.813493071 → 1.863474170 → 1.910067770 0.290824732 → 0.646261524 → 0.299833814 → 0.705946729 → 0.290824732 0.646261524 → 1.871357080 → 1.896273050
7 814 850:	0.000913826 → 0.000668324 → 0.000110180 → 0.000305019 → 0.000111695 0.000269460 → 0.000110180 → 0.000305019 → 0.000907055 → 0.000676184 0.000108333 → 0.000248012 → 0.000111695 → 0.000269460 → 0.000108333 0.000248012 → 0.000909991 → 0.000672775
8 816 818:	0.226663920 → 0.120274988 → 0.000000000 → 0.000000000 → 0.000000000 0.000000000 → 0.000000000 → 0.000000000 → 0.000000000 → 0.000000000 0.000000000 → 0.000000000 → 0.000000000 → 0.000000000 → 0.000000000 0.000000000 → 0.000000000 → 0.000000000
9 816 824:	0.933016091 → 0.682358549 → 0.112493704 → 0.311424399 → 0.114040672 0.275118915 → 0.112493704 → 0.311424399 → 0.926103155 → 0.690383609 0.110608325 → 0.253219614 → 0.114040672 → 0.275118915 → 0.110608325 0.253219614 → 0.929100301 → 0.686902765
10 818 820:	6.382378800 → 3.386690438 → 0.000000000 → 0.000000000 → 0.000000000 0.000000000 → 0.000000000 → 0.000000000 → 0.000000000 → 0.000000000 0.000000000 → 0.000000000 → 0.000000000 → 0.000000000 → 0.000000000 0.000000000 → 0.000000000 → 0.000000000

Line FB TB:	Ra → Xa → Rab → Xab → Rac Xac → Rba → Xba → Rb → Xb Rbc→ Xbc → Rca → Xca → Rbc Xbc → Rc → Xc
11 820 822:	1.821264480 → 0.966420075 → 0.000000000 → 0.000000000 → 0.000000000 0.000000000 → 0.000000000 → 0.000000000 → 0.000000000 → 0.000000000 0.000000000 → 0.000000000 → 0.000000000 → 0.000000000 → 0.000000000 0.000000000 → 0.000000000 → 0.000000000
12 824 826:	0.000000000 → 0.000000000 → 0.000000000 → 0.000000000 → 0.000000000 0.000000000 → 0.000000000 → 0.000000000 → 0.401632560 → 0.213118838 0.000000000 → 0.000000000 → 0.000000000 → 0.000000000 → 0.000000000 0.000000000 → 0.000000000 → 0.000000000
13 824 828:	0.076761363 → 0.056139195 → 0.009255114 → 0.025621596 → 0.009382386 0.022634661 → 0.009255114 → 0.025621596 → 0.076192620 → 0.056799435 0.009099999 → 0.020832956 → 0.009382386 → 0.022634661 → 0.009099999 0.020832956 → 0.076439202 → 0.056513058
14 828 830:	1.867859833 → 1.366053745 → 0.225207767 → 0.623458836 → 0.228304733 0.550776751 → 0.225207767 → 0.623458836 → 1.854020420 → 1.382119585 0.221433316 → 0.506935251 → 0.228304733 → 0.550776751 → 0.221433316 0.506935251 → 1.860020582 → 1.375151078
15 830 854:	0.047518939 → 0.034752835 → 0.005729356 → 0.015860988 → 0.005808144 0.014011933 → 0.005729356 → 0.015860988 → 0.047166860 → 0.035161555 0.005633333 → 0.012896592 → 0.005808144 → 0.014011933 → 0.005633333 0.012896592 → 0.047319506 → 0.034984274
16 832 858:	0.447774618 → 0.327478638 → 0.053988163 → 0.149459310 → 0.054730587 0.132035523 → 0.053988163 → 0.149459310 → 0.444456950 → 0.331330038 0.053083329 → 0.121525574 → 0.054730587 → 0.132035523 → 0.053083329 0.121525574 → 0.445895345 → 0.329659505
17 834 860:	0.184592802 → 0.135001398 → 0.022256345 → 0.061613838 → 0.022562405 0.054430971 → 0.022256345 → 0.061613838 → 0.183225110 → 0.136589118 0.021883332 → 0.050098298 → 0.022562405 → 0.054430971 → 0.021883332 0.050098298 → 0.183818081 → 0.135900449
18 834 842:	0.025587121 → 0.018713065 → 0.003085038 → 0.008540532 → 0.003127462 0.007544887 → 0.003085038 → 0.008540532 → 0.025397540 → 0.018933145 0.003033333 → 0.006944319 → 0.003127462 → 0.007544887 → 0.003033333 0.006944319 → 0.025479734 → 0.018837686
19 836 840:	0.078589015 → 0.057475843 → 0.009475474 → 0.026231634 → 0.009605777 0.023173582 → 0.009475474 → 0.026231634 → 0.078006730 → 0.058151803 0.009316666 → 0.021328978 → 0.009605777 → 0.023173582 → 0.009316666 0.021328978 → 0.078259183 → 0.057858607
20 836 862:	0.025587121 → 0.018713065 → 0.003085038 → 0.008540532 → 0.003127462 0.007544887 → 0.003085038 → 0.008540532 → 0.025397540 → 0.018933145 0.003033333 → 0.006944319 → 0.003127462 → 0.007544887 → 0.003033333 0.006944319 → 0.025479734 → 0.018837686
21 842 844:	0.123366476 → 0.090223706 → 0.014874290 → 0.041177565 → 0.015078835 0.036377134 → 0.014874290 → 0.041177565 → 0.122452425 → 0.091284806 0.014624999 → 0.033481536 → 0.015078835 → 0.036377134 → 0.014624999 0.033481536 → 0.122848718 → 0.090824558

Line FB TB:	Ra → Xa → Rab → Xab → Rac Xac → Rba → Xba → Rb → Xb Rbc→ Xbc → Rca → Xca → Rbc Xbc → Rc → Xc
22 844 846:	0.332632573 → 0.243269845 → 0.040105493 → 0.111026916 → 0.040657007 0.098083531 → 0.040105493 → 0.111026916 → 0.330168020 → 0.246130885 0.039433330 → 0.090276141 → 0.040657007 → 0.098083531 → 0.039433330 0.090276141 → 0.331236542 → 0.244889918
23 846 848:	0.048432765 → 0.035421159 → 0.005839536 → 0.016166007 → 0.005919839 0.014281393 → 0.005839536 → 0.016166007 → 0.048073915 → 0.035837739 0.005741666 → 0.013144603 → 0.005919839 → 0.014281393 → 0.005741666 0.013144603 → 0.048229497 → 0.035657049
24 850 816:	0.028328598 → 0.020718036 → 0.003415578 → 0.009455589 → 0.003462547 0.008353268 → 0.003415578 → 0.009455589 → 0.028118705 → 0.020961696 0.003358333 → 0.007688353 → 0.003462547 → 0.008353268 → 0.003358333 0.007688353 → 0.028209706 → 0.020856010
25 852 832:	0.000913826 → 0.000668324 → 0.000110180 → 0.000305019 → 0.000111695 0.000269460 → 0.000110180 → 0.000305019 → 0.000907055 → 0.000676184 0.000108333 → 0.000248012 → 0.000111695 → 0.000269460 → 0.000108333 0.000248012 → 0.000909991 → 0.000672775
26 854 856:	0.000000000 → 0.000000000 → 0.000000000 → 0.000000000 → 0.000000000 0.000000000 → 0.000000000 → 0.000000000 → 3.092438160 → 1.640944713 0.000000000 → 0.000000000 → 0.000000000 → 0.000000000 → 0.000000000 0.000000000 → 0.000000000 → 0.000000000
27 854 852:	3.365620238 → 2.461436371 → 0.405792664 → 1.123384977 → 0.411372961 0.992422101 → 0.405792664 → 1.123384977 → 3.340683565 → 2.490384751 0.398991636 → 0.913425894 → 0.411372961 → 0.992422101 → 0.398991636 0.913425894 → 3.351495013 → 2.477828484
28 858 864:	0.214734240 → 0.113944725 → 0.000000000 → 0.000000000 → 0.000000000 0.000000000 → 0.000000000 → 0.000000000 → 0.000000000 → 0.000000000 0.000000000 → 0.000000000 → 0.000000000 → 0.000000000 → 0.000000000 0.000000000 → 0.000000000 → 0.000000000
29 858 834:	0.532760412 → 0.389632746 → 0.064234896 → 0.177826077 → 0.065118229 0.157095326 → 0.064234896 → 0.177826077 → 0.528813065 → 0.394215126 0.063158329 → 0.144590632 → 0.065118229 → 0.157095326 → 0.063158329 0.144590632 → 0.530524462 → 0.392227534
30 860 836:	0.244905301 → 0.179110765 → 0.029528220 → 0.081745092 → 0.029934280 0.072215347 → 0.029528220 → 0.081745092 → 0.243090740 → 0.181217245 0.029033331 → 0.066467049 → 0.029934280 → 0.072215347 → 0.029033331 0.066467049 → 0.243877454 → 0.180303566
31 862 838:	0.000000000 → 0.000000000 → 0.000000000 → 0.000000000 → 0.000000000 0.000000000 → 0.000000000 → 0.000000000 → 0.442208970 → 0.327037905 0.000000000 → 0.000000000 → 0.000000000 → 0.000000000 → 0.000000000 0.000000000 → 0.000000000 → 0.000000000
32 832 890:	0.668399952 → 0.667149912 → 0.105050009 → 0.288950112 → 0.106500003 0.250749999 → 0.105050009 → 0.288950112 → 0.661900008 → 0.678449904 0.103300006 → 0.229550007 → 0.106500003 → 0.250749999 → 0.103300006 0.229550007 → 0.664699992 → 0.673550064

23

Transactive Energy Systems in Decentralized Autonomous Renewable Energy Communities

Riccardo Trevisan, Emilio Ghiani, Marco Galici, Susanna Mocci, and Fabrizio Pilo

University of Cagliari, Deptartment of Electrical and Electronic Engineering, Via Marengo, Cagliari, Italy

23.1 Introduction

The concept of transactive energy (TE) [1, 2] represents a novel strategy for managing the electrical power system that goes beyond the conventional hierarchical systems. TE systems encompass the active participation of local actors and their distributed energy resources (DERs), leveraging automated control technologies to dynamically balance supply and demand. This concept enhances the grid's energy efficiency, cost efficiency, safety, and reliability and also supports the diffusion of renewable energy communities (RECs). In RECs, value creation occurs as participants improve their energy efficiency performance and engage in local energy networks that comprise production, transmission/distribution, and consumption. These communities can be envisioned in the broader context of smart cities and the so-called positive energy districts (PEDs), which consist of interconnected smart buildings, renewable DERs, storage systems, and their social and commercial frameworks. RECs align with the objectives of the European Union (EU)'s Clean Energy Package [3] and are a key element of the Renewable Energy Directive (RED) II [4], contributing to the ambitious goal of decarbonizing the power system. More recently, EU bodies revised the RED II directive issuing the RED III [5] in which even more demanding objectives are set, including a significant increase in renewable energies in the European energy mix. In this context, RECs are even more challenged and challenging: Energy solutions are intertwined with various complexities, necessitating active engagement from end users. This involves stimulating them through innovative models of production and consumption that prioritize self-reliance and collaborative efforts.

Figure 23.1 schematically depicts a REC composed of a modern, intelligent energy distribution system, incorporating renewable energy generation and energy storage capabilities, along with smart metering and management technologies. The main actors, i.e., passive users, producers, and prosumers, are highlighted as well. The figure shows a centralized management system where informational flows are elaborated and used to devise the appropriate operational strategy. Efficiently administered either RECs or clusters can offer novel services that enhance the overall grid's performances. The success of RECs depends on the synergetic collaboration among consumers, prosumers, and energy producers, all striving for the shared objectives of social, economic, and environmental sustainability. Effective governance and operation of RECs demand interdisciplinary knowledge and a continuous multitiered strategic planning to foster new business

Smart Cyber-Physical Power Systems: Fundamental Concepts, Challenges, and Solutions, Volume 1, First Edition.
Edited by Ali Parizad, Hamid Reza Baghaee, and Saifur Rahman.

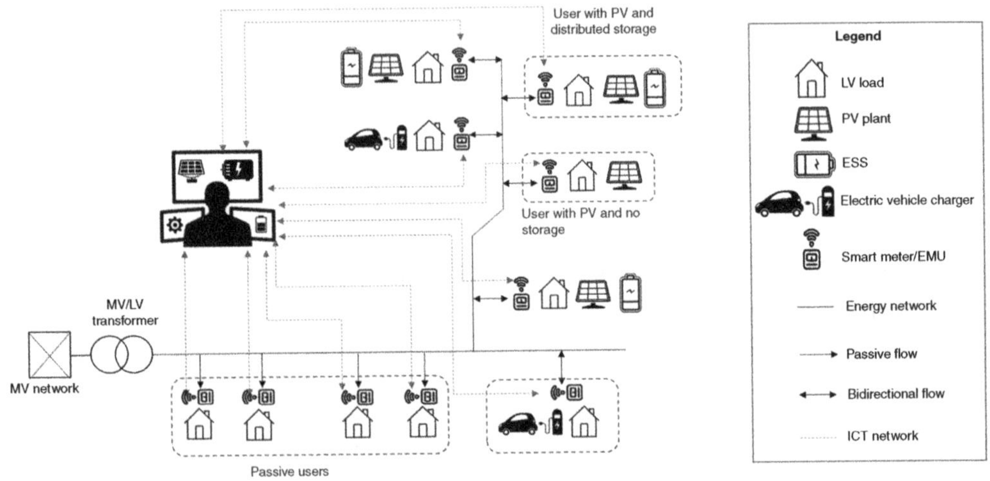

Figure 23.1 REC schematic representation.

models [6]. Nonetheless, the absence of a universal framework for RECs' development and management poses significant legal, technical, governance, and societal challenges that must be considered and tackled at the local level. To overcome these hurdles, innovative and inclusive strategies for RECs' diffusion and governance are emerging at the national level within each EU country [7, 8]. Challenges are not confined to one domain of knowledge or research field and require a multisectoral and holistic approach. Said challenges can be inferred as technical/technological, economic, social, regulatory, and market. Relatively to the first one, technical and technological challenges can be associated with the physics of the electrical system (i.e., the higher the penetration of DERs, the more challenges to operate the network in a reliable manner). Considering the smart grid architecture model (SGAM) [9], technical and technological challenges can be inferred to all the layers except for the business one; therefore, difficulties are associated with both hardware and software and range from the physical mean of transportation, up to the collection and transmission of data, passing by the definition of standard protocols. Other challenges (i.e., economic, social, regulatory, and market) can be tackled by the underlying business objectives and operational dynamics of the organization, which has to find its place in an uncertain and mutable scenario. An expertise in social dynamics is also required as members' interactions should be addressed as well. Literature highlights that the main social issues are associated with leadership, cooperation, and proximity [10–12] as well as the cultural background and social roots of the social fabric [13, 14]. The aforementioned points result in difficulties in reaching consensus among participants. Further economic challenges arise during the definition of the investment dynamics but can have aftereffects throughout the life of the project also due to instabilities attributable to the macroeconomic scenario. Said macroscenario is also responsible for regulatory policies, which often slow down entrepreneurial actions and, possibly, the flourishing of a relevant number of initiatives. Private initiatives require a favorable economic outlook; hence, instability and uncertainty must be harnessed from a market perspective.

It is thus clear that RECs perfectly suit all the hurdles envisioned in a TE system, combining economics, control mechanisms, the physics of the power system, and the inclusion of the users that should be able to actively take part in the local management of the energy resources according to the open-door principle.

23.2 RECs as DAOs

Blockchain's role could be pivotal in enabling local energy and resource markets and promoting a circular economy and has significant potential in many energy sector applications. It is often talked about as simply an innovation in data storage, although there are many aspects and applications that can be implemented to achieve a fully distributed, inclusive, transparent, secure, tamper-proof, and shared system. For instance, in [15], the authors highlight the increasing interest in using blockchain technology in the energy industry, with many developed countries already implementing pilot projects. Also, in [16], the authors discussed the applications of blockchain in power engineering, particularly in settlements on the electricity market and the management of electricity transactions. In [17], the authors suggest an application of blockchain as an energy open data ledger to track data regarding the energy footprint, while in [18], it is emphasized that blockchain can optimize energy management processes throughout the value chain in the increasingly decentralized energy system. Other applications in peer-to-peer trading and electric vehicle applications are discussed, respectively, in [19, 20]. A recently published whitepaper briefly identifies the main challenges and opportunities on the role of energy communities in optimizing the generation, consumption, and storage of energy within the community and trading energy and services locally and outside the community [21]. Several solutions can be found in the literature. For instance, in [22], the authors present TRANSAX, a blockchain based on the TE system concept that offers an efficient and secure market built on smart contracts and addresses implementing automated mechanisms in a distributed setting while ensuring safety, privacy, and market efficiency through a decentralized trading platform built on blockchain technology. In [23], the authors introduced a blockchain based on the TE framework that improves individual benefits and guarantees socially optimal performance, incentivizing prosumers to participate. Moreover, the authors of [24] propose a blockchain-based system that enables distributed peers to provide grid operation services, maintaining trustless reputation ratings and enforcing valid transactions through smart contracts. These papers collectively highlight the opportunities and benefits of using blockchain technology in TE systems but are all focused on highly specialized sectorial and limited use cases, lacking the holistic approach that is required in a TE system. Nevertheless, the academic contribution suggests that blockchain technology can facilitate the transition to a decentralized energy system and offer solutions for various energy-related challenges. The aforementioned resources, despite their claims to be realized from a TE system perspective, tackle just a singular aspect of a REC, whether it focused on a market perspective or a socio/governance one. In light of this lack, the authors believe that a wider, more comprehensive, and holistic approach is required toward the creation of distributed autonomous renewable communities (DARCs) [25] and that this new form of organization can use blockchain technology, particularly decentralized autonomous organizations (DAOs), to uniform and define standards, internal policies, best practices, etc., on both the socioeconomic perspective and the technical/technological one. To exemplify this idea, Figure 23.2 shows how technological innovation can support and push the socioeconomic aspects closely related to the REC life cycle.

Figure 23.2 highlights DLTs as the foundational technology that attributes verifiability, transparency, security, and accessibility. DLTs act as enabler for digital governance, local markets, local asset management, and coordinated controlling strategies, thus allowing fully distributed and decentralized management of the TE.

Historically, DAOs can be traced back to the early days of blockchain and cryptocurrency. They have been theorized to be particularly well-suited for open-source projects, community governance, and collective funding initiatives. The concept evolved from the desire to eliminate

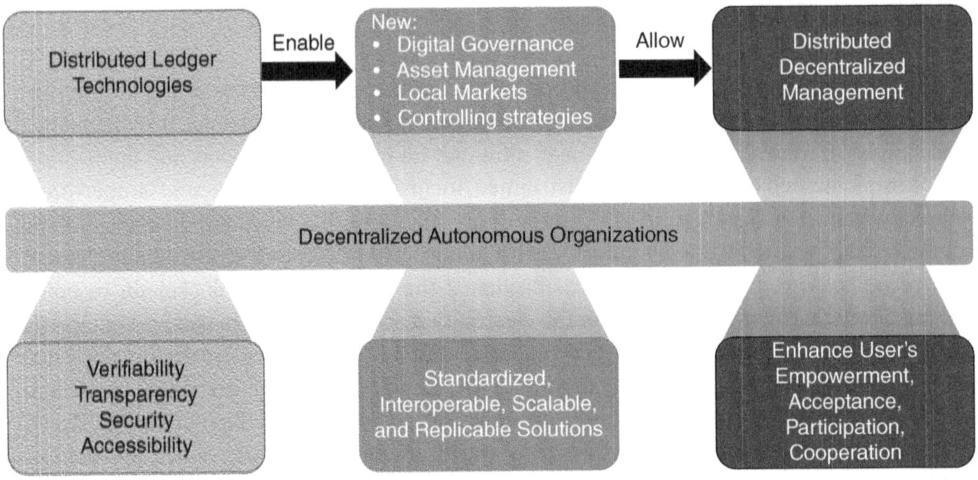

Figure 23.2 DAO contribution to REC.

the inefficiencies and human intermediaries and was formally proposed by Vitalik Buterin, cofounder of Ethereum, in 2013 [26]. Ethereum's distributed consensus and immutable smart contracts provide the ideal infrastructure for DAOs, where the rules are programmed into the code and cannot be altered once live, thus ensuring the organization operates as intended without centralized control [27]. Blockchain technology allows the execution of smart contracts, self-executing contracts with the terms of the agreement directly written into code [28, 29]. This technological backbone enables DAOs to operate without traditional management hierarchies, thus challenging conventional corporate governance models [30–32]. The first significant implementation of a DAO was "*The DAO*" on the Ethereum blockchain in 2016, which aimed to operate as a venture capital fund without a typical management structure [33–35]. Although it faced a significant setback due to a code exploit, the incident led to a surge in research exploring the security of decentralized governance models [36, 37].

Their applications are not limited to the digital realm; there are experiments in using DAOs for real-world asset management and community-led initiatives [38, 39]. The authors of [40] provide a review of the architecture, models, and mechanisms for DAOs. In [41], the DAO concept and applications are identified and the potential of DAOs to revolutionize traditional hierarchical management models and reduce organizational costs is highlighted. In [42], the characteristics and challenges of DAOs are further explored, emphasizing their decentralized governance structure and decision-making based on community consensus. Additionally, the authors of [43] discuss the promises and challenges of DAOs, focusing on decentralized governance and disintermediation. Collectively, academic literature suggests that DAOs have the potential to transform the organizational, technical, and economic facets or RECs and introduce a new era of democratic and distributed organizations, offering transparency, mutual trust, and collective decision-making through blockchain and smart contracts, putting automatization at the center. DAOs streamline membership dynamics, reduce barriers to entry and exit, and enable the replication of successful practices across different regions. Finally, in [44], the authors explore several DAO platforms that offer DAO creation as a service on the blockchain.

In this plethora of DAO applications, the energy sector represents a ripe domain for DAO [45], which nevertheless, as per the TE, is highly specialized and focused on one domain of knowledge.

DAOs can significantly enrich the SGAM framework by bringing in elements of decentralized governance, enhanced security, community engagement, and adaptive mechanisms, all of which are pivotal for the next generation of smart grid systems. To achieve all the abovementioned goals, a set of well thought and integrated smart contracts must be designed, and the suitable digital asset must be adopted.

The SGAM is a structured approach to conceptualize the functionalities and interoperability of a smart grid system. The design of SGAM is aimed at providing a standardized framework to ensure consistent and efficient development, deployment, and operation of smart grids. It involves multiple stakeholders and domains of the energy supply chain. Namely, these domains pertain to the social, virtual (or digital), and physical domains.

The physical domain includes power lines, substations, transformers, and smart devices representing the tangible components of the smart grid. The virtual domain surrounds the physical domain, and this implies the digital and information processing technologies that support the operation of the physical components. The social domain encapsulates both the physical and virtual domains and represents the stakeholders and their roles, social impacts, and the human elements of the smart grid. This model thus encompasses a wide range of actors and spans across various domains of the energy supply chain, and SGAM's core strengths lie in its interoperability, modularity, and adaptability. These attributes can be further enhanced by integrating SGAM with a DAO model, facilitating the development of the previously mentioned DARCs. Figure 23.3 shows the envisioned model integrating SGAM with the intersecting domains and the DAO model, thus realizing a governance framework for DARCs. The layout of the diagram is such that the different layers of the SGAM framework are wrapped around these domains, indicating the multifaceted

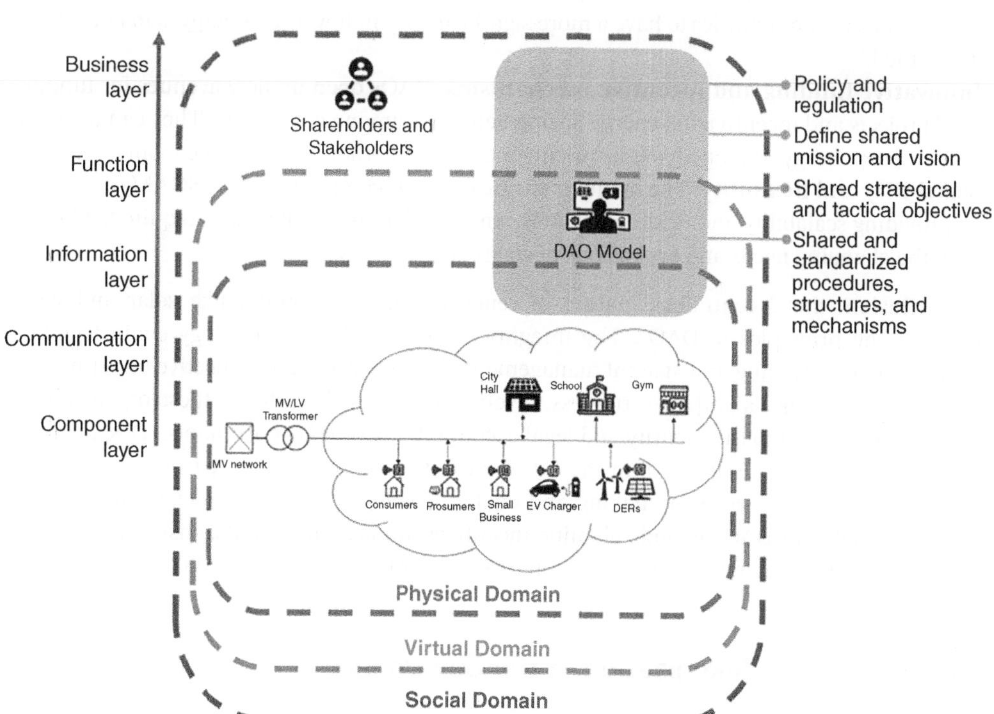

Figure 23.3 Conceptualization of a DAO model under a TE optic for REC using the SGAM.

approach to integrating business, functional, informational, communicational, and physical components to achieve a resilient and efficient smart grid system with a holistic approach.

RECs organized according to the SGAM principles with the help of DAOs can benefit from many multifacet positive contributions:

1) **Enhanced decentralization and autonomy:** By integrating DAOs into SGAM, the grid can operate more autonomously. DAOs facilitate decentralized decision-making and management, allowing for more localized control and responsiveness within the grid.
2) **Improved interoperability:** DAOs can improve interoperability within the smart grid by standardizing protocols and processes across different actors and systems. This is crucial in a multilayered structure like SGAM, where seamless interaction between various components is essential for optimal functioning.
3) **Greater transparency and trust:** Blockchain technology ensures transparency and trust in transactions and interactions within the smart grid. This can lead to increased confidence among stakeholders and potentially more collaboration and innovation.
4) **Data management and security:** DAOs, with their blockchain backbone, can handle data securely and efficiently. This capability is vital for the dynamic and complex data needs of a smart grid, ensuring data integrity and security.
5) **Adaptive governance and regulation compliance:** DAOs can offer adaptive governance mechanisms that evolve with changing regulatory requirements and technological advancements. This adaptability is crucial in the ever-evolving landscape of smart grid technology and energy regulation.
6) **Community participation and engagement:** DAOs encourage community participation in energy management decisions. This aspect can democratize energy governance, allowing end users and local communities to have a more significant say in how their energy is managed and distributed.
7) **Innovative funding and incentive mechanisms:** DAOs open up new avenues for funding REC projects and incentivizing energy-saving behaviors among consumers. They can facilitate peer-to-peer energy trading, dynamic pricing models, and other innovative economic models.
8) **Scalability and flexibility:** The modular nature of DAOs complements the SGAM framework by providing scalability and flexibility. DAOs can be scaled up or adapted as required, aligning with the changing needs and expansions of smart grid systems.

In conclusion, the decentralized nature of renewable energy sources like solar and wind aligns with the principles of DAOs. The integration of blockchain technology could facilitate peer-to-peer energy trading, transparent management of energy grids, and collective investment in renewable energy projects [10]. Nevertheless, several challenges characterize these organizations and are mainly related to the security and legal accountability of DAO governance. Despite these challenges, DAOs offer a glimpse into the future of organizational design, particularly in their potential to democratize decision-making and enable global collaboration. Sociotechnical challenges associated with DAOs include aligning incentives among participants, ensuring equitable participation, and managing dispute resolution in a decentralized context [46–48].

23.3 Toward the Tokenization of the Governance

DARC organizational governance can be realized according to the holacracy model [49], which is a system of organizational governance where authority and decision-making power are distributed throughout a holarchy of self-organizing teams rather than being centralized in a traditional management hierarchy. In a holacracy, power is more evenly spread across the organization, and participants have more autonomy and flexibility in their roles. This system is designed to increase agility, efficiency, transparency, innovation, and accountability within an organization. Such principles can be fulfilled with blockchain technology that is revolutionizing the way governance is conceptualized and executed as it is able to provide decentralized, secure, and transparent mechanisms, which are crucial in governance processes realized according to holacracy theory. The essence of blockchain in governance lies in its ability to foster trust among stakeholders, streamline bureaucratic processes, and enhance the accountability of governing bodies. The technology's application extends to various governance aspects, including voting systems, public record management, the definition of standards and shared rules among participants, and the enforcement of contracts and laws. By utilizing blockchain, governments can significantly reduce fraud, corruption, and inefficiencies, paving the way for a more participative and transparent governing process [50].

Blockchains can be broadly categorized into three types: public, private, and consortium blockchains [51]. Each type has distinct characteristics tailored to different governance needs. Public blockchains are completely open and decentralized, allowing anyone to participate in the consensus process. Private blockchains are controlled by a single organization or authority. Consortium blockchains represent a middle ground, operated by a group of organizations rather than a single entity. They combine the benefits of decentralized control with the management efficiency of private networks. This type is particularly useful for collaborative governance efforts among multiple organizations or government agencies. Consensus on blockchain can be reached with different protocols [52] that enable agreement among nodes. One of the most famous is the proof of work (PoW). In this algorithm, miners solve complex mathematical problems to add blocks. It is secure but energy-intensive. Another famous one is the proof of stake (PoS). Validators are chosen to create blocks based on the amount of cryptocurrency they hold and are willing to "stake." It is energy-efficient but requires trust. On the same line of the PoS is the delegated proof of stake (DPoS). Here, the elected delegates validate transactions according to the PoS protocol. It is faster but less decentralized. Finally, the proof of authority (PoA) allows the identified validators to confirm transactions. Despite the efficiency, it is centralized. The proof of space and time (PoST) is a middle ground even if it is still unproven at scale. Here, the miners prove they have allocated storage space over time. Each has strengths and weaknesses, impacting not only governance but security, scalability, and energy consumption as well.

Tokenization, in the context of blockchain and digital assets, refers to the process of converting rights to an asset into a digital token on a blockchain. The tokens can be native to a blockchain, for example, the Bitcoin one, or hosted on an existing blockchain via a smart contract. Tokens can represent real-world assets like real estate and stocks or even intangible assets like intellectual property or voting rights. Tokens are thus an enabling piece of technology toward innovative governance and operational processes. Tokenization can streamline various processes, enhance liquidity, and open up new investment opportunities by allowing fractional ownership of assets. This transformation enables more efficient, transparent, and secure governance processes. Tokenization can be applied to various governance areas such as asset management, service delivery, and voting mechanisms.

In asset management, tokenization allows for the digital representation of physical assets on the blockchain, facilitating efficient tracking and management. It enables fractional ownership of large assets or projects, democratizing investment opportunities in shared projects. This could lead to more community-driven development initiatives, where citizens have a direct stake in local projects. In services, tokens can be used as digital vouchers or credits, streamlining the distribution and redemption of benefits. In electoral/voting processes, tokenization can enhance the security and transparency of voting, reducing the risks of fraud and manipulation.

There is also a growing interest in tokenizing intangible assets like intellectual property, emission credits, or even community reputation scores. This broadens the scope of governance and public policy, providing innovative tools for managing resources and incentivizing desired behaviors. This can be used, for instance, in stimulating virtuous behaviors in demand response mechanisms. The NIST (National Institute of Technology) Internal Report 8301 report explores different categories of tokens in the context of blockchain networks:

- **Blockchain-based tokens:** These tokens exist on a blockchain and are often used to represent a variety of digital or physical assets. They include:
  - *Fungible tokens*: These are interchangeable and identical to each other, similar to fiat currency. One common standard for fungible tokens is the ERC-20 (Ethereum Requests for Comments) standard on the Ethereum blockchain.
  - *Nonfungible tokens* (NFTs): These are unique and not interchangeable. Each NFT represents a specific asset or a piece of data and has unique properties. NFTs have gained popularity in representing digital art, collectibles, and other unique items. The ERC-721 standard is commonly associated with NFTs.
  - *Semi-fungible tokens*: These combine properties of both fungible and NFTs. They can behave like fungible tokens until a certain condition is met, after which they act like NFTs.
- **Self-contained tokens:** These tokens are not reliant on a blockchain for their existence. They are used in closed systems and are often centralized. Examples include digital tokens used in online gaming platforms or loyalty points in commercial reward programs.

Each type of token has distinct characteristics and use cases. For instance, fungible tokens are suitable for use as digital currencies, in decentralized finance (DeFi) applications, or in any context where interchangeable value units are needed. NFTs are best suited for cases where uniqueness and provenance are important, such as in digital art, real estate, or unique digital identities. Semi-fungible tokens are useful in scenarios where an asset might need to change its nature over time, like a ticket that is fungible until it is used, after which it becomes a collectible (nonfungible).

Tokens can furthermore be classified into several categories, each serving specific purposes in governance processes [51]:

- **Utility tokens**: Essentially, digital coupons are a crucial component in blockchain-enabled governance ecosystems. They are designed not to represent an investment but to provide access to a service or product. For instance, a municipal government could issue utility tokens that residents use to pay for utilities, parking fees, or even access public libraries. The use of these tokens can streamline municipal services, making them more efficient and user-friendly. Moreover, utility tokens can play a role in incentivizing behaviors that align with public policy goals, such as tokens awarded for participating in recycling programs or community service activities [53].
- **Security tokens**: These are digital assets that represent ownership or a stake in an asset, such as real estate, stocks, or bonds. In governance, they can be instrumental in funding public projects or infrastructures. For example, a city could issue security tokens to finance the development of renewable energy facilities or public transportation systems. These tokens could entitle holders to a share of the revenue generated from these projects. This approach can not only democratize the financing of public projects but also foster a sense of ownership and investment in local communities.
- **Governance tokens**: These confer voting rights and are instrumental in decentralized decision-making processes. They can be particularly transformative in local governance and

community decision-making. By issuing governance tokens, a city or community can enable residents to vote on budget allocations, community projects, or local policy changes. This approach fosters a more engaged and participatory governance model, giving citizens a direct voice in matters that affect their daily lives. It can lead to more community-centric decision-making and increased transparency in how decisions are made.

- **Transactional tokens**: These are used as a medium of exchange within a specific ecosystem. In governance, they could revolutionize the way citizens interact with government services. For example, a government could issue transactional tokens that citizens use to pay taxes, fees, or fines. This system could simplify and secure transactions, reduce administrative costs, and increase efficiency. Furthermore, transactional tokens could also be used in government procurement processes, ensuring transparency and traceability in government spending. Each category of token has the potential to streamline and enhance different aspects of governance, contributing to more effective and citizen-centric governance models.

The application of blockchain and tokenization in governance presents numerous practical use cases, each demonstrating the potential to revolutionize aspects of public administration and citizen engagement. One example is smart cities. Integrating blockchain into smart city digital infrastructures can revolutionize urban governance. From traffic management to environmental monitoring, blockchain can provide a secure and efficient backbone for managing city services and citizen engagement. Other aspects that can be revolutionized are related to global health and managing digital identities enabling e-governance frameworks. The voting systems can be enhanced by blockchains, by means of the improvement on the integrity and transparency of electoral processes. By tokenizing votes, blockchain can ensure that each vote is tamper-proof and traceable, thereby reducing the risk of fraud. A practical example could be a local government using blockchain to conduct community votes on public projects or policy changes, ensuring a transparent and secure voting process [50]. Blockchain can streamline property registration and transfer processes, making them more transparent and efficient. Moreover, blockchain can transform public finance management, enhancing transparency in budgeting and expenditure tracking. Finally, blockchain can ensure the authenticity of products and transparency in the supply chain, in public procurement.

However, the idea of tokenization requires that these digital assets must be stored on digital wallets. The NIST report identifies two types of wallets, i.e., how tokens are stored, managed, and protected in the network.

- **Self-hosted wallets:** These are wallets where the user directly controls the private keys. They can be software-based (like mobile or desktop apps) or hardware-based (like Universal Serial Bus [USB] devices). Self-hosted wallets offer high levels of control and security but require the user to responsibly manage their private keys.
- **Custodial wallets:** In this case, a third-party service provider manages the private keys on behalf of the user. While this relieves the user from the responsibility of key management, it also means trusting the custodian to maintain security and accessibility.

23.3.1 Proposition of a 2-Token Governance Model for DARCs

This subsection explores a theoretical framework for a governance model functioning on a dual-token basis: (i) one for governance processes and (ii) the other for trading goods/services. The discussion intentionally does not provide an overly detailed implementation, recognizing the freedom for community members to make their own choices. It is a necessity to integrate best practices as outlined by the Token Taxonomy Framework (TTF) into the design of the

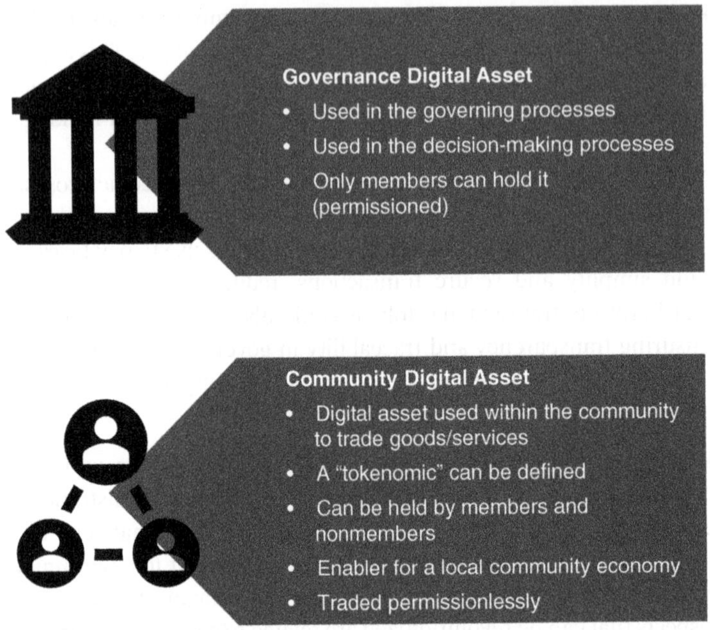

Figure 23.4 Two-token governance model.

community's digital assets. Figure 23.4 includes an infographic that illustrates the essential digital assets of the community.

To enhance the properties of governance and community tokens (CT) in the blockchain-based model, the TTF can provide valuable guidance [54]. According to TTF, the properties of a token could include:

- **Specification of token type:** It determines whether the token is fungible, nonfungible, or hybrid to suit different use cases.
- **Behavior definition:** The defined digital asset should have clear behavior, such as divisibility, transferability, and the ability to be minted or burned, in line with the community's governance structure.
- **Data attributes:** A token can hold or be linked to specific data attributes that are relevant to the community, such as emission data or proof of membership.
- **Interoperability:** A token should be designed to interact seamlessly with different blockchain systems and standards.
- **Educational clarity:** For community members to effectively participate in governance, the token's structure and capabilities should be easily understandable.
- **Legal compliance:** A token must adhere to regulatory requirements, ensuring that its creation, distribution, and use are legally sound.

The proposed theoretical framework introduces a governance digital asset designed for managing various processes, including engagement, decision-making, and project development. This token centralizes control among a select group of stakeholders, thereby promoting a sense of ownership and encouraging active participation. The influence each member has over the REC governance can be symbolized by the number of tokens they own. This system allows for the creation of different "membership tiers" or a more democratic governance structure, depending on the number of tokens allocated to each participant. The design of the governance token should reflect the REC's underlying governing principles. Whether this token takes the form of a NFT, a FT, or even a hybrid token, it should align with the REC's ethos. In its simplest form, the governance token can be realized as an NFT, which can also be representative of membership. In this case, due to the unique nature of this kind of asset, each member is assigned the same level of influence in the REC's activities. On the other hand, if implemented as an FT, a standard allocation process would need to be established, which could be advantageous in environments that are highly dynamic and flexible. Choosing a hybrid token architecture is appropriate in scenarios where specific use cases are envisioned as they can offer innovative benefits and are well suited to particular contexts. However, they also introduce operational challenges that must be thoughtfully addressed. Such tokens require careful management to ensure they effectively meet the community's needs and regulatory obligations. In the suggested framework, the governance token could be implemented either as a NFT or a FT, and it plays a crucial role in the decision-making process. The influence a member wields in these decisions is directly proportional to the number of governance tokens they possess, like in a PoS consensus algorithm. If the system is designed so that each member receives a single token, then each vote carries equal weight. In this scenario, NFTs are more suitable due to their unique characteristics, and they can also be used to symbolize membership within the REC. On the other hand, to facilitate a variable number of tokens per member, FTs are necessary. This last option allows for a reputation-based participation system, where members can earn more tokens as they contribute more to the community. However, given the potential complexity inherent in such governance mechanisms, it is vital to implement safeguards against the concentration of power. These checks and balances are essential to ensure fair and equitable governance within the REC, preventing any single member or group of members from gaining disproportionate influence. Moreover, the governance token can be used to permit access to a "closed" market accessible only for members. This can be used as an enabler for a local energy market (LEM).

The subdivision among governance and community token is required to allow only members of the community to take part in certain activities of the REC (governance token), without posing a limit to the REC economy by also allowing nonmembers to take part in the exchange of goods and services. To this end, a community token, envisioned as a utility token, is required. The community (utility) token can be traded permissionlessly between users in exchange of products and services. In this case, a secondary market is required, and rules must be defined. In this ecosystem, a utility token can be characterized by a tokenomic to govern supply, distribution, incentives, monetary policies, and utility cases. For instance, a utility token can be programmed so that a certain percentage of each transaction goes to a common fund, which can be used by the organization to achieve its objectives and goals (e.g., new installments and energy poverty policies). A well-thought monetary policy can be used to control variables like token supply, distribution rates, and incentivization to maintain a stable and growing economy within the blockchain ecosystem. Nevertheless, using a utility token can give rise to several issues, primarily related to token pricing, which can allow for speculative behaviors or be subjected to high volatility. Most cryptocurrencies are priced according to the classical rules of demand and supply, and in centralized markets (e.g., Binance, Crypto, and Coinbase), the order book mechanics is used for real-time trading. In DeFi, the usage of Automated Market Makers (AMMs) [55] is well established. Unlike traditional exchanges, which use an order book to match buy and sell orders, AMMs rely on a mathematical formula to price assets. Asset pricing is just the tip of the iceberg. Other issues are posed by technological and social barriers as the local social network might be reluctant to such a technology.

For both the governance and utility token, regulatory challenges might arise that can differ from country to country. Nevertheless, these innovative approaches to governance and local markets

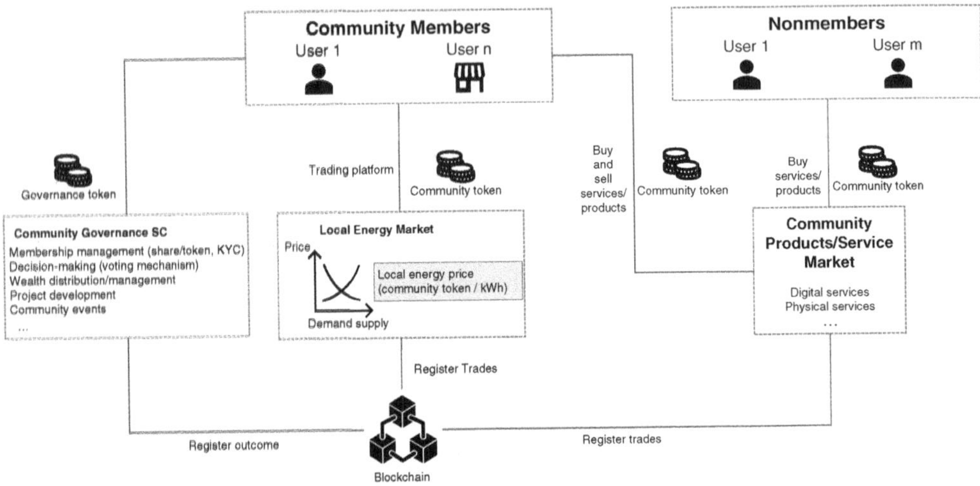

Figure 23.5 Two-token governance model: example of usages.

can open to fruitful research and innovative approaches to the development of energy policies at the local level, empowering the most vulnerable subjects ensuring mutual trust and transparency thanks to the blockchain technology.

A graphical representation of the devised framework where the use cases of the governance and community token is highlighted in Figure 23.5. In essence, this model outlines a self-sustaining, transparent, and decentralized community ecosystem underpinned by blockchain technology, where governance and market operations are tokenized to ensure fairness, efficiency, and member engagement.

23.3.2 Automated Market Makers: Enablers for Local Energy Markets

Since the advent of Bitcoin in 2008, the financial and energy landscape has been undergoing significant transformations driven by blockchain technology. Early discussions foresaw the disruptive potential of decentralized ledgers [56, 57]. However, despite the creation of numerous crypto tokens and a total cryptocurrency market capitalization exceeding 1.7 trillion as of early 2021, traditional financial intermediaries have faced substantial challenges from blockchain-based financial service providers [58]. In this regard, in 2020 a novel blockchain application known as DeFi emerged, employing open-source smart contracts to offer financial services without the need for traditional intermediaries. A pivotal innovation within DeFi is the rise of decentralized exchanges (DEXs), particularly those operating on the Ethereum blockchain. DEXs have emerged as a revolutionary concept in the blockchain space, utilizing smart contracts to facilitate peer-to-peer transactions of digital assets [59, 60]. Unlike traditional centralized exchanges (CEXs), DEXs operate without intermediaries, offering users a direct and secure means of transferring assets. The underlying technology is governed by smart contracts, ensuring noncustodial control where users retain ownership of their funds without reliance on custodians or depositories. Smart contracts, integral to the functioning of DEXs, are self-executing contracts stored on blockchains. These contracts contain predefined instructions, often written in solidity, especially when utilizing the Ethereum blockchain. Ethereum, a blockchain protocol featuring smart contract functionality, provides a platform for developing decentralized applications [60]. Smart contracts operate within the Ethereum Virtual Machine (EVM), designed to restrict access to the network, file system, or other processes running on the EVM.

DEXs have emerged as a response to the challenges faced by CEXs, including vulnerability to external attacks leading to the loss of user funds. Various approaches exist for building DEXs, with one method replicating the orderbook format of CEXs. On-chain order books, while providing high security, can be expensive due to transaction costs. Alternatively, off-chain order books record transactions off-chain and utilize the blockchain solely for settlement, offering a cost-effective solution but compromising on security and decentralization [61].

A pivotal advancement in the realm of DEXs is the adoption of AMMs [55]. AMMs revolutionize trading by employing liquidity pools (LPs), where traders interact directly with the pool to buy or sell assets. Mathematical formulas within the pool determine asset prices, allowing for efficient trading even with low liquidity. Commonly, DEXs adopt AMM through smart contracts, representing a paradigm shift from traditional order-book-based exchanges. Notably, Uniswap, the largest among these DEXs, swiftly ascended to become the fourth-largest cryptocurrency exchange by daily trading volume shortly after its launch in 2020 [62]. Transactions settle instantaneously upon confirmation and inclusion in the blockchain, eliminating counterparty risk. As a matter of fact, a trade on the AMM-based smart contract does not need counterparties. Additionally, the role of LP extends beyond professional market makers to include any token holder willing to deposit tokens and earn fees from trading activities. AMMs rely on algorithmic protocols to autonomously set prices and facilitate trades, eliminating the need for conventional order books. The fundamental components and procedures characterizing AMMs are outlined as follows:

- **Liquidity pools (LP):** AMMs operate through liquidity pools, consisting of tokens secured within smart contracts. Users create these pools by contributing equal values of two different tokens, establishing a specific token ratio that determines the asset prices within the pool.
- **Mathematical algorithms:** AMMs employ mathematical algorithms to ascertain **asset** prices within the liquidity pool. A prevalent algorithm is the constant product market-maker algorithm, often denoted as the $X \cdot Y = K$ formula.
- **Trading interface:** AMMs furnish a user-friendly trading interface, simplifying token transactions. Users are relieved from specifying prices or counterparties for their trades; the algorithm automatically determines the trade price.
- **Incentives:** To motivate users to contribute liquidity, AMMs provide incentives such as transaction fees and LP tokens. LP tokens are distributed to users contributing to the pool, representing a share of the pool's overall value. Holders of LP tokens receive a portion of the transaction fees collected from traders leveraging the AMM.
- **Smart contracts:** All operations within AMMs are executed through smart contracts, residing on a blockchain network. Smart contracts seamlessly carry out trades, adjust prices, and distribute transaction fees and LP tokens to participating users.

Instead of relying on an order book with all transactions in process, AMMs adopt the so-called liquidity pools. These pools in DEXs are pools of cryptocurrencies or, more generally, tradable assets. The liquidity pools managed by AMM's smart contract allow the exchange of these assets. It is possible to interact with these pools by adding or withdrawing additional assets. Each individual owning a certain quantity of the paired assets in the pool is identified as a LP. These actors play a crucial role in DeFi as they are responsible for providing capitals to the liquidity pool, price stability, and general trade efficiency. LPs are rewarded for their service in several ways: They can be awarded a trading fee; they can be offered with a high rate of return for the

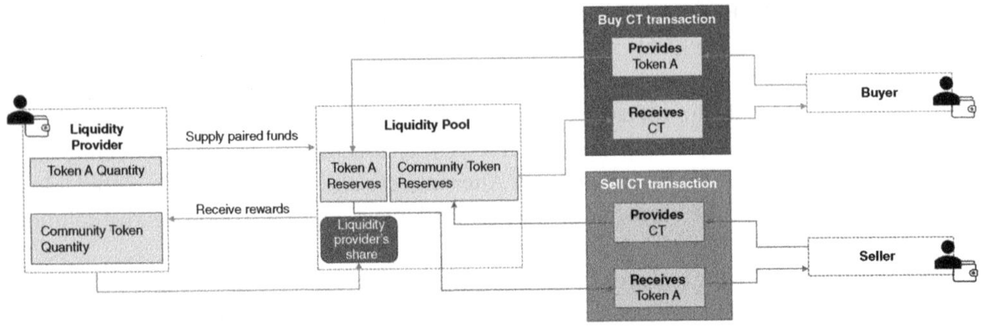

Figure 23.6 AMM functioning.

liquidity that they locked in the pool, or they can be awarded with the distribution of a third asset. Figure 23.6 provides a diagram that explains the mechanism of a buy transaction (blue) and a sell transaction (green) in the context of AMM, exemplified for the community token. On the left side of the picture, there is the LP that supplies the token pair to the pool. In the center, a simplified representation of the liquidity pool is provided. On the right side of the picture, two processes of buying and selling of the community token are depicted.

To interact with the pools, the investors (i.e., buyers and sellers) must follow well-defined rules. These rules, residing in smart contracts, are governed by mathematical algorithms that determine the inflation or the deflation of one of the paired assets. Therefore, while an investor does not require a counterpart (as they can directly interact with the AMM smart contract), it must instead implicitly adhere to well-defined rules that govern the liquidity pool. In other words, the investment operation, such as depositing a token A and withdrawing a token B, requires the satisfaction of a mathematical formula. The most well-known mathematical formula is that of the constant product (constant product market makers [CPMMs]). Namely, the product of the quantity of token A in pool A plus the investment multiplied by the quantity of token B in pool B minus the withdrawal due to the investment must satisfy a constant, usually denoted as k. To execute an investment, the investor will also need to pay a fee. This fee is generally used to pay network fees. In a more general way, AMMs allow to set a general fee percentage that can be used in an automated manner for a predefined intended use (i.e., save money for the community, repay LPs, and stabilize the CT price). Another crucial concept in the decentralized trading of asset is the so-called slippage. It refers to the difference between the expected price of a trade and the price at which the trade is executed and occurs due to the changes in a liquidity's pool price between the time a transaction is submitted and when it is executed. Usually, traders set a slippage tolerance to mitigate the risk of executing an unfavorable transaction. For instance, if slippage is set to 5%, it means that the trader is willing to accept a price variation for their transaction up to $\pm$5%, respectively, to the price at the time that the transaction is propagated to the network.

AMMs play a pivotal role in DEXs, revolutionizing the way assets are traded on blockchain platforms. One of the market-making protocols is the CPMM, but there is a general variety and research is very active on this end. Other market-making protocols are the constant sum market makers (CSMMs), constant mean market makers (CMMMs), and hybrid constant function market makers. These protocols use mathematical algorithms to facilitate peer-to-peer transactions, enhance liquidity, and determine prices autonomously. The CPMM is a foundational AMM model, expressed by the equation $X \cdot Y = K$. Here, X and Y represent tokens in the pool, both valued relative to each other, and K is their product. When creating the pool, it must have a 50:50 ratio of each asset. The equation must satisfy a condition considering reserves R_x and R_y for assets X and Y, respectively, along with a transaction fee f and amounts Δx and Δy being added and removed from the pool. The CPMM allows traders to adjust the reserves (Δy) when trading a certain amount of X (Δx). As trading occurs, the product $R_x \cdot R_y$ remains constant, although in practice, due to nonzero fees, this constant K tends to increase.

The CSMM uses the formula $X + Y = K$. Unlike CPMM, it allows for more than two assets to be stored in the pool and also allows for varying weights of each asset, deviating from the 50:50 ratio. The condition is $\sum_i R_i = K$, where R_i is the reserve of each asset in the pool. This model provides zero slippage but does not offer infinite liquidity. The reserve of the pool will not fall to zero, ensuring that the price movement when trades are made is not as drastic as in other AMMs.

The CMMM is an extension of CPMM, allowing more than two assets with varying weights. The equation is $\prod_i R_i^{w_i} = K$, where R_i is the reserve of each asset, w is the weight of each asset, and K is the constant. It ensures that the mean of the reserves remains constant.

The last protocol is the hybrid constant function market maker. Some projects use hybrid functions for specific characteristics based on the properties of the assets being traded. For instance, stable swap uses a hybrid of the constant product and constant sum functions [55]. The hybrid protocols aim at reducing slippage and address specific needs in the market.

The application of AMMs into blockchain and LEMs can create an active environment in which LEM's participants are able to manage the community platform through transactions. The combination of AMMs and LEMs can bring several benefits, but they also imply several drawbacks. For instance, AMMs streamline the trading process in LEMs by automating market functions, improving the whole market efficiency. As a matter of fact, this can lead to more efficient and faster transactions. In addition, AMMs operate on blockchain technology, enabling continuous and decentralized trading. This is particularly advantageous for LEMs where energy transactions may occur at any time. The decentralized nature of AMMs through blockchain aligns with the principles of LEMs, reducing the reliance on central authorities. This fosters a more democratic and inclusive energy market. Finally, AMMs can enhance liquidity in LEMs by providing a mechanism for participants to easily buy and sell energy assets without the need for a centralized authority.

However, the introduction of AMM into LEMs would increase the asset volatility, also resulting in an increased risk for the LPs. In the context of LEMs, this could lead to fluctuations in energy prices, making it challenging for participants to predict and plan effectively. Particularly, for this reason, a well-thought and resilient and robust monetary policy for the community token is highly suggested. Additionally, like any blockchain-based system, privacy concerns are on the spot. While blockchain itself is considered secure, privacy does not exist in the implementation of smart contracts that power the AMMs. Another barrier is provided by regulatory uncertainties. The regulatory environment for blockchain and decentralized technologies is still evolving. LEMs utilizing AMMs may face uncertainties regarding compliance and regulatory approval. Finally, participants may have limited control over the operation of AMMs, as these systems are often governed by predefined smart contracts. This lack of control could be a concern for entities accustomed to more centralized market structures.

In summary, while AMMs can offer significant advantages in terms of efficiency, decentralization, and liquidity in LEMs, they also pose challenges related to price volatility, security, regulatory issues, and a learning curve for participants. The successful integration of AMMs in LEMs would require addressing these challenges and ensuring a balance between automation and control.

23.4 Conclusions

This chapter comprehensively explored the transformative potential of distributed autonomous RECs, underpinned by blockchain and DAO technologies. As the energy sector evolves toward its decentralization and democratization, DARCs present a promising pathway to achieve inclusiveness and active participation. The integration of blockchain technology in these communities ensures transparent, secure, and efficient energy transactions, fostering trust and collaboration among participants. DAOs further enhance this model by offering innovative governance mechanisms, fostering community engagement and equitable decision-making. However, challenges in technical, social, and regulatory domains persist, necessitating continued research and innovative policy frameworks. The future of energy communities lies in harnessing these technologies, emphasizing sustainability, inclusivity, and resilience in energy systems. Refining these models, addressing scalability, and aligning with evolving regulatory landscapes are challenges that must be faced to ensure that DARCs contribute significantly to the global energy transition.

References

1 Gupta, N., Prusty, B.R., and Alrumayh, O. (2022). The role of transactive energy in the future energy industry: a critical review. *Energies* 15 (21): 8047.

2 NIST. (2017). Transactive energy overview. https://www.nist.gov: https://www.nist.gov/el/smart-grid-menu/hot-topics/transactive-energy-overview

3 European Commission. (2018). https://energy.ec.europa.eu/topics/energy-strategy/clean-energy-all-europeans-package_en.

4 European Parliament. (2018). Directive 2018/2001. https://eur-lex.europa.eu/legal-content/IT/TXT/PDF/?uri=CELEX

5 European Parliament. (2023). RED III. https://www.europarl.europa.eu/news/en/press-room/20230911IPR04926/meps-back-plans-to-boost-use-of-renewable-energy

6 Trevisan, R., Ghiani, E., and Pilo, F. (2023). Renewable energy communities in positive energy districts: a governance and realisation framework in compliance with the Italian regulation. *Smart Cities* 6 (1): 563–585.

7 Barchi, G. P. (2023). Residential renewable energy community: a techno-economic analysis of the Italian approach. *IEEE International Conference on Environment and Electrical Engineering and 2023 IEEE Industrial and Commercial Power Systems Europe (EEEIC/I&CPS Europe)* (pp. 1–6). IEEE.

8 D'Alpaos, C. and Andreolli, F. (2021). Renewable Energy Communities: The Challenge for New Policy and Regulatory Frameworks Design. *Smart Innovation, Systems and Technologies, Conference Paper. Vol 178. Springer*, 500–509.

9 International Electrotechnical Commission. (2023). https://syc-se.iec.ch/deliveries/sgam-basics/

10 Veelen, B.V. (2018). Negotiating energy democracy in practice: governance processes in community energy projects. *Environmental Politics* 27 (4): 644–655.

11 Simcock, N. (2016). Procedural justice and the implementation of community wind energy projects: a case study from South Yorkshire, UK. *Land Use Policy* 59: 467–477.

12 Vansintjan, D. (2022). https://www.rescoop.eu/uploads/rescoop/downloads/REScoop-Energy-Transition-to-Energy-Democracy-English.pdf

13 Avelino, F., Bosman, R., Frantzeskaki, N. et al. (2014). *The (Self-)Governance of Community Energy: Challenges & Prospects*. The Netherlands: Dutch Research Institute For Transitions (DRIFT).

14 Süsser, D., Doring, M., and Ratter, B.M.W. (2017). Harvesting energy: place and local entrepreneurship in community-based renewable energy transition. *Energy Policy* 101: 332–341.

15 Golosova, J., Romanovs, A., and Kunicinia, N. (2019). Review of the Blockchain Technology in the Energy Sector. *IEEE 7th IEEE Workshop on Advances in Information* (pp. 1–7). Latvia: Electronic and Electrical Engineering.

16 Zielińska, A. (2020). Application possibilities of blockchain technology in the energy sector. *6th International Conference – Renewable Energy Sources (ICoRES 2019)*, (pp. 1–6).

17 Galici, M.M. (2021). Energy Blockchain for Public Energy Communities. *Applied Sciences* 11 (8): 3457–3468.

18 Strüker, J.A. (2018). *Blockchain in the energy SectorBusiness Transformation Through Blockchain* (ed. H.B. Treiblmaier). Palgrave Macmillan.

19 Bao, J., He, D., Luo, M. et al. (2021). A Survey of Blockchain Applications in the Energy Sector. *IEEE Systems Journal* 15 (3): 3370–3381.

20 Andoni, M., Robu, V., Flynn, D. et al. (2019). Blockchain technology in the energy sector: a systematic review of challenges and opportunities. *Renewable and Sustainable Energy Reviews* 143–174.

21 Andoni, M., Norbu, S., Couraud, B. et al. (2022). https://eprints.gla.ac.uk/278207/1/278207.pdf

22 Eisele, S. (2020). Blockchains for transactive energy systems: opportunities, challenges, and approaches. *Computer* 53 (9): 66–76.

23 Yang, Q. and Wang, H. (2021). Blockchain-empowered socially optimal transactive energy system: framework and implementation. *IEEE Transactions on Industrial Informatics* 17 (5): 3122–3132.

24 Saxena, S., Farag, H.E.Z., Turesson, H. et al. (2019). Blockchain based transactive energy systems for voltage regulation. *arxiv Cryptography and Security* 3 (5): 646–656.

25 Trevisan, R., Mureddu, M., Ghiani, E. et al. (2023). Transactive Energy Systems in Decentralized Autonomous Renewable Energy Communities. *2023 PESGM*, (pp. 1–6). Florida.

26 Ethereum foundation blog. (2014). https://blog.ethereum.org/2014/05/06/daos-dacs-das-and-more-an-incomplete-terminology-guide

27 Ethereum. (n.d.). https://ethereum.org/en/dao/#:~:text=Ethereum%20and %20DAOs,rules%20it%20was%20programmed%20with

28 De Filippi, P. and Wright, A. (2018). *10 Blockchain of Things. Blockchain and the Law: The Rule of Code*, 156–170. Cambridge, MA and London, England: *Harvard University Press*.

29 De Filippi, P. and Hassan, S. (2018). *Blockchain Technology as a Regulatory Technology: From Code is Law to Law is Code. Computers and Society*.

30 Rozas, D., Tenorio-Fornes, A., and Hassan, S. (n.d.). When ostrom meets blockchain: exploring the potentials of blockchain for commons governance. *SAGE Journals*. doi: 10.2139/ssrn.3272329.

31 Leonhard, R. (2017). *Corporate Governance on Ethereum's Blockchain. SSRN*.

32 Hsieh, Y.Y., Vergne, J.-P., Anderson, P. et al. (2018). Bitcoin and the rise of decentralized autonomous organizations. *J Org Design* 7 (14).

33 American Cryptofed DAO. (2023). https://www.americancryptofed.org.

34 Decentraland DAO. (2023). https://dao.decentraland.org/en.

35 Uniswap. (2023). Uniswap decentralized exchange. https://uniswap.org/

36 Dupont, Q. (2017). Blockchain identities: notational technologies for control and management of abstracted entities. *Metaphilosophy* 48 (5): 634–653.

37 Garrod, J.Z. (2019). On the property of blockchains: comments on an emerging literature. *Economy and Society* 602–623.

38 Diallo, N. (2018). eGov-DAO: a better government using blockchain based decentralized autonomous organization. *International Conference on eDemocracy & eGovernment (ICEDEG)* 166–171.

39 Jentzsch, C. (2017). Decentralized Autonomous Organization to Automate Governance. DOCSLIB.ORG

40 Qin, R., Ding, W., Li, J. et al. (2023). Web3-Based Decentralized Autonomous Organizations and Operations: Architectures, Models, and Mechanisms. *IEEE Transactions on Systems, Man, and Cybernetics: Systems* 53 (4): 2073–2082.

41 Wang, S., Ding, W., Li, J. et al. (2019). Decentralized autonomous organizations: concept, model, and applications. *IEEE Transactions on Computational Social Systems* 6 (5): 870–878.

42 Schneider, B. B. (n.d.) Decentralized Autonomous Organizations. *Evolution, Challenges, and Opportunities.* PoEM Workshops.

43 Bellavitis, C. and Fisch, C. (2023). The rise of decentralized autonomous organizations (DAOs): a first empirical glimpse. *Journal of Business Research* 187–203.

44 El Faqir, Y., Arroyo, J., Hassan, S. et al. (2020). An overview of decentralized autonomous organizations on the blockchain. *Proceedings of the 16th International Symposium on Open Collaboration (OpenSym '20)* (pp. 1–8). New York: Association for Computing Machinery.

45 Mylrea, M. (2019). Distributed Autonomous Energy Organizations: Next-Generation Blockchain Applications for Energy Infrastructure. In: *Artificial Intelligence for the Internet of Everything* (ed. R.M.W. Lawless), 217–239. Academic Press.

46 Rodrigues, U.R. (2019). Law and the Blockchain. *IOWA Law Rev* 104 (2): 679–729.

47 Reijers, W. and Coeckelbergh, M. (2018). The blockchain as a narrative technology: investigating the social ontology and normative configurations of cryptocurrencies. *Philos. Technol* 31: 103–130.

48 Werbach, K. (2018). Trust, but verify: why the blockchain needs the Law. *Berkeley Technology Law Journal* 487–550.

49 Farkhondeh, M., Muller, B., Revue, M. et al. (2021). Holacracy: a new way of organizing? *Management Revue* 32 (4): 302–317.

50 Seebacher, S. S. (2017). Blockchain Technology as an Enabler of Service Systems: A Structured Literature Review. *Exploring Services Science, pp.* 12–23.

51 Lee, J.Y. (2019). A decentralized token economy: how blockchain and cryptocurrency can revolutionize business. *Business Horizons* 62 (6): 773–784.

52 Cachin, C. and Vukolic, M. (2017). Blockchain Consensus Protocols in the Wild. *Distributed, Parallel, and Cluster Computing.*

53 Artur, R., Varnavskiy, A., Gruzina, U. et al. (2018). Design of models for the tokenization of electric power industry basing on the blockchain technology. In: *Communication Papers of the Federated Conference on Computer Science and Information Systems*, vol. 17, 45–50.

54 Global Blockchain Business Council. (2023). https://gbbcouncil.org/interwork-alliance/token-taxonomy-framework/

55 Mohan, V. (2022). Automated market makers and decentralized exchanges: a DeFi primer. *Financ Innov* 8: 20.

56 Gan, R., Tsoukalas, G., and Netessine, S. (2021). Initial Coin Offerings, Speculation and Asset Tokenization. *Management Science* 914–931.

57 Cong, L.W. and He, Z. (2019). Blockchain Disruption and Smart Contracts. *The Review of Financial Studies* 32 (5): 1754–1797.

58 Gandal, N., Hamrick, J.T., Moore, T. et al. (2018). Price manipulation in the Bitcoin ecosystem. *Journal of Monetary Economics* 95: 86–96.

59 Lin, L.X. (2019). Deconstructing decentralized exchanges. *Stanford Journal of Blockchain Law & Policy* 58–77.

60 Gavin Zheng, L.G. (2021). *Ethereum Smart Contract Development in Solidity*. Berlin/Heidelberg: Springer.

61 Vikram Dhillon, D.M. (2017). *Blockchain Enabled Applications*. Berkeley, CA: Apress.

62 Didenko, A.N. (2022). *Decentralised Finance – A Policy Perspective*. CPA Australia Report.

24

Transactive Coordination Paradigm for Efficient Charging Management of Plug-in Electric Vehicles in Future Distribution Networks

Hossein Saber[1], Hossein Ranjbar[2], and Moein Moeini-Aghtaie[3]

[1] *Department of Electrical Engineering, Sharif University of Technology, Tehran, Iran*
[2] *School of Electrical and Mechanical Engineering, University of Adelaide, Adelaide, Australia*
[3] *Energy, Water and Environmental Institute, Sharif University of Technology, Tehran, Iran*

24.1 Introduction

Over the last decades, meeting the growing electricity demand imposed by electrification efforts in a clean, efficient, and reliable manner has become a primary objective for power system planners and operators. Therefore, the "smart grid" concept has been introduced concerning technological advances in data acquisition as well as demand-side communication and control [1]. The smart grid vision is reshaping the traditional view of distribution systems by enabling different coordination mechanisms to efficiently manage small-sized responsive devices. In this context, the transactive energy (TE) approach was established as a viable solution for coordinating a huge number of devices to meet operational requirements while considering economic values [2].

The definition of TE was first provided by the GridWise Architecture Council (GWAC) as "a set of economic and control mechanisms that allows the dynamic balance of supply and demand across the entire electrical infrastructure using value as a key operational parameter" [3]. In this definition, the term "value" implies the electricity price, which is a key parameter in energy trading influencing buying and selling decisions. Moreover, this definition declares that the TE approach includes the entire power system from the transmission network down to the microgrid level in the distribution network with a variety of customers. However, the aspect of the TE approach that has received more attention in recent years is its capability in distribution-level energy management.

Based on the GWAC's definition, the TE approach introduces a novel coordination mechanism that integrates economic and control aspects, enabling independent agents to engage in energy trading through an automated market platform. This objective is achieved through the development of a technology-driven infrastructure that establishes two-way communication links between sellers, buyers, and the market operator. The negotiations between market participants revolve around electricity prices and quantities. In this structure, mostly Internet technology is utilized to facilitate data transfer across all TE market participants, which successfully enables the briefly called Internet of Energy (IoE) structure field [4].

An important application of the TE is the optimal integration of a large number of electric vehicles (EVs) into distribution networks due to transportation electrification efforts. Besides the numerous benefits of EVs, the increasing growth of EVs particularly over the past decade has raised concerns about their potential negative impact on the distribution grid. These potential

Smart Cyber-Physical Power Systems: Fundamental Concepts, Challenges, and Solutions, Volume 1, First Edition.
Edited by Ali Parizad, Hamid Reza Baghaee, and Saifur Rahman.

challenges include overloading distribution feeders, accelerating transformer aging, voltage instability, and harmonic distortion [5]. In the literature, several solutions were proposed to efficiently address these challenges. One viable solution is TE-based charging management of EVs, which optimally determines the charging decisions of EVs considering the perspectives of EV owners and system operation. Using the TE approach, the EVs are enabled to actively participate in the local electricity market with less negative impacts on the grid functionality [6]. In this regard, the EVs can offer the charging flexibility and compete in an automated market environment to fulfill their charging requirements and support grid integration. In this structure, the EV charging stations (or parking lots) are equipped with Internet connectivity, allowing them to share information about plugged-in EVs with TE mechanisms through the IoE infrastructure.

An important concern when discussing TE-based charging management of EVs is the policy and market design [7]. The main challenges in designing an efficient TE model include the EV active participation model and the market-clearing mechanism. Since the TE model is a market platform based on automated controls, the individual EVs must be capable of computing their willingness to pay/accept in a user-friendly manner based on the EV owner-specified comfort and economic purposes. Thus, an efficient TE model requires a user-friendly algorithm to estimate the EVs' bids/offers, enabling them to participate in the local market trading on behalf of EV owners [8]. Moreover, regarding the market-clearing mechanisms, there are three different models, i.e., centralized, decentralized, and hybrid [9]. Each of these models has distinct advantages and limitations, making the choice of mechanism crucial for efficient coordination of EVs.

Considering the description above, this chapter focuses on the application of TE in the charging management of a large population of EVs in future distribution networks. To begin, an overview of the trends, opportunities, and challenges associated with transportation electrification is presented, offering insights into the forthcoming distribution network challenges. This brings the light to introducing different demand-side management approaches, aiming to clarify the advantages of TE models in addressing the trends and challenges of transportation electrification. Subsequently, worldwide examples of TE projects are presented, followed by a comprehensive explanation of the three-step design of an efficient TE market. This includes the modeling of active EV participation, clarification of market-clearing mechanisms, and the modeling of network constraints. Ultimately, concluding remarks and future directions are offered to enclose the key takeaways from the chapter.

24.2 Transportation Electrification

The transportation sector is one of the most significant contributors to greenhouse gas emissions. In 2021, it accounted for about 28% of global emissions [10]. This is due to the fact that most vehicles are powered by internal combustion engines with fossil fuels, which release pollutants into the atmosphere. The majority of transportation emissions come from road vehicles. For instance, a study in 2018 revealed that light-duty vehicles account for about 45% of transportation emissions in Queensland, while heavy-duty vehicles account for about 22% [11].

The good news is that there is a growing trend toward the electrification of transportation. This means that vehicles are powered by electricity instead of fossil fuels. In recent years, there has been a significant increase in the sales of EVs such that the global EV sales in 2022 reached 9.7 million, up from 2.1 million in 2018 (see Figure 24.1). Moreover, as illustrated in Figure 24.1, the share of EVs in total car sales jumped from 9% in 2021 to 14% in 2022, more than six times their share in 2018. Also, it is expected that EV market share will be more than 60% by 2030 [12].

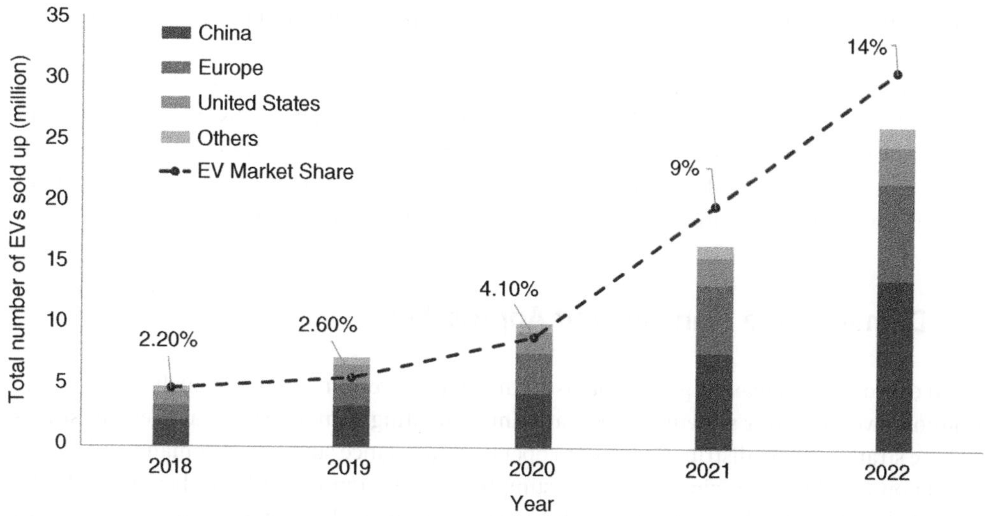

Figure 24.1 Global EV stock, 2018–2022.

The growth in EVs is driven by several factors including governmental incentives, decreasing battery costs, and increasing consumer demand for more sustainable transportation options [13]. For example, governmental incentives can be pointed to the Energy Improvement and Extension Act of 2008 in the United States [14].

With advancing technology and declining EV costs, the future holds a promising surge in EV adoption. This shift offers a multitude of environmental, economic, and social benefits for individuals, utility infrastructure, and governments. At present, the transportation system is heavily reliant on fossil fuel vehicles contributing around 31% of petroleum product consumption and nearly a quarter of global energy-related carbon dioxide (CO_2) emissions [13]. Embracing transportation electrification brings a compelling advantage in curbing emissions and air pollution by replacing fossil fuel–based cars with EVs, paving the way toward a greener and more sustainable future.

The benefits of transportation electrification extend beyond environmental gains, offering socioeconomic advantages for utilities, customers, and society as a whole. Studies reveal that transitioning to electric transportation can result in significant cost savings for customers, avoid premature deaths and environmental costs, create millions of net jobs, boost national gross domestic product (GDP), and generate substantial tax income for governments [15].

However, the electrification of transportation is not without any challenges. One challenge is the cost of EVs. EVs are still more expensive than traditional vehicles even considering subsidies [13]. For instance, the purchase price of the Hyundai i30 is around $24,000, while a similar electric model costs around $34,000. However, the cost of EVs is expected to decrease and make them comparable as the technology continues to improve. Another challenge is the lack of charging infrastructure and power supply capacity. Currently, home charging systems are mainly responsible for meeting most of the charging demand. However, it is not always feasible, especially in densely populated urban areas. In these areas, public charging stations provide a convenient and accessible way to charge EVs. However, the current number of public charging stations is not enough to support widespread EV adoption. For instance, in Norway, the ratio of EVs to public charging stations was 1.3 in 2011, and by 2022, this ratio had increased to 25 [13]. Therefore, as the number of EVs on the road increases, the need for public charging stations will become even more acute.

To ensure the continued growth of EV sales, it is imperative to invest in public charging infrastructure. This includes installing more charging stations by governments and private companies, as well as making them more accessible and affordable. By doing so, we can help to make EVs a more viable option for drivers everywhere. Moreover, the existing power distribution grid is not capable of meeting the demands of the large-scale penetration of EVs. Therefore, power system policymakers and planners should reinforce the power grid and develop charging energy management systems for maximizing the utilization of the existing energy infrastructures.

24.3 Demand-Side Management Approaches

As stated earlier, the increasing growth of distributed energy resources (DERs) such as EVs in the distribution level of power systems necessitates implementing demand-side management strategies. These strategies help distribution system operators to balance supply and demand and reduce the operation cost by optimizing energy consumption/generation of DERs. To better clarify the benefits of the TE method and categorize different approaches, the "smart energy management matrix" has been introduced as depicted in Figure 24.2 [16]. This matrix classifies demand-side management approaches into four general categories based on whether the local decisions are made centrally or locally and whether negotiations are based on one- or two-way communications. Thus, there are four approaches including incentive-based demand response (DR), centralized optimization, price-based DR, and transactive coordination. Among them, the TE approach that makes operational decisions locally based on two-way negotiation involving price and energy quantity is the most effective method with distinct and clear advantages.

24.3.1 Incentive-Based DR

This quadrant is the simplest and classical DR program that has demonstrated successful implementation across various regions worldwide for many decades. As shown in Figure 24.2, the decisions on local issues are made centrally based on one-way communications. In this approach, a group of consumers allow the local utility to switch off their appliances such as heating, ventilation, and air conditioning (HVAC) systems during peak demand hours. The incentive-based DR program can be classified into three distinct categories: direct load control, load as capacity resources, and interruptible loads. Although this approach is both simple and efficient, it fails to fully harness the response potential of consumers, as the state of responsive devices is not considered. Another drawback is that user preferences are not considered in this approach, which leads to interference with the autonomy of consumers [16, 17].

24.3.2 Centralized Optimization

The centralized optimization approach, similar to the incentive-based DR program, maintains central decision-making but sets itself apart by integrating two-way communications. Using this approach, a central optimization system oversees all available flexibility of DERs to optimally manage demand consumption and supply generation. In this regard, the optimizer requires all data associated with the device state to find the best solution for the whole system. Thus, as all relevant data are communicated to the central optimizer, this approach has the capability to completely unleash the response potential of each flexible device. Besides, since the optimizer directly controls the local devices, certain system reactions become evident when specific responses are

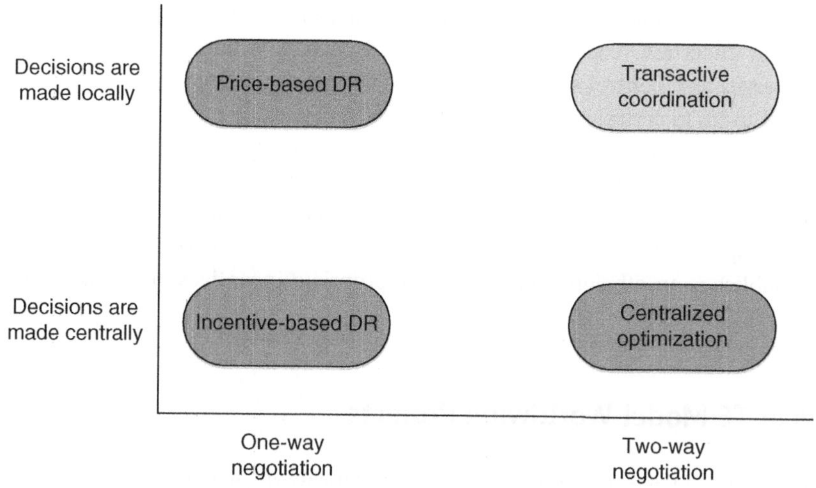

Figure 24.2 Smart energy management matrix.

triggered. In this approach, concerns about the autonomy of the incentive-based DR program remain. Moreover, due to the transfer of detailed local data to the central optimizer, an additional privacy concern arises. Further, the transfer of all local data to a central optimization engine imposes limitations on the scalability aspect of the approach. It is worth noting that as the number of responsive devices increases, both communication and optimization times experience nonlinear growth [16].

24.3.3 Price-Based DR

The price-based DR programs are based on one-way communication of dynamic price signals to manage the consumption/generation of DERs in distribution systems. In this approach, electricity prices are increased by the local utility with the expectation that consumers will respond by reducing their energy consumption. In this regard, the consumers can react to the received price signal and shift their energy consumption from peak demand hours to the durations with the lower price so that their energy expenses will be reduced. This approach offers several advantages: (i) There is lower system complexity due to the one-way communications, (ii) decisions are made locally leading to no concerns related to privacy or autonomy, and (iii) ease of implementation in regions with a wholesale electricity market, as day-ahead or intraday price profiles from this market are readily available. Despite all these advantages, the main drawback of this approach is related to the uncertainty of the system's reaction to the price signal [16, 18].

24.3.4 Transactive Coordination

In the transactive coordination (or TE) approach, the local decisions are made locally based on two-way communication. Here, the flexible devices participate in an automated market platform both within the distribution system and interacting with the representation of the bulk system. Similar to the price-based DR approach, the optimization of flexible devices is achieved through the local smart controllers; however, prior to price reaction, the local controller communicates the available flexibility along with the end users' conditions and requirements to an automated marketplace. In this regard, the consuming and producing devices communicate their bid prices (willingness to pay) and offer prices (willingness to accept). In this structure, the price-based DR approach transitions from uncertain reaction to market-based control, where a collaboratively derived dynamic price serves as a control signal to provide a certain system reaction. Thus, the TE approach enables end users to actively participate in the local electricity market to optimally schedule their flexible devices. That is the reason why this approach is also called market-based coordination. Moreover, this approach ensures the privacy of the customers, as the bids only convey data regarding energy quantities and electricity prices. In conclusion, the TE approach harnesses the complete response potential of flexible devices, offers higher certainty regarding the system reaction, establishes an efficient market with appropriate incentives, and addresses customers' privacy concerns [16, 19].

24.4 Examples of TE Model Worldwide Projects

There are several TE projects that have been implemented around the world, with a particular focus on the United States and Europe [20, 21]. Two of the most notable projects include the GridWise Olympic Peninsula Project and the American Electric Power's (AEP) gridSMART Demonstration Project [21]. These projects have demonstrated the potential of TE to improve the grid's operational condition. However, there are still several challenges that need to be addressed before TE can be widely implemented. These challenges include the need for common standards, the need for secure and reliable communication networks, and the need for consumer education. Despite these challenges, TE is a promising technology that has the potential to revolutionize the way we use and manage electricity. As these projects continue to mature and the challenges are addressed, TE is poised to become a key part of the future of the grid. These projects are reviewed in detail as follows.

24.4.1 GridWise Olympic Peninsula Project

The GridWise Olympic Peninsula project is the first proof-of-concept implemented TE project that was a field demonstration and test of advanced price signal-based control of DERs. It was sponsored by the US Department of Energy (DOE) and conducted by the Pacific Northwest National Laboratory (PNNL) in cooperation with several utilities and other organizations in the Olympic Peninsula region of Washington State [22].

The GridWise project revolutionized the energy landscape by introducing a sophisticated five-minute double auction market platform to effectively coordinate various types of energy resources. These resources included four sizable municipal water pumps, two backup diesel generators, and DR capabilities from electric water and space heating systems in 112 residential homes [22]. Through various contract agreements, participants gained invaluable insights into how their energy consumption behaviors adapted in response to price changes. The options offered included a transactive real-time price, a time-of-use rate with critical peak pricing, and a traditional flat rate.

In this local market platform, the utility submitted supply bids based on the wholesale energy price in the area. The bids from the diesel generators reflected their actual fixed and variable operation costs. Meanwhile, the municipal water pumps' bids were informed by water reservoir levels they regulated, ensuring efficient water management. Additionally, the residential DR equipment allowed households to specify their automatic price-response preferences, categorized by comfort settings ranging from nonprice responsive (most comfortable) to highly price responsive (greatest economy). During the five-minute market intervals, the clearing price for energy was determined and communicated to all market participants. Each participant's bidding equipment would then operate based on whether their bid was higher or lower than the market-clearing price, facilitating an efficient and dynamic energy exchange.

The results of the GridWise project were promising. The project demonstrated that it was possible to use TE to improve the efficiency and reliability of the grid. For instance, the project demonstrated successfully sustained peak load reductions of 15% throughout the year, with peak reductions reaching up to an impressive 50% for periods lasting up to three days. Remarkably, it showed the tremendous potential of TE in achieving multiple objectives, such as managing system peak loads and distribution network constraints, enabling wholesale price purchases by the utility, and facilitating cost savings for residential, commercial, and municipal energy consumption [21].

A considerable number of residential customers eagerly enrolled in a real-time pricing (RTP) plan, enabling them to save on their electricity bills. This plan offered pricing variations on a five-minute interval, while also granting customers the flexibility to use energy Internet technology. This technology allowed them to retain control over their preferences for comfort or savings, granting them the freedom to change their settings at any time. Moreover, the system seamlessly reflected their thermostat and appliance management choices automatically.

Advanced technology facilitating automated customer responses to real-time prices emerged as a powerful tool for managing short-term power fluctuations, especially in the presence of substantial clean wind generation. This technology effectively reduced carbon emissions without compromising comfort. Remarkably, the implementation of this unobtrusive technology came at a significantly low cost, rendering it far more economical and environmentally friendly than the conventional practice of ramping power plants up and down [22].

Therefore, the GridWise project was a significant milestone in the development of TE. The project helped to prove the concept of TE and demonstrated its potential to improve the grid. The project also helped to develop the technologies and standards needed for TE models. The lessons learned from the GridWise project are being used to develop new TE projects around the world. These projects are helping to pave the way for a more efficient, reliable, and resilient grid.

24.4.2 AEP gridSMART Demonstration Project

The other promising TE project is the AEP gridSMART Demonstration Project, which was built over the Olympic Peninsula project. This project presented a groundbreaking RTP component known as SMART Choice. Leveraging a sophisticated five-minute double auction market approach, SMART Choice expertly dispatched responsive loads within each of the four distribution circuits. To accommodate the preferences of individual households, cutting-edge software agents were employed to construct a comprehensive price flexibility curve, effectively coordinating control actions of specific devices, such as HVAC units, with the market system. Through this innovative approach, the project empowers households to make more informed energy choices, fostering a more efficient and responsive electricity consumption landscape [23].

The heart of the operation resided in a market-clearing engine, strategically located at the operations center. This powerful engine efficiently aggregated bids from all participating households, thereby forming a dynamic and price-sensitive demand curve for each distribution circuit. Utilizing this demand curve, the engine meticulously calculated both the clearing price and a supply bid, expertly incorporating the regional market operator's five-minute wholesale locational marginal price for electricity. Once the clearing price was determined, it was promptly broadcast back to the households, seamlessly integrated into the billing system, and meticulously executed in accordance with the tariff approved by the esteemed regulator, the Public Utility Commission of Ohio [21, 23]. This remarkable demonstration showcased the tangible benefits of the SMART Choice initiative, marking a significant step forward in revolutionizing the future of energy consumption and grid management.

During the late spring and summer of 2013, the RTP experiments unfolded on four feeders, encompassing approximately 200 actively participating households. The implementation of this transactive system reveals a negative correlation between energy consumption and electricity prices, validating the effectiveness of the RTP model. From a system impact perspective, simulations demonstrate that with a 35% penetration of RTP of households, a notable load reduction of approximately 5% can be achieved during a 3.5-hour system peak event. Furthermore, for a 2-hour local feeder peak event, the potential load reduction reaches nearly 8% [23].

In such scenarios, the billing system responded proactively. Households were provided with rebates, compensating for the discrepancy between the congested clearing price and the normal 5-minute real-time price. Furthermore, the billing system offered attractive incentive payments to households whose bids fell above the normal clearing price but below the congestion clearing price [16].

24.4.3 European Experience and Implementation

In addition to the aforementioned projects established in the United States, there are several other projects, such as Quartierstrom Walenstadt, the FUSION, and the GridFlex Heeten, that have been implemented in European countries [20]. The Quartierstrom project ran from 2019 to January 2020 in Walenstadt, Switzerland. It involved 37 households, including 27 prosumers equipped with a combined photovoltaic (PV) capacity of 280 kW and a lithium-ion battery storage capacity of 80 kWh [24]. The project facilitated P2P trading of solar energy within the neighborhood, enabled by blockchain technology. Participants could adjust their electricity purchase and selling prices using smart meters. The market operated on a decentralized mesh of smart meters, creating bids every 15 minutes based on users' preferences and current electricity consumption. The project successfully doubled the purchase of locally produced solar power, with 33% of electricity demand covered by solar power within the neighborhood. Participants found the concept of P2P local energy markets appealing and green, resulting in positive feedback and recommendations [20, 24].

The ongoing FUSION pilot in East Fife, Scotland, is designed for large consumers and producers [20]. It is based on the Universal Smart Energy Framework (USEF) and aims to alleviate grid congestion by activating flexibility bids, thereby reducing the need for grid infrastructure upgrades. The pilot uses high-level forecasts of the ScottishPower (SP) network to inform congestion pricing. Bids are based on USEF's D-programs, which are prognosis profiles taking grid topology into account. The goal is to extend the regular (green) operational mode of the power grid by managing capacity through local prosumer flexibility. FUSION does not allow participants to join the transmission system operator's (TSO) balancing markets but focuses on larger parties to optimize their participation in existing markets.

The GridFlex Heeten project, conducted from 2017 to 2020, aimed to alleviate grid congestion by shifting consumption to less crowded periods. It included one community of 47 households, with all participants actively involved [24]. The pilot tested two different network tariff models and utilized sea-salt batteries for excess energy storage. The market managed congestion through a price signal, encouraging participants to adjust their consumption patterns. The pilot found that altering transport tariffs alone had a noticeable effect on consumer behavior, but it did not drastically change behavior as transport tariffs constituted only a small portion of residential prosumers' energy bills in the Netherlands.

24.5 TE Paradigm in Charging Management of EVs

As stated earlier, with the proliferation in the deployment of small-sized DERs (particularly EVs) along with the technological advances in data acquisition and communication, the distribution network moves from a passive system to an active system. The integration of EVs into the distribution networks brings several benefits such as reducing carbon emissions and less dependency on fuel price. However, the arbitrary and uncontrolled penetration of EVs may impose operational challenges such as voltage violation and distribution transformer overloading. Integration of EVs (or EV charging infrastructures) into the wholesale electricity market is impossible. This is due to the small sizes of these resources compared to other market participants as well as the complexity and cost of communication and processing infrastructures that would be required to integrate them into the wholesale electricity market. The idea of local electricity trading in the context of TE can be a viable solution to address the potential challenges of increasing the penetration level of EVs into the distribution systems.

The aim of the study on TE-based charging management of EVs is to design a local electricity trading platform that enables the EVs to actively manage their charging demand considering the preferences of EV owners and system operators. In other words, the EVs are enabled to trade their charging and discharging flexibility in an efficient TE market platform. The designed TE market platform should have four main features: easy implementation, scalability, satisfying EV owners' preferences, and addressing network constraints. Easy implementation means that EV owners can adjust their bids and offers (in charging and discharging modes) to participate in the TE market in a user-friendly manner. In other words, in the TE market platform, the EV owners must be capable of individually determining their willingness to pay/accept based on their charging requirements and economic purposes. Further, the scalability feature refers to the ability of the designed TE platform to accommodate a large number of EVs. The significant obstacle that emerges as the number of EVs increases is communication and computation overheads. Thus, the system must handle the increased computational load and communication infrastructure demands that naturally arise when accommodating a large number of EVs.

Moreover, to encourage EV owners' participation in the designed TE platform, their preferences and requirements must be addressed. The main preferences of EV owners in charging management programs can be classified into three groups: reaching the desired state of charge (SOC) at departure, decreasing charging costs, and privacy-preserving. These preferences can be accommodated in the EVs' bidding/offering strategy as well as the market-clearing mechanism. Finally, aside from the abovementioned aspects, it is crucial to consider the technical aspect, and a TE model without considering the grid constraints would be impractical. Thus, the fourth primary factor that must be taken into account in the TE market design is the modeling distribution network constraints.

To design an efficient TE-based EV charging management framework with respect to the features mentioned above, three main steps should be undertaken. These steps include modeling EVs' active participation strategy, market-clearing mechanism, and modeling network constraints. For a better understanding of these steps, they are described as follows.

24.5.1 EVs' Active Participation Models

To coordinate the charging demand of EVs based on the TE approach, the EVs must be capable of determining their charging flexibility in a user-friendly manner. In previous studies, two different models for the participation of EVs in the TE market were proposed. In the first one, the EVs calculate their bid and offer prices/quantities to participate in the local market with the main

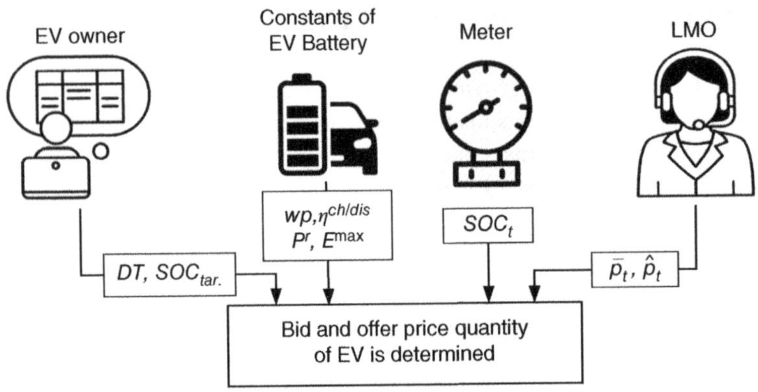

Figure 24.3 Algorithm for determining EV's bid and offer price/quantity.

objective of maximizing social welfare [25]. Figure 24.3 shows the algorithm that allows EVs to calculate their bid and offer prices/quantities considering their preferences. As seen in this figure, each EV calculates its real-time bid and offer price/quantity using four categories of data. In this regard, the local market operator (LMO) sends the mean and standard deviation of electricity price to the EV, and each EV owner specifies its departure time (DT) and target level of *SOC* at departure (SOC_{tar}). Moreover, the current SOC level of the EV battery (SOC_t) is measured, and the constant values associated with the EV battery, i.e., wear price (*wp*), charging and discharging efficiency ($\eta^{ch/dis}$), rated power (P^r), and energy capacity (E^{max}), are inputted to the algorithm. Considering these data, the bid and offer price/quantity of EV at time interval *t* is calculated as follows.

- **Bid Price:** The main goal of the EV owner regarding charging its EV battery is reaching the target SOC at departure and charging the battery at less cost. Thus, it is crucial to incorporate these two criteria in the EVs' bidding strategy. In this regard, in two conditions, the EV battery must be charged and act as an unresponsive load (bid price = ∞). The first condition is when the required time for reaching the target SOC (RT) is greater than or equal to the available time until departure (AT), and the second one is when the electricity price is in the minimum value during the plugged-in period. Based on these explanations, two priority indexes are defined for SOC and electricity price as follows [26]:

$$Pr_t^{SOC} = \begin{cases} \frac{RT_t}{AT_t} & RT_t < AT_t \\ 1 & RT_t \geq AT_t \end{cases} \tag{24.1}$$

$$Pr_t^{price} = \frac{\bar{p}_{\max} - \bar{p}_t}{\bar{p}_{\max} - \bar{p}_{\min}} \tag{24.2}$$

Considering the SOC and price priority indexes presented above, the charging priority index and the bid price of EV battery at time interval *t* are calculated as Eqs. (24.3) and (24.4), respectively [26]:

$$1 - Pr_t^{ch} = \sqrt{\left(1 - Pr_t^{SOC}\right)\left(1 - Pr_t^{price}\right)} \tag{24.3}$$

$$\pi_t^{ch} = \bar{p}_t + \sqrt{2}erf^{-1}\left(2Pr_t^{ch} - 1\right) \times \hat{p}_t \tag{24.4}$$

The EV's bid price (π_t^{ch}) is calculated using the concept of quantile function (inverse cumulative distribution function). In this regard, the inverse error function (erf^{-1}) is utilized to express the quantile function of the Gaussian distribution function. In this equation, the value of Pr_t^{ch} signifies the probability that the EV's bid price is greater than the clearing price during the time interval t. As an example, while $\bar{p}_t = \bar{p}_{\min}$ or $RT_t \geq AT_t$, the EV battery must be charged, and accordingly, the probability of charging EV will be 100%.

- **Bid Quantity:** The EV's bid quantity is the allowable charging power of the EV battery at time interval t. Thus, the EV's bid quantity is calculated as Eq. (24.5), which is the minimum of two values, i.e., rated power and possible charging power based on the residual capacity. In this equation, τ represents the duration of each time interval [26]:

$$P_t^{ch} = \min\left\{P^r, \frac{(SOC_{tar} - SOC_t) \times E^{\max}}{\eta^{ch}\tau}\right\} c \tag{24.5}$$

- **Offer Price:** The EVs with vehicle-to-grid (V2G) capabilities can discharge the stored energy in their batteries. In this regard, the V2G-capable EVs must calculate their offer price to actively participate in the TE market as generation units. Generally, generation units determine their offer price based on their imposed operation cost. Thus, to calculate the offer price of the EV in discharging mode, the cost imposed on the EV must be determined. This cost due to discharging a specific amount of power (P^{dis}) is calculated as follows [27]:

$$Cost^{dis} = Cost^{rech} + Cost^{w} \tag{24.6}$$

$$Cost^{rech} = \gamma \times \frac{P^{dis}\tau}{\eta^{ch}\eta^{dis}} \tag{24.7}$$

$$Cost^{w} = Cost^{w,dis} + Cost^{w,ch} \tag{24.8}$$

$$Cost^{w,dis} = wp \times \frac{P^{dis}\tau}{\eta^{dis}} \tag{24.9}$$

$$Cost^{w,ch} = wp \times \eta^{ch} \times P^{ch}\tau = wp \times \eta^{ch} \times \frac{P^{dis}\tau}{\eta^{ch}\eta^{dis}} = Cost^{w,dis} \tag{24.10}$$

$$Cost^{dis} = \left(\frac{\gamma}{\eta^{ch}\eta^{dis}} + \frac{2wp}{\eta^{dis}}\right) \times P^{dis}\tau \tag{24.11}$$

In Eq. (24.6), the cost imposed on the EV due to discharging ($Cost^{dis}$) is calculated by summation of the recharging cost ($Cost^{rech}$) and EV battery wear cost ($Cost^{w}$). As presented in Eq. (24.7), the recharging cost is equal to the base charging price (γ) multiplied by the required charging power to compensate for the removed energy during discharging. As expressed in Eq. (24.8), the EV battery wear cost is the summation of discharging and charging wear costs. Based on Eqs. (24.9) and (24.10), the discharging and charging wear costs are, respectively, formulated as multiplying battery wear price (wp) and energy removed and added through discharging and charging. In Eq. (24.10), we illustrated that the discharging and charging wear costs are the same. Based on Eqs. (24.6)–(24.10), the cost imposed on the EV due to discharging a specific power is calculated as Eq. (24.11). Thus, the offer price of EV is calculated as follows [27]:

$$\pi_t^{dis} = \frac{Cost^{dis}}{P^{dis}\tau} = \frac{\gamma}{\eta^{ch}\eta^{dis}} + \frac{2wp}{\eta^{dis}} \tag{24.12}$$

- **Offer Quantity:** The EV's offer quantity is the allowable discharging power of the EV battery at time interval t. Thus, the EV's offer quantity is calculated as Eq. (24.13), which is the minimum of two values, i.e., rated power and possible discharging power with respect to the minimum level of energy at EV battery ($E^{\min}$) [27]:

$$P_t^{dis} = \min\left\{P^r, \frac{\eta^{dis}(SOC_t \times E^{\max} - E^{\min})}{\tau}\right\} \tag{24.13}$$

Utilizing the above-described algorithm, all EVs calculate their bid and offer prices/quantities at each time interval. Then, the local market-clearing problem is solved to maximize social welfare at the corresponding time interval. After solving the market-clearing problem, the charging and discharging decisions of EVs as well as the exchanged electricity with the external grid are determined. Then, considering the determined charging and discharging power of EVs, the SOC level of each EV is updated for the next time interval. The proposed model is continued for the following time intervals.

In addition to the EVs' bidding/offering strategy presented above, the EVs can actively offer their charging flexibility based on the concept of the response curve [28]. In this structure, each EV owner upon arrival submits its DT, target SOC at departure, and response curve. The response curve represents the EV's required reimbursement from the system operator for their provided flexibility. In practice, each EV owner has a different required reimbursement for its response. Thus, each EV owner individually determines its response curve that allows the system operator to reduce the EV battery's charging power. Further, the EV owner will get reimbursed for the flexibility they provide. A sample of EV's response curve is depicted in Figure 24.4. As seen in this figure, each EV owner can readily determine its response curve by setting a single parameter, i.e., the required reimbursement for the maximum response ($\lambda^{\max}$) [29].

In each time interval, the EV can offer a maximum charging flexibility whose value is the minimum of two values, i.e., rated power and possible charging power based on the residual capacity. The maximum flexibility that the EV can offer ($\Delta P_t^{\max}$) is calculated as follows:

$$\Delta P_t^{\max} = \min\left\{ P^r, \frac{(SOC_{tar} - SOC_t)E^{\max}}{\eta^{ch}\tau} \right\} \tag{24.14}$$

Considering the EV's maximum offered flexibility at each time interval, the EV's response curve at time interval t can be mathematically expressed as follows:

$$\lambda_t = \begin{cases} k\Delta P_t & \Delta P_t < \Delta P_t^{\max} \\ k\Delta P_t^{\max} & \Delta P_t \geq \Delta P_t^{\max} \end{cases} \tag{24.15}$$

where the coefficient k represents the slope of the response curve, which is calculated as $\lambda^{\max}/P^r$. When all EVs have submitted their response curves, the system operator solves the local market-clearing problem. In this structure, the outputs of solving the market-clearing problem are the clearing price (λ_t^{cl}) and the decreased charging power of EVs (ΔP_t^{cl}) as illustrated in Figure 24.4. Further, the objective function of the optimization problem is minimizing the operation cost

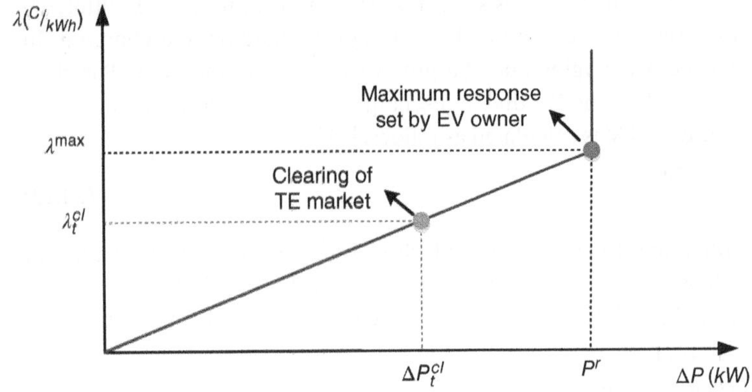

Figure 24.4 A sample of the response curve submitted by the EV owner.

or maximizing the total profit. The objective function includes several cost/revenue terms: cost/revenue of purchasing/selling electricity from/to the external grid, revenue for charging EV batteries, and reimbursements to the EVs for their provided flexibility. The revenue of the system operator for charging EVs is equal to the base charging price multiplied by the charged electricity, and the reimbursement to EVs for their response is calculated by multiplying the clearing price and decreased charging power.

After solving the local market-clearing problem and determining the clearing price and the decreased charging power of all EVs, the actual charging power of EV (P_t^{act-ch}) and its actual charging cost ($Cost_t^{act-ch}$) at time interval t are calculated as follows [29]:

$$P_t^{act-ch} = \Delta P_t^{\max} - \Delta P_t^{cl} \tag{24.16}$$

$$Cost_t^{act-ch} = Cost_t^{ch} - Reimb_t = \gamma P_t^{act-ch} - \lambda_t^{cl} \Delta P_t^{cl} \tag{24.17}$$

24.5.2 Market-clearing Mechanisms

In the electricity market, three models are available to clear the market and balance the supply and demand: centralized, decentralized, and hybrid. Figure 24.5 depicts the conceptual structure of these three market-clearing models. As seen in Figure 24.5(a), in the centralized model, a central entity acts as an intermediary between buyers and sellers in the market. It collects all the bids/offers (or response curves) from participants and matches them to determine the equilibrium price at which the market clears. Figure 24.5(b) shows that in the decentralized model, the market-clearing process occurs directly between buyers and sellers without the involvement of a central authority. This structure often occurs in P2P networks or decentralized marketplaces. The equilibrium price is determined by the collective actions of buyers and sellers. Finally, Figure 24.5(c) shows that the hybrid market-clearing model combines elements of both centralized and decentralized clearing models. In this structure, certain aspects of the market-clearing process are centralized, while others are decentralized. For instance, there may be a central authority responsible for setting certain rules or regulations and overseeing the market, while the actual trading and matching of orders occur through decentralized platforms.

Each of these market-clearing models has its pros and cons. As an example, although the centralized model is easy to implement, the growth in the number of EVs will pose a challenge to this model, as it necessitates handling an increasing amount of computational burden. Thus, the scalability of this model would be difficult in a local market with a large number of EVs. On the other hand, by applying the decentralized model, the computation load of the clearing problem will

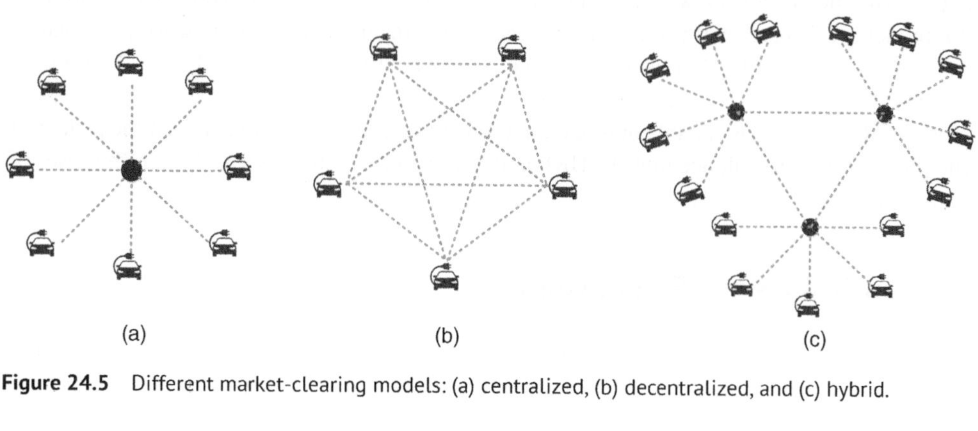

Figure 24.5 Different market-clearing models: (a) centralized, (b) decentralized, and (c) hybrid.

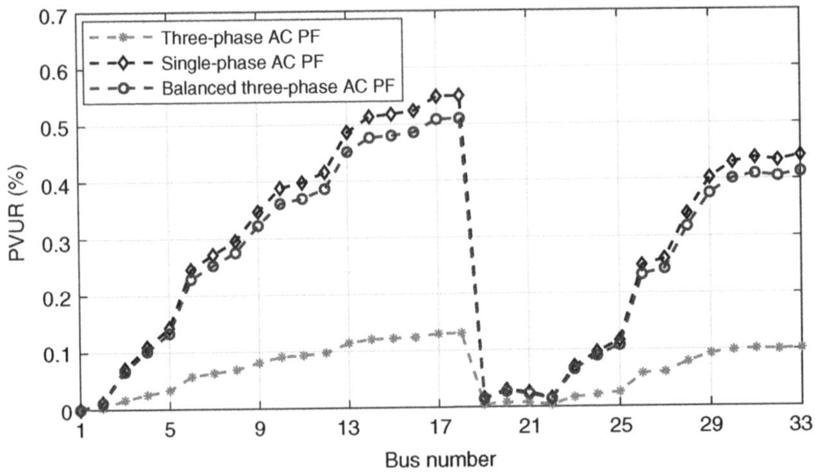

Figure 24.6 PVUR in all buses of test system with different power flow models.

be distributed among market participants. A notable drawback of the decentralized model stems from its implementation as an iterative price negotiation mechanism, which could potentially be impacted by the topology of the communication network, influencing its convergence rate [30].

24.5.3 Network Constraint Modeling

In designing any TE market platform, it is essential to take into account both economic and technical aspects, and neglecting the constraints associated with the distribution network would make the implementation of such a local market infeasible. Although EV owners are primarily focused on reaching the desired SOC at departure and decreasing their charging cost, the utilities are concerned with load shape and quality of service. The main operational constraints of distribution networks include nodal voltage limitation, transformer and distribution feeder capacity, and imbalance constraint. These operational constraints can be effectively integrated into the market-clearing problem using power flow (PF) equations. Among existing PF models, the direct current (DC) PF is unable to capture the voltage magnitude, power losses, and reactive power. Further, the single-phase alternating current (AC) PF cannot model the power and voltage imbalance among the three phases. It is the three-phase AC PF model that emerges as the optimal solution and can properly accommodate the network constraints in the local electricity market [27]. In addition, to better clarify the effectiveness of three-phase AC PF model in transactive coordination studies, this PF model is compared with other PF models. In this regard, the TE framework proposed in [27] was applied to the IEEE 33-bus test system considering different PF models, i.e., three-phase AC PF, single-phase AC PF, and balanced three-phase AC PF. As shown in Figure 24.6, since the three-phase AC PF model is capable of modeling the voltage and power imbalance in the proposed TE, the phase voltage unbalance rate (PVUR) of all buses in this model is less than in the cases of implementing other PF models.

24.6 Conclusions and Future Works

In this chapter, our focus was on the TE approach and its applicability in the efficient charging management of a large number of EVs in distribution networks. To begin, an overview of the trends, opportunities, and challenges associated with transportation electrification was presented that provided insights into the forthcoming distribution network challenges. Subsequently, different demand-side management approaches were introduced with the help of the smart energy management matrix, aiming to clarify the advantages of TE approach in addressing the trends and challenges of transportation electrification. This matrix classifies the demand-side management approaches into four classes, i.e., incentive-based DR, price-based DR, centralized optimization, and TE.

Examples of TE projects that were successfully implemented in the United States and Europe were examined in the next part of this chapter. These projects include the GridWise Olympic Peninsula Project and the AEP gridSMART Demonstration Project, which were implemented in the United States, and Quartierstrom Walenstadt, the FUSION, and the GridFlex Heeten, which were implemented in Europe. Following this, the core steps of designing an efficient TE-based charging management model were thoroughly explained, covering the modeling of active EV participation, market-clearing mechanisms, and network constraints. In this regard, two models were presented that enabled the EVs to determine their charging flexibility to participate in the real-time local electricity market. In the first model, the bid and offer price/quantity of each EV was calculated in a user-friendly manner, considering the economic and comfort preferences of EV owners. The algorithm introduced for bid and offer determination empowers EV owners to actively participate in the real-time local market, aiming to maximize social welfare. In the second model, the EV owners can submit their charging flexibility based on the concept of a response curve. In this model, the EVs will get reimbursed for the flexibility they provided in the real-time energy management program. Finally, the pros and cons of the centralized, decentralized, and hybrid market-clearing mechanisms and the capability of the TE-based approaches for resolving distribution network challenges via three-phase AC PF constraints were investigated.

For future directions, the blockchain-integrated EV charging management model and its advantages and limitations can be studied. The blockchain technology enables EVs to determine their optimal charging/discharging power using a P2P energy trading platform with secure, privacy-preserved, and anonymous transactions. Further, due to the increasing proliferation of renewable energy source (RES)-based DERs in the distribution networks, the uncertainties associated with their generations can be considered in the operation of TE market models. This integration opens the door to more robust and adaptable charging strategies, effectively harnessing the intermittent nature of RESs and enhancing the overall efficiency and reliability of the EV charging program. Finally, exploring the potential integration of artificial intelligence and machine learning algorithms could enhance the predictive capabilities of the TE-based charging management system. These advancements could enable real-time EV charging management based on evolving user behaviors, grid conditions, and environmental factors, leading to even more efficient and adaptive energy utilization.

References

1 Eltamaly, A.M., Alotaibi, M.A., Alolah, A.I., and Ahmed, M.A. (2021). A novel demand response strategy for sizing of hybrid energy system with smart grid concepts. *IEEE Access* 9: 20277–20294.

2 Pratt, A., Krishnamurthy, D., Ruth, M. et al. (2016). Transactive home energy management systems: the impact of their proliferation on the electric grid. *IEEE Electrification Magazine* 4 (4): 8–14.

3 R. B. Melton (2013). Gridwise transactive energy framework (draft version). Pacific Northwest National Lab. (PNNL), Richland, WA (United States).

4 Daneshvar, M., Pesaran, M., and Mohammadi-ivatloo, B. (2019). Transactive energy in future smart homes. In: *The Energy Internet* (ed. W. Su and A. Huang), 153–179. Elsevier.

5 Kasani, V.S., Tiwari, D., Khalghani, M.R. et al. (2021). Optimal coordinated charging and routing scheme of electric vehicles in distribution grids: real grid cases. *Sustainable Cities and Society* 73: 103081.

6 Ranjbar, H., Saber, H., and Sharifzadeh, M. (2022). Transactive charging control of electric vehicles considering voltage unbalance and transformers' loss of life. *IEEE Transactions on Smart Grid* 14 (3): 2226–2235.

7 Parag, Y. and Sovacool, B.K. (2016). Electricity market design for the prosumer era. *Nature Energy* 1 (4): 1–6.

8 Hoque, M.M., Khorasany, M., Razzaghi, R. et al. (2022). Network-aware coordination of aggregated electric vehicles considering charge-discharge flexibility. *IEEE Transactions on Smart Grid* 14 (3): 2125–2139.

9 Ahlqvist, V., Holmberg, P., and Tangerås, T. (2022). A survey comparing centralized and decentralized electricity markets. *Energy Strategy Reviews* 40: 100812.

10 Bashmakov, I., Nilsson, L.J., Acquaye, A. et al. (2022). *Climate Change 2022: Mitigation of Climate Change. Contribution of Working Group III to the Sixth Assessment Report of the Intergovernmental Panel on Climate Change, Chapter 11*. Berkeley, CA (United States): Lawrence Berkeley National Lab. (LBNL).

11 Queensland Government (2020). Transport sector greenhouse gas emissions, State of the Environment Report 2020, Queensland Government. [Online]. https://www.stateoftheenvironment.des.qld.gov.au/pollution/greenhouse-gas-emissions/transport-sector-greenhouse-gas-emissions (accessed October, 2024).

12 IEA (2022). By 2030 EVs represent more than 60% of vehicles sold globally, and require an adequate surge in chargers installed in buildings, IEA, vol. License: CC BY 4.0. [Online]. https://www.iea.org/reports/by-2030-evs-represent-more-than-60-of-vehicles-sold-globally-and-require-an-adequate-surge-in-chargers-installed-in-buildings (accessed October, 2024).

13 Ranjbar, H. and Sharifzadeh, M. (2022). Electrification of transportation: transition toward energy sustainability. In: *Industry 4.0 Vision for Energy and Materials: Enabling Technologies and Case Studies* (ed. M. Sharifzadeh), 269–296.

14 Energy improvement and extension act of 2008," 2008. [Online]. https://www.congress.gov/bill/110th-congress/house-bill/6049 (accessed October, 2024).

15 Baldwin, S., Myers, A., O'Boyle, M., and Wooley, D. (2021). Accelerating clean, electrified transportation by 2035: policy priorities. *Policy*.

16 Kok, K. and Widergren, S. (2016). A society of devices: integrating intelligent distributed resources with transactive energy. *IEEE Power and Energy Magazine* 14 (3): 34–45.

17 Tsui, K.M. and Chan, S.-C. (2012). Demand response optimization for smart home scheduling under real-time pricing. *IEEE Transactions on Smart Grid* 3 (4): 1812–1821.

18 Chen, Z., Wu, L., and Fu, Y. (2012). Real-time price-based demand response management for residential appliances via stochastic optimization and robust optimization. *IEEE Transactions on Smart Grid* 3 (4): 1822–1831.

19 Saber, H., Ranjbar, H., Fattaheian-Dehkordi, S. et al. (2022). Transactive energy management of V2G-capable electric vehicles in residential buildings: an MILP approach. *IEEE Transactions on Sustainable Energy* 13 (3): 1734–1743.

20 Doumen, S.C., Boff, D.S., Widergren, S.E., and Kok, J.K. (2023). Taming the wild edge of smart grid–lessons from transactive energy market deployments. *The Electricity Journal* 36 (2–3): 107253.

21 Lee, D., Hess, D.J., and Neema, H. (2020). The challenges of implementing transactive energy: a comparative analysis of experimental projects. *The Electricity Journal* 33 (10): 106865.

22 D. J. Hammerstrom, Ambrosio R., Carlon T. et al. 2008. Pacific northwest gridwise™ testbed demonstration projects; part I. Olympic peninsula project. Pacific Northwest National Lab.(PNNL), Richland, WA (United States).

23 S. E. Widergren, A. Somani, K. Subbarao et al. (2014). AEP Ohio gridSMART demonstration project real-time pricing demonstration analysis. Pacific Northwest National Lab. (PNNL), Richland, WA (United States).

24 J. K. Kok, A. van der Veen, S. Doumen, and P. C. Loonen, Transactive Energy in the Dutch Context, Topsector Energy, 2022. [Online]. https://pure.tue.nl/ws/portalfiles/portal/305313823/2022-TUe_TNO-Transactive_Energy-Survey_Report.pdf (accessed October, 2024).

25 Saber, H., Ehsan, M., Moeini-Aghtaie, M. et al. (2020). Network-constrained transactive coordination for plug-in electric vehicles participation in real-time retail electricity markets. *IEEE Transactions on Sustainable Energy* 12 (2): 1439–1448.

26 Saber, H., Ehsan, M., Moeini-Aghtaie, M. et al. (2022). Distributed transactive coordination of residential communities aiming at fulfilling households' preferences. *IEEE Access* 10: 122010–122021.

27 Saber, H., Ehsan, M., Moeini-Aghtaie, M. et al. (2022). A user-friendly transactive coordination model for residential prosumers considering voltage unbalance in distribution networks. *IEEE Transactions on Industrial Informatics* 18 (9): 5748–5759.

28 Wu, Q., Shahidehpour, M., Li, C. et al. (2019). Transactive real-time electric vehicle charging management for commercial buildings with PV on-site generation. *IEEE Transactions on Smart Grid* 10 (5): 4939–4950.

29 Saber, H., Ranjbar, H., Ehsan, M., and Anvari-Moghaddam, A. (2022). Transactive charging management of electric vehicles in office buildings: a distributionally robust chance-constrained approach. *Sustainable Cities and Society* 87: 104171.

30 Khorasany, M. (2020). *Market Design for Peer-to-peer Energy Trading in a Distribution Network with High Penetration of Distributed Energy Resources.* Queensland University of Technology.

25

Optimal Peer-to-Peer Energy Trading Using Machine Learning: Architecture, Strategies, and Algorithms

Nadya Noorfatima and Jaesung Jung

Department of Energy Systems Research, Ajou University, Suwon, South Korea

25.1 Introduction

Peer-to-peer (P2P) energy trading is a transactive energy concept that allows direct energy exchange between end-node customers with or without the ownership of distributed energy resources (DERs). In P2P energy trading, electricity customers perform transactions by considering the conditions of an electric power system. Based on a market approach, P2P energy trading can solve various power system operational problems, including distribution network congestion and DER integration. To enhance performance, the configuration and regulation of P2P energy trading are customized based on existing electricity market operations and their challenges.

Initially, two types of market models were adopted in P2P energy trading: centralized and decentralized models. To obtain the combined benefits of centralized and decentralized market models, hybrid P2P energy trading is developed. Furthermore, the operation of P2P energy trading significantly affects network reliability without appropriately compensating for the costs. The network cost allocation (NCA) method is implemented to compensate for the costs and alter the trading results such that they are more efficient with respect to the network's conditions. In this regard, the three principles of the NCA method, namely, those based on non-power flows, power flows, and game theory, are discussed herein.

P2P energy-trading operations have created a challenging frontier owing to the integration of renewables into power systems through the deregulated market. P2P energy trading is affected by specific challenges, but they are not limited to the volatility of load and generation capacities, increasing DER integration, and energy markets with incomplete information. Hence, the application of machine learning (ML) to P2P energy trading is considered to reduce the complexity of solving these problems. By identifying the challenges of P2P energy trading, ML techniques can be applied to P2P energy trading to address three aspects: forecasting, clustering, and decision-making. These applications can be extended to address other challenges in P2P energy trading.

The remainder of this chapter is organized into three sections: Section 25.2 presents the P2P energy trading architecture, which includes the components of the market, configuration models, and NCA methods. In Section 25.3, the application of ML to P2P energy trading is discussed based on the various principles of each aspect mentioned above. Section 25.4 summarizes the discussions of this chapter and presents the conclusions.

Smart Cyber-Physical Power Systems: Fundamental Concepts, Challenges, and Solutions, Volume 1, First Edition.
Edited by Ali Parizad, Hamid Reza Baghaee, and Saifur Rahman.

25.2 P2P Energy Trading Architecture

P2P energy trading aims to smoothen the integration of DERs into a distribution system via a market approach. However, P2P energy trading must be investigated comprehensively to define all the market components. The market components are introduced in this section as they maintain the smooth transactions of prosumers and consumers by preserving the economic and physical aspects. Regardless, Ableitner et al. [1] considered two types of components: physical and non-physical components.

The physical component refers to stakeholders, such as distribution system operators, market operators, aggregators, prosumers, and consumers. The members of a physical component can vary depending on the P2P energy trading configuration. The non-physical components include configuration models and pricing strategies. The configuration model determines the manner by which market stakeholders interact with each other in a virtual layer. In terms of the pricing strategy, despite the trading price from bid processes, cost allocation significantly affects the change in trading results depending on the network conditions, as investigated previously by Noorfatima et al. [2].

This section discusses P2P energy-trading configurations and NCA methods. P2P energy-trading configurations can be categorized into three types: centralized, decentralized, and hybrid. Subsequently, the NCA method, which can be classified into methods based on non-power flows, power flows, and game theory, is discussed.

25.2.1 Configuration Models

To construct an appropriate P2P energy trading operation, the configuration of the local energy market should be defined initially. The configuration model should be customized based on the characteristics of the local electricity operation or its purpose. Nevertheless, this section details the three most considered models by referring to Zhou et al. [3]: centralized, decentralized, and hybrid. The centralized operation of P2P energy trading is similar to that of a community-based local electricity market. Furthermore, decentralized P2P energy trading identifies a fully liberal electricity market. Meanwhile, the hybrid or semi-decentralized model reflects the combination of centralized and decentralized models, as it combines the merits of both. Centralized, decentralized, and hybrid configuration models are discussed as follows.

25.2.1.1 Centralized Model

The centralized P2P energy trading model adopts the design of conventional power electricity markets. This model is implemented to reduce the challenges of implementing a fully deregulated market. In this model, transactions within P2P energy trading are deregulated and requested by the operators. Thus, the participants cannot freely determine the amount of capacity they wish to offer, as it depends on requests from the market operator. The market operator can be represented by various institutions, including the aggregator, retailer, or distribution operation.

The equations for a centralized P2P energy-trading model can be written as follows:

Social Welfare

$$= \max\left(\sum_{l} U_l(P_l, \lambda_{l,j}, \lambda_{NUC_{l,j}}) + \sum_{j} U_j(P_j, \lambda_{i,j}, \lambda_{NUC_{i,j}})\right) \tag{25.1}$$

s.t. :

$$\sum_i P_i + \sum_j P_j = 0$$

$$P_i \geq 0; P_j \geq 0$$

where P_i and P_j denote the trading capacities of sellers i and buyer j, respectively; and $\lambda_{i,j}$ and $\lambda_{NUC_{i,j}}$ denote the confirmed trading price and network use cost of seller i and buyer j, respectively; U_i and U_j denote the utility function of seller i and buyer j. The interaction between prosumers and consumers in this model is illustrated in Figure 25.1. In the figure, the dashed line indicates the communication flow between the market operator and electricity customers. Meanwhile, the straight line with a light color represents the network that physically connects the customers.

The advantages and disadvantages of this model can be evaluated based on the aforementioned explanations. The centralized model features less complicated operations and can be easily implemented in the existing retail electricity market. A stable market is one of the characteristics of this model, which can be beneficial to the grid operator for estimating network usage arising from the P2P energy trading operation. However, using this model, the profit toward prosumers and consumers is limited and biased owing to the restricted regulation for performing transactions in P2P energy trading.

25.2.1.2 Decentralized Model

The decentralized P2P energy trading model depicts direct interactions between prosumers and consumers without any intervention from the regulators or market operators. In this model, P2P energy trading is performed via a platform that can facilitate energy transactions with several requirements.

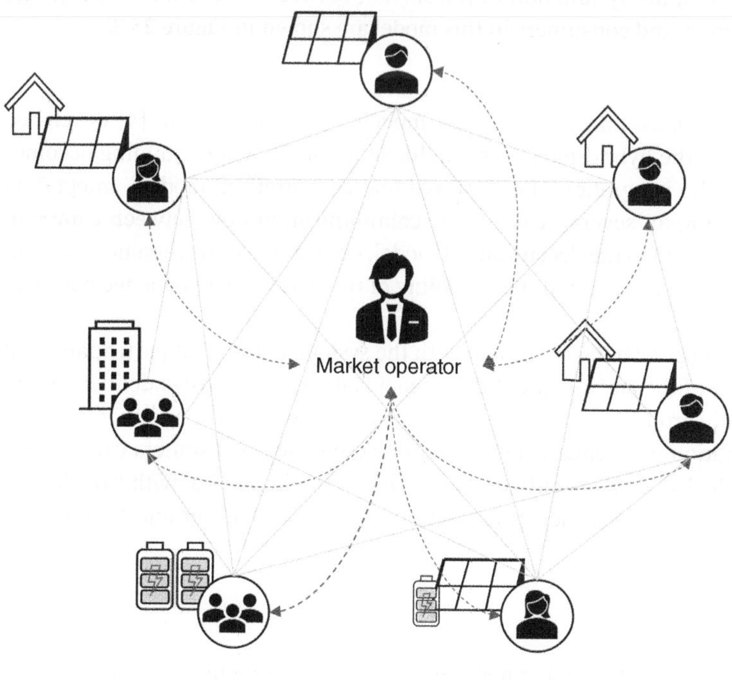

Figure 25.1 Centralized P2P energy trading market configuration with single market operator.

- Flexibility
- Security
- Usability
- Performance

These requirements are necessary owing to the following threats:

- Data leaks
- Unfair trading
- Network reliability issues

If this condition can be satisfied, then applying the decentralized P2P energy trading model can be beneficial, particularly for sellers and buyers. In decentralized P2P energy trading, sellers and buyers can maximize their benefits based on their preferences without any limits from the regulators. Therefore, a decentralized model can mitigate the disadvantages of the centralized model. To fully understand this model, equations for the decentralized P2P energy trading model can be formulated as follows:

$$\textbf{Total Profit} := \sum_{i}^{I}\sum_{j}^{J} U_{i,j}(P_i, P_j, \lambda_{i,j}, \lambda_{NUC_{i,j}}) \tag{25.2}$$

$$\text{s.t.}:$$

$$0 \leq \sum_{i} P_{i,j} \leq P_j$$

$$0 \leq \sum_{j} P_{i,j} \leq P_i$$

$$\lambda_i < \lambda_{i,j} < \lambda_j$$

where $U_{i,j}$ denotes the marginal utility function of transaction between seller i and buyer j. The interactions between prosumers and consumers in this model are shown in Figure 25.2.

25.2.1.3 Hybrid Model

To exploit the merits of centralized and decentralized models, Noorfatima et al. [4] developed hybrid P2P energy trading. In the hybrid model, the architecture of P2P energy trading becomes more complicated owing to the integration of centralized and decentralized models. Integrating both P2P models is challenging for several reasons: the communication flow between computation processes from the centralized to the decentralized model, or vice versa; convergence issues in the optimization process of both models; and the capability of the brain to acknowledge both P2P models.

Nevertheless, if these challenges can be mitigated, both the system and market participants can benefit from the model. One of the advantages of this model is its comprehensive deliberation of P2P energy trading operations from the perspectives of regulators and market participants. In the hybrid model, market operation can remain stable, and prosumers and consumers can actively participate in the market. The hybrid P2P configuration is shown in Figure 25.3, with two clusters presented to identify hierarchical transactions: transactions inside each cluster and transactions between clusters with the market operator.

25.2.2 Market Operation

In this section, the market operation is elaborated based on various principles under two main approaches: the trading strategy and auction mechanism. By considering game theory, the trading

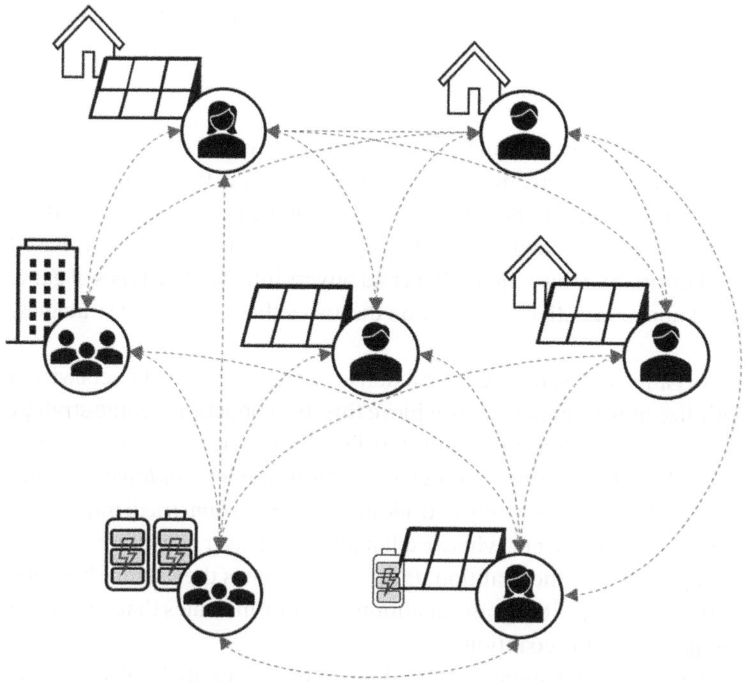

Figure 25.2 Decentralized P2P configuration with direct transactions between electricity customers.

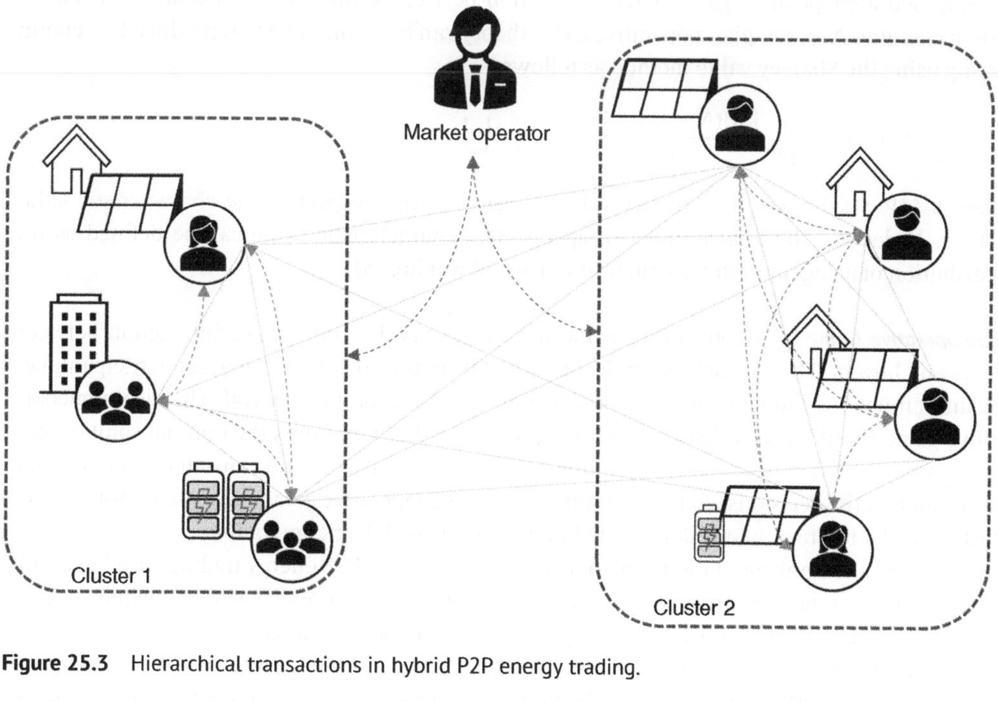

Figure 25.3 Hierarchical transactions in hybrid P2P energy trading.

strategy illustrates the interaction between sellers and buyers in achieving the market goals. The auction mechanism explains the established regulations for determining the P2P trading price.

25.2.2.1 Trading Strategy

Sellers and buyers can interact unrestrictedly to achieve the optimal benefits of P2P energy trading. Because P2P energy trading is considered a deregulated electricity market, the trading strategies of sellers and buyers can be difficult to comprehend. However, based on game theory, interactions between sellers and buyers can be identified and applied to improve the market efficiency. Based on the purpose of the game and the manner by which sellers and buyers interact, two types of game strategies exist: cooperative and noncooperative. This section discusses these two game strategies.

Cooperative Game A cooperative game represents the interaction between sellers and buyers within a community that aims to optimize mutual benefits. To achieve this, the cooperative game strategy evaluates the contribution of each agent toward the overall contribution. Cooperative game features certain characteristics such as additivity, symmetry, efficiency, and dummies. *Additivity* implies that a solution can be provided to all the players in any two additive games. *Symmetrical* implies the equal amounts should be paid to identical players, where exchanging any player between coalitions will not change the corresponding worth of the coalition. *Efficiency* evaluates the balance between the sum of the attributes and the total gain of the grand coalition. *Dummy* indicates that any player with no contribution has zero payoff for any coalition.

Cooperative game theory has been implemented in various applications in local electricity markets. Sharma and Abhyankar [5] developed a Shapley value-based cost coalition method to evaluate the actual contributions of prosumers in utilizing the distribution network, and Du et al. [6] proposed a cooperative game-based interaction between a multi-microgrid and distribution system operator. Accordingly, cooperative game theory can be formulated to reproduce P2P energy trading using the Shapley value method as follows:

$$SV_c(V(S)) = \sum_{s \subseteq N\{c\}}^{S} \frac{|S|!(N-|S|-1)!}{N!} \cdot \left(V\left(S \bigcup \{c\}\right) - V(S)\right) \tag{25.3}$$

where $SV_c(V(S))$ denotes the Shapley value of agent c with respect to the characteristic value of $V(S)$. N denotes the total number of agents. The characteristic value can be defined as the contribution of an agent to the overall profit or social welfare (S).

Noncooperative Game Noncooperative game theory depicts the competitive interaction between sellers and buyers. The interaction can be identified from the goal of each market participant, i.e., to gain as much individual benefit as possible, whether it is a profit or payoff. The ideal noncooperative game should have at least one Nash equilibrium that represents the optimal solution of a player. A decision is considered Nash equilibrium if a market participant can achieve an outcome by considering the decisions of other participants. A noncooperative game theory is appropriate for illustrating the interaction of a deregulated market, such as P2P energy trading.

Specifically, the implementation of a noncooperative game in P2P energy trading correlates with the form of the noncooperative game. The Stackelberg game is a noncooperative game strategy that is performed via an iterative game approach using a leader-follower algorithm. Noorfatima et al. [7] developed hierarchical P2P energy trading using a Stackelberg game to represent the operation of the P2P market. Furthermore, the non-cooperative game for P2P energy trading can be represented using a matrix game. The matrix game is a conventional method for representing the competition between market participants via a matrix-formed analysis. Within the matrix, each cell contains a payoff between row and column players. The matrix game shows the interaction between prosumers and consumers with incomplete information. Thus, in P2P energy trading, the noncooperative game theory can be formulated as follows:

$$\mathbf{OF} = \max\left(\sum_{i}^{I}\sum_{j}^{J}U_{i,j}\right) + \max\left(\sum_{j}^{J}\sum_{i}^{I}U_{j,i}\right) \tag{25.4}$$

$$U_{i,j} := P_{i,j}\cdot\lambda_{UP} - P_{i,j}\cdot\lambda_{i,j}\left(P_{g_i} - \sum_{j}^{J}P_{i,j}\right) \tag{25.5}$$

$$U_{j,i} := P_{i,j}\cdot\lambda_{j,i}\left(P_{d_j} - \sum_{i}^{I}P_{i,j}\right) - P_{i,j}\cdot\lambda_{UP} \tag{25.6}$$

$$\lambda_{i,j} = \lambda_{j,i} := \frac{\lambda_i\left(P_{g_i} - \sum_j^J P_{i,j}\right) + \lambda_j\left(P_{d_j} - \sum_i^I P_{i,j}\right)}{2} \tag{25.7}$$

s.t. :

$$0 \leq \sum_{i}^{I}P_{i,j} \leq P_j \tag{25.8}$$

$$0 \leq \sum_{j}^{J}P_{i,j} \leq P_i \tag{25.9}$$

where λ_{UP} denotes the uniform price.

25.2.2.2 Auction Mechanism

Auctions are a well-established method for determining product prices. In the electricity market, an auction mechanism is implemented to achieve the same purpose, i.e., to obtain the optimal price while maintaining the system's reliability by considering the generation capacity, load capacity, bidding prices, and physical constraints. Depending on the type of auction mechanism, the price can be formed as a single value or multiple values. This section details two types of auction mechanisms: uniform- and discriminatory-price auctions.

Uniform Price-based Auctions Uniform-price-based auctions represent a trading process to obtain a single optimal price. The optimality level of a uniform price is strictly dependent on various aspects, such as market efficiency, profitability, cost effectiveness, and attractiveness. However, these requirements may not be satisfied by only a single price. In P2P energy trading, particularly, the market can be publicly followed by various types of electricity customers with different degrees of market power. In a market with equal power, a uniform price-based auction can induce competitive behavior among market participants and thus increase the market efficiency. However, in a market with unbalanced power, a uniform-price-based auction can reduce the number of participants because the price is only beneficial to participants with higher market power.

Nevertheless, even in an electricity market with a high variability of price references, uniform price-based auctions are still an effective solution for maintaining price stability, which is one of the challenges of a market that utilizes intermittent energy resources by Bach [8]. Naturally, a uniform price-based model identifies an equilibrium point comprising the market-clearing price and capacity, which are extracted from the supply and demand curves of sellers and buyers, respectively. Under the uniform-price strategy, the operator unifies the bid prices from prosumers and consumers when the production cost curve and demand curve intersect, or $\frac{\lambda_m \leq \lambda_e \leq (\lambda_m + \lambda_n)}{2}$ if the condition $\lambda^+_{m,\varphi} \geq \lambda^-_{n,\varphi}$ is satisfied. Here, λ_m and λ_n represent the bid price of seller m and buyer n, respectively; $\lambda^+_{m,\varphi}$ and $\lambda^-_{n,\varphi}$ are the highest selling bid price and lowest buying bid price at phase φ, respectively; and λ_e represents the expected price at equilibrium point. The objective function of P2P energy trading with a uniform price-based auction is formulated as social welfare maximization, as follows:

$$\textbf{Social Welfare}^{\textbf{UP}}_{\textbf{i,j}} = \max \left\{ \int_{P_j^-}^{P_j^+} [\lambda_j(P_e) \cdot P_e] dP_e - \int_{P_i^-}^{P_i^+} [\lambda_i(P_e) \cdot P_e] dP_e \right\} \tag{25.10}$$

$$\begin{aligned} &\textbf{s.t.}\ : \\ &0 < P_e \leq P_i, P_j \\ &\lambda_j^- \leq \lambda_j(P_e) \leq \lambda_j^+ \\ &\lambda_i^- \leq \lambda_i(P_e) \leq \lambda_i^+ \\ &\lambda_e = \lambda_j(P_e) = \lambda_i(P_e) \end{aligned} \tag{25.11}$$

where P_e is the expected trading capacity at equilibrium. This section details the consumer and prosumer surpluses and the social welfare of the UP strategy. Consumer surplus is regarded as the accumulative product of λ_j, whereas the capacity of sellers aggregated on a certain level corresponds to λ_j. Similarly, the prosumer surplus is derived from λ_i and the trading capacity. Meanwhile, the trading capacity is determined from the optimization of social welfare. In the UP strategy, social welfare is defined as the difference between surpluses of the consumer and prosumer. Under this conflicting interest, the social welfare is maximized by selecting an optimal Q_e, and thus λ_e is achieved by satisfying the constraints.

Discriminatory Price-based Auctions Unlike uniform price-based auctions, the discriminatory price-based auction has been widely applied in P2P energy trading owing to its ability to attract broader types of sellers and buyers. Without disregarding the market power, the discriminatory price-based auction determines multiple price values and thus defines the market-clearing price as the average of various price values. In discriminatory price-based auctions, the market matches sellers with buyers by considering the profit maximization. In this case, individual profit maximization can be performed instead of the cumulative profit of the system. Heo et al. [9] proposed a discriminatory price-based application in P2P energy trading as a mixed-integer linear programming-based problem to optimize trading results. Based on the results, the sellers and buyers are paired not only by considering a similar price range but also by the possibility of obtaining high profits.

Despite being a potential auction mechanism for P2P energy trading, the equilibrium analysis of discriminatory price-based auctions should fully consider not only the economic but also the physical aspects of network reliability. The range of prices of sellers and buyers are described as $\lambda_i < \lambda_{i,j} \leq \frac{(\lambda_i + \lambda_j)}{2}$, with $P_{i,j}(\lambda_{i,j} < \lambda_i) = \emptyset$. λ_i and λ_j are the expected trading prices from cluster of sellers i and buyers j, respectively. If this condition is violated, then an empty set of matched trading capacities will be resulted. Furthermore, the discriminatory price-based strategy is performed at a higher stage where cluster-based participant trading is anticipated. Thus, the total benefit of P2P energy trading based on the discriminatory price-based strategy is formulated as follows:

$$\textbf{Total Profit} = \max \left\{ \int_{q_0}^{q_{max}} [(\lambda_j - \lambda_i) \cdot q]\, dq \right\} \tag{25.12}$$

s.t. :

$$q_0 = \begin{cases} P_i^-; \sum_j^J P_j \geq \sum_i^I P_i \\ P_j^+; \sum_j^J P_j < \sum_i^I P_i \end{cases}$$

$$q_{max} = \begin{cases} P_i^+; \sum_j^J P_j \geq \sum_i^I P_i \\ P_j^+; \sum_j^J P_j < \sum_i^I P_i \end{cases} \tag{25.13}$$

Based on the formulas above, the total profit of the discriminatory price-based auction strategy is defined as the product of the difference in the bid prices of prosumers and consumers toward the accumulative trading capacity. The lower and upper bounds of the trading capacity are determined based on two conditions: the prosumer and consumer markets. The seller's market occurs when the total demand surpasses the total supply. Meanwhile, the buyer's market occurs when the total available generation is greater than the total demand.

25.2.3 NCA

P2P energy trading, as a type of electricity market, can be affected by several factors, including the network cost. The network cost can regulate the market because it must satisfy strict criteria. This section focuses on determining the appropriate allocation method to calculate participants' contributions toward network costs, rather than calculating individual costs. However, capturing the actual contributions without bias is challenging. Therefore, Section 25.2.3.1–25.2.3.3 detail the principles and formulations of three primary types of NCAs, i.e., those based on non-power flows, power flows, and game theory. The classification of the NCA method is shown in Figure 25.4.

25.2.3.1 Non-Power Flow-Based

Non-power flow-based NCA methods formulate a network use cost (NUC) based on the magnitude of the agents, including the generators and loads in the network. This NCA method is widely used in pre-modern power systems owing to its simple and straightforward approach. In this section, two NCA methods are classified into the non-power flow-based NCA method, i.e., Postage Stamp and MW-Mile.

Postage Stamp The postage stamp method allocates the total network cost based on the magnitude of transacted power of each generator and demand. In P2P energy trading, the generator and demand are represented as seller and buyer, respectively. The transacted power can be measured at the time of the system's peak demand. The postage stamp is independent of the transmission distance and network configuration. This method assumes that the entire network is used regardless of the actual transaction capacity and facilities. Thus, it can send incorrect signals regarding

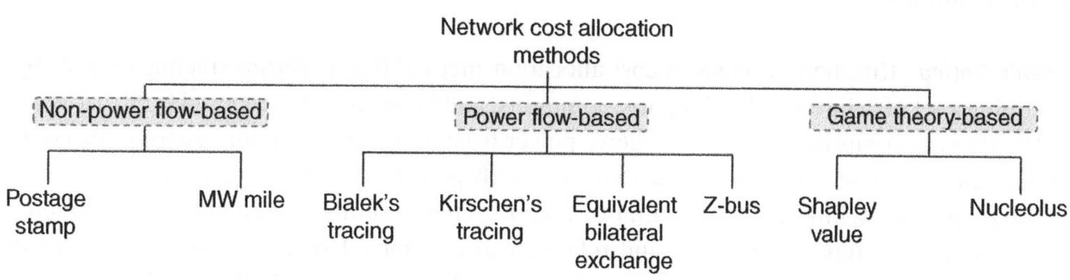

Figure 25.4 Classification of network cost allocation methods.

the actual network conditions. Nevertheless, it is widely used in European countries to allocate the transmission network to customers, as studied by Korab et al. [10]. It is characterized by a stable and predictable tariff, which is favorable for transmission owners and investors.

MW-Mile MW-Mile is a line-by-line method because it considers changes in MW transmission flows and transmission line lengths. In this method, the transacted-related flow in each branch is multiplied by the length of the transmission line. Subsequently, the values are accumulated over all transmission lines as they are used for the transaction. A previous study by Shirmohammadi et al. [11] considered two applications of the MW-Mile method: fixed values of generation and load that represent the power balance for the transaction, or every transaction identified based on the capacity variation and locations of generation and loads. The challenge with this method is to apply it to deregulated power systems because it does not consider counterflows. Counterflows typically occur in deregulated power systems such as P2P energy trading. This method is more suitable for application to systems operated by a utility or network owner. The MW-Mile method improves the technical quality of transmission services by maintaining easy regulation, charge continuity, and economic signals for dimensioning.

25.2.3.2 Power Flow-Based

Power flow-based NCA methods assume that the NUC is derived from the capacity flowing through specific lines owing to specific agents. The flowing capacity is traced from or toward a specific seller or buyer, respectively. This NCA method is favorable because of its capability in capturing actual economic signals with respect to the power flow condition. This section details power flow-based NCA methods, including Bialek's tracing, Kirschen's NCA, equivalent bilateral exchange (EBE), and Z-bus NCA.

Bialek's Tracing The tracing concept proposed by Bialek [12] is formulated based on the principle of Kirchoff's current law, as nodal inflows are shared proportionally among nodal outflows. Bialek's tracing method determines the contribution of individual generators or loads based on calculated topological distribution factors, which are derived from the relationship between nodal inflows and outflows. As nodal inflows are divided proportionally among nodal outflows, this method is also known as the proportional sharing method. The relationship between nodal inflows and outflows is represented by matrices comprising elements with binary numbers. This method has been proven suitable for deregulated electricity markets owing to its capability to calculate counterflows, including in both alternating current and direct current power flow calculations, as was studied by Hörsch et al. [13]. The key aspect of this method is that it will significantly affect the financing of market participants, as it yields an NUC with high volatility. Meanwhile, the network owner can benefit from this method, as its application can effectively regulate the market based on the actual network conditions.

Kirschen's Tracing Kirschen proposed a cost-allocation method that performs tracing analysis by determining a set of definitions for domains, commons, and links, as were proposed by Strbac et al. [14]. The domain comprises buses that receive power from a certain generator, whereas the common domain is a set of contiguous buses injected with power from the same set of generators. Based on graph theory, links connect commons via power flows, similar to branches. Furthermore, the NUC is derived from the absolute and relative contributions. However, Kirschen presented only the contribution of generators to the network cost, whereas the contribution of the loads was undefined. Therefore, Kirschen's tracing method is applied to P2P energy trading in this study by assuming negative generation loads. Despite weak mathematical evidence, Kirschen's tracing method offers several advantages. Incremental changes in injections are not limited and do not require linearization of the network model. Furthermore, Kirschen's tracing method can address the concerns of network owners, as it allocates a considerable amount of NUC to effectively solve congestion was proposed by Noorfatima et al. [15].

EBE As the name suggests, EBE allocates the contribution of sellers and buyers equivalently based on their bilateral exchanges. In particular, EBE calculates the distribution factors based on two parameters: the transacted capacity and network conductance. The principle of EBE is to provide a predefined fraction of each seller to each demand, and vice versa. In addition, the distribution factor is calculated by considering the line flow from a seller to a certain buyer through specific lines. This method requires power flow calculation results to obtain the line flow capacity and phase angles, as was studied by Galiana et al. [16]. Accordingly, the NUC of the seller and buyer can be calculated after obtaining all the distribution factors. The main feature of the EBE method is that it is independent of the slack bus, which is typically determined arbitrarily. Thus, the cost-allocation results remain stable and positive despite counterflows. This condition is preferred by P2P energy trading participants as it allows them to easily estimate the benefits of participating in the market.

Z-bus NCA The Z-bus NCA method was developed by Conejo et al. [17] to allocate the contribution of each seller and buyer in P2P energy trading by apportioning the active power flow among all nodal currents. The cost allocation of the Z-bus method requires a predefined impedance matrix that, when combined with current injections, can quantify the individual contribution of each current injection to deliver power throughout the distribution system. The implementation of a physical network for cost allocation induces the proximity effect, which implies regarding similar branches as either near or distant. This method is preferable for a deregulated electricity market with no access to actual network specifications because it can provide the electrical distance. The Z-bus NCA method has been proven to yield a positive correlation between electrical distance and the NUC allocated to sellers and buyers.

25.2.3.3 Game-Theory-Based

The game-theory-based NCA method was developed to address the dynamic possibilities of DGs supplying the grid. Game theory is the concept of modeling coalitions to obtain the objectives of either an individual or among rational decision-makers. In this section, the cost allocation methods are based on cooperative game theory such as Shapley value and nucleolus.

Shapley Value Sharma and Abhyankar [18] proposed the Shapley value, which determines the unbiased attribution of either gain or cost to all possible coalitions based on the average incremental cost when the agent is excluded from a subcoalition, including an empty subcoalition. The contribution of each player is calculated based on the marginal cost of the agent as they are joining, based on all possible orders. A possible joining order is defined as a permutation of the total number of agents in the coalition. The application of the Shapley value in cost allocation for P2P energy trading network usage is preferable because it assumes equal opportunity for all agents, whether in the best or worst coalitions. The application of the Shapley value to allocate network costs was proven by Benedek et al. [19], who developed an NCA method for distribution networks with distributed renewable generators. The main reason this method is preferable for P2P energy trading applications is that it provides a symmetrical cost-allocation solution. This feature benefits P2P energy trading operations, as the cost of the pricing mechanisms should not be biased or symmetrically represent the actual network cost.

Nucleolus Instead of allocating costs based on marginal contributions, nucleolus determines the cost allocated to each agent based on the dissatisfaction level of an agent toward all possible coalitions. This concept was originally proposed based on the idea that every game has only one nucleolus unless the core is empty, and that the nucleolus lies in the core. Therefore, the main objective of this method is to determine a set of charges ($Y = (y_1, y_2, \dots, y_x)$) that is not larger than the marginal contribution, which is the characteristic value ($v(S)$). The nucleolus is evaluated to determine the minimum dissatisfaction, which is the excess of the characteristic value ($v(s)$) and marginal contribution (y_x). The marginal contribution is determined as the division of the total cost without agent x and the total transacted capacity without agent x. Gao [20] applied a nucleolus for a system with a predefined fixed cost of the distribution network. In addition, this NCA method is an efficient solution for market operations that consider major alliances that can be represented as aggregated sellers or buyers.

25.3 ML Operation in P2P Energy Trading

ML is a branch of computer science that focuses on emphasizing the benefits of data processing to improve various fields involving computation. In P2P trading, ML can improve the trading process by considering physical and nonphysical aspects. By implementing ML, stakeholders can achieve optimal decisions based on conflicting aspects more efficiently. These conditions are enabled by adopting methods such as forecasting, clustering, and decision-making via ML approaches. In this section, ML operations and their applications in P2P energy trading are presented.

25.3.1 Forecasting

Forecasting using ML is performed to obtain values for a certain period in the future with reliable accuracy by acquiring less data for modeling. Initially, the forecasting method is based on a statistical process, which requires a significant amount of data to obtain reliable forecasting results. Makridakis et al. [21] showed that applying ML models to forecasting methods afforded significant improvements as compared with using the original statistical models. In addition, prior knowledge regarding the relationships between variables may not be required to obtain accurate predictions when using the ML approach.

These features are beneficial for P2P energy trading owing to several reasons. First, ML applies a preprocessing step to manage outliers, e.g., high- volatility data resulting from the generation capacity of renewable resources such as solar photovoltaic and wind turbines, and electricity market prices, which are typically involved in P2P energy trading. Second, P2P energy trading features limited historical data that can be exploited to obtain reliable forecasting results through arbitrary complex mappings and to support multiple inputs and outputs using more advanced ML-based forecasting methods.

To fully exploit these benefits, ML-based forecasting methods should be customized based on the purpose of P2P energy trading. This can be achieved by understanding the different types of ML-based forecasting methods and their mechanisms. This section details ML-based forecasting methods based on two categories: regression- and neural-based methods.

25.3.1.1 Regression-Based ML

Regression-based ML is a forecasting method that transforms a time-series prediction model into a regression model to generate useful data representations. To enable forecasting, regression-based

ML comprises some components that are crucial for achieving reliable prediction results. By varying the methods from data preparation to post-processing, the application of regression-based ML in P2P energy trading can be customized depending on the forecasting purpose. An electricity market price forecasting method that integrates a Gaussian regression model with a flower pollination algorithm was proposed by Sahoo et al. [22] to simultaneously predict the model's performance and forecast the energy price in competitive P2P energy trading. The forecasted results prove that regression-based ML presents low dependency toward missing data.

25.3.1.2 Neural Forecasting ML

Compared with the case of regression-based ML, forecasting using neural-based ML allows the direct processing of time series through neural network architectures to generate forecasted models. The simplest neural network is a linear model constructed using a bottom layer with predictor neurons, a top layer with forecast neurons, and no hidden layers. By adding a hidden layer, the neural network exhibits a nonlinear function that reduces the effects of extreme input values. Among various forecasting methods using neural networks, the long short-term memory (LSTM) is preferable for managing highly volatile data such as the load and generation capacity of P2P energy trading participants, as was studied by Kong et al. [23]. The LSTM calculates the forecasted value based on a modular operation, and the LSTM module is shown in Figure 25.5.

25.3.2 Clustering

Clustering has been widely used in various applications in the electricity market, such as locational marginal pricing and localizing DER operation. Direct grouping is the conventional clustering method for P2P energy trading operations based on customer locations. In ML, clustering constitutes unsupervised learning. ML-based clustering yields better results than the conventional clustering method owing to some of its features, i.e., it does not require initial information and can perform pattern discovery, thus providing a reliable representation of the data was studied by Aggarwal and Reddy [24]. In this case, the clustering method analyzes data patterns and categorizes them into several clusters. To achieve this, the clustering parameters are initially determined based on the clustering purpose.

Applying the ML-based clustering method with the appropriate parameter construction enhances the optimal operation of P2P energy trading. P2P energy trading typically focuses on either minimizing operational costs, maximizing individual profits, or network management.

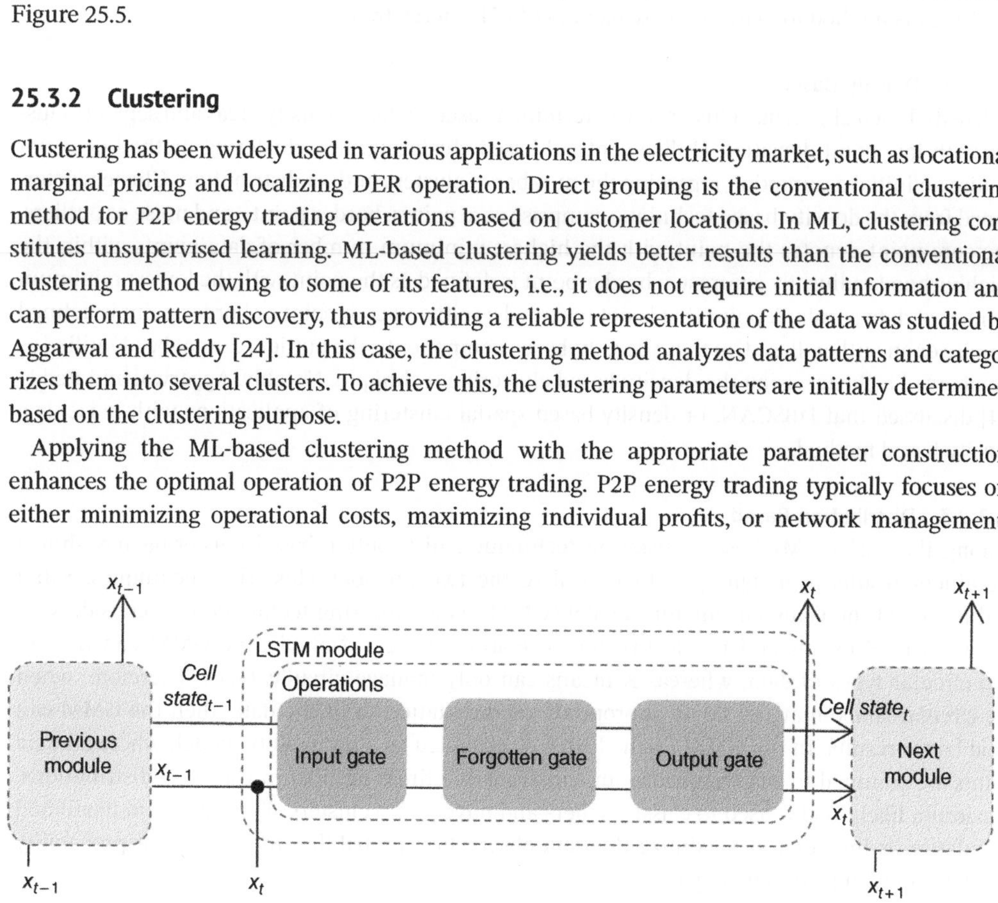

Figure 25.5 Module of long short-term memory forecasting technique.

The authors of Noorfatima et al. [7] applied Gaussian-based clustering by considering distance, load, and generation to minimize the operation cost and thus increase the profit of trading participants via an optimal analysis of the data pattern. The applications of ML-based clustering for other P2P energy trading purposes indicated improvements in terms of profit and network reliability. The appropriate ML-based clustering technique should be adopted based on the complexity level of each trading system. The categories of clustering techniques are presented next.

25.3.2.1 Centroid-Based

Centroid-based clustering identifies clusters based on the distance between a dataset and its data points. The number of data points is defined based on the number of clusters, which should be determined initially. Centroid-based clustering applies a non-hierarchical algorithm, which implies a straightforward and efficient clustering technique. Examples of clustering methods include K-means clustering, K-medoids clustering, and CLARANS were elaborated by Aggarwal and Reddy [24]. K-means clustering is the most widely used centroid-based clustering method, which allocates cluster members based on the Euclidean distance. K-medoid clustering applies partitioning around the medoid algorithm, which is defined as a point in the cluster with the minimum difference from all other points. CLARANS represents the clustering of large applications based on a randomized search. Noorfatima et al. [4] developed a dynamic clustering method using the K-means method to manage the scalability of P2P energy trading.

25.3.2.2 Density-Based

In density-based clustering, clusters are determined based on high-density areas and separate clusters with low point densities. Unlike centroid-based clustering, this clustering method does not require an initial assumption regarding the number of clusters or data distribution. Clusters determined from the density-based technique comprise core points, border points, and noise or outliers. The core point denotes the point with the highest minimum number of data points within the neighborhood radius. By contrast, a border point is defined as the point with the lowest minimum number of data points. A point that is not regarded as either a core point or border point is defined as noise or an outlier. Based on these density-based components, the outliers cannot be identified as members of a cluster using the density-based clustering technique. Notably, Aggarwal and Reddy [24] discussed that DBSCAN, or density-based spatial clustering of applications with noise, is a density-based method.

25.3.2.3 Distribution-Based

Among the various ML-based clustering techniques, distribution-based clustering has shown prominent results in assigning outliers, unlike the two previous clustering techniques. Sahbi [25] proposed the Gaussian mixture model (GMM) as a clustering technique that considers an extension of K-means and is effective for estimating clusters. Notably, the GMM can process non-circular types of data, whereas K-means can only manage circular types. Therefore, when the circular shape may not be an appropriate representation to fit the data, then the GMM can yield better results. In the GMM, each cluster is presented as a probability model, where all data points are assumed to be generated from a mixture of a finite number of Gaussian distributions. Maximum likelihood is used to classify sellers and buyers into clusters. The maximum likelihood calculation is intended to maximize the data observation probability from the joint probability distribution with respect to all parameters.

25.3.2.4 Hierarchical-Based

This clustering technique was developed to manage both quantitative and qualitative data. Hierarchical clustering adopts a recursive partitioning method to analyze the feature space, which is continually segregated based on a segregation criterion. In general, hierarchical clustering can be represented using a tree-learning method. Tree learning uses nodes to predict the value of each partition of the learned tree. Aggarwal and Reddy [24] explained that the segregation is continued until the final subsets, which are known as terminal or leaf nodes, are formed. Subsequently, the algorithm for hierarchical-based clustering is constructed via four primary steps: the construction of the proximity measure of the dissimilarity matrix, the visualization of data points as the terminal nodes, and the aggregation of the closest sets of clusters at each level; thus, the dissimilarity matrix is updated correspondingly. Some hierarchical-based clustering techniques, include single links, complete links, group averages, Ward's criterion, and decision trees, were studied by Rohlf [26], Murtagh and Legendre [27], and Gabidolla and Carreira-Perpiñán [28].

25.3.3 Decision-Making

P2P energy trading involves a local electricity market that can be directed by private parties. Consequently, the actual trading data may be difficult to obtain. Another technique can be implemented in ML to improve the P2P energy trading performance. ML-based decision-making techniques utilize historical data to arrive at better decisions. Therefore, even when real-time data are not provided, an optimal trading decision can be made by applying an ML-based decision-making technique. ML-based decision-making attempts to minimize the discrepancy between the actions performed by the operator. By considering the optimal policy, the actions performed can be improved accordingly.

25.3.3.1 Reinforcement-Learning (RL)-Based

Decision-making is a complicated process, but several pieces of evidence have proven that reinforcement-learning (RL) methods can be used to assess optimal decisions and thus achieve optimal results. RL-based decision-making is performed through a learning paradigm that sequentially optimizes decisions. However, in each sequence, the learning process can be improved significantly by learning the previous decisions. In RL, dynamic system transitions are evaluated with respect to several aspects, including the state condition, reward, and action. This model is widely known as the Markov decision process. Two types of RL algorithms exist: policy maximization and value maximization. In policy maximization, RL must update the policy for every learning sequence. Meanwhile, in value maximization, RL uses a fixed policy that is initially determined and attempts to improve the reward at every iteration.

Applying RL-based decision-making can improve the automation of allocation processes in P2P energy trading via a comprehensive analysis. Although P2P energy trading is performed in real-time, every decision made at one time can affect the trading results at other times. RL-based decision-making processes can be used to estimate future trading based on previous trading results. Wang et al. [29] applied RL to P2P energy trading to increase the trading capacity and reduce the energy cost. RL-based decision-making assists in power plant scheduling by considering the physical constraints of a distributed renewable energy system. Furthermore, Naseri et al. [30] proposed RL-based applications being used to solve dynamic retail pricing problems in transactive energy operations. In this case, the Q-learning model was developed to obtain optimal pricing without requiring complete knowledge regarding the system dynamics and uncertainties.

25.4 Simulation

25.4.1 Case Study: Hybrid P2P Energy Trading Using ML-Based Clustering

This section presents an algorithm for hybrid P2P energy trading using ML-based clustering. The details of the proposed method are available in the literature, Noorfatima et al. [7]. The hybrid P2P energy trading model comprises multiple aggregators. The algorithm for hybrid P2P energy trading using ML-based clustering is shown in Figure 25.6. A more extensive discussion of the proposed method is presented below.

25.4.1.1 Customer Classification Using GMM Clustering Method

The clustering process using the GMM considers three parameters: the generation capacity, load capacity, and participant location. GMM clustering is categorized as an unsupervised ML that specializes in classification processes with probability density estimations. In this case, to construct

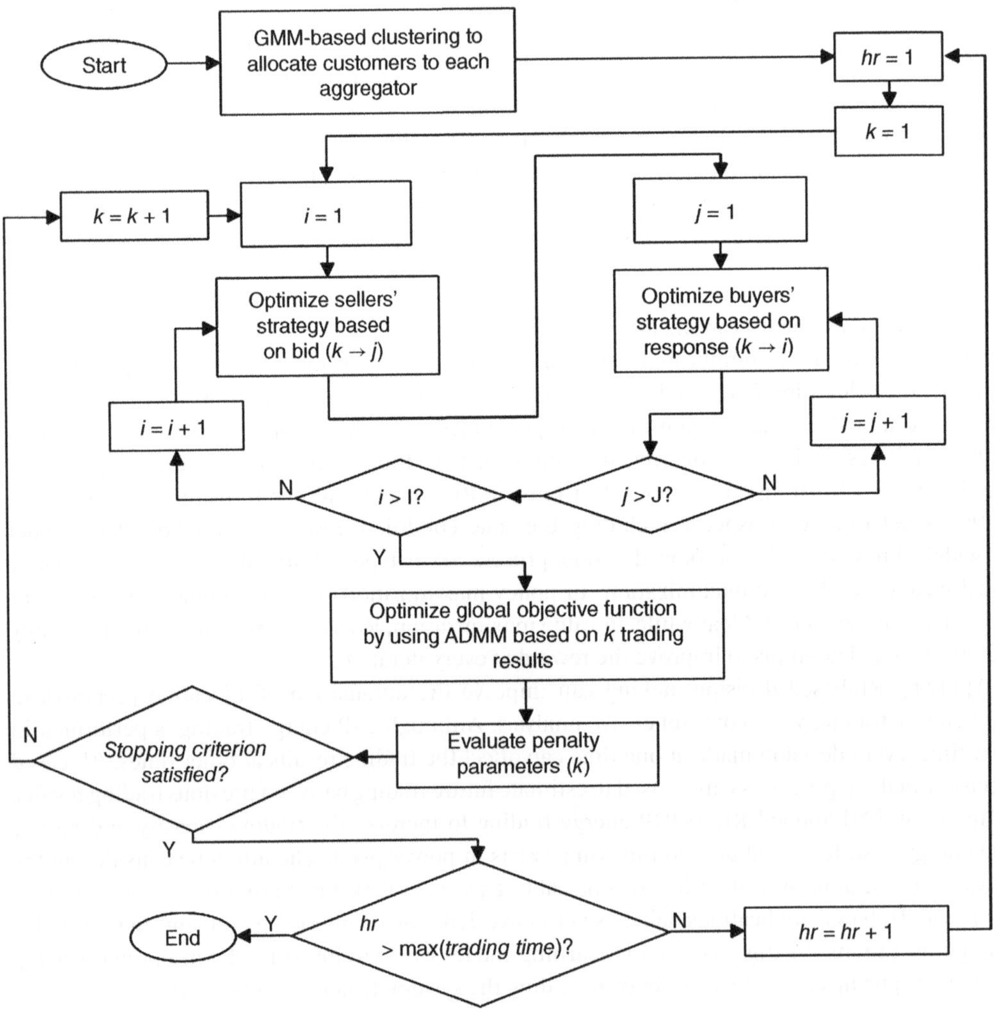

Figure 25.6 Algorithm of hybrid P2P energy trading with multi-aggregators.

the classification models for the sellers and buyers, the initial probabilities of the sellers ($\mathbb{P}^0_{i,s}$) and buyers ($\mathbb{P}^0_{j,b}$) are expressed as follows:

$$P^0_{i,s} = \sum_{(i=1)}^{I} \sum_{(s=1)}^{S} \left\| x_S^{P_s,r_s} - \sigma_i^{P_i,r_i} \right\|^2 \tag{25.14}$$

$$P^0_{j,b} = \sum_{(j=1)}^{J} \sum_{(b=1)}^{B} \left\| x_b^{P_b,r_b} - \sigma_j^{P_j,r_j} \right\|^2 \tag{25.15}$$

In addition, the dataset probability models of the sellers and buyers, $\mathbb{P}^s$ and $\mathbb{P}^b$, respectively, can be written as follows:

$$P^s \left(x_S^{\{P_s,r_s\}} | \theta_i \right) = \sum_{i=1}^{I} \varphi_i \cdot p \left(x_S^{\{P_s,r_s\}} | \theta_i \right) \tag{25.16}$$

$$P^b \left(x_b^{\{P_b,r_b\}} | \theta_j \right) = \sum_{j=1}^{J} \varphi_j \cdot p \left(x_b^{\{P_b,r_b\}} | \theta_j \right) \tag{25.17}$$

where θ_i and θ_j are Gaussian mixtures of aggregators i and j, respectively. Each *theta* comprises three components, i.e., μ, σ, and φ, which denote the mean, covariance, and mixture weights of the clusters toward each data point, respectively. Responsibility indicates the likelihood level of a data point toward the mixture components. After determining the probability function of each seller and buyer, the likelihood of all sellers and buyers is maximized. Sellers and buyers are allocated to the most appropriate aggregators based on their maximum likelihood.

25.4.1.2 Stackelberg Game Theory

Based on the clustering results, the interaction between the aggregators of sellers and buyers is represented using the Stackelberg game theory to obtain a possible trading settlement between the aggregators. The aggregators of sellers and buyers can be regarded as leaders and followers, respectively. Thus, to perform the Stackelberg game, convex sets of strategies and continuous utility functions of the aggregators of sellers and buyers are defined as follows:

1) **Strategy for aggregators of sellers**: The aggregators of sellers aim to maximize profit by obtaining the highest transaction margin from the aggregators of buyers. Thus, the utility function of the aggregators of sellers is formulated as shown in Eq. (1.5).
2) **Strategy of aggregators of buyers**: The objective of the aggregators of buyers is to maximize the profit by selecting the most profitable buyers that respond to the aggregators of the sellers' signals, i.e., the bid prices offered based on the aggregated generation capacity. Hence, the utility function is defined as the buyer's welfare function, which is formulated as shown in Eq. (1.6).

25.4.1.3 ADMM-Based Trading Optimization for Hybrid P2P

This section presents the alternating direction method of multipliers (ADMM) method for hybrid P2P energy trading using a multi-aggregator. ADMM with updating dual variables, as proposed by Nguyen [31], is applied not only to optimize general problems through a distributed approach but also to overcome the convergence issues of the Stackelberg game theory. The system operator aims to determine the optimal trading results for the entire system, which are formulated as shown in Eq. (1.2). To optimize the global objective function, the constraints of the system operator's objective function are formulated using the Karush–Kuhn–Tucker (KKT) conditions. The KKT conditions evaluate the satisfaction level of the constraints, which are derived from the total supply capacity of aggregators of sellers and the total demand capacity of the aggregators of buyers.

25.4.2 Simulation Results and Discussions

The performance of the ML-based clustering application in hybrid P2P energy trading was evaluated via an analytical comparison of various customer-allocation methods and an analytical comparison among the Stackelberg, conventional ADMM, and proposed adaptive methods.

1) **Simulation setup**: The experimental scenarios were established based on actual cases in the South Korean retail electricity market and were simulated on a 141-bus IEEE distribution network with randomly located DERs and loads, as shown in Figure 25.7 is cited from Khodr et al. [32]. Electricity tariff structures for various electricity customer categories based on South Korean power systems are available in KEPCO [33]. DER generation data was obtained from NREL [34].
2) **Comparison of total profit from various customer allocations**: To evaluate the significance of the proposed method, the total profit results of the three methods, the equally distributed, conventional GMM, and proposed methods, were compared. As shown in Table 25.1, the

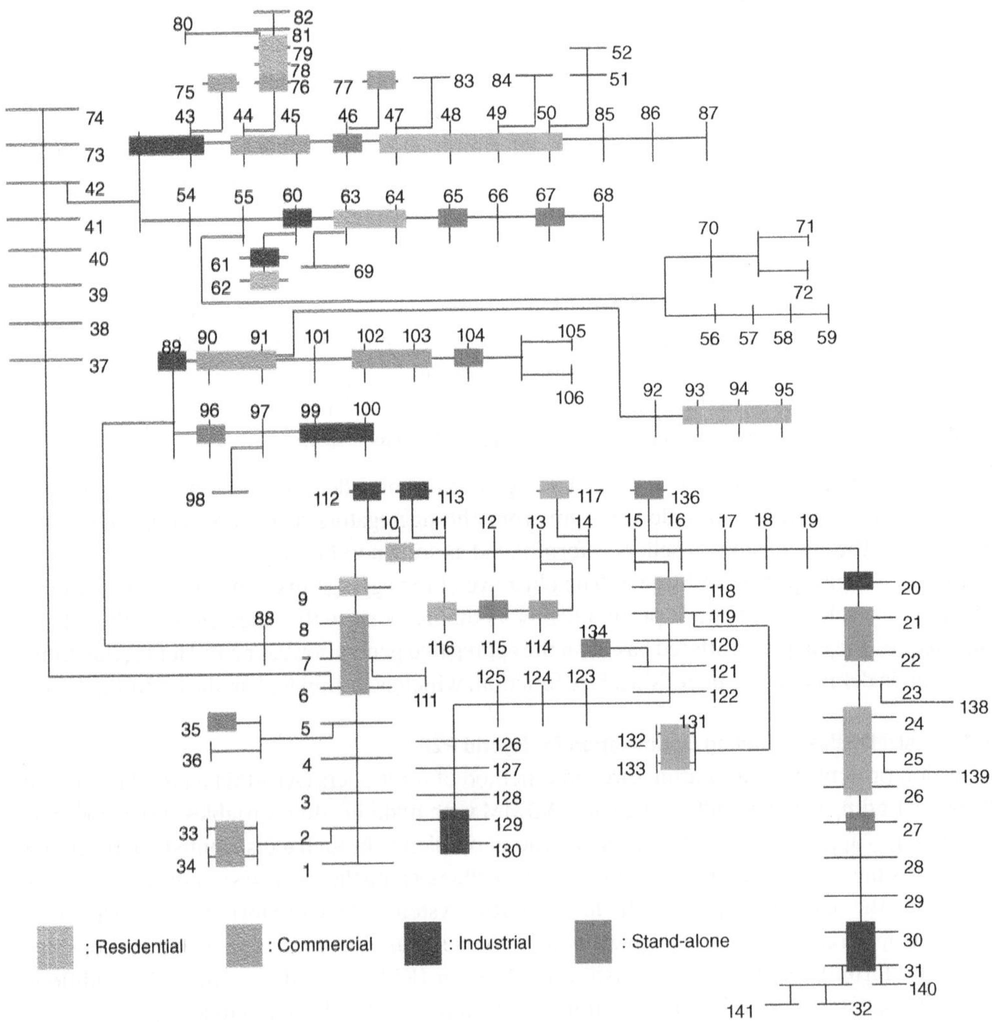

Figure 25.7 Diagram of 141-bus distribution system.

Table 25.1 Total profit comparison.

	Total profit (KRW)					
	Equally distributed		GMM		Proposed method	
No	Agg. of sellers	Agg. of buyers	Agg. of sellers	Agg. of buyers	Agg. of sellers	Agg. of buyers
1	304,257.80	63,193.33	90,212.34	414,392.56	522,058.89	301,707.76
2	423,419.70	54,950.46	480,282.11	287,520.11	861,841.81	48.69
3	948,506.62	63,193.33	74,433.99	244,141.40	448,513.24	414,058.89
4	131,298.98		0.00		57,462.81	
Total	1,807,483.11	181,337.12	644,928.43	946,054.06	1,889,876.76	715,815.34

proposed method yielded the highest total profit. Comparing the direct clustering approach of equally distributed clusters with the GMM clustering method, the total profit of the former was higher than that of the latter. However, the GMM clustering allocates more equally distributed profits among the aggregators of sellers and buyers. Furthermore, an additional optimization method should be incorporated to enhance the profitability of P2P trading. In this case, the ADMM was used to integrate the optimization process of the two layers of the hybrid P2P energy trading. Based on the results, the proposed method enhanced the profitability of the aggregators of buyers and maintained the profitability of the sellers as compared with the equally distributed method.

3) **Comparison between proposed and existing methods**: In P2P energy trading, the number of participants determines the market performance. Hence, a comparison of hybrid P2P energy trading with and without clustering was performed. As shown in Figure 25.8, if ML-based clustering is used, then the trading capacity is higher compared to the hybrid P2P

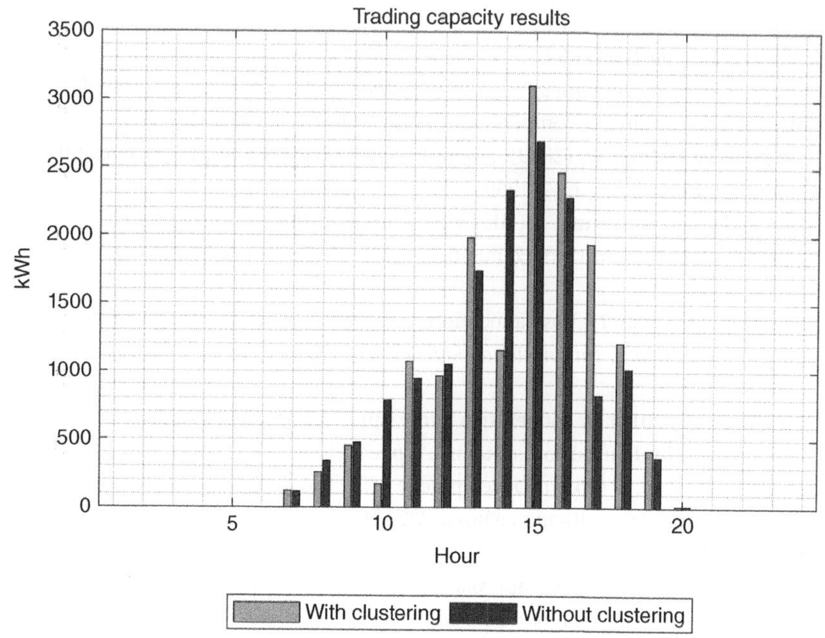

Figure 25.8 Comparison of trading amount.

energy trading without clustering. Because hybrid P2P energy trading concerns the sellers and buyers operated under an aggregated model, members must be selected for each cluster. By optimally allocating sellers and buyers to clusters, trading beneath each aggregator can be more efficient owing to the similarity among the members.

25.5 Conclusion

The increasing number of DERs, particularly in distribution networks with individual electricity customers, has encouraged researchers to focus on facilitating and managing them. However, P2P energy trading can potentially integrate DERs into a grid using a market approach. This chapter presents a theoretical approach relevant to compatible market configuration models and NCA method formulations for P2P energy trading applications. Despite recent efforts toward overcoming conflicts arising from market interactions and network reliability issues, significant discrepancies remain in some areas of P2P energy trading operations, which necessitates extensive research and investigation before they can be fully implemented in the real world.

Therefore, this chapter presents opportunities for applying ML methods to various P2P energy trading problems. To compare the improvements from using ML methods, the GMM clustering method applied to hybrid P2P energy trading was evaluated. The results showed that ML-based clustering improved the performance of hybrid P2P energy trading. Thus, ML is key for customer data analysis and pattern identification, as presented by ML-based clustering applications. In the future, other ML approaches, such as forecasting and decision-making, shall be applied to P2P energy trading to improve its performance.

References

1 Ableitner, L., Tiefenbeck, V., Meeuw, A. et al. (2020). User behavior in a real-world peer-to-peer electricity market. *Applied Energy* 270: 115061.

2 Noorfatima, N., Choi, Y., Onen, A., and Jung, J. (2022). Network cost allocation methods for pay-as-bid peer-to-peer energy trading: a comparison. *Energy Reports* 8: 14442–14463.

3 Zhou, Y., Wu, J., Long, C., and Ming, W. (2020). State-of-the-art analysis and perspectives for peer-to-peer energy trading. *Engineering* 6 (7): 739–753.

4 Noorfatima, N., Choi, Y., Lee, S., and Jung, J. (2022). Development of community-based peer-to-peer energy trading mechanism using Z-bus network cost allocation. *Frontiers in Energy Research* 10: 920885.

5 Sharma, S. and Abhyankar, A.R. (2017). Loss allocation of radial distribution system using Shapley value: a sequential approach. *International Journal of Electrical Power & Energy Systems* 88: 33–41.

6 Du, Y., Wang, Z., Liu, G. et al. (2018). A cooperative game approach for coordinating multi-microgrid operation within distribution systems. *Applied Energy* 222: 383–395.

7 Noorfatima, N., Nam, J., and Jung, J. (2023). Development of hybrid peer-to-peer energy trading for distribution system with multi-aggregators. *2023 IEEE Power & Energy Society General Meeting (PESGM)*, 1–5. IEEE.

8 Bach, D.X. (2023). Uniform purchasing price approach for Vietnam wholesale electricity market: modeling and discussing. *International Journal of Electrical Power & Energy Systems* 148: 109012.

9 Heo, K., Kong, J., Oh, S., and Jung, J. (2021). Development of operator-oriented peer-to-peer energy trading model for integration into the existing distribution system. *International Journal of Electrical Power & Energy Systems* 125: 106488.

10 Korab, R., Kocot, H., and Majchrzak, H. (2021). Fixed transmission charges based on the degree of network utilization. *Energies* 14 (3): 614.

11 Shirmohammadi, D., Gribik, P.R., Law, E.T.K. et al. (1989). Evaluation of transmission network capacity use for wheeling transactions. *IEEE Transactions on Power Systems* 4 (4): 1405–1413.

12 Bialek, J. (1996). Tracing the flow of electricity. *IEE Proceedings-Generation, Transmission and Distribution* 143 (4): 313–320.

13 Hörsch, J., Schäfer, M., Becker, S. et al. (2018). Flow tracing as a tool set for the analysis of networked large-scale renewable electricity systems. *International Journal of Electrical Power & Energy Systems* 96: 390–397.

14 Strbac, G., Kirschen, D., and Ahmed, S. (1998). Allocating transmission system usage on the basis of traceable contributions of generators and loads to flows. *IEEE Transactions on Power Systems* 13 (2): 527–534.

15 Noorfatima, N., Yang, Y., Jung, J., and Kim, J.-S. (2021). Congestion management by allocating network use cost for the small-scale DER aggregator market in South Korea. *Energies* 14 (12): 3524.

16 Galiana, F.D., Conejo, A.J., and Gil, H.A. (2003). Transmission network cost allocation based on equivalent bilateral exchanges. *IEEE Transactions on Power Systems* 18 (4): 1425–1431.

17 Conejo, A.J., Contreras, J., Lima, D.A., and Padilha-Feltrin, A. (2007). Zbus transmission network cost allocation. *IEEE Transactions on Power Systems* 22 (1): 342–349.

18 Sharma, S. and Abhyankar, A.R. (2016). Loss allocation for weakly meshed distribution system using analytical formulation of Shapley value. *IEEE Transactions on Power Systems* 32 (2): 1369–1377.

19 Benedek, M., Fliege, J., and Nguyen, T.-D. (2021). Finding and verifying the nucleolus of cooperative games. *Mathematical Programming* 190 (1-2): 135–170.

20 Gao, X. (2018). Fixed transmission cost allocation based on a nucleolus solution with economic incentive on power market. *IOP Conference Series: Materials Science and Engineering* 452: 032124.

21 Makridakis, S., Spiliotis, E., and Assimakopoulos, V. (2018). Statistical and machine learning forecasting methods: concerns and ways forward. *PLoS One* 13 (3): e0194889.

22 Sahoo, S., Swain, S., Dash, R. et al. (2021). Novel Gaussian flower pollination algorithm with IoT for unit price prediction in peer-to-peer energy trading market. *Energy Reports* 7: 8265–8276.

23 Kong, W., Dong, Z.Y., Jia, Y. et al. (2017). Short-term residential load forecasting based on LSTM recurrent neural network. *IEEE Transactions on Smart Grid* 10 (1): 841–851.

24 Aggarwal, C.C. and Reddy, C.K. (2018). *Data Clustering: Algorithms and Applications*, Chapman & Hall/CRC Data Mining and Knowledge Discovery Series. CRC Press. ISBN: 9781315360416. https://books.google.co.kr/books?id=cH50DwAAQBAJ.

25 Sahbi, H. (2008). A particular Gaussian mixture model for clustering and its application to image retrieval. *Soft Computing* 12 (7): 667–676.

26 Rohlf, F.J. (1982). 12 Single-link clustering algorithms. *Handbook of Statistics* 2: 267–284.

27 Murtagh, F. and Legendre, P. (2014). Ward's hierarchical agglomerative clustering method: which algorithms implement ward's criterion? *Journal of Classification* 31: 274–295.

28 Gabidolla, M. and Carreira-Perpiñán, M.Á. (2022). Optimal interpretable clustering using oblique decision trees. *Proceedings of the 28th ACM SIGKDD Conference on Knowledge Discovery and Data Mining*, 400–410.

29 Wang, J., Li, L., and Zhang, J. (2023). Deep reinforcement learning for energy trading and load scheduling in residential peer-to-peer energy trading market. *International Journal of Electrical Power & Energy Systems* 147: 108885.

30 Naseri, N., Talari, S., Ketter, W., and Collins, J. (2022). Dynamic retail market tariff design for an electricity aggregator using reinforcement learning. *Electric Power Systems Research* 212: 108560.

31 Nguyen, D.H. (2020). Optimal solution analysis and decentralized mechanisms for peer-to-peer energy markets. *IEEE Transactions on Power Systems* 36 (2): 1470–1481.

32 Khodr, H.M., Olsina, F.G., De Oliveira-De Jesus, P.M., and Yusta, J.M. (2008). Maximum savings approach for location and sizing of capacitors in distribution systems. *Electric Power Systems Research* 78 (7): 1192–1203.

33 KEPCO (2023). Korean electricity rate table. https://online.kepco.co.kr/PRM004D00 (accessed 16 October 2024).

34 NREL (2022). PVWatts®Calculator. https://pvwatts.nrel.gov/pvwatts.php (accessed 16 October 2024).

26

Optimal Peer-to-Peer Power Sharing in DC Islanded Microgrids

Rabia Khan and Noel N. Schulz

School of Electrical Engineering & Computer Science, Washington State University, Pullman, WA, USA

26.1 Introduction

The International Energy Agency (IEA) states that 770 million people (12% of the global population) are still suffering from energy poverty [1]. The United Nations (UN) has introduced sustainable development goals (SDG), where SDG-7 aims to provide clean, reliable, cheap, and sustainable energy to everyone by 2030 [2]. Approximately 85% of energy-deprived people are rural inhabitants who depend on kerosene, candles, and wood to fulfill their basic requirements for domestic lighting and heating [3]. The usage of these resources harms public health and the environment. Alternatively, these communities must be electrified to improve their socio-economic status and enhance human development, e.g., as education, health, agriculture, and employment opportunities [4, 5].

Extending the grid to remote rural regions is prohibitive because of (i) high infrastructure costs, (ii) significant transmission line losses, (iii) electricity theft, and (iv) unreliable service with frequent load shedding [6, 7]. Alternatively, there has been an increase in the deployment of stand-alone systems over the past decade due to their ease of installation [8]. However, individual solar home systems lack scalability and require complex financial resources. They are expensive and suboptimal compared to integrated microgrid (MG) systems, which have a better levelized cost of electricity (LCOE) and benefit from usage diversity. Therefore, this research focuses on DCMGs for their superior utility and higher electrification potential.

The integrated DCMG systems can be centralized or distributed, but the latter is more advantageous in scalability, modularity, higher efficiency, and life-cycle cost effectiveness [9–11]. Distributed generation and distributed storage (DGDS) MGs are considered the most efficient compared to (i) centralized generation centralized storage (CGCS), (ii) distributed generation centralized storage (DGCS), and (iii) centralized generation distributed storage (CGDS) systems [12]. Therefore, the research presented in this chapter is focused on DGDS MG architecture.

Most rural areas in Southeast Asia and Africa receive abundant sunlight, with solar irradiance of 5.5 $kWh/m^2/day$ and above [13], making solar a promising DC source compared to other renewable energy resources. Over the past decade, there has been a significant increase in the practical installation of solar PV-based DCMGs in these areas [14–16]. The reasons include (i) rapid decrease in the prices of solar panels and batteries, (ii) extensive market availability of DC loads, (iii) development

Smart Cyber-Physical Power Systems: Fundamental Concepts, Challenges, and Solutions, Volume 1, First Edition.
Edited by Ali Parizad, Hamid Reza Baghaee, and Saifur Rahman.

of power electronics resulting in high-efficiency DC–DC converters [17–19], and (iv) reduction of redundant conversion stages (AC–DC and vice versa) [20–22]. PV-battery-based islanded DC microgrids (IDCMGs) are a viable solution for rural electrification in remote areas, offering additional benefits of high operational efficiency, cost-effectiveness, and availability of DC sources and loads [4]. Therefore, the research presented in this chapter focuses on PV-battery-based IDCMG for cooperative rural electrification.

Nasir et al. proposed a scalable distributed DCMG architecture with power-sharing capability in [23–26]. The peer-to-peer energy-sharing approach is developed in [23]. However, the research presented in the literature focused solely on addressing distribution losses.

In contrast, our research utilizes the DGDS MG architecture with neighborhood-level power-sharing capability to model both distribution and conversion losses. In [24], the optimal planning of DCMGs was performed with constant converter losses. Nevertheless, subsequent research by Kolar et al. [27] and Gelani et al. [28] has demonstrated that the efficiency of a DC–DC converter varies with the output power. As a result, power electronic losses are not static throughout the operation; they depend on the percentage loading and the output power. Thus, a notable contribution of our work involves evaluating and optimizing all distribution and conversion losses, encompassing: (i) constant, (ii) linear, and (iii) quadratic losses.

The optimal power flow (OPF) algorithm is paramount for optimizing power system planning, operation, analysis, and scheduling. Its implementation is challenging due to the non-linear nature of power flow equations, rendering the optimization problem non-convex. Consequently, obtaining a globally optimal solution is intricate, although the problem can be approximated through relaxation to form a convex problem [29]. This relaxed problem can be verified by ensuring that the obtained solutions satisfy the nonlinear inequality constraints. The second-order cone program (SOCP) and semi-definite program (SDP) represent widely used relaxation techniques. In a previous study, the branch flow method (BFM) was employed [30, 31], and Low et al. demonstrated that the SOCP relaxation can be exact. Particularly, the optimal solution can be unique within DC networks, independent of system topologies and operational modes. They successfully solved the OPF for DCMGs. Our research enhances the BFM introduced in [31] to incorporate converter and distribution losses.

Given that many distributed DCMG systems are dominated by power electronics converters, considering only distribution losses while neglecting power electronic conversion losses may lead to inaccurate system-level analysis. Our approach addresses this gap by solving the OPF problem to optimize the power dispatch of distributed energy resources (DERs), thereby minimizing total distribution losses and maximizing power electronics efficiencies.

The key topics covered in this chapter are summarized below:

i) Development of a mathematical framework based on the improved BFM, which offers detailed modeling of distribution and power electronics conversion losses for IDCMGs aimed at rural electrification.
ii) Formulating a multi-objective optimization problem that seeks to maximize conversion efficiencies and simultaneously minimize distribution losses within the system. Converter losses (constant, linear, and quadratic) are incorporated into the constraints to enhance the accuracy of the designed OPF problem.
iii) Adaptation of the OPF problem to suit any DC system for creating a BFM and optimization of objective functions for various interconnection schemes, such as ring and radial, is essential for planning and designing IDCMGs for rural areas.

26.2 Modeling of Islanded DC Microgrid System

The IDCMG considered for rural electrification of a remote rural community is illustrated in Figure 26.1 (building block) and Figure 26.2 (architecture). The distributed architecture comprises multiple PV-battery-based users interconnected with each other. The objective is to establish an optimal power dispatch mechanism to enhance excess power-sharing among different users. For operational efficiency considerations, system losses are categorized as (i) power electronics converter losses and (ii) distribution losses. Additionally, certain other losses are assumed to be constant in this study, e.g., heat losses in solar panels and storage charging or discharging losses [32].

Figure 26.1 Nanogrid model with components.

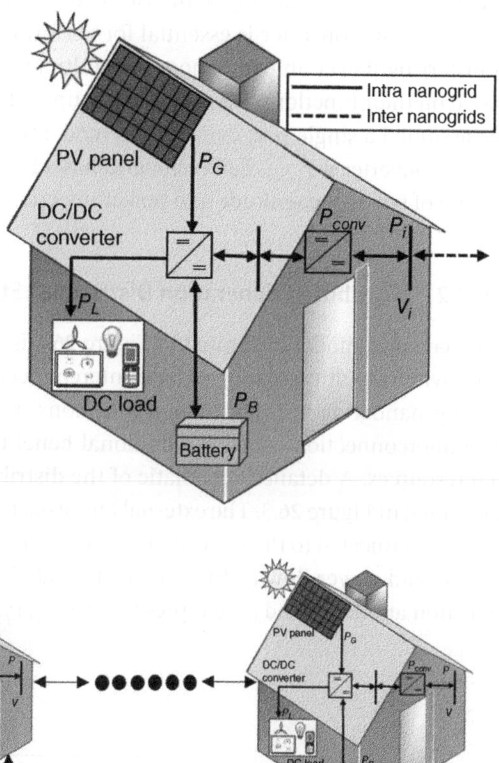

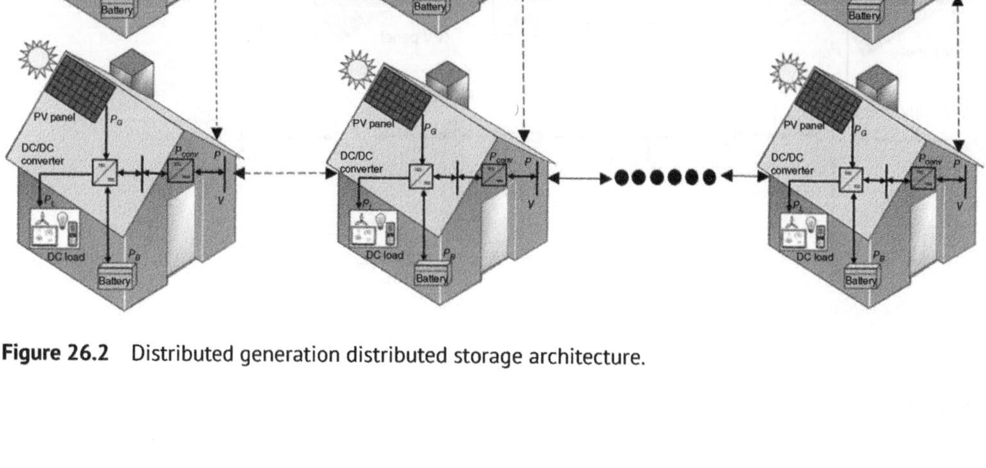

Figure 26.2 Distributed generation distributed storage architecture.

26.2.1 Nanogrid Model

In DGDS MG architecture of a village, several users are present. Each user is assumed to feature an independent rooftop solar panel, battery, and DC loads; this configuration is called a nanogrid. The nanogrids are designed to be self-sufficient and can operate both as prosumers and consumers. They can function independently or be integrated to form a MG [24]. This integration offers the advantage of resource-sharing and scalability. The nanogrid model with its components is shown in Figure 26.1. The power generated by each user is denoted as P_G and serves one of three purposes: (i) meeting internal load demands, (ii) being stored in the internal battery, or (iii) being shared with other nanogrids. The path of power channeling relies on resource availability and load demand. The power stored in the battery is represented as P_B with a corresponding state of charge (SOC). Due to the high market availability of the DC loads, each user is assumed to have DC loads with demand P_L. A DC–DC converter is essential for performing maximum power point tracking (MPPT) and transferring power among nanogrids. Multiple DC–DC converters are integrated within a system based on their functions, e.g., MPPT, step-up and step-down voltage, and power transfer. However, we assumed a single-port converter to model the converter losses explicitly. The power processed by the converter is P_{conv}. Each nanogrid has a DC external bus for inter-nanogrid interactions. The square of voltage magnitude and power injection at buses are represented by v and P, respectively.

26.2.2 Distributed Generation Distributed Storage Architecture

The nanogrid model illustrated in Figure 26.1 is a fundamental block for the DGDS MG architecture, where multiple nanogrids are interconnected, as depicted in Figure 26.2. The dotted lines among nanogrids indicate inter-connections, while the solid lines represent intra-connections. The interconnections offer an additional benefit of peer-to-peer power-sharing through individual resources. A detailed schematic of the distributed architecture, focusing on two nanogrids, is presented in Figure 26.3. The external bus of each nanogrid is referred to as a node. A node, denoted as "i" is connected to the node, denoted as "j" through a resistance r_{ij} with a corresponding current flow I_{ij} and power flow P_{ij} from node i to node j. The square magnitude of voltage and real power injection at nodes i and j are represented as v_i, v_j, P_i, and P_j.

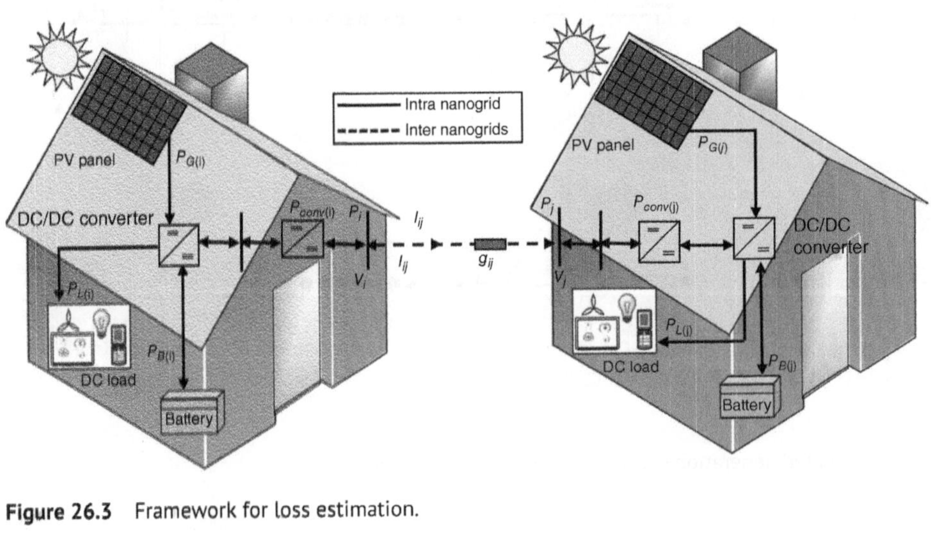

Figure 26.3 Framework for loss estimation.

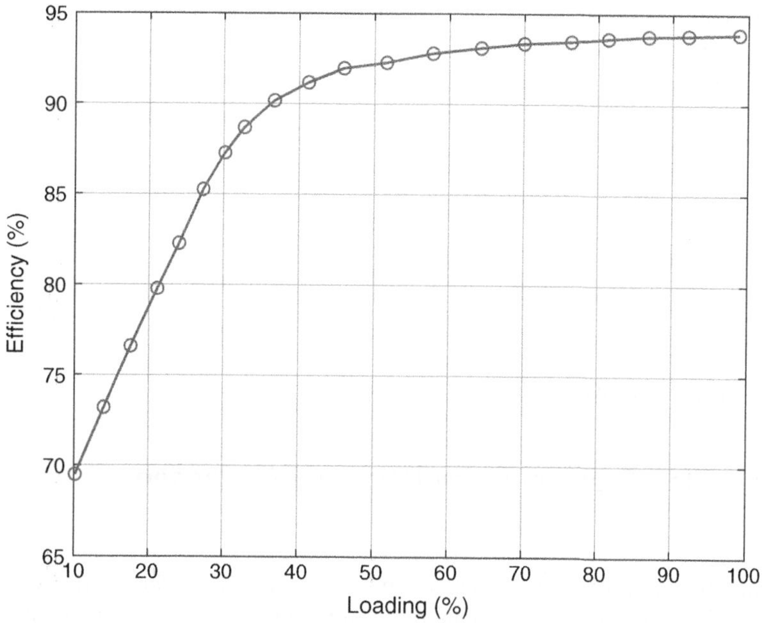

Figure 26.4 Loading vs. efficiency curve of a DC–DC converter.

26.2.3 Distribution Losses

Distribution losses depend on various factors, with voltage level, distance, and power distribution being the most significant [24]. The framework discussed in Section 26.3 employs the branch flow model for DC power systems that accounts for distribution loss calculation. For a DC system with N nodes, distribution losses are calculated using the square magnitude of current flow l_{ij} through distribution line resistance r_{ij}, as expressed in Eq. (26.1):

$$DL = \sum_{j:j\sim i} l_{ij} r_{ij} \quad \forall i \in \mathcal{N} \tag{26.1}$$

26.2.4 Converter Losses

The primary function of the DC–DC converter is to perform MPPT or step-up/down transfer. The power processed by each converter results in losses called power electronic losses. With advancements in power electronics, converters are designed to operate efficiently while minimizing these losses. Nonetheless, these losses are contingent upon the output power, whereby the converter's efficiency is a non-linear function of the ratio of output power to the rated capacity [27, 28, 33] (expressed as % loading), as defined by Eq. (26.2), as illustrated in Figure 26.4. The figure is obtained using grabbit tool in MATLAB, by obtaining a certain DC-DC converters' datasheet and ratings.

$$\eta_c = \sum_{x=0}^{K} k_x \left(\frac{P_0}{P_R} \right)^x \tag{26.2}$$

In Eq. (26.2), η_c denotes the converter's efficiency, P_0 and P_R represent output and rated powers, respectively, and k_x stands for the co-efficient of conversion efficiency. The value of x varies from 0 to K. In our work, we have only considered the low-order terms, assuming $K = 2$, so converter losses are modeled as (i) constant, (ii) linear, and (iii) quadratic. Higher-order terms are disregarded, leading to the simplified form of Eq. (26.2) as shown in Eq. (26.3). In order to map the converter losses and distribution losses on the same time scale and to achieve the combined objective of their minimization, we have assumed that converters are operating in the steady state with constant switching frequency. Therefore, the converters' dynamic characteristics are not considered within this work's scope.

$$\eta_c = k_0 + k_1\left(\frac{P_o}{P_R}\right)^1 + k_2\left(\frac{P_o}{P_R}\right)^2 \tag{26.3}$$

$$P_{conv_i} = \frac{1 - \sum_{x=0}^{K} k_x\left(\frac{P_o}{P_R}\right)^x}{\sum_{x=0}^{K} k_x\left(\frac{P_o}{P_R}\right)^x} * P_o \tag{26.4}$$

The total conversion losses (P_{conv_i}) are given by Eq. (26.4).

26.3 Optimal Power Flow Problem Formulation of DC Islanded Microgrid System

The formulation of BFM and the multi-objective optimization problem for the considered IDCMG system are presented in this section. Next, we elaborate on formulating the multi-objective optimization problem, aiming to minimize distribution losses and maximize conversion efficiencies within the system through optimal power dispatch. Moving forward, we introduce the SOCP relaxation, which involves transforming the non-convex quadratic equality constraints into inequality constraints. The proposed mathematical model is designed to solve the optimization problem on both the system level (distribution lines) and the device level (converters).

26.3.1 Branch Flow Model

The IDCMG system depicted in Figure 26.2 can be represented as a graph denoted as $\mathcal{G}(\mathcal{N}, \mathcal{E})$. This graph consists of $\mathcal{N}$ nodes, where $\mathcal{N} := \{1,2, \ldots, n\}$ and $\mathcal{E}$ edges, representing the connections between nodes. The terms "nodes" and "edges" are interchangeably used here to correspond to "buses" and "branches" within the power system. It can be radial or mesh, assuming $\mathcal{G}$ is connected. Nodes i and j are linked through an edge (i,j) and these nodes are indexed as 1,2, …, n, represented as $(i,j) \in \mathcal{E}$, with the condition that $i \sim j$ and $i < j$, denoted as $i \to j$.

In the DC power system each branch $(i,j) \in \mathcal{E}$ is characterized by an impedance $z_{ij} = r_{ij} + \iota x_{ij}$, which is purely resistive with no reactance, i.e., $x_{ij} = 0$. Consequently, $z_{ij} = r_{ij}$. Therefore, the admittance $y_{ij} = g_{ij}$, where $g_{ij} = 1/r_{ij}$. The apparent power $S_{ij} = P_{ij} + \iota Q_{ij}$ comprises solely real power flow (no reactive component), thus $S_{ij} = P_{ij}$. The current magnitude I_{ij} flows through the line. For each bus $(i) \in \mathcal{N}$, let V_i and P_i denote the magnitude of the voltage and net real power injection, respectively, as illustrated in Figure 26.1. In the context of a DCMG, all variables, i.e., V_i, P_i, I_{ij}, g_{ij}, and P_{ij} are real numbers.

The power flows in the power system are governed by three primary physical laws [30, 31]. These principles establish the branch-flow model for the IDCMG system.

$$\text{Ohm's Law;} \quad I_{ij} = g_{ij}(V_i - V_j) \quad \forall (i,j) \in \mathcal{E} \tag{26.5}$$

$$\text{Current Balance;} \quad I_i = \sum_{j:j\sim i} I_{ij} \quad \forall i \in \mathcal{N} \tag{26.6}$$

$$\text{Power Equation;} \quad P_i = V_i I_i \quad \forall i \in \mathcal{N} \tag{26.7}$$

26.3.2 Converter Efficiency and Distribution Loss Optimization

This chapter aims to reduce the distribution losses and maximize the converter's operational efficiencies within the system, achieving optimal power dispatch at each participating nanogrid and fostering optimal power sharing among nanogrids. An optimization problem (Eq. (26.8)) was previously introduced in the literature [31] for distribution loss minimization.

$$\text{OPF-1; Minimize:} \quad \sum_{j:j\sim i} l_{ij} r_{ij} \qquad \forall i \in \mathcal{N} \tag{26.8}$$

Our study extends the optimization problem by incorporating conversion efficiencies into distribution losses. The formulated multi-objective optimization problem (denoted as OPF-2) is presented as follows:

$$\text{OPF-2; Minimize:} \quad \sum_{j:j\sim i} l_{ij} r_{ij} - \sum_{i\in\mathcal{N}} \eta_i \qquad \forall i \in \mathcal{N} \tag{26.9}$$

Subject to:

$$P_{G_i}(t) - P_{L_i}(t) - P_{B_i}(t) - P_{conv_i}(t) = \sum_{j:i\to j} P_{ij}(t) \forall i \in \mathcal{N} \tag{26.10}$$

$$P_i(t) = \sum_{j:i\to j} P_{ij}(t) \qquad \forall i \in \mathcal{N} \tag{26.11}$$

$$P_{ij}(t) + P_{ji}(t) = r_{ij} * l_{ij}(t) \qquad i \to j \tag{26.12}$$

$$v_i(t) - v_j(t) = r_{ij} * (P_{ij}(t) - P_{ji}(t)) \qquad i \to j \tag{26.13}$$

$$v_i(t) * l_{ij}(t) = (P_{ij}(t))^2 \qquad i \sim j \tag{26.14}$$

$$l_{ij}(t) \leq (I_{ij,rated})^2 \qquad \forall (i,j) \in \mathcal{E} \tag{26.15}$$

$$P_{B_i}(t) - P_{L_i}(t) \leq C_B(t) * [SOC_i(t) - \underline{SOC_i}] \quad \forall i \in \mathcal{N} \tag{26.16}$$

$$P_{L_i}(t) - P_{B_i}(t) \leq C_B(t) * [\overline{SOC_i} - SOC_i(t)] \quad \forall i \in \mathcal{N} \tag{26.17}$$

$$SOC_i(t+1) = SOC_i(t) + 1/C_B(t) * [P_{B_i}(t)] \quad \forall i \in \mathcal{N} \tag{26.18}$$

Voltage limits at each node:

$$\underline{v_i} \leq v_i \leq \overline{v_i} \qquad \forall i \in \mathcal{N} \tag{26.19}$$

Power generation limits at each node:

$$\underline{P_{G_i}} \leq P_{G_i} \leq \overline{P_{G_i}} \qquad \forall i \in \mathcal{N} \tag{26.20}$$

SOC limits at each battery:

$$\underline{SOC_i} \leq SOC_i \leq \overline{SOC_i} \qquad \forall i \in \mathcal{N} \tag{26.21}$$

where P_{G_i} signifies the generated power, P_{Bi} represents the battery power, P_{Li} denotes the load power, $P_{convloss_i}$ encompass the total converter losses, and P_i stands for the active power flow injected at each node i. The terms $\underline{P_{G_i}}$ and $\underline{v_i}$ indicate the lower bounds of generated power and squared voltage magnitudes, respectively, while the $\overline{P_{G_i}}$ and $\overline{v_i}$ represent the corresponding upper bounds. The terms $\underline{SOC_i}$ and $\overline{SOC_i}$ denote lower and upper SOC bounds.

Algorithm 1: Optimal Power Dispatch Strategy

```
Data: Data Collection from all nodes: P_Gi, P_Bi, P_Li, SOC_i
∀i = 1, 2, 3, i, j,...N
Result: Optimal Power Dispatch at each node
initialization;
if P_Gi > P_Li & P_Gj < P_Lj then
    if P_Bi > P_Li & SOC_i(min) < SOC_i then
        if P_Bj < P_Lj & SOC_j(min) < SOC_j then
            P_ij > 0
        end
    end
else
    if P_Gi > P_Li & P_Gj < P_Lj then
        if P_Bi > P_Li & SOC_i(min) < SOC_i then
            if P_Bj < P_Lj & SOC_j(min) < SOC_j then
                P_ij > 0
            end
        end
    end
end
```

Figure 26.5 Proposed algorithm for optimal power dispatch.

The branch flow equations (26.12)–(26.14) involve v_i representing the voltage magnitude, and l_{ij}, signifying the squared current magnitudes, $v_i = |V_i|^2$ and $l_{ij} = |I_{ij}|^2$. Equation (26.12) corresponds to the power balance equation, while Eq. (26.14) captures the relationship between current flow, power flow, and voltage. Equations (26.12) and (26.13) are linear, but Eq. (26.14) introduces non-linearity due to its dependence on P, l, and v.

The optimal values for power dispatch at each bus, obtained by solving SOCP, are not necessarily the original problem's actual value. The feasible domain resulting from relaxing the constraints is larger than that of the actual problem. Accordingly, the optimal values are checked using the relaxed constraints while satisfying the nonlinear equality constraints. The feasibility and exactness of the conic relaxation are verified.

26.3.3 Proposed Algorithm

The pseudo-code for the algorithm for the power flow among nanogrids is represented in Figure 26.5. The power dispatch is determined considering P_{Gi}, P_{Bi}, P_{Li}, and SOC_i. The algorithm developed for optimal power dispatch strategy is independent of the SOC imbalance.

26.4 Results and Discussion

The algorithm's effectiveness is validated for static and dynamic load scenarios by comparing the results of the OPF solved using the Newton-Raphson Power Flow method modified for DC systems. The nonlinear OPF problems, namely, OPF-1 and OPF-2, are also solved. The relaxed SOCP problem is formulated and solved in MATLAB using the "fmincon" function from its optimization toolbox, ensuring the results' global optimality. The centralized optimization is performed for the multi-objective system, where state and control data from each participating user are received, and optimized data values are returned.

26.4.1 Test System

Due to the lack of a benchmark test system for IDCMGs, AC test systems are adapted with certain assumptions detailed below. The test system data is obtained from MATPOWER, a MATLAB toolbox. These test systems are modified to simulate DCMG conditions, involving adjustments to line reactance, setting reactive power flows to zero, and reducing line resistance values by 10% [30, 31]. Figure 26.2 illustrates the system diagram for the DGDS MG architecture with "n" number of users. The upper and lower bounds for voltage magnitudes are 1.05 per unit (p.u.) and 0.95 p.u., respectively.

26.4.2 Case Study

The case study for implementing our proposed optimization algorithm revolves around multiple independent nanogrids and a communal load. The fourteenth home is assumed as the communal load in the modified 14-user MG system. Figure 26.2 showcases the orientation of users utilized for implementing our proposed optimization algorithm. The architecture of the considered DC system, modified according to the IEEE 14-bus system, is displayed in Figure 26.6.

26.4.3 Static Loads

The optimization problem is solved to analyze the steady-state performance of the system. Results for (i) distribution loss optimization (OPF-1) and (ii) optimization for both distribution loss and conversion efficiency (OPF-2) in the modified 14-node system are demonstrated in Tables 26.1

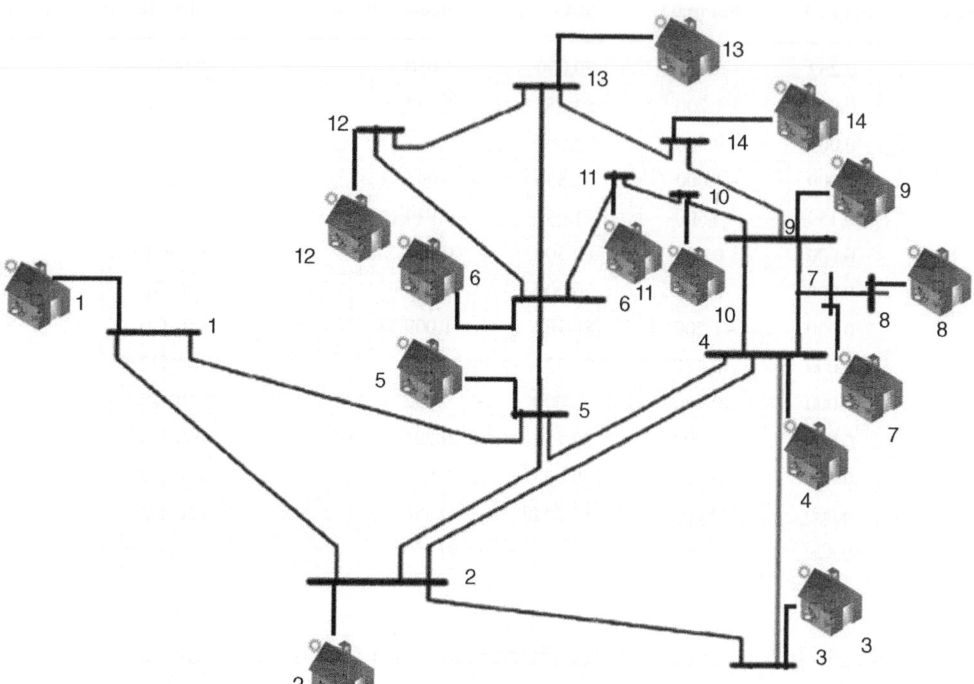

Figure 26.6 Modified 14 bus test system architecture as DC microgrid for rural areas.

Table 26.1 OPF-1 for modified 14 bus DC microgrid system.

Bus No.	Pi (p.u.)	PBi (p.u.)	SOCi (%)	Nodal voltage (p.u.)	Converter efficiency (%)
1	0.000	0.350	82.914	1.050	Non-Op
2	0.000	−1.200	70.000	1.050	Non-Op
3	0.000	0.500	84.165	1.050	Non-Op
4	0.000	−1.000	71.663	1.050	Non-Op
5	0.000	0.500	84.165	1.050	Non-Op
6	0.002	−1.002	71.647	1.050	77.691
7	0.000	0.300	82.497	1.050	Non-Op
8	0.000	−1.200	70.000	1.050	Non-Op
9	0.238	0.246	82.047	1.050	79.932
10	0.015	0.384	83.203	1.050	77.809
11	0.000	−1.200	70.000	1.050	Non-Op
12	0.002	0.496	84.137	1.050	77.692
13	0.183	−0.095	79.205	1.050	79.406
14	−0.429			1.019	81.752

Table 26.2 OPF-2 for modified 14 bus DC microgrid system.

Bus No.	Pi (p.u.)	PBi (p.u.)	SOCi (%)	Nodal voltage (p.u.)	Converter efficiency (%)
1	0.252	−0.392	70.000	1.050	80.063
2	0.000	−1.200	83.557	1.047	Non-Op
3	0.069	0.500	71.667	1.047	78.319
4	0.000	−1.000	82.836	1.043	Non-Op
5	0.150	0.500	71.667	1.044	79.089
6	0.000	0.400	82.500	0.995	Non-Op
7	0.000	0.300	70.000	1.009	Non-Op
8	0.000	−1.200	84.167	1.009	Non-Op
9	0.000	0.437	83.333	0.991	Non-Op
10	0.000	0.400	70.000	0.992	Non-Op
11	0.000	−1.200	84.158	0.993	Non-Op
12	0.000	0.325	80.833	0.992	Non-Op
13	0.000	0.100	89.6625	0.986	Non-Op
14	−0.429			0.956	81.752

and 26.2, respectively. The fourteenth user, assumed to be a community load without PV and battery, has its load demand met through power sharing from neighboring nanogrids. In Table 26.1, users 6, 9, 10, and 13 contribute power to the 14th user. However, in Table 26.2 for OPF-2, only users 1, 3, and 5 contribute to power-sharing with user 14. This suggests that OPF-2 concentrates

higher power scheduling at fewer users than OPF-1, resulting in fewer operating converters at any given time. The SOC at most non-power scheduling users is higher for OPF-2, indicating that batteries are generally at a higher SOC, consuming the available power instead of losses. Moreover, converters in OPF-2 operate more efficiently at each scheduling bus than those in OPF-1.

26.4.4 Dynamic Loads

Using time-based varying load and PV data is more practical and realistic. The optimization utilizes time-based data acquired from the National Renewable Energy Laboratories (NREL) [34]. Figure 26.7 demonstrates the data for PV and load multipliers, assuming an increase in demand during late afternoon hours, while the load requirement decreases during nighttime and early morning. Load and PV profiles depend highly on the selected site and its residents and can be adjusted accordingly.

26.4.4.1 Scheduled Power

The optimization problem explained in this chapter aims to minimize distribution losses, maximize the converter's efficiencies, and optimize power scheduling at each nanogrid. The power scheduling results for the modified 14-bus system are shown in Figures 26.8 and 26.9. During time instants $t = 1$ to $t = 5$ and $t = 21$ to $t = 24$, no PV generation occurs, and batteries meet the load demand. For OPF-1, two buses contribute to meeting the load demand at the 14th bus, whereas for OPF-2, only one bus is involved, albeit with a higher power value. Throughout the remaining time, OPF-2 schedules more power for fewer buses compared to OPF-1, which schedules power for more buses but at a lower magnitude as shown in Tables 26.3 and 26.4 for OPF-1 and OPF-2 respectively. Notably, the power scheduled for the community load remains the same due to the assumption that the community load lacks generation and storage in this study.

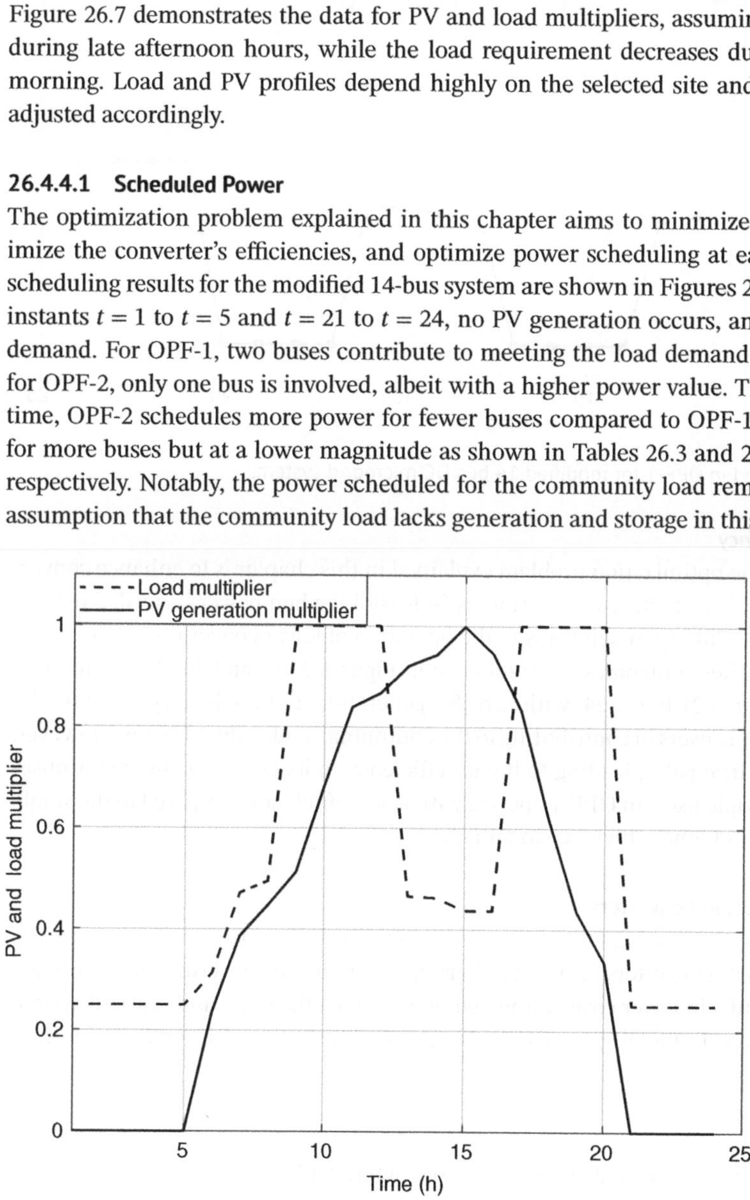

Figure 26.7 PV generation and load multipliers.

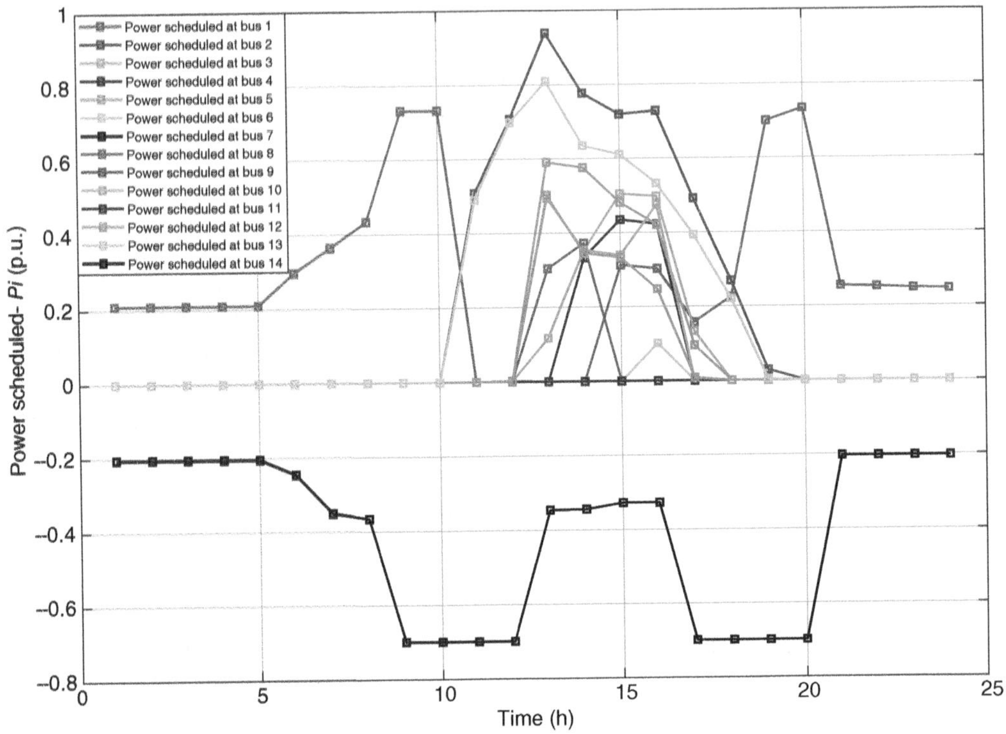

Figure 26.8 Power scheduled in OPF-1 for modified 14 bus DC microgrid system.

26.4.4.2 Converter Efficiency

The primary objective of the optimization problem explained in this chapter is to enhance conversion efficiency such that only a single user contributes (if feasible) when power is needed instead of involving multiple users. This approach leads to the operation of fewer converters, each operating at a higher efficiency. These outcomes are evident from Figures 26.10 and 26.11. During time instants $t = 1$ to $t = 5$ and $t = 21$ to $t = 24$, with zero PV generation and the battery fulfilling the load, OPF-1 involves multiple users in contributing to the communal load at the 14th bus. However, in OPF-2, only one user participates, leading to higher efficiency in its DC–DC converter. Consequently, converters of multiple users in OPF-1 operate with lower efficiency compared to the single user's converter operating at higher efficiency in OPF-2.

26.4.4.3 Number of Operating Converters

The optimization problem addressed in OPF-2 ensures power scheduling at users to minimize the necessary number of DC/DC converters for power-sharing. The reduced converter count, coupled with higher efficiency, results in lower costs, improved operational efficiency, and reduced system losses. The comparison of results for OPF-1 and OPF-2 in the modified 14-bus DC system is depicted in Figure 26.12.

26.4.4.4 Loss Evaluation

The objective of OPF-1 is to minimize distribution losses, while OPF-2 aims to minimize distribution losses and maximize conversion efficiencies. The total losses encompass distribution and conversion losses, which are calculated and compared for both OPF-1 and OPF-2.

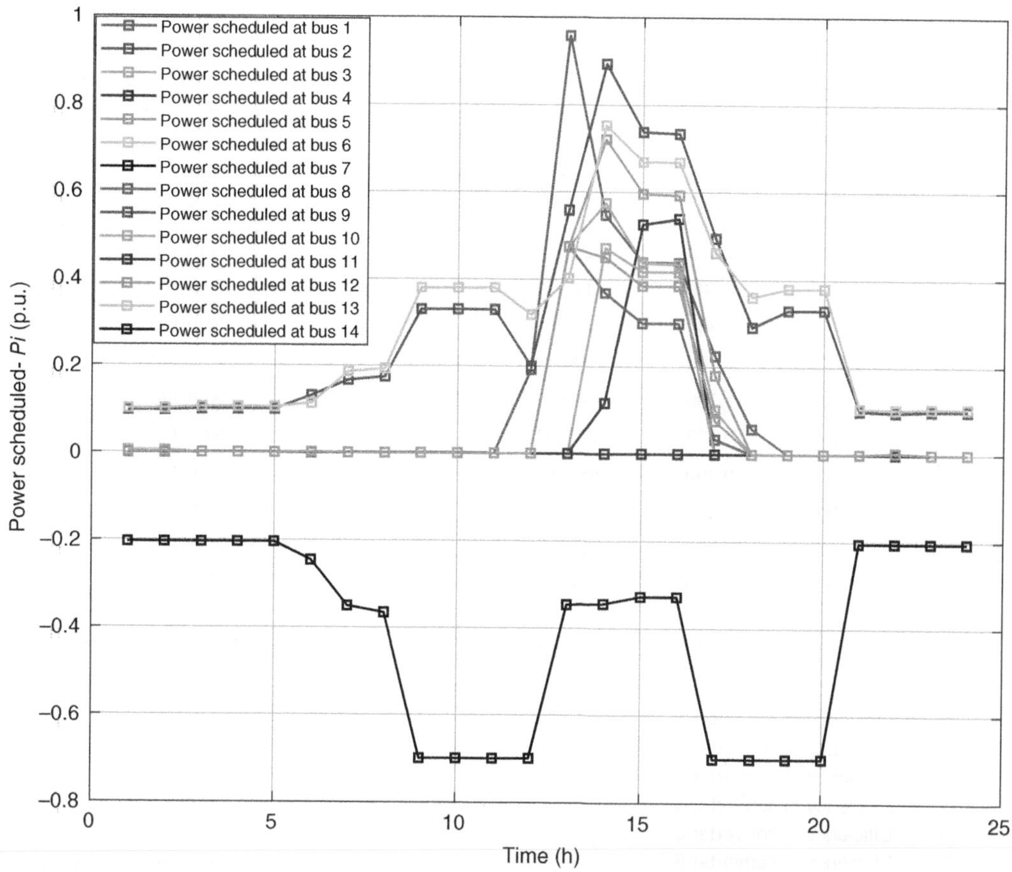

Figure 26.9 Power scheduled in OPF-2 for modified 14 bus DC microgrid system.

Table 26.3 OPF-1 with unbalanced SOCs.

Bus No.	Pi (p.u.)	PBi (p.u.)	SOCi (%)	Nodal voltage (p.u.)	Converter efficiency (%)
1	0.000	0.350	82.914	1.050	Non-Op
2	0.000	−1.200	50.000	1.050	Non-Op
3	0.000	0.500	64.165	1.050	Non-Op
4	0.000	−1.000	61.663	1.050	Non-Op
5	0.000	0.500	64.165	1.050	Non-Op
6	0.003	−1.003	71.644	1.050	77.700
7	0.000	0.300	62.498	1.050	Non-Op
8	0.000	−1.200	50.000	1.050	Non-Op
9	0.239	0.245	82.042	1.050	80.460
10	0.014	0.385	73.212	1.050	77.830
11	0.000	−1.200	50.000	1.050	Non-Op
12	0.003	0.496	84.136	1.050	77.700
13	0.183	−0.095	69.204	1.050	79.800
14	−0.429			1.019	82.750

Table 26.4 OPF-2 with unbalanced SOCs.

Bus No.	Pi (p.u.)	PBi (p.u.)	SOCi (%)	Nodal voltage (p.u.)	Converter efficiency (%)
1	0.252	−0.392	70.000	1.050	80.063
2	0.000	−1.200	83.557	1.047	Non-Op
3	0.069	0.500	71.667	1.047	78.319
4	0.000	−1.000	82.836	1.043	Non-Op
5	0.150	0.500	71.667	1.044	79.089
6	0.000	0.400	82.500	0.995	Non-Op
7	0.000	0.300	70.000	1.009	Non-Op
8	0.000	−1.200	84.167	1.009	Non-Op
9	0.000	0.437	83.333	0.991	Non-Op
10	0.000	0.400	70.000	0.992	Non-Op
11	0.000	−1.200	84.158	0.993	Non-Op
12	0.000	0.325	80.833	0.992	Non-Op
13	0.000	0.100	89.6625	0.986	Non-Op
14	−0.429			0.956	81.752

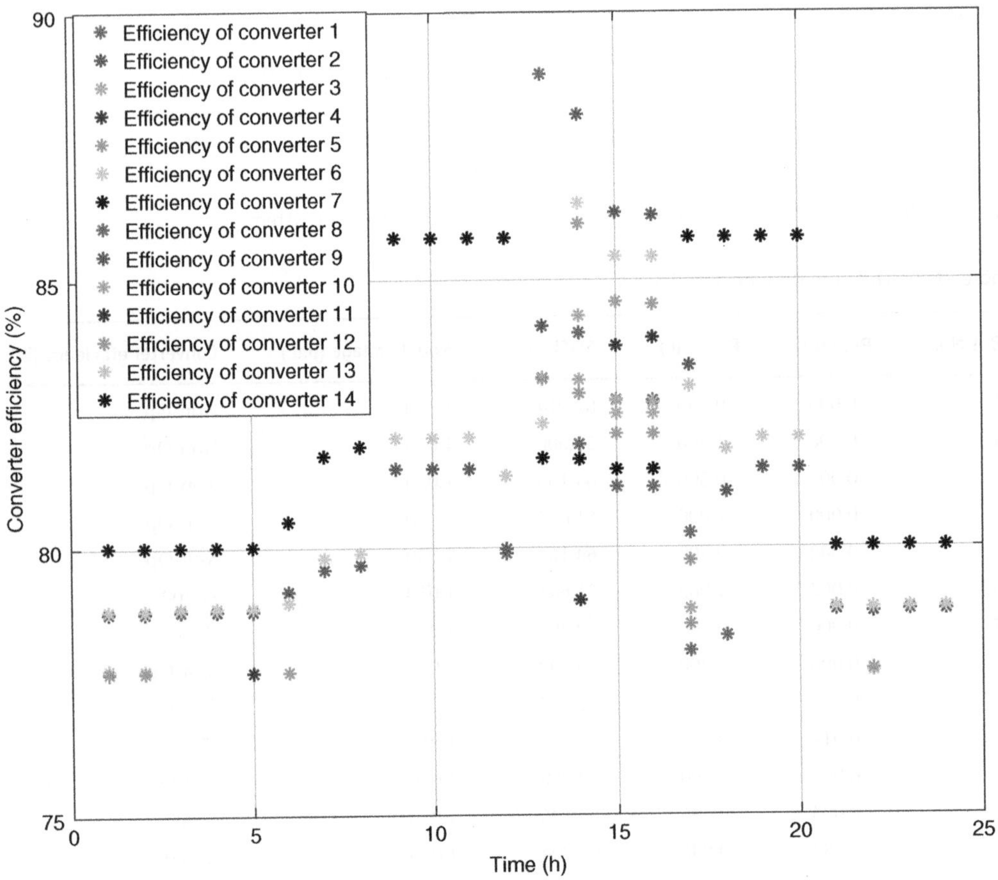

Figure 26.10 Converter efficiency in OPF-1 for 14 bus system.

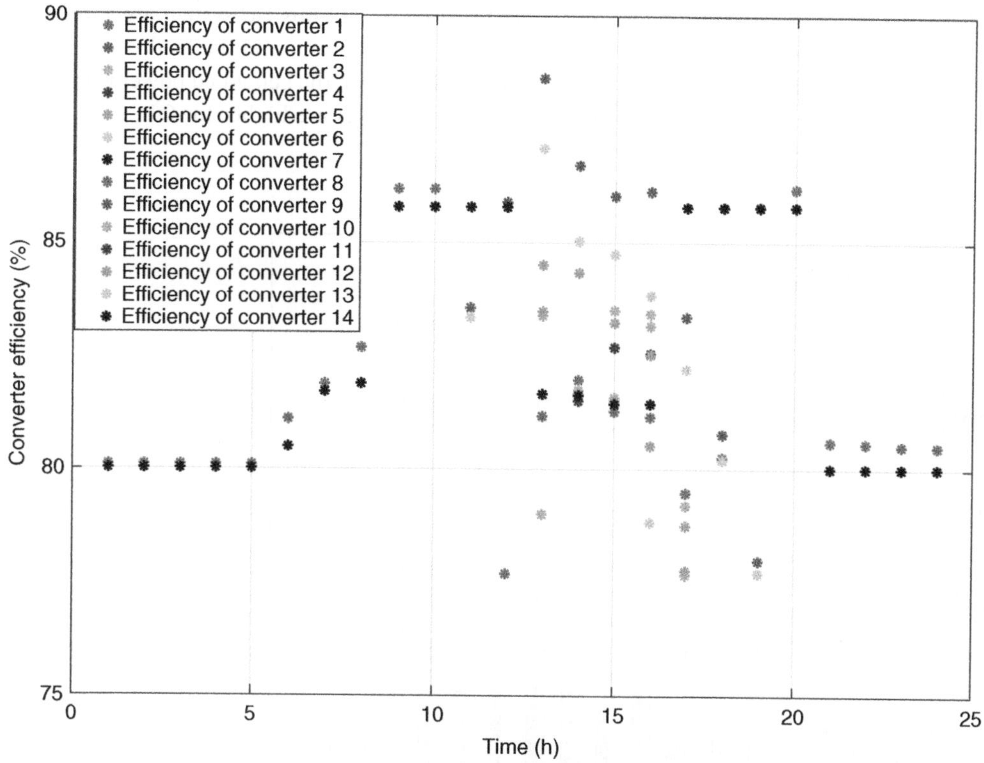

Figure 26.11 Converter efficiency in OPF-2 for 14 bus system.

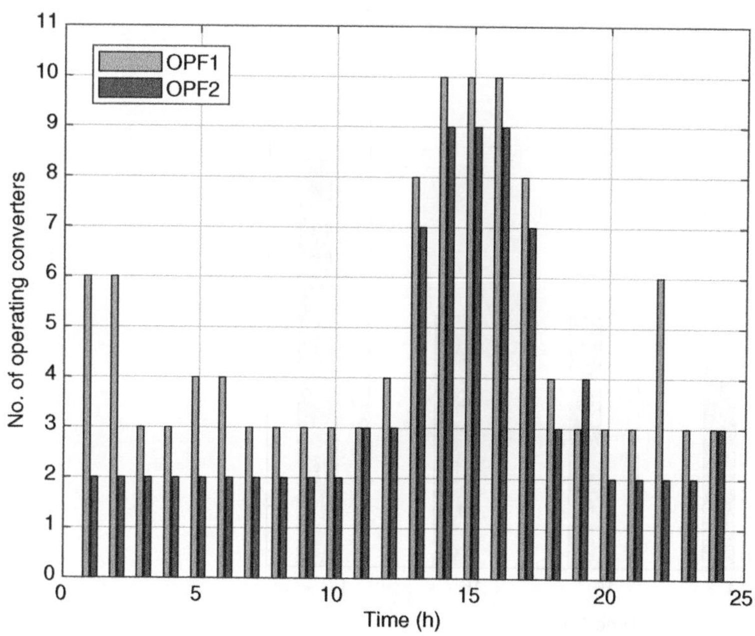

Figure 26.12 Comparison of operating converters in 14 bus system.

Distribution Losses The distribution loss results for the modified 14-bus DC system are shown in Figure 26.13. These losses are minimal when PV generation is absent, and the load is less. During daytime hours with high PV generation and load consumption, the generated power primarily charges batteries and fulfills the load demand. Distribution losses are higher for OPF-1 than for OPF-2.

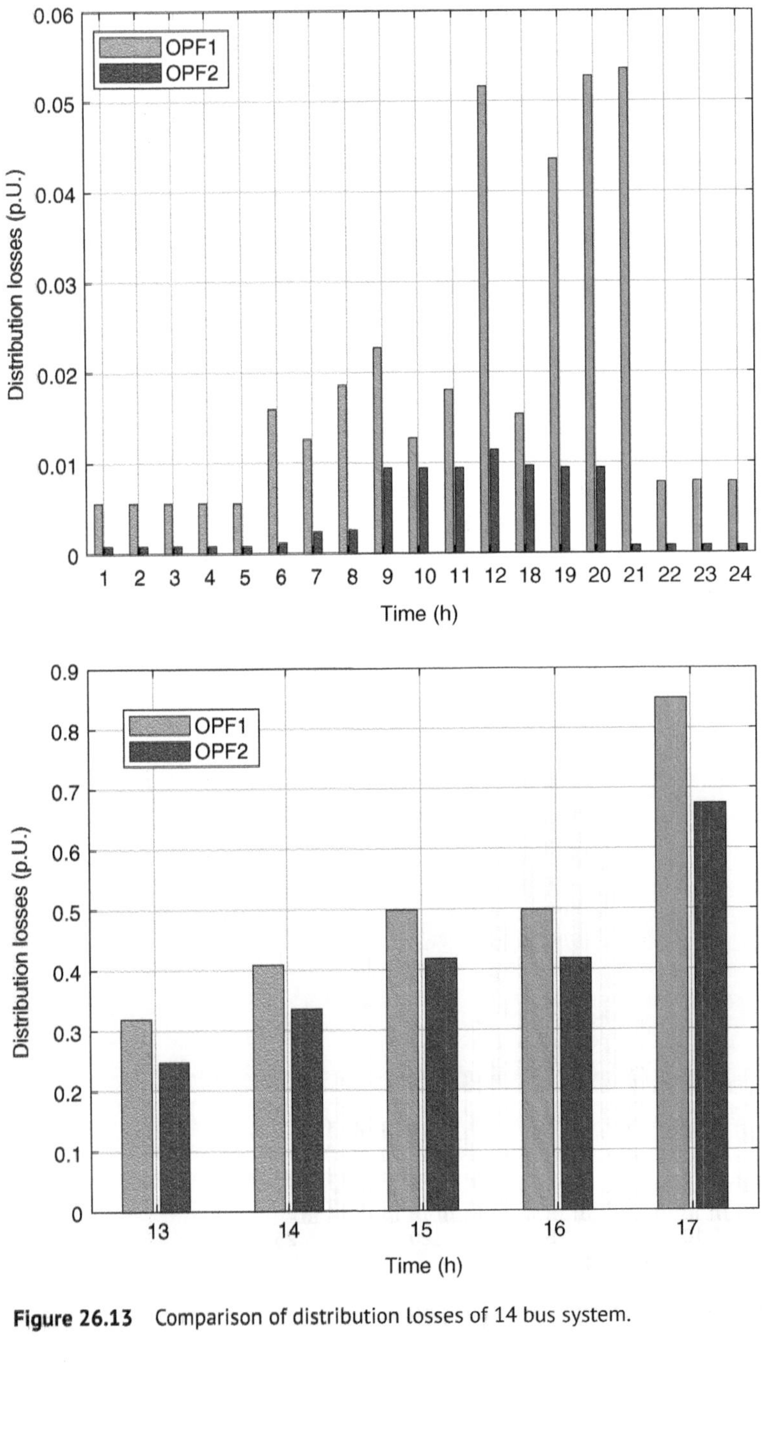

Figure 26.13 Comparison of distribution losses of 14 bus system.

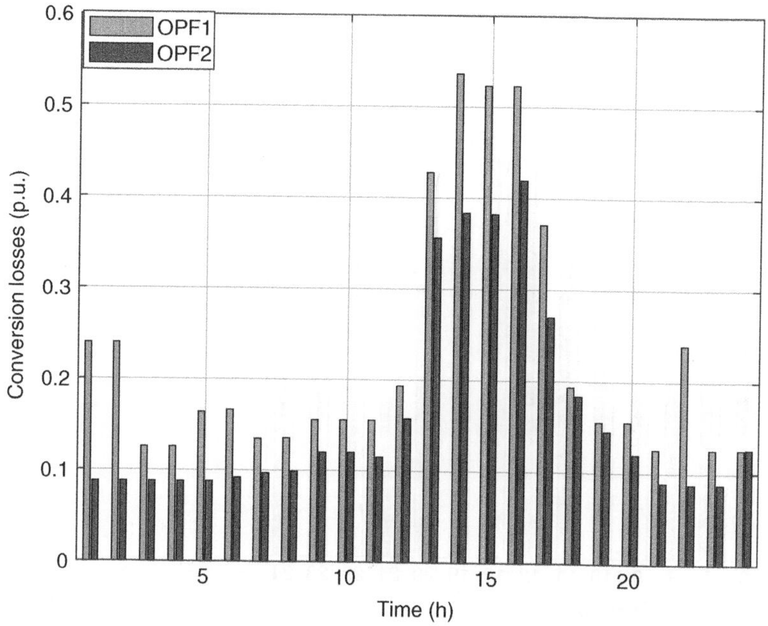

Figure 26.14 Comparison of conversion losses of 14 bus system.

Conversion Losses The results for conversion losses shown in Figure 26.14 are less when we don't have PV generation and the load is less. When both PV generation and load consumption are high during the daytime, the generation is mostly used to charge batteries besides meeting the load demand. The conversion losses are higher for OPF-1 than OPF-2.

Total Losses Figure 26.15 showcases the results for total losses in the modified 14-bus DC system. Losses are lower during nighttime and early morning but higher during daytime. Overall, OPF-1 results in higher total losses compared to OPF-2 throughout the 24-hour period, indicating that OPF-2 enhances efficiency and reduces total system losses.

26.5 Conclusion and Future Work

This chapter discusses the optimal power dispatch strategy for an islanded community consisting of multiple prosumers to minimize system losses and enhance efficiency. A mathematical framework is developed using a modified branch flow model using distribution and conversion losses. The non-linear optimization problem is relaxed using convex relaxation through second-order conic programming, and the relaxed solutions are determined to be feasible and exact. The optimization algorithm is implemented in the community of prosumers, capable of sharing energy with neighbors using the PV and battery as DERs. The proposed strategy is implemented for case studies of steady-state and time-varying (24-hour time period). The results indicate that system efficiency is improved by 6%, and the total losses are reduced by 4% compared to when only distribution losses are considered. The results demonstrate that incorporating converter losses in addition to distribution losses is essential for the system's optimal dispatch operation due to their high contribution to

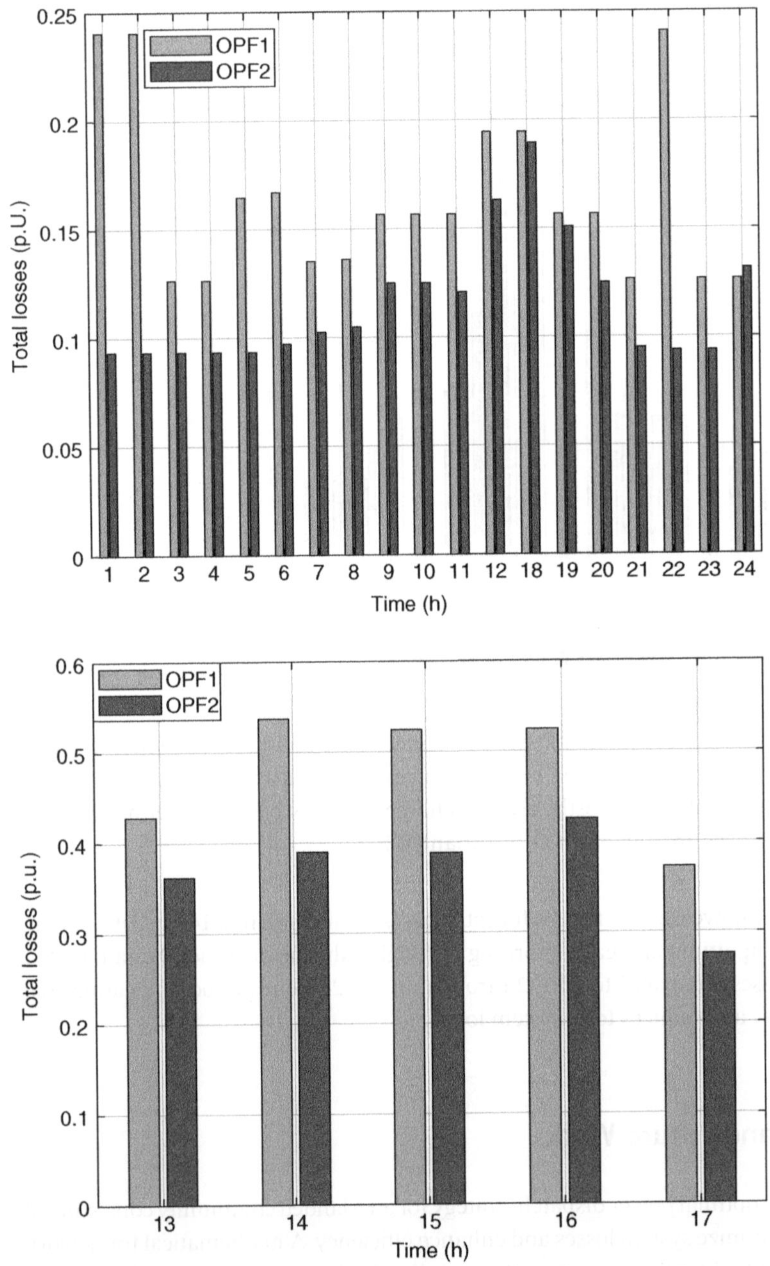

Figure 26.15 Comparison of total losses of 14 bus system.

system losses. In our study, the converters operate at higher efficiency, utilizing batteries at a high SOC to store energy or share it with other users.

Future work will extend the same optimization algorithm to grid-connected DC MGs and clusters of nanogrids. Although DC MGs are relatively small for remote rural areas, the potential integration of distributed renewable energy sources may necessitate the establishment of more DC MGs.

Acknowledgment

This material is based upon work supported by the U.S. Department of Energy under Award Number DE-IA0000025. The views and opinions of the authors expressed do not state or reflect those of the United States Government or any agency.

Bibliography

1 Küfeoğlu, S. (2022). SDG-7 affordable and clean energy. In: *Emerging Technologies: Value Creation for Sustainable Development*, 305–330. Springer.

2 United Nations (2015). *Transforming Our World: The 2030 Agenda for Sustainable Development*. New York: United Nations, Department of Economic and Social Affairs.

3 Hamza, M., Shehroz, M., Fazal, S. et al. (2017). Design and analysis of solar PV based low-power low-voltage DC microgrid architectures for rural electrification. *2017 IEEE Power & Energy Society General Meeting*, 1–5. IEEE.

4 Khan, R. and Schulz, N.N. (2020). Network loss analysis of low-voltage low-power DC microgrids for rural electrification. *2020 IEEE/PES Transmission and Distribution Conference and Exposition (T&D)*, 1–5. IEEE. doi: https://doi.org/10.1109/TD39804.2020.9299657.

5 Madduri, P.A., Poon, J., Rosa, J. et al. (2016). Scalable DC microgrids for rural electrification in emerging regions. *IEEE Journal of Emerging and Selected Topics in Power Electronics* 4 (4): 1195–1205.

6 Moksnes, N., Korkovelos, A., Mentis, D., and Howells, M. (2020). Corrigendum: Electrification pathways for Kenya–linking spatial electrification analysis and medium to long term energy planning (2017 Environ. Res. Lett. 12 095008). *Environmental Research Letters* 15 (12): 129501.

7 Moner-Girona, M., Bódis, K., Korgo, B. et al. (2017). Mapping the least-cost option for rural electrification in Burkina Faso. *European Commission Joint Research Centre*.

8 Khan, J. and Arsalan, M.H. (2016). Solar power technologies for sustainable electricity generation–a review. *Renewable and Sustainable Energy Reviews* 55: 414–425.

9 Inam, W., Strawser, D., Afridi, K.K. et al. (2015). Architecture and system analysis of microgrids with peer-to-peer electricity sharing to create a marketplace which enables energy access. *2015 9th International Conference on Power Electronics and ECCE Asia (ICPE-ECCE Asia)*, 464–469. IEEE.

10 Khan, R. (2024). Peer-to-peer power sharing in DC microgrids for rural electrification. PhD thesis. Washington State University.

11 Khan, R. and Schulz, N.N. (2023). Optimal peer-to-peer power dispatch in islanded DC clustered nanogrids for rural electrification. *2023 IEEE Power & Energy Society General Meeting (PESGM)*, 1–5. IEEE.

12 Khan, R., Schulz, N.N., and Nasir, M. (2019). Distribution loss analysis of DC microgrids for rural electrification. *2019 IEEE Global Humanitarian Technology Conference (GHTC)*, 1–8. IEEE.

13 Khan, R., Khan, A., and Zahra, A. (2019). Cost optimization of hybrid microgrid for rural electrification along western alignment of China-Pakistan economic corridor (CPEC) in Pakistan. *2019 IEEE Global Humanitarian Technology Conference (GHTC)*, 1–6. IEEE.

14 Balls, J.N. and Fischer, H.W. (2019). Electricity-centered clientelism and the contradictions of private solar microgrids in India. *Annals of the American Association of Geographers* 109 (2): 465–475.

15 Gelani, H.E., Dastgeer, F., Nasir, M. et al. (2021). AC vs. DC distribution efficiency: are we on the right path? *Energies* 14 (13): 4039.

16 Palit, D. (2013). Solar energy programs for rural electrification: experiences and lessons from South Asia. *Energy for Sustainable Development* 17 (3): 270–279.

17 Bardouille, P., Avato, P., Levin, J. et al. (2012). *From Gap to Opportunity: Business Models for Scaling Up Energy Access*. International Finance Corporation.

18 Ubilla, K., Jiménez-Estévez, G.A., Hernádez, R. et al. (2014). Smart microgrids as a solution for rural electrification: ensuring long-term sustainability through cadastre and business models. *IEEE Transactions on Sustainable Energy* 5 (4): 1310–1318.

19 Williams, N.J., Jaramillo, P., Taneja, J., and Ustun, T.S. (2015). Enabling private sector investment in microgrid-based rural electrification in developing countries: a review. *Renewable and Sustainable Energy Reviews* 52: 1268–1281.

20 Liu, J., Cui, B., Molzahn, D.K. et al. (2021). Optimal power flow in DC networks with robust feasibility and stability guarantees. *IEEE Transactions on Control of Network Systems* 9 (2): 904–916.

21 Madduri, P.A., Rosa, J., Sanders, S.R. et al. (2013). Design and verification of smart and scalable DC microgrids for emerging regions. *2013 IEEE Energy Conversion Congress and Exposition*, 73–79. IEEE.

22 Zhang, L., Wu, T., Xing, Y. et al. (2011). Power control of DC microgrid using DC bus signaling. *2011 26th Annual IEEE Applied Power Electronics Conference and Exposition (APEC)*, 1926–1932. IEEE.

23 Iqbal, S., Mehran, K., and Nasir, M. (2021). A novel approach of peer to peer energy sharing in DC microgrid with optimal distribution losses. *2021 IEEE PES Innovative Smart Grid Technologies Europe (ISGT Europe)*, 1–5. https://doi.org/10.1109/ISGTEurope52324.2021.9639947.

24 Nasir, M., Khan, H.A., Hussain, A. et al. (2017). Solar PV-based scalable DC microgrid for rural electrification in developing regions. *IEEE Transactions on Sustainable Energy* 9 (1): 390–399.

25 Nasir, M., Jin, Z., Khan, H.A. et al. (2018). A decentralized control architecture applied to DC nanogrid clusters for rural electrification in developing regions. *IEEE Transactions on Power Electronics* 34 (2): 1773–1785.

26 Nasir, M., Khan, H.A., Niazi, K.A.K. et al. (2019). Dual-loop control strategy applied to PV/battery-based islanded DC microgrids for swarm electrification of developing regions. *The Journal of Engineering* 2019 (18): 5298–5302.

27 Kolar, J.W., Krismer, F., Lobsiger, Y. et al. (2012). Extreme efficiency power electronics. *2012 7th International Conference on Integrated Power Electronics Systems (CIPS)*, 1–22. IEEE.

28 Gelani, H.E., Nasir, M., Dastgeer, F., and Hussain, H. (2017). Efficiency comparison of alternating current (AC) and direct current (DC) distribution system at residential level with load characterization and daily load variation. *Proceedings of the Pakistan Academy of Sciences: A. Physical and Computational Sciences* 54 (2): 111–118.

29 Farivar, M. and Low, S.H. (2013). Branch flow model: relaxations and convexification: Part I. *IEEE Transactions on Power Systems* 28 (3): 2554–2564. https://doi.org/10.1109/TPWRS.2013.2255317.

30 Gan, L. and Low, S.H. (2014). Optimal power flow in direct current networks. *IEEE Transactions on Power Systems* 29: 2892–2904.

31 Li, J., Liu, F., Wang, Z. et al. (2018). Optimal power flow in stand-alone DC microgrids. *IEEE Transactions on Power Systems* 33 (5): 5496–5506.

32 Maghami, M.R., Hizam, H., Gomes, C. et al. (2016). Power loss due to soiling on solar panel: a review. *Renewable and Sustainable Energy Reviews* 59: 1307–1316.

33 Nasir, M., Iqbal, S., Khan, H.A. et al. (2020). Sustainable rural electrification through solar PV DC microgrids—an architecture-based assessment. *Processes* 8 (11): 1417.

34 Zhang, Y. (2022). Solar Power Data for Integration Studies. *Technical Report*. NREL. https://www.nrel.gov/grid/solar-power-data.html.

27

Blockchain-Based Energy Trading Employing Hyperledger and Anomaly Detection Algorithms

Zejia Jing, Ali Parizad, and Saifur Rahman*

Advanced Research Institute (ARI), The Bradley Department of Electrical and Computer Engineering, Virginia Tech, Arlington, VA, USA

27.1 Introduction

27.1.1 Background

As more and more rooftop photovoltaic (PV) and demand response-enabled smart devices are installed on the distribution side of the power system, the traditional grid is transforming into a decentralized peer-to-peer (P2P) electricity trading network from a centralized network. In a P2P electricity trading network, it is necessary for participants submitting their bids with an anomaly detection system.

As more and more energy producers and consumers partake in hourly energy consumption reduction ("negawatt-hour") and electricity trading P2P, protecting these transactions and maintaining trust among participants becomes a critical concern. Blockchain technology has emerged as a promising encryption technology that can track and store transactions in distributed ledgers shared with each participant. It is promising to deploy blockchain technology in energy trading.

A two-layer scheme needs to be designed for the blockchain energy trading platform. One is the physical layer which includes traditional transmission and distribution lines, power electronic devices, relay protection devices, and end-user energy consumer facilities. Another layer is the financial layer, where the blockchain energy trading platform is to deploy. Building trust between the physical and financial layers is a big concern. As the physical layer, the power transmission network, and the financial layer, the blockchain-based electricity market, are separately deployed while working together in blockchain, anomaly detection needs to be considered. The relays and electricity protection devices are taking effect for the physical layer to prevent and detect the anomaly. While for the financial blockchain layer, it is necessary to consider anomaly detection in participants' bid quantities.

In this chapter, anomaly detection in blockchain-enabled P2P electricity trading is introduced to secure and guarantee trust in P2P networks. Experiments based on the Hyperledger blockchain platform demonstrate the benefits of blockchain-enabled electricity trading.

Smart Cyber-Physical Power Systems: Fundamental Concepts, Challenges, and Solutions, Volume 1, First Edition.
Edited by Ali Parizad, Hamid Reza Baghaee, and Saifur Rahman.

27.1.2 Contribution

The contributions of this chapter are listed as follows:

- **Practical Insights for Blockchain-Based Energy Trading Systems:** By focusing on rooftop PV energy systems and negawatts, the chapter addresses the growing importance of prosumers in energy markets, offering practical insights for their enhanced participation and decision-making in energy trading.
- **Development of an Anomaly Detection Framework Based on Predicted Building Hourly Energy Consumption and PV Energy:** An anomaly detection framework based on semi-supervised machine learning is proposed to help detect anomalies with a confidence interval of 95%.
- **Develop a Blockchain-Based Energy Trading Platform Demo using Hyperledger:** A blockchain-based transactive energy platform, including PV energy trading and negawatt-hour trading, is simulated based on the Hyperledger platform.

27.2 Literature Review

27.2.1 Electricity Market

The electricity market is the retail and wholesale market where electricity is bought, sold, and traded. Early introductions to the concept of energy markets and the privatization of power systems began in Chile in the early 1980s. The traditional electricity market models are all linear from generation to consumption. With the increasing installation of rooftop PV panels, more and more user-end energy consumers are transforming into consumers and producers, called prosumers. Moreover, more user-end smart electricity devices, such as the smart thermostat, smart inverter, lights, etc., are deployed on the user end. Energy consumers can also participate in the demand response regulation by reducing their energy consumption accordingly. In this way, the energy consumers play a more important and active role in the electricity market in generating distributed PV energy, energy storage, and negawatt-hours.

As the traditional electricity market is becoming more decentralized because of end-user participation, P2P energy trading attracts more and more attention.

27.2.1.1 Peer-to-Peer Electricity Market

There are two P2P energy trading application scenarios: one is PV energy trading, and another is negawatt-hour trading.

P2P energy trading is a new paradigm for power systems. People can generate their energy from renewable sources (RES) in homes, offices, and factories and share it locally. Moreover, P2P trading buys and sells energy between two or more connected parties, usually PV energy. Any excess energy can be transferred through secure platforms and sold to other users. Excess PV energy is returned to the grid at a smaller feed-in tariff rate. But as more and more people seek flexibility and control over how resources are allocated, this approach has become obsolete.

P2P energy trading platforms would allow consumers to share their surplus energy and control how it is distributed across the microgrid. Users who both sell and consume energy are called "producers." Even without solar panels, you can still buy energy from others.

The potential of these innovative technologies has many benefits, including:

a) Those without solar panels can still get renewable energy from their neighbors at a reasonable price, while those who sell their excess energy can get it at a higher price than the feed-in tariff they get from retailers.

b) Energy does not have to be transported from a central power plant, reducing the cost of transporting electricity. According to Aurora Energy, 41.1% of your electricity bill is used to manage and maintain the poles and wires that carry power from the generator to the customer's house.
c) The ability to create energy from RES has several advantages in itself.
d) Energy can be purchased from known sources (this allows you to choose the energy source, for example, specific community projects you might want to support).
e) It offers the option of dealing with other consumers and eliminating middlemen (electricity retailers).

27.2.1.2 Peer-to-Peer PV Energy Trading

As shown in the graph below, renewable energy, including PV, has a big benefit in the wholesale market and can be bought at a lower price. Thus, it is very promising for homeowners to sell their surplus PV energy to other energy consumers. For over one hundred years, electricity has been bought from the generator to consumers. While with the development of solar panels, consumers are acting as the role of power producers, and they can be called prosumers. They begin to have more opportunities, benefits, and interests in joining the electricity market by selling their surplus energy. This kind of energy trading network is called P2P PV energy trading, defined and described in the paper [1].

There is already some P2P energy trading project of PV energy implemented on the lab scale using blockchain technology [2]. One is known as the Brooklyn Microgrid project, named TransActive, a DER management microgrid project conducted by company LO3 [3]. The TransActive Grid they developed uses blockchain and Ethereum smart contracts to build the market for local renewable generators and buyers. The neighbors can sell and buy energy generated from residential solar panels without a centralization operator in the Brooklyn Microgrid project. The Electron [4] is another revolutionary blockchain open-source platform that enables gas and electricity metering and billing platform for energy suppliers. The Solar Coin is a famous blockchain application in energy trading as a new blockchain token, just like Bitcoin [5]. The generator can get one case coin when the solar panel generates one-megawatt power. At the same time, the value of the solar coin is still doubtable because it is given as a reward for free. Many research teams or institutes are beginning to work in this field, such as IERC T2 collaborative research project [6]. Some papers demonstrate the P2P PV energy trading structure, such as paper [7] focusing on flexible load control in PV energy trading.

27.2.1.3 Peer-to-Peer Negawatt-Hour Trading

Another promising P2P electricity trading is negawatt-hour trading, as shown in Figure 27.1. Increasing energy efficiency and trading on energy efficiency capability are also very important. The negawatt is a suppositional power representing the electrical power saved by energy

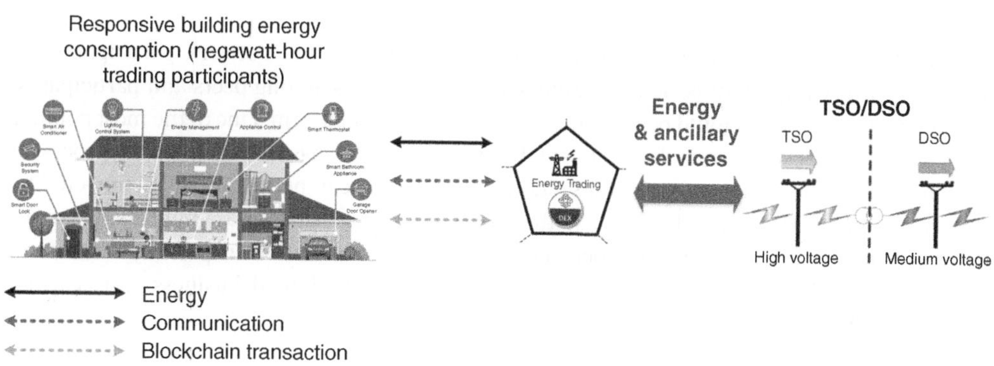

Figure 27.1 Negawatt-hour trading with responsive building hourly energy consumption.

conservation or energy efficiency increase. Amory Lovins first invented the term "Negawatt" in 1989 at the Green Energy Conference in Montreal [8]. Since then, people have realized that energy efficiency increases and energy savings as another kind of energy resource different from traditional energy resources [9]. The generation of negawatt-hour is predictable using the hourly energy consumption forecasting technique. The baseline of homeowners' or energy consumers' hourly energy consumption is pre-defined before negawatt-hour trading according to historical data of past years. The hourly energy consumption demand reduction can be predicted when adjusting AC set points.

With the development of hourly energy consumption forecasting techniques, people have begun to realize that the demand reduction can also be forecasted and traded as a kind of assessment among utilities, Demand Response Aggregators (DRAs), and homeowners. Thus, P2P trading enables customers to trade their hourly energy consumption reduction responsibility and ability in needs. It will benefit both homeowners and commercial building premises. When they receive a requirement for a load deduction signal from the utility, DRAs, or other homeowners, they can respond to this request as a negawatt-hour seller by turning up their setpoints and selecting acceptable temperature control bands based on their lifestyle preference and value proposition. By enabling negawatt-hour trading, if they are not willing to meet the contractual obligation to cut the load at the cycling period, they can join the P2P negawatt-hour trading network as a negawatt-hour buyer and buy hourly energy reduction capability from neighborhoods. This way, energy-saving capabilities, also called negawatt-hours, are traded from seller to buyer. An overview of negawatt-hour trading is shown in Figure 27.1.

Several previous papers have been published to discuss the P2P negawatt-hour trading issue. The journal [10] discusses the law of federal and state negawatt-hour trading in the US. And it summarized the ongoing negawatt-hours demand response efforts by FERCS to promote DRA programs in the electricity market. It also demonstrates the feasibility of a P2P negawatt-hour trading market. In papers [11, 12], the author proposed the price design of P2P distributed negawatt-hour trading in a real-time electricity market with energy storage systems. And authors in the [13] research on the daily market of P2P demand response, including PV panels, hearting appliances, refrigeration devices, and energy storage systems, using a proposed optimal bidding strategy. There are also papers like [14] demonstrating P2P trading case studies using different platforms, such as blockchain. An overview of all these papers will help understand the big picture of negawatt-hour trading with responsive building hourly energy consumption. In reality, the negawatt-hours aggregators will train and run the building model for negawatt-hour trading participants. Each building will be trained and simulated using different building models based on building types: school, hospital, office, etc.

27.2.2 Blockchain

While for P2P trading, because there is no central regulator, all the network participants act equally. It is a very important issue to guarantee trust and ledgers among peers and participants. Thus, energy transactions should be done through a very secure platform. One of the most popular ways is using the blockchain-enabled platform. Blockchain is a database technology that processes and stores information, such as asset transactions. These assets can be renewable energy credits, which can be traded through a database. With blockchain, all transactions are public and cannot be changed in any way once on the blockchain, resulting in complete transparency. Therefore, blockchain technology which is a great fit for this kind of distributed business case, can be deployed to enhance and benefit P2P energy trading.

Many papers are already researching this topic, such as in paper [15], a carbon market coupled with energy trading is demonstrated using blockchain. And in paper [16], the encryption theory using byzantine-based blockchain consensus is discussed in the case of energy trading. And in paper [17], energy trading based on the Ethereum network is demonstrated and simulated by conducting case study research.

The benefits of blockchain can be summarized as follows:

a) **Consensus:** In the network, all participants must agree on a consensus: the trading strategy or smart contract.
b) **Provenance:** All participants can know the information about assets' origin and change in their real-time ownership.
c) **Immutability:** No participant can make a change on the distributed ledger and database.
d) **Finality:** After transactions are done, a single, updated, shared ledger defines asset ownership.

27.2.2.1 Blockchain Revolution

Blockchain is a relatively new concept and technology. Satoshi Nakamoto first proposed and implemented it in Bitcoins [18]. Blockchain is the underlying digital foundation of Bitcoins. It is a network that records and tracks transactions in a shared, unchangeable ledger. In a traditional business network, every participant keeps their ledger, which is expensive, vulnerable, and easy to manipulate. While in the blockchain network, participants share a ledger and update peer to peer when a new transaction occurs. In this network, the participant is no longer a signal receiver and subscriber of the central system organizer. Participants play the role of both the publisher and subscriber. And all the databases and ledgers are synchronized among participants. At the same time, blockchain shows many pros in trust and efficiency among participants' exchange business.

Blockchain has several big pros compared to the traditional business network.

a) It is a distributed node network and has no central point failure risk.
b) It is manipulation proof, and the distributed database will remain very secure and not changeable.
c) It is a trusted network, and the network majority will collect and verify the behavioral data.
d) It is private, and the third party will not collect sensitive data.
e) It saves lots of transaction time by reducing the interaction time among multi-party from days to minutes.
f) It saves lots of cost by less oversight, fewer intermediaries, and less duplication effort.

27.2.2.2 Blockchain-Enabled Electricity Market

Blockchain technology for energy trading and the electricity market is new and very hot in research. Many papers have proposed or developed different algorithms and outlooks on future blockchain-based energy trading. Some papers are most representative in this field based on research on distributed energy resources and blockchain technology.

The paper [19] describes Switzerland's first real blockchain-based electricity market in design and technical specification. In the papers [20, 21], a blockchain framework that facilitates P2P energy trading in a microgrid is proposed in which both the microgrid control flow and power transaction flow are designed. The paper [22] demonstrates a multiagent structure that combines smart contracts and blockchain to facilitate P2P electricity trading without human regulation. Ethereum-based blockchain architecture is proposed with multi-agent system (MAS), blockchain network, and the device layer, including smart meters, PV panels, and batteries. A multi-agent architecture is deployed in Ethereum, and a non-cooperative game theory model is used. In the paper [23], blockchain and edge computing are used together to solve the challenge of the decentralized electricity market. The transactions layer is built in blockchain, and the electricity pricing algorithm is calculated in the edge computing terminal accordingly.

Authors in [24] discuss and simulate a proof-of-work cost architecture of demand-side blockchain. The optimal demand load management problem is investigated, studied, and compared. Results show the new architecture and management method save more electricity costs. The paper [25] proposes Elecbay, a three-dimension layer hierarchical architecture platform for P2P energy trading among prosumers and consumers. An associated bidding system is proposed using game theory and the Nash Equilibrium for P2P energy trading. The paper shows that P2P energy trading can reduce energy exchange between the utility grid and microgrid. In the paper [26], a blockchain-based trading algorithm is proposed and tested. The linear supply bidding mechanism is applied to the market and shows a competitive equilibrium that all participants maximize their profit.

27.2.2.3 Blockchain's Limitation in Electricity Trading

In a blockchain-based transactive energy platform, the consensus mechanisms and distributed ledger guarantee that transactions are executed and stored correctly. However, they cannot guarantee that the submitted transaction requests are reasonable and not malicious. For instance, an attacker could intentionally submit excessively high/low-priced bids or transactions not backed by actual energy resources, to disrupt the market. They might even hijack the IoT-based smart meters to send bogus measurements. Because the blockchain is agnostic to tangible real-world resources like "actual" energy, such an attack poses significant threats to a transactive energy platform. Thus, an added anomaly detection layer is needed.

27.2.3 Anomaly Detection

Anomaly detection in the trading network, also called outlier detection, is a way to detect abnormal data among many similar datasets. As summarized in [27], three anomaly types are classified.

a) **Point anomaly:** The single data point stays far from other data point distributions.
b) **Contextual anomaly:** It can be the noise in data, for instance, the background noise in a phone call.
c) **Collective anomaly:** Indicate there is a new phenomenon, usually because of the novelties in data.

27.2.3.1 Anomaly Detection Tools

Different methods are also used to detect anomalies. For point anomaly, classification-based anomaly detection techniques are most commonly used. There are five types of classification-based methods to detect anomalies which are Neural Networks Based [28], Bayesian Network Based [29], Support Vector Machines based [30, 31], and Rule-Based [32]. Besides the classification methods, other techniques are also developed to detect point anomalies, such as clustering-based anomaly detection [33], information-theoretic anomaly detection [34], nearest-neighbor-based anomaly detection [35], statistical anomaly detection [36], and spectral anomaly detection techniques [37].

As for the contextual, the most important step is to reduce the problem to point to an anomaly detection problem and solve it. The general step includes two ways, as summarized in the paper [38]. The first is identifying a context for each test data by taking advantage of the contextual attributes. And then calculate the point anomaly detection score for the test data within the context.

As for collective anomalies such as sequential anomalies, the general method is to change the original problem sequences to a finite feature space problem. Then use a point anomaly detection method in the finite feature space. The approach is developed for time-series data sets in the paper [39] for equal lengths of sequences and in the paper [40] for unequal lengths of sequences.

Usually, the reason for the anomaly is errors (which possibly happen in data entry, data measurement, data processing, experiment, and sampling), intentional injection, and natural factors. For example, one party could hijack smart meters to falsely provide high energy consumption measurements to artificially inflate demand so that its conspirators could sell energy at a higher price. Even though the symptoms of an attack might be of diverse forms, the number of ultimate goals of an attacker is limited: for example, it could be service/market disruption or selfish financial gains.

Among different anomaly detection tools, extreme value analysis, linear regression models, probabilistic and statistical modeling, and proximity-based models are the most commonly used. The extreme value analysis of one-dimensional data clarification is introduced in the paper [41]. This method gives a rough boundary of data. For example, the setpoint settings for a room HVAC are usually bounded as 66–82. That means all values not within this range will be seen as abnormal. This method may not work well if the data set is sparse. Another method is probabilistic model-based anomaly detection. In further probabilistic and statistical models, the parameter of the data model can be learned, and the form of a distribution is defined. This method may not work well because some data patterns do not follow a particular distribution. The third popular anomaly detection method is using the regression model. As mentioned in the book "Outlier analysis" [42], the proximity-based model is a very popular method for discovering data patterns.

27.2.3.2 Anomaly Detection for the Electricity Market

In an electricity trading market, the prosumer or participants may misbehave, cheat, or inject many promises that will depress the market. Then the participants may go and buy the electricity at a lower price and sell it at a higher price. Using the machine learning technique to identify the anomaly is very necessary. And the anomaly detection should also be able to predict system failures.

Some papers focus on end-user energy consumption anomaly detection in the electricity market[43–46]. The paper [43] identifies the malicious power consumption and anomaly detected by a waveform feature extraction model. The analysis of the target line loss analysis is used, and the SEEP algorithm is used to extract the model of power use data. The anomaly of electricity leakage, power theft, or other warnings can be detected by comparing the line loss and power consumption. The paper [44] introduces a method that helps identify abnormal days during DR event periods. By diagnosing and removing abnormal days such as vocation in energy consumption data, much better accuracy of DR baseline prediction can be achieved. The paper [45] proposes a trained deep neural network method to detect anomaly forecasting data in electricity time-series consumption. And in the paper [46] introduces an unsupervised method to systematically help anomaly detection in building energy consumption. This model utilizes the relationship between sensors and uncovers the true inter-device relationships to detect abnormal consumption. Some papers research the electricity market's physical parameter anomaly detection [47, 48]. Some papers research detecting anomalies in the price issue in the electricity market [49]. And some papers detect load forecasting anomalies in the electricity market [50].

In the paper [51], the author discusses electricity price anomaly detection and uses machine learning techniques to assist abnormal detection in the electricity market. It assumes that the original data has a certain kind of pattern which follows the market fundamentals. In contrast, the abnormal data has a special pattern, which deviates from market fundamentals. This paper uses supervised outlier detection, time series, and streaming outlier detection to identify the bid price and quantity anomaly. The author identifies the anomaly using the time series history data, and unusual data can be quickly caught. The methodology introduced by the paper mainly includes four steps:

a) Choose the input parameters and values that may impact electricity prices.
b) The machine learning technique selects the most relevant variables and finds the data patterns. These selected features will help get the data pattern more accurately and lower computational costs.
c) Using the selected feature as input and the expected price change as output. Those values with a high deviation from normal price expectations from the model will be anomalies. In the modeling part, kernel smoothing is used to find the internal mathematical relationship between price, generation capacity, and hourly energy consumption.
d) Output the anomalies and get people to investigate them for future analysis.

After the kernel model is set up, use the normal cases dataset as input to define the low and high boundaries. All the other data that fall out of this boundary will be anomalies. These four-steps machine learning-based anomaly detection method also inspired the method in this dissertation.

In the paper [52], the detection algorithms of electrical data that focus on abnormal trend detection are introduced and discussed. And the backtracking dynamic window model is used as the proposed algorithm. The electrical consumption data in different regions is time-series data. Thus, the anomaly detection techniques used in time series data will be used. Many papers are already discussing the different techniques of time series abnormal detection. In paper [53], an exception analysis method based on sequence similarity in time series CIoTA is discussed. Detecting trend anomalies is very useful for blockchain P2P trading. For instance, the PV panels are usually at peak power output during the daytime and back to zero at night. And PV energy output is usually higher in the summertime and lower in the winter. For energy consumption and demand response, energy consumption is usually at a peak in the daytime and the lowest point in the night hours. Besides, the energy consumption during the winter is usually much lower than in the summertime when most demand responses happen because the HVAC system is not running during the wintertime. And according to the participants' living and working schedules, the weekdays and weekend energy consumption data may differ a lot. These certain trends of anomaly detection of energy consumption will benefit the anomaly detection in blockchain P2P PV and demand response negawatt-hour trading.

In the paper [54], the way to detect collusive shill bidding in the case of online auctions is discussed. While for auctions of bidding period in the energy trading market, it is very important to avoid collusive shill bidding because it can harm the potential honest electricity or negawatt-hour buyer by driving up the final price for the electricity or negawatt-hour seller. The paper proposes a new algorithm called "Collusive Shill Bidding Detection (CSBD)" to calculate each participant's anomaly score Local Outlier Factor.

The above papers summarize current studies of anomaly detection tools used in electricity and other markets. While for the P2P electricity market based on a blockchain platform, how to deploy the anomaly detection tools in the blockchain network should be investigated.

27.2.3.3 Anomaly Detection for Blockchain-Enabled Peer-to-Peer Market

Blockchain is a distributed ledger technology that can make transactions secure and fair. While it cannot identify the input of the network is malicious or not. Some studies have addressed these issues. In the paper [55], two machine learning algorithms, Support Vector Machine (SVM) and K-means, are used to detect anomaly transactions in bitcoin blockchain cryptocurrency systems, and the result shows good accuracy. In the paper [56], as digitally signed transactions are stored in distributed ledgers in blockchain, a machine learning-based method is introduced to the automated signing process and protection from malicious blockchain transactions. The proposed method is to be deployed on top of blockchain technology. In the paper [57], a method to identify the healthiness of the blockchain network is proposed. The malicious node or blocks are detected using the data from the IoT blockchain by analyzing transaction generation interval, blockchain generation rate, and statistical analysis.

The paper [58] proposes a wastewater reuse control system. And blockchain technology is used to securely store data and incentivize human water reuse willingness by the token mechanism. To prevent data tampering, three machine learning-based anomaly detection algorithms: polynomial regression, DBSCAN (density-based spatial clustering of applications with noise), autoencoders, and long short-term memory (LSTM), are used to detect water reuse data frauds in the blockchain, and anomaly detection is used together in the wastewater system. And a case study lab implemented based on Hyperledger Fabric is simulated in the paper. The paper [59] introduces an anomaly detection tool used especially in blockchain systems named ADvISE. An Anomaly Detection tool for blockchain SystEms called ADvISE uses blockchain meta-data to collect unusual requests in the network. The structure of the ADvISE platform uses the history attacks information shared by peers in blockchain to protect against eclipse attacks. It means the malicious forks will be distributed to all peers in the community, and they will be warned and prepared for future attacks. This application is extremely efficient if the attackers inject harmful transactions in the same way on every peer. For general usage of the application, an additional layer to identify and prevent a potential attack is built by comparison of suspicious transactions and historical malicious sequences using machine learning or similar measurement methodology.

The paper [60] proposes a Blockchain Anomaly Detection solution named BAD. This paper is a continuous research study of the paper [59]. In this paper, the author leverages the past forks information collected. The information related to orphaned blocks and forks will not be discarded as other blockchain-based applications. These past orphaned blocks and forks will be taken into the analysis, collected, extended, and shared with all peers in the network. The recording information includes the start and detected time of the fork and the number and type of harmful transactions. The test result shows that preventing malicious transactions takes effect, and few malicious transactions are accepted.

27.2.4 Necessity of Blockchain-Based Peer-to-Peer Electricity Trading Framework Through Machine Learning-Based Anomaly Detection Technique

After reviewing the related literature, there are already a few papers published recently studying the crossing field of blockchain && anomaly detection, blockchain && electricity market, and anomaly detection && electricity market. While the interesting gap is there is no paper combining all three topics and research on the intersection among these three topics. As a review of the papers above, many papers have demonstrated ideas for anomaly detection. While until now, none of those articles combines blockchain technology, anomaly detection technique, and the electricity market. Thus, there is a big gap here. And it is a good research opportunity that allows future researchers to take a deeper insight into this area. The state of the art in this certain field is promising as more and more people begin to realize the importance and future of blockchain-based energy trading.

As to research necessity, because the blockchain-based transactive energy platform is distributed, however, existing anomaly detection methods require total visibility into the behavior of all participating entities. It is necessary and in great need to find the best solution without centralizing the system. Thus, this chapter proposes the importance and beginning research of an added anomaly detection infrastructure to the blockchain-based transactive energy platform to automatically detect and flag malicious transactions and manipulated measurements from compromised smart meters. The machine learning-based detection model can be used in anomaly detection of user malicious input in blockchain-based electricity trading. Moreover, the blockchain feature can also help detect anomalies by keeping track of historical trading data. And by implementing anomaly detection methods and blockchain architecture, a more secure, fair, and trusty P2P electricity trading network can be built and used by energy consumers and prosumers.

27.3 Anomaly Detection

Users may intentionally or unintentionally submit extremely high/low bids that do not match their solar panel capability or are not backed by substantial negawatt-hours and PV energy resources. Some anomalies occur because the participant's sensor is suffering from integrity errors. In contrast, some other anomalies occur because the participant maliciously submits extreme orders to benefit attackers themselves from market disruption. In both cases, anomalies should be detected by the algorithm and rejected by the market. Thus, based on the above negawatt-hours and PV energy forecasting model, a semi-supervised machine learning-based anomaly detection technique is developed and explained in this chapter.

The semi-supervised approach used in negawatt-hours and PV energy trading anomaly detection includes three following steps:

a) The model is first trained only on normal data (without any anomalies). This step is a purely supervised learning approach to get a model for building hourly energy consumption prediction and hourly solar irradiance prediction. The P2P energy trading system operator and negawatt-hours aggregators will train and run the model for building participants.
b) Building hourly energy consumption predicted result before and after setpoint adjustment is calculated, and their difference is the predicted negawatt-hours. The next hour's predicted PV energy is calculated using open source software for simulating solar power of photovoltaic energy systems (PVLIB).
c) Compare the predicted negawatt-hours and PV energy with user-submitted data point to predict whether the new data point is normal or not (based on the Root Mean Square Error[RMSE] value of the data in the training model). This step is based on knowledge without labeling data. Thus, it is a purely unsupervised learning process to classify any data point as an anomaly as long as it exceeds 95 percentile values.

These steps comprise the semi-supervised anomaly detection method called the predictive confidence level approach, as illustrated in [61, 62].

The advantages and disadvantages of this approach are summarized in Table 27.1.

Table 27.1 Pros and cons of predictive confidence level approach.

Pros	Cons
• Easy to understand and implement • Performance well when the model is accurate enough • Detect local outlier robustly	• Poor performance when the data is volatile and granularly • Need the model to be accurate and good enough

27.3.1 Mathematical Background

In statistics, the confidence interval for 95% is a multiplier of 1.96, as shown in Figure 27.2.

The overview of the semi-supervised anomaly detection method is shown in Figure 27.3. First, the deep neural network learning models such as artificial neural network (ANN), LSTM, gated recurrent units (GRU), and Convolutional Neural Network (CNN) train on past hours to predict the future 1–24 hours based on input parameters. Then upper and lower bounds can be calculated according to the standard deviation of residuals (RMSE instead) calculated according to historical prediction accuracy.

For a typical boundary with a 95% Predictive Confidence Level, the upper limit is

$$\text{predicted value} + 1.96 * \text{standard deviation}$$

the lower limit is

$$\text{predicted } value - 1.96 * standard \text{ deviation}$$

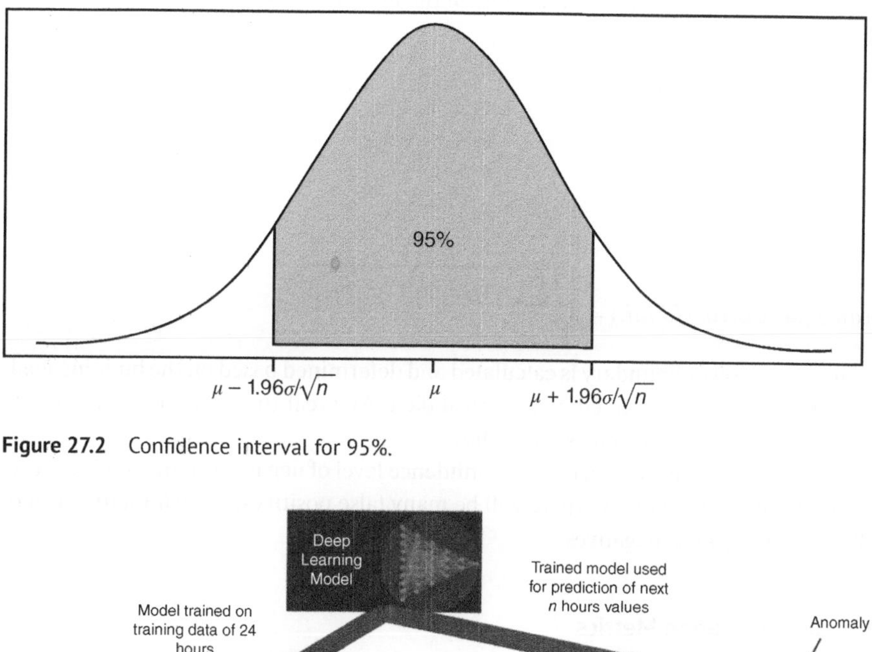

Figure 27.2 Confidence interval for 95%.

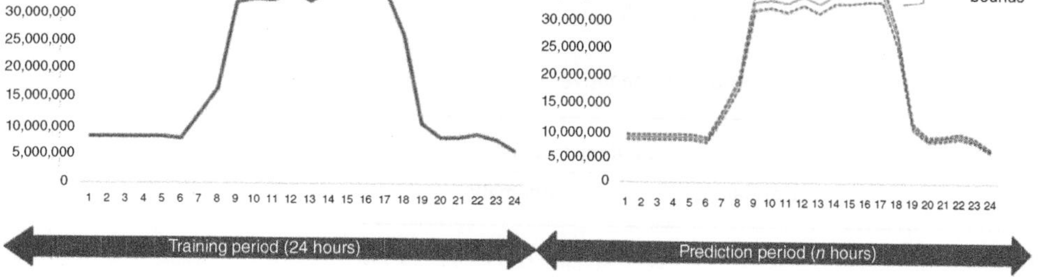

Figure 27.3 Semi-supervised anomaly detection architecture.

The standard deviation of the residuals is calculated using the formula:

$$Residual = y(actual) - y(pred)$$

$$S_{res} = \sqrt{\frac{\sum (y(actual) - y(pred))^2}{n-2}}$$

Where

S_{res} = Residual standard deviation

$y(actual)$ = Actual value

$y(pred)$ = Forecasted result

n = Data points in the dataset

In this dissertation, instead of S_{res}, the RMSE value of the model is used. That is, the upper limit is

$$predicted\ value + 1.96 * RMSE$$

the lower limit is

$$predicted\ value - 1.96 * RMSE$$

Where

$$Root\ Mean\ Square\ Error\ (RMSE) = \sqrt{\frac{\sum_{i=1}^{n} (y(actual) - y(pred))^2}{n}}$$

This way, the upper and lower boundary is calculated and determined based on the building load and PV energy forecasting models trained by historical data. Any real-time data point submitted by users outside the range will be treated as anomalies.

The multiplier 1.96 can be adjusted to get a new confidence level of negawatt-hours or PV energy submitted. If the limitation is too narrow, there will be many false positives; if the limitation is too wide, there will be too many false negatives.

27.3.2 Dataset and Evaluation Metrics

The performance metrics for anomaly detection effectiveness are summarized in Table 27.2.

$$TPR = True\ Positive\ Rate = \frac{TP}{TP + FN}$$

Table 27.2 Definition of true/false/positive/negative.

		Predicted value	
		Normal	**Anomaly**
Actual value	Normal	T.N. (True negative)	F.P. (False positive)
	Anomaly	F.N. (False negative)	T.P. (True positive)

TPR represents the number of samples predicted to be abnormal and actually abnormal as a proportion of the total abnormal number. The larger the *TPR* value, the better the performance

$$FPR = False\ Positive\ Rate = \frac{FP}{FP + TN}$$

FPR represents the metric that the number of samples that are predicted to be abnormal but are actually normal as a proportion of the actual normal total. The smaller the *FPR* value, the better the performance.

$$R = Recall = TPR = \frac{TP}{TP + FN}$$

The recall value is the same as the *TPR* value.

$$P = Precision = \frac{TP}{TP + FP}$$

Precision represents the metric that the number of samples that are predicted to be abnormal and are abnormal, accounting for the proportion of the total number of predicted anomalies. The larger the *P* value, the better the performance.

$$F1 = \frac{2 \times P \times R}{P + R}$$

*F*1 value is the weighted harmonic average of *P* and *R*. It represents the model's overall performance by considering both recall and precision metrics. The larger the *F*1 value, the better the performance.

$$A = Accuracy = \frac{TP + TN}{TP + TN + FP + FN}$$

Accuracy, different from precision, considers not only the abnormal class but also the normal class, that is, the number of all matching samples as a percentage of all samples. The larger the accuracy value, the better the performance.

27.3.3 Anomaly Injection Experiments

According to the definition in the paper [63], four patterns of anomalies in this session are used to evaluate the performance of negawatt-hours and PV energy anomaly detection models. The anomaly injection start and end times for two P2P energy trading scenarios are listed in Table 27.3.

The best forecast model RMSE from simulation and experimental results explained in Chapters 14 and 15 are summarized in Table 27.4.

27.3.3.1 Scaling Anomaly Injection

In this anomaly injection, the scaling parameter is employed to modify the PV energy/negawatt-hours bid quantity data y_t at time t to be submitted to P2P energy trading network as:

$$\hat{y}_t = y_t \times (1 + \lambda_s)$$

Table 27.3 Anomaly injection start time and end time.

	Negawatt-hour trading scenario 12:00–16:00 (HVAC Setpoint +2 °F)	**PV energy trading scenario 9:00–18:00 (Cloud cover < 3)**
Anomaly injection start time A_s	10/19/2017 0:00:00 a.m.	6/22/2013 2:00:00 a.m.
Anomaly injection end time A_e	12/31/2017 11:00:00 p.m.	9/24/2013 11:00:00 p.m.

Table 27.4 Forecast model RMSE summary.

	Building energy consumption forecasting model	PV energy forecasting model 9:00–18:00 (Cloud cover < 3)
RMSE	1154.750 kJ	10.97 Wh

Where,

y_t : Actual data before the anomaly injection
$\hat{y}_t$: Manipulated data after the anomaly injection
λ_s : Scaling parameter

Inspired by paper [64], scaling anomaly injection with $\lambda_s = 5$ is simulated to evaluate the model performance.

The PV energy data before and after scaling the anomaly injection is shown in Figure 27.4.

The PV energy data before and after scaling the anomaly injection in one typical summer week (June 22, 2013–June 28, 2013) is shown in Figure 27.5.

The negawatt-hour data before and after scaling the anomaly injection is shown in Figure 27.6.

The negawatt-hour data before and after scaling the anomaly injection in one typical summer week (October 19, 2017–October 25, 2017) is shown in Figure 27.7.

27.3.3.2 Simple Ramp Anomaly Injection

This anomaly injection gradually modified the PV energy/negawatt-hours bid quantity data y_t at time t to be submitted to P2P energy trading network by a ramp function

$$\hat{y}_t = y_t \times (1 + \lambda_r \times (t - A_s)), t = A_s, A_s + 1, \ldots, A_e$$

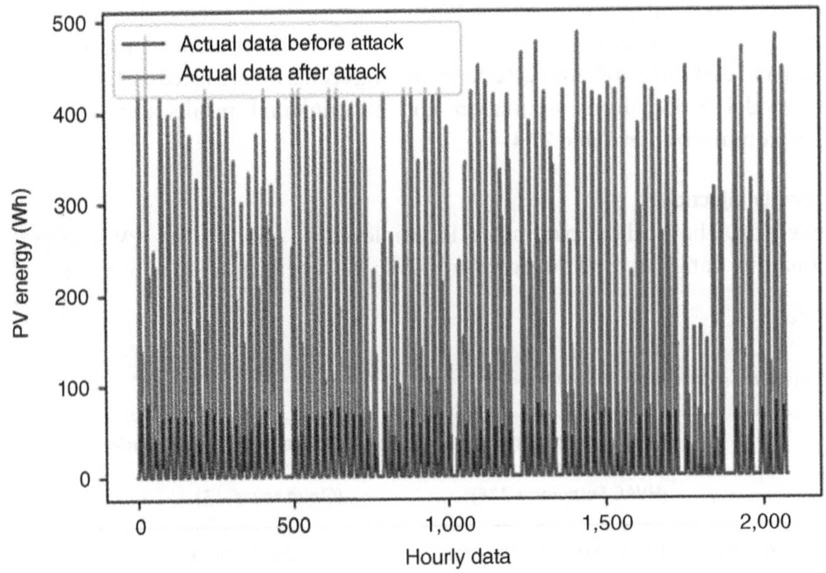

Figure 27.4 PV energy data before and after scaling the anomaly injection in the whole injection period.

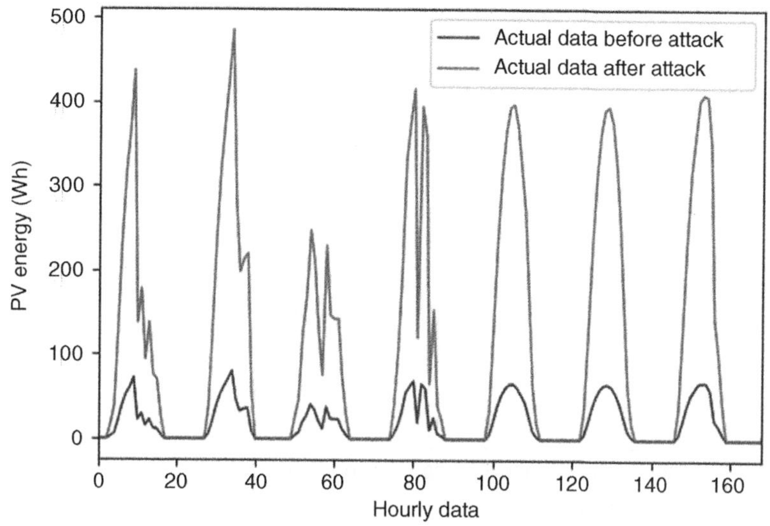

Figure 27.5 PV energy data before and after scaling the anomaly injection in one week.

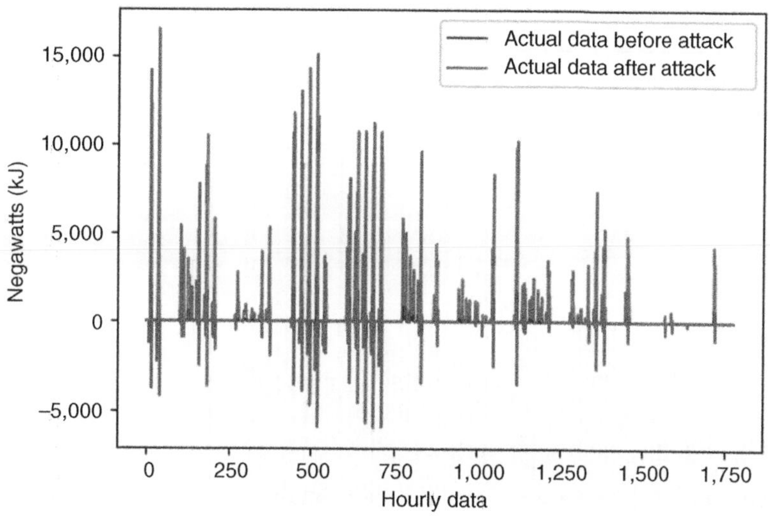

Figure 27.6 Negawatt-hour data before and after scaling the anomaly injection in the whole injection period.

Where,

y_t : Actual data before the anomaly injection
$\widehat{y}_t$: Manipulated data after the anomaly injection
λ_r : Ramp function parameter
A_s : The start time of the anomaly injection
A_e : The end time of the anomaly injection
t : Anomaly injection period

Inspired by paper [64], a simple ramp anomaly injection with $\lambda_r = 0.01$ is simulated to evaluate the model performance based on the given anomaly injection period.

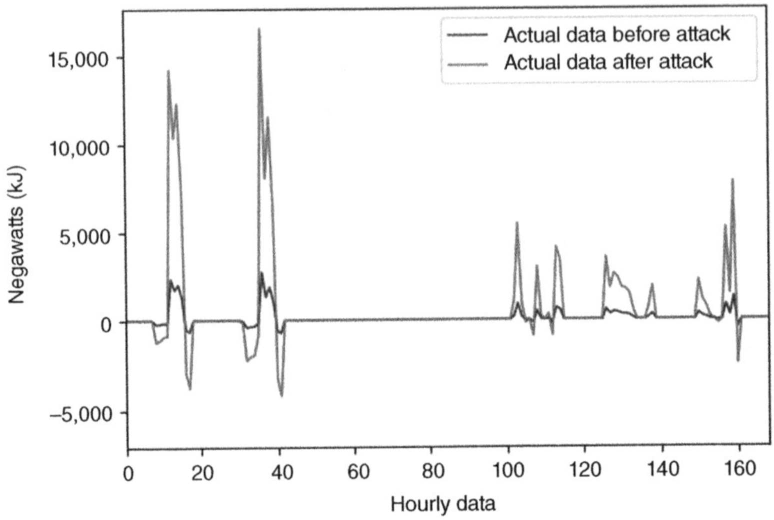

Figure 27.7 Negawatt-hour data before and after scaling the anomaly injection in one week.

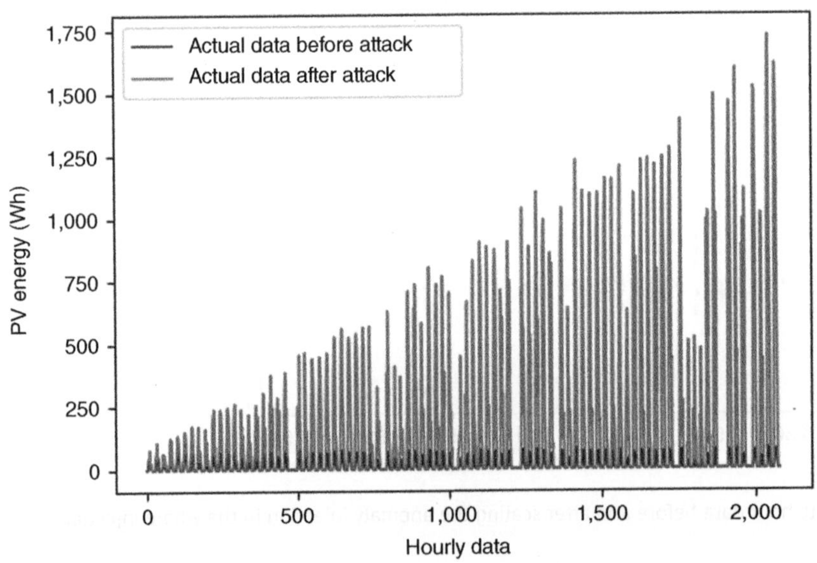

Figure 27.8 PV energy data before and after simple ramp anomaly injection in the whole injection period.

The PV energy data before and after a simple ramp anomaly injection is shown in Figure 27.8.

The PV energy data before and after a simple ramp anomaly injection in one typical summer week (June 22, 2013–June 28, 2013) is shown in Figure 27.9.

The negawatt-hour data before and after a simple ramp anomaly injection is shown in Figure 27.10.

The negawatt-hour data before and after a simple ramp anomaly injection in one typical summer week (October 19, 2017–October 25, 2017) is shown in Figure 27.11.

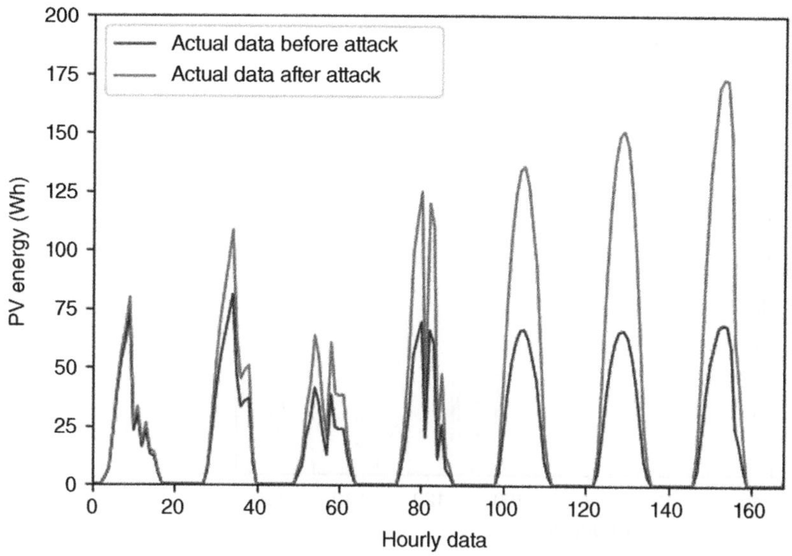

Figure 27.9 PV energy data before and after simple ramp anomaly injection in one week.

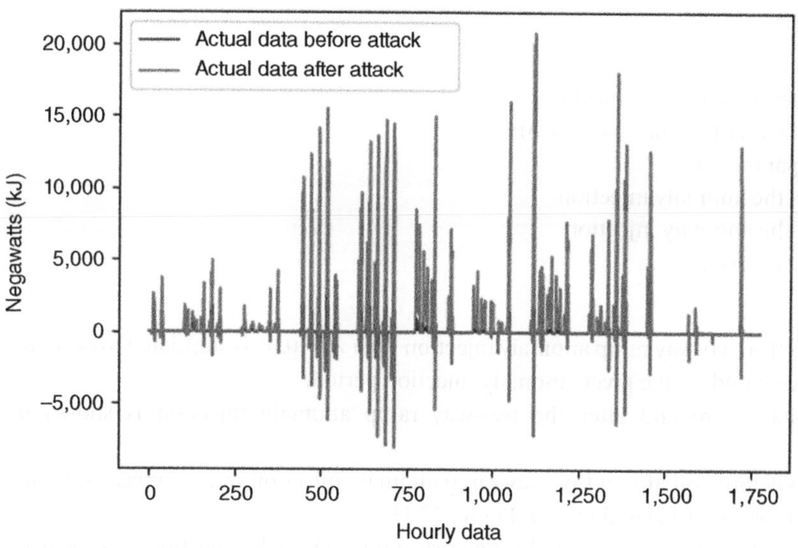

Figure 27.10 Negawatt-hour data before and after simple ramp anomaly injection in the whole injection period.

27.3.3.3 A Two-Way Ramp Anomaly Injection

In a two-way ramp anomaly injection, the malicious participants may increase the PV energy/negawatt-hours bid quantity data y_t first and then decrease it. In this case, the anomaly detection will be difficult and can be expressed by:

$$\hat{y}_t = y_t \times (1 + \lambda_r \times (t - A_s)), t = A_s, A_s + 1, \dots, \frac{A_s + A_e}{2}$$

$$\hat{y}_t = y_t \times (1 + \lambda_r \times (A_e - t)), t = \frac{A_s + A_e}{2} - 1, \dots, A_e$$

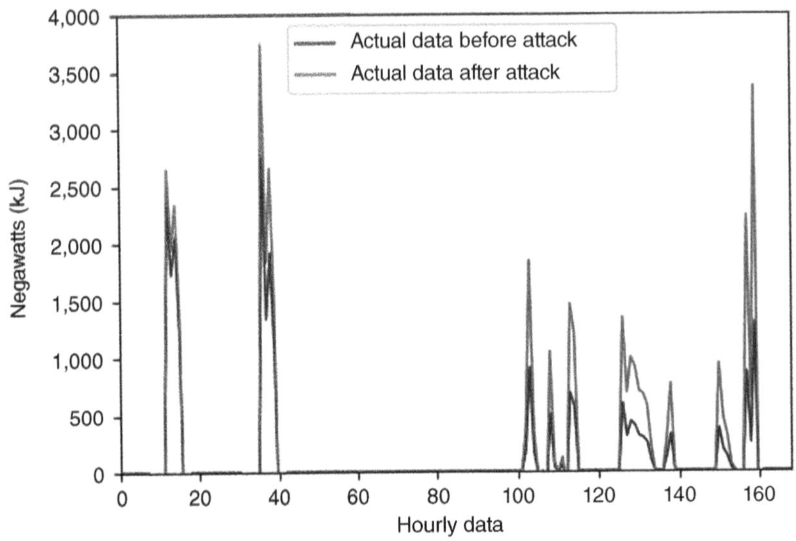

Figure 27.11 Negawatt-hour data before and after simple ramp anomaly injection in one week.

Where,

y_t : Actual data before the anomaly injection
$\hat{y}_t$: Manipulated data after the anomaly injection
λ_r : Ramp function parameter
A_s : The start time of the anomaly injection
A_e : The end time of the anomaly injection
t : Anomaly injection period

Inspired by paper [64], a two-way ramp anomaly injection with $\lambda_r = 0.01$ is simulated to evaluate the model performance based on the given anomaly injection period.

The PV energy data before and after the two-way ramp anomaly injection is shown in Figure 27.12.

The PV energy data before and after a two-way ramp anomaly injection in one typical summer week (June 22, 2013–June 28, 2013) is shown in Figure 27.13.

The negawatt-hour data before and after the two-way ramp anomaly injection is shown in Figure 27.14.

The negawatt-hour data before and after the two-way ramp anomaly injection in one typical summer week (October 19, 2017–October 25, 2017) is shown in Figure 27.15.

27.3.3.4 Random Anomaly Injection

Adding a random value to the PV energy/negawatt-hours bid quantity data y_t can be another type of anomaly injection and is given by:

$$\hat{y}_t = y_t + rand(a, b)$$

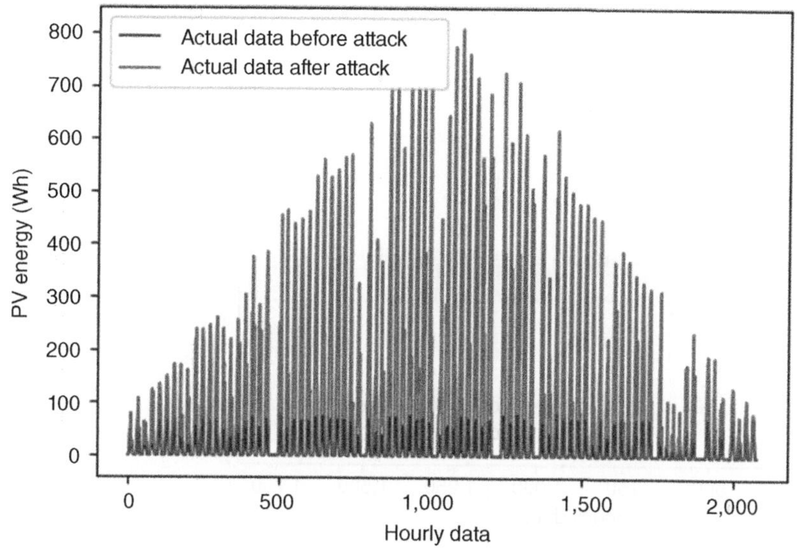

Figure 27.12 PV energy data before and after the two-way ramp anomaly injection in the whole injection period.

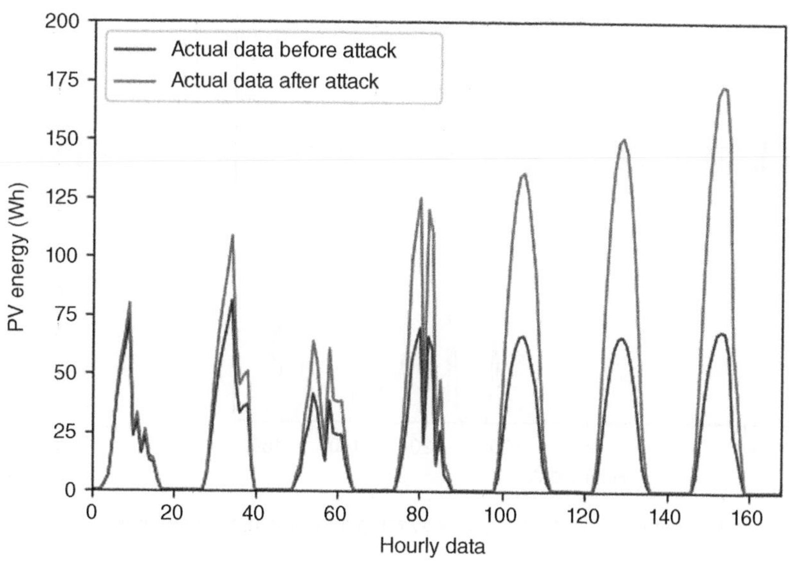

Figure 27.13 PV energy data before and after the two-way ramp anomaly injection in one week.

Where,

y_t	: Actual data before the anomaly injection
$\hat{y}_t$	: Manipulated data after the anomaly injection
rand	: Uniform random function
a and *b*	: Upper and lower bounds

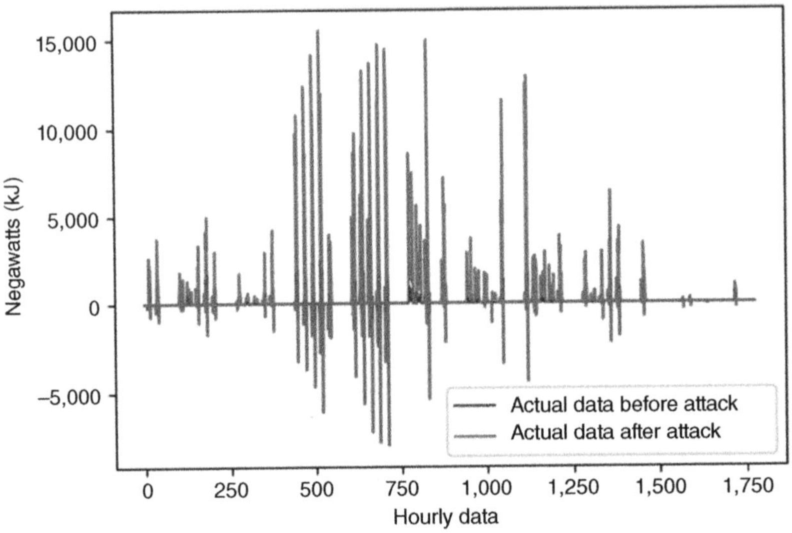

Figure 27.14 Negawatt-hour data before and after the two-way ramp anomaly injection in the whole injection period.

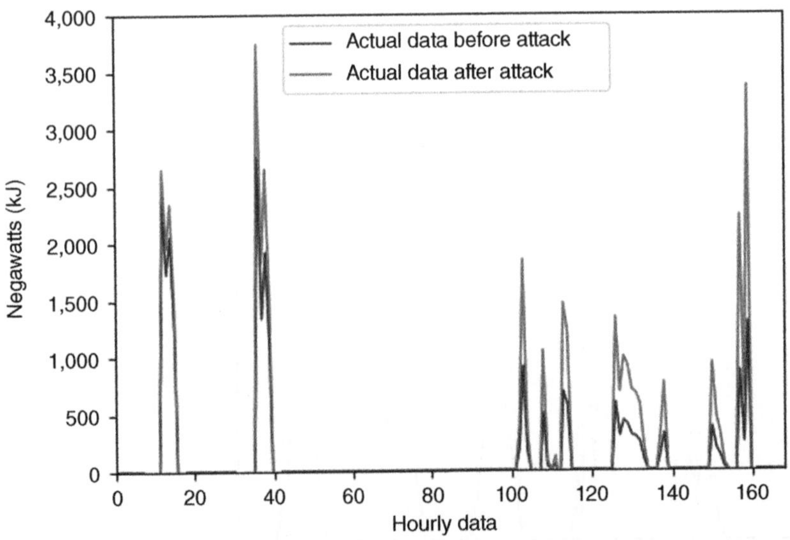

Figure 27.15 Negawatt-hour data before and after the two-way ramp anomaly injection in one week.

A random anomaly injection with $a = RMSE$ and $b = 2 * RMSE$ is simulated to evaluate the model performance based on the given anomaly injection period. As summarized in Table 27.4, the forecasted RMSE is 1154.750 kJ for negawatt-hours, and for PV energy, the RMSE value is 10.97 Wh.

The PV energy data before and after the random anomaly injection is shown in Figure 27.16.

The PV energy data before and after the random anomaly injection in one typical summer week (June 22, 2013–June 28, 2013) is shown in Figure 27.17.

The negawatt-hour data before and after the random anomaly injection is shown in Figure 27.18.

The negawatt-hours before and after the random anomaly injection in one typical summer week (October 19, 2017–October 25, 2017) are shown in Figure 27.19.

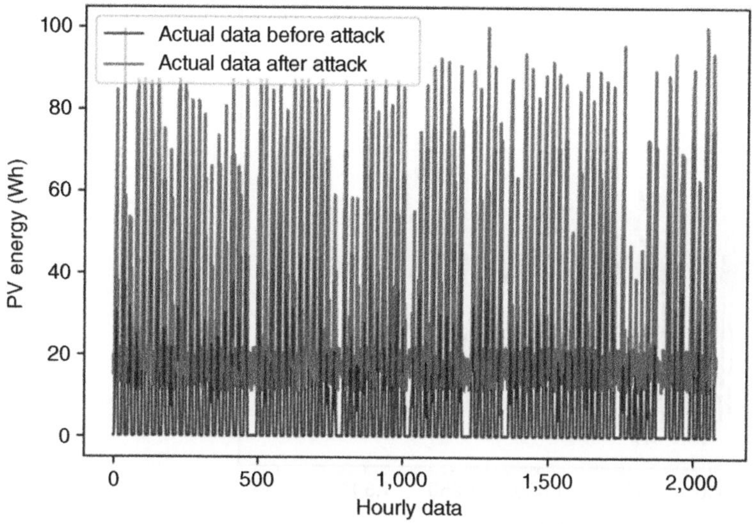

Figure 27.16 PV energy data before and after the random anomaly injection in the whole injection period.

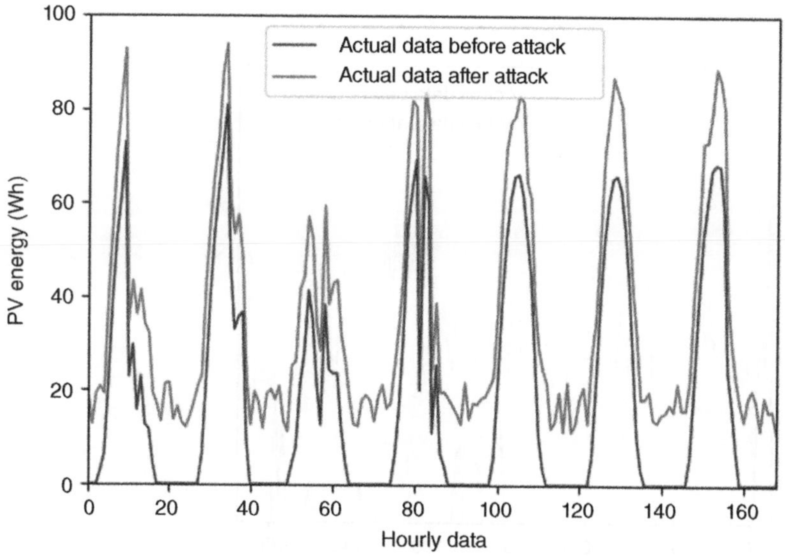

Figure 27.17 PV energy data before and after the random anomaly injection in one week.

27.3.4 Anomaly Detection Performance

The performance of the semi-supervised negawatt-hours anomaly detection technique facing four anomaly patterns is summarized in Tables 27.5–27.9.

The negawatt-hours anomaly detection performance is summarized in Table 27.9. The anomaly detection method's high precision value of around 97% indicates there will not be too many false alarms, and the operators will trust the system when the alarm happens. Besides, the false positive rate value of 1.50% is also satisfactory, indicating the customer will not be often disappointed if they submit normal data but are identified as abnormal by the system. The true positive rate is not that high, indicating that around 45–70% of the actual anomalies will be detected depending on

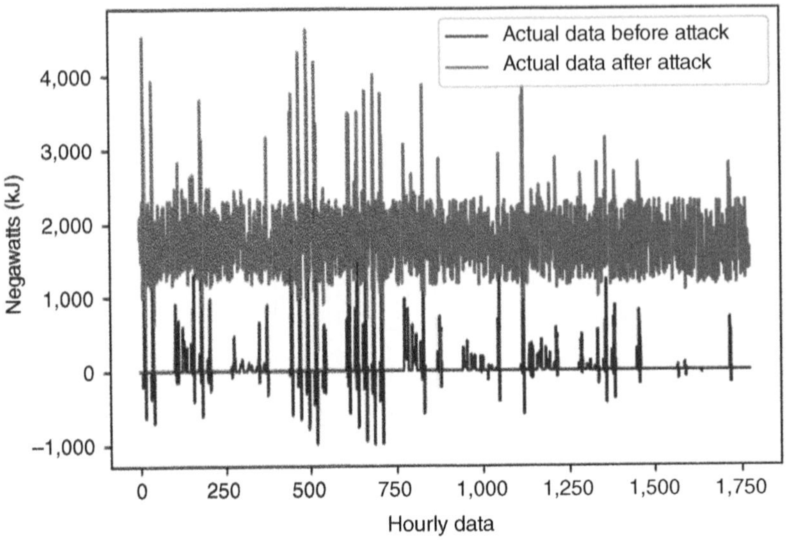

Figure 27.18 Negawatt-hour data before and after the random anomaly injection in the whole injection period.

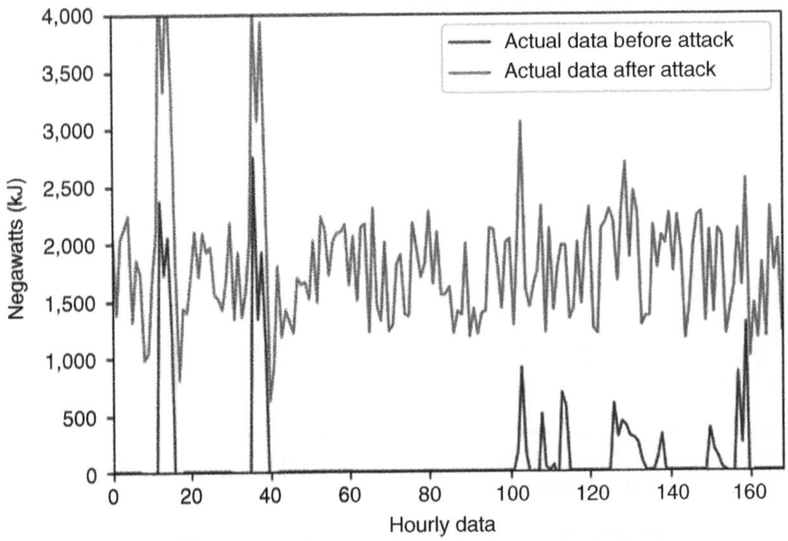

Figure 27.19 Negawatt-hour data before and after the random anomaly injection in one week.

Table 27.5 Negawatt-hours anomaly detection performance facing scaling anomaly injection.

Scaling anomalies		Negawatt-hours predicted value	
		Normal	**Anomaly**
Negawatt-hours actual value	Normal	True negative = 197	False positive = 3
	Anomaly	False negative = 108	True positive = 92

Table 27.6 Negawatt-hours anomaly detection performance facing simple ramp anomaly injection.

Simple ramp anomalies		Negawatt-hours predicted value	
		Normal	**Anomaly**
Negawatt-hours actual value	Normal	True negative = 197	False positive = 3
	Anomaly	False negative = 109	True positive = 91

Table 27.7 Negawatt-hours anomaly detection performance facing two-way ramp anomaly injection.

Two-way ramp anomalies		Negawatt-hours predicted value	
		Normal	**Anomaly**
Negawatt-hours actual value	Normal	True negative = 197	False positive = 3
	Anomaly	False negative = 117	True positive = 83

Table 27.8 Negawatt-hours anomaly detection performance facing random anomaly injection.

Random anomalies		Negawatt-hours predicted value	
		Normal	**Anomaly**
Negawatt-hours actual value	Normal	True negative = 197	False positive = 3
	Anomaly	False negative = 63	True positive = 137

Table 27.9 Negawatt-hours anomaly detection overall performance.

Negawatt-hours anomaly detection performance metrics	Scaling anomalies	Simple ramp anomalies	Two-way ramp anomalies	Random anomalies
True positive rate (TPR)	46.00%	45.50%	41.50%	68.50%
False positive rate (FPR)	1.50%	1.50%	1.50%	1.50%
Precision (P)	96.84%	96.81%	96.51%	97.86%
$F1$	62.37%	61.90%	58.04%	80.59%
Accuracy (A)	72.25%	72.00%	70.00%	83.50%

the anomaly injection types they face. The overall accuracy of the negawatt-hour trading anomaly detection will perform best when facing a random anomaly injection with an overall accuracy of 83.50%, while performance worst when facing a two-way ramp anomaly injection with an overall accuracy of 70.00%.

The performance of the semi-supervised PV energy anomaly detection technique facing four anomaly patterns is summarized in Tables 27.10–27.14.

The PV energy anomaly detection performance is summarized in Table 27.14. The anomaly detection method's high precision value of around 87–94% indicates there will not be too many false alarms, and the operators will trust the system when the alarm happens. Besides, the false positive rate value of 6.12% is also satisfactory, indicating the customer will not be often disappointed if they submit normal data but are identified as abnormal by the system. The true positive rate indicates that over 90% of the actual anomalies will be detected when facing scaling, simple, and two-way

Table 27.10 PV energy anomaly detection performance facing scaling anomaly injection.

Scaling anomalies		PV energy predicted value: Normal	PV energy predicted value: Anomaly
PV energy actual value	Normal	True negative = 230	False positive = 15
	Anomaly	False negative = 3	True positive = 242

Table 27.11 PV energy anomaly detection performance facing simple ramp anomaly injection.

Simple ramp anomalies		PV energy predicted value: Normal	PV energy predicted value: Anomaly
PV energy actual value	Normal	True negative = 230	False positive = 15
	Anomaly	False negative = 9	True positive = 236

Table 27.12 PV energy anomaly detection performance facing two-way ramp anomaly injection.

Two-way ramp anomalies		PV energy predicted value: Normal	PV energy predicted value: Anomaly
PV energy actual value	Normal	True negative = 230	False positive = 15
	Anomaly	False negative = 20	True positive = 225

Table 27.13 PV energy anomaly detection performance facing random anomaly injection.

Random anomalies		PV energy predicted value: Normal	PV energy predicted value: Anomaly
PV energy actual value	Normal	True negative = 230	False positive = 15
	Anomaly	False negative = 145	True positive = 100

Table 27.14 PV energy anomaly detection overall performance.

PV energy anomaly detection performance metrics	Scaling anomalies	Simple ramp anomalies	Two-way ramp anomalies	Random anomalies
True positive rate (TPR)	98.78%	96.33%	91.84%	40.82%
False positive rate (FPR)	6.12%	6.12%	6.12%	6.12%
Precision (P)	94.16%	94.02%	93.75%	86.96%
$F1$	96.41%	95.16%	92.78%	55.56%
Accuracy (A)	96.33%	95.10%	92.86%	67.35%

ramp anomaly injection. In contrast, around 40% of the actual anomalies will be detected when facing random anomaly injection. The overall accuracy of the PV energy trading anomaly detection will perform best when facing a scaling anomaly injection with an overall accuracy of 96.33%, while performance worst when facing a random anomaly injection with an overall accuracy of 67.35%.

To sum, this section introduces a semi-supervised machine learning model to predict and calculate the confidence interval of the expected value based on the standard deviation of residuals (which can be historical error rates, or RMSE). Any actual data point beyond this confidence interval is an anomaly. When the building's hourly energy consumption and PV energy forecast models are accurate enough, this method can efficiently detect anomalies, especially local outliers in time series trends.

27.4 Blockchain-Based Anomaly Detection Case Study

Blockchain technology has emerged as a promising encryption ledger technology to record energy-related transactions at the distributed level, ensuring P2P energy trading is more reliable and makes transactions undeniable.

In this dissertation, experiments based on the Hyperledger Composer blockchain platform are performed to demonstrate the benefits of blockchain-enabled negawatt-hours and PV energy trading. Once the blockchain-based energy trading platform runs on Hyperledger, the transaction record cannot be modified. Both the user balance and ledger held by peers are consistent and immutable. If a "block" of transactions wants to be appended to the blockchain, network participants must agree the transaction is valid through a process called consensus. This immutable feature makes the transaction more reliable and transparent to all participants.

Case studies are implemented and analyzed to demonstrate the use of blockchain platforms in P2P energy trading, and the anomaly detection feature is also demoed in this session.

27.4.1 Blockchain-Based Negawatt-Hour Trading Platform

The blockchain-based negawatt-hour trading platform was developed by me and published in the paper [14]. The summary of the blockchain-based negawatt-hours architecture is as follows.

Two negawatt-hour trading scenarios are proposed and simulated. One is the negawatt-hour trading between demand response aggregators and buildings. Another is the negawatt-hour trading among buildings to meet the contract that is not satisfied in the first scenario. A simulation platform built on Hyperledger Composer, including participants, assets, transactions, and transaction flows, is detailed, designed, and illustrated. This chapter shows how blockchain energy trading smart contracts and transaction flow work.

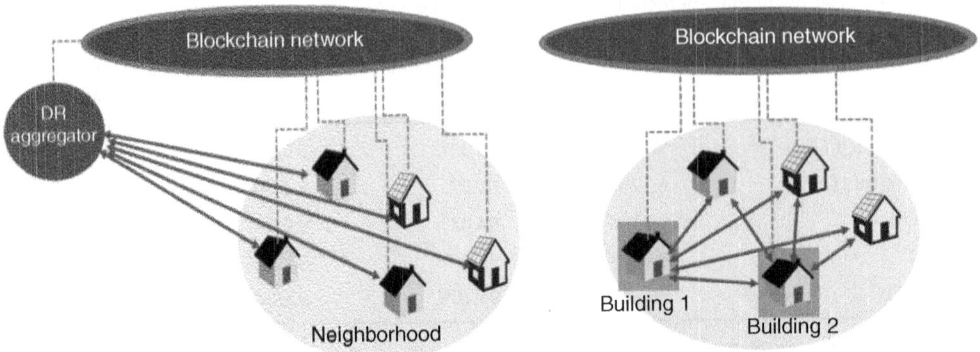

Figure 27.20 Negawatt-hour trading between DRA and buildings(left) and Peer-to-peer negawatt-hour trading (right).

Two negawatt-hour trading scenarios are proposed and simulated. One is the negawatt-hour trading between demand response aggregators (DRA) and buildings. Another is the negawatt-hour trading among buildings to meet the contract that is not satisfied in the first scenario, as shown in Figure 27.20.

A way of calculating market clearing price in blockchain energy trading is proposed as below:

$$MCP_t = \frac{\sum_{i=0}^{n} \beta_i \times kW_i}{\sum_{i=0}^{n} kW_i} \quad (27.1)$$

Where,

MCP_t : Market-clearing price at time t
n : Number of negawatt-hour sellers
i : Index of the negawatt-hour seller
β_i : Negawatt-hour seller's bid price
kW_i : Negawatt-hours ready to sell

In the first scenario, at the beginning of each hour, the business network accepts offers from the DRA aggregator by publishing "RequestnWBroadcast." This signal includes the necessary demand reduction quantity and the specified time frame, e.g., 50 kWh from 13:00 to 14:00. Once building owners receive the broadcasted message, they can respond with the negawatt amount (kWh) and available negawatt period. A negawatt offer from a building may consist of varying available negawatt amounts at different intervals, such as 6 kWh from 13:00 to 14:00 and 4 kWh from 13:00 to 14:00. The negawatt-hour sellers bid into the market by submitting "nWOffer." The market is cleared hourly by sorting the negawatt-hours offer amount at the "DRAofferPrice." Then in the next hour, the exchange of excess negawatt-hours occurs. Subsequently, following the hourly market clearance, the anticipated negawatts are transacted through the implementation of designated load restrictions, achievable via home/building energy management (HEM/BEM) software. The hourly bidding process and transaction flow are summarized in Figures 27.21 and 27.22.

In the second P2P scenario, the business network accepts offers from both negawatt-hour sellers and buyers at the beginning of each hour by publishing RequestP2PnWBroadcast in the blockchain. Then both negawatt-hour sellers and buyers bid into the market as blockchain participants. Building owners unable to fulfill the demand reduction mandated by the DRA inform other buildings of their negawatts contract deficiencies by disseminating a message to the blockchain network.

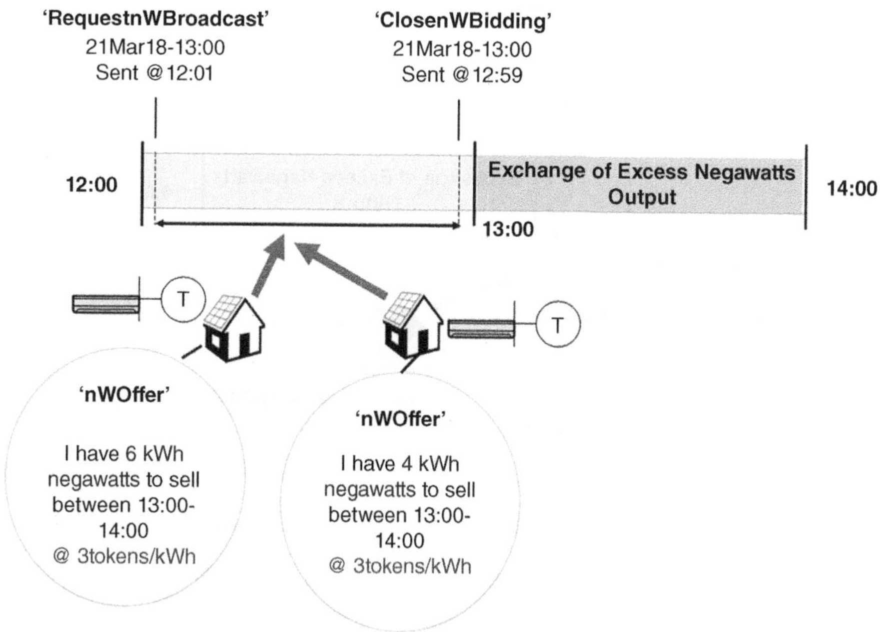

Figure 27.21 DRA negawatt-hour trading hourly bidding process.

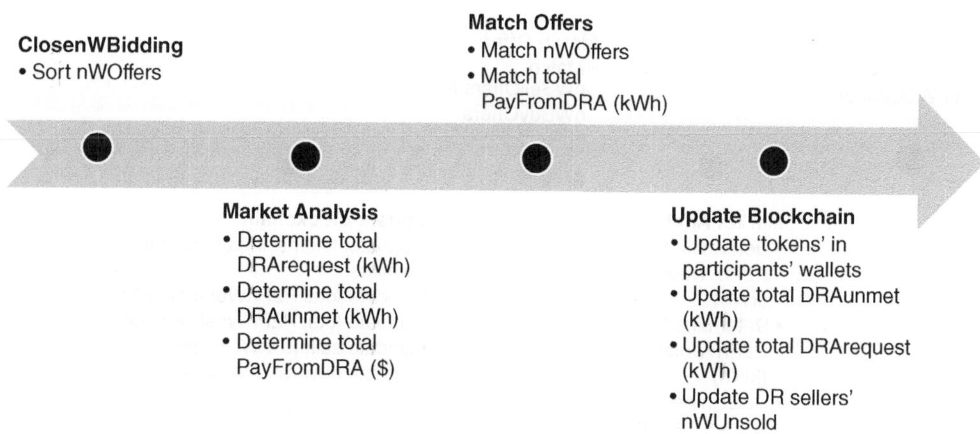

Figure 27.22 DRA negawatt-hour trading transaction flow in Hyperledger Composer.

This signal includes the required demand reduction from neighboring entities (kWh), the price (tokens/kWh) the building owner is prepared to pay, and the one-hour demand reduction interval, for instance, 10 kWh, 4 tokens/kWh, from 13:00 to 14:00. Building owners interested in selling their negawatts may answer to the broadcast message with the negawatt quantity (kWh), the bid price (token/kWh), and the duration of available negawatts. A single negawatt selling proposal may include varying negawatt quantities throughout distinct time intervals, for instance, 6 kWh at 3 tokens/kWh from 13:00 to 14:00, and 3 kWh at 8 tokens/kWh from 15:00 to 16:00. The market is cleared hourly according to the calculated MCP_t price in the smart contract. And the exchange of excess negawatt-hours occurs during the next hour. The hourly bidding process and transaction flow are summarized in Figures 27.23 and 27.24.

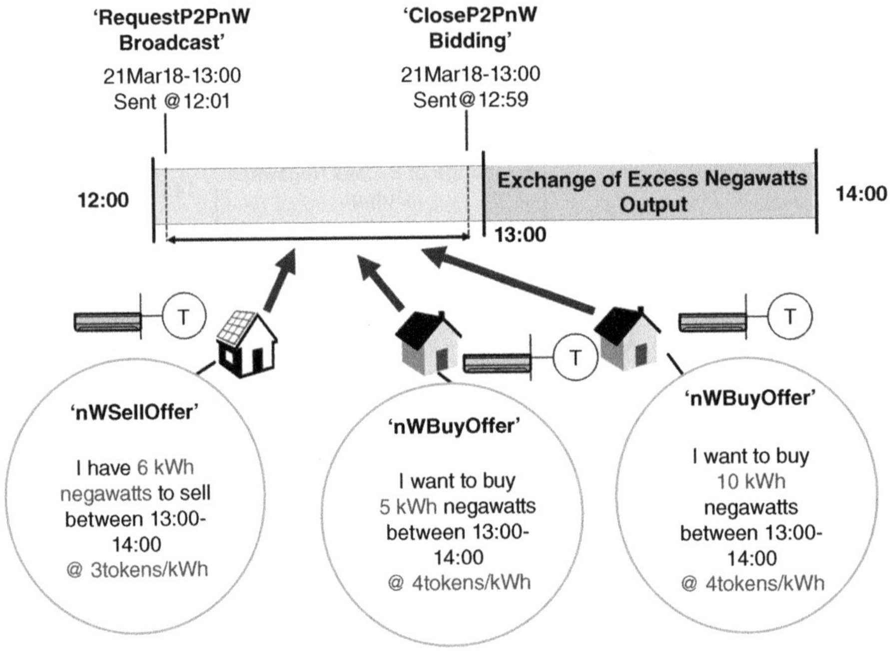

Figure 27.23 Peer-to-peer negawatt-hour trading hourly bidding process.

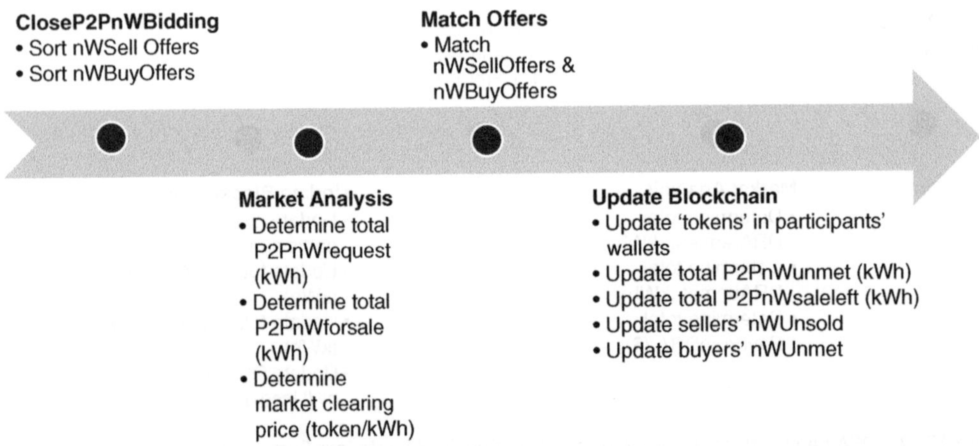

Figure 27.24 Peer-to-peer negawatt-hour trading transaction flow implemented in Hyperledger.

The chapter simulates an experimental negawatt-hour trading test case study in Hyperledger Composer. The result shows that blockchain-based negawatt-hour trading is very promising and efficient. Besides, it also demonstrates the feasibility of blockchain-based negawatt-hour trading. Five homeowners are included in this case study. The five homeowners with a beginning balance of 2000 tokens are willing to sell their demand response capability during the next trading hour.

In the case study, assuming there are five homeowners submitting offers to the blockchain network in response to the "RequestnWBroadcast" message with the "listingid" of "06091300" which means June 9th at 13:00. Their negawatt-hours available are broadcasted to all peers in the blockchain.

```
if (nWOffer.nWQuantity > 44 || nWOffer.nWQuantity < 36) {
  throw new Error('Input value is abnormal');
}
```

Figure 27.25 Code sample of negawatt-hours anomaly detection in Hyperledger Composer.

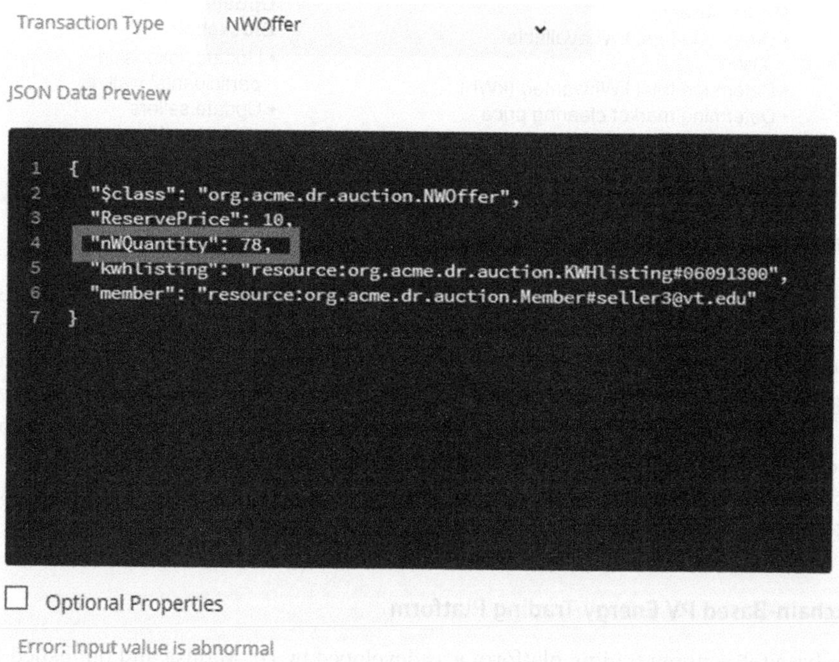

Figure 27.26 Error message showing the abnormal negawatt-hours is detected successfully.

It is assumed that all homeowners are willing to join the demand response and sacrifice their comfort by increasing the HVAC setpoint by 2°. As to seller 3's building, according to seller 3 building's negawatt-hours pattern, the electrical hourly energy consumption in the demand response peak period can be predicted and compared in two cases. One case is using constant HVAC. Another case is the HVAC setpoint adjustment case which means the HVAC setpoint is increased by 1.1 °C (2 °F) degrees from 12 p.m. to 4 p.m. Assuming the upper and lower bounds of the anomaly detection algorithm is 36 and 44 kWh, the negawatt-hours compacity of seller 3 should be between (36 and 44 kWh) if the homeowner are willing to join the demand response by increasing HVAC by 2° at noon. Any negawatt-hour quantity number out of this range will be abnormal.

In each negawatt-hour offer, the "nWQuantity" means the negawatt-hours available to sell by that negawatt-hour seller. One abnormal data of negawatt-hour quantity is submitted by seller 3 at 78 kWh, outside the normal margin of his negawatt-hours capacity. By coding in the Hyperledger Composer platform's script section, as shown in Figure 27.25, we can set up a negawatt-hour quantity limit for the offer submitted.

While if negawatt-hour seller 3 still wants to submit an offer with negawatt-hour quantity as 78 kWh, an error code will show up, as shown in Figure 27.26, saying "Error: Input value is abnormal."

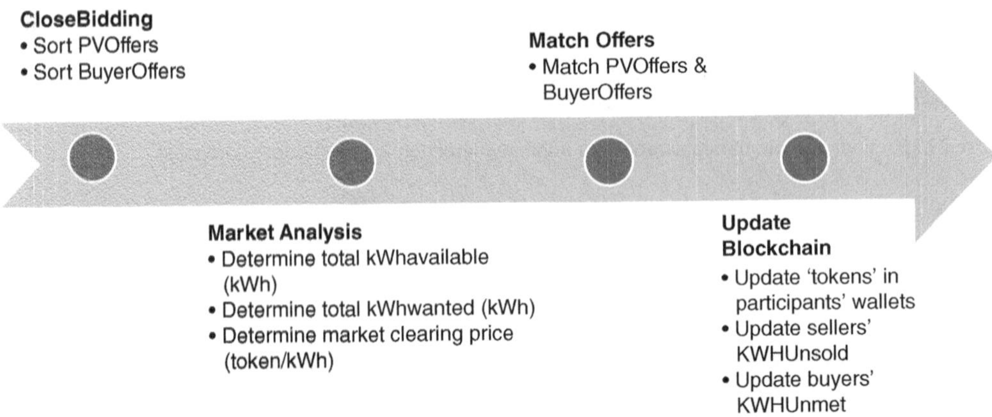

Figure 27.27 Peer-to-peer PV energy trading transaction flow implemented in Hyperledger Composer.

As illustrated in the anomaly detection section, the upper and lower bound numbers 44 and 36 will change according to the hourly energy consumption forecasting result of the machine learning forecasting model. And these bound numbers are directly from the difference between case I and case II building load forecasting results. Thus, anomaly detection features in a blockchain-based negawatt-hour trading platform can be achieved by linking hourly building energy consumption forecasting models and blockchain-based energy trading platforms.

Thus, in this case study, the abnormal offer by seller 3 is successfully detected among the five seller offers submitted.

27.4.2 Blockchain-Based PV Energy Trading Platform

The blockchain-based PV energy trading platform was developed by Dr. Manisa and published in the paper [65]. The implementation is also based on the Hyperledger Composer, and the design of participants, assets, transactions, and transaction flows is illustrated in the paper. The way how blockchain tracks the transaction record of PV energy output exchanges is also described and detailed in the paper. The hourly transaction flow of PV energy trading is summarized in Figure 27.27. In this dissertation, an added anomaly detection feature is developed, demoed, and tested based on Dr. Manisa's blockchain PV energy trading network.

Twelve homeowners are included in the experimental case study. The five homeowners with a beginning balance of 2000 tokens are willing to sell their surplus PV energy during the next trading hour. While at the same time, another seven homeowners with a beginning balance of 2000 tokens are willing to buy electricity during the next trading hour.

In this case study, assuming there are five homeowners submitting offers to the blockchain network in response to the "AcceptOfferBroadcast" message with the "listing of 14:00." Their kWhtobid available are broadcasted to all peers in the blockchain. The forecasted PV energy for the next hour is 50 kWh. The upper and lower anomaly boundary can be calculated from the anomaly detection section. Assuming the lower boundary is 40 kWh, the higher boundary is 60 kWh. All the normal kWhtobid PV sellers are within the margin (40 and 60 kWh). And the data that falls outside this margin will be detected as an anomaly.

The KWHlisting includes transaction offers details. In each PV offer, the "KWHtobid" means the kWh that PV sellers want to sell at least. And the "KWHavailable" means the PV available to sell by that PV seller. Five PV sellers join the market with four normal data by sellers 1, 2, 3, 5.

```
if (pvoffer.KWHavailable < 40 || pvoffer.KWHavailable > 60) {
  throw new Error('Input value is abnormal');
}
```

Figure 27.28 Code sample of PV energy seller's anomaly detection in Hyperledger Composer.

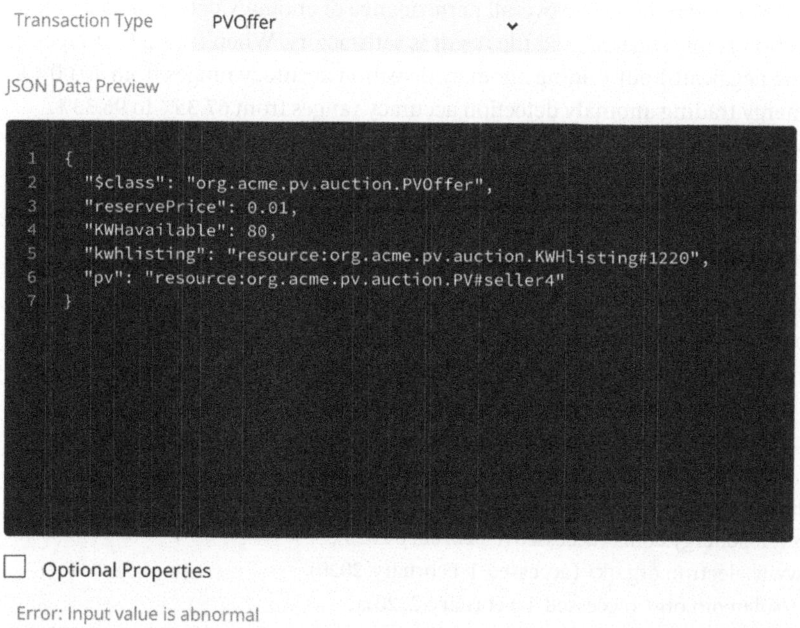

Figure 27.29 Error message showing the abnormal PV energy quantity is detected successfully.

And one abnormal data of available PV energy is submitted by seller 4 at 80 kWh, outside the normal margin.

By coding in the Hyperledger platform, as shown in Figure 27.28, we can set up a PV energy limit for the offer submitted below.

While if PV seller 4 still wants to submit an offer with a PV energy quantity of 80 kWh, after submission, an error code will show up, as shown in Figure 27.29, saying "Error: Input value is abnormal."

As illustrated in the anomaly detection section, the upper and lower bound numbers 40 and 60 kWh will change according to the model's hourly PV energy forecasting result. Thus, anomaly detection features in a blockchain-based PV energy trading platform can be achieved by linking the PV energy forecasting models and blockchain-based PV energy trading platforms.

In this case study, the abnormal offer by seller 4 is successfully detected among the five seller offers submitted.

27.5 Conclusion

This chapter has provided a semi-supervised machine learning architecture of anomaly detection tools used in negawatt-hour and PV energy trading. The prominence is given to the potential application benefits and research gaps in anomaly detection tools in P2P negawatt-hour and PV energy trading. This field's state of the art is promising as more people realize the importance and future of negawatt-hour and PV energy trading anomaly detection. The semi-supervised learning method introduced in this paper can detect malicious user input in negawatt-hour trading based on the accurate building hourly energy consumption model with setpoint adjustment. Besides, it can also be used to detect malicious user input in PV energy trading based on the accurate PV energy forecasting model on clear days. The overall performance of anomaly detection facing four types of anomaly injection is investigated, and the result is satisfactory. When facing four types of anomaly injections, the negawatt-hour trading anomaly detection accuracy ranges from 70.00% to 83.50%, and the PV energy trading anomaly detection accuracy ranges from 67.35% to 96.33%.

Besides, a blockchain-based P2P energy trading platform is demoed on the Hyperledger platform. A simple simulation shows the feasibility of anomaly detection in blockchain-based energy trading.

References

1 M. Pipattanasomporn, S. Rahman and M. Kuzlu (2019), "Blockchain-based Solar Electricity Exchange: Conceptual Architecture and Laboratory Setup," 2019 IEEE Power & Energy Society Innovative Smart Grid Technologies Conference (ISGT), Washington, DC, USA, 2019, pp. 1–5, https://doi.org/10.1109/ISGT.2019.8791663.

2 Zhang, C., Wu, J., Long, C., and Cheng, M. (2017). Review of existing peer-to-peer energy trading projects. *Energy Procedia* 105: 2563–2568. https://doi.org/10.1016/j.egypro.2017.03.737.

3 LO3 Energy. https://lo3energy.com/ (accessed 1 February 2020).

4 Electron. https://www.electron.org.uk/ (accessed 1 February 2020).

5 Solar Coin. https://solarcoin.org/ (accessed 1 February 2020).

6 IERC—International Energy Research Centre. https://www.ierc.ie/ (accessed 1 February 2020).

7 Liu, N., Yu, X., Wang, C. et al. (2017). Energy-sharing model with price-based demand response for microgrids of peer-to-peer prosumers. *IEEE Transactions on Power Apparatus and Systems* 32 (5): 3569–3583. https://doi.org/10.1109/TPWRS.2017.2649558.

8 Lovins, A.B. (1990). The negawatt revolution. *Across Board (NY)* XXVII (9): 18–23. [Online]. http://thewindway.com/pdf/E90-20_NegawattRevolution.pdf%0Afile:///Files/E5/E5CCF070-2847-4417-A027-14059BFAB958.pdf%0Ahttps://www.rmi.org/wp-content/uploads/2017/06/RMI_Negawatt_Revolution_1990.pdf.

9 Pentland, W. Beyond efficiency—the case for the 'virtual' negawatt. https://www.forbes.com/sites/williampentland/2012/03/12/beyond-energy-efficiency-the-case-for-the-virtual-negawatt/#29e41ab05d38 (accessed 2 February 2020).

10 Jacobs, S.B. (2015). Bypassing federalism and the administrative law of negawatts. *Iowa Law Review* 100 (3): 885–945. https://doi.org/10.2139/ssrn.2406684.

11 Tomic, S. (2012). Economic effects of trading watts and negawatts by agile customers in hierarchic energy markets. *9th International Conference on the European Energy Market, EEM 12*, pp. 1–6. https://doi.org/10.1109/EEM.2012.6254797.

12 Okawa, Y. and Namerikawa, T. (2017). Distributed optimal power management via negawatt trading in real-time electricity market. *IEEE Transactions on Smart Grid* 8 (6): 3009–3019. https://doi.org/10.1109/TSG.2017.2705291.

13 Liu, W., Wen, F., and Qi, D. (2020). Intraday residential demand response scheme based on peer-to-peer energy trading. *IEEE Transactions on Industrial Informatics* 16 (3): 1. https://doi.org/10.1109/tii.2019.2929498.

14 Jing, Z., Pipattanasomporn, M., and Rahman, S. (2019). Blockchain-based negawatt trading platform: conceptual architecture and case studies. *2019 IEEE PES GTD Grand International Conference and Exposition Asia (GTD Asia)* (May 2019), pp. 68–73. https://doi.org/10.1109/GTDASIA.2019.8715890.

15 Hua, W., and Sun, H. (2019). A blockchain-based peer-to-peer trading scheme coupling energy and carbon markets. *SEST 2019—International Conference on Smart Energy Systems and Technologies*, pp. 1–6. https://doi.org/10.1109/SEST.2019.8849111.

16 Sheikh, A., Kamuni, V., Asfia, U. et al. (2019). Secured energy trading using byzantine based blockchain consensus. *IEEE Access* 8: 1–1. https://doi.org/10.1109/access.2019.2963325.

17 Lee, B. S., Song, J. G., Moon, S. J., et al. (2020). Blockchain architectures for P2P energy trading between neighbors. *2019 International Conference on Information and Communication Technology Convergence (ICTC)*, pp. 1013–1017. https://doi.org/10.1109/ictc46691.2019.8939856.

18 Nakamoto, S. (2008). *Bitcoin: a peer-to-peer electronic cash system*. www.bitcoin.org (accessed 1 February 2020).

19 Wörner, A., Meeuw, A., Ableitner, L. et al. (2019). Trading solar energy within the neighborhood: field implementation of a blockchain-based electricity market. *Energy Informatics* 2 (S1): 1–12. https://doi.org/10.1186/s42162-019-0092-0.

20 Xue, L., Teng, Y., Zhang, Z., et al. (2018). Blockchain technology for electricity market in microgrid. *2017 2nd International Conference on Power and Renewable Energy (ICPRE)*, pp. 704–708. https://doi.org/10.1109/ICPRE.2017.8390625.

21 Li, H.A., and Nair, N.K.C. (2019). Cooperative control in an islanded microgrid under blockchain-based market operation. *2019 IEEE Innovative Smart Grid Technologies—Asia (ISGT Asia)*, pp. 2766–2771. https://doi.org/10.1109/ISGT-Asia.2019.8881229.

22 Mezquita, Y., Gazafroudi, A. S., Corchado, J. M., et al. (2019). Multi-agent architecture for peer-to-peer electricity trading based on blockchain technology. *ICAT 2019—27th International Conference on Information, Communication and Automation Technologies (ICAT) Proceedings*, pp. 1–6.

23 Chen, S. (2019). A framework of decentralized electricity market based on the collaborative mechanism of blockchain and edge computing. *2019 IEEE International Conference on Service Operations and Logistics, and Informatics (SOLI)*, pp. 219–223.

24 Dang, C., Zhang, J., Kwong, C.P., and Li, L. (2019). Demand side load management for big industrial energy users under blockchain-based peer-to-peer electricity market. *IEEE Transactions on Smart Grid* 10 (6): 6426–6435. https://doi.org/10.1109/TSG.2019.2904629.

25 Zhang, C., Wu, J., Zhou, Y. et al. (2018). Peer-to-peer energy trading in a microgrid. *Applied Energy* 220: 1–12. https://doi.org/10.1016/j.apenergy.2018.03.010.

26 Li, Nair, N.K.C. (2018). Blockchain-based microgrid market and trading mechanism. *Australasian Universities Power Engineering Conference, AUPEC 2018*, pp. 1–5. https://doi.org/10.1109/AUPEC.2018.8757870.

27 Santoyo, S. A brief overview of outlier detection techniques. https://towardsdatascience.com/a-brief-overview-of-outlier-detection-techniques-1e0b2c19e561 (accessed 1 February 2020).

28 De Stefano, C., Sansone, C., and Vento, M. (2000). To reject or not to reject: that is the question—an answer in case of neural classifiers. *IEEE Transactions on Systems, Man, and Cybernetics Part C: Applications and Reviews* 30 (1): 84–94. https://doi.org/10.1109/5326.827457.

29 Barbará, D., Wu, N., and Jajodia, S. (2001). Detecting novel network intrusions using bayes estimators, Proceedings of the 2001 SIAM International Conference on Data Mining, SDM 2001, Chicago, IL, USA, 2001, pp. 1–17. https://doi.org/10.1137/1.9781611972719.28.

30 Rätsch, G., Mika, S., Schölkopf, B., and Müller, K.R. (2002). Constructing boosting algorithms from SVMs: an application to one-class classification. *IEEE Transactions on Pattern Analysis and Machine Intelligence* 24 (9): 1184–1199. https://doi.org/10.1109/TPAMI.2002.1033211.

31 King, S.P., King, D.M., Anuzis, P. et al. (2002). *The Use of Novelty Detection Techniques for Monitoring High-Integrity Plant*, 221–226. Dept. Engineering Science Oxford University.

32 Tandon G. and Chan, P. K. (2007). Weighting versus pruning in rule validation for detecting network and host anomalies. *KDD '07: Proceedings of the 13th ACM SIGKDD international conference on Knowledge discovery and data mining*, pp. 697–706. https://doi.org/10.1145/1281192.1281267.

33 Ramadas, M., Ostermann, S., and Tjaden, B. (2003). Detecting anomalous network traffic with self-organizing maps. *Lecture Notes in Computer Science (including subseries Lecture Notes in Artificial Intelligence and Lecture Notes in Bioinformatics)* 2820: 36–54. https://doi.org/10.1007/978-3-540-45248-5_3.

34 Lin, J., Keogh, E., Fu, A., et al. (2005). Approximations to magic: Finding unusual medical time series. *18th IEEE Symposium on Computer-Based Medical Systems (CBMS'05)*, pp. 329–334. https://doi.org/10.1109/cbms.2005.34.

35 Hautamäki, V., Kärkkäinen, I., and Fränti, P. (2004). Outlier detection using k-nearest neighbour graph. *Proceedings of the 17th International Conference on Pattern Recognition, 2004. ICPR 2004*, vol. 3, pp. 430–433. https://doi.org/10.1109/ICPR.2004.1334558.

36 Yamanishi, K., Takeuchi, J.I., Williams, G., and Milne, P. (2004). On-line unsupervised outlier detection using finite mixtures with discounting learning algorithms. *Data Mining and Knowledge Discovery* 8 (3): 275–300. https://doi.org/10.1023/B:DAMI.0000023676.72185.7c.

37 Lakhina, A., Crovella, M., and Diot, C. (2005). Mining anomalies using traffic feature distributions. *Computer Communication Review* 35 (4): 217–228. https://doi.org/10.1145/1090191.1080118.

38 Xiuyao, S., Mingxi, W., Jermaine, C., and Ranka, S. (2007). Conditional anomaly detection. *IEEE Transactions on Knowledge and Data Engineering* 19 (5): 631–644. https://doi.org/10.1109/TKDE.2007.1009.

39 Blender, R., Fraedrich, K., and Lunkeit, F. (1997). Identification of cyclone-track regimes in the North Atlantic. *Quarterly Journal of the Royal Meteorological Society* 123 (539): 727–741. https://doi.org/10.1256/smsqj.53909.

40 Chan, P. K. and Mahoney, M. V. (2005). Modeling multiple time series for anomaly detection. *Fifth IEEE International Conference on Data Mining (ICDM'05)*, pp. 90–97. https://doi.org/10.1109/ICDM.2005.101.

41 Pickands, J. III, (1975). Statistical inference using extreme order statistics. *The Annals of Statistics* 3 (1): 119–131.

42 Aggarwal, C.C. (2016). *Outlier analysis*, vol. 9781461463. Nagoya, Japan: Springer.

43 Yijia, T. and Hang, G. (2016). Anomaly detection of power Consumption based on waveform feature recognition. *ICCSE 2016—11th International Conference on Computer Science & Education*, Nagoya, Japan, 2016, vol. 2, no. ICCSE, pp. 587–591. https://doi.org/10.1109/ICCSE.2016.7581646.

44 Zhang, Y., Chen, W., and Black, J. (2011). Anomaly detection in premise energy consumption data. *IEEE Power and Energy Society General Meeting*, pp. 1–8. https://doi.org/10.1109/PES.2011.6039858.

45 Liang, P., Yang, H.D., Chen, W.S. et al. (2018). Transfer learning for aluminium extrusion electricity consumption anomaly detection via deep neural networks. *International Journal*

of Computer Integrated Manufacturing 31 (4–5, 405): 396. https://doi.org/10.1080/0951192X.2017.1363410.

46 Fontugne, R., Ortiz, J., Tremblay, N., et al. (2013). Strip, bind, and search: a method for identifying abnormal energy consumption in buildings. *IPSN 2013—2013 ACM/IEEE International Conference on Information Processing in Sensor Networks*, pp. 129–140. https://doi.org/10.1145/2461381.2461399.

47 Ashok, A., Govindarasu, M., and Ajjarapu, V. (2018). Online detection of stealthy false data injection attacks in power system state estimation. *IEEE Transactions on Smart Grid* 9 (3): 1636–1646. https://doi.org/10.1109/TSG.2016.2596298.

48 Lin, Y., Abur, A., and Xu, H. (2019). Critical model parameters: a security vulnerability in electricity market operation. *2019 IEEE Milan PowerTech, PowerTech 2019*, pp. 1–6. https://doi.org/10.1109/PTC.2019.8810679.

49 Gao, X., Ye, N., and Song, J., et al. (2018). A method to inspect the implementation of electricity price based on deep learning variational autoencoder. *2018 2nd IEEE Conference on Energy Internet and Energy System Integration (EI2)*, pp. 1–5. https://doi.org/10.1109/EI2.2018.8582186.

50 Yue, M. (2018). An integrated anomaly detection method for load forecasting data under cyberattacks. *2017 IEEE Power & Energy Society General Meeting*, vol. 2018-January, pp. 1–5. https://doi.org/10.1109/PESGM.2017.8273964.

51 Zamani-Dehkordi, P., Rakai, L., and Zareipour, H. (2016). A data-driven method to detect the abnormal instances in an electricity market. *Proceedings—2015 IEEE 14th International Conference on Machine Learning and Applications, ICMLA 2015*, pp. 1050–1055. https://doi.org/10.1109/ICMLA.2015.63.

52 Zhou, A., Zhu, L., Qiu, H., et al. (2016). Detection of abnormal trends in electrical data. *Proceedings of 2015 IEEE International Conference on Progress in Informatics and Computing, PIC 2015*, pp. 247–251. https://doi.org/10.1109/PIC.2015.7489847.

53 Meiyu, S. (2012). Research on the discords detecting on time series based on distance and density. *Proceedings—2012 9th International Conference on Fuzzy Systems and Knowledge Discovery, FSKD 2012*, no. Fskd, pp. 1084–1088. https://doi.org/10.1109/FSKD.2012.6233997.

54 Majadi, N., Trevathan, J., and Bergmann, N. (2019). Collusive shill bidding detection in online auctions using Markov Random Field. *Electronic Commerce Research and Applications* 34 (January): 100831. https://doi.org/10.1016/j.elerap.2019.100831.

55 Sayadi, S., Ben Rejeb, S., and Choukair, Z. (2019). Anomaly detection model over blockchain electronic transactions. *2019 15th International Wireless Communications & Mobile Computing Conference (IWCMC)*, pp. 895–900. https://doi.org/10.1109/IWCMC.2019.8766765.

56 Podgorelec, B., Turkanović, M., and Karakatič, S. (2020). A machine learning-based method for automated blockchain transaction signing including personalized anomaly detection. *Sensors (Switzerland)* 20 (1): https://doi.org/10.3390/s20010147.

57 Song, J., Nang, J., and Jang, J. (2020). Design of anomaly detection and visualization tool for IoT blockchain. *2018 International Conference on Computational Science and Computational Intelligence (CSCI)*, pp. 1464–1465. https://doi.org/10.1109/csci46756.2018.00292.

58 Iyer, S., Thakur, S., Dixit, M., et al. (2020). Blockchain and anomaly detection based monitoring system for enforcing wastewater reuse. *2019 10th International Conference on Computing, Communication and Networking Technologies (ICCCNT)*, pp. 1–7. https://doi.org/10.1109/icccnt45670.2019.8944586.

59 Signorini, M., Pontecorvi, M., Kanoun, W., et al. (2018). Advise: Anomaly detection tool for blockchain systems. *2018 IEEE World Congress on Services (SERVICES)*, pp. 67–68. https://doi.org/10.1109/SERVICES.2018.00046.

60 Signorini, M., Pontecorvi, M., Kanoun, W., et al. [1807.03833] BAD: Blockchain Anomaly Detection. https://arxiv.org/abs/1807.03833? (accessed 2 February 2020).

61 Effective Approaches for Time Series Anomaly Detection | by Aditya Bhattacharya | Towards Data Science. https://towardsdatascience.com/effective-approaches-for-time-series-anomaly-detection-9485b40077f1 (accessed 8 July 2022).

62 Hyndman and R. J. and Athanasopoulos, G. (2018). Forecasting: principles and practice. *Otexts* online, open-access textbook.

63 Sridhar, S. and Govindarasu, M. (2014). Model-based attack detection and mitigation for automatic generation control. *IEEE Transactions on Smart Grid* 5 (2): 580–591. https://doi.org/10.1109/TSG.2014.2298195.

64 Akbarian, F., Ramezani, A., Hamidi-Beheshti, M.T., and Haghighat, V. (2020). Advanced algorithm to detect stealthy cyber attacks on automatic generation control in smart grid. *IET Cyber-Physical Systems: Theory & Applications* 5 (4): 351–358. https://doi.org/10.1049/IET-CPS.2019.0074.

65 Pipattanasomporn, M., Kuzlu, M., and Rahman, S. (Feb. 2019). A blockchain-based platform for exchange of solar energy: laboratory-scale implementation. *2018 International Conference and Utility Exhibition on Green Energy for Sustainable Development (ICUE)*, vol. 2018-October. https://doi.org/10.23919/ICUE-GESD.2018.8635679.

28

Optimal Coordination of VSC-Interfaced Subsystems to Safeguard the Frequency Performance of Cyber-Physical Power Systems

Georgios Giannakopoulos, Arcadio Perilla Guerra, José Luis Rueda Torres, and Peter Palensky

Intelligent Electrical Power Grids, Department of Electrical Sustainable Energy, Faculty of Electrical Engineering, Mathematics and Computer Science, Delft University of Technology, The Netherlands

28.1 Motivation and Scope of the Chapter

Integrated energy systems are increasingly dominated by the operation and control of power electronic interfaced (PEI) devices at generation, multi-energy storage, transmission, and distribution, and responsive demand [1, 2]. Simultaneously, the so-called voltage source converter technology (VSC) is positioning as the preferred option for implementation of PEI due to its versatile and flexible operation and control modes [1, 3]. The progressive phase-out of conventional alternating current technologies (e.g., synchronous generators) and their replacement by VSC-based PEI brings major changes in the dynamic behavior of the emerging forms of DC-AC system topologies [4]. For instance, reduced, limited, or absent control capabilities may entail unprecedented frequency or voltage deviations, which could happen in shorter time scales (e.g., less than a fraction of a second), when compared to typical deviations observed in conventional systems dominated by synchronous generators [4, 5]. This motivates a renewed emphasis and significant research efforts towards the development of models of multi-area and multi-energy HVDC-HVAC cyber-physical power system, which enable a trustworthy computer-aided investigation of new forms of dynamic stability phenomena [4, 6].

This chapter particularly focuses on the problem of the implications of altered active power balance on the dynamic frequency performance of cyber-physical power systems. To this aim, and in line with the current state-of-the-art, it is considered that the PEIs of new component additions may be equipped with emerging method(s) for fast frequency support (FFS) as a function of the underlying operation and control capabilities of the latest VSC technologies [7]. Hence, FFS is assumed as an available functionality of new key components like turbines (WTs) [8], solar-photovoltaic systems (PVs) [9], HVDC links [10] and power-to-gas conversion units (e.g., electrolysers) [11].

FFS working under the control principle of active power gradient is one of the promising options of effective mitigation of undesirable fast and significant dynamic frequency deviations [12]. This technique can be tuned for one or several disturbances affecting the active power balance. This can be done by considering two settings, i.e., the amount of active power (ΔP) to be quickly injected/absorbed, and the rate or gradient (APG) at which the active power is adjusted. Due

Smart Cyber-Physical Power Systems: Fundamental Concepts, Challenges, and Solutions, Volume 1, First Edition.
Edited by Ali Parizad, Hamid Reza Baghaee, and Saifur Rahman.

to the different dynamic properties of PEI distributed across a multi-area and multi-energy HVDC-HVAC cyber-physical power system, achieving an optimally effective FFS constitutes an open research challenge [12, 13]. For instance, the investigation discussed in [12] illustrates the limited effectiveness of uncoordinated tuning, taking as an example the challenges of calibrating the active power gradient based FFS applied to two-terminal VSC-based HVDC interconnector.

The risk of ineffective FFS is specially higher in systems with low headroom and adverse interplay of the resources guided by FFS under different forms of active power imbalances [13]. Recently, attempts to tune FFS systems with few selected PEIs have been conducted based on single parameter parametric sensitivities. Nevertheless, as shown in [13, 14], optimal effectiveness under diverse operating conditions and disturbances is not ensured.

Hence, proposing a suitable optimization-oriented problem statement is also challenge tackled in this chapter. The goal is to obtain an effective coordinated tuning of PEIs. To this end, and for illustrative purposes, the proposed multi-area and multi-energy HVDC-HVAC cyber-physical power system assumes FFS attached to VSC-HVDC links interconnecting the synchronous areas of the system as well as attached to proton exchange membrane (PEM) electrolysers. The example shown in the chapter considers the occurrence of a large-size active power imbalance.

The proposed optimization statement targets the enhancement of the overall frequency stability by properly and cooperatively mitigating the imbalance through controllable active power resources that are in service in the synchronous areas, also taking into account their own characteristics and limitations for FFS. The model is implemented by using library components of the software package known as DIgSILENT PowerFactory 2022 (DPF2022), whereas the optimization search process of the mean-variance mapping optimization (MVMO) algorithm, described in [15], is tailored and deployed in the chapter to find a near-to-optimal solution within a restricted computational budget (due to the need repetitively performing RMS simulations).

The reminder of the chapter is structured in the following sequence: The multi-area and multi-energy HVDC-HVAC cyber-physical power system is presented in Section 28.2. The proposed optimization problem statement is provided in Section 28.3. Section 28.4 concisely overviews the application of the MVMO algorithm. Findings from numerical simulations are given in Section 28.5. Concluding reflections and prospective future subsequent research aspects are summarized in Section 28.6.

28.2 The HVDC–HVAC Cyber-Physical Test Power System

The first version of the PST16 power system is presented in [15]. This system is modified and utilized in this chapter for the purpose of testing the proposed optimization statement on a futuristic multi-area and multi-energy HVDC-HVAC cyber-physical system. This entails using the model to perform the diverse RMS simulations of interest when tackling the challenging optimization search process. The topological layout of the multi-area and multi-energy HVDC-HVAC cyber-physical system is shown in Figure 28.1. It is worth pointing out that the weak HVAC tie-lines in the original system are replaced with three VSC-based HVDC links in different areas. Taking into account the forecasted hydrogen demand scenario shown in [16], 30% of the load demand of each area is replaced with 11 PEM electrolyzers. This is essentially done to investigate the importance and possible collateral impacts of the considered installation location with respect to disturbance altering the active power balance and overall dynamic frequency performance of the system. In this way, the proposed system basically allows investigating the effectiveness of an assumed form of FFS to be attached to the considered controllable active power sources, like the VSC-HVDC inter-area links

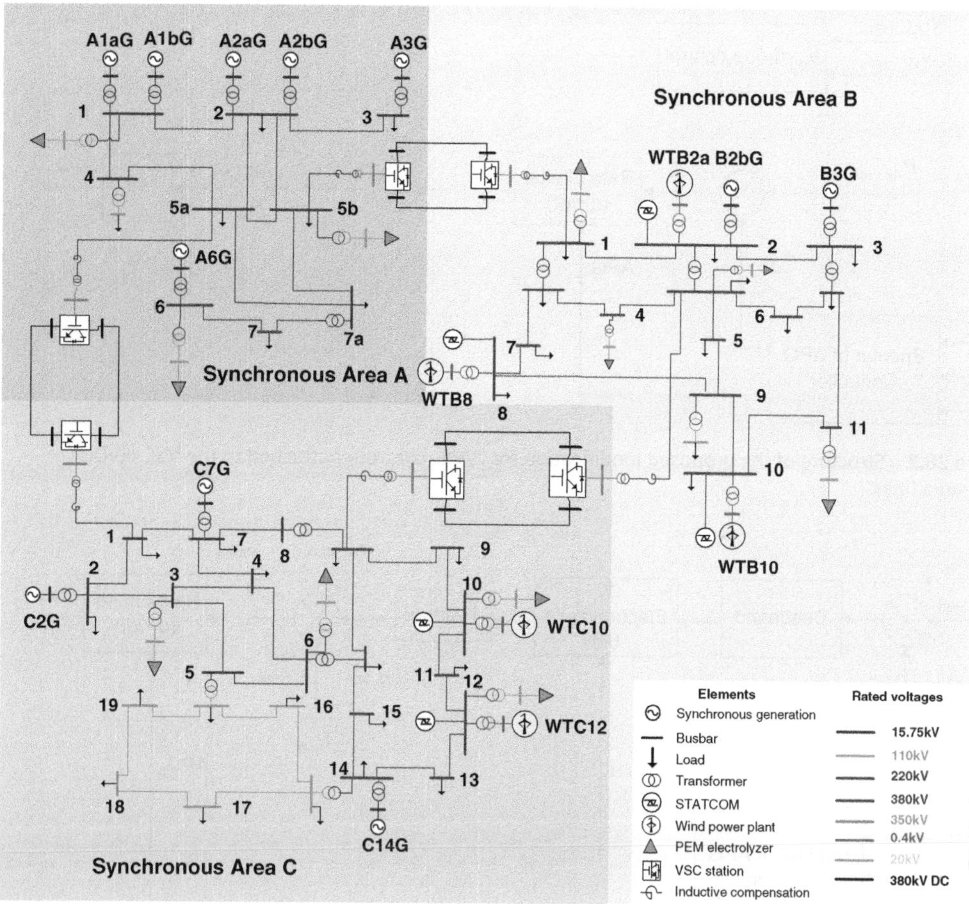

Figure 28.1 Single-line diagram of the proposed futuristic multi-area and multi-energy HVDC-HVAC cyber-physical system.

and the PEM electrolysers. Additional details about adopted component models and the parameters used in the multi-area and multi-energy HVDC-HVAC cyber-physical system are available in [17].

A model provided by DIgSILENT which is detailed in [17] and developed under the guidelines outlined in [18], is utilized for modeling the steady-state and dynamic performances of the VSC-HVDC inter-area links in DPF2022. The model comprises of two converter stations which exploit the current-vector control strategy proposed in [19]. On the DC side, the converters are configured in a bi-polar layout, and on the AC side, they are connected to the corresponding local synchronous area through conventional three-winding power transformers. Specific active and reactive power setpoints are provided to the HVDC links for recreating the power flow profiles of the AC tie lines that were used in the original PST16 system. These setpoints are used by the inverter stations to control the power output. The rectifiers operate based on the DC voltage and reactive power references and are responsible of maintaining the active power balance. Figure 28.2 shows the modified P/Vdc controller of the links, which enables FFS from the HVDC links. Basically, the APG control loop is introduced which controls the amount of power (ΔP) and the active power (APG) injection rate when a systemic frequency deviation is detected. As envisioned in this chapter, the parameter selection is determined based on the solution to a formally defined

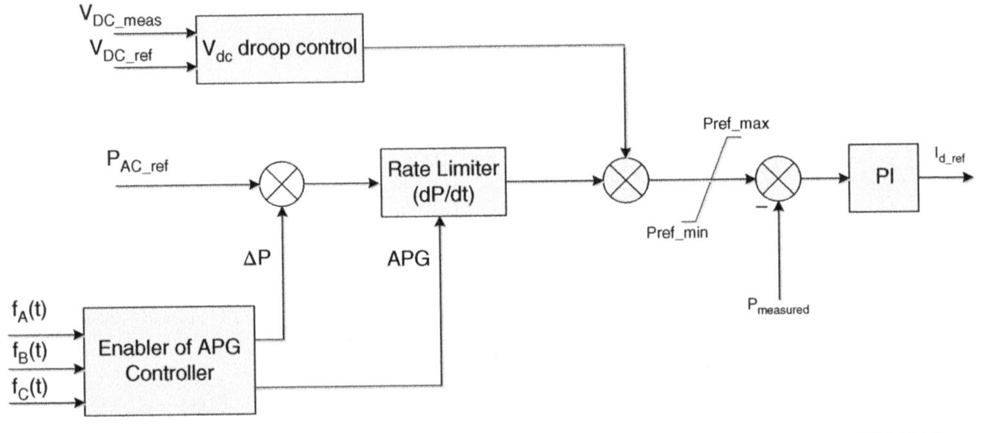

Figure 28.2 Structure of the proposed modification for P/Vdc controller attached to the VSC-HVDC inter-area links.

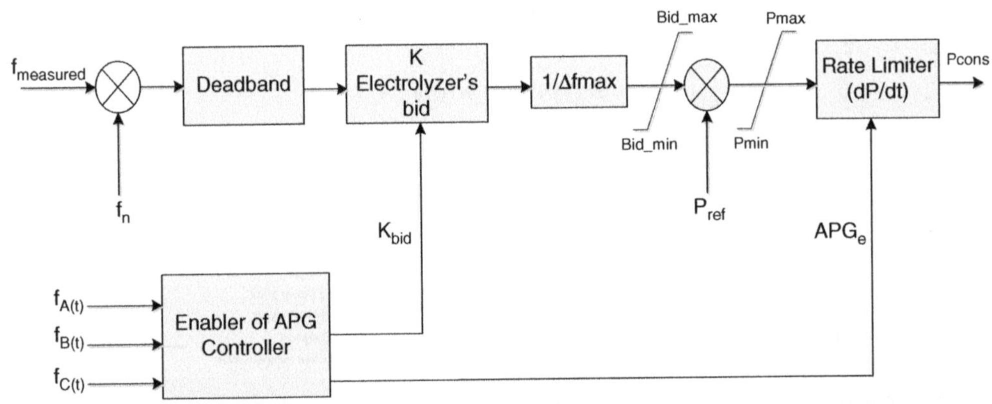

Figure 28.3 Proposed frequency controller for PEM electrolysers.

optimization problem. This is done by using the proposed formulation, which considers the time evolution of the dynamic frequency responses of each synchronous area considered in the optimization of the FFS attached to the involved controllable resources for active power support. This aspect is elaborated in Section 3. It is assumed that the VSC-HVDC links can adjust their power flows for FFS once in each simulation, with an activation delay of approximately 0.3 s after the occurrence of a disturbance altering the overall active power balance.

In this chapter, the modeling of PEM electrolysers has been done considering them as electromagnetically decoupled general loads, which are interconnected to the local grid through conventional power transformers. This assumption is motivated by a similar modeling approach reported in [20]. The electrolysers are activated for FFS, and their consumption is dynamically controlled in proportion to the measured systemic frequency deviation. The overall control structure considered in this chapter for the PEM electrolyzers is shown in Figure 28.3. The available bid and the ramp rate determine the tuning of the proposed frequency controller. The APG control block presented in Figure 28.3 is used for the dynamic active power adjustment. It operates as per the solution of the optimization problem under the proposed formulation, which is explained in detail in Section 28.3. The alternative representation of PEM electrolyzers, as discussed in [20], offers a versatile yet trustworthy model for numerical simulation-based power system stability assessments. This representation has a primary focus on the critical parameter of interest, which is the ramp-rate of the electrolyser's active power consumption.

28.3 Statement of the Optimization of FFC for PEI

A system with multiple synchronous areas that are electromagnetically decoupled through VSC-HVDC inter-area links entails a particular characteristic of the dynamic frequency response for each individual area. Hence, any active power imbalance occurring in one of the synchronous areas shall be in principle confined within the affected area [21]. In other terms, without the intervention of the VSC-HVDC inter-area links, the frequencies of the other areas remain constant at their corresponding nominal values. The VSC-HVDC inter-area links can, however, be controlled to dynamically alter the inter-area power exchange to provide FFS to the affected area. In such a case, the supporting synchronous area undergoes a deviation of frequency from the nominal value as a function of the defined form of controlled variation of the power exchange between the involved synchronous areas.

The representation in Figure 28.4 conveys that the objective function (OF) of the proposed problem statement is conceived to minimize the graphical area in the three-area multi-energy HVDC-HVAC cyber-physical system. The minimization is done among each pair of the frequency responses, of the synchronous areas that are participating in the frequency regulation. In this manner, the problem statement attempts to share the active power imbalance among the areas as per the kinetic energy characteristics and active power reserves. Therefore, it results in the improvement of the frequency response of the affected synchronous area, with very little impact on the supporting synchronous areas. The scheme tries to develop an artificially coupled frequency response, securing the best possible response for the complete system in a decoupled multi-area

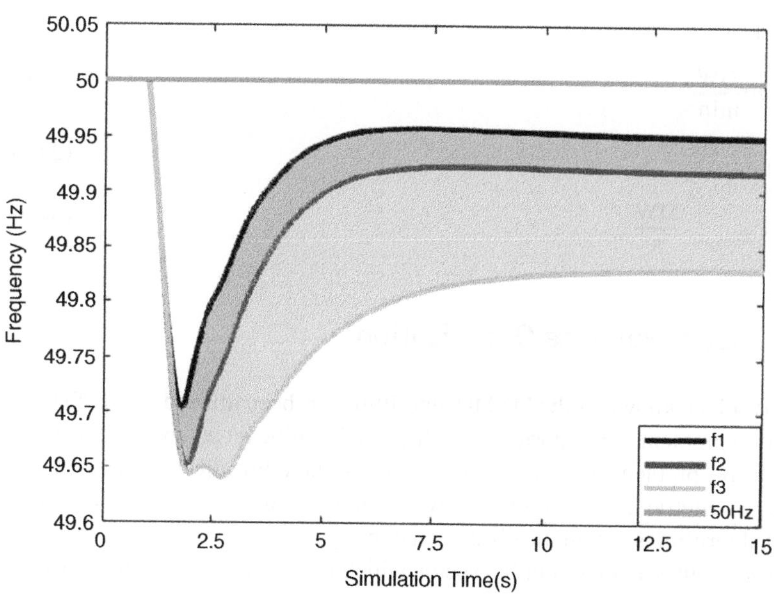

Figure 28.4 Example visualization of the optimal tuning of FFC.

HVDC-HVAC cyber-physical system. This is done by optimally dividing the imbalance across the active power reserves available for each area. The mathematical representation of the problem formulation is shown in Eqs (28.1–28.7).

Minimize:

$$OF(x) = \int_0^{t} [|f_A(\mathbf{x}, t) - f_B(\mathbf{x}, t)| + |f_B(\mathbf{x}, t) - f_C(\mathbf{x}, t)| + |f_A(x, t) - f_C(x, t)|]dt \tag{28.1}$$

$$\text{subjected to: } x_{min} \leq x \leq x_{max} \tag{28.2}$$

where $f_k(\mathbf{x},t)$ is the instantaneously measured frequency in the k-th area at instant t, on application of the candidate solution **x**. Two variables are present in the optimization vector x for each of the FFS elements that take part in the fast frequency regulation. In the multi-area and multi-energy HVDC-HVAC cyber-physical system, three VSC-HVDC inter-area links and eleven electrolyzers have been integrated. Thus, x constitutes a total of 28 optimization variables x.

The first parameter for the VSC-HVDC inter-area links indicates the APG ramp rate. The second parameter reflects the corresponding link's ΔP. The PEM electrolyzers are characterized by two primary parameters: the first one represents the electrolyser's bid size (Bid_e), while the second parameter signifies the rate of change in the electrolyser's consumption (APG_e). In Equation (28.3), the optimization vector x is observed. Additionally, each optimization variable is subject to specific bound constraints based on the technical limitations of each unit. For the VSC-HVDC inter-area links, the new setpoint should not exceed the rated capacity of the link in both directions [13, 14], and the ramp rate should not surpass the maximum rate that the link can handle [22]. Likewise, for the PEM electrolyzers, the available bid must not exceed 70% of their rated capacity, and the ramp rate should not exceed 0.5 pu/s, as elaborated in the references [23–25]. The assumed boundary conditions for these optimization variables are specified in Eqs. (28.4–28.7).

$$x = \begin{bmatrix} \Delta P_{HVDC_1}, APG_{HVDC_1}, \ldots, \Delta P_{HVDC_i}, APG_{HVDC_i}, K_{ebid1}, \\ APG_{e_1}, \ldots, K_{ebidj}, APG_{e_j} \end{bmatrix} \tag{28.3}$$

$$P_{refi} + \Delta P_i \leq |P_{rated_i}| \tag{28.4}$$

$$0\frac{GW}{min} \leq APGi \leq 60\frac{GW}{min} \tag{28.5}$$

$$0 \leq K_{bid_w} \leq 0.7P_{elec_{ratew}} \tag{28.6}$$

$$0 \leq APG_{e_w} \leq 0.5\, P_{elec_{rated_w}} \frac{MW}{s} \tag{28.7}$$

28.4 Solution by Mean–Variance Optimization

A powerful metaheuristic solver, known as the MVMO algorithm, has been utilized in this chapter for solving the optimization problem. The algorithm has displayed significant potential in the solution of computationally heavy problems related to power systems, including diverse statements of optimal power flow and parametric identification of power system dynamic equivalents [26]. The flowchart of the MVMO algorithm is schematically depicted in Figure 28.5.

Initially, the algorithm computes a random initial vector, which is within the research space that is defined by the boundary conditions assumed for each variable. The selected parameters are then

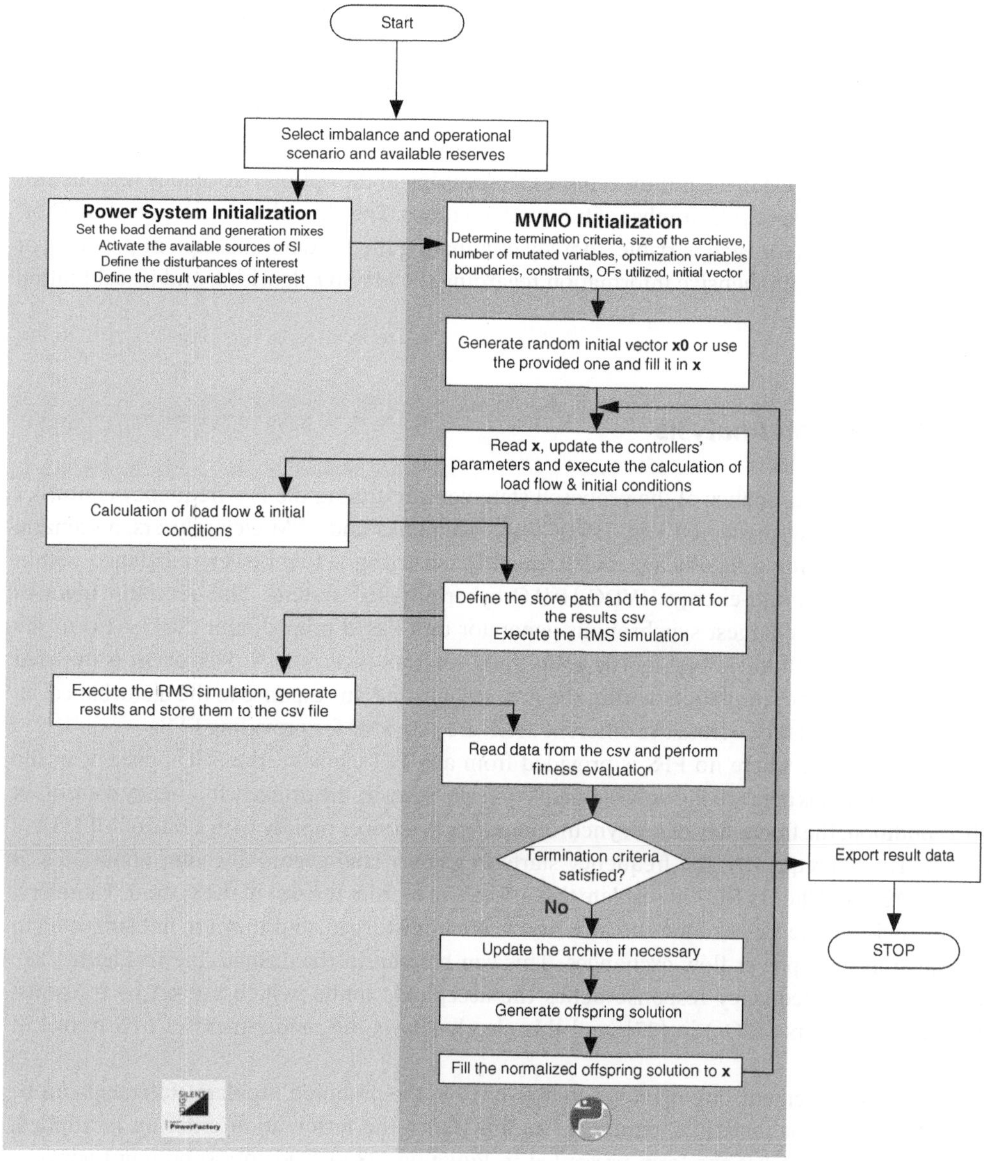

Figure 28.5 Solution procedure performed by MVMO.

applied to the corresponding system element in DPF2022, and an RMS simulation is performed for obtaining the time evolution of the dynamic frequency responses that are of interest for evaluating the systemic frequency performance. The latter ones are exploited to execute the fitness evaluation and calculation of the OF value. Numerous predefined fitness evaluations follow the same process. In this chapter, assuming a restricted computing budget due to time-consuming RMS simulations, 100 pre-defined fitness evaluations are considered, which act as the algorithm termination criterion. During these evaluations, MVMO generates potential offspring solution vectors based on the optimization variables of the best-performing parent solution achieved thus far. These offspring solutions are generated in accordance with evolution guidelines established by a mapping function for each variable. These guidelines take into consideration statistical data, including the mean and variance of each optimization variable as inputs. As a result, the most successful parent solutions and the statistical data pertaining to each variable are recorded in an archive of a predefined memory capacity. Eventually, once the termination criterion has been met, the solution vector contains the variables that have minimized the OF, resulting in the optimal frequency response for the system within the specified number of fitness evaluations. The algorithm operates within a normalized range of [0,1] for each variable, thereby eliminating the need for subsequent penalties or corrections. More comprehensive information regarding the MVMO algorithm and its calibration can be found in [26].

28.5 Simulation Analysis

For assessing the performance of the proposed statement for tuning the fast frequency controller (FFC) attached to the considered VSC-HVDC inter-area links and PEM electrolysers, a scenario is considered involving a significant and frequently occurring active power imbalance within the multi-area and multi-energy HVDC-HVAC cyber-physical system. This scenario involves the sudden loss of the largest synchronous generator unit-based power plant (SGU A1a) in the synchronous area with the lowest inertia conditions, referred to as area A. This event is initiated at the beginning of the simulation, within the first second, and, in response to the disturbance, all the PEI elements installed across the different areas are activated for providing FFS.

Under a condition where no FFS is provided from any PEI element, the SGUs present in the affected synchronous area are the only elements that can support the primary frequency control. In such a condition, the frequency of the synchronous area A reduces rapidly with a rate of −0.51Hz/s up to 48.8Hz. Subsequently, the frequency starts to recover, and eventually, after about 15 s, it settles at 49.55 Hz. The HVDC links isolate this affected area from the rest of the system. Therefore, the frequencies in the synchronous areas B and C are undisturbed, and they are not subjected to any change in their power flow exchanges. This can be seen in the dashed line in Figure 28.6. The affected area frequency response defies the acceptable limits, which are set by transmission system operators (TSOs) in [27], and this clearly shows the requirement of FFS from PEI units.

In case the PEI elements are operated to deliver FFS, the proposed problem statement can be effectively utilized for their FFC parameter tuning. Once the termination criterion is fulfilled, the solution vector x takes the form presented in Equation (28.8). The corresponding frequency response of the system is shown in Figure 28.6. It is observed that the affected synchronous area frequency considerably improves on sharing the imbalance among the available supporting synchronous areas. The initial rate of change of frequency (RoCoF) drops to −0.30 Hz/s from −0.51 Hz/s. This is solely because of the quick response from the PEM electrolyzers present in the affected synchronous area. Subsequently, the synchronous area A frequency decreases to 49.76 Hz and it settles at 49.93 Hz within a time duration of 5 s. The frequencies in the synchronous areas B and C fall to 49.92 and 49.93 Hz, respectively, and they eventually stabilize at 49.93 and 49.94 Hz, respectively. This operation validates the objective of the proposed statement, which is to minimize the frequency drop in the affected area, while also minimizing the impact on the active power

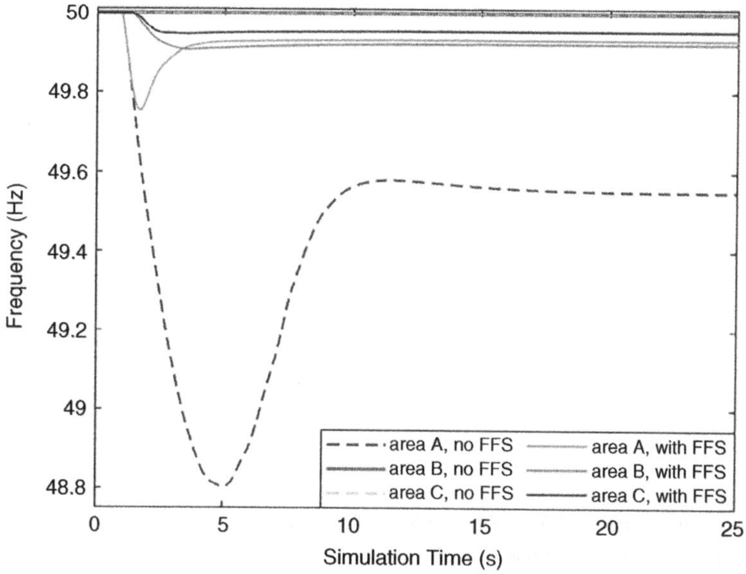

Figure 28.6 Dynamic frequecny evolution of the synchronous areas, with and without FFS.

supporting areas through appropriate tuning of FFC. As such, the formulation aims to establish uniform frequency responses across all interconnected synchronous areas by effectively utilizing the available active power reserves, thus resulting in frequency responses that fall within the acceptable operating limits as defined by TSOs in [27].

$$x = \begin{bmatrix} 932, 185, 796, 377, 412, 188.9, \ \ 95.4, 67.5, 38.7, 72.2, 21.4, 62.3, 242, 121, 85.6, 29.2, 186, \\ 36.7, 12.1, 16.4, 166, 74.5, 32.9, 24.7, 45.9, 94.5, 118, 53.5 \end{bmatrix} \tag{28.8}$$

Equation (28.8) also denotes that the electrolyzers play a considerably important role for the FFS. On the occurrence of an imbalance, the PEM electrolyzers present in synchronous area A are activated to offer rapid support and arrest the initial drop before the VSC-HVDC inter-area links are activated. The VSC-HVDC inter-area links A-B and A-C start changing their powers after 0.3 s. The power delivered to areas B and C is reduced such that support can be provided to the affected areas. On the other hand, the VSC-HVDC inter-area link B-C alters its power flow direction for balancing the impact on the synchronous areas B and C. In this manner, the optimization problem statement is able to achieve an optimal share of power among the affected synchronous areas as per their kinetic energy characteristics and active power reserves. The PEM electrolysers present in areas B and C are also activated for the purpose of FFS and mitigating the influence on the supporting areas, which lead to comparable post-disturbance frequency responses. Consequently, the formulation strives to achieve a coordinated frequency response within a multi-area system that is electromagnetically decoupled.

Figure 28.7 shows the performance of MVMO. The algorithm is able to effectively minimize the value of the proposed OF without the need of too many iterations. This significantly reduces the computational burden and produces high-quality results in terms of the frequency response.

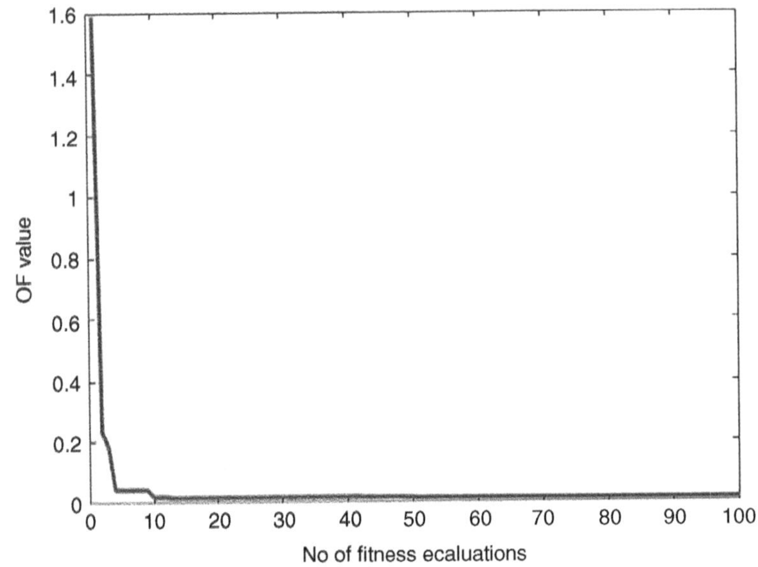

Figure 28.7 Resulting convergence of the optimization's objective function.

28.6 Concluding Reflections

The chapter presents a multi-area multi-energy HVDC-HVAC cyber-physical power system and an optimization problem statement for optimal calibration of the parameters of FFC attached to PEI elements. The system can be configured and is suitable for studying different penetration levels of PEI, which is essential for the design of new HVDC-HVDC architectures. Such investigations unlock the potential of new technological upgrades for an accelerated energy transition towards 100% stable and affordable sustainable and highly integrated energy systems.

The developed case study attempts to illustrate the implications of creating an artificially coupled fast active power-frequency support within the context of a PEI-isolated (i.e., electromagnetically decoupled) multi-area multi-energy HVDC-HVAC systems. Numerical analysis conducted by using DPF2022 shows that active power imbalances can be optimally and effectively tackled by the PEI elements equipped with FFC, which are deployed in the affected synchronous area, and with the support of PEIs with FFC like those attached to VSC-HVDC inter-area links and the PEIs with FFC from other synchronous areas. The proposed optimization problem statement allows defining FFS by other PEIs as a function of the available kinetic energy and active power headrooms without altering the intra-area active power balances of the external synchronous areas. Hence, by using the proposed optimization problem statement, it is possible to target the preservation of the time evolution of the dynamic frequency response in the involved synchronous areas in line with the operational limits pursued by a TSO. A powerful optimization solver is needed to tackle the optimization of FFCs of PEI within a limited computing budget. The MVMO seems an attractive option, as shown in this chapter. Future research will cover the study of other HVDC-HVAC architectures and optimization solvers.

References

1 Halley, A., Martins, N., Gomes, P. et al. (2018). Effects of increasing power electronics-based technology on power system stability: Performance and operations. *CIGRE Science & Engineering* 11: 5–17, June.

2 Khan, A., Hosseinzadehtaher, M., Shadmand, M.B. et al. (2020). On the Stability of the Power Electronics-Dominated Grid: A New Energy Paradigm. *IEEE Industrial Electronics Magazine* 14 (4): 65–78, Dec. https://doi.org/10.1109/MIE.2020.3002523.

3 Yuan, X., Hu, J., and Cheng, S. (2017). Multi-time scale dynamics in power electronics-dominated power systems. *Front. Mech. Eng.* 12: 303–311. https://doi.org/10.1007/s11465-017-0428-z.

4 N. Hatziargyriou, J. V. Milanović, C. Rahmann, V. Ajjarapu, C. Cañizares, I. Erlich, D. Hill, I. Hiskens, I. Kamwa, B. Pal, P. Pourbeik, J. J. Sanchez- Gasca, A. Stanković, T. Van Cutsem, V. Vittal and C. Vournas: "Definition and Classification of Power System Stability – Revised and Extended", *IEEE Transactions on Power Systems*, vol. 36, no. 4, pp. 3271-3281, https://doi.org/10.1109/TPWRS.2020.3041774, (2021).

5 Meegahapola, L., Sguarezi, A., Bryant, J.S. et al. (2020). Power System Stability with Power-Electronic Converter Interfaced Renewable Power Generation: Present Issues and Future Trends. *Energies* 13 (13): 13133441. https://doi.org/10.3390/en13133441.

6 Sewdien, V.N., Chatterjee, R., Val Escudero, M., and Van Putten, J. (2020). System Operational Challenges from the Energy Transition. *CIGRE Science & Engineering* 17: 5–19, Feb.

7 Maibach, P., Hernandez, A., Peiro, J. et al. (2021). Capabilities of Power Electronic Devices in Enabling the Energy Transition and Mitigating System Operational Challenges. *CIGRE Science & Engineering* 20: 125–136, Feb.

8 Rakhshani, E., Perilla, A., Torres, J.L.R. et al. (2020). FAPI Controller for Frequency Support in Low-Inertia Power Systems. *IEEE Open Access Journal of Power and Energy* 7: 276–286. https://doi.org/10.1109/OAJPE.2020.3010224.

9 Crăciun, B., Kerekes, T., Séra, D. et al. (2017). Power ramp limitation capabilities of large PV power plants with active power reserves. *IEEE Transactions on Sustainable Energy* 8 (2): 573–581.

10 Karaolanis, A., Perilla, A., Rueda, J.L. et al. (2018). Generic model of a VSC-based HVDC link for RMS simulations in PSS/E. *IFAC-PapersOnLine* 51 (28): 303–308. 10th IFAC Symposium on Control of Power and Energy Systems CPES.

11 Veerakumar, N., Ahmad, Z., Ebrahim Adabi, M. et al. (2020, 2020). *Fast Active Power - Frequency Support Methods in Large Scale Electrolyzers for Multi-Energy Systems.* The Hague, The Netherlands: IEEE PES Innovative Smart Grid Technologies Europe (ISGT-Europe).

12 Perilla Guerra, A.D., Rueda Torres, J.L., van der Meer, A.A. et al. (2017). *Influence of Active Power Gradient Control of an MMC-HVDC Link on Long-Term Frequency Stability*, 2017. Chicago, IL, USA: IEEE Power & Energy Society General Meeting.

13 Perilla Guerra, A.D., Gusain, D., Rueda Torres, J.L. et al. (2020). *Optimal Tuning of Active Power Control for Frequency Support in Multi-Energy Systems.* The Hague, The Netherlands: IEEE PES Innovative Smart Grid Technologies Europe (ISGT-Europe).

14 J.L. Rueda Torres, A.D. Perilla Guerra, E. Rakhshani. et al. (2019). MVMO-based tuning of active power gradient control of VSC-HVDC links for frequency support. *2nd International Conference on Smart Grid and Renewable Energy(SGRE)*, Doha, Qatar.

15 Gonzalez-Longatt, F.M. and Rueda Torres, J.L. (2014). *PowerFactory Applications for Power System Analysis*. Delft, The Netherlands: Springer.

16 Hydrogen Council (2017). *Hydrogen Scaling up – A Sustainable Pathway for the Global Energy Transition*. Hydrogen Council Tech. Rep.

17 DIgSILENT GmbH (2018). *"PowerFactory Seminar HVDC and Facts", DIgSILENT GmbH*. Gomaringen, Germany: Tech. Rep.

18 CIGRE Working Group B4.57 (2014). *Guide for the Development of Models for HVDC Converters in a HVDC Grid*. Paris, France: CIGRE Tech. Rep.

19 Trinh, N.T., Zeller, M., Wuerflinger, K., and Erlich, I. (2016). Generic Model of MMC-VSC-HVDC for Interaction Study With AC Power System. *IEEE Transactions on Power Systems* 31 (1): 27–34, Jan.

20 TSO 2020. 2019. Stability Analysis of an International Electricity System Connected to Regional Local Sustainable Gas Systems. TSO 2020 Activity 2 Final Report.

21 Kundur, P. (1994). *Power System Stability and Control*, 1ste. McGraw-Hill Professional.

22 Wang, W., Beddard, A., Barnes, M., and Marjanovic, O. (2014). Analysis of active power control for VSC HVDC. *IEEE Transactions on Power Delivery* 29 (4): 1978–1988.

23 Tuinema, B.W., Adabi, E., Ayivor, P.K.S. et al. (2020). Modelling of Large-sized Electrolysers for Real-Time Simulation and Study of the Possibility of Frequency Support by Electrolysers. *IET Generation, Transmission & Distribution* 14 (10): 1985–1992, 13 May.

24 Ayivor, P., Rueda Torres, J. L., van der Meijden, M.A.M.M. et al. (2018). Modelling of large size electrolyzer for electrical grid stability studies in real time digital simulation. *Proceedings of Energynautics 3rd International Hybrid Power Systems Workshop*, Tenerife, Spain.

25 Giannakopoulos, G. (April 2021). *Master Thesis: Optimal Tuning of the Active Power Gradient Control in Multi-Energy HVDC-HVAC Power Systems*. Delft, Netherlands: Delft University of Technonolgy.

26 J. L. Rueda Torres. and I. Erlich. (2018). Hybrid single parent-offspring MVMO for solving CEC2018 computationally expensive problems. *2018 IEEE World Congress on Computational Intelligence*, pp. 1-8, Rio de Janeiro Brazil.

27 European Commission, "Commission Regulation (EU) 2017/1485 of 2 August 2017 establishing a guideline on electricity transmission system operation", European Commission, Brussels (2017).

Index

Smart Cyber-Physical Power Systems: Fundamental Concepts, Challenges, and Solutions, Volume 1, First Edition.
Edited by Ali Parizad, Hamid Reza Baghaee, and Saifur Rahman.

C

d

e

u

IEEE Press Series on Power and Energy Systems

Series Editor: Ganesh Kumar Venayagamoorthy, Clemson University, Clemson, South Carolina, USA.

The mission of the IEEE Press Series on Power and Energy Systems is to publish leading-edge books that cover a broad spectrum of current and forward-looking technologies in the fast-moving area of power and energy systems including smart grid, renewable energy systems, electric vehicles and related areas. Our target audi- ence includes power and energy systems professionals from academia, industry and government who are interested in enhancing their knowledge and perspectives in their areas of interest.

1. *Electric Power Systems: Design and Analysis, Revised Printing*
   Mohamed E. El-Hawary
2. *Power System Stability*
   Edward W. Kimbark
3. *Analysis of Faulted Power Systems*
   Paul M. Anderson
4. *Inspection of Large Synchronous Machines: Checklists, Failure Identification, and Troubleshooting*
   Isidor Kerszenbaum
5. *Electric Power Applications of Fuzzy Systems*
   Mohamed E. El-Hawary
6. *Power System Protection*
   Paul M. Anderson
7. *Subsynchronous Resonance in Power Systems*
   Paul M. Anderson, B.L. Agrawal, and J.E. Van Ness
8. *Understanding Power Quality Problems: Voltage Sags and Interruptions*
   Math H. Bollen
9. *Analysis of Electric Machinery*
   Paul C. Krause, Oleg Wasynczuk, and S.D. Sudhoff
10. *Power System Control and Stability, Revised Printing*
    Paul M. Anderson and A.A. Fouad
11. *Principles of Electric Machines with Power Electronic Applications, Second Edition*
    Mohamed E. El-Hawary
12. *Pulse Width Modulation for Power Converters: Principles and Practice*
    D. Grahame Holmes and Thomas Lipo

13. *Analysis of Electric Machinery and Drive Systems, Second Edition*
Paul C. Krause, Oleg Wasynczuk, and S.D. Sudhoff
14. *Risk Assessment for Power Systems: Models, Methods, and Applications*
Wenyuan Li
15. *Optimization Principles: Practical Applications to the Operations of Markets of the Electric Power Industry*
Narayan S. Rau
16. *Electric Economics: Regulation and Deregulation*
Geoffrey Rothwell and Tomas Gomez
17. *Electric Power Systems: Analysis and Control*
Fabio Saccomanno
18. *Electrical Insulation for Rotating Machines: Design, Evaluation, Aging, Testing, and Repair*
Greg C. Stone, Edward A. Boulter, Ian Culbert, and Hussein Dhirani
19. *Signal Processing of Power Quality Disturbances*
Math H.J. Bollen and Irene Y.H. Gu
20. *Instantaneous Power Theory and Applications to Power Conditioning*
Hirofumi Akagi, Edson H. Watanabe, and Mauricio Aredes
21. *Maintaining Mission Critical Systems in a 24/7 Environment*
Peter M. Curtis
22. *Elements of Tidal-Electric Engineering*
Robert H. Clark
23. *Handbook of Large Turbo-Generator Operation and Maintenance, Second Edition*
Geoff Klempner and Isidor Kerszenbaum
24. *Introduction to Electrical Power Systems*
Mohamed E. El-Hawary
25. *Modeling and Control of Fuel Cells: Distributed Generation Applications*
M. Hashem Nehrir and Caisheng Wang
26. *Power Distribution System Reliability: Practical Methods and Applications*
Ali A. Chowdhury and Don O. Koval
27. *Introduction to FACTS Controllers: Theory, Modeling, and Applications*
Kalyan K. Sen and Mey Ling Sen
28. *Economic Market Design and Planning for Electric Power Systems*
James Momoh and Lamine Mili
29. *Operation and Control of Electric Energy Processing Systems*
James Momoh and Lamine Mili
30. *Restructured Electric Power Systems: Analysis of Electricity Markets with Equilibrium Models*
Xiao-Ping Zhang

31. *An Introduction to Wavelet Modulated Inverters*
S.A. Saleh and M.A. Rahman

32. *Control of Electric Machine Drive Systems*
Seung-Ki Sul

33. *Probabilistic Transmission System Planning*
Wenyuan Li

34. *Electricity Power Generation: The Changing Dimensions*
Digambar M. Tagare

35. *Electric Distribution Systems*
Abdelhay A. Sallam and Om P. Malik

36. *Practical Lighting Design with LEDs*
Ron Lenk and Carol Lenk

37. *High Voltage and Electrical Insulation Engineering*
Ravindra Arora and Wolfgang Mosch

38. *Maintaining Mission Critical Systems in a 24/7 Environment, Second Edition*
Peter Curtis

39. *Power Conversion and Control of Wind Energy Systems*
Bin Wu, Yongqiang Lang, Navid Zargari, and Samir Kouro

40. *Integration of Distributed Generation in the Power System*
Math H. Bollen and Fainan Hassan

41. *Doubly Fed Induction Machine: Modeling and Control for Wind Energy Generation Applications*
Gonzalo Abad, Jesús López, Miguel Rodrigues, Luis Marroyo, and Grzegorz Iwanski

42. *High Voltage Protection for Telecommunications*
Steven W. Blume

43. *Smart Grid: Fundamentals of Design and Analysis*
James Momoh

44. *Electromechanical Motion Devices, Second Edition*
Paul Krause Oleg, Wasynczuk, and Steven Pekarek

45. *Electrical Energy Conversion and Transport: An Interactive Computer-Based Approach, Second Edition*
George G. Karady and Keith E. Holbert

46. *ARC Flash Hazard and Analysis and Mitigation*
J.C. Das

47. *Handbook of Electrical Power System Dynamics: Modeling, Stability, and Control*
Mircea Eremia and Mohammad Shahidehpour

48. *Analysis of Electric Machinery and Drive Systems, Third Edition*
Paul C. Krause, Oleg Wasynczuk, S.D. Sudhoff, and Steven D. Pekarek

49. *Extruded Cables for High-Voltage Direct-Current Transmission: Advances in Research and Development*
Giovanni Mazzanti and Massimo Marzinotto

50. *Power Magnetic Devices: A Multi-Objective Design Approach*
S.D. Sudhoff

51. *Risk Assessment of Power Systems: Models, Methods, and Applications, Second Edition*
Wenyuan Li

52. *Practical Power System Operation*
Ebrahim Vaahedi

53. *The Selection Process of Biomass Materials for the Production of Bio-Fuels and Co-Firing*
Najib Altawell

54. *Electrical Insulation for Rotating Machines: Design, Evaluation, Aging, Testing, and Repair, Second Edition*
Greg C. Stone, Ian Culbert, Edward A. Boulter, and Hussein Dhirani

55. *Principles of Electrical Safety*
Peter E. Sutherland

56. *Advanced Power Electronics Converters: PWM Converters Processing AC Voltages*
Euzeli Cipriano dos Santos Jr. and Edison Roberto Cabral da Silva

57. *Optimization of Power System Operation, Second Edition*
Jizhong Zhu

58. *Power System Harmonics and Passive Filter Designs*
J.C. Das

59. *Digital Control of High-Frequency Switched-Mode Power Converters*
Luca Corradini, Dragan Maksimoviæ, Paolo Mattavelli, and Regan Zane

60. *Industrial Power Distribution, Second Edition*
Ralph E. Fehr, III

61. *HVDC Grids: For Offshore and Supergrid of the Future*
Dirk Van Hertem, Oriol Gomis-Bellmunt, and Jun Liang

62. *Advanced Solutions in Power Systems: HVDC, FACTS, and Artificial Intelligence*
Mircea Eremia, Chen-Ching Liu, and Abdel-Aty Edris

63. *Operation and Maintenance of Large Turbo-Generators*
Geoff Klempner and Isidor Kerszenbaum

64. *Electrical Energy Conversion and Transport: An Interactive Computer-Based Approach*
George G. Karady and Keith E. Holbert

65. *Modeling and High-Performance Control of Electric Machines*
John Chiasson

66. *Rating of Electric Power Cables in Unfavorable Thermal Environment*
George J. Anders

67. *Electric Power System Basics for the Nonelectrical Professional*
Steven W. Blume

68. *Modern Heuristic Optimization Techniques: Theory and Applications to Power Systems*
Kwang Y. Lee and Mohamed A. El-Sharkawi

69. *Real-Time Stability Assessment in Modern Power System Control Centers*
Savu C. Savulescu

70. *Optimization of Power System Operation*
Jizhong Zhu

71. *Insulators for Icing and Polluted Environments*
Masoud Farzaneh and William A. Chisholm

72. *PID and Predictive Control of Electric Devices and Power Converters Using MATLAB®/Simulink®*
Liuping Wang, Shan Chai, Dae Yoo, Lu Gan, and Ki Ng

73. *Power Grid Operation in a Market Environment: Economic Efficiency and Risk Mitigation*
Hong Chen

74. *Electric Power System Basics for Nonelectrical Professional, Second Edition*
Steven W. Blume

75. *Energy Production Systems Engineering Thomas*
Howard Blair

76. *Model Predictive Control of Wind Energy Conversion Systems*
Venkata Yaramasu and Bin Wu

77. *Understanding Symmetrical Components for Power System Modeling*
J.C. Das

78. *High-Power Converters and AC Drives, Second Edition*
Bin Wu and Mehdi Narimani

79. *Current Signature Analysis for Condition Monitoring of Cage Induction Motors: Industrial Application and Case Histories*
William T. Thomson and Ian Culbert

80. *Introduction to Electric Power and Drive Systems*
Paul Krause, Oleg Wasynczuk, Timothy O'Connell, and Maher Hasan

81. *Instantaneous Power Theory and Applications to Power Conditioning, Second Edition*
Hirofumi, Edson Hirokazu Watanabe, and Mauricio Aredes

82. *Practical Lighting Design with LEDs, Second Edition*
Ron Lenk and Carol Lenk

83. *Introduction to AC Machine Design*
Thomas A. Lipo

84. *Advances in Electric Power and Energy Systems: Load and Price Forecasting*
Mohamed E. El-Hawary

85. *Electricity Markets: Theories and Applications*
Jeremy Lin and Jernando H. Magnago

86. *Multiphysics Simulation by Design for Electrical Machines, Power Electronics and Drives*
Marius Rosu, Ping Zhou, Dingsheng Lin, Dan M. Ionel, Mircea Popescu, Frede Blaabjerg, Vandana Rallabandi, and David Staton

87. *Modular Multilevel Converters: Analysis, Control, and Applications*
Sixing Du, Apparao Dekka, Bin Wu, and Navid Zargari

88. *Electrical Railway Transportation Systems*
Morris Brenna, Federica Foiadelli, and Dario Zaninelli

89. *Energy Processing and Smart Grid*
James A. Momoh

90. *Handbook of Large Turbo-Generator Operation and Maintenance, 3rd Edition*
Geoff Klempner and Isidor Kerszenbaum

91. *Advanced Control of Doubly Fed Induction Generator for Wind Power Systems*
Dehong Xu, Dr. Frede Blaabjerg, Wenjie Chen, and Nan Zhu

92. *Electric Distribution Systems, 2nd Edition*
Abdelhay A. Sallam and Om P. Malik

93. *Power Electronics in Renewable Energy Systems and Smart Grid: Technology and Applications*
Bimal K. Bose

94. *Distributed Fiber Optic Sensing and Dynamic Rating of Power Cables*
Sudhakar Cherukupalli and George J. Anders

95. *Power System and Control and Stability, Third Edition*
Vijay Vittal, James D. McCalley, Paul M. Anderson, and A.A. Fouad

96. *Electromechanical Motion Devices: Rotating Magnetic Field-Based Analysis and Online Animations, Third Edition*
Paul Krause, Oleg Wasynczuk, Steven D. Pekarek, and Timothy O'Connell

97. *Applications of Modern Heuristic Optimization Methods in Power and Energy Systems*
Kwang Y. Lee and Zita A. Vale

98. *Handbook of Large Hydro Generators: Operation and Maintenance*
Glenn Mottershead, Stefano Bomben, Isidor Kerszenbaum, and Geoff Klempner

99. *Advances in Electric Power and Energy: Static State Estimation*
Mohamed E. El-hawary

100. *Arc Flash Hazard Analysis and Mitigation, Second Edition*
J.C. Das

101. *Maintaining Mission Critical Systems in a 24/7 Environment, Third Edition*
Peter M. Curtis

102. *Real-Time Electromagnetic Transient Simulation of AC-DC Networks*
Venkata Dinavahi and Ning Lin

103. *Probabilistic Power System Expansion Planning with Renewable Energy Resources and Energy Storage Systems*
Jaeseok Choi and Kwang Y. Lee

104. *Power Magnetic Devices: A Multi-Objective Design Approach, Second Edition*
Scott D. Sudhoff

105. *Optimal Coordination of Power Protective Devices with Illustrative Examples*
Ali R. Al-Roomi

106. *Resilient Control Architectures and Power Systems*
Craig Rieger, Ronald Boring, Brian Johnson, and Timothy McJunkin

107. *Alternative Liquid Dielectrics for High Voltage Transformer Insulation Systems: Performance Analysis and Applications*
Edited by U. Mohan Rao, I. Fofana, and R. Sarathi

108. *Introduction to the Analysis of Electromechanical Systems*
Paul C. Krause, Oleg Wasynczuk, and Timothy O'Connell

109. *Power Flow Control Solutions for a Modern Grid using SMART Power Flow Controllers*
Kalyan K. Sen and Mey Ling Sen

110. *Power System Protection: Fundamentals and Applications*
John Ciufo and Aaron Cooperberg

111. *Soft-Switching Technology for Three-phase Power Electronics Converters*
Dehong Xu, Rui Li, Ning He, Jinyi Deng, and Yuying Wu

112. *Power System Protection, Second Edition*
Paul M. Anderson, Charles Henville, Rasheek Rifaat, Brian Johnson, and Sakis Meliopoulos

113. *High Voltage and Electrical Insulation Engineering, Second Edition*
Ravindra Arora and Wolfgang Mosch

114. *Modeling and Control of Modern Electrical Energy Systems*
Masoud Karimi-Ghartemani

115. *Control of Power Electronic Converters with Microgrid Applications*
Armdam Ghosh and Firuz Zare

116. *Coordinated Operation and Planning of Modern Heat and Electricity Incorporated Networks*
Mohammadreza Daneshvar, Behnam Mohammadi-Ivatloo, and Kazem Zare

117. *Smart Energy for Transportation and Health in a Smart City*
Chun Sing Lai, Loi Lei Lai, and Qi Hong Lai

118. *Wireless Power Transfer: Principles and Applications*
Zhen Zhang and Hongliang Pang

119. *Intelligent Data Mining and Analysis in Power and Energy Systems: Models and Applications for Smarter Efficient Power Systems*
Zita Vale, Tiago Pinto, Michael Negnevitsky, and Ganesh Kumar Venayagamoorthy

120. *Introduction to Modern Analysis of Electric Machines and Drives*
Paul C. Krause and Thomas C. Krause

121. *Electromagnetic Analysis and Condition Monitoring of Synchronous Generators*
Hossein Ehya and Jawad Faiz

122. *Transportation Electrification: Breakthroughs in Electrified Vehicles, Aircraft, Rolling Stock, and Watercraft*
Ahmed A. Mohamed, Ahmad Arshan Khan, Ahmed T. Elsayed, and Mohamed A. Elshaer

123. *Modular Multilevel Converters: Control, Fault Detection, and Protection*
Fuji Deng, Chengkai Liu, and Zhe Chen

124. *Stability-Constrained Optimization for Modern Power System Operation and Planning*
Yan Xu, Yuan Chi, and Heling Yuan

125. *Interval Methods for Uncertain Power System Analysis*
Alfredo Vaccaro

126. *Practical Partial Discharge Measurement on Electrical Equipment*
Greg Stone, Andrea Cavallini, Glenn Behrmann, and Claudio Angelo Serafino

127. *Graph Database and Graph Computing for Power System Analysis*
Renchang Dai and Guangyi Liu

128. *The Power of Artificial Intelligence for the Next-Generation Oil and Gas Industry: Envisaging AI-inspired Intelligent Energy Systems and Environments*
Pethuru Raj Chelliah, Venkatraman Jayasankar, Mats Agerstam, B. Sundaravadivazhagan, and Robin Cyriac

129. *Microgrids: Theory and Practice*
Peng Zhang

130. *Smart Cyber-Physical Power Systems, Volume 1: Fundamental Concepts, Challenges, and Solutions*
Ali Parizad, Hamid Reza Baghaee, and Saifur Rahman

131. *Smart Cyber-Physical Power Systems, Volume 2: Solutions from Emerging Technologies*
Ali Parizad, Hamid Reza Baghaee, and Saifur Rahman

Printed and bound by CPI Group (UK) Ltd, Croydon, CR0 4YY

17/09/2026

14964313-0008